Lecture Notes in Computer Sc

Commenced Publication in 1973
Founding and Former Series Editors:
Gerhard Goos, Juris Hartmanis, and Jan van Leeuwen

Editorial Board

Advanced Research in Computing and Software Science

Subline of Lectures Notes in Computer Science

Subline Series Editors

Subline Advisory Board

Artur Czumaj   Kurt Mehlhorn
Andrew Pitts   Roger Wattenhofer (Eds.)

Automata, Languages, and Programming

39th International Colloquium, ICALP 2012
Warwick, UK, July 9-13, 2012
Proceedings, Part I

 Springer

Volume Editors

Artur Czumaj
University of Warwick
Department of Computer Science and
Centre for Discrete Mathematics and its Applications
Warwick, UK
E-mail: a.czumaj@warwick.ac.uk

Kurt Mehlhorn
Max-Planck-Institut für Informatik
Saarbrücken, Germany
E-mail: mehlhorn@mpi-inf.mpg.de

Andrew Pitts
University of Cambridge
Computer Laboratory
Cambridge, UK
E-mail: andrew.pitts@cl.cam.ac.uk

Roger Wattenhofer
ETH Zurich, Switzerland
E-mail: wattenhofer@ethz.ch

ISSN 0302-9743   e-ISSN 1611-3349
ISBN 978-3-642-31593-0   e-ISBN 978-3-642-31594-7
DOI 10.1007/978-3-642-31594-7
Springer Heidelberg Dordrecht London New York

Library of Congress Control Number: 2012940794

CR Subject Classification (1998): F.2, F.1, C.2, H.3-4, G.2, I.2

LNCS Sublibrary: SL 1 – Theoretical Computer Science and General Issues

Typesetting: Camera-ready by author, data conversion by Scientific Publishing Services, Chennai, India

Printed on acid-free paper

Springer is part of Springer Science+Business Media (www.springer.com)

Preface

This volume contains the papers presented at the 39th International Colloquium on Automata, Languages and Programming (ICALP 2012), held during July 9–13, 2012 at the University of Warwick, UK. ICALP is the main conference and annual meeting of the European Association for Theoretical Computer Science (EATCS) and first took place in 1972. This year the ICALP program consisted of three tracks:

- Track A: Algorithms, Complexity and Games
- Track B: Logic, Semantics, Automata and Theory of Programming
- Track C: Foundations of Networked Computation

In response to the call for papers, the three Program Committees received a total of 432 submissions: 248 for Track A, 105 for Track B, and 79 for Track C. Each submission was reviewed by three or more Program Committee members, aided by sub-reviewers. The committees decided to accept 123 papers for inclusion in the scientific program: 71 papers for Track A, 30 for Track B, and 22 for Track C. The selection was made by the Program Committees based on originality, quality, and relevance to theoretical computer science. The quality of the submissions was very high indeed, and many deserving papers could not be selected.

The EATCS sponsored awards for both a best paper and a best student paper (to qualify for which, all authors must be students) for each of the three tracks, selected by the Program Committees.

The best paper awards were given to the following papers:

Track A: Leslie Ann Goldberg and Mark Jerrum for their paper "The Complexity of Computing the Sign of the Tutte Polynomial (and Consequent #P-hardness of Approximation)"

Track B: Volker Diekert, Manfred Kufleitner, Klaus Reinhardt, and Tobias Walter for their paper "Regular Languages are Church-Rosser Congruential"

Track C: Piotr Krysta and Berthold Vöcking for their paper "Online Mechanism Design (Randomized Rounding on the Fly)"

The best student paper awards were given to the following papers:

Track A: jointly, Shelby Kimmel for her paper "Quantum Adversary (Upper) Bound" and Anastasios Zouzias for his paper "A Matrix Hyperbolic Cosine Algorithm and Applications"

Track B: Yaron Velner for his paper "The Complexity of Mean-Payoff Automaton Expression"

Track C: Leonid Barenboim for his paper "On the Locality of Some NP-Complete Problems"

In addition to the contributed papers, the conference included six invited lectures, by Gilles Dowek (INRIA Paris), Kohei Honda (Queen Mary London), Stefano Leonardi (Sapienza University of Rome), Daniel Spielman (Yale), Berthold Vöcking (RWTH Aachen University), and David Harel (The Weizmann Institute of Science). David Harel's talk was in honor of Alan Turing, since the conference was one of the Alan Turing Centenary Celebration events, celebrating the life, work, and legacy of Alan Turing.

The following workshops were held as satellite events of ICALP 2012 on July 8, 2012:

- Workshop on Applications of Parameterized Algorithms and Complexity (APAC)
- 4th International Workshop on Classical Logic and Computation (CL&C)
- Third Workshop on Realistic models for Algorithms in Wireless Networks (WRAWN)

We wish to thank all the authors who submitted extended abstracts for consideration, the members of the three Program Committees for their scholarly efforts, and all sub-reviewers who assisted the Program Committees in the evaluation process. We thank the sponsors Microsoft Research, Springer-Verlag, EATCS, and the Centre for Discrete Mathematics and its Applications (DIMAP) for their support, and the University of Warwick for hosting ICALP 2012. We are also grateful to all members of the Organizing Committee and to their support staff. The conference-management system EasyChair was used to handle the submissions, to conduct the electronic Program Committee meeting, and to assist with the assembly of the proceedings.

May 2012

Artur Czumaj
Kurt Mehlhorn
Andrew Pitts
Roger Wattenhofer

Organization

Program Committee

Susanne Albers	Humboldt-Universität zu Berlin, Germany
Albert Atserias	Universitat Politecnica de Catalunya, Spain
Andrei Brodnik	University of Ljubljana, Slovenia
Harry Buhrman	University of Amsterdam, The Netherlands
Bernard Chazelle	Princeton University, USA
James Cheney	University of Edinburgh, UK
Siu-Wing Cheng	HKUST, Hong Kong, SAR China
Bob Coecke	Oxford University, UK
Amin Coja-Oghlan	University of Warwick, UK
Pierre-Louis Curien	CNRS, France
Ugo Dal Lago	Università di Bologna, Italy
Benjamin Doerr	Max Planck Institute for Informatics, Germany
Stefan Dziembowski	University of Warsaw, Poland
Javier Esparza	Technische Universität München, Germany
Michal Feldman	Hebrew University, Israel
Antonio Fernandez Anta	Institute IMDEA Networks, Spain
Paola Flocchini	University of Ottawa, Canada
Fedor Fomin	University of Bergen, Norway
Pierre Fraigniaud	CNRS and University of Paris 7, France
Philippa Gardner	Imperial College London, UK
Philipp Gibbons	Intel Labs Pittsburgh, USA
Erich Graedel	RWTH Aachen University, Germany
Magnus Halldorsson	Reykjavik University, Iceland
Moritz Hardt	IBM Almaden Research, USA
Maurice Herlihy	Brown University, USA
Daniel Hirschkoff	ENS Lyon, France
Nicole Immorlica	Northwestern University, USA
Bart Jacobs	Radboud University Nijmegen, The Netherlands
Valerie King	University of Victoria, Canada/Microsoft Research SVC, USA
Bartek Klin	University of Warsaw, Poland
Elias Koutsoupias	University of Athens, Greece
Dariusz Kowalski	University of Liverpool, UK
Piotr Krysta	University of Liverpool, UK

Amit Kumar	IIT Delhi, India
Stefano Leonardi	Sapienza University of Rome, Italy
Pinyan Lu	Microsoft Research Asia, Shanghai, China
Nancy Lynch	MIT CSAIL, USA
Kazuhisa Makino	University of Tokyo, Japan
Dahlia Malkhi	Microsoft Research, Silicon Valley, USA
Laurent Massoulie	Thomson Research, Paris, France
Kurt Mehlhorn	Max Planck Institut fur Informatik Saarbrücken, Germany
Julian Mestre	University of Sydney, Australia
Aleksandar Nanevski	IMDEA-Software, Madrid, Spain
Flemming Nielson	Technical University of Denmark, Denmark
Rasmus Pagh	IT University of Copenhagen, Denmark
Alessandro Panconesi	Sapienza University of Rome, Italy
Rafael Pass	Cornell University, USA
Boaz Patt-Shamir	Tel Aviv University, Israel
Andrew Pitts	University of Cambridge, UK
Harald Raecke	Technische Universität München, Germany
Arend Rensink	University of Twente, The Netherlands
Peter Sanders	University of Karlsruhe, Germany
Davide Sangiorgi	University of Bologna, Italy
Piotr Sankowski	University of Warsaw, Poland
Saket Saurabh	The Institute of Mathematical Sciences, Chennai, India
Carsten Schuermann	IT University of Copenhagen, Denmark
Nicole Schweikardt	Goethe-Universität Frankfurt am Main, Germany
Angelika Steger	ETH Zurich, Switzerland
Andrzej Tarlecki	Warsaw University, Poland
Patrick Thiran	EPFL, Switzerland
Suresh Venkatasubramanian	University of Utah, USA
Bjorn Victor	Uppsala University, Sweden
Igor Walukiewicz	CNRS, LaBRI, France
Roger Wattenhofer	ETH Zurich, Switzerland
Thomas Wilke	University of Kiel, Germany
James Worrell	Oxford University, UK
Masafumi Yamashita	Kyushu University, Japan

Additional Reviewers

Aaronson, Scott
Abdalla, Michel
Abdullah, Amirali
Abu Zaid, Faried
Aceto, Luca
Adamczyk, Marek
Adler, Isolde
Agmon, Noa
Ahn, Jae Hyun
Alaei, Saeed
Albers, Susanne
Alessi, Fabio
Alglave, Jade
Almagor, Shaull
Altenkirch, Thorsten
Althaus, Ernst
Ambainis, Andris
Ambos-Spies, Klaus
Anand, S.
Andoni, Alexandr
Anh Dung, Phan
Anshelevich, Elliot
Antoniadis, Antonios
Arjona, Jordi
Arrighi, Pablo
Arvind, Vikraman
Asarin, Eugene
Asharov, Gilad
Aspnes, James
Ateniese, Giuseppe
Atig, Mohamed Faouzi
Azar, Yossi
Aziz, Haris
Bachrach, Yoram
Bae, Sang Won
Bansal, Nikhil
Barceló, Pablo
Bateni, Mohammadhossein
Batu, Tuğkan
Bauer, Andreas
Beame, Paul
Becchetti, Luca
Becker, Florent

Bei, Xiaohui
Benedikt, Michael
Bengtson, Jesper
Berenbrink, Petra
Bernáth, Attila
Berwanger, Dietmar
Bezhanishvili, Nick
Bhaskara, Aditya
Bhattacharyya, Arnab
Bhawalkar, Kshipra
Bidkhori, Hoda
Bienkowski, Marcin
Bienvenu, Laurent
Bierman, Gavin
Bille, Philip
Bilo', Vittorio
Bingham, Brad
Birget, Jean-Camille
Birkedal, Lars
Björklund, Henrik
Björklund, Johanna
Bloem, Roderick
Blondin Massé, Alexandre
Blumensath, Achim
Blömer, Johannes
Bonchi, Filippo
Bonifaci, Vincenzo
Bonsma, Paul
Boreale, Michele
Borgstrom, Johannes
Bosnacki, Dragan
Bosque, Jose Luis
Bouajjani, Ahmed
Boyle, Elette
Brach, Paweł
Brautbar, Michael
Bremer, Joachim
Briet, Jop
Bringmann, Karl
Broadbent, Anne
Brodal, Gerth Stølting
Bucciarelli, Antonio
Buchbinder, Niv

Bukh, Boris
Bulatov, Andrei
Cai, Leizhen
Caires, Luis
Calzavara, Stefano
Canetti, Ran
Capecchi, Sara
Carayol, Arnaud
Carbone, Marco
Caskurlu, Bugra
Celis, Laura Elisa
Censor-Hillel, Keren
Chakrabarti, Amit
Chakraborty, Sourav
Chakraborty, Tanmoy
Chan, Sze-Hang
Chatterjee, Krishnendu
Chatzigiannakis, Ioannis
Chekuri, Chandra
Chen, Ning
Chen, Xi
Childs, Andrew
Christodoulou, George
Chung, Kai-Min
Churchill, Martin
Chuzhoy, Julia
Ciancia, Vincenzo
Cicerone, Serafino
Cittadini, Luca
Clairambault, Pierre
Clemente, Lorenzo
Cohen, David
Colcombet, Thomas
Corbo, Jacomo
Cormode, Graham
Cornejo, Alejandro
Courcelle, Bruno
Cygan, Marek
Czerwinski, Wojciech
Czumaj, Artur
D'Souza, Deepak
Dadush, Daniel
David, Alexandre
De Leoni, Massimiliano
De Liguoro, Ugo

de Wolf, Ronald
Degorre, Aldric
Dell, Holger
Demangeon, Romain
Deng, Xiaotie
Deng, Yuxin
Deshpande, Amit
Devanur, Nikhil
Diekert, Volker
Dinitz, Michael
Dinsdale-Young, Thomas
Dittmann, Christoph
Downey, Rod
Doyen, Laurent
Drange, Pål
Drewes, Frank
Dreyer, Derek
Drucker, Andrew
Duflot, Marie
Dughmi, Shaddin
Durnoga, Konrad
Dyagilev, Kirill
Dyer, Martin
Efthymiou, Charilaos
Eisentraut, Christian
Elbassioni, Khaled
Ellen, Faith
Englert, Matthias
Epstein, Leah
Ergun, Funda
Fahrenberg, Uli
Feige, Uriel
Fekete, Sándor
Ferns, Norman
Filipiuk, Piotr
Finocchi, Irene
Fischer, Johannes
Fogarty, Seth
Forej, Vojtech
Fortnow, Lance
Fotakis, Dimitris
Frandsen, Gudmund Skovbjerg
Freydenberger, Dominik
Frieze, Alan
Fu, Hu

Fujito, Toshihiro
Fukunaga, Takuro
Gabbay, Murdoch
Gacs, Peter
Gal, Anna
Gamarnik, David
Gambin, Anna
Gamzu, Iftah
Ganguly, Sumit
Ganty, Pierre
Garcia Saavedra, Andres
García, Álvaro
Garg, Naveen
Gargano, Luisa
Gaspers, Serge
Gauwin, Olivier
Gavoille, Cyril
Gawrychowski, Pawel
Gay, Simon
Ge, Rong
Ghaffari, Mohsen
Ghica, Dan
Giachino, Elena
Gimbert, Hugo
Godskesen, Jens Chr.
Goerdt, Andreas
Goergiou, Chryssis
Goldberg, Leslie Ann
Goldhirsh, Yonatan
Golovach, Petr
Gorla, Daniele
Gottesman, Daniel
Goubault-Larrecq, Jean
Gould, Victoria
Grabmayer, Clemens
Grandoni, Fabrizio
Gravin, Nikolay
Grenet, Bruno
Griffin, Christopher
Grigorieff, Serge
Grohe, Martin
Gugelmann, Luca
Guha, Sudipto
Guillon, Pierre
Gupta, Anupam

Guruswami, Venkatesan
Gutin, Gregory
Gutkovas, Ramunas
Göller, Stefan
Haghpanah, Nima
Hague, Matthew
Hajiaghayi, Mohammadtaghi
Halevi, Shai
Halldorsson, Magnus M.
Hallgren, Sean
Hamano, Masahiro
Han, Xin
Hansen, Helle Hvid
Harwath, Frederik
Haviv, Ishay
Hay, David
Herbreteau, Frédéric
Hernich, André
Hertel, Philipp
Herzen, Julien
Hildebrandt, Thomas
Hodkinson, Ian
Hoefer, Martin
Hoenicke, Jochen
Hofheinz, Dennis
Hohenberger, Susan
Holzer, Markus
Horn, Florian
Hsu, Tsan-Sheng
Huang, Chien-Chung
Huang, Zengfeng
Hunter, Paul
Husfeldt, Thore
Høyer, Peter
Hüffner, Falk
Indyk, Piotr
Iosif, Radu
Ishii, Toshimasa
Ito, Takehiro
Jain, Abhishek
Jain, Rahul
Jancar, Petr
Jansen, Bart
Jansen, Klaus
Jeż, Łukasz

Jones, Neil
Jowhari, Hossein
Jurdzinski, Marcin
Jurdzinski, Tomasz
Kaiser, Lukasz
Kakimura, Naonori
Kaminski, Marcin
Kamiyama, Naoyuki
Kanté, Mamadou
Kapron, Bruce
Kari, Lila
Kartzow, Alexander
Kashefi, Elham
Katoen, Joost-Pieter
Katoh, Naoki
Katz, Jonathan
Kaufman, Tali
Kawachi, Akinori
Kawamura, Akitoshi
Kawase, Yasushi
Kazana, Tomasz
Kesner, Delia
Khabbazian, Majid
Kiefer, Stefan
Kijima, Shuji
Kirsten, Daniel
Kiwi, Marcos
Kiyomi, Masashi
Klasing, Ralf
Kleinberg, Robert
Kliemann, Lasse
Kobayashi, Yusuke
Koenemann, Jochen
Koenig, Barbara
Kolliopoulos, Stavros
Konrad, Christian
Kopczynski, Eryk
Kopparty, Swastik
Kortsarz, Guy
Korula, Nitish
Kosowski, Adrian
Koutavas, Vasileios
Kozen, Dexter
Kratsch, Dieter
Krauthgamer, Robert

Krcal, Jan
Krcal, Pavel
Kretinsky, Jan
Krishnaswamy, Ravishankar
Krugel, Johannes
Kucherov, Gregory
Kufleitner, Manfred
Kullmann, Oliver
Kupferman, Orna
Kuske, Dietrich
Kutzkov, Konstantin
Könemann, Jochen
Łącki, Jakub
Laird, Jim
Lammersen, Christiane
Land, Kati
Lasota, Slawomir
Laurent, Monique
Lee, Troy
Lelarge, Marc
Lengrand, Stephane
Leroux, Jerome
Levavi, Ariel
Ley-Wild, Ruy
Li, Shi
Lin, Anthony Widjaja
Lin, Huijia Rachel
Loff, Bruno
Lotker, Zvi
Lu, Chi-Jen
Lui, Edward
Luttenberger, Michael
Löffler, Maarten
M.S., Ramanujan
Macedonio, Damiano
Madet, Antoine
Madry, Aleksander
Maffei, Matteo
Magniez, Frederic
Mahajan, Meena
Maheshwari, Anil
Mahini, Hamid
Mahmoody, Mohammad
Mairesse, Jean
Makarychev, Konstantin

Malacaria, Pasquale
Malec, David
Malekian, Azarakhsh
Malkis, Alexander
Mandemaker, Jorik
Maneth, Sebastian
Maneva, Elitza
Manthey, Bodo
Marchetti-Spaccamela, Alberto
Mardare, Radu
Martens, Wim
Martin, Greg
Martini, Simone
Marx, Dániel
Massar, Serge
Matisziv, Tim
Mayr, Richard
Mazza, Damiano
McBride, Conor
Mcgregor, Andrew
Medina, Moti
Megow, Nicole
Mei, Alessandro
Mellies, Paul-Andre
Merro, Massimo
Methrayil Varghese, Praveen Thomas
Meyer, Roland
Michail, Othon
Milani, Alessia
Milius, Stefan
Mitra, Pradipta
Miyazaki, Shuichi
Mohar, Bojan
Momigliano, Alberto
Monaco, Gianpiero
Monmege, Benjamin
Montanari, Angelo
Moseley, Benjamin
Mostefaoui, Achour
Mosteiro, Miguel A.
Mozes, Shay
Mucha, Marcin
Mulligan, Dominic
Munagala, Kamesh
Møgelberg, Rasmus Ejlers

Nagarajan, Viswanath
Nandy, Subhas
Nanz, Sebastian
Navarra, Alfredo
Neis, Georg
Nesme, Vincent
Nestmann, Uwe
Neumann, Adrian
Nguyen, Huy
Nicolaou, Nicolas
Niehren, Joachim
Nielson, Hanne Riis
Nies, Andre
Nishide, Takashi
Nussbaum, Yahav
Nutov, Zeev
O Dunlaing, Colm
Obdrzalek, Jan
Obremski, Maciej
Onak, Krzysztof
Ono, Hirotaka
Ooshita, Fukuhito
Oren, Sigal
Orlandi, Alessio
Orlandi, Claudio
Oshman, Rotem
Otachi, Yota
Otto, Friedrich
Otto, Martin
Ouaknine, Joel
Oveis Gharan, Shayan
Padovani, Luca
Paes Leme, Renato
Pakusa, Wied
Panagiotou, Konstantinos
Parker, Matthew
Parrow, Joachim
Parys, Pawel
Patitz, Matthew
Paturi, Ramamohan
Paulusma, Daniel
Pavlovic, Dusko
Perdrix, Simon
Person, Yury
Peter, Ueli

Petit, Barbara
Philip, Geevarghese
Philippou, Anna
Phillips, Jeff
Pientka, Brigitte
Pietrzak, Krzysztof
Piliouras, Georgios
Pilipczuk, Marcin
Pilipczuk, Michal
Pin, Jean-Eric
Pinkau, Chris
Plandowski, Wojciech
Poll, Erik
Poplawski, Laura
Pountourakis, Emmanouil
Pretnar, Matija
Pruhs, Kirk
Puppis, Gabriele
Pyrga, Evangelia
Rabinovich, Roman
Radeva, Tsvetomira
Radoszewski, Jakub
Rafiey, Arash
Rafnsson, Willard
Raghavendra, Prasad
Rahmati, Zahed
Raman, Parasaran
Rao B.V., Raghavendra
Raskin, Jean-Francois
Rawitz, Dror
Reichardt, Ben
Reichman, Daniel
Renaud, Fabien
Reynier, Pierre-Alain
Riba, Colin
Ribichini, Andrea
Richard, Gaétan
Riveros, Cristian
Roland, Jérémie
Rosamond, Frances
Rosulek, Mike
Rybicki, Bartosz
Röglin, Heiko
Sachdeva, Sushant
Sadakane, Kunihiko

Saha, Chandan
Sahai, Amit
Salavatipour, Mohammad
Salvati, Sylvain
Samborski-Forlese, Julian
Sangnier, Arnaud
Sankur, Ocan
Santha, Miklos
Santhanam, Rahul
Santi, Paolo
Santos, Agustin
Sastry, Srikanth
Satti, Srinivasa Rao
Sauerwald, Thomas
Saxena, Nitin
Scafuro, Alessandra
Scarpa, Giannicola
Schaffner, Christian
Schieferdecker, Dennis
Schnoor, Henning
Schudy, Warren
Schwartz, Roy
Schweitzer, Pascal
Schwoon, Stefan
Scott, Philip
Segev, Danny
Seki, Shinnosuke
Senizergues, Geraud
Serre, Olivier
Seshadhri, C.
Seth, Karn
Sevilla, Andres
Sewell, Peter
Seyalioglu, Hakan
Shah, Chintan
Shapira, Asaf
Shen, Alexander
Shioura, Akiyoshi
Sikdar, Somnath
Silva, Alexandra
Silvestri, Riccardo
Singh, Mohit
Sivan, Balasubramanian
Skrzypczak, Michał
Smid, Michiel

Sobocinski, Pawel
Sommer, Christian
Soto, Jose
Spalek, Robert
Spieksma, Frits
Spirakis, Paul
Spöhel, Reto
Srinivasan, Aravind
Srinivasan, Srikanth
Srivastava, Gautam
Stacho, Juraj
Stark, Ian
Staton, Sam
Steurer, David
Stoddard, Greg
Stoelinga, Marielle I.A.
Straubing, Howard
Sun, He
Sun, Xiaoming
Sun, Xiaorui
Sviridenko, Maxim
Swamy, Chaitanya
Szegedy, Mario
Sørensen, Troels Bjerre
Tabareau, Nicolas
Takamatsu, Mizuyo
Takazawa, Kenjiro
Takimoto, Eiji
Tamaki, Suguru
Tamir, Tami
Tanaka, Keisuke
Tanigawa, Shin-Ichi
Taraz, Anusch
Tasson, Christine
Tautschnig, Michael
Telelis, Orestis
Terepeta, Michał
Terui, Kazushige
Tesson, Pascal
Thomas, Henning
Thraves, Christopher
Tokuyama, Takeshi
Toninho, Bernardo
Torenvliet, Leen
Toruńczyk, Szymon

Torán, Jacobo
Touili, Tayssir
Tredan, Gilles
Trevisan, Luca
Tulsiani, Madhur
Turrini, Andrea
Törmä, Ilkka
Uchizawa, Kei
Ueno, Kenya
Ueno, Shuichi
Ullman, Jon
Uno, Takeaki
Urzyczyn, Paweł
Vaidya, Nitin
Van Breugel, Franck
van Stee, Rob
Van Zuylen, Anke
Vassilvitskii, Sergei
Venema, Yde
Venkitasubramaniam,
    Muthuramakrishnan
Venturini, Rossano
Vereshchagin, Nikolay
Versari, Cristian
Verschae, Jose
Vidick, Thomas
Vijayaraghavan, Aravindan
Vilenchik, Dan
Vilenchik, Danny
Villard, Jules
Viola, Emanuele
Visconti, Ivan
Vishkin, Uzi
Vishnoi, Nisheeth
Vitanyi, Paul
Vondrak, Jan
Wachter, Björn
Wahlström, Magnus
Wang, Yajun
Ward, Justin
Wasowski, Andrzej
Weber, Tjark
Wee, Hoeteck
Wei, Zhewei
Weihmann, Jeremias

Weimann, Oren
Welch, Jennifer
Whittle, Geoff
Wiedijk, Freek
Wiese, Andreas
Williams, Ryan
Wimmer, Karl
Winzen, Carola
Wong, Prudence W.H.
Woodruff, David
Worrell, James
Wu, Yi
Wulff-Nilsen, Christian
Xia, Mingji
Xiao, David
Xie, Ning
Yamamoto, Masaki
Yamauchi, Yukiko
Yannakakis, Mihalis
Ye, Deshi
Yekhanin, Sergey
Yi, Ke

Yin, Yitong
Yokoo, Makoto
Yoshida, Yuichi
Young, Neal
Zadimoghaddam, Morteza
Zając, Michał
Zavattaro, Gianluigi
Zavou, Elli
Zdeborova, Lenka
Zelke, Mariano
Zhang, Chihao
Zhang, Fuyuan
Zhang, Jialin
Zhang, Jiawei
Zhang, Qin
Zhang, Shengyu
Zheng, Colin
Zheng, Jia
Zhong, Ning
Zhou, Yuan
Zohar, Aviv

Table of Contents – Part I

Track A – Algorithms, Complexity and Games

Table of Contents – Part II

Track C – Foundations of Networked Computation

Unsatisfiability Bounds for Random CSPs
from an Energetic Interpolation Method

Dimitris Achlioptas[1,2,3,*] and Ricardo Menchaca-Mendez[3]

[1] University of Athens, Greece
[2] CTI, Greece
[3] University of California, Santa Cruz, USA

Abstract. The interpolation method, originally developed in statistical physics, transforms distributions of random CSPs to distributions of much simpler problems while bounding the change in a number of associated statistical quantities along the transformation path. After a number of further mathematical developments, it is now known that, in principle, the method can yield rigorous unsatisfiability bounds if one "plugs in an appropriate functional distribution". A drawback of the method is that identifying appropriate distributions and plugging them in leads to major analytical challenges as the distributions required are, in fact, infinite dimensional objects. We develop a variant of the interpolation method for random CSPs on arbitrary sparse degree distributions which trades accuracy for tractability. In particular, our bounds only require the solution of a 1-dimensional optimization problem (which typically turns out to be very easy) and as such can be used to compute explicit rigorous unsatisfiability bounds.

1 Introduction

The problem of determining the satisfiability of Boolean formulas is central to the understanding of computational complexity. Moreover, it is of tremendous practical interest as it arises naturally in numerous settings. Random CNF formulas have emerged as a mathematically tractable vehicle for studying the performance of satisfiability algorithms and proof systems. For a given set of n Boolean variables, let B_k denote the set of all possible disjunctions of k distinct, non-complementary literals from its variables (k-clauses). A random k-SAT formula $F_k(n, m)$ is formed by selecting uniformly and independently m clauses from B_k and taking their conjunction. Such random formulas have been shown to be hard both for proof systems, e.g., in the seminal work of Chvátal-Szemérdi on resolution [7], and, more recently, for some of the most sophisticated satisfiability algorithms known [8].

More generally, in Random Constraint Satisfaction Problems (RCSPs) one has a set of n variables all with the same (small) domain D and a set of $m = rn$ constraints, for some constant $r > 0$, each of which binds a randomly selected subset of $O(1)$ variables. Canonical examples are finding large independent sets and colorings sparse random graphs, variations of satisfiability, and systems of random linear equations. We will be interested in random CSPs (RCSPs) from an asymptotic point of view, i.e., as the

---

* Research supported by NSF CCF-0546900, a Sloan Fellowship, and ERC grant 210743.

A. Czumaj et al. (Eds.): ICALP 2012, Part I, LNCS 7391, pp. 1–12, 2012.
© Springer-Verlag Berlin Heidelberg 2012

number of variables grows. In particular, we will say that a sequence of random events $\mathcal{E}_n$ occurs *with high probability (w.h.p.)* if $\lim \Pr[\mathcal{E}_n] = 1$. The ratio of constraints-to-variables, $r = m/n$, known as density, plays a fundamental role here as most interesting monotone properties are believed to exhibit 0-1 laws with respect to density. Perhaps the best known example is the satisfiability property for random k-CNF formulas. Let $g_k(n, r)$ denote the probability that $F_k(n, rn)$ is satisfiable.

Conjecture 1 (Satisfiability Threshold Conjecture). For each $k \geq 3$, there exists a constant r_k such that for any $\epsilon > 0$,

$$\lim_{n \to \infty} g_k(n, r_k - \epsilon) = 1, \quad \text{and} \quad \lim_{n \to \infty} g_k(n, r_k + \epsilon) = 0 .$$

The satisfiability threshold conjecture, which motivates our work, has attracted a lot of attention in computer science, mathematics and statistical physics [17,18,16]. At this point, neither the value, nor even the existence of r_k has been established. For $k = 3$, the best known bounds are $3.52 < r_3 < 4.49$, due to results in [9] and [13], respectively.

The last decade has seen a great deal of rigorous results on random CSPs, including a proliferation of upper and lower bounds for the satisfiability threshold of a number of problems. Equally importantly, random CSPs have been the domain of an extensive exchange of ideas between computer science and statistical physics [15], including the positing of the clustering phenomenon, establishing it rigorously, and relating it to algorithmic performance. In this work we take another step in this direction by taking a technique from mathematical physics, the *interpolation method* of Guerra [12], and using it to show how to derive end-to-end rigorous explicit upper bounds for the satisfiability threshold of a number of problems. To do so, we introduce a new version of the interpolation method that can be made computationally effective and give a new, much simpler, extension of the method to CSPs with arbitrary degree distributions.

Our method can be used to prove among other things the following result [5] regarding the satisfiability of mixtures of 2- and 3-clauses.

Theorem 1 ([5]). *Let F be a random CNF formula on n variables with $(1-\epsilon)n$ random 2-clauses, and $(1 + \epsilon)n$ random 3-clauses. W.h.p. F is unsatisfiable for $\epsilon = 10^{-4}$.*

Theorem 1, combined with the methods of [3], implies that a number of DPLL algorithms require exponential time on easily satisfiable random 3-CNF formulas. For example, ORDERED DLL requires exponential time for all $r \geq 2.78$, while GUC for all $r \geq 3.1$. Similar results hold for a host of other algorithms, including, for example, all algorithms analyzed in [2] and [1].

2 Motivation and Past Work

Perhaps the simplest possible upper bound on the satisfiability threshold comes from taking the union bound over all assignments $\sigma \in \{0, 1\}^n$ of the probability they satisfy a random formula $F = F_k(n, rn)$. That is,

$$\Pr[F_k(n, rn) \text{ is satisfiable}] \leq \sum_{\sigma} \Pr[\sigma \text{ satisfies } F_k(n, rn)] = \left[2(1 - 2^{-k})^r\right]^n \to 0 ,$$

for all $r > r_k^*$, where $2(1 - 2^{-k})^{r_k^*} = 1$. For example, $r_3^* = 5.19..$, but a long series of increasingly sophisticated results has culminated with the bound $r_3 < 4.48...$ by Díaz et al. [9]. At the same time, statistical physics results by Mertens et al. [14] give evidence that $r_3 < 4.26....$

2.1 Past Work on the Interpolation Method for Random CSPs

The *interpolation method* is a remarkable tool originally developed by Guerra and Toninelli [12] to deal with the Sherrignton Kirkpatrick model (SK) of statistical physics. Following their breakthrough, Franz and Leone [10], in a very important paper, applied the interpolation method to random k-SAT and random k-XOR-SAT to prove that certain expressions derived via the non-rigorous replica method of statistical physics for these problems can, in principle, be used to derive upper bounds for the satisfiability threshold of each problem. As we will see, though, doing so involves the solution of certain functional equations that appear beyond analytical penetration. In [20], Panchenko and Talagrand showed that the results of [10] can be derived in a simpler and uniform way, unifying the treatment of different levels of Parisi's Replica Symmetry Breaking.

A crucial ingredient in all the above proofs is a Poissonization device exploiting that in Erdős-Renyi (hyper)graphs the degrees of the vertices behave, essentially, like independent, Poisson random variables. Franz, Leone, and Tonnineli [11] extended the interpolation method to other degree sequences, but at the cost of introducing another level of complexity (multi-overlaps), thus placing the method even further out of reach in terms of explicit computations. In [19], Montanari gave a simpler method for dealing with degree sequences in the context of error-correcting codes, which proceeds by approximating the intended degree distribution "in chunks". This, unfortunately, requires the number of approximation steps to go to infinity (so that the chunk size goes to zero) in order to give results for the original problem.

Finally, in a recent paper, Bayati, Gamarnik and Tetali [6], showed that a combinatorial analogue of the interpolation method can be used to elegantly derive an approximate subadditivity property for a number of CSPs on Erdős-Renyi and regular random graphs. This allowed them to prove the *existence* of a number of limits in these problems. The simplicity of that approach, though, comes at the cost of losing the capacity to give bounds for the associated limiting quantities.

3 Highlights of the Interpolation Method on RCSPs

For simplicity of exposition we focus on the case where all constraints have the same arity $k \geq 2$. It is very easy to see that the proof goes through transparently for CSPs that are mixtures of constraints of different arities. Let $C_{k,n}$ denote the set of all possible k-constraints on n variables for the CSP at hand and let D denote the domain of each variable. So, for example, $C_{k,n}$ could contain all $2^k \binom{n}{k}$ clauses of length k on n variables, or all $\binom{n}{2} D!$ possible unique-games constraints on a graph with n vertices. A random CSP instance $I_k(n, r)$ is a conjunction of m constrains taken independently with replacement from the set $C_{k,n}$, where m is a *Poisson random variable* with mean $\mathbb{E}[m] = rn$. Note that in the more standard models of random CSPs m is fixed (not a random variable). Since, though, the standard deviation of the Poisson distribution is the square root of its mean we have $m = (1+o(1))rn$ w.h.p., thus not affecting any asymptotic results regarding densities. At the same time, along with the Poissonization of the variable degrees, this is key to the original development of the method. Eliminating the need for Poisson variable degrees and allowing arbitrary (sparse) degree sequence, as we do in Section 5, is part of the technical contribution in our work.

We shall work with the random variable $H_{n,r}(\sigma)$, known as the Hamiltonian, counting the number of unsatisfied constraints in the instance for each $\sigma \in D^n$. (The randomness of H being in the random choice of the instance). We will sometimes refer to $H_{n,r}(\sigma)$ as the energy function. The goal is to compute lower bounds for the following quantity as a function of $\beta \geq 0$,

$$f_r = f_r(\beta) = n^{-1}\mathbb{E}\left[\log\left(\sum_{\sigma \in D^n} \exp\left(-\beta H_{n,r}(\sigma)\right)\right)\right] . \tag{1}$$

For each fixed value of $\beta > 0$, the sum in $f_r(\beta)$ is dominated by those value assignment having energy (violated constraints) in some narrow window that depends on β. (The idea being that assignments violating more constraints are penalized too heavily to contribute significantly to the sum, while assignment violating even fewer constraints are too rare to have substantial contribution.) Thus, $f_r(\beta)$ effectively counts the number of assignments at each energy level is known as the *free entropy density*. Note that as β is increased $f_r(\beta)$ places more and more weight to assignments violating fewer constraints, recovering the number of solutions as $\beta \to \infty$ (writing $\beta = 1/T$ this is also known as the zero-temperature limit). Standard martingale arguments imply that if any finite $\beta > 0$ we have $\lim_{n\to\infty} f_r(\beta) < 0$, then w.h.p. no solutions exist. The goal of the interpolation method is to give negative upper bounds for $f_r(\beta)$ and since f_r is the free entropy, we refer to this as entropic interpolation.

Given $\sigma = (x_1, x_2, \ldots, x_n)$ we will write $H_{n,r}(\sigma)$ as the sum of m functions $\theta_a(x_{a_1}, \ldots, x_{a_k})$, one for each constraint. That is, $\theta_a(x_{a_1}, \ldots, x_{a_k}) = 1$ if the associated constraint is not satisfied and 0 otherwise. For example, for k-SAT, we take the domain of the variables to be $\{+1, -1\}^n$ and for each k-clause $c_a(x_{a_1}, \ldots, x_{a_k})$ we let

$$\theta_a(x_{a_1}, \ldots, x_{a_k}) = \prod_{j=1}^{k} \frac{1 + J_{a_j}x_{a_j}}{2} , \tag{2}$$

where $J_{a,j} \in \{+1, -1\}$ represents the sign of literal a_j in clause c_a: $+1$ if the literal is negated and -1 otherwise.

The basic object of the interpolation method is a modified energy function that interpolates between $H_{n,r}(\sigma)$ and the energy function of a dramatically simpler and fully tractable model. Specifically, for $t \in [0,1]$, let

$$\beta H_{n,r,t}(x_1, \ldots, x_n) = \sum_{m=1}^{m_t} \beta \theta_{a_m}(x_{a_{m,1}}, \ldots, x_{a_{m,k}}) + \sum_{i=1}^{n}\sum_{j=1}^{k_{i,t}} \log\left(\hat{v}_{i,j}(x_i)\right) , \tag{3}$$

where m_t is a Poisson random variable with mean $\mathbb{E}[m_t] = trn$, the $k_{i,t}$'s are i.i.d. Poisson random variables with mean $\mathbb{E}[k_{i,t}] = (1-t)kr$, and the functions $\hat{v}_{i,j}(\cdot)$ are i.i.d. random functions distributed as the function defined in (5) below.

Before delving into the meaning of the random functions $\hat{v}_{i,j}(\cdot)$, which are the heart of the method, let us first make a few observations about (3). First, note that (3) is simply the energy function of the original model when $t = 1$. On the other hand, when $t < 1$, we expect that $(1 - t)m$ of the original k-clauses will be replaced by k times as many functions each of which takes as input a single variable. A helpful way to think about this replacement is as a decombinatorialization of the energy function wherein k-ary functions are replaced by univariate, and therefore, independent functions. As one can imagine, for $t = 0$ the model is fully tractable. In particular, letting

$$f_r(t) = n^{-1}\mathbb{E}\left[\log\left(\sum_{\sigma \in D^n} \exp\left(-\beta H_{n,r,t}(\sigma)\right)\right)\right] , \qquad (4)$$

one can readily compute $f_r(0)$ since one can compute $H_{n,r,0}(\sigma)$ by examining one variable at a time. To relate the two models the plan is to give a lower bound for the change in f_r as t goes from 1 to 0, hence the name interpolation, thus bounding $f_r(1)$ by $f_r(0)$ plus a term depending on our bound on the derivative.

The main idea of the interpolation method is to select the (still mysterious) univariate functions $\hat{v}_{i,j}(\cdot)$ independently, from a probability distribution that reflects aspects of the geometry of the underlying solution space. The more accurate the reflection, the better the bound. One, of course, needs to guess this geometry and here is where the insights from statistical physics are most valuable. A beautiful aspect of the interpolation method is that it projects all information about the geometry of the solution space into a single object, a distribution γ as defined below. With that in mind, we now define the random univariate functions, *but without specifying the all-important distribution γ*. This is because the method gives a valid bound for *any* γ, i.e., the choice of γ affects the quality but not the validity of the derived bound.

Let $v(x)$ denote the density function of a random variable over D, where the probabilities $p_1, \ldots, p_{|D|}$ are themselves chosen at random from a distribution γ with support on the unit $(|D| - 1)$-dimensional simplex. Let $\hat{v}(x)$ be a random univariate function defined as follows

$$\hat{v}(x) = \sum_{y_1,\ldots,y_{k-1}} \exp\left(-\beta\theta(y_1, ..., y_{k-1}, x)\right) \prod_{j=1}^{k-1} v_j(y_j) , \qquad (5)$$

where $\theta(\cdot)$ is a random constraint-function and the functions $v_i(\cdot)$ are i.i.d. with the same distribution as $v(x)$.

To interpret the function in (5) it helps to think of its argument x as corresponding to a particular occurrence of a variable in a constraint c, e.g., a literal occurrence in a random k-clause. The idea is for (5) to simulate·the biases that this particular occurrence of x "feels" from its presence in c. To do this we replace c with a brand new random constraint (appearing as θ in (5)) containing $k-1$ new variables $y_1, \ldots, y_{k-1}$ which are "private" to θ, i.e., which will occur in no other constraint in the interpolating energy function. To simulate the statistical joint behavior of the $k - 1$ original variables in c due to their participation in clauses other than c, we assume that since the underlying random hypergraph is sparse, these $k - 1$ new variables are independent in the absence of θ, hence the product in (5). Finally, specifying the probability distribution γ governing the behavior of each ersatz variable is precisely what reflects our beliefs about the geometry of the space of solutions. Statistical physics considerations suggest candidate distributions as solutions to distributional equations.

To see how the geometry of the space of solutions enters the distribution γ, consider two dramatically different settings, precisely those separated by the so-called shattering (or clustering, or dynamical) transition. In one setting, the set of solutions has the property that if a solution is chosen uniformly at random, changing the value of any variable to any other value can be accommodated by changing, in expectation, the value of $O(1)$ other variables, i.e., by "local repair". In such a world, γ is a single density function over the $(|D| - 1)$-dimensional simplex. In contrast, after shattering occurs [4] the set of solutions consists of exponentially many clusters (connected components of solutions), separated by linear Hamming distance. In each cluster, a constant fraction of all variables take the same value in all solutions in the cluster, while all other variables are locally repairable. In this world, γ becomes a distribution over densities, the different densities corresponding to different clusters.

3.1 Why an Energetic Interpolation Method

What motivates our derivation of a different, so-called energetic, interpolation method is that dealing with the shattered case above leads to massive analytical obstacles, rendering the derivation of explicit, mathematically rigorous bounds problematic. In the realm of statistical mechanics, these are addressed via a numerical stochastic method known as population dynamics, used to derive the estimates in [14] for random k-SAT.

In contrast, we will see that the energetic approach leads to bounds which can be derived analytically, precisely because we dramatically collapse the information captured by γ. In particular, in our bounds γ will be specified by a single real number, while the bound itself is expressed by truncating an infinite sum to a finite one (at any desired degree of accuracy) and adding up the corresponding explicit terms, each involving the joint behavior of a finite number of Poisson random variables.

The reason this approach works is that in bivariate binary CSPs, such as random [MAX] 2-SAT, random MAX 2-LIN-2, and random $(2 + p)$-SAT, whenever a frozen variable appears in a constraint "the wrong way" (the freezing being due to its participation in other constraints) this necessarily causes the other variable in the constraint to also freeze. This percolative type of behavior causes the fraction of frozen variables to take off smoothly in such problems, a situation that can be captured by a simple model for the distribution γ if one focuses on states of lowest energy. This is precisely what we exploit in deriving our new upper bounds for these problems.

4 Energetic Interpolation for General CSPs

To develop an energetic interpolation method we replace the (far richer) free entropy density of the previous section with the following much simpler quantity

$$\xi_r = n^{-1}\mathbb{E}\left[\min_{\sigma \in D^n} H_{n,r}(\sigma)\right], \tag{6}$$

known as ground-state energy density, which simply tells us the fraction of violated constraints in the optimal (least-violating) assignments. By standard martingale arguments the random variable $\min_\sigma H_{n,r}(\sigma)$ concentrates around its expectation (consider the martingale exposing the constraints one by one and note that changing any one constraint cannot change its value by more than 1). Therefore, if $\liminf_{n\to\infty} \xi_r > 0$ we can conclude that the satisfiability threshold is upper bounded by r.

The univariate factors in the energy interpolation method are given as follows:

- For $1 \le j \le |D|$, let "j" denote the indicator function that the input is j, i.e., "j" is 1 if its input is j and 0 otherwise.
- Let "$*$" denote the function that assigns 0 to all elements of D.
- Let $\mathrm{h}(x)$ be a random function in $\{$"1"$, \ldots,$ "$|D|$"$,$ "$*$"$\}$ with $\Pr(\mathrm{h}(\cdot) =$ "$*$"$) = 1 - p$ and $\Pr(\mathrm{h}(\cdot) =$ "j"$) = p/|D|$.

The analogue of (5) is now

$$\hat{h}(x) = \min_{y_1,\ldots,y_{k-1}} \left\{ \theta(y_1, .., y_{k-1}, x) + \sum_{i=1}^{k-1} \mathrm{h}_i(y_i) \right\} , \tag{7}$$

where $\theta(\cdot)$ is a random constraint-function as before while the functions $\mathrm{h}_i(\cdot)$ are i.i.d. random functions distributed as $\mathrm{h}(x)$.

Observe that the energy interpolation method models all information about the geometry of the solution space into a single probability p, which can be interpreted as the probability that a variable picked at random will be frozen, i.e., have the same value in all optimal assignments. If that occurs for all $k - 1$ variables $y_1, .., y_{k-1}$ and they all happen to be frozen the wrong way as far as θ is concerned, then unless variable x takes the value desired by θ the function $\hat{h}(x)$ will evaluate to 1. When, at the end of the interpolation, we will have replaced all k-ary constraints with univariate random functions $\hat{h}$, the optimal overall assignment is simply found by assigning to each variable the value that makes the majority of its $\hat{h}$ functions evaluate to 0. The method delivers a valid bound for *any* choice of $p \in [0, 1]$ and the bound is then optimized by choosing the best value of p, i.e., performing a single-parameter search.

While we could give lower bounds on (6) for RCSPs defined on Erdős-Renyi (hyper)graphs by exploiting the same Poissonization device as in earlier works, we will instead show how to carry out the method in arbitrary sparse degree distributions.

5   The Interpolation Method on Sparse Degree Sequences

Let d_i denote the number of times variable i should appear in the random instance and let $L_i = \{l_{i,j}\}_{j=1}^{d_i}$ denote the set of occurrences corresponding to variable i. Note that the occurrences can be decorated so that, for example in k-SAT, we can specify how many of the L_i occurrences correspond to positive occurrences of the variables and how many to negative occurrences. It will be helpful to think of each occurrence as a piece of paper carrying the index of the underlying variable along with any desired decoration. To form a random instance with $m = rn$ constraints we simply choose a random permutation of the krn elements of $\mathcal{L} = \{L_i\}_{i=1}^n$ and consider the first k to specify the first constraint, the next k to specify the second constraint etc.

Consider now the following algorithm to build a random Hamiltonian composed of a mixture of k-ary constraint-factors of the desired CSP and of univariate functions as in (5). The algorithm has three inputs: The collection of occurrences $\mathcal{L}$, an integer t, and a sequence $x \in \{b,c\}^t$.

1. Set $H = \emptyset$, set $L = \mathcal{L}$, and set $j = 1$.
2. Select a random permutation π of the elements of L.
3. **While** $j \leq \min\{t, |L|\}$ **do:**
   (a) **If** $x_j = b$ **then**
       i.   Add a random univariate factor to H with argument $\pi(j)$.
       ii. $j \leftarrow j + 1$
   (b) **If** $x_t = c$ **then** with probability $1/k$
       i.   Add a random k-constraint to H on occurrences $\pi(j), \ldots, \pi(j+k-1)$.
       ii. $j \leftarrow j + k$

Let $\mathbb{H}(\mathcal{L}, x)$ denote the family of energy functions produced by the above algorithm. Observe that when $t = |\mathcal{L}|$ and $x = u \cdots u$, the energy functions produced by the algorithm have variable degree distribution given by $\mathcal{L}$ and consist of univariate factors only. On the other hand when $t = |\mathcal{L}|$ and $x = c \cdots c$ the resulting energy functions consist of $\tilde{m}$ energy constraint functions of arity k where $\tilde{m}$ is a Binomial random variable with km trials and probability of success $1/k$, conditioned on being at most m. In other words, w.h.p. the instance generated will have the desired degree sequence except for $o(n)$ variables (and, therefore, $o(n)$ constraints). Since we are interested in establishing a non-vanishing lower bound for (6) this will not affect any of our results.

The goal now is to relate the ground state energy of these two extreme cases. A key property, which will allow us to establish such relation, is that $\mathbb{H}(\mathcal{L}, x)$ is invariant under any permutation $\pi(x)$ of the elements in x.

Lemma 1. *For every sequence x, and every permutation π, the families $\mathbb{H}(\mathcal{L}, x)$ and $\mathbb{H}(\mathcal{L}, \pi(x))$ have the same distribution.*

Proof. The very first step of our construction is to take a uniformly random permutation of the elements of $\mathcal{L}$.

For any $\mathcal{L}$ and any $s \leq t$, since the order of the steps in x does not matter, let us write $\mathbb{H}(\mathcal{L}, t, s)$ to denote the distribution of energy functions generated by the algorithm when we take t steps in total, $t - s$ of which are additions of a univariate factor. Let

$$\xi_{\mathcal{L}}(t, s) = n^{-1} \mathbb{E}\left[\min_{\sigma \in D^n} H_{\mathcal{L},t,s}(\sigma)\right] .$$

Observe that if $t = km$ and $s = km$, then $\xi_{\mathcal{L}} = \xi_{\mathcal{L}}(km, km)$ corresponds to the original ground state energy, whereas $\xi_{\mathcal{L}}(km, 0)$ corresponds to the ground state energy of the model composed of univariate factors only.

Our lower bounds come from the following theorem.

Theorem 2. *For any choice of* $p \in [0, 1]$*, if* $m = rn$ *then*

$$\xi_{\mathcal{L}} \geq \xi_r(km, 0) - r(k-1)\mathbb{E}[h_c] - o(1) , \tag{8}$$

where

$$h_c = \min_{y_1, \ldots, y_k} \left\{ \theta(y_1, .., y_k) + \sum_{i=1}^{k} \mathbf{h}_i(y_i) \right\} .$$

To prove this we will prove that as s goes from t to 0, we can control the change of $\xi_r(t, s)$. Specifically,

Lemma 2. *If* $m = rn$ *then for any* $\epsilon > 0$*, all* $t \in [0, (kr - \epsilon)n]$*, and all* $1 \leq s \leq t$,

$$\mathbb{E}[\min\{H_{\mathcal{L}, t, s-1}(\sigma)\}] - (k-1)k^{-1}\mathbb{E}[h_c] \leq \mathbb{E}[\min\{H_{\mathcal{L}, t, s}(\sigma)\}] + o(1) .$$

Iteratively applying Lemma 2 so that we can increase the number of univariate factors from 0 to $t = (kr - \epsilon)n$ and letting $\epsilon \to 0$ yields Theorem 2.

Proof (Lemma 2). Let H_0 be an energy function from the family $\mathbb{H}(\mathcal{L}, t - 1, s - 1)$, that is, the energy function resulting from executing $t - 1$ steps of the algorithm where $s - 1$ of such steps correspond to adding a univariate factor. The key observation is that $H_{\mathcal{L}, t, s-1}(\sigma)$ and $H_{\mathcal{L}, t, s}(\sigma)$ can be obtained from H_0 by execution an additional step of the algorithm: $H_{\mathcal{L}, t, s-1}(\sigma)$ corresponds to the processing of a c symbol and $H_{\mathcal{L}, t, s}(\sigma)$ corresponds to the precessing of a u symbol.

We will show that conditional on any realization of H_0 we have

$$\mathbb{E}[\min\{H_{\mathcal{L}, t, s-1}(\sigma)\}|H_0] - (k-1)k^{-1}\mathbb{E}[h_c] \leq \mathbb{E}[\min\{H_{\mathcal{L}, t, s}(\sigma)\}|H_0] + o(1) . \tag{9}$$

That is, the proof reduces to comparing the effect of adding a single univariate factor to the effect of adding, with probability $1/k$, a single constraint. As one can imagine, the proof of (9) is problem specific. Below we prove it for random k-SAT and random Max-k-Lin-2. For all other random CSPs with binary domains the proof is very similar.

6 Applying Energetic Interpolation to Random CSPs

6.1 Random k-SAT

Let $C^* \subseteq \{0, 1\}^n$ be the set of optimal assignments in H_0. A variable x_i is frozen if its value is the same in all optimal assignments. The processing of a c symbol will increase the value of the minimum by at most 1 only if the following two conditions hold: 1) a new clause is added, which occurs with probability $1/k$, and 2) all the literals appearing in the new random factor correspond to a frozen variables. By the principle of deferred decisions we can think of the permutation π as generated on-the-fly, i.e., as we need occurrences to consume. Therefore, if the number of remaining occurrences is $\Omega(n)$ and f denotes the fraction of them that are associated with frozen variables corresponding to H_0, then

$$\mathbb{E}[\min\{H_{\mathcal{L}, t, s}(\sigma)\}|H_0] - \min\{H_0\} = k^{-1}2^{-k}f^k + O(1/n) ,$$

where the last term is due to the fact that we are selecting without replacement.

Similarly, the processing of a u symbol will increase the value of the minimum by 1 if the chosen literal correspond to a frozen variable x and x must take the opposite of its frozen value to minimize the added factor $\hat{h}(x)$. Thus the expected change is

$$\mathbb{E}\left[\min\{H_{\mathcal{L},t,s-1}(\sigma)\}|H_0\right] - \min\{H_0\} = 2^{-k}p^{k-1}f \ .$$

Finally,

$$\mathbb{E}\left[h_c\right] = \mathbb{E}\left[\min_{y_1,\ldots,y_k} \{\theta(y_1,..,y_k) + \sum_{i=1}^{k} \mathbf{h}_i(y_i)\}\right] = 2^{-k}p^k \ .$$

By combining the above equations and adding $-(k-1)k^{-1}2^{-k}p^k$ we get

$$\mathbb{E}\left[\min\{H_{\mathcal{L},t,s-1}(\sigma)\}|H_0\right] - (k-1)k^{-1}2^{-k}p^k - \mathbb{E}\left[\min\{H_{\mathcal{L},t,s}(\sigma)\}|H_0\right]$$
$$= k^{-1}2^{-k}\left(kp^{k-1}f - f^k - (k-1)p^k\right) + O(1/n) \ .$$

Finally, the polynomial $F(x,p) = kp^{k-1}x + x^k - (k-1)p^k \leq 0$ for all $0 \leq x, p \leq 1$. To see this last statement note that (i) $F(0,p), F(1,p), F(x,0), F(x,1) \leq 0$ and, (ii) the derivative of F with respect to p is 0 only when $p = x$, in which case $F(x,x) = 0$.

6.2   Random Max-k-Lin-2

The constraints in the Max-k-Lin-2 problem are chosen uniformly from the set of all $2n^k$ possible boolean equations on n variables, i.e., the k variables are chosen at random with replacement and the required parity is equally likely to be 0 or 1. Let $C^* \subseteq \{0,1\}^n$ be the set of optimal assignments in H_0. A variable x_i is frozen if its value is the same in all optimal assignments. The processing of a c symbol will increase the value of the minimum by at most 1 only if the following three conditions hold: 1) a new Boolean equation is added, which occurs with probability $1/k$, 2) all the literals appearing in the new random factor correspond to frozen variables and 3) the parity of the frozen variables is different from the one required by the new equation. As in the proof for random k-SAT above, if the number of remaining occurrences is $\Omega(n)$ and f denotes the fraction of them that are associated with frozen variables corresponding to H_0, then,

$$\mathbb{E}\left[\min\{H_{\mathcal{L},t,s}(\sigma)\}|H_0\right] - \min\{H_0\} = k^{-1}2^{-1}f^k + O(1/n) \ .$$

where the last term is due to the fact that we are selecting without replacement. Similarly, the processing of a c symbol can increase the value of the minimum by 1 if the chosen literal correspond to a frozen variable x and and x must take the opposite of its frozen value to minimize the added factor $\hat{h}(x)$. Thus the expected change is given by

$$\mathbb{E}\left[\min\{H_{\mathcal{L},t,s-1}(\sigma)\}|H_0\right] - \min\{H_0\} = 2^{-1}p^{k-1}f \ .$$

Finally,

$$\mathbb{E}\left[h_c\right] = \mathbb{E}\left[\min_{y_1,\ldots,y_k} \{\theta(y_1,..,y_k) + \sum_{i=1}^{k} \mathbf{h}_i(y_i)\}\right] = 2^{-1}p^k \ .$$

Combining the above equations and adding $-(k-1)k^{-1}2^{-1}p^k$ we get

$$\mathbb{E}\left[\min\{H_{\mathcal{L},t,s-1}(\sigma)\}|H_0\right] - (k-1)k^{-1}2^{-k}p^k - \mathbb{E}\left[\min\{H_{\mathcal{L},t,s}(\sigma)\}|H_0\right]$$
$$= k^{-1}2^{-1}\left(kp^{k-1}f - f^k - (k-1)p^k\right) + O(1/n) \ ,$$

where the r.h.s. of the equality entails the same polynomial as for random k-SAT.

7   Computing Explicit Energetic Interpolation Bounds for k-SAT

Applying Theorem 2 on a Poisson degree sequence we get that

$$\xi_r(0) = \mathbb{E}\left[\min_{x\in\{0,1\}}\left(\sum_{j=1}^{s}\hat{h}_j(x)\right)\right] \ , \tag{10}$$

where s is a Poisson random variable with mean kr, and the functions $\hat{h}_j(\cdot)$, i.e., random functions in $\{"0","1","*"\}$ with $\Pr(\hat{h}_j(\cdot) = "1") = \Pr(\hat{h}_j(\cdot) = "0") = 2^{-k}p^{k-1}$.

Let l_0, l_1, and l_* denote the number "0", "1", and "*" functions respectively among the $\hat{h}_j(\cdot)$ functions inside the summation of equation (10). Conditional on the value of s, the random vector (l_0, l_1, l_*) is distributed as a multinomial random vector and, therefore,

$$\xi_r(0) = \sum_{x=0}^{\infty}\sum_{l_0=0}^{x}\sum_{l_1=0}^{x-l_0}\min\{l_0,l_1\}\times\text{Poi}(kr,x)\text{Multi}(l_0,l_1,x-l_0-l_1) \ ,$$

where $\text{Multi}(\cdot,\cdot,\cdot)$ denotes the multinomial density function.

Changing the limits of all summations to infinity, does not change the value of $\xi_r(0)$, since $\text{Multi}(\cdot,\cdot,\cdot)$ evaluates to zero for negative numbers, hence, we can interchange the order of the summations to get

$$\xi_r(0) = \sum_{l_0=0}^{\infty}\sum_{l_1=0}^{\infty}\min\{l_0,l_1\}\times\sum_{x=0}^{\infty}\text{Poi}(kr,x)\text{Multi}(l_0,l_1,x-l_0-l_1) \ .$$

The last equation can be simplified by summing out the randomness in the Poisson random variable. The result is that l_0 and l_1 become two independent Poisson random variables with mean $\frac{k}{2^k}rp^{k-1}$. Thus,

$$\xi_r(0) = \sum_{l_0=0}^{\infty}\sum_{l_1=0}^{\infty}\min\{l_0,l_1\}\times\text{Poi}\left(\frac{k}{2^k}rp^{k-1},l_0\right)\times\text{Poi}\left(\frac{k}{2^k}rp^{k-1},l_1\right) \ ,$$

i.e., $\xi_r(0)$ is the expected value of the minimum of two independent Poisson random variables l_0, l_1 with mean $\lambda = \frac{k}{2^k}rp^{k-1}$. Finally, we note that

$$\mathbb{E}\left[\min\{l_0,l_1\}\right] = \sum_{i=0}^{\infty}i\left(2\text{Poi}(\lambda,i)\left(1 - \sum_{j=0}^{i-1}\text{Poi}(\lambda,j)\right) - (\text{Poi}(\lambda,i))^2\right) \ . \tag{11}$$

To compute a rigorous lower bound for (11) one now simply truncates the sum at any desired level of accuracy.

Acknowledgements. We are grateful to Andrea Montanari for a number of useful conversations.

References

1. Achioptas, D., Sorkin, G.: Optimal myopic algorithms for random 3-sat. In: Proceedings of the 41st Annual Symposium on Foundations of Computer Science 2000, pp. 590–600. IEEE (2000)
2. Achlioptas, D.: Lower bounds for random 3-sat via differential equations. Theoretical Computer Science 265(1-2), 159–185 (2001)
3. Achlioptas, D., Beame, P., Molloy, M.: A sharp threshold in proof complexity yields lower bounds for satisfiability search. Journal of Computer and System Sciences 68(2), 238–268 (2004)
4. Achlioptas, D., Coja-Oghlan, A.: Algorithmic barriers from phase transitions. In: 49th Annual IEEE Symp. on Foundations of Computer Science 2008, pp. 793–802 (2008)
5. Achlioptas, D., Menchaca-Mendez, R.: Exponential lower bounds for dpll algorithms on satisfiable random 3-cnf formulas (2012) (to appear in SAT 2012)
6. Bayati, M., Gamarnik, D., Tetali, P.: Combinatorial approach to the interpolation method and scaling limits in sparse random graphs. In: STOC 2010, pp. 105–114 (2010)
7. Chvatal, V., Szemeredi, E.: Many hard examples for resolution. Journal of the Association for Computing Machinery 35(4), 759–768 (1988)
8. Coja-Oghlan, A.: On belief propagation guided decimation for random k-sat. In: Proceedings of the Twenty-Second Annual ACM-SIAM Symposium on Discrete Algorithms, pp. 957–966. SIAM (2011)
9. Díaz, J., Kirousis, L., Mitsche, D., Pérez-Giménez, X.: On the satisfiability threshold of formulas with three literals per clause. Theoretical Computer Science 410(30-32), 2920–2934 (2009)
10. Franz, S., Leone, M.: Replica bounds for optimization problems and diluted spin systems. Journal of Statistical Physics 111(3), 535–564 (2003)
11. Franz, S., Leone, M., Toninelli, F.: Replica bounds for diluted non-poissonian spin systems. Journal of Physics A: Mathematical and General 36, 10967 (2003)
12. Guerra, F., Toninelli, F.: The thermodynamic limit in mean field spin glass models. Communications in Mathematical Physics 230(1), 71–79 (2002)
13. Kaporis, A., Kirousis, L., Lalas, E.: The probabilistic analysis of a greedy satisfiability algorithm. Random Structures & Algorithms 28(4), 444–480 (2006)
14. Mertens, S., Mézard, M., Zecchina, R.: Threshold values of random k-sat from the cavity method. Random Structures & Algorithms 28(3), 340–373 (2006)
15. Mezard, M., Montanari, A.: Information, physics, and computation. Oxford University Press, USA (2009)
16. Monasson, R., Zecchina, R.: Tricritical points in random combinatorics: the-sat case. Journal of Physics A: Mathematical and General 31, 9209 (1998)
17. Monasson, R., Zecchina, R.: Entropy of the K-satisfiability problem. Phys. Rev. Lett. 76, 3881–3885 (1996),
http://link.aps.org/doi/10.1103/PhysRevLett.76.3881
18. Monasson, R., Zecchina, R.: Statistical mechanics of the random k-satisfiability model. Phys. Rev. E 56, 1357–1370 (1997),
http://link.aps.org/doi/10.1103/PhysRevE.56.1357
19. Montanari, A.: Tight bounds for ldpc and ldgm codes under map decoding. IEEE Transactions on Information Theory 51(9), 3221–3246 (2005)
20. Panchenko, D., Talagrand, M.: Bounds for diluted mean-fields spin glass models. Probability Theory and Related Fields 130(3), 319–336 (2004)

The NOF Multiparty Communication Complexity of Composed Functions[*]

Anil Ada[1], Arkadev Chattopadhyay[2], Omar Fawzi[1], and Phuong Nguyen[2]

[1] School of Computer Science, McGill University
{aada,ofawzi}@cs.mcgill.ca
[2] Department of Computer Science, University of Toronto
{arkadev,pnguyen}@cs.toronto.edu

Abstract. We study the k-party 'number on the forehead' communication complexity of composed functions $f \circ g$, where $f : \{0,1\}^n \to \{\pm 1\}$, $g : \{0,1\}^k \to \{0,1\}$ and for $(x_1,\dots,x_k) \in (\{0,1\}^n)^k$, $f \circ g(x_1,\dots,x_k) = f(\dots,g(x_{1,i},\dots,x_{k,i}),\dots)$. We show that there is an $O(\log^3 n)$ cost simultaneous protocol for $\text{SYM} \circ g$ when $k > 1 + \log n$, SYM is any symmetric function and g is *any function*. Previously, an efficient protocol was only known for $\text{SYM} \circ g$ when g is symmetric and "compressible". We also get a non-simultaneous protocol for $\text{SYM} \circ g$ of cost $O((n/2^k)\log n + k\log n)$ for any $k \geq 2$.

In the setting of $k \leq 1 + \log n$, we study more closely functions of the form $\text{MAJORITY} \circ g$, $\text{MOD}_m \circ g$, and $\text{NOR} \circ g$, where the latter two are generalizations of the well-known and studied functions Generalized Inner Product and Disjointness respectively. We characterize the communication complexity of these functions with respect to the choice of g. As the main application of our results, we answer a question posed by Babai et al. (*SIAM Journal on Computing, 33:137–166, 2004*) and determine the communication complexity of $\text{MAJORITY} \circ \text{QCSB}_k$, where QCSB_k is the "quadratic character of the sum of the bits" function.

1 Introduction

The 'number on the forehead' (NOF) model of communication complexity was introduced by Chandra, Furst and Lipton [7] who used it to obtain branching program lower bounds. In this model, k players wish to evaluate a function $F : X_1 \times \cdots \times X_k \to \{\pm 1\}$ on a given input $(x_1,\dots,x_k)$. The input is distributed among the players in a way that Player i sees every x_j for $j \neq i$. This scenario is visualized as x_i being written on the forehead of Player i. In order to compute $F(x_1,\dots,x_k)$, the players communicate by means of broadcasting, according to a protocol which they have agreed upon beforehand. The goal is to compute $F(x_1,\dots,x_k)$ by communicating as few bits as possible. Note that for $k = 2$, this model is equivalent to the standard two player model introduced by Yao [24]. We are mainly interested in the case $X_i = \{0,1\}^n$ for all i. Here, every function can be trivially computed using $n+1$ bits of communication, and protocols of cost at most polylogarithmic in n are considered to be efficient. Deterministic, non-deterministic, randomized and quantum communication complexity models naturally manifest themselves in this setting. The overlap of information among the players

---

[*] A full version can be found online [1].

A. Czumaj et al. (Eds.): ICALP 2012, Part I, LNCS 7391, pp. 13–24, 2012.
© Springer-Verlag Berlin Heidelberg 2012

makes the NOF model interesting, powerful and fruitful in terms of applications. Apart from the aforementioned application in branching programs, this model also has important applications in circuit complexity, proof complexity and pseudorandom generators.

The class ACC^0 represents functions computable by polynomial-size, constant-depth circuits with unbounded fan-in AND, OR, NOT and MOD_m gates. Showing NP is not in ACC^0 is one of the frontiers in complexity theory. It is well known that a function in ACC^0 has a polylog(n) k-party deterministic communication complexity, where k is polylog(n) [12,6]. In fact the protocol is *simultaneous* where all the players, without interacting, speak once to an external referee who determines the output based only on the messages she receives. Proving that a function in NP requires super-polylogarithmic communication in the simultaneous model for polylogarithmic number of players would result in a major breakthrough. Currently no non-trivial lower bound is known for an explicit function for $k = \log n$ and this has proven to be a formidable barrier. Despite intense effort, even the 3 player model is far from being well understood and many important problems that are solved in the 2 player setting remain open in the 3 player setting. For example, in the 3 player setting, there is no known explicit function that is hard in the deterministic model but easy in the randomized model. On the other hand, the *equality* function is a canonical example of such a function in the 2 player setting.

Most of the well known and studied functions in the standard two party as well as the multiparty model have the following 'composed' structure. Let $f : \{0,1\}^n \to \{\pm 1\}$ be a function and $\overrightarrow{g} = (g_1, \ldots, g_n)$ be a vector of functions $g_i : \{0,1\}^k \to \{0,1\}$. Define $f \circ \overrightarrow{g}(x_1, \ldots, x_k) = f(\ldots, g_i(x_{1,i}, x_{2,i}, \ldots, x_{k,i}), \ldots)$, where $x_{j,i}$ denotes the ith coordinate of the n-bit string x_j. When all the g_i are the same, say g, we denote $f \circ \overrightarrow{g}$ by $f \circ g$. In this notation, the famous communication functions *generalized inner product, disjointness* and *hamming distance* can be written as $\text{GIP} = \text{MOD}_2 \circ \text{AND}$, $\text{DISJ} = \text{NOR} \circ \text{AND}$, and $\text{HD} = \text{THR}_t \circ \text{XOR}$ respectively. In an important paper [19], Razborov characterizes the 2 party communication complexity of $\text{SYM} \circ \text{AND}$ functions, where SYM denotes a symmetric function. Shi and Zhang [22] obtain a similar characterization for $\text{SYM} \circ \text{XOR}$ functions. Note that when $k = 2$, AND and XOR are the only interesting "inside functions" as other functions are either trivial or reduce to the case of AND or XOR.

In this paper, we study the multiparty communication complexity of composed functions with two goals in mind. The first goal is to better understand the power of $\log n$ and more players. The second and more general goal is to understand which combinations of the "inside" function g and the "outside" function f lead to hard communication problems and which combinations lead to easy communication problems. The focus of previous research has been on proving lower bounds for composed functions by selecting a "hard" outside function and a convenient inside function (see e.g. [20,23,15,8,5,16]). Our approach is to study composed functions without putting any restriction on g and obtain characterizations for the communication complexity of composed functions with respect to the choice of g. This *dual* approach is particularly interesting in the multiparty setting where the choice for g increases double exponentially in k.

First, we consider $\text{SYM} \circ g$ functions in the setting of $k > \log n$. This rich class contains many interesting functions and it is tempting to conjecture that some of these functions are candidates to break the $\log n$ barrier mentioned earlier. In particular, since the *majority* function $\text{MAJ} = \text{THR}_{n/2}$ is conjectured to be outside of ACC^0, it is of interest to try to determine the communication complexity of $\text{MAJ} \circ g$ for all g. For instance, Babai, Kimmel and Lokam [3] identified $\text{MAJ} \circ \text{MAJ}$ as a candidate function to be hard for more than $\log n$ many players. Later, in a significantly expanded version of [3], Babai et al. [2] show that $\text{MAJ} \circ \text{MAJ}$ has an efficient simultaneous protocol when $k > 1 + \log n$. Their upper bound in fact applies to $\text{SYM} \circ g$ where SYM is any symmetric function and g is any symmetric "compressible" function. Although the class of symmetric compressible functions contains natural functions like THR_t and MOD_m, this class is only a small portion of all symmetric functions as a random symmetric function is not compressible with high probability. Babai et al. [2] in fact identify QCSB, the *quadractic character of the sum of bits* function, as a symmetric inside function g for which their method fails. In this paper, we remove the symmetry and compressibility conditions on g and show that functions of the form $\text{SYM} \circ g$ are easy in the simultaneous model when $k > 1 + \log n$, for *any* choice of the inside function g.

Theorem 1 (Informal statement). For any g, there is a simultaneous deterministic k-party protocol for $\text{SYM} \circ g$ of cost $O(\log^3 n)$ when $k > 1 + \log n$. When $k > 1 + 2\log n$, the simultaneous protocol applies to $\text{SYM} \circ \vec{g}$ for any vector of functions $\vec{g}$. Furthermore, there is a deterministic protocol (non-simultaneous) for $\text{SYM} \circ \vec{g}$ of cost $O((n/2^k)\log n + k \log n)$ for any k.

The above result rules out functions of the form $\text{SYM} \circ g$ as candidates to break the $\log n$ barrier. Moreover, by the well known connections of the multiparty model with Ramsey theory [7], our $k + 1$ party protocol for $\text{NOR} \circ \text{XOR}$ gives the first non-trivial upper bound on the number of colors needed to color $(\mathbb{F}_2^n)^k$ so that no k dimensional *corner* is monochromatic. Although communication complexity bounds have been proven using Ramsey theory, no bounds on Ramsey numbers have been proven via communication complexity bounds before[1].

The insight for our (non-simultaneous) protocols is from the work of Grolmusz and Pudlák. Grolmusz [10] presented an efficient non-simultaneous protocol for the function $\text{SYM} \circ \text{AND}$ when $k \geq \log n$ players. Later, Pudlák [17] gave an elegant reinterpretation of Grolmusz's protocol. We obtain our simultaneous protocols when k is sufficiently large by extending [10,17] and combining it with a result of Babai et al [2].

In the setting of $k \leq \log n$, we study more closely functions of the form $\text{MAJ} \circ g$, $\text{MOD}_m \circ g$ and $\text{NOR} \circ g$, where the latter two are generalizations of arguably the most well known and studied functions GIP and DISJ respectively. We are able to obtain dichotomies, with respect to the choice of g, that characterize the communication complexity of $\text{MAJ} \circ g$, $\text{MOD}_m \circ g$ and $\text{NOR} \circ g$ for every g. Furthermore, our results show that these functions have polynomially related quantum and classical communication complexities[2]. It is worth noting that these characterizations are tightly connected to our upper bound result mentioned above. The upper bounds for these functions in the setting of $k \leq \log n$ use crucially the ideas developed for the upper bound for $\text{SYM} \circ g$ in the setting of $k > \log n$. Perhaps surprisingly, even our lower bounds for $\text{MOD}_m \circ g$ functions use these ideas as well. We state our results below.

---

[1] The details of this result are given in the full version of the paper.

[2] Note that by the work of [14], all our lower bounds hold in the quantum model, but we confine ourselves to the classical setting for simplicity.

Theorem 2 (Informal statement). Let $S_0 = \{y \in g^{-1}(1) \;:\; y \text{ has even weight}\}$ and $S_1 = \{y \in g^{-1}(1) \;:\; y \text{ has odd weight}\}$. If m divides $|S_0| - |S_1|$, $\text{MOD}_m \circ g$ has a simultaneous deterministic protocol of cost $O(k \log m)$. On the other hand, if m does not divide $|S_0| - |S_1|$, $\text{MOD}_m \circ g$ is a very hard function[3] in the randomized model, up to $\approx \frac{1}{2} \log n$ many players and m up to $n^{\frac{1}{2} - \delta}$ for a constant $\delta > 0$.

The first strong lower bounds in the NOF model were obtained by Babai, Nisan and Szegedy [4], who showed a very strong lower bound for the GIP function. Their proof is later refined by [9,18]. Grolmusz [11] extended the technique of [4] to show a lower bound for $\text{MOD}_m \circ \text{AND}$. We obtain our lower bound for $\text{MOD}_m \circ g$, where m is coprime to $|S_0| - |S_1|$, by extending the analysis of [9,18]. For other m for which $\text{MOD}_m \circ g$ is hard (i.e. m and $|S_0| - |S_1|$ are not coprime but m does not divide $|S_0| - |S_1|$), this analysis does not apply. In this case, we obtain the lower bound through a reduction to the previous case. This reduction uses ideas from our upper bound for $\text{SYM} \circ g$.

Theorem 3 (Informal statement). If $|S_0| = |S_1|$, then $\text{MAJ} \circ g$ has a k-party simultaneous deterministic protocol of cost $O(k \log n)$. If $|S_0| \neq |S_1|$, $\text{MAJ} \circ g$ is hard in the randomized bounded error model for k up to $\approx \frac{1}{2} \log n$

The above result is obtained by using our characterization for $\text{MOD}_m \circ g$. As immediate applications, we show for instance that $\text{MAJ} \circ \text{MAJ}$ and $\text{MAJ} \circ \text{XOR}$ are hard in the randomized model for k up to $\approx \frac{1}{2} \log n$.

Theorem 4 (Informal statement). $\text{NOR} \circ g$ is hard in the randomized bounded error model for k up to $\approx \frac{1}{2} \log n$ many players if and only if g has support size 1.

This result shows that the hardness of DISJ crucially relies on the fact that g has singleton support. The lower bound is obtained by a simple reduction and follows from the best known lower bound on $\text{DISJ} = \text{NOR} \circ \text{AND}$ [21]. An important ingredient in our upper bound is the use of our characterization for $\text{MOD}_m \circ g$.

As an application of our $\text{MAJ} \circ g$ characterization (Theorem 3) and our protocol for $\text{SYM} \circ g$ functions (Theorem 1), we answer an open question posed by Babai et al. [2] and determine the communication complexity of $\text{MAJ} \circ \text{QCSB}$. Recall that the techniques of Babai et al. fail for QCSB as it is a non-compressible function.

Corollary 1 (Informal statement). If $k \equiv 1 \mod 4$, $\text{MAJ} \circ \text{QCSB}_k$ has cost $O(k \log n)$ in the simultaneous deterministic model, and if $k \equiv 3 \mod 4$, the function is hard in the randomized model for up to $\approx \frac{1}{2} \log n$ many players with. For $k > 1 + \log n$, $\text{MAJ} \circ \text{QCSB}$ has cost $O(\log^3 n)$ in the simultaneous deterministic model.

2 Preliminaries

We refer the reader to [13] for details about the communication complexity models discussed in this paper. For $F : X_1 \times \cdots \times X_k \to \{\pm 1\}$, we denote by $\mathbf{D}_k(F)$, $\mathbf{D}_k^{\|}(F)$

---

[3] Here 'very hard' means that even if the error probability of the protocol is allowed to be exponentially close to 1/2, the function does not have an efficient protocol. Note that achieving error probability 1/2 is trivial for any function.

and $\mathbf{R}_k^{\varepsilon}(F)$ the k-party deterministic, simultaneous deterministic and randomized ε-error communication complexities of F respectively. A stronger model allowing quantum communication between the players can similarly be defined, and in fact, all the lower bounds in the randomized model that we prove here carry over to the quantum model using the results of [14].

A subset C_i of $X_1 \times \cdots \times X_k$ is a cylinder in the ith direction if membership in C_i does not depend on the ith coordinate, i.e., if $(x_1, \ldots, x_i, \ldots, x_k) \in C_i$, then $(x_1, \ldots, x_i', \ldots, x_k) \in C_i$ for every $x_i' \in X_i$. A cylinder intersection C is an intersection of k cylinders, one in each direction. It is well known that a k-party deterministic protocol for F of cost c partitions the input space into at most 2^c monochromatic (with respect to F's output) cylinder intersections. We identify a cylinder intersection $C \subseteq X_1 \times \cdots \times X_k$ with its characteristic function $C : X_1 \times \cdots \times X_k \to \{0,1\}$.

We define the discrepancy of $F : X_1 \times \cdots \times X_k \to \mathbb{C}$ under μ and with respect to a cylinder intersection C as $\mathrm{disc}_\mu(F,C) = \left| \mathbf{E}_{x \sim \mu} [F(x)C(x)] \right|$. The discrepancy of F under μ is $\mathrm{disc}_\mu(F) = \max_C \mathrm{disc}_\mu(F,C)$, where the maximum is over all possible cylinder intersections C. By the well-known discrepancy method:

$$\mathbf{R}_k^{\varepsilon}(F) \geq \log \left(\frac{1-2\varepsilon}{\mathrm{disc}_\mu(F)} \right). \tag{1}$$

In order to upper bound the discrepancy we will use the *cube measure*. Let μ be a product distribution over $X_1 \times \cdots \times X_k$, i.e., $\mu(x_1, \ldots, x_k) = \mu_1(x_1) \cdots \mu_k(x_k)$, where μ_i is a distribution over X_i. We define the cube measure of F under μ as

$$\mathcal{E}_\mu(F) = \mathbf{E}_{\substack{x_1^0, x_2^0, \ldots, x_k^0 \\ x_1^1, x_2^1, \ldots, x_k^1}} \left[\prod_{u \in \{0,1\}^k} C^{u_1 + \cdots + u_k}(F(x_1^{u_1}, \ldots, x_k^{u_k})) \right],$$

where in the expectation, x_i^0 and x_i^1 are distributed according to μ_i, and C denotes the complex conjugation operator: $C^b(z) = z$ if b is even, and $C^b(z) = \bar{z}$ otherwise. It is not difficult to verify that the cube measure is always a non-negative real number. In fact, the quantity $(\mathcal{E}_{\mathcal{U}}(F))^{1/2^k}$, where $\mathcal{U}$ is the uniform distribution, is known as the *hypergraph uniformity norm* and is a measure of "quasirandomness" of F. When $F(x_1, \ldots, x_k) = f(x_1 \oplus \cdots \oplus x_k)$, the hypergraph uniformity norm of F corresponds to Gowers uniformity norm of f over $\mathbb{F}_2^n$.

Lemma 1 ([9,18]). *Let $F : X_1 \times \cdots \times X_k \to \mathbb{C}$ be a complex valued function and μ_i a distribution over X_i. Define μ as the product of the μ_i. Then, $\mathrm{disc}_\mu(F) \leq (\mathcal{E}_\mu(F))^{1/2^k}$.*

In this paper $X_i = \{0,1\}^n$ for all i. We let $x = (x_1, \ldots, x_k)$ denote an input in $(\{0,1\}^n)^k$. Often we will view the input as a $k \times n$ dimensional matrix X, where the ith row of X is x_i. We reserve the variables x_i to denote an n-bit string whose j-th bit is denoted by $x_{i,j}$, and reserve the variables y_i to denote a single bit. Let $\mathcal{H}_k$ denote the k dimensional hypercube where the vertex set is $\{0,1\}^k$ and there is an edge between two vertices iff their Hamming distance is 1. Given an input in the $k \times n$ dimensional matrix form X, we associate each column of X with the corresponding vertex of $\mathcal{H}_k$. For each $v \in \{0,1\}^k$, define n_v as the number of times v occurs as a column of X.

3 Communication Complexity of Composed Functions

3.1 SYM ∘ g

A boolean function $f : \{0,1\}^n \to \{\pm 1\}$ is called *symmetric* if the output depends only on the Hamming weight of the input. In this section we present our deterministic protocol for $\mathrm{SYM} \circ \vec{g}$ where g is any function. Our result improves upon the result of Babai et al. [2], who give an efficient simultaneous protocol for $\mathrm{SYM} \circ g$, where g is symmetric and compressible, when $k > 1 + \log n$. First, we remove the symmetry and compressibility conditions on g and allow inside function(s) to be selected arbitrarily, and second, we provide a non-trivial protocol even when $k \le 1 + \log n$. We obtain our protocols in the non-simultaneous model by extending the ideas of Grolmusz [10] and Pudlák [17]. We combine this with a beautiful lemma of Babai et al. [2, Lemma 6.10] in order to make our protocols simultaneous:

Lemma 2 ([2]). *Suppose $k > 1 + \log n$ and let X be a $k \times n$ boolean matrix given as an input for a k-party communication problem. Let n_i be the number of columns of X with Hamming weight i. Then by communicating $O(k^2 \log n)$ bits, the players can compute n_i for all i in the simultaneous deterministic model.*

We note that in the following theorem, it will be clear from the proof that allowing different inner functions for different columns is important even to handle functions $f \circ g$ when the number of players $k \gg \log n$.

Theorem 1. *Let $f : \{0,1\}^n \to \{\pm 1\}$ be a symmetric function, $g : \{0,1\}^n \to \{0,1\}$ an arbitrary function, and $\vec{g} = (g_1, \ldots, g_n)$ a vector of n functions where $g_i : \{0,1\}^k \to \{0,1\}$ are arbitrary functions. Then,*

(a) $\mathbf{D}_k(f \circ \vec{g}) \le O((n/2^k) \log n + k \log n)$,
(b) *for $k > 1 + \log n$:* $\mathbf{D}_k^{\|}(f \circ g) \le O(\log^3 n)$,
(c) *for $k > 1 + 2 \log n$:* $\mathbf{D}_k^{\|}(f \circ \vec{g}) \le O(\log^3 n)$.

Proof. We first prove part (a). Fix an input for $f \circ \vec{g}$ given in $k \times n$ matrix form X. The protocol proceeds in two steps. In the first step, the players determine the column positions of some $u \in \mathcal{H}_k$. Later, they use this to compute the output of $f \circ \vec{g}$.

We now describe the first step. Let $X^{\ge 3}$ denote the $(k-2) \times n$ dimensional submatrix of X where the first two rows are deleted. Since $X^{\ge 3}$ has n columns and there are 2^{k-2} possible strings of length $k - 2$, the string $s \in \{0,1\}^{k-2}$ that appears the least number of times as a column of $X^{\ge 3}$ appears at most $n/2^{k-2}$ times. Without any communication, players 1 and 2 can agree on this string (breaking ties in say lexicographical order). Player 2, using at most $n/2^{k-2}$ bits of communication, can send player 1 the bits on player 1's forehead corresponding to the positions that string s appears. With this information, player 1 knows the positions of four vertices $00s$, $01s$, $10s$ and $11s$. Now player 1 can announce one of these vertices (call it u) and the column indices corresponding to u. The total cost is at most $O((n/2^k) \log n)$.

We proceed to step 2. Observe that the columns corresponding to u are taken care of, that is, we already know the value $g_j(u)$ where j is a column index corresponding to u. Let $S_j = g_j^{-1}(1)$. For $v \in \{0,1\}^k$, let $\mathbf{1}_j(v) = 1$ if v is in column j, and $\mathbf{1}_j(v) = 0$ otherwise. To compute the output of $f \circ \overrightarrow{g}$, it suffices to compute

$$\sum_j \sum_{v \in S_j} \mathbf{1}_j(v), \tag{2}$$

where the outer sum is over all column indices that do not correspond to u. Consider a shortest path from v to u: $v = w_1, w_2, \ldots, w_t = u$. Observe that since $\mathbf{1}_j(u) = 0$,

$$\mathbf{1}_j(v) = \sum_{i=1}^{t-1} (-1)^{i+1} (\mathbf{1}_j(w_i) + \mathbf{1}_j(w_{i+1})). \tag{3}$$

Each term $(\mathbf{1}_j(w_i) + \mathbf{1}_j(w_{i+1}))$ is known by some player because w_i and w_{i+1} differ only in one coordinate. To compute (2), each player announces her part of the sum. Since $\sum_j \sum_{v \in S_j} \mathbf{1}_j(v) \le n$, it suffices for players to send their part of the sum modulo $n+1$. Therefore this step of the protocol has cost at most $k \cdot \lceil \log(n+1) \rceil$. This completes the proof of part (a). Note that the second step of the protocol is simultaneous while the first step is not. When k is sufficiently large, we bypass the first step using Lemma 2.

We now prove part (c). Let $\ell = 2 + 2\log n$. Only the first ℓ players will speak. For each column j, the rows $\ell + 1$ to k naturally induce a function $g'_j : \{0,1\}^\ell \to \{0,1\}$; $g'_j(u) = g_j(u \cdot v)$ where $v \in \{0,1\}^{k-\ell}$ appears in column j from row $\ell + 1$ to k. Thus our task reduces to finding a protocol for $f \circ \overrightarrow{g'}$ with ℓ players. From now on we drop the superscript in g'_j and denote it by g_j.

As before we are interested in computing

$$\sum_{j=1}^n \sum_{v \in S_j} \mathbf{1}_j(v). \tag{4}$$

Let $\overrightarrow{0}$ be the all 0 vertex. Let $v \in S_j$ and let $v = w_1, \ldots, w_t = \overrightarrow{0}$ be a shortest path between v and $\overrightarrow{0}$. Then we have

$$\mathbf{1}_j(v) = \sum_{i=1}^{t-1} (-1)^{i+1} (\mathbf{1}_j(w_i) + \mathbf{1}_j(w_{i+1})) + (-1)^{|v|} \mathbf{1}_j(\overrightarrow{0}). \tag{5}$$

Substitute (5) into (4). Since the quantity in (4) is at most n, we can do arithmetic modulo $n+1$. As before, each term $(\mathbf{1}_j(w_i) + \mathbf{1}_j(w_{i+1}))$ in the sum is known to a player so the part of the sum involving these terms can be computed by the players using at most $\ell \cdot \lceil \log(n+1) \rceil$ bits. For each $j \in \{1, \ldots, n\}$, we group the terms involving $\mathbf{1}_j(\overrightarrow{0})$ when substituting (5) into (4) and let c_j be the coefficient of $\mathbf{1}_j(\overrightarrow{0})$ modulo $n+1$. We also need to compute $\sum_j c_j \mathbf{1}_j(\overrightarrow{0})$, which can be done as follows. From the original $\ell \times n$ input matrix X, we create a new matrix X' by duplicating the jth column c_j many times. Note that X' has at most n^2 columns so we can apply Lemma 2 on X' to compute the number of all 0 columns in X', which is exactly what we want. This step has cost $O(\log^3 n)$. So putting things together, we can compute (4) with at most $O(\log^3 n)$ bits of communication. The whole protocol is easily seen to be simultaneous. This completes the proof of part (c).

We conclude with the proof of part **(b)**. The strategy is exactly the same as above. We need to calculate $\sum_j c_j \mathbf{1}_j(\vec{0})$. Since all the g_j are the same, $c_j = c$ for all j for some c. So we want to compute $c\sum_j \mathbf{1}_j(\vec{0})$, which is precisely $cn_{\vec{0}}$. We can compute $n_{\vec{0}}$ using Lemma 2 when $k > 1 + \log n$. So putting things together, we can compute (4) using at most $O(k^2 \log n)$ bits of communication. Given part **(c)**, we are done.

3.2 $\mathrm{MOD}_m \circ g$

For $(y_1, y_2, \ldots, y_n) \in \{0,1\}^n$, let $\mathrm{MOD}_m(y_1, y_2, \ldots, y_n) = -1$ iff $\sum_{j=1}^n y_j = 0 \mod m$. In this section we show that the complexity of $\mathrm{MOD}_m \circ g$ is determined by the quantity $\big||S_0| - |S_1|\big|$, where S_i is the subset of the support of g that consists of all inputs whose Hamming weight has parity i. Part **(b)** of Theorem 2 is important because it will be used to derive the lower bound in the next subsection.

Theorem 2. *Let $m \geq 2$ be an integer. The function $\mathrm{MOD}_m \circ g$ satisfies:*

(a) *If m divides $|S_0| - |S_1|$, then $\mathbf{D}_k^{\|}(\mathrm{MOD}_m \circ g) \leq k\lceil \log m\rceil$.*
(b) *Otherwise, $\mathbf{R}_k^\varepsilon(\mathrm{MOD}_m \circ g) \geq \frac{5n}{m^2 4^k} + \log(1 - 2\varepsilon) - (k+1)\lceil \log m\rceil - 1$.*

Before sketching the proof, we first state a lemma which is essential for our protocols here and in the next subsection.

Lemma 3. *Let $S_0 = \{u_1, \ldots, u_r\}$ and $S_1 = \{v_1, \ldots, v_r\}$ be subsets of the vertices of $\mathcal{H}_k$ such that for each i, the distance between u_i and v_i is odd. Then $\sum_{i=1}^r n_{u_i} + \sum_{i=1}^r n_{v_i} \mod m$ can be computed in the simultaneous model using at most $k \cdot \lceil \log m\rceil$ bits. Similarly, if the distance between u_i and v_i is even for each i, $\sum_{i=1}^r n_{u_i} - \sum_{i=1}^r n_{v_i} \mod m$ can be computed in the simultaneous model using at most $k \cdot \lceil \log m\rceil$ bits.*

Proof. Using the notation in the proof of Theorem 1, note that we are interested in computing $\sum_{i=1}^r \sum_{j=1}^n \mathbf{1}_j(u_i) + \mathbf{1}_j(v_i) \mod m$. Each term $(\mathbf{1}_j(u_i) + \mathbf{1}_j(v_i))$ can be written as a telescoping sum as in (3). Each term in the telescoping sum is known by a player. Since we can do arithmetic modulo m, the desired value can be computed with each player sending their part of the sum modulo m. So the total cost is $k \cdot \lceil \log m\rceil$. The second part holds similarly. □

Proof (Proof Sketch of Theorem 2). **Part (a):** Suppose m divides $|S_0| - |S_1|$ and assume without loss of generality that $|S_0| \geq |S_1|$. We choose (arbitrarily) a subset $S_0' \subseteq S_0$ of size $|S_1|$. As the distance between an element of S_0' and an element of S_1 is odd, we can compute $\sum_{v \in S_0'} n_v + \sum_{v \in S_1} n_v \mod m$ using Lemma 3. For the remaining elements in $S_0 - S_0'$, we pair them with $\vec{0}$. Hence, using Lemma 3 once again, we can compute $(|S_0| - |S_1|)n_{\vec{0}} + \sum_{v \in S_0 - S_0'} n_v \equiv \sum_{v \in S_0 - S_0'} n_v \mod m$. Thus, we have computed $\sum_{v \in S_0 \cup S_1} n_v \mod m$, from which the output of $\mathrm{MOD}_m \circ g$ is determined. Observe that the sums $\sum_{v \in S_0'} n_v + \sum_{v \in S_1} n_v \mod m$ and $\sum_{v \in S_0 - S_0'} n_v \mod m$ need not be computed separately and that we can compute $\sum_{v \in S_0 \cup S_1} n_v \mod m$ in one shot using $k\lceil \log m\rceil$ bits.
Part (b), Case 1: We consider two cases, depending on whether m and $|S_0| - |S_1|$ are coprime or not. The first case is when m and $|S_0| - |S_1|$ are coprime. The proof makes

use of the characterization of the MOD_m function in terms of exponential sums. Fix $2 \leq m \in \mathbb{N}$ and $0 \leq a, b \leq m - 1$. Let $\omega = e^{2\pi i/m}$ be an m-th root of unity. The function $\mathrm{EXP}_m^{a,b}$ is defined as $\mathrm{EXP}_m^{a,b}(y_1, y_2, \ldots, y_n) = \omega^{a((\sum_{j=1}^{n} y_j) - b)}$.

The strategy is as follows. Define $f_m(y_1, \ldots, y_n) = \sum_j y_j \mod m$. First we show that for any cylinder intersection, the fraction of points x in the cylinder intersection that satisfy $f_m \circ g(x) = b$ is roughly (with exponentially small error) $1/m$ for all $b \in \{0, 1, \ldots, m - 1\}$. This step uses an estimate of the cube measure of $\mathrm{EXP}_m^{a,b} \circ g$ under the uniform distribution.

Lemma 4. *Assume m and $|S_0| - |S_1|$ are coprime. For any $a \in \{1, 2, \ldots, m - 1\}$ and $b \in \{0, 1, \ldots, m - 1\}$, $\mathcal{E}_{\mathcal{U}}(\mathrm{EXP}_m^{a,b} \circ g) \leq e^{-8n/(m^2 2^k)}$.*

It is perhaps remarkable that one can obtain a bound on the cube measure as a function of only $|S_0| - |S_1|$ and m. A proof of this lemma can be found in the full version [1, Lemma 3.5]. Define the distribution μ that puts equal weight to all x with $f_m \circ g(x) = 0$ and $f_m \circ g(x) = 1$. All other points get 0 weight. It will easily follow that $\mathrm{disc}_\mu(\mathrm{MOD}_m \circ g)$ is exponentially small and thus the desired lower bound is achieved using the discrepancy method (Inequality (1)).

Part (b), Case 2: It is not hard to show that the above analysis of Case 1 cannot work when m and $|S_0| - |S_1|$ are not coprime. Thus, to get the complete characterization, we need new ideas. For this, we construct a reduction to Case 1 using insights from the protocol of Theorem 1. We can assume without loss of generality that $|S_0| - |S_1| > 0$. Let $1 < d = \gcd(m, |S_0| - |S_1|)$, and let $m = dq$ and $|S_0| - |S_1| = dr$, where q and r are coprime integers. Because m does not divide $|S_0| - |S_1|$, $q \geq 2$. Our strategy is to use a protocol for $\mathrm{MOD}_m \circ g$ in order to construct a protocol for $\mathrm{MOD}_q \circ g'$ for some function g' for which we can apply the lower bound proved in Case 1.

We start by partitioning the set S_0 into sets $S_0', T_1, \ldots, T_d$ with $|S_0'| = |S_1|$ and $|T_1| = \cdots = |T_d| = r$. Let g' be the function whose support is T_1. Note that the support of g' has size r and consists only of inputs of even Hamming weight. So we can apply the lower bound of Case 1 to $\mathrm{MOD}_q \circ g'$.

Using a protocol for $\mathrm{MOD}_m \circ g$, we will construct a protocol for $\mathrm{MOD}_q \circ g'$ as follows. Fix an input $X \in \{0, 1\}^{k \times n'}$ in matrix form. Recall that for each $v \in \{0, 1\}^k$, n_v denotes the number of occurrences of v as a column of X. First, using Lemma 3 we can compute $\sum_{v \in S_0' \cup S_1} n_v \mod m$ using $k \lceil \log m \rceil$ bits of communication. Again using Lemma 3, for any $\ell \in \{2, \ldots, d\}$, the difference $\sum_{v \in T_\ell} n_v - \sum_{v \in T_1} n_v \mod m$ can also be computed at a cost of $k \lceil \log m \rceil$ bits. As a result, we can compute

$$\sum_{v \in S_0' \cup S_1} n_v + \sum_{\ell=2}^{d} \left(\sum_{v \in T_\ell} n_v - \sum_{v \in T_1} n_v \right) \equiv \sum_{v \in S} n_v - d \sum_{v \in T_1} n_v \mod m.$$

Let $s = s(X)$ denote this number. Observe that $\sum_{v \in T_1} n_v \equiv 0 \mod q$ iff $d \sum_{v \in T_1} n_v \equiv 0 \mod m$. So $\sum_{v \in T_1} n_v \equiv 0 \mod q$ iff $\sum_{v \in S} n_v \equiv s \mod m$. The latter can be determined by running the protocol for $\mathrm{MOD}_m \circ g$ on the input which is obtained from X (viewed as an $k \times n'$ array) by appending $m - s$ columns all of which belong to S.

In short, the protocol for $\mathrm{MOD}_q \circ g'$ on inputs from $(\{0, 1\}^{n'})^k$ consists of two steps: First, the players compute s. Then they simulate the protocol for $\mathrm{MOD}_m \circ g$ on the input of size $(\{0,1\}^n)^k$ specified above, where $n = n' + (m - s)$. A lower bound for $\mathrm{MOD}_m \circ g$ then follows from Case 1.

3.3 MAJ $\circ g$

For each $n \geq 1$, the *majority* function $\mathrm{MAJ}^n : \{0,1\}^n \to \{-1,1\}$ is defined as $\mathrm{MAJ}^n(y_1,\ldots,y_n) = -1$ iff $\sum_i y_i \geq n/2$. When no confusion arises we drop the superscript n from MAJ^n. It is not difficult to show that $\mathrm{MAJ} \circ g$ cannot be much easier than $\mathrm{SYM} \circ g$:

Proposition 1. *Let* $g : \{0,1\}^k \to \{0,1\}$ *be a boolean function and* $f : \{0,1\}^n \to \{-1,1\}$ *be a symmetric function on* n *variables. For any* $\varepsilon \geq 0$, $\mathbf{R}_k^{\varepsilon'}(f \circ g) \leq \mathbf{R}_k^\varepsilon(\mathrm{MAJ}^{2n} \circ g) \cdot \lceil \log(n+1) \rceil$, *where* $\varepsilon' = \varepsilon \lceil \log(n+1) \rceil$.

We can combine Proposition 1 with our lower bounds for $\mathrm{MOD}_m \circ g$ functions (Theorem 2) to obtain a characterization for the complexity of $\mathrm{MAJ} \circ g$ for every g.

Theorem 3. *Let* $g : \{0,1\}^k \to \{0,1\}$ *be a boolean function and* S *be its support. The function* $\mathrm{MAJ} \circ g$ *satisfies:*

- *If* $|S_0| = |S_1|$, *then* $\mathbf{D}_k^{\|}(\mathrm{MAJ} \circ g) \leq k \cdot \lceil \log(n+1) \rceil$.
- *Otherwise,* $\mathbf{R}_k^{1/3}(\mathrm{MAJ} \circ g) \geq \Omega\left(\frac{n}{(k \log k)^2 \cdot 4^k \log n \log \log n} \right)$.

Theorem 3 can be used to determine the communication complexity of a class of functions considered by Babai et al. [2]. For an odd prime k, define the function $\mathrm{QCSB}_k :$ $\{0,1\}^k \to \{0,1\}$ by $\mathrm{QCSB}_k(y_1,\ldots,y_k) = 1$ if and only if $y_1 + \cdots + y_k$ is a quadratic residue modulo k. Recall that $z \in \mathbb{F}_k$ is a quadratic residue if there exists $a \in \mathbb{F}_k$ such that $z = a^2$. The authors of [2] prove that QCSB_k is not 'compressible', so their protocol for $k > 1 + \log n$ does not apply for $\mathrm{SYM} \circ \mathrm{QCSB}_k$. They leave as an open question the problem of finding good bounds for the communication complexity of the function $\mathrm{MAJ} \circ \mathrm{QCSB}_k$. The following corollary completely determines the hardness of this function for any number of players, except in the range between $\approx \frac{1}{2} \log n$ and $\log n$.

Corollary 1 (Answers Babai et al. [2]). *Let* k *be an odd prime.*

- *If* $k \equiv 1 \mod 4$, *then* $\mathbf{D}_k^{\|}(\mathrm{MAJ} \circ \mathrm{QCSB}_k) \leq O(k \log n)$.
- *If* $k \equiv 3 \mod 4$, *then* $\mathbf{R}_k^{1/3}(\mathrm{MAJ} \circ \mathrm{QCSB}_k) \geq \Omega\left(\frac{n}{(k \log k)^2 4^k \log n \log \log n} \right)$.
- *If* $k > 1 + \log n$, *then* $\mathbf{D}_k^{\|}(\mathrm{MAJ} \circ \mathrm{QCSB}_k) \leq O(\log^3 n)$.

3.4 NOR $\circ g$

We obtain a simple and perhaps surprising characterization for the k-player randomized communication complexity of $\mathrm{NOR} \circ g$, where $\mathrm{NOR}(y_1,\ldots,y_n) = -1$ iff $(y_1,\ldots,y_n) = (0,\ldots,0)$. In a very recent paper, Sherstov [21] significantly improves on the bounds of [15],[8] and [5] on the multiparty bounded error communication complexity of disjointness: $\mathbf{R}_k^{1/3}(\mathrm{DISJ}) \geq \Omega\left(\frac{n}{4^k} \right)^{1/4}$. First we observe that this lower bound applies - via a simple reduction - to NOR $\circ$ g when g's support size is 1. We complement this with an efficient randomized protocol for NOR $\circ$ g when g's support size is more than one. The main ingrediants of the upper bound is Theorem 2 together with random sampling.

Theorem 4. *Let* $g : \{0,1\}^k \rightarrow \{0,1\}$ *be a boolean function and* $S = \{y \in \{0,1\}^k : g(y) = 1\}$ *be its support.*

- *If* $|S| = 1$, $\mathbf{R}_k^{1/3}(\text{NOR} \circ g) \geq \Omega \left(\frac{n}{4^k}\right)^{1/4}$,
- *Otherwise,* $\mathbf{R}_k^{\varepsilon}(\text{NOR} \circ g) \leq O(k)$ *for a constant* ε.

4 Conclusion

The most well-studied communication problems like GIP, set-disjointness have a composed structure with an outer function f and an inner function g. Recently, this structure has been exploited by several authors to prove hardness in the NOF model. A natural question that arises is what combination of f and g results in hardness. Almost all previous work focused on fixing the inner function g with a convenient property that allows one to prove hardness for a range of outer functions f. In this work, we address the dual and natural problem of studying families of functions that arise from varying the inner function g. We obtain complete characterizations of hard and easy functions in three of these families: MAJ $\circ$ g, MOD$_m$ $\circ$ g and NOR $\circ$ g. Our characterizations show that hard functions in each of these families, somewhat unexpectedly, exhibit simple and elegant structure.

A key component of our characterization is a new simultaneous protocol for SYM $\circ$ g that is efficient for every g, when the number of players is more than $\log n$. This rules out the possibility of composing a symmetric function with any inner function to take us past the $\log n$ barrier for proving strong lower bounds. To the best of our knowledge, such an impossibility was not known before. In particular, Babai et. al., ten years ago, posed an open problem of determining the communication complexity of the function MAJ $\circ$ QCSB$_k$, where QCSB$_k$ is the quadratic residuosity function. Combining our protocol for SYM $\circ$ g with our characterization of MAJ $\circ$ g, we are able to completely answer this question. While this may sound as a setback to the hope of going past the $\log n$ barrier, it highlights the importance of considering *block composition* where the inner function acts on a block of columns rather than one column as presented in this paper. We end this discussion by pointing out an open problem: Is there an inner function g that acts on two columns such that MAJ $\circ$ g is hard for more than $\log n$ players?

References

1. Ada, A., Chattopadhyay, A., Fawzi, O., Nguyen, P.: The NOF Multiparty Communication Complexity of Composed Functions. Technical report, In Electronic Colloquium on Computational Complexity (ECCC) TR11–155 (2011)
2. Babai, L., Gál, A., Kimmel, P.G., Lokam, S.V.: Communication complexity of simultaneous messages. SIAM Journal on Computing 33, 137–166 (2004)

3. Babai, L., Kimmel, P.G., Lokam, S.V.: Simultaneous Messages vs. Communication. In: Mayr, E.W., Puech, C. (eds.) STACS 1995. LNCS, vol. 900, pp. 361–372. Springer, Heidelberg (1995)
4. Babai, L., Nisan, N., Szegedy, M.: Multiparty protocols, pseudorandom generators for logspace, and time-space trade-offs. J. Comput. Syst. Sci. 45(2), 204–232 (1992)
5. Beame, P., Huynh-Ngoc, D.-T.: Multiparty communication complexity and threshold circuit size of AC^0. In: Proceedings of the 2009 50th Annual IEEE Symposium on Foundations of Computer Science, FOCS 2009, pp. 53–62. IEEE Computer Society, Washington, DC (2009)
6. Beigel, R., Tarui, J.: On ACC. Computational Complexity 4, 350–366 (1994)
7. Chandra, A.K., Furst, M.L., Lipton, R.J.: Multi-party protocols. In: Proceedings of the Fifteenth Annual ACM Symposium on Theory of Computing, STOC 1983, pp. 94–99. ACM, New York (1983)
8. Chattopadhyay, A., Ada, A.: Multiparty communication complexity of disjointness. Technical report. In: Electronic Colloquium on Computational Complexity (ECCC) TR08–002 (2008)
9. Chung, F.R.K., Tetali, P.: Communication complexity and quasi randomness. SIAM Journal on Discrete Mathematics 6(1), 110–123 (1993)
10. Grolmusz, V.: The BNS lower bound for multi-party protocols is nearly optimal. Information and Computation 112, 51–54 (1994)
11. Grolmusz, V.: Separating the communication complexities of MOD m and MOD p circuits. In: Proceedings of the 33rd Annual Symposium on Foundations of Computer Science (FOCS), pp. 278–287 (1995)
12. Håstad, J., Goldmann, M.: On the power of small-depth threshold circuits. Computational Complexity 1, 610–618 (1991)
13. Kushilevitz, E., Nisan, N.: Communication Complexity. Cambridge university press (1997)
14. Lee, T., Schechtman, G., Shraibman, A.: Lower bounds on quantum multiparty communication complexity. In: Proceedings of the 24th Annual IEEE Conference on Computational Complexity (CCC), pp. 254–262 (2009)
15. Lee, T., Shraibman, A.: Disjointness is hard in the multiparty number-on-the-forehead model. Computational Complexity 18, 309–336 (2009)
16. Lee, T., Zhang, S.: Composition Theorems in Communication Complexity. In: Abramsky, S., Gavoille, C., Kirchner, C., Meyer auf der Heide, F., Spirakis, P.G. (eds.) ICALP 2010. LNCS, vol. 6198, pp. 475–489. Springer, Heidelberg (2010)
17. Pudlák, P.: Personal communication (2006)
18. Raz, R.: The BNS-Chung criterion for multi-party communication complexity. Computational Complexity 9(2), 113–122 (2000)
19. Razborov, A.: Quantum communication complexity of symmetric predicates. Izvestiya: Mathematics 67(1), 145–159 (2003)
20. Sherstov, A.A.: The pattern matrix method for lower bounds on quantum communication. In: Proceedings of the 40th Symposium on Theory of Computing (STOC), pp. 85–94 (2007)
21. Sherstov, A.A.: The multiparty communication complexity of set disjointness. Technical report, In Electronic Colloquium on Computational Complexity (ECCC) TR11–145 (2011)
22. Shi, Y., Zhang, Z.: Communication complexities of symmetric XOR functions. Quantum Information and Computation 9, 255–263 (2009)
23. Shi, Y., Zhu, Y.: Quantum communication complexity of block-composed functions. Quantum Information and Computation 9, 444–460 (2009)
24. Yao, A.C.: Some complexity questions related to distributive computing (preliminary report). In: Proceedings of the Eleventh Annual ACM Symposium on Theory of Computing, pp. 209–213. ACM Press, New York (1979)

Quantum Strategies Are Better Than Classical in Almost Any XOR Game[*]

Andris Ambainis[1], Artūrs Bačkurs[1], Kaspars Balodis[1], Dmitrijs Kravčenko[1], Raitis Ozols[1], Juris Smotrovs[1], and Madars Virza[2]

[1] Faculty of Computing, University of Latvia, Raina bulv. 19, Riga, LV-1586, Latvia
{andris.ambainis,Juris.Smotrovs}@lu.lv,
{abackurs,kbalodis,kdmitry}@gmail.com, raitis.ozols@inbox.lv
[2] Computer Science and Artificial Intelligence Laboratory, Massachusetts Institute of Technology, 32 Vassar Street, Cambridge, MA 02139, USA
madars@gmail.com

Abstract. We initiate a study of random instances of nonlocal games. We show that quantum strategies are better than classical for almost any 2-player XOR game. More precisely, for large n, the entangled value of a random 2-player XOR game with n questions to every player is at least 1.21... times the classical value, for $1 - o(1)$ fraction of all 2-player XOR games.

1 Introduction

Quantum mechanics is strikingly different from classical physics. In the area of information processing, this difference can be seen through quantum algorithms which can be exponentially faster than conventional algorithms [27,25] and through quantum cryptography which offers degree of security that is impossible classically [5].

Another information-theoretic way of seeing the difference between quantum mechanics and the classical world is through non-local games. An example of a non-local game is the CHSH (Clauser-Horne-Shimony-Holt) game [10]. This is a game played by two players against a referee. The two players cannot communicate but can share common randomness or a common quantum state that is prepared before the beginning of the game. The referee sends an independent uniformly random bit to each of the two players. Each player responds by sending one bit back to the referee. Players win if $x \oplus y = i \wedge j$ where i, j are the bits that the referee sent to the player and x, y are players' responses. The maximum winning probability that can be achieved is 0.75 classically and $\frac{1}{2} + \frac{1}{2\sqrt{2}} = 0.85...$ quantumly.

There are several reasons why non-local games are interesting. First, CHSH game provides a very simple example to test the validity of quantum mechanics. If we have implemented the referee and the two players A, B by devices so that

---

[*] Supported by ESF project 2009/0216/1DP/1.1.1.2.0/09/APIA/VIAA/044 and FP7 FET-Open project QCS. Full version available as arXiv preprint arXiv:1112.3330.

A. Czumaj et al. (Eds.): ICALP 2012, Part I, LNCS 7391, pp. 25–37, 2012.

there is no communication possible between A and B and we observe the winning probability of 0.85..., there is no classical explanation possible. Second, non-local games have been used in device-independent cryptography [1,26].

Some non-local games show big gaps between the classical and the quantum winning probabilities. For example, Buhrman et al. [8] construct a 2-player quantum game where the referee and the players send values $x, y, i, j \in \{1, \ldots, n\}$ and the classical winning probability is $\frac{1}{2} + \Theta(\frac{1}{\sqrt{n}})$ while the quantum winning probability is 1. In contrast, Almeida et al. [2] construct a non-trivial example of a game in which quantum strategies provide no advantage at all.

Which of those is the typical behaviour? In this paper, we study this question by looking at random instances of non-local games.

More specifically, we study two-party XOR games with uniform distribution of inputs. This is a subclass of non-local games with 2 players, where the referee chooses inputs $i \in \{1, 2, \ldots, n\}$, $j \in \{1, 2, \ldots, k\}$ uniformly at random and sends them to the players. The players reply by sending bits x and y. The rules of the game are specified by an $n \times k$ matrix A whose entries are $+1$ and -1. To win, the players must produce x and y with $x = y$ if $A_{ij} = 1$ and x and y with $x \neq y$ if $A_{ij} = -1$.

We consider the case when the matrix A that specifies the rules of the game is chosen randomly against all ± 1-valued $n \times k$ matrices A. For the case when $n = k$, we show that

- The maximum winning probability p_q that can be achieved by a quantum strategy is $\frac{1}{2} + \frac{1 \pm o(1)}{\sqrt{n}}$ with a probability $1 - o(1)$;
- The maximum winning probability p_{cl} that can be achieved by a classical strategy satisfies

$$\frac{1}{2} + \frac{0.6394... - o(1)}{\sqrt{n}} \leq p_{cl} \leq \frac{1}{2} + \frac{0.8325... + o(1)}{\sqrt{n}}$$

with a probability $1 - o(1)$.

In the literature on non-local games, one typically studies the difference between the winning probability p_q (p_{cl}) and the losing probability $1 - p_q$ $(1 - p_{cl})$: $\Delta_q = 2p_q - 1$ $(\Delta_{cl} = 2p_{cl} - 1)$. The advantage of quantum strategies is then evaluated by the ratio $\frac{\Delta_q}{\Delta_{cl}}$. For random XOR games, our results imply that

$$1.2011... < \frac{\Delta_q}{\Delta_{cl}} < 1.5638...$$

for almost all games. Our computer experiments suggest that, for large n, $\frac{\Delta_q}{\Delta_{cl}} \approx 1.305....$ For comparison, the biggest advantage that can be achieved in any 2-player XOR game is equal to Grothendieck's constant K_G [14] about which we know that [16,23,6]

$$1.67696.... \leq K_G \leq 1.7822139781...$$

Thus, the quantum advantage in random XOR games is comparable to the maximum possible advantage for this class of non-local games.

We find this result quite surprising. Quantum-over-classical advantage usually makes use of a structure that is present in the computational problem (such as the algebraic structure that enables Shor's quantum algorithm for factoring [25]). Such structure is normally not present in random computational problems.

The methods that we use to prove our results are also quite interesting. The upper bounds are easy in both classical and quantum case but both lower bounds are fairly sophisticated. The lower bound for the entangled value requires proving a new version of Marčenko-Pastur law [19] for random matrices.

The classical value of random XOR games is equal to a natural quantity ($l_\infty \to l_1$ norm of a random matrix) that might be interesting for other purposes. The lower bound for it requires a subtle argument that reduces lower-bounding the classical value to analyzing a certain random walk.

Related Work. Junge and Palazuelos [17] and Briet and Vidick [7] have constructed non-local games with a big gap between the quantum (entangled) value and the classical value, via randomized constructions. The difference between this paper and [7,17] is as follows. The goal of [7,17] was to construct a big gap between the entangled value and the classical value of a non-local game and the probability distribution on non-local games and inputs was chosen so that this goal would be achieved.

Our goal is to study the behaviour of non-local games in the case when the conditions are random. We therefore choose a natural probability distribution on non-local games (without the goal of optimizing the quantum advantage) and study it. The surprising fact is that a substantial quantum advantage still exists in such setting.

2   Technical Preliminaries

We use $[n]$ to denote the set $\{1, 2, \ldots, n\}$.

In a 2-player XOR game, we have two players A and B playing against a referee. Players A and B cannot communicate but can share common random bits (in the classical case) or an entangled quantum state (in the quantum case). The referee randomly chooses values $i \in \{1, \ldots, n\}$ and $j \in \{1, \ldots, n\}$ and sends them to A and B, respectively. Players A and B respond by sending answers $x \in \{0, 1\}$ and $y \in \{0, 1\}$ to the referee.

Players win if answers x and y satisfy some winning condition $P(i, j, x, y)$. For XOR games, the condition may only depend on the parity $x \oplus y$ of players' responses. Then, it can be written as $P(i, j, x \oplus y)$.

For this paper, we also assume that, for any i, j, exactly one of $P(i, j, 0)$ and $P(i, j, 1)$ is true. Then, we can describe a game by an $n \times n$ matrix $(A_{ij})_{i,j=1}^n$ where $A_{ij} = 1$ means that, given i and j, players must output x, y with $x \oplus y = 0$ (equivalently, $x = y$) and $A_{ij} = -1$ means that players must output x, y with $x \oplus y = 1$ (equivalently, $x \neq y$).

Let $p_{S,win}$ be the probability that the players win if they use a strategy S and $p_{S,los} = 1 - p_{S,win}$ be the probability that they lose. We will be interested in the difference $\Delta_S = p_{S,win} - p_{S,los}$ between the winning and the losing probabilities. The *classical value* of a game, Δ_{cl}, is the maximum of Δ_S over all classical strategies S. The *entangled value* of a game, Δ_q, is the maximum of Δ_S over all quantum strategies S.

Let p_{ij} be the probability that the referee sends question i to player A and question j to player B. Then [11, section 5.3], the classical value of the game is equal to

$$\Delta_{cl} = \max_{u_1,\ldots,u_n \in \{-1,1\}} \max_{v_1,\ldots,v_n \in \{-1,1\}} \sum_{i,j=1}^{n} p_{ij} A_{ij} u_i v_j. \tag{1}$$

In the quantum case, Tsirelson's theorem [9] implies that

$$\Delta_q = \max_{u_i : \|u_i\|=1} \max_{v_j : \|v_j\|=1} \sum_{i,j=1}^{n} p_{ij} A_{ij} \langle u_i, v_j \rangle \tag{2}$$

where the maximization is over all tuples of unit-length vectors $u_1, \ldots, u_n \in R^d$, $v_1, \ldots, v_n \in R^d$ (in an arbitrary number of dimensions d).

We will assume that the probability distribution on the referee's questions i, j is uniform: $p_{ij} = \frac{1}{n^2}$ and study Δ_{cl} and Δ_q for the case when A is a random Bernoulli matrix (i.e., each entry A_{ij} is $+1$ with probability $1/2$ and -1 with probability $1/2$, independently of other entries).

Other probability distributions on referee's questions can be considered, as well. For example, one could choose y_{ij} to be normally distributed random variables with mean 0 and variance 1 and take $p_{ij} = \frac{|y_{ij}|}{\sum_{i,j=1}^{n} |y_{ij}|}$. Or, more generally, one could start with y_{ij} being i.i.d. random variables from some arbitrary distribution D and define p_{ij} in a similar way.

Most of our results are still true in this more general setting (with mild assumptions on the probability distribution D). Namely, Theorem 1 and the upper bound part of Theorem 4 remain unchanged. The only exception is the lower bound part of Theorem 4 which relies on the fact that the probability distribution p_{ij} is uniform. It might be possible to generalize our lower bound proof to other distributions D but the exact constant in such generalization of our lower bound could depend on the probability distribution D.

3 Quantum Upper and Lower Bound

Theorem 1. *For a random 2-player XOR game with n inputs for each player,*

$$\Delta_q = \frac{2 \pm o(1)}{\sqrt{n}}$$

with probability $1 - o(1)$.

Proof. Because of (2), proving our theorem is equivalent to showing that

$$\max_{\|u_i\|=\|v_j\|=1} \sum_{i=1}^{n}\sum_{j=1}^{n} A_{ij}\langle u_i, v_j\rangle = (2 \pm o(1))n^{3/2}$$

holds with probability $1 - o(1)$.

For the upper bound, we rewrite this expression as follows. Let u be a vector obtained by concatenating all vectors u_i and v be a vector obtained by concatenating all v_j. Since $\|u_i\| = \|v_j\| = 1$, we have $\|u\| = \|v\| = \sqrt{n}$. We have

$$\sum_{i=1}^{n}\sum_{j=1}^{n} A_{ij}\langle u_i, v_j\rangle = \langle u, (A \otimes I)v\rangle \leq \|u\| \cdot \|A \otimes I\| \cdot \|v\| \leq \|A\|n.$$

By known results on operator norms of random matrices [30], $\|A\| = (2+o(1))\sqrt{n}$ with a high probability.

For the lower bound, we note that

$$\max_{\|u_i\|=\|v_j\|=1} \sum_{i=1}^{n}\sum_{j=1}^{n} A_{ij}\langle u_i, v_j\rangle = \max_{\|u_i\|\leq 1,\|v_j\|\leq 1} \sum_{i=1}^{n}\sum_{j=1}^{n} A_{ij}\langle u_i, v_j\rangle.$$

We have

Theorem 2 (Marčenko-Pastur law, [19]). *Let A be a $n \times n$ random matrix whose entries A_{ij} are independent random variables with mean 0 and variance 1. Let $C \in [0,2]$. With probability $1 - o(1)$, the number of singular values λ of A that satisfy $\lambda \geq C\sqrt{n}$ is $(f(C) - o(1))n$ where*

$$f(C) = \frac{1}{2\pi} \int_{x=C^2}^{4} \sqrt{\frac{4}{x} - 1}\, dx.$$

Let $\lambda_1, \ldots, \lambda_m$ be the singular values of A that satisfy $\lambda_i \geq (2 - \epsilon)\sqrt{n}$. With high probability, we have $m \in [(f(2 - \epsilon) - o(1))n, (f(2 - \epsilon) + o(1))n]$. We now assume that this is the case.

Let l_i and r_i be the corresponding left and right singular vectors: $Ar_i = \lambda_i l_i$. (Here, we choose l_i and r_i so that $\|l_i\| = \|r_i\| = 1$ for all i.) Let l_{ij} and r_{ij} be the components of l_i and r_i: $l_i = (l_{ij})_{j=1}^{n}$ and $r_i = (r_{ij})_{j=1}^{n}$.

We define u_j and v_j in a following way:

$$u_j = (l_{ij})_{i=1}^{m}, \quad v_j = (r_{ij})_{i=1}^{m}.$$

We have

$$\sum_{i=1}^{n}\sum_{j=1}^{n} A_{ij}\langle u_i, v_j\rangle = \sum_{i=1}^{n}\sum_{j=1}^{n}\sum_{k=1}^{m} A_{ij}l_{ki}r_{kj}$$

$$= \sum_{k=1}^{m}\langle l_k, Ar_k\rangle = \sum_{k=1}^{m}\lambda_k \geq (2 - \epsilon)m\sqrt{n}. \tag{3}$$

Since $\|l_i\| = \|r_i\| = 1$ and the vectors u_i and v_j are obtained by rearranging the entries of l_i and r_i, we have

$$\sum_{i=1}^{n} \|u_i\|^2 = \sum_{i=1}^{n} \|l_i\|^2 = m$$

and, similarly, $\sum_i \|v_i\|^2 = m$. If u_i and v_i all were of the same length, we would have $\|u_i\|^2 = \|v_i\|^2 = \frac{m}{n}$. Then, replacing u_i and v_i by $u_i' = \frac{u_i}{\|u_i\|}$ and $v_i' = \frac{v_i}{\|v_i\|}$ would increase each vector $\sqrt{\frac{n}{m}}$ times and result in

$$\sum_{i=1}^{n} \sum_{j=1}^{n} A_{ij} \langle u_i', v_j' \rangle \geq (2 - \epsilon) n^{3/2}.$$

To deal with the general case, we will show that almost all u_i and v_i are of roughly the same length. Then, a similar argument will be used. The key to our proof is a new modification of Marčenko-Pastur law.

Theorem 3 (Modified Marčenko-Pastur law). *Let A be an $n \times n$ random matrix whose entries A_{ij} are independent random variables with mean 0 and variance 1. Let $C \in [0,2]$. Let e_i be the ith vector of the standard basis. Let P_C be the projector on the subspace spanned by the right singular vectors with singular values at least $C\sqrt{n}$. Then,*

$$Pr\left[\left| \|P_C e_i\|^2 - f(C) \right| > \epsilon \right] = O\left(\frac{1}{n} \right)$$

with the big-O constant depending on C and ϵ.

The same result also holds for the left singular vectors.

Proof. The proof is given in the full version of the paper.   □

We now complete the proof, assuming the modified Marčenko-Pastur law. Since P_C is spanned by the right singular vectors $r_1, \ldots, r_m$, we have

$$\|P_C e_i\|^2 = \sum_{j=1}^{m} \langle r_j, e_i \rangle^2 = \sum_{j=1}^{m} r_{ji}^2 = \|v_i\|^2. \tag{4}$$

Therefore, the modified Marčenko-Pastur law means that

$$Pr[\|v_i\|^2 > f(2 - \epsilon) + \delta] = O\left(\frac{1}{n} \right).$$

Thus, the expected number of $i \in \{1, \ldots, n\}$ for which $\|v_i\|^2 > f(2 - \epsilon) + \delta$ is $O(1)$. We now apply the following transformations to vectors v_i:

1. For each v_i with $\|v_i\|^2 > f(2 - \epsilon) + \delta$ (or u_i with $\|u_i\|^2 > f(2 - \epsilon) + \delta$), we replace it by the zero vector $\vec{0}$;

2. We replace each v_i by

$$v_i' = \frac{v_i}{\sqrt{f(2-\epsilon)+\delta}}$$

and similarly for u_i.

After the first step $\|v_i\|^2 \leq f(2-\epsilon)+\delta$ for all i. Hence, after the second step, $\|v_i'\|^2 \leq 1$ for all i.

We now bound the effect of those two steps on the sum

$$\sum_{i=1}^{n}\sum_{j=1}^{n} A_{ij}\langle u_i, v_j\rangle.$$

Because of (3), the initial value of this sum is at least

$$(2-\epsilon)m\sqrt{n} \geq (2-\epsilon)(f(2-\epsilon) - o(1))n^{3/2}. \tag{5}$$

Because of (4), $\|v_j\|^2 = \|P_C e_j\|^2 \leq \|e_j\|^2 = 1$. Similarly, $\|u_i\|^2 \leq 1$. Hence, $|\langle u_i, v_j\rangle| \leq 1$ and replacing one v_j (or u_i) by 0 changes the sum by at most $\sum_{i=1}^{n}|A_{ij}| = n$. Replacing $O(1)$ v_j's (or u_i's) changes it by $O(n)$. Since the sum (5) is of the order $\Theta(n^{3/2})$, this is a lower order change.

Replacing v_i's by v_i''s (and u_i's by similarly defined u_i''s) increases each inner product $\langle u_i, v_j\rangle$ $\frac{1}{f(2-\epsilon)+\delta}$ times and achieves

$$\sum_{i=1}^{n}\sum_{j=1}^{n} A_{ij}\langle u_i', v_j'\rangle \geq \frac{(2-\epsilon)(f(2-\epsilon) - o(1))}{f(2-\epsilon)+\delta}n^{3/2}.$$

Since this can be achieved for any fixed $\epsilon > 0$ and $\delta > 0$, we get that

$$\max_{\|u_i'\|\leq 1, \|v_j'\|\leq 1}\sum_{i=1}^{n}\sum_{j=1}^{n} A_{ij}\langle u_i', v_j'\rangle \geq (2 - o(1))n^{3/2}.$$

□

4   Classical Upper and Lower Bound

In the classical case, we have to estimate

$$\Delta_{cl} = \max_{u_1,\ldots,u_n \in \{-1,1\}} \max_{v_1,\ldots,v_n \in \{-1,1\}} \sum_{i,j=1}^{n} A_{ij} u_i v_j. \tag{6}$$

There are several ways how one can interpret this expression and several contexts in which similar quantities have been studied before:

1. (6) is equal to the $l_\infty \to l_1$ norm of A (denoted $\|A\|_{\infty\to1}$). It is known that, for a random matrix A, $\|A\|_{\infty\to1} = \Theta(n\sqrt{n})$ (e.g., from [21] or [18]), but the exact constant under Θ is not known.

2. One can also interpret (6) combinatorially, as a problem of "unbalancing lights" [3]. In this interpretation, $n \times n$ matrix represents an array of lights, with each light being "on" ($A_{ij} = 1$) or "off" ($A_{ij} = -1$). We are allowed to choose a row or a column and switch all lights in this row or column. The task is to maximize the difference between the number of lights that are on and the number of lights that are off. It is known that for any $n \times n$ matrix A with ± 1 entries, (6) is at least $\sqrt{\frac{2}{\pi}} n^{3/2}$ [3, p.19]. We are not aware of any work on evaluating (6) for a random matrix A in this context.

3. In the context of statistical physics, there has been substantial work on determining the order of

$$\max_{u_1,\ldots,u_n \in \{-1,1\}} \sum_{i,j=1}^{n} A_{ij} u_i u_j \tag{7}$$

when A_{ij} is a symmetric Gaussian matrix (each $A_{ij} = A_{ji}$ is an independent Gaussian random variable with mean 0 and variance 1). It is known that (7) is equal to $(1.527\ldots + o(1))n^{3/2}$ with probability $1 - o(1)$. This was first discovered in [24,22] and rigorously proven by Talagrand [29].

The quantities (6) and (7) are of similar flavour but are not identical and there is no clear relation between them.

Theorem 4. *For a random 2-player XOR game, its classical value Δ_{cl} satisfies*

$$\frac{1.2789\ldots}{\sqrt{n}} \leq \Delta_{cl} \leq \frac{2\sqrt{\ln 2} + o(1)}{\sqrt{n}} = \frac{1.6651\ldots + o(1)}{\sqrt{n}}$$

with probability $1 - o(1)$.

This is equivalent to

$$1.2789\ldots n^{3/2} \leq \|A\|_{\infty \to 1} \leq 1.6651\ldots n^{3/2}$$

for a Bernoulli random matrix A.

In computer experiments, the ratio $\frac{\|A\|_{\infty \to 1}}{n^{3/2}}$ grows with n and reaches $1.4519\ldots$ for $n = 26$. By fitting a formula $an^{3/2} + bn$ where the leading term is of the order $n^{3/2}$ and the largest correction term is of the order n to the data, we obtained that

$$\|A\|_{\infty \to 1} \approx 1.53274\ldots n^{3/2} - 0.472806\ldots n.$$

Figure 1 shows the fit. Curiously, the constant in front of $n^{3/2}$ is very close to the constant $1.527\ldots$ for the sum (7). We do not know whether this is a coincidence or there is some connection between the asymptotic behaviour of the two sums.

Proof. The upper bound follows straightforwardly from Chernoff bounds (and is similar to the argument in [18] which provides an upper bound on (6) which holds with probability $1 - O(1/c^n)$). We use the following form of Chernoff inequality:

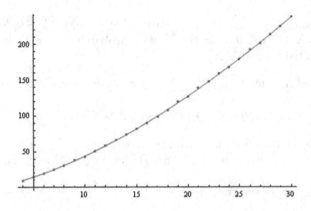

Fig. 1. $\|A\|_{\infty \to 1}$, for random $n \times n$ matrices A

Theorem 5. *[3, p.263] Let $X_1, \ldots, X_n$ be independent random variables with $Pr[X_i = 1] = Pr[X_i = -1] = \frac{1}{2}$ and let $X = X_1 + \ldots + X_n$. Then,*

$$Pr[X \geq a] < e^{-\frac{a^2}{2n}}.$$

Let $x_1, \ldots, x_n \in \{-1, 1\}$ and $y_1, \ldots, y_n \in \{-1, 1\}$ be arbitrary. If $A_{ij} \in \{-1, 1\}$ are uniformly random, then $A_{ij}x_iy_j \in \{-1, 1\}$ are also uniformly random. Hence, $\sum_{i,j} A_{ij}x_iy_j$ is a sum of n^2 uniformly random values from $\{-1, 1\}$. By Theorem 5,

$$Pr\left[\sum_{i,j} A_{ij}x_iy_j > Cn^{\frac{3}{2}}\right] < e^{-\frac{\left(Cn^{\frac{3}{2}}\right)^2}{2n^2}} = \frac{1}{e^{\frac{C^2 n}{2}}}.$$

By taking $C = 2\sqrt{\ln 2} + 2\frac{\sqrt{\ln n}}{\sqrt{n}}$, we can ensure that this probability is less than $\frac{1}{2^{2n}n^2}$. Then, by the union bound, the probability that $\sum_{i,j} A_{ij}x_iy_j > Cn^{\frac{3}{2}}$ for some choice of x_i's and y_j's is less than $2^{2n}\frac{1}{2^{2n}n^2} = \frac{1}{n^2}$.

We now prove the lower bound[1]. We first show

Lemma 1. *Let A be an $n \times n$ random Bernoulli matrix. Then,*

$$E_A\left[\max_{u_i, v_j \in \{-1, 1\}} \sum_{i,j} u_i v_j A_{ij}\right] \geq (1.2789\ldots - o(1))n^{3/2}.$$

---

[1] This lower bound is not necessary for proving the advantage of quantum strategies which follows by combining the classical upper bound and the quantum lower bound. But it is interesting for two other reasons. First, it is necessary to show that, for a random XOR game, $\frac{A_q}{A_{cl}}$ is less than Grothendiek's constant. Second, as discussed at the beginning of this section, the classical value is equal to a natural quantity that comes up in several other settings.

Let $X = \max_{u_i, v_j \in \{-1,1\}} \sum_{i,j} u_i v_j A_{ij}$. By Lemma 1, $E[X] \geq (1.2789... - o(1))n^{3/2}$. To prove that $X \geq (1.2789... - o(1))n^{3/2}$ with probability $1 - o(1)$, we show that X is concentrated around $E[X]$.

Lemma 2. *Let* $X = \max_{u_i, v_j \in \{-1,1\}} \sum_{i,j} u_i v_j A_{ij}$ *for a random* $n \times n$ *matrix* A. *Then,*

$$Pr\left[|X - E[X]| \geq an\right] < 2e^{-a^2/8}.$$

We then apply Lemma 2 with $a = \log n$ (or with $a = f(n)$ for any other $f(n)$ that has $f(n) \to \infty$ when $n \to \infty$ and $f(n) = o(\sqrt{n})$) and combine it with Lemma 1.

It remains to prove the two lemmas.

Proof (of Lemma 1). Let A be a random ± 1 matrix. We choose u_i and v_j, according to Algorithm 1.

Because of the last step, we get that

$$\sum_{i=1}^{n}\sum_{j=1}^{n} u_i v_j A_{ij} = \sum_{j=1}^{n} |S_{n,j}|.$$

Each of $S_{n,j}$ is a random variable with an identical distribution. Hence,

$$E\left[\sum_{i=1}^{n}\sum_{j=1}^{n} u_i v_j A_{ij}\right] = \sum_{j=1}^{n} E|S_{n,j}| = nE|S_{n,1}|. \tag{8}$$

---

1. Set $u_1 = 1$.
2. For each $k = 2, \ldots, n$ do:
   (a) For each $j = 1, \ldots, n$, compute $S_{k-1,j} = \sum_{i=1}^{k-1} A_{ij} u_i$.
   (b) Let $a_k = (Z(S_{k-1,1}), Z(S_{k-1,2}), ..., Z(S_{k-1,n}))$ where $Z(x) = 1$ if $x > 0$, $Z(x) = -1$ if $x < 0$ and $Z(x) = 1$ or $Z(x) = -1$ with equal probability $\frac{1}{2}$ if $x = 0$.
   (c) Let $b_k = (A_{k1}, A_{k2}, ..., A_{kn})$.
   (d) Let $u_k \in \{+1, -1\}$ be such that a_k and $u_k b_k$ agree in the maximum number of positions.
3. For each $j = 1, \ldots, n$, let v_j be such that $v_j S_{n,j} \geq 0$ where $S_{n,j} = \sum_{i=1}^{n} A_{ij} u_i$.

Algorithm 1. Algorithm for choosing u_i and v_j for a given matrix A

We now consider a random walk with a reflecting boundary. The random walk starts at position 0. If it is at the position 0, it always moves to the position 1. If it is at the position $i > 0$, it moves to the position $i+1$ with probability $\frac{1}{2} + \frac{\epsilon}{2}$ and position $i - 1$ with probability $\frac{1}{2} - \frac{\epsilon}{2}$. Let K_i^ϵ be the position of the walker after i steps.

Lemma 3. $|S_{n,1}| = K_n^\epsilon$ *for some* $\epsilon = (1 + o(1))\sqrt{\frac{2}{\pi n}}$.

Proof. $b_i = (A_{i1}, \ldots, A_{in})$ is a vector consisting of random ± 1's that is independent of a_i. Hence, the expected number of agreements between a_i and $u_i b_i$ is $(\frac{1}{2} + \frac{\epsilon}{2})n$ where $\epsilon = (1 + o(1))\sqrt{\frac{2}{\pi n}}$ [3, p.21]. Moreover, the probability of a_i and $u_i b_i$ agreeing in location j is the same for all j.

Hence, if $|S_{i-1,1}| > 0$, we have $|S_{i,1}| = |S_{i-1,1}| + 1$ with probability $\frac{1}{2} + \frac{\epsilon}{2}$ and $|S_{i,1}| = |S_{i-1,1}| - 1$ with probability $\frac{1}{2} - \frac{\epsilon}{2}$. If $|S_{i-1,1}| = 0$, then we always have $|S_{i,1}| = 1$. $\qquad\square$

Lemma 4. *For a random walk with a reflecting boundary and* $\epsilon = \frac{\alpha}{\sqrt{n}}$, *we have* $E[K_n^\epsilon] \geq (f(\alpha) - o(1))\sqrt{n}$ *where*

$$f(\alpha) = \frac{1}{2}\left(e^{-\frac{\alpha^2}{2}}\sqrt{\frac{2}{\pi}} + \alpha + \left(\frac{1}{\alpha} + \alpha\right)\mathrm{Erf}\left(\frac{\alpha}{\sqrt{2}}\right)\right).$$

Proof. The proof is given in the full version of the paper. $\qquad\square$

By combining (8) and Lemmas 3 and 4, the probability of winning minus the probability of losing in the classical case of a random XOR game is at least

$$f\left(\sqrt{\frac{2}{\pi}}\right)\sqrt{n} \cdot n \cdot \frac{1}{n^2} = \frac{2 + 2e^{-1/\pi} + (2 + \pi)\mathrm{Erf}\left(\frac{1}{\sqrt{\pi}}\right)}{2\sqrt{2\pi}}n^{-\frac{1}{2}}$$

$$= 1.2789076012442957...n^{-\frac{1}{2}}.$$

$\qquad\square$

Proof (of Lemma 2). Let

$$f(A_{11}, A_{12}, \ldots, A_{nn}) = \max_{u_i, v_j \in \{-1,1\}} \sum_{i,j} u_i v_j A_{ij}.$$

Then, changing one A_{ij} from $+1$ to -1 (or from -1 to $+1$) changes $\sum_{i,j} u_i v_j A_{ij}$ by at most 2. This means that $f(A_{11}, \ldots, A_{nn})$ changes by at most 2 as well. In other words, f is 2-Lipschitz. By applying Azuma's inequality [20, p. 303-305] with $c = 2$, $t = n^2$, $\lambda = \frac{a}{2}$, we get

$$Pr\left[|f(A_{11}, \ldots, A_{nn}) - E[f(A_{11}, \ldots, A_{nn})]| \geq an\right] < 2e^{-a^2/8}.$$

$\qquad\square$

5   Conclusion

We showed that quantum strategies are better than classical for random instances of XOR games. We expect that similar results may be true for other classes of non-local games.

A possible difficulty with proving them is that the mathematical methods for analyzing other classes of non-local games are much less developed. There is a well developed mathematical framework for studying XOR games [9,11,31] which we used in our paper. But even with that, some of our proofs were quite involved. Proving a similar result for a less well-studied class of games would be even more difficult.

Acknowledgments. We thank Assaf Naor, Oded Regev, Stanislaw Szarek and several anonymous referees for useful comments and references to related work.

References

1. Acin, A., Brunner, N., Gisin, N., Massar, S., Pironio, S., Scarani, V.: Device-independent security of quantum cryptography against collective attacks. Physical Review Letters 98, 230501 (2007)
2. Almeida, M.L., Bancal, J.-D., Brunner, N., Acin, A., Gisin, N., Pironio, S.: Guess your neighbour's input: a multipartite non-local game with no quantum advantage. Physical Review Letters 104, 230404 (2010), also arXiv:1003.3844
3. Alon, N., Spencer, J.: The Probabilistic Method. Wiley (2000)
4. Bai, Z., Silverstein, J.: Spectral Analysis of Large Dimensional Random Matrices. Springer (2010)
5. Bennett, C.H., Brassard, G.: Quantum Cryptography: Public key distribution and coin tossing. In: Proceedings of the IEEE International Conference on Computers, Systems, and Signal Processing, Bangalore, p. 175 (1984)
6. Braverman, M., Makarychev, K., Makarychev, Y., Naor, A.: The Groethendieck constant is strictly smaller than Krivine's bound. In: Proceedings of FOCS 2011, pp. 453–462 (2011)
7. Briet, J., Vidick, T.: Explicit lower and upper bounds on the entangled value of multiplayer XOR games, arxiv: 1108.5647
8. Buhrman, H., Regev, O., Scarpa, G., de Wolf, R.: Near-optimal and explicit Bell inequality violations. In: Proceedings of Complexity 2011, pp. 157–166 (2011); also arxiv: 1012.5403
9. Cirel'son, B. (Tsirelson): Quantum generalizations of Bell's inequality. Letters in Mathematical Physics 4, 93–100 (1980)
10. Clauser, J., Horne, M., Shimony, A., Holt, R.: Physical Review Letters 23, 880–884 (1969)
11. Cleve, R., Höyer, P., Toner, B., Watrous, J.: Consequences and limits of non-local strategies. In: Proceedings of CCC 2004, pp. 236–249 (2004); also quant-ph/0404076
12. Davidson, K., Szarek, S.: Local operator theory, random matrices and Banach spaces. In: Johnson, W.B., Lindenstrauss, J. (eds.) Handbook on the Geometry of Banach Spaces, vol. 1, pp. 317–366. Elsevier (2001)
13. Graham, R.L., Knuth, D.E., Patashnik, O.: Concrete Mathematics, 2nd edn. Addison-Wesley, Reading (1994)
14. Grothendieck, A.: Resume de la theorie metrique des produits tensoriels topologiques. Boletim Sociedade De Matematico de Sao Paulo 8, 1–79 (1953)
15. Grover, L.K.: A fast quantum mechanical algorithm for database search. In: Proceedings of STOC 1996, pp. 212–219 (1996)

16. Krivine, J.-L.: Sur la constante de Grothendieck. Comptes Rendus de l'Académie des Sciences, Series A-B 284, A445–A446 (1977)
17. Junge, M., Palazuelos, C.: Large violation of Bell inequalities with low entanglement. Communications in Mathematical Physics 306(3), 695–746 (2011); arXiv:1007.3043
18. Linial, N., Mendelson, S., Schechtman, G., Shraibman, A.: Complexity measures of sign matrices. Combinatorica 27(4), 439–463 (2007)
19. Marčenko, V.A., Pastur, L.A.: Distribution of eigenvalues for some sets of random matrices. Math. USSR Sbornik 1, 457–483 (1967)
20. Mitzenmacher, M., Upfal, E.: Probability and Computing. Randomized Algorithms and Their Analysis. Cambridge University Press (2005)
21. Montero, A.M., Tonge, A.M.: The Schur multiplication in tensor algebras. Studia Math. 68(1), 1–24 (1980)
22. Parisi, G.: The order parameter for spin glasses: a function on the interval 0-1. Journal of Physics A: Mathemathical and General 13, 1101–1112 (1980)
23. Reeds, J.A.: A new lower bound on the real Grothendieck constant (1991) (unpublished manuscript), http://www.dtc.umn.edu/reedsj/bound2.dvi
24. Sherrington, D., Kirkpatrick, S.: Infinite ranged models of spin glasses. Physical Review B 17, 4384–4403 (1978)
25. Shor, P.W.: Algorithms for quantum computation: Discrete logarithms and factoring. In: FOCS 1994, pp. 124–134. IEEE (1994)
26. Silman, J., Chailloux, A., Aharon, N., Kerenidis, I., Pironio, S., Massar, S.: Fully distrustful quantum cryptography. Physical Review Letters 106, 220501 (2011)
27. Simon, D.R.: On the power of quantum computation. In: FOCS 1994, pp. 116–123. IEEE (1994)
28. Stanley, R.: Enumerative Combinatorics, vol. 2. Cambridge University Press (1999)
29. Talagrand, M.: The generalized Parisi formula. Comptes Rendus de l'Académie des Sciences, Series I 337, 111–114 (2003)
30. Tao, T.: Topics in Random Matrix Theory, Draft of a book, http://terrytao.files.wordpress.com/2011/02/matrix-book.pdf
31. Wehner, S.: Tsirelson bounds for generalized Clauser-Horne-Shimony-Holt inequalities. Physical Review A 73, 022110 (2006)

Efficient Submodular Function Maximization under Linear Packing Constraints[*]

Yossi Azar[1] and Iftah Gamzu[2]

[1] Blavatnik School of Computer Science, Tel-Aviv Univ., Israel
azar@tau.ac.il
[2] Computer Science Division, The Open Univ.,
and Blavatnik School of Computer Science, Tel-Aviv Univ., Israel
iftah.gamzu@cs.tau.ac.il

Abstract. We study the problem of maximizing a monotone submodular set function subject to linear packing constraints. An instance of this problem consists of a matrix $A \in [0,1]^{m \times n}$, a vector $b \in [1, \infty)^m$, and a monotone submodular set function $f : 2^{[n]} \to \mathbb{R}_+$. The objective is to find a set S that maximizes $f(S)$ subject to $A x_S \leq b$, where x_S stands for the characteristic vector of the set S. A well-studied special case of this problem is when f is linear. This special linear case captures the class of packing integer programs.

Our main contribution is an efficient combinatorial algorithm that achieves an approximation ratio of $\Omega(1/m^{1/W})$, where $W = \min\{b_i/A_{ij} : A_{ij} > 0\}$ is the width of the packing constraints. This result matches the best known performance guarantee for the linear case. One immediate corollary of this result is that the algorithm under consideration achieves constant factor approximation when the number of constraints is constant or when the width of the constraints is sufficiently large. This motivates us to study the large width setting, trying to determine its exact approximability. We develop an algorithm that has an approximation ratio of $(1-\epsilon)(1-1/e)$ when $W = \Omega(\ln m/\epsilon^2)$. This result essentially matches the theoretical lower bound of $1 - 1/e$. We also study the special setting in which the matrix A is binary and k-column sparse. A k-column sparse matrix has at most k non-zero entries in each of its column. We design a fast combinatorial algorithm that achieves an approximation ratio of $\Omega(1/(W k^{1/W}))$, that is, its performance guarantee only depends on the sparsity and width parameters.

1 Introduction

Let $f : 2^{[n]} \to \mathbb{R}$ be a set function, where $[n] = \{1, 2, \ldots, n\}$. The function f is called *submodular* if and only if $f(S) + f(T) \geq f(S \cup T) + f(S \cap T)$, for all $S, T \subseteq [n]$. An alternative definition of submodularity is through the property of decreasing marginal values. Given a function $f : 2^{[n]} \to \mathbb{R}$ and a set $S \subseteq [n]$, the function f_S is defined by $f_S(j) = f(S \cup \{j\}) - f(S)$. The value $f_S(j)$ is called the incremental marginal value of

---

[*] This research was supported in part by the Israeli Centers of Research Excellence (I-CORE) program (Center No.4/11), the Israel Science Foundation (grant No. 1404/10), and by the Google Inter-university center. Due to space limitations, some proofs are omitted from this extended abstract. We refer the reader to the full version of this paper (available online at http://arxiv.org/abs/1007.3604), in which all missing details are provided.

A. Czumaj et al. (Eds.): ICALP 2012, Part I, LNCS 7391, pp. 38–50, 2012.
© Springer-Verlag Berlin Heidelberg 2012

element j to the set S. The *decreasing marginal values* property requires that $f_S(j)$ is non-increasing function of S for every fixed j. Formally, it requires that $f_S(j) \geq f_T(j)$ for all $S \subseteq T$. Since the amount of information necessary to convey an arbitrary submodular function may be exponential, we assume a value oracle access to the function. A *value oracle* for the function f allows us to query about the value of $f(S)$ for any set S. Throughout the rest of the paper, whenever we refer to a submodular function, we shall also imply a *normalized* and *monotone* function. Specifically, we assume that a submodular function f also satisfies $f(\emptyset) = 0$ and $f(S) \leq f(T)$ whenever $S \subseteq T$.

In this paper, we focus our attention on the problem (or rather class of problems) of maximizing a monotone submodular set function subject to linear packing constraints. Formally, the input of this problem consists of a matrix $A \in [0,1]^{m \times n}$, a vector $b \in [1, \infty)^m$, and a monotone submodular set function $f : 2^{[n]} \to \mathbb{R}_+$. The objective is to find a set S that maximizes $f(S)$ subject to $Ax_S \leq b$, where x_S stands for the characteristic vector of the set S. We note that the domain restrictions on the entries of A and b are without loss of generality since arbitrary non-negative packing constraints can be reduced to the above form by first eliminating any element j for which there is some constraint i such that $A_{ij} > b_i$, and then scaling the input (see, e.g., the discussion in [26]). A well-studied special setting of our problem is when the objective function f is *linear*, namely, there is a weight vector $c \in \mathbb{R}_+^n$ such that $f(S) = \sum_{j \in S} c_j$. This special setting captures the class of packing integer programs, which models many fundamental combinatorial optimization problems, including maximum independent set, hypergraph matching, and disjoint paths.

Previous Work. There has been a long line of research on maximizing monotone submodular functions subject to matroid and knapsack constraints. Arguably, the most classic scenario is maximizing a submodular function subject to a cardinality constraint, that is, $\max\{f(S) : |S| \leq k\}$. It is known that a simple greedy algorithm achieves an approximation ratio of $1 - 1/e$ for this problem [23]. Furthermore, this result is optimal in two different ways: (i) given only oracle access to f, one cannot attain a better approximation ratio without asking exponentially many value queries [22], and (ii) even if f has a compact representation, it is still NP-hard to obtain a better approximation result [11]. The greedy approach and its variants has been shown to be useful in additional constraint structures [15,19,6,18]. One relevant setting is maximizing a monotone submodular function under a knapsack constraint [30]. A knapsack constraint is essentially a single packing constraint, and may be viewed as the weighted analog of a cardinality constraint. Sviridenko [27] demonstrated that a greedy algorithm with partial enumeration achieves an approximation guarantee of $1 - 1/e$ for this problem.

Another approach that has been proven effective in handling submodular function maximization under different constraint structures is based on approximately solving a continuous fractional relaxation of the problem, followed by pipage or randomized rounding. The pipage rounding technique was originally developed by Ageev and Sviridenko [1], and was adapted to submodular maximization scenarios by Calinescu, Chekuri, Pál and Vondrák [5]. Vondrák [28] utilized the continuous relaxation approach to achieve a tight $(1 - 1/e)$-approximation for maximizing a monotone submodular function subject to a matroid constraint, and Kulik, Shachnai and Tamir [20] used this approach to attain a $(1 - \epsilon)(1 - 1/e)$-approximation for maximizing a monotone submodular function under a constant number of packing constraints. Later on, Chekuri, Vondrák and Zenklusen [8] presented a dependent randomized rounding scheme that can be utilized to extend those results for maximizing a monotone submodular function subject to one matroid and constant number of packing constraints. Recently, Feldman, Naor and Schwartz [14] presented a new unified continuous relaxation approach that finds approximate fractional solutions in both monotone and non-monotone scenarios.

Our Contribution. Our main result is an efficient multiplicative updates algorithm for maximizing a monotone submodular function subject to any number of linear packing constraints. The approximation ratio of our algorithm matches the best known performance guarantee for the special case when the objective function f is linear, which is achieved using the randomized rounding technique [25,24,26]. More precisely, let $W = \min\{b_i/A_{ij} : A_{ij} > 0\}$ be the *width* of the packing constraints, we attain the following result.

Theorem 1. *There is a deterministic polynomial-time algorithm that attains an approximation guarantee of $\Omega(1/m^{1/W})$ for maximizing a monotone submodular function under linear packing constraints.*

It is worth noting that our combinatorial algorithm is deterministic and efficient. Moreover, our technique is different than the two leading approaches used in the past for submodular maximization, namely, the greedy approach and the continuous relaxation approach. Our algorithm is based on a multiplicative updates method (see, e.g., [31,16,2,4]). This method is known to be fruitful for approximately solving problems that can be cast as linear and integer programs. Nevertheless, the analysis of these algorithms relies heavily on primal-dual results, which are not applicable in our submodular setting. We believe that this new approach may be suitable for other submodular optimization problems. We also like to remark that a comparable approximation guarantee may be obtained using the continuous relaxation approach applied with randomized rounding [7]. However, in contrast with that approach, our algorithm is deterministic, efficient and combinatorial. In particular, the continuous relaxation approach runs in polynomial-time but is very far from being practical.

One immediate corollary of Theorem 1 is that the algorithm under consideration achieves a constant factor approximation when the number of constraints is constant or when the width of the packing constraints is sufficiently large, say $W = \Omega(\ln m)$. This motivates us to study the large width setting, trying to determine its exact approximability. The following theorem summarizes our result in this context.

Theorem 2. *There is a deterministic polynomial-time algorithm that achieves an approximation ratio of $(1 - \epsilon)(1 - 1/e)$ for maximizing a monotone submodular function subject to linear packing constraints when $W = \Omega(\ln m/\epsilon^2)$, for any fixed $\epsilon > 0$.*

We note that this result almost matches the theoretical lower bound of $1 - 1/e$, which already holds for maximizing a monotone submodular function subject to a cardinality constraint [23,11]. Specifically, the large width setting captures the hard instances of that problem. We remark that the $(1 - 1/e)$-approximation in the submodular setting stands in contrast with a $(1 + \epsilon)$-approximation which can be achieved by randomized rounding when the objective function is linear and the width is sufficiently large.

We also study the interesting special setting of the problem in which the constraints matrix is binary, namely, $A \in \{0,1\}^{m \times n}$ instead of $A \in [0,1]^{m \times n}$. We demonstrate how to fine-tune our algorithm and its analysis to achieve an improved approximation guarantee of $\Omega(1/m^{1/(W+1)})$. We like to emphasize that this result is optimal unless P = ZPP. Recently, Bansal et al. [3] considered the special case of maximizing a submodular function under k-*column sparse* packing constraints. In this setting, the constraints matrix has at most k non-zero entries in each column. They developed an algorithm whose approximation ratio only depends on the sparsity and width parameters of the input matrix. Specifically, they presented a $\Omega(1/k^{1/W})$-approximation algorithm that employs the continuous relaxation approach in conjunction with randomized rounding and alteration. We make a first step towards attaining their performance guarantee in a deterministic and efficient way. We present a fast combinatorial algorithm for the binary k-column sparse setting whose approximation ratio only depends on the sparsity and width parameters of the input matrix. The following theorem outlines this result.

Theorem 3. *There is a deterministic polynomial-time algorithm that achieves an approximation guarantee of $\Omega(1/(Wk^{1/W}))$ for maximizing a monotone submodular function under binary packing constraints.*

Other Related Work. The problem of maximizing a non-monotone submodular function without any structural constraints is known to be both NP-hard and APX-hard since it generalizes the maximum cut problem. Feige, Mirrokni and Vondrák [12] developed an algorithm whose approximation ratio is 0.4. This result was iteratively improved by Oveis Gharan and Vondrák [17], and then by Feldman, Naor and Shwartz [13] to a ratio of 0.42. Lee, Mirrokni, Nagarajan and Sviridenko [21] presented a $(1/4 - \epsilon)$-approximation algorithm for non-monotone submodular maximization subject to a constant number of packing constraints. This result was iteratively improved by Chekuri, Vondrák and Zenklusen [9], and then by Feldman, Naor and Shwartz [14] to a ratio of $1/e - \epsilon$. Vondrák [29], and very recently, Dobzinski and Vondrák [10] developed general approaches to derive inapproximability results in the value oracle model.

2   Submodular Maximization with Linear Packing Constraints

In this section, we develop a multiplicative updates algorithm for the problem and analyze its performance. An important input parameter of our algorithmic template is an update factor. This parameter plays an essential role in achieving the desired approximation guarantees in the two settings of interest. We first consider the general problem, and demonstrate that there is an update factor for which our algorithm attains an approximation ratio of $\Omega(1/m^{1/W})$. In particular, this implies that the algorithm achieves constant factor approximation for input instances that have a large width, e.g., instances with $W = \Omega(\ln m)$. This motivates us to study this large width setting, trying to determine its exact approximability. We match (up to a disparity of ϵ) the theoretical lower bound of $1 - 1/e$ using a different update factor and a refined analysis.

2.1   The Algorithm

The multiplicative updates algorithm, formally described below, maintains a collection of weights that are updated in a multiplicative way. Informally, these weights capture the extent to which each constraint is close to be violated under a given solution. The algorithm is built around one main loop. In each iteration of that loop, the algorithm extends the current solution with a non-selected element that minimizes a normalized sum of the weights. When the loop terminates, the algorithm returns the resulting solution in case it is feasible; otherwise, either the last selected element or the resulting solution without that element is returned, depending on their value. Recall that $f_S(j) = f(S \cup \{j\}) - f(S)$ is the incremental marginal value of element j to the set S, and x_S is the characteristic vector of the set S.

Algorithm 1. Multiplicative Updates

Input: A collection of linear packing constraints defined by $A \in [0,1]^{m \times n}$ and $b \in [1, \infty)^m$, a monotone submodular set function $f : 2^{[n]} \to \mathbb{R}_+$, an update factor $\lambda \in \mathbb{R}_+$

Output: A subset of $[n]$

1: $S \leftarrow \emptyset$
2: **for** $i \leftarrow 1$ to m **do** $w_i \leftarrow 1/b_i$ **end for**
3: **while** $\sum_{i=1}^m b_i w_i \leq \lambda$ and $S \neq [n]$ **do**
4:      Let $j \in [n] \setminus S$ be the element with minimal $\sum_{i=1}^m A_{ij} w_i / f_S(j)$
5:      $S \leftarrow S \cup \{j\}$
6:      **for** $i \leftarrow 1$ to m **do** $w_i \leftarrow w_i \lambda^{A_{ij}/b_i}$ **end for**
7: **end while**
8: **if** $Ax_S \leq b$ **then** return S
9: **else if** $f(S \setminus \{j\}) \geq f(\{j\})$ **then** return $S \setminus \{j\}$
10: **else** return $\{j\}$ **end if**

2.2 Analysis

In the remainder of this section, we analyze the performance of the algorithm. We begin by establishing several lemmas that hold independently of the value of the update factor. Later on, we consider specific update factors, and study their effect on the approximation ratio of the algorithm. For ease of presentation, it would be convenient to first introduce some notation and terminology:

– Let $S^* \subseteq [n]$ be a solution that maximizes the submodular function subject to the linear packing constraints, with value of $f(S^*)$.
– Let S_t be the solution at the end of iteration t of the algorithm, and note that $S_0 = \emptyset$ indicates the solution at the beginning of the algorithm. Moreover, let $\gamma(t)$ denote the element selected at iteration t of the algorithm, and let $\delta_t = f(S_t) - f(S_{t-1})$ be its incremental marginal value to the solution. Finally, let w_{it} be the value of w_i at the end of iteration t of the algorithm, and remark that $w_{i0} = 1/b_i$ is the value of w_i at the beginning of the algorithm.
– Let $\Lambda_t = \sum_{i=1}^m b_i w_{it}$ and $\alpha_t = \sum_{i=1}^m A_{i\gamma(t)} w_{i(t-1)} / \delta_t$. Notice that the algorithm may proceed to iteration $t + 1$ only if $\Lambda_t \leq \lambda$, and that $\Lambda_0 = m$. Also note that α_t is the value which gave rise to the selection of element $\gamma(t)$ at iteration t.

Correctness. We prove that the algorithm outputs a feasible solution. This is achieved by demonstrating that the returned solution respects the packing constraints.

Lemma 1. *The algorithm outputs a feasible solution.*

Approximation. We turn to analyze the approximation guarantee of the algorithm. We begin by establishing a generic algebraic bound applicable for any monotone submodular function and any arbitrary sequence of element additions.

Proposition 1. *Given a submodular function* $f : 2^{[n]} \to \mathbb{R}_+$, *a set collection* $S_0 \subseteq S_1 \subseteq \cdots \subseteq S_t \subseteq [n]$, *and a set* $S^* \subseteq [n]$ *satisfying* $f(S^*) > f(S_t)$ *then*

$$\sum_{\ell=1}^{t} \frac{f(S_\ell) - f(S_{\ell-1})}{f(S^*) - f(S_{\ell-1})} \le \ln \left(\frac{f(S^*) - f(S_0)}{f(S^*) - f(S_t)} \right) .$$

We continue by bounding the value of the optimal solution using the main parameters of the algorithm at the end of iteration ℓ.

Proposition 2. $f(S^*) \le f(S_\ell) + \Lambda_\ell/\alpha_{\ell+1}$ *in every iteration* ℓ.

Proof. The element selected at iteration $\ell+1$ minimizes the term $\sum_{i=1}^{m} A_{ij} w_{i\ell}/f_{S_\ell}(j)$ with respect to every $j \in [n]\backslash S_\ell$. This clearly implies that $\alpha_{\ell+1} \le \sum_{i=1}^{m} A_{ij} w_{i\ell}/f_{S_\ell}(j)$ for every j under consideration. Rearranging the terms in this inequality, we can bound the marginal value of each element $j \in [n] \setminus S_\ell$ with respect to S_ℓ, and obtain that $f_{S_\ell}(j) \le \sum_{i=1}^{m} A_{ij} w_{i\ell}/\alpha_{\ell+1}$. Let $J^* = \{j : j \in S^* \text{ and } j \notin S_\ell\}$ be the set of elements selected by the optimal solution, but not selected by the algorithm up to the end of iteration ℓ. Note that $J^* \subseteq [n] \setminus S_\ell$, and notice that

$$f(S^*) \le f(S^* \cup S_\ell) \le f(S_\ell) + \sum_{j \in J^*} f_{S_\ell}(j) ,$$

where the first inequality follows from the monotonicity of f, and the last inequality holds as a result of its submodularity. Specifically, the latter inequality is obtained using the decreasing marginal values property. We now focus on bounding the above right-hand side term. For this purpose, we utilize the bound derived earlier on the marginal values of the elements in $[n] \setminus S_\ell$, and attain

$$\sum_{j \in J^*} f_{S_\ell}(j) \le \sum_{j \in J^*} \sum_{i=1}^{m} \frac{A_{ij} w_{i\ell}}{\alpha_{\ell+1}} = \sum_{i=1}^{m} \frac{w_{i\ell}}{\alpha_{\ell+1}} \sum_{j \in J^*} A_{ij} \le \sum_{i=1}^{m} \frac{b_i w_{i\ell}}{\alpha_{\ell+1}} = \frac{\Lambda_\ell}{\alpha_{\ell+1}} ,$$

where the last inequality follows by recalling that the elements in J^* are a subset of the elements in the optimal solution, and thus, constitute a feasible solution respecting all constraints. As a result, $\sum_{j \in J^*} A_{ij} \le b_i$. ∎

We next prove that the algorithm attains an approximation guarantee of $\Omega(1/m^{1/W})$ when the update factor is $\lambda = e^W m$. Recall that $W = \min\{b_i/A_{ij} : A_{ij} > 0\}$ is the width of the constraints.

Lemma 2. *The algorithm archives* $\Omega(1/m^{1/W})$-*approximation by using* $\lambda = e^W m$.

Proof. Suppose the main loop terminates after t iterations. Notice that when the loop terminates either $S_t = [n]$ or $\sum_{i=1}^{m} b_i w_{it} > e^W m$. In the former case, one can easily infer that the returned solution is $1/2$-approximation to the optimal solution. Specifically, if S_t is returned by the algorithm then the outcome is clearly optimal since S_t consists of all elements, and if one of $S_t \setminus \{j\}$ or $\{j\}$ is returned then the value of the solution is a $1/2$-approximation since $\max\{f(S_t \setminus \{j\}), f(\{j\})\} \geq (f(S_t \setminus \{j\}) + f(\{j\}))/2 \geq f(S_t)/2$, where the last inequality uses the submodularity of f. In fact, one can easily validate that the above analysis also holds in case that $f(S_t) \geq f(S^*)$, which can happen since S_t may be infeasible. Hence, in the remainder of the proof, we shall assume that $f(S^*) > f(S_t)$ and that the loop terminates with $\Lambda_t = \sum_{i=1}^{m} b_i w_{it} > e^W m$.

We concentrate on upper bounding the value of Λ_t. For this purpose, we analyze the change in $\sum_{i=1}^{m} b_i w_i$ along the loop iterations. Observe that for any $\ell = 1, \ldots, t$,

$$\Lambda_\ell = \sum_{i=1}^{m} b_i w_{i\ell} = \sum_{i=1}^{m} b_i w_{i(\ell-1)} \cdot \left(e^W m\right)^{A_{i\gamma(\ell)}/b_i}$$

$$\leq \sum_{i=1}^{m} b_i w_{i(\ell-1)} \cdot \left(1 + \frac{eW m^{1/W} A_{i\gamma(\ell)}}{b_i}\right)$$

$$= \sum_{i=1}^{m} b_i w_{i(\ell-1)} + eW m^{1/W} \sum_{i=1}^{m} A_{i\gamma(\ell)} w_{i(\ell-1)}$$

$$= \Lambda_{\ell-1} + eW m^{1/W} \alpha_\ell \delta_\ell .$$

The first inequality follows by plugging $a = em^{1/W}$ and $y = W A_{i\gamma(\ell)}/b_i$ to the inequality $a^y \leq 1 + ay$, which is known to be valid for any $a \in \mathbb{R}_+$ and $y \in [0,1]$, and the last equality results from the definition of α_ℓ. By Proposition 2, we know that $\alpha_\ell \leq \Lambda_{\ell-1}/(f(S^*) - f(S_{\ell-1}))$ in case $f(S^*) > f(S_{\ell-1})$. The latter condition clearly holds since $f(S^*) > f(S_t)$ by previous assumption, and $f(S_t) \geq f(S_{\ell-1})$ for any ℓ under consideration. Therefore,

$$\Lambda_\ell \leq \Lambda_{\ell-1} \cdot \left(1 + \frac{eW m^{1/W} \delta_\ell}{f(S^*) - f(S_{\ell-1})}\right) \leq \Lambda_{\ell-1} \cdot \exp\left(\frac{eW m^{1/W} \delta_\ell}{f(S^*) - f(S_{\ell-1})}\right) ,$$

where the last inequality holds since $1 + y \leq e^y$. The resulting recursive definition can be used, in conjunction with the base case $\Lambda_0 = m$, to upper bound Λ_t by

$$\Lambda_t \leq \Lambda_0 \cdot \prod_{\ell=1}^{t} \exp\left(\frac{eW m^{1/W} \delta_\ell}{f(S^*) - f(S_{\ell-1})}\right) = m \cdot \exp\left(eW m^{1/W} \sum_{\ell=1}^{t} \frac{f(S_\ell) - f(S_{\ell-1})}{f(S^*) - f(S_{\ell-1})}\right) .$$

Recall that we assumed that the loop terminated with $\Lambda_t > e^W m$. This lower bound on Λ_t can be utilized, together with the upper bound on Λ_t, to yield

$$1 \leq em^{1/W} \sum_{\ell=1}^{t} \frac{f(S_\ell) - f(S_{\ell-1})}{f(S^*) - f(S_{\ell-1})} \leq em^{1/W} \ln\left(\frac{f(S^*) - f(S_0)}{f(S^*) - f(S_t)}\right) ,$$

where the last inequality is due to the Proposition 1. We note that $f(S_0) = 0$ since f is normalized and $S_0 = \emptyset$. Subsequently, one can obtain that $1 - 1/\exp(1/em^{1/W}) \leq f(S_t)/f(S^*)$ using simple algebraic manipulations. This can be further simplified to $1/(em^{1/W} + 1) \leq f(S_t)/f(S^*)$ by reutilizing the fact that $1 + y \leq e^y$. Notice that this proves that the algorithm archives $\Omega(1/m^{1/W})$-approximation since the value of the returned solution is at least $f(S_t)/2$. This follows from arguments similar to those presented at the beginning of the proof.   ∎

We are now ready to complete the proof of the first main result of the paper. We note that this result matches the best known approximation guarantee for the case that the objective function f is linear, achievable using the randomized rounding technique.

Proof of Theorem 1. By Lemma 1 and Lemma 2, we know that when the algorithm uses an update factor of $\lambda = e^W m$, it constructs a feasible solution which approximates the optimal solution within a factor of $\Omega(1/m^{1/W})$.   ∎

One immediate corollary of this theorem is that the algorithm under consideration attains a constant approximation guarantee when the number of constraints is constant or when the width is sufficiently large, say $W = \Omega(\ln m)$. In particular, one can reexamine the analysis presented in the proof of Lemma 2, and deduce that the approximation ratio of the algorithm approaches $1/(2e + 2)$ for sufficiently large W's. A natural followup question is whether one can improve upon this result. In what follows, we demonstrate that we can beat this approximation ratio by a careful selection of the update factor. We present a refined analysis that proves an approximation ratio of $(1 - \epsilon)(1 - 1/e)$ when $W = \Omega(\ln m/\epsilon^2)$. In particular, our analysis avoids the two-factor loss due to the max-selection in the last two lines of the algorithm.

Lemma 3. *The algorithm achieves an approximation ratio of* $(1 - 4\epsilon)(1 - 1/e)$ *by using* $\lambda = e^{\epsilon W}$ *when* $W \geq \max\{\ln m/\epsilon^2, 1/\epsilon\}$ *for any fixed* $\epsilon > 0$.

We are now ready to complete the proof of the second principal result of the paper. We note that this result almost matches the theoretical lower bound of $1 - 1/e$, which already holds for maximizing a monotone submodular function subject to a cardinality constraint [23,11]. In particular, our large width setting captures the hard instances of the latter problem as this problem can be solved in polynomial-time when $W = O(1/\epsilon)$ by enumerating over all sets of size at most W.

Proof of Theorem 2. Given an instance of the problem in which $W = \Omega(\ln m/\epsilon^2)$ for any fixed $\epsilon > 0$, Lemma 1 and Lemma 3 guarantee that employing the algorithm with an update factor of $\lambda = e^{\epsilon W/4}$ results in a feasible solution that approximates the optimal solution within a factor of $(1 - \epsilon)(1 - 1/e)$.   ∎

3   Submodular Maximization with Binary Packing Constraints

We consider the special setting of monotone submodular maximization under binary packing constraints, namely, when $A \in \{0, 1\}^{m \times n}$ instead of $A \in [0, 1]^{m \times n}$. Note that we may assume without loss of generality that $b \in \mathbb{N}_+^m$ since each vector entry can be rounded down to the nearest integer without any consequences whatsoever. This natural setting has been considered in the past for linear objective functions. Similarly to the general linear case, the randomized rounding technique attains the best known approximation guarantee in this case as well. In particular, it achieves an approximation ratio of $\Omega(1/m^{1/(W+1)})$. This outcome is also known to be optimal unless P = ZPP [6].

We can demonstrate that our multiplicative updates approach from Section 2 can be utilized to obtain the above-mentioned improved approximation guarantee for the underlying setting. This requires a fine-tuning of the algorithm and its analysis.

We next develop a different multiplicative updates algorithm for the special setting in which the constraints matrix is k-column sparse. In this case, the number of 1-value entries in each column of the input matrix is at most k. We prove that our algorithm achieves an approximation guarantee that does not depend on the number of rows m, but only depends on the sparsity parameter k and width parameter W. More precisely, we establish that the algorithm attains an approximation ratio of $\Omega(1/(Wk^{1/W}))$.

3.1 The Algorithm

The multiplicative updates algorithm, formally described below, maintains a collection of weights that capture the extent to which each constraint is close to be violated under a given solution. The algorithm is built around one main loop. In each iteration of that loop, the algorithm considers a remaining element whose marginal contribution to the current solution is maximal, and adds it to the solution set if its corresponding sum of weights is sufficiently small. In any case, the element under consideration is removed from the list of remaining elements. When the loop terminates, the algorithm returns the resulting solution.

Algorithm 2. Column Sparse Multiplicative Updates

Input: A collection of linear packing constraints defined by $A \in \{0,1\}^{m \times n}$ and $b \in \mathbb{N}_+^m$, a monotone submodular set function $f : 2^{[n]} \to \mathbb{R}_+$, an update factor $\lambda \in \mathbb{R}_+$

Output: A subset of $[n]$

1: $S \leftarrow \emptyset, R \leftarrow [n]$
2: **for** $i \leftarrow 1$ to m **do** $w_i \leftarrow 0$ **end for**
3: **while** $R \neq \emptyset$ **do**
4:    Let $j \in R$ be the element with maximal $f_S(j)$
5:    **if** $\sum_{i=1}^{m} A_{ij} w_i < (\lambda - 1)$ **then** $S \leftarrow S \cup \{j\}$
6:    $R \leftarrow R \setminus \{j\}$
7:    **for** $i \leftarrow 1$ to m **do** $w_i \leftarrow \lambda^{\sum_{j \in S} A_{ij}/b_i} - 1$ **end for**
8: **end while**
9: return S

3.2 Analysis

In what follows, we analyze the performance of the algorithm. We begin by establishing an algebraic bound applicable for any monotone submodular function and any solution set of elements, attained by an algorithm that considers the elements in a greedy fashion. Note that our algorithm indeed considers the elements in such fashion. We define the *greedy elements sequence* $\mathcal{E}(f, S) = \langle e_1, \ldots, e_n \rangle$ of a submodular function f and a set S as the ordered sequence of elements considered by a greedy process whose outcome is S. Specifically, the greedy process is initialized with $R_0 = [n]$ and $S_0 = \emptyset$. Then , it runs for n steps, where in each step t, it considers the element $e_t \in R_{t-1}$ that has a maximum marginal value with respect to the current solution set S_{t-1}, and adds it to the solution set S_t of the next step if $e_t \in S$. In any case, the element e_t is removed from R_{t-1} to obtain the set R_t of remaining elements for the next step. With this definition in mind, let $E_t = \{e_1, \ldots, e_t\}$ be the set of first t elements in the sequence.

Proposition 3. *Given a submodular function* $f : 2^{[n]} \to \mathbb{R}_+$, *a set* $S \subseteq [n]$, *their greedy elements sequence* $\mathcal{E}(f, S) = \langle e_1, \ldots, e_n \rangle$, *and another set* $S^* \subseteq [n]$ *satisfying* $|S \cap E_t| \geq \alpha \cdot |S^* \cap E_t|$ *for every* $t \in [n]$ *and a parameter* $\alpha \leq 1$, *it holds that* $f(S) \geq (\alpha/(\alpha+1)) \cdot f(S^*)$.

Proof. Let us assume without loss of generality that the greedy process goes over the elements according to the order 1 to n, namely, $E_1 = \{1\}, E_2 = \{1, 2\}$, and so on. We note that this assumption is valid since one can appropriately rename the elements. Furthermore, let $S = \{a_1, \ldots, a_{|S|}\}$ and $S^* = \{b_1, \ldots, b_{|S^*|}\}$ be the respective elements of S and S^* sorted in an increasing order. Let us suppose that $1/\alpha$ is integral. We emphasize that this assumption is merely for simplicity of presentation, as we demonstrate later. We match between each element of S and $1/\alpha$ distinct elements from S^*. Specifically, each element a_t is matched to the elements set $S_t^* = \{b_{(t-1)/\alpha+1}, \ldots, b_{t/\alpha}\}$. Notice that every element of S^* is matched to an element of S; else, it must be that $|S^*| > |S|/\alpha$, but this contradicts the fact that $|S| = |S \cap E_n| \geq \alpha \cdot |S^* \cap E_n| = \alpha|S^*|$. We next argue that each $a_t \leq b_{(t-1)/\alpha+1}$. As a result, we attain that each

$$f_{S \cap E_{a_t-1}}(a_t) \geq f_{S \cap E_{a_t-1}}(b_{(t-1)/\alpha+1}), \ldots, f_{S \cap E_{a_t-1}}(b_{t/\alpha}) .$$

The last inequality holds since we known that when the element a_t was considered by the greedy process, all the elements of S_t^* were still available, and therefore, their marginal value with respect to the solution $S \cap E_{a_t-1}$ was no more than the marginal value of the element a_t. Consequently,

$$f(S^*) \leq f(S) + \sum_{b \in S^* \setminus S} f_S(b) = f(S) + \sum_{t=1}^{\lceil \alpha |S^*| \rceil} \sum_{b \in S_t^*} f_S(b)$$

$$\leq f(S) + \frac{1}{\alpha} \sum_{t=1}^{|S|} f_{S \cap E_{a_t-1}}(a_t) = \left(1 + \frac{1}{\alpha}\right) f(S) ,$$

where both inequalities hold by the submodularity of f. For the purpose of establishing the previously mentioned argument, suppose by way of contradicting that there is some t for which $a_t > b_{(t-1)/\alpha+1}$. Let us concentrate on the elements set $E_{(t-1)/\alpha+1}$. Notice that $|S \cap E_{(t-1)/\alpha+1}| \leq t-1$, whereas $|S^* \cap E_{(t-1)/\alpha+1}| = (t-1)/\alpha+1$. This implies that $|S \cap E_{(t-1)/\alpha+1}| < \alpha \cdot |S^* \cap E_{(t-1)/\alpha+1}|$, a contradiction. We conclude by noting that our assumption that $1/\alpha$ is integral can be easily neglected. Specifically, one need to modify that proof in such a way that a fractional part of an element from S^* may be matched to an element form S. Then, notice that at most two fractional parts of an element of T are matched to elements of S, and those elements must appear before the element of S^* in the greedy elements sequence. ∎

We now turn to establish our main result for the special setting of maximizing a monotone submodular function under k-column sparse packing constraints.

Proof of Theorem 3. We first claim that the algorithm outputs a feasible solution, that is, a solution that respects the packing constraints. Suppose by way of contradiction that ℓ is the first element that is added to S and induces a violation in some constraint i at iteration t of the main loop. Note that $A_{i\ell} = 1$. Let S_t be the solution at the end of iteration t, and notice that $\sum_{j \in S_t} A_{ij} = b_i + 1$ since all the entries of A are binary. This implies that $w_i = \lambda - 1$ at the beginning of the iteration in which ℓ was considered, and thus, $\sum_{i=1}^{m} A_{i\ell} w_i \geq \lambda - 1$. Inspecting the selection rule, one can infer that ℓ could not have been selected.

We next prove that the algorithm attains an approximation ratio of $\Omega(1/(Wk^{1/W}))$ when the update factor is $\lambda = k+1$. Recall that W is the width of the constraints, which is equal to $\min\{b_i\}$ in our case. Similarly to before, we denote by $S^* \subseteq [n]$ a solution that maximizes the submodular function subject to the linear packing constraints. Let $\langle e_1, \ldots, e_n \rangle$ be the ordered sequence of elements considered by our algorithm, and note that it is essentially the greedy elements sequence $\mathcal{E}(f, S)$. Moreover, let $E_t = \{e_1, \ldots, e_t\}$ be the set of first t elements in that sequence, $S_t^* = S^* \cap E_t$ be the elements of E_t in the optimal solution, $S_t = S \cap E_t$ be the elements of E_t in our algorithm's solution, and $w_{it} = \lambda^{\sum_{j \in S_t} A_{ij}/b_i} - 1$ be the value of w_i at the end of iteration t of the algorithm. We prove the two following propositions:

Proposition 4. *For every $t \in \{0, \ldots, n\}$,*

$$|S_t| \geq \frac{\sum_{i=1}^{m} b_i w_{it}}{W \lambda^{1/W}(k + \lambda - 1)} .$$

Proposition 5. *For every $t \in \{0, \ldots, n\}$,*

$$|S_t^*| \leq |S_t| + \frac{\sum_{i=1}^{m} b_i w_{it}}{\lambda - 1} .$$

We can now utilize the above propositions and get that for every $t \in \{0, \ldots, n\}$,

$$|S_t^*| \leq |S_t| + \frac{\sum_{i=1}^{m} b_i w_{it}}{\lambda - 1} \leq |S_t| + \frac{W \lambda^{1/W}(k + \lambda - 1)}{\lambda - 1}|S_t| = \left(1 + 2W\lambda^{1/W}\right) \cdot |S_t| ,$$

where the last equality holds as $\lambda = k+1$. Therefore, we can employ Proposition 3 with $\alpha = 1/(1 + 2W\lambda^{1/W})$, and attain that our algorithm's solution approximates the optimal solution to within a factor of $\alpha/(\alpha+1) = 1/(2+2W\lambda^{1/W}) = \Omega(1/(Wk^{1/W}))$. ∎

Acknowledgments. The authors thank Chandra Chekuri, Ilan Cohen, Gagan Goel, and Jan Vondrák for valuable discussions on topics related to the subject of this study.

References

1. Ageev, A.A., Sviridenko, M.: Pipage rounding: A new method of constructing algorithms with proven performance guarantee. J. Comb. Optim. 8(3), 307–328 (2004)
2. Azar, Y., Regev, O.: Combinatorial algorithms for the unsplittable flow problem. Algorithmica 44(1), 49–66 (2006)
3. Bansal, N., Korula, N., Nagarajan, V., Srinivasan, A.: On k-Column Sparse Packing Programs. In: Eisenbrand, F., Shepherd, F.B. (eds.) IPCO 2010. LNCS, vol. 6080, pp. 369–382. Springer, Heidelberg (2010)

4. Briest, P., Krysta, P., Vöcking, B.: Approximation techniques for utilitarian mechanism design. In: 37th STOC, pp. 39–48 (2005)
5. Calinescu, G., Chekuri, C., Pál, M., Vondrák, J.: Maximizing a Submodular Set Function Subject to a Matroid Constraint (Extended Abstract). In: Fischetti, M., Williamson, D.P. (eds.) IPCO 2007. LNCS, vol. 4513, pp. 182–196. Springer, Heidelberg (2007)
6. Chekuri, C., Khanna, S.: On multidimensional packing problems. SIAM J. Comput. 33(4), 837–851 (2004)
7. Chekuri, C., Vondrák, J.: Personal Communication (2010)
8. Chekuri, C., Vondrák, J., Zenklusen, R.: Dependent randomized rounding via exchange properties of combinatorial structures. In: 51st FOCS, pp. 575–584 (2010)
9. Chekuri, C., Vondrák, J., Zenklusen, R.: Submodular function maximization via the multilinear relaxation and contention resolution schemes. In: 43rd STOC, pp. 783–792 (2011)
10. Dobzinski, S., Vondrák, J.: From query complexity to computational complexity. In: 44th STOC (2012)
11. Feige, U.: A threshold of ln n for approximating set cover. J. ACM 45(4), 634–652 (1998)
12. Feige, U., Mirrokni, V.S., Vondrák, J.: Maximizing non-monotone submodular functions. In: 48th FOCS, pp. 461–471 (2007)
13. Feldman, M., Naor, J(S.), Schwartz, R.: Nonmonotone Submodular Maximization via a Structural Continuous Greedy Algorithm (Extended Abstract). In: Aceto, L., Henzinger, M., Sgall, J. (eds.) ICALP 2011 Part I. LNCS, vol. 6755, pp. 342–353. Springer, Heidelberg (2011)
14. Feldman, M., Naor, J., Schwartz, R.: A unified continuous greedy algorithm for submodular maximization. In: 52nd FOCS, pp. 570–579 (2011)
15. Fisher, M.L., Nemhauser, G.L., Wolsey, L.A.: An analysis of approximations for maximizing submodular set functions II. Math. Program. Study 8, 73–87 (1978)
16. Garg, N., Könemann, J.: Faster and simpler algorithms for multicommodity flow and other fractional packing problems. SIAM J. Comput. 37(2), 630–652 (2007)
17. Gharan, S.O., Vondrák, J.: Submodular maximization by simulated annealing. In: 22nd SODA, pp. 1098–1116 (2011)
18. Goundan, P.R., Schulz, A.S.: Revisiting the greedy approach to submodular set function maximization (2007) (manuscript)
19. Khuller, S., Moss, A., Naor, J.: The budgeted maximum coverage problem. Inf. Process. Lett. 70(1), 39–45 (1999)
20. Kulik, A., Shachnai, H., Tamir, T.: Maximizing submodular set functions subject to multiple linear constraints. In: 20th SODA, pp. 545–554 (2009)
21. Lee, J., Mirrokni, V.S., Nagarajan, V., Sviridenko, M.: Maximizing nonmonotone submodular functions under matroid or knapsack constraints. SIAM J. Discrete Math. 23(4), 2053–2078 (2010)
22. Nemhauser, G.L., Wolsey, L.A.: Best algorithms for approximating the maximum of a submodular set function. Math. Operations Research 3(3), 177–188 (1978)
23. Nemhauser, G.L., Wolsey, L.A., Fisher, M.L.: An analysis of approximations for maximizing submodular set functions I. Math. Program. 14, 265–294 (1978)
24. Raghavan, P.: Probabilistic construction of deterministic algorithms: Approximating packing integer programs. Journal of Computer and System Sciences 37(2), 130–143 (1988)
25. Raghavan, P., Thompson, C.D.: Randomized rounding: a technique for provably good algorithms and algorithmic proofs. Combinatorica 7(4), 365–374 (1987)
26. Srinivasan, A.: Improved approximation guarantees for packing and covering integer programs. SIAM J. Comput. 29(2), 648–670 (1999)
27. Sviridenko, M.: A note on maximizing a submodular set function subject to a knapsack constraint. Oper. Res. Lett. 32(1), 41–43 (2004)

28. Vondrák, J.: Optimal approximation for the submodular welfare problem in the value oracle model. In: 40th STOC, pp. 67–74 (2008)
29. Vondrák, J.: Symmetry and approximability of submodular maximization problems. In: 50th FOCS, pp. 651–670 (2009)
30. Wolsey, L.A.: Maximising real-valued submodular functions: Primal and dual heuristics for location problems. Math. Operations Research 7(3), 410–425 (1982)
31. Young, N.E.: Randomized rounding without solving the linear program. In: 6th SODA, pp. 170–178 (1995)

Polynomial-Time Isomorphism Test for Groups with No Abelian Normal Subgroups

(Extended Abstract⋆)

László Babai[1,⋆⋆], Paolo Codenotti[2], and Youming Qiao[3,⋆⋆⋆]

[1] University of Chicago
laci@cs.uchicago.edu
[2] University of Minnesota
paolo@ima.umn.edu
[3] Institute for Theoretical Computer Science,
Institute for Interdisciplinary Information Sciences,
Tsinghua University
jimmyqiao86@gmail.com

Abstract. We consider the problem of testing isomorphism of groups of order n given by Cayley tables. The trivial $n^{\log n}$ bound on the time complexity for the general case has not been improved upon over the past four decades. We demonstrate that the obstacle to efficient algorithms is the presence of abelian normal subgroups; we show this by giving a polynomial-time isomorphism test for groups without nontrivial abelian normal subgroups. This concludes a project started by the authors and J. A. Grochow (SODA 2011). Two key new ingredient are: (a) an algorithm to test permutational isomorphism of permutation groups in time, polynomial in the order and simply exponential in the degree; (b) the introduction of the "twisted code equivalence problem," a generalization of the classical code equivalence problem by admitting a group action on the alphabet. Both of these problems are of independent interest.

Keywords: Group Isomorphism, Permutational Isomorphism, Code Equivalence.

1 Introduction, Main Results

The isomorphism problem for groups asks to determine whether or not two groups of order n, given by their Cayley tables (multiplication tables), are isomorphic. As pointed out long ago [7,15], if G is generated by k elements then isomorphism can be decided, and all isomorphisms listed, in time $n^{k+O(1)}$. Since

---

⋆ See http://people.cs.uchicago.edu/~laci/papers for the full version of this paper.

⋆⋆ This collaboration was supported in part by László Babai's NSF Grants CCF-0830370 and CCF 1017781. The views expressed in this paper are those of the authors and do not necessarily reflect the views of the NSF.

⋆⋆⋆ Youming's work was supported in part by the National Basic Research Program of China Grant 2011CBA00300, 2011CBA00301, the National Natural Science Foundation of China Grant 61033001, 61061130540, 61073174.

A. Czumaj et al. (Eds.): ICALP 2012, Part I, LNCS 7391, pp. 51–62, 2012.
© Springer-Verlag Berlin Heidelberg 2012

$k \leq \log_2 n$ for all groups, this gives an $n^{\log_2 n + O(1)}$-time algorithm for all groups and a polynomial-time algorithm for finite simple groups (because the latter are generated by 2 elements [20,1]). In spite of considerable attention to the problem over the decades, no general bound with a sublogarithmic exponent has been obtained. While the abelian case is easy ($O(n)$ by [9]), just one step away from the abelian case lurk the most notorious cases: nilpotent groups of class 2 (groups G such that the quotient $G/Z(G)$ is abelian, where $Z(G)$ is the center of G). No complete structure theory of such groups is known; recent work in this direction by James B. Wilson [22,23] commands attention. Recently, special classes of non-nilpotent solvable groups have been considered [11,16,4].

The group isomorphism problem is of great importance to computational group theory; heuristic methods (e. g., Cannon and Holt [6]) have been implemented in the Magma and GAP computational algebra systems for groups are given as permutation groups, represented by sets of generators. In this context the isomorphism problem is graph-isomorphism hard and therefore no subexponential ($\exp(n^{o(1)})$) worst-case algorithm can currently be expected (n is now the size of the permutation domain), while efficient practical algorithms remain a possibility.

While class-2 nilpotent groups have long been recognized as the chief bottleneck in the group isomorphism problem, this intuition has never been formalized. The ultimate formalization would be to reduce the general case to this case. As a first step, we consider a significant class without a complete structure theory at the opposite end of the spectrum: *groups without abelian normal subgroups*. Following Robinson [17], we call such groups *semisimple*[1].

In 2010, J. A. Grochow and the present authors started a project to test isomorphism of semisimple groups [3]. In that paper we observed, based on an elementary analysis of the group structure, that this problem can be solved in time $n^{\log \log n}$, and using a combination of additional structure theory and combinatorial/algorithmic techniques, we gave a polynomial time algorithm assuming the boundedness of certain parameters. The main result of the present paper, stated next, concludes the project.

Theorem 1. *Isomorphism of semisimple groups given by their Cayley tables can be decided in polynomial time.*

We note that semisimplicity of a given group can be decided in polynomial time (trivially for groups given by Cayley tables and nontrivially even for permutation groups given by generators [14]).

Our second main result concerns permutational isomorphism.

Definition 1. *Two permutation groups $G, H \leq S_k$ are permutationally isomorphic if $\exists \pi \in S_k$ such that $G^\pi = H$, where $G^\pi := \{\pi^{-1}\sigma\pi \mid \sigma \in G\}$.*

---

[1] We note that various authors use the term 'semisimple group' in several different meanings (see e. g. [21]). The definition we use conforms to the general practice in algebra that an algebraic structure (ring, algebra, inner product space, etc.) is semisimple if its 'radical' is trivial; each concept of 'radical' then corresponds to a notion of semisimplicity. In our case, the 'radical' is the solvable radical, i. e., the largest solvable normal subgroup.

Theorem 2. *Permutational Isomorphism of permutation groups $G, H \leq S_k$, given by lists of generators, can be decided in time $\mathrm{poly}(|G|)\, c^k$, for an absolute constant c.*

The proofs of these two main results are intertwined. First we solve the permutational isomorphism problem for the special case of *transitive groups* in a stronger sense (see Section 1.1); then we use this to solve our main problem, isomorphism of semisimple groups, via "twisted code equivalence." The general case of the permutational isomorphism problem follows via a simple reduction (cf. [3, Theorem 7.2]).

The two main ingredients that support Theorem 1 and 2 are: *permutational isomorphism of transitive permutation groups* and *twisted code equivalence*. We explain them in the next section.

1.1 Technical Ingredients

Theorem 3. *There exists an absolute constant c such that for all pairs of transitive permutation groups $G, H \leq S_k$ (a) the number of permutational automorphisms of G is at most $|G|\, c^k$; (b) we can list the set of all permutational isomorphisms of G and H in time $|G|\, c^k$.*

The proof involves a detailed group-theoretic study of the structure of transitive permutation groups. (See Section 4 for an outline of the proof.)

Another key ingredient is the concept of "twisted code equivalence," a generalization of the code equivalence problem, and a problem of independent interest. A *code* of length ℓ over a finite alphabet Γ is a subset of Γ^A for some set A with $|A| = \ell$. An *equivalence* of the codes $\mathcal{A} \subseteq \Gamma^A$ and $\mathcal{B} \subseteq \Gamma^B$ is a bijection $A \to B$ that takes $\mathcal{A}$ to $\mathcal{B}$. If $|\Gamma| = 2$ then the code is a Boolean function or hypergraph, so the code equivalence problem is a generalization of the hypergraph isomorphism problem. Slightly simplifying and extending Luks's C^ℓ dynamic programming algorithm for hypergraph isomorphism [12] to treat code equivalence, in [3] we gave an algorithm to test equivalence of codes of length ℓ over an alphabet Γ in time $(c|\Gamma|)^{2\ell}$, for an absolute constant c. In the present paper we introduce a generalization of this problem by allowing to permute the symbols by some group W acting on Γ, independently for each coordinate.

Definition 2 (Twisted code equivalence). *Let $\mathcal{A} \subseteq \Gamma^A$, $\mathcal{B} \subseteq \Gamma^B$ be codes of length ℓ over Γ. Let a group W act on Γ. Given a bijection $\pi : A \to B$ and a function $w : B \to W$, the pair (π, w) is a W-twisted equivalence of the codes $\mathcal{A}$ and $\mathcal{B}$ if by applying π to the coordinates of each codeword in $\mathcal{A}$ and then applying $w(b)$ to the entry in position b for each $b \in B$ we obtain the code $\mathcal{B}$.*

Generalizing the algorithm from [3] and improving its data management, we obtain the following result, proved in Section 3. Our algorithm uses a coset intersection subroutine and operates on rather large alphabets. In our case, W will have a low-degree faithful permutation representation, and we take advantage of this.

Theorem 4. *Let Γ be an alphabet, $W \leq \mathrm{Sym}(\Gamma)$, and assume a faithful permutation representation of W of degree d is given. The set of W-twisted equivalences of two codes $\mathcal{A}, \mathcal{B} \subseteq \Gamma^\ell$ can be found in time $c^{d\ell}\mathrm{poly}(|\mathcal{A}|, |W|, |\Gamma|)$.*

In the special case where $W = \{\mathrm{id}\}$, twisted code equivalence is simply code equivalence and the theorem gives a running time of $c^\ell \mathrm{poly}(|\Gamma|, |\mathcal{A}|)$. This improvement in the dependence on $|\Gamma|$ over the previous bound of $(c|\Gamma|)^{2\ell}$ from [3] is critical to our main result.

1.2 Strategy for the Main Result

Our approach is motivated by the Babai-Beals filtration of groups [2] (see [3, Section 7.5]). Specifically, the *socle* of a group is the product of its minimal normal subgroups. The socle of a semisimple group G is the direct product of nonabelian simple groups, and G acts on the set of simple factors of the socle by conjugation, producing a permutation group of degree at most $\log_{60}|G|$. (60 is the order of A_5, the smallest nonabelian simple group.)

First we observe that isomorphism of groups that are direct products of simple groups can be tested in polynomial time. So we can assume that our semisimple groups G and H have isomorphic socles. The second observation is that an isomorphism of the socles extends in at most one way to an isomorphism of G and H. Moreover, given an isomorphism of the socles, we can find the unique extension if it exists (Observation 8). Our next step is to identify isomorphic simple factors of the socles with a canonical copy. From now on we look for only those isomorphisms that respect the specific identification. We note that the number of identifications to consider is polynomially bounded ([3, Lemma 4.1]). We look at the conjugation action of the groups G and H on the set of simple factors of their socles. The orbits of this action correspond to the minimal normal subgroups. Our alphabets will be the isomorphism types of these actions under identification-preserving isomorphisms; these can be computed using our algorithm for transitive permutational isomorphism (Theorem 3). The problem then reduces to twisted code equivalence over these alphabets. The twisting groups consist of the identification-preserving automorphisms of the alphabets; they can be represented as permutation groups acting on the set of simple factors, thus giving a small value of d for the application of Theorem 4.

1.3 Organization of the Paper

We first present the two technical ingredients: twisted code equivalence in Section 3, and permutational isomorphism of transitive groups in Section 4. We outline the proof of the main result, Theorem 1, in Section 5. Detailed proofs, and in some cases the detailed statements, appear in the full version of this paper.

2 Group-Theoretic Preliminaries

For a function $f : X \to Y$, we write x^f for the image of $x \in X$.

General group theory. For groups H, G, we write $H \leq G$ to say that H is a subgroup of G. Given groups G and H, $\mathrm{ISO}(G, H)$ denotes the set of $G \to H$ isomorphisms; $\mathrm{Aut}(G) = \mathrm{ISO}(G, G)$. The set $\mathrm{ISO}(G, H)$ is either empty or a coset of $\mathrm{Aut}(G)$. For $g \in G$, *conjugation by g* means the map $g : G \to G$ defined by $x \mapsto x^g := g^{-1}xg$. For $S \subseteq G$ and $g \in G$ we set $S^g = \{s^g \mid s \in S\}$.

Permutation groups. Let $\mathrm{Sym}(\Omega)$ denote the symmetric group acting on the set Ω, the group of all permutations of Ω. We write S_k for $\mathrm{Sym}([k])$ where $[k] = \{1, \ldots, k\}$. Permutation groups of *degree* k are subgroups of $\mathrm{Sym}(\Omega)$ with $|\Omega| = k$. A homomorphism $\varphi \colon G \to \mathrm{Sym}(\Omega)$ is called a *permutation representation* of G of degree $|\Omega|$; such a homomorphism defines a G-action $x \mapsto x^\pi := x^{\varphi(\pi)}$ on Ω ($x \in \Omega, \pi \in G$). We say that φ is *faithful* if it is injective. Let $\mathrm{Alt}(\Omega) \leq \mathrm{Sym}(\Omega)$ denote the *alternating group*. Let $G \leq \mathrm{Sym}(\Omega)$. The *orbit* of $x \in \Omega$ is the set $x^G := \{x^\pi : \pi \in G\}$. The *length* of an orbit is its size. A permutation group $G \leq \mathrm{Sym}(\Omega)$ is *transitive* if $x^G = \Omega$ for some (any) $x \in \Omega$. The *stabilizer* G_x of a point $x \in \Omega$ is $G_x = \{\pi \in G \mid x^\pi = x\}$.

Given a G-action on Ω, a nonempty set $B \subseteq \Omega$ is a *block of imprimitivity* (or simply "a block") if $(\forall \pi \in G)(B^\pi = B$ or $B \cap B^\pi = \emptyset)$. A transitive action is *primitive* if all blocks are trivial (the singletons or the whole domain Ω), and *imprimitive* otherwise. Let $G \leq \mathrm{Sym}(\Omega)$ and $H \leq \mathrm{Sym}(\Delta)$. The *wreath product* $G \wr H$ is a permutation group acting on $\Omega \times \Delta$ viewed as $|\Delta|$ copies of Ω. $|\Delta|$ copies of G act independently on each copy of Ω and H permutes the copies. This defines the *standard action* of this group. In its *product action*, the same group acts on Ω^Δ, such that the copies of G act on each coordinate and H permutes the coordinates.

Algorithms for permutation groups. For the purposes of computation, a permutation group $G \leq \mathrm{Sym}(\Omega)$ will be represented by a list of generators. Many basic computational tasks for permutation groups, including membership testing, finding the order of a group, finding pointwise stabilizers of subsets of the domain, finding blocks of imprimitivity, can be performed in polynomial time ([19,8,10], cf. [18]).

3 Twisted Code Equivalence

Let $\mathrm{EQ}_W(\mathcal{A}, \mathcal{B})$ denote the set of W-twisted equivalences of the codes $\mathcal{A}$ and $\mathcal{B}$. (See Section 1.1 for the definitions.) Note that this is either empty or a coset of the group $\mathrm{EQ}_W(\mathcal{A}, \mathcal{A}) \leq W \wr \mathrm{Sym}(\mathcal{A})$.

In this section we prove the following, more precise version of Theorem 4.

Theorem 5. *Let $\mathcal{A} \subseteq \Gamma^A$ and $\mathcal{B} \subseteq \Gamma^B$ be codes of length ℓ. Let $W \leq \mathrm{Sym}(\Gamma)$, and assume we are given a faithful permutation representation of W of degree d. Then $\mathrm{EQ}_W(\mathcal{A}, \mathcal{B})$ can be found in time $O(2^{\ell(d+1)} |W| |\Gamma| |\mathcal{A}|^2 \log |\mathcal{A}|)$.*

Proof. For a subset $U \subseteq A$ we call a function $y \colon U \to \Gamma$ a "partial string over A." The set U is the domain of y, denoted $\mathrm{dom}(y)$, and $|U|$ the *length* of y. For a partial string y over A, let $\mathcal{A}_y$ be the set of strings in $\mathcal{A}$ that are extensions of y. We make analogous definitions for $\mathcal{B}$.

We construct a dynamic programming table with an entry for each pair (y, z) of partial strings, y over A and z over B, of equal length such that $\mathcal{A}_y \neq \emptyset$, and $\mathcal{B}_z \neq \emptyset$. For each such pair (y, z), we store the set $I(y, z)$ of W-twisted equivalences of the restriction $\mathcal{A}_y^*$ of $\mathcal{A}_y$ to $A \backslash \mathrm{dom}(y)$ with the restriction $\mathcal{B}_z^*$ of $\mathcal{B}_z$ to $B \backslash \mathrm{dom}(z)$. Note that the $I(y, z)$ are either empty or cosets of the groups $\mathrm{EQ}_W(\mathcal{A}_y^*, \mathcal{A}_y^*) \leq W \wr \mathrm{Sym}(A \backslash \mathrm{dom}(y))$, and hence they can be stored efficiently.

We start with full strings $y \in \mathcal{A}$, $z \in \mathcal{B}$ and work our way down to $\mathrm{dom}(y) = \mathrm{dom}(z) = \emptyset$, at which point we shall have constructed all $\mathcal{A} \to \mathcal{B}$ twisted W-equivalences. When y, z are full strings, we have $|\mathcal{A}_y| = 1$, $|\mathcal{B}_z| = 1$, and $I(y, z)$ is trivial.

Let y, z be proper partial strings of length h, and assume we have constructed $I(y', z')$ for all y', z' of length greater than h. To construct $I(y, z)$ we augment the domain of y by one index $r \in A \backslash \mathrm{dom}(y)$, and the domain of z by one index, $s \in B \backslash \mathrm{dom}(z)$. We fix r, and make all possible choices of $s \in B \backslash \mathrm{dom}(z)$. For each $s \in B \backslash \mathrm{dom}(z)$ and $\sigma \in W$, we will separately find the set of all elements of $I(y, z)$ that move s to r, and act on the symbol in that position by σ.

More formally, for $\gamma \in \Gamma$, and $r \in A \backslash \mathrm{dom}(y)$, let $y(r, \gamma)$ be the partial string extending y by γ at position r, and let $\sigma \in W$, $s \in B \backslash \mathrm{dom}(z)$. Given some $\pi \colon (A \backslash \mathrm{dom}(y) \backslash \{r\}) \to (B \backslash \mathrm{dom}(z) \backslash \{s\})$, let $\pi^* \colon (A \backslash \mathrm{dom}(y)) \to (B \backslash \mathrm{dom}(z))$ extend π by sending r to s; and for $w \colon (B \backslash \mathrm{dom}(z) \backslash \{s\}) \to W$, let $w^* \colon (B \backslash \mathrm{dom}(z)) \to W$ extend w by sending s to σ. Let $I^*(y, z \, ; \, r \xrightarrow{(\gamma, \sigma)} s)$ be the set of all (π^*, w^*) for $(\pi, w) \in I(y(r, \gamma), z(s, \gamma^{\sigma^{-1}}))$. Then

$$I(y, z) = \bigcup_{s \in B} \bigcup_{\sigma \in W} \bigcap_{\gamma \in \Gamma : \mathcal{A}_{y(r, \gamma)} \neq \emptyset} I^*(y, z \, ; \, r \xrightarrow{(\gamma, \sigma)} s).$$

If $z(s, \gamma^{\sigma^{-1}}) \in \mathcal{B}$, then we can look up the value of $I(y(r, \gamma), z(s, \gamma^{\sigma^{-1}}))$ in the table and use that to compute the corresponding I^*. If for some σ and γ, $z(s, \gamma^{\sigma^{-1}}) \notin \mathcal{B}$, then $I(y(r, \gamma), z(s, \gamma^{\sigma^{-1}}))$ is empty and so is I^*.

Analysis. We consider $\ell |\mathcal{A}|$ partial strings of $\mathcal{A}$ (all prefixes), and $2^\ell |\mathcal{A}|$ partial strings of $\mathcal{B}$ (we can assume $|\mathcal{A}| = |\mathcal{B}|$, otherwise we reject equivalence). So the number of table entries we store is at most $2^\ell \ell |\mathcal{A}|^2$. The cost of computing each dynamic programming entry is $\ell |\Gamma| |W|$ coset intersection operations, and $\ell |\Gamma| |W|$ times the cost of checking whether some $\mathcal{A}_{y'}$ is empty. Standard techniques allow us to compute the I^* and paste cosets together in polynomial time. The cost of coset intersection is $O(2^{\ell d})$ [12]. (Stronger bounds for coset intersection are available but not needed here.) The cost of checking if $\mathcal{A}_{y'}$ is empty is $\log |\mathcal{A}|$ if we add a preprocessing step to sort $\mathcal{A}$. □

We shall need a simple generalization of this result to multiple alphabets. (As before, we refer to the full version for all missing details.)

4   Permutational Isomorphism for Transitive Groups

4.1   Further Group-Theoretic Preliminaries

Permutational Isomorphism. If $G \leq \mathrm{Sym}(\Omega)$ and $H \leq \mathrm{Sym}(\Delta)$ are permutation groups, a bijection $\pi \colon \Omega \to \Delta$ is a *permutational isomorphism* from G to H if $G^\pi = H$. We denote the set of all $G \to H$ permutational isomorphisms by $\mathrm{PISO}(G, H)$; $\mathrm{PAut}(G) := \mathrm{PISO}(G, G)$. We say G and H are *permutationally isomorphic* if $\mathrm{PISO}(G, H) \neq \emptyset$. Each $\pi \in \mathrm{PISO}(G, H)$ induces an isomorphisim $\widehat{\pi} \colon G \to H$. Let $\widehat{\mathrm{PISO}}(G, H) := \{\widehat{\pi} \mid \pi \in \mathrm{PISO}(G, H)\}$.

Proposition 1. *Given* $G, H \leq S_k$ *and* $f \in S_k$, *we can decide in* $\mathrm{poly}(k)$ *time whether or not* $f \in \mathrm{PISO}(G, H)$.   Proof: use membership testing.

Bounds on primitive groups. Let $G \leq S_k$. We call G a *giant*[2] if $k \geq 7$ and $G = S_k$ or A_k. These two groups are far larger than any other primitive group.

Lemma 1. *Let* $G \leq S_k$ *be primitive and let* $H \leq S_k$. *(a) If* G *is non-giant then* $|\mathrm{PAut}(G)| \leq \exp(\widetilde{O}(\sqrt{k}))$. *(The tilde hides polylog factors.)*
(b) If G *is non-giant then we can list* $\mathrm{PISO}(G, H)$ *in time* $\exp(\widetilde{O}(\sqrt{k}))$.
(c) We can find the coset $\mathrm{PISO}(G, H)$ *in quasipolynomial time* $(\exp(\mathrm{polylog}(k)))$.

The proof requires the classification of finite simple groups via Cameron's classification of the large primitive groups [5].

Structure trees. For a transitive group $G \leq \mathrm{Sym}(\Omega)$, a *$G$-invariant tree* is a rooted tree whose set of leaves is Ω and to which the G-action extends as tree automorphisms. (Such extension is necessarily unique.) A G-invariant tree is a *structure tree* for G if every internal node has at least 2 children, and for every internal node u of the tree, the action of the stabilizer G_u on the set of children of u is primitive. The following observation will allow us to list all structure trees of a transitive group.

Lemma 2. *Let* $G \leq \mathrm{Sym}(\Omega)$ *be a transitive permutation group of degree* k. *Then (a) G has at most* $k^{2 \log k}$ *structure trees; (b) we can list all structure trees of G in time* $O(k^{2 \log k + O(1)})$.

Subdirect products, diagonals. Given groups $G_1, \ldots, G_r$, we write $\pi_j \colon \prod_{i=1}^r G_i \to G_j$ for the projection map of the direct product onto the j-th factor. A *subdirect product* of the G_i is a subgroup $H \leq \prod_{i=1}^r G_i$ such that $\pi_j(H) = G_j$ for each j. A particularly important example of subdirect products is a diagonal. Let $V_1, \ldots, V_r$ be isomorphic groups, $(\forall i)(V_i \cong T)$. A *diagonal* of $(V_1, \ldots, V_r)$ is an embedding $\phi : T \hookrightarrow \prod_{i=1}^r V_i$ such that $\mathrm{Im}(\phi)$ is a subdirect product of the V_i. Its image is denoted $\mathrm{diag}(V_1 \times \cdots \times V_r)$. The *standard diagonal* of T^r is the map $\Delta : t \mapsto (t, \ldots, t)$.

---

[2] We remark that S_k and A_k are primitive for $k \geq 3$. We look at $k \geq 5$ and $k \neq 6$ since A_k is simple when $k \geq 5$; and $\mathrm{Aut}(A_k) \cong S_k$ when $k \geq 4$, $k \neq 6$.

For permutation groups, analogously, we can define permutational diagonals. Let $V_i \leq \mathrm{Sym}(\Omega_i)$ (for $i \in [r]$) be permutation groups, permutationally isomorphic to $T \leq \mathrm{Sym}(\Xi)$. (In particular, for every i, $|\Omega_i| = |\Xi|$.) A *permutational diagonal* of $\prod_{i=1}^{r} V_i$ is a list of permutational isomorphisms $\phi_i \in \mathrm{PISO}(T, V_i)$ (so $\widehat{\phi}_i : T \to V_i$ is the corresponding isomorphism). Let $\widehat{\phi} : T \to \prod V_i$ be defined by $t \mapsto (t^{\widehat{\phi}_1}, \ldots, t^{\widehat{\phi}_r})$. We use $\mathrm{pdiag}(V_1, \times \cdots \times V_r)$ to denote the image $\widehat{\phi}(T)$ for some permutational diagonal of $(V_1, \ldots, V_r)$.

For example, given $G \leq \mathrm{Sym}(\Omega)$, consider the induced action of G^r on Ω^r. The standard diagonal T of G^r acting on $\{(\omega, \ldots, \omega) \mid \omega \in \Omega\}$ is a permutational diagonal defined by the identity bijections.

Fact 6. *Let $G \leq H_1 \times \cdots \times H_m$ be a subdirect product, where each H_i is a simple group. Then G is a direct product of diagonals.*

Fact 7. *Let $G \leq \mathrm{Alt}(\Omega_1) \times \cdots \times \mathrm{Alt}(\Omega_m)$, where $(\forall i)(|\Omega_i| \geq 5, \neq 6)$ be a subdirect product of alternating groups. Then G is a direct product of permutational diagonals.*

4.2   Transitive Groups: Outline of the Proof of Theorem 3

Let $G \leq \mathrm{Sym}(\Omega)$ and $H \leq \mathrm{Sym}(\Delta)$ be transitive permutation groups of degree $k = |\Omega| = |\Delta|$. Our job is to list all their permutational isomorphisms in time $c^k |G|$. Our strategy is to fix a structure tree of G and work by induction on its depth. The base case is when G is primitive; this is settled by Lemma 1.

We call a structure tree T of G and a structure tree U of H *compatible* if their depth is the same and, for every ℓ, the primitive groups arising on level ℓ in G and H (as actions of the stabilizers of a node on level ℓ on the children of that node) are permutationally isomorphic.

By Lemma 2, we can try all structure trees of H that are compatible with the chosen structure tree of G. So we may assume we have fixed structure trees T and U on G and H, resp., and we are looking for permutational isomorphisms respecting them. Assume these trees have depth $d \geq 1$ (the root is level 0). Let G^* denote the action of G on $T(d-1)$: the set of nodes at level $d-1$; define H^* analogously. Assume by induction that the set $\mathrm{PISO}(G^*, H^*)$ is available. Now, for each $\pi \in \mathrm{PISO}(G^*, H^*)$ we wish to list the set $\mathrm{PISO}(G, H, \pi)$ of extensions of π to elements of $\mathrm{PISO}(G, H)$.

For $i \in T(d-1)$ let Ω_i denote the set of children of i and let $G(i)$ denote the action of G_i, the stabilizer of i, restricted to Ω_i. Note that $\{\Omega_1, \ldots, \Omega_m\}$ is a maximal system of imprimitivity for G, where $m = |T(d-1)|$. For $j \in U(d-1)$, define Δ_j and $H(j)$ analogously. Since T and U (the structure trees) are compatible, all the $G(i)$ and $H(j)$ are permutationally isomorphic primitive groups. Let K be the pointwise stabilizer of $T(d-1)$ in G, and L the corresponding stabilizer in H. So $K \lhd G$ is the kernel of the restriction homomorphism $G \to G^*$; analogously for $L \lhd H$. Let $K(i)$ be the restriction of K to Ω_i; and $L(j)$ the restriction of L to Δ_j. We now distinguish two cases.

Case 1. $G(i)$ is not a giant. In this case, by Lemma 1, the number of permutational isomorphisms between $G(i)$ and $H(i^\pi)$ is at most c^h where $h = k/m = |\Omega_i|$. We can combine these in $(c^h)^m = c^k$ ways, settling this case.

Case 2. $G(i)$ is a giant. This case is more technical. For simplicity, in this case we shall assume $K(i) = \mathrm{Alt}(\Omega_i)$ in this outline. The basic observation is that $K(i) \lhd G(i)$ and therefore either $K(i) = \{\mathrm{id}\}$ (Case 2a) or $K(i)$ is a giant (Case 2b).

Case 2a. $K(i) = \{\mathrm{id}\}$. If this is true for some i, it is true for all. Therefore, $K = \{\mathrm{id}\}$, so G embeds in G^*, i.e., π has at most one extension. Let us call this unique extension φ if it exists. Let $x \in \Omega_i$; we wish to find $y = x^\varphi \in \Delta_{i^\pi}$. One can show that if such a y exists then it is unique, and to find such a y it is sufficient to consider the stabilizer G_x and look for a corresponding stabilizer H_y for some $y \in \Delta_{i^\pi}$.

Case 2b. $K(i) = \mathrm{Alt}(\Omega_i)$. Clearly, $\mathrm{PISO}(G, H) \subseteq \mathrm{PISO}(K, L)$. While this is always true, under Case 2b we shall also see that the set $\mathrm{PISO}(K, L; \pi)$ has "affordable size," so we can list $\mathrm{PISO}(K, L; \pi)$ and check each of its elements (Proposition 1). Assuming $h \geq 5$, $h \neq 6$, we can bound the number of extensions of π by $2^m |K|$, and list them in time $O(2^m |K| k)$.

5   Semisimple Group Isomorphism: The Proof of Theorem 1

5.1   The Framework

Recall our strategy from Section 1.2. If G is semisimple then $\mathrm{Soc}(G)$ is the direct product of its minimal normal subgroups; and each minimal normal subgroup is the direct product of isomorphic nonabelian simple groups. Both of these decompositions are unique. The conjugation action of G permutes the simple factors of the socle; this defines a permutation representation $G \to S_k$ where k is the number of simple factors of the socle. The kernel of this representation is denoted $\mathrm{Pker}(G)$; this is the first characteristic subgroup in the BB chain [2]. So $G/\mathrm{Pker}(G)$ is a permutation group of degree $k \leq \log_{60} |G|$. The orbits of this permutation representation correspond to the minimal normal subgroups of G; in particular, it is transitive exactly if G has a unique minimal normal subgroup (namely, the socle).

Lumping together isomorphic minimal normal subgroups, we obtain the product decomposition $\mathrm{Soc}(G) = \prod_{i=1}^d \prod_{j=1}^{z_i} N_{i,j} \cong \prod_{i=1}^d K_i^{z_i}$, where the $N_{i,j}$ are the minimal normal subgroups and $(\forall i, j)(N_{i,j} \cong K_i)$. The K_i are pairwise non-isomorphic. Let $N_{ij} = \prod_{h=1}^{t_i} V_{ijh}$ be the decomposition of N_{ij} into simple factors. Let $T_i \cong V_{ijh}$ be a canonical copy of the simple factors of $K_i \cong N_{ij}$.

Our first trick is to fix an isomorphism between T_i and each V_{ijh}, i.e., a diagonal of $\prod_{j,h} V_{ijh}$ for each i. Since T_i is generated by two elements, we have $|\mathrm{Aut}(T_i)| \leq |T_i|^2$, and therefore the number of diagonals we need to consider is at most $|\mathrm{Soc}(G)|^2$. Let ϕ and ψ be diagonals of $\mathrm{Soc}(G)$ and $\mathrm{Soc}(H)$, resp., with respect to their factorizations into simple factors. We write $\mathrm{ISOds}(G, H; \phi, \psi)$ for the set of $G \to H$ isomorphisms which respect these diagonals.

Observation 8 ([3, Corollary 3.1]) *If G and H are semisimple groups then (a) any isomorphism* $f : \mathrm{Soc}(G) \to \mathrm{Soc}(H)$ *extends in at most one way to a* $\bar{f} : G \to H$ *isomorphism; and, (b) given f we can decide if* $\bar{f}$ *exists, and find it if it does, in polynomial time.*

Part (a) of the observation follows from a well-known fact [17, Claim 3.3.19, page 90] (cf. [6, sec. 3.1], [3, Lemma 3.1]). It follows that an isomorphism $\chi \in$ ISOds$(G, H; \phi, \psi)$ is uniquely determined by the permutational isomorphism it induces between $G/\mathrm{Pker}(G)$ and $H/\mathrm{Pker}(H)$.

5.2   Semisimple Groups with a Unique Minimal Normal Subgroup

Corollary 1. *Let G and H be semisimple groups with a unique minimal normal subgroup. (a)* $|\mathrm{Aut}(G)| \leq |G|^{O(1)}$, *and (b) we can list* ISO(G, H) *in time* $|G|^{O(1)}$.

Proof. Fix a diagonal ϕ of $\mathrm{Soc}(G)$ and consider all diagonals ψ of $\mathrm{Soc}(H)$; in each case, we shall compute ISOds$(G, H; \phi, \psi)$. Because simple groups have 2 generators, the latter needs to be performed $\leq |\mathrm{Soc}(G)|^2 \leq |G|^2$ times (cf. [3, Lemma 4.1]).

By part (a) of Observation 8, every isomorphism $\chi \in$ ISOds$(G, H; \phi, \psi)$ is uniquely determined by the permutational isomorphism it induces between $G/\mathrm{Pker}(G)$ and $H/\mathrm{Pker}(H)$. Therefore the number of automorphisms respecting ϕ is at most the number of permutational automorphisms of $G/\mathrm{Pker}(G)$, which in turn is at most $c^k|G/\mathrm{Pker}(G)|$, by part (a) of Theorem 3. Since $k = O(\log |G|)$, this proves part (a). For part (b), we apply the algorithm in part (b) of Theorem 3 to the transitive permutation groups $G/\mathrm{Pker}(G)$ and $H/\mathrm{Pker}(H)$ and list all $\pi \in$ PISO$(G/\mathrm{Pker}(G), H/\mathrm{Pker}(H))$. For each such π, let f be the isomorphism of $\mathrm{Soc}(G)$ and $\mathrm{Soc}(H)$ determined by π and the diagonals ϕ, and ψ. For each such f, apply part (b) of Observation 8 to check whether f extends to an isomorphism $\bar{f} : G \to H$.   □

5.3   All Semisimple Groups

In this subsection we outline the reduction of testing isomorphism of the semisimple groups G, H to twisted code equivalence. Since isomorphism of the socles and their direct decomposition to minimal normal subgroups are easy to test, we may assume that $\mathrm{Soc}(G) = \mathrm{Soc}(H)$ and they have the same decomposition into minimal normal subgroups, using the notation from Section 5.1. The conjugation action of G on $\mathrm{Soc}(G)$ embeds G into $\prod_i \prod_j \mathrm{Aut}(N_{ij})$; we call this embedding α, and let $G^* = \alpha(G)$; we define the embedding β and $H^* = \beta(H)$ analogously. We view G^* and H^* as permutation groups acting on $\mathrm{Soc}(G)$.

Having fixed diagonals as in Section 5.1, we are only interested in those permutational isomorphisms $G^* \to H^*$ which act on the socles by permuting the copies of the K_i and within them, permuting the copies of T_i, respecting their standard diagonals. Let X be the set of these $G^* \to H^*$ permutational isomorphisms (given as a coset). If we knew X, by [17, Claim 3.3.19, page 90], we can recover the coset of $G \to H$ isomorphisms respecting the diagonals by the formula $\alpha X \beta^{-1}$. We now describe how to find X.

Let G_{ij}^* denote the restriction of G^* to N_{ij}, and similarly H_{ij}^* the restriction of H^* to N_{ij}. By Cor. 1, we can compute the isomorphism types of the G_{ij}^* and H_{ij}^* under permutational isomorphisms in X. Let $\Gamma_1, \ldots, \Gamma_r$ be representatives of these isomorphism types. These will be our alphabets. For each i, j, pick an arbitrary isomorphism φ_{ij} of this type between G_{ij}^* and the corresponding representative, and analogously for H^*.

We create codes $\overline{G}$ and $\overline{H}$ over the alphabets Γ_i as follows. Let $\sigma \in G^*$, and let $\sigma_{ij} \in G_{ij}^*$ denote the restriction of σ to N_{ij}. The string associated with σ is $\overline{\sigma} = (\sigma_{ij}^{\varphi_{ij}})_{ij}$. Then $\overline{G} = \{\overline{\sigma} \mid \sigma \in G^*\}$. Define $\overline{H}$ analogously.

For $\ell \in [r]$, let W_ℓ be the group of automorphisms of Γ_ℓ induced by permutational automorphisms that preserve the standard diagonals. These automorphisms are determined by the permutation they induce on the set of simple factors. Thus if Γ_ℓ is acting on a copy of $K_i (= T_i^{t_i})$ then W_ℓ has a faithful permutation representation of degree t_i.

Let $\chi : G^* \to H^*$ be a permutational isomorphism respecting the standard diagonals (i.e., $\chi \in X$). So χ induces a permutation $\pi(\chi)$ of the minimal normal subgroups. $\pi = \pi(\chi)$ determines χ up to elements of the W_ℓ applied to each letter of the codes $\overline{G}$ and $\overline{H}$. In other words, the set X of $G^* \to H^*$ isomorphisms we are looking for corresponds exactly to the set of (W_ℓ)-twisted equivalences of $\overline{G}$ and $\overline{H}$.

Analysis. Fixing diagonals will only add a factor of $|\mathrm{Soc}(G)|^2 \leq |G|^2$. The algorithm for groups with a unique minimal normal subgroup (part (b) of Corollary 1) takes polynomial time. The length of the codes is the number of minimal normal subgroups, which is $O(\log|G|)$. The alphabets Γ_ℓ are subgroups of $G^* \cong G$ and hence $(\forall \ell)(|\Gamma_\ell| \leq |G|)$. The groups W_ℓ have polynomial size by part (a) of Corollary 1, and faithful permutation representations of degree t_i. Therefore the permutation group where we will perform coset intersection will have a permutation representation of degree $k = \sum_{i,j} t_i$, the number of simple factors of $\mathrm{Soc}(G)$, which is $O(\log|G|)$. Finally the size of the codes themselves is the order of the groups. Therefore the total running time of the twisted code equivalence algorithm is polynomial. □

6   Comparison with Prior Work

A 2003 paper by Cannon and Holt [6] describes a practical method to test isomorphism of permutation groups. Sec. 3 of their paper is dedicated to semisimple groups, underlining the significance of this class. Naturally, our framework is based on the same simple structural observations regarding the socle as theirs (Sec. 5.1); the most notable common element is part (a) of Obs. 8. After these initial observations, the two algorithms diverge in accordance with their very different goals: [6] describes heuristic algorithms with no performance guarantees and with reference to programs that use backtracking which would count as illegal steps for us; [6] reports practical efficiency. We devise algorithms which take time, polynomial in the *order* of the group, a prohibitive cost in their context.

References

1. Aschbacher, M., Guralnick, R.: Some applications of the first cohomology group. J. Algebra 90(2), 446–460 (1984)
2. Babai, L., Beals, R.: A polynomial-time theory of black-box groups I. In: Groups St Andrews 1997 in Bath. LMS Lect. Notes, vol. 260, pp. 30–64. Cambr. U. Press (1999)
3. Babai, L., Codenotti, P., Grochow, J.A., Qiao, Y.M.: Code equivalence and group isomorphism. In: Proc. 22nd SODA, pp. 1395–1408 (2011)
4. Babai, L., Qiao, Y.M.: Polynomial-time isomorphism test for groups with abelian Sylow towers. In: 29th STACS, pp. 453–464 (2012)
5. Cameron, P.J.: Finite permutation groups and finite simple groups. Bull. London Math. Soc. 13(1), 1–22 (1981)
6. Cannon, J.J., Holt, D.F.: Automorphism group computation and isomorphism testing in finite groups. J. Symb. Comput. 35, 241–267 (2003)
7. Felsch, V., Neubüser, J.: On a programme for the determination of the automorphism group of a finite group. In: Proc. Conf. on Computational Problems in Algebra, Oxford, 1967, pp. 59–60. Pergamon Press (1970)
8. Furst, M.L., Hopcroft, J., Luks, E.M.: Polynomial-time algorithms for permutation groups. In: Proc. 21st FOCS, pp. 36–41. IEEE Comp. Soc. (1980)
9. Kavitha, T.: Linear time algorithms for abelian group isomorphism and related problems. J. Comput. Syst. Sci. 73(6), 986–996 (2007)
10. Knuth, D.E.: Efficient representation of perm groups. Combinat. 11, 57–68 (1991)
11. Le Gall, F.: Efficient isomorphism testing for a class of group extensions. In: 26th STACS, pp. 625–636 (2009)
12. Luks, E.M.: Hypergraph isomorphism and structural equivalence of boolean functions. In: Proc. 31st ACM STOC, pp. 652–658. ACM Press (1999)
13. Luks, E.M., Miyazaki, T.: Polynomial-time normalizers for permutation groups with restricted composition factors. In: 13th ISAAC, pp. 176–183 (2002)
14. Luks, E.M., Seress, Á.: Computing the Fitting subgroup and solvable radical for small-base permutation groups in nearly linear time. In: Workshop on Groups and Computation II, DIMACS Series in DMTCS, pp. 169–181 (1991)
15. Miller, G.L.: On the $n^{\log n}$ isomorphism technique. In: 10th STOC, pp. 51–58 (1978)
16. Qiao, Y.M., Sarma, J.M.N., Tang, B.: On isomorphism testing of groups with normal Hall subgroups. In: Proc. 28th STACS, pp. 567–578 (2011)
17. Robinson, D.J.S.: A Course in the Theory of Groups, 2nd edn. Springer (1996)
18. Seress, Á.: Permutation Group Algorithms. Cambridge Univ. Press (2003)
19. Sims, C.C.: Computation with permutation groups. In: Petrick, S.R. (ed.) Proc. 2nd Symp. Symb. Algeb. Manip., pp. 23–28. ACM Press (1971)
20. Steinberg, R.: Generators for simple groups. Canad. J. Math. 14, 277–283 (1962)
21. Suzuki, M.: Group Theory II. Springer (1986)
22. Wilson, J.B.: Decomposing p-groups via Jordan algebras. J. Algebra 322, 2642–2679 (2009)
23. Wilson, J.B.: Finding central decompositions of p-groups. J. Group Theory 12, 813–830 (2009)

Clustering under Perturbation Resilience

Maria Florina Balcan and Yingyu Liang

School of Computer Science, Georgia Institute of Technology
ninamf@cc.gatech.edu, yliang39@gatech.edu

Abstract. Motivated by the fact that distances between data points in many real-world clustering instances are often based on heuristic measures, Bilu and Linial [6] proposed analyzing objective based clustering problems under the assumption that the optimum clustering to the objective is preserved under small multiplicative perturbations to distances between points. In this paper, we provide several results within this framework. For separable center-based objectives, we present an algorithm that can optimally cluster instances resilient to $(1 + \sqrt{2})$-factor perturbations, solving an open problem of Awasthi et al. [2]. For the k-median objective, we additionally give algorithms for a weaker, relaxed, and more realistic assumption in which we allow the optimal solution to change in a small fraction of the points after perturbation. We also provide positive results for min-sum clustering which is a generally much harder objective than k-median (and also non-center-based). Our algorithms are based on new linkage criteria that may be of independent interest.

Keywords: clustering, perturbation resilience, k-median, min-sum.

1 Introduction

Problems of clustering data from pairwise distance information are ubiquitous in science. A common approach for solving such problems is to view the data points as nodes in a weighted graph (with the weights based on the given pairwise information), and then to design algorithms to optimize various objective functions such as k-median or min-sum. For example, in the k-median clustering problem the goal is to partition the data into k clusters C_i, giving each a center c_i, in order to minimize the sum of the distances of all data points to the centers of their cluster. In the min-sum clustering approach the goal is to find k clusters C_i that minimize the sum of all intra-cluster pairwise distances. Yet unfortunately, for most natural clustering objectives, finding the optimal solution to the objective function is NP-hard. As a consequence, there has been substantial work on approximation algorithms [1,5,7,8,9] with both upper and lower bounds on the approximability of these objective functions on worst case instances.

Recently, Bilu and Linial [6] suggested an exciting, alternative approach aimed at understanding the complexity of clustering instances which arise in practice. Motivated by the fact that distances between data points in clustering instances are often based on a heuristic measure, they argue that interesting instances should be resilient to small perturbations in these distances. In particular, if small perturbations can cause the optimal clustering for a given objective to change drastically, then that probably is not

A. Czumaj et al. (Eds.): ICALP 2012, Part I, LNCS 7391, pp. 63–74, 2012.

a meaningful objective to be optimizing. They specifically define an instance to be α-perturbation resilient for an objective Φ if perturbing pairwise distances by multiplicative factors in the range $[1, \alpha]$ does not change the optimum clustering under Φ. They consider in detail the case of max-cut clustering and give an efficient algorithm to recover the optimum when the instance is resilient to perturbations on the order of $O(\sqrt{n\Delta})$ where n is the number of points and Δ is the maximum degree of the graph. They also give an efficient algorithm for instance of unweighted max-cut that is resilient to perturbations on the order of $O(\frac{n}{\delta})$ where δ is the minimum degree of the graph.

Two important questions raised by the work of Bilu and Linial [6] are: (1) the degree of resilience needed for their algorithm to succeed is quite high: can one develop algorithms for important clustering objectives that require much less resilience? (2) the resilience definition requires the optimum solution to remain *exactly* the same after perturbation: can one succeed under weaker conditions? In the context of *separable center-based* objectives such as k-median and k-center, Awasthi et al. [2] partially address the first question and show that an algorithm based on the single-linkage heuristic can efficiently find the optimal clustering for α-perturbation-resilient instances for $\alpha = 3$. They also conjecture it to be NP-hard to beat 3 and prove beating 3 is NP-hard for a related notion.

In this work, we address both questions raised by Bilu and Linial [6] and additionally improve over Awasthi et al. [2]. First, for separable center-based objectives we design a polynomial time algorithm for finding the optimum for instances resilient to perturbations of value $\alpha = 1 + \sqrt{2}$, thus beating the previously best known factor of 3 of Awasthi et al. [2]. Second, for k-median, we consider a weaker, relaxed, and more realistic notion of perturbation-resilience where we allow the optimal clustering of the perturbed instance to differ from the optimal of the original in a small ϵ fraction of the points. This is arguably a more natural though also more difficult condition to deal with. We give positive results for this case as well, showing for somewhat larger values of α that we can still achieve a near-optimal clustering. We additionally give positive results for min-sum clustering which is a generally much harder objective than k-median (and also non-center-based). For example, the best known guarantee for min-sum clustering on worst-case instances is an $O(\delta^{-1} \log^{1+\delta} n)$-approximation in time $n^{O(1/\delta)}$ due to [5]; by contrast, the best guarantee known for k-median is factor $3 + \epsilon$ due to [1].

Our results are achieved by carefully deriving structural properties of perturbation-resilience. At a high level, all the algorithms we introduce work by first running appropriate linkage procedures to produce a tree, and then running dynamic programming to retrieve the best k-clustering in the tree. To ensure that (under perturbation resilience) the tree output in the first step has a low-cost pruning, we derive new linkage procedures (closure linkage and approximate closure linkage) which are of independent interest.

Our Results: We provide several results for clustering perturbation-resilient instances in the metric space for separable center-based objectives and for the min-sum objective.

In Section 3 we improve on the bounds of Awasthi et al. [2] for α-perturbation resilient instances for separable center-based objectives, giving an algorithm that efficiently [1] finds the optimum for $\alpha = 1 + \sqrt{2}$. Commonly used separable center-based

---

[1] For clarity, efficient means polynomial in n (number of points) and k (number of clusters).

objectives, such as k-median, are NP-hard to even approximate, yet we can recover the exact solution for perturbation resilient instances. Our algorithm is based on a new linkage procedure using a new notion of distance (closure distance) between sets that may be of independent interest.

In Section 4 we consider the more challenging and more general notion of (α, ϵ)-perturbation resilience for k-median, where we allow the optimal solution after perturbation to be ϵ-close to the original. We provide an efficient algorithm which for $\alpha > 2 + \sqrt{7}$ produces $(1 + O(\epsilon/\rho))$-approximation to the optimum, where ρ is the fraction of the points in the smallest cluster. The key property we derive and exploit is that, except for ϵn bad points, most points are α closer to their own center than to any other center. Using this, we then design an approximate version of the closure linkage criterion that allows us to carefully eliminate the noise introduced by the bad points and construct a tree with a low-cost pruning that is a good approximation to the optimum.

In Section 5 we provide the first efficient algorithm for optimally clustering α-min-sum perturbation resilient instances. Our algorithm is based on an appropriate modification of average linkage that exploits the structure of such instances.

Due to the lack of space we only provide sketches for most proofs in this paper. Full proofs appear in the long version of the paper [4]. In the long version, we also provide sublinear-time algorithms, showing algorithms that can return an implicit clustering from only access to a small random sample.

2   Notation and Preliminaries

In a clustering instance, we are given a set S of n points in a finite metric space, and we denote $d : S \times S \to \mathbb{R}_{\geq 0}$ as the distance function. Φ denotes the objective function over a partition of S into $k < n$ clusters which we want to optimize, i.e. Φ assigns a score to every clustering. The optimal clustering w.r.t. Φ is denoted as $\mathcal{C} = \{C_1, C_2, \ldots, C_k\}$, and its cost is denoted as $\mathcal{OPT}$. The core concept we study in this paper is the perturbation resilience notion introduced by Bilu and Linial [6]. Formally:

Definition 1. *A clustering instance (S, d) is α-perturbation resilient to an objective Φ if for any $d' : S \times S \to \mathbb{R}$ s.t. $\forall p, q \in S, d(p, q) \leq d'(p, q) \leq \alpha d(p, q)$, there is a unique optimal clustering C' for Φ under d' that equals the optimal clustering C under d.*

In this paper, we focus on center-based and min-sum objectives. For center-based objectives, we consider *separable center-based objectives* defined by Awasthi et al. [2].

Definition 2. *A clustering objective is **center-based** if the solution can be defined by partitioning S into k clusters $\mathcal{P} = \{P_1, P_2, \ldots, P_k\}$ and assigning a set of centers $\mathbf{p} = \{p_1, p_2, \ldots, p_k\} \subseteq S$ for the clusters. Such an objective is **separable** if it furthermore satisfies the following two conditions: 1) The objective function value of a given clustering is either a (weighted) sum or the maximum of the individual cluster scores; 2) Given a proposed single cluster, its score can be computed in polynomial time.*

For example, for the k-median objective which we study substantially, the objective is $\Phi(\mathcal{P}, \mathbf{p}) = \sum_{i=1}^{k} \sum_{p \in P_i} d(p, p_i)$. Other examples of center-based objectives include k-means for which $\Phi(\mathcal{P}, \mathbf{p}) = \sum_{i=1}^{k} \sum_{p \in P_i} d^2(p, p_i)$, and k-centers for which

$\Phi(\mathcal{P}, \mathbf{p}) = \max_{i=1}^{k} \max_{p \in P_i} d(p, p_i)$. The centers in the optimal solution are denoted as $\mathbf{c} = \{c_1, \ldots, c_k\}$. Clearly, in an optimal solution, each point is assigned to its nearest center. In such cases, the objective is denoted as $\Phi(\mathbf{c})$.

We also consider a different type of objective function: the *min-sum objective*. For this objective, S is partitioned into k clusters $\mathcal{P} = \{P_1, P_2, \ldots, P_k\}$, and the goal is to minimize $\Phi(\mathcal{P}) = \sum_{i=1}^{k} \sum_{p,q \in P_i} d(p, q)$.

In Section 4 we consider a generalization of Definition 1 where we allow a small difference between the original and the new optimum after perturbation. Formally:

Definition 3. *Let $\mathcal{C}$ be the optimal k-clustering and $\mathcal{C}'$ be another k-clustering of a set of n points. We say $\mathcal{C}'$ is ϵ-**close** to $\mathcal{C}$ if $\min_{\sigma \in S_k} \sum_{i=1}^{k} |C_i \setminus C'_{\sigma(i)}| \le \epsilon n$, where σ is a matching between indices of clusters of $\mathcal{C}'$ and those of $\mathcal{C}$.*

Definition 4. *A clustering instance (S, d) is (α, ϵ)-**perturbation resilient** to an objective Φ if for any $d' : S \times S \to \mathbb{R}$ s.t. $\forall p, q \in S, d(p, q) \le d'(p, q) \le \alpha d(p, q)$, the optimal clustering $\mathcal{C}'$ for Φ under d' is ϵ-close to the optimal clustering $\mathcal{C}$ under d.*

For simplicity, we use shorthand $d(A, B) = \sum_{p \in A} \sum_{q \in B} d(p, q)$ and $d(p, B) = d(\{p\}, B)$. Also, we will sometimes assume that $\min_i |C_i|$ and ϵn is known. (Otherwise, we can simply search over the n possible different values for each parameter.)

3 α-Perturbation Resilience for Center-Based Objectives

In this section we show that, for $\alpha \ge 1 + \sqrt{2}$, if the clustering instance is α-perturbation resilient for separable center-based objectives, then we can efficiently find the optimal clustering. This improves on the $\alpha \ge 3$ bound of Awasthi et al. [2] and stands in sharp contrast to the NP-Hardness results on worst-case instances. Our algorithm succeeds for an even weaker property, the α-center proximity, introduced in Awasthi et al. [2].

Definition 5. *A clustering instance (S, d) satisfies the α-**center proximity** property if for any optimal cluster $C_i \in \mathcal{C}$ with center c_i, $C_j \in \mathcal{C} (j \ne i)$ with center c_j, any point $p \in C_i$ satisfies $\alpha d(p, c_i) < d(p, c_j)$.*

Lemma 1. *([2]) Any clustering instance that is α-perturbation resilient to separable center-based objectives also satisfies the α-center proximity.*

The proof follows by constructing a specific perturbation that blows up all the pairwise distances within C_i by a factor of α. By α-perturbation resilience, the optimal clustering remains the same, which then implies the desired result. In this section, we prove our results for α-center proximity. The results also hold for α-perturbation resilience since it implies α-center proximity. We begin with some key properties.

Lemma 2. *For any points $p \in C_i$ and $q \in C_j (j \ne i)$ in the optimal clustering of an α-center proximity instance, when $\alpha \ge 1 + \sqrt{2}$, we have:*
(1) $d(c_i, q) > d(c_i, p)$,   (2) $d(p, c_i) < d(p, q)$.

Proof. (1) By Lemma 1, $d(q, c_i) > \alpha d(q, c_j)$. By triangle inequality, $d(c_i, c_j) \leq d(q, c_j) + d(q, c_i) < (1 + \frac{1}{\alpha})d(q, c_i)$. Also, $d(p, c_j) > \alpha d(p, c_i)$ and thus $d(c_i, c_j) \geq d(p, c_j) - d(p, c_i) > (\alpha - 1)d(p, c_i)$. The result follows by these inequalities.
(2) It also follows from triangle inequality. The proof appears in [2]. □

Lemma 2 implies that for any optimal cluster C_i, the ball of radius $\max_{p \in C_i} d(c_i, p)$ around the center c_i contains *only* points from C_i, and moreover, points inside the ball are each closer to the center than to any point outside the ball. Inspired by this structural property, we define the notion of closure distance between two sets as the radius of the minimum ball that covers the sets and has some margin from points outside the ball. We show that any (strict) subset of an optimal cluster has smaller closure distance to another subset in the same cluster than to any subset or union of other clusters. Using this, we will be able to define an appropriate linkage procedure that produces a tree on subsets that will all be laminar with respect to the optimal clusters. This will then allow us to extract from the tree the optimal solution using dynamic programming. We now define the notion of closure distance and then present our algorithm.

Definition 6. *Let* $\mathbb{B}(p, r) = \{q : d(q, p) \leq r\}$. *The **closure distance** $d_S(A, A')$ between two disjoint non-empty subsets A and A' of point set S is the minimum $d \geq 0$ such that there is a point $c \in A \cup A'$ satisfying the following requirements:*
(1) coverage: the ball $\mathbb{B}(c, d)$ *covers A and A', i.e.* $A \cup A' \subseteq \mathbb{B}(c, d)$;
(2) margin: points inside $\mathbb{B}(c, d)$ *are closer to the center c than to points outside,*
   i.e. $\forall p \in \mathbb{B}(c, d), q \notin \mathbb{B}(c, d)$, *we have* $d(c, p) < d(p, q)$.

Note that for any A, A', $d_S(A, A') = d_S(A', A) \leq \max_{p, q \in S} d(p, q)$, and it can be computed in polynomial time.

Algorithm 1. Separable center-based objectives, α perturbation resilience

Input: Data set S, distance function $d(\cdot, \cdot)$ on S.
Phase 1: Begin with n singleton clusters.
- Repeat till only one cluster remains: merge clusters C, C' which minimize $d_S(C, C')$.
- Let T be the tree with single points as leaves and internal nodes corresponding to the merges.

Phase 2: Apply dynamic programming on T to get the minimum cost pruning $\tilde{C}$.
Output: Clustering $\tilde{C}$.

Theorem 1. *For* $(1 + \sqrt{2})$-*center proximity instances, Algorithm 1 outputs the optimal clustering in polynomial time.*

The proof follows from the following key property of the Phase 1 of Algorithm 1.

Theorem 2. *For* $(1 + \sqrt{2})$-*center proximity instances, Phase 1 of Algorithm 1 constructs a binary tree such that the optimal clustering is a pruning of this tree.*

Proof. We prove correctness by induction. In particular, assume that our current clustering is *laminar* to the optimal clustering – that is, for each cluster A in our current clustering and each C in the optimal clustering, we have either $A \subseteq C$, or $C \subseteq A$ or $A \cap C = \varnothing$. This is clearly true at the start. To prove that the merge steps preserve the laminarity, we need to show the following: if A is a strict subset of an optimal cluster C_i, A' is a subset of another optimal cluster or the union of one or more other clusters, then there exists B from $C_i \setminus A$ in the current clustering, such that $d_S(A, B) < d_S(A, A')$.

Let $d = \max_{p \in C_i} d(c_i, p), p^* = \arg\max_{p \in C_i} d(c_i, p)$. We first prove that there is a cluster $B \subseteq C_i \setminus A$ in the current clustering such that $d_S(A, B) \leq d$. There are two cases. First, if $c_i \notin A$, then define B to be the cluster in the current clustering that contains c_i. By induction, $B \subseteq C_i \setminus A$. Then we have $d_S(B, A) \leq d$ since there is $c_i \in B$, and (1) for any $p \in A \cup B$, $d(c_i, p) \leq d$, (2) for any $p \in S$ satisfying $d(c_i, p) \leq d$, and any $q \in S$ satisfying $d(c_i, q) > d$, by Lemma 2 we know $p \in C_i$ and $q \notin C_i$, and thus $d(c_i, p) < d(p, q)$. Second, if $c_i \in A$, we pick any $B \subseteq C_i \setminus A$ and a similar argument gives $d_S(A, B) \leq d$.

As a second step, we need to show that $d < \hat{d} = d_S(A, A')$. There are two cases: the center for $d_S(A, A')$ is in A or in A'. In the first case, there is a point $c \in A$ such that c and $\hat{d}$ satisfy the requirements of the closure distance. Pick a point $q \in A'$, and suppose C_j is the optimal cluster that contains q. As $d(c, q) \leq \hat{d}$, and by Lemma 2 $d(c_j, q) < d(c, q)$, we must have $d(c_j, c) \leq \hat{d}$ (otherwise it violates the second requirement of closure distance). Then we have $d = d(p^*, c_i) < d(p^*, c_j)/\alpha \leq (d + d(c_i, c) + d(c, c_j))/\alpha$ from Lemma 1 and triangle inequality. Since $d(c_i, c) < d(c, c_j)/\alpha$, we can combine the above inequalities and compare d and $d(c, c_j)$, and when $\alpha \geq 1 + \sqrt{2}$ we have $d < d(c, c_j) \leq \hat{d}$.

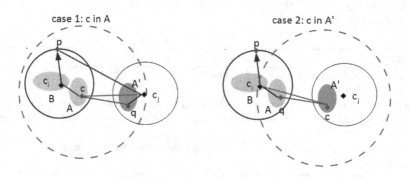

Fig. 1. Illustration for comparing d and $d_S(A, A')$ in Theorem 2

Now consider the second case, when there is a point $c \in A'$ such that c and $\hat{d}$ satisfy the requirements of the closure distance. Pick a point $q \in A$. We have $\hat{d} \geq d(c, q)$ from the first requirement, and $d(c, q) > d(c_i, q)$ by Lemma 2. Then from the second requirement $d(c_i, c) \leq \hat{d}$. So by Lemma 2, $d = d(c_i, p^*) < d(c_i, c) \leq \hat{d}$. □

Note: Our factor of $\alpha = 1 + \sqrt{2}$ beats the NP-hardness *lower bound* of $\alpha = 3$ of [2] for center proximity instances. The reason is that the lower bound requires the addition of Steiner points that can act as centers but are not part of the data to be clustered (though the upper bound of [2] does not allow such Steiner points). One can also show a lower bound for center proximity instances without Steiner points. In particular one can show that for any $\epsilon > 0$, solving $(2 - \epsilon)$-center proximity k-median instances is NP-hard [10].

4 (α, ϵ)-Perturbation Resilience for the k-Median Objective

In this section we consider a natural relaxation of the α-perturbation resilience, the (α, ϵ)-perturbation resilience, that requires the optimal clustering after perturbation to be ϵ-close to the original. We show that for (α, ϵ)-perturbation resilient instances, with $\alpha > 2 + \sqrt{7}$ and $\epsilon = O(\epsilon' \rho)$ where ρ is the fraction of the points in the smallest cluster, we can in polynomial time output a clustering that provides a $(1 + \epsilon')$-approximation to the optimum. Thus this improves over the best worst-case approximation guarantees known when $\epsilon' \leq 2$ and also beats the lower bound of $(1 + 2/e)$ on the best approximation achievable on worst case instances for metric k-median [9] when $\epsilon' \leq 1/e$.

The key idea is to understand and leverage the structure implied by (α, ϵ)-perturbation resilience. We show that perturbation resilience implies that there exists only a small fraction of points that are bad in the sense that their distance to their own center is not α times smaller than their distance to any other centers in the optimal solution. We then use this bounded number of bad points in our clustering algorithm.

4.1 Structure of (α, ϵ)-Perturbation Resilience

To understand (α, ϵ)-perturbation resilience, we need to consider the difference between the optimal clustering $\mathcal{C}$ under d and the optimal clustering $\mathcal{C}'$ under d', defined as $\min_{\sigma \in \mathcal{S}_k} \sum_{i=1}^{k} |C_i \setminus C'_{\sigma(i)}|$. Without loss of generality, we assume in this subsection that $\mathcal{C}'$ is indexed so that the argmin σ is the identity, and the difference is $\sum_{i=1}^{k} |C_i \setminus C'_i|$. We denote by c'_i the center of C'_i.

In the following we call a point *good* if it is α times closer to its own center than to any other center in the optimal clustering; otherwise we call it *bad*. Let B_i be the set of bad points in C_i. That is, $B_i = \{p : p \in C_i, \exists j \neq i, \alpha d(c_i, p) > d(c_j, p)\}$. Let $G_i = C_i \setminus B_i$ be the good points in cluster C_i. Let $B = \cup_i B_i$ and $G = \cup_i G_i$. We show that under perturbation resilience we do not have too many bad points. Formally:

Theorem 3. *Suppose the clustering instance is* (α, ϵ)*-perturbation resilient to* k*-median and* $\min_i |C_i| > \frac{6\alpha}{\alpha - 1} \epsilon n$. *Then* $|B| \leq \epsilon n$.

Here we describe a proof sketch of the theorem. In the full version we provide the detailed proof, and also point out that the bound in Theorem 3 is an optimal bound for the bad points in the sense that for any $\alpha > 1$ and $\epsilon < \frac{1}{5}$, we can construct an (α, ϵ)-perturbation resilient 2-median instance which has ϵn bad points.

Proof Sketch of [Theorem 3] The main idea is to construct a specific perturbation that forces certain selected bad points to move from their original optimal clusters. For technical reasons, we only perturb a selected subset of bad points, and show that they move out after perturbation. Then the (α, ϵ)-perturbation resilience leads to a bound on the number of selected bad points, which can also be proved to be a bound on all the bad points. The selected bad points $\hat{B}_i$ in cluster C_i are defined by arbitrarily selecting $\min(\epsilon n + 1, |B_i|)$ points from B_i. Let $\hat{B} = \cup_i \hat{B}_i$. For $p \in \hat{B}_i$, let $c(p) = \arg\min_{c_j, j \neq i} d(p, c_j)$ denote its second nearest center; for $p \in C_i \setminus \hat{B}_i$, $c(p) = c_i$. The perturbation we consider blows up all distances by a factor of α except for those distances between p and $c(p)$. Formally, we define d' as $d'(p, q) = d(p, q)$ if $p = c(q)$ or $q = c(p)$, and $d'(p, q) = \alpha d(p, q)$ otherwise.

The key challenge in proving a bound on the selected bad points is to show that $c_i' = c_i$ for all i, i.e., the optimal centers do not change after the perturbation. Then in the optimum under d' each point p is assigned to the center $c(p)$, and therefore the selected bad points ($\hat{B}$) will move from their original optimal clusters. By (α, ϵ)-perturbation resilience property we get an upper bound on the number of selected bad points.

Suppose C_i' is obtained by adding point set A_i and removing point set M_i from C_i, i.e. $A_i = C_i' \setminus C_i$, $M_i = C_i \setminus C_i'$. At a high level, we prove that $c_i = c_i'$ for all i as follows. We first show that for each cluster, its new center is close to its old center, roughly speaking since the new and old clusters have a lot in common (Claim 1). We then show if $c_i' \neq c_i$ for some i, then the weighted sum of the distances $\sum_{1 \leq i \leq k} |C_i| d(c_i, c_i')$ should be large (Claim 2). However, this contradicts Claim 1, so $c_i' = c_i$ for all i.

Claim 1. *For each i, $d(c_i, (C_i \cap C_i') \setminus \hat{B}_i) \geq \frac{\alpha+2}{\alpha+1} \frac{|C_i|}{3} d(c_i, c_i')$.*

Proof Sketch: The key idea is that under d', c_i' is the optimal center, so it has no more cost than c_i on C_i'. Since $\hat{B}_i \setminus M_i$ and A_i are small compared to $(C_i \cap C_i') \setminus \hat{B}_i$, c_i' cannot save much on $\hat{B}_i \setminus M_i$ and A_i, thus it cannot have much more cost on $(C_i \cap C_i') \setminus \hat{B}_i$ than c_i. Then c_i' is close to $(C_i \cap C_i') \setminus \hat{B}_i$, and so is c_i, then c_i' is close to c_i. Formally, we have $d'(c_i', C_i') \leq d'(c_i, C_i')$. We divide C_i' into $(C_i \cap C_i') \setminus \hat{B}_i$, $\hat{B}_i \setminus M_i$ and A_i, and move terms on $(C_i \cap C_i') \setminus \hat{B}_i$ to one side (the cost more than c_i on $(C_i \cap C_i') \setminus \hat{B}_i$), the rest terms to another side (the cost saved on $\hat{B}_i \setminus M_i$ and A_i). After translating from d' to d, we apply triangle inequality and obtain the claim. □

Claim 2. *Let $I_i = 1$ if $c_i \neq c_i'$ and $I_i = 0$ otherwise. Then we have*
$$\sum_{1 \leq i \leq k} I_i d(c_i, (C_i \cap C_i') \setminus \hat{B}_i) \leq \sum_{1 \leq i \leq k} \frac{|C_i|}{3} d(c_i, c_i').$$

Proof Sketch: The key idea is that the clustering that under d' assigns points in $C_i' \setminus \hat{B}_i$ to c_i and points p in $\hat{B}_i \setminus M_i$ to $c(p)$, saves much cost on $(C_i \cap C_i') \setminus \hat{B}_i$ compared to the optimal clustering $\{C_i'\}$ under d', if $c_i' \neq c_i$. Then $\{C_i'\}$ must save this cost on other parts of points. So $\{c_i'\}$ should be near these points and $\{c_i\}$ should be far away, and the weighted sum of the distances between $\{c_i'\}$ and $\{c_i\}$ should be large. Formally, $\sum_i d'(c_i', C_i') \leq \sum_i [d'(c_i, C_i' \setminus \hat{B}_i) + \sum_{p \in \hat{B}_i \setminus M_i} d'(c(p), p)]$ since $\{c_i'\}$ are the optimal centers for C_i' under d'. By dividing C_i' into A_i, $\hat{B}_i \setminus M_i$ and $(C_i \cap C_i') \setminus \hat{B}_i$, and by the fact $\alpha \sum_i d(c_i, C_i) \leq \alpha \sum_i d(c_i', C_i)$ since c_i are the optimal centers, we can show that $\{C_i'\}$ should save as much as approximately $(\alpha - 1) \sum_i d(c_i, (C_i \cap C_i') \setminus \hat{B}_i)$ cost on points other than $(C_i \cap C_i') \setminus \hat{B}_i$. Then the result follows by triangle inequality. □

These claims lead to $\sum_{1 \leq i \leq k} |C_i| d(c_i, c_i') [1 - (\alpha + 2) I_i / (\alpha + 1)] \geq 0$. If $I_i = 0$, then $d(c_i, c_i') = 0$; if $I_i = 1$, the coefficient of $d(c_i, c_i')$ is negative. So the left hand side is at most 0. Then all terms equal 0, i.e. $d(c_i, c_i') = 0 (1 \leq i \leq k)$. Then points in $\hat{B}_i$ will move to other clusters after perturbation, which means that $\hat{B}_i \subseteq M_i$, thus $\hat{B} \subseteq \cup_i M_i$. Then $|\hat{B}| \leq |\cup_i M_i| \leq \epsilon n$. In particular, $|\hat{B}_i| \leq \epsilon n$ for any i. Then $|B_i| \leq \epsilon n$, otherwise $|\hat{B}_i|$ would be $\epsilon n + 1$. So $\hat{B}_i = B_i$, and $\hat{B} = B$ and $|B| = |\hat{B}| \leq \epsilon n$. $\qquad\square$

4.2   Approximating the Optimal Clustering

Since (α, ϵ)-perturbation resilient instances have at most ϵn bad points, we can show that for $\alpha > 4$ such instances satisfy the ϵ-strict separation property (the property that after eliminating an ϵ fraction of the points, the remaining points are closer to points in their own cluster than to other points in different clusters). Therefore, we could use the algorithms in [3] to output a tree with a pruning ϵ-close to the optimal clustering. However, this pruning might not have a small cost and it is not clear how to retrieve a small cost clustering from the tree constructed by these generic algorithms. Here we design a new algorithm for obtaining a good approximation for (α, ϵ)-perturbation resilient instances. This algorithm first uses a novel linkage procedure based on an approximate version of the closure condition in Section 3 to construct a tree, and then processes the tree to output a desired clustering. We first define the approximate closure condition.

Definition 7. *Suppose C' is a clustering of S and $p, q \in S$.*
Let $U_{p,q}$ denote the set of clusters that are nearly contained in the ball $\mathbb{B}(p, d(p,q))$,
i.e. $U_{p,q} = \{C | C \in C', |C \setminus \mathbb{B}(p, d(p,q))| \leq \epsilon n, C \cap \mathbb{B}(p, d(p,q)) \neq \emptyset\}$.
*The ball $\mathbb{B}(p, d(p,q))$ satisfies the **approximate closure** condition with respect to C' if*
$|\cup_{C \in U_{p,q}} C| \geq \min_i |C_i| - \epsilon n$ and the following conditions are satisfied:
(1) approximate coverage: it covers most of $U_{p,q}$, i.e. $|\cup_{C \in U_{p,q}} C_i \setminus \mathbb{B}(p, d(p,q))| \leq \epsilon n$;[2]
(2) approximate margin: after removing a few points outside the ball, points inside
are closer to each other than to points outside, i.e. $\exists E \subseteq S \setminus \mathbb{B}(p, d(p,q)), |E| \leq \epsilon n$,
s.t. $\forall p_1, p_2 \in \mathbb{B}(p, d(p,q)), q_1 \in S \setminus \mathbb{B}(p, d(p,q)) \setminus E$, we have $d(p_1, p_2) < d(p_1, q_1)$.

We are now ready to present our main algorithm for the (α, ϵ)-perturbation resilient instances, Algorithm 2. Informally, it starts with singleton points in their own clusters. It then checks in increasing order of $d(p, q)$ whether the ball $\mathbb{B}(p, d(p,q))$ satisfies the approximate closure condition, and if so it merges all the clusters nearly contained within $\mathbb{B}(p, d(p,q))$. As we show below, the tree produced has a pruning that respects the optimal clustering. However, this pruning may contain more than k-clusters, so in the second phase, we clean the tree so that there is a pruning with k-clusters that coincides with the optimal clustering on the good points. Finally we run dynamic programming to get the minimum cost pruning, which provides a good approximation to the optimum.

Our main result in this section is Theorem 4, which follows from Lemma 3 for Phase 1 of the algorithm and Lemma 4 for Phase 2.

Theorem 4. *For (α, ϵ)-perturbation resilient instances to k-median, if $\alpha > 2 + \sqrt{7}$ and $\epsilon \leq \rho/8$ where $\rho = \min_i |C_i|/n$, then in polynomial time, Algorithm 2 outputs a tree $\tilde{T}$ that contains a pruning ϵ-close to the optimal clustering. Moreover, if $\epsilon \leq \rho\epsilon'/8$ where $\epsilon' \leq 1$, the clustering produced is a $(1 + \epsilon')$-approximation to the optimum.*

---

[2] Note that in the definition of $U_{p,q}$, each cluster in it has at most ϵn points outside $\mathbb{B}(p, d(p,q))$. But the approximate coverage is stronger: $U_{p,q}$, as a whole, can have at most ϵn outside.

Algorithm 2. k-median, (α, ϵ) perturbation resilience

Input: Data set S, distance function $d(\cdot, \cdot)$ on S, $\min_i |C_i|$, $\epsilon > 0$
Phase 1: Initialize C' to be the clustering with each singleton point being a cluster.
• Sort all the pairwise distances $d(p, q)$. For $d(p, q)$ in ascending order,
•   If $\mathbb{B}(p, d(p, q))$ satisfies approximate closure condition and $|U_{p,q}| > 1$, merge $U_{p,q}$.
• Construct the tree T with points as leaves and internal nodes corresponding to the merges.
Phase 2: If a node has only singleton points as children, delete his children; get T'.
• Assign any singleton node p to the non-singleton leaf of smallest median distance; get $\tilde{T}$.
Phase 3: Apply dynamic programming on $\tilde{T}$ to get the minimum cost pruning $\tilde{C}$.
Output: Clustering $\tilde{C}$, (optional) tree $\tilde{T}$.

Lemma 3. *If $\alpha > 2 + \sqrt{7}$, $\epsilon \le \rho/8$, then the tree T contains nodes $N_i(1 \le i \le k)$ such that $N_i \setminus B = C_i \setminus B$.*

Proof Sketch: For each i, we let $q_i^* = \arg\max_{q \in C_i \setminus B} d(c_i, q)$. The proof follows from two key facts: (1) If $C' \setminus B$ is laminar to $C \setminus B$ right before checking some $d(p, q)$, and $U_{p,q}$ contains both good points from C_i and $C_j (i \ne j)$, then $d(c_i, q_i^*)$ and $d(c_j, q_j^*)$ are checked before $d(p, q)$. (2) If $C' \setminus B$ is laminar to $C \setminus B$ right before checking $d(c_i, q_i^*)$, we have that right after checking $d(c_i, q_i^*)$ there is a cluster containing all the good points in cluster i and no other good points.

Consider any merge step s.t. $U_{p,q}$ contains good points from both C_i and $C_j (j \ne i)$. Fact (1) implies both $d(c_i, q_i^*)$ and $d(c_j, q_j^*)$ must have been checked, and then fact (2) implies all good points in C_i and C_j respectively have already been merged. So the laminarity is always satisfied. Then the lemma follows from fact (2).

We now prove fact (1). Suppose that there exist good points from C_i and C_j in $U_{p,q}$. From the laminarity assumption, the fact that clusters in $U_{p,q}$ have only ϵn points outside $\mathbb{B}(p, d(p, q))$ and $|B| \le \epsilon n$, we can show there exist good points $p_i \in C_i$ and $p_j \in C_j$ in $\mathbb{B}(p, d(p, q))$. When $\alpha > 2 + \sqrt{7}$ we can show $d(c_i, q_i^*) < d(p_i, p_j)/2$, and by triangle inequality $d(p_i, p_j)/2 \le d(p, q)$, so $d(p, q) > d(c_i, q_i^*)$. The same argument leads to $d(p, q) > d(c_j, q_j^*)$. So $d(c_i, q_i^*)$ and $d(c_j, q_j^*)$ are checked before $d(p, q)$.

We now prove fact (2). It is sufficient to show that $\cup_{C \in U_{c_i, q_i^*}} C \setminus B = C_i \setminus B$ and U_{c_i, q_i^*} satisfies the approximate closure condition. First, U_{c_i, q_i^*} contains no good points outside C_i by fact (1). Second, any C containing good points from C_i is in U_{c_i, q_i^*}. By fact (1), C has no good points outside C_i. Since $\mathbb{B}(c_i, d(c_i, q_i^*))$ contains all good points in C_i, C has only bad points outside the ball, so $C \in U_{c_i, q_i^*}$. We finally show U_{c_i, q_i^*} satisfies the approximate closure condition. Since in addition to all good points in C_i, $\cup_{C \in U_{c_i, q_i^*}} C$ can only contain bad points, it has at most ϵn points outside $\mathbb{B}(c_i, d(c_i, q_i^*))$, so approximate coverage condition is satisfied. And we can show for $\alpha > 2 + \sqrt{7}$, $2d(c_i, q_i^*)$ is smaller than the distance between any point in $\mathbb{B}(c_i, d(c_i, q_i^*))$ and any good point outside C_i. Then let $E = B \setminus \mathbb{B}(c_i, d(c_i, q_i^*))$, approximate margin condition is satisfied. We also have $|\cup_{C \in U_{c_i, q_i^*}} C| \ge |C_i \setminus B| \ge \min_i |C_i| - \epsilon n$. $\square$

Lemma 4. *If $\alpha > 2 + \sqrt{7}$, $\epsilon \le \epsilon' \rho/8$ where $\epsilon' \le 1$, then $\tilde{C}$ is a $(1 + \epsilon')$-approximation.*

Proof Sketch: By Lemma 3, T has a pruning $\mathcal{P}$ that contains $N_i(1 \le i \le k)$ and possibly some bad points, such that $N_i \setminus B = C_i \setminus B$. Therefore, each non-singleton leaf in T' has only good points from one optimal cluster and has more good points than bad points. This implies that each singleton good point in T' is assigned to a leaf that has good points from its own optimal cluster.

So after Phase 2, $\mathcal{P}$ in T becomes $\mathcal{P}' = \{N'_i\}$ in $\tilde{T}$ such that $N'_i \setminus B = C_i \setminus B$. It is sufficient to prove the cost of $\mathcal{P}'$ approximates $\mathcal{OPT}$, i.e. to bound the increase of cost caused by a bad point $p_j \in C_j$ ending up in $N'_i (i \neq j)$. There are two cases: p_j belongs to a non-singleton leaf node in T' or p_j is a singleton in T'. In either case, we can find $K = (\min_i |C_i| - \epsilon n)/2 - \epsilon n$ good points p_{it} from C_i in the leaf in which p_j ends up in $\tilde{T}$, and K good points p_{js} from C_j in any other leaf containing only good points from C_j, such that $d(p_j, p_{it}) \leq d(p_j, p_{js})$. Then $d(p_j, c_i) - d(p_j, c_j)$ can be bounded by

$$\frac{1}{K}\left\{\sum_{1 \leq t \leq K}[d(p_j, p_{it}) + d(p_{it}, c_i)] - \sum_{1 \leq s \leq K}[d(p_j, p_{js}) - d(p_{js}, c_j)]\right\} \leq \frac{1}{K}\mathcal{OPT}.$$

As $|B| \leq \epsilon n$, the cost of $\mathcal{P}'$ is $\leq (1 + \frac{\epsilon n}{K})\mathcal{OPT}$. Setting $\epsilon' \geq \frac{\epsilon n}{K}$ gives the lemma. $\square$

We note that approximate margin condition in the Definition 7 can be verified in $O(n^3)$ time by enumerating $p_1, p_2 \in \mathbb{B}(p, d(p, q)), q_1 \notin \mathbb{B}(p, d(p, q))$, and checking if there are no more than ϵn such q_1 that there exist p_1, p_2 violating the condition. So the algorithm runs in polynomial time.

5 α-Perturbation Resilience for the Min-Sum Objective

In this section we provide an efficient algorithm for clustering α-perturbation resilient instances for the min-sum k-clustering problem (Algorithm 3). We use the following notations: $d_{avg}(A, B) = d(A, B)/(|A||B|)$ and $d_{avg}(p, B) = d_{avg}(\{p\}, B)$.

Theorem 5. *For* $(3\frac{\max_i |C_i|}{\min_i |C_i| - 1})$-*perturbation resilient instances to min-sum, Algorithm 3 outputs the optimal min-sum k-clustering in polynomial time.*

Algorithm 3. Min-sum, α perturbation resilience

Input: Data set S, distance function $d(\cdot, \cdot)$ on S, $\min_i |C_i|$.
Phase 1: Connect each point with its $\frac{1}{2}\min_i |C_i|$ nearest neighbors.
• Initialize the clustering $\mathcal{C}'$ with each connected component being a cluster.
• Repeat till one cluster remains in $\mathcal{C}'$: merge clusters C, C' that minimize $d_{avg}(C, C')$.
• Let T be the tree with components as leaves and internal nodes corresponding to the merges.

Phase 2: Apply dynamic programming on T to get the minimum cost pruning $\tilde{\mathcal{C}}$.
Output: Output $\tilde{\mathcal{C}}$.

Proof Sketch: First we show that the α-perturbation resilience property implies that for any two optimal clusters C_i and C_j and any $A \subseteq C_i$, we have $\alpha d(A, C_i \setminus A) < d(A, C_j)$. This follows by considering the perturbation where $d'(p, q) = \alpha d(p, q)$ if $p \in A, q \in C_i \setminus A$ and $d'(p, q) = d(p, q)$ otherwise, and using the fact that the optimum does not change after the perturbation. This can be used to show that when $\alpha > 3\frac{\max_i |C_i|}{\min_i |C_i|-1}$ we have: (1) for any optimal clusters C_i and C_j and any $A \subseteq C_i$, $A' \subseteq C_j$ s.t. $\min(|C_i \setminus A|, |C_j \setminus A'|) > \min_i |C_i|/2$ we have $d_{avg}(A, A') > \min\{d_{avg}(A, C_i \setminus A), d_{avg}(A', C_j \setminus A')\}$; (2) for any point p in the optimal cluster C_i, twice its average distance to points in $C_i \setminus \{p\}$ is smaller than the distance to any point in other optimal cluster C_j. Fact (2) implies that for any point $p \in C_i$ its $|C_i|/2$ nearest neighbors are in the same optimal cluster, so the leaves of the tree T are laminar to the optimum clustering. Fact (1) can be used to show that the merges preserve the laminarity with the optimal clustering, so the minimum cost pruning of T will be the optimal clustering, as desired. See the full version for the details.   □

6   Discussion and Open Questions

In this work, we advance the line of research on perturbation resilience in clustering in multiple ways. For α-perturbation resilient instances, we improve on the known guarantees for center-based objectives and give the first analysis for min-sum. Furthermore, for k-median, we analyze and give the first algorithmic guarantees known for a relaxed but more challenging condition of (α, ϵ)-perturbation resilience, where an ϵ fraction of points are allowed to move after perturbation. We also give sublinear-time algorithms for k-median and min-sum under perturbation resilience in the long version.

A natural direction for future investigation is to explore whether one can take advantage of smaller perturbation factors for perturbation resilient instances in Euclidian spaces. More broadly, it would be interesting to explore other ways in which perturbation resilient instances behave better than worst case instances (e.g., natural algorithms converge faster).

Acknowledgments. This work was supported by NSF grant CCF-0953192, by AFOSR grant FA9550-09-1-0538, by a Microsoft Research Faculty Fellowship, and by a Google Research award.

References

1. Arya, V., Garg, N., Khandekar, R., Meyerson, A., Munagala, K., Pandit, V.: Local search heuristics for k-median and facility location problems. SIAM J. Comput. 33(3) (2004)
2. Awasthi, P., Blum, A., Sheffet, O.: Center-based clustering under perturbation stability. Inf. Process. Lett. 112(1-2), 49–54 (2012)
3. Balcan, M.F., Gupta, P.: Robust hierarchical clustering. In: COLT (2010)
4. Balcan, M.F., Liang, Y.: Clustering under Perturbation Resilience. CoRR, abs/1112.0826 (2011)
5. Bartal, Y., Charikar, M., Raz, D.: Approximating min-sum -clustering in metric spaces. In: STOC (2001)
6. Bilu, Y., Linial, N.: Are stable instances easy? In: Innovations in Computer Science (2010)
7. Charikar, M., Guha, S., Tardos, É., Shmoys, D.B.: A constant-factor approximation algorithm for the k-median problem. J. Comput. Syst. Sci. 65(1) (2002)
8. de la Vega, W.F., Karpinski, M., Kenyon, C., Rabani, Y.: Approximation schemes for clustering problems. In: STOC (2003)
9. Jain, K., Mahdian, M., Saberi, A.: A new greedy approach for facility location problems. In: STOC (2002)
10. Reyzin, L.: Data stability in clustering: A closer look. CoRR, abs/1107.2379 (2011)

Secretary Problems with Convex Costs*

Siddharth Barman, Seeun Umboh, Shuchi Chawla, and David Malec

University of Wisconsin–Madison
{sid,seeun,shuchi,dmalec}@cs.wisc.edu

Abstract. We consider online resource allocation problems where given a set of requests our goal is to select a subset that maximizes a value minus cost type of objective. Requests are presented online in random order, and each request possesses an adversarial value and an adversarial size. The online algorithm must make an irrevocable accept/reject decision as soon as it sees each request. The "profit" of a set of accepted requests is its total value minus a convex cost function of its total size. This problem falls within the framework of secretary problems. Unlike previous work in that area, one of the main challenges we face is that the objective function can be positive or negative, and we must guard against accepting requests that look good early on but cause the solution to have an arbitrarily large cost as more requests are accepted. This necessitates new techniques. We study this problem under various feasibility constraints and present online algorithms with competitive ratios only a constant factor worse than those known in the absence of costs for the same feasibility constraints. We also consider a multi-dimensional version of the problem that generalizes multi-dimensional knapsack within a secretary framework. In the absence of feasibility constraints, we present an $O(\ell)$ competitive algorithm where ℓ is the number of dimensions; this matches within constant factors the best known ratio for multi-dimensional knapsack secretary.

1 Introduction

We study online resource allocation problems under a natural profit objective: a single server accepts or rejects requests for service so as to maximize the total value of the accepted requests minus the cost imposed by them on the system. This model captures, for example, the optimization problem faced by a cloud computing service accepting jobs, a wireless access point accepting connections from mobile nodes, or an advertiser in a sponsored search auction deciding which keywords to bid on. In many of these settings, the server must make accept or reject decisions in an online fashion as soon as requests are received without knowledge of the quality of future requests. We design online algorithms with the goal of achieving a small competitive ratio—ratio of the algorithm's performance to that of the best possible (offline optimal) solution.

* This work was supported in part by NSF awards CCF-0643763 and CNS-0905134. A full version [6] of this paper can be found at http://arxiv.org/abs/1112.1136.

A. Czumaj et al. (Eds.): ICALP 2012, Part I, LNCS 7391, pp. 75–87, 2012.
© Springer-Verlag Berlin Heidelberg 2012

A classical example of online decision making is the secretary problem. Here a company is interested in hiring a candidate for a single position; candidates arrive for interview in *random order*, and the company must accept or reject each candidate following the interview. The goal is to select the best candidate with as high a probability as possible. What makes the problem challenging is that each interview merely reveals the rank of the candidate relative to the ones seen previously, but not the ones following. Nevertheless, Dynkin [12] showed that it is possible to succeed with constant probability using the following algorithm: unconditionally reject the first $1/e$ fraction of the candidates; then hire the next candidate that is better than all of the ones seen previously. Dynkin showed that as the number of candidates goes to infinity, this algorithm hires the best candidate with probability approaching $1/e$ and in fact this is the best possible.

More general resource allocation settings may allow picking multiple candidates subject to a certain feasibility constraint. We call such a problem a generalized secretary problem (GSP) and use $(\Phi, \mathcal{F})$ to denote an instance of the problem. Here $\mathcal{F}$ denotes a feasibility constraint that the set of accepted requests must satisfy (e.g. the size of the set cannot exceed a given bound), and Φ denotes an objective function that we wish to maximize. As in the classical setting, we assume that requests arrive in random order; the feasibility constraint $\mathcal{F}$ is known in advance but the quality of each request, in particular its contribution to Φ, is only revealed when the request arrives. Recent work has explored variants of the GSP where Φ is the sum over the accepted requests of the "value" of each request. For such a sum-of-values objective, constant factor competitive ratios are known for various kinds of feasibility constraints including cardinality constraints [18,20], knapsack constraints [4], and certain matroid constraints [5].

In many settings, the linear sum-of-values objective does not adequately capture the tradeoffs that the server faces in accepting or rejecting a request, and feasibility constraints provide only a rough approximation. Consider, e.g., a wireless access point accepting connections. Each accepted request improves resource utilization and brings value to the access point. However as the number of accepted requests grows the access point performs greater multiplexing of the spectrum, and must use more and more transmitting power in order to maintain a reasonable connection bandwidth for each request. The power consumption and its associated cost are *non-linear* functions of the total load on the access point. This directly translates into a value minus cost type of objective where the cost is an increasing function of the total size of accepted requests.

Our goal then is to accept a set A out of a universe U of requests such that the "profit" $\pi(A) = v(A) - \mathsf{C}(s(A))$ is maximized; here $v(A)$ is the total value of all requests in A, $s(A)$ is the total size, and C is a known increasing convex cost function[1].

Note that when the cost function takes on only the values 0 and ∞ it captures a knapsack constraint, and therefore the problem $(\pi, 2^U)$ (i.e. where the feasibility

---

[1] Convexity is crucial in obtaining any non-trivial competitive ratio—if the cost function were concave, the only solutions with a nonnegative objective function value may be to accept everything or nothing.

constraint is trivial) is a generalization of the knapsack secretary problem [4]. We further consider objectives that generalize the ℓ-dimensional knapsack secretary problem. Here, we are given ℓ different (known) convex cost functions C_i for $1 \leq i \leq \ell$, and each request is endowed with ℓ sizes, one for each dimension. The profit of a set is given by $\pi(A) = v(A) - \sum_{i=1}^{\ell} C_i(s_i(A))$ where $s_i(A)$ is the total size of the set in dimension i.

We consider the profit maximization problem under various feasibility constraints. For single-dimensional costs, we obtain online algorithms with competitive ratios within a constant factor of those achievable for a sum-of-values objective with the same feasibility constraints. For ℓ-dimensional costs, in the absence of any constraints, we obtain an $O(\ell)$ competitive ratio. We remark that this is essentially the best approximation achievable even in the offline setting: Dean et al. [10] show an $\Omega(\ell^{1-\epsilon})$ hardness for the simpler ℓ-dimensional knapsack problem under a standard complexity-theoretic assumption. For the multi-dimensional problem with general feasibility constraints, our competitive ratios are worse by a factor of $O(\ell^5)$ over the corresponding versions without costs. Improving this factor is a possible avenue for future research.

We remark that the profit function π is a submodular function. Recently several works [14,7,17] have looked at secretary problems with submodular objective functions and developed constant competitive algorithms. However, all of these works make the crucial assumption that the objective is always nonnegative; it therefore does not capture π as a special case. In particular, if Φ is a monotone increasing submodular function (that is, if adding more elements to the solution cannot decrease its objective value), then to obtain a good competitive ratio it suffices to show that the online solution captures a good fraction of the optimal solution. In the case of [7] and [17], the objective function is not necessarily monotone. Nevertheless, nonnegativity implies that the universe of elements can be divided into two parts, over each of which the objective essentially behaves like a monotone submodular function in the sense that adding extra elements to a good subset of the optimal solution does not decrease its objective function value. In our setting, in contrast, adding elements with too large a size to the solution can cause the cost of the solution to become too large and therefore imply a negative profit, even if the rest of the elements are good in terms of their value-size tradeoff. As a consequence we can only guarantee good profit when no "bad" elements are added to the solution, and must ensure that this holds with constant probability. This necessitates designing new techniques.

Our Techniques. In the absence of feasibility constraints (see Section 3), we note that it is possible to classify elements as "good" or "bad" based on a threshold on their value to size ratio (a.k.a. density) such that any large enough subset of the good elements provides a good approximation to profit; the optimal threshold is defined according to the offline optimal fractional solution. Our algorithm learns an estimate of this threshold from the first few elements (that we call the sample) and accepts all the elements in the remaining stream that cross the threshold. Learning the threshold from the sample is challenging. First, following the intuition about avoiding all bad elements, our estimate must be conservative, i.e. exceed the true threshold, with constant probability. Second, the optimal threshold for the sample can differ significantly from the optimal threshold for the entire stream and is therefore not a good candidate for our estimate. Our key observation is that the optimal profit over the sample is a much better behaved random variable and is, in particular, sufficiently concentrated; we use this observation to carefully pick an estimate for the density threshold.

With general feasibility constraints, it is no longer sufficient to merely classify elements as good and bad: an arbitrary feasible subset of the good elements is not necessarily a good approximation. Instead, we decompose the profit function into two parts, each of which can be optimized by maximizing a certain sum-of-values function (see Section 4). This suggests a reduction from our problem to two different instances of the GSP with sum-of-values objectives. The catch is that the new objectives are not necessarily non-negative and so previous approaches for the GSP don't work directly. We show that if the decomposition of the profit function is done with respect to a good density threshold and an extra filtering step is applied to weed out bad elements, then the two new objectives on the remaining elements are always non-negative and admit good solutions. At this point we can employ previous work on GSP with a sum-of-values objective to obtain a good approximation to one or the other component of profit. We note that while the exposition in Section 4 focuses on a matroid feasibility constraint, the results of that section extend to any downwards-closed feasibility constraint that admits good offline and online algorithms with a sum-of-values objective[2].

In the multi-dimensional setting (discussed in Section 5), elements have different sizes along different dimensions. Therefore, a single density does not capture the value-size tradeoff that an element offers. Instead we can decompose the value of an element into ℓ different values, one for each dimension, and define densities in each dimension accordingly. This decomposes the profit across dimensions as well. Then, at a loss of a factor of ℓ, we can approximate the profit objective along the "best" dimension. The problem with this approach is that a solution that is good (or even best) in one dimension may in fact be terrible with respect to the overall profit, if its profit along other dimensions is negative. Surprisingly we show that it is possible to partition values across dimensions in such a way that there is a *single* ordering over elements in terms of their value-size tradeoff that is respected in each dimension; this allows us to prove that a solution that is good in one dimension is also good in other dimensions. We present an $O(\ell)$ competitive algorithm for the unconstrained setting based on this approach in Section 5, and defer a discussion of the constrained setting to the full version of the paper.

Related Work. The classical secretary problem has been studied extensively; see [15,16] and [24] for a survey. Recently a number of papers have explored variants of the GSP with a sum-of-values objective. Hajiaghayi et al. [18] considered the variant where up to k secretaries can be selected (a.k.a. the k-secretary

---

[2] We obtain an $O(\alpha^4 \beta)$ competitive algorithm where α is the best offline approximation and β is the best online competitive ratio for the sum-of-values objective.

problem) in a game-theoretic setting and gave a strategyproof constant-competitive mechanism. Kleinberg [20] later showed an improved $1 - O(1/\sqrt{k})$ competitive algorithm for the classical setting. Babaioff et al. [4] generalized this to a setting where different candidates have different sizes and the total size of the selected set must be bounded by a given amount, and gave a constant factor approximation. In [5] Babaioff et al. considered another generalization of the k-secretary problem to matroid feasibility constraints. A matroid is a set system over U that is downwards closed (that is, subsets of feasible sets are feasible), and satisfies a certain exchange property (see [22] for a comprehensive treatment). They presented an $O(\log r)$ competitive algorithm, where r is the rank of the matroid, or the size of a maximal feasible set. This was subsequently improved to a $O(\sqrt{\log r})$-competitive algorithm by Chakraborty and Lachish [8]. Several papers have improved upon the competitive ratio for special classes of matroids [1,11,21]. Bateni et al. [7] and Gupta et al. [17] were the first to (independently) consider non-linear objectives in this context. They gave online algorithms for non-monotone nonnegative submodular objective functions with competitive ratios within constant factors of the ratios known for the sum-of-values objective under the same feasibility constraint. Other versions of the problem that have been studied recently include: settings where elements are drawn from known or unknown distributions but arrive in an adversarial order [9,19,23], versions where values are permuted randomly across elements of a non-symmetric set system [25], and settings where the algorithm is allowed to reverse some of its decisions at a cost [2,3].

2 Notation and Preliminaries

We consider instances of the generalized secretary problem represented by the pair $(\pi, \mathcal{F})$, and an implicit number n of requests or elements that arrive in an online fashion. U denotes the universe of elements. $\mathcal{F} \subseteq 2^U$ is a known downwards-closed feasibility constraint. Our goal is to accept a subset of elements $A \subseteq U$ with $A \in \mathcal{F}$ such that the objective function $\pi(A)$ is maximized. For a given set $T \subseteq U$, we use $O^*(T) = \text{argmax}_{A \in \mathcal{F} \cap 2^T}\, \pi(A)$ to denote the optimal solution over T; O^* is used as shorthand for $O^*(U)$.

We now describe the function π. In the single-dimensional cost setting, each element $e \in U$ is endowed with a value $v(e)$ and a size $s(e)$. Values and sizes are integral and are a priori unknown. The size and value functions extend to sets of elements as $s(A) = \sum_{e \in A} s(e)$ and $v(A) = \sum_{e \in A} v(e)$. The "profit" of a subset is given by $\pi(A) = v(A) - C(s(A))$ where C is a non-decreasing convex function of size, $C : \mathbb{Z}^+ \to \mathbb{Z}^+$, satisfying $C(0) = 0$. The following quantities will be useful in our analysis:

- The *density* of an element, $\rho(e) := v(e)/s(e)$. We assume w.l.o.g that densities of elements are unique and let e_γ denote the unique element with density γ.
- The *marginal cost* function, $c(s) := C(s) - C(s - 1)$. Note that this is non-decreasing.

- The *inverse marginal cost* function, $\bar{s}(\rho)$, is defined to be the maximum size for which $c(s) \leq \rho$.
- The *density prefix* for a given density γ and a set T, $P_\gamma^T := \{e \in T : \rho(e) \geq \gamma\}$, and the partial density prefix, $\bar{P}_\gamma^T := P_\gamma^T \setminus \{e_\gamma\}$. We use P_γ and $\bar{P}_\gamma$ as shorthand for P_γ^U and $\bar{P}_\gamma^U$ respectively.

We will sometimes find it useful to discuss fractional relaxations of the offline problem of maximizing π subject to $\mathcal{F}$. To this end, we extend the definition of subsets of U to allow for fractional membership. We use αe to denote an α-fraction of element e; this has value $v(\alpha e) = \alpha v(e)$ and size $s(\alpha e) = \alpha s(e)$. We compute the cost $C(s)$ for a non-integral size s by piecewise linear interpolation. We say that a fractional subset A is feasible if its support $\mathrm{supp}(A)$ is feasible. Note that when the feasibility constraint can be expressed as a set of linear constraints, this relaxation is more restrictive than the natural linear relaxation.

Note that since cost is a convex non-decreasing function of size, it may at times be more profitable to accept a fraction of an element rather than the whole. That is, $\mathrm{argmax}_\alpha \pi(\alpha e)$ may be strictly less than 1. For such elements, $\rho(e) < c(s(e))$. We use $\mathbb{F}$ to denote the set of all such elements: $\mathbb{F} = \{e \in U : \mathrm{argmax}_\alpha \pi(\alpha e) < 1\}$, and $\mathbb{I} = U \setminus \mathbb{F}$ to denote the remaining elements. Our solutions will generally approximate the optimal profit from $\mathbb{F}$ by running Dynkin's algorithm for the classical secretary problem; most of our analysis will focus on $\mathbb{I}$. Let $F^*(T)$ denote the optimal (feasible) fractional subset of $T \cap \mathbb{I}$ for a given set T. Then $\pi(F^*(T)) \geq \pi(O^*(T \cap \mathbb{I}))$. We use F^* as shorthand for $F^*(U)$, and let s^* be the size of this solution.

In the multi-dimensional setting each element has an ℓ-dimensional size $s(e) = (s_1(e), \ldots, s_\ell(e))$. The cost function is composed of ℓ different non-decreasing convex functions, $C_i : \mathbb{Z}^+ \to \mathbb{Z}^+$. The cost of a set of elements is defined to be $C(A) = \sum_i C_i(s_i(A))$ and the profit of A is its value minus its cost: $\pi(A) = v(A) - C(A)$.

3   Unconstrained Profit Maximization

We begin by developing an algorithm for the unconstrained version of the generalized secretary problem with $\mathcal{F} = 2^U$, which already exhibits some of the challenges of the general setting. Note that this setting captures as a special case the knapsack secretary problem of [4] where the goal is to maximize the total value of a subset of size at most a given bound. In fact in the offline setting, the generalized secretary problem is very similar to the knapsack problem. If all elements have the same (unit) size, then the optimal offline algorithm orders elements in decreasing order of value and picks the largest prefix in which each element contributes a positive marginal profit. When element sizes are different, a similar approach works: we order elements by density, and note that either a prefix of this ordering or a single element is a good approximation (much like the greedy 2-approximation for knapsack). The full version of this paper [6] provides a detailed analysis of the algorithm as well as other missing proofs.

Precisely, we show that $|O^* \cap \mathbb{F}| \leq 1$, and we can therefore focus on approximating π over the set $\mathbb{I}$. Furthermore, let $\mathcal{A}(U)$ denote the greedy subset obtained by considering elements in $\mathbb{I}$ in decreasing order of density and picking the largest prefix where every element has nonnegative marginal profit. The following lemma implies that either $\mathcal{A}(U)$ or the single best element is a 3-approximation to O^*.

Lemma 1. *We have that* $\pi(O^*) \leq \pi(F^*) + \max_{e \in U} \pi(e) \leq \pi(\mathcal{A}(U)) + 2 \max_{e \in U} \pi(e)$. *Therefore the greedy offline algorithm achieves a 3-approximation for* $(\pi, 2^U)$.

The offline greedy algorithm suggests an online solution as well. In the case where a single element gives a good approximation, we can use the classical secretary algorithm to get a good competitive ratio. In the other case, to get good competitive ratio, we only need to estimate the smallest density, say ρ^-, in the prefix of elements that the offline greedy algorithm picks, and then accept every element that exceeds this threshold.

We pick an estimate for ρ^- by observing the first few elements of the stream U. Note that it is important for our estimate of ρ^- to be no smaller than ρ^-. In particular, if there are many elements with density just below ρ^-, and our algorithm uses a density threshold less than ρ^-, then the algorithm may be fooled into mostly picking elements with density below ρ^- (since elements arrive in random order), while the optimal solution picks elements with densities far exceeding ρ^-. We now describe how to pick an overestimate of ρ^- which is not too conservative, that is, such that there is still sufficient profit in elements whose densities exceed the estimate.

In the remainder of this section, we assume that every element has profit at most $\frac{1}{k_1+1}\pi(O^*)$ for an appropriate constant k_1, to be defined later. (If this does not hold, the classical secretary algorithm obtains an expected profit of at least $\frac{1}{e(k_1+1)}\pi(O^*)$). Then Lemma 1 implies $\pi(F^*) \geq (1 - 1/(k_1+1))\pi(O^*)$, $\max_{e \in U}\pi(e) \leq (1/k_1)\pi(F^*)$, and $\pi(\mathcal{A}(U)) \geq (1 - 1/k_1)\pi(F^*)$.

We divide the stream U into two parts X and Y, where X is a random subset of U. Our algorithm unconditionally rejects elements in X and extracts a density threshold τ from this set. Over the remaining stream Y, it accepts an element if and only if its density is at least τ and if it brings in non-negative marginal profit. Under the assumption of small element profits, we can use a concentration lemma of Feige et al. [13] to show that $\pi(X \cap \mathcal{A}(U))$ is concentrated and is a large enough fraction of $\pi(O^*)$. This implies that with high probability $\pi(X \cap \mathcal{A}(U))$ (which is a prefix of $\mathcal{A}(X)$) is a significant fraction of $\pi(\mathcal{A}(X))$. Therefore we attempt to identify $X \cap \mathcal{A}(U)$ by looking at profits of prefixes of X. We will need the following lemma about $\mathcal{A}()$.

Lemma 2. *For any set S, consider subsets $A_1, A_2 \subseteq \mathcal{A}(S)$. If $A_1 \supseteq A_2$, then $\pi(A_1) \geq \pi(A_2)$. That is, π is monotone-increasing when restricted to $\mathcal{A}(S)$ for all $S \subseteq U$.*

We define two good events. E_1 asserts that $X \cap \mathcal{A}(U)$ has high enough profit. Our final output is the set P_τ^Y. E_2 asserts that the profit of P_τ^Y is a large enough

---

Algorithm 1. Online algorithm for single-dimensional $(\pi, 2^U)$

1: With probability $1/2$ run the classic secretary algorithm to pick the single most profitable element else execute the following steps.
2: Draw k from Binomial$(n, 1/2)$.
3: Select the first k elements to be in the sample X. Reject these elements.
4: Let τ be largest density such that $\pi(P_\tau^X) \geq \beta \left(1 - \frac{1}{k_1}\right) \pi(F^*(X))$ for constants β and k_1 to be specified later.
5: Initialize selected set $O \leftarrow \emptyset$.
6: **for** $i \in Y = U \setminus X$ **do**
7:     **if** $\pi(O \cup \{i\}) - \pi(O) \geq 0$ and $\rho(i) \geq \tau$ and $i \notin \mathbb{F}$ **then**
8:         $O \leftarrow O \cup \{i\}$
9:     **end if**
10: **end for**

---

fraction of the profit of P_τ. Recall that $\mathcal{A}(U)$ is a density prefix, say P_{ρ^-}, and so $X \cap \mathcal{A}(U) = P_{\rho^-}^X$. Let E_1 denote the event that $\pi(P_{\rho^-}^X) > \beta \, \pi(P_{\rho^-})$, where β is a constant to be specified later. Conditioned on E_1, we have $\pi(P_{\rho^-}^X) > \beta(1 - 1/k_1) \pi(F^*) \geq \beta(1 - 1/k_1) \pi(F^*(X))$. Note that threshold τ, as selected by Algorithm 1, is the largest density such that $\pi(P_\tau^X) \geq \beta(1 - 1/k_1) \pi(F^*(X))$. Therefore, E_1 implies $\tau \geq \rho^-$, and we have the following lemma.

Lemma 3. *Conditioned on E_1, $O = P_\tau \cap Y \subseteq \mathcal{A}(U)$.*

On the other hand, $P_\tau^X \subseteq P_\tau \subset \mathcal{A}(U)$ along with Lemma 2 implies

$$\pi(P_\tau) \geq \pi(P_\tau^X) \geq \beta(1 - 1/k_1) \pi(F^*(X)) \geq \beta(1 - 1/k_1) \pi(P_{\rho^-}^X) \geq \beta^2 (1 - 1/k_1)^2 \pi(F^*)$$

where the second inequality is by the definition of τ, the third by optimality and the last is obtained by applying E_1 and $\mathcal{A}(U) \geq (1 - 1/k_1) F^*$.

We define ρ^+ to be the largest density such that $\pi(P_{\rho^+}) \geq \beta^2 (1 - 1/k_1)^2 \pi(F^*)$. Then $\rho^+ \geq \tau$, which implies $P_{\rho^+} \subseteq P_\tau$ and the following lemma.

Lemma 4. *Event E_1 implies $O \supseteq Y \cap P_{\rho^+}$.*

Based on the above lemma, we define event $E_2 : \pi(P_{\rho^+}^Y) \geq \beta' \pi(P_{\rho^+})$, for an appropriate constant β'. Conditioned on events E_1 and E_2, and using Lemma 2 again, we get $\pi(O) \geq \pi(P_{\rho^+}^Y) \geq \beta'\beta^2(1 - 1/k_1)^2 \pi(F^*)$. To wrap up the analysis, we show that E_1 and E_2 are high probability events.

Lemma 5. *If no element of U has profit more than $\frac{1}{113}\pi(O^*)$, then $\Pr[E_1 \wedge E_2] \geq 0.52$, where $\beta = 0.262$ and $\beta' = 0.094$.*

Putting everything together we get the following theorem.

Theorem 1. *Algorithm 1 achieves a competitive ratio of 616 for $(\pi, 2^U)$ using $k_1 = 112$ and $\beta = 0.262$.*

Proof. If there exists an element with profit at least $\frac{1}{113}\pi(O^*(U))$, the classical secretary algorithm (Step 1) gives a competitive ratio of $\frac{1}{113e} \geq \frac{1}{308}$. Otherwise, using Lemma 5, with $\beta' = 0.094$, we have $\mathbb{E}[\pi(O)] \geq \mathbb{E}[\pi(O) \mid E_1 \wedge E_2]\Pr[E_1 \wedge E_2] \geq 0.52\beta'\beta^2(1-1/k_1)^2\pi(F^*) \geq 0.52\beta'\beta^2(1-1/k_1)^2(1-1/(k_1+1))\pi(O^*) \geq \frac{1}{307}\pi(O^*)$. Since we flip a fair coin to decide whether to output the result of running the classical secretary algorithm, or output the set O, we achieve a $2\max\{308, 307\} = 616$-approximation to $\pi(O^*)$ in expectation (over the coin flip). □

4   Matroid-Constrained Profit Maximization

We now extend the algorithm of Section 3 to the setting $(\pi, \mathcal{F})$ where $\mathcal{F}$ is a matroid constraint. In particular, $\mathcal{F}$ is the set of all independent sets of a matroid over U. We skip a precise definition of matroids and will only use the following facts: $\mathcal{F}$ is a downward closed feasibility constraint and there exist an exact offline and an $O(\sqrt{\log r})$-competitive online algorithm for $(\Phi, \mathcal{F})$, where Φ is a sum-of-values objective and r is the rank of the matroid. The algorithms and detailed proofs for this section are given in the full version [6] of the paper.

In the unconstrained setting, we showed that there always exists either a density prefix or a single element with near-optimal profit. So in the online setting it sufficed to determine the density threshold for a good prefix. In constrained settings this is no longer true, and we need to develop new techniques. Our approach is to develop a general reduction from the π objective to two different sum-of-values type objectives over the same feasibility constraint. This allows us to employ previous work on the $(\Phi, \mathcal{F})$ setting; we lose only a constant factor in the competitive ratio. We will first describe the reduction in the offline setting and then extend it to the online algorithm using techniques from Section 3.

Decomposition of π. For a given density γ, we define the *shifted density function* $h_\gamma()$ over sets as $h_\gamma(A) := \sum_{e \in A}(\rho(e) - \gamma)s(e)$ and the *fixed density function* $g_\gamma()$ over sizes as $g_\gamma(s) := \gamma s - \mathsf{C}(s)$. For a set A we use $g_\gamma(A)$ to denote $g_\gamma(s(A))$. It is immediate that for any density γ we can split the profit function as $\pi(A) = h_\gamma(A) + g_\gamma(A)$. In particular $\pi(O^*) = h_\gamma(O^*) + g_\gamma(O^*)$. Our goal will be to optimize the two parts separately and then return the better of them.

Note that the function h_γ is a sum of values function where the value of an element is defined to be $(\rho(e) - \gamma)s(e)$. Its maximizer is a subset of P_γ, the set of elements with nonnegative shifted density $\rho(e) - \gamma$. In order to ensure that the maximizer of h_γ, say A, also obtains good profit, we must ensure that $g_\gamma(A)$ is nonnegative, and therefore $\pi(A) \geq h_\gamma(A)$. This is guaranteed for a set A as long as $s(A) \leq \bar{s}(\gamma)$.

Likewise, the function g_γ increases as a function of size s as long as s is at most $\bar{s}(\gamma)$, and decreases thereafter. Therefore, in order to maximize g_γ, we merely need to find the largest (in terms of size) feasible subset of size no more than $\bar{s}(\gamma)$. As before, if we can ensure that for such a subset h_γ is nonnegative (e.g. if the set is a subset of P_γ), then the profit of the set is no smaller than its g_γ value. This motivates the following definition of "bounded" subsets:

Definition 1. *Given a density γ a subset $A \subseteq U$ is said to be γ-bounded if $A \subseteq P_\gamma$ and $s(A) \leq \bar{s}(\gamma)$.*

Proposition 1. *For any γ-bounded set A, $\pi(A) \geq h_\gamma(A)$ and $\pi(A) \geq g_\gamma(A)$.*

For a density γ, we define $H_\gamma := \mathrm{argmax}_{H \in \mathcal{F}, H \subseteq P_\gamma}\, h_\gamma(H)$ along with $G_\gamma := \mathrm{argmax}_{G \in \mathcal{F}, G \subseteq \bar{P}_\gamma}\, s(G)$.

Following our observations above, both H_γ and G_γ can be determined efficiently (in the offline setting) using standard matroid maximization. However, we must ensure that the two sets are γ-bounded. Further, in order to compare the performance of G_γ against O^*, we must ensure that its size is at least a constant fraction of the size of O^*.

We show in the full version of this paper that there exists a density ρ^- for which H_γ and G_γ satisfy these properties. The following is our main claim of this section.

Lemma 6. *There exists a density ρ^- such that for any density $\gamma > \rho^-$, $\pi(O^*(\bar{P}_\gamma)) \leq \pi(H_\gamma) + \pi(G_\gamma)$. Furthermore, $\pi(O^*) \leq \pi(H_{\rho^-}) + \pi(G_{\rho^-}) + 2\max_{e \in U} \pi(e)$.*

This lemma immediately gives us an offline approximation algorithm for $(\pi, \mathcal{F})$: for every element density γ, we find the sets H_γ and G_γ; we then output the best (in terms of profit) of these sets or the best individual element. We obtain the following theorem:

Theorem 2. *The algorithm outlined above 4-approximates $(\pi, \mathcal{F})$ in the offline setting.*

The Online Setting. Our online algorithm, as in the unconstrained case, uses a sample X from U to obtain an estimate τ for the density ρ^-. Then with equal probability it applies the online algorithm for $(h_\tau, \mathcal{F})$ on the remaining set $Y \cap P_\tau$ or the online algorithm for $(s, \mathcal{F})$ (in order to maximize g_τ) on $Y \cap P_\tau$. The algorithm is described in detail in the full version of the paper. Lemma 6 indicates that it should suffice for τ to be larger than ρ^- while ensuring that $\pi(O^*(P_\tau))$ is large enough. As in Section 3 we define the density ρ^+ as the upper limit on τ, and claim that τ satisfies the required properties w.h.p.

Theorem 3. *If there exists an α-competitive algorithm for the matroid secretary problem $(\Phi, \mathcal{F})$ where Φ is a sum-of-values objective, then the online algorithm outlined above achieves a competitive ratio of $O(\alpha)$ for the problem $(\pi, \mathcal{F})$.*

5 Multi-dimensional Profit Maximization

In this section, we consider the GSP with a multi-dimensional profit objective. Recall that in this setting each element e has ℓ different sizes $s_1(e), \ldots, s_\ell(e)$, and the cost of a subset is defined by ℓ different convex functions $C_1, \ldots, C_\ell$. The profit function is defined as $\pi(A) = v(A) - \sum_i C_i(s_i(A))$.

As in the single-dimensional setting, we partition U into two sets $\mathbb{I}$ and $\mathbb{F}$ with $\mathbb{F} = \{e \in U : \text{argmax}_\alpha \, \pi(\alpha e) < 1\}$. We show in the full version that, as before, an optimal solution cannot contain too many elements of $\mathbb{F}$: $|O^* \cap \mathbb{F}| \leq \ell$. We therefore devote the remainder of this section to approximating π over $\mathbb{I}$. Here we focus on the unconstrained problem $(\pi, 2^U)$, but our results extend to constrained settings as well [6].

Our high level approach is to distribute the value of each element across the ℓ dimensions, thereby defining densities and decomposing profit across dimensions appropriately. We do this in such a way that a maximizer of the ith dimensional profit for some dimension i gives us a good overall solution (albeit at a cost of a factor of ℓ).

Formally, let $\rho : U \to \mathbb{R}^\ell$ denote an ℓ-dimensional vector function $\rho(e) = (\rho_1(e), \ldots, \rho_\ell(e))$ that satisfies $\sum_i \rho_i(e) s_i(e) = v(e)$ for all e. We set $v_i(e) = \rho_i(e) s_i(e)$ and $\pi_i(A) = v_i(A) - C_i(s_i(A))$ and note that $\pi(A) = \sum_i \pi_i(A)$. Let F_i^* denote the maximizer of π_i over $\mathbb{I}$. Then, $\pi(F^*) \leq \sum_i \pi_i(F_i^*)$.

Given this observation, it is natural to try to obtain an approximation to π by solving for F_i^* for all i and rounding the best one. This does not immediately work: even if $\pi_i(F_i^*)$ is very large, $\pi(F_i^*)$ could be negative because of the profit of the set being negative in other dimensions. Instead, with each element we can associate a density vector ρ such that for any two elements e and e' either $\rho(e)$ component wise dominates $\rho(e')$ or vice versa. We call such vectors *proper* densities. We show that under ρ the best set F_i^* indeed gives an $O(\ell)$ approximation to $O^*(\mathbb{I})$.

Proper density vectors induce a single ordering over elements, say, $e_1, \ldots, e_n$. Note that each F_i^* is a (fractional) prefix of this sequence. Let F_1^* be the shortest prefix and let $\mathcal{A} = \{e_1, \ldots, e_{k_1}\}$ denote the integral part of F_1^*. $\mathcal{A}$ satisfies the inequality $\pi(F^*) \leq \ell(\pi(\mathcal{A}) + \max_e \pi(e))$. Thus $\mathcal{A}$ or the single best element gives us an offline $O(\ell)$-approximation for $(\pi, 2^U)$ in the multi-dimensional setting.

The Online Setting. Note that proper densities essentially define a 1-dimensional manifold in ℓ-dimensional space. We can therefore hope to apply our online algorithm from Section 3 to this setting. However, there is a caveat: the algorithm from Section 3 uses the offline algorithm as a subroutine on the sample X to estimate the threshold τ; naïvely replacing the subroutine by the $O(\ell)$ approximation described above leads to an $O(\ell^2)$ competitive online algorithm[3]. In order to improve the competitive ratio to $O(\ell)$ we pick the threshold τ more carefully. (The online algorithm is described in detail in the full version of the paper).

Via a similar argument as for Theorem 1, we get

Theorem 4. *Algorithm 1 with the modifications outlined above is $O(\ell)$ competitive for $(\pi, 2^U)$ where π is a multi-dimensional profit function.*

---

[3] Note the $(1 - 1/k_1)^2$ factor in the final competitive ratio in Theorem 1; this factor is due to the use of the offline subroutine in determining τ.

References

1. Babaioff, M., Dinitz, M., Gupta, A., Immorlica, N., Talwar, K.: Secretary problems: weights and discounts. In: SODA 2009 (2009)
2. Babaioff, M., Hartline, J., Kleinberg, R.: Selling banner ads: Online algorithms with buyback. In: Fourth Workshop on Ad Auctions (2008)
3. Babaioff, M., Hartline, J.D., Kleinberg, R.D.: Selling ad campaigns: Online algorithms with cancellations. In: EC 2009 (2009)
4. Babaioff, M., Immorlica, N., Kempe, D., Kleinberg, R.: A knapsack secretary problem with applications. In: Approximation, Randomization, and Combinatorial Optimization. Algorithms and Techniques, pp. 16–28 (2007)
5. Babaioff, M., Immorlica, N., Kleinberg, R.: Matroids, secretary problems, and online mechanisms. In: SODA 2007 (2007)
6. Barman, S., Umboh, S., Chawla, S., Malec, D.L.: Secretary problems with convex costs. CoRR, abs/1112.1136 (2011)
7. Bateni, M.H., Hajiaghayi, M.T., Zadimoghaddam, M.: Submodular secretary problem and extensions. In: Approximation, Randomization, and Combinatorial Optimization. Algorithms and Techniques, pp. 39–52 (2010)
8. Chakraborty, S., Lachish, O.: Improved competitive ratio for the matroid secretary problem. In: SODA 2012 (2012)
9. Chawla, S., Hartline, J.D., Malec, D.L., Sivan, B.: Multi-parameter mechanism design and sequential posted pricing. In: STOC 2010 (2010)
10. Dean, B.C., Goemans, M.X., Vondrák, J.: Adaptivity and approximation for stochastic packing problems. In: SODA 2005 (2005)
11. Dimitrov, N.B., Plaxton, C.G.: Competitive Weighted Matching in Transversal Matroids. In: Aceto, L., Damgård, I., Goldberg, L.A., Halldórsson, M.M., Ingólfsdóttir, A., Walukiewicz, I. (eds.) ICALP 2008, Part I. LNCS, vol. 5125, pp. 397–408. Springer, Heidelberg (2008)
12. Dynkin, E.B.: The optimum choice of the instant for stopping a Markov process. Soviet Math. Dokl 4(627-629) (1963)
13. Feige, U., Flaxman, A.D., Hartline, J.D., Kleinberg, R.D.: On the Competitive Ratio of the Random Sampling Auction. In: Deng, X., Ye, Y. (eds.) WINE 2005. LNCS, vol. 3828, pp. 878–886. Springer, Heidelberg (2005)
14. Feldman, M., Naor, J., Schwartz, R.: Improved competitive ratios for submodular secretary problems. In: Approximation, Randomization, and Combinatorial Optimization. Algorithms and Techniques, pp. 218–229 (2011)
15. Ferguson, T.: Who solved the secretary problem. Statist. Sci. 4(3), 282–289 (1989)
16. Freeman, P.R.: The secretary problem and its extensions: a review. International Statistical Review 51(2), 189–206 (1983)
17. Gupta, A., Roth, A., Schoenebeck, G., Talwar, K.: Constrained non-monotone submodular maximization: Offline and secretary algorithms. Internet and Network Economics, 246–257 (2010)
18. Hajiaghayi, M.T., Kleinberg, R., Parkes, D.C.: Adaptive limited-supply online auctions. In: EC 2004 (2004)
19. Kennedy, D.P.: Prophet-type inequalities for multi-choice optimal stopping. Stochastic Processes and their Applications 24(1), 77–88 (1987)
20. Kleinberg, R.: A multiple-choice secretary algorithm with applications to online auctions. In: SODA 2005 (2005)

21. Korula, N., Pál, M.: Algorithms for Secretary Problems on Graphs and Hypergraphs. In: Albers, S., Marchetti-Spaccamela, A., Matias, Y., Nikoletseas, S., Thomas, W. (eds.) ICALP 2009. LNCS, vol. 5556, pp. 508–520. Springer, Heidelberg (2009)
22. Oxley, J.: Matroid Theory. Oxford University Press (1992)
23. Samuel-Cahn, E.: Comparison of threshold stop rules and maximum for independent nonnegative random variables. The Annals of Probability 12 (1984)
24. Samuels, S.: Secretary problems. In: Handbook of Sequential Analysis, pp. 381–405. Marcel Dekker (1991)
25. Soto, J.A.: Matroid Secretary Problem in the Random Assignment Model. In: SODA 2011 (2011)

Nearly Simultaneously Resettable Black-Box Zero Knowledge

Joshua Baron[1], Rafail Ostrovsky[2], and Ivan Visconti[3]

[1] UCLA, Los Angeles, CA, USA 90095
jwbaron@math.ucla.edu, rafail@cs.ucla.edu
[2] Università di Salerno, 84084 Fisciano (SA) - Italy
visconti@dia.unisa.it

Abstract. An important open question in Cryptography concerns the possibility of achieving secure protocols even in the presence of physical attacks. Here we focus on the case of proof systems where an adversary forces the honest player to re-use its randomness in different executions. In 2009, Deng, Goyal and Sahai [1] constructed a *simultaneously resettable* non-black-box zero-knowledge argument system that is secure against resetting provers *and* verifiers.

In this work we study the case of the black-box use of the code of the adversary and show a nearly simultaneously resettable black-box zero-knowledge proof systems under standard assumptions. Compared to [1], our protocol is a *proof* (rather then just argument) system, but requires that the resetting prover can reset the verifier up to a bounded number of times (which is unavoidable for black-box simulation), while the verifier can reset the prover an arbitrary polynomial number of times. The main contribution of our construction is that the round complexity is independent of the above bound. To achieve our result, we construct a constant-round nearly simultaneously resettable coin-flipping protocol that we believe is of independent interest.

Keywords: Reset attacks, Black-box simulation.

1 Introduction

In this work, we study the feasibility of achieving efficient zero-knowledge proof systems in the presence of physical attacks. Specifically, we examine the role of the black-box use of the code of the adversary with respect to simultaneously resettable proof systems. Such proof systems are of interest as examples of proof systems that are secure under very relaxed constraints on the re-use of the same randomness in multiple executions. In the case of *resettable zero knowledge (rZK)*, a malicious verifier may cheat against an honest prover who must use the same random tape polynomially many times. Further, *resettably sound* zero knowledge constrains the randomness used by the verifier: a malicious prover may try to cheat against an honest verifier who must use the same random tape polynomially many times. The former property was introduced and instantiated

A. Czumaj et al. (Eds.): ICALP 2012, Part I, LNCS 7391, pp. 88–99, 2012.

by Canetti, Goldreich, Goldwasser and Micali [2]; the later property was introduced by Micali and Reyzin [3] and later instantiated by Barak, Goldreich, Goldwasser and Lindell [4]. Because rZK is a generalization of concurrent zero knowledge (cZK) [5,6,7,8,9], every rZK proof system is a cZK proof system. A question opened by [4] and resolved by [1] is *"Does there exist a resettably sound rZK proof system for all $\mathcal{NP}$?"*. [1] answered this question in the affirmative, but they required a construction with a security proof that required a non-black-box simulator strategy, which utilize the specific strategy of a cheating verifier in its specification. Currently, there is no known practical protocol that relies on a non-black-box[1] simulation strategy, while for instance there *do* exist efficient constructions for cZK and concurrent non-malleable zero knowledge that rely on black-box simulation strategies [10,11], which work against any verifier strategy.

It is proved in [4] that resettably sound black-box zero-knowledge arguments can be constructed for languages in $\mathcal{BPP}$. Instead, we study whether there exist *t-bounded resettably sound rZK proof systems with black-box simulation*, and more in general, with only a black-box use of the code of the adversary (i.e., both the simulation and the proof of soundness do not rely on non-black-box uses of the code of the adverary). Such proof systems are rZK but also allow a malicious prover to conduct at most $t(n)$ resets against an honest verifier, where t is any fixed polynomial and n is the security parameter. Such a security setting has practical applications (indeed, in [12] it has been considered for the case of e-passports) because real hardware that may be reset to break security, such as smart cards or stateless devices, have certain wear costs; after enough resets, the hardware may simply break, a simple counter may run out, or built-in battery may become depleted. Our black-box construction is also of theoretical interest, and moreover may lead to more efficient near-simultaneously rZK protocols. Indeed while all known non-black-box constructions based on standard assumptions are inefficient, there are in several cases alternative efficient black-box constructions [10,11]. Further, unlike [4], we obtain unconditional soundness, a property that is hard to achieve when the simulator is non-black-box.

We remark that constructing t-resettably sound rZK proof systems for any language $L \in \mathcal{NP}$ with black-box simulation is quickly accomplished if round complexity is allowed to be t-dependent. For any t, take any rZK proof system with black-box simulation and repeat it sequentially with independent randomness $t + 1$ times; a verifier then accepts only if he accepts for each of the $t + 1$ protocol runs. What we desire is to construct a t-resettably sound rZK proof system where the round complexity is *t-independent*.

1.1   Overview of Our Contribution

For all $\mathcal{NP}$ and for any polynomial t, we construct a t-resettably sound rZK proof system with black-box use only of the code of the adversary and round complexity $O(n^{\epsilon})$ for security parameter n and for any constant $\epsilon > 0$. We require

---

[1] We ignore controversial non-black-box extraction assumptions.

standard assumptions as the existence of enhanced trapdoor permutations and collision-resistant hash functions.

We re-examinine the rZK protocol of [2]. Their protocol involves an adaptation of the cZK construction of [6]. We first give a high-level description of the protocol of [2]. V commits to a set of n^ϵ strings of length n. Then, in the next $2n^\epsilon$ rounds, P first commits to an n-bit string and V subsequently decommits to the next string, eventually decommitting to the entire set. The protocol concludes by P giving a rWI proof that either $x \in L$ or that at least one of the strings committed by P is identical to the subsequent decommitment of V. Clearly, such a protocol is not t-resettably sound (for any $t \geq 1$), as a resetting prover can simply obtain one of V's committed strings, rewind a round and then commit to that string. However, we use this protocol as a basis to construct our protocol.

We think of the initial commitment by V to n^ϵ strings of length n as a *database*. The idea for our protocol is that V should commit to a database of $poly(n,t)$ strings of length n; then, in each of the next $2n^\epsilon$ rounds, P asks for n entries of V's database, which V then reveals. Finally, P provides a t-resettably sound, rWI proof that either $x \in L$ or that P can commit to a large constant fraction of V's database. The idea is that even if P was able to successfully ask for $tn^{\epsilon+1}$ indices of the database, P would still not know a large constant fraction of the database; in this way, the protocol will be t-resettably sound.

We overcome several challenges to accomplish such a protocol. First, we require the simulator to discover significantly more indices of V's database than a t-resetting P^* possibly could. We note that we can modify the (black-box) simulator strategy given in [2]; there, at each prover commitment phase, the simulator executes an independent look-ahead subprotocol call to t discover the string that V would decommit to. In fact, these look-ahead subprotocol calls are independent from one round to the next. We take advantage of this independence by having our simulator execute *polynomially* many look-ahead subprotocol calls for each round and proving that such a strategy produces more than half of V's database. On the other hand, we will show that for suitable parameters, a t-resetting P^* will only be able to discover at most $1/16$ of V's database except with negligible probability. Therefore, our protocol starts by V committing to a large database followed by $2n^\epsilon$ rounds where V decommits to the n (distinct) random indices in each round that P asks for. Finally, P commits to a guess of the *entire* database and proves that either $x \in L$ or that a large constant fraction of the guess correctly corresponds to V's database.

However to have a meaningful statement for the proof given by P, it seems that V should reveal the entire database, but this exposes again V to reset attacks. Therefore, a second challenge is that V will reveal a small fraction of the database and P will prove that it committed to a large portion of this fraction. The challenge of such a strategy is that a cheating V^* might skew the distribution of what it reveals to be used for the rWI proof at the conclusion of our protocol such that the simulator might not discover enough entries of the database. Therefore, we require a special coin-flipping protocol by which V reveals n uniformly random elements of its database, at which point P proves, using a t-resettably sound rWI proof, that either $x \in L$ or that P's database guess, committed to before the final proof, contains a correct guess for at least $1/4$ of the n indices that V last decommitted to. Since the n revealed indices are uniformly distributed over V's database, only if P's initial database guess matches at least $1/4$ of V's database will P be able to give a correct WI proof without using the witness for $x \in L$ (except with negligible probability).

A first attempt for such a coin-flipping protocol between left player P_L and right player P_R might be for P_L to a apply pseudorandom function family, f_*^{PRF}, with a previously committed seed, s, on input a random string sent by P_R. The string sent by P_R would be constructed by applying a $(t+1)$-wise independent hash function on the transcript thus far (P_R cannot use a pseudorandom function because P_L^* is unbounded and could distinguish the output). However, a cheating P_L^* might commit to a seed in such a way that even on random input, the output of the pseudorandom function has skewed distribution. Once can also try to modify the protocol by having P_R hash and output the pair (R, R') and send it to P_L, who then computes $f_s^{PRF}(R) \oplus R'$. However, cheating P_R^* could then simply rewind, keep R fixed and send whatever R' he wished. Instead, we solve our problem as follows: P_R hashes to obtain the triple (R, R', r'), computes a (statistically hiding) commitment to R', denoted c, using randomness r', and sends (R, c) to P_L. Then, using previous committed seed s, P_L applies f_s^{PRF} to the concatenation of R and c, which also binds the output of the pseudorandom function to R' before R' has been revealed. Finally, P_R opens the decommitment of R' to P_L, and both set the random string τ to be the sum of the output of the pseudorandom function and R'. We remark that an adversary (resetting or not) may always guess $O(\log n)$ bits of the coin-flipping output; however, since the output length is n, an adversary will have only a negligible chance of correctly guessing a constant fraction of the coin-flipping output.

A final note is that while P_R^* may construct R and R' as he wishes, it is very important that a cheating P_L^* formats his pseudorandom function outputs correctly; otherwise, upon discovering P_R's R and R', P_L^* could simply rewind and lie about the output of f_s^{PRF}. Therefore, P_L must send a rWI proof that either $x \in L$ or the function output was formatted correctly. In this way, only in the case that P_L^* is cheating with $x \notin L$ the correct formatting will need to be assured; we can bootstrap the witness for $x \in L$ to make the rWI proof of consistency witness hiding.

A third challenge is that our coin-flipping protocol makes the larger protocol not admissible. In particular, the simulator in [2] was given in the so-called hybrid model, where a cheating V^* is somewhat constrained. Therefore to prove rZK for our protocol, we must demonstrate that our protocol is not meaningfully different enough from the protocol of [2] even though their simulator no longer applies to our setting. To accomplish this task, we prove a theorem based on the observation that the only place where the simulators might differ are where they play identically to the honest prover against the verifier but using a different witness. We therefore construct a variant of our own protocol that more explicitly models the protocol of [2] but is only rZK (and *not* bounded resettably sound). We then construct the simulator for the intermediate protocol. Finally, we prove that the existence of a simulator for this intermediate protocol implies the existence of a simulator for the protocol that we desire. We believe that such a proof strategy is of independent interest.

We note that our protocol has communication complexity that is dependent of t; finding a protocol using standard assumptions with communication complexity independent of t is an interesting open question[2], as is improving the round complexity of our construction.

1.2   Other Related Work

The notion of rZK was first introduced by [2]; later constructions of black-box rZK protocols have better round complexity. In [7] it is shown a rZK proof system with black-box simulation and $\omega(\log^2 n)$ rounds. The protocol improved the round complexity of [6], by examining a static simulator rewind schedule and showing that such a schedule produced a successful single extraction, except with negligible probability. It is not clear how to adapt such a scheduling strategy to the polynomial many successful extractions that we require. The results of [9] can also be used to construct a black-box rZK proof system. Their protocol requires $\tilde{O}(\log n)$ rounds and also build upon the protocol of [6]. The simulator strategy of [9] relies on a careful analysis of the random tapes used by the simulator throughout its run together with the oblivious simulator strategy of [7] to obtain a single successful extraction, while our approach relies on segmenting the simulator of [2] and running various of its subprotocols in parallel to obtain polynomially many successful look-aheads. Finding compatibility between the two approaches is an interesting open question.

The first resettably sound (non-black-box) zero-knowledge argument was constructed by [4]. Deng and Lin [13] constructed a zero-knowledge argument secure in a weakened notion of simultaneous resettability: both cheating prover and verifier can reset the other polynomially many times, but can only reset a particular party with a fixed random tape (e.g., an incarnation) a bounded number of times. Their protocol requires only a constant number of rounds and also required non-black-box simulation in the proof.

The construction of a simultaneously rZK argument was first provided by [1] and requires polynomial round complexity. Their protocol relies on the prover initially committing to his challenges for the extraction stage and using the resettably sound zero-knowledge argument of [4] to prove that either $x \in L$ or that the decommitted challenges are correct. Because the protocol heavily relies on the non-black-box zero-knowledge argument of [4], the simulator used for the security proof is non-black-box. Recently, it has been shown in [19] how to obtain a constant-round resettably sound resettable witness indistinguishable argument of knowledge.

We note that all the protocols listed here are in the standard model. In particular, [2,4,13,14,15,16] also provide constructions in the bare public-key model.

---

[2] Using less standard assumptions like complexity leveraging, constructing a protocol t-independent communication complexity seems to be more easily to accomplish.

In addition, [17] give a resettable black-box *statistical* zero-knowledge proof for several non-trivial languages, but they do not have a construction for all $\mathcal{NP}$. As such research is incomparable to our work, we do not consider it here.

2 Preliminaries, Definitions and Tools

We denote by n the security parameter throughout our discussion., by $[m]$ the set $\{1, ..., m\}$, by $x|y$ the concatenation of x and y, by U_n the uniform distribution over $\{0,1\}^n$ and by f_*^{PRF} a pseudorandom function family, where $f_s^{PRF}(x)$ is the evaluation of the function specified by seed s at x.

We will denote by C the PPT committer and by R the PPT receiver. We use the standard notions of statistically (respectively, computationally) binding and computationally (respectively, statistically) hiding. When statistical hiding or binding is discussed for a commitment scheme, the not mentioned property is assumed to hold with computational security.

We denote by $(P_1(x), P_2(y))$ the interactive protocol between party P_1 with input x and party P_2 with input y; moreover, we denote the sequential composition of protocols $\pi_i = (P_1^i, P_2^i)$ and $\pi_j = (P_1^j, P_2^j)$ by $(\pi_i, \pi_j) = ((P_1^i, P_1^j), (P_2^i, P_2^j))$.

We refer the reader to [2] for the definition of rZK (and witness indistinguishable) proof systems, and the definition of admissible proof systems as well as the hybrid model. Our definition of t-bounded resettable soundness follows from the definition in [4] for resettable soundness, except that a malicious prover P^* has a bound of $t(n)$ many resets he can execute against a verifier V. We omit the formal definition here due to lack of space.

We will utilize the following construction[3] of Dwork and Naor [18].

Theorem 1 (zaps). *If enhanced trapdoor permutations exist, then for every language L there exists a two-round simultaneously resettable WI proof system.*

3 Black-Box rZK with t-Resettable Soundness

Before giving the exact protocol specification for our candidate construction $\Pi = (P, V)$, we first outline its crucial steps. We consider our protocol as the composition of three subprotocols, π_0, π_1, and π_2, for two reasons. The first reason is that the purpose of each of the subprotocols is distinct and so discussing them separately is natural. The second reason is that in order to prove rZK of Π, we will construct another protocol that will rely on the first two subprotocols but will require a different third subprotocol, π_2'. In what follows, fix a language $L \in \mathcal{NP}$, let n be the security parameter, let $\epsilon > 0$ be any constant, and let t be the polynomial resetting bound of the prover.

---

[3] In fact, zaps are not inherently resettable WI, though they are resettably sound, as noted by [4]. However, when the prover's zap message is computed using a random tape that is a pseudorandom function applied to V's initial message and P's random tape, as is done here, zaps are rWI. We will therefore refer to zaps here and implicitly assume that their instantiation in our protocol constructions utilize the appropriate PRF-random tape construction.

Table 1. Outline of protocol $\Pi = (\pi_0, \pi_1, \pi_2)$. CHCom is statistically binding, while SHCom is statistically hiding.

π_0 : *Setup Phase*
1) P sets up SHCom
2) V constructs DB, $\|DB\| = 16tn(n^\epsilon + 1)$
3) V sets up zap, V sends SHCom(DB)
π_1 : *Extraction Phase*
1) n^ϵ Iterations:
i) P asks V for n indices of DB
ii) V decommits to the n indices
π_2 : *rWI Proof Phase*
1) P guesses DB as γ; P sends PBCom(γ)
2) P and V jointly generate n random indices, τ, of DB (coin flipping)
3) V decommits to the indices τ of DB
4) P sends zap for "$x \in L$ or γ agrees with at least $1/4$ of the n values of DB in positions τ"

Table 2. Outline of nearly re-settable coin-flipping subprotocol with output τ. f_*^{PRF} is a pseudorandom function family. (P executes P_L and V executes P_R.)

Coin Flipping Subprotocol
1) P_L sends CHCom(s) to P_R
2) a) P_R applies $(t + 1)$-wise independent hash function h to transcript to obtain (R, R', r')
b) P_R computes $c \leftarrow$ SHCom(R') using randomness r'
c) P_R sends R, c to P_L
3) P_L computes $r \leftarrow f_s^{PRF}(R\|c)$, and sends r to P_R
4) P_L sends zap for "$x \in L$ or r formatted correctly"
5) P_R decommits R'
6) P_L and P_L output $\tau = r \oplus R'$

For subprotocol π_0, P and V instantiate the proof system. P sends the setup message for a 2-round statistically hiding, computationally binding commitment scheme. V then constructs an ordered database, DB, consisting of $16tn(n^\epsilon + 1)$ random distinct strings of length n, and sends a statistically hiding commitment of DB to P. V also sends the setup message used by P to execute zap proofs. At this point, P applies a pseudorandom function ($f_*^{P,1}$ with seed chosen using P's initial random tape) to the current transcript and uses the output as his random tape for the rest of the protocol.

For subprotocol π_1, for each of the n^ϵ sequential iterations, P asks for a random sequence of n entries of DB, which V then decommits to. Note that for this subprotocol, a resetting P can discover at most $tn^{1+\epsilon}$ entries of DB. We note that this protocol is very similar to the protocol in [2]; where their protocol requires $O(n)$-length (random) P commitments, our protocol requires $O(n \log n)$-length random index requests, where both protocols require corresponding V decommitments.

For subprotocol π_2, P guesses V's database and commits to the guess (which we call γ) using a non-interactive perfectly binding commitment scheme. P and V then attempt to jointly compute an $(n \log |DB|)$-length random string as follows: P commits to a seed s using a non-interactive perfectly binding commitment scheme. V uses a $(t + 1)$-wise independent hash function h with input the transcript (of π_1 and π_2 thus far) to output a random triple (R, R', r'). V then computes c, a statistically hiding commitment to R' using randomness r', and sends R and c to P. P sends back, using the PRF family $f_*^{P,2}$, $r = f_s^{P,2}(R\|c)$. P proves using a zap that either $x \in L$ or that r is properly formed from R,

c and the commitment of s. Note that P's commitment to s earlier in π_2 can now be viewed as a partial commitment to a random string. V then decommits to R' and sets $\tau = R' \oplus r$ as the set of n indices of DB. V decommits to the n indices of DB corresponding to τ. Therefore, a resetting P^* can discover at most $tn(n^\epsilon + 1)$ entries, or $1/16$ of DB, including protocol π_1. P provides a zap that either $x \in L$ or that $1/4$ of the entries of P's guess γ corresponds to the final n decommitments from DB (that correspond to τ). We denote this language by Λ_2. We note that only in the case of $x \notin L$ we have that the property that τ is distributed randomly is important; we use this fact to "bootstrap" well-formedness of τ in the unbounded, cheating P^* case.

Theorem 2. *Assuming the existence of enhanced trapdoor permutations and collision-resistant hash functions, protocol Π is a t-bounded resettably sound rZK proof system for L.*

The proof that Π is rZK will follow from proofs that Π is t-resettably sound and complete (Lemma 3), that (π_0, π_2) is rWI (Theorem 4) and that there exists a specific, simpler protocol Π' that is also rZK (Theorem 6). We will then prove that since Π' is both rZK and sufficiently similar to Π, in a manner we define as *near compatible*, Π is also rZK (Lemma 5).

Lemma 3. *Π is t-resettably sound and t-resettably complete for L.*

Completeness follows from the completeness of the zap protocols. Resettable soundness follows from the rWI of zap proofs and the security of the coin-tossing protocol; since P^* cannot discover $1/4$ of the database DB except with negligible probability (due to the statistical hiding property of V's commitment scheme), P^* cannot find a correct witness for language Λ_2 except with negligible probability due to the distribution of the coin-tossing protocol output.

In what follows, we will need that the sequential composition (π_0, π_2) is rWI for languages Λ_2 and L.

Theorem 4. *Assuming the existence of enhanced trapdoor permutations and collision-resistant hash functions, protocol (π_0, π_2) is rWI for Λ_2 and for L.*

For lack of space, we omit the proof of Theorem 4. The intuition for the proof is that rWI holds due to the rWI of zap proofs, the security of the respective commitment schemes, and the security of the coin-flipping protocol.

3.1 From a rZK Proof System to a New rZK Proof System

We now outline how we prove rZK of protocol Π by constructing another protocol where rZK is easier to prove. We note that this definition may likely be generalized, but we only detail properties that will apply in our case for simplicity. It is important to note that the "simpler" rZK protocol does not need to be t-resettably sound; since the purpose here is to prove rZK, resettable soundness is not required. Due to lack of space, we omit here the precise definition of *near-compatible* protocols.

The idea of our transformation stems from the idea of constructing a rZK proof system for L from a rWI proof system; see the constructions of [2,7,9]. What generally occurs is that first the setup of the rWI proof is executed; then a so-called extraction protocol is executed, where a cheating prover learns nothing, but a simulator learns some secret s. Finally, a rWI proof is completed for the language "$x \in L$ or the secret s has been learned"; in the specific case of [2], the "secret" was that the prover had committed to a string before the verifier had decommitted to that same string, while in our case, the prover commits to a largely correct guess of the database previously committed to by the verifier.

At a high level, we say that a protocol $\Pi = (\pi_0, \pi_1, \pi_2)$, which is the protocol that we wish to prove rZK, is near-compatible to a protocol $\Pi' = (\pi_0, \pi_1, \pi_2')$ if the following holds. Fix a language $L \in \mathcal{NP}$. Then (π_0, π_1, π_2') is rZK for L. (π_0, π_1, π_2) is an interactive proof for L, and (π_0, π_2) must be rWI so that it does not reveal to the verifier whether the transcript is generated by using the genuine witness of a real prover or by fake witness belonging to a simulator[4]. Finally, we wish that the extraction stage, π_1, is essential for the simulator to complete both (π_0, π_2') and (π_0, π_2) but extraneous for the honest prover.

Lemma 5. *(Informal) Fix a language L. Let (π_0, π_1, π_2) be near-compatible to (π_0, π_1, π_2'). Let (π_0, π_1, π_2') be rZK with a simulator that plays honestly for π_0 and π_2' and such that any witness extracted by the simulator is, except with negligible probability, a valid witness for (π_0, π_1, π_2) (with the same messages sent for π_0 and π_1). Then (π_0, π_1, π_2) is rZK.*

For lack of space, the formal version of Lemma 5 in omitted. The intuition for the proof of Lemma 5 is that by definition of near-compatible protocols and by the lemma statement, if simulator Sim$'$ for (π_0, π_1, π_2') is able to extract a witness to complete the protocol, then so is simulator Sim for (π_0, π_1, π_2) that acts identically to Sim$'$ for π_1 and honestly for π_0 and π_2. This is because both simulators act identically for the rounds where extraction occurs. Further, (π_0, π_2) being rWI implies that V^* cannot distinguish whether the transcript is generated by a real prover using a witness for $x \in L$ or by a simulator using an extracted witness.

4   An Admissible, Near-Compatible rZK Proof System

Here we outline an admissible rZK proof system that has the same initialization phase and extraction phase as protocol Π but with a simplified end stage in order to make the proof of rZK easier. In particular, (π_0, π_1, π_2') is not constructed to be t-resettably sound, and therefore the verifier can eventually reveal the entire DB.

---

[4] Some additional technical properties specified in the precise definition: it is enough for our purposes that the setup phase, π_0, consists of one round of messages sent by P followed by a round of messages sent by V. In order to prove the lemma, we will require security reductions that will need limited access to the prover's random tape; therefore, P's message for π_0 must be public coin.

For lack of space, we will only sketch π'_2. π'_2 has the same inputs as π_2 above. π'_2 also begins like π_2: first, prover commits his guess γ using a statistically binding commitment scheme and sends it to V. Then, however, V decommits the *entire* DB. P then executes a zap that either $x \in L$ or that γ corresponds to 1/4 of DB; we denote this new language by Λ_3.

Note that by construction, (π_0, π_1, π'_2) satisfies the respective properties of near-compatibility. Further, (π_0, π_1, π'_2) is admissible since the verifier, after its initial message, only sends decommitments.

Since (π_0, π_1, π'_2) and (π_0, π_1, π_2) are near-compatible, the only remaining subtlety is to note that the witness extraction property of Lemma 5 holds. But this is indeed the case because the simulator, will extract a set of entries from V that correspond to at least 1/2 of V's DB except with negligible probability. Since, for π_2, the entries chosen for Λ_2 are selected uniformly at random from DB, if the simulator knows 1/2 of DB, then the simulator will know 1/4 of the entries selected for Λ_2 except with negligible probability.

Theorem 6. *Assuming the existence of enhanced trapdoor permutations and collision-resistant hash functions, protocol (π_0, π_1, π'_2) is zero knowledge in the hybrid model (i.e, hZK) for L.*

Since the protocol (π_0, π_1, π'_2) is hZK and already in the form needed to transform zero-knowledge proofs secure in the hybrid model to zero-knowledge proofs secure in the resettably model, Theorem 6 implies the following[5].

Corollary 7. (π_0, π_1, π'_2) *is rZK for L.*

We would like to contrast the protocol (π_0, π_1, π'_2) with that given in [2]. As noted in the high-level outline of Π in Section 3, the extraction stage of [2] and the subprotocol π_1 are very similar. Indeed, π'_2 is a natural extension of the protocol in [2] because in both their protocol and ours, DB is revealed and the rWI proof incorporates the whole DB. In order to prove Theorem 6, we will need the fact that (π_0, π'_2) is rWI for Λ_3 and for L.

Lemma 8. *Assuming the existence of enhanced trapdoor permutations and collision-resistant hash functions, then protocol (π_0, π'_2) is rWI for Λ_3 and for L.*

4.1 High-Level Simulator Strategy in the Proof of Theorem 6

In [2], the high level strategy of the simulator was that it would try to "look ahead" to try to figure out the verifier's commitment ahead of time, but otherwise

---

[5] We note that the simulator does not change from Theorem 6 to Corollary 7. The reason is that the proof in [2] that takes a hZK protocol and proves that it is rZK does not change the simulator; rather, it proves that for every hybrid adversary there exists a corresponding adversary that however is still simulatable. In particular, if the simulator given here in the hybrid model only rewinds during π_1 and otherwise plays honestly, so does the simulator in the full rZK model.

play honestly for all other (non-extraction stage) rounds. This is also true for the simulator here: the simulator would like to discover as many DB decommitments as possible and otherwise plays honestly. The most important difference between the rZK simulator here and the simulator in [2] is that their protocol only requires 1 successful look-ahead to proceed, while our protocol requires polynomially many successful look-aheads to proceed.

To construct our simulator, we will use a nearly identical strategy as the simulator from [2] except that we will execute the individual (main-thread level) look-aheads $|DB|$ many times in parallel. Namely, the simulator Sim′ in the extraction stage, π_1, attempts to discover half of DB; if this has not occurred at the end of the extraction stage, then the simulator simply aborts and fails to complete. One of the main inefficiencies of the [2] simulator is that it computes a distinct look-ahead subprotocol run (embedded in the subroutine NextProverMsg, which then unfolds recursively, see details in [2]) at each of the n^ϵ round iterations of the extraction stage. The idea of their simulator is that if the simulator makes a distinct look-ahead subprotocol run at each round, which in turn consists of polynomially many look-ahead attempts, then except with negligible probability, the simulator will be able to extract one "secret". Since the look-ahead subprotocol success probability is independent from one round to the next, the strategy of our simulator is that instead of making one independent look-ahead subprotocol run at each round, we make $poly(n,t) = |DB|$ independent calls at each (main-thread) iteration of the extraction stage[6]. By a union bound, our simulator will also fail to extract only with negligible probability. A subtlety is that $|DB|$ successful look-aheads might not reveal as much of $|DB|$ as desired. However, because the prover messages in π_1 consist of n randomly chosen indices, V^* is unable to both complete the protocol with $P/\text{Sim}′$ and sufficiently control the distribution of the prover messages that V^* chooses to proceed with.

We omit the full simulator specification and proof here due to lack of space.

Acknowledgments. The work of the second author is supported in part by NSF grants 0830803, 09165174, 1065276, 1118126 and 1136174, US-Israel BSF grant 2008411, OKAWA Foundation Research Award, IBM Faculty Research Award, Xerox Faculty Research Award, B. John Garrick Foundation Award, Teradata Research Award, and Lockheed-Martin Corporation Research Award. This material is based upon work supported by the Defense Advanced Research Projects Agency through the U.S. Office of Naval Research under Contract N00014-11-1-0392. The views expressed are those of the author and do not reflect the official policy or position of the Department of Defense or the U.S. Government. The work of the third author has been done while visiting UCLA and is supported in part by the European Commission through the FP7 programme under contract 216676 ECRYPT II.

---

[6] We make these independent calls only at the main-thread level of the recursion; in this manner, simulator run-time does not expand exponentially.

References

1. Deng, Y., Goyal, V., Sahai, A.: Resolving the simultaneous resettability conjecture and a new non-black-box simulation strategy. In: FOCS 2009, pp. 251–260 (2009)
2. Canetti, R., Goldreich, O., Goldwasser, S., Micali, S.: Resettable zero-knowledge (extended abstract). In: STOC 2000, pp. 235–244. ACM (2000)
3. Micali, S., Reyzin, L.: Soundness in the Public-Key Model. In: Kilian, J. (ed.) CRYPTO 2001. LNCS, vol. 2139, pp. 542–565. Springer, Heidelberg (2001)
4. Barak, B., Goldreich, O., Goldwasser, S., Lindell, Y.: Resettably-sound zero-knowledge and its applications. In: FOCS 2001, pp. 116–125 (2001)
5. Dwork, C., Naor, M.,, S.: Concurrent zero-knowledge. In: STOC 1998, pp. 409–418. ACM (1998)
6. Richardson, R., Kilian, J.: On the Concurrent Composition of Zero-Knowledge Proofs. In: Stern, J. (ed.) EUROCRYPT 1999. LNCS, vol. 1592, pp. 415–431. Springer, Heidelberg (1999)
7. Kilian, J., Petrank, E.: Concurrent and resettable zero-knowledge in polylogarithmic rounds. In: STOC 2001, pp. 560–569. ACM (2001)
8. Canetti, R., Kilian, J., Petrank, E., Rosen, A.: Black-box concurrent zero-knowledge requires $\tilde{\Omega}(\log n)$ rounds. In: STOC 2001, pp. 570–579. ACM, USA (2001)
9. Prabhakaran, M., Rosen, A., Sahai, A.: Concurrent zero knowledge with logarithmic round-complexity. In: FOCS 2002, pp. 366–375. IEEE Computer Society (2002)
10. Micciancio, D., Petrank, E.: Simulatable Commitments and Efficient Concurrent Zero-Knowledge. In: Biham, E. (ed.) EUROCRYPT 2003. LNCS, vol. 2656, pp. 644–645. Springer, Heidelberg (2003)
11. Ostrovsky, R., Pandey, O., Visconti, I.: Efficiency Preserving Transformations for Concurrent Non-Malleable Zero Knowledge. In: Micciancio, D. (ed.) TCC 2010. LNCS, vol. 5978, pp. 535–552. Springer, Heidelberg (2010)
12. Blundo, C., Persiano, G., Sadeghi, A.R., Visconti, I.: Improved Security Notions and Protocols for Non-Transferable Identification. In: Jajodia, S., Lopez, J. (eds.) ESORICS 2008. LNCS, vol. 5283, pp. 364–378. Springer, Heidelberg (2008)
13. Deng, Y., Lin, D.: Instance-Dependent Verifiable Random Functions and Their Application to Simultaneous Resettability. In: Naor, M. (ed.) EUROCRYPT 2007. LNCS, vol. 4515, pp. 148–168. Springer, Heidelberg (2007)
14. Crescenzo, G., Persiano, G., Visconti, I.: Constant-Round Resettable Zero Knowledge with Concurrent Soundness in the Bare Public-Key Model. In: Franklin, M. (ed.) CRYPTO 2004. LNCS, vol. 3152, pp. 237–253. Springer, Heidelberg (2004)
15. Crescenzo, G., Persiano, G., Visconti, I.: Improved Setup Assumptions for 3-Round Resettable Zero Knowledge. In: Lee, P.J. (ed.) ASIACRYPT 2004. LNCS, vol. 3329, pp. 530–544. Springer, Heidelberg (2004)
16. Scafuro, A., Visconti, I.: On Round-Optimal Zero Knowledge in the Bare Public-Key Model. In: Pointcheval, D., Johansson, T. (eds.) EUROCRYPT 2012. LNCS, vol. 7237, pp. 153–171. Springer, Heidelberg (2012)
17. Garg, S., Ostrovsky, R., Visconti, I., Wadia, A.: Resettable Statistical Zero Knowledge. In: Cramer, R. (ed.) TCC 2012. LNCS, vol. 7194, pp. 494–511. Springer, Heidelberg (2012)
18. Dwork, C., Naor, M.: Zaps and their applications. In: FOCS 2000, pp. 283–293. IEEE Computer Society (2000)
19. Cho, C., Ostrovsky, R., Scafuro, A., Visconti, I.: Simultaneously Resettable Arguments of Knowledge. In: Cramer, R. (ed.) TCC 2012. LNCS, vol. 7194, pp. 530–547. Springer, Heidelberg (2012)

Complexity of Complexity and Maximal Plain versus Prefix-Free Kolmogorov Complexity

Bruno Bauwens*

Instituto de Telecomunicações Faculdade de Ciência da Universidade do Porto

Abstract. Peter Gacs showed [2] that for every n there exists a bit string x of length n whose plain complexity $C(x)$ has almost maximal conditional complexity relative to x, i.e., $C(C(x)|x) \geq \log n - \log^{(2)} n - O(1)$. Here $\log^2(i) = \log\log i$ etc. Following Elena Kalinina [4], we provide a game-theoretic proof of this result; modifying her argument, we get a better (and tight) bound $\log n - O(1)$. We also show the same bound for prefix-free complexity.

Robert Solovay's showed [11] that infinitely many strings x have maximal plain complexity but not maximal prefix-free complexity (among the strings of the same length); i.e. for some c: $|x| - C(x) \leq c$ and $|x| + K(|x|) - K(x) \geq \log^{(2)} |x| - c \log^{(3)} |x|$. Using the result above, we provide a short proof of Solovay's result. We also generalize it by showing that for some c and for all n there are strings x of length n with $n - C(x) \leq c$, and $n + K(n) - K(x) \geq K(K(n)|n) - 3K(K(K(n)|n)|n) - c$. This is very close to the upperbound $K(K(n)|n) + O(1)$ proved by Solovay.

Introduction

Plain Kolmogorov complexity $C(x)$ of a binary string x was defined in [5] as the minimal length of a program that computes x. (See the preliminaries or [3,6,10] for the details.) It was clear from the beginning (see, e.g., [13]) that complexity function is not computable: no algorithm can compute $C(x)$ given x. In [2,3,6] a stronger non-uniform version of this result was proven: for every n there exists a string x of length n such that conditional complexity $C(C(x)|x)$, i.e., the minimal length of a program that maps x to $C(x)$, is at least $\log n - O(\log^{(2)} n)$. (If complexity function were computable, this conditional complexity would be bounded.)

In Section 1 we revisit this classical result and improve it a bit by removing the $\log^{(2)} n$ term. No further improvement is possible because $C(n) \leq n + O(1)$, therefore $C(C(n)|x) \leq \log n + O(1)$ for all x. We use a game technique that was developed by Andrej Muchnik (see [9,8,12]) and turned out to be useful in many cases. Recently

* Full version: www.bcomp.be/papers/compcompfull.pdf. Supported by the Portuguese science foundation FCT (SFRH/BPD/75129/2010), and partially supported by the project CSI^2 (PTDC/EIAC/099951/2008). The author is grateful to Elena Kalinina and (Nikolay) Kolia Vereshchagin for giving the text [4]. The author is also grateful to (Alexander) Sasha Shen for his very generous help: for reading earlier texts on these results, for discussion, for providing a clear exposition of section 1 and some parts of section 2, and for his permission to publish it (with small modifications).

A. Czumaj et al. (Eds.): ICALP 2012, Part I, LNCS 7391, pp. 100–108, 2012.
© Springer-Verlag Berlin Heidelberg 2012

Elena Kalinina (in her master thesis [4]) used it to provide a proof of Gacs' result. We use a more detailed analysis of essentially the same game to get a better bound.

For some c, a bit string x is C-random if $n - C(x) \leq c$. Note that $n + O(1)$ is the smallest upper bound for $C(x)$. A variant of plain complexity is prefix-free or self-delimiting complexity, which is defined as the shortest program that produces x on a Turing machine with binary input tape, i.e. without blanc or terminating symbol. (See the preliminaries or [3,6,10] for the details.) The smallest upper bound for $K(x)$ for strings of length n is $n + K(n) + O(1)$. For some c, the string x is defined to be K-random if $n + K(n) - K(x) \leq c$.

Robert Solovay [11,1] observed that K-random strings are also C-random strings (for some $c' \leq O(c)$), but not vice versa. Moreover, he showed that some c and infinitely many x satisfy $|x| - C(x) \leq c$ and

$$|x| + K(|x|) - K(x) \geq \log^{(2)} |x| - c \log^{(3)} |x|.$$

He also showed that for C-random x the left-hand side of the equation is upper-bounded by $K(K(n)|n) + O(1)$, which is bounded by $\log^{(2)} n + O(1)$. Later Joseph Miller [7] and Alexander Shen [9] generalized this, by showing that every co-enumerable set (i.e., the complement is enumerable) containing strings of every length, also contains infinitely many x such that the above equation holds. (Note that the set of C-random strings is co-enumerable but the set of K-random strings not.)

In Section 2 we provide a short proof for Solovay's result using the improved version of Gacs' theorem. Then we generalize it by showing that for some c and every n there are strings x of length n with $n - C(x) \leq c$ and

$$n + K(n) - K(x) \geq K(K(n)|n) - 3K(K(K(n)|n) |n) - c.$$

This is very close to the upperbound $K(K(n)|n) - O(1)$, which was shown by Solovay [11]. By the improved version of Gacs' result, we can choose n such that $K(K(n)|n) = \log^{(2)} n + O(1)$. For such n we obtain Solovay's theorem with the $c \log^{(3)} |x|$ term replaced by a $O(1)$ constant.

Preliminaries: Let U be a Turing machine. The *plain (Kolmogorov) complexity* relative to U is defined by

$$C_U(x|y) = \min \{|p| : U(p,y) = x\} .$$

If the machine U is prefix-free (i.e., for every p, y such that $U(p,y)$ halts, there is no prefix q of p such that $U(q,y)$ halts) then we write $K_U(x|y)$ rather than $C_U(x|y)$, and refer to it as *prefix-free (Kolmogorov) complexity* relative to U. There exist plain and prefix-free Turing machines U and V for which $C_U(x|y)$ and $K_V(x|y)$ are minimal within an $O(1)$ constant. We fix such machines and omit the indexes U,V. If y is the empty string we use the notation $C(x)$ and $K(x)$.

1 Complexity of Complexity Can Be High

Theorem 1. *There exist some constant c such that for every n there exists a string x of length n such that $C(C(x)|x) \geq \log n - c$.*

To prove this theorem, we first define some game and show a winning strategy for the game. (The connection between the game and the statement that we want to prove will be explained later.)

1.1 The Game

Game G_n has parameter n and is played on a rectangular board divided into cells. The board has 2^n columns and n rows numbered $0, 1, \ldots, n-1$ (the bottom row has number 0, the next one has number 1 and so on, the top row has number $n-1$), see Fig. 1.

Initially the board is empty. Two players: White and Black, alternate their moves. At each move, a player can pass or place a pawn (of his color) on the board. The pawn can not be moved or removed afterwards. Also Black may blacken some cell instead. Let us agree that White starts the game (though it does not matter).

The position of the game should satisfy some restrictions; the player who violates these restrictions, loses the game immediately. Formally the game is infinite, but since the number of (non-trivial) moves is a priori bounded, it can be considered as finite, and the winner is determined by the last (limit) position on the board.

Restrictions: (1) each player may put at most 2^i pawns in row i (thus the total number of black and white pawns in a row can be at most $2^i + 2^i$); (2) in each column Black may blacken at most half of the cells.

We say that a white pawn is *dead* if either it is on a blackened cell or has a black pawn in the same column strictly below it.

Winning rule: Black wins if he killed all white pawns, i.e., if each white pawn is dead in the final position.

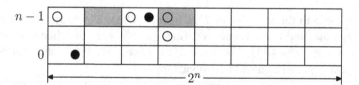

Fig. 1. Game board

For example, if the game ends in the position shown at Fig. 1, the restrictions are not violated (there are $3 \le 2^2$ white pawns in row 2 and $1 \le 2^1$ white pawn in row 1, as well as $1 \le 2^2$ black pawn in row 2 and $1 \le 2^0$ black pawn in row 0). Black loses because the white pawn in the third column is not dead: it has no black pawn below and the cell is not blackened. (There is also one living pawn in the fourth column.)

1.2 How White Can Win

The strategy is quite simple. White starts by placing a white pawn in an upper row of some column and waits until Black kills it, i.e., blackens the cell or places a black pawn below. In the first case White puts her pawn one row down and waits again. Since Black has no right to make all cells in a column black (at most half may be blackened), at some point he will be forced to place a black pawn below the white pawn in this column. After that White switches to some other column. (The ordering of columns is not important; we may assume that White moves from left to right.)

Note that when White switches to a next column, it may happen that there is a black pawn in this column or some cells are already blackened. If there is already a black pawn, White switches again to the next column; if some cell is blackened, White puts her pawn in the topmost white (non-blackened) cell.

This strategy allows White to win. Indeed, Black cannot place his pawns in all the columns due to the restrictions (the total number of his pawns is $\sum_{i=0}^{n-1} 2^i = 2^n - 1$, which is less than the number of columns). White also cannot violate the restriction for the number of her pawns on some row i: all dead pawns have a black pawns strictly below them, so the number of them on row i is $\sum_{j=0}^{i-1} 2^j = 2^i - 1$, hence White can put an additional pawn.

In fact we may even allow Black to blacken all the cells except one in each column, and White will still win, but this is not needed (and the $n/2$ restriction will be convenient later).

1.3   Proof of Gacs' Theorem

Let us show that for each n there exists a string x of length n such that $C(C(x|n)|x) \geq \log n - O(1)$. Note that here $C(x|n)$ is used instead of $C(x)$; the difference between these two numbers is $O(\log n)$ since n can be described by $\log n$ bits, so the difference between the complexities of these two numbers is $O(\log \log n)$.

Consider the following strategy for Black (assuming that the columns of the table are indexed by strings of length n):

- Black blackens the cell in column x and row i as soon as he discovers that $C(i|x) < \log n - 1$. (The constant 1 guarantees that less than half of the cells will be blackened.) Note that Kolmogorov complexity is an upper semicomputable function, and Black approximates it from above, so more and more cells are blackened.
- Black puts a black pawn in a cell (x, i) when he finds a program of length i that produces x with input n (this implies that $C(x|n) \leq i$). Note that there are at most 2^i programs of length i, so Black does not violate the restriction for the number of pawns on any row i.

Let White play against this strategy (using the strategy described above). Since the strategy is computable, the behavior of White is also computable. One can construct a decompressor V for the strings of length n as follows: each time White puts a pawn in a cell (x, i), a program of length i is assigned to x. By White's restriction, no more than 2^i programs need to be assigned. By universality, a white pawn on cell (x, i) implies that $C(x|n) \leq i + O(1)$. If White's pawn is alive in column x, there is no black pawn below, so $C(x|n) \geq i$, and therefore $C(x|n) = i + O(1)$. Moreover, for a winning pawn, the cell (x, i) is not blackened, so $C(i|x) \geq \log n - 1$. Therefore, $C(C(x|n)|x) \geq \log n - O(1)$.

Remark: the construction also guarantees that $C(x|n) \geq n/2 - O(1)$ for that x. (Here the factor $1/2$ can be replaced by any $\alpha < 1$ if we change the rules of the game accordingly.) Indeed, according to white's strategy, he always plays in the highest non-black cell of some column, and at most half of the cells in a column can be blackened, therefore no white pawns appear in the lower half of the board.

1.4   Modified Game and the Proof of Theorem 1

Now we need to get rid of the condition n and show that for every n there is some x such that $C(C(x)|x) \geq \log n - O(1)$. Imagine that White and Black play simultaneously all the games G_n. Black blackens the cell (x,i) in game $G_{|x|}$ when he discovers that $C(i|x) < \log n - 1$, as he did before, and puts a black pawn in a cell (x,i) when he discovers an *unconditional* program of length i for x. If Black uses this strategy, he satisfies the stronger restriction: the total number of pawns in row i *on all boards* is bounded by 2^i.

Assume that White uses the described strategy on each board. What can be said about the total number of white pawns in row i? The dead pawns have black pawns strictly below them and hence the total number of them does not exceed $2^i - 1$. On the other hand, there is at most one live white pawn on each board. We know also that in G_n white pawns never appear below row $n/2 - 1$, so the number of live white pawns does not exceed $2i + O(1)$. Therefore we have $O(2^i)$ white pawns on the i-th row in total.

For each n there is a cell (x,i) in G_n where White wins in G_n. Hence, $C(x) < i + O(1)$ (because of property just mentioned and the computability of White's behavior), $C(x) \geq i$ and $C(i|x) \geq \log n - 1$ (by construction of Black's strategies and the winning condition). Theorem 1 is proven.

1.5   Version for Prefix Complexity

Theorem 2. *There exist some constant c such that for every n there exists a string x of length n such that $C(K(x)|x) \geq \log n - c$ and $K(x) \geq n/2$. This also implies that $K(K(x)|x) \geq \log n - c$.*

The proof of $C(K(x)|x) \geq \log n - c$ goes in the same way. Black places a pawn in cell (i,x) if some program of length i for a prefix-free (unconditional) machine computes x (and hence $K(x) \leq i$); White uses the same strategy as described above. The sum of 2^{-i} for all black pawns is less than 1 (Kraft-inequality); some white pawns are dead, i.e., strictly above black ones, and for each column the sum of 2^{-j} where j is the row number, does not exceed $\sum_{j>i} 2^{-j} < 2^{-i}$. Hence the corresponding sum for all dead white pawns is less than 1; for the rest the sum is bounded by $\sum_n 2^{-n/2+1}$, so the total sum is bounded by a constant, and we conclude that for x in the winning column the row number is $K(x) + O(1)$, and this cell is not blackened.

2   Strings with Maximal Plain and Non-maximal Prefix-Free Complexity

In this section we compare two measures of non-randomness. Let x be a string of length n; we know that $C(x) \leq n + O(1)$, and the difference $n - C(n)$ measures how "nonrandom" x is. Let us call it C-deficiency of x. On the other hand, $K(x) \leq n + K(n) + O(1)$, so $n + K(n) - K(x)$ also measures "nonrandomness" in some other way; we call this quantity K-deficiency of x.

The following proposition means that K-random strings (for which K-deficiency is small; they are also called "Chaitin random") are always C-random (C-deficiency is small; such strings are also called "Kolmogorov random").

Proposition 1 (Solovay [11]). $|x| + K(|x|) - K(x) \leq c$ *implies* $|x| - C(x) \leq O(c)$.

Proof. We use a result of Levin: for every string u

$$K(u \mid C(u)) = C(u) + O(1),$$

and, on the other hand, for any positive or negative integer number c:

$$K(u \mid i) = i + c,$$

implies $C(u) = i + O(c)$.[1]

Let $n = |x|$. Notice that

$$n + K(n) \leq K(x) - c = K(x, n) - O(c) \leq K(x \mid n) + K(n) - O(c).$$

Hence, $K(x \mid n) \geq n - O(c)$, thus $K(x \mid n) = n + O(c)$ and thus: $C(x) = n + O(c)$.

R. Solovay showed that the reverse statement is not always true: a C-random string may be not K-random. However, as the following result shows, the K-deficiency still can be bounded for C-random strings:

Proposition 2 (Solovay [11]). *For any x of length n the inequality $C(x) \geq n - c$ implies:*

$$n + K(n) - K(x) \leq K(K(n) \mid n) + O(c).$$

Note that $K(K(n) \mid n) \leq \log^{(2)} n + O(1)$.

Proof. The proof uses another result of Levin [2,3,6]: for all u, v we have the additivity property

$$K(u, v) = K(u) + K(v \mid u, K(u)) + O(1).$$

To prove Proposition 2, notice that $n = C(x) = K(x \mid C(x)) = K(x \mid n)$ with $O(c)$-precision. By additivity we have: $K(x) = K(n, x) = K(n) + K(x \mid n, K(n))$. Putting these observations together, we get

$$n + K(n) - K(x) = K(x \mid n) + K(n) - (K(n) + K(x \mid n, K(n))) + O(c)$$
$$= K(x \mid n) - K(x \mid n, K(n)) + O(c). \tag{1}$$

Observe that $K(x \mid n) \leq K(x \mid n, K(n)) + K(K(n) \mid n) + O(1)$, hence the K-deficiency is bounded by $K(K(n) \mid n) + O(c)$.

---

[1] Textbooks like [6, Lemma 3.1.1] mention only the first statement. To show the second, note that the function $i \mapsto K(x \mid i)$ maps numbers at distance c to numbers at distance $O(\log c)$, hence, the fixed point $C(x)$ must be unique within an $O(1)$ constant. Furthermore, for any i, the fixed point must be within distance $O(|i - K(u \mid i)|)$ from i, hence $|C(u) - i| \leq O(|i - K(u \mid i)|) = O(c)$.

The following theorem shows that for all n the bound $K(K(n)|n)$ for K-deficiency for C-random strings can almost be achieved. The error is at most $O(\log K(K(n)|n))$.

Theorem 3. *For some c and all n there are strings x of length n such that $n - C(x) \leq c$, and*

$$n + K(n) - K(x) \geq K(K(n)|n) - 3K(K(K(n)|n)|n) - c.$$

By corollary, infinitely many C-random strings have K-deficiency $\log^{(2)}|x| + O(1)$. Indeed, for n such that $K(K(n)|n) = \log^{(2)}n + O(1)$, we have $K(K(K(n)|n)|n) \leq O(1)$, and hence, a slightly stronger statement than proved by Solovay [11] is obtained.

Corollary 1. *There exists a constant c and infinitely many x such that $|x| - C(x) \leq c$ and $|x| + K(|x|) - K(x) \geq \log^{(2)}|x| - c$.*

Before proving Theorem 3, we prove the corollary directly.

Proof. First we choose n, the length of string x. It is chosen in such a way that $K(K(n)|n) = \log^{(2)}n + O(1)$ and $K(n) \geq (\log n)/2$ (Theorem 2). (So the bound of Proposition 2 is not an obstacle.) We know already (see equation 1) that for a string x with C-deficiency c the value of K-deficiency is $O(c)$-close to $K(x|n) - K(x|n, K(n))$. This means that adding $K(n)$ in the condition should decrease the complexity, so let us include $K(n)$ in x somehow. We also have to guarantee maximal C-complexity of x. This motivates the following choice:

- choose r of length $n - \log^{(2)}n$ such that $K(r|n, K(n)) \geq |r|$. Note that this implies $K(r|n, K(n)) = |r| + O(1)$, since the length of r is determined by the condition.
- Let $x = \langle K(n) \rangle r$, the concatenation of $K(n)$ (in binary) with r. Note that $\langle K(n) \rangle$ has at most $\log^{(2)}n + O(1)$ bits for every n, and by choice of n has at least $\log^{(2)}n - 1$ bits, hence $|x| = n + O(1)$.

As we have seen (looking at equation (1)), it is enough to show that $K(x|K(n), n) \leq n - \log^{(2)}n$ and $K(x|n) \geq n$ (the latter equality implies $C(x) = n$); all the equalities here and below are up to $O(1)$ additive term.

- Knowing n, we can split x in two parts $\langle K(n) \rangle$ and r. Hence, $K(x|K(n), n) = K(K(n), r|n, K(n))$, and this equals $K(r|n, K(n))$, i.e., $n - \log^{(2)}n$ by choice of r.
- To compute $K(x|n)$, we use additivity:

$$K(x|n) = K(K(n), r|n) = K(K(n)|n) + K(r|K(n), K(K(n)|n), n).$$

By choice of n, we have $K(K(n)|n) = \log^{(2)}n$, and the last term simplifies to $K(r|K(n), \log^{(2)}n, n)$, and this equals $K(r|K(n), n) = n - \log^{(2)}n$ by choice of r. Hence $K(x|n) = \log^{(2)}n + (n - \log^{(2)}n) = n$.

Remark 1: One can ask how many strings exist that satisfy the conditions of Corollary 1. By Proposition 2, the length n of such a string must satisfy $K(K(n)|n) \geq \log^{(2)}n - O(1)$. By Theorem 2, there is at least one such an n for every length of n

in binary. Hence such n, can be found within exponential intervals. One can also ask for such n, how many strings x of length n satisfy the conditions of Corollary 1. By a theorem of Chaitin [6], there are at least $O(2^{n-k})$ strings with K-deficiency k, hence we can have at most $O(2^{n-\log^{(2)} n})$ such strings. It turns out that indeed at least a fraction $1/O(1)$ of them satisfy the conditions of Corollary 1. To show this, note that in the proof Theorem 3, every different r of length $|n| - \log^{(2)} n + O(1)$ leads to the construction of a different x. For such r we essentially need $K(r|n, K(n)) \geq |r| - O(1)$, and hence there are $O(2^{n-\log^{(2)} n})$ of them.

Proof of Theorem 3. In the proof above, in order to obtain a large value $K(x|n) - K(x|n, K(n))$, we incorporated $K(n)$ in a direct way (as $\langle K(n)\rangle$) in x. To show that $C(x) = K(x|n) + O(1)$ is large we essentially used that the length of $\langle K(n)\rangle$ equals $K(K(n)|n) + O(1)$. For general n, this trick does not work anymore, but we can use a shortest program for $K(n)$ given n (on a plain machine). For every n we can construct x as follows:

- let q be a shortest program that computes $K(n)$ from n on a *plain* machine (if there are several shortest programs, we choose the one with shortest running time). Note that $|q| = C(K(n)|n) + O(1) = C(q|n) + O(1)$ (remind that by adding some fixed instructions, a program can print itself, and that a shortest program is always incompressible, thus up to $O(1)$ constants: $|q| \geq C(K(n)|n) \geq C(q|n) \geq |q|$), by Levin's result (conditional version), the last term also equals $K(q|n, |q|) + O(1)$;
- let r have length $n - |q|$, such that $K(r|n, K(n), q) \geq |r|$. Note that this implies $K(r|n, K(n), q) = |r| + O(1)$, (since the length of r is determined by the condition).
- We define x as the concatenation qr.

We show that $C(x) = n + O(1)$ and that the K-deficiency is at least $|q| - K(|q||n) + O(1)$. To show that this implies the theorem, we need that

$$K(K(n)|n) - 3K(K(K(n)|n)|n) \leq C(K(n)|n) - K(C(K(n)|n)|n) + O(1),$$

which is for $a = K(n)$ the conditioned version of Lemma 1:

$$K(a|n) - 3K(K(a|n)|n) \leq C(a|n) - K(C(a|n)|n) + O(1).$$

Following the same structure as the proof above, it remains to show that $K(x|K(n), n) \leq n - |q| + K(|q||n)$ and $K(x|n) \geq n$ (the latter equality implies $C(x) = n$); all the equalities here and below are up to $O(1)$ additive term.

- Knowing $|q|$, we can split x in two parts q and r. Hence, $K(x|K(n), n, |q|) = K(q, r|n, K(n), |q|)$. Given $n, K(n), |q|$ we can search for a program of length $|q|$ that on input n outputs $K(n)$; the one with shortest computation time is q. Hence, $K(q, r|n, K(n), |q|) = K(r|n, K(n), |q|)$, i.e., $n - |q|$ by choice of r, and therefore $K(x|K(n), n) \leq n - |q| + K(|q||n)$.
- To compute $K(x|n)$, we use additivity:

$$K(x|n) \geq K(x|n, |q|) = K(q, r|n, |q|) = K(q|n, |q|) + K(r|q, K(q|n, |q|), n).$$

By choice of q we have $K(q|n, |q|) = |q|$. The last term is $K(r|q, |q|, n)$ which equals $K(r|q, n) = n - |q|$ by choice of r. Hence, $K(x|n) \geq |q| + (n - |q|) = n$. $\square$

Lemma 1. $K(a) - 3K(K(a)) \leq C(a) - K(C(a)) + O(1)$

Proof. Note that $K(a) - C(a) \leq K(C(a))$. Indeed, any program for a plain machine can be considered as a program for a prefix-free machine conditional to it's length. Hence, we can transform a plain program p to a prefix-free program by adding a description of $|p|$ of length $K(|p|)$ to p. Hence it remains to show $2K(C(a)) \leq 3K(K(a)) + O(1)$. Solovay [11,1] showed that

$$K(a) - C(a) = K(K(a)) + O(K(K(K(a)))),$$

hence,

$$|K(K(a)) - K(C(a))| \leq O(\log K(K(a))).$$

References

1. Downey, R.G., Hirschfeldt, D.R.: Algorithmic Randomness and Complexity. Theory and Applications of Computability. Springer (2010)
2. Gács, P.: On the symmetry of algorithmic information. Soviet Math. Dokl. 15(5), 1477–1480 (1974)
3. Gács, P.: Lecture notes on descriptional complexity and randomness (1988-2011), http://www.cs.bu.edu/faculty/gacs/papers/ait-notes.pdf
4. Kalinina, E.: Some applications of the method of games in Kolmogorov complexity. Master thesis, Moscow State University (2011)
5. Kolmogorov, A.N.: Three approaches to the quantitative definition of information. Problemy Peredachi Informatsii 1(1), 3–11 (1965)
6. Li, M., Vitányi, P.M.B.: An Introduction to Kolmogorov Complexity and Its Applications. Springer, New York (2008)
7. Miller, J.S.: Contrasting plain and prefix-free complexities, Preprint, http://www.math.wisc.edu/~jmiller/downloads.html
8. Muchnik, A.: On the basic structures of the descriptive theory of algorithms. Soviet Math. Dokl. 32, 671–674 (1985)
9. Muchnik, A.A., Mezhirov, I., Shen, A., Vereshchagin, N.: Game interpretation of Kolmogorov complexity (2010), arxiv:1003.4712v1
10. Shen, A.: Algorithmic Information theory and Kolmogorov complexity. Technical report TR2000-034. Uppsala University (2000)
11. Solovay, R.: Draft of a paper (or series of papers) on Chaitin's work, unpublished notes, 215 pages (1975)
12. Vereshchagin, N.: Kolmogorov complexity and Games. Bulletin of the European Association for Theoretical Computer Science 94, 51–83 (2008)
13. Zvonkin, A.K., Levin, L.A.: The complexity of finite objects and the development of the concepts of information and randomness by means of the theory of algorithms. Russian Math. Surveys 25(6(156)), 83–124 (1970)

On Quadratic Programming with a Ratio Objective[*]

Aditya Bhaskara, Moses Charikar,
Rajsekar Manokaran, and Aravindan Vijayaraghavan[**]

Department of Computer Science, Princeton University
Center for Computational Intractability
{bhaskara,moses,rajsekar,aravindv}@cs.princeton.edu

Abstract. Quadratic Programming (QP) is the well-studied problem of maximizing over $\{-1, 1\}$ values the quadratic form $\sum_{i \neq j} a_{ij} x_i x_j$. QP captures many known combinatorial optimization problems, and assuming the Unique Games conjecture, Semidefinite Programming (SDP) techniques give optimal approximation algorithms. We extend this body of work by initiating the study of Quadratic Programming problems where the variables take values in the domain $\{-1, 0, 1\}$. The specific problem we study is

$$\text{QP-Ratio}: \quad \max_{\{-1,0,1\}^n} \frac{\sum_{i \neq j} a_{ij} x_i x_j}{\sum x_i^2}$$

This is a natural relative of several well studied problems (in fact Trevisan introduced a normalized variant as a stepping stone towards a spectral algorithm for Max Cut Gain). Quadratic ratio problems are good testbeds for both algorithms and complexity because the techniques used for quadratic problems for the $\{-1, 1\}$ and $\{0, 1\}$ domains do not seem to carry over to the $\{-1, 0, 1\}$ domain. We give approximation algorithms and evidence for the hardness of approximating these problems.

We consider an SDP relaxation obtained by adding constraints to the natural eigenvalue (or SDP) relaxation for this problem. Using this, we obtain an $\tilde{O}(n^{1/3})$ approximation algorithm for QP-ratio. We also give a $\tilde{O}(n^{1/4})$ approximation for bipartite graphs, and better algorithms for special cases.

As with other problems with ratio objectives (e.g. uniform sparsest cut), it seems difficult to obtain inapproximability results based on **P $\neq$ NP**. We give two results that indicate that QP-Ratio is hard to approximate to within any constant factor: one is based on the assumption that random instances of Max k-AND are hard to approximate, and the other makes a connection to a ratio version of Unique Games. We also give a natural distribution on instances of QP-Ratio for which an n^{ε} approximation (for ε roughly $1/10$) seems out of reach of current techniques.

---

[*] The full version of the paper [BCMV11] can be accessed at http://arxiv.org/abs/1101.1710

[**] The first, second and fourth authors are supported by NSF AF 0916218 and CCF 0832797. The third author is support by NSF CCF 0832797.

A. Czumaj et al. (Eds.): ICALP 2012, Part I, LNCS 7391, pp. 109–120, 2012.

1 Introduction

Semidefinite programming techniques have proved very useful for quadratic optimization problems (i.e. problems with a quadratic objective) over $\{0,1\}$ variables or $\{\pm1\}$ variables. Such problems admit natural SDP relaxations and beginning with the seminal work of Goemans and Williamson [GW95], sophisticated techniques have been developed for exploiting these SDP relaxations to obtain approximation algorithms. For a large class of constraint satisfaction problems, a sequence of exciting results [KKMO07, KV05, Aus07] culminating in the work of Raghavendra [Rag08], shows that in fact, such SDP based algorithms are optimal (assuming the Unique Games Conjecture).

In this paper, we initiate a study of a quadratic programming problem (QP-Ratio) with variables in $\{0, \pm1\}$.

$$\text{QP-Ratio}: \quad \max_{\{-1,0,1\}^n} \frac{\sum_{i \neq j} a_{ij} x_i x_j}{\sum x_i^2} \tag{1}$$

An alternate phrasing of the ratio-quadratic programming problem is the following: the goal is to select a subset of non-zero variables S and assign them values in $\{\pm1\}$ so as to maximize the ratio of the quadratic programming objective $\sum_{i<j\in S} a_{i,j} x_i x_j$ to the size of S. This can be viewed as an outlier version of quadratic programming, where the variables corresponding to outliers must be set to 0, and the goal is to maximize the solution quality on the rest. Note that the numerator itself is the quadratic programming objective $\sum_{i<j} a_{i,j} x_i x_j$, and can be maximized by setting all variables to be ±1. However, the denominator term in the objective makes it worthwhile to set variables to 0.

Variants of this problem are well known: Restricting to $\{\pm1\}$ variables results in a problem with an $O(\log n)$ approximation [NRT99, CW04]. On the other hand, restricting to non-negative variable values (when the $a_{i,j}$ are non-negative) yields a polynomial time solvable problem. In fact, ratio objectives like this have been studied in several contexts and algorithms to optimize them are often useful subroutines in designing approximation algorithms.

Despite these connections, QP-Ratio seems to fall outside the realm of our current understanding on both the algorithmic and inapproximability fronts. One of the goals of our work is to enhance (and understand the limitations of) the SDP toolkit for approximation algorithms by applying it to this natural problem. On the hardness side, the issues that come up are akin to those arising in other problems with a ratio/expansion flavor, where conventional techniques in inapproximability have been ineffective.

A normalized version of the QP-Ratio objective arose in recent work of Trevisan [Tre09] on computing Max Cut Gain using eigenvalue techniques. The idea here is to use the eigenvector to come up with a 'good' partial assignment, and recurse. Crucial to this procedure is a quantity called the *GainRatio* defined for a graph; this is a special case of Normalized QP-Ratio where $a_{ij} = -1$ for edges, and 0 otherwise.

1.1 Our Results

Algorithms. We first study mathematical programming relaxations for QP-Ratio. The main difficulty in obtaining such relaxations is imposing the constraint that the variables take values $\{-1, 0, 1\}$. Capturing this using convex constraints is the main challenge in obtaining good algorithms for the problem.

We consider a semidefinite programming (SDP) relaxation obtained by adding constraints to the natural eigenvalue relaxation, and round it to obtain an $\tilde{O}(n^{1/3})$ approximation algorithm. A natural, interesting special case is bipartite instances of QP-Ratio, where the support of a_{ij} is the adjacency matrix of a bipartite graph (akin to bipartite instances of quadratic programming, also known as the Grothendieck problem). For bipartite instances, we obtain an $\tilde{O}(n^{1/4})$ approximation and an almost matching SDP integrality gap of $\Omega(n^{1/4})$.

Techniques. The main challenge in semi-definite programming(SDP) based approaches for ratio quadratic programs is the situation that vectors in the SDP solution could have very different lengths. To overcome this, we strengthen the SDP by having additional constraints in the SDP, and come up with new rounding techniques to move to SDP solutions with a smaller range of lengths.

We can take further advantage of these additional inequalities in the case of bipartite graphs to get an improved $n^{1/4}$ approximation. Here, we combine the previous SDP rounding with a different rounding scheme, which performs well when the vector solution has equal contributions from many length scales.

Inapproximability. Complementing our algorithmic result for QP-Ratio, we show hardness results for the problem. We first show that there is no PTAS for the problem assuming $P \neq NP$. We also provide evidences that it is hard to approximate to within any constant factor.

In section 3.2 we rule out constant factor approximation algorithms for QP-Ratio assuming Feige's hypothesis [Fei02] that random instances of k-AND are hard to distinguish from 'well-satisfiable' instances. Even the strongest known SDP relaxations ($\Omega(n)$ rounds of the Lasserre hierarchy) cannot refute this conjecture [Tul09]. We also show in section 3.3 a reduction from a ratio version of the well-known Unique Games problem to QP-Ratio. We think that ratio version of Unique Games is an interesting problem worthy of study that could shed light on the complexity of other ratio optimization questions. The technical challenge in our reduction is to develop the required fourier-analytic machinery to tackle PCP-based reductions to ratio problems.

As with other ratio problems like Sparsest Cut, there is a big gap in the approximation guarantees and inapproximability results. We suspect that the problem is in fact hard to approximate to an n^ε factor for some $\varepsilon > 0$. In Section 3.1, we decribe a natural distribution over instances which we believe are hard to approximate up to polynomial factors.

Normalized QP-Ratio. Our original motivation to study quadratic ratio problems was the related GainRatio problem studied in Trevisan [Tre09]. We give a sharp contrast between the strengths of different relaxations and disprove Trevisan's conjecture that the eigenvalue approach towards Max Cutgain is as powerful as the SDP-based approach[CW04]. See Section 4 for details.

2   Algorithms for QP-Ratio

We start with the most natural relaxation for QP-Ratio (1) :

$$\max \frac{\sum_{i,j} A_{ij} x_i x_j}{\sum_i x_i^2} \text{ subject to } x_i \in [-1, 1]$$

(instead of $\{0, \pm 1\}$). The solution to this is precisely the largest eigenvector of A (scaled such that entries are in $[-1, 1]$). However it is easy to construct instances with a large integrality gap of $\Omega(\sqrt{n})$ (for example, the instance given by the adjacency matrix of a n-vertex star graph).

We show that SDP relaxations give more power in expressing the constraints $x_i \in \{0, \pm 1\}$. Consider the following relaxation:

$$\max \sum_{i,j} A_{ij} \cdot \langle \mathbf{w}_i, \mathbf{w}_j \rangle \text{ subject to } \sum_i \|\mathbf{w}_i\|^2 = 1, \text{ and}$$

$$|\langle \mathbf{w}_i, \mathbf{w}_j \rangle| \le \mathbf{w}_i^2 \text{ for all } i, j \tag{2}$$

It is easy to see that this is indeed a relaxation: start with an integer solution $\{x_i\}$ with k non-zero x_i, and set $\mathbf{v}_i = (x_i/\sqrt{k}) \cdot \mathbf{v}_0$ for a fixed unit vector $\mathbf{v}_0$.

Without constraint (2), the SDP relaxation is equivalent to the eigenvalue relaxation given above. Roughly speaking, equation (2) tries to impose the constraint that non-zero vectors are of equal length. In the example of the n-vertex star, this relaxation has value equal to the true optimum. In fact, for any instance with $A_{ij} \ge 0$ for all i, j, this relaxation is exact [Cha00]. [1]

In the remainder of the section, we describe a simple $\tilde{O}(n^{1/3})$ rounding algorithm, which shows that the additional constraints (2) indeed help.

2.1   An $\tilde{O}(n^{1/3})$ Rounding Algorithm

Consider an instance of QP-Ratio defined by $A_{(n \times n)}$. Let $\mathbf{w}_i$ be an optimal solution to the SDP, and let the objective value be denoted sdp.

Since the problem is the same up to scaling the A_{ij}, let us assume that $\max_{i,j} |A_{ij}| = 1$. There is a trivial solution which attains a value $1/2$ (if i, j are indices with $|A_{ij}| = 1$, set x_i, x_j to be ± 1 appropriately, and the rest of the x's to 0). Now, since we are aiming for an $\tilde{O}(n^{1/3})$ approximation, we can assume that sdp $> n^{1/3}$.

As alluded to earlier (and as can be seen in the gap example), the difficulty is when most of the contribution to sdp is from non-zero vectors with very different lengths. The idea of the algorithm will be to move to a situation in which this does not happen. First, we show that if the vectors indeed have roughly equal length, we can round well. Roughly speaking, the algorithm uses the lengths $\|\mathbf{v}_i\|$ to determine whether to pick i, and then uses the ideas of [CW04] or [NRT99] applied to the vectors $\frac{\mathbf{v}_i}{\|\mathbf{v}_i\|}$.

---

[1] We consider other SDP relaxations that can be writing by viewing the $\{0, \pm 1\}$ as a 3-alphabet CSP, and show a $\Omega(\sqrt{n})$-integrality gap in the full version. It is interesting to see if lift and project methods starting with this relaxation can be useful.

Lemma 1. *Given a vector solution $\{\mathbf{v}_i\}$, with $\|\mathbf{v}_i\|^2 \in [\tau/\Delta, \tau]$ for some $\tau > 0$ and $\Delta > 1$, we can round it to obtain an integer solution with cost at least $\mathsf{sdp}/(\sqrt{\Delta}\log n)$.*

Proof. Starting with $\mathbf{v}_i$, we produce vectors $\mathbf{w}_i$ each of which is either 0 or a unit vector, such that

$$\text{If } \frac{\sum_{i,j} A_{ij}\langle \mathbf{v}_i, \mathbf{v}_j\rangle}{\sum_i \mathbf{v}_i^2} = \mathsf{sdp}, \text{ then } \frac{\sum_{i,j} A_{ij}\langle \mathbf{w}_i, \mathbf{w}_j\rangle}{\sum_i \mathbf{w}_i^2} \geq \frac{\mathsf{sdp}}{\sqrt{\Delta}}.$$

Stated this way, we are free to re-scale the $\mathbf{v}_i$, thus we may assume $\tau = 1$. Now note that once we have such $\mathbf{w}_i$, we can throw away the zero vectors and apply the rounding algorithm of [CW04] (with a loss of an $O(\log n)$ approximation factor), to obtain a $0, \pm 1$ solution with value at least $\mathsf{sdp}/(\sqrt{\Delta}\log n)$.

So it suffices to show how to obtain the $\mathbf{w}_i$. Let us set (recall we assumed $\tau = 1$)

$$\mathbf{w}_i = \begin{cases} \mathbf{v}_i/\|\mathbf{v}_i\|, \text{ with prob. } \|\mathbf{v}_i\| \\ 0 \text{ otherwise} \end{cases}$$

(this is done independently for each i). Note that the probability of picking i is proportional to the length of $\mathbf{v}_i$ (as opposed to the typically used square lengths, [CMM06] say). Since $A_{ii} = 0$, we have

$$\frac{\mathbf{E}\left[\sum_{i,j} A_{ij}\langle \mathbf{w}_i, \mathbf{w}_j\rangle\right]}{\mathbf{E}\left[\sum_i \mathbf{w}_i^2\right]} = \frac{\sum_{i,j} A_{ij}\langle \mathbf{v}_i, \mathbf{v}_j\rangle}{\sum_i |\mathbf{v}_i|} \geq \frac{\sum_{i,j} A_{ij}\langle \mathbf{v}_i, \mathbf{v}_j\rangle}{\sqrt{\Delta}\sum_i \mathbf{v}_i^2} = \frac{\mathsf{sdp}}{\sqrt{\Delta}}. \quad (3)$$

The above proof only shows the existence of vectors $\mathbf{w}_i$ which satisfy the bound on the ratio. The proof can be made constructive using the method of conditional expectations, by setting the variables one by one, and use the fact that if $c, d > 0$ and $\frac{a+b}{c+d} > \theta$, then either $\frac{a}{c} > \theta$ or $\frac{b}{d} > \theta$.

Let us define the 'value' of a set of vectors $\{\mathbf{w}_i\}$ to be $\mathsf{val} := \frac{\sum A_{ij}\langle \mathbf{w}_i, \mathbf{w}_j\rangle}{\sum_i \mathbf{w}_i^2}$. The $\mathbf{v}_i$ we start will have $\mathsf{val} = \mathsf{sdp}$.

Lemma 2. *We can move to a set of vectors such that (a) val is at least $\mathsf{sdp}/2$, (b) each non-zero vector $\mathbf{v}_i$ satisfies $\mathbf{v}_i^2 \geq 1/n$, (c) vectors satisfy (2), and (d) $\sum_i \mathbf{v}_i^2 \leq 2$.*

The proof is by showing that very small vectors can either be enlarged or thrown away (proof in full version). The next lemma also gives an upper bound on the lengths – this is where the constraints (2) are crucial. It uses equation 2 to upper bound the contribution from each vector – hence large vectors can not contribute much in total, since they are few in number (see the full version for details).

Lemma 3. *Suppose we have a solution of value Bn^ρ and $\sum_i \mathbf{v}_i^2 \leq 2$. We can move to a solution with value at least $Bn^\rho/2$, and $\mathbf{v}_i^2 < 16/n^\rho$ for all i.*

Theorem 1. *Suppose A is an $n \times n$ matrix with zero's on the diagonal. Then there exists a polynomial time $O(n^{1/3}\log n)$ approximation algorithm for the QP-Ratio problem defined by A.*

Proof. As before, let us rescale and assume $\max i, j |A_{ij}| = 1$. Now if $\rho > 1/3$, Lemmata 2 and 3 allow us to restrict to vectors satisfying $1/n \leq \mathbf{v}_i^2 \leq 4/n^\rho$, and using Lemma 1 gives the desired $\tilde{O}(n^{1/3})$ approximation; if $\rho < 1/3$, then the trivial solution of $1/2$ is an $\tilde{O}(n^{1/3})$ approximation.

We now describe an integrality gap of roughly $n^{1/4}$, as it highlights the issues that arise in getting better approximations.

2.2   Integrality Gap Instance

Consider a complete bipartite graph on L, R, with $|L| = n^{1/2}$, and $|R| = n$. The edge weights are set to ± 1 uniformly at random. Denote by B the $n^{1/2} \times n$ matrix of edge weights (rows indexed by L and columns by R). A standard Chernoff bound argument shows (see the full version for a proof):

Lemma 4. *With high probability over the choice of B,* opt $\leq \sqrt{\log n} \cdot n^{1/4}$.

Let us now exhibit an SDP solution with value $n^{1/2}$. Let $\mathbf{v}_1, \mathbf{v}_2, \ldots, \mathbf{v}_{\sqrt{n}}$ be mutually orthogonal vectors, with each $\mathbf{v}_i^2 = 1/2n^{1/2}$. We assign these vectors to vertices in L. Now to the jth vertex in R, assign the vector $\mathbf{w}_j$ defined by $\mathbf{w}_j = \sum_i B_{ij} \frac{\mathbf{v}_i}{\sqrt{n}}$.

It is easy to check that this assignment satisfies the SDP constraints , and attains a value $\Omega(n^{1/2})$. This gives a gap of $\Omega(n^{1/4})$.

This gap instance can be seen as a collection of $n^{1/2}$ stars (vertices in L are the 'centers'). $O(\sqrt{n})$ different coordinates allow us to satisfy the constraints (2).

This gap instance is bipartite. In such instances it turns out that there is a better rounding algorithm with a ratio $\tilde{O}(n^{1/4})$.

2.3   The Bipartite Case

In this section, we prove the following theorem:

Theorem 2. *When A is bipartite (i.e. the adjacency matrix of a weighted bipartite graph), there is a (tight upto logarithmic factor) $O(n^{1/4} \log^2 n)$ approximation algorithm for QP-Ratio .*

Bipartite instances of QP-Ratio can be seen as the ratio analog of the Grothendieck problem [AN06]. The algorithm works by rounding the semidefinite program relaxation from section 2. As before, let us assume $\max_{i,j} a_{ij} = 1$ and consider a solution to the SDP (2). To simplify the notation, let u_i and v_j denote the vectors on the two sides of the bipartition. Suppose the solution satisfies:

$$(1) \sum_{(i,j) \in E} a_{ij} \langle u_i, v_j \rangle \geq n^\alpha, \qquad (2) \sum_i u_i^2 = \sum_j v_j^2 = 1.$$

If the second condition does not hold, we scale up the vectors on the smaller side, losing at most a factor 2. Further, we can assume from Lemma 2 that the squared lengths u_i^2, v_j^2 are between $\frac{1}{2n}$ and 1. Let us divide the vectors $\{u_i\}$ and $\{v_j\}$ into $\log n$ groups based on their squared length. There must exist two levels (for the u and v's respectively) whose contribution to the objective is at least $n^\alpha/\log^2 n$.[2] Let L denote the set of indices corresponding to these u_i, and R denote the same for v_j. Thus we have $\sum_{i\in L, j\in R} a_{ij}\langle u_i, v_j\rangle \geq n^\alpha/\log^2 n$. We may assume, by symmetry that $|L| \leq |R|$. Now since $\sum_j v_j^2 \leq 1$, we have that $v_j^2 \leq 1/|R|$ for all $j \in R$. Also, let us denote by A_j the $|L|$-dimensional vector consisting of the values a_{ij}, $i \in L$. Thus

$$\frac{n^\alpha}{\log^2 n} \leq \sum_{i\in L, j\in R} a_{ij}\langle u_i, v_j\rangle \leq \sum_{i\in L, j\in R} |a_{ij}| \cdot v_j^2 \leq \frac{1}{|R|}\sum_{j\in R}\|A_j\|_1. \qquad (4)$$

We will construct an assignment $x_i \in \{+1, -1\}$ for $i \in L$ such that $\frac{1}{|R|} \cdot \sum_{j\in R}|\sum_{i\in L} a_{ij}x_i|$ is 'large'. This suffices, because we can set $y_j \in \{+1, -1\}$, $j \in R$ appropriately to obtain the value above for the objective (this is where it is crucial that the instance is bipartite – there is no contribution due to other y_j's while setting one of them).

Lemma 5. *There exists an assignment of $\{+1, -1\}$ to the x_i such that*

$$\sum_{j\in R}|\sum_{i\in L} a_{ij}x_i| \geq \frac{1}{24}\sum_{j\in R}\|A_j\|_2$$

Furthermore, such an assignment can be found in polynomial time.

Proof. The intuition is the following: suppose $X_i, i \in L$ are i.i.d. $\{+1, -1\}$ random variables. For each j, we would expect (by random walk style argument) that $\mathbf{E}[|\sum_{i\in L} a_{ij}X_i|] \approx \|A_j\|_2$, and thus by linearity of expectation, $\mathbf{E}[\sum_{j\in R}|\sum_{i\in L} a_{ij}X_i|] \approx \sum_{j\in R}\|A_j\|_2$. Thus the existence of such x_i follows. This can in fact be formalized using the following lemma (please refer to full version for the proof)

Lemma 6. *Let $b_1, \ldots, b_n \in \mathbb{R}$ with $\sum_i b_i^2 = 1$, and let $X_1, \ldots, X_n$ be i.i.d. $\{+1, -1\}$ r.v.s. Then*

$$\mathbf{E}[|\sum_i b_i X_i|] \geq 1/12.$$

We also make this lemma constructive (please see the appended full version for details).

Proof (Proof of Theorem 2.). By Lemma 5 and Eq (4), there exists an assignment to x_i, and a corresponding assignment of $\{+1, -1\}$ to y_j such that the value of the solution is at least

$$\frac{1}{|R|} \cdot \sum_{j\in R}\|A_j\|_2 \geq \frac{1}{|R|\,|L|^{1/2}}\sum_{j\in R}\|A_j\|_1 \geq \frac{n^\alpha}{|L|^{1/2}\log^2 n}. \qquad \text{[By Cauchy Schwarz]}$$

---

[2] Such a clean division into levels can only be done in the bipartite case – in general there could be negative contribution from 'within' the level.

Now if $|L| \leq n^{1/2}$, we are done because we obtain an approximation ratio of $O(n^{1/4} \log^2 n)$. On the other hand if $|L| > n^{1/2}$ then we must have $\|u_i\|_2^2 \leq 1/n^{1/2}$. Since we started with u_i^2 and v_i^2 being at least $1/2n$ (Lemma 2) all the vector squared lengths are within a factor $O(n^{1/2})$ of each other. Thus by Lemma 1 we obtain an approximation ratio of $O(n^{1/4} \log n)$. This completes the proof.

2.4   Algorithms for Special Cases

We obtain better approximation algorithms for QP-Ratio in restricted settings. We defer the proofs to the full version of the paper.

- When A is positive semi-definite ($A \succeq 0$), we can round the eigenvector to get an $O(\log^2 n)$ approximation for QP-Ratio. In independent work, [DKS11] recently showed an $O(\sqrt{\log n})$ approximation to QP-Ratio when A is psd.
- When $OPT \geq \varepsilon D_{\max}$ (where $D_{\max} = \max_i \sum_i |a_{ij}|$ is the maximum degree), we can find a solution of value $e^{-O(1/\varepsilon)} D_{\max}$ using techniques from [Tre09].

3   Hardness of Approximating QP-Ratio

Given that our algorithmic techniques give only an $n^{1/3}$ approximation in general, and the natural relaxations do not seem to help, it is natural to ask how hard we expect the problem to be. Our results in this direction are as follows: we show that the problem is APX-hard i.e., there is no PTAS unless $P = NP$ (see Appendix B for details). Next, we show that there cannot be a constant factor approximation assuming that Max k-AND is hard to approximate 'on average' (related assumptions are explored in [Fei02]). Our reduction therefore gives a (fairly) natural *hard distribution* for the QP-Ratio problem.

3.1   Candidate Hard Instances

To reconcile the large gap between our upper bounds and lower bounds, we describe a natural distribution on instances we do not know how to approximate to a factor better than n^δ (for some fixed $\delta > 0$).

Let $\mathcal{G}$ denote a bipartite random graph with vertex sets V_L of size n and V_R of size $n^{2/3}$, left degree n^δ for some small δ (say 1/10) [i.e., each edge between V_L and V_R is picked i.i.d. with prob. $n^{-(9/10)}$]. Next, we pick a random (planted) subset P_L of V_L of size $n^{2/3}$ and random assignments $\rho_L : P_L \mapsto \{+1, -1\}$ and $\rho_R : V_R \mapsto \{+1, -1\}$. For an edge between $i \in P_L$ and $j \in V_R$, $a_{ij} := \rho_L(i)\rho_R(j)$. For all other edges we assign $a_{ij} := \pm 1$ independently at random.

The optimum value of such a *planted* instance is roughly n^δ, because the assignment of ρ_L, ρ_R (and assigning 0 to $V_L \setminus P_L$) gives a solution of value n^δ. However, for $\delta < 1/6$, we do not know how to find such a planted assignment: simple counting and spectral approaches do not seem to help.

Making progress on such instances appears to be crucial to improving the algorithm or the hardness results. In fact, the instances produced by the reduction from Random k-AND are similar in essence. We also note the similarity to other problems which are beyond current techniques, such as the Planted Clique and Planted Densest Subgraph problems [BCC+10].

3.2   Reduction from Random k-AND

We start out by quoting the assumption we use.

Conjecture 1 (Hypothesis 3 in [Fei02]). For some constant $c > 0$, for every k, $\exists \Delta_0$, such that for every $\Delta > \Delta_0$, there is no polynomial time algorithm that, on most k-AND formulas with n-variables and $m = \Delta n$ clauses, outputs 'typical', but never outputs 'typical' on instances with $m/2^{c\sqrt{k}}$ satisfiable clauses.

The reduction to QP-Ratio is as follows: Given a k-AND instance on n variables $X = \{x_1, x_2, \ldots x_n\}$ with m clauses $C = \{C_1, C_2, \ldots C_m\}$, and a parameter $0 < \alpha < 1$, let $A = \{a_{ij}\}$ denote the $m \times n$ matrix such that a_{ij} is $1/m$ if variable x_j appears in clause C_i as is, a_{ij} is $-1/m$ if it appears negated and 0 otherwise.

Let $f : X \to \{-1, 0, 1\}, g : C \to \{-1, 0, 1\}$ denote functions which correspond to assignments. Let $\mu_f = \sum_{i \in [n]} |f(x_i)|/n$ and $\mu_g = \sum_{j \in m} |g(C_j)|/m$. Let

$$\vartheta(f, g) = \frac{\sum_{ij} a_{ij} f(x_i) g(C_j)}{\alpha \mu_f + \mu_g}. \tag{5}$$

Observe that if we treat $f(), g()$ as variables, we obtain an instance of QP-Ratio (we describe how to get rid of the weighting in the denominator in the full version) We pick $\alpha = 2^{-c\sqrt{k}}$ and Δ a large enough constant so that Conjecture 1 and the rest of the proofs work. The completeness follows from the natural assignment (proof in full version).

Lemma 7 (Completeness). *If α fraction of the clauses in the k-AND instance can be satisfied, then there exists function f, g such that θ is at least $k/2$.*

Soundness: We will show that for a typical random k-AND instance (i.e., with high probability), the maximum value $\vartheta(f, g)$ can take is at most $o(k)$.

Let the maximum value of ϑ obtained be ϑ_{max}. We first note that there exists a solution f, g of value $\vartheta_{max}/2$ such that the equality $\alpha \mu_f = \mu_g$ holds[3] – so we only need consider such assignments.

Now, the soundness argument is two-fold: if only a few of the vertices (X) are picked ($\mu_f < \frac{\alpha}{400}$) then the expansion of small sets guarantees that the value is small . On the other hand, if many vertices (and hence clauses) are picked, then we claim that for every assignment to the variables (every f), only a small fraction ($2^{-\omega(\sqrt{k})}$) of the clauses contribute more than $k^{7/8}$ to the numerator. These lemmas shows together show a gap of k vs $k^{7/8}$ assuming Hypothesis 1. The complete proof is included in the full version of the paper. Since we can pick k to be arbitrarily large, we can conclude that QP-Ratio is hard to approximate to any constant factor.

3.3   Reductions from Ratio versions of CSPs

Here we ask: is there a reduction from a *ratio version* of Label Cover to QP-Ratio? For this to be useful we must also ask: is the (appropriately defined) ratio

---

[3] if $\alpha \mu_f > \mu_g$, we can pick more constraints such that the numerator does not decrease (by setting $g(C_j) = \pm 1$ in a greedy way so as to not decrease the numerator) till $\mu_{g'} = \alpha \mu_f$, while losing a factor 2. Similarly for $\alpha \mu_f < \mu_g$, we pick more variables.

version of Label Cover hard to approximate? The answer to the latter question is yes (see the full version for details and proof that Ratio-LabelCover is hard to approximate to any constant factor). Unfortunately, we do not know how to reduce from Ratio-LabelCover.

Here, we present a reduction starting from a ratio version of *Unique Games* to QP-Ratio (inspired by [ABH+05], who give a reduction from Label Cover to Quadratic Programming, without the ratio). However, we do not know whether it is hard to approximate for the parameters we need. While it seems related to *Partial Unique Games* introduced by [RS10], they have an added size constraint that at least α fraction of vertices should be labeled, which enables a reduction from *Unique Games with Small-set Expansion*. However, a key point to note is that we do not need 'near perfect' completeness, as in typical UG reductions.

We hope the Fourier analytic tools we use to analyze the ratio objective could find use in other PCP-based reductions to ratio problems. Informally, Ratio UG is a *Unique Label Cover* problem $\mathcal{U}\big(G(V,E),[R],\{\pi_e|e \in E\}\big)$ where we only ask for a partial labeling $(L : V \to [R]\cup\{\bot\})$. The objective value is the ratio of the number of satisfied constraints to the number of labeled variables (please see the full version for details). We reduce Ratio UG to the following useful intermediate problem:

QP-Intermediate. Given $A_{(n\times n)}$ with $A_{ii} \leq 0$, maximize $\frac{x^T Ax}{\sum_i |x_i|}$ s.t. $x_i \in [-1,1]$.

Now given an instance $\Upsilon = (V,E,\Pi)$ of Ratio UG, with alphabet $[R]$ and a regular graph (V,E), we associate 2^R variables to each vertex, which are denoted $f_u(x)$, indexed by $x \in \{-1,1\}^R$. The intended solution to each vertex is either the long code corresponding to the label, or $f_u = 0$ (for each x). Now,

- Fourier coefficients $(\widehat{f_u}(S))$ are linear forms in the variables $f_u(x)$.
- For $(u,v) \in E$, $T_{uv} =^{\text{def}} \sum_i \widehat{f_u}(\{i\})\widehat{f_v}(\{\pi_{uv}(i)\})$. [It is 1 if edge is satisfied]
- For $u \in V$, $L(u) =^{\text{def}} \sum_{S:|S|\neq 1} \widehat{f_u}(S)^2$. [Penalizes f_u that are not dictators]

The instance of QP-Intermediate we consider is (here $\|f_u\|_1$ denotes $\mathbf{E}_x[|f_u(x)|]$)

$$\mathcal{Q} := \max \; \frac{\mathbf{E}_{(u,v)\in E}T_{uv} - \eta\mathbf{E}_u L(u)}{\mathbf{E}_u|f_u|_1}, \quad \text{where } \eta \text{ will be picked large enough.}$$

For a function f, we define the 'linear' and the 'non-linear' parts to be

$$f^{=1} := \sum_i \widehat{f}(i)\chi(\{i\}) \quad \text{and} \quad f^{\neq 1} := f - f^{=1} = \sum_{|S|\neq 1} \widehat{f}(S)\chi(S).$$

The choice of η will ensure that for *each* u, $\|f_u^{\neq 1}\|_2^2$ is *tiny*.

A key step in the analysis is the following: if a boolean function f is 'nearly linear', then it must also be *spread out* [i.e. $\|f\|_2 \approx \|f\|_1$]. This helps us deal with the main issue in a reduction with a ratio objective – showing we cannot have a large numerator along with a very small value of $\|f\|_1$ (the denominator). Morally, this is similar to a statement that a boolean function with a *small support* cannot have all its Fourier mass on the *linear* Fourier coefficients. Please refer to the full version for the complete proof.

4   Normalized QP-Ratio

Given any symmetric matrix A, the normalized QP-Ratio problem aims to find the best $\{-1, 0, 1\}$ assignment which maximizes the following:

$$\max_{\{-1,0,1\}^n} \frac{\sum_{i \neq j} a_{ij} x_i x_j}{\sum d_i x_i^2}. \quad \text{where } d_i = \sum_j |a_{ij}| \text{ are "the degrees"} \quad (6)$$

Note that when the degrees d_i are all equal ($d_i = d \; \forall i$), this is the same as QP-Ratio upto a scaling. In the non-regular case, the normalized objective tends to penalize picking vertices of high degree in the solution.

Let us consider the natural eigenvalue relaxation below. This is also the maximum eigenvalue of $D^{-1/2} A D^{1/2}$ where D is the diagonal matrix of degrees.

$$\text{Eigenval } \lambda(A) = \max_{x \in [-1,1]^n} \frac{x^t A x}{\sum_i d_i x_i^2} = \frac{2 \sum_{i \neq j} a_{ij} x_i x_j}{\sum_{i \neq j} |a_{ij}|(x_i^2 + x_j^2)} \quad (7)$$

GainRatio and Trevisan[Tre09]'s Conjecture
The special case when $A = -Adj(G)$ where $Adj(G)$ is the matrix of edge weights is called *GainRatio* of G. [Tre09] studied in the context of a purely spectral algorithm for Max CutGain: he gave an algorithm for GainRatio based on the above eigenvalue relaxation (7), and used this as a subroutine to obtain algorithms for Max CutGain. His randomized rounding technique showed that if the eigenvalue is ε, the GainRatio is at least $e^{-O(1/\varepsilon)}$. This also gives an algorithm for Normalized QP-Ratio with a similar guarantee (we defer to the full version for details). Trevisan[Tre09] also *conjectures* a better dependence:

$$\text{GainRatio} = \Omega\left(\frac{\lambda(A)}{\log(1/\lambda(A))}\right).$$

This would give an spectral algorithm which matches the SDP-based algorithm of [CW04] (as for Max Cut [Tre09]). We show that this conjecture is false, and describe an instance for which

$$\text{GainRatio} = O\left(\exp\left(-1/\lambda(A)^{1/4}\right)\right).$$

This shows that the eigenvalue based approach is necessarily 'exponentially' weaker than an SDP-based one. Roughly speaking, SDPs are stronger because they can enforce vectors to be all of equal length, while this cannot be done in an eigenvalue relaxation. The description of the instance and proof of the exponential integrality gap can be found in the full version.

References

[ABH+05]   Arora, S., Berger, E., Hazan, E., Kindler, G., Safra, M.: On non-approximability for quadratic programs. In: FOCS 2005: Proceedings of the 46th Annual IEEE Symposium on Foundations of Computer Science, pp. 206–215. IEEE Computer Society, Washington, DC (2005)

[AN06]   Alon, N., Naor, A.: Approximating the cut-norm via grothendieck's in-
            equality. SIAM J. Comput. 35, 787–803 (2006)
[Aus07]   Austrin, P.: Towards sharp inapproximability for any 2-csp. In: Proceed-
            ings of the 48th Annual IEEE Symposium on Foundations of Computer
            Science, pp. 307–317. IEEE Computer Society, Washington, DC (2007)
[BCC+10]   Bhaskara, A., Charikar, M., Chlamtac, E., Feige, U., Vijayaraghavan,
            A.: Detecting high log-densities: an $o(n^{1/4})$ approximation for densest
            k-subgraph. In: STOC 2010: Proceedings of the 42nd ACM Symposium
            on Theory of Computing, pp. 201–210. ACM, New York (2010)
[BCMV11]   Bhaskara, A., Charikar, M., Manokaran, R., Vijayaraghavan, A.: On
            quadratic programming with a ratio objective. CoRR, abs/1101.1710
            (2011)
[Cha00]   Charikar, M.: Greedy Approximation Algorithms for Finding Dense Com-
            ponents in a Graph. In: Jansen, K., Khuller, S. (eds.) APPROX 2000.
            LNCS, vol. 1913, pp. 84–95. Springer, Heidelberg (2000)
[CMM06]   Charikar, M., Makarychev, K., Makarychev, Y.: Near-optimal algorithms
            for unique games. In: Proceedings of the Thirty-Eighth Annual ACM Sym-
            posium on Theory of Computing, STOC 2006, pp. 205–214. ACM, New
            York (2006)
[CW04]   Charikar, M., Wirth, A.: Maximizing quadratic programs: Extending
            grothendieck's inequality. In: FOCS 2004: Proceedings of the 45th An-
            nual IEEE Symposium on Foundations of Computer Science, pp. 54–60.
            IEEE Computer Society, Washington, DC (2004)
[DKS11]   Deshpande, A., Kannan, R., Srivastava, N.: Zero-one rounding of singular
            vectors. Manuscript (2011)
[Fei02]   Feige, U.: Relations between average case complexity and approximation
            complexity. In: Proceedings of the 34th annual ACM Symposium on The-
            ory of Computing (STOC 2002), pp. 534–543. ACM Press (2002)
[GW95]   Goemans, M.X., Williamson, D.P.: Improved approximation algorithms
            for maximum cut and satisfiability problems using semidefinite program-
            ming. J. ACM 42(6), 1115–1145 (1995)
[KKMO07]   Khot, S., Kindler, G., Mossel, E., O'Donnell, R.: Optimal inapproximabil-
            ity results for max-cut and other 2-variable csps? SIAM J. Comput. 37(1),
            319–357 (2007)
[KV05]   Khot, S., Vishnoi, N.K.: The unique games conjecture, integrality gap
            for cut problems and embeddability of negative type metrics into ℓ_1. In:
            FOCS 2005, pp. 53–62 (2005)
[NRT99]   Nemirovski, A., Roos, C., Terlaky, T.: On maximization of quadratic form
            over intersection of ellipsoids with common center. Mathematical Pro-
            gramming 86, 463–473 (1999), doi:10.1007/s101070050100
[Rag08]   Raghavendra, P.: Optimal algorithms and inapproximability results for
            every CSP? In: STOC 2008, pp. 245–254 (2008)
[RS10]   Raghavendra, P., Steurer, D.: Graph expansion and the unique games
            conjecture. In: STOC 2010: Proceedings of the 42nd ACM Symposium on
            Theory of Computing, pp. 755–764. ACM, New York (2010)
[Tre09]   Trevisan, L.: Max cut and the smallest eigenvalue. In: STOC 2009: Pro-
            ceedings of the 41st annual ACM Symposium on Theory of Computing,
            pp. 263–272. ACM, New York (2009)
[Tul09]   Tulsiani, M.: Csp gaps and reductions in the lasserre hierarchy. In: Pro-
            ceedings of the 41st Annual ACM Symposium on Theory of Computing,
            STOC 2009, pp. 303–312. ACM, New York (2009)

De-amortizing Binary Search Trees[*]

Prosenjit Bose[1,**], Sébastien Collette[2,***],
Rolf Fagerberg[3,†], and Stefan Langerman[2,‡]

[1] School of Computer Science, Carleton University
[2] Département d'Informatique, Université Libre de Bruxelles
[3] Department of Mathematics and Computer Science, University of Southern Denmark
jit@scs.carleton.ca, {secollet,stefan.langerman}@ulb.ac.be,
rolf@imada.sdu.dk

Abstract. We present a general method for de-amortizing essentially any Binary Search Tree (BST) algorithm. In particular, by transforming Splay Trees, our method produces a BST that has the same asymptotic cost as Splay Trees on any access sequence while performing each search in $O(\log n)$ worst case time. By transforming Multi-Splay Trees, we obtain a BST that is $O(\log \log n)$ competitive, satisfies the scanning theorem, the static optimality theorem, the static finger theorem, the working set theorem, and performs each search in $O(\log n)$ worst case time. Transforming OPT proves the existence of an $O(1)$-competitive offline BST algorithm which performs at most O(log n) BST operations between each access to the keys in the input sequence. Finally, we obtain that if there is an $O(1)$-competitive online BST algorithm, then there is also one that performs every search in $O(\log n)$ operations worst case.

1 Introduction

Over half a century since the discovery of rotation-based Binary Search Trees, their exact performance is still not fully understood. The very first works on BST focused on maintaining $O(\log n)$ height during insertions and deletions [1,18], or guaranteeing better average case bounds for searches with known distributions [22].

By introducing splay trees [23], Sleator and Tarjan proposed an alternate view of the problem, where instead of looking at the cost of individual searches, it is the entire cost of a sequence of accesses which is bounded, using amortized analysis.

The purpose of this article is to show that the two approaches are not exclusive—i.e., that it is possible to combine the good amortized performances of self-adjusting and other adaptive BST with strong worst case guarantees for individual searches.

The BST Model. Central to the line of work originating from the splay tree paper [23] is the *BST model*. This is because competitive analysis (see below) of online BST algorithms requires lower bounds on the optimal offline algorithm, which again requires a

[*] Full version available at http://arxiv.org/abs/1111.1665
[**] Research supported in part by NSERC.
[***] Chargé de recherches du F.R.S.-FNRS.
[†] Partially supported by the Danish Council for Independent Research, Natural Sciences.
[‡] Maître de recherches du F.R.S.-FNRS.

A. Czumaj et al. (Eds.): ICALP 2012, Part I, LNCS 7391, pp. 121–132, 2012.
© Springer-Verlag Berlin Heidelberg 2012

precise model of computation. All existing lower bounds [10,13,25] use one of several existing, asymptotically equivalent, variants of this model. In order to describe accurately our results, we choose one specific BST model, which we now describe. In line with previous work, we do not consider insertions and deletions. Hence, our BST model consists of a binary search tree T containing the n distinct keys, which wlog. may be taken to be $\{1, 2, \ldots, n\}$ with their natural order. The position of a finger, initially at the root of T, is maintained, and the following two *BST operations*, each of unit cost, are allowed: 1) moving the finger from a node to its parent or to one of its children, and 2) performing a rotation between the node pointed to by the finger and its parent.

Given the current tree T and the current finger position, an *access* to a key x is a list of BST operations (finger movements and rotations), during which the finger position is at the node containing x at least once.

For an input sequence $S = \langle s_1, s_2, \ldots, s_m \rangle$ of keys to be accessed, a BST algorithm $\mathcal{A}$ that *realizes* S returns a list $\mathcal{A}(S)$ of BST operations for accessing the keys $s_1, s_2, \ldots$ in that order—that is, where S is a subsequence of the sequence of keys pointed to by the finger during the execution of $\mathcal{A}(S)$. An *offline* algorithm $\mathcal{A}$ is given the entire sequence S and the starting tree T as input and then outputs the sequence of operations $\mathcal{A}(S)$, while an *online* algorithm is fed the keys from S one by one and must output the BST operations for the access of one key before the next key is given. More formally, A is online if $\mathcal{A}(S)$ is a prefix of $\mathcal{A}(S')$ whenever S is a prefix of S'. The *cost* of $\mathcal{A}(S)$ is the number of BST operations it contains.

Note that the model, as all the standard variants of the BST model used in competitive analysis of online BST algorithms, only requires the algorithm to list the BST operations $\mathcal{A}(S)$ to be performed (see, e.g, [25]). In particular, the model does not restrict how those operations are generated, what auxiliary memory is used in order to generated them, or even how much time is used to generate them.

Of course, real-world implementations of practical BST algorithms have some sensible limits on their time and space usage. In fact, almost all BST implementations in the literature besides adhering to the standard BST model described above also have the following additional features: they work in the pointer machine model, use no more space than the tree itself plus $O(1)$ words of balance information in each node of the tree and $O(1)$ extra working variables, and generate their access sequence $\mathcal{A}(S)$ in time proportional to the BST model cost of $\mathcal{A}(S)$. We in this version of the paper show how to de-amortize BST algorithms with a method working in the standard BST model. In the full version we show how to extend the method to maintain the additional features just listed, should the BST algorithm being de-amortized have these.

Denote by OPT the best offline algorithm, that is, $OPT(S)$ is a shortest possible list of operations that realizes S. An algorithm $\mathcal{A}$ (online or offline) is $f(n)$-*competitive* if we have $\mathcal{A}(S) = O(f(n) \cdot OPT(S))$ for all sequences S. It is *dynamically optimal* if it is $O(1)$-competitive.

Prior Works. The study of self-adjusting BSTs to minimize the overall cost over a sequence of accesses was initiated by Allen and Munro [2] with their analysis of the move-to-root and the simple exchange heuristics, and then by Sleator and Tarjan with the introduction of Splay Trees [23], which they conjectured to be dynamically optimal. They show how the running time of Splay Trees can be upper bounded in several ways as a function of the access sequence. They prove the *balance theorem* (accesses run in $O(\log n)$ amortized), the *static optimality theorem* (any sequence of accesses runs within a constant factor of the time to run it on the best possible static tree for that sequence; in particular it reaches the entropy bound), the static finger theorem (access x runs in $O(\log d(x, f))$, where $d(x, f)$ is the number of keys between the query item x and any fixed *finger* element f), the *working set theorem* (access x runs in time $O(\log w(x))$ where $w(x)$ is the number of distinct elements accessed since the previous access to x), and the *scanning theorem* (accessing all nodes in symmetric order takes time $O(n)$). They also conjectured the *dynamic finger theorem* (access to y runs in amortized $O(\log d(x, y))$ where x is the previous item in the access sequence), which was subsequently proved by Cole [9,8]. All bounds above are amortized.

On another front, Wilber [25] gave a formal analysis of several variants of the BST model, providing equivalence reductions between them, and provided two lower bounds on the number of operations that any BST algorithm must perform for a given sequence. In particular, he proved that the bit reversal sequence requires $\Omega(\log n)$ amortized operations per access. These lower bounds were recently generalized in [13,10]. Splay Trees were also shown to be *key independent optimal* [20], that is, they are $O(1)$-competitive if the order of the keys is arbitrary or random, and that they are $O(1)$-competitive with respect to a wide class of balanced BST algorithms [15].

New bounds have been designed: the *queueish* bound (opposite of the working set bound: the number of elements *not* accessed since the last access to x) was shown not to be achievable by any BST algorithm [21]. Recent papers have attempted to engineer a BST that satisfies the unified property, a bound that implies both the dynamic finger and the working set bound [19,3]. The *skip-splay trees* [12] perform each access within a multiplicative factor $O(\log \log n)$ of the unified bound, amortized. The *layered working set trees* [7] are BSTs that achieve the working set bound worst case. By combining it with the skip-splay structure, the authors show how to achieve the unified bound, amortized, with an additive cost of $O(\log \log n)$.

The first significant breakthrough on the competitive analysis of BST algorithms came with the invention of *tango trees* [11], the first provably $O(\log \log n)$-competitive BST. This result was subsequently improved independently by the *multi-splay trees* [24] and the *chain-splay trees* [16] which both offer the additional guarantee of performing each access in $O(\log n)$ amortized time. Further properties of multi-splay trees were proved in [14], where they were shown to satisfy static optimality, the static finger property, the working set property, and key-independent optimality. They further satisfy the dequeue property which is not known to be satisfied by splay trees.

In recent years, the question was raised as to whether the good amortized properties could be reconciled with the $O(\log n)$ worst case bounds satisfied by well balanced trees such as AVL or red-black trees. Such results were known for static trees [5], however recent works gave indication that strong balance constraints at every node forces the working set bound to be an amortized lower bound, thus forbidding any such tree to have stronger properties such as the dynamic finger property [4] (the proof was given for self-adjusting skip-lists and B-trees, however the proofs can easily be adapted to BST with balance constraints at every node). However, it remained open whether relaxing the balance condition to just bounding the height of the tree would be compatible with obtaining better amortized performances. In [6], a BST based on Tango trees [11] is engineered to be both $O(\log \log n)$-competitive and guarantees $O(\log n)$ worst case access time for each access. However, this structure is unlikely to possess all the other desirable properties of Splay Trees.

Our Results. In this paper we show that it is possible to automatically transform *any* BST algorithm into one that provides worst case time guarantees per access while keeping the same asymptotic amortized running times. Our core result shows how to keep a BST balanced while losing only a constant factor in the running time:

- Any BST algorithm $\mathcal{A}$ on tree T can be transformed into a BST algorithm $\mathcal{A}'$ on a tree T' such that for any access sequence S, $|\mathcal{A}'(S)| = O(|\mathcal{A}(S)|)$, while the depth of T' is always $O(\log n)$. If $\mathcal{A}$ is online, so is $\mathcal{A}'$.

Using this, we then show how to de-amortize the BST and answer each query in $O(\log n)$ worst case cost:

- Any BST algorithm $\mathcal{A}$ on tree T can be transformed into a BST algorithm $\mathcal{A}''$ on a tree T'' such that for any access sequence S, $|\mathcal{A}''(S)| = O(|\mathcal{A}(S)|)$ and each access to a node is performed in $O(\log n)$ operations worst case. If $\mathcal{A}$ is online, so is $\mathcal{A}''$.

Finally, we in the full version show that we can extend the method to maintain the additional features of real-world online BST algorithms described above. In particular, we have that if $\mathcal{A}$ works in the pointer machine model, with working space being $O(1)$ words of information in the nodes and $O(1)$ global working variables, and computes each access to a key in time proportional to the number of BST operation of the access, then so does our final algorithm.

Applying this transformation to Splay Trees, we obtain a BST that executes every sequence within a constant factor of the Splay Tree and thus satisfies the scanning theorem, the working set property, static optimality, the key-independent optimality, the static finger property, the dynamic finger property, and that performs each access in $O(\log n)$ worst case. Applying it to Multi-Splay Trees, we obtain a BST that is $O(\log \log n)$ competitive, satisfies the scanning theorem, the working set property, static optimality, the key-independent optimality, the static finger property, and performs each search in $O(\log n)$ worst case time. Applying it to OPT proves the existence of an $O(1)$-competitive offline BST algorithm which performs at most O(log n) BST operations between each access to the keys in the input sequence. Furthermore, if there is an $O(1)$-competitive online BST algorithm, then there is also one that performs every search in $O(\log n)$ operations worst case.

Overview of Paper. On a high level, our construction works by performing a heavy-path decomposition of the tree of the original algorithm $\mathcal{A}$, and then during $\mathcal{A}$'s BST operations maintain each heavy-path as a constant number of structures accessed in a stack-like fashion. The remaining and most technical ingredient is a method for maintaining such stack structures as trees in the BST model, while fulfilling a weight-based balance criterion that ensures the total composition of the stack representations of the heavy-paths to be a balanced tree. In the paper, these ingredients are covered in the reverse order of above.

2   Pop-Tarts

We start by implementing a stack using a balanced BST. We differentiate internal nodes, which always have two children, and leaves which have no children (leaves can also be seen as empty pointers). In order to fit the stack data structure in the BST model, we assume that nodes to be pushed onto the stack appear as the parent of the root of the current stack, and that nodes are pushed onto the stack in decreasing key order (that is, after the push operation the old stack is the right child of the newly inserted node, and its left child is a leaf). Our later application of the stack structure fulfils these assumptions. An empty stack is composed of one leaf. The structure will maintain the invariant that the left child of the root is always a leaf, to allow for easy pop operations. After each push or pop operation, the structure is allowed to perform a sequence of operations in the BST model (finger movements and rotations), and at the end of the sequence, the finger is back at the root. Leaves can have a weight associated to them, and we use the convention that internal nodes all have weight 1 (it would not be difficult to generalize these structures to support arbitrary internal weights, however this is not necessary for our application).

A BST implementing a stack in this manner we call a *Pop-tart*[1]. A pop-tart is *good* if push and pop operations are performed in $O(1)$ amortized time and $O(\log n)$ worst-case time. It is *crazy good* [17] if it is good and the depth of every leaf of weight w is $O(\log(W/w))$, where W is the total weight of all leaves in the pop-tart, or $O(\log n)$ for an unweighted pop-tart with n leaves[2].

In the remainder of this section, we will describe three pop-tart structures. The first two lay down ground concepts that will be used to construct the third pop-tart (Chocolate), which is always crazy good.

Vanilla Pop-Tart. Implementing a good pop-tart is easy. In fact, performing no BST operations after each push or pop operation will produce a linear tree with exactly $O(1)$ time per operation. This elementary implementation is called *Vanilla Pop-Tart*. A vanilla pop-tart will be crazy good if the weight of each pushed leaf is always larger than the total weight of all other leaves in the pop-tart.

Lemma 1. *The Vanilla Pop-Tart is crazy good if nodes are added in decreasing key order and new leaves have weight larger or equal to the total weight of all other leaves in the pop-tart. That is, it uses $O(1)$ time per push and pop operation and the depth of a leaf of weight w is at most $1 + \log W/w$ where W is the total weight of all leaves in the pop-tart.*

Proof. The proof is by induction. If the pop-tart contains one leaf, then it is at depth 0, this covers the base case. Assume by induction that the lemma is true for the right subtree of the root, which is of total weight W'. Then the left child of the root is the last added leaf and it has weight at least W', thus, $W \geq 2W'$. The left child of the root is

---

[1] Pop-Tarts are a line of crazy good [17] breakfast products. that *pop* out of the toaster, which reminds us of popping a stack. Pop-tart is a trademark of the Kellogg Company.

[2] We slightly abuse the big-Oh notation and write $O(\log(W/w))$ to mean a function which is smaller than $c \log(W/w) + d$ for some constants c and d.

at depth $1 \leq 1 + \log W/w$. Any other leaf in the tree by induction is at depth at most $2 + \log W'/w \leq 1 + \log W/w$.

Cherry Pop-Tart. We now describe the *Cherry Pop-Tart*, which is a crazy good pop-tart if all leaves have weight 1. Although Cherry Pop-tarts are not used explicitly in this paper, they serve as a warm up, introducing some key concepts needed to define the Chocolate Pop-tart structure, which is used later.

The algorithm used is a variant of a 2-4 tree implemented as a BST. On a high level, it may be viewed as reversing edges on the leftmost path in a red-black tree, and then having a permanent finger at the leftmost internal node (effectively making it the root of the BST).

In greater detail: The Cherry Pop-tart is a BST with the nodes on the right path of the tree grouped into layers. A layer consists of consecutive nodes on the right path, and the left subtrees of these nodes are called *crumbs*. The right child of the last node in the layer is the top node of the next layer (except for the last layer, where it is the original leaf of the initial empty stack). By definition of BSTs, the layers are linearly ordered, that is, all keys in a layer are smaller than the keys in the next layer.

We number the layers as follows: the layer containing the root is layer 0, the next one along the right path is layer 1, and so on. We maintain the invariants that each layer has between 1 and 3 nodes on the right path (hence that many crumbs), and that the crumbs pointed to by layer i (called i-*crumbs*) are perfectly balanced trees containing exactly 2^i leaves.

The invariant is true for a pop-tart containing one node: that node is layer 0 and it points to one 0-crumb (containing one leaf). When a new node is pushed as the parent of the root, it is added to layer 0. Layer 0 therefore has one more node and one more 0-crumb. Either the new layer 0 still has no more than 3 crumbs, maintaining the invariant, or layer 0 now has 4 0-crumbs (each composed of exactly one leaf). In this case, we perform a left rotation between the last two nodes of the layer. This replaces the last two nodes of the layer with one node whose left pointer points to a 1-crumb. We now move that node from layer 0 to layer 1. See Figure 1. Again, the reconfiguration could either stop there or ripple down further. In general, as a node is added as the parent of the first node in layer i, either layer i still has no more than 3 i-crumbs, or we preform a rotation on the node between the last two crumbs, forming a $(i + 1)$-crumb with twice as many leaves which is inserted into layer $i + 1$. A pop operation works symmetrically.

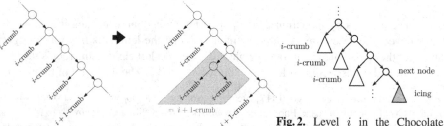

Fig. 1. Restoring the Cherry Pop-tart invariant at level i

Fig. 2. Level i in the Chocolate Pop-tart

Lemma 2. *The Cherry Pop-Tart is crazy good if nodes are added in decreasing key order and all leaves have weight 1. That is, it uses $O(1)$ amortized time and $O(\log n)$ worst case time per push and pop operation and its tree has height $O(\log n)$.*

Proof. To show that a push or pop operation has amortized cost $O(1)$, we assign a potential of 0 to layers with 2 nodes, and a potential of 1 to layers with 1 or 3 nodes. A push or pop operation has actual cost proportional to the number of layers that had to be readjusted to restore the invariant. Each readjusted layer had a potential of 1 before the operation (i.e., had 3 nodes before a push or 1 node before a pop) and of 0 after the operation (i.e., has 2 nodes exactly). Therefore, the decrease of potential pays exactly for the readjustments. The insertion or deletion in the last layer possibly increases its potential by 1, which is the amortized cost of the operation. Therefore, this pop-tart is good.

Since layer i has at least one i-crumb containing 2^i leaves, a pop-tart with n leaves has at most $\log n$ layers, each having crumbs of height $O(\log n)$, thus the total height of the tree is $O(\log n)$. So in the unweighted case, this pop-tart is crazy good.

Chocolate Pop-Tart. Again, the structure will be decomposed into a sequence of layers whose nodes form a right path and point to crumbs. This time, the right path of the ith layer will be composed of 1 to 3 *regular* nodes whose left child is an i-crumb, then a *next* node whose left child points to the next layer and whose right child points to a subtree called the *icing*. This will be called the *structural invariant*. See Figure 2. The icing is itself a stack, implemented using a Vanilla Pop-tart (that is, a simple linear tree), whose leaves will be *frozen*[3] subtrees of the chocolate pop-tart. In order for the icing to be crazy good, we will ensure that the nodes (frosted subtrees) pushed onto it will always be at least as heavy as the total weight of the icing. The subtrees to be frosted and pushed into the icing of level i will always be the next node and the entire subtree rooted at the top node of level $i + 1$. Therefore, we maintain the invariant that the total weight of layer $i + 1$ (that is, the the total weight of the subtree rooted at the topmost node of that layer) is smaller than the total weight of the icing of layer i (*thick icing invariant*). If violated, layer $i + 1$ will be frosted and pushed into the icing, to maintain the invariant.

The last layer, say, layer i, is incomplete: it is composed of 0 to 3 regular nodes, has no pointer to the next layer, and always contains an icing as its rightmost subtree. It can only have 0 regular nodes if the icing contains exactly one element (which is always an i-crumb).

As before, when a new node is pushed onto the i-layer (starting with $i = 0$), either the i-layer has at most 3 regular nodes, in which case we are done, or it contains 4 regular nodes and we need to restore the structural invariant. We start by performing a left rotation between the two lowest regular nodes in the layer, creating an $(i+1)$-crumb. We have two cases to consider. If the ith layer is not the last one, then we perform a left rotation between the next node and the lowest regular node, to move the new $(i + 1)$-crumb and its node to the $(i + 1)^{th}$ layer. On the other hand, if the ith layer is the last one, then it has no next node. Then the lowest regular node becomes a next node which

---

[3] or *frosted*.

points to the new $(i + 1)^{th}$ layer. That $(i + 1)^{th}$ layer contains 0 regular nodes, no next node and an icing which contains the $(i + 1)$-crumb as its only leaf.

Having done this, there are again two cases to consider: if the total weight of the subtree rooted at the (new) top node of the $(i+1)^{th}$ layer is smaller than the total weight of the icing of the ith layer, then we proceed with the insertion of the $(i + 1)$-crumb, by restoring the structural invariant if necessary, and so on. Otherwise, we restore the thick icing invariant by frosting the $(i + 1)^{th}$ layer without modifying it further (even if it contains now 4 regular nodes), and push it and its parent node (the next node of the ith layer) into the icing of the ith layer. The ith layer then becomes the last layer. It has no next node and two regular nodes.

The deletion operation is symmetric: when the first regular node of the ith layer is deleted, either the layer still has at least one regular node left, in which case we are done, or we have to restore the structural invariant. If i is not the last layer, we pull two nodes and their associated i-crumbs from the $(i + 1)^{th}$ layer (by performing two right rotations and possibly recursively restoring the invariant in the $(i + 1)^{th}$ layer). If the $(i + 1)^{th}$ layer is only composed of an icing (which then contains one frosted $(i + 1)$-crumb), we defrost the icing, perform a right rotation, transforming the next layer into two regular nodes pointing to i-crumbs and the ith layer becomes the last one. On the other hand, if i is the last layer, then we pop a frosted subtree from the icing (unless it contains only one leaf), and perform a right rotation to turn the frosted subtree into one regular node and a next node, the latter pointing to the new, unfrosted, $(i + 1)^{th}$ layer and to the remaining icing.

Lemma 3. *The Chocolate Pop-Tart is crazy good if nodes are added in decreasing key order and new leaves are added with arbitrary weights. That is, it uses $O(1)$ amortized time per push and pop operation and the depth of a leaf of weight w is $O(\log W/w)$ where W is the total weight of all leaves in the pop-tart.*

Proof. We first show that the Chocolate Pop-tart is good, that is, it uses $O(1)$ amortized time per push and pop operation. For this, we assign a potential of 0 to layers with 2 regular nodes, and a potential of 1 to all other layers. A push operation will cause a bunch of reconfigurations in successive layers, that end in either adding a crumb to a layer that does not overflow, or pushing an element in the icing of a layer. Either case costs $O(1)$ amortized. As in the case of Cherry Pop-tarts, it is easily verified that every layer that overflows had 3 regular nodes before, and thus a potential of 1, and two regular nodes after, so a potential of 0 (except possibly for the last rearranged layer). Likewise, during a pop operation, the potential of a rearranged layer (except the last one) goes from 1 to 0 since the number of regular nodes it contains goes from 1 to 2. Thus, the decrease of potential of a layer during a push or a pop pays for its rearrangement, while the amortized cost of $O(1)$ pays for the potential increase and the rearrangement in the last node and the push in the icing if it occurs.

It now remains to prove that the depth of a node of weight w is $O(\log W/w)$. The proof will be by induction on the layer number. Consider the subtree rooted at the first node of the ith layer and let W_i be the total weight of that subtree. Assume by induction that at any moment in the algorithm, any leaf of weight w has depth $i + 6 + 7 \log W_i/w$ starting from the root of the ith layer. We want to show that in the subtree rooted at the first node of the $(i-1)^{th}$ layer, any leaf of weight w has depth $(i-1)+6+7 \log W_{i-1}/w$.

Obviously, the hypothesis is true for an ith layer that contains only an icing with one frosted i-crumb, since all its leaves are at distance i; this covers the base case.

For a $(i-1)^{th}$ layer, we consider the leaves located (i) in $(i-1)$-crumbs pointed by regular nodes, (ii) in the ith layer if it exists, and (iii) in the icing of the $(i-1)^{th}$ layer. Any leaf of type (i) is at distance $\leq 3 + i - 1$ which is small enough. For type (ii) leaves, notice that as long as i-crumbs are being moved from the $(i-1)^{th}$ layer to the ith layer without being frosted and pushed to the icing, $W_i \leq W_{i-1}/2$. Therefore, for any leaf of weight w in the subtree of the ith layer, the depth of that leaf is at most

$$4 + i + 6 + 7 \log W_i/w \leq 10 + i + 7 \log W_{i-1}/w - 7 \leq (i-1) + 4 + 7 \log W_{i-1}/w$$

which is below the desired bound.

Finally for case (iii), since the icing of the $(i-1)^{th}$ layer is implemented as a Vanilla pop-tart and the frosted subtrees are pushed with (total) weights always larger than all other leaves (frosted subtrees) in the icing, the icing is crazy good, that is, a frosted subtree of total weight W will have its root at depth at most $5 + \log W_{i-1}/W$. Let p be the parent of the frosted subtree containing the node of weight w, let W_p be the weight of the subtree rooted at p. The depth of p is at most $4 + \log W_{i-1}/W_p$ since the left child of every node on the right path of the icing contains at least half of the weight of that node. Every frosted subtree has its first node whose left pointer points to a possibly heavy i-crumb, and whose right pointer points to what used to be the ith layer at some point in time. Let W' be the weight of that i^{th} layer. Then $W' \leq W_p/2$ otherwise the ith layer would have been frosted earlier. By induction, a leaf of weight w in this former i^{th} layer must have depth no more than $4 + \log W_{i-1}/W_p + 2 + i + 6 + 7 \log W'/w$

$$\leq 12 + i + \log W_{i-1}/W_p + 7 \log W_p/w - 7 \leq (i-1) + 6 + 7 \log W_{i-1}/w$$

which is the desired bound. A leaf in the i-crumb pointed by the left pointer of the root node of the frosted subtree has weight at most W_p, and its depth is

$$4 + \log W_{i-1}/W_p + 2 + i \leq (i-1) + 6 + 7 \log W_{i-1}/w.$$

This completes the induction proof. For $i = 0$, we have that any leaf of weight w has depth at most $6 + 7 \log W/w$, so the chocolate pop-tart is crazy good for arbitrary weights.

Note that all pop-tarts described in this section can also be flipped to maintain elements pushed in increasing order. If the cherry or chocolate pop-tarts need to be implemented in a real-world BST, $O(1)$ extra bits of information in each node is sufficient for storing the function of that node (regular, next, icing, crumb).

3 Simulation

We now show how to efficiently simulate any BST algorithm while keeping the tree of logarithmic height. The method will work for trees with weighted nodes as well. Let w_i be the weight of the node with key i and let $W = \sum_{i=1}^{n} w_i$. For unweighted trees, set $w_i = 1$ and $W = n$. We represent the tree T of the original BST algorithm using a heavy path decomposition. To construct this decomposition, we denote every edge of T as either *solid* or *dotted*. For each non-leaf node, the edge to its child with largest total subtree weight (or the left child, in case of a tie) is a solid edge, and the edge to its other child is dotted. The solid edges form *heavy paths* connected together by dotted edges.

We simulate the original BST algorithm as follows: When its finger is at the root of T, each heavy path is implemented using a pair of weighted pop-tarts: a heavy path from node y to node x (with y an ancestor of x) is a sequence of nodes that can be decomposed into the subsequence $L(y, x)$ of nodes smaller than x on the path, and the subsequence $R(y, x)$ of nodes larger than x on the path. Note that $L(y, x)$ is increasing, and $R(y, x)$ is decreasing. In our simulation, the end of the path x does not change, but y can move up or down along the path to the root. As y moves up, the new nodes are added to $L(y, x)$ in decreasing order, or to $R(y, x)$ in increasing order.

The sequences $L(y, x)$ and $R(y, x)$ will each be stored in the weighted chocolate pop-tart structure described in the previous section, and these two pop-tarts will be left and right children of x, respectively, see Fig. 3. Each node on the path is connected

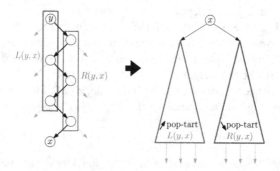

Fig. 3. Representing a heavy path with Pop-tarts

via a dotted edge to a subtree which will be considered as a leaf in the pop-tart, whose weight is exactly the total weight of all the nodes in that subtree. The subtrees contained in those leaves will be structured in the same manner, recursively. The nodes in the tree will contain two extra bits, one to determine if the edge to its parent node is solid or dotted, and another to determine if the next node on its heavy path is in $L(y, x)$ or $R(y, x)$.

When the finger f is not at the root r of the tree, the path from the finger to the the root is also represented as a pair of pop-tarts in a similar way, but this time upside-down (see Fig. 4). Thus, as f walks down, the elements of $L(r, f)$ are added in increasing order, and the elements of $R(r, f)$ are added in decreasing order. Hence, finger movements in the original BST algorithm can be implemented using one push and one pop operation by transferring a node from one pop-tart to the other using $O(1)$ rotations. Likewise, rotations in the original BST algorithm only involve the first few nodes on the pop-tarts linked from the finger, and thus can be implemented in $O(1)$ rotations and push/pop operations. Note that the finger in the tree maintained by our simulation always stays at the root.

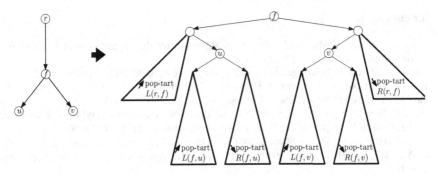

Fig. 4. Representing the finger in general position

Any path from the root to a node x of weight w uses at most $\log W/w$ dotted edges. Further, let $W_1, W_2, \ldots, W_k$ be the total weights of the successive heavy paths (along with their descendants) on the path from the root to x. By Lemma 3, the ith heavy path will be stored at depth $O(\log(W_{i-1}/W_i))$ in the pop-tart of the $(i-1)^{th}$ heavy path, and node x will be at depth $O(\log(W_k/w))$ in the pop-tart of the last heavy path. Thus, the total depth of x in the tree is bounded by a telescoping sum that sums up to $O(\log(W/w))$. Clearly, if $\mathcal{A}$ is online, so is $\mathcal{A}'$. We obtain:

Theorem 1. *Given a BST algorithm $\mathcal{A}$ with a starting tree T, there is a BST algorithm $\mathcal{A}'$ with a starting tree T' such that $|\mathcal{A}'(S)| = O(|\mathcal{A}(S)|)$, and such that the depth of a node i in T' is always $O(\log(W/w_i))$ and the finger is always at the root of T'. If $\mathcal{A}$ is online, so is $\mathcal{A}'$.*

We note that $O(1)$ extra bits per node is sufficient for storing the structure of the original tree and the function of each node in the simulation: each node needs to indicate whether a child is part of the same heavy path or not, and for all nodes on the path from r to f, a bit is needed to determine if the next node on the path is stored in $L(r, f)$ or in $R(r, f)$.

4   De-amortization

Theorem 2. *For any BST algorithm $\mathcal{A}$ with a starting tree T there is a BST algorithm $\mathcal{A}''$ with a starting tree T'' such that for any access sequence S, $|\mathcal{A}''(S)| = O(|\mathcal{A}(S)|)$ and each access to a node is performed in $O(\log n)$ operations worst case. If $\mathcal{A}$ is online, so is $\mathcal{A}''$.*

Proof. Using Theorem 1, transform $\mathcal{A}$ and T into $\mathcal{A}'$ and T' such that the depth of node i in T' is always $c \log n$ for some constant c. Algorithm $\mathcal{A}'$ is then modified as follows: while running the sequence of operations in $\mathcal{A}'(S)$, every time $c \log n$ operations from the original $\mathcal{A}'(S)$ sequence have been performed without accessing the next unaccessed element of the input sequence, access this element by moving the finger to it and back (thereby inserting $\leq 2c \log n$ extra BST operations into the sequence at this point). Thus every access is performed in worst case $3c \log n$, and the total cost of the sequence is the same within a factor 3. If $\mathcal{A}$ is online, so is $\mathcal{A}''$.

References

1. Adel'son-Vel'skii, G.M., Landis, E.M.: An algorithm for the organization of information. Soviet. Math. 3, 1259–1262 (1962)
2. Allen, B., Munro, I.: Self-organizing binary search trees. JACM 25(4), 526–535 (1978)
3. Badoiu, M., Cole, R., Demaine, E.D., Iacono, J.: A unified access bound on comparison-based dynamic dictionaries. Theor. Comput. Sci. 382(2), 86–96 (2007)
4. Bose, P., Douïeb, K., Langerman, S.: Dynamic optimality for skip lists and B-trees. In: Proc. of the ACM-SIAM Symposium On Discrete Algorithms, pp. 1106–1114 (2008)
5. Bose, P., Douïeb, K.: Should Static Search Trees Ever Be Unbalanced? In: Cheong, O., Chwa, K.-Y., Park, K. (eds.) ISAAC 2010. LNCS, vol. 6506, pp. 109–120. Springer, Heidelberg (2010)
6. Bose, P., Douïeb, K., Dujmović, V., Fagerberg, R.: An O(log log n)-Competitive Binary Search Tree with Optimal Worst-Case Access Times. In: Kaplan, H. (ed.) SWAT 2010. LNCS, vol. 6139, pp. 38–49. Springer, Heidelberg (2010)
7. Bose, P., Douïeb, K., Dujmović, V., Howat, J.: Layered Working-Set Trees. In: López-Ortiz, A. (ed.) LATIN 2010. LNCS, vol. 6034, pp. 686–696. Springer, Heidelberg (2010)
8. Cole, R.: On the dynamic finger conjecture for splay trees. Part II: the proof. SIAM J. Computing 30(1), 44–85 (2000)
9. Cole, R., Mishra, B., Schmidt, J., Siegel, A.: On the dynamic finger conjecture for splay trees. Part I: splay sorting log n-block sequences. SIAM J. Computing 30(1), 1–43 (2000)
10. Demaine, E.D., Harmon, D., Iacono, J., Kane, D., Pătraşcu, M.: The geometry of binary search trees. In: Proc. of the 20th Annual ACM-SIAM Symposium on Discrete Algorithms, New York, January 4-6, pp. 496–505 (2009)
11. Demaine, E.D., Harmon, D., Iacono, J., Patrascu, M.: Dynamic optimality - almost. SIAM J. Comput. 37(1), 240–251 (2007)
12. Derryberry, J., Sleator, D.D.: Skip-Splay: Toward Achieving the Unified Bound in the BST Model. In: Dehne, F., Gavrilova, M., Sack, J.-R., Tóth, C.D. (eds.) WADS 2009. LNCS, vol. 5664, pp. 194–205. Springer, Heidelberg (2009)
13. Derryberry, J., Sleator, D.D., Wang, C.C.: A lower bound framework for binary search trees with rotations. Technical Report CMU-CS-05-187. Carnegie Mellon University (November 2005)
14. Derryberry, J., Sleator, D.D., Wang, C.C.: Properties of multi-splay trees. Technical Report CMU-CS-09-180. Carnegie Mellon University (November 2009)
15. Georgakopoulos, G.F.: Splay trees: a reweighing lemma and a proof of competitiveness vs. dynamic balanced trees. Journal of Algorithms 51(1), 64–76 (2004)
16. Georgakopoulos, G.F.: Chain-splay trees, or, how to achieve and prove log log n-competitiveness by splaying. Inf. Process. Lett. 106, 37–43 (2008)
17. Gold Effie Award, http://www.effie.org/winners/showcase/2006/256
18. Guibas, L.J., Sedgewick, R.: A dichomatic framework for balanced trees. In: Proc. 19th Ann. IEEE Symp. on Theory of Computing, pp. 8–21 (1978)
19. Iacono, J.: Alternatives to splay trees with $O(\log n)$ worst-case access times. In: Proc. 12th ACM-SIAM Sympos. Discrete Algorithms, pp. 516–522 (2001)
20. Iacono, J.: Key-independent optimality. Algorithmica 42(1), 3–10 (2005)
21. Iacono, J., Langerman, S.: Queaps. Algorithmica 42(1), 49–56 (2005)
22. Knuth, D.E.: Optimum binary search trees. Acta Inf. 1, 14–25 (1971)
23. Sleator, D.D., Tarjan, R.E.: Self-adjusting binary trees. JACM 32, 652–686 (1985)
24. Wang, C.C., Derryberry, J., Sleator, D.D.: O(log log n)-competitive dynamic binary search trees. In: Proc. of the Seventeenth Annual ACM-SIAM Symposium on Discrete Algorithm, pp. 374–383. ACM, New York (2006)
25. Wilber, R.: Lower bounds for accessing binary search trees with rotations. SIAM J. Computing 18(1), 56–67 (1989)

Efficient Sampling Methods
for Discrete Distributions

Karl Bringmann[1] and Konstantinos Panagiotou[2]

[1] Max Planck Institute for Informatics
Campus E1.4, 66123 Saarbrücken, Germany
[2] Department of Mathematics, University of Munich,
Theresienstraße 39, 80333 Munich, Germany

Abstract. We study the fundamental problem of the exact and efficient generation of random values from a finite and discrete probability distribution. Suppose that we are given n distinct events with associated probabilities $p_1, \ldots, p_n$. We consider the problem of sampling a subset, which includes the ith event independently with probability p_i, and the problem of sampling from the distribution, where the ith event has a probability proportional to p_i. For both problems, we present on two different classes of inputs – sorted and general probabilities – efficient preprocessing algorithms that allow for asymptotically optimal querying, and prove almost matching lower bounds for their complexity.

1 Introduction

Generating random variables from finite and discrete distributions has long been an important building block in many applications. For example, in computer simulations usually a huge number of random decisions based on prespecified or dynamically changing distributions is made. In this work we consider two fundamental computational problems, namely sampling *independent events* and sampling *from a distribution*, on two different classes of inputs, sorted and unsorted probabilities. As we will see, there is a rich interplay in designing efficient algorithms that solve these different variants.

Our results are valid in the classical RealRAM model [1, 9] of computation. In particular, we will assume that the following operations take constant time:

- Accessing the content of any memory cell.
- Generating a uniformly distributed real number in the interval $[0, 1]$.
- Performing any basic arithmetical operation involving real numbers like addition, multiplication, division, comparison, truncation, and evaluating any fundamental function like exp and log.

Whether our results can be generalized to more realistic machine models is an interesting question for future work.

In the remainder, we will abbreviate $[n] = \{1, \ldots, n\}$ and we will write $\ln x$ for the natural logarithm of x and $\log x$ for the binary logarithm of x. Finally, we will write rand() for a uniform random number in $[0, 1]$.

A. Czumaj et al. (Eds.): ICALP 2012, Part I, LNCS 7391, pp. 133–144, 2012.
© Springer-Verlag Berlin Heidelberg 2012

1.1 Subset Sampling

We consider n independent events with indicator random variables $X_1, \ldots, X_n$, and $\Pr[X_i = 1] = p_i$. For shortcut we write $\mu = \mu_{\mathbf{p}} = \sum_{i=1}^n p_i = \mathbb{E}[\sum_{i=1}^n X_i]$ and $\mathbf{p} = (p_1, \ldots, p_n)$. Consider the random variable $X = X_{\mathbf{p}} = \{i \in [n] \mid X_i = 1\}$, which is the set of all events that occurred.

We concern ourselves with the problem of sampling X. We study this problem on two different classes of input sequences, sorted and general (i.e., not necessarily sorted) sequences; dependent on the class under consideration we call the problem SORTEDSUBSETSAMPLING or UNSORTEDSUBSETSAMPLING.

A *single-sample algorithm* for SORTEDSUBSETSAMPLING or UNSORTEDSUBSETSAMPLING gets input $\mathbf{p}$ and outputs a set $S \subseteq [n]$ that has the same distribution as X. When we speak of "input $\mathbf{p}$" we mean that the algorithm gets to know n and can access every p_i in constant time. This can be achieved by storing all p_i's in an array, but also, e.g., by a constant depth arithmetic circuit computing p_i given i. In particular, the algorithm does not know the number of i's with $p_i = 0$ (i.e., the input format is not sparse).

Such an algorithm cannot run faster than $\mathcal{O}(1 + \mu)$, as its expected output size is μ and any algorithm requires a running time of $\Omega(1)$. This runtime, however, is in general not achievable, as our results below make more precise. Hence, we consider a *preprocessing-query* variant of the problem, where we want to be able to answer queries in the optimal expected runtime of $\mathcal{O}(1 + \mu)$ after a certain preprocessing.

In the preprocessing-query variant we consider the interplay of two algorithms. First, the *preprocessing algorithm* P gets $\mathbf{p}$ as input and computes some auxiliary data $D = D(\mathbf{p})$. Second, the *query algorithm* Q gets input $\mathbf{p}$ and D, and samples X, i.e., for any $S \subseteq [n]$ we have $\Pr[Q(\mathbf{p}, D) = S] = \Pr[X_{\mathbf{p}} = S]$. Here $\Pr$ goes only over the random choices of Q, so that, after running the preprocessing once, running the query algorithm multiple times generates multiple independent samples. Note that if the preprocessing time is p and the query time is q, then we can generate a single sample of X in time $p + q$, so the single-sample variant of the problem is also solved by the preprocessing-query variant. In this paper we will not consider single-sample algorithms any further, because our constructed preprocessing-query algorithms are already for a *single* query as efficient as the best single-sample algorithm we can devise. This holds for all problem variants we consider.

The single-sample variant of UNSORTEDSUBSETSAMPLING can be solved trivially in time $\mathcal{O}(n)$; we just toss a biased coin for every p_i. A classic algorithm solves this problem for $p_1 = \ldots = p_n = p$ in the optimal expected time $\mathcal{O}(1 + \mu)$, see e.g. the monographs [2] by Devroye and [5] by Knuth, where also many other cases are discussed. Indeed, observe that the index i_1 of the first sampled element is geometrically distributed, i.e., $\Pr[i_1 = i] = (1 - p)^{i-1}p$. Such a random value can be generated by setting $i_1 = \lfloor \frac{\log \text{rand}()}{\log(1-p)} \rfloor$. Moreover, after having sampled the index of the first element, we iterate the process starting at $i_1 + 1$ to sample the second element, and so on, until we arrive for the first time at an index $i_k > n$. In [13] the "orthogonal" problem is considered, where we want to uniformly sample a fixed number of elements from a stream of objects.

In this paper we generalize the algorithm for equal probabilities as far as possible. More precisely, we ask whether the optimal query time $\mathcal{O}(1 + \mu)$ is achievable for larger classes of inputs and how much preprocessing is needed. We obtain the following answers.

Theorem 1. SORTEDSUBSETSAMPLING *can be solved in* $\mathcal{O}(\log n)$ *preprocessing time and* $\mathcal{O}(1 + \mu)$ *expected query time. Moreover, the bound on the preprocessing time is nearly tight, as the sum of preprocessing and query time is* $\Omega\left(\frac{\log n}{\log \log n}\right)$ *for any such algorithm, as* $n \to \infty$ *and* $\mu = \mu(n) \geqslant (\log n)^{-\mathcal{O}(1)}$.

Note that all our lower bounds only hold for algorithms that work for all n and all sorted sequences $p_1, \ldots, p_n$. They are worst-case bounds over the input sequence **p** and asymptotic in n. For particular instances **p** there can be faster algorithms. Due to space limitations, the proof of the lower bound of Theorem 1 is not included in this extended abstract.

To avoid any confusion, note that we mean worst-case bounds whenever we speak of *(running) time* and expected bounds whenever we speak of *expected (running) time*. The next result addresses the case where the probabilities are not necessarily sorted.

Theorem 2. UNSORTEDSUBSETSAMPLING *can be solved in* $\mathcal{O}(n)$ *preprocessing time and* $\mathcal{O}(1 + \mu)$ *expected query time. Moreover, this is optimal, as even any single-sample algorithm for* UNSORTEDSUBSETSAMPLING *needs time* $\Omega(n)$.

Both positive results in the previous theorems depend highly on each other. In particular, as it is demonstrated in Section 3, we prove them by repeatedly reducing the instance size n and switching from the one problem variant to the other.

The problem of UNSORTEDSUBSETSAMPLING was considered also recently in the two papers [11, 12], where algorithms with linear preprocessing time and suboptimal query time $\mathcal{O}(\log n + \mu)$ were designed. Thus, our results improve upon these running times, and provide accompanying and (almost) matching lower bounds.

1.2   Proportional Sampling

In the previous section we considered the problem of sampling subsets. Here we will focus on a slightly different and more classical problem. Given $\mathbf{p} = (p_1, \ldots, p_n) \in \mathbb{R}_{\geqslant 0}^n$, we define a random variable $Y = Y_\mathbf{p}$ that takes values in $[n]$ such that $\Pr[Y = i] = p_i/\mu$, where again $\mu = \sum_{i=1}^n p_i$. We call the problem of sampling Y SORTEDPROPORTIONALSAMPLING or UNSORTEDPROPORTIONAL-SAMPLING, if we consider it on sorted or general input sequences, respectively.

As previously, we consider two variations of the problem. In the *single-sample* variant we are given **p** and we want to compute an output that has the same distribution as Y. Moreover, in the *preprocessing-query* variant we have a pre-computation algorithm that, given **p**, computes some auxiliary data D, and a query algorithm that is given $\mathbf{p}$ and D and has an output with the same distribution as Y; where the results of multiple calls to the query algorithm are independent.

In this setting, we no longer output μ elements. So, it could be that the optimal expected query time reduces to $\mathcal{O}(1)$. For sorted sequences, this optimal query time can be indeed achieved after a relatively small preprocessing time, as the next result shows.

Theorem 3. SORTEDPROPORTIONALSAMPLING *can be solved in* $\mathcal{O}(\log n)$ *preprocessing time and* $\mathcal{O}(1)$ *expected query time.*

For general input sequences, this problem can be solved by the technique known as *pairing* or *aliasing* [5, 14]. This result is not new, but will be used in the proofs of Theorem 1 and Theorem 2, so we include it for completeness.

Theorem 4. UNSORTEDPROPORTIONALSAMPLING *can be solved in* $\mathcal{O}(n)$ *preprocessing time and* $\mathcal{O}(1)$ *query time. Moreover, this is optimal, as any single-sample algorithm for* UNSORTEDPROPORTIONALSAMPLING *needs time* $\Omega(n)$.

The fundamental problem of the exact and efficient generation of random values from discrete and continuous distributions has been studied extensively in the literature. Knuth and Yao investigated in their seminal work [6] the power of several restricted devices, like finite-state machines; the articles [3, 15] provide a further refined treatment of the topic. However, their results are not directly comparable to ours, since they do not make any assumption on the sequence of probabilities, and use unbiased coin flips as the only source of randomness, but cannot guarantee efficient precomputation on general sequences. Furthermore, Hagerup, Mehlhorn and Munro [4] and Matias, Vitter and Ni [7] provided algorithms for a dynamic version of UNSORTEDPROPORTIONALSAMPLING, where the probabilities may change over time. In particular, under certain mild conditions their results guarantee the same bounds as in Theorem 4.

The rest of the paper is structured as follows. In the following section we will show Theorem 4. Section 3 contains the proofs of Theorems 1 and 2, while Section 4 is devoted to the proof of Theorem 3. We discuss relaxations to our input model and possible extensions in Section 5.

2   Proportional Sampling on Unsorted Probabilities

In this section we consider UNSORTEDPROPORTIONALSAMPLING and prove Theorem 4. The upper bound can be reached by the old technique known as *pairing* or *aliasing* [14]; see also Mihai Pătraşcu's blog [10] for a nice explanation. Basically, we use $\mathcal{O}(n)$ preprocessing to distribute the probabilities of all elements over n urns such that any urn contains probability mass of at most two elements. For querying we choose an urn uniformly at random and choose a random one of the two included elements according to their probability mass in the urn, which gives $\mathcal{O}(1)$ worst-case querying time.

The lower bound for Theorem 4 is provided by the following lemma, which reduces UNSORTEDPROPORTIONALSAMPLING to searching in an unordered array.

Moreover, the same proof yields the lower bound of Theorem 2 for UNSORTED-SUBSETSAMPLING.

Lemma 1. *Any single-sample algorithm for* UNSORTEDPROPORTIONALSAMPLING *needs* $\Omega(n)$ *expected time. Moreover, any single-sample algorithm for* UNSORTEDSUBSETSAMPLING *needs* $\Omega(n)$ *expected time.*

Proof. Consider the instances $\mathbf{p}^{(k)} = (p_1^{(k)}, \ldots, p_n^{(k)})$ with $p_i^{(k)} = \delta_{ik}$, where δ_{ik} is the Kronecker delta. Any sampling algorithm for UNSORTEDPROPORTIONAL-SAMPLING returns k on instance $\mathbf{p}^{(k)}$ with probability 1. This cannot be done better than with linear search for k, and randomness does not help, either. With varying μ, no better bound is possible, either: Simply set $p_i^{(k)} = \mu\delta_{ik}$.

Observe that on the same instance any sampling algorithm for UNSORTED-SUBSETSAMPLING returns $\{k\}$ with probability 1. This needs runtime $\Omega(n)$ for the same reasons. With varying μ, no better bound is possible, either: Set the first $s := \lceil \mu-1 \rceil$ probabilities p_i to values that sum up to $\mu-1$, and let $p_i^{(k)} = \delta_{ik}$ for $s < i \leqslant n$. Then we still need runtime $\Omega(n - \mu)$ for searching k. As we also need runtime $\Omega(\mu)$ for outputting the result, the claim follows. $\square$

3   Subset Sampling

In this section we consider SORTEDSUBSETSAMPLING and UNSORTEDSUBSET-SAMPLING and prove Theorems 1 and 2. An interesting interplay between both of these problems will be revealed on the way.

We begin with a first algorithm for unsorted probabilities that has a quite large preprocessing time, but will be used for a base case later. The algorithm uses Theorem 4, which we proved in the preceding section.

Lemma 2. UNSORTEDSUBSETSAMPLING *can be solved in* $\mathcal{O}(n^2)$ *preprocessing time and* $\mathcal{O}(1 + \mu)$ *expected query time.*

Proof. For $i \in [n]$ let X_i be the smallest sampled element which is at least i, or ∞, if no such element is sampled. X_i is a random variable with $\Pr[X_i = j] = p_j \prod_{i \leqslant k < j}(1 - p_k)$ and $\Pr[X_i = \infty] = \prod_{i \leqslant k \leqslant n}(1 - p_k)$. These probabilities can be computed in time $\mathcal{O}(n)$ for any i, i.e., in time $\mathcal{O}(n^2)$ for all i. After having computed the distribution of the X_i's, we execute, for each $i \in [n]$, the preprocessing of Theorem 4, see the beginning of Section 2, which allows us to quickly sample X_i later on. This preprocessing costs in total $\mathcal{O}(n^2)$.

For querying, we start at $i = 1$ and iteratively sample the smallest element $j \geqslant i$ (i.e., sample X_i), output j, and start over with $i = j+1$. This is done until $j = \infty$ or $i = n+1$. Note that any sample of X_i can be computed in $\mathcal{O}(1)$ time with our preprocessing, so that sampling $S \subseteq [n]$ will be done in time $\mathcal{O}(1+|S|)$. The expected runtime is, thus, $\mathcal{O}(1 + \mu)$. $\square$

After having this base case, we turn towards reductions between SORTEDSUB-SETSAMPLING and UNSORTEDSUBSETSAMPLING. First, we give an algorithm for UNSORTEDSUBSETSAMPLING, that reduces the problem to SORTEDSUBSET-SAMPLING. For this, we roughly sort the probabilities so that we get good upper bounds for each probability. Then these upper bounds will be a sorted instance. After querying from this sorted instance, we use rejection (see, e.g., [5]) to sample with the original probabilities.

Lemma 3. *Assume that* SORTEDSUBSETSAMPLING *can be solved in* $p(n, \mu)$ *pre-processing time and* $q(n, \mu)$ *expected query time, where* p *and* q *are monotonically increasing in* n *and* μ. *Then* UNSORTEDSUBSETSAMPLING *can be solved in* $\mathcal{O}(n + p(n, 2\mu + 1))$ *preprocessing time and* $\mathcal{O}(1 + \mu + q(n, 2\mu + 1))$ *expected query time.*

Proof. Let $p_1, \ldots, p_n$ be an input sequence to UNSORTEDSUBSETSAMPLING. For preprocessing, we permute the input **p** so that it is approximately sorted, by putting it into buckets $B_k := \{i \in [n] \mid 2^{-k} \geqslant p_i \geqslant 2^{-k-1}\}$, for $k \in \{0, 1, \ldots, L\}$, and $B_L := \{i \in [n] \mid 2^{-L} \geqslant p_i\}$, where $L = \lceil \log n \rceil$. For each $i \in B_k$ we set $\overline{p}_i := 2^{-k}$, which is an upper bound on p_i. We sort the probabilities $\overline{p}_i$, $i \in [n]$, descendingly using bucket sort with the buckets B_k, yielding $\overline{p}'_1 \geqslant \ldots \geqslant \overline{p}'_n$. In this process we store the original index $\mathrm{ind}(i)$ corresponding to $\overline{p}'_i$, so that we can find $p_{\mathrm{ind}(i)}$ corresponding to $\overline{p}'_i$ in constant time. Then we run the preprocessing of SORTEDSUBSETSAMPLING on $\overline{p}'_1, \ldots, \overline{p}'_n$. Note that

$$\overline{\mu} := \sum_{i=1}^{n} \overline{p}'_i = \sum_{i=1}^{n} \overline{p}_i \leqslant \sum_{i=1}^{n} \max\left\{ 2p_i, \frac{1}{n} \right\} \leqslant 2\mu + 1.$$

For querying, we query $\overline{p}'_1, \ldots, \overline{p}'_n$ using SORTEDSUBSETSAMPLING, yielding $S' \subseteq [n]$. We compute $S := \{\mathrm{ind}(i) \mid i \in S'\}$. Each $i \in S$ was sampled with probability $\overline{p}_i \geqslant p_i$. We use rejection to get this probability down to p_i. For this, we generate for each $i \in S$ a random number rand() and check whether it is smaller than or equal to $\frac{p_i}{\overline{p}_i}$. If this is not the case, we delete i from S. Note that we have thus sampled i with probability p_i, and all elements are sampled independently, so that we can return S.   □

We also give a reduction in the other direction, solving SORTEDSUBSETSAM-PLING by UNSORTEDSUBSETSAMPLING.

Lemma 4. *Assume that* UNSORTEDSUBSETSAMPLING *can be solved in* $p(n, \mu)$ *preprocessing time and* $q(n, \mu)$ *expected query time, where* p *and* q *are monotonically increasing in* n *and* μ. *Then* SORTEDSUBSETSAMPLING *can be solved in* $\mathcal{O}(\log n + p(1 + \log n, 2\mu))$ *preprocessing time and* $\mathcal{O}(1 + \mu + q(1 + \log n, 2\mu))$ *expected query time.*

Proof. Let $p_1, \ldots, p_n$ be an input sequence to SORTEDSUBSETSAMPLING. We consider blocks $B_k = \{i \in [n] \mid 2^k \leqslant i < 2^{k+1}\}$, with $k \in \{0, \ldots, L\}$ and $L := \lfloor \log n \rfloor$. For $i \in B_k$ we let $\overline{p}_i := p_{2^k}$, which is an upper bound on p_i. We will first sample with respect to the probabilities $\overline{p}_i$ - call the sampled elements *potential* - and then use rejection. For this, let X_k be an indicator random variable for the event that we sample *at least one* potential element in B_k. Then $q_k := \Pr[X_k = 1] = 1 - (1 - p_{2^k})^{|B_k|}$. Moreover, let Y_k be a random variable for the first potential element in block B_k minus 2^k. Let $Y_k = \infty$, if no element in B_k is sampled as a potential element. Then $\Pr[Y_k = i] = p_{2^k}(1 - p_{2^k})^i$ for $i \in \{0, \ldots, |B_k| - 1\}$, and $\Pr[Y_k = \infty] = \Pr[X_k = 0] = 1 - q_k$. We calculate

$$\Pr[Y_k = i \mid X_k] = \frac{\Pr[Y_k = i]}{\Pr[X_k]} = \frac{p_{2^k}}{q_k}(1 - p_{2^k})^i.$$

Since this is a geometric distribution, we can sample from it in constant time as sketched in the introduction; see also [5].

Now, for preprocessing, we compute the probabilities q_k, which can be done in time $\mathcal{O}(\log n)$ (as $a^b = \exp(b \log a)$ can be computed in constant time on a Real RAM), and run the preprocessing of UNSORTEDSUBSETSAMPLING on them. Note that the q_k are in general unsorted.

For querying, we query the blocks B_k that contain potential elements using the query algorithm for UNSORTEDSUBSETSAMPLING. Then for each block B_k that contains a potential element, we sample all potential elements in this block. Note that the first of the potential elements in B_k is distributed as $\Pr[Y_k = i \mid X_k]$, which is geometric, so we can sample it in constant time, while all further potential elements are distributed as Y_k (but only on the remainder of the block), which is still geometric. After thus sampling potential elements $\overline{S}$, we reject each potential element with the right probability: We keep each $i \in \overline{S}$ only if rand() $\leqslant \frac{p_i}{\overline{p}_i}$. This yields a correctly distributed sample.

Let $\overline{\mu} := \sum_{i=1}^{n} \overline{p}_i$. The overall query time is at most $q(1 + \log n, \overline{\mu}) + \mathcal{O}(1 + |\overline{S}|)$ when sampling potential elements $\overline{S}$. As the expected value of $|\overline{S}|$ is $\overline{\mu}$, all we need to show in order to finish the proof is $\overline{\mu} \leqslant 2\mu$. For this, note that $\overline{p}_i \leqslant p_{\lceil i/2 \rceil}$. This yields

$$\overline{\mu} = \sum_{i=1}^{n} \overline{p}_i \leqslant \sum_{i=1}^{n} p_{\lceil i/2 \rceil} \leqslant 2 \sum_{i=1}^{n} p_i = 2\mu.$$

$\square$

Next, we put above three lemmas together to prove the upper bounds of Theorems 1 and 2.

Proof (Theorem 2, upper bound). To solve UNSORTEDSUBSETSAMPLING, we use the reduction Lemma 3 and then Lemma 4, followed by the base case Lemma 2. This reduces the instance size from n to $\mathcal{O}(\log n)$, so that preprocessing costs $\mathcal{O}(n)$ for the invocation of the first lemma, $\mathcal{O}(\log n)$ for the second, and $\mathcal{O}(\log^2 n)$ for the third. Note that μ is increased by constant factors only, so that we indeed get the optimal query time $\mathcal{O}(1 + \mu)$.

$\square$

Proof (Theorem 1, upper bound). To solve SORTEDSUBSETSAMPLING, we use the reductions Lemma 4, Lemma 3, and Lemma 4 again, followed by the base case Lemma 2. This reduces the instance size from n to $\mathcal{O}(\log n)$ and further down to $\mathcal{O}(\log \log n)$, while μ is increased by constant factors only. For precomputation this yields a runtime of $\mathcal{O}(\log n)$ from Lemmas 4 and 3, $\mathcal{O}(\log \log n)$ from the second invocation of Lemma 4, and $\mathcal{O}(\log^2 \log n)$ from the base case Lemma 2, summing up to $\mathcal{O}(\log n)$. The query time is the optimal expected time $\mathcal{O}(1 + \mu)$. □

4   Proportional Sampling on Sorted Probabilities

We prove Theorem 3 in this section, i.e., we show how to solve SORTEDPRO-PORTIONALSAMPLING in $\mathcal{O}(\log n)$ preprocessing time and $\mathcal{O}(1)$ expected query time. We do this by first considering the special case of $\frac{1}{2} \leqslant \mu \leqslant 1$, so that we have a (nearly) proper probability distribution. Lemma 7 shows how to reduce SORTEDPROPORTIONALSAMPLING to SORTEDSUBSETSAMPLING in this special case. Then we reduce the general case with arbitrary μ to the special case.

4.1   Special Case $1/2 \leqslant \mu \leqslant 1$

We first fix some notation for this section. Let $\mathbf{p}$ be an instance to SORTEDPRO-PORTIONALSAMPLING with $\mu = \mu_{\mathbf{p}}$ in the range $[\frac{1}{2}, 1]$. Instead of $\mathbf{p}$ we consider $\mathbf{p}' = (p'_1, \ldots, p'_n)$ with $p'_i := \frac{p_i}{1+p_i}$. Note that $\mathbf{p}'$ ist still sorted and $\mu' := \sum_{i=1}^n p'_i$ is in the range $[\frac{\mu}{2}, \mu]$, thus in the range $[\frac{1}{4}, 1]$.

Let $Y = \text{SORTEDPROPORTIONALSAMPLING}(\mathbf{p})$ be the random variable denoting proportional sampling on input $\mathbf{p}$, and $X = \text{SORTEDSUBSETSAMPLING}(\mathbf{p}')$ be the random variable denoting subset sampling on input $\mathbf{p}'$. Then conditioned on sampling exactly one element $X = \{i\}$, this element i is distributed exactly as Y, as formulated by the following lemma.

Lemma 5. *With the definitions and assumptions of this section we have for all* $i \in [n]$
$$\Pr[X = \{i\} \mid |X| = 1] = \Pr[Y = i].$$

Proof. Bayes' rule and straightforward calculation give

$$\Pr[X = \{i\} \mid |X| = 1] = \Pr[X = \{i\}] / \Pr[|X| = 1]$$

$$= \left(\frac{p'_i}{1 - p'_i} \prod_{k=1}^n (1 - p'_k) \right) \bigg/ \left(\sum_{j=1}^n \frac{p'_j}{1 - p'_j} \prod_{k=1}^n (1 - p'_k) \right)$$

$$= \left(\frac{p'_i}{1 - p'_i} \right) \bigg/ \left(\sum_{j=1}^n \frac{p'_j}{1 - p'_j} \right)$$

Plugging in the definition of p'_i yields

$$\Pr[X = \{i\} \mid |X| = 1] = p_i / \sum_{j=1}^n p_j = \Pr[Y = i].$$

□

Moreover, the probability of sampling exactly one element is large, as shown in the following lemma. Note that this bound is not best possible but sufficient for our purposes.

Lemma 6. *With the definitions and assumptions of this section we have*

$$\Pr[|X| = 1] \geqslant 1/8.$$

Proof. Clearly,

$$\Pr[|X| = 1] = \sum_{j=1}^{n} \frac{p_j'}{1 - p_j'} \prod_{k=1}^{n} (1 - p_k').$$

Assume there is no p_i' greater than $1/2$. Then we have $1 - p_i' \geqslant 4^{-p_i'}$ for all $i \in [n]$, so we get

$$\Pr[|X| = 1] \geqslant \sum_{j=1}^{n} p_j' \cdot \prod_{k=1}^{n} 4^{-p_k'} = \mu' \cdot 4^{-\sum_{k=1}^{n} p_k'} = \mu' \cdot 4^{-\mu'} \geqslant \frac{1}{8}.$$

Otherwise, there is exactly one $p_{i*}' > 1/2$, as $\mu' \leqslant 1$. Then $1 - p_k' \geqslant 4^{-p_k'}$ holds for all $k \in [n]$, $k \neq i^*$, which yields

$$\Pr[|X| = 1] \geqslant \Pr[X = \{i^*\}] = p_{i*}' \prod_{\substack{1 \leqslant k \leqslant n \\ k \neq i^*}} (1 - p_k') \geqslant \frac{1}{2} \prod_{\substack{1 \leqslant k \leqslant n \\ k \neq i^*}} 4^{-p_k'}$$

$$\geqslant \frac{1}{2} 4^{-\sum_{k=1}^{n} p_k'} = \frac{1}{2} 4^{-\mu'} \geqslant \frac{1}{8}.$$

$\square$

We put these facts together to show the following result.

Lemma 7. *Assume that* SORTEDSUBSETSAMPLING *can be solved in* $p(n, \mu)$ *preprocessing time and* $q(n, \mu)$ *expected query time, where* p *and* q *are monotonically increasing in* n *and* μ. *Then* SORTEDPROPORTIONALSAMPLING *on instances with* $\frac{1}{2} \leqslant \mu \leqslant 1$ *can be solved in* $\mathcal{O}(p(n, 1))$ *preprocessing time and* $\mathcal{O}(q(n, 1))$ *expected query time.*

Proof. For preprocessing, given input $\mathbf{p}$, we run the preprocessing of SORTEDSUBSETSAMPLING on input $\mathbf{p}'$. This does not mean that we compute the vector $\mathbf{p}'$ beforehand, but if the preprocessing algorithm of SORTEDSUBSETSAMPLING reads the i-th input value, we compute $p_i' = \frac{p_i}{1+p_i}$ on the fly, so that this needs runtime $\mathcal{O}(p(n, 1))$. It allows to sample X later on in expected runtime $\mathcal{O}(q(n, 1))$ using the same trick of computing $\mathbf{p}'$ on the fly.

For querying, we repeatedly sample X until we sample a set S of size one. Returning the unique element of S results in a proper sample according to SORTEDPROPORTIONALSAMPLING by Lemma 5. Moreover, by Lemma 6 and the fact that sampling X needs expected time $\mathcal{O}(q(n, 1))$ after our preprocessing, we need expected query time $\mathcal{O}(q(n, 1))$.

$\square$

4.2   General Case

Lemma 8. *Assume that* SortedProportionalSampling *on instances with* $\frac{1}{2} \leqslant \mu \leqslant 1$ *can be solved in* $p(n)$ *preprocessing time and* $q(n)$ *expected query time. Then* SortedProportionalSampling *(for general instances) can be solved in* $\mathcal{O}(\log n + p(n))$ *preprocessing time and* $\mathcal{O}(q(n))$ *expected query time.*

Proof. We need to compute a good upper bound $\overline{\mu} \geqslant \mu$. For this we reuse an idea of the proof of Lemma 4: For $i \in [n]$ let 2^k be the largest power of two less than or equal to i, and set $\overline{p}_i := p_{2^k}$. Then $\overline{\mu} := \sum_{i=1}^n \overline{p}_i \geqslant \sum_{i=1}^n p_i = \mu$, and we have $\overline{p}_i \leqslant p_{\lceil i/2 \rceil}$, so that

$$\overline{\mu} = \sum_{i=1}^n \overline{p}_i \leqslant \sum_{i=1}^n p_{\lceil i/2 \rceil} \leqslant 2 \sum_{i=1}^n p_i = 2\mu.$$

Hence, $\overline{\mu}$ is indeed a good upper bound on μ. Moreover, $\overline{\mu}$ can be computed in time $\mathcal{O}(\log n)$, as

$$\overline{\mu} = \sum_{k=0}^{\lfloor \log n \rfloor} p_{2^k}(\min\{2^{k+1} - 1, n\} - 2^k + 1).$$

Now, for preprocessing, we compute $\overline{\mu}$ and consider $\mathbf{p}' = (p'_1, \ldots, p'_n)$ with $p'_i := \frac{p_i}{\overline{\mu}}$. Since $\overline{\mu} \geqslant \mu \geqslant \frac{\overline{\mu}}{2}$ we have $\mu' := \sum_{i=1}^n p'_i$ in the range $[\frac{1}{2}, 1]$. Thus, we can run the preprocessing of SortedProportionalSampling (on instances with bounded μ) on $\mathbf{p}'$. We do this without computing the whole vector $\mathbf{p}'$. Instead, if the preprocessing algorithm reads the i-th input value, we compute $p'_i = \frac{p_i}{\overline{\mu}}$ on the fly. This way we need a runtime of $\mathcal{O}(\log n + p(n))$.

For querying, we query according to $\mathbf{p}'$ within expected runtime $\mathcal{O}(q(n))$, where we again compute values of $\mathbf{p}'$ on the fly as needed. As we want to sample proportional to the input probabilities, a sample with respect to $\mathbf{p}'$ has the same distribution as a sample with respect to $\mathbf{p}$, so that we simply return the sample we have.   □

Proof (Theorem 3). To solve SortedProportionalSampling we take Lemmas 8 and 7 and Theorem 1 together.   □

5   Relaxations

In this section we describe some natural relaxations for the input model studied so far in this paper.

Large Deviations for the Running Times. The query runtimes in Theorems 1, 2 and 3 are, in fact, not only small in expectation, but they are also concentrated, i.e., they satisfy large deviation estimates in the following sense. Let t be the expected runtime bound and T the actual runtime. Then

$$\Pr[T > kt] = e^{-\Omega(k)},$$

where the asymptotics are with respect to k. This is shown rather straightforwardly along the lines of our proofs of these theorems. The fundamental reason for this is that the size of the random set X is concentrated. Indeed, let X_i be an indicator random variable for the i-th element as above. Then for any $a > 1$ we obtain along the lines of the proof of the Chernoff bound

$$\Pr[|S| > k(\mu + 1)] = \Pr[a^{\sum_{i=1}^{n} X_i} > a^{k(\mu+1)}] \leqslant \frac{\mathbb{E}[a^{\sum_{i=1}^{n} X_i}]}{a^{k(\mu+1)}}.$$

Then, the independence of the X_i's implies that

$$\Pr[|S| > k(\mu + 1)] \leqslant \frac{\prod_{i=1}^{n} \mathbb{E}[a^{X_i}]}{a^{k(\mu+1)}}$$

$$= \frac{\prod_{i=1}^{n} (ap_i + (1 - p_i))}{a^{k(\mu+1)}} \leqslant \exp((a - 1)\mu - k(\mu + 1) \ln a).$$

Setting $a = k + 1$ yields

$$\Pr[|S| > k(\mu + 1)] \leqslant \exp(k\mu - k(\mu + 1) \log(k + 1)) \leqslant (k + 1)^{-k},$$

for $k \geqslant 2$, as claimed.

Unimodular Input. Many natural distributions **p** are not sorted, but unimodular, meaning that p_i is monotonically increasing for $1 \leqslant i \leqslant m$ and monotonically decreasing for $m \leqslant i \leqslant n$ (or the other way round). Knowing m, we can run the algorithms developed in this paper on both sorted halves, and combine the return values, which gives an optimal query algorithm for unimodular inputs. Alternatively, if we have strong monotonicity, we can search for m in time $\mathcal{O}(\log n)$ using ternary search, which does not increase our precomputation time.

This can be naturally generalized to k-modular inputs, where the monotonicity changes k times.

Approximate Input In some applications it may be costly to compute the probabilities p_i exactly, but we are able to compute approximations $\overline{p}_i(\varepsilon) \geqslant p_i \geqslant \underline{p}_i(\varepsilon)$, with relative error at most ε, where the cost of computing these approximations depends on ε. We can still guarantee optimal query time, if the costs of computing these approximations are small enough, see e.g. [8].

Indeed, we can surely sample a superset $\overline{S}$ with respect to the probabilities $\overline{p}_i(1)$. Then we want to use rejection, i.e., for each element $i \in \overline{S}$ we want to compute a random number $r := \text{rand}()$ and delete i from $\overline{S}$ if $r \cdot \overline{p}_i(1) > p_i$, to get a sample set S. This check can be performed as follows. We initialize $k := 1$. If $r \cdot \overline{p}_i(1) > \overline{p}_i(2^{-k})$ we delete i from $\overline{S}$. If $r \cdot \overline{p}_i(1) \leqslant \underline{p}_i(2^{-k})$ we keep i and are done. Otherwise, we increase k by 1. This method needs an expected number of $\mathcal{O}(1)$ rounds of increasing k; the probability of needing k rounds is $\mathcal{O}(2^{-k})$. Hence, if the cost of computing $\overline{p}_i(\varepsilon)$ and $\underline{p}_i(\varepsilon)$ is $\mathcal{O}(\varepsilon^{-c})$ with $c < 1$, the expected overall cost is constant, and we get an optimal expected query time of $\mathcal{O}(1 + \mu)$.

References

[1] Borodin, A., Munro, I.: The computational complexity of algebraic and numeric problems. American Elsevier Publishing Co., Inc., New York (1975)

[2] Devroye, L.: Nonuniform random variate generation. Springer, New York (1986)

[3] Flajolet, P., Saheb, N.: The complexity of generating an exponentially distributed variate. Journal of Algorithms 7(4), 463–488 (1986)

[4] Hagerup, T., Mehlhorn, K., Munro, J.I.: Maintaining Discrete Probability Distributions Optimally. In: Lingas, A., Carlsson, S., Karlsson, R. (eds.) ICALP 1993. LNCS, vol. 700, pp. 253–264. Springer, Heidelberg (1993)

[5] Knuth, D.E.: The Art of Computer Programming. Seminumerical Algorithms, 3rd edn., vol. 2. Addison-Wesley Publishing Co, Reading (2009)

[6] Knuth, D.E., Yao, A.C.: The complexity of nonuniform random number generation. In: Algorithms and Complexity (Proc. Sympos.), pp. 357–428. Carnegie-Mellon Univ., Pittsburgh (1976)

[7] Matias, Y., Vitter, J.S., Ni, W.C.: Dynamic generation of discrete random variates. Theory of Computing Systems 36(4), 329–358 (2003)

[8] Nacu, Ş., Peres, Y.: Fast simulation of new coins from old. The Annals of Applied Probability 15(1A), 93–115 (2005)

[9] Preparata, F.P., Shamos, M.I.: Computational Geometry. Texts and Monographs in Computer Science. Springer, New York (1985)

[10] Pătraşcu, M.: Webdiarios de motocicleta, sampling a discrete distribution (2011), `infoweekly.blogspot.com/2011/09/sampling-discrete-distribution.html`

[11] Tsai, M.-T., Wang, D.-W., Liau, C.-J., Hsu, T.-s.: Heterogeneous Subset Sampling. In: Thai, M.T., Sahni, S. (eds.) COCOON 2010. LNCS, vol. 6196, pp. 500–509. Springer, Heidelberg (2010)

[12] Tsai, M.T., Wang, D.W., Liau, C.J., Hsu, T.S.: Heterogeneous subset sampling (submitted for publication, 2012)

[13] Vitter, J.S.: Random sampling with a reservoir. ACM Trans. Math. Softw. 11(1), 37–57 (1985)

[14] Walker, A.J.: New fast method for generating discrete random numbers with arbitrary distributions. Electronic Letters 10, 127–128 (1974)

[15] Yao, A.C.: Context-free grammars and random number generation. In: Combinatorial algorithms on words (Maratea, 1984), vol. 12, pp. 357–361. Springer (1985)

Approximation Algorithms for Online Weighted Rank Function Maximization under Matroid Constraints

Niv Buchbinder[1,*], Joseph (Seffi) Naor[2,**], R. Ravi[3,***], and Mohit Singh[4]

[1] Computer Science Dept., Open University of Israel
niv.buchbinder@gmail.com
[2] Computer Science Dept., Technion, Haifa, Israel
naor@cs.technion.ac.il
[3] Tepper School of Business, Carnegie Mellon University, Pittsburgh, PA
ravi@cmu.edu
[4] Microsoft Research, Redmond and School of Computer Science,
McGill University, Montreal, Quebec, Canada
mohitsinghr@gmail.com

Abstract. Consider the following online version of the submodular maximization problem under a matroid constraint. We are given a set of elements over which a matroid is defined. The goal is to incrementally choose a subset that remains independent in the matroid over time. At each time, a new weighted rank function of a different matroid (one per time) over the same elements is presented; the algorithm can add a few elements to the incrementally constructed set, and reaps a reward equal to the value of the new weighted rank function on the current set. The goal of the algorithm as it builds this independent set online is to maximize the sum of these (weighted rank) rewards. As in regular online analysis, we compare the rewards of our online algorithm to that of an offline optimum, namely a single independent set of the matroid that maximizes the sum of the weighted rank rewards that arrive over time. This problem is a natural extension of two well-studied streams of earlier work: the first is on online set cover algorithms (in particular for the max coverage version) while the second is on approximately maximizing submodular functions under a matroid constraint.

In this paper, we present the first randomized online algorithms for this problem with poly-logarithmic competitive ratio. To do this, we employ the LP formulation of a scaled reward version of the problem. Then we extend a weighted-majority type update rule along with uncrossing properties of tight sets in the matroid polytope to find an approximately optimal fractional LP solution. We use the fractional solution values as probabilities for a online randomized rounding algorithm. To show that our rounding produces a sufficiently large reward independent set, we prove and use new covering properties for randomly rounded fractional solutions in the matroid polytope that may be of independent interest.

* Supported in part by ISF grant 954/11 and by BSF grant 2010426.
** Supported in part by the Google Inter-university center for Electronic Markets and Auctions, by ISF grant 954/11, and by BSF grant 2010426.
*** Supported in part by NSF award CCF-1143998.

A. Czumaj et al. (Eds.): ICALP 2012, Part I, LNCS 7391, pp. 145–156, 2012.

1   Introduction

Making decisions in the face of uncertainty is the fundamental problem addressed by online computation [5]. In many planning scenarios, a planner must decide on the evolution of features to a product without knowing the evolution of the demand for these features from future users. Moreover, any features initially included must be retained for backward compatibility, and hence leads to an online optimization problem: given a set of features, the planner must phase the addition of the features, so as to maximize the value perceived by a user at the time of arrival. Typically, users have diminishing returns for additional features, so it is natural to represent their utility as a submodular function of the features that are present (or added) when they arrive. Furthermore, the set of features that are thus monotonically added, are typically required to obey some design constraints. The simplest are of the form that partition the features into classes and there is a restriction on the number of features that can be deployed in each class. A slight extension specifies a hierarchy over these classes and there are individual bounds over the number of features that can be chosen from each class. We capture these, as well as other much more general restrictions on the set of deployed features, via the constraint that the chosen features form an independent set of a *matroid*. Thus, our problem is to monotonically construct an independent set of features (from a matroid over the features) online, so as to maximize the sum of submodular function values (users) arriving over time and evaluated on the set of features that have been constructed so far.

This class of online optimization problems generalizes some early work of Awerbuch et al. [2]. They considered a set-cover instance, in which the restriction is to choose at most k sets with the goal of maximizing the coverage of the elements as they arrive over time. In this setting there is no gain from an element which is covered later than its arrival time. This is precisely the online maximization version of the well-studied maximum coverage problem. Even this special case of our problem already abstracts problems in investment planning, strategic planning, and video-on-demand scheduling.

1.1   Problem Setting, Main Result and Techniques

In our setting, we are given a universe of elements E, $|E| = m$, and a matroid $\mathcal{M} = (E, \mathcal{I}(\mathcal{M}))$ whose independent sets characterize the limitations on which sets of elements can be chosen. At every time step i, $1 \leq i \leq n$, a client arrives with a non-negative monotone submodular function $f_i : 2^E \to \mathbb{Z}_+$ representing her welfare function. The objective is to maintain a monotonically increasing set $F \in \mathcal{I}(\mathcal{M})$ over time; that is, the set F_{i-1} of elements (at time $i-1$) can only be augmented to F_i after seeing f_i at time step i. The welfare of client i is then $f_i(F_i)$, and our objective is to maximize $\sum_{i=1}^{n} f_i(F_i)$. We compare our performance to the offline optimum $\max_{O \in \mathcal{I}(\mathcal{M})} \sum_{i=1}^{n} f_i(O)$.

We are concerned with the case that each of the submodular functions f_i is a weighted rank function of a matroid $\mathcal{N}_i$[1], i.e., $f_i(S) = \max_{I \subseteq S, I \in \mathcal{I}(\mathcal{N}_i)} \sum_{e \in I} w_{i,e}$

---

[1] Matroid $\mathcal{N}_i$ is defined on the same set of elements as $\mathcal{M}$.

where $w_i : E \to \mathbb{R}_+$ is an arbitrary weight function. This class of submodular functions is very broad and includes all the examples discussed above; Furthermore, we believe it captures the difficulty of general submodular functions even though we have not yet been able to extend our results to the general case. Nevertheless, there are submodular functions which are not weighted rank functions of a matroid, for example, multi-set coverage function [8].

Theorem 1. *There exists a randomized polynomial time algorithm which is* $O\left(\log^2 n \log m \log f_{ratio}\right)$*-competitive, for the online submodular function maximization problem under a matroid constraint over m elements, when each* f_i, $1 \le i \le n$*, is a weighted rank function of a matroid and* $f_{ratio} = 2\frac{\max_{i,e} f_i(\{e\})}{\min_{i,e | f_i(\{e\}) \neq 0} f_i(\{e\})}$*. In other words, the algorithm maintains monotonically increasing independent sets* $F_i \in \mathcal{I}(\mathcal{M})$ *such that*

$$E\left[\sum_{i=1}^n f_i(F_i)\right] \ge \Omega\left(\frac{1}{\log^2 n \log m \log f_{ratio}}\right) \cdot \max_{O \in \mathcal{I}(\mathcal{M})} \sum_{i=1}^n f_i(O).$$

Our result should be contrasted with the lower bound proved in [2][2].

Lemma 1. *Any randomized algorithm for the submodular maximization problem under a matroid constraint is* $\Omega(\log n \log(m/r))$*-competitive, where r is the rank of the matroid. This lower bound holds even for uniform matroids and when all* f_i *are unweighted rank functions.*

We note that the $O(\log m)$ factor in our analysis can be slightly improved to an $O(\log(m/r))$ factor with a more careful analysis. A lower bound of $\Omega(\log f_{ratio})$ also follows even when the functions f_i are linear (see, for example, [6]).

Main Techniques. To prove our results, we combine techniques from online computation and combinatorial optimization. The first step is to formulate an integer linear programming formulation for the problem. Unfortunately, the natural linear program is not well-suited for the online version of the problem. Thus, we formulate a different linear program in which we add an extra constraint that each element e *contributes* roughly the same value to the objective of the optimal solution. While this may not be true in general, we show that an approximate optimal solution satisfies this requirement.

We note that the online setting we study is quite different from the online packing framework studied by [6] and leads to new technical challenges. In particular, there are two obstacles in applying the primal-dual techniques in [6] to our setting. First, the linear formulation we obtain (which is natural for our problem) is not a strict packing LP and contains negative variables (see Section 3). Second, the number of packing constraints is exponential, and hence the techniques of [6] would give a linear competitive factor rather than a polylogarithmic one. Nevertheless, we present in Section 3 an online algorithm which

---

[2] The lower bounds in [2] even apply to a special case of a uniform matroid and very restricted submodular functions.

gives a fractional solution to the linear program having a large objective value. One of the crucial ingredients is the *uncrossing* property of tight sets for any feasible point in the matroid polytope.

To obtain an integral solution, we perform in Section 4 a natural randomized rounding procedure to select fractionally chosen elements. But, we have to be careful to maintain that the selected elements continue to form an independent set. The main challenge in the analysis is to tie the performance of the randomized algorithm to the performance of the fractional algorithm. As a technical tool in our proof, we show in Lemma 9 that randomly rounding a fractional solution in the matroid polytope gives a set which can be covered by $O(\log n)$ independent sets with high probability. This lemma may be of independent interest and similar in flavor to the results of Karger [15] who proved a similar result for packing bases in the randomly rounded solution.

Some of the proofs are excluded here due to space considerations and appear in the full version of the paper [7].

1.2   Related Results

Maximizing monotone submodular function under matroid constraints has been a well studied problem and even many special cases have been studied widely (see survey by Goundan and Schulz [14]). Fisher, Nemhauser and Wolsey [13] gave a $(1 - \frac{1}{e})$-approximation for a uniform matroid and showed that the greedy algorithm gives a $\frac{1}{2}$-approximation. This was improved by Calinescu *at al* [8] and Vondrák [21] who gave a $(1 - \frac{1}{e})$-approximation for the general problem. They also introduced the *multi-linear extension* of a submodular function and used pipage rounding introduced by Ageev and Sviridenko [1]. The facility location problem was introduced by Cornuejols et al. [10] and was the impetus behind studying the general submodular function maximization problem subject to matroid constraints. The submodular welfare problem can be cast as a submodular maximization problem subject to a matroid constraint (the reduction appears in Fisher *et al.* [13]), and the problem has been extensively studied [19,17,18,16]. The result of Vondrák [21] implies a $(1 - \frac{1}{e})$-approximation for the problem. Despite the restricted setting of our benefit functions, we note that recent work in welfare maximization in combinatorial auctions [11] has focused on precisely the case when the valuations are matroid rank sums (MRS) that we consider in our model.

A special case of our online problem was studied by Awerbuch et al. [2]. They studied an online variant of the max-coverage problem, where given n sets covering m elements, the elements arrive one at a time, and the goal is to pick up to k sets online to maximize coverage. They obtained a randomized algorithm which is $O(\log n \log(m/k))$-competitive for the problem and proved that this is optimal in their setting. Our results generalize both the requirement on the cardinality of the chosen sets to arbitrary matroid constraints, and the coverage functions of the arriving elements to monotone submodular functions that are weighted rank functions of matroids.

Another closely related problem with a different model of uncertainty was studied by Babaioff et al. [3]. They studied a setting in which elements of a matroid arrive in an online fashion and the goal is to construct an independent set which is competitive with the maximum weight independent set. They considered the random permutation model which is a non-adversarial setting, and obtained an $O(\log k)$-competitive algorithm for general matroids, where k is the rank of the matroid, and constant competitive ratio for several interesting special cases. Recently, Bateni et al. [4] studied the same model where the objective function is a submodular function (rather than linear).

Chawla et al. [9] study Bayesian optimal mechanism design to maximize expected revenue for a seller while allocating items to agents who draw their values for the items from a known distribution. Their development of agent-specific posted price mechanisms when the agents arrive in order, and the items allocated must obey matroid feasibility constraints, is similar to our setting. In particular, we use the ideas about certain ordering of matroid elements (Lemma 7 in their paper) in the proof that our randomized rounding algorithm give sufficient profit.

2 Preliminaries

Given a set E, a function $f : 2^E \to \mathbb{R}_+$ is called *submodular* if for all sets $A, B \subseteq E$, $f(A) + f(B) \geq f(A \cap B) + f(A \cup B)$. Given set E and a collection $\mathcal{I} \subseteq 2^E$, $\mathcal{M} = (E, \mathcal{I}(\mathcal{M}))$ is a *matroid* if (i) for all $A \in \mathcal{I}$ and $B \subset A$ implies that $B \in \mathcal{I}$ and (ii) for all $A, B \in \mathcal{I}$ and $|A| > |B|$ then there exists $a \in A \setminus B$ such that $B \cup \{a\} \in \mathcal{I}$. Sets in $\mathcal{I}$ are called *independent sets* of the matroid $\mathcal{M}$. The rank function $r : 2^E \to R^+$ of matroid $\mathcal{M}$ is defined as $r(S) = \max_{T \in \mathcal{I}: T \subseteq S} |T|$. A basic property of matroids is the fact that the rank function of any matroid is submodular.

We also work with weighted rank functions of a matroid, defined as $f(S) = \max_{I \subseteq S, I \in \mathcal{I}(\mathcal{M})} \sum_{e \in I} w_e$ for some weight function $w : 2^E \to \mathbb{R}_+$. Given any matroid $\mathcal{M}$, we define the matroid polytope to be the convex hull of independent sets $P(\mathcal{M}) = conv\{\mathbf{1}_I : I \in \mathcal{I}\} \subseteq \mathbb{R}^{|E|}$. Edmonds [12] showed that $P(\mathcal{M}) = \{x \geq 0 : x(S) \leq r(S) \ \forall S \subseteq E\}$. We also use the following fact about fractional points in the matroid polytope (The proof follows from standard uncrossing arguments. See, e.g., Schrijver [20], Chapter 40).

Fact 2. *Given a matroid $\mathcal{M} = (E, \mathcal{I}(\mathcal{M}))$ with rank function r and feasible point $x \in P(\mathcal{M})$, let $\tau = \{S \subseteq E : x(S) = r(S)\}$. Then, τ is closed under intersection and union and there is a single maximal set in τ.*

3 Linear Program and the Fractional Algorithm

We now give a linear program for the online submodular function maximization problem and show how to construct a feasible fractional solution online which is $O(\log m \log n \log f_{ratio})$-competitive. Before we give the main theorem, we first

$$LP_1: \quad \max \ \textstyle\sum_{i=1}^{n} \sum_{e \in E} z_{i,e}$$

s.t.

$$\forall S \subseteq E \ \ \textstyle\sum_{e \in S} x_e \leq r(S) \tag{1}$$

$$\forall 1 \leq i \leq n, S \subseteq E \ \textstyle\sum_{e \in S} z_{i,e} \leq r_i(S) \tag{2}$$

$$\forall 1 \leq i \leq n, e \in E \quad z_{i,e} \leq x_e \tag{3}$$

$$\forall 1 \leq i \leq n, e \in E \quad z_{i,e}, x_e \geq 0$$

Fig. 1. LP for maximizing a sum of (unweighted) rank functions subject to matroid constraint

formulate a natural LP. Let $O \subseteq E$ denote the optimal solution having value $\sum_{i=1}^{n} f_i(O)$. Since each f_i is the weighted rank function of matroid $\mathcal{N}_i$, we have that $f_i(O) = w_i(O_i) = \sum_{e \in O_i} w_{i,e}$ where $O \supseteq O_i \in \mathcal{I}(\mathcal{N}_i)$. For the sake of simplicity, we assume that $w_{i,e} = 1$ (In the full version we show that this assumption can be removed with an additional loss of an $O(\log f_{ratio})$ factor in the competitive ratio). Observe that in this case, $f_i(S) = r_i(S)$, where r_i is the rank function of matroid $\mathcal{N}_i$ for any set $S \subseteq E$.

We next formulate a linear program where x_e is the indicator variable for whether $e \in O$ and $z_{i,e}$ is the indicator variable for whether $e \in O_i$. Since $O \in \mathcal{I}(\mathcal{M})$ and $O_i \in \mathcal{I}(\mathcal{N}_i)$, we have that $x \in P(\mathcal{M})$ and $z_i \in P(\mathcal{N}_i)$ as represented by constraints (1) and constraints (2), respectively in Figure 1.

We prove the following theorem.

Theorem 3. *There exists a polynomial time algorithm $\mathcal{A}$ that constructs a feasible fractional solution (x, z) online to LP_1 which is $O(\log n \log m)$-competitive. That is, algorithm $\mathcal{A}$ maintains a monotonically increasing solution (x, z) such that $\sum_{i=1}^{n} \sum_{e \in E} z_{i,e} = \Omega(\frac{\sum_{i=1}^{n} f_i(O)}{\log n \log m})$ where O is an optimal integral solution.*

To prove Theorem 3, instead of working with the natural linear program LP_1, we formulate a different linear program. The new linear program is indexed by an integer α and places the constraints that each $e \in O$ occurs in $[\frac{\alpha}{2}, \alpha]$ different O_i's as represented by constraints (7) and (8). The parameter α will be defined later.

The next lemma, whose proof is omitted, shows that if we pick $O(\log n)$ different values of α then the sum of the integer solutions to the linear programs $LP_2(\alpha)$ perform as well as the optimal solution.[3]

Lemma 2. *Let OPT denote the value of an optimal integral solution to linear program LP_1 and let OPT_α denote the value of an optimal integral solution to the linear program $LP_2(\alpha)$ for each $\alpha \in \{1, 2, 4, \ldots, 2^{\lceil \log n \rceil}\}$. Then $OPT \leq \sum_{\alpha \in \{1,2,4,\ldots,2^{\lceil \log n \rceil}\}} OPT_\alpha$.*

Using the above lemma, a simple averaging argument shows that for some guess α, the optimal integral solution to $LP_2(\alpha)$ is within a $\log n$ factor of the

---

[3] We assume that the algorithm knows the value of n. In the full version of the paper we show how to deal with an unknown n losing an additional small factor.

$$LP_2(\alpha): \quad \max \ \sum_{i=1}^{n} \sum_{e \in E} z_{i,e}$$

s.t.

$$\forall S \subseteq E \ \sum_{e \in S} x_e \leq r(S) \tag{4}$$

$$\forall 1 \leq i \leq n, S \subseteq E \ \sum_{e \in S} z_{i,e} \leq r_i(S) \tag{5}$$

$$\forall 1 \leq i \leq n, e \in E \quad z_{i,e} \leq x_e \tag{6}$$

$$\forall e \in E \ \sum_{i=1}^{n} z_{i,e} \leq \alpha x_e \tag{7}$$

$$\forall e \in E \ \sum_{i=1}^{n} z_{i,e} \geq \frac{\alpha x_e}{2} \tag{8}$$

$$\forall 1 \leq i \leq n, e \in E \quad z_{i,e}, x_e \geq 0$$

Fig. 2. A restricted LP for the submodular function maximization subject to matroid constraint

optimal integral solution to LP_1. Hence, we construct an algorithm which first guesses α and then constructs an approximate fractional solution to $LP_2(\alpha)$.

3.1   Online Algorithm for a Fractional LP Solution

Given a fractional solution x, a set $S \subseteq E$ tight is called *tight* if $x(S) = r(S)$.

Guessing Algorithm:

– Guess a value $\alpha \in_R \{1, 2, 4 \ldots, n\}$.
– Run AlgG with value α.

AlgG:

– Initialize $x_e \leftarrow 1/m^2$ (where $m = |E|$), set $z_{i,e} = 0$ for each i, e.
– When function f_i arrives, order the elements arbitrarily.
– For each element e in order:
– If $\forall S | e \in S$, $x(S) < r(S)$ and $z_i(S) < r_i(S) - 1/2$:

$$x_e \leftarrow \min\left\{ x_e \cdot \exp\left(\frac{8 \log m}{\alpha}\right), \ \min_{S | e \in S}\{r(S) - x(S \setminus \{e\})\} \right\} \tag{9}$$

$$z_{i,e} \leftarrow x_e/2 \tag{10}$$

Using an independence oracle for each of the matroids $\mathcal{N}_i$, the above conditions can be checked in polynomial time by a reduction to submodular function minimization (See Schrijver [20], Chapter 40) and therefore the running time of the algorithm is polynomial. Note that the fractional algorithm is carefully designed. For example, it is very reasonable to greedily update the value of $z_{i,e}$ even when the value x_e is not updated by the algorithm(of course, ensuring that $z_i \in P(\mathcal{N}_i)$). While such an algorithm does give the required guarantee on the performance of the fractional solution, it is not clear how to round such a solution to an integral solution. In particular, our algorithm for finding a fractional solution is tailored so as to allow us to use the values later on as rounding probabilities in a randomized algorithm.

Before we continue, we define some helpful notation regarding the online algorithm. Let $x_{i,e}(\alpha)$ be the value of the variable x_e after the arrival of user i for some guess α. Let $\Delta x_{i,e}(\alpha)$ be the change in the value of x_e when user i arrives. Let $x_e(\alpha)$ be the value of x_e at the end of the execution. Similarly, let $z_{i,e}(\alpha)$ be the value of $z_{i,e}$ at the end of the execution. We start with the following lemma that follows from the update rule (9). The proof is omitted.

Lemma 3. *For any element $e \in E$, and guess α,*

$$\sum_{i=1}^{n} z_{i,e}(\alpha) \geq \frac{\alpha}{48 \log m} \left(x_e(\alpha) - \frac{1}{m^2} \right),$$ (11)

where $x_e(\alpha)$ is the value at the end of the execution of AlgG.

Next we prove that the solution produced by AlgG is almost feasible with respect to an optimal solution to $LP_2(\alpha)$.

Lemma 4 (Feasibility Lemma). *Let $(x(\alpha), z(\alpha))$ be the fractional solution generated by AlgG at the end of the sequence. Then, it satisfies all constraints of $LP_2(\alpha)$ except constraints (8).*

Proof. We prove that the solution is feasible.
Matroid constraints (4). Clearly, the algorithm never violates the matroid constraints by choosing the minimum of the two terms in (9).
Constraints (5) and constraints (6). $z_{i,e} \leftarrow x_{i,e}(\alpha)/2 \leq x_e(\alpha)/2$, thus constraints (6) hold. Finally, the algorithm only updates $z_{i,e}$ if for all $S|e \in S$, $z_i(S) < r_i(S) - 1/2$. Since by the above observations $z_{i,e} \leq x_e(\alpha)/2 \leq 1/2$, we never violate constraints (5) after the update.
Constraints (7). This constraint follows since

$$\sum_{i=1}^{n} z_{i,e} = \sum_{i:\Delta x_{i,e}>0} x_{i,e}(\alpha)/2 \leq x_e(\alpha) |\{i : \Delta x_{i,e} > 0\}|$$

However, after α augmentations, $x_e(\alpha) \geq \frac{1}{m^2} \exp\left(\frac{8\log m}{\alpha} \cdot \alpha\right) > 1$. Thus, x_e must be in a tight set and so by design we never update x_e and any $z_{i,e}$.

In order to evaluate the performance of the algorithm we first show that the *size* of the solution returned by the algorithm is large as compared to the optimal integral solution. Later in Lemma 6, we relate the objective value of the solution to its size. This lemma uses crucially the properties of the matroid. The proof is omitted.

Lemma 5 (Large Fractional Size). *Let $(x^*(\alpha), z^*(\alpha))$ be an optimal integral solution to $LP_2(\alpha)$. Let $(x(\alpha), z(\alpha))$ be the fractional solution generated by AlgG at the end of the sequence. Then, we have $\sum_{e \in E} x_e(\alpha) \geq \frac{1}{16} \sum_{e \in E} x_e^*(\alpha)$.*

Finally, we prove a lemma bounding the performance of the algorithm.

Lemma 6. *For any guess value α, the algorithm maintains a fractional solution to $LP_2(\alpha)$ such that:*

$$\sum_{e \in E} \sum_{i=1}^{n} z_{i,e}(\alpha) = \Omega\left(\frac{OPT_\alpha}{\log m}\right),$$

where OPT_α is objective of an optimal integral solution to $LP_2(\alpha)$.

Proof. Let (x^*, z^*) denote the optimal integral solution to $LP_2(\alpha)$. If $x_e^* = 0$ for each e, then the lemma follows immediately. We have the following

$$\sum_{e \in E} \sum_{i=1}^{n} z_{i,e}(\alpha) \geq \frac{\alpha}{48 \log m} \sum_{e \in E} \left(x_e(\alpha) - \frac{1}{m^2}\right) \quad (Lemma\ 3)$$

$$\geq \frac{\alpha}{48 \log m} \sum_{e \in E} \left(\frac{x_e^*(\alpha)}{16} - \frac{1}{m^2}\right) \quad (Lemma\ 5)$$

$$= \Omega\left(\frac{1}{\log m} \sum_{e \in E} \sum_{i=1}^{n} z_{i,e}^*(\alpha)\right)$$

where the last equality follows since in $LP_2(\alpha)$ for each element $\sum_{i=1}^{n} z_{i,e}^*(\alpha) \leq \alpha x_e^*$ and $\sum_{e \in E} x_e^* \geq 1$. This completes the proof of Lemma.

Finally, we get our main theorem.

Theorem 4. *The online algorithm for the fractional LP solution (of LP_1) is $O(\log m \log n)$-competitive.*

Proof. The proof follows by combining Lemma (4), Lemma (6), Lemma (2) and the observation that there are $O(\log n)$ possible values of α, where each is guessed with probability $\Omega(1/\log n)$.

4   Randomized Rounding Algorithm

In this section we present a randomized algorithm for the unweighted problem which is $O(\log^2 n \log m)$-competitive when each submodular function f_i is a rank function of a matroid. The algorithm is based on the fractional solution designed in Section 3. Although our rounding scheme is extremely simple, the proof of its correctness involves carefully matching the performance of the rounding algorithm with the performance of the fractional algorithm. Indeed, here the fact that $LP_2(\alpha)$ has extra constraints not present in LP_1 is used very crucially.

Theorem 5. *The expected profit of the randomized algorithm is $\Omega\left(\frac{OPT}{\log m \log^2 n}\right)$.*

The randomized algorithm follows the following simple rounding procedure.

Matroid Randomized Rounding Algorithm:

- $F \leftarrow \emptyset$.
- Guess the value $\alpha \in_R \{1, 2, 4 \ldots, n\}$.
- Run AlgG with value α.
  - Whenever x_e increases by Δx_e, if $F \cup \{e\} \in \mathcal{I}(\mathcal{M})$ then $F \leftarrow F \cup \{e\}$ with probability $\frac{\Delta x_e}{4}$.

In order to prove our main theorem, we prove several crucial lemmas. The main idea is to tie the performance of the randomized algorithm to the performance of the fractional solution that is generated. In the process we lose a factor of $O(\log n)$. We first introduce some notation. All of the following notation is with respect to the execution of the online algorithm for a fixed value of α and we omit it from the notation. Let F_i denote the solution formed by the randomized algorithm at the end of iteration i and let F denote the final solution returned by the algorithm. Let Y_e^i denote the indicator random variable of the event that element e has been selected *till* iteration i. Let ΔY_e^i denote the indicator random variable that element e is selected in iteration i. Let $y_e^i = Pr[Y_e^i = 1]$ and $\Delta y_e^i = Pr[\Delta Y_e^i = 1]$. Finally, let y_e denote the probability element e is in the solution at the end of the execution. Recall that $x_{i,e}$ denotes the value of the variable x_e in the fractional solution after iteration i and let x_e denote the fractional value of element e at the end of the execution of the fractional algorithm, and let $\Delta x_{i,e}$ be the change in the value of e in iteration i.

Since the algorithm tosses a coin for element e in iteration i with probability $\Delta x_{i,e}/4$, therefore the probability that an element e is included in the solution till iteration i is at most $x_{i,e}/4$. Our first lemma states that the expected number of elements chosen by the algorithm is at least half that amount in expectation and is comparable to the total size of the fractional solution. Thus, Lemma 7 plays the role of Lemma 5 in the analysis of the randomized algorithm. The proof is omitted.

Lemma 7. *Let F be the solution returned by the randomized rounding algorithm, then $E[|F|] = \sum_{e \in E} y_e \geq \frac{\sum_{e \in E} x_e}{8}$.*

Our second lemma relates the change in the probability an element is chosen to the change in the fractional solution. This lemma shows that a crucial property of the exponential update rule for the fractional solution is also satisfied by the integral solution. The proof is omitted.

Lemma 8. *For each element e and iteration i, $\frac{\Delta y_e^i}{y_e^i} \leq \frac{\Delta x_{i,e}}{x_{i,e}} \leq \frac{24 \log m}{\alpha}$.*

We next prove a general lemma regarding randomized rounding in any matroid polytope. The proof of the lemma utilizes a lemma proved in Chawla et al. [9] and it is omitted.

Lemma 9. *Given a matroid $\mathcal{N} = (E, \mathcal{I})$ and a solution z such that for all $S \subseteq E$, $z(S) \leq r(S)/2$, construct a set F by including in $e \in F$ with probability*

z_e for each $e \in E$ independently. Then, with high probability $(1 - \frac{1}{m^2 n^2})$, F can be covered by $O(\log m + \log n)$ independent sets where $m = |N|$.

We now prove a relation between the profit accrued by the algorithm at iteration i, denoted by the random variable $r_i(F_i)$, and the events that a particular set of elements are chosen in the solution. For any i, Let H_i denote the set of elements such that $z_{i,e} > 0$. Note that $z_{i,e} > 0$ if and only if $\Delta x_{i,e} > 0$.

Lemma 10. $\sum_{i=1}^{n} E[r_i(F_i)] \geq \frac{1}{c \log n} \sum_{i=1}^{n} \sum_{e \in H_i} y_e^i$, where c is some constant.

Now we have all the ingredients for proving Theorem 5.

Proof of Theorem 5: We prove that the expected profit of the algorithm with guess α is at least $\Omega\left(\frac{OPT_\alpha}{\log m \log n}\right)$. Since each α is guessed with probability $1/\log n$, and the value of OPT is the sum over all values α, we get the desired value. The expected profit of the algorithm when we guess α is at least.

$$\sum_{i=1}^{n} E[f_i(F_i)] \geq \frac{1}{c \log n} \sum_{i=1}^{n} \sum_{e \in H_i} y_e^i \qquad (Lemma\ 10)$$

$$\geq \sum_{i=1}^{n} \sum_{e \in H_i} \frac{\alpha}{c' \log m \log n} \Delta y_e^i \qquad (Lemma\ 8)$$

$$= \sum_{e \in E} \frac{\alpha}{c' \log m \log n} y_e \qquad (\sum_{i:e \in H_i} \Delta y_e^i = y_e)$$

$$\geq \sum_{e \in E} \frac{\alpha}{8 c' \log m \log n} x_e \qquad (Lemma\ 7)$$

$$= \Omega\left(\frac{\alpha \cdot n_\alpha}{\log m \log n}\right) = \Omega\left(\frac{OPT_\alpha}{\log m \log n}\right). \ (Lemma\ 5)$$

$\square$

References

1. Ageev, A.A., Sviridenko, M.: Pipage rounding: A new method of constructing algorithms with proven performance guarantee. J. Comb. Optim. 8(3), 307–328 (2004)
2. Awerbuch, B., Azar, Y., Fiat, A., Leighton, T.: Making commitments in the face of uncertainty: how to pick a winner almost every time (extended abstract). In: STOC 1996: Proceedings of the Twenty-Eighth Annual ACM Symposium on Theory of Computing, pp. 519–530 (1996)
3. Babaioff, M., Immorlica, N., Kleinberg, R.: Matroids, secretary problems, and online mechanisms. In: ACM-SIAM Symposium on Discrete Algorithms, pp. 434–443 (2007)
4. Bateni, M., Hajiaghayi, M., Zadimoghaddam, M.: Submodular Secretary Problem and Extensions. In: Serna, M., Shaltiel, R., Jansen, K., Rolim, J. (eds.) APPROX and RANDOM 2010. LNCS, vol. 6302, pp. 39–52. Springer, Heidelberg (2010)
5. Borodin, A., El-Yaniv, R.: Online computation and competitive analysis. Cambridge University Press (1998)
6. Buchbinder, N., Naor, J.: The design of competitive online algorithms via a primal-dual approach. Foundations and Trends in Theoretical Computer Science 3(2-3), 93–263 (2009)
7. Buchbinder, N., Naor, J. (Seffi)., Ravi, R., Singh, M.: Approximation Algorithms for Online Weighted Rank Function Maximization under Matroid Constraints (2012), http://arxiv.org/abs/1205.1477

8. Calinescu, G., Chekuri, C., Pál, M., Vondrák, J.: Maximizing a Submodular Set Function Subject to a Matroid Constraint (Extended Abstract). In: Fischetti, M., Williamson, D.P. (eds.) IPCO 2007. LNCS, vol. 4513, pp. 182–196. Springer, Heidelberg (2007)
9. Chawla, S., Hartline, J.D., Malec, D.L., Sivan, B.: Multi-parameter mechanism design and sequential posted pricing. In: ACM Symposium on Theory of Computing, pp. 311–320 (2010)
10. Cornuejols, G., Fisher, M.L., Nemhauser, G.L.: Location of bank accounts to optimize float: An analytic study of exact and approximate algorithms. Management Science 23(8), 789–810 (1977)
11. Dughmi, S., Roughgarden, T., Yan, Q.: From convex optimization to randomized mechanisms: toward optimal combinatorial auctions. In: ACM Symposium on Theory of Computing, pp. 149–158 (2011)
12. Edmonds, J.: Submodular functions, matroids, and certain polyhedra. In: Proceedings of the Calgary International Conference on Combinatorial Structures and their Application, pp. 69–87. Gordon and Breach, New York (1969)
13. Fisher, M.L., Nemhauser, G.L., Wolsey, L.A.: An analysis of approximations for maximizing submodular set functions - part ii. Mathematical Programming 14, 265–294 (1978)
14. Goundan, P.R., Schulz, A.S.: Revisiting the greedy approach to submodular set function maximization (January 2009) (preprint)
15. Karger, D.R.: Random sampling and greedy sparsification for matroid optimization problems. Mathematical Programming 82, 99–116 (1998)
16. Khot, S., Lipton, R.J., Markakis, E., Mehta, A.: Inapproximability results for combinatorial auctions with submodular utility functions. Algorithmica 52(1), 3–18 (2008)
17. Lehmann, B., Lehmann, D.J., Nisan, N.: Combinatorial auctions with decreasing marginal utilities. In: ACM Conference on Electronic Commerce, pp. 18–28 (2001)
18. Mirrokni, V.S., Schapira, M., Vondrák, J.: Tight information-theoretic lower bounds for welfare maximization in combinatorial auctions. In: ACM Conference on Electronic Commerce, pp. 70–77 (2008)
19. Nemhauser, G.L., Wolsey, L.A.: Best Algorithms for Approximating the Maximum of a Submodular Set Function. Mathematics of Operations Research 3(3), 177–188 (1978)
20. Schrijver, A.: Combinatorial optimization - polyhedra and efficiency. Springer (2005)
21. Vondrak, J.: Optimal approximation for the submodular welfare problem in the value oracle model. In: STOC 2008: Proceedings of the 40th Annual ACM Symposium on Theory of Computing, pp. 67–74 (2008)

Improved LP-Rounding Approximation Algorithm for k-level Uncapacitated Facility Location

Jaroslaw Byrka and Bartosz Rybicki

Institute of Computer Science, University of Wroclaw
Joliot-Curie 15, PL-50-383 Wrocław
jby@ii.uni.wroc.pl, rybicki.bartek@gmail.com

Abstract. We study the k-level uncapacitated facility location problem, where clients need to be connected with paths crossing open facilities of k types (levels). In this paper we give an approximation algorithm that for any constant k, in polynomial time, delivers solutions of cost at most α_k times OPT, where α_k is an increasing function of k, with $\lim_{k \to \infty} \alpha_k = 3$.

Our algorithm rounds a fractional solution to an extended LP formulation of the problem. The rounding builds upon the technique of iteratively rounding fractional solutions on trees (Garg, Konjevod, and Ravi SODA'98) originally used for the group Steiner tree problem.

We improve the approximation ratio for k-UFL for all $k \geq 3$, in particular we obtain the ratio equal 2.02, 2.14, and 2.24 for $k = 3, 4$, and 5.

1 Introduction

In k-level facility location problem we have a set C of clients and a set $F = \bigcup_{i=1}^{k} F_i$ of facilities (locations to potentially open a facility). Facilities are of k different types (levels), e.g., for $k = 3$ one may think of these facilities as shops, warehouses and factories. Each set F_i contains facilities on level i. Each facility i has cost of opening it f_i and for each $i, j \in C \cup F$ there is distance $c_{i,j} \geq 0$ which satisfies the triangle inequality. The task is to connect each client to an open facility at each level, i.e., for each client j it needs to be connected with a path $p_j = (j, i_1, i_2, \cdots, i_{k-1}, i_k)$, where i_l is an open facility at level l. We aim at minimizing the total cost of opening facilities (at all levels) plus the total connection cost, i.e., the sum of the lengths of clients paths.

1.1 Related Work and Our Results

The studied k-level UFL, generalizes the standard 1-level UFL, for which Guha and Khuller [8] showed a 1.463-hardness of approximation. This was recently improved by Krishnaswamy and Sviridenko [9] who showed 1.539-hardness for two levels ($k = 2$) and 1.61-hardness for general k, which demonstrates that multilevel facility location is strictly harder to approximate than the single level variant for which the current best known approximation ratio is 1.488 by Li [10].

A. Czumaj et al. (Eds.): ICALP 2012, Part I, LNCS 7391, pp. 157–169, 2012.
© Springer-Verlag Berlin Heidelberg 2012

The first constant factor approximation algorithm for $k = 2$ is due to Shmoys, Tardos, and Aardal [2], who gave a 3.16-apx. algorithm. For general k, the first constant factor approximation algorithm was the 3-apx. algorithm by Aardal, Chudak, and Shmoys [1].

As it was naturally expected that the problem is easier for smaller number of levels, Ageev, Ye, and Zhang [3] gave an algorithm which reduces an instance of the k-level problem into a pair of instances of the $(k - 1)$-level problem and of the single level problem. By this reduction they obtained 2.43-apx. for $k = 2$ and 2.85-apx. for $k = 3$. This was later improved by Zhang [12], who got 1.77-apx for $k = 2$, 2.53-apx [1] for $k = 3$, and 2.81-apx for $k = 4$. Byrka and Aardal [4] have then improved the approximation ratio for $k = 3$ to 2.492.

Zhang [12] predicted the existence of an algorithm that for any fixed k has approximation ratio strictly smaller than 3. In this paper we give such an algorithm, which is a natural generalization of LP-rounding algorithms for single level UFL. Our new LP-rounding algorithm improves the currently best known approximation ratio for k-level UFL for any $k > 2$. The ratios we obtain for $k \leq 10$ are summarized in the following table.

k	1	2	3	4	5	6	7	8	9	10
previous best	1.49	1.77	2.50	2.81	3	3	3	3	3	3
our alg. (no scaling)	1.74	2.07	2.26	2.38	2.47	2.53	2.59	2.63	2.66	2.69
our alg. (with scaling)	1.58	1.85	2.02	2.14	2.24	2.31	2.37	2.42	2.46	2.50

In this paper we describe the simpler variant (with no scaling) in full detail. The application of the scaling and filtering techniques from UFL is straightforward but technical. It turns out that the analysis analogous to the one in [5] gives best approximation when applied to the version of our algorithm with scaling. In Section 5, we briefly discuss the application of scaling to our algorithm.

1.2   The Main Idea behind Our Algorithm

The 3-approximation algorithm of Aardal, Chudak, and Shmoys, rounds a fractional solution to the standard path LP-relaxation of the studied problem by clustering clients around so-called cluster centers. Each cluster center gets a direct connection, while all the other clients only get a 3-hop connection via their centers. In the single level UFL problem, Chudak and Shmoys observed that by randomly opening facilities one may obtain an improved algorithm using the fact that for each client, with at least some fixed probability, he gets an open facility within a 1-hop path distance. While in the single level problem independently sampling facilities to open is sufficient, the multilevel variant requires coordinating the process of opening facilities across levels.

The key idea behind our solution relies on an observation that the optimal integral solution has a form of a forest, while the fractional solution to the standard LP-relaxation may not have this structure. We start by modifying

---

[1]   This value of γ deviates slightly from the value 2.51 given in the paper. The original argument contained a minor calculation error.

the instance and hence the LP, so that we enforce the forest structure also for the fractional solution of the relaxation. Having the hierarchical structure of the trees, we then use the technique of Garg, Konjevod, and Ravi [7], to first round the top of the tree, and then only consider the descendant edges if the parent edge is selected. This approach naturally leads to sampling trees (not opening lower level facilities if their parent facilities are closed), but to eventually apply the technique to a location problem, we need to make it compatible with clustering. To this end we must ensure that all cluster centers get a direct 1-hop path service. This we obtain by a specific modification of the rounding algorithm, which ensures opening exactly one direct path for each cluster center, while preserving the necessary randomness for all the other clients. It is only possible because cluster centers do not share top level facilities, and in rounding a single tree we only care about at most one cluster center. In Section 3.2 we propose a token-passing based rounding procedure which has exactly the desired properties.

2   Extended LP Formulation

To describe our new LP we first describe a process of splitting vertices of the input graph into a number of (polynomially many for fixed k) copies of each potential facility location.

Graph Modification. Our idea is to have a graph in which each facility f on level j may only be connected to a single facility on level $j + 1$. Since we do not know a priori to which facility on level $j+1$ facility f is connected in the optimal solution, we will introduce multiple copies of f, one for each possible parent on level $j + 1$.

To be more precise, we let F' denote the original set of facilities, and we construct the new set of facilities denoted by F. Nothing will change for facilities in set F'_k, so $F_k = F'_k$. For each facility $f \in F'_{k-1}$ we have $|F_k|$ facilities each connected with different facility in set F_k. So the cardinality of the set F_{k-1} is equal to $|F_k| \cdot |F'_{k-1}|$. In general: for each $i = 1, 2, \ldots, k - 1$ set F_i has $|F_{i+1}|$ copies of each element in set F'_i and each copy is connected with a different element in the set F_{i+1}, so $|F_i| = |F_{i+1}| \cdot |F'_i|$. Observe that so created copies of facilities at level l are in one to one correspondence with paths $(i_l, i_{l+1}, \ldots, i_k)$ on original facilities on levels $l, l+1, \ldots, k$. We will use such paths on the original facilities as names for the facilities in the extended instance.

The distance between any two copies i^1, i^2 of element i is equal to zero and the cost of opening facility i^1 and i^2 is the same and equal to f_i. If i'_1 is a copy of i_1 and i'_2 is a copy of i_2 then $c_{i'_1 i'_2} = c_{i_1 i_2}$. Distance between copy of facility i and client c is equal to c_{ci}. Set C of clients will stay unchanged.

Connection and Service Cost. P_C is the set of paths (in the above described graph), which start in some client and end in a facility at level k. P_j is the set of facilities at level j in the extended instance, or alternatively the set of paths on

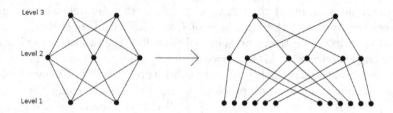

Fig. 1. Figure presets graph before (left part) and after (right part) modification. As you can see vertices in the highest level do not change.

original facilities which start in a facility at level j and end in a facility at level k. Now we define the cost of path p denoted by c_p. For $p = (c, i_1, i_2, \cdots i_k) \in P_C$ we have $c_p = c_{c,i_1} + c_{i_1,i_2} + \ldots + c_{i_{k-1},i_k}$ and for $p = (i_j, i_{j+1}, \cdots i_k) \in P_j$ we have $c_p = f_{i_j}$. So if $p \in P_C$ then c_p is a service cost (i.e., the length of path p), and if $p \in P_j$ then c_p is the cost of opening the first facility on this path. $P = P_C \cup \bigcup_{j=1}^{k} P_j$.

2.1   The LP

$$min \sum_{p \in P} x_p c_p \tag{1}$$

$$\sum_{p \in P_C : j \in p} x_p \geq 1 \quad \forall_{j \in C} \tag{2}$$

$$x_{(i_{l+1},i_{l+2},\ldots i_k)} - x_{(i_l,i_{l+1},\ldots i_k)} \geq 0 \quad \forall_{p=(i_l,i_{l+1},\ldots i_k) \in P_l, l < k} \tag{3}$$

$$x_q - \sum_{p=(j,\ldots i_l,i_{l+1}\ldots i_k) \in P_C} x_p \geq 0 \quad \forall_{j \in C} \forall_{q=(i_l,i_{l+1},\ldots i_k) \in P \setminus P_C} \tag{4}$$

$$x_p \geq 0 \quad \forall_{p \in P} \tag{5}$$

The natural interpretation of the above LP is as follows. Inequality (2) states that each client is assigned to at least one path. Inequality (3) encodes that opening of a lower level facility implies opening of its unique higher level facility. The most complicated inequality (4) for a client $j \in C$ and a facility $i_l \in F_l$, imposes that the opening of i_l must be at least the total usage of it by client j.

Let $p \sqsupseteq q$ denote that p is suffix of q. The dual program to the above LP is:

$$max \sum_{j \in C} v_j \tag{6}$$

$$v_j - \sum_{q \in P_1 : q \sqsupseteq p} y_p - \sum_{q \in P \setminus P_C : q \sqsupseteq p} w_{j,q} \leq c_p \quad \forall_j \forall_{p \in P_C} \tag{7}$$

$$\sum_{q \in P_{k-1}:p \sqsupseteq q} y_q + \sum_{j \in C} w_{j,p} \le c_p \qquad \forall_{p \in P_k} \qquad (8)$$

$$\sum_{q \in P_{l-1}:p \sqsupseteq q} y_p - \sum_{q \in P_{l+1}:q \sqsupseteq p} y_p + \sum_{j \in C} w_{j,p} \le c_p \qquad \forall_{l \in \{1,\dots,k-1\}} \forall_{p \in P_l} \qquad (9)$$

$$v_j, y_p, w_{j,q} \ge 0 \qquad \forall_{j,p,q} \qquad (10)$$

Lemma 1. *Let x and (v, y, w) be optimal solutions to the above primal and dual linear programs, respectively. For any $p \in P_C$, if $x_p > 0$, then $c_p \le v_j$, where j is the client connected by the path c_p.*

Proof. Using (8) we can write following complementary slackness condition:

$$x_p \left(c_p - v_j + \sum_{q \in P_1:q \sqsupseteq p} y_p + \sum_{q \in P \setminus P_C:q \sqsupseteq p} w_{j,q} \right) = 0 \; \forall_{j \in C} \; \forall_{p \in P_C:j \in p}$$

We are interested in p for which $x_p > 0$, so

$$c_p + \sum_{q \in P_1:q \sqsupseteq p} y_p + \sum_{q \in P \setminus P_C:q \sqsupseteq p} w_{j,q} = v_j$$

From (10) we know that each variable in dual program is non-negative, so we can write that $x_p > 0$ implies $c_p \le v_j$. □

Let P^j denote the set of paths beginning in client j, which have positive chance to open. Define $d^{av}(j) = C_j^* = \sum_{p \in P^j} c_p x_p^*$, $d^{max}(j) = \max_{p \in P^j:x_p > 0} c_p \le v_j$, and $F_j^* = v_j^* - C_j^*$. Of course $F^* = \sum_{j \in C} F_j^*$ and $C^* = \sum_{j \in C} C_j^*$.

3   Algorithm

The approximation algorithm that we propose has the following structure:

1: formulate and solve the extended LP (1)-(5);
2: scale up facility opening by $\gamma \ge 1$
   (optional, only to improve the approximation ratio)
3: cluster clients;
4: round facility opening (tree by tree);
5: connect each client j with a closest open connection path $p \in P^j$.

It starts by solving the above described extended LP which, by contrast to the LP used in [1], enforces the fractional solution to have a forest like structure. The step 3. can be interpreted as an adaptation of (by now standard) LP-rounding techniques used for (single level) facility location. Step 4. is an almost direct application of a method from [7]. The final connection step 5. is straightforward, the algorithm simply connects each client via a shortest path of open facilities.

For the clarity of presentation we first only describe the algorithm without scaling which achieves a slightly weaker approximation ratio. We will now present steps 3. and 4. in more detail.

3.1   Clustering

Like in LP-rounding algorithms for UFL, we will partition clients into disjoint clusters and for each cluster center select a single client which will be called the center of this cluster.

Please recall that the solution x^* we obtain by solving LP (1)-(5) gives us (possibly fractional) weights on paths. Paths $p \in P_c$ we interpret as connections from clients to open facilities, while other (shorter) paths from $P \setminus P_c$ encode the (fractional) opening of facilities, which has a structure of a forest (i.e., every facility from a lower level is assigned only to a single facility at a higher level).

Observe that if two client paths $p_1, p_2 \in P_c$ share at least one facility, then they must also end in the same facility at the highest k-th level. For a client j and a k-th level facility i we will say j is fractionally connected to i in x^* if and only if there exists path $p \in P_c$ of the form $(j, \dots, i)$ with $x_p > 0$. Two clients are called neighbors if they are fractionally connected to the same k-th level facility.

The clustering is done as follows. Consider all clients to be initially unclustered. While there remains at least one unclustered client do the following:

- select an unclustered client j that minimizes $d^{av}(j) + d^{max}(j)$,
- create cluster containing j and all its yet unclustered neighbors,
- call j the center of the new cluster;

The procedure is known (see e.g., [6]) to provide good clustering, i.e., no two cluster centers are neighbors and the distance from each client to his cluster center is not too big.

3.2   Randomized Facility Opening

We will now give details on how the algorithm decides which facilities to open. Recall that the facility opening part of the fractional solution can be interpreted as a set of trees rooted in top level facilities and having leaves in level-1 facilities.

We will start by describing how a single tree is rounded. For the clarity of presentation we will change the notation and denote the set of vertices (facilities) of such tree by V, and we will use x_v to denote the fractional opening of $v \in V$ in the initial fractional solution x^*. We will also use y_v to denote how much a cluster center uses v. Please note, that for each of the trees of the fractional solution there is at most one cluster center client j using this tree. If the tree we are currently rounding is not used by any cluster center, then we set all $y_v = 0$. If cluster center j uses the tree, then for each facility v in the tree, we set $y_v = \sum_{p \in P^j : v \in p} x_p$, i.e., y_v is the sum over the connection paths p of j crossing v of the extent the fractional solution uses this path.

Let $p(v)$ denote the parent node of v for all (not-root) nodes, and let $C(v)$ denote the set of children nodes of v for all nodes except on the lowest level. Observe, that x and y satisfy:

1. if v is not a leaf, then $y_v = \sum_{u \in C(v)} y_u$;
2. if v is not the root node, then $x_v \leq x_{p(v)}$;
3. for all $v \in V$ we have $x_v \geq y_v$.

The following procedure will be used to round both the fractional x into an integral $\hat{x}$ and the fractional y into an integral $\hat{y}$. The procedure will visit each node of the tree exactly once. For certain nodes it will be run in a 'with a token' mode and for some others it will be run 'without a token'. It will be initiated in the root node and will recursively execute itself on a subset of lower level nodes. Initially $\hat{x}_v$ and $\hat{y}_v$ are set to 0 for all nodes v, and unless indicated otherwise a node does not have a token.

Procedure $ROUND(v)$

1: **if** v has a token **then**
2:      $\hat{x}_v = 1 \; \hat{y}_v = 1$
3:      **if** v is not a leaf **then**
4:          select a single node $u \in C(v)$
             taking each $i \in C(v)$ with probability equal $\frac{y_i}{y_v}$
5:          give the token to the node u
6:          **for** $i \in C(v)$ **do**
7:              ROUND(i)
8:          **end for**
9:      **end if**
10: **else**
11:     **if** v is the root node **then**
12:         $x_{pred} = 1$
13:     **else**
14:         $x_{pred} = x_{p(v)}$
15:     **end if**
16:     toss a coin that comes up "heads" with probability $\frac{x_v - y_v}{x_{pred} - y_v}$
17:     **if** it is "heads" **then**
18:         $\hat{x}_v = 1$
19:         **if** v is not a leaf **then**
20:             **for** $i \in C(v)$ **do**
21:                 ROUND(i)
22:             **end for**
23:         **end if**
24:     **end if**
25: **end if**

Now we briefly describe what the algorithm does. Suppose that we are in node v which is not a leaf. If v has a token then we set $\hat{x}_v = \hat{y}_v = 1$, choose one son (each son i has probability $\frac{y_i}{y_u}$) and give him a token. Make recursive call on each son. If v doesn't have a token then with probability $\frac{x_v - y_v}{x_{pred} - y_v}$ (x_{pred} is 1 if v is a root or $x_{p(v)}$ otherwise) set $\hat{x}_v = 1$ and make recursive call on each son. If v is a leaf we don't choose a son to give him a token and don't make a recursive calls on sons. We execute the above procedure on the root of the tree, possibly assigning the token to the root node just before the execution. Observe, that an execution of the procedure $ROUND(v)$ on a root of the tree brings the token to a single leaf of the tree if and only if it starts with the token at the root node. In case of the token, the $\hat{y}_v$ variables will record the path of the token, and hence will form a single path from the root to a leaf.

Consider a procedure that first with probability y_r gives the token to the root r of the tree and then executes $ROUND(r)$. We will argue that this procedure preserves marginals when used to round x into $\hat{x}$ and y into $\hat{y}$.

Lemma 2. $E[\hat{y}_v] = y_v$ for all $v \in V$.

Proof. By induction on the distance of v from the root r. $E[\hat{y}_r]$ is just the probability that we started with a token in r, hence it is y_r. For a non-root node v, by inductive assumption, his parent node $u = p(v)$ has $E[\hat{y}_u] = y_u$. The probability of $\hat{y}_v = 1$ can be written as:

$$Pr[\hat{y}_v = 1] = Pr[\hat{y}_v = 1|\hat{y}_u = 1] \cdot Pr[\hat{y}_u = 1] + Pr[\hat{y}_v = 1|\hat{y}_u = 0] \cdot Pr[\hat{y}_u = 0]$$
$$= \frac{y_v}{y_u} \cdot y_u + 0 = y_v.$$

$\square$

Lemma 3. $E[\hat{x}_v] = x_v$ for all $v \in V$.

Proof. By Lemma 2 is is now sufficient to show, that $E[\hat{x}_v - \hat{y}_v] = x_v - y_v$ for all $v \in V$. Observe that $\hat{x}_v - \hat{y}_v$ is always either 0 or 1, hence $E[\hat{x}_v - \hat{y}_v] = Pr[\hat{x}_v = 1, \hat{y}_v = 0]$.

The proof is again by induction on the distance of v from the root node r. Clearly, $E[\hat{x}_r - \hat{y}_r] = Pr[\hat{x}_r = 1, \hat{y}_r = 0] = Pr[\hat{x}_r = 1|\hat{y}_r = 0] \cdot Pr[\hat{y}_r = 0] = \frac{x_r - y_r}{1 - y_r} \cdot (1 - y_r) = x_r - y_r$.

For a non-root node v, by inductive assumption, his parent node $u = p(v)$ has $E[\hat{x}_u] = x_u$. Note that $\hat{y}_v = 1$ implies $\hat{x}_u = 1$. Hence, by Lemma 2, $Pr[\hat{x}_u = 1, \hat{y}_v = 0] = x_u - y_v$. The probability of $\hat{x}_v = 1$ and $\hat{y}_v = 0$ can be written as:

$$Pr[\hat{x}_v = 1, \hat{y}_v = 0] = Pr[\hat{x}_v = 1, \hat{x}_u = 1, \hat{y}_v = 0]$$
$$= Pr[\hat{x}_v = 1|\hat{x}_u = 1, \hat{y}_v = 0] \cdot Pr[\hat{x}_u = 1, \hat{y}_v = 0]$$
$$= \frac{x_v - y_v}{x_u - y_v} \cdot (x_u - y_v) = x_v - y_v.$$

$\square$

To round the entire fractional solution we run the above described single tree rounding procedure as follows:

1. For each cluster center j, put a single token on the root node of one of the trees he is using in the fractional solution. Every single tree is selected with probability equal the fractional connection of j to this tree.
2. Execute the $ROUND(.)$ procedure on the root of each tree.

By the construction of the rounding procedure, every single cluster center, since he had placed his token on a tree, will have one of his paths open so that he can directly connect via this path. Moreover, by Lemma 2 the probability of opening a particular connection path $p \in P^j$ for him (as indicated by variables $\hat{y}$) is exactly equal the weight x_p the fractional solution assigns to this path. Hence, his expected connection cost is exactly his fractional cost.

To bound the expected connection cost of the other (non-center) clients is slightly more involved and will be discussed in the following section.

4   Analysis

Let us first comment on the running time of the algorithm. The algorithm first solves a linear program of size $O(n^k)$, where n is the maximal number of facilities on a single level. For fixed k it is of polynomial size, hence may be directly solved by the ellipsoid algorithm. The rounding of facility openings is by traversing trees whose total size is again bounded by $O(n^k)$. Finally each client can try each of his at most $O(n^k)$ possible connecting paths and see which of them is the closest open one.

Every client j will find an open connecting path to connect with, since he is a part of a cluster, and the client j' who is the center of this cluster certainly has a good open connecting path. Client j may simply use (the facility part of) the path of cluster center j', which by the triangle inequality will cost him at most the distance $c_{j,j'}$ more than it costs j'.

In fact a slightly stronger bound on the expected length of the connection path of j is easy to derive. We use the following bound, which is analogous to the Chudak ans Shmoys [6] argument for UFL.

Lemma 4. *For a non central client $j \in C$, if all paths from P^j are closed, then the expected connection cost of client j is*

$$E[C_j] \leq 2d^{max}(j) + d^{av}(j).$$

Again like in the work of Chudak ans Shmoys [6], the crux of our improvement lies in the fact that with a certain probability the quite expensive 3-hop path guaranteed by the above lemma will not be necessary, because j will happen to have a shorter direct connection. The main part of the analysis which will now follow is to evaluate the probability of this lucky event.

We will use the following technical lemma .

Lemma 5. *For $c, d > 0$, $\sum_i x_i = c$ and $\forall_i\, x_i \geq 0$ we have*

$$\prod_{i=1}^{n}(1 - x_i + x_i d) \leq (1 - \frac{c}{n} + \frac{cd}{n})^n.$$

Suppose a (non-center) client is connected with a flow of value z to a tree in the fractional solution. Suppose further that this flow saturates all the fractional openings on this tree, then the following function $f_k(z)$ gives a lower bound on the probability that at least one path of this tree will be open as a result of the rounding routine. Function $f_k(z)$ is defined recursively. For $k = 1$ it is just equal to fractional opening, i.e., $f_1(z) = z$. For $k \geq 2$ it is $f_k(z) = z \cdot min_z(1 - \prod_{i=1}^{n}(1 - f_{k-1}(\frac{z_i}{z})))$[**]. It is a product of the probability of opening the root node, and the (recursively bounded) probability that at least one of the subtrees has an open path, conditioned on the root being open.

The following lemma displays the structure of $f_k(.)$.

Lemma 6. *Inequality $f_k(x) \geq x \cdot (1 - c)$ implies $f_{k+1}(x) \geq x \cdot (1 - e^{c-1})$.*

Proof. Note that $f_1(x) \geq x$ and $f_2(x) \geq x(1 - \frac{1}{e})$, so base follows. Now we show induction step. Suppose that $f_k(x) \geq x \cdot (1 - c)$ then $f_{k+1}(x) = x \cdot (1 - max_x \prod_{i=1}^{n}(1 - f_k(\frac{x_i}{x}))) \geq x \cdot (1 - max_x \prod_{i=1}^{n}(1 - \frac{x_i}{x} + \frac{x_i}{x} \cdot c)) = x(1 - (1 - \frac{1}{n} + \frac{c}{n})^n) \mapsto x(1 - e^{c-1})$. Last equality base on Lemma 5, but we have to put $\overline{x_i} = \frac{x_i}{x}$. $\square$

Since a single client j may not use the full opening (capacity) of the tree he is using, a more direct and accurate estimate of his probability of getting a path would be the following function $f_k(x, z)$ which depends on both the opening of the root node x and the fractional usage of the tree by j given as z.

$$f_k(x, z) = \begin{cases} x & \text{when } k = 1, \\ x \cdot min_{x,z}(1 - (\prod_{i=1}^{n}(1 - f_{k-1}(\frac{x_i}{x}, \frac{z_i}{x})))) & \text{otherwise.} \end{cases}$$

Fortunately enough, we may inductively prove the following lemma, which states that the worst case for our analysis is when the tree capacity is saturated by the connectivity flow of a client.

Lemma 7. *If $1 \geq x \geq z \geq 0$ then $f_k(x, z) \geq f_k(z)$*

Consider now a single client j who is fractionally connected to a number of trees with a total weight of his connection paths equal γ (you may think he sends a total flow of value γ through these trees, from leaves to roots). Now, to bound the probability of at least one of these paths getting opened by the rounding procedure, we introduce function $F_k(\gamma)$ defined as follows. $F_k(\gamma) = 1 - max_\gamma \prod_{i=1}^{n}(1 - f_k(x_i))$. That function is one minus the biggest chance that no tree gives route from root to leaf, using the previously defined $f_k(.)$ function to express the success probability on a single tree.

Now we can give an analogue of Lemma 6 but for $F_k(\gamma)$.

Lemma 8. *Inequality $F_k(\gamma) \geq 1 - e^{(c-1)\gamma}$ implies $F_{k+1}(\gamma) \geq 1 - e^{(e^{c-1}-1)\gamma}$.*

[**] For notational convenience we use max_x (min_x) to denote $max_{x_1+...+x_n=x, x_i>0}$ ($min_{x_1+...+x_n=x, x_i>0}$).

Proof. Suppose that $f_k(x) \geq x(1-c)$. Note that $F_k(\gamma) = 1 - max_\gamma \prod_{i=1}^n (1 - f_k(x_i)) \geq 1 - max_\gamma \prod_{i=1}^n (1 - x_i + x_i c)) = 1 - (1 - \frac{\gamma}{n} + \frac{\gamma}{n}c)^n \mapsto 1 - e^{(c-1)\gamma}$.
(Last equality base on Lemma 5). Leading observation is that in the last equality there is no requirement for positive constant c - we can replace it with any other positive constant and equality will be still true. Using Lemma 6 we know that $f_{k+1}(x) \geq x(1 - e^{c-1})$. The only difference in the way we evaluate $F_{k+1}(\gamma)$ is the replacement of constant c by other constant e^{c-1}, so the equality for $F_k(\gamma)$ implies the equality for $F_{k+1}(\gamma)$, and hence the lemma holds.   □

We are now ready to combine our arguments into a bound on the expected total cost of the algorithm.

Theorem 1. *Expected total cost of the algorithm is at most* $(3 - 2F_k(1))OPT$.

Proof. Note first that by Lemma 3, the probability of opening of each single facility equals its fractional opening, and hence the expected facility opening cost is exactly the fractional opening cost F^*.

Consider client $j \in C$ which is a cluster center. He randomly chooses one of the paths from set P^j. Expected connection cost for client j is $E[C_j] = d^{av}(j) = \sum_{p \in P^j} c_p x_p = C_j^*$. Suppose now $j \in C$ is not a cluster center. As discussed above, the chance that at least one path from P^j is open is not less than $F_k(1)$. Suppose that at least one path from P^j is open. Each path from that set has proportional probability to open, so the expected length of the chosen path is equal to $d^{av}(j)$. If there is no open paths in set P^j, client j will use path $p' \in P^{j'}$ which was chosen by his cluster center $j' \in C$, but j has to pay extra for the distance to the center. In this case, by Lemma 4 we have $E[C_j] \leq 2d^{max}(j) + d^{av}(j)$.

The total cost of the algorithm can be bounded by the following expression:

$$F^* + \sum_{j \in D}(F_k(1)C_j^* + (1 - F_k(1))(2d^{max}(j) + d^{av}(j))) \leq$$

$$F^* + \sum_{j \in D}(F_k(1)C_j^* + (1 - F_k(1))(2(C_j^* + F_j^*) + C_j^*)) =$$

$$(3 - 2F_k(1))(F^* + C^*)$$

Note that $F_k(1) > 0$ for each k, so expected total cost of algorithm is strictly less than three times the optimum cost.   □

5   How to Apply Scaling

By means of scaling up facility opening variables before rounding, just like in the case of 1-level UFL, we gain on the connectivity cost in two ways. First of all, the probability for j of connecting to one of his fractional facilities via a shorter 1-hop path increases, decreasing the usage of the longer backup paths. The second effect is that in clustering clients may ignore the furthest of their fractionally used facilities, hence filtering the solution and reducing the lengths of the 3-hop connections. In fact, if the scaling factor is sufficient, which is the case for our application, we eventually do not need the dual program to upper bound the length of a fractional connection with a dual variable. All this is well studied for UFL (see, e.g., [5]), but would require a few pages to present in detail.

All we need in order to use the techniques from UFL is to give bounds on the probability of opening a connection to specific groups of facilities as a function of the scaling parameter γ. So the probability of connecting j to one of his close facilities (total opening equal 1 after scaling) will be at least $F_k(1)$. The probability of connecting j to either a close or a distant facility (total opening equal γ after scaling) will be at least $F_k(\gamma)$. The probability of using the backup 3-hop path via the cluster center will be at most $1 - F_k(\gamma)$. To obtain the approximation ratios claimed in the table in Section 1.1, it remains to plug in these numbers to the analysis in [5], and for each value of k find the optimal value for the scaling parameter γ. A complete description of the algorithm with scaling will appear in the full version of this paper.

Acknowledgements. We thank Karen Aardal for insightful discussions on the k-UFL problem. We also thank Thomas Rothvoss for teaching us the "rounding on trees" technique form [7]. Research supported by FNP HOMING PLUS/2010-1/3 grant.

References

1. Aardal, K., Chudak, F., Shmoys, D.: A 3-Approximation Algorithm for the k-Level Uncapacitated Facility Location Problem. Inf. Process. Lett. 72(5-6), 161–167 (1999)
2. Aardal, K., Tardos, E., Shmoys, D.: Approximation Algorithms for Facility Location Problems (Extended Abstract). In: STOC 1997, pp. 265–274 (1997)
3. Ageev, A., Ye, Y., Zhang, J.: Improved Combinatorial Approximation Algorithms for the k-Level Facility Location Problem. SIAM J. Discrete Math. 18(1), 207–217 (2004)
4. Byrka, J., Aardal, K.: An Optimal Bifactor Approximation Algorithm for the Metric Uncapacitated Facility Location Problem. SIAM J. Comput. 39(6), 2212–2231 (2010)
5. Byrka, J., Ghodsi, M., Srinivasan, A.: LP-rounding algorithms for facility-location problems CoRR, abs/1007.3611 (2010)
6. Chudak, F., Shmoys, D.: Improved Approximation Algorithms for the Uncapacitated Facility Location Problem. SIAM J. Comput. 33(1), 1–25 (2003)
7. Garg, N., Konjevod, G., Ravi, R.: A polylogarithmic approximation algorithm for the group Steiner tree problem. In: SODA 1998, pp. 253–259 (1998)
8. Guha, S., Khuller, S.: Greedy strikes back: Improved facility location algorithms
9. Krishnaswamy, R., Sviridenko, M.: Inapproximability of the multi-level uncapacitated facility location problem. SODA 2012 31(1), 718–734 (1999); Journal of Algorithms 31(1), 228–248 (1999)

10. Li, S.: A 1.488 Approximation Algorithm for the Uncapacitated Facility Location Problem. In: Aceto, L., Henzinger, M., Sgall, J. (eds.) ICALP 2011, Part II. LNCS, vol. 6756, pp. 77–88. Springer, Heidelberg (2011)
11. Sviridenko, M.: An Improved Approximation Algorithm for the Metric Uncapacitated Facility Location Problem. In: Cook, W.J., Schulz, A.S. (eds.) IPCO 2002. LNCS, vol. 2337, pp. 240–257. Springer, Heidelberg (2002)
12. Zhang, J.: Approximating the two-level facility location problem via a quasi-greedy approach. In: SODA 2004, pp. 808–817 (2004)

Testing Coverage Functions

Deeparnab Chakrabarty[1] and Zhiyi Huang[2,*]

[1] Microsoft Research, Bangalore
dechakr@microsoft.com
[2] University of Pennsylvania
hzhiyi@cis.upenn.edu

Abstract. A *coverage function* f over a ground set $[m]$ is associated with a universe U of weighted elements and m sets $A_1, \ldots, A_m \subseteq U$, and for any $T \subseteq [m]$, $f(T)$ is defined as the total weight of the elements in the union $\cup_{j \in T} A_j$. Coverage functions are an important special case of submodular functions, and arise in many applications, for instance as a class of utility functions of agents in combinatorial auctions.

Set functions such as coverage functions often lack succinct representations, and in algorithmic applications, an access to a value oracle is assumed. In this paper, we ask whether one can test if a given oracle is that of a coverage function or not. We demonstrate an algorithm which makes $O(m|U|)$ queries to an oracle of a coverage function and completely reconstructs it. This gives a polytime tester for *succinct* coverage functions for which $|U|$ is polynomially bounded in m. In contrast, we demonstrate a set function which is "far" from coverage, but requires $2^{\tilde{\Theta}(m)}$ queries to distinguish it from the class of coverage functions.

1 Introduction

Submodular set functions are set functions $f : 2^{[m]} \mapsto \mathbb{R}$ defined over a ground set $[m]$ which satisfy the property: $f(S \cap T) + f(S \cup T) \leq f(S) + f(T)$. These are arguably the most extensively studied set functions, and arise in various fields such as combinatorial optimization, computer science, electrical engineering, economics, etc. In this paper, we focus on a particular class of submodular functions, called coverage functions.

Coverage functions arise out of families of sets over a universe. Given a universe U and sets $A_1, \cdots, A_m \subseteq U$, the coverage of a collection of sets $T \subseteq [m]$ is the number of elements in the union $\bigcup_{j \in T} A_j$. More generally, each element $i \in U$ has a weight $w_i \geq 0$, inducing the function $f : 2^{[m]} \mapsto \mathbb{R}_{\geq 0}$:

$$\forall T \subseteq [m] : f(T) = w \left(\bigcup_{i \in T} A_i \right)$$

with the usual notation of $w(S) := \sum_{i \in S} w_i$. A set function is called a *coverage function* iff f is induced by a set system as described above. In the definition above, the size of the universe U of the inducing set system can be arbitrarily large. We call a coverage function *succinct* if $|U|$ is bounded by a fixed polynomial in m.

---

* The second author is supported in part by the ONR MURI Grant N000140710907.

A. Czumaj et al. (Eds.): ICALP 2012, Part I, LNCS 7391, pp. 170–181, 2012.

Coverage functions arise in many applications (plant location [5], machine learning [10]); an important one being that in combinatorial auctions [11,3]. Utilities of agents are often modeled as coverage functions – agents are thought to have certain requirements (the universe U) and the items being auctioned (the A_i's) fulfill certain subsets of these. Many auction mechanisms take advantage of the specific property of these utility functions; a notable one is the recent work of Dughmi, Roughgarden and Yan [6] who give $O(1)$-approximate truthful mechanisms when utilities of agents are coverage. (Such a result is not expected for general submodular functions [7].)

In general, set functions have exponentially large (in m) description, and algorithmic applications often assume access to a value oracle which returns $f(T)$ on being queried a subset $T \subseteq [m]$. Efficient algorithms making only polynomially many queries to this oracle, exploit the coverage property of the underlying function to ensure correctness. This raises the question we address in this paper:

Can one test, in polynomial time, whether the oracle at hand is indeed that of a coverage (or a 'close' to coverage) function?

It is easy to see that the parenthesized qualification in the above question is necessary. Using property testing parlance [9,8], we say a function is ε-far from coverage if it needs to be modified in ε-fraction of the points to make it a coverage function.

Our first result (Theorem 2) is a reconstruction algorithm which makes $O(m|U|)$ queries to a value oracle of a *true* coverage function and reconstructs the coverage function, that is, deduces the underlying set system $(U; A_1, \ldots, A_m)$ and weights of the elements in U. Such an algorithm can be used distinguish coverage functions with those which are ε-far from being coverage (Corollary 1). In particular, for succinct coverage functions, the answer to the above question is yes.

Our second result illustrates why the testing question may have a negative answer for general coverage functions. We show that certifying 'non-coverageness' requires exponentially many queries. To explain this, let us first consider a certificate of a non-submodularity. By definition, for any non-submodular function f, there must exist sets $S, T, S \cup T$, and $S \cap T$ such that $f(S) + f(T) < f(S \cup T) + f(S \cap T)$. Therefore, four queries (albeit non-deterministic) to a value oracle of f certifies non-submodularity of f. In contrast, we exhibit non-coverage functions for which *any* certificate needs to query the function at exponentially many sets (Corollary 2).

In fact, from just the definition of coverage functions it is not *a priori* clear what a certificate for coverageness should be. In Section 1.1, we show that a particular linear transformation (the W-transform) of set functions can be used: we show a function f is coverage iff all its W-coefficients are non-negative. This motivates a new notion of distance to coverageness which we call W-distance: a set function has W-distance ε if at least an ε-fraction of the W-coefficients are negative. This notion of distance captures the density of certificates to non-coverageness. Our lower bound results show that testing coverage functions against this notion of distance is infeasible: we construct set functions with W-distance at least $1 - e^{-\Theta(m)}$ which require $2^{\Theta(m)}$ queries to distinguish them from coverage functions (Corollary 3).

How is the usual notion of distance to coverage related to the W-distance? We show in Section 4 that there are functions which are far in one notion but close in the other. Nonetheless, we believe that the functions we construct for our lower bounds also have large (usual) distance to coverage functions. We prove this assuming a conjecture on the number of roots of certain multilinear polynomials; we also provide some partial evidence for this conjecture.

Related Work. The work most relevant to, and indeed which inspired this paper, is that by Seshadhri and Vondrák [16], where the authors address the question of testing general submodular set functions. The authors focus on a particular simple testing algorithm, the "square tester", which samples a random set R, $i, j \notin R$ and checks whether or not $f(R, i, j) + f(R) \leq f(R, i) + f(R, j)$. [16] show that $\varepsilon^{-\tilde{O}(\sqrt{m})}$ random samples are sufficient to distinguish submodular functions from those ε-far from submodularity, and furthermore, at least $\varepsilon^{-4.8}$ samples are necessary. Apart from the obvious problem of closing this rather large gap, the authors of [16] suggest tackling special, well-motivated cases of submodularity. In fact, the question of testing coverage functions was specifically raised by Seshadhri in [15] (attributed to N. Nisan).

It is instructive to compare our results with that of [16]. Firstly, although coverage functions are a special case of submodular functions, the sub-exponential time tester of [16] *does not* imply a tester for coverage functions. This is because a function might be submodular but far from coverage; in fact, the function f^* in our lower bound result is submodular. Given our result that there are no small certificates of non-coverageness, we believe testing coverageness is *harder* than testing submodularity.

A recent relevant paper is that of Badanidiyuru et. al. [1]. Among other results, [1] shows that any coverage function f can be arbitrarily well approximated by a succinct coverage function. More precisely, if f is defined via $(U; A_1, \ldots, A_m)$ with weights w, then for any $\varepsilon > 0$, there exists another coverage function f' defined via $(U'; A'_1, \ldots, A'_m)$ with weights w' such that $f'(T)$ is within $(1 \pm \varepsilon)f(T)$ such that $|U'| = \text{poly}(m, 1/\varepsilon)$. This, in some sense shows that succinct coverage functions capture the essence of coverage functions. Unfortunately, this 'sketch' is found using random sampling on the universe U and it is open whether this can be obtained via polynomially many queries to an oracle for f.

1.1   The W-Transform: Characterizing Coverage Functions

Given a set function $f : 2^{[m]} \mapsto \mathbb{R}_{\geq 0}$, we define the W-transform $w : 2^{[m]} \setminus \emptyset \mapsto \mathbb{R}$ as

$$\forall S \in 2^{[m]} \setminus \emptyset, \qquad w(S) = \sum_{T : S \cup T = [m]} (-1)^{|S \cap T| + 1} f(T) \tag{1}$$

We call the resulting set $\{w(S) : S \subseteq [m]\}$ the W-coefficients of f. The W-coefficients are unique; this follows since the $(2^m - 1) \times (2^m - 1)$ matrix M defined as $M(S, T) = (-1)^{|S \cap T| + 1}$ if $S \cup T = [m]$ and 0 otherwise, is full rank[1]. Inverting we get the *unique* evaluation of f in terms of its W-coefficients.

$$\forall T \subseteq [m], \quad f(T) = \sum_{S \subseteq [m] : S \cap T \neq \emptyset} w(S) \tag{2}$$

---

[1] One can check $M^{-1}(S, T) = 1$ if $S \cap T \neq \emptyset$ and 0 otherwise.

If f is a coverage function induced by the set system $(U; A_1, \cdots, A_m)$, then the function $w(S)$ precisely is the size of $\bigcap_{i \in S} A_i$ and is hence non-negative. This follows from the inclusion-exclusion principle. Indeed the non-negativity of the W-coefficients is a characterization of coverage functions.

Theorem 1. *A set function $f : 2^{[m]} \mapsto \mathbb{R}_{\geq 0}$ is coverage iff all its W-coefficients are non-negative.*

Proof. Suppose that f is a function with all W-coefficients non-negative. Consider a universe U consisting of $\{S : S \subseteq [m]\}$ with weight of element S being $w(S)$, the Sth W-coefficient of f. Given U, for $i = 1 \ldots m$, define $A_i := \{S \subseteq [m] : i \in S\}$. For any $T \subseteq [m]$, $\bigcup_{i \in T} A_i = \{S \subseteq [m] : S \cap T \neq \emptyset\}$. From (2) we get $f(T) = w \left(\bigcup_{i \in T} A_i \right)$ proving that f is a coverage function.

Suppose f is a coverage function. By definition, there exists $(U; A_1, \ldots, A_m)$ with non-negative weights on elements in U such that $f(T) = w \left(\bigcup_{i \in T} A_i \right)$. Each element in $S \in U$ corresponds to a subset of $[m]$ defined as $\{i : S \in A_i\}$. We may assume each element of U corresponds to a unique subset; if more than one elements have the same incidence structure, we may merge them into one element with weight equalling sum of both the weights. This transformation doesn't change the function value and keeps the weights non-negative. Furthermore, we may also assume every subset on $[m]$ is an element of U by giving weights equal to 0; this doesn't change the function value either. In particular, $|U|$ may be assumed to be 2^m. As before, one can check that for any $T \subseteq [m]$, $f(T) = \sum_{S : S \cap T \neq \emptyset} w(S)$. From (2) we get that these are the W-coefficients of f, and are hence non-negative. $\square$

From the second part of the proof above, note that the positive W-coefficients of a coverage function f correspond to the elements in the universe U. Let $\{S : w(S) > 0\}$ be the support of a coverage function f. Note that succinct coverage functions are precisely those with polynomial support size.

One can use Theorem 1 to certify non-coverageness of a function f: one of its W-coefficients $w(S)$ must be negative, and the function values in the summand of (1) certifies it. Observe, however, that this certificate can be exponentially large. In Section 3 we'll show this is inherent in *any* certificate of coverageness. The W-transformation also motivates the following notion of distance to coverage functions.

Definition 1. *The W-distance of a function f from coverage functions is the fraction of its negative W-coefficients.*

Comparison with Fourier Transformation. Readers who are familiar with the analysis of Boolean functions might find (1) similar to the Fourier transformation. Indeed, if we sum over all T in the summation of (1) instead of only over the T s.t. $S \cup T = [m]$, then it becomes the Fourier transformation. However, it is worth pointing out that due to this subtle change, the W-transformation behaves quite differently to the representation by Fourier basis. In particular, unlike the Fourier basis, the basis of the W-transform is not orthonormal with respect to the usual notion of inner product.

2 Reconstructing Succinct Coverage Functions

Given a coverage function f, suppose $\{S_1, \ldots, S_n\}$ is the support of f. That is, these are the sets in the W-transform of f with $w(S_i) > 0$, and all the other sets have weight 0. We now give an algorithm to find these sets and weights using $O(mn)$ queries. As a corollary, we will obtain a polynomial time algorithm for testing succinct coverage functions where $n = \text{poly}(m)$.

The procedure is iterative. The algorithm maintains a partition of $2^{[m]}$ at all times, and for each part in the partition, stores the total weight of the all the sets contained in the part. We start of with the trivial partition containing all sets whose weight is given by $f([m])$. In each iteration, these partitions are refined; for instance, in the first iteration we divide the partition into sets containing a given element i and those that don't contain the element i. The total weights of the first collection can be found by querying $f(\{i\})$. Any time the sum of a part evaluates to 0, we discard it and subdivide it no more[2]. After m iterations, the remaining n parts give the support sets and their weights. To describe formally, we introduce some notation.

Given a vector $\mathbf{x} \in \{0, 1\}^k$ we associate a subset of $[k]$ containing the elements i iff $\mathbf{x}(i) = 1$. At times, we abuse notation and use the vector to imply the subset. Let $\mathcal{F}(\mathbf{x}) := \{S \subseteq [m] : S \cap [k] = \mathbf{x}\}$, that is, subsets of $[m]$ which "match" with the vector $\mathbf{x}$ on the first k elements. Note that $|\mathcal{F}(\mathbf{x})| = 2^{m-k}$, and $\{\mathcal{F}(\mathbf{x}) : \mathbf{x} \in \{0, 1\}^k\}$ is a partition of $2^{[m]}$; if $k = 0$, then $\mathcal{F}(\mathbf{x})$ is the trivial partition consisting of all subsets of $[m]$. Given $\mathbf{x} \in \{0, 1\}^k$, we let $\mathbf{x} \oplus 0$ be the $(k + 1)$ dimensional vector with $\mathbf{x}$ appended with a 0. Similarly, define $\mathbf{x} \oplus 1$. At the kth iteration, the algorithm maintains the partition $\{\mathcal{F}(\mathbf{x}) : \mathbf{x} \in \{0, 1\}^k\}$ and the total weight of subsets in each $\mathcal{F}(\mathbf{x})$. In the subsequent iteration refines each partition $\mathcal{F}(\mathbf{x})$ into $\mathcal{F}(\mathbf{x} \oplus 0)$ and $\mathcal{F}(\mathbf{x} \oplus 1)$. However, if a certain weight of a part of the partition evaluates to 0, then the algorithm does not need to refine that part any further since all the weights of that subset *must* be zero. The algorithm terminates in m iterations making $O(mn)$ queries. We now give the refinement procedure. In what follows, we say a vector $\mathbf{y} \le \mathbf{x}$ if they are of the same dimension and $\mathbf{y}(i) = 1 \Rightarrow \mathbf{x}(i) = 1$. We say $\mathbf{y} < \mathbf{x}$ if $\mathbf{y} \le \mathbf{x}$ and $\mathbf{y} \ne \mathbf{x}$.

Claim. The procedure **Refine** returns the correct weights of the refinement.

Proof. It suffices to show that $\Delta_{\mathbf{x}_i} = w\left(\mathcal{F}(\mathbf{x}_i \oplus 1)\right) = \sum_{S:S\cap[k]=\mathbf{x}_i, k+1 \in S} w(S)$. The RHS equals

$$\sum_{S:S\cap[k] \subseteq \mathbf{x}_i,\ k+1 \in S} w(S) - \sum_{\mathbf{y} < \mathbf{x}_i} \sum_{S\cap[k]=\mathbf{y}, k+1 \in S} w(S). \qquad (3)$$

The first term above equates to

$$\sum_{\substack{S:S\cap[k]\setminus\mathbf{x}_i=\emptyset, \\ k+1 \in S}} w(S) = \sum_{S:S\cap([k]\setminus\mathbf{x}_i \cup k+1) \ne \emptyset} w(S) - \sum_{S:S\cap([k]\setminus\mathbf{x}_i) \ne \emptyset} w(S) = F_i^1 - F_i^0$$

---

[2] Familiar readers will observe the similarity of our algorithm and the Goldreich-Levin algorithm to compute 'large' Fourier coefficients (see, for instance, [13] for an exposition).

Procedure Refine

1: **Input:** $0 \le k \le m$, $\{w(\mathcal{F}(\mathbf{x})) > 0 : \mathbf{x} \in \{0,1\}^k\}$
2: **Output:** $\{w(\mathcal{F}(\mathbf{x} \oplus 0)), w(\mathcal{F}(\mathbf{x} \oplus 1)) : \mathbf{x} \in \{0,1\}^k\}$
3: Order $\{\mathbf{x} : w(\mathcal{F}(\mathbf{x})) > 0\}$ by increasing number of 1's breaking ties arbitrarily. Call the order $\{\mathbf{x}_1, \dots, \mathbf{x}_N\}$;
4: **for** $i = 1 \to N$ **do**
5:     Query $f([k] \setminus \mathbf{x}_i) = F_i^0$ and $f(([k] \setminus \mathbf{x}_i) \cup k + 1) = F_i^1$.
6:     Define $\Delta_{\mathbf{x}_i} := F_i^1 - F_i^0 - \sum_{j<i} \Delta_{\mathbf{x}_j}$.
7:     $w(\mathcal{F}(\mathbf{x}_i \oplus 1)) = \Delta_{\mathbf{x}_i}$; $w(\mathcal{F}(\mathbf{x} \oplus 0)) = w(\mathcal{F}(\mathbf{x}_i)) - \Delta_{\mathbf{x}_i}$
8: **end for**

Procedure Recover Coverage

1: **Input:** Value oracle to coverage function f,
2: **Output:** $\{S_1, \dots, S_n\}$ with $w(S_i) > 0$.
3: Initialize $k = 0$, $\mathbf{x}$ to be the empty vector, and list L to contain $\mathbf{x}$. Let $w(\mathcal{F}(\mathbf{x})) = f([m])$.
4: **for** $k = 1 \to m$ **do**
5:     Run **Refine** on each $\mathbf{x}$ in list L and remove it.
6:     Add $\mathbf{x} \oplus 0$ and $\mathbf{x} \oplus 1$ to L only if the weights evaluate to positive.
7: **end for**
8: For each $\mathbf{x} \in \{0,1\}^m$ in L, return corresponding set and weight calculated by the **Refine** procedure.

Note that the summation $\sum_{S \cap [k] = \mathbf{y}, k+1 \in S} w(S)$ equals 0 if $w(\mathcal{F}(\mathbf{y})) = \sum_{S \cap [k] = \mathbf{y}} w(S)$ equals zero since $w(S) \ge 0$ for all S. Therefore, the second term in (3) is precisely $\sum_{j<i} w(\mathcal{F}(\mathbf{x}_j \oplus 1))$. If $i = 1$, then this is 0; for other i this equates to $\sum_{j<i} \Delta_{\mathbf{x}_j}$ by induction. □

Theorem 2. *Given value oracle access to a coverage function f with positive weight sets $\{S_1, \dots, S_n\}$, the procedure* **Recover Coverage** *returns the correct weights with $O(mn)$ queries to the oracle.*

Proof. Whenever a certain $w(\mathcal{F}(\mathbf{x}))$ evaluates to 0, we can infer that $w(S) = 0$ for all $S \in \mathcal{F}(\mathbf{x})$ since f is a coverage function. It is also clear that the algorithm terminates in m steps since the partition refines to singleton sets. The number of oracle accesses is proportional (twice) to the number of calls to the **Refine** subroutine. The latter is at most mn since in each iteration the number of parts remaining is at most the number of parts remaining in the end. □

Corollary 1. *Given any n, there exists a $O(mn + \epsilon^{-1})$ time tester which will return* YES *for coverage functions having W-support size at most n, and return* NO *with $\Omega(1)$ probability for functions that are ϵ-far from the set of coverage functions with W-support at most n.*

Proof. Run the reconstruction algorithm described above. If we get a set with negative weight, return NO. If we succeed, then if f is truly a coverage function, we have derived the unique weights. We sample $O(\varepsilon^{-1})$ random sets and compare the value of our computed function with that of the oracle; if the function is ε-far from coverage, then we will catch it with probability $O(1)$.   □

Theorem 3. *Reconstructing coverage functions on m elements with W-support size n requires at least $\Omega(mn/\log n)$ probes.*

Proof. Consider the bipartite graphs with m and n vertices on the A and U side. Let the weight be 1 on all vertices in U. Each non-isomorphic (on permutation of the U vertices) maps to a different coverage function over the A side: the neighborhood of a vertex $A_i \in A$ is precisely the elements it contains. Note each such graph corresponds to a way of allocating n identical balls (U-side vertices) into 2^m different bins (different choice of set of adjacent A-side vertices). This number is at least $\binom{2^m+n-1}{n-1} \geq \left(\frac{2^m}{n}\right)^{n-1}$.

Hence, we need at least $\Omega(mn)$ bits of information. Notice that each probe of function value only provides $O(\log n)$ bits of information since the function value is always an integer between 0 and n, we get the lower bound in Theorem 3.   □

3 Testing Coverage Functions Is Hard?

In this section we demonstrate a set function whose W-distance to coverage functions is 'large', but it takes exponentially many queries to distinguish from coverage functions. In particular, the function has W-coefficients $w(S) = -1$ if $|S| > k := k(m)$, and $w(S) = N$ if $|S| \leq k$, where N is a positive integer and $k(m)$ is a growing function of m, which will be precisely determined later. Let this function be called f^*.

Firstly, observe that from (1) it follows that $w(S)$ can be precisely determined by querying the $2^{|S|}$ sets in $\{T : T \cup S = [m]\} = \{\overline{S} \cup X : X \subseteq S\}$. It follows that f^* can be distinguished from coverage using 2^{k+1} queries.

In this section we show an almost tight lower bound: Any tester which makes less than 2^k queries cannot distinguish f^* from a coverage function. Our bound is information theoretic and holds even if the tester has infinite computation power. More precisely, we show that given the value of f^* on a collection of sets $\mathcal{J}$ with $|\mathcal{J}| < 2^k$, there exists a *coverage* function f which has the same values on the sets in $\mathcal{J}$.

Theorem 4. *There exists a coverage function consistent with the queries of f^* on $\mathcal{J}$ if $|\mathcal{J}| < 2^k$.*

Corollary 2. *Any certificate of non-coverageness of f^* must be of size at least 2^k.*

Setting $k(m) = m/4$, we get f^* has W-distance at least $(1 - e^{-\Theta(m)})$, giving us:

Corollary 3. *Any tester distinguishing between coverage functions and functions of W-distance as large as $(1 - e^{-\Theta(m)})$ needs at least $2^{\Theta(m)}$ queries.*

We give a sketch of the proof before diving into the details. Suppose a tester queries the collection $\mathcal{J}$. We first observe that the existence of a coverage function consistent with the queries in $\mathcal{J}$ can be expressed as a set of linear inequalities. Using Farkas' lemma, we get a certificate of the *non-existence* of such a completion. This certificate, at a high level, corresponds to an assignment of values on the m-dimensional hypercube satisfying certain linear constraints. We show that if the parameter N is properly chosen, most of these assignments can be assumed to be 0. In the next step we use this property to show that unless the size of $|\mathcal{J}| \geq 2^k$, all the assignments need to be 0 which contradicts the Farkas linear constraints, thereby proving the existence of the coverage function consistent with $\mathcal{J}$.

3.1 Consistent Coverage Functions and Farkas Lemma

Recall, from Theorem 1, a function $f : 2^{[m]} \mapsto \mathbb{R}_{\geq 0}$ is coverage iff it satisfies

$$\forall S \subseteq [m] : \qquad \sum_{T:S\cup T=[m]} (-1)^{|S\cap T|+1} f(T) \geq 0$$

$$\forall T \subseteq [m] : \qquad f(T) \geq 0$$

Let $\mathcal{J}$ be the collection of sets on which the function f^* has been queried. Define

$$b(S) := \sum_{T\in\mathcal{J}:S\cup T=[m]} (-1)^{|S\cap T|} f^*(T)$$

Therefore, if we can find assignments $f : 2^{[m]} \setminus \mathcal{J} \mapsto \mathbb{R}_{\geq 0}$ satisfying:

$$\forall S \subseteq [m] : \qquad \sum_{T\notin\mathcal{J}:S\cup T=[m]} (-1)^{|S\cap T|+1} f(T) \geq b(S) \qquad (4)$$

$$\forall T \notin \mathcal{J} : \qquad f(T) \geq 0 \qquad (5)$$

we can complete the queries on $\mathcal{J}$ to a coverage function. Applying Farkas' lemma (see for instance [2]), we see that there is *no* feasible solution to (4), (5) if and only if there is a feasible solution $\alpha : 2^{[m]} \mapsto \mathbb{R}_{\geq 0}$ satisfying:

$$\sum_{S\subseteq[m]} \alpha(S)b(S) > 0 \qquad (6)$$

$$\forall T \notin \mathcal{J} : \qquad \sum_{S:S\cup T=[m]} (-1)^{|S\cap T|+1} \alpha(S) \leq 0 \qquad (7)$$

$$\forall S \subseteq [m] : \qquad \alpha(S) \geq 0 \qquad (8)$$

Now we define the parameter N for the function f^*; let N be any integer larger than $(2m)!$. Note that this makes the values doubly exponential, but we are interested in the power of an all powerful tester. In the next lemma we show that one can assume there is a feasible solution to (6), (7), and (8) with half of the $\alpha(S)$'s set to 0.

Lemma 1. *If there exists α satisfying* (6), (7), *and* (8), *then we may assume* $\alpha_S = 0$ *for all S such that $|S| \leq k$.*

Intuitively, what this lemma says is that the constraint (4) for sets of size $\leq k$ should not help in catching the function not being coverage. This is because the true function values satisfies the constraints with huge 'redundancy': $\sum_{T:S\cup T=[m]} (-1)^{|S\cap T|+1} f^*(T) = N \gg 0$. Formally, we can prove the lemma as follows.

Proof. Suppose there is an α satisfying (6), (7), and (8). Then, by scaling we may assume that

$$\sum_{S\subseteq[m]} \alpha(S) = 1 \qquad (9)$$

Equivalently, there is a positive valued solution to the LP $\{\max \sum_{S \subseteq [m]} b(S)\alpha(S) :$ (7), (8), (9)$\}$. Choose α to be a basic feasible optimal solution. Such a solution makes 2^m of the inequalities in (7), (8), and (9) tight, and therefore by Cramer's rule, each of the non-zero $\alpha(S) \geq \frac{1}{(2^m)!}$ since all coefficients are $\{-1, 0, 1\}$.

Now we show that if α is basic feasible and $N > (2^m)!$, then we must have that $\alpha(S) = 0$ for all S such that $|S| \leq k$. We first note that $\forall S \subseteq [m]$:

$$w(S) = \sum_{T:S\cup T=[m]} (-1)^{|S\cap T|+1} f^*(T) = \sum_{T\notin \mathcal{J}:S\cup T=[m]} (-1)^{|S\cap T|+1} f^*(T) - b(S) \ .$$

Therefore, $\sum_{S \subseteq [m]} \alpha(S)b(S) > 0$ and the above equality imply that

$$\sum_{T\notin \mathcal{J}} \sum_{S:S\cup T=[m]} \alpha(S)(-1)^{|S\cap T|+1} f^*(T) - \sum_{S\subseteq [m]} \alpha(S)w_S = \sum_{S\subseteq [m]} \alpha(S)b(S) > 0 \ .$$

But by (7), $\sum_{S\subseteq [m]} \alpha(S)(-1)^{|S\cap T|+1} \leq 0$ for all $T \notin \mathcal{J}$, and $f^*(T) \geq 0$ for all $T \subseteq [m]$. So we have that $\sum_{S\subseteq [m]} \alpha(S)w(S) < 0$. Assume for contradiction that there exists S_0, $|S_0| \leq k$ such that $\alpha_{S_0} \neq 0$. From the earlier discussion we know that $\alpha_{S_0} \geq \frac{1}{(2^m)!}$. Therefore, we have $\sum_{S\subseteq [m]} \alpha(S)w(S) \geq \frac{1}{(2^m)!} N - \sum_{S\subseteq [m]:|S|\leq k} \alpha(S) > 1 - 1 = 0$, a contradiction. The latter inequality follows from (9) and our assumption that $N > (2^m)!$. $\qquad \square$

3.2 Nullity of Farkas Certificate

In the following discussion, we assume without loss of generality $\alpha(S) = 0$ for all S, $|S| \leq k$. We will work with the following linear function of the α's. For a set T, define

$$g(T) := \sum_{S:S\cup T=[m]} (-1)^{|S\cap T|+1} \alpha(S)$$

From (7), we get $g(T) \leq 0$ for all $T \notin \mathcal{J}$. Inverting, we get

$$\alpha(S) = \sum_{T:T\cap S\neq \emptyset} g(T) = G - \sum_{T\subseteq \overline{S}} g(T), \quad \text{where } G := \sum_{T\subseteq [m]} g(T) \quad (10)$$

We now show that if $\alpha(S) = 0$ for all $|S| \leq k$, then $g(T)$ *must* be > 0 for at least 2^k sets T. This will imply $|\mathcal{J}| \geq 2^k$.

Lemma 2. *If $\alpha(S) = 0$ for all $|S| \leq k$, then $g(T) > 0$ for at least 2^k subsets $T \subseteq [m]$.*

Proof. Let S^* be any minimal set with $\alpha(S^*) > 0$. Note that $|S^*| \geq k + 1$. From (10), we get $\hat{G} := \alpha(S^*) = G - \sum_{T\subseteq \overline{S^*}} g(T) > 0$. Consider any $i \in S^*$. By minimality, we have $\alpha(S^* \setminus i) = 0$, giving us

$$0 = G - \sum_{T\subseteq \overline{S^*\setminus i}} g(T) = G - \sum_{T\subseteq \overline{S^*}} g(T) - \sum_{T\subseteq \overline{S^*}} g(T \cup i)$$

Therefore for all $i \in S^*$, $\sum_{T\subseteq \overline{S^*}} g(T \cup i) = \hat{G} > 0$. By induction, we can extend the above calculation to any subset $X \subseteq S^*$,

$$\sum_{T\subseteq \overline{S^*}} g(T \cup X) = (-1)^{|X|+1} \hat{G} \quad (11)$$

Note that the summands in (11) are disjoint for different sets X, and furthermore, whenever $|X|$ is odd, the sum is > 0 implying at least one of the summands must be positive for each odd subset $X \subseteq S^*$. This proves the lemma since $|S^*| = k + 1$.

Proof of (11): Let's denote the sum $\sum_{T \subseteq \overline{S^*}} g(T \cup X)$ as $h(X)$. So $\hat{G} = G - h(\emptyset)$, and by induction, $h(Y) = (-1)^{|Y|+1} \hat{G}$ for every proper subset of X. Now, $\alpha(S^* \setminus X) = 0$ gives us

$$0 = G - \sum_{T \subseteq \overline{S^* \setminus X}} g(T) = G - \sum_{Y \subseteq X} h(Y)$$

Rearranging, $h(X) = G - \sum_{Y \subsetneq X} h(Y) = \hat{G} - \sum_{i=1}^{|X|-1} \binom{|X|}{i}(-1)^{i+1}\hat{G} = (-1)^{|X|+1}\hat{G}$ $\square$

Theorem 4. Suppose there is no consistent completion, implying α's satisfying (6), (7) and (8). By Lemma 1 and Lemma 2, we get that if (7) holds, then $|\mathcal{J}| \geq 2^k$. $\square$

4 W-Distance and Usual Distance

We first note that the two notions are unrelated; in particular, we show two functions each "far" in one notion, but "near" in the other. The proofs of the following two lemmas can be found in the full version of the paper [4].

Lemma 3. *There is a function with W-distance $1 - e^{-\Theta(m)}$ whose distance to coverage is $e^{-\Theta(m)}$.*

Lemma 4. *There is a function with W-distance $O(m^2/2^m)$ whose distance to coverage is $\Omega(1)$.*

Despite the fact that the two notions are incomparable, we argue that the lower bound example of Section 3 is in fact also far from coverage (with proper choice of $k(m)$) in the usual notion of distance, under a reasonable conjecture about the property of multilinear polynomials. Unfortunately, we are unable to prove this conjecture and leave it as an open question.

Conjecture 1. For any m-variate multilinear polynomials $f(x) = \sum_{S \subseteq [m]} \lambda_S \prod_{i \in S} x_S$ with $\lambda_S < 0$ for all $|S| > k$, has at most $O(k2^m/\sqrt{m})$ zeroes on the hypercube $\{0, 1\}^m$.

In fact, we conjecture that the maximum number of zeros is achieved when the $k + 1$ layers of function values in the "middle of the hypercube" are zero, that is, $f(x) = 0$ iff. $(m - k)/2 \leq \|x\|_1 \leq (m + k)/2$. At the end of this section, we present some evidence for this conjecture by giving a proving it for symmetric functions, that is, when $f(x_1, \ldots, x_m) = f(x_{\sigma(1)}, \ldots, x_{\sigma(m)})$ for any permutation σ of $[m]$. We now show that the conjecture implies f^* is far from coverage in the usual notion of distance.

Lemma 5. *Assuming Conjecture 1, with $k(m) = o(\sqrt{m})$, f^* is $1 - o(1)$ far from coverage.*

Remark 1. Theorem 4 implies that f^* requires superpolynomial queries to test as long as we have $k(m) = \omega(\log m)$.

Proof. Consider the coverage function f' that is closest to f^* in the usual notion of distance. Let w', w^* be the W-coefficients of f', f^*. Define the function $\Delta f := f' - f^*$ and let $\Delta w := w' - w^*$. By linearity of W-transformation, we get that Δw are the W-coefficients for Δf. Therefore,

$$\Delta f(T) = \sum_{S: T \cap S \neq \emptyset} \Delta w(S) = \sum_{S \subseteq [m]} \Delta w(S)(1 - \mathbf{1}_{T \cap S = \emptyset}) .$$

Consider the following binary vector representation of $S \subseteq [m]$: $\mathbf{x} \in \{0,1\}^m$ such that $\mathbf{x}_i = 0$ iff. $i \in S$. Using this, the function Δf can be interpreted as $\Delta f(\mathbf{x}) = W - \sum_{S \subseteq [m]} w(S) \prod_{i \in S} x_i$. We are using here the fact that $T \cap S = \emptyset$ is equivalent to $S \subseteq \overline{T}$. By our choice of w^* and the assumption that $w'(S) \geq 0$ for all S, we have $\Delta w(S) \geq 1$ for all $|S| > k$. From Conjecture 1, we get that at most $O(k/\sqrt{m})$-fraction of the function values of Δf are zeroes. So f' is at least $1 - O(k/\sqrt{m})$ far from f^*. The lemma follows since $k = o(\sqrt{m})$. □

Support for Conjecture 1: Proof for Symmetric Functions. Since f is symmetric, each λ_S is equal for sets of the same cardinality. Let λ_j denote the value of λ_S when $|S| = j$. Then f is equivalent to the function $g : [m] \mapsto \mathbb{R}$

$$g(i) = f(\mathbf{x} : \|\mathbf{x}\|_1 = i) = \sum_{j=0}^m \sum_{S:|S|=j} \lambda_j \prod_{i \in S} x_i = \sum_{j=0}^m \lambda_j \binom{i}{j} .$$

By our assumption, $\lambda_j < 0$ for all $j > k$. Hence, all the high order derivatives (at least $k + 1$-th order) of f are negative. Intuitively, since the high order derivatives of g are negative, there are at most $k + 1$ sign-changes of $g(i)$. Therefore, there are at most $k + 1$ different i's such that $g(i) = 0$. This implies the conjecture for symmetric functions.

Acknowledgements. The authors wish to thank C. Seshadhri, Jan Vondrák, Sampath Kannan, Jim Geelen and Mike Saks for very fruitful conversations. DC especially thanks Sesh for illuminating conversations over the past few years, and Mike for asking insightful questions.

References

1. Badanidiyuru, A., Dobzinski, S., Fu, H., Kleinberg, R., Nisan, N., Roughgarden, T.: Sketching valuation functions. In: SODA (2012)
2. Bertsimas, D., Tsitsiklis, J.: Introduction to linear optimization. Athena Scientific Belmont, MA (1997)
3. Blumrosen, L., Nisan, N.: Combinatorial auctions. Algorithmic Game Theory (2007)
4. Chakrabarty, D., Huang, Z.: Testing coverage functions. Arxiv (2012)
5. Cornuejols, G., Fisher, M., Nemhauser, G.: Location of bank accounts to optimize float: An analytic study of exact and approximate algorithms. Management Science, 789–810 (1977)
6. Dughmi, S., Roughgarden, T., Yan, Q.: From convex optimization to randomized mechanisms: toward optimal combinatorial auctions. In: STOC, pp. 149–158. ACM (2011)
7. Dughmi, S., Vondrák, J.: Limitations of randomized mechanisms for combinatorial auctions. In: FOCS (2011)
8. Goldreich, O.: Combinatorial property testing (a survey). Randomization Methods in Algorithm Design 43, 45–59 (1999)

9. Goldreich, O., Goldwasser, S., Ron, D.: Property testing and its connection to learning and approximation. Journal of the ACM (JACM) 45(4), 653–750 (1998)
10. Krause, A., McMahan, H., Guestrin, C., Gupta, A.: Robust submodular observation selection. Journal of Machine Learning Research 9, 2761–2801 (2008)
11. Lehmann, B., Lehmann, D., Nisan, N.: Combinatorial auctions with decreasing marginal utilities. Games and Economic Behavior 55(2), 270–296 (2006)
12. Mahler, K.: Introduction to p-adic numbers and their functions (1973)
13. O'Donnell, R.: Chapter 3.5 highlight: The Goldreich-Levin algorithm. In: Analysis of Boolean Functions (2012), http://analysisofbooleanfunctions.org/
14. Robert, A.: A course in p-adic analysis, vol. 198. Springer (2000)
15. Seshadhri, C.: Open problems 2: Open problems in data streams, property testing, and related topics (2011),
   http://www.cs.umass.edu/~mcgregor/papers/11-openproblems.pdf
16. Seshadhri, C., Vondrák, J.: Is submodularity testable? In: ICS (2011)

Sparse Fault-Tolerant Spanners for Doubling Metrics with Bounded Hop-Diameter or Degree

T.-H. Hubert Chan, Mingfei Li, and Li Ning

The University of Hong Kong

Abstract. We study *fault-tolerant* spanners in doubling metrics. A subgraph H for a metric space X is called a k-vertex-fault-tolerant t-spanner $((k,t)$-VFTS or simply k-VFTS), if for any subset $S \subseteq X$ with $|S| \leq k$, it holds that $d_{H \setminus S}(x,y) \leq t \cdot d(x,y)$, for any pair of $x, y \in X \setminus S$.

For any doubling metric, we give a basic construction of k-VFTS with stretch arbitrarily close to 1 that has optimal $O(kn)$ edges. In addition, we also consider bounded hop-diameter, which is studied in the context of fault-tolerance for the first time even for Euclidean spanners. We provide a construction of k-VFTS with bounded hop-diameter: for $m \geq 2n$, we can reduce the hop-diameter of the above k-VFTS to $O(\alpha(m,n))$ by adding $O(km)$ edges, where α is a functional inverse of the Ackermann's function.

Finally, we construct a fault-tolerant single-sink spanner with bounded maximum degree, and use it to reduce the maximum degree of our basic k-VFTS. As a result, we get a k-VFTS with $O(k^2 n)$ edges and maximum degree $O(k^2)$.

1 Introduction

A metric space (X, d) can be represented by a complete graph $G = (X, E)$, where the edge weight $w(e)$ on an edge $e = \{x, y\}$ is $d(x, y)$. A t-*spanner* of X, is a weighted subgraph $H = (X, E')$ of G that preserves all pairwise distance within a factor of t, i.e., $d_H(x, y) \leq t \cdot d(x, y)$ for all $x, y \in X$. Here, $d_H(x, y)$ denotes the shortest-path distance between x and y in H. The factor t is called the *stretch* of H. A path between x and y in H with length at most $t \cdot d(x, y)$ is called a t-*spanner path*. Spanners have been studied extensively since the mid-eighties (see [2,8,1,16,12,3,4,10] and the references therein; also refer to [15] for an excellent survey).

Spanners are important structures, as they enable approximation of a metric space in a much more economical form. One natural requirement is that spanners should be sparse, ideally with the number of edges being linear in the number of points in the metric space. In addition, for some applications, it might also be required that a spanner should have small maximum degree, or a small *hop-diameter*, i.e., every pair of points x and y should be connected by a t-spanner path with small number of edges.

In many applications of spanners, we want our spanner to be robust to failures, meaning that even when some of the points in the spanner fail, the remaining

A. Czumaj et al. (Eds.): ICALP 2012, Part I, LNCS 7391, pp. 182–193, 2012.

part is still a t-spanner. Formally, given $1 \leq k \leq n - 2$, a spanner H of X is called a k-vertex-fault-tolerant t-spanner ((k, t)-VFTS or simply k-VFTS if the stretch t is clear from context), if for any subset $S \subseteq X$ with $|S| \leq k$, $H \setminus S$ is a t-spanner for $X \setminus S$.

The notion of fault-tolerant spanners was introduced by Levcopoulos et al. [13] in the context of *Euclidean spanners* (the special case when X is a finite subset of low dimensional Euclidean space and $d(x, y) = ||x - y||_2$). They presented an algorithm that constructs Euclidean ($k, 1 + \epsilon$)-VFTS with $O(k^2 n)$ edges. This result was later improved by Lukovszki in [14]. They provided two constructions of ($k, 1 + \epsilon$)-VFTS: one with optimal $O(kn)$ edges, and the other with $O(k^2 n)$ edges and maximum degree $O(k^2)$. There has also been research on the trade-off between maximum degree and weight in fault-tolerant Euclidean spanners [13,7], and fault-tolerant spanners for general graphs [6,9].

In this paper, we study fault-tolerant spanners for *doubling metrics*. The *doubling dimension* of a metric space (X, d), denoted by $\dim(X)$ (or dim when the context is clear), is the smallest value ρ such that every ball in X can be covered by 2^ρ balls of half the radius [11]. A metric space is called *doubling*, if its doubling dimension is bounded by some constant. Doubling dimension is a generalization of Euclidean dimension to arbitrary metric spaces, as the space $\mathbb{R}^T$ equipped with ℓ_p-norm has doubling dimension $\Theta(T)$ [11]. Spanners for doubling metrics have been studied in [12,3,4,10,16].

1.1 Our Results and Techniques

Basic spanner with small number of edges. Our first result is a construction of ($k, 1 + \epsilon$)-VFTS for doubling metrics with $O(kn)$ edges. Note that the size is optimal up to a constant factor [14].

Theorem 1 (($k, 1 + \epsilon$)-**VFTS with** $O(kn)$ **Edges**). *Let (X, d) be a doubling metric with n points and let $0 < \epsilon < \frac{1}{2}$ be a constant. Given $1 \leq k \leq n - 2$, there exists a ($k, 1 + \epsilon$)-VFTS of X with $\epsilon^{-O(\dim)} \cdot kn$ edges.*

Our technique of the basic k-VFTS construction is an extension of that in [3]. Specifically, we give a k-fault-tolerant version of the hierarchical nets and net-trees used in [3], which guarantee that even under the failure of at most k points, for any functioning point x and level i, there exists a net-tree in which there is a path with length at most $O(2^i)$ between x and some level-i net point. We also add cross edges between net points at the same level that are reasonably close. Then, we show that under the failure of at most k points, a ($1 + \epsilon$)-spanner path between x and y can be formed by first climbing from x to some net point x' at an appropriate level i, then going along a cross edge $\{x', y'\}$, and finally going from y' down to y.

The upper bound on the number of edges in the spanner is established in a way similar to [3] by carefully assigning a direction to each edge, and then showing that the out-degree of each point is bounded by $O(k)$. However, we simplify and improve the analysis because in our case cross edges can be added between arbitrarily close points.

Spanners with small hop-diameters. We also consider the hop-diameter, which is studied in the context of fault-tolerance for the first time even for Euclidean spanners. The k-vertex-fault-tolerant hop-diameter is defined as follows.

Definition 1 (k-Vertex-Fault-Tolerant Hop-Diameter). *Let H be a (k,t)-VFTS for the metric space (X, d). The k-vertex-fault-tolerant hop-diameter (or simply hop-diameter) of H is at most D, if for any set of points $S \subseteq X$ with $|S| \leq k$, there exists a t-spanner path in $H \setminus S$ with at most D edges (hops) between every pair of $x, y \in X$.*

We show that by adding a few extra edges to our basic k-VFTS, we can significantly reduce its hop-diameter.

Theorem 2 ($(k, 1 + \epsilon)$-VFTS with Small Hop-Diameter). *Let $m \geq 2n$. We can add $O(km)$ extra edges to the spanner in Theorem 1 to get a $(k, 1 + \epsilon)$-VFTS with hop-diameter at most $O(\alpha(m, n))$, where α is the functional inverse of Ackermann's function.*

The technique of reducing the hop-diameter is similar to that in [4]. Let H be the spanner in Theorem 1. Recall that when the points in S fail, the $(1 + \epsilon)$-spanner path in $H \setminus S$ between any two points of x and y is the concatenation of a path P_1 in some net-tree T_1, the cross edge $\{x', y'\}$ and a path P_2 in some net-tree T_2. We add edges to net-trees to shortcut the paths P_1 and P_2, and hence obtain a spanner with small hop-diameter.

Spanners with bounded degree. We also give a construction of (k,t)-VFTS with bounded maximum degree. This is achieved with a sacrifice of increasing the number of edges. The result matches the state-of-the-art result of the bounded-degree Euclidean $(k, 1 + \epsilon)$-VFTS in [14].

Theorem 3 ($(k, 1 + \epsilon)$-VFTS with Bounded Degree). *Let (X, d) be a doubling metric with n points and let $0 < \epsilon < 1$ be a constant. Given $1 \leq k \leq n - 2$, there exists a $(k, 1 + \epsilon)$-VFTS with $\epsilon^{-O(\mathrm{dim})} \cdot k^2 n$ edges and maximum degree $\epsilon^{-O(\mathrm{dim})} \cdot k^2$.*

In [3], it is shown how to reduce the maximum degree of net-tree based spanners for doubling metrics. This is achieved by replacing some cross edges with inter-level edges. As a result, the end points of a replaced cross edge $\{u, v\}$ are connected by a path $\{u, w_1, w_2, \ldots, w_i, v\}$, with approximately the same length. However, in the context of fault-tolerance, some of the w_j's might fail, and it is unclear how to make this procedure resilient to failures.

However, we note that the degree-reduction techniques similar to those in [1,14] can be applied. Recall that the edges of the k-VFTS in Theorem 1 can be directed such that the out-degree of each point is bounded by $O(k)$. To reduce the in-degrees, for each point x, we replace the star consisting of x and edges going into x with a k-vertex-fault-tolerant single-sink spanner that approximately preserves the distances to x, and has maximum degree $O(k)$.

We show how to construct k-vertex-fault-tolerant single-sink spanners with bounded degree for doubling metrics. The construction for those in Euclidean space is based on Θ-graphs [1,14]. We provide a novel technique called *ring-partition*, which can be seen as a replacement for the Θ-graph in doubling metrics. Given a specific root point from X, the metric space X is partitioned into rings centered at the root with geometrically increasing radii, and each ring is further partitioned into small clusters. From each cluster, we select some *portals*, which are connected to the root by short paths. In addition, points in each cluster are connected to the portals with short paths as well. As a result, every point's distance to the root is approximately preserved.

1.2 Preliminaries

For any positive integer m, we denote $[m] := \{1, 2, \ldots, m\}$.

Throughout this paper, let (X, d) be a metric space with n points, $1 \leq k \leq n-2$ be an integer representing the number of faults allowed, and let $0 < \epsilon < \frac{1}{2}$ be a constant. Without loss of generality, we also assume that the minimum inter-point distance of X is strictly greater than 1. We denote $\Delta := \max_{x,y \in X} d(x, y)$ as the *diameter* of X.

Suppose $r > 0$. The ball of radius r centered at x is $B(x, r) := \{y \in X : d(x, y) \leq r\}$. We say that a cluster $C \subseteq X$ has radius at most r, if there exists $x \in C$ such that $C \subseteq B(x, r)$. Let $r_2 > r_1 > 0$. The ring of inner radius r_1 and outer radius r_2 centered at x is $R(x, r_1, r_2) := B(x, r_2) \setminus B(x, r_1)$.

A set $Y \subseteq X$ is an *r-cover* for X if for any point $x \in X$ there is a point $y \in Y$ such that $d(x, y) \leq r$. A set Y is an *r-packing* if for any pair of distinct points $y, y' \in Y$, it holds that $d(y, y') > r$. We say that a set $Y \subseteq X$ is an *r-net* for X if Y is both an r-cover for X and an r-packing. Note that if X is finite, an r-net can be constructed greedily.

By recursively applying the definition of doubling dimension, we can get the following key proposition [11].

Proposition 1 (Nets Have Small Size). *Let $R \geq 2r > 0$ and let $Y \subseteq X$ be an r-packing contained in a ball of radius R. Then, $|Y| \leq (\frac{R}{r})^{2\dim}$.*

2   Basic Construction of Sparse Fault-Tolerant Spanners

In this section, we extend the $O(n)$-edge spanner construction in [4] and build a k-VFTS with $O(kn)$ edges. We construct $k + 1$ sequences of hierarchical nets and assign each sequence with a distinct "color". Then based on the hierarchy of each color, we extract a net-tree. We show some properties similar to those in [3], and in addition we show that the fault-tolerance property can be established.
Fault-Tolerant Hierarchical Nets. We color each point in X with one of $k+1$ colors and let X_c be the set of points with color c. For each color $c \in [k + 1]$, we build a sequence of hierarchical nets of $\ell := \lceil \log_2 \Delta \rceil$ levels, $X_c = N_0^c \supseteq N_1^c \supseteq \cdots \supseteq N_\ell^c$. We denote by $N_i := \cup_{c \in [k+1]} N_i^c$ the set of all level-i net points. Let $r_i := 2^i$ be the *distance scale* of level i. Fault-tolerant hierarchical nets should satisfy the following properties:

1. *Packing.* For each $0 \leq i \leq \ell$ and $c \in [k+1]$, N_i^c is an r_i-packing;
2. *Covering.* For any $1 \leq i \leq \ell$, if $x \in X$ is not a net point in N_i, then for each color $c \in [k+1]$, there exists a net point $y_c \in N_i^c$ such that $d(y_c, x) \leq r_i$.

Construction. The hierarchical nets can be constructed in a top-down approach. Initially, each N_ℓ^c consists of a distinct point in X. Note that $k \leq n-2$ and hence the initialization is well defined. Also, the single point in N_ℓ^c is colored with c and points not included in any cluster N_ℓ^c stay uncolored.

Suppose all nets on level $i+1$ have been built and we construct the level-i nets as follows. For c from 1 to $k+1$, let U_c be the set of uncolored points when we start to build N_i^c, i.e., after finishing the construction of $N_i^1, N_i^2, \ldots, N_i^{c-1}$. We initialize $N_i^c := N_{i+1}^c$, and extend N_{i+1}^c to get N_i^c by greedily adding points in U_c to N_i^c such that the resulting N_i^c is an r_i-net for U_c; we color the points in $N_i^c \cap U_c$ with color c.

Note that the packing property and the covering property follow directly from the net construction.

Fault-Tolerant Net-Trees. For each color $c \in [k+1]$, we define a *net-tree* T_c, which spans all nodes in X except nodes in the highest level N_ℓ with colors different from c. The construction is given in Algorithm 1. It follows from the construction that all internal nodes of T_c have color c, and all points excluding $N_\ell \cup X_c$ are leaves of T_c.

---

1 Initialize T_c to be the only point in N_ℓ^c, which is the root of T_c;
2 **for** $i = \ell - 1$ *to* 0 **do**
3   **for** *each point* $x \in N_i \setminus N_{i+1}$ **do**
4   let $y_c \in N_{i+1}^c$ be a point such that $d(y_c, x) \leq r_{i+1}$ (such a point exists by the covering property of fault-tolerant hierarchical nets);
5   add x to T_c and set y_c as its parent in T_c by adding the edge $\{y_c, x\}$ to T_c;
6   **end**
7 **end**
8 **return** T_c;

Algorithm 1. Construction of net-tree T_c for color c

For any $c \in [k+1]$, a path $P = \{x_0, x_1, \ldots, x_i\}$ is called a *c-path*, if all edges on P are contained in T_c. Note that for a c-path $P = \{x_0, x_1, \ldots, x_i\}$, any point x_j with $0 < j < i$ has degree at least 2 in T_c and hence is an internal node. Thus, its color must be c. The length of P is defined as $\text{length}(P) := \sum_{j=1}^i d(x_{i-1}, x_i)$. The following lemma shows that for any point $x \notin N_\ell$, any color $c \in [k+1]$ and any $0 \leq i \leq \ell$, there is a c-path from x to some node in N_i with length at most $2r_i$.

Lemma 1 (Climbing Path in Net-tree). *Let T_c be a net-tree obtained above and let $x \in X \setminus N_\ell$ be a non-root point. For any $0 \leq i \leq \ell$, there exists a c-path P_i*

starting from x and ending at some net point $x_i \in N_i$, such that $\mathsf{length}(P_i) \leq 2 \cdot r_i$. *In addition, x_i also has color c if $x_i \neq x$.*

Proof. Let $i^*(x)$ be the largest i such that $x \in N_i$. For $0 \leq i \leq i^*(x)$, let $P_i = \{x\}$ and $x_i = x$. The conclusion holds trivially for P_i.

Now suppose $i^*(x) < i \leq \ell$ and we use induction on i. The base case is for $i = i^*(x) + 1$. We let x_i be x's parent in T_c and let $P_i := \{x, x_i\}$. Note that P_i is a c-path. By the construction of T_c, we know that $x_i \in N_i^c$ and $d(x, x_i) \leq r_i$. Hence, x_i has color c and $\mathsf{length}(P_i) = d(x, x_i) \leq r_i \leq 2 \cdot r_i$.

Suppose $i > i^*(x) + 1$ and there exists an x_{i-1} with color c, and a c-path $P_{i-1} = \{x, \ldots, x_{i-1}\}$ such that $\mathsf{length}(P_{i-1}) \leq 2 \cdot r_{i-1}$. If $x_{i-1} \in N_i$, let $x_i = x_{i-1}$ and $P_i = P_{i-1}$. In this case, we have $\mathsf{length}(P_i) = \mathsf{length}(P_{i-1}) \leq 2 \cdot r_{i-1} \leq 2 \cdot r_i$. Other properties of P_i follow directly from the properties of P_{i-1}, and hence P_i is a c-path.

Otherwise, $x_{i-1} \notin N_i$ and we know that x_{i-1} is not the root of T_c and let x_i be x_{i-1}'s parent in T_c. By our construction of net-trees, $x_i \in N_i^c$ and hence has color c; in addition, $d(x_{i-1}, x_i) \leq r_i$. We let $P_i = P_{i-1} \oplus x_i$, which is formed by appending x_i to the end of P_{i-1}. Note that the edge $\{x_{i-1}, x_i\} \in T_c$ and hence $P_i \subseteq T_c$ is a c-path. Also, $\mathsf{length}(P_i) = \mathsf{length}(P_{i-1}) + d(x_{i-1}, x_i) \leq 2 \cdot r_{i-1} + r_i = 2 \cdot r_i$. $\qquad\square$

Fault-Tolerant Spanners. We have added inter-level edges in the net-trees. Now we add edges connecting net points at the same level to achieve small stretch. Define $\gamma := 4 + \frac{32}{\epsilon}$. For any $0 \leq i \leq \ell$, we call $\{x, y\}$ a *cross edge* at level i, iff x and y are both in N_i, and $d(x, y) \leq \gamma \cdot r_i$. (An edge can be a cross edge at more than one level.) We construct a spanner H by taking the union of all edges in the net-trees of all colors, and all cross edges at all levels, and claim that H is a k-VFTS with stretch at most $1 + \epsilon$.

Lemma 2 (Fault-Tolerant Stretch). *Let $S \subseteq X$ be any set with $|S| \leq k$. For any $x, y \in X \setminus S$, $d_{H \setminus S}(x, y) \leq (1 + \epsilon) \cdot d(x, y)$.*

Proof. Fix $x \neq y \in X \setminus S$ and suppose $r_i < d(x, y) \leq r_{i+1}$. Let q be some integer such that $\frac{8}{2^q} \leq \epsilon < \frac{16}{2^q}$, say $q := \lceil \log \frac{8}{\epsilon} \rceil$.

If $i \leq q - 1$, then $d(x, y) < r_{i+1} \leq 2^q < \frac{16}{\epsilon} < \gamma \cdot r_0$. Hence, $\{x, y\}$ is a cross edge at level 0 and $d_{H \setminus S}(x, y) = d(x, y)$.

Now suppose $i \geq q$ and let $j := i - q \geq 0$. Since $|S| \leq k$, there exists some $c \in [k+1]$ such that $X_c \cap S = \emptyset$. We first show that there exist x' and y' such that $d_{H \setminus S}(x, x') \leq 2 \cdot r_j$ and $d_{H \setminus S}(y, y') \leq 2 \cdot r_j$. Note that either the node x is in N_j, or $i^*(x) < j$ and hence $x \notin N_\ell$. In the latter case, by Lemma 1, there exists an $x_j \in N_j$ with color c and a c-path P connecting x and x_j. We let $x' = x$ and $x' = x_j$ in respective cases. If $x' = x$, then $d_{H \setminus S}(x, x') = d(x, x') = 0 \leq 2 \cdot r_j$. Otherwise, since P is a c-path and x' has color c, no point on P is contained in S and hence $P \subseteq H \setminus S$. Therefore, we have $d_{H \setminus S}(x, x') \leq 2 \cdot r_j$. Similarly, we can choose some $y' \in N_j \setminus S$ such that $d_{H \setminus S}(y, y') \leq 2 \cdot r_j$.

Note that $d(x, x') \leq d_{H \setminus S}(x, x') \leq 2 r_j$ and $d(y, y') \leq d_{H \setminus S}(y, y') \leq 2 r_j$. Hence, we have $d(x', y') \leq d(x', x) + d(x, y) + d(y, y') \leq 2 r_j + r_{i+1} + 2 r_j <$

$(4 + \frac{32}{\epsilon}) \cdot r_j = \gamma \cdot r_j$. It follows that $\{x', y'\}$ is a cross edge at level j and thus $d_{H \backslash S}(x', y') = d(x', y')$.

Note that $d(x', y') \leq d(x', x) + d(x, y) + d(y, y') \leq 4r_j + d(x, y)$, and $d_{H \backslash S}(x, y) \leq d_{H \backslash S}(x, x') + d_{H \backslash S}(x', y') + d_{H \backslash S}(y', y)$. Hence we conclude that $d_{H \backslash S}(x, y) \leq 8 \cdot r_j + d(x, y) = \frac{8}{2^q} \cdot r_i + d(x, y) \leq (1 + \epsilon) \cdot d(x, y)$. □

Remark. Note that in the proof, there exists a color c such that no node in S has color c. Also, the spanner path in $H \backslash S$ is the concatenation of a path $P_1 = \{x, \ldots, x'\}$, a cross edge $\{x', y'\}$, and a path $P_2 = \{y', \ldots, y\}$, such that each of P_1 and P_2 is either a c-path or a trivial path with only one point. This property is useful in our later construction of k-VFTS with small hop-diameter.
Bounding the Number of Edges. We show that the number of edges in H is $O(kn)$. We actually show a stronger result: we can direct the edges in H in a way such that every point has out-degree $O(k)$.

Lemma 3 (Bounding the Number of Edges). *Let H be the $(k, 1 + \epsilon)$-VFTS we construct above. Then, the number of edges in H is $\epsilon^{-O(\dim)} \cdot kn$. Moreover, the edges of H can be directed such that the out-degree of each point is bounded by $\epsilon^{-O(\dim)} \cdot k$.*

Proof. Note that the edges of H come from two sources: the net-trees and cross edges; we bound them separately.

In any net-tree T_c, we direct the edge $\{x, p(x)\}$ from x to $p(x)$, where $p(x)$ is the parent of x in T_c. Note that every point has out-degree at most 1 in each tree, and hence the out-degree due to the net-tree edges is bounded by $k + 1$.

Now we bound the out-degree due to the cross edges. Recall that $i^*(x)$ is the maximum i such that $x \in N_i$. Given an edge $\{x, y\}$, we direct it from x to y if $i^*(x) < i^*(y)$, and direct it arbitrarily if $i^*(x) = i^*(y)$.

Fix $x \in X$. We bound the number of edges coming out of x and going into some point with a fixed color c. Let $i := i^*(x)$. For any directed edge (x, y) such that y has color c, we know that $i^*(y) \geq i^*(x) = i$ and hence $y \in N_i^c$. Note that the existence of a cross edge $\{x, y\}$ implies that $d(x, y) \leq \gamma \cdot r_i$. Also note that N_i^c is an r_i-packing. Then, by Proposition 1, the number of such edges is $\gamma^{O(\dim)} = (4 + \frac{32}{\epsilon})^{O(\dim)}$. Since there are at most $k + 1$ colors, the number of cross edges coming out of x is bounded by $(k + 1) \cdot (4 + \frac{32}{\epsilon})^{O(\dim)} = \epsilon^{-O(\dim)} \cdot k$.

The upper bound on the number of edges follows directly from the analysis of out-degree. □

3   Achieving Small Hop-Diameter

In this section, we show that a technique similar to that in [4] can be used to reduce the hop-diameter of our basic k-VFTS.

Let T be a tree metric with n nodes. It is shown in [5,4] that for $m \geq 2n$, we can add m edges to T to obtain a spanner R, such that for the unique tree path P between x and y in T, there is a path P' in R that connects x and y via at most $O(\alpha(m, n))$ nodes on P (in the same order), where $\alpha(\cdot, \cdot)$ is defined below. By the triangle inequality, $\text{length}(P') \leq \text{length}(P)$.

Definition 2 (Ackermann's Function [17]). *Let $A(i,j)$ be a function defined for integers $i,j \geq 0$ as the following.*

$$A(0,j) = 2j \qquad\qquad\qquad \text{for } j \geq 0$$
$$A(i,0) = 0, A(i,1) = 2 \qquad\qquad \text{for } i \geq 1$$
$$A(i,j) = A(i-1, A(i,j-1)) \qquad \text{for } i \geq 1, j \geq 2$$

Define the function $\alpha(m,n) := \min\{i | i \geq 1, A(i, 4 \lceil \frac{m}{n} \rceil) > \log_2 n\}$.

Adding Edges to Reduce Hop-diameter. Let H be the $(k, 1 + \epsilon)$-VFTS constructed in Section 2. Now we show how to add edges to H to reduce the hop-diameter. For each net-tree T_c, we use the technique in [5] to add m edges to T_c to get a spanner R_c such that between any two points in T_c, there is a path between them in R_c with $O(\alpha(m,n))$ hops which preserves their original path distance in T_c. Let H' denote the spanner constructed by taking the union of all edges in R_c's for all colors, and all cross edges at all levels. Hence, H' has $k(m + \epsilon^{-O(\dim)} \cdot n)$ edges. We prove in the following lemma that H' has small hop-diameter.

Lemma 4 (Bounded Hop-Diameter). *For $m \geq 2n$, the spanner H' constructed above has $k(m + \epsilon^{-O(\dim)} \cdot n)$ edges. Let $S \subseteq X$ be a set with $|S| \leq k$. For any pair $x, y \in X \setminus S$, there exists a path between x and y in $H' \setminus S$ with $O(\alpha(m,n))$ hops, and the path has length at most $(1 + \epsilon) \cdot d(x,y)$.*

Proof. In Lemma 2, we have proved that there is a color c such that no point in S has color c, and the $(1 + \epsilon)$-spanner path in $H \setminus S$ connecting x and y is a concatenation of a path P_1, a cross edge and a path P_2, where each of P_1 and P_2 is either a c-path or a trivial path consisting of only one point. Note that if P_1 is not a trivial path, it can be substituted by a path P_1' in $R_c \setminus S \subseteq H' \setminus S$ consisting of $O(\alpha(m,n))$ hops. Similarly, P_2 can also be substitued by a path P_2' with $O(\alpha(m,n))$ hops. The new spanner path connecting x and y in $H' \setminus S$ after the substitution has length at most $(1 + \epsilon) \cdot d(x,y)$, as $\mathsf{length}(P_1') \leq \mathsf{length}(P_1)$ and $\mathsf{length}(P_2') \leq \mathsf{length}(P_2)$. $\square$

4   Achieving Bounded Degree

4.1   Fault-Tolerant Single-Sink Spanners

Our technique of reducing degrees in fault-tolerant spanners is based on single-sink spanners. Given a point $v \in X$, a spanner H for X is a *k-vertex-fault-tolerant v-single-sink t-spanner* $((k, t, v)$-VFTssS$)$, if for any subset $S \subseteq X \setminus \{v\}$ with $|S| \leq k$, and any point $x \in X \setminus S$, it holds that $d_{H \setminus S}(v, x) \leq t \cdot d(v, x)$. Here, t is called the *root-stretch* of H. In this section, we show a construction of a $(k, 1 + \epsilon, v)$-VFTssS with maximum degree $O(k)$. Throughout this section, we assume a point $v \in X$ is given. Without loss of generality, we assume $0 < \epsilon < \frac{1}{9}$ is a constant and build a $(k, 1 + 9\epsilon, v)$-VFTssS. Our construction is based on a technique called *ring-partition*.

Ring-Partition. Let $\ell = \left\lceil \log_{\frac{1}{\epsilon}} \Delta \right\rceil$ and $r_i = \frac{1}{\epsilon^i}$ with $i \in [\ell]$. For convenience, let $r_0 = 1$ (recall that we assume inter-point distances are larger than 1). Consider the rings, denoted by $R_1, \ldots, R_\ell$, where $R_i := R(v, r_{i-1}, r_i)$. For convenience, let $R_0 := \{v\}$. The rings are pairwise disjoint and their union covers X. For each $i \in [\ell]$, we build an ϵr_{i-1}-net N_i for R_i. By Proposition 1, N_i contains at most $(\frac{r_i}{\epsilon r_{i-1}})^{2\dim} = \epsilon^{-4\dim}$ points. We denote this upper bound by $\Gamma := \left\lceil \epsilon^{-4\dim} \right\rceil$ and then we have $|N_i| \le \Gamma$. Let $N := \cup_{i>0} N_i$ be the set of net points. Then, for each net point $y \in N_i$, we construct a *net cluster* C_y, such that a point $x \in X$ is in C_y iff x is in R_i, and among all points in N_i, y is the closest one to x (breaking ties arbitrarily). For each $y \in N$, we arbitrarily choose $k + 1$ *portals* $Q_y \subseteq C_y$ (if $|C_y| < k+1$, we let $Q_y = C_y$). Let $Q_i := \cup_{y \in N_i} Q_y$ be the portals in R_i, and and $Q := \cup_{y \in N} Q_y$ be the set of all portals. Note that $|Q_i| \le (k+1) \cdot |N_i| \le \Gamma \cdot (k+1)$.

We construct a $(k, 1 + 9\epsilon, v)$-VFTssS in two stages. First, we add edges to connect portals to the root and obtain a $(k, 1 + \epsilon, v)$-VFTssS for $Q \cup \{v\}$. Then, we add edges to connect the points in each cluster with their portals by short paths.

Connecting Portals to the Root. Assign each point $q \in Q$ a unique identifier $\mathsf{id}(q) \in [|Q|]$, such that for any $q \in R_i$ and $q' \in R_j$ with $i < j$, it holds that $\mathsf{id}(q) < \mathsf{id}(q')$; also let $\mathsf{id}(v) = 0$. In other words, points closer to the root v have smaller identifiers. We divide the points in Q into groups of size $k + 1$. Specifically, let $A_j := \{q \in Q | (j-1) \cdot (k+1) + 1 \le \mathsf{id}(q) \le j \cdot (k+1)\}$ for $j \ge 1$. The edges to connect portals with the root are added as follows.

- For $q \in Q$ with $1 \le \mathsf{id}(q) \le (2\Gamma + 1) \cdot (k+1)$, add an edge $\{q, v\}$. Let E_0 denote the set of such edges.
- For $j > 2\Gamma + 1$, we add an edge between every point in A_j and every point in $A_{j-2\Gamma-1}$, and let E_j denote the set of such edges, i.e. $E_j := \{\{x, y\} | x \in A_j \text{ and } y \in A_{j-2\Gamma-1}\}$.

Define $\tilde{E} := E_0 \cup (\cup_{j>2\Gamma+1} E_j)$ and $\tilde{H} := (Q \cup \{v\}, \tilde{E})$. Note that the degree of v in $\tilde{H}$ is at most $(2\Gamma + 1) \cdot (k+1) = \epsilon^{-O(\dim)} \cdot k$, and the degree of any point $q \in Q$ is at most $2(k+1) = O(k)$.

Lemma 5 (Edges Connect Portals at least Two Levels Apart). *Let $q \in Q$, and let $\{q, x\} \in \tilde{E}$ be an edge with $\mathsf{id}(x) < \mathsf{id}(q)$. Then, either $x = v$ or there exist $i > 2$ and $0 < i' \le i - 2$ such that $q \in Q_i$ and $x \in Q_{i'}$.*

Proof. We first consider $\{q, x\}$ with $q \in Q_1 \cup Q_2$. Since $|Q_1| \le \Gamma \cdot (k+1)$ and $|Q_2| \le \Gamma \cdot (k+1)$, it holds that $\mathsf{id}(q) \le 2\Gamma \cdot (k+1) \le (2\Gamma+1) \cdot (k+1)$. Hence, q is connected to the root and $x = v$.

Now consider the case that $q \in Q_i$ for some $i > 2$. If $x = v$, then we are done. Now suppose $x \ne v$. Then, we know that $q \in A_j$ for some $j > 2\Gamma + 1$ and $x \in A_{j-2\Gamma-1}$. Note that there are exactly $2\Gamma \cdot (k+1)$ points in $A_{j-2\Gamma} \cup \cdots \cup A_{j-1}$, and hence $\mathsf{id}(x) < \mathsf{id}(q) - 2\Gamma \cdot (k+1)$. On the other hand, since there are at most $2\Gamma \cdot (k+1)$ points in $Q_i \cup Q_{i-1}$, x cannot be in $Q_i \cup Q_{i-1}$. Hence, $x \in Q_{i'}$ for some $i' \le i - 2$. □

Lemma 6. *Let $S \subseteq X \setminus \{v\}$ be a set of at most k points. Then, for any point $q \in Q \setminus S$, $d_{\tilde{H}\setminus S}(v, q) \leq (1 + 3\epsilon) \cdot d(v, q)$.*

Proof. We use induction on $\mathrm{id}(q)$. For $q \in Q \setminus S$ with $1 \leq \mathrm{id}(q) \leq (2\Gamma+1) \cdot (k+1)$, we know that q is connected to v in $\tilde{H}$ and hence $d_{\tilde{H}}(v, q) = d(v, q)$.

Now suppose $\mathrm{id}(q) > (2\Gamma + 1) \cdot (k + 1)$ and for any point $q' \in Q \setminus S$ with $\mathrm{id}(q') < \mathrm{id}(q)$, it holds that $d_{\tilde{H}\setminus S}(v, q') \leq (1 + 3\epsilon)d(v, q')$. Let $j > 2\Gamma + 1$ be such that $q \in A_j$. From the construction of $\tilde{H}$, we know that for any point $p \in A_{j-2\Gamma-1}$, there exists an edge $\{p, q\} \in \tilde{H}$. Since $A_{j-2\Gamma-1}$ contains $k + 1$ points, there exists a functioning point $p^* \notin S$ in $A_{j-2\Gamma-1}$. Note that $\mathrm{id}(p^*) < \mathrm{id}(q)$. Hence, we have

$$d_{\tilde{H}\setminus S}(v, p^*) \leq (1 + 3\epsilon)d(v, p^*) \tag{1}$$

by induction hypothesis.

Note that $\mathrm{id}(q) > (2\Gamma+1) \cdot (k+1)$ implies $q \in Q_i$ for some $i > 2$. By Lemma 5, we have $p^* \in Q_{i'} \subseteq R_{i'}$ for some $0 < i' \leq i - 2$. Hence,

$$d(v, p^*) \leq r_{i'} = \epsilon r_{i'+1} \leq \epsilon r_{i-1} < \epsilon d(v, q) \tag{2}$$

By the triangle inequality, $d_{\tilde{H}\setminus S}(v, q) \leq d_{\tilde{H}\setminus S}(v, p^*) + d(p^*, q) \leq d_{\tilde{H}\setminus S}(v, p^*) + d(p^*, v) + d(v, q)$, which by (1) and (2) is at most $(1 + \epsilon(2 + 3\epsilon))d(v, q) \leq (1 + 3\epsilon)d(v, q)$, where the last inequality holds when $\epsilon \leq \frac{1}{3}$. $\square$

Connecting Points in Clusters to Portals. Fix $i \in [\ell]$ and a point $y \in N_i$. Recall that C_y denotes the net cluster centered at y, whose radius is at most $r := \epsilon \cdot r_{i-1}$, and Q_y is the set of portals for C_y. We call the portals in Q_y r-portals since they are portals for clusters with radius at most r. We define a procedure $\mathsf{Add}(C_y, Q_y, r)$ which adds edges to connect points in C_y with portals in Q_y.

1. *Sub-clustering.* We return immediately if $C_y = Q_y$. Suppose $C_y \neq Q_y$. We build an $\frac{r}{2}$-net $\widehat{N}$ for $C_y \setminus Q_y$. Recall that r is an upper bound on C_y's radius. By Proposition 1, $|\widehat{N}| \leq 4^{\mathrm{dim}}$. Then for each node $z \in \widehat{N}$, we construct a cluster $\widehat{C}_z$, such that a point $x \in C_y$ is in $\widehat{C}_z$ iff x is in $C_y \setminus Q_y$, and among all points of $\widehat{N}$, z is the closest one to x (breaking ties arbitrarily).
2. *Connecting sub-portals.* For each sub-cluster $\widehat{C}_z$, we arbitrarily select $k + 1$ sub-portals $\widehat{Q}_z$ (called $\frac{r}{2}$-portals) in $\widehat{C}_z$ (select all points if $|\widehat{C}_z| < k + 1$). Then, for each sub-portal in $\widehat{Q}_z$ and each portal in Q_y, we add an edge between them. Note that since $\widehat{N} \leq 4^{\mathrm{dim}}$, and thus each portal $q \in Q_y$ is connected with at most $4^{\mathrm{dim}} \cdot (k + 1) = 2^{O(\mathrm{dim})} \cdot k$ sub-portals.
3. *Recursion.* For every $z \in \widehat{N}$, recursively call $\mathsf{Add}(\widehat{C}_z, \widehat{Q}_z, \frac{r}{2})$.

Let $\widehat{H}_y$ be the resulting spanner returned by $\mathsf{Add}(C_y, Q_y, r)$. We have the following lemma.

Lemma 7. *Let $S \subseteq X$ be a set of at most k points. For any $x \in C_y \setminus (Q_y \cup S)$ and any r-portal $q \in Q_y \setminus S$, it holds that $d_{\widehat{H}_y \setminus S}(x, q) \leq 2r$.*

Proof. Suppose that x is an $\frac{r}{2^i}$-portal. Note that x is connected to $k+1$ distinct $\frac{r}{2^{i-1}}$-portals, and at least one of them must be functioning (i.e., not in S). We let x_{i-1} be such a $\frac{r}{2^{i-1}}$-portal. Using this argument, we can find a sequence of portals $x = x_i, x_{i-1}, x_{i-2}, \ldots, x_0 = q$, such that for all $0 \le j \le i$, $x_j \notin S$ is an $\frac{r}{2^j}$-portal. In addition, for $j \in [i]$, $\{x_j, x_{j-1}\} \in \widehat{H}_y$ and $d(x_j, x_{j-1}) \le \frac{r}{2^{j-1}}$. Hence, $d_{\widehat{H}_y \setminus S}(x, q) \le \sum_{j=1}^i \frac{r}{2^{j-1}} \le 2r$. □

Obtaining the $(k, 1+9\epsilon, v)$-VFTssS. Our final $(k, 1+9\epsilon, v)$-VFTssS, denoted by H_v, is the union of $\tilde{H}$ and $\widehat{H}_y$'s for all $y \in N$. Note that the degree of v is bounded by $\epsilon^{-O(\dim)} \cdot k$ and the degree of any other point in H_v is bounded by $2^{O(\dim)} \cdot k$. It remains to show that H_v has root-stretch at most $1+9\epsilon$ under the failure of at most k points.

Lemma 8. *Let $S \subseteq X \setminus \{v\}$ be a set of at most k points. For any $x \in X \setminus S$, $d_{H_v}(v, x) \le (1+9\epsilon) \cdot d(v, x)$.*

Proof. Suppose $x \ne v$. Otherwise the conclusion holds trivially. Let $y \in N$ be the net point covering x, i.e., $x \in C_y$, and let R_i be the ring that contains C_y.

If x is a portal for C_y, then by Lemma 6, we know that $d_{\tilde{H} \setminus S}(v, x) \le (1 + 3\epsilon)d(v, x)$ and hence $d_{H_v \setminus S}(v, x) \le d_{\tilde{H} \setminus S}(v, x) \le (1 + 9\epsilon)d(v, x)$.

Otherwise, $Q_y \ne C_y$ and hence $|Q_y| = k + 1$. Therefore, there must be some $q \in Q_y$ which is functioning. Let $r := \epsilon r_{i-1}$ be an upper bound on the radius of C_y. From the construction of the ring-partition, we know that $d(q, x) \le r$ and $r \le \epsilon \cdot d(v, x)$. By Lemma 7, it holds that $d_{\widehat{H}_y \setminus S}(q, x) \le 2r$. Hence, $d_{H_v \setminus S}(v, x) \le d_{H_v \setminus S}(v, q) + d_{H_v \setminus S}(q, x) \le d_{\tilde{H} \setminus S}(v, q) + d_{\widehat{H}_y \setminus S}(q, x) \le (1 + 3\epsilon)d(v, q) + 2r \le (1 + 3\epsilon)(d(v, x) + d(x, q)) + 2r \le (1 + 3\epsilon)d(v, x) + 6r \le (1 + 9\epsilon)d(v, x)$. □

4.2 $(k, 1+\epsilon)$-VFTS with Bounded Degree

Now we construct a $(k, 1+\epsilon)$-VFTS with bounded degree as follows. We first construct a basic $(k, 1 + \frac{\epsilon}{3})$-VFTS for X with $O(kn)$ edges, and denote it by H_0. Recall that the edges of H_0 can be directed such that the out-degree of each point in H_0 is $\epsilon^{-O(\dim)} \cdot k$. Denote an edge $\{x, y\}$ by (x, y) if it is directed from x to y in H_0. For any point $x \in X$, we let $N_{in}(x) := \{y \in X | (y, x) \in H_0\}$, and build a $(k, 1 + \frac{\epsilon}{3}, x)$-VFTssS H_x for $N_{in}(x) \cup \{x\}$. We take the spanner $H := \cup_{x \in X} H_x$, and show that H is a $(k, 1+\epsilon)$-VFTS with maximum degree $O(k^2)$.

Lemma 9. *For $0 < \epsilon \le \frac{1}{2}$, H is a $(k, 1+\epsilon)$-VFTS, in which the degree of any point $x \in X$ is $\epsilon^{-O(\dim)} \cdot k^2$. Consequently, H has $\epsilon^{-O(\dim)} \cdot k^2 n$ edges.*

Proof. We first prove that H is a $(k, 1+\epsilon)$-VFTS. Let $S \subseteq X$ be a set of at most k points. Since H_0 is a $(k, 1 + \frac{\epsilon}{3})$-VFTS, for any $x, y \in X \setminus S$, there exists a $(1 + \frac{\epsilon}{3})$-spanner path P_0 in $H_0 \setminus S$ between x and y.

For each edge $\{u, v\} \in P_0$, suppose the edge is directed as (u, v). Then, since H_v is a $(k, 1 + \frac{\epsilon}{3}, v)$-VFTssS and u, v are both funcitoning, there is a $(1 + \frac{\epsilon}{3})$-spanner path $P_{uv} \subseteq H_v \setminus S$ between u and v. Let P denote the concatenation of P_{uv}'s for all edges $\{u, v\} \in P_0$. Then, P is contained in $H \setminus S$ and is a spanner path between x and y with stretch at most $(1 + \frac{\epsilon}{3})^2 \le 1 + \epsilon$.

Next we bound the degree of an arbitrary point $x \in X$ in H. The edges incident to x in H are contained in H_x and H_y's such that the edge $\{x, y\}$ is directed from x to y in H_0. Note that the number of H_y's involving x is bounded by the out-degree of x in H_0, which is $\epsilon^{-O(\dim)} \cdot k$. Also recall that the degree of x in H_x is $\epsilon^{-O(\dim)} \cdot k$ and the degree of x in each H_y is $2^{O(\dim)} \cdot k$. Hence, we conclude that the degree of x in H is $\epsilon^{-O(\dim)} \cdot k^2$.

The upper bound on the number of edges follows directly from the degree analysis.   □

References

1. Arya, S., Das, G., Mount, D.M., Salowe, J.S., Smid, M.H.M.: Euclidean spanners: short, thin, and lanky. In: STOC 1995, pp. 489–498 (1995)
2. Callahan, P.B., Kosaraju, S.R.: Faster algorithms for some geometric graph problems in higher dimensions. In: SODA 1993, pp. 291–300 (1993)
3. Chan, H.T.-H., Gupta, A., Maggs, B.M., Zhou, S.: On hierarchical routing in doubling metrics. In: SODA 2005, pp. 762–771 (2005)
4. Chan, T.-H.H., Gupta, A.: Small hop-diameter sparse spanners for doubling metrics. Discrete & Computational Geometry 41(1), 28–44 (2009)
5. Chazelle, B.: Computing on a free tree via complexity-preserving mappings. Algorithmica 2, 337–361 (1987)
6. Chechik, S., Langberg, M., Peleg, D., Roditty, L.: Fault-tolerant spanners for general graphs. In: STOC 2009, pp. 435–444 (2009)
7. Czumaj, A., Zhao, H.: Fault-tolerant geometric spanners. Discrete & Computational Geometry 32(2), 207–230 (2004)
8. Das, G., Narasimhan, G.: A fast algorithm for constructing sparse euclidean spanners. In: Symposium on Computational Geometry, pp. 132–139 (1994)
9. Dinitz, M., Krauthgamer, R.: Fault-tolerant spanners: better and simpler. In: PODC 2011, pp. 169–178 (2011)
10. Gottlieb, L.-A., Roditty, L.: An Optimal Dynamic Spanner for Doubling Metric Spaces. In: Halperin, D., Mehlhorn, K. (eds.) ESA 2008. LNCS, vol. 5193, pp. 478–489. Springer, Heidelberg (2008)
11. Gupta, A., Krauthgamer, R., Lee, J.R.: Bounded geometries, fractals, and low-distortion embeddings. In: FOCS 2003, pp. 534–543 (2003)
12. Har-Peled, S., Mendel, M.: Fast construction of nets in low dimensional metrics, and their applications. In: Symposium on Computational Geometry, pp. 150–158 (2005)
13. Levcopoulos, C., Narasimhan, G., Smid, M.H.M.: Efficient algorithms for constructing fault-tolerant geometric spanners. In: STOC 1998, pp. 186–195 (1998)
14. Lukovszki, T.: New Results on Fault Tolerant Geometric Spanners. In: Dehne, F., Gupta, A., Sack, J.-R., Tamassia, R. (eds.) WADS 1999. LNCS, vol. 1663, pp. 193–204. Springer, Heidelberg (1999)
15. Narasimhan, G., Smid, M.H.M.: Geometric spanner networks. Cambridge University Press (2007)
16. Solomon, S., Elkin, M.: Balancing Degree, Diameter and Weight in Euclidean Spanners. In: de Berg, M., Meyer, U. (eds.) ESA 2010. LNCS, vol. 6346, pp. 48–59. Springer, Heidelberg (2010)
17. Tarjan, R.E.: Efficiency of a good but not linear set union algorithm. J. ACM 22(2), 215–225 (1975)

A Dependent LP-Rounding Approach
for the k-Median Problem*

Moses Charikar and Shi Li

Department of Computer Science, Princeton University, Princeton NJ 08540, USA

Abstract. In this paper, we revisit the classical k-median problem. Using the standard LP relaxation for k-median, we give an efficient algorithm to construct a probability distribution on sets of k centers that matches the marginals specified by the optimal LP solution. Analyzing the approximation ratio of our algorithm presents significant technical difficulties: we are able to show an upper bound of 3.25. While this is worse than the current best known $3 + \epsilon$ guarantee of [2], because: (1) it leads to 3.25 approximation algorithms for some generalizations of the k-median problem, including the k-facility location problem introduced in [10], (2) our algorithm runs in $\tilde{O}(k^3 n^2 / \delta^2)$ time to achieve $3.25(1+\delta)$-approximation compared to the $O(n^8)$ time required by the local search algorithm of [2] to guarantee a 3.25 approximation, and (3) our approach has the potential to beat the decade old bound of $3 + \epsilon$ for k-median.

We also give a 34-approximation for the knapsack median problem, which greatly improves the approximation constant in [13]. Using the same technique, we also give a 9-approximation for matroid median problem introduced in [11], improving on their 16-approximation.

Keywords: Approximation, k-Median Problem, Dependent Rounding.

1 Introduction

In this paper, we present a novel LP rounding algorithm for the metric k-median problem which achieves approximation ratio 3.25. For the k-median problem, we are given a finite metric space $(\mathcal{F} \cup \mathcal{C}, d)$ and an integer $k \geq 1$, where $\mathcal{F}$ is a set of facility locations and $\mathcal{C}$ is a set of clients. Our goal is to select k facilities to open, such that the total connection cost for all clients in $\mathcal{C}$ is minimized, where the connection cost of a client is its distance to its nearest open facility. When $\mathcal{F} = \mathcal{C} = X$, the set of points with the same nearest open facility is known as a cluster and thus the sum measures how well X can be partitioned into k clusters. The k-median problem has numerous applications, starting from clustering to data mining [3], to assigning efficient sources of supplies to minimize the transportation cost([12,16]).

The problem is NP-hard and has received a lot of attention ([15], [6], [7], [10], [1]). The best known approximation factor is $3 + \epsilon$ approximation due to [2].

---

* A full version of this paper is available at the authors' web pages.

A. Czumaj et al. (Eds.): ICALP 2012, Part I, LNCS 7391, pp. 194–205, 2012.
ⓒ Springer-Verlag Berlin Heidelberg 2012

Jain et al. [9] proved that the k-median problem is $1 + 2/e \approx 1.736$-hard to approximate.

Our algorithm (like several previous ones) for the k-median problem is based on the following natural LP relaxation:

$$\text{LP(1)} \qquad \min \quad \sum_{i \in \mathcal{F}, j \in \mathcal{C}} d(i,j) x_{i,j} \qquad \text{s.t.}$$

$$\sum_{i \in \mathcal{F}} x_{i,j} = 1, \quad \forall j \in \mathcal{C}; \qquad x_{i,j} \leq y_i, \quad \forall i \in \mathcal{F}, j \in \mathcal{C};$$

$$\sum_{i \in \mathcal{F}} y_i \leq k; \qquad x_{i,j}, y_i \in [0,1], \quad \forall i \in \mathcal{F}, j \in \mathcal{C}.$$

It is known that the LP has an integrality gap of 2. On the positive side, [1] showed that the integrality gap is at most 3 by giving an exponential time rounding algorithm.

Very recently, Kumar [13] gave a (large) constant-factor approximation algorithm for a generalization of the k-median problem, which is called knapsack median problem. In the problem, each facility $i \in \mathcal{F}$ has an opening cost f_i and we are given a budget M. The goal is to open a set of facilities such that their total opening cost is at most M, and minimize the total connection cost. When $M = k$ and $f_i = 1$ for every facility $i \in \mathcal{F}$, the problem becomes k-median.

Krishnaswamy et al. [11] introduced another generalization of k-median, called matroid-median problem. In the problem, the set of open facilities has to form an independent set of some given matroid. [11] gave a 16-approximation for this problem, assuming there is a separation oracle for the matroid polytope.

1.1 Our Results

We give a simple and efficient rounding procedure. Given an LP solution, we open a set of k facilities from some distribution and connect each client j to its closest open facility, such that the expected connection cost of j is at most 3.25 times its fractional connection cost. This leads to a 3.25 approximation for k-median. Though the provable approximation ratio is worse than that of the current best algorithm, we believe the algorithm (and particularly our approach) is interesting for the following reasons:

Firstly, our algorithm is more efficient than the $3 + \epsilon$-approximation algorithm with the same approximation guarantee. The bottleneck of our algorithm is solving the LP, for which we can apply Young's fast algorithm for the k-median LP [17].

Secondly, our approach has the potential to beat the decade old $3 + \epsilon$-approximation algorithm for k-median. In spite of the simplicity of our algorithm, we are unable to exploit its full potential due to technical difficulties in the analysis. Our upper bound of 3.25 is not tight. The algorithm has some parameters which we have instantiated for ease of analysis. It is possible that the algorithm with these specific choices gives an approximation ratio strictly better than 3; further there is additional room for improvement by making a judicious choice of algorithm parameters.

The distribution of solutions produced by the algorithm satisfies marginal conditions and negative correlation. Consequently, the algorithm can be easily extended to solve the k-median problem with facility costs and the k-median problem (called k-facility location problem) with multiple types of facilities, both introduced in [10]. The techniques of this paper yield a factor 3.25 algorithm for the two generalizations.

Based on our techniques for the k-median problem, we give a 34-approximation algorithm for the knapsack median problem, which greatly improves the constant approximation given by [13].(The constant was 2700.) Following the same line of the algorithm, we can give a 9-approximation for the matroid-median problem, improving on the 16-approximation in [11].

2 The Approximation Algorithm for the k-Median Problem

Our algorithm is inspired by the $6\frac{2}{3}$-approximation for k-median by [7] and the clustered rounding approach of Chudak and Shmoys [8] for facility location as well as the analysis of the 1.5-approximation for UFL problem by [4]. In particular, we are able to save the additive factor of 4 that is lost at the beginning of the $6\frac{2}{3}$-approximation algorithm by [7], using some ideas from the rounding approaches for facility location.

We first give with a high level overview of the algorithm. A simple way to match the marginals given by the LP solution is to interpret the y_i variables as probabilities of opening facilities and sample independently for each i. This has the problem that with constant probability, a client j could have no facility opened close to j. In order to address this, we group fractional facilities into bundles, each containing a total fractional of between $1/2$ and 1. At most one facility is opened in each bundle and the probability that some facility in a bundle is picked is exactly the volume, i.e. the sum of y_i values for the bundle.

Creating bundles reduces the uncertainty of the sampling process. E.g. if the facilities in a bundle of volume $1/2$ are sampled independently, with probability $e^{-1/2}$ in the worst case, no facility will be open; while sampling the bundle as a single entity reduces the probability to $1/2$. The idea of creating bundles alone does not reduce the approximation ratio to a constant, since still with some non-zero probability, no nearby facilities are open.

In order to ensure that clients always have an open facility within expected distance comparable to their LP contribution, we pair the bundles. Each pair now has at least a total fraction of 1 facility and we ensure that the rounding procedure always picks one facility in each pair. The randomized rounding procedure makes independent choices for each pair of bundles and for fractional facilities that are not in any bundle. This produces k facilities in expectation. We get exactly k facilities by replacing the independent rounding by a dependent rounding procedure with negative correlation properties so that our analysis need only consider the independent rounding procedure. (The technique of dependent rounding was used in [5] to approximate the fault-tolerant facility location problem.)

Now we proceed to give more details. We solve LP(1) to obtain a fractional solution (x, y). By splitting one facility into many if necessary, we can assume $x_{i,j} \in \{0, y_i\}$. We remove all facilities i from $\mathcal{C}$ with $y_i = 0$. Let $\mathcal{F}_j = \{i \in \mathcal{F} : x_{i,j} > 0\}$. So, instead of using x and y, we shall use $(y, \{\mathcal{F}_j | j \in \mathcal{C}\})$ to denote a solution.

For a subset of facilities $\mathcal{F}' \subseteq \mathcal{F}$, define $\mathsf{vol}(\mathcal{F}') = \sum_{i \in \mathcal{F}'} y_i$ to be the *volume* of $\mathcal{F}'$. So, $\mathsf{vol}(\mathcal{F}_j) = 1, \forall j \in \mathcal{C}$. W.L.O.G, we assume $\mathsf{vol}(\mathcal{F}) = k$. Denote by $d(j, \mathcal{F}')$ the average distance from j to $\mathcal{F}'$ w.r.t weights y, i.e, $d(j, \mathcal{F}') = \sum_{i \in \mathcal{F}'} y_i d(j, i) / \mathsf{vol}(\mathcal{F}')$. Define $d_{av}(j) = \sum_{i \in \mathcal{F}_j} y_i d(i, j)$ to be the connection cost of j in the fractional solution. For a client j, let $B(j, r)$ denote the set of facilities that have distance strictly smaller than r to j.

Our rounding algorithm consists of 4 phases, which we now describe.

2.1   Filtering Phase

We begin our algorithm with a filtering phase, where we select a subset $\mathcal{C}' \subseteq \mathcal{C}$ of clients. $\mathcal{C}'$ has two properties: (1) The clients in $\mathcal{C}'$ are far away from each other. With this property, we can guarantee that each client in $\mathcal{C}'$ can be assigned an exclusive set of facilities with large volume. (2) A client in $\mathcal{C} \backslash \mathcal{C}'$ is close to some client in $\mathcal{C}'$, so that its connection cost is bounded in terms of the connection cost of its neighbour in $\mathcal{C}'$. So, $\mathcal{C}'$ captures the connection requirements of $\mathcal{C}$ and also has a nice structure. After this filtering phase, our algorithm is independent of the clients in $\mathcal{C} \backslash \mathcal{C}'$. Following is the filtering phase.

Initially, $\mathcal{C}' = \emptyset, \mathcal{C}'' = \mathcal{C}$. At each step, we select the client $j \in \mathcal{C}''$ with the minimum $d_{av}(j)$, breaking ties arbitrarily, add j to $\mathcal{C}'$ and remove j and all j's that $d(j, j') \le 4 d_{av}(j')$ from $\mathcal{C}''$. This operation is repeated until $\mathcal{C}'' = \emptyset$.

Lemma 1. *(1) For any $j, j' \in \mathcal{C}', j \ne j', d(j, j') > 4 \max\{d_{av}(j), d_{av}(j')\}$;*
(2) For any $j' \in \mathcal{C} \backslash \mathcal{C}'$, there is a client $j \in \mathcal{C}'$ such that $d_{av}(j) \le d_{av}(j'), d(j, j') \le 4 d_{av}(j')$.

We leave the proof of the lemma to the full version of the paper.

2.2   Bundling Phase

Since clients in $\mathcal{C}'$ are far away from each other, each client $j \in \mathcal{C}'$ can be assigned a set of facilities with large volume. To be more specific, for a client $j \in \mathcal{C}'$, we define a set $\mathcal{U}_j$ as follows. Let $R_j = \frac{1}{2} \min_{j' \in \mathcal{C}', j' \ne j} d(j, j')$ be half the distance of j to its nearest neighbour in $\mathcal{C}'$, and $\mathcal{F}'_j = \mathcal{F}_j \cap B(j, 1.5 R_j)$ to be the set of facilities that serve j and are at most $1.5 R_j$ away.[1] A facility i which belongs to

---

[1] It is worthwhile to mention the motivation behind the choice of the scalar 1.5 in the definition of $\mathcal{F}'_j$. If we were only aiming at a constant approximation ratio smaller than 4, we could replace 1.5 with 1, in which case the analysis is simpler. On the other hand, we believe that changing 1.5 to ∞ would give the best approximation, in which case the algorithm also seems cleaner (since $\mathcal{F}'_j = \mathcal{F}_j$). However, if the scalar were ∞, the algorithm is hard to analyze due to some technical reasons. So, the scalar 1.5 is selected so that we don't lose too much in the approximation ratio and yet the analysis is still manageable.

at least one $\mathcal{F}'_j$ is *claimed* by the nearest $j \in \mathcal{C}'$ such that $i \in \mathcal{F}'_j$, breaking ties arbitrarily. Then, $\mathcal{U}_j \subseteq \mathcal{F}_j$ is the set of facilities claimed by j.

Lemma 2. *The following two statements are true:*
(1) $1/2 \le \mathsf{vol}(\mathcal{U}_j) \le 1, \forall j \in \mathcal{C}'$, *and (2)* $\mathcal{U}_j \cap \mathcal{U}_{j'} = \emptyset, \forall j, j' \in \mathcal{C}', j \ne j'$.

Proof. Statement 2 is trivial; we only consider the first one. Since $\mathcal{U}_j \subseteq \mathcal{F}'_j \subseteq \mathcal{F}_j$, we have $\mathsf{vol}(\mathcal{U}_j) \le \mathsf{vol}(\mathcal{F}_j) = 1$. For a client $j \in \mathcal{C}'$, the closest client $j' \in \mathcal{C}' \setminus \{j\}$ to j has $d(j, j') > 4d_{av}(j)$ by lemma 1. So, $R_j > 2d_{av}(j)$ and the facilities in $\mathcal{F}_j$ that are at most $2d_{av}(j)$ away must be claimed by j. The set of these facilities has volume at least $1 - d_{av}(j)/(2d_{av}(j)) = 1/2$. Thus, $\mathsf{vol}(\mathcal{U}_j) \ge 1/2$.

The sets $\mathcal{U}_j$'s are called *bundles*. Each bundle $\mathcal{U}_j$ is treated as a single entity in the sense that at most 1 facility from it is open, and the probability that 1 facility is open is exactly $\mathsf{vol}(\mathcal{U}_j)$. From this point, a bundle $\mathcal{U}_j$ can be viewed as a single facility with $y = \mathsf{vol}(\mathcal{U}_j)$, except that it does not have a fixed position. We will use the phrase "opening the bundle $\mathcal{U}_j$" the operation that opens 1 facility randomly from $\mathcal{U}_j$, with probabilities $y_i / \mathsf{vol}(\mathcal{U}_j)$.

2.3 Matching Phase

Next, we construct a matching $\mathcal{M}$ over the bundles (or equivalently, over $\mathcal{C}'$). If two bundles $\mathcal{U}_j$ and $\mathcal{U}_{j'}$ are matched, we sample them using a joint distribution. Since each bundle has volume at least $1/2$, we can choose a distribution such that with probability 1, at least 1 bundle is open.

We construct the matching $\mathcal{M}$ using a greedy algorithm. While there are at least 2 unmatched clients in $\mathcal{C}'$, we choose the closest pair of unmatched clients $j, j' \in \mathcal{C}'$ and match them.

2.4 Sampling Phase

Following is our sampling phase.
1: **for** each pair $(j, j') \in \mathcal{M}$ **do**
2:    With probability $1 - \mathsf{vol}(\mathcal{U}_{j'})$, open $\mathcal{U}_j$; with probability $1 - \mathsf{vol}(\mathcal{U}_j)$, open $\mathcal{U}_{j'}$; and with probability $\mathsf{vol}(\mathcal{U}_j) + \mathsf{vol}(\mathcal{U}_{j'}) - 1$, open both $\mathcal{U}_j$ and $\mathcal{U}_{j'}$;
3: **end for**
4: If some $j \in \mathcal{C}'$ is not matched in $\mathcal{M}$, open $\mathcal{U}_j$ randomly and independently with probability $\mathsf{vol}(\mathcal{U}_j)$;
5: For each facility i not in any bundle $\mathcal{U}_j$, open it independently with probability y_i.

After we selected the open facilities, we connect each client to its nearest open facility. Let C_j denote the connection cost of a client $j \in \mathcal{C}$. Our sampling process opens k facilities in expectation, since each facility i is open with probability y_i. It does not always open k facilities as we promised. In the full version of the paper, we shall prove the following lemma:

Lemma 3. *There is a rounding procedure in which we always open k facilities and the probability that i is open is exactly y_i. The $\mathbb{E}[C_j]$ in this procedure is at most the $\mathbb{E}[C_j]$ in the rounding procedure we described. Moreover, the events that facilities are open are negatively-correlated; that is, for every set S of facilities,*

$$\Pr[\text{all facilities in } S \text{ are open}] \leq \prod_{i \in S} y_i.$$

By Lemma 3, it suffices to consider the rounding procedure we described. We shall outline the proof of the 3.25 approximation ratio for the above algorithm in section 3. As a warmup, we conclude this section with a much weaker result:

Lemma 4. *The algorithm gives a constant approximation for k-median.*

Proof. It is enough to show that the ratio between $\mathbb{E}[C_j]$ and $d_{av}(j)$ is bounded, for any $j \in C$. Moreover, it suffices to consider a client $j \in C'$. Indeed, if $j \notin C'$, there is a client $j_1 \in C'$ such that $d_{av}(j_1) \leq d_{av}(j), d(j, j_1) \leq 4d_{av}(j)$, by the second property of lemma 1. So $\mathbb{E}[C_j] \leq \mathbb{E}[C_{j_1}] + 4d_{av}(j)$. Thus, the ratio for j is bounded by the ratio for j_1 plus 4. So, it suffices to consider j_1.

W.L.O.G, assume $d_{av}(j_1) = 1$. Let j_2 be the client in $C' \setminus \{j_1\}$ that is closest to j_1. Consider the case where j_1 is not matched with j_2 (this is worse than the case where they are matched). Then, j_2 must be matched with another client, say $j_3 \in C'$, before j_1 is matched, and $d(j_2, j_3) \leq d(j_1, j_2)$. The sampling process guarantees that there must be a open facility in $\mathcal{U}_{j_2} \cup \mathcal{U}_{j_3}$. It is true that j_2 and j_3 may be far away from j_1. However, if $d(j_1, j_2) = 2R$ (thus, $d(j_1, j_3) \leq 4R, d_{av}(j_2), d_{av}(j_3) \leq R/2$), the volume of $\mathcal{U}_{j_1}$ is at least $1 - 1/R$. That means with probability at least $1 - 1/R$, j_1 will be connected to a facility that serves it in the fractional solution; only with probability $1/R$, j_1 will be connected to a facility that is $O(R)$ away. This finishes the proof.

3 Outline of the Proof of the 3.25-Approximation Ratio

If we analyze the algorithm as in the proof of lemma 4, an additive factor of 4 is lost at the first step. This additive factor can be avoided,[2] if we notice that there is a set $\mathcal{F}_j$ of facilities of volume 1 around j. Hopefully with some probability, some facility in $\mathcal{F}_j$ is open. It is not hard to show that this probability is at least $1 - 1/e$. So, only with probability $1/e$, we are going to pay the additive factor of 4. Even if there are no open facilities in $\mathcal{F}_j$, the facilities in $\mathcal{F}_{j_1}$ and $\mathcal{F}_{j_2}$ can help to reduce the constant.

A natural style of analysis is: focus on a set of "potential facilities", and consider the expected distance between j and the closest open facility in this set. An obvious candidate for the potential set is $\mathcal{F}_j \cup \mathcal{F}_{j_1} \cup \mathcal{F}_{j_2} \cup \mathcal{F}_{j_3}$. However, we are unable to analyze this complicated system.

Instead, we will consider a different potential set. Observing that $\mathcal{U}_{j_1}, \mathcal{U}_{j_2}, \mathcal{U}_{j_3}$ are disjoint, the potential set $\mathcal{F}_j \cup \mathcal{U}_{j_1} \cup \mathcal{U}_{j_2} \cup \mathcal{U}_{j_3}$ is much more tractable. Even with

---

[2] This is inspired by the analysis for the facility location problem in [8,4,14].

this simplified potential set, we still have to consider the intersection between $\mathcal{F}_j$ and each of $\mathcal{U}_{j_1}$, $\mathcal{U}_{j_2}$ and $\mathcal{U}_{j_3}$. Furthermore, we tried hard to reduce the approximation ratio at the cost of complicating the analysis(recall the argument about the choice of the scalar 1.5). With the potential set $\mathcal{F}_j \cup \mathcal{U}_{j_1} \cup \mathcal{U}_{j_2} \cup \mathcal{U}_{j_3}$, we can only prove a worse approximation ratio. To reduce it to 3.25, different potential sets are considered for different bottleneck cases.

W.L.O.G, we can assume $j \notin \mathcal{C}'$, since we can think of the case $j \in \mathcal{C}'$ as $j \notin \mathcal{C}'$ and there is another client $j_1 \in \mathcal{C}'$ with $d(j, j_1) = 0$. We also assume $d_{av}(j) = 1$. Let $j_1 \in \mathcal{C}'$ be the client such that $d_{av}(j_1) \leq d_{av}(j) = 1, d(j, j_1) \leq 4d_{av}(j) = 4$. Let j_2 be the closest client in $\mathcal{C}' \setminus \{j_1\}$ to j_1, thus $d(j_1, j_2) = 2R_{j_1}$. Then, either j_1 is matched with j_2, or j_2 is matched with a different client $j_3 \in \mathcal{C}'$, in which case we will have $d(j_2, j_3) \leq d(j_1, j_2) = 2R_{j_1}$. We only consider the second case. Readers can verify this is indeed the bottleneck case.

For the ease of notation, define $2R := d(j_1, j_2) = 2R_{j_1}, 2R' := d(j_2, j_3) \leq 2R, d_1 := d(j, j_1), d_2 := d(j, j_2)$ and $d_3 := d(j, j_3)$.

At the top level, we divide the analysis into two cases : the case $2 \leq d_1 \leq 4$ and the case $0 \leq d_1 \leq 2$. (Notice that we assumed $d_{av}(j) = 1$ and thus $0 \leq d_1 \leq 4$.) For some technical reason, we can not include the whole set $\mathcal{F}_j$ in the potential set for the former case. Instead we only include a subset $\mathcal{F}'_j$ (notice that $j \notin \mathcal{C}'$ and thus $\mathcal{F}'_j$ was not defined before). $\mathcal{F}'_j$ is defined as $\mathcal{F}_j \cap B(j, d_1)$.

The case $2 \leq d_1 \leq 4$ is further divided into 2 sub-cases : $\mathcal{F}'_j \cap \mathcal{F}'_{j_1} \subseteq \mathcal{U}_{j_1}$ and $\mathcal{F}'_j \cap \mathcal{F}'_{j_1} \not\subseteq \mathcal{U}_{j_1}$. Thus, we will have 3 cases, and the proof of the approximation ratios appear in the full paper.

1. $2 \leq d_1 \leq 4, \mathcal{F}'_j \cap \mathcal{F}'_{j_1} \subseteq \mathcal{U}_{j_1}$. In this case, we consider the potential set $\mathcal{F}'' = \mathcal{F}'_j \cup \mathcal{F}'_{j_1} \cup \mathcal{U}_{j_2} \cup \mathcal{U}_{j_3}$. Notice that $\mathcal{F}'_j = \mathcal{F}_j \cap B(j, d_1)$, $\mathcal{F}'_{j_1} = \mathcal{F}_{j_1} \cap B(j_1, 1.5R)$. In this case, $E[C_j] \leq 3.243$.

2. $2 \leq d_1 \leq 4, \mathcal{F}'_j \cap \mathcal{F}'_{j_1} \not\subseteq \mathcal{U}_{j_1}$. In this case, some facility i in $\mathcal{F}'_j \cap \mathcal{F}'_{j_1}$ must be claimed by some client $j' \neq j_1$. Since $d(j, i) \leq d_1, d(j_1, i) \leq 1.5R$, we have

$$d(j, j') \leq d(j, i) + d(j', i) \leq d(j, i) + d(j_1, i) \leq d_1 + 1.5R.$$

If $j' \notin \{j_2, j_3\}$, we can include $\mathcal{U}_{j'}$ in the potential set and thus the potential set is $\mathcal{F}'' = \mathcal{F}'_j \cup \mathcal{F}'_{j_1} \cup \mathcal{U}_{j_2} \cup \mathcal{U}_{j_3} \cup \mathcal{U}_{j'}$. If $j \in \{j_2, j_3\}$, then we know j and j_2, j_3 are close. So, we either have a "larger" potential set, or small distances between j and j_2, j_3. Intuitively, this case is unlikely to be the bottleneck case. In this case, we show $E[C_j] \leq 3.189$.

3. $0 \leq d_1 \leq 2$. In this case, we consider the potential set $\mathcal{F}'' = \mathcal{F}_j \cup \mathcal{U}_{j_1} \cup \mathcal{U}_{j_2} \cup \mathcal{U}_{j_3}$. In this case, $E[C_j] \leq 3.25$.

3.1   Running Time of the Algorithm

We now analyze the running time of our algorithm in terms of $n = |\mathcal{F} \cup \mathcal{C}|$. The bottleneck of the algorithm is solving the LP. Indeed, the total running time for rounding is $O(n^2)$.

To solve the LP, we use the $(1 + \epsilon)$ approximation algorithm for the fractional k-median problem in [17]. The algorithm gives a fractional solution which opens

$(1 + \epsilon)k$ facilities with connection cost at most $1 + \epsilon$ times the fractional optimal in time $O(kn^2 \ln(n/\epsilon)/\epsilon^2)$. We set $\epsilon = \delta/k$ for some small constant δ. Then, our rounding procedure will open k facilities with probability $1 - \delta$ and $k + 1$ facilities with probability δ. The expected connection cost of the integral solution is at most $3.25(1 + \delta/k)$ times the fractional optimal. Conditioned on the rounding procedure opening k facilities, the expected connection cost is at most $3.25(1 + \delta/k)/(1 - \delta) \leq 3.25(1 + O(\delta))$ times the optimal fractional value.

Theorem 1. *For any $\delta > 0$, there is a $3.25(1 + \delta)$-approximation algorithm for k-median problem that runs in $\tilde{O}\left((1/\delta^2)k^3n^2\right)$ time.*

3.2 Generalization of the Algorithm to Variants of k-Median

The distribution of k open facilities produced by our algorithm satisfies marginal conditions. That is, the probability that a facility i is open is exactly y_i. This allows our algorithm to be extended to some variants of the k-median problem.

The first variant is called k-facility location problem, which is a common generalization of k-median and UFL introduced in [10]. In the problem, we are given set $\mathcal{F}$ of facilities, set $\mathcal{C}$ of clients, metric $(d, \mathcal{F} \cup \mathcal{C})$, opening cost f_i for each facility $i \in \mathcal{F}$ and an integer k. The goal is to open at most k facilities and connect each client to its nearest open facility so as to minimize the sum of the opening cost and the connection cost. The best known approximation ratio for the k-facility location problem was $2 + \sqrt{3} + \epsilon$, due to Zhang [18]. For this problem, the LP is the same as LP(1), except that we add a term $\sum_{i \in \mathcal{F}} f_i y_i$ to the objective function. After solving the LP, we use our rounding procedure to obtain an integer solution. The expected opening cost of the solution is exactly the fractional opening cost in the LP solution, while the expected connection cost is at most 3.25 times the fractional connection cost. This gives a 3.25-approximation for the problem, improving the $2 + \sqrt{3} + \epsilon$-approximation.

Another generalization introduced in [10] is the k-median problem with t types of facilities. The goal of the problem is to open at most k facilities and connect each client to *one facility of each type* so as to minimize the total connection cost. Our techniques yield a 3.25 approximation for this problem as well. We first solve the natural LP for this problem. Then, we apply the rounding procedure to each type of facilities. The only issue is that the number of open facilities of some type in the LP solution might not be an integer. This can be handled using the techniques in the proof of Lemma 3.

4 Approximation Algorithms for Knapsack-Median and Matroid-Median

The LP for knapsack-median is the same as LP (1), except that we change the constraint $\sum_{i \in \mathcal{F}} y_i \leq k$ to the knapsack constraint $\sum_{i \in \mathcal{F}} f_i y_i \leq M$.

As shown in [13], the LP has unbounded integrality gap. To amend this, we do the same trick as in [13]. Suppose we know the optimal cost OPT for the instance. For a client j, let L_j be its connection cost. Then, for some other client j', its connection cost is at least $\max\{0, L_j - d(j, j')\}$. This suggests

$$\sum_{j' \in C} \max\{0, L_j - d(j, j')\} \leq \mathsf{OPT}. \tag{1}$$

Thus, knowing OPT, we can get an upper bound L_j on the connection cost of j: L_j is the largest number such that the above inequality is true. We solve the LP with the additional constraint that $x_{i,j} = 0$ if $d(i,j) > L_j$. Then, the LP solution, denoted by LP, must be at most OPT. By binary searching, we find the minimum OPT so that $\mathsf{LP} \leq \mathsf{OPT}$. Let $\left(x^{(1)}, y^{(1)}\right)$ be the fractional solution given by the LP. We use $\mathsf{LP}_j = d_{av}(j) = \sum_{i \in F} d(i,j)x_{i,j}^{(1)}$ to denote the contribution of the client j to LP.

Then we select a set of filtered clients C' as we did in the algorithm for the k-median problem. For a client $j \in C$, let $\pi(j)$ be a client $j' \in C'$ such that $d_{av}(j') \leq d_{av}(j), d(j, j') \leq 4d_{av}(j)$. Notice that for a client $j \in C'$, we have $\pi(j) = j$. This time, we can not save the additive factor of 4; instead, we move the connection demand on each client $j \notin C'$ to $\pi(j)$. For a client $j' \in C'$, let $w_{j'} = \left|\pi^{-1}(j')\right|$ be its connection demand. Let $\mathsf{LP}^{(1)} = \sum_{j' \in C', i \in F} w_{j'} x_{i,j'} d(i,j') = \sum_{j' \in C'} w_{j'} d_{av}(j')$ be the cost of the solution $\left(x^{(1)}, y^{(1)}\right)$ to the new instance. For a client $j \in C$, let $\mathsf{LP}_j^{(1)} = d_{av}(\pi(j))$ be the contribution of j to $\mathsf{LP}^{(1)}$. (The amount $w_{j'} d_{av}(j')$ is evenly spread among the $w_{j'}$ clients in $\pi^{-1}(j')$.) Since $\mathsf{LP}_j = d_{av}(j) \leq d_{av}(\pi(j)) \leq \mathsf{LP}_j^{(1)}$, we have $\mathsf{LP}^{(1)} \leq \mathsf{LP}$.

For any client $j \in C'$, let $2R_j = \min_{j' \in C', j' \neq j} d(j, j')$, if $\mathsf{vol}(B(j, R_j)) \leq 1$; otherwise let R_j be the smallest number such that $\mathsf{vol}(B(j, R_j)) = 1$. ($\mathsf{vol}(S)$ is defined as $\sum_{i \in S} y_i^{(1)}$.) Let $B_j = B(j, R_j)$ for the ease of notation. If $\mathsf{vol}(B_j) = 1$, we call B_j a full ball; otherwise, we call B_j a partial ball. Notice that we always have $\mathsf{vol}(B_j) \geq 1/2$. Notice that $R_j \leq L_j$ since $x_{i,j}^{(1)} = 0$ for all facilities i with $d_{i,j} > L_j$.

We find a matching $\mathcal{M}$ over the partial balls as in Section 2: while there are at least 2 unmatched partial balls, match the two balls B_j and $B_{j'}$ with the smallest $d(j, j')$. Consider the following LP.

$$\mathsf{LP(2)} \qquad \min \quad \sum_{j' \in C'} w_{j'} \left(\sum_{i \in B_{j'}} d(i, j')y_i + \left(1 - \sum_{i \in B_{j'}} y_i\right) R_{j'} \right)$$

$$\sum_{i \in B_{j'}} y_i = 1, \quad \forall j' \in C', B_{j'} \text{ full}; \qquad \sum_{i \in B_{j'}} y_i \leq 1, \quad \forall j' \in C', B_{j'} \text{ partial};$$

$$\sum_{i \in B_j} y_i + \sum_{i \in B_{j'}} y_i \geq 1, \quad \forall(B_j, B_{j'}) \in \mathcal{M}; \qquad \sum_{i \in F} f_i y_i \leq M;$$

$$y_i \geq 0, \quad \forall i \in F$$

Let $y^{(2)}$ be an optimal *basic solution* of LP (2) and let $\mathsf{LP}^{(2)}$ be the value of LP(2). For a client $j \in C$ with $\pi(j) = j'$, let $\mathsf{LP}_j^{(2)} = \sum_{i \in B_{j'}} d(i, j')y_i + \left(1 - \sum_{i \in B_{j'}} y_i\right) R_{j'}$ be the contribution of j to $\mathsf{LP}^{(2)}$. Then we prove

Lemma 5. $LP^{(2)} \leq LP^{(1)}$.

Proof. It is easy to see that $y^{(1)}$ is a valid solution for $LP(2)$. By slightly abusing the notations, we can think of $LP^{(2)}$ is the cost of $y^{(1)}$ to $LP(2)$. We compare the contribution of each client $j \in \mathcal{C}$ with $\pi(j) = j'$ to $LP^{(2)}$ and to $LP^{(1)}$. If $B_{j'}$ is a full ball, j' contributes the same to $LP^{(2)}$ and as to $LP^{(1)}$. If $B_{j'}$ is a partial ball, j' contributes $\sum_{i \in \mathcal{F}_{j'}} d(i,j')y_i^{(1)}$ to $LP^{(1)}$ and $\sum_{i \in B_{j'}} d(i,j')y_i^{(1)} + (1 - \sum_{i \in B_{j'}} y_i^{(1)})R_{j'}$ to $LP^{(2)}$. Since $B_{j'} = B(j', R_{j'}) \subseteq \mathcal{F}_{j'}$ and $\text{vol}(\mathcal{F}_{j'}) = 1$, the contribution of j' to $LP^{(2)}$ is at most that to $LP^{(1)}$. So, $LP^{(2)} \leq LP^{(1)}$.

Notice that $LP(2)$ only contains y-variables. We show that any basic solution y^* of $LP(2)$ is almost integral. In particular, we prove the following lemma in the full version of the paper:

Lemma 6. *Any basic solution y^* of $LP(2)$ contains at most 2 fractional values. Moreover, if it contains 2 fractional values $y_i^*, y_{i'}^*$, then $y_i^* + y_{i'}^* = 1$ and either there exists some $j \in \mathcal{C}'$ such that $i, i' \in B_j$ or there exists a pair $(B_j, B_{j'}) \in \mathcal{M}$ such that $i \in B_j, i' \in B_{j'}$.*

Let $y^{(3)}$ be the integral solutin obtained from $y^{(2)}$ as follows. If $y^{(2)}$ contains at most 1 fractional value, we zero-out the fractional value. If $y^{(2)}$ contains 2 fractional values $y_i^{(2)}, y_{i'}^{(2)}$, let $y_i^{(3)} = 1, y_{i'}^{(3)} = 0$ if $f_i \leq f_{i'}$ and let $y_i^{(3)} = 0, y_{i'}^{(3)} = 1$ otherwise. Notice that since $y_i^{(2)} + y_{i'}^{(2)} = 1$, this modification does not increase the budget. Let SOL be the cost of the solution $y^{(3)}$ to the original instance.

We leave the proof of the following lemma to the full version of the paper.

Lemma 7. $\sum_{i \in B(j', 5R_{j'})} y_i^{(2)} \geq 1$ *and* $\sum_{i \in B(j', 5R_{j'})} y_i^{(3)} \geq 1$. *i.e, there is an open facility (possibly two facilities whose opening fractions sum up to 1) inside $B(j', 5R_{j'})$ in both the solution $y^{(2)}$ and the solution $y^{(3)}$.*

Lemma 8. $\mathsf{SOL} \leq 34 \mathsf{OPT}$.

Proof. Let $\tilde{i}$ be the facility that $y_{\tilde{i}}^{(2)} > 0, y_{\tilde{i}}^{(3)} = 0$, if it exists; let $\tilde{j}$ be the client that $\tilde{i} \in B_{\tilde{j}}$.

Now, we focus on a client $j \in \mathcal{C}$ with $\pi(j) = j'$. Then, $d(j, j') \leq 4d_{av}(j) = 4\mathsf{LP}_j$. Assume that $j' \neq \tilde{j}$. Then, to obtain $y^{(3)}$, we did not move or remove an open facility from $B_{j'}$. In other words, for every $i \in B_{j'}, y_i^{(3)} \geq y_i^{(2)}$. In this case, we show

$$\mathsf{SOL}_{j'} \leq \sum_{i \in B_{j'}} d(i, j')y_i^{(2)} + (1 - \sum_{i \in B_{j'}} y_i^{(2)}) \times 5R_{j'}.$$

If there is no open facility in $B_{j'}$ in $y^{(3)}$, then there is also no open facility in $B_{j'}$ in $y^{(2)}$. Then, by Lemma 7, $\mathsf{SOL}_{j'} = 5R_{j'} = $ right-side. Otherwise, there is exactly one open facility in $B_{j'}$ in $y^{(3)}$. In this case, $\mathsf{SOL}_{j'} = \sum_{i \in B_{j'}} d(j', i)y_i^{(3)} \leq$ right-side since $y_i^{(3)} \geq y_i^{(2)}$ and $d(i, j') \leq 5R_{j'}$ for every $i \in B_{j'}$.

Observing that the right side of the inequality is at most $5\mathsf{LP}_j^{(2)}$, we have $\mathsf{SOL}_j \le 4\mathsf{LP}_j + \mathsf{SOL}_{j'} \le 4\mathsf{LP}_j + 5\mathsf{LP}_j^{(2)}$.

Now assume that $j' = \tilde{j}$. Since there is an open facility in $B(j', 5R_{j'})$ by Lemma 7, we have $\mathsf{SOL}_j \le 4\mathsf{LP}_j + 5R_{j'}$. Consider the set $\pi^{-1}(j')$ of clients. Notice that we have $R_{j'} \le L_{j'}$ since $x_{i,j'}^{(1)} = 0$ for facilities i such that $d(i, j') > L_{j'}$. Also by Inequality (1), we have $\sum_{j \in \pi^{-1}(j')}(R_{j'} - d(j, j')) \le \sum_{j \in \pi^{-1}(j')}(L_{j'} - d(j, j')) \le \mathsf{OPT}$. Then, since $d(j, j') \le 4\mathsf{LP}_j$ for every $j \in \pi^{-1}(j')$, we have

$$\sum_{j \in \pi^{-1}(j')} \mathsf{SOL}_j \le \sum_j (4\mathsf{LP}_j + 5R_{j'}) \le 4\sum_j \mathsf{LP}_j + 5\sum_j R_{j'}$$

$$\le 4\sum_j \mathsf{LP}_j + 5\left(\mathsf{OPT} + \sum_j d(j, j')\right) \le 24\sum_j \mathsf{LP}_j + 5\mathsf{OPT},$$

where the sums are all over clients $j \in \pi^{-1}(j')$. Summing up all clients $j \in \mathcal{C}$, we have

$$\mathsf{SOL} = \sum_{j \in \mathcal{C}} \mathsf{SOL}_j = \sum_{j \notin \pi^{-1}(\tilde{j})} \mathsf{SOL}_j + \sum_{j \in \pi^{-1}(\tilde{j})} \mathsf{SOL}_j$$

$$\le \sum_{j \notin \pi^{-1}(\tilde{j})} (4\mathsf{LP}_j + 5\mathsf{LP}_j^{(2)}) + 24 \sum_{j \in \pi^{-1}(\tilde{j})} \mathsf{LP}_j + 5\mathsf{OPT}$$

$$\le 24\sum_{j \in \mathcal{C}} \mathsf{LP}_j + 5\sum_{j \in \mathcal{C}} \mathsf{LP}_j^{(2)} + 5\mathsf{OPT} \le 24\mathsf{LP} + 5\mathsf{LP}^{(2)} + 5\mathsf{OPT} \le 34\mathsf{OPT},$$

where the last inequality follows from the fact that $\mathsf{LP}^{(2)} \le \mathsf{LP}^{(1)} \le \mathsf{LP} \le \mathsf{SOL}$. Thus, we proved

Theorem 2. *There is an efficient 34-approximation algorithm for the knapsack-median problem.*

It is not hard to change our algorithm so that it works for the matroid median problem. The analysis for the matroid median problem is simpler, since $y^{(2)}$ will already be an integral solution. We leave the proof of the following theorem to the full version of the paper.

Theorem 3. *There is an efficient 9-approximation algorithm for the matroid median problem, assuming there is an efficient oracle for the input matroid.*

References

1. Archer, A., Rajagopalan, R., Shmoys, D.B.: Lagrangian Relaxation for the k-Median Problem: New Insights and Continuity Properties. In: Di Battista, G., Zwick, U. (eds.) ESA 2003. LNCS, vol. 2832, pp. 31–42. Springer, Heidelberg (2003)
2. Arya, V., Garg, N., Khandekar, R., Meyerson, A., Munagala, K., Pandit, V.: Local search heuristic for k-median and facility location problems. In: Proceedings of the Thirty-Third Annual ACM Symposium on Theory of Computing, STOC 2001, pp. 21–29. ACM, New York (2001), http://doi.acm.org/10.1145/380752.380755

3. Bradley, P.S., Fayyad, U.M., Mangasarian, O.L.: Mathematical programming for data mining: Formulations and challenges. INFORMS Journal on Computing 11, 217–238 (1998)
4. Byrka, J.: An Optimal Bifactor Approximation Algorithm for the Metric Uncapacitated Facility Location Problem. In: Charikar, M., Jansen, K., Reingold, O., Rolim, J.D.P. (eds.) RANDOM 2007 and APPROX 2007. LNCS, vol. 4627, pp. 29–43. Springer, Heidelberg (2007)
5. Byrka, J., Srinivasan, A., Swamy, C.: Fault-Tolerant Facility Location: A Randomized Dependent LP-Rounding Algorithm. In: Eisenbrand, F., Shepherd, F.B. (eds.) IPCO 2010. LNCS, vol. 6080, pp. 244–257. Springer, Heidelberg (2010)
6. Charikar, M., Guha, S.: Improved combinatorial algorithms for the facility location and k-median problems. In: Proceedings of the 40th Annual IEEE Symposium on Foundations of Computer Science, pp. 378–388 (1999)
7. Charikar, M., Guha, S., Tardos, É., Shmoys, D.B.: A constant-factor approximation algorithm for the k-median problem (extended abstract). In: Proceedings of the Thirty-First Annual ACM Symposium on Theory of Computing, STOC 1999, pp. 1–10. ACM, New York (1999), http://doi.acm.org/10.1145/301250.301257
8. Chudak, F.A., Shmoys, D.B.: Improved approximation algorithms for the uncapacitated facility location problem. SIAM J. Comput. 33(1), 1–25 (2004)
9. Jain, K., Mahdian, M., Saberi, A.: A new greedy approach for facility location problems. In: Proceedings of the Thiry-Fourth Annual ACM Symposium on Theory of Computing, STOC 2002, pp. 731–740. ACM, New York (2002), http://doi.acm.org/10.1145/509907.510012
10. Jain, K., Vazirani, V.V.: Approximation algorithms for metric facility location and k-median problems using the primal-dual schema and lagrangian relaxation. J. ACM 48(2), 274–296 (2001)
11. Krishnaswamy, R., Kumar, A., Nagarajan, V., Sabharwal, Y., Saha, B.: The matroid median problem. In: Proceedings of ACM-SIAM Symposium on Discrete Algorithms, pp. 1117–1130 (2011)
12. Kuehn, A.A., Hamburger, M.J.: A heuristic program for locating warehouses, vol. 9(9), pp. 643–666 (July 1963)
13. Kumar, A.: Constant factor approximation algorithm for the knapsack median problem. In: Proceedings of the Twenty-Third Annual ACM-SIAM Symposium on Discrete Algorithms, SODA 2012, pp. 824–832. SIAM (2012), http://dl.acm.org/citation.cfm?id=2095116.2095182
14. Li, S.: A 1.488 Approximation Algorithm for the Uncapacitated Facility Location Problem. In: Aceto, L., Henzinger, M., Sgall, J. (eds.) ICALP 2011, Part II. LNCS, vol. 6756, pp. 77–88. Springer, Heidelberg (2011)
15. Lin, J.H., Vitter, J.S.: Approximation algorithms for geometric median problems. Inf. Process. Lett. 44, 245–249 (1992), http://portal.acm.org/citation.cfm?id=152566.152569
16. Manne, A.: Plant location under economies-of-scale-decentralization and computation. Managment Science (1964)
17. Young, N.E.: K-medians, facility location, and the chernoff-wald bound. In: Proceedings of the Eleventh Annual ACM-SIAM Symposium on Discrete Algorithms, SODA 2000, pp. 86–95. Society for Industrial and Applied Mathematics, Philadelphia, PA, USA (2000), http://portal.acm.org/citation.cfm?id=338219.338239
18. Zhang, P.: A New Approximation Algorithm for the k-Facility Location Problem. In: Cai, J.-Y., Cooper, S.B., Li, A. (eds.) TAMC 2006. LNCS, vol. 3959, pp. 217–230. Springer, Heidelberg (2006), http://dx.doi.org/10.1007/11750321_21

Node-Weighted Network Design in Planar and Minor-Closed Families of Graphs*

Chandra Chekuri, Alina Ene, and Ali Vakilian

Dept. of Computer Science, University of Illinois, Urbana, IL 61801, USA
{chekuri,ene1,vakilia2}@illinois.edu

Abstract. We consider *node-weighted* network design in planar and minor-closed families of graphs. In particular we focus on the edge-connectivity survivable network design problem (EC-SNDP). The input consists of a node-weighted undirected graph $G = (V, E)$ and integral connectivity requirements $r(uv)$ for each pair of nodes uv. The goal is to find a minimum node-weighted subgraph H of G such that, for each pair uv, H contains $r(uv)$ edge-disjoint paths between u and v. Our main result is an $O(k)$-approximation algorithm for EC-SNDP where $k = \max_{uv} r(uv)$ is the maximum requirement. This improves the $O(k \log n)$-approximation known for node-weighted EC-SNDP in general graphs [15]. Our algorithm and analysis applies to the more general problem of covering a proper function with maximum requirement k. Our result is inspired by, and generalizes, the work of Demaine, Hajiaghayi and Klein [5] who gave constant factor approximation algorithms for node-weighted Steiner tree and Steiner forest problems (and more generally covering 0-1 proper functions) in planar and minor-closed families of graphs.

1 Introduction

Network design is an important area of discrete optimization with several practical applications. Moreover, the clean optimization problems that underpin the applications have led to fundamental theoretical advances in combinatorial optimization, algorithms and mathematical programming. In this paper we consider a class of problems that can be modeled as follows. Given an undirected graph $G = (V, E)$ find a subgraph H of *minimum weight/cost* such that H satisfies certain desired *connectivity* properties. A common cost model is to assign a non-negative weight $w(e)$ to each $e \in E$ and the weight/cost of H is simply the total weight of edges in it. A number of well-studied problems can be cast as special cases. Examples include polynomial-time solvable problems such as the minimum spanning tree (MST) problem when H is required to connect all the nodes of G, and the NP-Hard Steiner tree problem where H is required to connect only a given subset $S \subseteq V$ of terminals. A substantial generalization of these problems is the *survivable network design problem* which is defined as follows. The input, in addition to G, consists of an integer requirement function $r(uv)$ for each (unordered) pair of nodes uv in G; the goal is to find a minimum-weight subgraph H that contains $r(uv)$ edge-disjoint paths between u and v for each pair uv. This problem is called the

---

* The authors are partially supported by NSF grant CCF-1016684.

A. Czumaj et al. (Eds.): ICALP 2012, Part I, LNCS 7391, pp. 206–217, 2012.

edge-connectivity SNDP (EC-SNDP) to distinguish from more general problems such as Elem-SNDP and VC-SNDP that require the paths to be element and vertex disjoint respectively. SNDP arises naturally in the design of fault-tolerant networks, and various special cases have been extensively studied. Algorithmic approaches for SNDP and related problems are based on solving a larger class of abstract network design problems such as covering proper and skew-supermodular cut-requirement functions that we describe formally later.

Node Weights: The cost of a network is dependent on the application. In connectivity problems, as we remarked, a common model is the edge-weight model. A more general problem is obtained when each node v of G has a weight $w(v)$ and the weight of H is the total weight of the nodes in H[1]. Node weights are relevant in several applications, in particular telecommunication networks, where they can model the cost of setting up routing and switching infrastructure at a given node. There have also been several recent applications in wireless network design [17,16] where the weight function is closely related to that of node weights. We refer the reader to [5] for some additional applications of node weights to network formation games.

The node-weighted versions of network design problems often turn out to be strictly harder to approximate than their corresponding edge-weighted versions. For instance the Steiner tree problem admits a 1.39-approximation for edge-weights [2], however, Klein and Ravi [12] showed, via a simple reduction from the Set Cover problem, that the node-weighted Steiner tree problem on n nodes is hard to approximate to within an $\Omega(\log n)$-factor unless $P = NP$. They also described a $(2 \log k)$-approximation where k is the number of terminals. A more dramatic difference emerges if we consider SNDP. Jain gave a 2-approximation for EC-SNDP with edge-weights [10]. The best known approximation for EC-SNDP with node-weights is $O(k \log n)$ [15] where $k = \max_{uv} r(uv)$ is the maximum connectivity requirement. Nutov [15] gives evidence, via a reduction from the k-densest-subgraph problem, that for the node-weighted problem a dependence on k in the approximation ratio is necessary.

Demaine, Hajiaghayi and Klein [5] considered the approximability of the node-weighted Steiner tree problem in planar graphs. In an interesting result, they adapted the well-known primal-dual algorithm for the edge-weighted problem [1,7] to the node-weighted problem and showed that it gives a 6-approximation in planar graphs. Demaine et al. also showed that their algorithm works for a more general class of 0-1-valued proper functions (first considered by Goemans and Williamson [7]) that includes several other problems such as the Steiner forest problem ([14] claims an improved $9/4$ approximation for the Steiner forest problem). Their analysis also shows that one obtains a constant factor approximation (the algorithm is the same) for any minor-closed family of graphs where the constant depends on the family. In addition to their theoretical value, these results have the potential to be useful in practice since in many real-world networks the underlying graph G is either planar or has very few crossings.

---

[1] For many problems of interest, including Steiner tree and SNDP, the version with weights on both edges and nodes can be reduced to the version with only node weights; sub-divide an edge e by placing a new node v_e and set the weight of v_e to be that of e.

Our Results: In this paper we consider node-weighted network design problems in planar graphs for higher connectivity. In particular we consider EC-SNDP and show that the insights in [5] can be used to develop improved approximation algorithms for this more general problem as well. However, the results require non-trivial technical work that we explain after we state the results. The algorithm works for any graph but the ratio is constant for planar graphs and more generally graphs from any minor-closed family; we articulate the precise dependence of the ratio on the family in later sections.

Our main result is the following.

Theorem 1. *There is an $O(k)$-approximation for node-weighted EC-SNDP in planar graphs where k is the maximum requirement.*

The above theorem extends to a more general problem that we describe now. An integer valued set function $f : 2^V \rightarrow \mathbb{Z}_+$ on the vertex set of G is said to be proper if it satisfies the following conditions: (i) f is symmetric, that is, $f(S) = f(V - S)$ for all S, and (ii) f is maximal, that is, $f(A \cup B) \leq \max\{f(A), f(B)\}$ for any two disjoint sets A, B. Given a proper function f on V (by a value oracle) and a graph G on V, the f-covering problem is to find a subgraph H of minimum weight such that $|\delta_H(S)| \geq f(S)$ for all S^2. EC-SNDP is a special case of this problem [18]. We obtain an $O(k)$-approximation for the node-weighted version of this problem in planar graphs where $k = \max_S f(S)$.

Overview of Technical Ideas and Contribution: The two main algorithmic approaches for SNDP are the following. The first is the augmentation approach pioneered by Williamson et al. [18] in which the required network is built in k phases. At the end of the first $(i - 1)$ phases the connectivity of a pair uv is at least $\min\{r(uv), i - 1\}$. Thus the i'th phase is required to increase the connectivity of some of the pairs by 1 by adding additional edges; the advantage of this approach is that we now work with a 0-1 covering problem. On the other hand the covering problem is no longer so simple. The function that we need to cover falls into the more general class of *uncrossable* functions: A requirement function $f : 2^V \rightarrow \{0, 1\}$ is uncrossable if for any sets $A, B \subseteq V$, $f(A) = f(B) = 1$ implies $f(A \cap B) = f(A \cup B) = 1$ or $f(A - B) = f(B - A) = 1$. Williamson et al. [18] showed that a primal-dual algorithm achieves a 2-approximation for the edge-weighted version of covering uncrossable functions. Nutov [15] gave an $O(\log n)$-approximation for the node-weighted case. These results for uncrossable functions, when combined with the augmentation framework, give a $2k$ and an $O(k \log n)$ approximation for the edge-weighted and node-weighted versions of EC-SNDP in general graphs[3]. The second approach for SNDP is the powerful iterated rounding technique pioneered by Jain which led to a 2-approximation for EC-SNDP [10] and also for covering a certain class of skew-supermodular functions[4]. Iterated rounding does not quite apply to node-weighted problems for various technical reasons.

---

[2] We work with node-induced subgraphs H of G in which case H may not contain all the nodes of a set $S \subset V$. In that case $\delta_H(S)$ denotes the edges of H with exactly one endpoint in S.

[3] The approximation for the edge-weighted version can be improved to $2H_k$ by doing the augmentation in the reverse [6].

[4] A function $f : 2^V \rightarrow \mathbb{Z}$ is skew-supermodular if for all $A, B \subseteq V$, $f(A) + f(B) \leq \max\{f(A \cap B) + f(A \cup B), f(A - B) + f(B - A)\}$. A skew-supermodular function f with $f(A) \leq 1$ for all A gives rise to an uncrossable function.

We follow the augmentation approach. Demaine et al. adapted the primal-dual algorithm for edge-weighted 0-1-proper functions to the node-weighted case. The novel technical ingredient in their analysis is to understand properties of *node-minimal* feasible solutions instead of edge-minimal feasible solutions. For the most part, problems captured by 0-1-proper functions are very similar to the Steiner forest problem, a canonical problem in this class. In this setting it is possible to visualize and understand node-minimal solutions through connected components and basic reachability properties. In the augmentation approach for higher-connectivity, as we remarked, the problem in each phase is no longer that of covering a proper function but belongs to the richer class of covering uncrossable functions. The primal-dual analysis for this class of functions is more subtle and abstract [18] and proceeds via uncrossing arguments and laminar witness families.

Our main technical contribution is understanding properties of node-minimal feasible solutions for uncrossable functions. We refer the reader to Theorem 3 in Section 3 for the precise statement; the theorem holds for general graphs (not just planar graphs) and may have other applications. We remark on a crucial aspect of our algorithm and analysis. Why do our results only apply for covering proper functions and not the more general class of skew-supermodular functions? For the node-weighted problem of covering an arbitrary uncrossable function there is no natural covering LP relaxation. However, we observe that the particular uncrossable functions that arise in the augmentation framework for a proper function (including EC-SNDP) have certain additional connectivity properties that allow for an LP relaxation and the primal-dual approach. We obtain a constant factor approximation in each phase and this results in an $O(k)$-approximation overall where k is the maximum requirement.

As in [5] we use planarity only in one step of the analysis where we argue about the average degree of a certain graph that is a minor of the original graph; this is the reason that the algorithm and analysis generalize to any minor-closed family. In this paper, in the interest of clarity and exposition, we have not attempted to optimize the constants in the approximation.

Extensions: Our ideas for EC-SNDP can be extended to give an $O(k)$ approximation for node-weighted Elem-SNDP in planar graphs. We again use the augmentation approach but for Elem-SNDP we use a primal-dual algorithm and analysis with respect to the setpair relaxation [11,3]. There are however some non-trivial differences and the generalization is not immediate. An improved algorithm for node-weighted VC-SNDP in planar graphs follows from a generic reduction of VC-SNDP to Elem-SNDP [4]. A longer version of this paper will discuss these extensions.

Other Related Work: There is extensive literature on network design but due to space limitations we are unable to discuss it in detail. We refer the reader to [8] for a survey on primal-dual based algorithms for network design, and to recent surveys [13,9] for an overview of the known approximation results and references to related work.

Organization: Section 2 describes our algorithm based on the augmentation approach and the primal-dual algorithm for each phase of the augmentation. The analysis is done by assuming the main technical theorem on a node-minimal augmentation of the uncrossable requirement functions that arise in the augmentation framework. We state and prove this theorem in Section 3. Some of the proofs are omitted in this version. A longer version with detailed proofs as well as the claimed extensions will be made available on arXiv and the authors' web pages in the near future.

2   Algorithm for Node-Weighted EC-SNDP and Proper Functions

We start by defining the node-weighted EC-SNDP problem formally. The input consists of an undirected node-weighted graph $G = (V, E)$ (weight of node v is denoted by $w(v)$) and a requirement function $r(uv)$ for each pair of nodes. The goal is to find a minimum node-weighted subgraph H of G such that H contains $r(uv)$ edge-disjoint paths for each pair uv. We use k to denote the maximum requirement. A node u is called a *terminal* if there is some node v such that $r(uv) > 0$. Since any feasible solution has to contain all terminals, we can assume without loss of generality that the weight of every terminal is zero. We define a function $f : 2^V \to \mathbb{Z}_+$ where $f(S) = \max\{r(uv) \mid u \in S, v \notin S\}$. It is well-known that f is a proper function. By Menger's theorem, solving node-weighted EC-SNDP is equivalent to finding a minimum node-weight subgraph H such that $|\delta_H(S)| \geq f(S)$ for all $S \subset V$. (Recall that $\delta_H(S)$ is the set of all edges of H with exactly one endpoint in S.) Our algorithm and analysis extend to the problem of finding a node-weighted subgraph to cover a given proper function. For an arbitrary proper function f we call a node v a terminal if $f(\{v\}) > 0$; maximality of f implies that S contains a terminal if $f(S) > 0$. Again, we can assume without loss of generality that terminals have zero weight, since they are included in any feasible solution.

We alert the reader that, in order to cover the function f, we need to pick a set of *edges*. But since the weights are (only) on the nodes, we pay for a set of nodes and we can use any of the edges in the graph induced by the nodes in order to cover the function. More precisely, our goal is to select a minimum-weight node-induced subgraph $H = G[X]$ that covers f, where X is a subset of nodes of G. We will always assume that X contains the terminals.

As we mentioned, our algorithm for covering f uses the augmentation framework introduced in [18]. Let $f_p : 2^V \to \mathbb{Z}$ be the function such that $f_p(S) = \min\{f(S), p\}$ for each set S. If f is a proper function then f_p is also a proper function. The algorithm performs k phases: for $1 \leq p \leq k$, at the end of phase p, the algorithm has a subgraph H_p that covers f_p. In phase p the algorithm starts with H_{p-1} that covers f_{p-1} and adds some additional nodes to obtain H_p that covers f_p. We can express the underlying optimization problem in phase p as follows.

It is convenient to assume that all of the vertices of H_{p-1} have zero weight; since we have already paid for the nodes, we can set their weight to zero at the beginning of phase p. Let $G'_p = (V, E(G) - E(H_{p-1}))$. (We emphasize that G'_p has all of the nodes of G and that the terminals and vertices of $V(H_{p-1})$ have zero weight.) Our goal is to select a minimum-weight subgraph H of G'_p that covers the following 0-1 function $h_p : 2^V \to \{0, 1\}$. For each set S, we have $h_p(S) = 1$ iff $f(S) \geq p$ and $|\delta_{H_{p-1}}(S)| = p - 1$. The function h_p is known to be an uncrossable function [18]; note that it may no longer be a proper function. We use a primal-dual algorithm to cover h_p in the graph G'_p. A 2-approximation exists for this covering problem for the edge-weighted problem and an $O(\log n)$-approximation for the node-weighted case [15]. We show that the primal-dual algorithm achieves an $O(1)$-approximation for the node-weighted case in planar graphs, however, we emphasize that it only applies for the specific uncrossable functions that arise from proper functions as above; in particular it is important that the chosen subgraphs at the end of each phase are node-induced. We describe and analyze the primal-dual algorithm below and point out the place where we need this restriction.

2.1 A Primal-Dual Algorithm for the Augmentation Problem

In the following, we fix a phase p of the augmentation framework. Let $h = h_p$ and $G' = G'_p$. Recall that all of the terminals and the vertices selected in the first $p-1$ phases have zero weight. In the following, we use $\Gamma_{G'}(S)$ to denote the set of all vertices v such that $v \notin S$ but there is an edge $uv \in E(G')$ such that $u \in S$. We use a primal-dual algorithm in order to select a subgraph H of G' that covers h. The primal and dual LPs that we use are described below. We remark that the primal LP has unbounded integrality gap for an arbitrary uncrossable function[5]. However, the function h that arises from a proper function f in the augmentation framework has additional properties that allow us to avoid such examples.

Primal:	Dual:
$\min \sum\limits_{v \in V} x(v)w(v)$	$\max \sum\limits_{S \subseteq V} y(S)h(S)$
s.t. $\sum\limits_{v \in \Gamma_{G'}(S)} x(v) \geq h(S) \quad \forall S \subseteq V$	s.t. $\sum\limits_{S:v \in \Gamma_{G'}(S)} y(S) \leq w(v) \quad \forall v \in V$
$x(v) \geq 0 \quad \forall v \in V$	$y(S) \geq 0 \quad \forall S \subseteq V$

We omit the constraint $x(v) \leq 1$ in the primal since h is a 0-1 function.

The primal-dual algorithm is a "standard" one in that it is the natural adaptation to the node-weighted setting (as done in [5]) of the primal-dual algorithm for edge-weighted network design formalized by Goemans and Williamson [7]. The algorithm selects a set $X \subseteq V(G')$ of nodes such that the graph $G'[X]$ covers h. Initially, X consists of all vertices that have zero weight. We also maintain a feasible dual solution $\mathbf{y}$ that is implicitly initialized to zero. We proceed in iterations. Consider iteration i and let X_{i-1} be the set of nodes selected in the first $i-1$ iterations; the set X_0 consists of all zero-weight vertices. A set S is *violated* with respect to X_{i-1} iff $h(S) = 1$ and $\delta_{G'[X_{i-1}]}(S) = \emptyset$. A set S is a *minimal violated set* with respect to X_{i-1} iff S is a violated set and no proper subset of S is violated. Let $\mathcal{C}_i$ denote the collection of all minimal violated sets with respect to X_{i-1}. As shown in [18], no two minimal violated sets of an uncrossable function can intersect; further the collection of minimal violated sets for h arising from proper functions can be computed in polynomial time. Moreover, Lemma 1 below shows that the sets in $\mathcal{C}_i$ are subsets of X_{i-1}. If $\mathcal{C}_i$ is empty,

---

[5] A simple example is a function h such that there is a single set S such that $h(S) = 1$. Each vertex in S has weight 1, and each vertex in $V - S$ has weight 0. The optimum solution has value 1 since at least one node in S has to be picked but the optimum LP value is 0; note that the value is 0 even if we have integrality constraints.

$G'[X_{i-1}]$ covers h and we return $G'[X_{i-1}]$. Otherwise, we increase the dual variables $\{y(S)\}_{S \in C_i}$ uniformly until a dual constraint for a vertex v becomes tight, i.e., we have $\sum_{S:v \in \Gamma_{G'}(S)} y(S) = w(v)$; we add v to X. Note that, since the components of C_i are contained in X_{i-1}, for each minimal violated component $C \in C_i$, none of the vertices in $\Gamma_{G'}(C)$ are in X_{i-1} and thus it is possible to increase the dual variables $\{y(S)\}_{S \in C_i}$.

Finally we perform a *reverse-delete* step. Let X be the set of vertices selected by the primal-dual algorithm. We select a subset Y of X as follows. We start with $Y = X$. We order the vertices of Y in the reverse of the order in which they were selected by the primal-dual algorithm. Let v be the current vertex. If $G'[Y - v]$ is a feasible cover for h, we remove v from Y.

The primal-dual algorithm described above is not well-defined for an arbitrary uncrossable function h but the following property holds for those that arise from proper functions. Using the following lemma, we can show that the algorithm is well-defined and it outputs a cover of h in polynomial time.

Lemma 1. *Every minimal violated component $C \in C_i$ is a subset of X_{i-1}.*
Proof: Consider $C \in C_i$ and suppose $C \not\subseteq X_{i-1}$. Let $C' = C \cap X_{i-1}$. We observe that $f_p(C \setminus C') = 0$ since all the terminals are in X_{i-1}. Since f_p is maximal, we have $f_p(C) \leq \max\{f_p(C'), f_p(C \setminus C')\} = \max\{f_p(C'), 0\} = f_p(C')$. Since $C \in C_i$, we have $f_p(C) = p$ and $|\delta_{G[X_{i-1}]}(C)| = p - 1$. Therefore $f_p(C') \geq f_p(C) = p$. Additionally, $\delta_{G[X_{i-1}]}(C) = \delta_{G[X_{i-1}]}(C')$, since $G[X_{i-1}]$ does not have any edges incident to vertices in $V \setminus X_{i-1}$. It follows that C' is violated with respect to X_{i-1}, which contradicts the minimality of C.   □

Now we turn our attention to the analysis of the primal-dual algorithm. In the following, we show that the algorithm achieves an $O(1)$ approximation for the augmentation problem when the graph G is from a minor-closed family of graphs $\mathcal{G}$; the constant depends on the family $\mathcal{G}$.

Theorem 2. *If G is a graph from a minor-closed family of graphs $\mathcal{G}$, the weight of the set Y is $O(\text{OPT}_h)$, where OPT_h is the optimum solution to the LP relaxation for covering h.*

The dual variables are grown uniformly in each iteration and the standard primal-dual analysis [7] gives a condition under which the approximation ratio can be upper bounded. This is encapsulated in the lemma below.

Lemma 2. *Let $B_i = Y - X_{i-1}$. Suppose there exists a γ such that, for each iteration i of the primal-dual algorithm, $\sum_{C \in C_i} |B_i \cap \Gamma_{G'}(C)| \leq \gamma |C_i|$. Then the weight of Y is at most γOPT_h, where OPT_h is the value of an optimal solution to the LP relaxation.*

The content of the above lemma is the following. Consider the minimal violated sets in C_i. The set $B_i = Y - X_{i-1}$ forms a *node-minimal* set that together with X_{i-1} covers h (minimality follows from the reverse delete step). We are interested in γ, the "average degree"[6] of the components in C_i, with respect to nodes in B_i. In general graphs γ can

---

[6] Here we are abusing the term slightly and we refer to the ratio $\sum_{C \in C_i} |B_i \cap \Gamma_{G'}(C)|/|C_i|$ as the average degree of the components in C_i. One can view the ratio as the average degree of the components if we shrink each of the components in C_i to a single vertex and we remove parallel edges.

be $\Omega(n)$ in the worst case which does not give a useful bound. However, planar graphs are sparse. Thus one can bound the average degree if one can bound the number of *nodes* in B_i that are adjacent to components in C_i. This was done in [5] for 0-1 proper functions but the case of uncrossable functions is more involved and it is our main technical contribution. Theorem 3 is stated in a general and useful form and proved in Section 3. Assuming the theorem, we finish the analysis as follows. The following lemma upper bounds the number of nodes in B_i that are adjacent to components in C_i.

Lemma 3. *Let B'_i be the set of all vertices $u \in B_i$ such that $u \in \Gamma_{G'}(C)$ for some component $C \in C_i$. We have $|B'_i| \leq 4|C_i|$.*

In order to take advantage of the fact that planar and minor-closed graphs are sparse, we need the following technical ingredient. The proof of Lemma 4 follows from the maximality of f_p and it is similar to the proof of Lemma 1.

Lemma 4. *For each component $C \in C_i$, the graph $G[C]$ is connected.*

In order to finish the average degree argument, we shrink each component $C \in C_i$ into a single node and we use Lemma 3 and the fact that, for a graph K from a minor-closed family $\mathcal{G}$ there is a constant c' that depends only on the family such that $|E(K)| \leq c'|V(K)|$.

Lemma 5. *Let $B_i = Y - X_{i-1}$. If G is a graph from a minor-closed family of graphs $\mathcal{G}$, we have $\sum_{C \in C_i} |B_i \cap \Gamma_{G'}(C)| \leq c|C_i|$, where c is a constant that depends only on the family $\mathcal{G}$.*

Theorem 2 follows from Lemma 2 and Lemma 5. Theorem 2 together with the augmentation framework gives an $O(k)$-approximation for finding a minimum node-weighted subgraph to cover a proper function with maximum requirement k. The result for EC-SNDP is a special case of this result.

Remark 1. For planar graphs, we get a 10-approximation for the augmentation problem and a $10k$-approximation for the EC-SNDP problem. Demaine et al. [5] get a 6-approximation for planar graphs when $k = 1$, and thus our ratio is slightly weaker. Our analysis in Lemma 5 could be tightened in several ways. We believe that the analysis in Theorem 3 and consequently Lemma 3 can be improved to obtain a factor of 3 instead of 4. The analysis uses the maximality of f but not symmetry and hence our results hold for a larger class of functions than proper functions.

3   Proof of Theorem 3

Let $G = (V, E)$ be a graph. Let $h : 2^V \rightarrow \{0, 1\}$ be a requirement function. A set S is *violated* if $h(S) = 1$. A set C is a *minimal violated component* of h if C is violated and no proper subset of C is violated. Let H be a subgraph of G. The graph H is a *feasible cover* for h if, for any set $S \subseteq V$ such that $h(S) = 1$, there is at least one edge of H leaving S; in other words, $|\delta_H(S)| \geq h(S)$. We say that H is a *node-minimal* feasible cover for h if, for any vertex $v \in V(H)$, $H - v$ is not a feasible cover for h.

Now we are ready to state our main theorem.

Theorem 3. *Let* $h : 2^V \rightarrow \{0,1\}$ *be an uncrossable function. Let* C *be the minimal violated components of* h. *Let* H *be a node-minimal feasible cover for* h. *Let* X *be the set of all vertices* $v \in V(H)$ *such that* v *is not in the union of the components in* C *and there is an edge of* H *connecting* v *to a component of* C. *Then* $|X| \leq 4|C|$.

We devote the rest of this section to the proof of Theorem 3. A basic property of uncrossable functions [18] is stated below.

Lemma 6. *Let* h *be an uncrossable function. The minimal violated components of* h *are disjoint. Moreover, if* S *is a violated set and* C *is a minimal violated component,* S *and* C *do not properly intersect.*

We start with a high-level overview of the proof. The main idea is to pick a subset M of the edges of H such that M is an *edge-minimal* feasible cover for h. Such a minimal cover has nice properties that were pointed out and used in the analysis for edge-weighted problems [18]. More precisely, for each edge $e \in M$, we can pick a "witness set" that is a violated set such that e is the only edge of M that is leaving the set. Moreover, we can pick a family of witness sets, one for each edge of M, such that the family is laminar[7]. This laminar family can be used to upper bound the number of edges of M that are incident to the components of C.

We are interested in analyzing a node-minimal cover H which is not necessarily edge-minimal; there can be a node u that is adjacent to components in C but it is possible that an edge-minimal cover M does not contain any of the edges connecting u to components of C. Thus we cannot use the witness family to count such vertices. We address this issue by counting them separately using a witness family for a *different* set of edges.

We now turn our attention to the formal proof of the theorem. We refer to the vertices in X as *critical* vertices. We refer to edges connecting a critical vertex to a component $C \in C$ as *red edges*, and we refer to all other edges of H as *blue edges*.

We define two subsets of edges F_1 and F_2 as follows. We start with $F_1 = E(H)$ and we remove some of the edges as follows. We order the *blue edges* arbitrarily. We consider the blue edges in this order. Let e be the current edge. If $F_1 - e$ is a feasible solution for h, we remove e from F_1. At the end of this process, each red edge is in F_1 and each blue edge in F_1 is necessary to cover h. We refer to critical vertices that are incident to at least one blue edge of F_1 as *regular* vertices; critical vertices that are not regular are referred to as *special* vertices. As we will see shortly, we can use the blue edges in F_1 to upper bound the number of regular vertices.

In order to count the special vertices, we pick a subset F_2 of F_1 as follows. We start with $F_2 = F_1$. We consider the *red edges* of F_2 in some order. Let e be the current edge. If $F_2 - e$ is a feasible cover, we remove e from F_2. We can use the red edges in F_2 to upper bound the number of special vertices. Since H is a node-minimal cover for h, each special vertex is incident to at least one red edge of F_2.

Note that F_2 is an edge-minimal feasible cover for h while F_1 is a feasible cover but is not necessarily edge-minimal. The difficulty is in counting the regular vertices via F_1. We consider the regular and special vertices separately. Theorem 3 follows from the following two lemmas.

---

[7] A set family $\mathcal{F}$ is laminar iff no two sets in $\mathcal{F}$ properly intersect.

Lemma 7. *The number of regular vertices is at most* $2|\mathcal{C}|$.

Lemma 8. *The number of special vertices is at most* $2|\mathcal{C}|$.

Our counting arguments are based on the laminar witness family approach of Williamson et al. More precisely, we define a witness set as follows.

Definition 1. *Let* F *be a set of edges. A set* $S_e \subseteq V$ *is an* F-***witness set*** *for an edge* e *iff* $h(S_e) = 1$ *and* $\delta_F(S_e) = \{e\}$.

An F-witness set S_e is a violated set; from Lemma 6 it follows that for each component $C \in \mathcal{C}, C \subseteq S_e$ or $C \cap S_e = \emptyset$.

Recall that a family of sets $\mathcal{L}$ is *laminar* if no two sets in $\mathcal{L}$ properly intersect; differently said, for any two sets $A, B \in \mathcal{L}$, either A and B are disjoint or one is contained in the other. The following lemma follows from [18].

Lemma 9 ([18]). *Let* F *be a feasible cover for an uncrossable function* h. *Let* $M \subseteq F$ *be a subset of* F *such that, for each edge* $e \in M$, $F - e$ *is not a feasible cover for* h. *There is a laminar family* $\mathcal{L} = \{S_e \mid e \in M\}$ *such that* S_e *is an* F-*witness set for* e.

Our approach is to use laminar witness families for the blue edges of F_1 and the red edges of F_2 in order to count the regular and special vertices. Before we turn our attention to the counting arguments, we describe some properties of laminar witness families that we need.

We can associate a forest $\mathcal{F}$ with a laminar set family $\mathcal{L}$ as follows. The forest $\mathcal{F}$ has a node ν_S for each set $S \in \mathcal{L}$. We add an edge between ν_A and ν_B iff A is the smallest set in $\mathcal{L}$ that contains B. Let $\mathcal{L} = \{S_e \mid e \in M\}$ be a laminar F-witness family for a set $M \subseteq F$ of edges. Let $\mathcal{T}$ be the tree associated with $\mathcal{L} \cup \{V\}$; we root $\mathcal{T}$ at the node ν_V.

We define the following bijection between the edges of the tree $\mathcal{T}$ and the edges of M. Let e be an edge of M and let S_e be the witness set for e. The node ν_{S_e} has a parent ν_A in $\mathcal{T}$, and we associate the edge $e \in M$ with the edge (ν_A, ν_{S_e}) of $\mathcal{T}$. We say that the edge e *corresponds* to the edge (ν_A, ν_{S_e}). A node ν_S of $\mathcal{T}$ *owns* a vertex $v \in V$ iff S is the smallest set in $\mathcal{L} \cup \{V\}$ that contains v.

Proposition 1. *Let* $\mathcal{L} = \{S_e \mid e \in M\}$ *be a laminar* F-*witness family for a set* $M \subseteq F$ *of edges. Let* $\mathcal{T}$ *be the tree associated with* $\mathcal{L} \cup \{V\}$. *For each leaf* ν_S *of* $\mathcal{T}$ *there is a distinct component* $C \in \mathcal{C}$ *such that* $C \subseteq S$.

The following simple observation plays a crucial role in our counting argument.

Proposition 2. *Let* $\mathcal{L} = \{S_e \mid e \in M\}$ *be a laminar* F-*witness family for a set* $M \subseteq F$ *of edges. Let* $\mathcal{T}$ *be the tree associated with* $\mathcal{L} \cup \{V\}$. *Let* ν_S *be a node of* $\mathcal{T}$ *and let* e *be an edge of* $F \setminus M$. *Either both endpoints of* e *are contained in* S *or neither endpoint of* e *is contained in* S. *In particular, the endpoints of* e *are owned by the same node of* $\mathcal{T}$.

The following lemma was proved in [18].

Lemma 10 ([18]). *Let $\mathcal{L} = \{S_e \mid e \in M\}$ be a laminar F-witness family for a set $M \subseteq F$ of edges. Let $\mathcal{T}$ be the tree associated with $\mathcal{L} \cup \{V\}$. Let e be an edge of M and let (ν_A, ν_{S_e}) be the edge of $\mathcal{T}$ corresponding to e, where S_e is the witness set for e and ν_A is the parent of ν_{S_e}. Then ν_A owns one endpoint of e and ν_{S_e} owns the other endpoint of e.*

Counting Argument for Regular Vertices. Let $\mathcal{L}_{F_1} = \{S_e \mid e$ is a blue edge in $F_1\}$ be a laminar F_1-witness family for the blue edges in F_1 that is guaranteed by Lemma 9. Let $\mathcal{T}_{F_1}$ be the tree associated with the family $\mathcal{L}_{F_1} \cup \{V\}$; we view $\mathcal{T}_{F_1}$ as a rooted tree whose root is the node corresponding to V.

Recall that each regular vertex u is incident to a red edge ur; the edge ur is in F_1, since F_1 contains all the red edges. Additionally, u is incident to a blue edge $ub \in F_1$. Since r is contained in a minimal component of $\mathcal{C}$, it follows from Proposition 2 that the node of $\mathcal{T}_{F_1}$ that owns u also owns a component $C_u \in \mathcal{C}$. Our approach is to charge each regular vertex u in its subtree; more precisely, we charge u to a component $C \in \mathcal{C}$ that is owned by a node in the subtree rooted at the node that owns u and C_u.

We charge each regular vertex u as follows. Recall that there is a blue edge $ub \in F_1$ that is incident to u. Let ν_A and ν_B be the nodes of $\mathcal{T}_{F_1}$ that own u and b, respectively. By Lemma 10, one of ν_A, ν_B is the parent of the other.

Suppose that ν_A is the parent of ν_B. Since each leaf owns a component of $\mathcal{C}$ (from Proposition 1), there is a descendant of ν_B (possibly ν_B itself) that owns a component of $\mathcal{C}$. Let ν_S be the closest such descendant, i.e., a descendant whose distance to ν_B is minimized. (If there are several descendants whose distance to ν_B is minimum, we pick one of them arbitrarily.) We charge u to one of the components of $\mathcal{C}$ that ν_S owns; we refer to this charge as a *subtree charge* (since u is charged in a subtree rooted at a child of the node ν_A that owns u). Since a regular vertex v and its component C_v are owned by the same node of the tree, the components C_v serve as sentinels that ensure that there is at most one subtree charge to each component of $\mathcal{C}$.

Suppose that ν_A is a child of ν_B. We charge u to the component C_u; we refer to this charge as a *parent charge* (since the charge corresponds to the tree edge connecting the node ν_A that owns C to its parent). Since each node has at most one parent edge, there is at most one parent charge to each component of $\mathcal{C}$.

Proposition 3. *There is at most one subtree charge to each component $C \in \mathcal{C}$.*

Proposition 4. *There is at most one parent charge to each component $C \in \mathcal{C}$.*

Proof of Lemma 7: Each component of $\mathcal{C}$ is charged at most twice and thus the number of regular vertices is at most $2|\mathcal{C}|$.   □

Counting Argument for Special Vertices. Recall that F_2 is an edge-minimal cover of h. Moreover, a critical vertex v is special only if there is an edge $e \in F_2$ (in fact a red edge) such that e connects v to a minimal violated component C. Thus, the total number of special vertices is upper bounded by $\sum_{C \in \mathcal{C}} |\delta_{F_2}(C)|$. Williamson et al. [18] show that for any edge-minimal cover of an uncrossable function this is upper bounded by $2|\mathcal{C}|$. Thus we can upper bound the number of special vertices by $2|\mathcal{C}|$ which proves Lemma 8. We remark that some of the regular vertices are counted in this step as well, but this can only help us.

References

1. Agrawal, A., Klein, P., Ravi, R.: When trees collide: An approximation algorithm for the generalized Steiner problem on networks. SIAM Journal on Computing 24(3), 440–456 (1995)
2. Byrka, J., Grandoni, F., Rothvoß, T., Sanità, L.: An improved LP-based approximation for Steiner tree. In: Proc. of ACM STOC 2010, pp. 583–592 (2010)
3. Cheriyan, J., Vempala, S., Vetta, A.: Network design via iterative rounding of setpair relaxations. Combinatorica 26(3), 255–275 (2006)
4. Chuzhoy, J., Khanna, S.: An $O(k^3 \log n)$-approximation algorithm for vertex-connectivity survivable network design. In: Proc. of FOCS, pp. 437–441. IEEE (2009)
5. Demaine, E.D., Hajiaghayi, M., Klein, P.N.: Node-Weighted Steiner Tree and Group Steiner Tree in Planar Graphs. In: Albers, S., Marchetti-Spaccamela, A., Matias, Y., Nikoletseas, S., Thomas, W. (eds.) ICALP 2009. LNCS, vol. 5555, pp. 328–340. Springer, Heidelberg (2009)
6. Goemans, M.X., Goldberg, A.V., Plotkin, S., Shmoys, D.B., Tardos, E., Williamson, D.P.: Improved approximation algorithms for network design problems. In: Proc. of ACM-SIAM SODA, pp. 223–232 (1994)
7. Goemans, M.X., Williamson, D.P.: A general approximation technique for constrained forest problems. SIAM Journal on Computing 24, 296 (1995)
8. Goemans, M.X., Williamson, D.P.: The primal-dual method for approximation algorithms and its application to network design problems. In: Approximation Algorithms for NP-Hard Problems, pp. 144–191. PWS Publishing Co. (1996)
9. Gupta, A., Könemann, J.: Approximation algorithms for network design: A survey. Surveys in Operations Research and Management Science 16(1), 3–20 (2011)
10. Jain, K.: A factor 2 approximation algorithm for the generalized Steiner network problem. Combinatorica 21(1), 39–60 (2001)
11. Jain, K., Mandoiu, I., Vazirani, V.V., Williamson, D.P.: A primal-dual schema based approximation algorithm for the element connectivity problem. Journal of Algorithms 45(1), 1–15 (2002)
12. Klein, P., Ravi, R.: A nearly best-possible approximation algorithm for node-weighted Steiner trees. Journal of Algorithms 19(1), 104–115 (1995)
13. Kortsarz, G., Nutov, Z.: Approximating minimum cost connectivity problems. In: Gonzalez, T.F. (ed.) Handbook on Approximation Algorithms and Metaheuristics. Chapman and Hall/CRC (2007)
14. Moldenhauer, C.: Primal-Dual Approximation Algorithms for Node-Weighted Steiner Forest on Planar Graphs. In: Aceto, L., Henzinger, M., Sgall, J. (eds.) ICALP 2011. LNCS, vol. 6755, pp. 748–759. Springer, Heidelberg (2011)
15. Nutov, Z.: Approximating Steiner networks with node-weights. SIAM Journal of Computing 39(7), 3001–3022 (2010)
16. Nutov, Z.: Approximating Steiner network activation problems. In: Proc. of LATIN (2012)
17. Panigrahi, D.: Survivable network design problems in wireless networks. In: Proc. of ACM-SIAM SODA (2011)
18. Williamson, D.P., Goemans, M.X., Mihail, M., Vazirani, V.V.: A primal-dual approximation algorithm for generalized Steiner network problems. Combinatorica 15(3), 435–454 (1995)

Computing the Visibility Polygon
of an Island in a Polygonal Domain*

Danny Z. Chen and Haitao Wang**

Department of Computer Science and Engineering
University of Notre Dame, Notre Dame, IN 46556, USA
{dchen,hwang6}@nd.edu

Abstract. Given a set $\mathcal{P}$ of h pairwise-disjoint polygonal obstacles of totally n vertices in the plane, we study the problem of computing the (weakly) visibility polygon from a polygonal obstacle P^* (an island) in $\mathcal{P}$. We give an $O(n^2 h^2)$ time algorithm for it. Previously, the special case where P^* is a line segment was solved in $O(n^4)$ time, which is worst-case optimal. In addition, when all obstacles in $\mathcal{P}$ (including P^*) are convex, our algorithm runs in $O(n + h^4)$ time.

1   Introduction

Given a set $\mathcal{P}$ of h pairwise-disjoint polygonal obstacles of totally n vertices in the plane, the space minus the interior of all obstacles is called the *free space*. Two points are *visible* to each other if the open line segment joining them lies in the free space. Two objects are *visible* to each other if a point of one object is visible to a point of the other object (this is often called *weakly visible* in the literature; we use *visible* when there is no confusion from the context). Consider a polygonal obstacle $P^* \in \mathcal{P}$ (an *island*). The *(weak) visibility polygon/region* of P^* (or *from P^**), denoted by $Vis(P^*)$, is the set of points in the plane visible to P^*. In this paper, we present an $O(n^2 h^2)$ time algorithm for computing $Vis(P^*)$. When all obstacles in $\mathcal{P}$ (including P^*) are convex, referred to as the *convex version*, we give an $O(n + h^4)$ time solution for computing $Vis(P^*)$.

Visibility problems have been studied extensively (e.g., [1,2,4,7,8,10,11,13,14,16,17,18]). Linear time algorithms were given for computing the visibility polygon inside a simple polygon P from a single point [7,13,14,16], from a line segment [10], and from another simple polygon [8] contained in the polygon P. For the problem versions on a polygonal domain $\mathcal{P}$ as defined above, if P^* is a single point, Suri and O'Rourke [18] and Asano et al. [1] presented $O(n \log n)$ time algorithms for computing $Vis(P^*)$; later, Heffernan and Mitchell [11] gave an $O(n + h \log h)$ time algorithm. If P^* is a line segment, Suri and O'Rourke [18] presented an $O(n^4)$ time algorithm and showed that this is optimal in the worst case. To our best knowledge, no result for the general problem of computing $Vis(P^*)$ in $\mathcal{P}$ when P^* is an arbitrary simple polygon was known before.

* This research was supported in part by NSF under Grant CCF-0916606.
** Corresponding author.

A. Czumaj et al. (Eds.): ICALP 2012, Part I, LNCS 7391, pp. 218–229, 2012.
© Springer-Verlag Berlin Heidelberg 2012

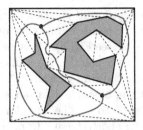 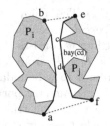 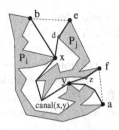

Fig. 1. [6]Illustrating a triangulation of the free space among two obstacles and the corridors (with red solid curves). There are two junction triangles indicated by the large dots inside them, connected by three solid (red) curves. Removing the two junction triangles results in three corridors.

Fig. 2. [6]Illustrating an open hourglass (left) and a closed hourglass (right) with a corridor path connecting the apices x and y of the two funnels. The dashed segments are diagonals. The paths $\pi(a, b)$ and $\pi(e, f)$ are shown with thick solid curves. A bay $bay(\overline{cd})$ with gate $\overline{cd}$ (left) and a canal $canal(x, y)$ with gates $\overline{xd}$ and $\overline{yz}$ (right) are also indicated.

Our $O(n^2h^2)$ time algorithm for computing $Vis(P^*)$ with P^* being a simple polygon improves (for small h) the $O(n^4)$ time solution [18] for the special case where P^* is a line segment, and actually solves a more general problem.

Our approach generalizes Suri and O'Rourke's algorithm [18], which we call the *SO algorithm*, by exploiting a corridor structure of the obstacles in $\mathcal{P}$. Corridor structures have been used in solving shortest path problems (e.g., [5,6,12,15]). In Section 2, we review the corridor structure of $\mathcal{P}$ and introduce some concepts and observations. In Section 3, we present our algorithm. The convex version is discussed in Section 4. Due to the space limit, some details are omitted and can be found in the full version of the paper.

2 Preliminaries

We review the corridor structure [15] and define some new concepts, e.g., the *ocean* $\mathcal{M}$, *bays*, and *canals*, etc. For simplicity, we assume all obstacles in $\mathcal{P}$ are contained in a rectangle $\mathcal{R}$ (see Fig. 1). We also use $\mathcal{R}$ to denote the space inside the rectangle, and use $\mathcal{F}$ to denote the free space in $\mathcal{R}$.

Denote by $Tri(\mathcal{F})$ a triangulation of $\mathcal{F}$. Let $G(\mathcal{F})$ denote the (planar) dual graph of $Tri(\mathcal{F})$. The degree of each node in $G(\mathcal{F})$ is at most three. Based on $G(\mathcal{F})$, we compute a planar 3-regular graph, denoted by G^3 (the degree of every node in G^3 is three), possibly with loops and multi-edges, as follows. First, we remove every degree-one node from $G(\mathcal{F})$ along with its incident edge; repeat this process until no degree-one node remains in the graph. Second, remove every degree-two node from $G(\mathcal{F})$ and replace its two incident edges by a single edge; repeat this process until no degree-two node remains. The resulting graph is G^3 (see Fig. 1). The resulting graph G^3 has $h + 1$ faces, $2h - 2$ nodes, and $3h - 3$ edges [15]. Every node of G^3 corresponds to a triangle in $Tri(\mathcal{F})$, which is called a *junction triangle* (see Fig. 1). The removal of the nodes for all junction triangles from G^3 results in $O(h)$ *corridors*, each of which corresponds to one edge of G^3.

The boundary of a corridor C consists of four parts (see Fig. 2): (1) A boundary portion of an obstacle $P_i \in \mathcal{P}$, from a point a to a point b; (2) a diagonal of a junction triangle from b to a boundary point e on an obstacle $P_j \in \mathcal{P}$ ($P_i = P_j$ is possible); (3) a boundary portion of the obstacle P_j from e to a point f; (4) a diagonal of a junction triangle from f to a. The corridor C is a simple polygon. Let $\pi(a,b)$ (resp., $\pi(e,f)$) be the shortest path from a to b (resp., e to f) inside C. The region H_C bounded by $\pi(a,b), \pi(e,f)$, and the two diagonals $\overline{be}$ and $\overline{fa}$ is called an *hourglass*, which is *open* if $\pi(a,b) \cap \pi(e,f) = \emptyset$ and *closed* otherwise (see Fig. 2). If H_C is open, then both $\pi(a,b)$ and $\pi(e,f)$ are convex chains and are called the *sides* of H_C; otherwise, H_C consists of two "funnels" and a path $\pi_C = \pi(a,b) \cap \pi(e,f)$ joining the two apices of the two funnels, called the *corridor path* of C. Each funnel side is also a convex chain. We compute the hourglass for each corridor. After $Tri(\mathcal{F})$ is produced, the total time for computing all hourglasses is $O(n)$.

Let $\mathcal{M}$ be the union of the $O(h)$ junction triangles, open hourglasses, and funnels. We call the space $\mathcal{M}$ the *ocean*, and $\mathcal{M} \subseteq \mathcal{F}$. Since the sides of open hourglasses and funnels are convex, the boundary $\partial \mathcal{M}$ of $\mathcal{M}$ consists of $O(h)$ convex chains with totally $O(n)$ vertices; further, there are $O(h)$ reflex vertices on $\partial \mathcal{M}$. This implies that the complementary region $\mathcal{R} \setminus \mathcal{M}$ consists of a set of polygons bounded by $O(h)$ convex chains with $O(h)$ reflex vertices. Thus, $\mathcal{R} \setminus \mathcal{M}$ can be partitioned into a set $\mathcal{P}'$ of $O(h)$ pairwise interior-disjoint convex polygons of totally $O(n)$ vertices [15]. If we view the convex polygons in $\mathcal{P}'$ as obstacles, then the ocean $\mathcal{M}$ is the free space with respect to $\mathcal{P}'$. A point on $\partial \mathcal{M}$ must be on the boundary of a convex obstacle in $\mathcal{P}'$. The set $\mathcal{P}'$ can be obtained easily in $O(n + h \log h)$ time. It should be pointed out that our algorithms given later can be applied to $\mathcal{M}$ directly without having to explicitly partition $\mathcal{R} \setminus \mathcal{M}$ into convex polygons in $\mathcal{P}'$. But for ease of exposition, we always discuss our algorithms on $\mathcal{P}'$. Next, we examine the other free space of $\mathcal{F}$ than $\mathcal{M}$, i.e., $\mathcal{F} \setminus \mathcal{M}$, which consists of two types of regions: *bays* and *canals*, defined below.

Consider the hourglass H_C of a corridor C. We first discuss the case when H_C is open (see Fig. 2). H_C has two sides. Let $S_1(H_C)$ be an arbitrary side of H_C. The obstacle vertices on $S_1(H_C)$ all lie on the same obstacle, say $P \in \mathcal{P}$. Let c and d be any two consecutive vertices on $S_1(H_C)$ such that the line segment $\overline{cd}$ is not an edge of P (see the left figure in Fig. 2, with $P = P_j$). The free region enclosed by $\overline{cd}$ and a boundary portion of P between c and d is called the *bay* of $\overline{cd}$ and P, denoted by $bay(\overline{cd})$, which is a simple polygon. We call $\overline{cd}$ the *bay gate* of $bay(\overline{cd})$, which is a common edge of $bay(\overline{cd})$ and $\mathcal{M}$.

If the hourglass H_C is closed, let x and y be the two apices of its two funnels. Consider two consecutive vertices c and d on a side of a funnel such that $\overline{cd}$ is not an obstacle edge. If neither c nor d is a funnel apex, then c and d must lie on the same obstacle and the segment $\overline{cd}$ also defines a bay with that obstacle. However, if c or d is a funnel apex, say, $c = x$, then c and d may lie on different obstacles. If they lie on the same obstacle, then they also define a bay; otherwise, we call $\overline{xd}$ the *canal gate* at $x = c$ (see Fig. 2). Similarly, there is also a canal gate at the other funnel apex y, say $\overline{yz}$. Let P_i and P_j be the two obstacles bounding the hourglass H_C. The free region enclosed by P_i, P_j, and the two canal gates $\overline{xd}$ and $\overline{yz}$ that contains the corridor path of H_C is the *canal* of H_C, denoted by $canal(x, y)$, which is also a simple polygon.

Note that the bays and canals together constitute the space $\mathcal{F} \setminus \mathcal{M}$. While the total number of all bays is $O(n)$, the total number of all canals is $O(h)$ since each canal corresponds to a corridor and the number of corridors is $O(h)$.

The fact that each bay has only one gate allows us to process a bay easily. Intuitively, an observer outside a bay cannot see any point outside the bay "through" its gate. But, each canal has two gates, which could possibly cause trouble. The next lemma discovers an important property that an observer outside a canal cannot see any point outside the canal through the canal (and its two gates); we call it *the opaque property* of canals.

Lemma 1. (The Opaque Property) *For any canal, suppose a line segment $\overline{pq}$ is in $\mathcal{F}$ (i.e., p is visible to q) such that neither p nor q is in the canal. Then, $\overline{pq}$ cannot contain any point of the canal that is not on its two gates.*

Proof. A sketch. Assume to the contrary that $\overline{pq}$ contains a point t in a canal such that t is not on either gate of the canal. Let $canal(x, y)$ be the canal as shown in Fig. 2. Since $\overline{pq}$ travels through the canal without intersecting any obstacle, the shortest paths $\pi(a, b)$ and $\pi(e, f)$ do not intersect. Thus, the hourglass defined by the corridor containing $canal(x, y)$ is open. But this means the corridor cannot contain a canal, incurring contradiction.

For a line segment $\overline{pq}$ in the free space $\mathcal{F}$, consider extending $\overline{pq}$ along both directions of the line containing $\overline{pq}$ until it first hits an obstacle (or goes to infinity) in each direction; the two points on obstacles (or at infinity) first hit by extending $\overline{pq}$ are called the *extension ends* of $\overline{pq}$.

3   Our Algorithm

This section presents our algorithm for computing $Vis(P^*)$. We first compute the corridor structure of $\mathcal{P}$ and obtain the convex polygonal obstacle set $\mathcal{P}'$. Denote by ∂P^* the boundary of P^*. Clearly, a point p is in $Vis(P^*)$ if and only if p is visible to a point on ∂P^*. With respect to $\mathcal{P}'$, we partition the edges of ∂P^* into three types. Some edges of ∂P^* may be in a bay (resp., canal), and we call these edges of ∂P^* the *Type-I edges* (resp., *Type-II edges*). Other edges of ∂P^* lie on $\partial \mathcal{M}$ and thus on the boundaries of some convex obstacles in $\mathcal{P}'$, and we call these edges of ∂P^* the *Type-III edges*. Clearly, $Vis(P^*)$ is the union of the visibility regions of the Type-I, Type-II, and Type-III edges. We first give some observations on the three types of edges.

3.1   Observations

The Type-I Edges. We begin with the Type-I edges. Suppose a bay $bay(\overline{cd})$ contains some Type-I edges. Since the boundary of $bay(\overline{cd})$ except its gate $\overline{cd}$

lies on the same obstacle of $\mathcal{P}$, all edges on the boundary of $bay(\overline{cd})$ except its gate are Type-I edges. The following observation is self-evident.

Observation 1. *If a bay B contains Type-I edges, then every point in B is visible to the Type-I edges in B; further, a point outside B is visible to the Type-I edges in B if and only if it is visible to the gate of B.*

Let α denote the set of the Type-I edges in $bay(\overline{cd})$. By Observation 1, the visibility region of α in $\mathcal{F}$ is the union of $bay(\overline{cd})$ and the visibility region of $\overline{cd}$ in the space $\mathcal{F} \setminus bay(\overline{cd})$. In other words, besides $bay(\overline{cd})$, computing the visibility region of α is reduced to computing the visibility region of $\overline{cd}$ in $\mathcal{F} \setminus bay(\overline{cd})$. If a bay contains any Type-I edges, we call its gate an *illumination gate*.

The Type-II Edges. Consider a canal $canal(x,y)$ with two gates $\overline{xd}$ and $\overline{yz}$ (see Fig. 2) such that $canal(x,y)$ contains some Type-II edges. Recall that the boundary of $canal(x,y)$ consists of $\overline{xd}$, a boundary portion of an obstacle $P_i \in \mathcal{P}$, $\overline{yz}$, and a boundary portion of an obstacle $P_j \in \mathcal{P}$. Then, one of P_i and P_j must be P^*. Denote by α the set of Type-II edges in $canal(x,y)$.

Recall that $P_i = P_j$ is possible. If $P_i = P_j = P^*$, then α consists of all edges of $canal(x,y)$ except its two gates. Due to the opaque property in Lemma 1, we have the next lemma, similarly to Observation 1.

Lemma 2. *If $P_i = P_j = P^*$, then every point in $canal(x,y)$ is visible to α and a point outside $canal(x,y)$ is visible to α if and only if the point is visible to a gate of $canal(x,y)$.*

Proof. A sketch. Since $canal(x,y)$ is a simple polygon and α contains all edges of $canal(x,y)$ except $\overline{xd}$ and $\overline{yz}$, every point in $canal(x,y)$ is visible to α.

Consider any point p outside $canal(x,y)$ that is visible to α, say, at a point $q \in \alpha$. Then $\overline{pq} \subset \mathcal{F}$. Since p is outside $canal(x,y)$ and q is in $canal(x,y)$, $\overline{pq}$ must intersect a gate of $canal(x,y)$. Therefore, p is visible to that gate.

Consider any point p outside $canal(x,y)$ that is visible to a gate, say $\overline{xd}$, of $canal(x,y)$. Then, there must be a point $q \in \overline{xd}$ such that $\overline{pq} \in \mathcal{F}$. By extending $\overline{pq}$ along the direction from p to q, the first point on obstacles of $\mathcal{P}$ hit by this extension of $\overline{pq}$ must be on α, which implies that p is visible to α. □

By Lemma 2, when $P_i = P_j = P^*$, the visibility region of α is the union of $canal(x,y)$ and the visibility region of the two gates of $canal(x,y)$ in $\mathcal{F} \setminus canal(x,y)$. If $P_i = P_j = P^*$, then the two gates of $canal(x,y)$ are also called *illumination gates*.

We then discuss the case of $P_i \neq P_j$. In this case, only one of P_i and P_j is P^*, say $P_i = P^*$, and the edge sequence of α has two endpoints at the two gates of $canal(x,y)$ respectively (e.g., they are x and y in Fig. 2). Each such endpoint of α is on a side of a funnel and thus lies on $\partial \mathcal{M}$. For simplicity of discussion, we assume that each such endpoint of α is on $\partial \mathcal{M}$ but not on α. In other words, α does not include these two endpoints. With this assumption, we mean that these two endpoints of α are considered as part of the Type-III edges, and the visibility region of the two endpoints of α will be found when we compute the visibility region of the Type-III edges.

Let $Vis(\alpha, \mathcal{F})$ (resp., $Vis(\alpha, canal(x, y))$) denote the visibility region of α in $\mathcal{F}$ (resp., $canal(x, y)$). We will show below that in this case, $Vis(\alpha, \mathcal{F})$ is either $Vis(\alpha, canal(x, y))$, or the union of $Vis(\alpha, canal(x, y))$ and the visibility region of one or both gates of $canal(x, y)$ in $\mathcal{F}$ (not in $\mathcal{F} \setminus canal(x, y)$). For example, in the situation of Fig. 2 (with $P_i = P^*$), $Vis(\alpha, \mathcal{F}) = Vis(\alpha, canal(x, y))$.

Recall that x and y are the two funnel apices of $canal(x, y)$. For its gate $\overline{dx}$, if the vertex $d \in P^*$, then we call $\overline{dx}$ an *illumination gate*, and we also assume that $\overline{dx}$ does not include its two end vertices since they are on $\partial \mathcal{M}$ and are treated as part of the Type-III edges. Similarly, if $z \in P^*$, then $\overline{yz}$ is an (open) *illumination gate*. Note that $canal(x, y)$ may have zero, one, or two illumination gates (e.g., the example in Fig. 2 has zero illumination gates with $P_i = P^*$). The following lemma (with proof omitted) characterizes the visibility region $Vis(\alpha, \mathcal{F})$.

Lemma 3. *The visibility region $Vis(\alpha, \mathcal{F})$ is the union of $Vis(\alpha, canal(x, y))$ and the visibility region of the (open) illumination gates (if any) of $canal(x, y)$ in $\mathcal{F}$.*

By Lemma 3, to find $Vis(\alpha, \mathcal{F})$, it suffices to compute $Vis(\alpha, canal(x, y))$ and the visibility region of the illumination gates of $canal(x, y)$ (if any) in $\mathcal{F}$ and then take their union. Note that we can compute $Vis(\alpha, canal(x, y))$ in linear time in terms of the number of vertices of $canal(x, y)$ [8].

The Type-III Edges and Illumination Gates. For the Type-III edges (which lie on $\partial \mathcal{M}$), since the illumination gates of all bays/canals also lie on $\partial \mathcal{M}$, our algorithm computes the visibility region of the union of all Type-III edges and illumination gates as a whole. Let Υ denote the union of all Type-III edges and illumination gates. The following lemma (with proof omitted) is due to the fact that the number of canals is $O(h)$.

Lemma 4. Υ *consists of $O(h)$ convex chains on $\partial \mathcal{M}$.*

3.2   Computing Vis(P*)

Let $Vis_1(P^*)$ denote the union of the visibility regions of the Type-I edges inside their corresponding bays and the visibility regions of the Type-II edges inside their corresponding canals which are characterized by Lemma 2 (i.e., each point in $Vis_1(P^*)$ is either in a bay that contains Type-I edges or in a canal with $P_i = P_j = P^*$). Let $Vis_2(P^*)$ denote the union of the visibility regions of the Type-II edges inside their corresponding canals which are characterized by Lemma 3 (i.e., $P_i = P^*$ or $P_j = P^*$, but $P_i \neq P_j$). Let $Vis_3(P^*)$ denote the visibility region of Υ in $\mathcal{F}$. Then by Observation 1 and Lemmas 2 and 3, we have $Vis(P^*) = Vis_1(P^*) \cup Vis_2(P^*) \cup Vis_3(P^*)$.

We first compute $Vis_1(P^*)$ and $Vis_2(P^*)$. By Observation 1 and Lemma 2, $Vis_1(P^*)$ is the union of all bays/canals each of which either (as a bay) contains some Type-I edges or (as a canal) satisfies Lemma 2. $Vis_2(P^*)$ can be computed in totally $O(n)$ time by using the algorithm in [8] since no two different canals intersect in their interior. Henceforth, we focus on computing $Vis_3(P^*)$.

When computing $Vis_3(P^*)$ in $\mathcal{F}$, we can ignore all bays/canals involved in $Vis_1(P^*)$ since they are entirely in $Vis_1(P^*)$ (i.e., we treat these bays/canals as being inside the obstacles defining them). But, the canals involved in $Vis_2(P^*)$ cannot be ignored since some points in such a canal $canal(x, y)$ that are not visible to the Type-II edges in $canal(x, y)$ can still be visible to other boundary portions of P^*. In the discussion below, we assume all bays/canals contained in $Vis_1(P^*)$ have been ignored. In other words, each bay considered below contains no Type-I edges, and if a canal contains some Type-II edges, then only one of the two obstacles defining the canal is P^*.

Consider an (open) illumination gate $\overline{xd}$ of a canal $canal(x, y)$. We may view $\overline{xd}$ as having two "sides", one (called the *canal side*) facing the inside of $canal(x, y)$ and the other (called the *ocean side*) facing the ocean $\mathcal{M}$. A point in $\mathcal{F}$ visible to $\overline{xd}$ must be visible to a side of $\overline{xd}$. In other words, the visibility region of $\overline{xd}$ in $\mathcal{F}$ is the union of the visibility regions of its two sides. By Lemma 1, the visibility polygon of the canal side of $\overline{xd}$ is a subset of $canal(x, y)$. By the definition of illumination gates of the canal $canal(x, y)$, the obstacle on which the vertex d lies is P^*. Let α be the set of the Type-II edges contained in $canal(x, y)$. We can show that the visibility polygon of the canal side of $\overline{xd}$ is a subset of the visibility polygon of α in $canal(x, y)$, which is contained in $Vis_2(P^*)$. The details are omitted. Thus, when computing $Vis_3(P^*)$, we can ignore the visibility polygon of the canal side of $\overline{xd}$. For this purpose, we view $\overline{xd}$ as two obstacle edges that are close infinitely to each other, and these two edges connect the two obstacles that define $canal(x, y)$ into one obstacle. Further, the edge of $\overline{xd}$ that is adjacent to $\mathcal{M}$ is still viewed as an edge in Υ, i.e., it is used to compute the visibility region of the ocean side of $\overline{xd}$ later; but, the other edge of $\overline{xd}$ is not treated as an edge of Υ. For the other gate $\overline{yz}$ of $canal(x, y)$, if $\overline{yz}$ is not an illumination gate, then the canal $canal(x, y)$ now becomes a bay with $\overline{yz}$ as its gate. Otherwise, we do the same thing on $\overline{yz}$, and the interior of $canal(x, y)$ can be ignored in computing $Vis_3(P^*)$. We process each illumination gate of a canal in this way. Processing all illumination gates of the canals (which satisfy Lemma 3) takes $O(n)$ time. Then, we obtain a new obstacle set, and all illumination gates of the canals satisfying Lemma 3 now become obstacle edges in Υ.

Further, we view all illumination gates of the bays/canals contained in $Vis_1(P^*)$ also as obstacle edges in Υ (since we ignore these bays/canals). Hence, Υ now has only obstacle edges and no more illumination gates. WLOG, we still use $\mathcal{P}$, $\mathcal{P}'$, and $\mathcal{M}$ to denote such structures built on the new obstacle set. Note that $\mathcal{P}'$ still consists of $O(h)$ convex polygons of totally $O(n)$ vertices and Υ still consists of $O(h)$ convex chains on $\partial\mathcal{M}$. Below, we compute $Vis_3(P^*)$ of Υ.

In addition to the corridor structure, the efficiency of our algorithm is also due to the property of Υ in Lemma 4. For this, we generalize the SO algorithm [18] for a single line segment to convex chains.

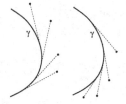
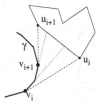

Fig. 3. The segments in $S(\gamma)$ are partitioned in two sorted lists (left and right)

Fig. 4. Illustrating the case when u_i and u_{i+1} are on the same obstacle edge

We first compute a set S' of line segments in which each segment $\overline{uv}$ connects two mutually visible obstacle vertices (u and v) and has an extension end on Υ. The segments of S' adjacent to each obstacle vertex are sorted by their slopes. The set S' can be computed in $O(n^2)$ time [9] and $|S'| = O(n^2)$. By applying the SO algorithm, $Vis_3(P^*)$ may be computed in $O(|S'|^2)$ time. Since Υ may have $\Omega(n)$ edges, $|S'| = \Omega(n^2)$ is possible. Our idea is to extend the SO algorithm and use a subset S of S' with $|S| = O(nh)$, as follows.

We compute the set S by removing some segments from S'. For each segment $\overline{uv} \in S'$, if one of u and v is on a convex chain $\gamma \in \Upsilon$ and $\overline{uv}$ is not tangent to γ, then we remove $\overline{uv}$ from S' (if one of u and v is an endpoint of γ, then $\overline{uv}$ is considered to be tangent to γ). Further, if $\overline{uv}$ is an edge of Υ, we also remove it from S' (alternatively, we could keep such $\overline{uv}$ and modify the following algorithm accordingly although the running time may be the same). The remaining segments in S' constitute the set S, which can be obtained in $O(n^2)$ time.

For each obstacle vertex v, denote by $S(v)$ the set of segments in S that are incident to v; the segments in $S(v)$ are sorted by their slopes (this is done when constructing S). For each obstacle vertex $v \notin \Upsilon$, we perform a rotational sweeping on v using $S(v)$; this is done exactly as in the SO algorithm, in $O(|S(v)|)$ time. The sweeping on all obstacle vertices not on Υ *generates* $O(|S|)$ triangles altogether, in $O(n + |S|)$ time. Below, we discuss the vertices on Υ.

For the vertices on Υ, we do not sweep each such vertex individually. Instead, we perform, for each convex chain $\gamma \in \Upsilon$, a rotational sweeping on γ as a whole. For this, all segments of S adjacent to γ are maintained in two cyclically sorted lists. Specifically, let $S(\gamma)$ be the set of segments in S adjacent to γ. Thus, for each segment $\overline{uv} \in S(\gamma)$, either u or v is on γ (say, $v \in \gamma$). According to our construction of S, $\overline{uv}$ must be tangent to γ at v. If we view $\overline{vu}$ as a segment directed from v to u, we say $\overline{vu}$ is a *left segment* (resp., *right segment*) of $S(\gamma)$ if γ is on the left (resp., right) of the directed $\overline{vu}$. We partition $S(\gamma)$ into two lists: one list contains the left segments of $S(\gamma)$ and the other list contains the right segments of $S(\gamma)$ (see Fig. 3); each list is sorted by the slopes of its segments.

Using each list of $S(\gamma)$ (say, the left segment list), we perform a sweeping on γ by rotating a ray that keeps its origin on γ and is tangent to γ at its origin, as follows. Suppose the rotating ray currently contains a segment $\overline{v_i u_i}$ in the list with $v_i \in \gamma$ (i.e., the origin of the ray is at v_i and the direction of the ray is from v_i to u_i), and $\overline{v_{i+1} u_{i+1}}$ is the next segment in the list to be encountered with

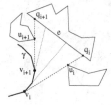 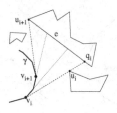

Fig. 5. Illustrating the case when u_i and u_{i+1} are not on the same obstacle edge and e is incident to neither u_i nor u_{i+1}. Both q_i and q_{i+1} must be on e.

Fig. 6. Illustrating the case when u_i and u_{i+1} are not on the same obstacle edge and e is incident to u_{i+1}. The point q_i must be on e.

$v_{i+1} \in \gamma$ (see Fig. 4). Note that $v_i = v_{i+1}$ is possible. Let $\gamma(v_i, v_{i+1})$ denote the portion of γ from v_i counterclockwise to v_{i+1}. If u_i and u_{i+1} are on the same obstacle edge (see Fig. 4), then the sweeping generates a region that is bounded by $\overline{v_i u_i}$, $\overline{u_i u_{i+1}}$, $\overline{v_{i+1} u_{i+1}}$, and $\gamma(v_i, v_{i+1})$. Further, if $v_i \neq v_{i+1}$, then we extend each edge of $\gamma(v_i, v_{i+1})$ into this region until it hits $\overline{u_i u_{i+1}}$, thus partitioning the region into g triangles where g is the number of edges on $\gamma(v_i, v_{i+1})$. If u_i and u_{i+1} are not on the same obstacle edge, then let q_i (resp., q_{i+1}) be the extension end of $\overline{v_i u_i}$ (resp., $\overline{v_{i+1} u_{i+1}}$) from v_i to u_i (resp., v_{i+1} to u_{i+1}). Note that during the sweeping from $\overline{v_i u_i}$ to $\overline{v_{i+1} u_{i+1}}$, the *hitting end* of the rotating ray (i.e., the end of the ray that is not its origin and hits an obstacle) must be moving on a single obstacle edge e (or at infinity). In other words, during the above sweeping, the rotating ray must be hitting e continuously such that the portion of the rotating ray between its origin on γ and its intersection with e lies entirely in the free space. Note that the vertex u_i (resp., u_{i+1}) is incident to two obstacle edges. Depending on whether e is one of the obstacle edges incident to u_i or u_{i+1}, there are several cases. If e is incident to neither u_i nor u_{i+1}, then the two extension ends q_i and q_{i+1} must be on e (see Fig. 5). In this case, the sweeping generates a region bounded by $\overline{v_i q_i}$, $\overline{q_i q_{i+1}}$, $\overline{v_{i+1} q_{i+1}}$, and $\gamma(v_i, v_{i+1})$, and triangulates this region in a similar way if $v_i \neq v_{i+1}$ (see Fig. 5). If e is incident to u_{i+1} (see Fig. 6), then q_i is on e, and the sweeping generates a region bounded by $\overline{v_i q_i}$, $\overline{q_i u_{i+1}}$, $\overline{u_{i+1} v_{i+1}}$, and $\gamma(v_i, v_{i+1})$ and triangulates this region similarly if $v_i \neq v_{i+1}$ (see Fig. 6). The case when e is incident to u_i can be handled similarly.

Since each edge of γ introduces at most two additional triangles (one for each list of $S(\gamma)$) and Υ has $O(n)$ edges, the entire sweeping algorithm on Υ generates $O(n + |S|)$ triangles. In summary, the sweeping on all obstacle vertices generates totally $O(n + |S|)$ triangles. The proof of Lemma 5 is omitted.

Lemma 5. $Vis_3(P^*)$ *is the union of all triangles generated by the sweeping algorithm.*

Therefore, we have obtained $Vis_3(P^*)$ as the union of $O(n + |S|)$ triangles; computing the union takes $O((n + |S|)^2)$ time [18]. Below, we prove $|S| = O(nh)$.

We define three subsets S_1, S_2, and S_3 of S. A segment $\overline{uv} \in S$ is in S_1 if and only if $\overline{uv}$ is contained in a bay/canal and $\overline{uv}$ is not a gate of that bay/canal. A segment $\overline{uv} \in S$ is in S_2 if and only if $\overline{uv}$ is tangent to a convex obstacle of $\mathcal{P}'$ at u or v, or $\overline{uv}$ is an edge of an obstacle in $\mathcal{P}'$. For S_3, a segment $\overline{uv} \in S$ is in S_3 if and only if $\overline{uv}$ has an endpoint on Υ. By our construction of S, if $\overline{uv}$ has an endpoint (say u) on Υ, then $\overline{uv}$ must be tangent to the convex chain of Υ that contains u. Since Υ has $O(h)$ convex chains, $|S_3| = O(nh)$ holds. Note that $S_2 \cap S_3 \neq \emptyset$ is possible. We can prove $|S_1| = O(n)$, $|S_2| = O(nh)$, and $S = S_1 \cup S_2 \cup S_3$. The proofs are omitted. Hence, $|S| = O(nh)$ follows. In summary, we can compute $Vis_3(P^*)$ as the union of $O(nh)$ triangles in $O(n^2h^2)$ time.

It remains to compute $Vis(P^*) = Vis_1(P^*) \cup Vis_2(P^*) \cup Vis_3(P^*)$. Since $Vis_1(P^*)$ and $Vis_2(P^*)$ consist of simple polygons of totally $O(n)$ complexity, we triangulate them into $O(n)$ triangles in $O(n)$ time [3]. Thus, we have $Vis(P^*)$ as the union of $O(nh)$ triangles, which can be computed in $O(n^2h^2)$ time [18].

Theorem 1. *If P^* is a simple polygon, then the visibility polygon of P^* among the obstacles in $\mathcal{P}$ can be computed in $O(n^2h^2)$ time.*

4   The Convex Version

We sketch our $O(n + h^4)$ time algorithm for the convex version. A complete description is in our full paper. Our algorithm generalizes the SO algorithm [18].

We call a line segment tangent to two convex obstacles of $\mathcal{P}$ at the two endpoints of the segment a *bitangent*. A bitangent is *free* if it lies entirely in the free space $\mathcal{F}$. The total number of free bitangents of $\mathcal{P}$ is $O(h^2)$. In the following discussion, unless otherwise specified, a bitangent always refers to a free one.

We first compute all bitangents of the obstacles in $\mathcal{P}$ as well as their extension ends. We retain only those bitangents each of which either has an extension end on P^* or has an endpoint on P^*, and let $\mathcal{B}$ denote the resulting bitangent set. For each obstacle $A \in \mathcal{P}$, let $\mathcal{B}(A)$ be the set of bitangents in $\mathcal{B}$ each of which has an endpoint on ∂A. We further partition $\mathcal{B}(A)$ into two subsets $\mathcal{B}_l(A)$ and $\mathcal{B}_r(A)$, as follows. Consider a bitangent $\overline{uv} \in \mathcal{B}(A)$ and assume u is on A. If $A = P^*$, then $\overline{uv}$ is in $\mathcal{B}_l(P^*)$ (resp., $\mathcal{B}_r(P^*)$) if P^* is on the left (resp., right) of $\overline{uv}$ directed from u to v. If $A \neq P^*$, $\overline{uv}$ has an extension, say a, on P^*. Then, $\overline{uv}$ is in $\mathcal{B}_l(A)$ (resp., $\mathcal{B}_r(A)$) if A is on the left (resp., right) of $\overline{au}$ directed from a to u. Further, the bitangents in $\mathcal{B}_l(A)$ (resp., $\mathcal{B}_r(A)$) are sorted cyclically around A by their slopes. In other words, when rotating a line tangent to A counterclockwise around A, the rotating line encounters the bitangents of $\mathcal{B}_l(A)$ (resp., $\mathcal{B}_r(A)$) in this sorted order. Clearly, $|\mathcal{B}| = O(h^2)$.

For each obstacle $A \in \mathcal{P}$, we perform a rotational sweeping on the vertices along ∂A and the bitangents of $\mathcal{B}_r(A)$ (resp., $\mathcal{B}_l(A)$), as follows.

Let the bitangents of $\mathcal{B}_r(A)$ be $\{t_1, t_2, \ldots, t_g\}$ sorted counterclockwise around A. For each $1 \leq i \leq g$, let $t_i = \overline{v_iu_i}$ with v_i lying on A, and the two extension ends of t_i be p_i and q_i with p_i lying on P^* (if t_i has an endpoint on P^*, then let p_i be that endpoint). Let $l(t_i)$ denote the line containing t_i. We rotate a line l tangent to A counterclockwise around A starting at the point v_1. When l encounters a

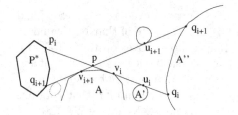 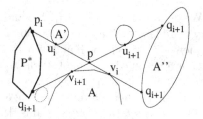

Fig. 7. Illustrating the situation when $v_i \in \overline{p_i u_i}$ and A' is below the line $l(t_i)$ ($t_i = \overline{v_i u_i}$)

Fig. 8. Illustrating the situation when $u_i \in \overline{p_i v_i}$ and A' is above the line $l(t_i)$ ($t_i = \overline{v_i u_i}$)

bitangent in $\mathcal{B}_r(A)$, two regions, called *pseudo-triangles* will be generated, such that each point in the two pseudo-triangles is visible to $\mathcal{P}^*$. We will show later that $Vis(P^*)$ is the union of such pseudo-triangles for all obstacles in $\mathcal{P}$.

Suppose the rotating line l currently overlaps with $l(t_i)$ for $t_i \in \mathcal{B}_r(A)$. Note that either $v_i \in \overline{p_i u_i}$ (see Fig. 7) or $u_i \in \overline{p_i v_i}$ (see Fig. 8) is possible. We only discuss the former case and the other case is similar.

If $v_i \in \overline{p_i u_i}$, then let A' be the convex obstacle in $\mathcal{P}$ on which u_i lies. The obstacle A' can be either above or below the line $l(t_i)$. We only discuss the case where A' is below $l(t_i)$ (see Fig. 7) and the other case is similar.

At this moment, the rotating line l overlaps with $l(t_i)$. We continue rotating l counterclockwise until encountering t_{i+1}. Then, we produce two *pseudo-triangles* (defined below). Let p be the intersection of $l(t_i)$ and $l(t_{i+1})$ ($p = v_i = v_{i+1}$ is possible). Let A'' be the convex obstacle in $\mathcal{P}$ on which q_i lies. It is possible that u_{i+1} is also on A'', in which case t_{i+1} is tangent to A'' at u_{i+1}. If u_{i+1} is not on A'', then q_{i+1} must be on A'' (see Fig. 7). Suppose $u_{i+1} \notin A''$. Then the first pseudo-triangle (refer to Fig. 7) is bounded by $\overline{pq_i}$, $\overline{pq_{i+1}}$, and the boundary portion of A'' between q_i and q_{i+1}. The second pseudo-triangle is bounded by $\overline{pv_i}$, $\overline{pv_{i+1}}$, and the boundary portion of A between v_i and v_{i+1}. (When $p = v_i = v_{i+1}$, the second pseudo-triangle is degenerated to $\emptyset$.) If $u_{i+1} \in A''$, then the first pseudo-triangle is bounded by $\overline{pq_i}$, $\overline{pu_{i+1}}$, and the boundary portion of A'' between q_i and u_{i+1}, and the second pseudo-triangle is the same as above. Thus, each pseudo-triangle is bounded by two line segments and a convex chain on the boundary of an obstacle in $\mathcal{P}$. Note that when a pseudo-triangle is found, only its two sides need to be output explicitly (i.e., its base is represented implicitly). In this way, each pseudo-triangle is output in $O(1)$ time and is represented in $O(1)$ space.

After we perform rotational sweeping on all obstacles in $\mathcal{P}$ as above, we can obtain $O(h^2)$ pseudo-triangles. We can show that the visibility polygon $Vis(P^*)$ is the union of all generated pseudo-triangles and their union can be computed in $O(n + h^4)$ time. The details are omitted.

References

1. Asano, T., Asano, T., Guibas, L., Hershberger, J., Imai, H.: Visibility of disjoint polygons. Algorithmica 1(1), 49–63 (1986)

2. Atallah, M., Chen, D., Wagener, H.: An optimal parallel algorithm for the visibility of a simple polygon from a point. Journal of the ACM 38(3), 516–533 (1991)
3. Chazelle, B.: Triangulating a simple polygon in linear time. Discrete & Computational Geometry 6, 485–524 (1991)
4. Chazelle, B., Guibas, L.: Visibility and intersection problems in plane geometry. Discrete Comput. Geom. 4, 551–589 (1989)
5. Chen, D., Wang, H.: Computing shortest paths among curved obstacles in the plane (2011), arXiv:1103.3911
6. Chen, D.Z., Wang, H.: A Nearly Optimal Algorithm for Finding L_1 Shortest Paths Among Polygonal Obstacles in the Plane. In: Demetrescu, C., Halldórsson, M.M. (eds.) ESA 2011. LNCS, vol. 6942, pp. 481–492. Springer, Heidelberg (2011)
7. ElGindy, H., Avis, D.: A linear algorithm for computing the visibility polygon from a point. Journal of Algorithms 2(2), 186–197 (1981)
8. Ghosh, S.: Computing the visibility polygon from a convex set and related problems. Journal of Algorithms 12, 75–95 (1991)
9. Ghosh, S., Mount, D.: An output-sensitive algorithm for computing visibility graphs. SIAM Journal on Computing 20(5), 888–910 (1991)
10. Guibas, L., Hershberger, J., Leven, D., Sharir, M., Tarjan, R.: Linear-time algorithms for visibility and shortest path problems inside triangulated simple polygons. Algorithmica 2(1-4), 209–233 (1987)
11. Heffernan, P., Mitchell, J.: An optimal algorithm for computing visibility in the plane. SIAM Journal on Computing 24(1), 184–201 (1995)
12. Inkulu, R., Kapoor, S.: Planar rectilinear shortest path computation using corridors. Computational Geometry: Theory and Applications 42(9), 873–884 (2009)
13. Joe, B.: On the correctness of a linear-time visibility polygon algorithm. International Journal of Computer Mathematics 32, 155–172 (1990)
14. Joe, B., Simpson, R.: Corrections to Lee's visibility polygon algorithm. BIT 27, 458–473 (1987)
15. Kapoor, S., Maheshwari, S., Mitchell, J.: An efficient algorithm for Euclidean shortest paths among polygonal obstacles in the plane. Discrete and Computational Geometry 18(4), 377–383 (1997)
16. Lee, D.: Visibility of a simple polygon. Computer Vision, Graphics, and Image Processing 22(2), 207–221 (1983)
17. Lee, D., Lin, A.: Computing the visibility polygon from an edge. Computer Vision, Graphics, and Image Processing 34, 594–606 (1986)
18. Suri, S., O'Rourke, J.: Worst-case optimal algorithms for constructing visibility polygons with holes. In: Proc. of the 2nd Annual Symposium on Computational Geometry, pp. 14–23 (1986)

Directed Subset Feedback Vertex Set
Is Fixed-Parameter Tractable*

Rajesh Chitnis[1,**], Marek Cygan[2,***], Mohammadtaghi Hajiaghayi[1,†],
and Dániel Marx[3,‡]

[1] Department of Computer Science, University of Maryland at College Park, USA
{rchitnis,hajiagha}@cs.umd.edu
[2] IDSIA, University of Lugano, Switzerland
marek@idsia.ch
[3] Computer and Automation Research Institute, Hungarian Academy of Sciences
(MTA SZTAKI), Budapest, Hungary
dmarx@cs.bme.hu

Abstract. Given a graph G and an integer k, the FEEDBACK VERTEX SET (FVS) problem asks if there is a vertex set T of size at most k that hits all cycles in the graph. Bodlaender (WG '91) gave the first fixed-parameter algorithm for FVS in undirected graphs. The fixed-parameter tractability status of FVS in directed graphs was a long-standing open problem until Chen et al. (STOC '08) showed that it is fixed-parameter tractable by giving an $4^k k! n^{O(1)}$ algorithm. In the subset versions of this problems, we are given an additional subset S of vertices (resp. edges) and we want to hit all cycles passing through a vertex of S (resp. an edge of S). Indeed both the edge and vertex versions are known to be equivalent in the parameterized sense. Recently the SUBSET FEEDBACK VERTEX SET in undirected graphs was shown to be FPT by Cygan et al. (ICALP '11) and Kakimura et al. (SODA '12). We generalize the result of Chen et al. (STOC '08) by showing that SUBSET FEEDBACK VERTEX SET in directed graphs can be solved in time $2^{2^{O(k)}} n^{O(1)}$, i.e., FPT parameterized by size k of the solution. By our result, we complete the picture for feedback vertex set problems and their subset versions in undirected and directed graphs.

The technique of random sampling of important separators was used by Marx and Razgon (STOC '11) to show that UNDIRECTED MULTICUT is FPT and was generalized by Chitnis et al. (SODA '12) to directed graphs to show that DIRECTED MULTIWAY CUT is FPT. In this paper we give a general family of problems (which includes DIRECTED MULTIWAY CUT and DIRECTED SUBSET FEEDBACK VERTEX SET among others) for which we can do random sampling

* A full version of the paper is available at http://arxiv.org/pdf/1205.1271v1.pdf

** Supported in part by NSF CAREER award 1053605, ONR YIP award N000141110662, DARPA/AFRL award FA8650-11-1-7162 and a University of Maryland Research and Scholarship Award (RASA).

*** Supported in part by ERC Starting Grant NEWNET 279352, NCN grant N206567140 and Foundation for Polish Science.

† Supported in part by NSF CAREER award 1053605, ONR YIP award N000141110662, DARPA/AFRL award FA8650-11-1-7162 and a University of Maryland Research and Scholarship Award (RASA). The author is also with AT&T Labs–Research.

‡ Supported by ERC Starting Grant PARAMTIGHT (No. 280152).

A. Czumaj et al. (Eds.): ICALP 2012, Part I, LNCS 7391, pp. 230–241, 2012.

of important separators and obtain a set which is disjoint from a minimum solution and covers its "shadow". We believe this general approach will be useful for showing the fixed-parameter tractability of other problems in directed graphs.

1 Introduction

The FEEDBACK VERTEX SET (FVS) problem has been one of the most extensively studied problems in the parameterized complexity community. Given a graph G and an integer k, it asks if there is a set T of size at most k which hits all cycles in G. FVS in both undirected and directed graphs was shown to be NP-hard by Karp [18]. A generalization of the FVS problem is the SUBSET FEEDBACK VERTEX SET (SFVS) problem: given a subset $S \subseteq V$ (resp. $S \subseteq E$), find a set T of size at most k such that T hits all cycles passing through a vertex of S (resp. an edge of S). It is easy to see that $S = V$ (resp. $S = E$) gives the FVS problem.

As compared to undirected graphs, FVS behaves differently on digraphs. In particular the trick of replacing each edge of an undirected graph G by arcs in both directions does not work: every feedback vertex set of the resulting digraph is a vertex cover of G and vice versa. Any other simple transformation does not seem possible either and thus the directed and undirected versions are very different problems. This is reflected in the best known approximation ratio for the directed versions as compared to the undirected problems: FVS in undirected graphs has an 2-approximation [1] while FVS in directed graphs has an $O(\log |V| \log \log |V|)$-approximation [13,24]. For SFVS in undirected graphs there is an 8-approximation [14] while the best-known approximation in directed graphs is $O(\min\{\log |V| \log \log |V|, \log^2 |S|\})$ [13].

Rather than finding approximate solutions in polynomial time, one can look for exact solutions in time that is superpolynomial, but still better than the running time obtained by brute force solutions. In both the directed and the undirected versions of the feedback vertex set problems, brute force can be used to check in time $n^{O(k)}$ if a solution of size at most k exists: one can go through all sets of size at most k. Thus the problem can be solved in polynomial time if the optimum is assumed to be small. In the undirected case, we can do significantly better: since the first FPT algorithm for FVS in undirected graphs by Bodlaender [3] almost 21 years ago, there have been a number of papers [2,5,6,17] giving faster algorithms and the current fastest algorithm runs in $O^*(3^k)$ time [10] (the O^* notation hides all factors which are polynomial in size of input). That is, undirected FVS is fixed-parameter tractable parameterized by the size of the cutset we remove. Recall that a problem is *fixed-parameter tractable* (FPT) with a particular parameter p if it can be solved in time $f(p)n^{O(1)}$, where f is an arbitrary function depending only on p; see [12,15,22] for more background. For digraphs, the fixed-parameter tractability status of FVS was a long-standing open problem (almost 16 years) until Chen et al. [7] resolved it by giving an $O^*(4^k k!)$ algorithm. This was recently generalized by Bonsma and Lokshtanov [4] who gave a $O^*(47.5^k k!)$ algorithm for FVS in mixed graphs, i.e., graphs having both directed and undirected edges.

In the more general SUBSET FEEDBACK VERTEX SET problem, given an additional subset S of vertices and we want to find a set T of size at most k that hits all cycles passing through a vertex of S. In the edge version we are given a subset $S \subseteq E(G)$ and we want to hit all cycles passing through an edge of S. The vertex and edge versions are indeed known to be equivalent in the parameterized sense in both undirected and directed graphs. Recently Cygan et al. [11] and independently Kakimura et al. [16] have shown that SUBSET FEEDBACK VERTEX SET in undirected graphs is FPT parameterized by the size of the solution. Our main result is that SUBSET FEEDBACK VERTEX SET in digraphs is also fixed-parameter tractable parameterized by the size of the solution:

Theorem 1. (main result) SUBSET FEEDBACK VERTEX SET *(SUBSET-DFVS) in directed graphs can be solved in* $O^*(2^{2^{O(k)}})$ *time.*

Our Techniques. As a first step, we use the standard technique of *iterative compression* [23] to argue that it is sufficient to solve the compression version of SUBSET-DFVS, where we assume that a solution T of size $k+1$ is given in the input and we have to find a solution of size k. Our algorithm for the compression problem is inspired by the algorithm of Marx and Razgon [21] for undirected MULTICUT and Chitnis et al. [8] for DIRECTED MULTIWAY CUT. We define the "shadow" of a solution X as those vertices that are disconnected from T (in either direction) after the removal of X. Our goal is to ensure that there is a solution whose shadow is empty, as finding such a shadowless solution can be a significantly easier task. For this purpose, we use the technique of "random sampling of important separators," which was introduced in [21] for undirected graphs and was generalized to directed graphs in [8]. We present this approach here in generic way that can be used for the following general family of problems:

Finding an $\mathcal{F}$-transversal for some T-connected $\mathcal{F}$

Input : A directed graph $G = (V,E)$, a positive integer k, a set $T \subseteq V$ and a set $\mathcal{F} = \{F_1, F_2, \ldots, F_q\}$ of subgraphs such that $\mathcal{F}$ is T-connected, i.e., $\forall\, i \in [q]$ each vertex of F_i can reach some vertex of T by a walk completely contained in F_i and is reachable from some vertex of T by a walk completely contained in F_i.
Parameter : k
Question : Does there exist an $\mathcal{F}$-transversal $W \subseteq V$ with $|W| \leq k$, i.e., a set W such that $F_i \cap W \neq \emptyset$ for every $i \in [q]$?

It is easy to see that the above family includes DIRECTED MULTIWAY CUT (take T as the set of terminals and $\mathcal{F}$ as the set of all walks between different terminals) and the compression version of SUBSET-DFVS (take T as the solution that we want to compress and $\mathcal{F}$ as set of all S-closed-walks). For this family of problems, we can invoke the random sampling of important separators technique and obtain a set which is disjoint from a minimum solution and covers its shadow. Given such a set, we can use (some problem specific variant of) the "torso operation" to find an equivalent instance that has a shadowless solution. Therefore, we can focus on the simpler task of finding a shadowless solution. We believe this will be a useful opening step in the design of FPT algorithms for other transversal and cut problems on digraphs.

In the case of undirected MULTICUT [21], if there was a shadowless solution, then the problem could be reduced to an FPT problem called ALMOST 2SAT. In the case of DIRECTED MULTIWAY CUT [8], if there was a solution whose shadow is empty, then the problem could be reduced to the undirected version which was known to be FPT.

For SUBSET-DFVS, the situation is a bit more complicated. As mentioned above, we first use the technique of iterative compression to reduce the problem to an instance where we are given a solution T and we want to find a disjoint solution of size at most k. We define the "shadows" with respect to the solution T that we want to compress whereas in [8], the shadows were defined with respect to the terminal set T. The "torso" operation we define in this paper is specific to the SUBSET-DFVS problem and differs from the one defined in [8]. Even after ensuring that there is a solution T' whose shadow is empty, we are not done unlike in [8]. We then analyze the structure of the graph $G \setminus T'$ and use "pushing" to branch on some important separators. Then for each branch, we need to do the whole process of random sampling of important separators to find a solution whose shadow is empty. This is followed again by branching on important separators. We repeat this two-step process until the budget k becomes zero.

2   Preliminaries

Observe, that a directed graphs contains no cycles if and only if it contains no closed-walks, for this reason throughout the article we use the term closed-walks, since it is sometimes easier to show a closed walk and avoid discussion whether it is a simple cycle or not. A feedback vertex set is a set of vertices that hits all the closed-walks of the graph.

Definition 2. (feedback vertex set) *Let G be a directed graph. A set $T \subseteq V(G)$ is a* feedback vertex set *of G if $G \setminus T$ does not contain any closed-walks.*

This gives rise to the DIRECTED FEEDBACK VERTEX SET (DFVS) problem where we are given a directed graph G and we want to find if G has a feedback vertex set of size at most k. DFVS was shown to be FPT by Chen et al. [7], closing a long-standing open problem in the parameterized complexity community.

In this paper we consider a generalization of the DFVS problem where given a set $S \subseteq V(G)$, we ask if there exists a vertex set of size $\leq k$ that hits all closed-walks passing through S.

SUBSET DIRECTED FEEDBACK VERTEX SET (SUBSET-DFVS)
Input : A directed graph $G = (V, E)$, a set $S \subseteq V(G)$ and a positive integer k.
Parameter : k
Question : Does there exist a set $T \subseteq V(G)$ with $|T| \leq k$ such that $G \setminus T$ has no closed walk containing a vertex of S?

It is easy to see that SUBSET-DFVS is a generalization of DFVS by setting $S = V(G)$. We also define an equivalent variant of SUBSET-DFVS where the set S is a subset of edges. First we define a special type of closed-walks:

Definition 3. *(S-closed-walk) Let $G = (V, E)$ be a digraph and $S \subseteq E(G)$. A closed walk (starting and ending at same vertex) C in G is said to be a S-closed-walk if it contains an edge from S.*

EDGE SUBSET DIRECTED FEEDBACK VERTEX SET (EDGE-SUBSET-DFVS)
Input : A directed graph $G = (V, E)$, a set $S \subseteq E(G)$ and a positive integer k.
Parameter : k
Question : Does there exist a set $T \subseteq V(G)$ with $|T| \leq k$ such that $G \setminus T$ has no S-closed-walks?

2.1 Iterative Compression

We now use the technique of *iterative compression* introduced by Reed et al. [23]. It has been used to obtain faster FPT algorithms for various problems [6,7,21]. In the first step we transform the SUBSET-DFVS problem into the following problem:

SUBSET-DFVS REDUCTION
Input : A directed graph $G = (V, E)$, a set $S \subseteq E(G)$, a positive integer k and a set $T \subseteq V$ such that $G \setminus T$ has no S-closed-walks .
Parameter : $k + |T|$
Question : Does there exist a set $T' \subseteq V(G)$ with $|T'| \leq k$ such that $G \setminus T'$ has no S-closed-walks?

Lemma 4. [⋆][1] *(power of iterative compression)* SUBSET-DFVS *can be solved by* $O(n)$ *calls to an algorithm for the* SUBSET-DFVS REDUCTION *problem.*

Now we transform the SUBSET-DFVS REDUCTION problem into the following problem whose only difference is that the subset feedback vertex set in the output must be disjoint from the one in the input:

DISJOINT SUBSET-DFVS REDUCTION
Input : A directed graph $G = (V, E)$, a set $S \subseteq E(G)$, a positive integer k and a set $T \subseteq V$ such that $G \setminus T$ has no S-closed-walks.
Parameter : $k + |T|$
Question : Does there exist a set $T' \subseteq V(G)$ with $|T'| \leq k$ such that $T \cap T' = \emptyset$ and $G \setminus T'$ has no S-closed-walks?

Lemma 5. [⋆] *(adding disjointness)* SUBSET-DFVS REDUCTION *can be solved by* $O(2^{|T|})$ *calls to an algorithm for the* DISJOINT SUBSET-DFVS REDUCTION *problem.*

From Lemmas 4 and 5, an FPT algorithm for DISJOINT SUBSET-DFVS REDUCTION translates into an FPT algorithm for SUBSET-DFVS with an additional blowup factor of $O(2^{|T|}n)$.

3   Covering the Shadow of a Solution

The purpose of this section is to present the "random sampling of important separators" technique used in [8] for DIRECTED MULTIWAY CUT in a generalized way that applies to SUBSET-DFVS as well. The technique consists of two steps:

---

[1] The proofs of the results labeled with ⋆ have been deferred to the full version of the paper.

1. First find a set Z *small* enough to be disjoint from a solution X (of size $\leq k$) but *large* enough to cover the "shadow" of X.
2. Then define a "torso" operation which uses the set Z to reduce the problem instance in such a way that X becomes a shadowless solution.

In this section, we define a general family of problems for which Step 1 can be efficiently performed. The general technique to execute Step 1 is very similar to what was done for DIRECTED MULTIWAY CUT [8] and so we defer most of the proofs to the full version of the paper. In Section 4, we show how Step 2 can be done for the specific problem of DISJOINT SUBSET-DFVS REDUCTION. First we start by defining shadows:

Definition 6. (**separator**) *Let $G = (V,E)$ be a directed graph. Given two disjoint non-empty sets $X,Y \subseteq V$ we call a set W of vertices as an $X - Y$ separator if W is disjoint from $X \cup Y$ and there is no walk from X to Y in $G \setminus W$. A set W is a minimal $X - Y$ separator if no proper subset of W is an $X - Y$ separator.*

Definition 7. (**shadow**) *Let G be graph and $W \subseteq V(G)$. Then for $v \in V(G)$ we say that v is in the "forward shadow" $f_{G,T}(W)$ of W (with respect to T), if W is a $T - \{v\}$ separator in G. Similarly, we say that v is in the "reverse shadow" $r_{G,T}(W)$ of W (with respect to T), if W is a $\{v\} - T$ separator in G.*

That is, we can imagine T as a light source with light spreading on the directed edges. The forward shadow of W is the set of vertices that remain dark if the set W blocks the light. In the reverse shadow, we imagine that light is spreading on the edges backwards. We abuse the notation slightly and write $v - T$ separator instead of $\{v\} - T$ separator. We also drop G and T from the subscript if they are clear from the context. Note that W itself is not in the shadow of W (as a $T - v$ or $v - T$ separator needs to be disjoint from T and v), that is, W and $f_{G,T}(W) \cup r_{G,T}(W)$ are disjoint.

Let $G = (V,E)$ be a directed graph and $T \subseteq V(G)$. Consider $\mathcal{F} = \{F_1, F_2, \ldots, F_q\}$ which is a set of subgraphs of G. We define the following property:

Definition 8. (*T-connected*) *Let $\mathcal{F} = \{F_1, F_2, \ldots, F_q\}$ be a set of subgraphs of G. Then $\mathcal{F}$ is said to be T-connected if $\forall i \in [q]$, each vertex of the subgraph F_i can reach some vertex of T by a walk completely contained in F_i and is reachable from some vertex of T by a walk completely contained in F_i.*

For a set $\mathcal{F}$ of subgraphs of G, a transversal is a set of vertices which hits each subgraph in $\mathcal{F}$. We note that the subgraphs in $\mathcal{F}$ are given implicitly to us.

Definition 9. (*$\mathcal{F}$-transversal*) *Let $\mathcal{F} = \{F_1, F_2, \ldots, F_q\}$ be a set of subgraphs of G. Then W is said to be an $\mathcal{F}$-transversal if $\forall i \in [q]$ we have $F_i \cap W \neq \emptyset$.*

The main theorem of this section is the following:

Theorem 10. [⋆](**randomized covering of the shadow**) *Let $T \subseteq V(G)$. In $O^*(4^k)$ time, we can construct a set $Z \subseteq V(G)$ such that for any set of subgraphs $\mathcal{F}$ which is T-connected, if there exists an $\mathcal{F}$-transversal of size $\leq k$, then the following holds with probability $2^{-2^{O(k)}}$: there is an $\mathcal{F}$-transversal X of size $\leq k$ satisfying*

1. $X \cap Z = \emptyset$.
2. Z covers the shadow of X.

We also prove the following derandomized version of Theorem 10:

Theorem 11. [$\star$](**deterministic covering of the shadow**) *Let* $T \subseteq V(G)$. *In* $O^*(2^{2^{O(k)}})$ *time, we can construct a set* $\{Z_1, Z_2, \ldots, Z_t\}$ *where* $t = 2^{2^{O(k)}} \log^2 n$ *such that for any set of subgraphs* $\mathcal{F}$ *which is* T-connected, *if there exists an* $\mathcal{F}$-transversal *of size* $\leq k$, *then there is an* $\mathcal{F}$-transversal X *of size* $\leq k$ *such that for at least one* $i \in [t]$ *we have*

1. $X \cap Z_i = \emptyset$.
2. Z_i covers the shadow of X.

In DIRECTED MULTIWAY CUT, T was the set of terminals and the set $\mathcal{F}$ was the set of all walks from one vertex of T to another vertex of T. In SUBSET-DFVS , the set T is the solution that we want to compress and $\mathcal{F}$ is the set of all closed S-walks passing through some vertex of T.

We say that an $\mathcal{F}$-transversal T' is *shadowless* if $f(T') \cup r(T') = \emptyset$. Note that if T' is a shadowless solution, then in the graph $G \setminus T'$, each vertex is reachable from some vertex of T and can reach some vertex of T. In Section 5 we will see how we can make progress in DISJOINT SUBSET-DFVS REDUCTION if there exists a shadowless solution. So we would like to transform the instance in such a way that ensures the existence of a shadowless solution, by taking the torso (Section 4) and make progress by using the BRANCH algorithm from Section 5.

4   Reducing the Instance by Torso

We use the algorithm of Theorem 11 to construct a set Z of vertices that we want to get rid of. The second ingredient of our algorithm is an operation that removes a set of vertices without making the problem any easier. This transformation can be conveniently described using the operation of taking the *torso* of a graph. From this point onwards in the paper, we do not follow [8]. In particular, the *torso* operation is problem-specific. For DISJOINT SUBSET-DFVS REDUCTION, we define it as follows:

Definition 12. (**torso**) *Let* (G, S, T, k) *be an instance of* DISJOINT SUBSET-DFVS RE-DUCTION *and* $C \subseteq V(G)$. *The graph* $\text{torso}(G, C)$ *has vertex set* C *and there is (directed) edge* (a, b) *in* $\text{torso}(G, C)$ *if there is an* $a \rightarrow b$ *walk in* G *whose internal vertices are not in* C. *Furthermore, we add the edge* (a, b) *to* S *if there is an* $a \rightarrow b$ *walk in* G *which contains an edge from* S *and whose internal vertices are not in* C.

In particular, if $a, b \in C$ and (a, b) is a directed edge of G, then $\text{torso}(G, C)$ contains (a, b) as well. Thus $\text{torso}(G, C)$ is a supergraph of the subgraph of G induced by C. The following lemma shows that the *torso* operation preserves S-closed-walks inside C.

Lemma 13. [$\star$] (**torso preserves** S-**closed-walks**) *Let* G *be a directed graph and* $C \subseteq V(G)$. *Let* $G' = \text{torso}(G, C), v \in C$ *and* $W \subseteq C$. *Then* $G \setminus W$ *has an* S-closed-walk passing through v *if and only if* $G' \setminus W$ *has an* S-closed-walk passing through v.

If we want to remove a set Z of vertices, then we create a new instance by taking the torso on the *complement* of Z:

Definition 14. *Let $I = (G,S,T,k)$ be an instance of* DISJOINT SUBSET-DFVS RE-DUCTION *and $Z \subseteq V(G) \setminus T$. The reduced instance $I/Z = (G',S,T,p)$ is defined as*

- $G' = \text{torso}(G,V(G) \setminus Z)$
- *S is modified as specified in Definition 12.*

The following lemma states that the operation of taking the torso does not make the DISJOINT SUBSET-DFVS REDUCTION problem easier for any $Z \subseteq V(G) \setminus T$ in the sense that any solution of the reduced instance I/Z is a solution of the original instance I. Moreover, if we perform the torso operation for a Z that is large enough to cover the shadow of some solution T^* and also small enough to be disjoint from T^*, then T^* becomes a shadowless solution for the reduced instance I/Z.

Lemma 15. [$\star$] **(creating a shadowless instance)** *Let $I = (G,S,T,k)$ be an instance of* DISJOINT SUBSET-DFVS REDUCTION *and $Z \subseteq V(G) \setminus T$.*

1. *If I is a no-instance, then the reduced instance I/Z is also a no-instance.*
2. *If I has solution T' with $f_{G,T}(T') \cup r_{G,T}(T') \subseteq Z$ and $T' \cap Z = \emptyset$, then T' is a shadowless solution of I/Z.*

For every Z_i in the output of Theorem 11, we use the torso operation to remove the vertices in Z_i. We prove that this procedure is safe by showing the following:

Lemma 16. [$\star$] *Let $I = (G,S,T,k)$ be an instance of* DISJOINT SUBSET-DFVS RE-DUCTION. *Let the sets in the output of Theorem 11 be $Z_1,Z_2,\ldots,Z_t$. For every $i \in [t]$, let G_i be the reduced instance G/Z_i.*

1. *If I is a no-instance, then G_i is also a no-instance for every $i \in [t]$.*
2. *If I is a yes-instance, then there exists a solution T^* of I which is a shadowless solution of some G_j for some $j \in [t]$.*

5   Finding a Shadowless Solution

Consider an instance (G,S,T,k) of DISJOINT SUBSET-DFVS REDUCTION. First, let us assume that from each vertex of T, we can reach an edge of S, since otherwise we can clearly remove such a vertex from the set T, without violating the assumption that $G \setminus T$ has no S-closed walk. Next, we branch on all $2^{2^{O(k)}} \log^2 n$ choices for Z taken from $\{Z_1,Z_2,\ldots,Z_t\}$ (given by Theorem 11) and build a reduced instance I/Z for each choice of Z. By Lemma 15, if I is a no-instance then I/Z_j is a no-instance for each $j \in [t]$. If I is a yes-instance, then by Lemma 16, there is at least one $i \in [t]$ such that I has a solution T' which is a solution, and in fact a shadowless solution, for the reduced instance I/Z_i.

So for the reduced instance I/Z_i we know that each vertex in $G \setminus T'$ can reach some vertex of T and can be reached from a vertex of T. Since T' is a solution for the instance (G,S,T,k) of DISJOINT SUBSET-DFVS REDUCTION, we know that $G \setminus T'$ does

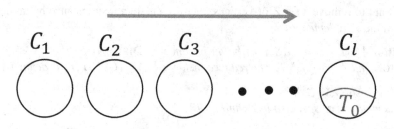

Fig. 1. We arrange the strong components of $G \setminus T'$ in topological order so that the only possible direction of edges between the strong components is as shown by the blue arrow. We will claim later that the last component C_ℓ must contain a non-empty subset T_0 of T and further that no edge of S can be present within C_ℓ. This allows us to make some progress as we shall see in Theorem 21

not have any S-closed-walks. Consider a topological ordering say $C_1, C_2, \ldots, C_\ell$ of the strong components of $G \setminus T'$, i.e., there can be an edge from C_i to C_j only if $i < j$. We illustrate this in Figure 1.

Definition 17. *(starting points of S) Let S^- be the set of starting points of edges in S, i.e., $S^- = \{u \mid (u,v) \in S\}$.*

Lemma 18. [$\star$] *(properties of C_ℓ) Let C_ℓ be the last strong component in the topological ordering of $G \setminus T'$ (refer to Figure 1). Then*

1. *C_ℓ contains a non-empty subset T_0 of T.*
2. *No edge of S is present within C_ℓ.*
3. *S^- is disjoint from C_ℓ.*

Since T_0 is the subset of T present in C_ℓ and only edges between strong components can be from left to right, we have that there are no $T_0 - (T \setminus T_0)$ walks in $G \setminus T'$. Along with the third claim of Lemma 18, this implies that the solution T' contains a $T_0 - (S^- \cup (T \setminus T_0))$ separator. We now define a special type of separators:

Definition 19. **(important separator)** *Let G be a digraph and let $X, Y \subseteq V$ be two disjoint non-empty sets. A minimal $X - Y$ separator W is called an* important $X - Y$ *separator if there is no $X - Y$ separator W' with $|W'| \leq |W|$ and $R^+_{G \setminus W}(X) \subset R^+_{G \setminus W'}(X)$, where $R^+_A(X)$ is the set of vertices reachable from X in A.*

For any $X, Y \subseteq V(G)$, the following lemma (proved in [8]) gives an upper bound the number of important $X - Y$ separators of size at most k:

Lemma 20. [$\star$]**(number of important separators)** *Let $X, Y \subseteq V(G)$ be disjoint sets in a directed graph G. Then for every $k \geq 0$ there are at most 4^k important $X - Y$ separators of size at most k. Furthermore, we can enumerate all these separators in time $O^*(4^k)$.*

By "pushing", we have the following theorem:

Algorithm 1. BRANCH

Input: An instance $I = (G, S, T, k)$ of DISJOINT SUBSET-DFVS REDUCTION.
Output: A new set of $2^{O(k+|T|)}$ instances of DISJOINT SUBSET-DFVS REDUCTION where the budget k is reduced.

1: **for** every non-empty subset T_0 of T: **do**
2:     Use Lemma 20 to enumerate all the at most 4^k important $T_0 - (S^- \cup (T \setminus T_0))$ separators of size at most k.
3:     Let the important separators be $\mathcal{B} = \{B_1, B_2, \ldots, B_m\}$.
4:     **for** each $i \in [m]$ **do**
5:         Create a new instance $I_{T_0, i} = (G \setminus B_i, S, T, k - |B_i|)$ of DISJOINT SUBSET-DFVS REDUCTION.

Theorem 21. [⋆] *(pushing) Either T' contains an important $T_0 - (S^- \cup (T \setminus T_0))$ separator or there is another solution T'' of the instance (G, S, T, k) such that $|T''| \le |T'|$ and T'' contains an important $T_0 - (S^- \cup (T \setminus T_0))$ separator.*

Theorem 21 tells us that there is always a minimum solution which contains an important $T_0 - (S^- \cup (T \setminus T_0))$ separator where T_0 is a non-empty subset of T. This gives $2^{|T|} - 1$ choices for T_0. For each guess of T_0 we enumerate all the at most 4^k important $T_0 - (S^- \cup (T \setminus T_0))$ separators of size at most k in time $O^*(4^k)$ as given by Lemma 20. This gives the following natural branching algorithm:

6   FPT Algorithm for DISJOINT SUBSET-DFVS REDUCTION

Lemma 16 and the BRANCH algorithm together combine to give a *bounded-search-tree* FPT algorithm for DISJOINT SUBSET-DFVS REDUCTION as follows:

FPT Algorithm for SUBSET-DFVS

Step 1: At the first step, for a given instance $I = (G, S, T, k)$, use Theorem 11 to obtain a set of instances $\{Z_1, Z_2, \ldots, Z_t\}$ where $2^{2^{O(k)}} \log^2 n$ and Lemma 16 implies

- If I is a no-instance, then all the reduced instances $G_j = G/Z_j$ are no-instances for all $j \in [t]$
- If I is a yes-instance, then there is at least one $i \in [t]$ such that there is a solution T^* for I which is a shadowless solution for the reduced instance $G_i = G/Z_i$.

So at this step we branch into $2^{2^{O(k)}} \log^2 n$ directions.

Step 2 : For each of the instances obtained from the above step, we run the BRANCH algorithm to obtain a set of $2^{O(k+|T|)}$ instances where in each case either the answer is NO, or the budget k is reduced.

We then repeatedly perform Steps 1 and 2. Note that for every instance, one execution of steps 1 and 2 gives rise to $2^{2^{O(k)}} \log^2 n$ instances such that for each instance, either we know that the answer is NO or the budget k has decreased, because we have assumed that from each vertex of T one can reach the set S^-, and hence each important separator is non-empty. Therefore, considering a level as an execution of Step 1 followed by Step 2, the height of the search tree is at most k. Each time we branch into at most $2^{2^{O(k)}} \log^2 n$ directions (as $|T|$ is at most $k+1$). Hence the total number of nodes in the search tree is $\left(2^{2^{O(k)}} \log^2 n\right)^k$.

Lemma 22. [*] *For every n and $k \le n$, we have* $(\log n)^k \le (2k \log k)^k + \frac{n}{2^k}$

So the total number of nodes in the search tree is $\left(2^{2^{O(k)}} \log^2 n\right)^k = \left(2^{2^{O(k)}}\right)^k (\log^2 n)^k = (2^{2^{O(k)}})(\log^2 n)^k \le (2^{2^{O(k)}})\left((2k \log k)^k + \frac{n}{2^k}\right)^2 \le 2^{2^{O(k)}} n^2$. We then check the leaf nodes and see if there are any S-closed-walks left even after the budget k has become zero. If the graph at least one of the leaf nodes is S-closed-walk free, then the given instance is a yes-instance. Otherwise it is a no-instance. This gives an $O^*(2^{2^{O(k)}})$ algorithm for DIS-JOINT SUBSET-DFVS REDUCTION. By Lemma 4, we have an $O^*(2^{2^{O(k)}})$ algorithm for the SUBSET-DFVS problem.

7   Conclusion and Open Problems

In this paper we gave the first fixed-parameter algorithm for DIRECTED SUBSET FEED-BACK VERTEX SET parameterized by the size of the solution. Our algorithm used various tools from the FPT world such as iterative compression, bounded-depth search trees, random sampling of important separators, etc. We also gave a general family of problems for which we can do random sampling of important separators and obtain a set which is disjoint from a minimum solution and covers its shadow. We believe this general approach will be useful for deciding the fixed-parameter tractability status of other problems in digraphs where we do not know that much techniques unlike undirected graphs.

The next natural question is whether SUBSET-DFVS has a polynomial kernel or can we rule out such a possibility under some standard assumptions? The recent developments [9,19,20] in the field of kernelization may be useful in answering this question. Another question is to try and reduce the complexity of our algorithm to single exponential. In the field of exact exponential algorithms, Razgon gave a $O(1.9977^n)$ algorithm for DFVS. It would be interesting to break the trivial $2^n n^{O(1)}$ barrier for SUBSET-DFVS.

References

1. Bafna, V., Berman, P., Fujito, T.: A 2-approximation algorithm for the undirected feedback vertex set problem. SIAM J. Discrete Math. 12(3), 289–297 (1999).
2. Becker, A., Bar-Yehuda, R., Geiger, D.: Randomized algorithms for the loop cutset problem. J. Artif. Intell. Res. (JAIR) 12, 219–234 (2000)
3. Bodlaender, H.L.: On disjoint cycles. In: WG, pp. 230–238 (1991)

4. Bonsma, P., Lokshtanov, D.: Feedback Vertex Set in Mixed Graphs. In: Dehne, F., Iacono, J., Sack, J.-R. (eds.) WADS 2011. LNCS, vol. 6844, pp. 122–133. Springer, Heidelberg (2011)

5. Cao, Y., Chen, J., Liu, Y.: On Feedback Vertex Set New Measure and New Structures. In: Kaplan, H. (ed.) SWAT 2010. LNCS, vol. 6139, pp. 93–104. Springer, Heidelberg (2010)

6. Chen, J., Fomin, F.V., Liu, Y., Lu, S., Villanger, Y.: Improved algorithms for feedback vertex set problems. J. Comput. Syst. Sci. 74(7), 1188–1198 (2008)

7. Chen, J., Liu, Y., Lu, S., O'Sullivan, B., Razgon, I.: A fixed-parameter algorithm for the directed feedback vertex set problem. In: STOC 2008, pp. 177–186 (2008)

8. Chitnis, R.H., Hajiaghayi, M., Marx, D.: Fixed-parameter tractability of directed multiway cut parameterized by the size of the cutset. In: SODA 2012, pp. 1713–1725 (2012)

9. Cygan, M., Kratsch, S., Pilipczuk, M., Pilipczuk, M., Wahlström, M.: Clique cover and graph separation: New incompressibility results. CoRR abs/1111.0570 (2011)

10. Cygan, M., Nederlof, J., Pilipczuk, M., Pilipczuk, M., van Rooij, J.M.M., Wojtaszczyk, J.O.: Solving connectivity problems parameterized by treewidth in single exponential time. In: FOCS 2011, pp. 150–159 (2011)

11. Cygan, M., Pilipczuk, M., Pilipczuk, M., Wojtaszczyk, J.O.: Subset Feedback Vertex Set is Fixed-Parameter Tractable. In: Aceto, L., Henzinger, M., Sgall, J. (eds.) ICALP 2011. LNCS, vol. 6755, pp. 449–461. Springer, Heidelberg (2011)

12. Downey, R.G., Fellows, M.R.: Parameterized Complexity, 530 p. Springer (1999)

13. Even, G., Naor, J., Schieber, B., Sudan, M.: Approximating Minimum Feedback Sets and Multi-Cuts in Directed Graphs (Extended Summary). In: IPCO 1995. LNCS, vol. 920, pp. 14–28. Springer, Heidelberg (1995)

14. Even, G., Naor, J., Zosin, L.: An 8-approximation algorithm for the subset feedback vertex set problem. SIAM J. Comput. 30(4), 1231–1252 (2000)

15. Flum, J., Grohe, M.: Parameterized Complexity Theory, 493 p. Springer (2006)

16. Kakimura, N., Kawarabayashi, K., Kobayashi, Y.: Erdös-pósa property and its algorithmic applications: parity constraints, subset feedback set, and subset packing. In: SODA 2012, pp. 1726–1736 (2012)

17. Kanj, I.A., Pelsmajer, M.J., Schaefer, M.: Parameterized Algorithms for Feedback Vertex Set. In: Downey, R.G., Fellows, M.R., Dehne, F. (eds.) IWPEC 2004. LNCS, vol. 3162, pp. 235–247. Springer, Heidelberg (2004)

18. Karp, R.M.: Reducibility among combinatorial problems. In: Complexity of Computer Computations, pp. 85–103 (1972)

19. Kratsch, S., Wahlström, M.: Representative sets and irrelevant vertices: New tools for kernelization. CoRR abs/1111.2195 (2011)

20. Kratsch, S., Wahlström, M.: Compression via matroids: a randomized polynomial kernel for odd cycle transversal. In: SODA 2012, pp. 94–103 (2012)

21. Marx, D., Razgon, I.: Fixed-parameter tractability of multicut parameterized by the size of the cutset. In: STOC 2011, pp. 469–478 (2011)

22. Niedermeier, R.: Invitation to Fixed-Parameter Algorithms. Oxford University Press (2006)

23. Reed, B.A., Smith, K., Vetta, A.: Finding odd cycle transversals. Oper. Res. Lett. (2004)

24. Seymour, P.D.: Packing directed circuits fractionally. Combinatorica 15(2), 281–288 (1995)

Max-Cut Parameterized above
the Edwards-Erdős Bound*

Robert Crowston[1], Mark Jones[1], and Matthias Mnich[2]

[1] Royal Holloway, University of London, UK
{robert,markj}@cs.rhul.ac.uk
[2] Cluster of Excellence, Saarbrücken, Germany
m.mnich@mmci.uni-saarland.de

Abstract. We study the boundary of tractability for the MAX-CUT problem in graphs. Our main result shows that MAX-CUT above the Edwards-Erdős bound is fixed-parameter tractable: we give an algorithm that for any connected graph with n vertices and m edges finds a cut of size
$$\frac{m}{2} + \frac{n-1}{4} + k$$
in time $2^{O(k)} \cdot n^4$, or decides that no such cut exists.

This answers a long-standing open question from parameterized complexity that has been posed a number of times over the past 15 years.

Our algorithm is asymptotically optimal, under the Exponential Time Hypothesis, and is strengthened by a polynomial-time computable kernel of polynomial size.

Keywords: Algorithms and data structures, maximum cuts, combinatorial bounds, fixed-parameter tractability.

1 Introduction

The study of cuts in graphs is a fundamental area in theoretical computer science, graph theory, and polyhedral combinatorics, dating back to the 1960s. A *cut* of a graph is an edge-induced bipartite subgraph, and its *size* is the number of edges it contains. Finding cuts of maximum size in a given graph was one of Karp's famous 21 NP-complete problems [18]. Since then, the MAX-CUT problem has received considerable attention in the areas of approximation algorithms, random graph theory, combinatorics, parameterized complexity, and others; see the survey [26].

As a fundamental NP-complete problem, the computational complexity of MAX-CUT has been intensively scrutinized. We continue this line of research and explore the boundary between tractability and hardness, guided by the question: *Is there a dichotomy of computational complexity of MAX-CUT that depends on the size of the maximum cut?*

---

* Due to space constraints, several proofs and details were omitted. A full version of the paper can be found at http://arxiv.org/abs/1112.3506.

A. Czumaj et al. (Eds.): ICALP 2012, Part I, LNCS 7391, pp. 242–253, 2012.

This question was already studied by Erdős [11] in the 1960s, who gave a randomized polynomial-time algorithm that in any n-vertex graph with m edges finds a cut of size at least $m/2$. Erdős [11, 12] also (erroneously) conjectured that the value $m/2$ can be raised to $m/2 + \varepsilon m$ for some $\varepsilon > 0$; only much later it was shown [15, 24] that finding cuts of size $m/2 + \varepsilon m$ is NP-hard for every $\varepsilon > 0$. Furthermore, the MAX-CUT GAIN problem—maximize the gain compared to a random solution that cuts $m/2$ edges—does not allow constant approximation [19] under the Unique Games Conjecture, and the best one can hope for is to cut a $1/2 + \Omega(\varepsilon/\log(1/\varepsilon))$ fraction of edges in graphs in which the optimum is $1/2 + \varepsilon$ [6].

However, the lower bound $m/2$ *can* be increased, by a sublinear function: Edwards [9, 10] in 1973 proved that a cut of size

$$m/2 + \frac{1}{8}\left(\sqrt{8m+1} - 1\right) \tag{1}$$

always exists, and for connected graphs this can be further increased to

$$m/2 + (n-1)/4, \tag{2}$$

which is always at least as large as (1). Thus, any graph with n vertices, m edges and t connected components has a cut of size at least $m/2 + (n-t)/4$. The lower bound (2) is famously known as the *Edwards-Erdős bound*, and it is tight for cliques of every odd order n.

The bound has been proved several times ([4, 7, 13, 24, 25]), with some proofs yielding polynomial-time algorithms to attain it. As (2) is tight for infinitely many non-isomorphic graphs, and finding maximum cuts is NP-hard, finding cuts beyond (2) requires a new approach: a *fixed-parameter algorithm*, that for any connected graph with n vertices and m edges, and integer $k \in \mathbb{N}$, finds a cut of size at least $m/2 + (n-1)/4 + k$ (if such exists) in time $f(k) \cdot n^c$, where f is an arbitrary function dependent only on k and c is an absolute constant independent of k. The point here is to confine the combinatorial explosion to the (small) parameter k. But at first sight, it is not even clear how to find a cut of size $m/2 + (n-1)/4 + k$ in time $n^{f(k)}$, for an arbitrary function f.

In 1997, Mahajan and Raman [22] gave a fixed-parameter algorithm for the variant of this problem with Erdős' lower bound $m/2$, and showed how to decide existence of a cut of size $m/2 + k$ in time $2^{O(k)} \cdot n^{O(1)}$. Their result was strengthened by Bollobás and Scott [4] who replaced $m/2$ by the stronger bound (1). It remained an open question ([7, 14, 22, 23, 27]) whether this result could be strengthened further by replacing (1) with the stronger bound (2).

Main Results

We settle the computational complexity of MAX-CUT above the Edwards-Erdős bound (2).

Theorem 1. *There is an algorithm that computes, for any connected graph G with n vertices and m edges and any integer $k \in \mathbb{N}$, in time $2^{O(k)} \cdot n^4$ a cut of G with size at least $m/2 + (n-1)/4 + k$, or decides that no such cut exists.*

Theorem 1 answers a question posed several times over the past 15 years ([22, 23] [7, 14, 27]). In particular, instances with $k = O(\log m)$ can be solved in polynomial time, thereby enlarging the realm of tractability.

The running time of our algorithm is likely to be optimal, as the following theorem shows.

Theorem 2. *No algorithm can find cuts of size $m/2 + (n-1)/4 + k$ in time $2^{o(k)} \cdot n^{O(1)}$ given a connected graph with n vertices and m edges, and integer $k \in \mathbb{N}$, unless the Exponential Time Hypothesis fails.*

The Exponential Time Hypothesis was introduced by Impagliazzo and Paturi [17], and states that n-variable SAT formulas cannot be solved in subexponential time.

Fixed-parameter tractability of MAX-CUT above Edwards-Erdős bound $m/2 + (n-1)/4$ implies the existence of a so-called *kernelization*, which efficiently transforms any instance (G, k) into an equivalent instance (G', k'), the *kernel*, whose size $g(k) = |G'| + k'$ itself depends on k only. Alas, the size $g(k)$ of the kernel for many fixed-parameter tractable problems is enormous, and in particular many fixed-parameter tractable problems do not admit kernels of size *polynomial* in k unless coNP $\subseteq$ NP/poly [3]. We prove the following.

Theorem 3. *There is a polynomial-time algorithm that transforms any connected graph $G = (V, E)$ with integer $k \in \mathbb{N}$ to a connected graph $G' = (V', E')$ of order $O(k^5)$, such that G has a cut of size $|E|/2 + (|V| - 1)/4 + k$ if and only if G' has a cut of size $|E'|/2 + (|V'| - 1)/4 + k'$, for some $k' \leq k$.*

Bollobás and Scott [4] proved fixed parameter tractability for the weighted version of Max Cut parameterized above (1). They give a $2^{O(k^4)} + n + w(G)$ time algorithm to find a cut of weight $w(G)/2 + \frac{1}{8}(\sqrt{8w(G) + 1} - 1) + k$ if such a cut exists, or else an optimal cut, where w is an edge-weighting on the graph G. The proof in [4] can easily be seen to give a kernel of size $O(k^4)$ (although it is not described as such in [4], as kernelization has only recently begun to attract significant attention). We improve this to a kernel of size $O(k^3)$.

Theorem 4. *There is a linear-time algorithm that, for any integer $k \in \mathbb{N}$, transforms any connected graph $G = (V, E)$ with edge-weighting w to a connected graph $G' = (V', E')$ of size $O(k^3)$ with edge-weighting w', such that G has a cut of size $w(G)/2 + \frac{1}{8}(\sqrt{8w(G) + 1} - 1) + k$ if and only if G' has a cut of size $w'(G')/2 + \frac{1}{8}(\sqrt{8w'(G') + 1} - 1) + k'$, for some $k' \leq k$.*

The proof is a slight modification of the proof of Theorem 22 in Bollobás and Scott [4]. Note that Theorems 1 and 3 only hold for unweighted graphs; the weighted versions remain open. Due to space contraints, the proofs of Theorems 2 and 4 are omitted.

Our Techniques and Related Work

Our results are based on algorithmic as well as combinatorial arguments. To prove Theorem 1, we design a Turing reduction to a generalization of the problem on block graphs, for which we show how to to solve the problem efficiently. Theorem 2 is established by a combinatorial reduction. Theorem 3 is proven by a careful analysis of random cuts via the probabilistic method, whereas the proof of Theorem 4 is through a characterization of graphs for which the lower bound is nearly tight, and weighted graph decompositions as "edge-sums".

A number of the standard approaches that have been developed for "above-guarantee" parameterizations of other problems are unavailable for this problem. The most common approach is to use probabilistic analysis of a random variable whose expected value corresponds to a solution matching the guarantee. However, there is no simple randomized procedure known giving a cut of size $m/2 + (n-1)/4$. Another approach is to make use of approximation algorithms that give a factor-c approximation, when the problem is parameterized above the bound $c \cdot n$; here, n is the maximum value of the objective function. But there is no such approximation algorithm for this problem.

Our paper also differs in its use of reduction rules. Most reduction rules for above-guarantee problems remove certain subgraphs of constant size, on which the bound is tight. But for our problem the bound is tight on cliques. Thus our reduction rules remove maximal cliques from the graph, which may contain a large fraction of the vertices in G. Moreover, rather than using reduction rules to reduce to an equivalent instance, which is then solved quickly, our reduction rules do not produce an equivalent instance. Instead, they either reduce to a 'yes"-instance or we can determine useful restrictions on the structure of the original instance, which can then be used to solve the original instance in fixed-parameter tractable time.

2   Preliminaries

We use standard graph theory terminology and notation. Given a graph G, let $V(G)$ be the vertices of G and let $E(G)$ be the edges of G. For disjoint sets $S, T \subseteq V(G)$, let $E(S,T)$ denote the set of edges in G with one vertex in S and one vertex in T. For $S \subseteq V(G)$, let $G[S]$ denote the subgraph induced by the vertices of S, and let $G - S$ denote the graph $G[V(G) \setminus S]$. We say that G has a *cut of size* t if there exists an $S \subseteq V(G)$ such that $|E(S, V(G) \setminus S)| = t$. The graph G is *connected* if any two of its vertices are connected by a path, and it is *2-connected* if $G - v$ is connected for every $v \in V(G)$. A *connected component of* G is a connected subgraph G' of G that is maximal with respect to vertex inclusion, and we often identify G' with its vertex set $V(G')$.

We study the following formulation of MAX-CUT parameterized above Edwards-Erdős bound:

MAX-CUT above Edwards-Erdős (MAX-CUT-AEE)
Instance: A connected graph G with $n = |V(G)|$ vertices and $m = |E(G)|$ edges, and an integer $k \in \mathbb{N}$.

Parameter: k.

Question: Does G have a cut of size at least $\frac{m}{2} + \frac{n-1}{4} + \frac{k}{4}$?

We ask for a cut of size $\frac{m}{2} + \frac{n-1}{4} + \frac{k}{4}$, rather than the more usual $\frac{m}{2} + \frac{n-1}{4} + k$, so that we may treat k as an integer at all times. Note that this does not affect the existence of a fixed-parameter algorithm or polynomial-size kernel. A pair (G, k) is called a *"yes"-instance* if G has a cut of size at least $\frac{m}{2} + \frac{n-1}{4} + \frac{k}{4}$, and a *"no"-instance* otherwise.

An *assignment* or *coloring* on G is a function $\alpha : V(G) \to \{\text{red}, \text{blue}\}$, and an edge is *cut* or *satisfied* by α if one of its vertices is colored red and the other vertex is colored blue. Note that a graph has a cut of size t if and only if it has an assignment that satisfies at least t edges. A *partial assignment* on G is a function $\alpha : X \to \{\text{red}, \text{blue}\}$, where X is a subset of $V(G)$.

A *clique* in G is a set of vertices $X \subseteq V(G)$ any two of which are adjacent in G. A *block* in G is a maximal subgraph of G which is 2-connected. A *block graph* is a connected graph in which every block is a clique. Observe that a complete graph is a block graph, and a graph formed by identifying together one vertex each from two disjoint block graphs is also a block graph. (For the purposes of this paper we count an isolated vertex as a block graph.)

3 Fixed-Parameter Algorithm for Max-Cut above the Edwards-Erdős Bound

In this section, we prove Theorem 1. To this end, we prove the following lemma, which also forms the basis of our kernel in Theorem 3.

Lemma 1. *Given a connected graph G with n vertices and m edges and an integer k, we can in polynomial time decide that either G has a cut of size at least $\frac{m}{2} + \frac{n-1}{4} + \frac{k}{4}$, or find a set S of at most $3k$ vertices in G such that each component of $G - S$ is a block graph.*

The algorithm starts by applying the following rules to the given connected graph G. These rules are such that if an instance (G', k') is reduced from (G, k) and (G', k') is a "yes"-instance, then (G, k) is also a "yes"-instance. The converse does not necessarily hold – a "yes"-instance may be reduced to a "no"-instance. In some rules, we *mark* certain vertices, that will be collected in the set S of Lemma 1.

Rule 1. *Let G be a connected graph with $v \in V(G), X \subseteq V(G)$ such that X is a connected component of $G - v$ and $X \cup \{v\}$ is a clique. Then remove all vertices in X and incident edges. Reduce k by 1 if $|X|$ is odd, otherwise leave k the same. Do not mark any vertices.*

Rule 2. *Let G be a connected graph reduced by Rule 1 with $v \in V(G)$ such that for all connected components X of $G - v$, except possibly one, X is a clique. Then remove v and all incident edges, and all vertices in X and incident edges, for every connected component X of $G - v$ which is a clique. Mark v, and reduce k by $2t - 1$, where t is the number of connected components of $G - v$ removed. (Only apply this rule if $t \geq 1$.)*

Rule 3. *Let G be a connected graph with $x, y \in V(G)$ such that $\{x, y\} \notin E(G)$, and for all connected components X of $G - \{x, y\}$, except possibly one, $X \cup \{x\}$ and $X \cup \{y\}$ are cliques. Then remove $\{x, y\} \cup X$ for any clique X satisfying these conditions. Mark x and y, and reduce k by $3t - 2$, where t is the number of connected components of $G - \{x, y\}$ removed. (Only apply this rule if $t \geq 1$.)*

Rule 4. *Let G be a connected graph with $a, b, c \in V(G)$ such that $\{a, b\}, \{b, c\} \in E(G), \{a, c\} \notin E(G)$, and $G - \{a, b, c\}$ is connected. Then mark a, b, c, and remove a, b, c and incident edges, and reduce k by 1.*

These rules can be applied exhaustively in polynomial time, as each rule reduces the number of vertices in G, and for each rule we can check for any applications of that rule by trying every set of at most three vertices in $V(G)$, and examining the connected components of the graph when those vertices are removed.

Lemma 2. *Let (G, k) and (G', k') be instances of MAX-CUT-AEE such that (G', k') is reduced from (G, k) by an application of Rules 1, 2, 3 and 4. Then G' is connected, and if (G', k') is a "yes"-instance of MAX-CUT-AEE then so is (G, k).*

Proof. First, we show that G' is connected. For Rule 1, observe that for $s, t \in V(G) \setminus X$, no path between s and t passes through X, so $G - X$ is connected. For Rules 2 and 3, observe that we remove some vertices together with all but at most one of the connected components in the resulting graph, so we are left with a single component. For Rule 4, the conditions explicitly state that we only apply the rule if the resulting graph is connected.

Second, we prove separately for each rule the following claim, in which n' denotes the number of vertices and m' the number of edges removed by the rule.

> Any assignment to the vertices of G' can be extended to an assignment on G that cuts an additional $\frac{m'}{2} + \frac{n'}{4} + \frac{k-k'}{4}$ edges. $\hfill (\star)$

The proofs for Rules 2 and 3 are omitted due to length; their proofs are similar to Rule 1 but more complicated.

Rule 1: Since v is the only vertex connecting X to the rest of the graph, any assignment to G' can be extended to one which is optimal on $X \cup \{v\}$. (Indeed, let α be an optimal coloring of $G[X \cup \{v\}]$, and let α' be the α with all colors reversed. Both α and α' are optimal colorings of $G[X \cup \{v\}]$, and one of these will agree with the coloring we are given on G' since the only overlap is v.) Observe that $n' = |X|$ and $m' = \frac{|X|(|X|+1)}{2}$, since the edges we remove form a clique including v, and all vertices in the clique except v are removed.

If $|X|$ is even then the maximum cut of the clique $X \cup \{v\}$ has size $\frac{|X|}{2}(\frac{|X|}{2} + 1) = \frac{|X|(|X|+2)}{4} = \frac{|X|(|X|+1)}{4} + \frac{|X|}{4} = \frac{m'}{2} + \frac{n'}{4}$, which is what we require as k is unchanged in this case.

If $|X|$ is odd then the maximum cut of the clique $X \cup \{v\}$ has size $\frac{(|X|+1)}{2} \frac{(|X|+1)}{2} = \frac{|X|(|X|+2)}{4} + \frac{1}{4} = \frac{m'}{2} + \frac{n'}{4} + \frac{1}{4}$, which is what we require as we reduce k by 1 in this case.

Rule 4: Observe that $n' = 3$ and $m' = 2 + |E(G', \{a, b, c\})|$. Consider two colorings α, α' of $\{a, b, c\}$: $\alpha(a) = \alpha(c) =$ red, $\alpha(b) =$ blue, and $\alpha'(a) = \alpha'(c) =$ blue, $\alpha'(b) =$ red. Both these colorings satisfy edges $\{a, b\}$ and $\{b, c\}$, and at least one of them will satisfy at least half the edges between $\{a, b, c\}$ and G'. Therefore, the number of satisfied edges incident with $\{a, b, c\}$ is at least $2 + \frac{|E(G', \{a, b, c\})|}{2} = \frac{m'}{2} + \frac{n'}{4} + \frac{1}{4}$.

This concludes the proof of the claim $(\star)$.

We now know that any assignment on G' can be extended to an assignment on G that cuts an additional $\frac{m'}{2} + \frac{n'}{4} + \frac{k-k'}{4}$ edges. Hence, if G' has a cut of size $\frac{|E(G')|}{2} + \frac{|V(G')|-1}{4} + \frac{k'}{4}$, then G has a cut of size $\frac{m-m'}{2} + \frac{n-n'-1}{4} + \frac{k'}{4} + \frac{m'}{2} + \frac{n'}{4} + \frac{k-k'}{4} = \frac{m}{2} + \frac{n-1}{4} + \frac{k}{4}$. Therefore, if (G', k') is a "yes"-instance then so is (G, k). $\qquad\square$

Lemma 3. *To any connected graph G with at least one edge, at least one of Rules 1–4 applies.*

Proof. The full proof is omitted due to length; we give an outline.

Suppose that G is reduced by Rules 1, 2, and 3. We show that there exist $a, b, c \in V(G)$ such that $\{a, b\}, \{b, c\} \in E(G)$ but $\{a, c\} \notin E(G)$ and $G - \{a, b, c\}$ is connected, that is, Rule 4 applies.

Observe that if G does not contain a set of vertices a, b, c such that $\{a, b\}, \{b, c\} \in E(G)$ and $\{a, c\} \notin E(G)$, then G is a clique and so Rule 1 applies. So such a set a, b, c must exist, and our only problem is if every such set a, b, c disconnects the graph. Assuming G is not a clique, it is possible to find a set of vertices a, b, c such that $\{a, b\}, \{b, c\} \in E(G)$, $\{a, c\} \notin E(G)$, and at most one component of $G \backslash \{a, b, c\}$ is not a clique. Reduction Rules 1, 2, and 3 impose restrictions on the edges between a, b, c and the clique components of $G \backslash \{a, b, c\}$ (for example, every clique component in $G \backslash \{a, b, c\}$ must be adjacent to at least two vertices of a, b, c), and we can use these restrictions to find a set of vertices satisfying Reduction Rule 4. $\qquad\square$

Lemma 4. *Let G be a connected graph and let $S \subseteq V(G)$ be the set of vertices that are marked after applying Rules 1–4 exhaustively to G; then every component of $G - S$ is a block graph.*

Proof. The complete proof is omitted due to length; we give an outline.

We proceed by induction on the number of applications of a reduction rule. By Lemma 3, a graph to which Rules 1, 2, 3 and 4 do not apply contains no edges, and is therefore a block graph. (In fact by Lemma 2, such a graph is also connected, and therefore consists of a single vertex.) This handles the base case. For the inductive step, it can be shown that for each reduction rule, if G' is reduced by an application of that rule from G, and $G' - S$ is a block graph, then so is $G - S$. $\qquad\square$

Putting Lemmas 2 and 4 together, we can now prove Lemma 1.

Proof (Proof of Lemma 1). Apply Rules 1, 2, 3 and 4 exhaustively, and let S be the set of vertices which are marked after doing this. By Lemma 4, every connected component in $G - S$ is a block graph. Therefore, if $|S| < 3k$ we are done. It remains to show that if $|S| \geq 3k$ then G has an assignment that satisfies at least $\frac{m}{2} + \frac{n-1}{4} + \frac{k}{4}$ edges.

So suppose that $|S| \geq 3k$. Let (G', k') be the instance obtained from (G, k) by exhaustively applying Rules 1, 2, 3 and 4. Observe that every time k is reduced, at most three vertices are marked. Therefore since at least $3k$ vertices are marked, we have $k' \leq 0$. But since the Edwards-Erdős bound holds for all connected graphs, (G', k') is a "yes"-instance. Therefore, by Lemma 2, (G, k) is a "yes"-instance, as required.   □

We now show that, for a given assignment to S, we can efficiently find an optimal extension to $G - S$. For this, we consider the following generalisation of MAX-CUT where each vertex has an associated weight for each part of the partition. These weights may be taken as an indication of how much we would like the vertex to appear in each part.

MAX-CUT-WITH-WEIGHTED-VERTICES

Instance: A graph G with weight functions $w_0 : V(G) \rightarrow \mathbb{N}_0$ and $w_1 : V(G) \rightarrow \mathbb{N}_0$, and an integer $t \in \mathbb{N}$.

Question: Does there exist an assignment $f : V \rightarrow \{0, 1\}$ such that $\sum_{xy \in E} |f(x) - f(y)| + \sum_{f(x)=0} w_0(x) + \sum_{f(x)=1} w_1(x) \geq t$?

Now MAX-CUT is the special case of MAX-CUT-WITH-WEIGHTED-VERTICES in which G is connected and $w_0(x) = w_1(x) = 0$ for all $x \in V(G)$.

Lemma 5. MAX-CUT-WITH-WEIGHTED-VERTICES *is solvable in polynomial time when G is a block graph.*

Proof. The full proof is omitted due to length; we give an outline.

For each vertex x let $\varepsilon(x) = w_1(x) - w_0(x)$. If G is a clique, we can solve the problem in polynomial time by numbering the vertices $x_1, \ldots x_n$ such that if $i < j$ then $\varepsilon(x_i) \geq \varepsilon(x_j)$. Then there is an optimal assignment in which x_i is assigned 1 for every $i \leq t$, and x_i is assigned 0 for every $i > t$, for some $0 \leq t \leq n$. We therefore find the optimal assignment by trying each value of t.

We may use this approach to reduce the problem when G is not a clique. Find a block C for which only one vertex $r \in V(C)$ is adjacent to any vertex in $V(G) \backslash V(C)$. Then we may reduce the problem by removing $V(C) \backslash r$, and changing $w_i(r)$ to the value of the optimal solution on C for which $f(r) = i$.   □

We are ready to prove Theorem 1, and show that MAX-CUT-AEE is fixed-parameter tractable.

Proof (of Theorem 1). By Lemma 1, we can in polynomial time either decide that G has an assignment that satisfies at least $\frac{m}{2} + \frac{n-1}{4} + \frac{k}{4}$ edges, or find a set S of at most $3k$ vertices in G such that $G - S$ is a block graph. So assume we have found such an S. Then we transform our instance into at most 2^{3k} instances of MAX-CUT-WITH-WEIGHTED-VERTICES, such that the answer to our original instance is "yes" if and only if the answer to at least one of the instances of MAX-CUT-WITH-WEIGHTED-VERTICES is "yes", and in each MAX-CUT-WITH-WEIGHTED-VERTICES instance the graph is a block graph. As each of these instances can be solved in polynomial time by Lemma 5, we have a fixed-parameter tractable algorithm.

For every possible coloring of the vertices in S, we construct one of the instances of MAX-CUT-WITH-WEIGHTED-VERTICES as follows. For every vertex $x \in G - S$, let $w_0(x)$ equal the number of vertices in S adjacent to x which are colored blue, and let $w_1(x)$ equal the number of vertices in S adjacent to x which are colored red. Then remove the vertices of S from G. By Lemma 1, each component of the resulting graph G' is a block graph. Let m' be the number of edges in $G - S$, let n' be the number of vertices in $G - S$, and let p be the number of edges within S satisfied by the assignment to S. Then for an assignment to the vertices of $G - S$, the total number of satisfied edges in G would be exactly $\sum_{xy \in E(G-S)} |f(x) - f(y)| + p + \sum_{f(x)=0} w_0(x) + \sum_{f(x)=1} w_1(x)$, where $f : V(G) \setminus S \to \{0,1\}$ is such that $f(x) = 0$ if x is colored red, and $f(x) = 1$ if x is colored blue. Thus, the assignment to S can be extended to one that cuts at least $\frac{m}{2} + \frac{n-1}{4} + \frac{k}{4}$ edges in G if and only if the instance of MAX-CUT-WITH-WEIGHTED-VERTICES is a "yes"-instance with $t = \frac{m}{2} + \frac{n-1}{4} + \frac{k}{4} - p$. $\square$

4   Polynomial Kernel for Max-Cut above Edwards-Erdős

In this section, we prove Theorem 3. By Lemma 1, in polynomial time we can either decide that (G, k) is a "yes"-instance, or find a set S of vertices in G such that $|S| < 3k$ and $G - S$ is a block graph. In what follows we assume we have found such a set S.

Observe that we can find all blocks in $G - S$ in polynomial time. Indeed, if X is a clique on at least 2 vertices then any vertex not in X which is adjacent to two or more members of X is part of a block containing X, and there is only one such block. Therefore, we can find all the blocks by expanding greedily from each edge in $G - S$.

Let $C_1, \ldots, C_{n^*}$ be the blocks in $G - S$. Let J be the set of vertices in $G - S$ which occur in two or more blocks. For each $i \in \{1, \ldots, n^*\}$ let $A_i = C_i - J$.

We first apply the following reduction rules.

Rule 5. *Let G be a connected graph with $v \in V(G), X \subseteq V(G)$ such that X is a connected component of $G - v$ and $X \cup \{v\}$ is a clique. Then remove all vertices in X and incident edges. Reduce k by 1 if $|X|$ is odd, otherwise leave k the same.*

Note that Reduction Rule 5 is the same as Reduction Rule 1, which we used to find the set S.

Rule 6. *Suppose there exists a vertex $x \in G - S$ and a set of vertices $X \subseteq V(G) \setminus S$ such that $X \cup \{x\}$ is a clique and X is a connected component of $G - (S \cup \{x\})$, and there is exactly one vertex $s \in S$ which is adjacent to X, and $X \cup \{s\}$ is a clique. Then remove all but one vertex from X, and incident edges. Reduce k by 1 if $|X|$ is even, otherwise leave k the same.*

Rule 7. *Suppose there exist vertex sets $X, Y \subseteq G - S$ such that X and Y are maximal odd cliques, with vertices $x \in X, y \in Y, \{z\} = X \cap Y$, such that x, z are the only vertices in X adjacent to a vertex in $G - X$, and y, z are the only vertices in Y adjacent to a vertex in $G - Y$. Then remove all vertices in $(X \cup Y) - \{x, y, z\}$ and incident edges, and add new vertices u, v, and edges such that $\{x, y, z, u, v\}$ is a clique. Do not change k.*

Rule 8. *Suppose for some block C_i in $G - S$, there exists $X \subseteq A_i$ such that $|X| > \frac{|A_i| + |J| + |S|}{2}$ and for all $x, y \in X$, x and y have exactly the same neighbors in S. Then remove any two vertices from X and incident edges. Do not change k.*

The complete proof of Theorem 3 is omitted due to length; instead we give a brief outline.

We first show that Reduction Rules 5, 6, 7 and 8 are valid; we then assume G is reduced by these rules. We show that for any "no"-instance, the number of blocks in $G \backslash S$ is $O(k^2)$, and for each block, the number of vertices in that block is $O(k^3)$.

We show the bound on the number of blocks by a probabilistic argument. Suppose a block C_i contains at most one vertex in J. For such a graph, given a random coloring on S, we can expect to achieve at least an extra $\frac{1}{4}$ above the Edwards-Erdős bound in the graph consisting of edges between vertices in A_i and between A_i and S. So if the number of blocks is big enough we have a "yes"-instance. Otherwise, we have a bound on the number of blocks with one vertex in J, and using this we can limit the total number of blocks in $G \backslash S$.

To limit the number of vertices in a block C_i, we show that if there there exist $X_1, X_2 \subseteq C_i$ such that X_1, X_2 are large enough and some $s \in S$ is adjacent to all vertices in X_1 and no vertices in X_2, then by coloring S to exploit this distinction we can satisfy enough edges in $E(S, C_i)$ to ensure that we have a "yes"-instance. It then follows that if C_i is large enough but G is a "no"-instance then a large number of vertices in C_i have exactly the same neighbourhood in S, and we would have an application of Rule 8.

Putting the limits on the number of vertices in a block together with the number of block in $G \backslash S$ gives us a bound of $O(k^5)$ on the number of vertices in G.

5   Discussion and Open Problems

We showed fixed-parameter tractability of MAX-CUT parameterized above the Edwards-Erdős bound $m/2 + (n-1)/4$, and thereby resolved an open question from [7, 14, 22, 23, 27]. Furthermore, we showed that the problem has a kernel with $O(k^5)$ vertices and the "edge version" of the bound admits a kernel of size $O(k^3)$. We have not attempted to optimize running time or kernel size, and indeed we conjecture that MAX-CUT has a kernel with $O(k^3)$ vertices and the edge version admits a linear kernel.

It remains an open problem whether the weighted version of MAX-CUT above the Edwards-Erdős bound is fixed-parameter tractable; our conjecture is that this problem is also fixed-parameter tractable with a polynomial kernel.

The problem MAX-BISECTION is a variant of MAX-CUT in which we seek a cut such that the number of vertices in both sides of the bipartition is as equal as possible. The the tight lower bound on the bisection size in terms of m is $m/2$. Fixed-parameter tractability of MAX-BISECTION above $m/2$ was shown by Gutin and Yeo [14]. An improved bound lower bound in terms of m and n is $mn/2(n-1)$. It is an open question whether MAX-BISECTION parameterized above $mn/2(n-1)$ is fixed-parameter tractable.

Acknowledgment. We thank Tobias Friedrich and Gregory Gutin for help with the presentation of the results. Part of this research has been supported by an International Joint Grant from the Royal Society.

References

[1] Alon, N.: Bipartite subgraphs. Combinatorica 16(3), 301–311 (1996)

[2] Andersen, L.D., Grant, D.D., Linial, N.: Extremal k-colourable subgraphs. Ars Combin. 16, 259–270 (1983)

[3] Bodlaender, H.L., Downey, R.G., Fellows, M.R., Hermelin, D.: On problems without polynomial kernels. J. Comput. System Sci. 75(8), 423–434 (2009)

[4] Bollobás, B., Scott, A.: Better bounds for Max Cut. In: Bollobás, B. (ed.) Contemporary Combinatorics. Bolyai Society Mathematical Studies, vol. 10, pp. 185–246 (2002)

[5] Cai, L., Juedes, D.: On the existence of subexponential parameterized algorithms. J. Comput. System Sci. 67(4), 789–807 (2003)

[6] Charikar, M., Wirth, A.: Maximizing quadratic programs: Extending Grothendieck's inequality. In: Proc. FOCS 2004, pp. 54–60 (2004)

[7] Crowston, R., Fellows, M.R., Gutin, G., Jones, M., Rosamond, F., Thomassé, S., Yeo, A.: Simultaneously satisfying linear equations over $\mathbb{F}_2$: MaxLin2 and Max-r-Lin2 parameterized above average. In: Proc. FSTTCS 2011. LIPICS, vol. 13, pp. 229–240 (2011)

[8] Downey, R.G., Fellows, M.R.: Parameterized complexity. Monographs in Computer Science (1999)

[9] Edwards, C.S.: Some extremal properties of bipartite subgraphs. Canad. J. Math. 25, 475–485 (1973)

[10] Edwards, C.S.: An improved lower bound for the number of edges in a largest bipartite subgraph. In: Recent Advances in Graph Theory, pp. 167–181 (1975)

[11] Erdős, P.: On some extremal problems in graph theory. Israel J. Math. 3, 113–116 (1965)

[12] Erdős, P.: On even subgraphs of graphs. Mat. Lapok 18, 283–288 (1967)

[13] Erdős, P., Gyárfás, A., Kohayakawa, Y.: The size of the largest bipartite subgraphs. Discrete Maths 177, 267–271 (1997)

[14] Gutin, G., Yeo, A.: Note on maximal bisection above tight lower bound. Inform. Process. Lett. 110(21), 966–969 (2010)

[15] Haglin, D.J., Venkatesan, S.M.: Approximation and intractability results for the maximum cut problem and its variants. IEEE Trans. Comput. 40(1), 110–113 (1991)

[16] Hofmeister, T., Lefmann, H.: A combinatorial design approach to Max Cut. Random Structures and Algorithms 9, 163–175 (1996)

[17] Impagliazzo, R., Paturi, R.: On the complexity of k-SAT. J. Comput. System Sci. 62(2), 367–375 (2001)
[18] Karp, R.M.: Reducibility among combinatorial problems. In: Complexity of Computer Computations (Proc. Sympos., IBM Thomas J. Watson Res. Center, Yorktown Heights, N.Y., 1972), pp. 85–103 (1972)
[19] Khot, S., O'Donnell, R.: SDP gaps and UGC-hardness for Max-Cut-Gain. Theory Comput. 5, 83–117 (2009)
[20] Lehel, J., Tuza, Z.: Triangle-free partial graphs and edge covering theorems. Discrete Math. 39(1), 59–65 (1982)
[21] Locke, S.C.: Maximum k-colorable subgraphs. J. Graph Theory 6(2), 123–132 (1982)
[22] Mahajan, M., Raman, V.: Parameterizing above guaranteed values: MaxSat and MaxCut. Technical Report TR97-033. Electronic Colloquium on Computational Complexity (1997), http://eccc.hpi-web.de/report/1997/033/
[23] Mahajan, M., Raman, V., Sikdar, S.: Parameterizing above or below guaranteed values. J. Comput. System Sci. 75(2), 137–153 (2009)
[24] Ngọc, N.V., Tuza, Z.: Linear-time approximation algorithms for the max cut problem. Combin. Probab. Comput. 2(2), 201–210 (1993)
[25] Poljak, S., Turzík, D.: A polynomial algorithm for constructing a large bipartite subgraph, with an application to a satisfiability problem. Canad. J. Math. 34(3), 519–524 (1982)
[26] Poljak, S., Tuza, Z.: Maximum cuts and large bipartite subgraphs. In: Combinatorial Optimization, New Brunswick, NJ, 1992-1993. DIMACS Ser. Discrete Math. Theoret. Comput. Sci., vol. 20, pp. 181–244 (1995)
[27] Sikdar, S.: Parameterizing from the Extremes: Feasible Parameterizations of some NP-optimization problems. PhD thesis, The Institute of Mathematical Sciences, Chennai, India (2010)

Clique Cover and Graph Separation: New Incompressibility Results*

Marek Cygan[1,**], Stefan Kratsch[2,***], Marcin Pilipczuk[3,†],
Michał Pilipczuk[4,‡], and Magnus Wahlström[5]

[1] IDSIA, University of Lugano, Switzerland
marek@idsia.ch
[2] Utrecht University, Utrecht, the Netherlands
s.kratsch@uu.nl
[3] Institute of Informatics, University of Warsaw, Poland
malcin@mimuw.edu.pl
[4] Department of Informatics, University of Bergen, Norway
michal.pilipczuk@ii.uib.no
[5] Max-Planck-Institute for Informatics, Saarbrücken, Germany
wahl@mpi-inf.mpg.de

Abstract. The field of kernelization studies polynomial-time preprocessing routines for hard problems in the framework of parameterized
complexity. In this paper we show that, unless $NP \subseteq coNP/poly$ and
the polynomial hierarchy collapses up to its third level, the following
parameterized problems do not admit a polynomial-time preprocessing
algorithm that reduces the size of an instance to polynomial in the parameter:

- EDGE CLIQUE COVER, parameterized by the number of cliques,
- DIRECTED EDGE/VERTEX MULTIWAY CUT, parameterized by the
  size of the cutset, even in the case of two terminals,
- EDGE/VERTEX MULTICUT, parameterized by the size of the cutset,
- and k-WAY CUT, parameterized by the size of the cutset.

The existence of a polynomial kernelization for EDGE CLIQUE COVER
was a seasoned veteran in open problem sessions. Furthermore, our results complement very recent developments in designing parameterized
algorithms for cut problems by Marx and Razgon [STOC'11], Bousquet
et al. [STOC'11], Kawarabayashi and Thorup [FOCS'11] and Chitnis
et al. [SODA'12].

---

* The full version of this paper is available online [1].
** Partially supported by ERC Starting Grant NEWNET 279352 and Foundation for
Polish Science.
*** Supported by the Netherlands Organization for Scientific Research (N.W.O.),
project "KERNELS: Combinatorial Analysis of Data Reduction".
† Partially supported by NCN grant N206567140 and Foundation for Polish Science.
‡ Partially supported by European Research Council (ERC) grant "Rigorous Theory
of Preprocessing", reference 267959.

1   Introduction

In order to cope with the NP-hardness of many natural combinatorial problems, various algorithmic paradigms such as brute-force, approximation, or heuristics are applied. However, while the paradigms are quite different, there is a commonly used opening move of first applying polynomial-time preprocessing routines, before making sacrifices in either exactness or runtime. The aim of the field of kernelization is to provide a rigorous mathematical framework for analyzing such preprocessing algorithms. One of its core features is to provide quantitative performance guarantees for preprocessing via the framework of parameterized complexity, a feature easily seen to be infeasible in classical complexity (cf. [2]).

In the framework of parameterized complexity an instance x of a parameterized problem comes with an integer parameter k, formally, a parameterized problem $Q \subseteq \Sigma^* \times \mathbb{N}$ for some finite alphabet Σ. We say that a problem is *fixed parameter tractable* (*FPT*) if there exists an algorithm solving any instance (x, k) in time $f(k)\text{poly}(|x|)$ for some (usually exponential) computable function f. It is known that a problem is FPT iff it is kernelizable. A *kernelization algorithm* (*kernel* for short) for a problem Q is a polynomial time preprocessing routine that takes an instance (x, k) and in time polynomial in $|x| + k$ produces an equivalent instance (x', k') (i.e., $(x, k) \in Q$ iff $(x', k') \in Q$) such that $|x'| + k' \leq g(k)$ for some computable function g. The function g is the *size of the kernel*, and if it is polynomial, we say that Q admits a polynomial kernel. If g is small, after preprocessing even an exponential-time brute-force algorithm might be feasible. Therefore small kernels, with g being linear or polynomial, are of big interest.

Although polynomial kernels for a wide range of problems have been developed for the last few decades (see the surveys of Guo and Niedermeier [3] and Bodlaender [4]), a framework for proving kernelization lower bounds was discovered only three years ago by Bodlaender et al. [5], with the backbone theorem proven by Fortnow and Santhanam [6]. The crux of the framework is the following idea of a composition. Assume we are able to combine in polynomial time an arbitrary number of instances $x_1, x_2, \ldots, x_t$ of an NP-complete problem L into a single instance (x, k) of a parameterized problem $Q \in NP$ such that $(x, k) \in Q$ if and only if one of the instances x_i is in L, while k is bounded polynomially in $\max_i |x_i|$. If such a *composition* algorithm was pipelined with a polynomial kernel for the problem Q, we would obtain an OR-distillation of the NP-complete language L: the resulting instance is of size polynomial in $\max_i |x_i|$, possibly significantly smaller than t, but encodes a disjunction of all input instances x_i (i.e., an OR-distillation is a compression of the logical OR of the instances). As proven by Fortnow and Santhanam [6], existence of such an algorithm would imply NP $\subseteq$ coNP/poly, which is known to cause a collapse of the polynomial hierarchy to its third level [7,8].

The astute reader may have noticed that the above description of a composition is actually using the slightly newer notion of a cross-composition [9]. This generalization of the original lower bound framework will be the main ingredient of our proofs. The framework of kernelization lower bounds was also extended by Dell and van Melkebeek [10] to allow excluding kernels of particular exponent in the polynomial. Recently, Dell and Marx [11] and, independently, Hermelin and Wu [12] simplified this approach and applied it to various packing problems.

The aforementioned (cross-)composition algorithm is sometimes called an *OR-composition*, as opposed to an *AND-composition*, where we require that the output instance (x, k) is in Q if and only if *all* input instances belong to L. Various problems have been shown to be AND-compositional, with the most important example being the problem of determining whether an input graph has treewidth no larger than the parameter [5]. It is conjectured [5] that no NP-complete problem admits an AND-distillation, which would be the result of pipelining an AND-composition with a polynomial kernel. However, it is now a major open problem in the field of kernelization to support this claim with a proof based on a plausible complexity assumption.

Although the framework of kernelization lower bounds has been applied successfully multiple times over the last three years, there are still many important problems where the existence of a polynomial kernel is widely open. The reason for this situation is that an application of the idea of a composition (or appropriate reductions, see [13]) is far from being automatic. To obtain a composition algorithm, usually one needs to carefully choose the starting language L (for example, the choice of the starting language is crucial for compositions of Dell and Marx [11], and the core idea of the composition algorithms for connectivity problems in degenerate graphs [14] is to use GRAPH MOTIF as a starting point) or invent sophisticated gadgets to merge the instances (for example, the colors and IDs technique introduced by Dom et al. [15] or the idea of an instance selector, used mainly for structural parameters [9,16]).

Our results. The main contribution of this paper is a proof of non-existence of polynomial kernels for four important problems.

Theorem 1. *Unless $NP \subseteq coNP/poly$, EDGE CLIQUE COVER, parameterized by the number of cliques, as well as DIRECTED MULTIWAY CUT, MULTICUT and k-WAY CUT, parameterized by the size of the cutset, do not admit polynomial kernelizations.*

The common theme of our compositions is a very careful choice of starting problems. Not only do we select particular NP-complete problems, but we also restrict instances given as the input, to make them satisfy certain conditions that allow designing cross-compositions. Each time we constrain the set of input instances of an NP-complete problem we need to prove that the problem remains NP-complete. Even though this paper is about negative results, in our constructions we use intuition derived from the design of parameterized algorithms techniques, including iterative compression (in case of EDGE CLIQUE COVER) introduced by Reed et al. [17] and important separators (in case of MULTICUT) defined by Marx [18].

For the three cut problems listed in Theorem 1 our kernelization hardness results complement very recent developments in the design of algorithm parameterized by the size of the cutset [19,20,21,22]. In this extended abstract we give some motivation and related work for each of the four problems, and informally describe the compositions algorithms. For detailed proofs, as well as a more thorough description of motivation and related work, see the full version [1].

Edge clique cover. In the EDGE CLIQUE COVER problem the goal is to cover the edges of an input graph G with at most k cliques all of which are subgraphs of G. This problem, NP-complete even in very restricted graph classes, is also known as COVERING BY CLIQUES (GT17), INTERSECTION GRAPH BASIS (GT59) [23] and KEYWORD CONFLICT [24]. It has multiple applications in various areas in practice, such as computational geometry [25], applied statistics [26,27], and compiler optimization [28].

From the point of view of parameterized complexity, EDGE CLIQUE COVER was extensively studied by Gramm et al. [29]. A simple kernelization algorithm is known that reduces the size of the graph to at most 2^k vertices; the best known fixed-parameter algorithm is a brute-force search on the 2^k-vertex kernel. The question of a polynomial kernel for EDGE CLIQUE COVER, probably first verbalized by Gramm et al. [29], was repeatedly asked in the parameterized complexity community, for example on the last Workshop on Kernels (WorKer, Vienna, 2011). We show that EDGE CLIQUE COVER is both AND- and OR-compositional (i.e., both an AND- and an OR-composition algorithm exist for some NP-complete input language L), thus the existence of a polynomial kernel would both cause a collapse of the polynomial hierarchy as well as violate the AND-conjecture. To the best of our knowledge, this is the first natural parameterized problem that is known to admit both an AND- and an OR-composition.

Multicut and directed multiway cut. With MULTICUT and DIRECTED MULTIWAY CUT we move on to the family of graph separation problems. The central problems of this area are two natural generalizations of the $s - t$ cut problem, namely MULTIWAY CUT and MULTICUT. In the first problem we are given a graph G with designated terminals and we are to delete at most p edges (or vertices, depending on the variant) so that the terminals remain in different connected components. In the MULTICUT problem we consider a more general setting where the input graph contains terminal *pairs* and we need to separate all pairs of terminals. The graph separation problems became one of the most important subareas in parameterized complexity after Marx introduced the concept of important separators [18]. This technique turns out to be very robust, and is now a key ingredient in fixed-parameter algorithms for several problems.

Although the picture of the fixed-parameter tractability of the graph separation problems becomes more and more complete, very little is known about polynomial kernelization. Very recently, Kratsch and Wahlström came up with a genuine application of matroid theory to graph separation problems. They were able to obtain randomized polynomial kernels for ODD CYCLE TRANSVERSAL [30], ALMOST 2-SAT, and MULTIWAY CUT and MULTICUT restricted to a bounded number of terminals, among others [31]. We are not aware of any other results on kernelization of the graph separation problems.

We prove that DIRECTED MULTIWAY CUT, even in the case of two terminals, as well as MULTICUT, parameterized by the size of the cutset, are OR-compositional, thus a polynomial kernel for any of these two problems would cause a collapse of the polynomial hierarchy. In fact, in the full version [1] we give two OR-compositions for MULTICUT: the constructions are very different and the presented gadgets may inspire lower bounds for similar problems.

The k-way cut problem. The last part of this work is devoted to another generalization of the *s-t* cut problem, but of a bit different flavor. The k-WAY CUT problem is defined as follows: given an undirected graph G and integers k and s, remove at most s edges from G to obtain a graph with at least k connected components. This problem has applications in numerous areas of computer science, such as finding cutting planes for the traveling salesman problem, clustering-related settings (e.g., VLSI design) or network reliability [32]. In general, k-WAY CUT is NP-complete [33] but solvable in polynomial time for fixed k: a long line of research led to a deterministic algorithm running in time $O(mn^{2k-2})$ [34]. The dependency on k in the exponent is probably unavoidable: from the parameterized perspective, the k-WAY CUT problem parameterized by k is $W[1]$-hard [35]. Moreover, the node-deletion variant is also $W[1]$-hard when parameterized by s [18]. Somewhat surprisingly, in 2011 Kawarabayashi and Thorup presented a fixed-parameter algorithm for (edge-deletion) k-WAY CUT parameterized by s [22]. In this paper we complete the parameterized picture of the edge-deletion k-WAY CUT problem parameterized by s by showing that it is OR-compositional and, therefore, a polynomial kernel is unlikely to exist.

2   Preliminaries

We here informally summarize the kernelization lower bounds framework; see the full version [1] for formal definitions.

We use the cross-composition technique due to Bodlaender et al. [9]. Let L be a (classical) language, and Q be a parameterized one. We first split instances of L into equivalence classes of a polynomially-computable relation (called *polynomial equivalence relation*) that partitions all instances of size n into $n^{O(1)}$ equivalence classes (e.g., we may partition the input graphs according to the number of their vertices and edges). Within each equivalence class, we exhibit a polynomial-time *cross-composition* algorithm that, given t instances $x_1, x_2, \ldots, x_t$ of L, produces one instance (x^*, k^*) of Q such that k^* is bounded polynomially in $\max_i |x_i|$ and $(x^*, k^*) \in Q$ iff at least one instance x_i belongs to L. If such a composition is pipelined with a polynomial kernelization algorithm for Q, we obtain a very efficient *distillation* algorithm for the language L: by the result of Fortnow and Santhanam [6], L belongs to coNP/poly. Thus, if L is NP-complete, we obtain NP $\subseteq$ coNP/poly and a collapse of the polynomial hierarchy.

If we assume that the output instance $(x^*, k^*) \in Q$ iff *all* input instances x_i belong to L, we obtain an *AND-cross-composition*. It is conjectured (the so-called *AND-conjecture* [5]) that no NP-complete problem admits an AND-distillation, obtained by pipelining such an AND-cross-composition and a polynomial kernel.

3   Clique Cover

EDGE CLIQUE COVER
Input: An undirected graph G and an integer k.
Task: Does there exist a set of k subgraphs of G, such that each subgraph is a clique and each edge of G is contained in at least one of these subgraphs?

In this section we present both a cross-composition and an AND-cross-composition of EDGE CLIQUE COVER parameterized by k. We start with the AND-cross-composition since the construction we present is also used in the cross-composition.

3.1   AND-Cross-Composition

Theorem 2. EDGE CLIQUE COVER *AND-cross-composes to* EDGE CLIQUE COVER *parameterized by* k.

Proof (sketch). For the equivalence relation we take a relation that puts two instances (G_1, k_1), (G_2, k_2) of EDGE CLIQUE COVER into the same equivalence class iff $k_1 = k_2$ and the number of vertices in G_1 is equal to the number of vertices in G_2. Therefore, in the rest of the proof we assume that we are given a sequence $(G_i, k)_{i=0}^{t-1}$ of EDGE CLIQUE COVER instances that are in the same equivalence class (to avoid confusion we number everything starting from zero in this proof). Let n be the number of vertices in each of the instances. By adding isolated vertices in the instances and duplicating some instances we may ensure that $n = 2^{h_n}$ and $t = 2^{h_t}$ for some integers h_n and h_t.

Now we construct an instance (G^*, k^*), where k^* is polynomial in $n + k + h_t$. Initially as G^* we take a disjoint union of graphs G_i for $i = 0, \ldots, t - 1$ with added edges between every pair of vertices from G_a and G_b for $a \neq b$. Next, in order to cover all the edges between different instances with few cliques we introduce the following construction. Let us assume that the vertex set of G_i is $V_i = \{v_0^i, \ldots, v_{n-1}^i\}$. For each $0 \leq a < n$, for each $0 \leq b < n$ and for each $0 \leq r < h_t$ we add to G^* a vertex $w(a, b, r)$ which is adjacent to exactly one vertex in each V_i, that is v_j^i where $j = (a + b\lfloor \frac{i}{2^r} \rfloor) \bmod n$. By W we denote the set of all added vertices $w(a, b, r)$. As the new parameter k^* we set $k^* = |W| + k = n^2 h_t + k$.

Note that W is an independent set of non-isolated vertices in G^*. As for each $w \in W$ the set $N_{G^*}[w]$ induces a clique, we may assume that an optimal clique cover of G^* contains $|W|$ cliques $N_{G^*}[w]$ for $w \in W$. We observe that these cliques cover no edges of G_i for any $0 \leq i < t$ while covering all other edges of G^*. Thus the remaining k cliques need to induce solutions for all input instances.   □

As a consequence we obtain the following result.

Corollary 1. *There is no polynomial kernel for the* EDGE CLIQUE COVER *problem parameterized by* k *unless the AND-conjecture fails.*

3.2   Cross-Composition

In this section we show a cross-composition to EDGE CLIQUE COVER, which we obtain by extending the AND-cross-composition gadgets from the previous section. We cross-compose from a strengthened variant of the EDGE CLIQUE COVER problem, proven to be NP-complete in the full version.

COMPRESSION CLIQUE COVER
Input: An undirected graph G, an integer k and a set $\mathcal{C}$ of $k + 1$ cliques in G covering all edges of G.
Task: Does there exist a set of k subgraphs of G, such that each subgraph is a clique and each edge of G is contained in at least one of the subgraphs?

Theorem 3. COMPRESSION CLIQUE COVER *cross-composes to* EDGE CLIQUE COVER *parameterized by* k.

Proof (sketch). Similarly as in the proof of Theorem 2, we assume that we are given a sequence $(G_i, k, \mathcal{C}_i)_{i=0}^{t-1}$ of COMPRESSION CLIQUE COVER instances with $|V(G_i)| = n$ for each $0 \leq i < t$, and $n = 2^{h_n}$, $t = 2^{h_t}$.

We extend the construction from Theorem 2 by adding h_t gadgets D_j, for $0 \leq j < h_t$. Each gadget D_j is a 6-vertex clique with a perfect matching removed and a vertex set partitioned into two halves L_j and R_j such that in each of the three non-edges of D_j one endpoint is in L_j and the second in R_j. For each instance $(G_i, k, \mathcal{C}_i)$ we connect the vertices of G_i to L_j if the j-th bit of the index i equals zero and to R_j otherwise. Moreover, we add simplicial vertices similar to the vertices $w(a, b, r)$ to cover the edges connecting the gadgets D_j with the graphs G_i. The requested number of cliques is: one for each simplicial vertex, four for each gadget D_j, and additional k cliques.

The key observation is that there are only two reasonable ways to cover the edges of the gadget D_j (see Figure 1). We first choose three triangles to cover the edges between the halves L and R. These three triangles contain vertices both from L and R and, therefore, cannot contain any other vertex outside D_j. The fourth clique contains the entire set L or the entire set R and may contain other vertices in the instances G_i connected to the chosen set (L or R).

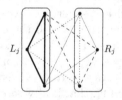

Fig. 1. One of the two optimal ways to cover the edges of a gadget D_j with four cliques

Thus, each gadgets D_j *grants* an extra clique to all instances with j-th bit set to zero or one. We infer that there is exactly one instance G_i left where the edges need to be covered by the remaining k cliques.   □

Corollary 2. *There is no polynomial kernel for the* EDGE CLIQUE COVER *problem parameterized by* k *unless* $NP \subseteq coNP/poly$.

4   Directed Multiway Cut

DIRECTED EDGE (VERTEX) MULTIWAY CUT

Input: A digraph $G = (V, A)$, a set of terminals $T \subseteq V$ and an integer p.
Task: Does there exist a set S of at most p arcs in A (p vertices in $V \setminus T$), such that in $G \setminus S$ there is no path between any pair of terminals in T?

It is well known that the edge- and vertex-deletion variants are equivalent (cf. [20]). Further, in the node-deletion variant we may assume that a set $V^\infty \supseteq T$ is given, and the solution cutset needs to be disjoint with V^∞: for any $v \in V^\infty \setminus T$, we can replace v with a clique on $p + 1$ vertices. Hence we show a cross-composition to DIRECTED VERTEX MULTIWAY CUT with a set of undeletable vertices V^∞. We start from the following restricted variant of DIRECTED VERTEX MULTIWAY CUT, proven to be NP-complete in the full version.

PROMISED DIRECTED VERTEX MULTIWAY CUT

Input: A digraph $G = (V, A)$, two terminals $T = \{s_1, s_2\}$, a set of forbidden vertices $V^\infty \supseteq T$ and an integer p. Moreover, after removing any set of at most $p/2$ vertices of $V \setminus V^\infty$, both an $s_1 s_2$-path and an $s_2 s_1$-path remain.
Task: Does there exist a set S of at most p vertices in $V \setminus V^\infty$, such that in $G \setminus S$ there is no $s_1 s_2$-path nor $s_2 s_1$-path?

Theorem 4. PROMISED DIRECTED VERTEX MULTIWAY CUT *cross-composes into* DIRECTED VERTEX MULTIWAY CUT *with two terminals, parameterized by the size of the cutset* p.

Proof (sketch). By choosing an appropriate equivalence relation, we assume that we are given a sequence $I_i = (G_i, T_i = \{s_1^i, s_2^i\}, V_i^\infty, p)_{i=1}^t$ of PROMISED DIRECTED VERTEX MULTIWAY CUT instances. As the graph G' we take the disjoint union of all the graphs G_i and for each $i = 1, \ldots, t - 1$, in G' we identify the vertices s_2^i and s_1^{i+1}. Let $I' = (G', \{s_1^1, s_2^t\}, \bigcup_{i=1}^t V_i^\infty, p)$ be an instance of DIRECTED VERTEX MULTIWAY CUT. To see the correctness of this cross-composition, observe that the crucial assumption that in the input instances a $p/2$-cut cannot separate s_1 from s_2 in any direction ensures that a p-cut in G' can make any significant separation only in one input instance, and in this instance it needs to separate both s_1^1 from s_2^t and s_2^t from s_1^1.   □

Corollary 3. *Both* DIRECTED VERTEX MULTIWAY CUT *and* DIRECTED EDGE MULTIWAY CUT *do not admit a polynomial kernel when parameterized by* p *unless* $NP \subseteq coNP/poly$, *even in the case of two terminals.*

5   Multicut

EDGE (VERTEX) MULTICUT

Input: An undirected graph $G = (V, E)$, a set of pairs of terminals $\mathcal{T} = \{(s_1, t_1), \ldots, (s_k, t_k)\}$ and an integer p.
Task: Does there exist a set $S \subseteq E$ ($S \subseteq V$) such that no connected component of $G \setminus S$ contains both vertices s_i and t_i, for some $1 \leq i \leq k$?

It is easy to see that the vertex version of the MULTICUT problem is at least as hard as the edge version. In order to show a cross-composition into the MUL-TICUT problem parameterized by p we consider the following restricted variant of the MULTIWAY CUT problem with three terminals, which we prove to be NP-complete in the full version.

PROMISED MULTIWAY CUT
Input: An undirected graph $G = (V, E)$, a set of three terminals $T = \{s_1, s_2, s_3\} \subseteq V$ and an integer p. An instance satisfies: (i) $\deg(s_1) = \deg(s_2) = \deg(s_3) = d > 0$, (ii) for each $j = 1, 2, 3$ and any non-empty set $X \subseteq V \setminus T$ we have $|\delta(X \cup \{s_j\})| > d$, and (iii) $d \le p < 2d$.
Task: Does there exist a set S of at most p edges in E, such that in $G \setminus S$ there is no path between any pair of terminals in T?

Condition (i) ensures that degrees of all the terminals are equal, whereas condition (ii) guarantees that the set of edges incident to a terminal s_j is the only minimum size s_j–$(T \setminus \{s_j\})$ cut. Having both (i) and (ii), condition (iii) verifies whether an instance is not a trivially YES- or NO-instance, because by (i) and (ii) there is no solution of size less than d and removing all the edges incident to two terminals always gives a solution of size at most $2d$.

Theorem 5. PROMISED MULTIWAY CUT *cross-composes into* EDGE MULTI-CUT *parameterized by the size of the cutset p.*

Proof (sketch). By choosing an appropriate relation and duplicating some input instances, we assume that we are given a sequence of an *odd* number of PROMISED MULTIWAY CUT instances with equal cutset size p and terminal degree d.

We arrange the instances as on Figure 2: the empty and full circles are the terminals of the input instances, and the multiple edges are of multiplicity d. The empty circles are the terminals of the constructed EDGE MULTICUT instance: we request to separate a terminal from the two terminals that lie on the opposite side of the circle (de-noted by dashed lines on Figure

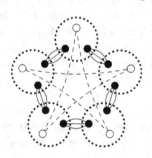

Fig. 2. Cross-composition for MULTICUT

2). The properties of PROMISED MULTIWAY CUT ensures that the only way to obtain a solution of size $p' = d + p$ is to solve one input instance and to cut an opposite edge of multiplicity d.   □

Corollary 4. MULTICUT *does not admit a polynomial kernel when parameter-ized by p unless $NP \subseteq coNP/poly$.*

A second (significantly different) cross-composition for MULTICUT is presented in the full version [1].

6 k-Way Cut

---

k-WAY CUT
Input: An undirected connected graph G and integers k and s.
Task: Does there exist a set X of at most s edges in G such that $G \setminus X$ has at least k connected components?

Note that in the problem definition we assume that the input graph is connected and, therefore, for $k > s + 1$ the input instances are trivial. However, if we are given an instance (G, k, s) where G has $c > 1$ connected components, we can easily reduce it to the connected version: we add to G a complete graph on $s + 2$ vertices (so that no two vertices of the complete graph can be separated by a cut of size s), connect one vertex from each connected component of G to all vertices of the complete graph, and decrease k by $c - 1$. Thus, by restricting ourselves to connected graphs G we do not make the problem easier.

The main result of this section is that k-WAY CUT, parameterized by s, does not admit a polynomial kernel (unless NP $\subseteq$ coNP/poly). We show a cross-composition from the CLIQUE problem, well-known to be NP-complete.

CLIQUE
Input: An undirected graph G and an integer ℓ.
Task: Does G contain a clique on ℓ vertices as a subgraph?

Theorem 6. CLIQUE *cross-composes to* k-WAY CUT *parameterized by* s.

Proof (sketch). By defining the polynomial equivalence relation appropriately, in the designed cross-composition we may assume that we are given t instances (G_i, ℓ) for $1 \le i \le t$ of the CLIQUE problem with $|V(G_i)| = n$ and $|E(G_i)| = m$ for all $1 \le i \le t$.

We consider a weighted version of the k-WAY CUT problem where each edge may have a positive integer weight and the cutset X needs to be of total weight at most s. We use three weights: light, medium and heavy; all weights in our construction are polynomial in n and m. The weighted version can be reduced to the unweighted one by replacing each vertex v by a huge clique H_v and connecting cliques H_u and H_v with the number of edges equal to the weight of the edge uv.

The construction is as follows. The input instances have light edges. In each input instance, between every two vertices we add an additional edge of medium weight. Moreover, we introduce a root vertex r and connect it to each vertex of the input instances with a heavy edge. We ask for $k = n - \ell + 1$ connected components created by cutting $n - \ell$ heavy edges, $\binom{n}{2} - \binom{\ell}{2}$ medium edges and $m - \binom{\ell}{2}$ light edges.

The heavy edges ensure that in any solution, after removal of the cutset we have one large connected component and $n - \ell$ isolated vertices. The only way to cut only $\binom{n}{2} - \binom{\ell}{2}$ medium edges is to cut $n - \ell$ isolated vertices from one input instance. The budget for light edges forces us to leave from this particular input instance a clique of size ℓ in the large connected component. $\square$

Corollary 5. k-WAY CUT *parameterized by* s *does not admit a polynomial kernel unless* $NP \subseteq coNP/poly$.

7   Conclusion and Open Problems

We have shown that four important parameterized problems do not admit a kernelization algorithm with a polynomial guarantee on the output size unless $NP \subseteq coNP/poly$ and the polynomial hierarchy collapses. We would like to mention here some open problems very closely related to our work.

- The OR-composition for DIRECTED MULTIWAY CUT in the case of two terminals excludes the existence of a polynomial kernel for most graph separation problems in directed graphs. There are two important cases not covered by this result: one is the MULTICUT problem in directed acyclic graphs, and the second is DIRECTED MULTIWAY CUT with deletable terminals.
- Both our OR-compositions for MULTICUT use a number of terminal pairs that is linear in the number of input instances. Is MULTICUT parameterized by both the size of the cutset and the number of terminal pairs similarly hard to kernelize?

Acknowledgements. We would like to thank Jakub Onufry Wojtaszczyk for some early discussions on the kernelization of the graph separation problems.

References

1. Cygan, M., Kratsch, S., Pilipczuk, M., Pilipczuk, M., Wahlström, M.: Clique cover and graph separation: New incompressibility results. CoRR abs/1111.0570 (2011)
2. Harnik, D., Naor, M.: On the compressibility of NP instances and cryptographic applications. SIAM J. Comput. 39(5), 1667–1713 (2010)
3. Guo, J., Niedermeier, R.: Invitation to data reduction and problem kernelization. SIGACT News 38(1), 31–45 (2007)
4. Bodlaender, H.L.: Kernelization: New Upper and Lower Bound Techniques. In: Chen, J., Fomin, F.V. (eds.) IWPEC 2009. LNCS, vol. 5917, pp. 17–37. Springer, Heidelberg (2009)
5. Bodlaender, H.L., Downey, R.G., Fellows, M.R., Hermelin, D.: On problems without polynomial kernels. J. Comput. Syst. Sci. 75(8), 423–434 (2009)
6. Fortnow, L., Santhanam, R.: Infeasibility of instance compression and succinct PCPs for NP. J. Comput. Syst. Sci. 77(1), 91–106 (2011)
7. Cai, J., Chakaravarthy, V.T., Hemaspaandra, L.A., Ogihara, M.: Competing provers yield improved Karp-Lipton collapse results. Inf. Comput. 198(1), 1–23 (2005)
8. Yap, C.K.: Some consequences of non-uniform conditions on uniform classes. Theor. Comput. Sci. 26, 287–300 (1983)
9. Bodlaender, H.L., Jansen, B.M.P., Kratsch, S.: Cross-composition: A new technique for kernelization lower bounds. In: STACS 2011, pp. 165–176 (2011)
10. Dell, H., van Melkebeek, D.: Satisfiability allows no nontrivial sparsification unless the polynomial-time hierarchy collapses. In: STOC 2010, pp. 251–260 (2010)
11. Dell, H., Marx, D.: Kernelization of packing problems. In: SODA 2012, pp. 68–81 (2012)
12. Hermelin, D., Wu, X.: Weak compositions and their applications to polynomial lower bounds for kernelization. In: SODA 2012, pp. 104–113 (2012)

13. Bodlaender, H.L., Thomassé, S., Yeo, A.: Kernel bounds for disjoint cycles and disjoint paths. Theor. Comput. Sci. 412(35), 4570–4578 (2011)
14. Cygan, M., Pilipczuk, M., Pilipczuk, M., Wojtaszczyk, J.O.: Kernelization Hardness of Connectivity Problems in d-Degenerate Graphs. In: Thilikos, D.M. (ed.) WG 2010. LNCS, vol. 6410, pp. 147–158. Springer, Heidelberg (2010)
15. Dom, M., Lokshtanov, D., Saurabh, S.: Incompressibility through Colors and iDs. In: Albers, S., Marchetti-Spaccamela, A., Matias, Y., Nikoletseas, S., Thomas, W. (eds.) ICALP 2009. LNCS, vol. 5555, pp. 378–389. Springer, Heidelberg (2009)
16. Bodlaender, H.L., Jansen, B.M.P., Kratsch, S.: Preprocessing for Treewidth: A Combinatorial Analysis Through Kernelization. In: Aceto, L., Henzinger, M., Sgall, J. (eds.) ICALP 2011. LNCS, vol. 6755, pp. 437–448. Springer, Heidelberg (2011)
17. Reed, B.A., Smith, K., Vetta, A.: Finding odd cycle transversals. Oper. Res. Lett. 32, 299–301 (2004)
18. Marx, D.: Parameterized graph separation problems. Theor. Comput. Sci. 351(3), 394–406 (2006)
19. Bousquet, N., Daligault, J., Thomassé, S.: Multicut is FPT. In: STOC 2011, pp. 459–468 (2011)
20. Chitnis, R., Hajiaghayi, M., Marx, D.: Fixed-parameter tractability of directed multiway cut parameterized by the size of the cutset. In: SODA 2012, pp. 1713–1725 (2012)
21. Marx, D., Razgon, I.: Fixed-parameter tractability of multicut parameterized by the size of the cutset. In: STOC 2011, pp. 469–478 (2011)
22. Kawarabayashi, K., Thorup, M.: The minimum k-way cut of bounded size is fixed-parameter tractable. In: FOCS 2011, pp. 160–169 (2011)
23. Garey, M.R., Johnson, D.S.: Computers and Intractability: A Guide to the Theory of NP-Completeness. W.H. Freeman, New York (1979)
24. Kellerman, E.: Determination of keyword conflict. IBM Technical Disclosure Bulletin 16(2), 544–546 (1973)
25. Agarwal, P.K., Alon, N., Aronov, B., Suri, S.: Can visibility graphs be represented compactly? Discrete & Computational Geometry 12, 347–365 (1994)
26. Gramm, J., Guo, J., Hüffner, F., Niedermeier, R., Piepho, H.P., Schmid, R.: Algorithms for compact letter displays: Comparison and evaluation. Computational Statistics & Data Analysis 52(2), 725–736 (2007)
27. Piepho, H.P.: An algorithm for a letter-based representation of all-pairwise comparisons. Journal of Computational and Graphical Statistics 13(2), 456–466 (2004)
28. Rajagopalan, S., Vachharajani, M., Malik, S.: Handling irregular ILP within conventional VLIW schedulers using artificial resource constraints. In: CASES 2000, pp. 157–164 (2000)
29. Gramm, J., Guo, J., Hüffner, F., Niedermeier, R.: Data reduction and exact algorithms for clique cover. ACM Journal of Experimental Algorithmics 13 (2008)
30. Kratsch, S., Wahlström, M.: Compression via matroids: a randomized polynomial kernel for odd cycle transversal. In: SODA 2012, pp. 94–103 (2012)
31. Kratsch, S., Wahlström, M.: Representative sets and irrelevant vertices: New tools for kernelization. In: CoRR abs/1111.2195 (2011)
32. Burlet, M., Goldschmidt, O.: A new and improved algorithm for the 3-cut problem. Oper. Res. Lett. 21(5), 225–227 (1997)
33. Goldschmidt, O., Hochbaum, D.S.: A polynomial algorithm for the k-cut problem for fixed k. Math. Oper. Res. 19(1), 24–37 (1994)
34. Thorup, M.: Minimum k-way cuts via deterministic greedy tree packing. In: STOC 2008, pp. 159–166 (2008)
35. Downey, R.G., Estivill-Castro, V., Fellows, M.R., Prieto, E., Rosamond, F.A.: Cutting up is hard to do: the parameterized complexity of k-cut and related problems. Electr. Notes Theor. Comput. Sci. 78, 209–222 (2003)

The Inverse Shapley Value Problem

Anindya De[1,*], Ilias Diakonikolas[1,**], and Rocco Servedio[2,***]

[1] UC Berkeley
{anindya,ilias}@cs.berkeley.edu
[2] Columbia University
rocco@cs.columbia.edu

Abstract. For f a weighted voting scheme used by n voters to choose between two candidates, the n *Shapley-Shubik Indices* (or *Shapley values*) of f provide a measure of how much control each voter can exert over the overall outcome of the vote. Shapley-Shubik indices were introduced by Lloyd Shapley and Martin Shubik in 1954 [SS54] and are widely studied in social choice theory as a measure of the "influence" of voters. The *Inverse Shapley Value Problem* is the problem of designing a weighted voting scheme which (approximately) achieves a desired input vector of values for the Shapley-Shubik indices. Despite much interest in this problem no provably correct and efficient algorithm was known prior to our work.

We give the first efficient algorithm with provable performance guarantees for the Inverse Shapley Value Problem. For any constant $\epsilon > 0$ our algorithm runs in fixed poly(n) time (the degree of the polynomial is independent of ϵ) and has the following performance guarantee: given as input a vector of desired Shapley values, if any "reasonable" weighted voting scheme (roughly, one in which the threshold is not too skewed) approximately matches the desired vector of values to within some small error, then our algorithm explicitly outputs a weighted voting scheme that achieves this vector of Shapley values to within error ϵ. If there is a "reasonable" voting scheme in which all voting weights are integers at most poly(n) that approximately achieves the desired Shapley values, then our algorithm runs in time poly(n) and outputs a weighted voting scheme that achieves the target vector of Shapley values to within error $\epsilon = n^{-1/8}$.

1 Introduction

In this paper we consider the common scenario in which each of n voters must cast a binary vote for or against some proposal. What is the best way to design such a voting scheme? [1] If it is desired that each of the n voters should have the same "amount of

---

[*] Research supported by NSF award CCF-1118083.

[**] Research supported by a Simons Postdoctoral Fellowship.

[***] Research supported in part by NSF awards CCF-0915929 and CCF-1115703.

[1] Throughout the paper we consider only *weighted voting schemes,* in which the proposal passes if a weighted sum of yes-votes exceeds a predetermined threshold. Weighted voting schemes are predominant in voting theory and have been extensively studied for many years, see [EGGW07, ZFBE08] and references therein. In computer science language, we are dealing with *linear threshold functions* (henceforth abbreviated as *LTFs*) over n Boolean variables.

A. Czumaj et al. (Eds.): ICALP 2012, Part I, LNCS 7391, pp. 266–277, 2012.
© Springer-Verlag Berlin Heidelberg 2012

power" over the outcome, then a simple majority vote is the obvious solution. However, in many scenarios it may be the case that we would like to assign different levels of voting power to the n voters – perhaps they are shareholders who own different amounts of stock in a corporation, or representatives of differently sized populations. In such a setting it is much less obvious how to design the right voting scheme; indeed, it is far from obvious how to correctly quantify the notion of the "amount of power" that a voter has under a given fixed voting scheme. As a simple example, consider an election with three voters who have voting weights 49, 49 and 2, in which a total of 51 votes are required for the proposition to pass. While the disparity between voting weights may at first suggest that the two voters with 49 votes each have most of the "power," any coalition of two voters is sufficient to pass the proposition and any single voter is insufficient, so the voting power of all three voters is in fact equal.

Many different *power indices* (methods of measuring the voting power of individuals under a given weighted voting scheme) have been proposed over the course of decades. These include the Banzhaf index [Ban65], the Deegan-Packel index [DP78], the Holler index [Hol82], and others (see the extensive survey of de Keijzer [dK08]). Perhaps the best known, and certainly the oldest, of these indices is the *Shapley-Shubik index* [SS54], which is also known as the index of *Shapley values* (we shall henceforth refer to it as such). Informally, the Shapley value of a voter i among the n voters is the fraction of all $n!$ orderings of the voters in which she "casts the pivotal vote" (see [Rot88] for much more on Shapley values). We shall work with the Shapley values throughout this paper.

Given a particular weighted voting scheme (i.e. an n-variable linear threshold function), standard sampling-based approaches can be used to efficiently obtain highly accurate estimates of the n Shapley values (see also the works of [Lee03, BMR+10]). However, the *inverse* problem is much more challenging: given a vector of n desired values for the Shapley values, how can one design a weighted voting scheme that (approximately) achieves these Shapley values? This problem, which we refer to as the *Inverse Shapley Value Problem*, is quite natural and has received considerable attention; various heuristics and exponential-time algorithms have been proposed, e.g. [APL07, FWJ08, dKKZ10, Kur11], but prior to our work no provably correct and efficient algorithms were known.

Our Results. We give the first efficient algorithm with provable performance guarantees for the Inverse Shapley Value Problem. Our results apply to "reasonable" voting schemes; roughly, we say that a weighted voting scheme is "reasonable" if fixing a tiny fraction of the voting weight does not already determine the outcome, i.e. if the threshold of the linear threshold function is not too extreme. This seems to be a plausible property for natural voting schemes. Roughly speaking, we show that if there is any reasonable weighted voting scheme that approximately achieves the desired input vector of Shapley values, then our algorithm finds such a weighted voting scheme. Our algorithm runs in fixed polynomial time in n, the number of voters, for any constant error parameter $\epsilon > 0$. In a bit more detail, our first main theorem, stated informally, is as follows (see Section 5 for Theorem 3 which gives a precise theorem statement):

Main Theorem (Arbitrary Weights, Informal Statement). *There is a poly(n)-time algorithm with the following properties: The algorithm is given any constant accuracy* *parameter $\epsilon > 0$ and any vector of n real values $\tilde{a}(1), \ldots, \tilde{a}(n)$. The algorithm has the following performance guarantee: if there is any monotone increasing reasonable LTF $f(x)$ whose Shapley values are very close to the given values $\tilde{a}(1), \ldots, \tilde{a}(n)$, then with very high probability the algorithm outputs $v \in \mathbb{R}^n$, $\theta \in \mathbb{R}$ such that the linear threshold function $h(x) = \text{sign}(v \cdot x - \theta)$ has Shapley values ϵ-close to those of f.*

Our second main theorem gives an even stronger guarantee if there is a weighted voting scheme with small weights (at most $\text{poly}(n)$) whose Shapley values are close to the desired values. For this problem we give an algorithm which achieves $1/\text{poly}(n)$ accuracy in $\text{poly}(n)$ time. An informal statement of this result is (see Section 5 for Theorem 4 which gives a precise theorem statement):

Main Theorem (Bounded Weights, Informal Statement). *There is a $\text{poly}(n, W)$-time algorithm with the following properties: The algorithm is given a weight bound W and any vector of n real values $\tilde{a}(1), \ldots, \tilde{a}(n)$. The algorithm has the following performance guarantee: if there is any monotone increasing reasonable LTF $f(x) = \text{sign}(w \cdot x - \theta)$ whose Shapley values are very close to the given values $\tilde{a}(1), \ldots, \tilde{a}(n)$ and where each w_i is an integer of magnitude at most W, then with very high probability the algorithm outputs $v \in \mathbb{R}^n$, $\theta \in \mathbb{R}$ such that the linear threshold function $h(x) = \text{sign}(v \cdot x - \theta)$ has Shapley values $n^{-1/8}$-close to those of f.*

Discussion and Our Approach. At a high level, the Inverse Shapley Value Problem that we consider is similar to the "Chow Parameters Problem" that has been the subject of several recent papers [Gol06, OS08, DDFS12]. The Chow parameters are another name for the n Banzhaf indices; the Chow Parameters Problem is to output a linear threshold function which approximately matches a given input vector of Chow parameters. (To align with the terminology of the current paper, the "Chow Parameters Problem" might perhaps better be described as the "Inverse Banzhaf Problem.")

Let us briefly describe the approaches in [OS08] and [DDFS12] at a high level for the purpose of establishing a clear comparison with this paper. Each of the papers [OS08, DDFS12] combines structural results on linear threshold functions with an algorithmic component. The structural results in [OS08] deal with anti-concentration of affine forms $w \cdot x - \theta$ where $x \in \{-1, 1\}^n$ is uniformly distributed over the Boolean hypercube, while the algorithmic ingredient of [OS08] is a rather straight-forward brute-force search. In contrast, the key structural results of [DDFS12] are geometric statements about how n-dimensional hyperplanes interact with the Boolean hypercube, which are combined with linear-algebraic (rather than anti-concentration) arguments. The algorithmic ingredient of [DDFS12] is more sophisticated, employing a boosting-based approach inspired by the work of [TTV08, Imp95].

Our approach combines aspects of both the [OS08] and [DDFS12] approaches. Very roughly speaking, we establish new structural results which show that linear threshold functions have good anti-concentration (similar to [OS08]), and use a boosting-based approach derived from [TTV08] as the algorithmic component (similar to [DDFS12]). However, this high-level description glosses over many "Shapley-specific" issues and complications that do not arise in these earlier works; below we describe two of the main challenges that arise, and sketch how we meet them in this paper.

First Challenge: Establishing Anti-concentration with Respect to Non-Standard Distributions. The Chow parameters (i.e. Banzhaf indices) have a natural definition in terms of the uniform distribution over the Boolean hypercube $\{-1, 1\}^n$. Being able to use the uniform distribution with its many nice properties (such as complete independence among all coordinates) is very useful in proving the required anti-concentration results that are at the heart of [OS08]. In contrast, it is not *a priori* clear what is (or even whether there exists) the "right" distribution over $\{-1, 1\}^n$ corresponding to the Shapley values. In this paper we derive such a distribution μ over $\{-1, 1\}^n$, but it is much less well-behaved than the uniform distribution (it is supported on a proper subset of $\{-1, 1\}^n$, and it is not even pairwise independent). Nevertheless, we are able to establish anti-concentration results for affine forms $w \cdot x - \theta$ corresponding to linear threshold functions under the distribution μ as required for our results. This is done by showing that any linear threshold function can be expressed with "nice" weights, and establishing anti-concentration for any "nice" weight vector by carefully combining anti-concentration bounds for p-biased distributions across a continuous family of different choices of p (see Section 3 for details).

Second Challenge: Using Anti-concentration to Solve the Inverse Shapley Problem. The main algorithmic ingredient that we use is a procedure from [TTV08]. Given a vector of values $(\mathbf{E}[f(x)x_i])_{i=1,\dots,n}$ (correlations between the unknown linear threshold function f and the individual input variables), it efficiently constructs a bounded function $g : \{-1, 1\}^n \to [-1, 1]$ which closely matches these correlations, i.e. $\mathbf{E}[f(x)x_i] \approx \mathbf{E}[g(x)x_i]$ for all i. Such a procedure is very useful for the Chow parameters problem, because the Chow parameters correspond precisely to the values $\mathbf{E}[f(x)x_i]$ – i.e. the degree-1 Fourier coefficients of f – with respect to the uniform distribution. (This correspondence is at the heart of Chow's original proof [Cho61] showing that the exact values of the Chow parameters suffice to information-theoretically specify any linear threshold function; anti-concentration is used in [OS08] to extend Chow's original arguments about degree-1 Fourier coefficients to the setting of approximate reconstruction.)

For the inverse Shapley problem, there is no obvious correspondence between the correlations of individual input variables and the Shapley values. Moreover, without a notion of "degree-1 Fourier coefficients" for the Shapley setting, it is not clear why anti-concentration statements with respect to μ should be useful for approximate reconstruction. We deal with both these issues by developing a notion of the *degree-1 Fourier coefficients of f with respect to distribution* μ and relating these coefficients to the Shapley values; see Section 2. [2] Armed with this notion, we prove a key result (Lemma 6) saying that if the LTF f is anti-concentrated under distribution μ, then any bounded

---

[2] We actually require two related notions: one is the "coordinate correlation coefficient" $\mathbf{E}_{x \sim \mu}[f(x)x_i]$, which is necessary for the algorithmic [TTV08] ingredient, and one is the "Fourier coefficient" $\hat{f}(i) = \mathbf{E}_{x \sim \mu}[f(x)L_i]$, which is necessary for Lemma 6. We define both notions and establish the necessary relations between them in Section 2.

We note that Owen [Owe72] has given a characterization of the Shapley values as a weighted average of p-biased influences (see also [KS06]). However, this is not as useful for us as our characterization in terms of "μ-distribution" Fourier coefficients, because we need to ultimately relate the Shapley values to anti-concentration with respect to μ.

function g which closely matches the degree-1 Fourier coefficients of f must be close to f in ℓ_1-measure with respect to μ. (This is why anti-concentration with respect to μ is useful for us.) From this point, exploiting properties of the [TTV08] algorithm, we can pass from g to an LTF whose Shapley values closely match those of f.

Organization. Because of space constraints most proofs are deferred to the full version. In Section 2 we define the distribution μ and the notions of Fourier coefficients and "coordinate correlation coefficients," and the relations between them, that we will need. At the end of that section we prove a crucial lemma, Lemma 6, which says that anti-concentration of affine forms and closeness in Fourier coefficients together suffice to establish closeness in ℓ_1 distance. Section 3 proves that "nice" affine forms have the required anti-concentration, and Section 4 describes the algorithmic tool from [TTV08] that lets us establish closeness of coordinate correlation coefficients. Section 5 puts the pieces together to prove our main theorems.

2 Reformulation of Shapley-Shubik Indices

Given $f : \{-1,1\}^n \to \{-1,1\}$, we will denote by $\tilde{f}(i)$ the i-th Shapley value of f. The original definition of Shapley values is somewhat cumbersome to work with. In this section we derive alternate characterizations of Shapley values in terms of "Fourier coefficients" and "coordinate correlation coefficients" and establish various technical results relating Shapley values and these coefficients; these technical results will be crucially used in the proof of our main theorems.

There is a particular distribution μ that plays a central role in our reformulations. We start by defining this distribution μ and introducing some relevant notation, and then give our results. Because of space constraints all proofs are deferred to the full version.

The Distribution μ. Let us define $\Lambda(n) := \sum_{0<k<n} \frac{1}{k} + \frac{1}{n-k}$; clearly we have $\Lambda(n) = \Theta(\log n)$, and more precisely we have $\Lambda(n) \le 2\log n$. We also define $Q(n,k)$ as $Q(n,k) := \frac{1}{k} + \frac{1}{n-k}$ for $0 < k < n$, so we have $\Lambda(n) = Q(n,1) + \cdots + Q(n, n-1)$.

For $x \in \{-1,1\}^n$ we write $\mathrm{wt}(x)$ to denote the number of 1s in x. We define the set B_n to be $B_n := \{x \in \{-1,1\}^n : 0 < \mathrm{wt}(x) < n\}$, i.e. $B_n = \{-1,1\}^n \setminus \{1, -1\}$.

The distribution μ is supported on B_n and is defined as follows: to make a draw from μ, sample $k \in \{1, \ldots, n-1\}$ with probability $Q(n,k)/\Lambda(n)$. Choose $x \in \{-1,1\}^n$ uniformly at random from the kth "weight level" of $\{-1,1\}^n$, i.e. from $\{-1,1\}^n_{=k} := \{x \in \{-1,1\}^n : \mathrm{wt}(x) = k\}$.

Useful Notation. For $i = 0, \ldots, n$ we define the "coordinate correlation coefficients" of a function $f : \{-1,1\}^n \to \mathbb{R}$ (with respect to μ) as:

$$f^*(i) := \mathbf{E}_{x \sim \mu}[f(x) \cdot x_i] \tag{1}$$

(here and throughout the paper x_0 denotes the constant 1).

Later in this section we will define an orthonormal set of linear functions L_0, $L_1, \ldots, L_n : \{-1,1\}^n \to \mathbb{R}$. We define the "Fourier coefficients" of f (with respect to μ) as:

$$\hat{f}(i) := \mathbf{E}_{x \sim \mu}[f(x) \cdot L_i(x)]. \tag{2}$$

An Alternative Expression for the Shapley Values. We start by expressing the Shapley values in terms of the coordinate correlation coefficients:

Lemma 1. *Given* $f : \{-1, 1\}^n \to [-1, 1]$, *for each* $i = 1, \ldots, n$ *we have* $\tilde{f}(i) = \frac{f(1) - f(-1)}{n} + \frac{\Lambda(n)}{2} \cdot \left(f^*(i) - \frac{1}{n} \sum_{j=1}^n f^*(j) \right)$.

Construction of a Fourier Basis for Distribution μ. For all $x \in B_n$ we have that $\mu(x) > 0$, and consequently we know that the functions $1, x_1, \ldots, x_{n+1}$ form a basis for the subspace of linear functions from $B_n \to \mathbb{R}$. By Gram-Schmidt orthogonalization, we can obtain an orthonormal basis $L_0, \ldots, L_n$ for this subspace, i.e. one that satisfies $\langle L_i, L_i \rangle_\mu = 1$ for all i and $\langle L_i, L_j \rangle_\mu = 0$ for all $i \neq j$.

We now give explicit expressions for these basis functions. We start by defining $L_0 : B_n \to \mathbb{R}$ as $L_0 : x \mapsto 1$. Next, by symmetry, we can express each L_i as

$$L_i(x) = \alpha(x_1 + \ldots + x_n) + \beta x_i.$$

Using the orthonormality properties it is straightforward to solve for α and β. The following Lemma gives the values of α and β:

Lemma 2. *For the choices* $\alpha = \frac{1}{n} \cdot \left(\sqrt{\frac{\Lambda(n)}{n\Lambda(n) - 4(n-1)}} - \frac{\sqrt{\Lambda(n)}}{2} \right)$, $\beta = \frac{\sqrt{\Lambda(n)}}{2}$, *the set* $\{L_i\}_{i=0}^n$ *is an orthonormal set of linear functions under the distribution* μ.

We note for later reference that $\alpha = -\Theta\left(\frac{\sqrt{\log n}}{n} \right)$ and $\beta = \Theta(\sqrt{\log n})$.

Relating the Shapley Values to the Fourier Coefficients. The next lemma gives a useful expression for $\hat{f}(i)$ in terms of $\tilde{f}(i)$:

Lemma 3. *Let* $f : \{-1, 1\}^n \to [-1, 1]$ *be any function. Then for each* $i = 1, \ldots, n$ *we have* $\hat{f}(i) = \frac{2\beta}{\Lambda(n)} \cdot \left(\tilde{f}(i) - \frac{f(1) - f(-1)}{n} \right) + \frac{1}{n} \cdot \sum_{j=1}^n \hat{f}(j)$.

Bounding Shapley Distance in Terms of Fourier Distance. Recall that the Shapley distance $d_{\text{Shapley}}(f, g)$ between $f, g : \{-1, 1\}^n \to [-1, 1]$ is defined as $d_{\text{Shapley}}(f, g) := \sqrt{\sum_{i=1}^n (\tilde{f}(i) - \tilde{g}(i))^2}$. We define the *Fourier distance* between f and g as $d_{\text{Fourier}}(f, g) := \sqrt{\sum_{i=0}^n (\hat{f}(i) - \hat{g}(i))^2}$.

Our next lemma shows that if the Fourier distance between f and g is small then so is the Shapley distance.

Lemma 4. *Let* $f, g : \{-1, 1\}^n \to [-1, 1]$. *Then,* $d_{\text{Shapley}}(f, g) \leq \frac{4}{\sqrt{n}} + \frac{\Lambda(n)}{2\beta} \cdot d_{\text{Fourier}}(f, g)$.

Bounding Fourier Distance by "Correlation Distance." The following lemma will be useful for us since it lets us upper bound Fourier distance in terms of the distance between vectors of correlations with individual variables:

Lemma 5. *Let* $f, g : \{-1, 1\}^n \to \mathbb{R}$. *Then we have* $d_{\text{Fourier}}(f, g) \leq O(\sqrt{\log n}) \cdot \sqrt{\sum_{i=0}^n (f^*(i) - g^*(i))^2}$.

From Fourier Closeness to ℓ_1-Closeness. An important technical ingredient in our work is the notion of an affine form $\ell(x)$ having "good anti-concentration" under distribution μ; we now give a precise definition to capture this.

Definition 1 (Anti-concentration). *Fix $w \in \mathbb{R}^n$ and $\theta \in \mathbb{R}$, and let the affine form $\ell(x)$ be $\ell(x) := w \cdot x - \theta$. We say that $\ell(x)$ is (δ, κ)-anti-concentrated under μ if $\mathbf{Pr}_{x \sim \mu}[|\ell(x)| \leq \delta] \leq \kappa$.*

The next lemma plays a crucial role in our results. It essentially shows that for $f = \text{sign}(w \cdot x - \theta)$, if the affine form $\ell(x) = w \cdot x - \theta$ is anti-concentrated, then *any* bounded function $g : \{-1, 1\}^n \to [-1, 1]$ that has $d_{\text{Fourier}}(f, g)$ small must in fact be close to f in ℓ_1 distance under μ.

Lemma 6. *Let $f : \{-1, 1\}^n \to \{-1, 1\}$, $f = \text{sign}(w \cdot x - \theta)$ be such that $w \cdot x - \theta$ is (δ, κ)-anti-concentrated under μ (for some $\kappa \leq 1/2$), where $|\theta| \leq \|w\|_1$. Let $g : \{-1, 1\}^n \to [-1, 1]$ be such that $d_{\text{Fourier}}(f, g) \leq \rho$. Then we have*

$$\mathbf{E}_{x \sim \mu}[|f(x) - g(x)|] \leq (4\|w\|_1 \sqrt{\rho})/\delta + 4\kappa.$$

3   A Useful Anti-concentration Result

In this section we prove an anti-concentration result for monotone increasing η-reasonable affine forms under the distribution μ. Note that even if k is a constant the result gives an anti-concentration probability of $O(1/\log n)$; this will be crucial in the proof of our first main result in Section 5.

Theorem 1. *Let $L(x) = w_0 + \sum_{i=1}^n w_i x_i$ be a monotone increasing η-reasonable affine form, so $w_i \geq 0$ for $i \in [n]$ and $|w_0| \leq (1 - \eta) \sum_{i=1}^n |w_i|$. Let $k \in [n], 0 < \zeta < 1/2, k \geq 2/\eta$ and $r \in \mathbb{R}_+$ be such that $|S| \geq k$, where $S := \{i \in [n] : |w_i| \geq r\}$. Then*

$$\mathbf{Pr}_{x \sim \mu}[|L(x)| < r] = O\left(\frac{1}{\log n} \cdot \frac{1}{k^{1/3 - \zeta}} \cdot \left(\frac{1}{\zeta} + \frac{1}{\eta}\right)\right).$$

This theorem essentially says that under the distribution μ, the random variable $L(x)$ falls in the interval $[-r, r]$ with only a very small probability. Such theorems are known in the literature as "anti-concentration" results, but almost all such results are for the uniform distribution or for other product distributions, and indeed the proofs of such results typically crucially use the fact that the distributions are product distributions.

In our setting, the distribution μ is not even a pairwise independent distribution, so standard approaches for proving anti-concentration cannot be directly applied. Instead, we exploit the fact that μ is a *symmetric* distribution; a distribution is symmetric if the probability mass it assigns to an n-bit string $x \in \{-1, 1\}^n$ depends only on the number of 1's of x (and not on their location within the string). This enables us to perform a somewhat delicate reduction to known anti-concentration results for biased product distributions. Our proof adopts a point of view which is inspired by the combinatorial proof of the basic Littlewood-Offord theorem (under the uniform distribution on the hypercube) due to Benjamini et. al. [BKS99]. The proof is given in the full version.

4   A Useful Algorithmic Tool

In this section we describe a useful algorithmic tool arising from recent work in computational complexity theory. The main result we will need is the following theorem of [TTV08] (the ideas go back to [Imp95] and were used in a different form in [DDFS12]):

Theorem 2. *[TTV08] Let X be a finite domain, μ be a samplable probability distribution over X, $f : X \to [-1,1]$ be a bounded function, and $\mathcal{L}$ be a finite family of Boolean functions $\ell : X \to \{-1,1\}$. There is an algorithm* **Boosting-TTV** *with the following properties: Suppose* **Boosting-TTV** *is given as input a list $(a_\ell)_{\ell \in \mathcal{L}}$ of real values and a parameter $\xi > 0$ such that $|\mathbf{E}_{x \sim \mu}[f(x)\ell(x)] - a_\ell| \leq \xi/16$ for every $\ell \in \mathcal{L}$. Then* **Boosting-TTV** *outputs a function $h : X \to [-1,1]$ with the following properties:*

(i) $|\mathbf{E}_{x \sim \mu}[\ell(x)h(x) - \ell(x)f(x)]| \leq \xi$ *for every $\ell \in \mathcal{L}$;*
(ii) $h(x)$ *is of the form $h(x) = P_1(\frac{\xi}{2} \cdot \sum_{\ell \in \mathcal{L}} w_\ell \ell(x))$ where the w_ℓ's are integers whose absolute values sum to $O(1/\xi^2)$.*

The algorithm runs for $O(1/\xi^2)$ iterations, where in each iteration it estimates $\mathbf{E}_{x \sim \mu}[h'(x)\ell(x)]$ to within additive accuracy $\pm \xi/16$. Here each h' is a function of the form $h'(x) = P_1(\frac{\xi}{2} \cdot \sum_{\ell \in \mathcal{L}} v_\ell \ell(x))$, where the v_ℓ's are integers whose absolute values sum to $O(1/\xi^2)$.

We note that Theorem 2 is not explicitly stated in the above form in [TTV08]; in particular, neither the time complexity of the algorithm nor the fact that it suffices for the algorithm to be given "noisy" estimates a_ℓ of the values $\mathbf{E}_{x \sim \mu}[f(x)\ell(x)]$ is explicitly stated in [TTV08]. So for the sake of completeness, in the full version we state the algorithm in full and sketch a proof of correctness of this algorithm using results that are explicitly proved in [TTV08].

5   Our Main Results

In this section we combine ingredients from the previous subsections and prove our main results, Theorems 3 and 4.

Our first main result gives an algorithm that works if *any* monotone increasing η-reasonable LTF has approximately the right Shapley values:

Theorem 3. *There is an algorithm* **IS** *(for* **Inverse-Shapley***) with the following properties.* **IS** *is given as input an accuracy parameter $\epsilon > 0$, a confidence parameter $\delta > 0$, and n real values $\tilde{a}(1), \ldots, \tilde{a}(n)$; its output is a pair $v \in \mathbb{R}^n, \theta \in \mathbb{R}$. Its running time is $\mathrm{poly}(n, 2^{\mathrm{poly}(1/\epsilon)}, \log(1/\delta))$. The performance guarantees of* **IS** *are the following:*

1. *Suppose there is a monotone increasing η-reasonable LTF $f(x)$ such that $d_{\mathrm{Shapley}}(a, f) \leq 1/\mathrm{poly}(n, 2^{\mathrm{poly}(1/\epsilon)})$. Then with probability $1 - \delta$ algorithm* **IS** *outputs $v \in \mathbb{R}^n, \theta \in \mathbb{R}$ which are such that the LTF $h(x) = \mathrm{sign}(v \cdot x - \theta)$ has $d_{\mathrm{Shapley}}(f, h) \leq \epsilon$.*

2. *For any input vector* $(\tilde{a}(1), \ldots, \tilde{a}(n))$, *the probability that* IS *outputs* $v \in \mathbb{R}^n, \theta \in \mathbb{R}$ *such that the LTF* $h(x) = \text{sign}(v \cdot x - \theta)$ *has* $d_{\text{Shapley}}(f, h) > \epsilon$ *is at most* δ.

Proof. We first note that we may assume $\epsilon > n^{-c}$ for a constant $c > 0$ of our choosing, for if $\epsilon \leq n^{-c}$ then the claimed running time is $2^{\Omega(n^2 \log n)}$. In this much time we can easily enumerate all LTFs over n variables (by trying all weight vectors with integer weights at most n^n; this suffices by [MTT61]) and compute their Shapley values exactly, and thus solve the problem. So for the rest of the proof we assume that $\epsilon > n^{-c}$.

It will be obvious from the description of IS that property (2) above is satisfied, so the main job is to establish (1). Before giving the formal proof we first describe an algorithm and analysis achieving (1) for an idealized version of the problem. We then describe the actual algorithm and its analysis (which build on the idealized version).

Recall that the algorithm is given as input ϵ, δ and $\tilde{a}(1), \ldots, \tilde{a}(n)$ that satisfy $d_{\text{Shapley}}(a, f) \leq 1/\text{poly}(n, 2^{\text{poly}(1/\epsilon)})$ for some monotone increasing η-reasonable LTF f. The idealized version of the problem is the following: we assume that the algorithm is also given the two real values $f^*(0), (f^*(1) + \ldots + f^*(n))/n$. It is also helpful to note that since f is monotone and η-reasonable (and hence is not a constant function), it must be the case that $f(1) = 1$ and $f(-1) = -1$.

The algorithm for this idealized version is as follows: first, using Lemma 1, the values $\tilde{f}(i), i = 1, \ldots, n$ are converted into values $a^*(i)$ which are approximations for the values $f^*(i)$. Each $a^*(i)$ satisfies $|a^*(i) - f^*(i)| \leq 1/\text{poly}(n, 2^{O(\text{poly}(1/\epsilon))})$. The algorithm sets $a^*(0)$ to $f^*(0)$. Next, the algorithm runs Boosting-TTV with the following input: the family $\mathcal{L}$ of Boolean functions is $\{1, x_1, \ldots, x_n\}$; the values $a^*(0), \ldots, a^*(n)$ comprise the list of real values; μ is the distribution; and the parameter ξ is set to $1/\text{poly}(n, 2^{\text{poly}(1/\epsilon)})$. (We note that each execution of Step 3 of Boosting-TTV, namely finding values that closely estimate $\mathbf{E}_{x \sim \mu}[h_t(x) x_i]$ as required, is easily achieved using a standard sampling scheme; details in the full version.) Boosting-TTV outputs an LBF $h(x) = P_1(v \cdot x - \theta)$; the output of our overall algorithm is the LTF $h'(x) = \text{sign}(v \cdot x - \theta)$.

Let us analyze this algorithm for the idealized scenario. By Theorem 2, the output function h that is produced by Boosting-TTV is an LBF $h(x) = P_1(v \cdot x - \theta)$ that satisfies $\sqrt{\sum_{j=0}^{n}(h^*(j) - f^*(j))^2} = 1/\text{poly}(n, 2^{\text{poly}(1/\epsilon)})$. Given this, Lemma 5 implies that $d_{\text{Fourier}}(f, h) \leq \rho := 1/\text{poly}(n, 2^{\text{poly}(1/\epsilon)})$.

At this point, we have established that h is a bounded function that has $d_{\text{Fourier}}(f, h) \leq 1/\text{poly}(n, 2^{\text{poly}(1/\epsilon)})$. We would like to apply Lemma 6 and thereby assert that the ℓ_1 distance between f and h (with respect to μ) is small. To see that we can do this, we first claim (see full version for details) that since f is a monotone increasing η-reasonable LTF, it has a representation as $f(x) = \text{sign}(w \cdot x + w_0)$ whose weights satisfy the following property: for any choice of $\zeta > 0$, after rescaling all the weights, the largest-magnitude weight has magnitude 1, and the $k := \Theta_{\zeta, \eta}(1/\epsilon^{6+2\zeta})$ largest-magnitude weights each have magnitude at least $r := 1/(n \cdot k^{O(k)})$. (Note that since $\epsilon \geq n^{-c}$ we indeed have $k \leq n$ as required.) Given this, Theorem 1 implies that the affine form $L(x) = w \cdot x + w_0$ satisfies

$$\mathbf{Pr}_{x \sim \mu}[|L(x)| < r] \leq \kappa := \epsilon^2/(1024 \log(n)), \tag{3}$$

i.e. it is (r, κ)-anticoncentrated with $\kappa = \epsilon^2/(1024 \log(n))$. Thus we may indeed apply Lemma 6, and it gives us that

$$\mathbf{E}_{x \sim \mu}[|f(x) - h(x)|] \leq \frac{4\|w\|_1 \sqrt{\rho}}{r} + 4\kappa \leq \epsilon^2/(128 \log n). \qquad (4)$$

Now let $h' : \{-1, 1\}^n \to \{-1, 1\}$ be the LTF defined as $h'(x) = \text{sign}(v \cdot x - \theta)$ (recall that h is the LBF $P_1(v \cdot x - \theta)$). Since f is a $\{-1, 1\}$-valued function, it is clear that for every input x in the support of μ, the contribution of x to $\mathbf{Pr}_{x \sim \mu}[f(x) \neq h'(x)]$ is at most twice its contribution to $\mathbf{E}_{x \sim \mu}[|f(x) - h(x)|]$. Thus we have that $\mathbf{Pr}_{x \sim \mu}[f(x) \neq h'(x)] \leq \epsilon^2/(64 \log n)$. By a standard argument, we obtain that $d_{\text{Fourier}}(f, h') \leq \epsilon/(4\sqrt{\log n})$. Finally, Lemma 4 gives that $d_{\text{Shapley}}(f, h') \leq 4/\sqrt{n} + \sqrt{\Lambda(n)} \cdot \epsilon/(4\sqrt{\log n}) < \epsilon/2$. So indeed the LTF $h'(x) = \text{sign}(v \cdot x - \theta)$ satisfies $d_{\text{Shapley}}(f, h') \leq \epsilon/2$ as desired.

Now we turn from the idealized scenario to actually prove Theorem 3, where we are not given the values of $f^*(0)$ and $(f^*(1) + \ldots + f^*(n))/n$. To get around this, we note that $f^*(0), (f^*(1) + \ldots + f^*(n))/n \in [-1, 1]$. So the idea is that we will run the idealized algorithm repeatedly, trying "all" possibilities (up to some prescribed granularity) for $f^*(0)$ and for $(f^*(1) + \ldots + f^*(n))/n$. At the end of each such run we have a "candidate" LTF h'; we use a simple procedure **Shapley-Estimate** to estimate $d_{\text{Shapley}}(f, h')$ to within additive accuracy $\pm\epsilon/10$, and we output any h' whose estimated value of $d_{\text{Shapley}}(f, h')$ is at most $8\epsilon/10$.

We may run the idealized algorithm $\text{poly}(n, 2^{\text{poly}(1/\epsilon)})$ times without changing its overall running time (up to polynomial factors). Thus we can try a net of possible guesses for $f^*(0)$ and $(f^*(1) + \ldots + f^*(n))/n$ which is such that one guess will be within $\pm 1/\text{poly}(n, 2^{\text{poly}(1/\epsilon)})$ of the the correct values for both parameters. It is straightforward to verify that the analysis of the idealized scenario given above is sufficiently robust that when these "good" guesses are encountered, the algorithm will with high probability generate an LTF h' that has $d_{\text{Shapley}}(f, h') \leq 6\epsilon/10$. A straightforward analysis of running time and failure probability shows that properties (1) and (2) are achieved as desired, and Theorem 3 is proved.   □

For any monotone η-reasonable target LTF f, Theorem 3 constructs an output LTF whose Shapley distance from f is at most ϵ, but the running time is exponential in $\text{poly}(1/\epsilon)$. We now show that if the target monotone η-reasonable LTF f has integer weights that are at most W, then we can construct an output LTF h with $d_{\text{Shapley}}(f, h) \leq n^{-1/8}$ running in time $\text{poly}(n, W)$; this is a far faster running time than provided by Theorem 3 for such small ϵ. (The "1/8" is chosen for convenience; it will be clear from the proof that any constant strictly less than 1/6 would suffice.)

Theorem 4. *There is an algorithm ISBW (for Inverse-Shapley with Bounded Weights) with the following properties. ISBW is given as input a weight bound $W \in \mathbb{N}$, a confidence parameter $\delta > 0$, and n real values $\tilde{a}(1), \ldots, \tilde{a}(n)$; its output is a pair $v \in \mathbb{R}^n, \theta \in \mathbb{R}$. Its running time is $\text{poly}(n, W, \log(1/\delta))$. The performance guarantees of ISBW are the following:*

1. *Suppose there is a monotone increasing η-reasonable LTF $f(x) = \text{sign}(u \cdot x - \theta)$, where each u_i is an integer with $|u_i| \leq W$, such that $d_{\text{Shapley}}(a, f) \leq 1/\text{poly}(n, W)$.*

Then with probability $1 - \delta$ *algorithm* ISBW *outputs* $v \in \mathbb{R}^n$, $\theta \in \mathbb{R}$ *which are such that the LTF* $h(x) = \text{sign}(v \cdot x - \theta)$ *has* $d_{\text{Shapley}}(f, h) \leq n^{-1/8}$.

2. *For any input vector* $(\tilde{a}(1), \ldots, \tilde{a}(n))$, *the probability that* IS *outputs* v, θ *such that the LTF* $h(x) = \text{sign}(v \cdot x - \theta)$ *has* $d_{\text{Shapley}}(f, h) > n^{-1/8}$ *is at most* δ.

Proof. Let $f(x) = \text{sign}(u \cdot x - \theta)$ be as described in the theorem statement. We may assume that each $|u_i| \geq 1$ (by scaling all the u_i's and θ by $2n$ and then replacing any zero-weight u_i with 1). Next we observe that for such an affine form $u \cdot x - \theta$, Theorem 1 immediately yields the following corollary:

Corollary 1. *Let* $L(x) = \sum_{i=1}^{n} u_i x_i - \theta$ *be a monotone increasing η-reasonable affine form. Suppose that* $u_i \geq r$ *for all* $i = 1, \ldots, n$. *Then for any* $\zeta > 0$, *we have*

$$\mathbf{Pr}_{x \sim \mu}[|L(x)| < r] = O\left(\frac{1}{\log n} \cdot \frac{1}{n^{1/3 - \zeta}} \cdot \left(\frac{1}{\zeta} + \frac{1}{\eta}\right)\right).$$

With this anti-concentration statement in hand, the proof of Theorem 4 closely follows the proof of Theorem 3. The algorithm runs Boosting-TTV with $\mathcal{L}$, $a^*(i)$ and μ as before but now with ξ set to $1/\text{poly}(n, W)$. The LBF h that Boosting-TTV outputs satisfies $d_{\text{Fourier}}(f, h) \leq \rho := 1/\text{poly}(n, W)$. We apply Corollary 1 to the affine form $L(x) := \frac{u}{\|u\|_1} \cdot x - \frac{\theta}{\|u\|_1}$ and get that for $r = 1/\text{poly}(n, W)$, we have

$$\mathbf{Pr}_{x \sim \mu}[|L(x)| < r] \leq \kappa := \epsilon^2/(1024 \log n) \tag{5}$$

where now $\epsilon := n^{-1/8}$, in place of Equation (3). Applying Lemma 6 we get that

$$\mathbf{E}_{x \sim \mu}[|f(x) - h(x)|] \leq \frac{4\|w\|_1 \sqrt{\rho}}{r} + 4\kappa \leq \epsilon^2/(128 \log n)$$

analogous to (4). The rest of the analysis goes through exactly as before, and we get that the LTF $h'(x) = \text{sign}(v \cdot x - \theta)$ satisfies $d_{\text{Shapley}}(f, h') \leq \epsilon/2$ as desired. The rest of the argument is unchanged so we do not repeat it. □

Acknowledgement. We thank Christos Papadimitriou for helpful conversations.

References

[APL07] Aziz, H., Paterson, M., Leech, D.: Efficient algorithm for designing weighted voting games. In: IEEE Intl. Multitopic Conf., pp. 1–6 (2007)

[Ban65] Banzhaf, J.: Weighted voting doesn't work: A mathematical analysis. Rutgers Law Review 19, 317–343 (1965)

[BKS99] Benjamini, I., Kalai, G., Schramm, O.: Noise sensitivity of Boolean functions and applications to percolation. Inst. Hautes Études Sci. Publ. Math. 90, 5–43 (1999)

[BMR+10] Bachrach, Y., Markakis, E., Resnick, E., Procaccia, A., Rosenschein, J., Saberi, A.: Approximating power indices: theoretical and empirical analysis. Autonomous Agents and Multi-Agent Systems 20(2), 105–122 (2010)

[Cho61] Chow, C.K.: On the characterization of threshold functions. In: Proc. 2nd FOCS 1961, pp. 34–38 (1961)

[DDFS12] De, A., Diakonikolas, I., Feldman, V., Servedio, R.: Near-optimal solutions for the Chow Parameters Problem and low-weight approximation of halfspaces. To appear in STOC (2012)

[dK08] de Keijzer, B.: A survey on the computation of power indices (2008), http://www.st.ewi.tudelft.nl/~tomas/theses/DeKeijzerSurvey.pdf

[dKKZ10] de Keijzer, B., Klos, T., Zhang, Y.: Enumeration and exact design of weighted voting games. In: AAMAS 2010, pp. 391–398 (2010)

[DP78] Deegan, J., Packel, E.: A new index of power for simple n-person games. International Journal of Game Theory 7, 113–123 (1978)

[EGGW07] Elkind, E., Goldberg, L.A., Goldberg, P.W., Wooldridge, M.: Computational complexity of weighted voting games. In: AAAI 2007, pp. 718–723 (2007)

[FWJ08] Fatima, S., Wooldridge, M., Jennings, N.: An Anytime Approximation Method for the Inverse Shapley Value Problem. In: AAMAS 2008, pp. 935–942 (2008)

[Gol06] Goldberg, P.: A Bound on the Precision Required to Estimate a Boolean Perceptron from its Average Satisfying Assignment. SIDMA 20, 328–343 (2006)

[Hol82] Holler, M.J.: Forming coalitions and measuring voting power. Political Studies 30, 262–271 (1982)

[Imp95] Impagliazzo, R.: Hard-core distributions for somewhat hard problems. In: Proc. 36th FOCS 1995, pp. 538–545 (1995)

[KS06] Kalai, G., Safra, S.: Threshold phenomena and influence. In: Computational Complexity and Statistical Physics, pp. 25–60. Oxford University Press (2006)

[Kur11] Kurz, S.: On the inverse power index problem. Optimization (2011), doi:10.1080/02331934.2011.587008

[Lee03] Leech, D.: Computing power indices for large voting games. Management Science 49(6) (2003)

[MTT61] Muroga, S., Toda, I., Takasu, S.: Theory of majority switching elements. J. Franklin Institute 271, 376–418 (1961)

[OS08] O'Donnell, R., Servedio, R.: The Chow Parameters Problem. In: Proc. 40th STOC 2008, pp. 517–526 (2008)

[Owe72] Owen, G.: Multilinear extensions of games. Management Science 18(5), 64–79 (1972); Part 2, Game theory and Gaming

[Rot88] Roth, A.E. (ed.): The Shapley value. University of Cambridge Press (1988)

[SS54] Shapley, L., Shubik, M.: A Method for Evaluating the Distribution of Power in a Committee System. American Political Science Review 48, 787–792 (1954)

[TTV08] Trevisan, L., Tulsiani, M., Vadhan, S.: Regularity, Boosting and Efficiently Simulating every High Entropy Distribution. Technical Report 103, ECCC, 2008. Conference version in Proc. CCC (2009)

[ZFBE08] Zuckerman, M., Faliszewski, P., Bachrach, Y., Elkind, E.: Manipulating the quota in weighted voting games. In: AAAI, pp. 215–220 (2008)

Zero-One Rounding of Singular Vectors

Amit Deshpande[1], Ravindran Kannan[1], and Nikhil Srivastava[2]

[1] Microsoft Research India
{amitdesh,kannan}@microsoft.com
[2] Center for Computational Intractability, Princeton
ns@cs.princeton.edu

Abstract. We propose a generic and simple technique called *dyadic rounding* for rounding real vectors to zero-one vectors, and show its several applications in approximating singular vectors of matrices by zero-one vectors, cut decompositions of matrices, and norm optimization problems. Our rounding technique leads to the following consequences.

1. Given any $A \in \mathbb{R}^{m \times n}$, there exists $z \in \{0,1\}^n$ such that

$$\frac{\|Az\|_q}{\|z\|_p} \geq \Omega\left(p^{1-\frac{1}{p}}(\log n)^{\frac{1}{p}-1}\right) \|A\|_{p \to q},$$

where $\|A\|_{p \to q} = \max_{x \neq 0} \|Ax\|_q / \|x\|_p$. Moreover, given any vector $v \in \mathbb{R}^n$ we can round it to a vector $z \in \{0,1\}^n$ with the same approximation guarantee as above, but now the guarantee is with respect to $\|Av\|_q / \|Av\|_p$ instead of $\|A\|_{p \to q}$. Although stated for $p \mapsto q$ norm, this generalizes to the case when $\|Az\|_q$ is replaced by *any* norm of z.

2. Given any $A \in \mathbb{R}^{m \times n}$, we can efficiently find $z \in \{0,1\}^n$ such that

$$\frac{\|Az\|}{\|z\|} \geq \frac{\sigma_1(A)}{2\sqrt{2 \log n}},$$

where $\sigma_1(A)$ is the top singular value of A. Extending this, we can efficiently find *orthogonal* $z_1, z_2, \ldots, z_k \in \{0,1\}^n$ such that

$$\frac{\|Az_i\|}{\|z_i\|} \geq \Omega\left(\frac{\sigma_k(A)}{\sqrt{k \log n}}\right), \quad \text{for all } i \in [k].$$

We complement these results by showing that they are almost tight.

3. Given any $A \in \mathbb{R}^{m \times n}$ of rank r, we can approximate it (under the Frobenius norm) by a sum of $O(r \log^2 m \log^2 n)$ cut-matrices, within an error of at most $\|A\|_F / \text{poly}(m,n)$. In comparison, the Singular Value Decomposition uses r rank-1 terms in the sum (but not necessarily cut matrices) and has zero error, whereas the cut decomposition lemma by Frieze and Kannan in their algorithmic version of Szemerédi's regularity partition [9,10] uses only $O(1/\epsilon^2)$ cut matrices but has a large $\epsilon \sqrt{mn} \|A\|_F$ error (under the cut norm). Our algorithm is deterministic and more efficient for the corresponding error range.

Keywords: rounding, matrix norms, singular value decomposition, cut decomposition.

A. Czumaj et al. (Eds.): ICALP 2012, Part I, LNCS 7391, pp. 278–289, 2012.
© Springer-Verlag Berlin Heidelberg 2012

1   Introduction

In most combinatorial optimization problems, once we come up with the right relaxation, solving the relaxation is often routine compared to the final rounding. Several sophisticated and clever rounding techniques are known that round real solutions to integer or zero-one solutions [19]. These rounding techniques often exploit the structure of the problem at hand (e.g., graph problems) as well as its corresponding relaxation (e.g., linear and semidefinite programs). In this paper, however, we propose a generic and simple rounding scheme that rounds any unit vector to a normalized vector in $\{0,1\}^n$, works for a wide range of problems, and has applications in cut decompositions of matrices and norm optimization problems.

The Singular Value Decomposition (SVD) decomposes any given matrix into a sum of rank-1 matrices, where the number of terms used in the decomposition is equal to the rank of the given matrix [11]. Often in practice, when the data is given as a matrix (e.g., document-term matrix, DNA microarray data), its rows and columns have special meanings as objects or attributes or features. When we use the singular value decomposition and its several analogs such as Principal Component Analysis (PCA) in practice, our actual intent is to find out the most important objects, attributes or features rather than just reducing the dimensionality by picking out a small number of important directions. The usual SVD or PCA fail to do this because the singular vectors are often real vectors and they correspond to linear combinations of objects, attributes or features, which are meaningless in practice [14]. Thus, it is desirable to have analogs of singular value decomposition that round the singular vectors to zero-one vectors.

One particular matrix decomposition that fits the above requirement is the cut decomposition of matrices. In cut decomposition, we decompose a given matrix into a sum of rank-1 matrices of a special type, known as cut-matrices. A cut-matrix is a rank-1 matrix obtained by taking an outer product of two zero-one vectors. Among the notable theoretical applications of cut decompositions are the algorithmic version of Szemerédi's regularity lemma [18] as well as the approximation schemes for dense constraint satisfaction problems due to Frieze and Kannan [9,10] and Alon et al. [1]. The cut decomposition lemma of Frieze and Kannan that lies at the heart of these results is a randomized algorithm to decomposes any given matrix (approximately) into a constant number of cut-matrices. While the approximation error in the Frieze-Kannan cut decomposition is large, it is still negligible for their applications, may it be the regularity lemma or dense constraint satisfaction problems. In this paper, we come up with a different *deterministic* cut decomposition algorithm that is based on the singular value decomposition and dyadic rounding of the singular vectors, and uses only polylogarithmically more number of terms in the sum than the SVD while keeping the error polynomially small.

The top singular value of a matrix is the same as its spectral or $\|\cdot\|_{2\mapsto 2}$ norm [11]. This can be generalized to p-to-q norm of a matrix, which is defined as $\max_{\|x\|_p=1} \|Ax\|_q$, where $\|\cdot\|_p$ and $\|\cdot\|_q$ denote the ℓ_p and ℓ_q-norms, respectively. Several natural problems and their relaxations can be expressed using matrix

p-to-q norms (see [6] for a survey), e.g., ℓ_p-Grothendieck problem [2], subspace approximation problem [7], condition number estimation [12], robust optimization [17], and spectral relaxations of graph cuts. Our rounding technique for the singular vectors generalizes to p-to-q norm optimization and even beyond to a larger class of norm optimization problems. To illustrate this, we give tensor norm optimization as an example of its generality.

2   Our Results

Our *dyadic rounding* technique leads to the following consequences.

1. Given any $A \in \mathbb{R}^{m \times n}$, there exists $z \in \{0,1\}^n$ such that

$$\frac{\|Az\|_q}{\|z\|_p} \geq \Omega \left(p^{1-\frac{1}{p}} (\log n)^{\frac{1}{p}-1} \right) \|A\|_{p \mapsto q},$$

   where $\|A\|_{p \mapsto q} = \max_{x \neq 0} \|Ax\|_q / \|x\|_p$. Moreover, given any vector $v \in \mathbb{R}^n$ we can round it to a vector $z \in \{0,1\}^n$ with the same approximation guarantee as above, but now with respect to $\|Av\|_q / \|Av\|_p$ instead of $\|A\|_{p \mapsto q}$. Thus, our rounding can be combined with the known algorithms [6,16,18,5] for computing or approximating p-to-q norms of matrices to finally get a zero-one solution while losing only a small $(\log n)^{1/q}$ factor.

2. *A special case of the above:* Given any $A \in \mathbb{R}^{m \times n}$, we can efficiently find $z \in \{0,1\}^n$ such that

$$\frac{\|Az\|}{\|z\|} \geq \frac{\sigma_1(A)}{2\sqrt{2 \log n}},$$

   where $\sigma_1(A)$ is the top singular value of A. Extending this, we can efficiently find *orthogonal* $z_1, z_2, \ldots, z_k \in \{0,1\}^n$ such that

$$\frac{\|Az_i\|}{\|z_i\|} \geq \Omega \left(\frac{\sigma_k(A)}{\sqrt{k \log n}} \right), \quad \text{for all } i \in [k].$$

   We complement these results by showing that they are almost tight.

3. Given any $A \in \mathbb{R}^{m \times n}$ of rank r, we can approximate it (under the Frobenius norm) by a sum of $O(r \log^2 m \log^2 n)$ cut-matrices, within an error of at most $\|A\|_F / \text{poly}(m, n)$. Our algorithm runs in time $O(T_{\text{svd}})$, where T_{svd} be the running time of the Singular Value Decomposition (SVD). In comparison, the singular value decomposition uses r rank-1 terms in the sum (but not necessarily cut matrices) and has zero error, whereas the cut decomposition lemma by Frieze and Kannan in their algorithmic version of Szemerédi's regularity partition [9,10] uses only $O(1/\epsilon^2)$ cut matrices but has a large $\epsilon \sqrt{mn} \|A\|_F$ error (under the cut norm), and runs in time $2^{O(1/\epsilon^2)}$. Notice that the cut norm of any m by n matrix is at most $\sqrt{mn}$ times its Frobenius norm (which can be shown by Cauchy-Schwarz inequality), so our upper bound for the approximation error under the Frobenius norm also applies to the cut norm.

4. Given any k-dimensional tensor $A \in (R^n)^{\otimes k}$, there exist $x_1, x_2, \ldots, x_k \in \{0,1\}^n$ such that

$$\frac{|A(x_1, \ldots, x_k)|}{\|x_1\| \cdots \|x_k\|} \geq \Omega\left((\log n)^{-k/2}\right) \max_{\|x_1\| = \ldots = \|x_k\| = 1} |A(x_1, \ldots, x_k)|.$$

3   Related Work

The problem of rounding the top singular vector to zero-one vector was considered by Bollobas and Nikiforov [4] in the context of 'discrepency of graphs'. They considered the slightly different formulation

$$\sigma_1(A) = \max_{x,y \in \mathbb{R}^n} \frac{x^T A y}{\|x\| \|y\|}$$

and showed that for Hermitian A there are always $x', y' \in \{0,1\}^n$ which come within a factor of $O(\log n)$ of achieving this optimum. By applying their theorem to adjacency matrices of graphs (minus the trivial top singular vector) they disproved a conjecture of Fan Chung which asserted that the rounding could be done to within a constant factor, thus providing a strong converse to the expander mixing lemma. This was refined by Bilu and Linial [3] to show that

$$\max_{x,y \in \mathbb{R}^n} \frac{x^T (A - (d/n)J)y}{\|x'\| \|y'\|} \leq O(\log d) \max_{x', y' \in \{0,1\}^n} \frac{x^T (A - (d/n)J)y}{\|x'\| \|y'\|}$$

where A is the adjacency matrix of a d-regular graph and J is the all 1's matrix. This is indeed a converse to the expander mixing lemma, since

$$\frac{|x^T (A - (d/n)J)y|}{\|x'\| \|y'\|} = |E(S,T) - (d/n)|S||T||$$

for x', y' indicator vectors of sets S, T. They used a randomized bucketing technique for the coordinates based on powers of 2 but their results apply only to matrices whose row and column sums are bounded by d.

Another special case appears in a recent manuscript of Matoušek [15] (see Lemma 7), where the given matrix is a vector of all 1's and the rounding gives a vector whose non-zero coordinates are almost equal, i.e., within a factor 2 of each other. Brubaker and Vempala [6] also used indicator decomposition for tensor norms to get $\Omega\left((\log n)^{-k}\right)$ guarantee, which we improve to $\Omega\left((\log n)^{-k/2}\right)$. Recently, it was pointed out to us that the dyadic rounding of only the top singular vector also appeared in [13] in a different context. Our dyadic rounding is also different from the binary expansion method of Beck and Spencer [8] – our method is deterministic and we do not round digit-by-digit.

4   Preliminaries and Notation

For a vector $v \in \mathbb{R}^n$ its ℓ_p-norm is defined as $\|v\|_p = \left(\sum_{i=1}^n |v_i|^p\right)^{1/p}$. When we use $\|v\|$ without any subscript, it should be considered as $\|v\|_2$. For a matrix

$A \in \mathbb{R}^{m \times n}$, its p-to-q norm is defined as

$$\|A\|_{p \mapsto q} = \max_{\|x\|_p = 1} \|Ax\|_q .$$

In particular, $\|A\|_{2 \mapsto 2}$ is known as the spectral norm or the operator norm of A, and the vector x that achieves this maximum is called the top (right) singular vector of A.

Given any matrix $A \in \mathbb{R}^{m \times n}$ of rank r, there exist non-negative real numbers $\sigma_1 \geq \sigma_2 \geq \ldots \sigma_r \geq 0$, an orthonormal system of vectors $u_1, \ldots, u_r \in \mathbb{R}^m$ and another orthonormal system of vectors $v_1, v_2, \ldots, v_r \in \mathbb{R}^n$ such that

$$A = \sum_{i=1}^{r} \sigma_i u_i v_i^T .$$

This is also known as the Singular Value Decomposition (SVD) of A. In other words, the Singular Value Decomposition decomposes A into a sum of r rank-1 matrices.

The Frobenius norm of a matrix $A \in \mathbb{R}^{m \times n}$ is defined as the ℓ_2-norm of it when thought of as a vector of length mn, i.e.,

$$\|A\|_F = \left(\sum_{ij} A_{ij}^2 \right)^{1/2} .$$

Using the Singular Value Decomposition, one can show that $\|A\|_F^2 = \sum_{i=1}^{r} \sigma_i^2$.

The cut-norm of a matrix $A \in \mathbb{R}^{m \times n}$ is defined as

$$\|A\|_C = \max_{I \subseteq [m], J \subseteq [n]} \left| \sum_{i \in I, j \in J} A_{ij} \right| .$$

By Cauchy-Schwarz inequality, we have $\|A\|_C \leq \sqrt{|I||J|} \|A\|_F \leq \sqrt{mn} \|A\|_F$.

5 Dyadic Rounding of Vectors

Here we state the *dyadic rounding* lemma that is at the core of our results.

Lemma 1. *(Dyadic rounding lemma) Given any $A \in \mathbb{R}^{m \times n}$, there exists a vector $z \in \{0,1\}^n$ such that*

$$\frac{\|Az\|_q}{\|z\|_p} \geq \Omega \left(p^{1 - \frac{1}{p}} (\log n)^{\frac{1}{p} - 1} \right) \|A\|_{p \mapsto q}, \quad \text{where } \|A\|_{p \mapsto q} = \max_{z \neq \bar{0}} \frac{\|Az\|_q}{\|z\|_p} .$$

Proof. Let $v = \operatorname{argmax}_{z \neq \bar{0}} \|Az\|_q / \|z\|_p$ and $\|v\|_p = 1$, without loss of generality. In the first step, we find a constant factor approximation x to v using a small grid of size $n^{-1/p}$. In the second step, we divide this vector into two vectors, call them

x_{pos} and x_{neg}, containing the positive and negative coordinates of x, respectively. We show that one of x_{pos} and x_{neg} gives a constant factor approximation to x, and therefore, to v. Finally in the third step, using the fact that the coordinates of our new vector are bounded integer multiples of the grid size, we divide them into $O(\log n)$ parts based on powers of 2 (which gives the name *dyadic rounding*), and write our vector as a linear combination of $O(\log n)$ vectors from $\{0,1\}^n$. One of these (up to scaling) is the vector z that we are looking for.

Here is a formal proof. Let $v = \text{argmax}_{z \neq 0}$ and $\|v\|_p = 1$, without loss of generality. We can write v as a convex combination $v = \sum_t \alpha_t x_t$, where $x_t \in \left(n^{-1/p}\mathbb{Z}\right)^n$, $\|x_t\|_p \leq 2$ and $\alpha_t \geq 0$ for all t, and $\sum_t \alpha_t = 1$. By triangle inequality, $\|Av\|_q \leq \sum_t \alpha_t \|Ax_t\|_q$, so there must exist some t such that $\|Ax_t\|_q \geq \|Av\|_q$. We proceed with this particular $x_t \in \left(n^{-1/p}\mathbb{Z}\right)^n$. Let $x_t = x_{\text{pos}} + x_{\text{neg}}$, where

$$(x_{\text{pos}})_i = \begin{cases} (x_t)_i & \text{if } (x_t)_i > 0 \\ 0 & \text{otherwise} \end{cases} \qquad (x_{\text{neg}})_i = \begin{cases} (x_t)_i & \text{if } (x_t)_i < 0 \\ 0 & \text{otherwise} \end{cases}$$

By triangle inequality, $\|Ax_t\|_q \leq \|Ax_{\text{pos}}\|_q + \|Ax_{\text{neg}}\|_q$ and $\|x_{\text{pos}}\|_p^p + \|x_{\text{neg}}\|_p^p = \|x_t\|_p^p \leq 2^p$. Define

$$y = \text{argmax}_{x \in \{x_{\text{pos}}, x_{\text{neg}}\}} \|Ax\|_q .$$

Then $\|Ay\|_q \geq \|Ax_t\|_q /2 \geq \|Av\|_q /2$ and $\|y\|_p \leq 2$, and all the non-zero coordinates of y are integer multiples of $n^{-1/p}$ upper bounded by 2 and have the same sign. Now we can write $y = \sum_{j=0}^{O(p^{-1}\log n)} z_j$, where

$$(z_j)_i = \begin{cases} 0 & \text{if binary expansion of } |y_i|\, n^{1/p} \text{ does not contain } 2^j \\ \text{sign}(y_i) 2^j n^{-1/p} & \text{if } |y_i|\, n^{1/p} > 0 \text{ and its binary expansion contains } 2^j \end{cases}$$

Therefore,

$$\|Ay\|_q \leq \sum_{j=0}^{O(p^{-1}\log n)} \|Az_j\|_q$$

$$= \sum_{j=0}^{O(p^{-1}\log n)} \frac{\|Az_j\|_q}{\|z_j\|_p} \|z_j\|_p$$

$$\leq \left(\sum_{j=0}^{O(p^{-1}\log n)} \left(\frac{\|Az_j\|_q}{\|z_j\|_p} \right)^{p/(p-1)} \right)^{(p-1)/p} \left(\sum_{j=0}^{O(p^{-1}\log n)} \|z_j\|_p^p \right)^{1/p}$$

by Hölder's inequality

$$\leq \left(\sum_{j=0}^{O(p^{-1}\log n)} \left(\frac{\|Az_j\|_q}{\|z_j\|_p} \right)^{p/(p-1)} \right)^{(p-1)/p} \|y\|_p ,$$

where the last inequality works because $\sum_{j=0}^{O(p^{-1}\log n)}(z_j)_i^p$ is subsumed by y_i^p, for each i. Thus, by averaging, there must exist some j such that

$$\frac{\|Az_j\|_q}{\|z_j\|_p} \geq \Omega\left(p^{1-\frac{1}{p}}(\log n)^{\frac{1}{p}-1}\right) \frac{\|Ay\|_q}{\|y\|_p} \geq \Omega\left(p^{1-\frac{1}{p}}(\log n)^{\frac{1}{p}-1}\right) \frac{\|Av\|_q}{\|v\|_p}.$$

Remark: Observe that the above proof works even when $\|Az\|_q$ is replaced by *any* norm of z. Moreover, this proof can can be made algorithmic using the following simple idea.

Proposition 1. *Given any $v \in [0,1]^n$, we can efficiently find $x_1, x_2, \ldots, x_{n+1} \in \{0,1\}^n$ such that $v = \sum_{t=1}^{n+1} \alpha_t x_t$, with $\sum_{t=1}^{n+1} \alpha_t = 1$ and $\alpha_t \geq 0$ for all t.*

Proof. We prove this by induction on the number of coordinates. Let v_i be the maximum coordinate of v. Then $w = (1/v_i)v$ is still in $[0,1]^n$ and v is a convex combination of $\bar{0}$ and w. Now w has its i-th coordinates as 1, so by induction hypothesis w can be written as a convex combination $w = \sum_{t=1}^{n} \beta_t x_t$, where $x_t \in \{0,1\}^n$ and all x_t have their i-coordinate as 1. Putting these two together, v can be written as a convex combination of $\bar{0}$ and $x_1, \ldots, x_n$.

5.1 Rounding Singular Vectors

Here are some immediate corollaries of Lemma 1. We skip the proofs as they are essentially identical to that of Lemma 1. For spectral or $\|\cdot\|_{2\to2}$ norm we get slightly better constants as follows.

Corollary 2. *Given any $A \in \mathbb{R}^{m\times n}$, there exists a vector $z \in \{0,1\}^n$ such that*

$$\frac{\|Az\|}{\|z\|} \geq \frac{\sigma_1(A)}{2\sqrt{2\log n}},$$

and such a vector z can be found in polynomial time.

Using a vector instead of matrix, we get the next corollary, which says that the set of normalized zero-one vectors is a weak ϵ-net for S^{n-1} with $\epsilon = 2 - \frac{1}{\sqrt{2\log n}}$.

Corollary 3. *Given any $a \in \mathbb{R}^n$, there exists a vector $z \in \{0,1\}^n$ such that*

$$\langle a, z\rangle \geq \frac{\|a\|\,\|z\|}{2\sqrt{2\log n}}.$$

This special case is actually *equivalent* to corollary 2 for matrices: given any matrix A with top singular vector u satisfying $\|Au\| = \sigma_1(A)u$, we simply round u to $z \in \{0,1\}^n$ and observe that

$$\|Az\|^2 = z^T A^T A z \geq z^T(\sigma_1(A)uu^T)z \geq \sigma_1(A)\frac{\|z\|}{2\sqrt{2\log n}}, \quad \text{as in Corollary 2.}$$

We prove an almost matching tightness result, and our rounding also generalizes to give $\Omega\left((\log n)^{-k/2}\right)$ guarantee for tensor norm optimization, which improves an earlier $\Omega\left((\log n)^{-k}\right)$ guarantee by Brubaker and Vempala [6]. We defer the proofs of both these to the full version.

6   Rounding Multiple Singular Vectors Simultaneously

Our dyadic rounding does not preserve orthogonality property when applied to the top k singular vectors simultaneously. But surprisingly, we can get around it to show that there exist k *orthogonal* zero-one vectors such that all of them are at least as good as the k-th singular vector.

Theorem 4. *(Multiple rounding with orthogonality constraint)* *Given any* $A \in \mathbb{R}^{m \times n}$, *there exist vectors* $x_1, x_2, \ldots, x_k \in \{0,1\}^n$ *such that*

$$\|Ax_i\| \geq \frac{\sigma_k(A)}{2\sqrt{2(2k-1)\log n}} \|x_i\|, \text{ for all } i \in [k], \text{ and } \langle x_i, x_j \rangle = 0 \text{ for } i \neq j.$$

This proof is also constructive and these vectors can be found efficiently.

Proof. Let $A^T A = \sum_{j=1}^n \sigma_j^2 v_j v_j^T$ be the singular value decomposition of $A^T A \in \mathbb{R}^{n \times n}$, with $\sigma_1 \geq \sigma_2 \geq \ldots \geq \sigma_n \geq 0$, and let $B = \sigma_k^2 \sum_{j=1}^k v_j v_j^T$. Since $A^T A \succeq B \succ 0$, we have $\|Ax\|^2 = x^T A^T A x \geq x^T B x$, and it suffices to find orthogonal vectors $x_1, x_2, \ldots, x_k \in \mathbb{R}^n$ such that

$$x_i^T B x_i = \Omega\left(\frac{\sigma_k^2}{k \log n}\right) \|x_i\|^2, \text{ for all } i \in [k].$$

However,

$$x_i^T B x_i = \sigma_k^2 \sum_{j=1}^k x_i^T v_j v_j^T x_i = \sigma_k^2 \sum_{j=1}^k \langle v_j, x_i \rangle^2 = \sigma_k^2 \|Vx_i\|^2,$$

where $V \in \mathbb{R}^{k \times n}$ be a matrix with $v_1, v_2, \ldots, v_k$ as its rows. Thus, it suffices to find orthogonal vectors $x_1, x_2, \ldots, x_k \in \mathbb{R}^n$ such that

$$\|Vx_i\| = \Omega\left(\frac{1}{\sqrt{k \log n}}\right) \|x_i\|, \text{ for all } i \in [k].$$

To prove this, we first divide the columns of V into k disjoint parts to get column submatrices $C_1, C_2, \ldots, C_k$ such that $\|C_j\|_F^2 = \Omega(1)$, for all $j \in [k]$. Let $a_1, a_2, \ldots, a_n$ be the squared lengths of the columns of V. Then $\sum_{i=1}^n a_i = \|V\|_F^2 = k$ and moreover, $a_i \in [0,1]$, for all $i \in [n]$, since the columns of V are in isotropic position. Therefore, using Lemma 2 we can partition the columns into k disjoint sets $P_1, P_2, \ldots, P_k$ such that

$$\sum_{i \in P_j} a_i \geq \frac{1}{2 - 1/k}, \text{ for all } j \in [k].$$

Now we can define matrices $C_1, C_2, \ldots, C_k \in \mathbb{R}^{k \times n}$ as

$$(C_j)_{pq} = \begin{cases} V_{pq} & \text{for } q \in P_j \\ 0 & \text{otherwise} \end{cases}$$

i.e., C_j is a matrix that keeps all the columns of V that are in P_j and makes all the others zero. Thus,

$$\sigma_1 \left(C_j\right)^2 \geq \frac{1}{k} \|C_j\|_F^2 \geq \frac{1}{2k-1}, \text{ for all } j \in [k].$$

Now by Corollary 2, for each C_j we can find a vector $x_j \in \{0,1\}^n$ such that

$$\|Vx_j\| = \|C_j x_j\| \geq \frac{\sigma_1\left(C_j\right)}{2\sqrt{2\log n}} \|x_j\| \geq \frac{1}{2\sqrt{2(2k-1)\log n}}, \text{ for all } j \in [k],$$

and since C_j use disjoint subsets of columns from V, we also have $\langle x_i, x_j \rangle = 0$, for $i \neq j$. This completes the proof.

Lemma 2. *(Claim B.1 in [7] restated) Let $a_1, a_2, \ldots, a_n \in [0,1]$ be such that $\sum_{i=1}^n a_i = k$. Then we can partition $[n]$ into k parts as $P_1 \uplus P_2 \uplus \ldots \uplus P_k$ that satisfy*

$$\sum_{i \in P_j} a_i \geq \frac{1}{2-1/k}, \text{ for all } j \in [k].$$

Moreover, this can be done by a greedy algorithm that considers a_i's in their decreasing order and then puts them into k bins one by one, where each time the bin chosen is the one with the least sum of a_i's thrown in it so far.

We show that the dependence on k in Theorem 4 cannot be improved by more than a logarithmic factor. The tight example is a random matrix, and the lower bound is essentially a consequence of the fact that orthogonal zero-one vectors must have disjoint supports. The proof is deferred to full version.

Proposition 5. *Suppose $k \leq n/2$ and $A = \frac{1}{\sqrt{n}}G$ where $G_{k \times n} = (g_{ij})$ has i.i.d. standard Gaussian entries. Then with high probability, $\sigma_k(A) = \Omega(1)$, but for every mutually orthogonal $z_1, \ldots, z_k \in \{0,1\}^n$, there is some i with*

$$\frac{\|Az_i\|}{\|z_i\|} \leq O\left(\sqrt{\frac{\log k}{k}}\right).$$

7   Rounding SVD to Cut Decompositio

Singular Value Decomposition (SVD) decomposes any matrix of rank r into a sum of r rank-1 matrices. Often these rank-1 matrices are not very meaningful for specific applications in practice related to real world data as well as in theory when the matrix has some underlying graph structure.

Frieze and Kannan came up with a notion of cut decomposition of matrices where any matrix can be written as a sum of a small number of cut-matrices, up to a small error of approximation (under the cut norm). Cut-matrices are rank-1 matrices obtained by taking outer product of zero-one vectors, i.e., xy^T with $x \in \{0,1\}^m$ and $y \in \{0,1\}^n$. This has found applications in their algorithmic version of the famous Szemerédi's regularity lemma in graph theory [18,9,10]. Cut decompositions have also been useful in efficient approximation schemes for dense constraint satisfaction problems [10,1]. The cut decomposition lemma of Frieze and Kannan says,

Theorem 6. *(Frieze-Kannan cut decomposition) Given any $A \in \mathbb{R}^{m \times n}$ and $\epsilon, \delta \in (0,1)$, we can find $t = O(1/\epsilon^2)$ matrices $M_1, M_2, \ldots, M_t$ such that each $M_i = \gamma_i x_i y_i^T$, for some $x_i \in \{0,1\}^m, y_i \in \{0,1\}^n$, with*

$$\sum_i \gamma_i^2 \leq \frac{27 \|A\|_F^2}{mn} \quad and \quad \|A - (M_1 + \cdots + M_t)\|_C \leq \epsilon \sqrt{mn} \|A\|_F.$$

Moreover, $M_1, \ldots, M_t$ can be found by a randomized algorithm that runs in time $2^{\tilde{O}(1/\epsilon^2)}/\delta^2$ and succeeds with probability at least $1 - \delta$.

Compare this with our new cut decomposition based on the singular value decomposition and dyadic rounding of singular vectors. This can be thought of as a cut decomposition that uses only polylogarithmically more terms in the decomposition that the SVD while keeping the error polynomially small. (The error can be made an arbitrarily small as an inverse polynomial in m and n).

Theorem 7. *Given any $A \in \mathbb{R}^{m \times n}$ with rank$(A) = r$, we can find cut-matrices $M_1, M_2, \ldots, M_t$, that is, each $M_i = \gamma_i x_i y_i^T$, for some $x_i \in \{0,1\}^m, y_i \in \{0,1\}^n$ and $\gamma_i \in \mathbb{R}$, such that $t = O(r \log^2 m \log^2 n)$ and*

$$\|A - (M_1 + \cdots + M_t)\|_F \leq \frac{\|A\|_F}{poly(m,n)}.$$

The coefficient length for this cut decomposition $\left(\sum_i \gamma_i^2\right)^{1/2}$ is $O(\|A\|_F \log m \log n)$, and there is a deterministic $O(T_{svd})$ time algorithm to find such cut matrices $M_1, M_2, \ldots, M_t$. (Remark: We can improve the $\log^2$ bounds to $\log$ for a weaker error guarantee of $\epsilon \|A\|_F$, for any constant $\epsilon > 0$.)

The core of the proof is the following simple discretization argument that allows us to write an arbitrary vector as a sum of a few zero-one vectors.

Corollary 8. *Given any $a \in \mathbb{R}^n$, and $\epsilon > 0$, there exist vectors $z_1, z_2, \ldots, z_t \in \{0,1\}^n$ and constants $\alpha_1, \alpha_2, \ldots, \alpha_t \in \mathbb{R}$ such that $t = O(\log n + \log(1/\epsilon))$ and $\left\| a - \sum_{j=1}^t \alpha_j z_j \right\| \leq \epsilon \|a\|$. Note that we can take ϵ to be as small as $1/poly(n)$ and t will still be $O(\log n)$.*

Proof. Assume $\|a\| = 1$ and as in the proof of Lemma 1 consider a fine grid $\left(\frac{\epsilon}{\sqrt{n}}\mathbb{Z}\right)^n$. Round a to the nearest grid point, which is at distance at most ϵ. Since the largest integer involved is of size $\sqrt{n}/\epsilon$, this grid point can be written as a weighted sum of at most $\log(\sqrt{n}/\epsilon)$ binary vectors.

We are now ready to finish the proof of the cut decomposition, Theorem 7.

Proof. Let $A = \sum_{i=1}^{r} \sigma_i u_i v_i^T$ be the singular value decomposition of A, where $\sigma_1 \geq \sigma_2 \geq \ldots \geq \sigma_r \geq 0$ with $\text{rank}(A) = r$, and $\{u_i \in \mathbb{R}^m\}_i$, $\{v_i \in \mathbb{R}^n\}_i$ form orthonormal systems of vectors. By Corollary 8, we know that there exist vectors $x_j(u_i) \in \{0,1\}^m$ and $y_j(v_i) \in \mathbb{R}^n$, and constants $\alpha_j(u_i), \beta_j(v_i) \in \mathbb{R}$ such that

$$\left\| u_i - \sum_{j=1}^{O(\log^2 m)} \alpha_j(u_i) x_j(u_i) \right\| \leq \frac{1}{\text{poly}(m)}, \qquad \text{and}$$

$$\left\| v_i - \sum_{j=1}^{O(\log^2 n)} \beta_j(v_i) y_j(v_i) \right\| \leq \frac{1}{\text{poly}(n)},$$

for all i. Therefore,

$$\left\| \sigma_i u_i v_i^T - \sigma_i \sum_{j_1=1}^{O(\log^2 m)} \sum_{j_2=1}^{O(\log^2 n)} \alpha_{j_1}(u_i) \beta_{j_2}(v_i) x_{j_1}(u_i) y_{j_2}(v_i)^T \right\|_F^2 \leq \frac{\sigma_i^2}{\text{poly}(m,n)},$$

for all i. Hence,

$$\left\| A - \sum_{i=1}^{r} \sum_{j_1=1}^{O(\log^2 m)} \sum_{j_2=1}^{O(\log^2 n)} \sigma_i \alpha_{j_1}(u_i) \beta_{j_2}(v_i) x_{j_1}(u_i) y_{j_2}(v_i)^T \right\|_F^2 \leq \frac{\sum_i \sigma_i^2}{\text{poly}(m,n)}$$

$$\leq \frac{\|A\|_F^2}{\text{poly}(m,n)}.$$

Notice that all the rank-1 matrices used in the sum are cut matrices, and therefore we have actually obtained a collection of $t = O(r \log^2 m \log^2 n)$ cut matrices $M_1, M_2, \ldots, M_t$ such that

$$\|A - (M_1 + \ldots + M_t)\|_F \leq \frac{\|A\|_F}{\text{poly}(m,n)}.$$

Using the relation between the cut norm and the Frobenius norm described in 4 we get

$$\|A - (M_1 + \ldots + M_t)\|_C \leq \sqrt{mn} \, \|A - (M_1 + \ldots + M_t)\|_F \leq \frac{\|A\|_F}{\text{poly}(m,n)}.$$

The *coefficient length* for this cut decomposition can be bounded by

$$\sum_{i=1}^{r} \sum_{j_1=1}^{O(\log^2 m)} \sum_{j_2=1}^{O(\log^2 n)} \sigma_i^2 \alpha_{j_1}(u_i)^2 \beta_{j_2}(v_i)^2 \leq \|A\|_F^2 \log^2 m \log^2 n.$$

8   Open Problems

Spectral algorithms are used for finding good cuts in graphs or finding important features via Principal Component Analysis (PCA). The sparsest cut problem can be rewritten in the form of $\max_{x \in \{-1,1\}^n, x \perp \bar{1}} x^T K x / x^T L x$, where K and L are the Laplacians of the complete graph and the given graph, respectively. It would be interesting to extend our techniques to *relative* singular value problems involving objective functions $\|Ax\| / \|Bx\|$ instead of $\|Ax\| / \|x\|$, with potential applications in designing faster algorithms for graph cuts or feature selection.

References

1. Alon, N., de la Vega, F., Kannan, R., Karpinski, M.: Random sampling and approximation of Max-CSPs. Journal of Computer and System Sciences 67, 212–243 (2003)
2. Alon, N., Naor, A.: Approximating the cut-norm via Grothendieck's inequality. SIAM Journal on Computing (SICOMP) 35(4), 787–803 (2006)
3. Bilu, Y., Linial, N.: Lifts, discrepancy and nearly optimal spectral gaps. Combinatorica 26, 495–519 (2006)
4. Bollobas, B., Nikiforov, V.: Graphs and hermitian matrices: discrepancy and singular values. Discrete Mathematics 285 (2004)
5. Boyd, D.W.: The power method for p-norms. Linear Algebra and Its Applications 9, 95–101 (1974)
6. Charles Brubaker, S., Vempala, S.S.: Random Tensors and Planted Cliques. In: Dinur, I., Jansen, K., Naor, J., Rolim, J. (eds.) APPROX 2009. LNCS, vol. 5687, pp. 406–419. Springer, Heidelberg (2009)
7. Deshpande, A., Tulsiani, M., Vishnoi, N.: Algorithms and hardness for subspace approximation. In: ACM-SIAM Symposium on Discrete Algorithms, SODA 2011 (2011)
8. Doerr, B.: Roundings Respecting Hard Constraints. In: Diekert, V., Durand, B. (eds.) STACS 2005. LNCS, vol. 3404, pp. 617–628. Springer, Heidelberg (2005)
9. Frieze, A., Kannan, R.: The regularity lemma and approximation schemes for dense problems. In: IEEE Symposium on Foundations of Computing (FOCS 1996), pp. 12–20 (1996)
10. Frieze, A., Kannan, R.: Quick approximation to matrices and applications. Combinatorica 19(2), 175–220 (1999)
11. Golub, G., van Loan, C.: Matrix Computations. Johns Hopkins University Press (1996)
12. Nicholas, J.: Higham. Estimating the matrix p-norm. Numerische Mathematik 62, 511–538 (1992)
13. Kasiviswanathan, S.P., Rudelson, M., Smith, A., Ullman, J.: The price of privately releasing contingency tables and the spectra of random matrices with correlated rows. In: STOC 2010, pp. 775–784 (2010)
14. Mahoney, M., Drineas, P.: CUR matrix decompositions for improved data analysis. Proceedings of the National Academy of Sciences USA 106, 697–702 (2009)
15. Matoušek, J.: The determinant bound for discrepancy is almost tight (2011), http://arxiv.org/PS_cache/arxiv/pdf/1101/1101.0767v2.pdf
16. Nesterov, Y.: Semidefinite relaxation and nonconvex quadratic optimization. Optimization Methods and Software 9, 141–160 (1998)
17. Steinberg, D.: Computation of matrix norms with applications to robust optimization. Research thesis. Technion – Israel University of Technology (2005)
18. Szemerédi, E.: Regular partitions of graphs. Problèmes combinatoires et théorie des graphes (Colloq. Internat. CNRS), Paris 260, 399–401 (1976)
19. Vazirani, V.: Approximation Algorithms. Springer (2001)

Label Cover Instances with Large Girth and the Hardness of Approximating Basic k-Spanner*

Michael Dinitz[1,**], Guy Kortsarz[2,***], and Ran Raz[3,†]

[1] Weizmann Institute of Science
mdinitz@cs.cmu.edu
[2] Department of Computer Science, Rutgers, Camden
guyk@camden.rutgers.edu
[3] Weizmann Institute of Science
ran.raz@weizmann.ac.il

Abstract. We study the well-known *Label Cover* problem under the additional requirement that problem instances have large girth. We show that if the girth is some k, the problem is roughly $2^{(\log^{1-\epsilon} n)/k}$ hard to approximate for all constant $\epsilon > 0$. A similar theorem was claimed by Elkin and Peleg [ICALP 2000] as part of an attempt to prove hardness for the basic k-spanner problem, but their proof was later found to have a fundamental error. Thus we give both *the first* non-trivial lower bound for the problem of Label Cover with large girth as well as the first full proof of strong hardness for the basic k-spanner problem, which is both the simplest problem in graph spanners and one of the few for which super-logarithmic hardness was not known. Assuming $NP \not\subseteq BPTIME(2^{polylog(n)})$, we show (roughly) that for every $k \geq 3$ and every constant $\epsilon > 0$ it is hard to approximate the basic k-spanner problem within a factor better than $2^{(\log^{1-\epsilon} n)/k}$. This improves over the previous best lower bound of only $\Omega(\log n)/k$ from [17]. Our main technique is subsampling the edges of 2-query PCPs, which allows us to reduce the degree of a PCP to be essentially equal to the soundness desired. This turns out to be enough to basically guarantee large girth.

1 Introduction

In this paper we deal with 2-query probabilistically checkable proofs (PCPs) and variants of the *Label Cover* problem. Label Cover was originally defined by Arora and Lund in their early survey on hardness of approximation [2]. Since then, Label Cover has been widely used as a starting point when proving hardness of approximation, as it corresponds naturally to 2-query probabilistically

* Full version available at http://arxiv.org/abs/1203.0224
** Research supported in part by an Israel Science Foundation grant #452/08, a US-Israel BSF grant #2010418, and by a Minerva grant.
*** Research supported in part by NSF award number 0829959.
† Research supported by an Israel Science Foundation grant and by the I-CORE Program of the Planning and Budgeting Committee and the Israel Science Foundation.

A. Czumaj et al. (Eds.): ICALP 2012, Part I, LNCS 7391, pp. 290–301, 2012.
© Springer-Verlag Berlin Heidelberg 2012

checkable proofs and one-round two-prover interactive proof systems. Certain reductions from Label Cover, though, require special properties of the Label Cover instances. So then the question becomes: is Label Cover still hard even when restricted to these instances? For example, the famous *Unique Games Conjecture* of Khot [16] can be thought of as a conjecture that Label Cover is still difficult to approximate when the relation on each edge is required to be a bijection. A different type of requirement is on the structure of the Label Cover graph rather than on the allowed relations; Elkin and Peleg [11] showed that if Label Cover (actually, a slight variant known as Min-Rep) is hard even on graphs with large girth then the *basic k-spanner* problem is also hard to approximate. They then gave a proof that Label Cover is indeed hard to approximate on large-girth graphs, but unfortunately this proof was later found to have a flaw (as Elkin and Peleg point out in [13, Section 1.3]). In this paper we give a completely new proof that Label Cover and Min-Rep are hard to approximate on large-girth graphs, and thus prove strong hardness for basic *k*-spanner. Our argument is based on subsampling edges of a 2-query PCP/Label Cover instance. Subsampling of 2-query PCPs and Label Cover instances has been done before for other reasons (see e.g. [15]), but we show that the sampling probability can be set low enough to destroy most small cycles while still preserving hardness, and this technique was not previously used in this context. Remaining cycles can then be removed deterministically.

1.1 Label Cover and Probabilistically Checkable Proofs

A *probabilistically checkable proof* (PCP) is a string (proof) together with a verifier algorithm that checks the proof (probabilistically). There are several important parameters of a PCP, including the following:

1. Completeness (c): the minimum probability that the verifier accepts a correct proof. All of the PCPs in this paper have completeness 1.
2. Soundness (or error) (s): the maximum probability that the verifier accepts a proof of an incorrect claim.
3. Queries: the number of queries that the verifier makes to the proof. In this paper we will study the case when the verifier only makes 2 queries.
4. Size: the number of positions in the proof (i.e. the length).
5. Alphabet (Σ): the set of symbols that the proof is written in. We will only be concerned with PCPs for which $|\Sigma|$ is at most polynomial in the size of the PCP, so we will assume this throughout.

For this paper we will be concerned with 2-query PCPs, which are the special case when the verifier is only allowed to query two positions of the proof. We will also assume (without loss of generality) that these two queries are to different parts of the proof, i.e. there is some set of positions A that can be read by the first query and some other set of positions B that can be read by the second query with $A \cap B = \emptyset$.

For this type of PCP, there are two natural graphs that represent it. The first (and simpler), which is sometimes called the *supergraph*, is a bipartite graph

(A, B, E) in which there is a vertex for every position of the proof and an edge between two positions if there is a possibility that the verifier might query those two positions. By our assumption, this graph is bipartite. We also assume that the verifier chooses its query uniformly at random from these edges, which is in fact the case in the PCPs that we will use (in particular in the PCP for Max-3SAT(5) obtained by parallel repetition). Vertices of this graph will sometimes be called *supervertices*, and edges will be *superedges*.

The second graph can be thought of as an expansion of the supergraph in which the test the verifier does is explicitly contained in the graph. This graph is also bipartite, with vertex set $(A \times \Sigma_A, B \times \Sigma_B)$, where Σ_A is the alphabet used in A positions of the proof and similarly for Σ_B. There is an edge between vertices (a, α) and (b, β) if the verifier might query a and b together (i.e. (a, b) is an edge in the supergraph) and if, upon such queries, the verifier will accept the proof if it sees values α in position a and β in position b. We call this graph the Min-Rep graph. This is related to the work in [17].

There is a natural correspondence between these types of PCPs and the optimization problem of Label Cover. In Label Cover we are given a bipartite graph $G = (A, B, E)$, alphabets Σ_A and Σ_B, and for every edge $e \in E$ a nonempty relation $\pi_e \subseteq \Sigma_A \times \Sigma_B$. The goal is to find assignments $\gamma_A : A \to \Sigma_A$ and $\gamma_B : B \to \Sigma_B$ in order to maximize the number of edges $e = (a, b)$ for which $(\gamma_A(a), \gamma_B(b)) \in \pi_e$ (in which case we say that the edge is satisfied or covered). It is easy to see that the existence of PCPs for NP-hard problems implies that Label Cover is hard to approximate: in particular, if we use the supergraph and set the relation to be answers on which the verifier accepts, then it is hard to distinguish instances in which at least a c fraction of the edges are satisfiable (a valid proof) from instances in which at most an s fraction of the edges are satisfiable (an invalid proof). The exact nature of the hardness assumption is based on the size of the PCP: if it has size $m(n)$ (where n is the size of the original problem instance) then the hardness assumption is that NP is not contained in DTIME($poly(m(n))$) (for deterministic PCP constructions and approximation algorithms) or BPTIME($poly(m(n))$) (for randomized PCP constructions or approximation algorithms). We let Label Cover$_k$ be the Label Cover problem with the additional restriction that the girth of G is larger than k.

We now describe a slight variant of Label Cover known as Min-Rep (originally defined by Kortsarz [17]) that has been useful for proving hardness of approximation for network design problems such as spanners. It can be thought of as a minimization version of Label Cover with the additional property that the alphabets are represented explicitly as vertices in a graph. Consider the Min-Rep graph $G' = (A \times \Sigma_A, B \times \Sigma_B, E')$. For every $i \in A$ let $A_i = \{(i, \alpha) \in A \times \Sigma_A\}$ be the set of vertices in the Min-Rep graph corresponding to vertex i in the Label Cover graph, and similarly for $i \in B$ let $B_i = \{(i, \beta) \in B \times \Sigma_B\}$. Our goal is to choose a set S of vertices of G' of minimum size so that for every $(i, j) \in E$ there are vertices (i, α) and (j, β) in S that are joined by an edge in E'. Such a set is called a *REP-cover*, and the vertices in it are called *representatives*. Less formally, we can think of Min-Rep as being the problem of assigning to every vertex in the supergraph a *set* of labels/representatives (unlike Label Cover, in which only a single label is assigned) so that for every superedge (a, b) there is at least one label assigned to a and at least one label assigned to b that satisfy the relation $\pi_{(a,b)}$. Note that in the Min-Rep graph the number of vertices is $|A| \cdot |\Sigma_A| + |B| \cdot |\Sigma_B|$, which means that the size of a Min-Rep instance might be larger than the size of the associated Label Cover instance. The *supergirth* of a Min-Rep graph is just the girth of the supergraph, i.e. the girth of the associated Label Cover instance. As with Label Cover, we let Min-Rep$_k$ be the Min-Rep problem with the additional restriction that the supergirth is larger than k.

Two parameters of PCPs/Label Cover that will be important for us are the *degree* and the *girth*. The degree of a PCP is the maximum degree of a vertex in the supergraph / associated Label Cover instance. Similarly, the girth of a PCP is the girth in the supergraph / associated Label Cover instance (recall that the girth of a graph is the length of the smallest cycle).

1.2   The Basic k-Spanner Problem and Previous Work

The *basic k-spanner problem*, also called the *minimum size k-spanner problem*, is the second main subject in this paper. In this problem we are given an undirected, unweighted graph G and are asked to find the subgraph $G' = (V, E')$ with the minimum number of edges with the property that

$$\frac{dist_{G'}(u, v)}{dist_G(u, v)} \leq k, \text{ for every two vertices } u, v \in V. \tag{1}$$

In the above, the distance between two vertices is just the number of edges in the shortest path between the two vertices, and $dist_H$ is the distance function on a graph H. Any subgraph G' satisfying (1) is called a *k-spanner* of G, and our goal is to find the k-spanner with the fewest edges. Elkin and Peleg [11] proved that if Min-Rep$_{k+1}$ is hard to approximate, then the basic k-spanner problem is also hard to approximate.

The concept of graph spanners was first invented by [19] in a geometric context. To the best of our knowledge the spanner problem on general graphs was first invented indirectly by Peleg and Upfal [22] in their work on small routing tables. This problem was first explicitly defined in [21,20].

Spanners appear in remarkably many applications; the following examples are certainly not exhaustive. Peleg and Upfal [22] showed an application of spanners to maintaining small routing tables. For further applications toward this subject see [24]. In [21] a relation between sparse spanners and synchronizers for distributed networks was found. In [8,14] applications of spanners to parallel, distributed, and streaming algorithms for shortest paths are described. For applications of spanners to distance oracles see [5,25]. For applications of spanners in property testing and related subjects see [7].

There is also quite a bit known about approximating spanners. For the basic k-spanner problem, a seminal paper of Althöfer et al. [1] shows that every undirected graph on n vertices has a $(2k - 1)$-spanner with at most $n^{1+1/k}$ edges (for any integer $k \geq 1$), which immediately implies an $n^{1/\lfloor (k+1)/2 \rfloor}$ approximation to basic k-spanner when $k \geq 3$. For $k = 2$ no nontrivial absolute bounds are possible, but in terms of approximation there is an $O(\log n)$ approximation [18], which is known to be the best possible [17]. There are also many spanner variants that have been studied, such as the directed k-spanner problem [9,6] and the client-server k-spanner problem [12]. Essentially all variants are known to be hard to approximate to better than $2^{\log^{1-\epsilon} n}$ (see [17,13]), leaving the hardness of the basic version a tantalizing question.

1.3   Our Results

All of our results hold for large n, so throughout this paper we will assume that n is sufficiently large. Our first result is on the hardness of approximating Label Cover with large girth:

Theorem 1. *Assuming $NP \not\subseteq BPTIME(2^{polylog(n)})$, for any constant $\epsilon > 0$ and for $3 \leq k \leq \log^{1-2\epsilon} n$ there is no polynomial-time algorithm that approximates Label Cover$_k$ to a factor better than $2^{(\log^{1-\epsilon} n)/k}$.*

We also show how to adapt this hardness from Label Cover to Min-Rep, which then gives us the strong hardness for basic k-spanner that was originally claimed by [11].

Theorem 2. *Assuming $NP \not\subseteq BPTIME(2^{polylog(n)})$, for any constant $\epsilon > 0$ and for $3 \leq k \leq \log^{1-2\epsilon} n$ there is no polynomial-time algorithm that approximates Min-Rep$_k$ to a factor better than $2^{(\log^{1-\epsilon} n)/k}$.*

Theorem 3. *Assuming $NP \not\subseteq BPTIME(2^{polylog(n)})$, for any constant $\epsilon > 0$ and for $3 \leq k \leq \log^{1-2\epsilon} n$ there is no polynomial-time algorithm that approximates the basic k-spanner problem to a factor better than $2^{(\log^{1-\epsilon} n)/k}$.*

1.4   The Error in [11] and Our Techniques

To the best of our knowledge the question answered in Theorem 2 regarding the hardness of Min-Rep with large supergirth was first presented in ICALP 2000 by Elkin and Peleg [11]. In [11] the authors tried to create Min-Rep instances with large supergirth that are also hard to approximate as follows. They started with a 3-Sat(5) instance and associated supergraph, where the supergraph has clauses and variables as vertices, with an edge between a clause and a variable if the variable is in the clause. They then showed how to change the instance to force this graph to have large girth, without losing much in the gap. They then applied the parallel repetition theorem [23] and claimed to boost the hardness while maintaining large supergirth. This reduction is incorrect (as Elkin and Peleg acknowledge in [13]), as non-cycles such as paths in the original graph become simple cycles after parallel repetition is applied. In fact the supergirth in the construction of [11] is 4, no matter what the initial supergirth before parallel repetition is, and thus [11] does not imply any hardness whatsoever for the large supergirth Min-Rep problem. For the interested reader, in the conference version of [11] it is Lemma 13 which is incorrect.

Our main idea is to apply parallel repetition *first*, boosting the gap, and then randomly sample superedges to sparsify the supergraph. It turns out, perhaps surprisingly, that to a certain extent these random choices *do not decrease the gap*. This may seem non-intuitive at first as usually the gap is closely related to superdegree and a smaller superdegree would imply a smaller gap. This may have been one of the obstacles in finding a lower bound for Min-Rep with large supergirth. However, it turns out that it is possible to keep the gap despite the smaller superdegree.

The hardness that we derive this way is actually for Label Cover$_k$ and not for Min-Rep$_k$. The standard reduction from Label Cover to Min-Rep [17] entails duplications of many super vertices. This is needed in order to ensure regularity in the Min-Rep graph, which is used to ensure that removing a μ fraction of the supervertices will imply a removal of at most a μ fraction of the superedges. In [17] this duplication is done after the parallel repetition step, as the supergirth was not an important quantity. However, such duplications add many cycles of length 4 in the supergraph. We handle this difficulty by performing duplication *before* we apply parallel repetition.

Regarding the hardness of basic k-spanner, in [11] a reduction is given from Min-Rep$_{k+1}$ to the basic k-spanner problem for $k \geq 3$. While this reduction is correct, since the hardness proof for large supergirth Min-Rep in [11] is incorrect the reduction does not actually imply any hardness for basic k-spanner.

The actual situation before our paper is as follows. No hardness whatsoever was known for the Min-Rep$_k$ problem; our hardness result comes 12 years after this question was first posed. Regarding lower bounds for the basic k-spanner problem, this question was first raised in [17] in APPROX 1998. The best lower bound known (before our paper) was $\Omega(\log n)/k$. The improved hardness we give comes 14 years after this question was first posed.

2   Sampling Lemma for 2-Query PCPs

We begin with our general 2-query PCP sampling lemma. We remark that similar subsampling techniques have been used before (notably by Goldreich and Sudan [15] to give almost-linear size PCPs), but we specialize the technique with an eye towards giving a tradeoff between the soundness and the girth. Since we will only be concerned with regular PCPs, we will phrase it for the special case when the supergraph has $|A| = |B| = n/2$ and is regular with degree d. We will assume without loss of generality that $|\Sigma_A| \geq |\Sigma_B|$. Given such a PCP verifier (i.e. Label Cover instance) $G = (A, B, E)$, let G_α be the verifier/instance that we get by sub-sampling the edges with probability $p_\alpha = \frac{\alpha \log |\Sigma_A|}{d}$, i.e. every edge $e \in E$ is included in G_α independently with probability p_α.

Lemma 1. *Let $G = (A, B, E)$ be a 2-query PCP verifier/Label Cover instance with completeness 1 and soundness s in which $|A| = |B| = n/2$, every vertex has* degree d, and $|\Sigma_A| \geq |\Sigma_B|$. Let $1 \leq \alpha \leq 1/s$. Then with high probability G_α is a PCP verifier with completeness 1 and soundness at most $4e/\alpha$.

Proof. It is obvious that G_α has completeness 1 with probability 1, since any valid labeling/proof of G is also valid for G_α. To bound the soundness, first fix a proof / labeling. We know that in the original verifier, at most an s fraction of the edges are satisfied. After sampling, the expected number of satisfied edges is at most

$$p_\alpha s |E| \leq \frac{|E| \log |\Sigma_A|}{d} = \frac{n}{2} \log |\Sigma_A|.$$

Since the sampling decisions are independent we can apply a Chernoff bound (see e.g. [10, Theorem 1.1]), giving us that the probability that more than $en \log |\Sigma_A|$ edges are satisfied is at most $2^{-en \log |\Sigma_A|} = |\Sigma_A|^{-en}$. But the total number of possible proofs is at most $|\Sigma_A|^{n/2} |\Sigma_B|^{n/2} \leq |\Sigma_A|^n$. So by a union bound, the probability that any labeling satisfies more than $en \log |\Sigma_A|$ edges is at most $|\Sigma_A|^{-(e-1)n} \leq 2^{-n}$, which is negligible. But the expected total number of edges after sampling is $p_\alpha |E| = \frac{n}{2}\alpha \log |\Sigma_A|$, and so another Chernoff bound implies that with high probability the number of edges after sampling is at least $(n/4)\alpha \log |\Sigma_A|$. Thus with high probability no proof is accepted with probability more than $(en \log |\Sigma_A|)/((n/4)\alpha \log |\Sigma_A|) = 4e/\alpha$. □

Lemma 1 shows that we can sample edges so that the average degree is about $\alpha \log |\Sigma|$ without hurting the soundness too much (in particular, the soundness becomes basically $1/\alpha$). Note that if we set $\alpha = 1/s$ this allows us to reduce the average degree to basically $(1/s) \log |\Sigma_A|$ (a possibly significantly reduction) without affecting the soundness by more than a constant. We would like to claim that this lets us increase the girth, but at this point we will merely prove that any edge is *unlikely* to be in short cycles. Later we will deterministically remove edges that take part in short cycles, but since that might destroy approximate-regularity (which is necessary for our reduction to Min-Rep) we put it off until later.

Lemma 2. *Fix an edge $(u,v) \in G$. Conditioned on $(u,v) \in G_\alpha$, the probability that (u,v) is in a cycle in G_α of length at most k is at most $\frac{2(\alpha \log |\Sigma_A|)^{k-1}}{d}$.*

Proof. Let $4 \leq k' \leq k$ (note that no edge is in a cycle of length less than 4 in any bipartite graph, including G). Fix a cycle of length k' in G that contains (u,v). After conditioning on (u,v) surviving the sampling, the probability that all of the other edges in the cycle are also in G_α is $p_\alpha^{k'-1} = \left(\frac{\alpha \log |\Sigma_A|}{d}\right)^{k'-1}$. On the other hand, we know from the degree bound in G that the number of k'-cycles containing (u,v) is at most $d^{k'-2}$. So a union bound implies that the probability that (u,v) is in a k'-cycle in G_α is at most $\frac{(\alpha \log |\Sigma_A|)^{k'-1}}{d}$. Now we take a union bound over all $4 \leq k' \leq k$ to get that the total probability that (u,v) is in a cycle of length at most k is at most $\sum_{k'=4}^{k} \frac{(\alpha \log |\Sigma_A|)^{k'-1}}{d} \leq \frac{2(\alpha \log |\Sigma_A|)^{k-1}}{d}$ as claimed (assuming without loss of generality that $\alpha \log |\Sigma_A| \geq 2$). □

It is easy to see that subsampling preserves approximate regularity. This is made formal in the next lemma.

Lemma 3. *If $\alpha \geq 16 \log n$ then with probability at least $1 - 2/n$ the degree of every vertex in G_α is between $\frac{1}{2}\alpha \log |\Sigma_A|$ and $2\alpha \log |\Sigma_A|$.*

3 Label Cover and Min-Rep with Large (Super)Girth

In this section we show that Label Cover and Min-Rep are both hard to approximate, even when restricted to instances with large (super)girth. We start with a PCP verifier with supergraph G and Min-Rep graph H, and then use the previously described random sampling technique to get a new supergraph G_α and Min-Rep graph H_α. We now remove from G_α any edge that is in a cycle of length at most k, giving us a new supergraph G'_α and Min-Rep graph H'_α (where an edge $((a, \delta), (b, \beta))$ from H is in H'_α if (a, b) remains as an edge in G'_α). These instances will form our reduction.

We say that an edge $(i, j) \in G_\alpha$ is *bad* if it is not in G'_α, i.e. it is part of a cycle of length of at most k in G_α.

Lemma 4. *Let $16 \log n \leq \alpha \leq \min\{1/s, d^{1/k}/\log |\Sigma_A|\}$. Then with probability larger than $2/3$ the number of bad edges is at most $O\left(\frac{n(\alpha \log |\Sigma_A|)^k}{d}\right) \leq O(n)$*

Proof. Lemma 2 and Markov's inequality imply that with probability at least $3/4$, at most a $\frac{8(\alpha \log |\Sigma_A|)^{k-1}}{d}$ fraction of the edges are bad. With high probability the total number of edges in G_α is $\Theta(n\alpha \log |\Sigma_A|)$, so this means that the number of bad edges is at most $O\left(\frac{n(\alpha \log |\Sigma_A|)^k}{d}\right)$. By our choice of α, this is at most $O(n)$. □

Theorem 4. *If there is no (randomized) polynomial time algorithm that can distinguish between instances of Label Cover in which $|A| = |B| = n/2$ and all vertices have degree d where all edges are satisfiable and instances where at most an $s \leq 1/(16 \log n)$ fraction of the edges are satisfiable, then there is some constant c so that for $16 \log n \leq \alpha \leq \min\{1/s, d^{1/k}/\log |\Sigma_A|\}$ there is no (randomized) polynomial time algorithm that can distinguish between instances of Label Cover$_k$ in which all edges are satisfiable and instances in which at most a c/α fraction of the edges are satisfiable.*

Proof. If there is a labeling that satisfies all edges of G, then clearly the same labeling satisfies all edges of G'_α. On the other hand, suppose that only an s fraction of the edges of G can be satisfied. By Lemma 4, the number of bad edges is at most $O(n)$, so the total number of edges in G'_α is still $\Theta(n\alpha \log |\Sigma_A|)$.

Fix any labeling of G'_α, and suppose that it satisfies a β fraction of the edges of G'_α. Then even if it would have satisfied all of the bad edges, the number of edges of G_α that it satisfies is at most $\beta \cdot \Theta(n\alpha \log |\Sigma_A|) + O(n)$. By Lemma 1 this must be at most $(4e/\alpha) \cdot \Theta(n\alpha \log |\Sigma_A|)$, and thus for some constant c we have that $\beta \leq c/\alpha$.

Thus if we could distinguish between the case when every edge of G'_α can be satisfied and the case when at most a c/α fraction can be satisfied, we could distinguish between the case when every edge of G can be satisfied and the case when at most an s fraction can be satisfied. □

We now reduce to Min-Rep$_k$. We first show how the size of the minimum REP-cover of H_α depends on G. We will then show that, similar to Label Cover, we can deterministically remove small cycles to get the instance H'_α with large supergirth that preserves this gap.

Lemma 5. *Let* $16 \log n \leq \alpha \leq 1/s$. *If there is a valid labeling of G then the Min-Rep instance H_α has a REP-cover of size n (where n is the number of vertices in the supergraph). Otherwise, with high probability the smallest REP-cover has size at least* $\Omega(n\sqrt{\alpha})$.

Proof. If there is a valid labeling of G then by Lemma 1 there is a valid labeling of G_α (since the completeness remains 1), and thus there is a REP-cover of H_α of size n as claimed. On the other hand, suppose that at most an s fraction of the edges of G can be satisfied. Then since the soundness of G_α is at most $4e/\alpha$ by Lemma 1, any labeling of G_α satisfies at most a $4e/\alpha$ fraction of the edges. Suppose that there is a REP-cover of H_α of size less than $n\sqrt{\alpha}/(36\sqrt{3e})$. We will show that this implies that there is a labeling of G_α that satisfies more than a $4e/\alpha$ fraction of the edges, giving a contradiction and proving the lemma.

Suppose that the smallest REP-cover for H_α has size βn. This means that the *average* number of representatives/labels assigned to each vertex in G_α by this cover is β. To analyze the labeling that covers the most edges, we analyze the random labeling obtained by choosing for each vertex a label uniformly at random from the set of labels it is assigned by the REP-cover. Let $A' \subseteq A$ be the set of vertices in A that receive at most 18β labels in this REP-cover, and define $B' \subseteq B$ analogously. Note that $|A'| \geq (8/9)|A|$ and similarly $|B'| \geq (8/9)|B|$, since otherwise the total number of representatives in the REP-cover is larger than $(1/9) \cdot (n/2) \cdot (18\beta) = \beta n$, contradicting our assumption on the size of the REP-cover. With high probability the fraction of edges of G_α that touch a vertex of $A \setminus A'$ is at most $\frac{(1/9) \cdot (2\alpha \log |\Sigma_A|)}{(1/9) \cdot (2\alpha \log |\Sigma_A|) + (8/9)((\alpha \log |\Sigma_A|)/2)} = 1/3$, and similarly for the fraction of edges that touch $B \setminus B'$ (where we used the approximate regularity from Lemma 3). So if we consider the subgraph of G_α induced by $A' \cup B'$ we still have at least $1/3$ of the edges of G_α.

Now let $(a, b) \in A' \times B'$ be such an edge. Since we started with a REP-cover, there is at least one representative assigned to a and one representatives assigned to b that satisfy the relation $\pi_{(a,b)}$. Since both endpoints have at most 18β representatives in the REP-cover, the probability that these two representatives are the assigned labels is at least $1/(18\beta)^2$. Thus by linearity of expectations we expect that at least $1/(3(18\beta)^2) = 1/(972\beta^2)$ fraction of the edges are covered by our random labeling, so there exists some labeling that covers at least that many. If $\beta \leq \frac{\sqrt{\alpha}}{36\sqrt{3e}}$ then this is at least $4e/\alpha$, giving a contradiction. Thus the smallest REP-cover has size at least $(n\sqrt{\alpha})/(36\sqrt{3e})$, proving the lemma. □

We will now get rid of small cycles by using the instance H'_α.

Theorem 5. *If there is no (randomized) polynomial time algorithm that can distinguish between instances of Label Cover in which $|A| = |B| = n/2$ and all vertices have degree d where all edges are satisfiable and instances where at most an $s \le 1/(16 \log n)$ fraction of the edges are satisfiable, then there is some constant c so that for $16 \log n \le \alpha \le \min\{1/s, d^{1/k}/\log|\Sigma_A|\}$ there is no (randomized) polynomial time algorithm that can distinguish between instances of Min-Rep$_k$ where the smallest REP-cover has size at most n and instances where the smallest REP-cover has size at least $n\sqrt{\alpha}/c$ (here n is the size of the supergraph).*

Proof. If there is a labeling that satisfies all edges of G, then clearly choosing that labeling gives a valid REP-cover of H'_α of size at most n. For the other case, suppose that any labeling of G satisfies at most an s fraction of the edges. Then by Lemma 5, with high probability the smallest REP-cover of H_α has size at least $\Omega(n\sqrt{\alpha})$. By Lemma 4, the number of bad edges is at most $O(n)$. Removing any particular edge (in particular a bad edge) can only decrease the size of the optimal REP-cover by at most 2, so if we remove all bad edges (getting H'_α) we are left with an instance with supergirth larger than k in which the smallest REP-cover has size at least $\Omega(n\sqrt{\alpha}) - O(n) = \Omega(n\sqrt{\alpha})$. By construction the supergirth is greater than k, so this proves the theorem. $\square$

Now we define and analyze the PCP / Label Cover instances to which we will apply Theorems 4 and 5. Recall that Max-3SAT(5) is the problem of finding an assignment to variables of a 3-CNF formula that maximizes the number of satisfied clauses, with the additional property that every variable appears in exactly 5 clauses of the formula. We begin with the standard Label Cover instance for Max-3SAT(5) (see for example [2]). The graph (A, B, E) has $|B| = n'$ and $|A| = 5n'/3$ (where n' is the number of variables in the instance), and every vertex in A has degree 3 and every vertex in B has degree 5. Vertices in A correspond to clauses and vertices in B correspond to variables. The alphabet sizes are $|\Sigma_A| = 7$ and $|\Sigma_B| = 2$. The PCP Theorem [3,4] implies that the gap for these instances is constant, i.e. it is hard to distinguish the case when all edges are satisfiable from the case in which $1/(1 + \epsilon)$ fraction of the edges are satisfiable, for some constant ϵ.

Now we take three copies of A, call them A_1, A_2, A_3, and let A' be their union (so $|A'| = 5n'$). Similarly we take five copies of B to get B_i for $i \in [5]$, and take their union to be B'. Now between each A_i and each B_j we put a copy of the original edge set E (which we will call E_{ij}), giving us edge set E'. Note that $|B'| = |A'| = 5n'$ and every vertex has degree 15. Obviously if the original instance has all edges satisfiable then that is still true of this instance. On the other hand, suppose in the original instance at most $1/(1 + \epsilon)$ of the edges are satisfiable. Then fix any labeling of A' and B'. For any i, j this induces some labeling of A_i and B_j, which we know can only satisfy $1/(1 + \epsilon)$ of the edges in E_{ij}. Since this is true for all i, j, the total fraction of satisfied edges is at most $1/(1 + \epsilon)$.

We now apply parallel repetition ℓ times. Now each side has size $(5n')^\ell$, the degree is $d = 15^\ell$, and the alphabet sizes are $|\Sigma_A| = 7^\ell$ and $|\Sigma_B| = 2^\ell$. By the parallel repetition theorem [23], unless $NP \subseteq BPTIME(n^{O(\ell)})$ it is hard to distinguish between the case when all edges are satisfiable and when at most a $2^{-\ell/c}$ fraction are satisfiable for some constant c. We can apply Theorem 4 to this construction to get the following hardness result.

Theorem 6. *Assuming $NP \not\subseteq BPTIME(2^{polylog(n)})$, for any constant $\epsilon > 0$ and $3 \leq k \leq \log^{1-2\epsilon} n$ there is no polynomial time algorithm that can approximate Label Cover$_k$ to a factor better than $2^{(\log^{1-\epsilon} n)/k}$.*

Proof. Set $\ell = \log^{1/\epsilon} n'$, so the size of the Label Cover instance is $n = (5n')^{\log^{1/\epsilon} n'}$ and $\ell^\epsilon = \log n'$. Note that $\log n = \Theta(\ell \log n') = \Theta(\ell^{1+\epsilon})$, so $\ell = \Theta((\log n)^{1/(1+\epsilon)})$. Let $\alpha = \min\{2^{\ell/c}, 15^{\ell/k}/\ell \log 7\}$. Assuming that k is at most $\log^{(1/(1+\epsilon))-\gamma} n$ for some constant $\gamma > 0$ implies that $\alpha \geq 16 \log n$, so applying Theorem 4 to this construction implies that, assuming $NP \not\subseteq BPTIME(n^{O(\ell)})$, there is no polynomial time algorithm that can distinguish between instances of Label Cover$_k$ in which all edges are satisfiable and instances in which at most a c/α fraction are satisfiable (for some constant c). Using a smaller ϵ to change $1/(1 + \epsilon)$ to $1 - \epsilon$, as well as to get rid of lower order terms, gives the theorem.   □

On the other hand, if we apply Theorem 5 to this construction then we get the following theorem:

Theorem 7. *Assuming $NP \not\subseteq BPTIME(2^{polylog(n)})$, for any constant $\epsilon > 0$ and $3 \leq k \leq \log^{1-2\epsilon} n$ there is no polynomial time algorithm that can distinguish between instances of Min-Rep$_k$ that have a REP-cover of size at most $\tilde{n}$ and instances in which the smallest REP-cover has size at least $2^{(\log^{1-\epsilon} n)/k} \cdot \tilde{n}$, where n is the size of the Min-Rep graph and $\tilde{n}$ is the size of the supergraph. Thus there is no polynomial time algorithm that can approximate Min-Rep$_k$ to a factor better than $2^{(\log^{1-\epsilon} n)/k}$.*

Proof. Analogous to Theorem 6.   □

Elkin and Peleg [11, Section 6] showed how to reduce Min-Rep$_{k+1}$ to basic k-spanner while losing only a factor of k. When combined with Theorem 7 it yields the following theorem (the loss of k can be fixed by choosing a smaller constant ϵ).

Theorem 8. *Assuming $NP \not\subseteq BPTIME(2^{polylog(n)})$, for any constant $\epsilon > 0$ and $3 \leq k \leq \log^{1-2\epsilon} n$ there is no polynomial time approximation algorithm for the basic k-spanner problem with ratio less than $2^{(\log^{1-\epsilon} n)/k}$.*

References

1. Althöfer, I., Das, G., Dobkin, D., Joseph, D., Soares, J.: On sparse spanners of weighted graphs. Discrete Comput. Geom. 9(1), 81–100 (1993)
2. Arora, S., Lund, C.: Hardness on Approximation. In: Hochbaum, D. (ed.) Approximation Algorithms for NP-Hard Problems, ch. 10. PWS Publishing (1996)
3. Arora, S., Lund, C., Motwani, R., Sudan, M., Szegedy, M.: Proof verification and the hardness of approximation problems. J. ACM 45(3), 501–555 (1998)
4. Arora, S., Safra, S.: Probabilistic checking of proofs: A new characterization of np. J. ACM 45(1), 70–122 (1998)
5. Baswana, S., Sen, S.: Approximate distance oracles for unweighted graphs in expected $o(n^2)$ time. ACM Transactions on Algorithms 2(4), 557–577 (2006)
6. Berman, P., Bhattacharyya, A., Makarychev, K., Raskhodnikova, S., Yaroslavtsev, G.: Improved Approximation for the Directed Spanner Problem. In: Aceto, L., Henzinger, M., Sgall, J. (eds.) ICALP 2011. LNCS, vol. 6755, pp. 1–12. Springer, Heidelberg (2011)
7. Bhattacharyya, A., Grigorescu, E., Jung, K., Raskhodnikova, S., Woodruff, D.P.: Transitive-closure spanners. In: SODA 2009, pp. 932–941 (2009)
8. Cohen, E.: Polylog-time and near-linear work approximation scheme for undirected shortest paths. J. ACM 47(1), 132–166 (2000)
9. Dinitz, M., Krauthgamer, R.: Directed spanners via flow-based linear programs. In: STOC 2011, pp. 323–332 (2011)
10. Dubhashi, D., Panconesi, A.: Concentration of Measure for the Analysis of Randomized Algorithms. Cambridge University Press, New York (2009)
11. Elkin, M., Peleg, D.: Strong Inapproximability of the Basic k-Spanner Problem. In: Welzl, E., Montanari, U., Rolim, J.D.P. (eds.) ICALP 2000. LNCS, vol. 1853, pp. 636–647. Springer, Heidelberg (2000)
12. Elkin, M., Peleg, D.: Approximating k-spanner problems for $k > 2$. Theor. Comput. Sci. 337(1-3), 249–277 (2005)
13. Elkin, M., Peleg, D.: The hardness of approximating spanner problems. Theory Comput. Syst. 41(4), 691–729 (2007)
14. Feigenbaum, J., Kannan, S., McGregor, A., Suri, S., Zhang, J.: Graph distances in the data-stream model. SIAM J. Comput. 38(5), 1709–1727 (2008)
15. Goldreich, O., Sudan, M.: Locally testable codes and pcps of almost-linear length. J. ACM 53, 558–655 (2006)
16. Khot, S.: On the unique games conjecture. In: FOCS 2005, p. 3 (2005)
17. Kortsarz, G.: On the hardness of approximating spanners. Algorithmica 1444 (1998)
18. Kortsarz, G., Peleg, D.: Generating sparse 2-spanners. Journal of Algorithms 17(2), 222–236 (1994)
19. Levcopoulos, C., Lingas, A.: There are planar graphs almost as good as the complete graphs and almost as cheap as minimum spanning trees. Algorithmica 8(3), 251–256 (1992)
20. Peleg, D., Schaffer, A.: Graph spanners. J. Graph Theory 13, 99–116 (1989)
21. Peleg, D., Ullman, J.D.: An optimal synchronizer for the hypercube. SIAM J. Comput. 18(4), 740–747 (1989)
22. Peleg, D., Upfal, E.: A trade-off between space and efficiency for routing tables. J. ACM 36(3), 510–530 (1989)
23. Raz, R.: A parallel repetition theorem. SIAM Journal on Computing 27(3), 763–803 (1998)
24. Thorup, M., Zwick, U.: Compact routing schemes. In: SPAA 2001, pp. 1–10 (2001)
25. Thorup, M., Zwick, U.: Approximate distance oracles. J. ACM 52(1), 1–24 (2005)

Space-Constrained Interval Selection[*]

Yuval Emek[1], Magnús M. Halldórsson[2,**], and Adi Rosén[3,***]

[1] ETH Zurich, Zurich, Switzerland
emek@tik.ee.ethz.ch
[2] ICE-TCS, School of Computer Science, Reykjavik University, Iceland
mmh@ru.is
[3] CNRS and Université Paris Diderot, France
adiro@lri.fr

Abstract. We study streaming algorithms for the interval selection problem: finding a maximum cardinality subset of disjoint intervals on the line. A deterministic 2-approximation streaming algorithm for this problem is developed, together with an algorithm for the special case of proper intervals, achieving improved approximation ratio of 3/2. We complement these upper bounds by proving that they are essentially best possible in the streaming setting: it is shown that an approximation ratio of $2 - \epsilon$ (or $3/2 - \epsilon$ for proper intervals) cannot be achieved unless the space is linear in the input size. In passing, we also answer an open question of Adler and Azar [1] regarding the space complexity of constant-competitive randomized preemptive online algorithms for the same problem.

1 Introduction

In this paper we consider the *interval selection* problem, namely, finding a maximum cardinality subset of disjoint intervals from a given collection of intervals on the real line. It is well known that this problem has a simple optimal algorithm in the classical setting when the complete set of intervals is given to the algorithm [15]. Here we study this problem in the *streaming* model [17,23], where the input is given to the algorithm as a stream of items (intervals in our case), one at a time, and the algorithm has a limited memory that precludes storing the whole input. Yet, the algorithm is still required to output a feasible solution, with a good approximation ratio.

The motivation for the streaming model stems from applications of managing very large data sets, such as biological data (DNA sequencing), network traffic data, and more. Although some function of the whole data set is to be computed, it is impossible to store the whole input. Depending on the setting, different variants of the streaming model have been considered in the literature, such as the classical streaming model [17] or the so-called *semi-streaming* model

---

[*] Refer to [9] for a full version of this extended abstract.
[**] Research partially supported by grant 90032021 from the Icelandic Research Fund.
[***] Research partially supported by ANR projects QRAC and ALADDIN.

A. Czumaj et al. (Eds.): ICALP 2012, Part I, LNCS 7391, pp. 302–313, 2012.
© Springer-Verlag Berlin Heidelberg 2012

[12]. Common to all of them is the fact that the space used by the streaming algorithm is linear in some natural upper bound on the size of the output it returns (sometimes, a multiplicative polylogarithmic overhead is allowed).

In many problems considered in the streaming literature, the size of the output is fully determined by some parameter of the input, and thus, one would typically express the space complexity as a function of this parameter (cf. [4,13]). However, in other problems, the size of the output cannot be a priori expressed that way as it depends on the given instance; in such settings it is natural to seek a streaming algorithm whose space complexity is not much larger than the output size of the given instance (cf. [16]). Clearly, as long as the computational model of the streaming algorithm is based on a Turing machine with no distinction between the working tape and the output tape, the size of the output is an inherent lower bound on the required space.

In this paper, we consider a setting where the algorithm is given a stream of real-line intervals, each one defined by its two endpoints, and the goal is to compute a maximum cardinality subset of disjoint intervals (or an approximation thereof). This problem finds many applications, e.g., in resource allocation problems, and it has been extensively studied in the online and offline settings in many variants. We seek algorithms with a good upper bound on the space they use for a given instance, expressed in terms of the size of the output for that specific instance. Typically, we seek algorithms that use space which is at most linear in the size of the output and yet guarantee a good approximation ratio.

Related Work. The offline interval selection problem corresponds to finding a maximum independent set in an interval graph. An optimal greedy algorithm was discovered early [15] and has since been a staple of algorithms textbooks [8,18]. It should be noted that the input can be given in (at least) two different ways: as an intersection graph with the nodes corresponding to the intervals, or as a set of intervals given by their endpoints. This distinction makes little difference in the traditional offline setting, where switching between these representations can be done efficiently. However, it can be important in access- or resource-constrained settings. We choose to study the interval selection problem assuming the latter representation — that is, the input is given as a set of intervals — since we believe that it makes more sense in applications related to the online and streaming settings (most previous works on online interval selection make the same assumption).

The study of space-constrained algorithms goes back at least to the 1980 work of Munro and Paterson on selection and sorting [22]. More recently, the streaming model was developed to capture the processing of massive data-sets that arise in practice [23]. Most streaming algorithms deal with the approximate computation of various statistics, or "heavy hitters", as exemplified by the celebrated paper of Alon, Matias, and Szegedy [4].

A number of classic graph theoretic problems have been treated in the streaming setting, for example, matching problems [20,11], diameter and shortest paths [12,13], min-cut [3], and graph spanners [13]. These were mostly studied under the *semi-streaming* model, introduced by Feigenbaum et al. [12]; in this model, the algorithm is allowed to use $n \log^{O(1)}(n)$ space on an n-vertex graph (i.e., $\log^{O(1)}(n)$ bits per vertex). Closest to our problem, the independent set problem in general sparse graphs (and hypergraphs) was studied in the streaming setting by Halldórsson et al. [16]. Geometric streaming algorithms have also been appearing in recent years, especially dealing with extent and ranges, such as [2].

There is a plethora of literature on interval selection in the online setting. Some papers capture the problem as a call admission problem on a linear network, with the objective of maximizing the number (or weight) of accepted calls. Awerbuch et al. [5] present a strongly $\lceil \log N \rceil$-competitive algorithm for the problem, where N is the number of nodes on the line (corresponding to the number of possible interval endpoints). This yields an $O(\log \Delta)$-competitive algorithm for the weighted case, where Δ is the ratio between the longest to the shortest interval. On the negative side, Awerbuch et al. [5] establish a lower bound of $\Omega(\log N)$ on the competitive ratio of randomized non-preemptive online interval selection algorithms. In the context of the real line, this immediately implies that such algorithms cannot have competitive ratio that does not depend on the length of the input. In fact, Bachmann et al. [6] recently showed that the competitive ratio of randomized non-preemptive online algorithms for interval selection on the real line must be linear in the number of intervals in the input. Preemptive online scheduling has a lower bound of $\Omega(\log \Delta / \log \log \Delta)$ in the weighted case [7]. In comparison, much better results are possible for preemptive online algorithms in the unweighted setting: Adler and Azar [1] devise a 16-competitive algorithm. One way of easing the task of the algorithm is to assume arrival by time, i.e., the intervals arrive in order of left endpoints. This has been treated for different weighted problems [19,24,21,14,10].

Our Results. We give tight results for the interval selection problem in the streaming setting. Our main positive result is a deterministic 2-approximation streaming algorithm that uses space linear in the size of the output (Sect. 3). This is complemented by a matching lower bound (Sect. 4), stating that an approximation ratio of $2 - \epsilon$ cannot be obtained by any randomized streaming algorithm with space significantly smaller than the size of the input (which is much larger than the size of the output). The special case of proper interval collections (i.e., collections of intervals with no proper containments) is also considered, for which a deterministic 3/2-approximation streaming algorithm that uses space linear in the output size is presented (deferred to [9]); a matching lower bound on the approximation ratio is established (Sect. 4) for streams of unit intervals (a special case of proper intervals). The upper bounds are extended to *multiple-pass* streaming algorithms: we show that an approximation ratio $1 + 1/(2p - 1)$ can be obtained in p passes over the input (deferred to [9]).

In passing, we also answer an open question posed by Adler and Azar [1] in the context of randomized preemptive online algorithms for the interval selection problem. Adler and Azar point out that the decisions made by their online algorithm depend on the whole history (i.e., the input seen so far) and that natural attempts to remove this dependency seem to fail. Consequently, they write (using the term "active call" for an interval in the solution maintained by the online algorithm) that *"it seems very interesting to find out whether there exist constant-competitive algorithms where each decision depends only on the currently active calls and maybe on additional bounded information".* We answer this question affirmatively by slightly modifying our main algorithm to achieve a randomized preemptive online algorithm that admits constant competitive ratio (slightly improving on that of [1]) and uses space linear in the size of the optimal solution, rather than the size of the input, as the algorithm of Adler and Azar does (deferred to [9]).

2   Preliminaries

We think of the real line $\mathbb{R}$ as stretching from left to right so that an *interval* I contains all points between its left *endpoint* left(I) and its right endpoint right(I), where left$(I) <$ right(I). Each endpoint can be either *open* (exclusive) or *closed* (inclusive). A *half-open* interval has a closed left endpoint and an open right endpoint. (This is, perhaps, the natural interval type to use in most resource allocation applications.) Observe that the assumption that left$(I) <$ right(I) implies that every interval contains an open set (in the topological sense) and that half-open intervals are always well defined.

The interval related notions of *intersection, disjointness,* and *containment* follow the standard view of an interval as a set of points. Two intervals I, J *properly* intersect if they intersect without containment; I properly contains J if I contains J and J does not contain I. An interval collection $\mathcal{I}$ is said to be *proper* (and the intervals in the collection, *proper* intervals) if no two intervals in $\mathcal{I}$ exhibit proper containment. The *load* of $\mathcal{I}$ is defined to be $\max_{p\in\mathbb{R}} |\{I \in \mathcal{I} \mid p \in I\}|$.

The *interval selection* problem asks for a maximum cardinality subset of pairwise disjoint intervals out of a given set S of intervals. In the streaming model, the input interval set S is considered to be an ordered set (a.k.a. a *stream*) and the intervals arrive one by one according to that order. The intervals are specified by their endpoints, where each endpoint is represented by a bit string of length b (the same b for all endpoints). This may potentially provide a streaming algorithm with the edge of knowing in advance some bounds on the number of intervals that will arrive and on the number of intervals that can be placed between two existing intervals. However, our algorithms do not take advantage of this extra information and our lower bounds show that it is essentially useless. An optimal solution to a given instance S of the interval selection problem is denoted by $\mathrm{Opt}(S)$.

We may sometimes talk about *segments*, rather than intervals, when we want to emphasize that the entities under consideration are not part of the input. Given a set $\mathcal{I}$ of intervals, a *component* (or *connected component*) of $\mathcal{I}$ is a maximal continuous segment in $\bigcup_{I\in\mathcal{I}} I$.

3   The Main Algorithm

Overview. Given a stream S of intervals, our algorithm maintains a collection $A \subseteq S$, referred to as the *actual* intervals, from which the output $\mathrm{Alg}(S) =$

$\text{Opt}(A)$ is taken. It also maintains a collection V of *virtual* intervals, where each virtual interval is the intersection of two actual intervals that existed in A at some point. The role of the virtual intervals is to filter out undesired intervals from joining A: an arriving interval $I \in S$ joins A if and only if it does not contain any currently maintained virtual or actual interval.

Our algorithm is designed to guarantee that each interval $I \in S$ leaves a *trace* in either A or V, namely, there exists some $J \in A \cup V$ such that $J \subseteq I$. Moreover, if $I, I' \in A$ properly intersect, then $I \cap I' \in V$. This essentially means that an arriving interval is rejected if and only if it contains some previous interval of S or the intersection of two properly intersecting previous intervals in S that have belonged to A.

Following that, it is not difficult to show that the load of the interval collection A is at most 2. Based on a careful analysis of the structure of the (connected) components in A and the locations of the virtual intervals within these components and between them, we can argue that $|V| \le |A|$. This immediately yields the desired upper bound on the space of our algorithm as $|A| \le 2 \cdot |\text{Opt}(A)|$. The bound on the approximation ratio essentially stems from the observation that $|\text{Opt}(S)| \le |\text{Opt}(A \cup V)|$ (a direct corollary of the fact that each interval in S leaves a trace in $A \cup V$) and from the invariant that each actual interval contains at most 2 virtual intervals.

It is interesting to point out that our algorithm is in fact a deterministic preemptive online algorithm that maintains a load-2 interval collection (the collection A). Since the main result of Adler and Azar [1] also relies on such an algorithm, one may wonder if the two algorithms can be compared. Actually, the algorithm of Adler and Azar bases its rejection (and preemption) decisions on similar conditions: an arriving interval is rejected if and only if it contains some previous interval of S or the intersection of two properly intersecting intervals in A. (Adler and Azar use a different terminology, but the essence is very similar.) The difference lies in the latter condition: whereas the algorithm of Adler and Azar considers only the properly intersecting intervals that are currently in A, our algorithm also (implicitly) considers properly intersecting intervals that belonged to A in the past and were preempted since. This seemingly small difference turns out to be crucial as it facilitates our algorithm to use much less memory, thus giving rise to an interesting phenomena: by remembering extra information (i.e., intersecting intervals that belonged to A in the past and are not in A anymore), we actually end up using less memory.

The Algorithm. Consider a stream $S = (I_1, \ldots, I_n)$ of intervals on the real line. It will be convenient to assume that all endpoints are distinct, i.e., $\{\text{left}(I), \text{right}(I)\} \cap \{\text{left}(J), \text{right}(J)\} = \emptyset$ for every two intervals $I, J \in S$. Unless stated otherwise, we will also assume that the intervals mentioned in this section are closed on both endpoints. These two assumptions are lifted in [9].

Our algorithm, denoted `Alg`, maintains a collection $A \subseteq S$ of *actual* intervals and a collection V of *virtual* intervals, where each virtual interval is realized by endpoints of intervals in S. That is, the virtual interval $I \in V$ satisfies $\{\text{left}(I), \text{right}(I)\} \subseteq \{\text{left}(J), \text{right}(J) \mid J \in S\}$. The algorithm initially sets

$A, V \leftarrow \emptyset$. Then, upon arrival of a new interval $I \in S$, Alg proceeds according to the policy presented in Algorithm 1.

Algorithm 1. The policy of Alg upon arrival of an interval $I \in S$

1: **if** $\exists J \in A \cup V$ s.t. $J \subseteq I$ **then**
2:     reject I and halt
3: $A \leftarrow A \cup \{I\}$
4: **for all** $J \in A$ s.t. $J \supseteq I$ **do**
5:     $A \leftarrow A - \{J\}$
6: **for all** $J \in V$ s.t. $J \supseteq I$ **do**
7:     $V \leftarrow V - \{J\}$
8: **for** $p \in \{\text{left}(I), \text{right}(I)\}$ **do**
9:     **if** $\exists J \in V$ s.t. $p \in J$ **then**
10:        $V \leftarrow V - \{J\} \cup \{I \cap J\}$
11:    **else if** $\exists J \in A$ s.t. $p \in J$ **then**
12:        $V \leftarrow V \cup \{I \cap J\}$
13: **for all** $J \in A$ and $K \in V$ **do**
14:    **if** $\text{left}(J) < \text{left}(K) < \text{right}(K) < \text{right}(J)$ **then**
15:        $A \leftarrow A - \{J\}$

Analysis (sketch). We provide here a sketch of the analysis; the detailed version is deferred to [9]. Throughout, we let $1 \leq t \leq n$ denote the time at which Alg completed processing interval $I_t \in S$; time $t = 0$ denotes the beginning of the execution. We refer to the period between time $t-1$ and time t as *round* t. The stream prefix $(S_1, \ldots, S_t)$ is denoted by S_t. The collections A and V at time t are denoted by A_t and V_t, respectively, although, when t is clear from the context, we may omit the subscript.

Lemma 1 lies at the core of our analysis: it states that each interval in S leaves some trace in either A or V. This will be employed later on to argue that $\text{Alg}(S)$ is not much smaller than $\text{Opt}(S)$.

Lemma 1. *For every interval $I_t \in S$ and for every time $t' \geq t$, there exists some interval $\rho \in A_{t'} \cup V_{t'}$ such that $\rho \subseteq I_t$.*

Lemma 2 — the main lemma regarding the updating phase in lines 8–12 and the resulting structure of the interval collections A and V — states seven invariants maintained by our algorithm; these invariants are proved simultaneously by induction on t, essentially by straightforward analysis of the policy presented in Algorithm 1.

Lemma 2. *For any round $1 \leq t \leq n$, the updating phase satisfies the following two properties:*
(P1) If ρ is added to V in round t, then $\rho \in V_t$.
(P2) If ρ and σ are added to V in round t, then $\rho \cap \sigma = \emptyset$.
Moreover, for any time $0 \leq t \leq n$, the interval collections A and V satisfy the following five properties:

(P3) For every $\rho \in A$ and $\sigma \in V$, if $\rho \cap \sigma \neq \emptyset$, then $\sigma \subset \rho$ with a common endpoint.
(P4) For every $\rho, \sigma \in A$, if $\rho \cap \sigma \neq \emptyset$, then $\rho \cap \sigma \in V$.
(P5) Every point $p \in \mathbb{R}$ is contained in at most 1 virtual interval.
(P6) Every point $p \in \mathbb{R}$ is contained in at most 2 actual intervals.
(P7) There do not exist two actual intervals $\rho, \sigma \in A$ such that $\rho \subseteq \sigma$.

We employ Lemma 2 in order to understand the structure of the components of A and their relations with the intervals in V. To that end, fix some time t and consider an arbitrary component C formed as the union of the actual intervals $\rho_1, \ldots, \rho_k \in A_t$. We denote the leftmost and rightmost points in (the segment) C by left(C) and right(C), respectively. Assume without loss of generality that left(ρ_i) < left(ρ_{i+1}) for every $1 \leq i \leq k - 1$. Lemma 2(P6) and (P7) then guarantees that

$$\text{left}(\rho_{i-1}) < \text{left}(\rho_i) < \text{right}(\rho_{i-1}) < \text{left}(\rho_{i+1}) < \text{right}(\rho_i) < \text{right}(\rho_{i+1})$$

for every $2 \leq i \leq k-1$. By Lemma 2(P4), we conclude that $\rho_i \cap \rho_{i+1} \in V_t$ for every $1 \leq i \leq k-1$, while Lemma 2(P3) implies that the segment [left(ρ_2), right(ρ_{k-1})] does not intersect with any other virtual interval in V_t. The segment C possibly contains two more virtual intervals at time t: an interval $\sigma_\ell \subseteq$ [left(ρ_1), left(ρ_2)) and an interval $\sigma_r \subseteq$ (right(ρ_{k-1}), right(ρ_k)], but then Lemma 2(P3) guarantees that left(σ_ℓ) = left(ρ_1) = left(C) and right(σ_r) = right(ρ_k) = right(C). An illustration of a component is provided in Fig. 1. There may also exist virtual intervals in between the components of A, but Lemma 3, to be stated soon, essentially shows that their number and structure are fairly limited.

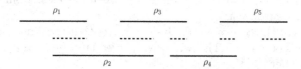

Fig. 1. A component C of A. The solid lines depict the actual interval ρ_i, $i = 1, \ldots, 5$; the dashed lines depict the virtual intervals contained in C.

Let Ψ_t denote the collection of the components of A_t and consider two adjacent components $C_\ell, C_r \in \Psi_t$, where C_ℓ is to the left of C_r. We say that the pair (C_ℓ, C_r) is *solid* at time t if at most one virtual interval in V_t intersects with the segment [right(C_ℓ), left(C_r)]. Lemma 3 states that the pair (C_ℓ, C_r) is always solid.

Lemma 3. *At every time $0 \leq t \leq n$, all pairs of adjacent components in Ψ_t are solid. Moreover, no virtual interval intersects with the segment $(-\infty, \text{left}(C_\ell)]$ nor with the segment $[\text{right}(C_r), +\infty)$, where C_ℓ and C_r are the leftmost and rightmost components in Ψ_t, respectively.*

Lemma 4 is established by combining Lemma 1 and Lemma 3 with a careful accounting of the virtual intervals.

Lemma 4. $|\mathtt{Alg}(S_t)| \geq |\mathtt{Opt}(S_t)|/2$ *at every time* $0 \leq t \leq n$.

Corollary 1. $|\mathtt{Alg}(S)| \geq |\mathtt{Opt}(S)|/2$.

It remains to bound the space of our algorithm, showing that it is linear in the length of the bit string representing $\mathtt{Alg}(S)$. At each time t, the space of $\mathtt{Alg}$ is linear in the length of the bit strings representing A_t and V_t. As $\mathtt{Opt}(S_t)/2 \leq \mathtt{Alg}(S_t) \leq \mathtt{Opt}(S_t)$ for every $0 \leq t \leq n$, and since $\mathtt{Opt}(S_t)$ is non-decreasing with t, it is sufficient to show that $|A_t| + |V_t| = O(|\mathtt{Alg}(S_t)|) = O(|\mathtt{Opt}(A_t)|)$. By Lemma 2(P6), we know that the actual intervals in A_t can be 2-colored such that if two intervals belong to the same color class, then they do not intersect. Thus, $|A_t| \leq 2 \cdot |\mathtt{Opt}(A_t)|$ at every time t. On the other hand, Lemma 3 implies that if we count the actual and virtual intervals by scanning the real line from left to right, then the number of virtual intervals never exceeds that of the actual intervals. Therefore, $|V_t| \leq |A_t|$ which concludes our analysis.

4   Lower Bound(s)

In this section we establish lower bounds on the approximation ratio of randomized streaming algorithms for the interval selection problem, establishing the following two theorems.

Theorem 1 (Lower bound for general intervals). *For every real* $\epsilon > 0$, *integers* $k_0, n_0 > 0$, *and subexponential (respectively, sublinear) function* $s :$ $\mathbb{N} \to \mathbb{N}$, *there exist* $k_0 \leq k \leq c \cdot k_0$, *where* c *is a universal constant,* $n > n_0$, *and an interval stream* S *such that (1)* $|S| = n$; *(2)* $|\mathtt{Opt}(S)| = k$; *and (3)* $\mathtt{Alg}(S) < k(1/2 + \epsilon)$ *for any randomized interval selection streaming algorithm* $\mathtt{Alg}$ *with space* $s(kb)$ *(resp., space* $s(nb)$*), where* b *is the length of the bit strings representing the endpoints of* S.

Theorem 2 (Lower bound for unit intervals). *For every real* $\epsilon > 0$, *integers* $k, n_0 > 0$, *and subexponential (respectively, sublinear) function* $s : \mathbb{N} \to \mathbb{N}$, *there exist* $n > n_0$, *and a unit interval stream* S *such that (1)* $|S| = n$; *(2)* $|\mathtt{Opt}(S)| = k$; *and (3)* $\mathtt{Alg}(S) < k(2/3 + \epsilon)$ *for any randomized proper interval selection streaming algorithm* $\mathtt{Alg}$ *with space* $s(kb)$ *(resp., space* $s(nb)$*), where* b *is the length of the bit strings representing the endpoints of* S.

Our lower bounds are proved by designing a random interval stream S for which every deterministic algorithm performs badly on expectation; the assertion then follows by Yao's principle. (Our construction uses half-open intervals, but this can be easily altered.) Note that under the setting used by our lower bounds, the algorithm is required to output a collection $\mathcal{C}$ of disjoint intervals, and the quality of the solution is then determined to be the cardinality of $\mathcal{C} \cap S$. In other words, the algorithm is allowed to output non-existing intervals (that is, intervals that never arrived in the input), but it will not be credited for them. This, obviously, can only increase the power of the algorithm.

The (k,n)-Gadget. Fix some positive integer m whose role is to bound the space of the algorithm. Our lower bounds rely on the following framework, characterized by the parameters $k, n \in \mathbb{Z}_{>0}$, denoted a (k,n)-*gadget*. Consider an extensive form two-player zero-sum game played between the algorithm (MAX) and the adversary (MIN), depicted by a sequence of k *phases*. Informally, in each phase t, the adversary chooses a permutation $\pi_t \in P_n$, where P_n is the collection of all permutations on n elements, and an index $i_t \in [n]$. The algorithm observes π_t (but not i_t) and produces a *memory image* M_t, i.e., a bit string of length m. The index i_t is handed to the algorithm after the memory image is produced. At the end of the last phase the algorithm tries to *recover* $\pi_t(i_t)$ for $t = 1, \ldots, k$: it outputs some $i_t^* \in [n]$ based on the memory image M_t, index i_t, and all other memory images and indices. For each t such that $i_t^* = \pi_t(i_t)$, the algorithm scores a (positive) point.

More formally, the adversarial strategy is depicted by the choices of the permutations π_t and the indices i_t for $t = 1, \ldots, k$. We commit the adversary to make those choices uniformly at random (so, the adversary reveals its mixed strategy), namely, $\pi_t \in_r P_n$ and $i_t \in_r [n]$ for every t, where all the random choices are independent. The strategy of the algorithm is depicted by the function sequences $\{f_t\}_{t=1}^k$ and $\{g_t\}_{t=1}^k$, where $f_t : P_n \times (\{0,1\}^m \times [n])^{t-1} \to \{0,1\}^m$ and $g_t : \{0,1\}^m \times [n] \times (\{0,1\}^m \times [n])^{k-1} \to [n]$. Let Γ_0 be the empty string and recursively define[1] $\Gamma_t = \Gamma_{t-1} \circ f_t(\pi_t, \Gamma_{t-1}) \circ i_t$. The payoff of the algorithm is the number of indices t, $1 \leq t \leq k$, such that

$$g_t\left(f_t(\pi_t, \Gamma_{t-1}), i_t, \{f_{t'}(\pi_{t'}, \Gamma_{t'-1}), i_{t'}\}_{t' \neq t}\right) = \pi_t(i_t) .$$

In the language of the aforementioned informal description, the role of the function f_t is to produce the memory image M_t based on the permutation π_t and all previous memory images and indices (whose concatenation is given by Γ_{t-1}). The role of the function g_t is to recover $\pi_t(i_t)$ based on the memory image M_t, index i_t, and all other memory images and indices.

Note that the memory images $M_{t'}$ and indices $i_{t'}$, $t' \neq t$, do not contain any information on the permutation π_t on top of that contained in M_t. In particular, the entropy in $\pi_t(i_t)$ given M_t, i_t, and $\{M_{t'}, i_{t'}\}_{t' \neq t}$ is equal to the entropy in $\pi_t(i_t)$ given M_t and i_t. Therefore, it will be convenient to decompose the domain of the function $g_t : \{0,1\}^m \times [n] \times (\{0,1\}^m \times [n])^{k-1} \to [n]$ so that the $(\{0,1\}^m \times [n])^{k-1}$-part determines which function $\hat{g}_t : \{0,1\}^m \times [n] \to [n]$ is chosen, and then this function $\hat{g}_t$ is used to produce i_t^* based on M_t and i_t. Similarly, we decompose the domain of the function $f_t : P_n \times (\{0,1\}^m \times [n])^{t-1} \to \{0,1\}^m$ so that the $(\{0,1\}^m \times [n])^{t-1}$-part determines which function $\hat{f}_t : P_n \to \{0,1\}^m$ is chosen, and then this function $\hat{f}_t$ is used to produce M_t based on π_t.

We now turn to bound the expected payoff of the algorithm as a function of k, m, and n. The key ingredient in this context is the following lemma, which is essentially a well known fact in slightly different settings; a proof is provided in [9] for completeness.

---

[1] We use the notation $u \circ v$ to denote the concatenation of the string u to string v.

Lemma 5. *For every real $\alpha > 0$ and integer $n_0 > 0$, there exists an integer $n > n_0$ such that for every two functions $\hat{f} : P_n \to \{0,1\}^m$ and $\hat{g} : \{0,1\}^m \times [n] \to [n]$, where $m = \alpha n \log n$, we have $\mathbb{P}_{\pi \in_r P_n, i \in_r [n]}(\hat{g}(\hat{f}(\pi), i) = \pi(i)) < 2\alpha$.*

Corollary 2. *For every real $\alpha > 0$ and integers $k, n_0 > 0$, there exists an integer $n > n_0$ such that if $m \leq \alpha n \log n$, then the expected payoff of the algorithm (MAX) player in a (k,n)-gadget is smaller than $2\alpha k$.*

The (n, π)-Stack. We now turn to implement a (k,n)-gadget via a carefully designed interval stream. As a first step, we introduce the (n, π)-*stack* construction. Given an integer $n > 0$ and a permutation $\pi \in P_n$, an (n, π)-stack deployed in the segment $[x, y)$, $x < y$, is a collection of n intervals $J_1, \ldots, J_n$ satisfying: (1) all intervals J_i are half open; (2) all intervals J_i have the same length $\mathrm{right}(J_i) - \mathrm{left}(J_i) = \lambda n$, where $\lambda = \frac{y-x}{2n-1/2}$; and (3) $\mathrm{left}(J_i) = x + \lambda(i-1) + \epsilon \pi(i)$ for every $i \in [n]$, where $\epsilon = \lambda/(2n)$. Note that this deployment ensures that $\mathrm{left}(J_n) < \mathrm{right}(J_1)$, hence the half open segment $[\mathrm{left}(J_n), \mathrm{right}(J_1))$ is contained in J_i for every $i \in [n]$. Moreover, the union of the intervals in the stack does not necessarily cover the whole segment $[x, y)$; it is always contained in $[x, y)$, though. The structure of an (n, π)-stack is illustrated in Fig. 2.

Fig. 2. The relative locations of the intervals in an (n, π)-stack for $n = 4$. The left and right endpoints of interval J_i are located in the segments depicted by the bidirectional arrows whose length is $\lambda/2$. The exact location within this segment is determined by $\pi(i)$. In the construction of the 2-lower bound for general intervals, the bold rectangles correspond to the segments in which the stacks (or auxiliary intervals) identified with the left and right children of the current node are deployed assuming that the good interval is interval J_2 (these segments do not intersect with the segments corresponding to the bidirectional arrows).

The (k,n)-gadget is implemented by introducing k stacks, each corresponding to one phase, and some *auxiliary* intervals; the stack corresponding to phase t is referred to as stack t. The permutation π used in the construction of stack t is π_t. The index i_t will dictate the choice of one *good* interval out of the n intervals in that stack. What exactly makes this interval good will be clarified soon; informally, the algorithm has no incentive to output an interval in a stack unless this interval is good.

The k stacks are used both by the construction of the 2-lower bound for general interval streams and by that of the (3/2)-lower bound for unit intervals. The difference between the two constructions lies in the manner in which these stacks are deployed in the real line, and in the addition of the auxiliary intervals. The details of the 2-lower bound are provided here; those of the (3/2)-lower bound are deferred to [9].

A 2-Lower Bound for General Intervals. The interval stream that realizes the (k, n)-gadget for the 2-lower bound for general intervals is constructed as follows. Assume that $k = 2^\kappa - 1$ for some positive integer κ and consider a perfect binary tree T of depth κ. The k stacks are identified with the internal nodes of T so that stack t precedes stack $t + 1$ in a pre-order traversal of T. (In other words, if stack t is identified with node u and stack t' is identified with a child of u, then $t < t'$.) In addition to the intervals in the stacks, we also introduce $2^\kappa = k + 1$ auxiliary intervals which are identified with the leaves of T; these auxiliary intervals arrive last in the stream. We say that an interval J is *assigned* to node $u \in T$ if J belongs to the stack identified with u or if u is a leaf and J is the auxiliary interval identified with it.

The deployment of the stacks and the auxiliary intervals in $\mathbb{R}$ is performed as follows. Stack 1 (identified with T's root) is deployed in $[0, 1)$. Given the deployment of stack t identified with internal node $u \in T$ in the segment $[x, y)$, we deploy the stacks identified with the left and right children of u in the segments $\sigma_\ell = [x + \lambda(i_t - 3/2), x + \lambda(i_t - 1))$ and $\sigma_r = [x + \lambda(i_t + n - 1/2), x + \lambda(i_t + n))$, respectively, where recall that $\lambda = \frac{y-x}{2n-1/2}$. If the children of u are leaves in T, then we deploy auxiliary intervals in those two segments instead of stacks, that is, one auxiliary interval in σ_ℓ and one in σ_r. Refer to Fig. 2 for an illustration.

The key observation regarding the choice of σ_ℓ and σ_r is that

$$\text{left}(J_{i_t - 1}) \leq \text{left}(\sigma_\ell) < \text{right}(\sigma_\ell) \leq \text{left}(J_{i_t}) \quad \text{and}$$
$$\text{right}(J_{i_t}) \leq \text{left}(\sigma_r) < \text{right}(\sigma_r) \leq \text{right}(J_{i_t + 1}).$$

This implies that: (1) the good interval in the stack identified with node $u \in T$ does not intersect with any interval assigned to a descendant of u in T; and (2) a non-good interval in the stack identified with node $u \in T$ contains every interval assigned to a descendant of either the left child of u or the right child of u in T.

The best response of the algorithm would clearly include all the auxiliary intervals in the output, hence it can include an interval J_i of stack t in the output only if it is the good interval of that stack, namely, $i = i_t$. For that purpose, the algorithm has to recover the exact locations of the endpoints of J_{i_t} that implicitly encode $\pi_t(i_t)$. Observing that the endpoints in this construction can be represented by bit strings of length $\log(n) \cdot \log(k)$, Theorem 1 follows by Corollary 2.

References

1. Adler, R., Azar, Y.: Beating the logarithmic lower bound: Randomized preemptive disjoint paths and call control algorithms. J. Scheduling 6(2), 113–129 (2003)
2. Agarwal, P.K., Sharathkumar, R.: Streaming algorithms for extent problems in high dimensions. In: SODA 2010, pp. 1481–1489 (2010)
3. Ahn, K.J., Guha, S.: Graph Sparsification in the Semi-Streaming Model. In: Albers, S., Marchetti-Spaccamela, A., Matias, Y., Nikoletseas, S., Thomas, W. (eds.) ICALP 2009. LNCS, vol. 5556, pp. 328–338. Springer, Heidelberg (2009)
4. Alon, N., Matias, Y., Szegedy, M.: The space complexity of approximating the frequency moments. J. Comput. Syst. Sci. 58(1), 137–147 (1999)

5. Awerbuch, B., Bartal, Y., Fiat, A., Rosén, A.: Competitive non-preemptive call control. In: SODA 1994, pp. 312–320 (1994)
6. Bachmann, U.T., Halldórsson, M.M., Shachnai, H.: Online Selection of Intervals and t-Intervals. In: Kaplan, H. (ed.) SWAT 2010. LNCS, vol. 6139, pp. 383–394. Springer, Heidelberg (2010)
7. Canetti, R., Irani, S.: Bounding the power of preemption in randomized scheduling. SIAM J. Comput. 27(4), 993–1015 (1998)
8. Cormen, T.H., Leiserson, C.E., Rivest, R.L., Stein, C.: Introduction to Algorithms, 3rd edn. MIT Press and McGraw-Hill (2009)
9. Emek, Y., Halldórsson, M., Rosén, A.: Space-constrained interval selection (2012), http://arxiv.org/abs/1202.4326
10. Epstein, L., Levin, A.: Improved randomized results for the interval selection problem. Theor. Comput. Sci. 411(34-36), 3129–3135 (2010)
11. Epstein, L., Levin, A., Mestre, J., Segev, D.: Improved approximation guarantees for weighted matching in the semi-streaming model. In: STACS 2010, pp. 347–358 (2010)
12. Feigenbaum, J., Kannan, S., McGregor, A., Suri, S., Zhang, J.: On graph problems in a semi-streaming model. Theor. Comput. Sci. 348, 207–216 (2005)
13. Feigenbaum, J., Kannan, S., McGregor, A., Suri, S., Zhang, J.: Graph distances in the data-stream model. SIAM J. Comput. 38(5), 1709–1727 (2008)
14. Fung, S.P.Y., Poon, C.K., Zheng, F.: Improved Randomized Online Scheduling of Unit Length Intervals and Jobs. In: Bampis, E., Skutella, M. (eds.) WAOA 2008. LNCS, vol. 5426, pp. 53–66. Springer, Heidelberg (2009)
15. Gavril, F.: Algorithms for minimum coloring, maximum clique, minimum covering by cliques, and maximum independent set of a chordal graph. SIAM J. Comput. 1(2), 180–187 (1972)
16. Halldórsson, B.V., Halldórsson, M.M., Losievskaja, E., Szegedy, M.: Streaming Algorithms for Independent Sets. In: Abramsky, S., Gavoille, C., Kirchner, C., Meyer auf der Heide, F., Spirakis, P.G. (eds.) ICALP 2010. LNCS, vol. 6198, pp. 641–652. Springer, Heidelberg (2010)
17. Henzinger, M.R., Raghavan, P., Rajagopalan, S.: Computing on data streams. In: AMS-DIMACS Series. Special Issue on Computing on Very Large Datasets (1998)
18. Kleinberg, J., Tardos, E.: Algorithm Design. Addison-Wesley (2005)
19. Lipton, R.J., Tomkins, A.: Online interval scheduling. In: SODA 1994, pp. 302–311 (1994)
20. McGregor, A.: Finding Graph Matchings in Data Streams. In: Chekuri, C., Jansen, K., Rolim, J.D.P., Trevisan, L. (eds.) APPROX 2005 and RANDOM 2005. LNCS, vol. 3624, pp. 170–181. Springer, Heidelberg (2005)
21. Miyazawa, H., Erlebach, T.: An improved randomized on-line algorithm for a weighted interval selection problem. J. of Scheduling 7(4), 293–311 (2004)
22. Munro, J.I., Paterson, M.: Selection and sorting with limited storage. Theor. Comput. Sci. 12, 315–323 (1980)
23. Muthukrishnan, S.: Data streams: Algorithms and applications. Foundations and Trends in Theoretical Computer Science 1(2) (2005)
24. Woeginger, G.J.: On-line scheduling of jobs with fixed start and end times. Theor. Comput. Sci. 130(1), 5–16 (1994)

Polynomial Time Algorithms
for Branching Markov Decision Processes
and Probabilistic Min(Max) Polynomial
Bellman Equations

Kousha Etessami[1], Alistair Stewart[1], and Mihalis Yannakakis[2]

[1] School of Informatics, University of Edinburgh
kousha@inf.ed.ac.uk, stewart.al@gmail.com
[2] Department of Computer Science, Columbia University
mihalis@cs.columbia.edu

Abstract. We show that one can approximate the least fixed point solution for a multivariate system of monotone probabilistic max (min) polynomial equations, in time polynomial in both the encoding size of the system of equations and in $\log(1/\epsilon)$, where $\epsilon > 0$ is the desired additive error bound of the solution. (The model of computation is the standard Turing machine model.)

These equations form the Bellman optimality equations for several important classes of *infinite-state* Markov Decision Processes (MDPs). Thus, as a corollary, we obtain the first polynomial time algorithms for computing to within arbitrary desired precision the *optimal value* vector for several classes of infinite-state MDPs which arise as extensions of classic, and heavily studied, purely stochastic processes. These include both the problem of maximizing and minimizing the *termination (extinction) probability* of multi-type branching MDPs, stochastic context-free MDPs, and 1-exit Recursive MDPs. We also show that we can compute in P-time an ϵ-optimal policy for any given desired $\epsilon > 0$.

1 Introduction

[1] Markov Decision Processes (MDPs) are a fundamental model for stochastic dynamic optimization and optimal control, with applications in many fields. They extend purely stochastic processes (Markov chains) with a controller (an agent) who can partially affect the evolution of the process, and seeks to optimize some objective. For many important classes of MDPs, the task of computing the *optimal value* of the objective, starting at any state of the MDP, can be rephrased as the problem of solving the associated *Bellman optimality equations* for that MDP model. In particular, for finite-state MDPs where, e.g., the objective is to maximize (or minimize) the probability of eventually reaching some target state, the associated Bellman equations are *max-(min-)linear* equations, and we know

---

[1] A full version of this paper is available at arxiv.org/abs/1202.4798. Research partially supported by NSF Grant CCF-1017955.

A. Czumaj et al. (Eds.): ICALP 2012, Part I, LNCS 7391, pp. 314–326, 2012.
© Springer-Verlag Berlin Heidelberg 2012

how to solve such equations in P-time using linear programming (see, e.g., [15]). The same holds for a number of other classes of finite-state MDPs.

In many important settings however, the state space of the processes of interest, both for purely stochastic processes, as well as for controlled ones (MDPs), is not finite, even though the processes can be specified in a finite way. For example, consider *multi-type branching processes* (BPs) [13], a classic probabilistic model with applications in many areas (biology, physics, etc.). A BP models the stochastic evolution of a population of entities of distinct types. In each generation, every entity of each type T produces a set of entities of various types in the next generation according to a given probability distribution on offsprings for the type T. In a *Branching Markov Decision Process* (BMDP) [14, 16], there is a controller who can take actions that affect the probability distribution for the sets of offsprings for each entity of each type. For both BPs and BMDPs, the state space consists of all possible populations, given by the number of entities of the various types, so there are an infinite number of states. From the computational point of view, the usefulness of such infinite-state models hinges on whether their analysis remains tractable.

In recent years there has been a body of research aimed at studying the computational complexity of key analysis problems associated with MDP extensions (and, more general stochastic game extensions) of important classes of finitely-presented but countably *infinite-state* stochastic processes, including controlled extensions of classic multi-type branching processes (i.e., BMDPs), *stochastic context-free grammars*, and discrete-time *quasi-birth-death processes*. In [11] a model called *recursive Markov decision processes* (RMDP) was studied that is in a precise sense more general than all of these; it forms the MDP extension of *recursive Markov chains* [12] (equivalently, *probabilistic pushdown systems* [7]), and can be viewed as the extension of finite-state MDPs with recursion.

A central analysis problem for all of these models, which forms the key to a number of other analyses, is the problem of computing their *optimal termination (extinction) probability*. For example, in the setting of multi-type Branching MDPs (BMDPs), these key quantities are the maximum (minimum) probabilities, over all control strategies (or policies), that starting from a single entity of a given type, the process will eventually reach extinction (i.e., the state where no entities have survived). From these quantities, one can compute the optimum probability for any initial population, as well as other quantities of interest.

One can indeed form Bellman optimality equations for the optimal extinction probabilities of BMDPs, and for a number of related important infinite-state MDP models. However, it turns out that these optimality equations are no longer max/min *linear* but rather are max/min *polynomial* equations ([11]). Specifically, the Bellman equations for BMDPs with the objective of maximizing (or minimizing) extinction probability are multivariate systems of monotone probabilistic max (or min) polynomial equations, which we call **max/minPPSs**, of the form $x_i = P_i(x_1, \ldots, x_n)$, $i = 1, \ldots, n$, where each $P_i(x) \equiv \max_j q_{i,j}(x)$ (respectively $P_i(x) \equiv \min_j q_{i,j}(x)$) is the max (min) over a finite number of probabilistic polynomials, $q_{i,j}(x)$. A *probabilistic polynomial*, $q(x)$, is a multi-variate polynomial where the monomial coefficients and constant term of $q(x)$ are all non-negative and sum to ≤ 1. We write these equations in vector form as $x = P(x)$. Then $P(x)$ defines a mapping $P : [0,1]^n \to [0,1]^n$ that is monotone, and thus (by Tarski's theorem) has a *least fixed point* in $[0,1]^n$. The equations $x = P(x)$, can have more than one solution, but it turns out that the optimal value vector for the corresponding BMDP is precisely the least fixed point (LFP) solution vector $q^* \in [0,1]^n$, i.e., the (coordinate-wise) least non-negative solution ([11]).

Already for pure stochastic multi-type branching processes (BPs), the extinction probabilities may be irrational values. The problem of *deciding* whether the extinction probability of a BP is $\geq p$, for a given probability p is in PSPACE ([12]), and likewise, deciding whether the optimal extinction probability of a BMDP is $\geq p$ is in PSPACE ([11]). These PSPACE upper bounds appeal to decision procedures for the existential theory of reals for solving the associated (max/min)PPS equations. However, already for BPs, it was shown in [12] that this quantitative *decision* problem is already at least as hard as the *square-root sum* problem, as well as a (much) harder and more fundamental problem called *PosSLP*, which captures the power of unit-cost exact rational arithmetic. It is a long-standing open problem whether either of these decision problems is in NP, or even in the polynomial time hierarchy (see [1, 12] for more information on these problems). Thus, such *quantitative decision problems* are unlikely to have P-time algorithms, even in the purely stochastic setting, so we can certainly not expect to find P-time algorithms for the extension of these models to the MDP setting. On the other hand, it was shown in [12] and [11], that for both BPs and BMDPs the *qualitative* decision problem of deciding whether the optimal extinction probability $q_i^* = 0$ or whether $q_i^* = 1$, can be solved in P-time.

Despite decades of theoretical and practical work on computational problems like extinction for multi-type branching processes, and equivalent termination problems stochastic context-free grammars, until recently it was not even known whether one could obtain *any* non-trivial *approximation* of the extinction probability of a purely stochastic multi-type branching processes (BP) in P-time. The extinction probabilities of pure BPs are the LFP of a system of probabilistic polynomial equations (PPS), without max or min. In recent work [9], we provided the first polynomial time algorithm for computing (i.e., approximating) to within any desired additive error $\epsilon > 0$ the LFP of a given PPS, and hence the extinction probabilities of a purely stochastic BP, in time polynomial in both the encoding size of the PPS (or the BP) and in $\log(1/\epsilon)$. The algorithm works in the standard Turing model of computation. Our algorithm was based on an approach using Newton's method that was first introduced and studied in [12]. In [12] the approach was studied for more general systems of *monotone* polynomial equations (MPSs), and it was subsequently further studied in [6].

Note that unlike PPSs and MPSs, min/maxPPSs which define Bellman equations for BMDPs are not differentiable functions (only piecewise differentiable). So it is not even clear how to apply a Newton-type method toward solving them.

In this paper we provide the first polynomial time algorithms for approximating the LFP of both maxPPSs and minPPSs, and thus the first polynomial time algorithm for computing (to within any desired additive error) the optimal value vector for BMDPs with the objective of maximizing or minimizing their extinction probability. Our approach is based on a *generalized Newton's method* (GNM), that extends Newton's method in a natural way to the setting of max/minPPSs, where each iteration requires the computation of the least (greatest) solution of a max- (min-) linear system of equations, both of which we show can be solved using linear programming. Our approach also makes crucial use of the P-time algorithms in [11] for *qualitative* analysis of max/min BMDPs, which allow us to remove variables x_i where the LFP is $q_i^* = 1$ or where $q_i^* = 0$. Our algorithms have the nice feature that they are relatively simple, although the analysis of their correctness and time complexity is rather involved.

We furthermore show that we can compute ϵ-optimal (pure) strategies (policies) for both maxPPSs and minPPSs, for *any* given desired $\epsilon > 0$, in time polynomial in both the encoding size of the max/minPPS and in $\log(1/\epsilon)$. This result is at first glance rather surprising, because there are only a bounded number of distinct pure policies for a max/minPPS, and computing an optimal policy is PosSLP-hard. The proof of this result involves an intricate analysis of bounds on the norms of certain matrices associated with (max/min)PPSs.

Finally, we consider *Branching simple stochastic games* (BSSGs), which are two-player turn-based stochastic games, where one player wants to maximize and the other to minimize the extinction probability. The *value* of these games (which are determined) is characterized by the LFP solution of associated min-maxPPSs which combine both min and max operators (see [11]). We observe that our results easily imply a FNP upper bound for ϵ-approximating the *value* of BSSGs and computing ϵ-optimal strategies for them.

Related Work: We have already mentioned some related results. BMDPs and related processes have been studied previously in both operations research (e.g. [14, 16, 4]) and computer science (e.g. [11, 5, 2]), but no efficient algorithms were known for (approximate) computation of the relevant optimal probabilities and policies; the best known upper bound was PSPACE [11]. In [11] we introduced RMDPs, a recursive extension of MDPs. We showed that for general RMDPs, the problem of computing the optimal termination probabilities, even within any nontrivial approximation, is undecidable. However, for the important class of 1-exit RMDPs (1-RMDP), the optimal probabilities can be expressed by min (or max) PPSs, and in fact the problems of computing (approximately) the LFP of a min/maxPPS and the termination probabilities of a max/min 1-RMDP, or BMDP, are all polynomially equivalent. We furthermore showed in [11] that there are always pure, memoryless optimal policies for both maximizing and minimizing 1-RMDPs (and for the more general turn-based stochastic games).

In [10], 1-RMDPs with different objectives were studied, namely optimizing the total expected reward in a setting with positive rewards. In that setting, things are much simpler: Bellman equations turn out to be max/min-linear, optimal values are rational, and can be computed *exactly* in P-time, using LP.

A work more closely related to this paper is [5] by Esparza, Gawlitza, Kiefer, and Seidl. They studied more general monotone min-maxMPSs, i.e., systems of monotone polynomial equations that include both min and max operators, and they presented two different iterative analogs of Newton's methods for approximating the LFP of a min-maxMPS, $x = P(x)$. Their methods are related to ours, but differ in key respects. Both of their methods use certain piece-wise linear functions to approximate the min-maxMPS in each iteration, which is also what we do to solve each iteration of our generalized Newton's method. However, the precise nature of their piece-wise linearizations, as well as how they solve them, differ in important ways from ours, even when they are applied in the context of maxPPSs or minPPSs. They show, working in the unit-cost *exact* arithmetic model, that using their methods one can compute j "valid bits" of the LFP (i.e., compute the LFP within relative error at most 2^{-j}) in $k_P + c_P \cdot j$ iterations, where k_P and c_P are terms that depend in *some* way on the input system, $x = P(x)$. However, they give no constructive upper bound on k_P, and their upper bounds on c_P are exponential in the number n of variables of $x = P(x)$. Note that MPSs are more difficult: even without min/max operators, it is PosSLP-hard to approximate their LFP within any nontrivial constant additive error $c < 1/2$, even for MPSs arising from Recursive Markov Chains [12]. We note that for MPSs and maxMPSs, computing their LFP can be formulated as a *geometric programming* problem, but this does not yield a P-time algorithm for approximating the LFP.

Proofs and details omitted due to space. See the full version of this paper [8].

2   Definitions and Background

For an n-vector of variables $x = (x_1, \ldots, x_n)$, and a vector $v \in \mathbb{N}^n$, we use the shorthand notation x^v to denote the monomial $x_1^{v_1} \ldots x_n^{v_n}$. Let $\langle \alpha_r \in \mathbb{N}^n \mid r \in R \rangle$ be a multi-set of n-vectors of natural numbers, indexed by the set R. We say that a multi-variate polynomial $P_i(x) = \sum_{r \in R} p_r x^{\alpha_r}$, is **monotone** if $p_r \geq 0$ for all $r \in R$. If in addition, $\sum_{r \in R} p_r \leq 1$, then we call $P_i(x)$ a **probabilistic polynomial**. A probabilistic (respectively, **monotone**) **polynomial system of equations**, $x = P(x)$, which we call a **PPS** (resp., a **MPS**), is a system of n equations, $x_i = P_i(x)$, in n variables $x = (x_1, x_2, \ldots, x_n)$, where $P_i(x)$ is a probabilistic (resp., monotone) polynomial for all i.

A **maximum-minimum probabilistic polynomial system of equations**, $x = P(x)$, called a **max-minPPS**, is a system of n equations in n variables $x = (x_1, x_2, \ldots, x_n)$, where for all $i \in \{1, 2, \ldots, n\}$, either:

- Max-polynomial: $P_i(x) = \max\{q_{i,j}(x) : j \in \{1, \ldots, m_i\}\}$, Or:
- Min-polynomial: $P_i(x) = \min\{q_{i,j}(x) : j \in \{1, \ldots, m_i\}\}$

where each $q_{i,j}(x)$ is a probabilistic polynomial. We call such a system a **maxPPS** (respectively, a **minPPS**) if for every i, $i \in \{1, \ldots, n\}$, $P_i(x)$ is a Max-polynomial (respectively, a Min-polynomial). Note that we can view a PPS in n variables as a maxPPS, or as a minPPS, where $m_i = 1$ for every i. For computational purposes we assume that all the coefficients are rational and that the polynomials are given in sparse form, i.e., by listing only the nonzero terms, with the coefficient and the nonzero exponents of each term given in binary. We let $|P|$ denote the total bit encoding length of a system $x = P(x)$ under this representation.

We use **max/minPPS** to refer to a system of equations that is either a maxPPS or a minPPS. While [11] also considered max-minPPSs which contain *both* max and min equations, our primary focus will be on systems that contain just one or the other. (But we get results about max-minPPSs as a corollary.)

As shown in [11], any max-minPPS, $x = P(x)$, has a **least fixed point** (**LFP**) solution, $q^* \in [0,1]^n$, i.e., $q^* = P(q^*)$ and if $q = P(q)$ for some $q \in [0,1]^n$ then $q^* \leq q$ (coordinate-wise inequality). As observed in [12, 11], q^* may contain irrational values, even for PPSs. This paper gives P-time algorithms for computing q^* to within arbitrary precision for both maxPPSs and minPPSs.

We define a **policy** for a max/minPPS, $x = P(x)$, to be a function $\sigma :$ $\{1, ... n\} \rightarrow \mathbb{N}$ such that $1 \leq \sigma(i) \leq m_i$. A policy σ induces a PPS $x = P_\sigma(x)$ where $(P_\sigma)_i(x) = q_{i,\sigma(i)}$. We use q^*_σ to denote the LFP solution vector for the PPS $x = P_\sigma(x)$. For a maxPPS, $x = P(x)$, a policy σ^* is called **optimal** if for all other policies σ, $q^*_{\sigma^*} \geq q^*_\sigma$. For a minPPS $x = P(x)$ a policy σ^* is optimal if for all other policies σ, $q^*_{\sigma^*} \leq q^*_\sigma$. For a max/minPPS with LFP q^*, a policy σ is ϵ**-optimal** for $\epsilon > 0$ if $\|q^*_\sigma - q^*\|_\infty \leq \epsilon$. A non-trivial fact is that optimal policies always exist and that they actually attain the LFP q^* of the max/minPPS:

Theorem 1 ([11]). *For any max/minPPS, $x = P(x)$, there always exists an optimal policy σ^*, and furthermore $q^* = q^*_{\sigma^*}$.*[2]

As discussed in the introduction, PPSs can be used to capture central probabilities of interest for several basic stochastic models, including Multi-type Branching Processes (BPs), while maxPPSs and minPPSs can be similarly used to capture the central optimum probabilities for corresponding stochastic optimization (MDP) models. In particular, a **Branching Markov Decision Process** (BMDP) consists of a finite set $V = \{T_1, \ldots, T_n\}$ of types, a finite set A_i of actions for each type, and a finite set $R(T_i, a)$ of probabilistic rules for each type T_i and action $a_i \in A_i$. Each rule $r \in R(T_i, a)$ has the form $T_i \xrightarrow{p_r} \alpha_r$, where α_r is a finite multi-set whose elements are in V, $p_r \in (0,1]$ is the probability of the rule, and the sum of probabilities of rules in $R(T_i, a)$ is 1: $\sum_{r \in R(T_i, a)} p_r = 1$.

Intuitively, a BMDP describes the stochastic evolution of entities of given types in the presence of a controller that can influence the evolution. Starting from an initial population (i.e. set of entities of given types) X_0 at time (generation) 0, a sequence of populations $X_1, X_2, \ldots$ is generated, where X_k is obtained from X_{k-1} as follows. First the controller selects for each entity of X_{k-1} an available action for that type of entity; then a rule is chosen independently and simultaneously for every entity of X_{k-1}, probabilistically according to the probabilities of the rules for the type of the entity and the selected action, and the entity is replaced by a new set of entities with the types specified by the right-hand side of the rule. The process is repeated as long as the current population X_k is nonempty, and terminates if and when X_k becomes empty. The objective of the controller is either to minimize the probability of termination (i.e., extinction of the population), in

---

[2] The theorem in [11] is more general, applying to 1-exit Recursive Simple Stochastic Games, and shows that also for max-minPPSs, both the max and the min player have optimal policies that attain the LFP q^*.

which case the process is a minBMDP, or to maximize it, in which case it is a maxBMDP. At each stage, k, the controller is allowed in principle to select the actions for the entities of X_k based on the entire history, may use randomization (a mixed strategy), and may make distinct choices for entities of the same type. However, it turns out that none of these flexibilities increase the controller's power: there is always an optimal pure, memoryless, strategy that always uses the same action for all entities of the same type ([11]).

For each type T_i of a minBMDP (respectively, maxBMDP), let q_i^* be the minimum (resp. maximum) probability of termination if the initial population consists of a single entity of type T_i. From the given minBMDP (maxBMDP) we can construct a minPPS (resp. maxPPS) $x = P(x)$ whose LFP is precisely the vector q^* of optimal termination (extinction) probabilities (see Theorem 20 in the full version of [11]): The min/max polynomial $P_i(x)$ for each type T_i contains one polynomial $q_{i,j}(x)$ for each action $j \in A_i$, with $q_{i,j}(x) = \sum_{r \in R(T_i,j)} p_r x^{\alpha_r}$.

It is convenient to put max/minPPSs in the following simple form. A maxPPS in **simple normal form (SNF)**, $x = P(x)$, is a system of n equations in n variables $x_1, x_2, ...x_n$ where each $P_i(x)$ for $i = 1, 2, ...n$ is in one of three forms:

- Form L: $P_i(x) = a_{i,0} + \sum_{j=1}^n a_{i,j} x_j$, where $a_{i,j} \geq 0$ for all j, and $\sum_{j=0}^n a_{i,j} \leq 1$
- Form Q: $P_i(x) = x_j x_k$ for some j, k
- Form M: $P_i(x) = \max\{x_j, x_k\}$ for some j, k

SNF form for minPPSs is analogous: just replace max with min in "Form M".

For a max/minPPS in SNF form, for simplicity in notation, when we refer to a policy, σ, if $P_i(x)$ has form M, say $P_i(x) \equiv \max\{x_j, x_k\}$, we will often use $\sigma(i) = k$ to mean σ chooses x_k among the two choices x_j and x_k.

Proposition 1. *Every max/minPPS, $x = P(x)$, can be transformed in P-time to an "equivalent" max/minPPS, $y = Q(y)$ in SNF form, such that $|Q| \in O(|P|)$. More precisely, the variables x are a subset of the variables y, the LFP of $x = P(x)$ is the projection of the LFP of $y = Q(y)$, and an optimal policy (respectively, ϵ-optimal policy) for $x = P(x)$ can be obtained in P-time from an optimal (resp., ϵ-optimal) policy of $y = Q(y)$.*

Thus from now on, and for the rest of this paper *we assume, without loss of generality, that all max/minPPSs are in SNF normal form*. We now summarize some of the key prior results on PPSs and max/minPPSs.

Proposition 2 ([11]). *There is a P-time algorithm that, given a minPPS or maxPPS, $x = P(x)$, over n variables, with LFP $q^* \in \mathbb{R}_{\geq 0}^n$, determines for every $i = 1, \ldots, n$ whether $q_i^* = 0$ or $q_i^* = 1$ or $0 < q_i^* < 1$.*

Thus, given a max/minPPS we can find in P-time all the variables x_i such that $q_i^* = 0$ or $q_i^* = 1$, remove them and their corresponding equations $x_i = P_i(x)$, and substitute their values on the RHS of the remaining equations. This yields a new max/minPPS, $x' = P'(x')$, where its LFP solution, q'^*, is $\mathbf{0} < q'^* < \mathbf{1}$, which corresponds to the remaining coordinates of q^*. Thus, it suffices to focus our attention to systems whose LFP is strictly between 0 and 1.

The problem of *deciding* whether a coordinate q_i^* of the LFP is $\geq 1/2$ (or whether $q_i^* \geq r$ for any other given bound $r \in (0,1)$) is at least as hard as the square-root-sum and the PosSLP problems, even for PPS (without the min and max operator) [12], and hence it is highly unlikely that it is in P-time. The problem of *approximating* the LFP of a PPS in P-time was solved recently in [9], by using Newton's method, after elimination of the variables with $0/1$ values.

Definition 1. *For a PPS $x = P(x)$ we use $P'(x)$ to denote the Jacobian matrix of partial derivatives of $P(x)$, i.e., $P'(x)_{i,j} := \frac{\partial P_i(x)}{\partial x_j}$. For a point $x \in \mathbb{R}^n$, if $(I - P'(x))$ is non-singular, then we define one Newton iteration at x via the operator: $\mathcal{N}(x) = x + (I - P'(x))^{-1}(P(x) - x)$. Given a max/minPPS, x=P(x), and a policy σ, we use $\mathcal{N}_\sigma(x)$ to denote the Newton operator of the PPS $x = P_\sigma(x)$; i.e., if $(I - P'_\sigma(x))$ is non-singular at a point $x \in \mathbb{R}^n$, then $\mathcal{N}_\sigma(x) = x + (I - P'_\sigma(x))^{-1}(P_\sigma(x) - x)$.*

Theorem 2 ([9]). *Let $x = P(x)$ be a PPS with rational coefficients in SNF form which has LFP $0 < q^* < 1$. If we conduct iterations of Newton's method as follows: $x^{(0)} := 0$, and for $k \geq 0$: $x^{(k+1)} := \mathcal{N}(x^{(k)})$, then the Newton operator is defined at all steps, and for any $j > 0$, $\|q^* - x^{(j+4|P|)}\|_\infty \leq 2^{-j}$ where $|P|$ is the bit encoding length of the system $x = P(x)$.*

Furthermore, there is an algorithm (based on suitable rounding of Newton iterations) which, given a PPS, $x = P(x)$, and given a positive integer j, computes a rational vector $v \in [0,1]^n$, such that $\|q^* - v\|_\infty \leq 2^{-j}$, and which runs in time polynomial in $|P|$ and j in the standard Turing model of computation.

The proof of this theorem involved various technical lemmas on PPSs and Newton's method, several of which we need to strengthen for this paper. To prove the P-time upper bounds in [9], an inductive step of the following form was used:

Lemma 1 ([9]). *Let $x = P(x)$ be a PPS in SNF with $0 < q^* < 1$. For any $0 \leq x \leq q^*$ and $\lambda > 0$, the operator $\mathcal{N}(x)$ is defined, $\mathcal{N}(x) \leq q^*$, and if $q^* - x \leq \lambda(1 - q^*)$ then $q^* - \mathcal{N}(x) \leq \frac{\lambda}{2}(1 - q^*)$.*

Our goal is to define an iteration $I(x)$ for max/minPPSs that has similar properties to the Newton operator for PPS, i.e., that can be computed efficiently for a given x and for which we can prove properties similar to Lemma 1.

3 Generalizing Newton's Method Using Linear Programming

If a max/minPPS, $x = P(x)$, has no equations of form Q, it amounts to precisely the Bellman equations for a finite-state MDP with the objective of maximizing/minimizing reachability probabilities. It is well known that we can compute the exact (rational) optimal values for such MDPs, and thus the exact LFP, q^*, for such a max(min)-linear systems, using linear programming (see, e.g., [15, 3]).

Computing the LFP of max/minPPSs is clearly a generalization of this problem to the infinite-state setting of branching and recursive MDPs. If we have no equations of form M, we have a PPS, which we can solve in P-time using Newton's method, as shown recently in [9]. An iteration of Newton's method works by approximating the system of equations by a linear system. For a maxPPS(or minPPS), we will define an analogous "approximate" system of equations that we have to solve in each iteration of **"Generalized Newton's Method"** (GNM) which has both linear equations and equations involving the max (or min) function. We will show that we can solve the equations that arise from each iteration of GNM using linear programming. We will then show that a polynomial (in fact, linear) number of iterations are enough to approximate the desired LFP solution, and that it suffices to carry out the computations with polynomial precision.

We begin by expressing the max/min linear equations that should be solved by one iteration of what will become GNM, applied at a point y. Recall that we assume w.l.o.g. throughout that max/minPPSs and PPSs are in SNF.

Definition 2. *For a max/minPPS, $x = P(x)$, with n variables, the **linearization of $P(x)$ at a point y** $\in \mathbb{R}^n$, is a system of max/min linear functions denoted by $P^y(x)$, which has the following form: if $P_i(x)$ has form L or M, then $P_i^y(x) = P_i(x)$, and if $P_i(x)$ has form Q, i.e., $P_i(x) = x_j x_k$ for some j,k, then $P_i^y(x) = y_j x_k + x_j y_k - y_j y_k$.*

Consider a PPS, $x = P_\sigma(x)$, obtained by fixing policy σ for a max/minPPS, $x = P(x)$, and define $P_\sigma^y(x) := (P_\sigma)^y(x)$. Note than the linearization $P^y(x)$ only changes equations of form Q, and fixing policy σ only changes equations of form M, so these operations commute and $P_\sigma^y(x) \equiv (P_\sigma)^y(x) = (P^y)_\sigma(x)$.

Lemma 2. *Let $x = P(x)$ be a PPS. For $y \in \mathbb{R}^n$, let $(P^y)'(x)$ be the Jacobian of $P^y(x)$. Then for any $x \in \mathbb{R}^n$, $(P^y)'(x) = P'(y)$ and $P^y(x) = P(y) + P'(y)(x-y)$.*

An iteration of Newton's method on $x = P_\sigma(x)$ at a point y solves a system of linear equations that can be expressed in terms of $P_\sigma^y(x)$. The next lemma establishes this fact in part *(i)*. Part *(ii)* provides conditions under which we are guaranteed to be doing "at least as well" as one such Newton iteration.

Lemma 3. *Suppose that the matrix inverse $(I - P_\sigma'(y))^{-1}$ exists and is nonnegative, for some policy σ, and some $y \in \mathbb{R}^n$. Then*

(i) $\mathcal{N}_\sigma(y)$ is defined, and it is the unique point $a \in \mathbb{R}^n$ such that $P_\sigma^y(a) = a$.
(ii) For all $x \in \mathbb{R}^n$: if $P_\sigma^y(x) \geq x$ then $x \leq \mathcal{N}_\sigma(y)$; if $P_\sigma^y(x) \leq x$, then $x \geq \mathcal{N}_\sigma(y)$.

We shall now define distinct iteration operators for a maxPPS and a minPPS, both of which we shall refer to with the overloaded notation $I(x)$. These operators will serve as the basis for GNM applied to maxPPSs and minPPSs, respectively.

Definition 3. *For a maxPPS, $x = P(x)$, with LFP q^*, such that $0 < q^* < 1$, and for a real vector y such that $0 \leq y \leq q^*$, we define the operator $I(y)$ to be the unique optimal solution, $a \in \mathbb{R}^n$, to the following mathematical program: Minimize: $\sum_i a_i$; Subject to: $P^y(a) \leq a$.*
For a minPPS, $x = P(x)$, with LFP q^*, such that $0 < q^* < 1$, and for a real vector y such that $0 \leq y \leq q^*$, we define the operator $I(y)$ to be the unique

optimal solution $a \in \mathbb{R}^n$ to the following mathematical program:
Maximize: $\sum_i a_i$;   Subject to: $P^y(a) \geq a$.

A priori, it is not even clear if these "definitions" of $I(y)$ are well-defined. We shall prove the following central claim separately for maxPPSs and minPPSs:

Theorem 3. *Let $x = P(x)$ be a max/minPPS, with LFP q^*, with $0 < q^* < 1$. For any $0 \leq y \leq q^*$:*

1. *$I(y)$ is well-defined, and $I(y) \leq q^*$, and:*
2. *For any $\lambda > 0$, if $q^* - y \leq \lambda(1 - q^*)$ then $q^* - I(y) \leq \frac{\lambda}{2}(1 - q^*)$.*

The next proposition observes that linear programming can be used to compute an iteration of the operator, $I(y)$, for both maxPPSs and minPPSs.

Proposition 3. *Given a max/minPPS, $x = P(x)$, with LFP q^*, and given a rational vector y, $0 \leq y \leq q^*$, the constrained optimization problem (i.e., mathematical program) "defining" $I(y)$ can be described by a LP whose encoding size is polynomial (in fact, linear) in both $|P|$ and the encoding size of the rational vector y. Thus, we can compute the (unique) optimal solution $I(y)$ to such an LP (assuming it exists, and is unique) in P-time.*

GNM for maxPPSs: For a maxPPS, $x = P(x)$, we know by Theorem 1 that there exists an optimal policy, τ, such that $q^* = q^*_\tau \geq q^*_\sigma$ for any policy σ.

Lemma 4. *If $x = P(x)$ is a maxPPS, with LFP solution $0 < q^* < 1$, and y is a real vector with $0 \leq y \leq q^*$, then $x = P^y(x)$ has a least fixed point solution, denoted μP^y, with $\mu P^y \leq q^*$. Furthermore, the operator $I(y)$ is well-defined, $I(y) = \mu P^y \leq q^*$, and for any optimal policy τ, $I(y) = \mu P^y \geq \mathcal{N}_\tau(y)$.*

To prove this lemma, we argue that the LP defining $I(y)$ is: (1) feasible, because q^* is a feasible point, and (2) is bounded from below, in particular by $\mathcal{N}_\tau(y)$ (which we show is defined), where τ is any optimal policy. Hence the LP has an optimal solution. We then argue that any optimal solution a must satisfy all the constraints with equality, $P^y(a) = a$, and that the coordinate-wise minimum of two optimal solution vectors is also optimal. This implies that there is a unique optimal solution, $I(y)$, satisfying $\mathcal{N}_\tau(y) \leq I(y) = \mu P^y \leq q^*$. We then show, using Lemma 1 for P_τ, that part (ii) of Theorem 3 also holds for maxPPSs.

GNM for minPPSs: Our proof for minPPSs is different, because it turns out we can not use the same argument based on LPs to prove that $I(y)$ is well-defined. Fortunately, for minPPSs we can show that $(I - P_\sigma(y))^{-1}$ exists and is non-negative for *any* policy σ, at those points y of interest, and thus $\mathcal{N}_\sigma(y)$ is defined. And we can use this to show that there is *some* policy, σ, such that $I(y)$ is equivalent to an iteration of Newton's method at y after fixing that policy σ:

Lemma 5. *Given a minPPS, $x = P(x)$, with LFP $0 < q^* < 1$, and a vector y with $0 \leq y \leq q^*$, there is some policy σ such that $P^y(\mathcal{N}_\sigma(y)) = \mathcal{N}_\sigma(y)$.*

We establish the existence of such a policy using a policy improvement argument. Note that policy improvement may not run in P-time, and we do not claim it does. We only use policy improvement to prove the existence of such a policy σ.

After identifying the policy σ of Lemma 5, we show that its Newton iterate $\mathcal{N}_\sigma(y)$: (1) is coordinate-wise minimum over all policies, (2) is the unique fixed point of $x = P^y(x)$, and (3) is $\leq q^*$. Using these properties then we argue that the LP that defines $I(y)$ has a unique solution, which is precisely this $\mathcal{N}_\sigma(y)$.

In the maxPPS case, we had an iteration at least as good as iterating with an optimal policy. Here we have an iteration that is at least as bad! Nevertheless, we show it is good enough. For maxPPSs, Theorem 3 (ii) followed by using Lemma 1. For minPPSs, we crucially need a stronger result than Lemma 1.

Lemma 6. *If $x = P(x)$ is a PPS and we are given $x, y \in \mathbb{R}^n$ with $0 \leq x \leq y \leq P(y) \leq 1$, and if the following conditions hold: $\lambda > 0$ and $y - x \leq \lambda(\mathbf{1} - y)$, and $(I - P'(x))^{-1}$ exists and is non-negative, then $y - \mathcal{N}(x) \leq \frac{\lambda}{2}(\mathbf{1} - y)$.*

Applying this to $x = P_\sigma(x)$, x, and $y := q^*$, yields Theorem 3 for minPPSs.

A P-time Algorithm (in the Turing Model) for Max/minPPSs: In [9] we gave a P-time algorithm, in the standard Turing model of computation, for approximating the LFP of a PPS, $x = P(x)$, using Newton's method. The proof in [9] uses induction based on the "halving lemma", Lemma 1. For the base case of the induction, the key property shown in [9] is that if the LFP q^* of a PPS is < 1 then $1 - q_i^* \geq 2^{-4|P|}$ for all i. From this, we easily derive:

Lemma 7. *If $0 < q^* < 1$ is the LFP of a max/minPPS, $x = P(x)$, then $1 - q_i^* \geq 2^{-4|P|}$ for all i.*

Theorem 3 (ii) provides suitable "halving lemmas" for max/minPPSs. Using it, we can now give a P-time algorithm, in the Turing model, for approximating the LFP, q^*, for a max/minPPS, by carrying out iterations of our *Generalized Newton's Method* using the same rounding technique as in [9]. Specifically, first find and remove the variables x_i with value $q_i^* = 0$ or 1, and then use the following algorithm with rounding parameter h:

Start with $x^{(0)} := 0$; For each $k \geq 0$ compute $x^{(k+1)}$ from $x^{(k)}$ as follows:

1. Calculate $I(x^{(k)})$ by solving the following LP:
   Minimize: $\sum_i x_i$; *Subject to:* $P^{x^{(k)}}(x) \leq x$, if $x = P(x)$ is a maxPPS, or:
   Maximize: $\sum_i x_i$; *Subject to:* $P^{x^{(k)}}(x) \geq x$, if $x = P(x)$ is a minPPS.
2. For each coordinate $i = 1, 2, ...n$, set $x_i^{(k+1)}$ to be the maximum (non-negative) multiple of 2^{-h} which is $\leq \max\{0, I(x^{(k)})_i\}$. (In other words, we round $I(x^{(k)})$ down to the nearest 2^{-h} and ensure it is non-negative.)

Theorem 4. *Given any max/minPPS, $x = P(x)$, with LFP $0 < q^* < 1$, if we use the above algorithm with rounding parameter $h = j + 2 + 4|P|$, then the iterations are all defined, and for every $k \geq 0$ we have $0 \leq x^{(k)} \leq q^*$, and furthermore after $h = j + 2 + 4|P|$ iterations we have: $\|q^* - x^{(j+2+4|P|)}\|_\infty \leq 2^{-j}$.*

Corollary 1. *Given any max/minPPS, $x = P(x)$, with LFP q^*, and given any integer $j > 0$, there is an algorithm that computes a rational vector v with $\|q^* - v\|_\infty \le 2^{-j}$, in time polynomial in $|P|$ and j.*

Computing an ϵ-Optimal Policy in P-time: In the full paper ([8]) we show that for max/minPPSs we can compute a ϵ-optimal (pure, memoryless) policy in time polynomial in $|P|$ and $\log(1/\epsilon)$. The proof requires, among other things, an intricate analysis of norm bounds for key matrices associated with max/minPPSs. We again must deal separately with minPPSs and maxPPSs (the latter is harder). We also observe in the full paper that computing an optimal policy is PosSLP-hard, and hence likely not in P-time.

Approximating the Value of BSSGs in FNP: In the full paper ([8]), we observe that an easy corollary of our results is an FNP upper bound on approximating the *value* of 2-player *Branching simple stochastic games* (BSSG), whose values correspond to the LFP of max-minPPSs. See the full version [8].

References

[1] Allender, E., Bürgisser, P., Kjeldgaard-Pedersen, J., Miltersen, P.B.: On the complexity of numerical analysis. SIAM J. Comput. 38(5), 1987–2006 (2009)

[2] Brázdil, T., Brozek, V., Forejt, V., Kucera, A.: Reachability in recursive markov decision processes. Inf. Comput. 206(5), 520–537 (2008)

[3] Courcoubetis, C., Yannakakis, M.: Markov decision processes and regular events. IEEE Trans. on Automatic Control 43(10), 1399–1418 (1998)

[4] Denardo, E., Rothblum, U.: Totally expanding multiplicative systems. Linear Algebra Appl. 406, 142–158 (2005)

[5] Esparza, J., Gawlitza, T., Kiefer, S., Seidl, H.: Approximative Methods for Monotone Systems of Min-Max-Polynomial Equations. In: Aceto, L., et al. (eds.) ICALP 2008, Part I. LNCS, vol. 5125, pp. 698–710. Springer, Heidelberg (2008)

[6] Esparza, J., Kiefer, S., Luttenberger, M.: Computing the least fixed point of positive polynomial systems. SIAM J. on Computing 39(6), 2282–2355 (2010)

[7] Esparza, J., Kučera, A., Mayr, R.: Model checking probabilistic pushdown automata. Logical Methods in Computer Science 2(1), 1–31 (2006)

[8] Etessami, K., Stewart, A., Yannakakis, M.: Polynomial Time Algorithms for Branching Markov Decision Processes and Probabilistic Min(Max) Polynomial Bellman Equations. Preprint of the full version of this paper on ArXiv: 1202.4798

[9] Etessami, K., Stewart, A., Yannakakis, M.: Polynomial-time algorithms for multitype branching processes and stochastic context-free grammars. In: Proc. 44th ACM STOC 2012 (2012), full preprint on ArXiv:1201.2374

[10] Etessami, K., Wojtczak, D., Yannakakis, M.: Recursive Stochastic Games with Positive Rewards. In: Aceto, L., et al. (eds.) ICALP 2008, Part I. LNCS, vol. 5125, pp. 711–723. Springer, Heidelberg (2008)

[11] Etessami, K., Yannakakis, M.: Recursive Markov Decision Processes and Recursive Stochastic Games. In: Caires, L., et al. (eds.) ICALP 2005. LNCS, vol. 3580, pp. 891–903. Springer, Heidelberg (2005), see full version at, http://homepages.inf.ed.ac.uk/kousha/j_sub_rmdp_rssg.pdf; which includes also the results of our paper: Etessami, K., Yannakakis, M.: Efficient Qualitative Analysis of Classes of Recursive Markov Decision Processes and Simple Stochastic Games. In: Durand, B., Thomas, W. (eds.) STACS 2006. LNCS, vol. 3884, pp. 634–645. Springer, Heidelberg (2006)

[12] Etessami, K., Yannakakis, M.: Recursive Markov chains, stochastic grammars, and monotone systems of nonlinear equations. Journal of the ACM 56(1) (2009)

[13] Harris, T.E.: The Theory of Branching Processes. Springer (1963)

[14] Pliska, S.: Optimization of multitype branching processes. Management Sci. 23(2), 117–124 (1976/1977)

[15] Puterman, M.L.: Markov Decision Processes. Wiley (1994)

[16] Rothblum, U., Whittle, P.: Growth optimality for branching Markov decision chains. Math. Oper. Res. 7(4), 582–601 (1982)

Succinct Indices for Range Queries
with Applications to Orthogonal Range Maxima[*]

Arash Farzan[1], J. Ian Munro[2], and Rajeev Raman[3]

[1] Max-Planck-Institut für Informatik, Saarbücken, Germany
[2] University of Waterloo, Canada
[3] University of Leicester, UK

Abstract. We consider the problem of preprocessing N points in 2D, each endowed with a priority, to answer the following queries: given a axis-parallel rectangle, determine the point with the largest priority in the rectangle. Using the ideas of the *effective entropy* of range maxima queries and *succinct indices* for range maxima queries, we obtain a structure that uses $O(N)$ words and answers the above query in $O(\lg N \lg \lg N)$ time. This a direct improvement of Chazelle's result from 1985 [10] for this problem – Chazelle required $O(N/\epsilon)$ words to answer queries in $O((\lg N)^{1+\epsilon})$ time for any constant $\epsilon > 0$.

1 Introduction

Range searching is one of the most fundamental problems in computer science with important applications in areas such as computational geometry, databases and string processing. The input is a set of N points in general position in $\mathbb{R}^d$ (we focus on the case $d = 2$), where each point is associated with *satellite* data, and an aggregation function defined on the satellite data. We wish to preprocess the input to answer queries of the following form efficiently: given any 2D axis-aligned rectangle R, return the value of the aggregation function on the satellite data of all points in R. Researchers have considered range searching with respect to diverse aggregation functions such as emptiness checking, counting, reporting, minimum/maximum, etc. [10,12,17]. In this paper, we consider the problem of *range maximum* searching (the minimum variant is symmetric), where the satellite data associated with each point is a numerical *priority*, and the aggregation function is "arg max", i.e., we want to report the point with the maximum priority in the given query rectangle. This aggregation function is *the* canonical one to study, among those that do not admit inverses [10].

Our primary concern is to obtain *linear-space* data structures, namely those that occupy $O(N)$ words, and we seek to minimize query time subject to this constraint. The space usage is a fundamental concern in geometric data structures due to very large data volumes; indeed, space usage is a main reason why range searching data structures like quadtrees, which have poor worst-case query performance, are preferred in many practical applications over data structures such

---

[*] Work done while Farzan was employed by, and Raman was visiting, MPI.

A. Czumaj et al. (Eds.): ICALP 2012, Part I, LNCS 7391, pp. 327–338, 2012.
© Springer-Verlag Berlin Heidelberg 2012

Table 1. Space/time tradeoffs for 2D range maximum searching in the word RAM

Citation	Size (in words)	Query time
Chazelle'88 [10]	$O(N \lg^\epsilon N)$	$O(\lg N)$
Chan et al.'10 [8]	$O(N \lg^\epsilon N)$	$O(\lg \lg N)$
Karpinski et al.'09 [15]	$O\left(N(\lg \lg N)^{O(1)}\right)$	$O((\lg \lg N)^2)$
Chazelle'88 [10]	$O(N \lg \lg N)$	$O(\lg N \lg \lg N)$
Chazelle'88 [10]	$O(\frac{1}{\epsilon}N)$	$O(\lg^{1+\epsilon} N)$
NEW	$O(N)$	$O(\lg N \lg \lg N)$

as range trees, which have asymptotically optimal query performance. Space efficient solutions to range searching date to the work of Chazelle [10] over a quarter century ago, and Nekrich [18] gives a nice survey of much of this work. Recently there has been a flurry of activity on various aspects of space-efficient range reporting, and for some aggregation functions there has even been attention given to the constant term within the space usage [5,18].

We now formalize the problem studied by our paper, as well as those of [10,8,15]. We assume input points are in *rank space*: the x-coordinates of the n points are $\{0,\ldots,N-1\} = [N]$, and the y-coordinates are given by a permutation $\upsilon : [N] \to [N]$, such that the points are $(i, \upsilon(i))$ for $i = 0,\ldots,N-1$. The priorities of the points are given by another permutation π such that $\pi(i)$ is the priority of the point $(i, \upsilon(i))$. The reduction to rank space can be performed in $O(\lg N)$ time with a linear space structure even if the original and query points are points in $\mathbb{R}^2$ [12,10]. The query rectangle is specifed by two points from $[N] \times [N]$ and includes the boundaries (see Fig. 1(R)). Analogous to previous work, we also assume the word-RAM model with word size $\Theta(\lg N)$ bits[1].

Range maximum searching is a well-studied problem (Table 1). Chazelle [10] gave a few space/time tradeoffs covering a broad spectrum. To the best of our knowledge, the solution with the lowest query time that uses only $O(N)$ words is that of Chazelle [10], who gave a data structure of size $O(\frac{1}{\epsilon}N)$ words with query time $O(\lg^{1+\epsilon} N)$ for any fixed $\epsilon > 0$. More recent results on the range maximum problem are as follows. Karpinski *et al.* [15] studied the problem of 3D five-sided range emptiness queries which is closely related to range maximum searching in 2D. As observed in [8], their solution yields a query time of $(\lg \lg N)^{O(1)}$ with an index of size $N(\lg \lg N)^{O(1)}$ words. Chan *et al.* [8] currently give the best query time of $O(\lg \lg N)$, but this is at the expense of using $O(N \lg^\epsilon N)$ words, for any fixed $\epsilon > 0$. However, there has been no improvement in the running time for linear-space data structures. In this paper, we improve Chazelle's long-standing result by giving a data structure of $O(N)$ words and reducing the query time from polylogarithmic to "almost" logarithmic, namely, $O(\lg N \lg \lg N)$.

Although our primary focus is on 4-sided queries, which specify a rectangle that is bounded from all sides, we also need to consider 2-sided and 3-sided queries, which are "open" on two and one side respectively (thus a 2-sided query

---

[1] $\lg x = \log_2 x$.

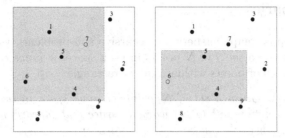

Fig. 1. 2-sided and 4-sided range maximum queries. The numbers with the points represent their priorities, and the unshaded points are the answers.

is specified by a single point (i, j)—see Fig. 1(L)—and a 3-sided query by two points (i, j) and (k, l) where either $i = k$ or $j = l$. Our solution recursively divides the points into horizontal and vertical slabs, and a query rectangle is decomposed into smaller 2-sided, 3-sided, and 4-sided queries. A key intermediate result is the data structure for 2-sided queries. The 2-sided sub-problems are partitioned into smaller sub-problems, which are stored in a "compressed" format that is "decompressed" at query time. The "compression" uses the idea that to answer 2-sided range maxima queries on a problem of size m, one need not store the entire total order of priorities using $\Theta(m \lg m)$ bits: $O(m)$ bits suffice, i.e., the *effective entropy* [13] of 2-sided queries is low. This does not help immediately, since $\Theta(m \lg m)$ bits are needed to store the coordinates of the points comprising these sub-problems. To overcome this bottleneck, the data structures for these sub-problems are *succinct indices* [2]: they do not store the coordinates, and instead obtain them from a global data structure during "decompression". We solve 3-sided and 4-sided subqueries by recursion, or by using succinct indices for range maximum queries on matrices [20,6]. When recursing, we cannot afford the space required to store the structures for rank space reduction for each such subproblem: a further key idea is to use a single global structure to achieve this.

By reusing ideas from this 2-sided result, we obtain two stand-alone results on succinct indices for 2-sided range maxima queries. We show that given N points in rank space, together with priorities, it is possible to answer 2-sided range maxima queries in just $O(N)$ additional bits (i.e excluding point coordinate information) in $(\lg N)^{O(1)}$ time, assuming the index can access point coordinates via an orthogonal range reporting [12,8] queries. This result has been recently used in a context where the input is a low-discrepancy point set whose coordinates need not be stored at all, and can be generated "on the fly". A different index for the *permuted-point* model of Bose et al. [4] uses $O(N)$ additional bits and answers 2-sided range maxima queries in only $O(\lg \lg N)$ time.

The paper is organized as follows. We first describe some building blocks used in Section 2. Section 3 is devoted to our main result, and Section 4 describes the succinct index results. Many proofs have been omitted from this extended abstract and may be found in [11].

2 Preliminaries

In order to support mapping between recursive sub-problems, we use the following primitives on a set S of N points in rank space. A *range counting* query reports the *number* of points within a query rectangle:

Lemma 1 ([14]). *Given a set of N points in rank space in two dimensions, there is a data structure with $O(N)$ words of space that supports range counting queries in $O(\lg N / \lg \lg N)$ time.*

A *range reporting* structure supports the operation of listing the coordinates of all points within a query rectangle. We use the following consequence of a result of Chan *et al.* [8]:

Lemma 2 ([8]). *Given a set of N points in rank space in two dimensions, there is a data structure with $O(N)$ words of space that supports range reporting queries in $O\left((1+k)\lg^{1/3} N\right)$ time where k is the number of points reported.*

The *range selection* problem is as follows: given an input array A of size N, to preprocess it so that given a query (i, j, k), with $1 \le i \le j \le N$, we return an index i_1 such that $A[i_1]$ is the k-th smallest of the elements in the subarray $A[i], A[i+1], \ldots, A[j]$.

Lemma 3 ([7]). *Given an array of size N, there is a data structure with $O(N)$ words of space that supports range selection queries in $O(\lg N / \lg \lg N)$ time.*

3 The Data Structure

In this section we show our main result:

Theorem 1. *Given N points in two-dimensional rank space, and their priorities, there is a data structure that occupies $O(N)$ words of space and answers range maximum queries in $O(\lg N \lg \lg N)$ time.*

We first give an overview of the data structure. We begin by storing all the points (using their input coordinates) once each in the structures of Lemmas 1 and 2. We also store an instance of the data structure of Lemma 3 once each for the arrays X and Y, where $X[i] = \nu(i)$ and $Y[i] = \nu^{-1}(i)$ for $i \in N$ (X stores the y-coordinates of the points in order of increasing x-coordinate, and Y the x-coordinates in order of increasing y-coordinate). These four "global" data structures use $O(N)$ words of space in all.

We recursively decompose the problem à la Afshani *et al.* [1]. Let n be the recursive problem size (initially $n = N$). Given a problem of size n, we divide the problem into n/k mutually disjoint horizontal *slabs* of size n by k, and n/k mutually disjoint vertical slabs of size k by n. A horizontal and vertical slab intersect in a *square* of size $k \times k$. We recurse on each horizontal or vertical slab: observe that each horizontal or vertical slab has exactly k points in it, and is treated as a problem of size k—i.e. it is logically comprised of two permutations

v and π on $[k]$ (Fig. 2(L); Sec. 3.1). Given a slab in a problem of size n containing k points, we need to map coordinates in the slab (which in one dimension will be from $[n]$) down to $[k] \times [k]$ in order to view the slab as a problem of size k— and back again. This mapping is *not* explicitly stored, and is achieved through *slab-rank* and *slab-select* operations (Sec. 3.2).

The given query rectangle is decomposed into a number of disjoint recursive 2-sided, 3-sided and 4-sided queries. In addition to queries that reach slabs at the bottom of the recursion, many other queries do not generate further recursive problems. Such queries are called *terminal*, and the problems (or data structures) that involve answering terminal queries are also called terminal. Each terminal query produces some *candidate* points: the set of all candidate points must contain the final answer. To achieve the space bound, we require that all terminal problems of size n—except those at the bottom of the recursion—use space $O(n\sqrt{\lg n})$ bits (Sec. 3.1). Terminal 3- and 4-sided problems are handled by the results of [20,6]. Terminal 2-sided problems reduce the range maximum query to planar point location, but the space bound precludes an explicit representation. Instead, the data structures are *succinct indices*—the points that comprise them are accessed by means of queries to a single global data structure. Using the key insight that $O(n)$ bits suffice to encode the priority information needed to answer 2-sided queries in a problem of size n (Sec. 3.3), we store parts of the planar sub-division in the recursive problems in a compressed form, relevant parts of which are recomputed at query time (Sec. 3.4).

3.1   A Recursive Formulation and Its Space Usage

The recursive structure is as follows. Let $L = \lg N$, and consider a recursive problem of size n (at the top level $n = N$). We assume wlog that N and n are powers of 2, as are a number of expressions which represent the size of recursive problems (if not, replace real-valued parameters $x \geq 1$ by $2^{\lfloor \lg x \rfloor}$ or $2^{\lceil \lg x \rceil}$ without affecting the asymptotic complexity). Unless we have reached the bottom of the recursion, we partition the input range $[n] \times [n]$ into mutually disjoint vertical slabs of width $k = \sqrt{nL}$ and also into mutually disjoint horizontal slabs of height $k = \sqrt{nL}$, which intersect in $O(n/L)$ $k \times k$ squares. We need to answer 2-sided, 3-sided or 4-sided queries on this problem, which we do as follows (see Fig. 2(R)):

- A 2-sided query is terminal and generates one candidate.
- A 3-sided query results in at most one recursive 3-sided query on a slab, plus up to three terminal problems, each generating one candidate: at most two 2-sided queries on slabs, and at most one *square-aligned* 3-sided query (a square-aligned query exactly covers a rectangular sub-array of squares).
- A 4-sided query either results in a recursive 4-sided query on a slab or results at most one square-aligned 4-sided query (generating one candidate), plus up to four recursive 3-sided queries in slabs.

Since each 3-sided query only generates one recursive 3-sided query, $O(r) = O(\lg \lg N)$ recursive problems are solved generating $O(r)$ candidates.

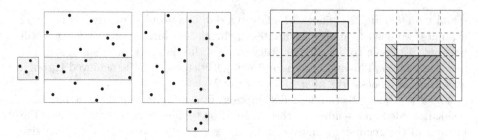

Fig. 2. The recursive decomposition of the input (L) and queries (R). In (R) shaded problems are terminal problems.

The data structures associated with the current recursive problem are:

- for problems at the bottom of the recursion, we store an instance of Chazelle's data structure which uses $O(n \lg n)$ bits of space and has query time $O((\lg n)^2)$.
- For 2-sided queries (which are terminal) in non-terminal problems we use the data structure with space usage $O(n\sqrt{\lg n})$ bits described in Sec. 3.4.
- For 3- and 4-sided square-aligned queries, we store an $n/k \times n/k$ matrix that contains the (top-level) coordinates and priority of the maximum point (if any) in each square. This matrix uses $O(n/L \cdot L) = O(n)$ bits. We then use the data structure of [20,6] for answering 2D range maximum queries on the elements in the above matrix; this also uses $O(n)$ bits.

Finally, each recursive problem has $O(\lg N) = O(L)$ bits of "header" information, containing, e.g., the bounding box of the problem in the top-level coordinate system. Ignoring the header information, the space usage is given by:

$$S(n) = 2\sqrt{n/L}S(\sqrt{nL}) + O(n\sqrt{\lg n}),$$

which after r levels of recursion becomes:

$$S(N) = 2^r \frac{N^{1-1/2^r}}{L^{1-1/2^r}} S(N^{1/2^r} L^{1-1/2^r}) + O\left(2^r N \sqrt{\lg(N^{1/2^r} L^{1-1/2^r})}\right).$$

The recursion is terminated for the first level r where $2^r \geq \lg N/\lg\lg N$. At this level, the problems are of size $O((\lg N)^2)$ and $\Omega((\lg N)^{1.5})$ and the second term in the space usage becomes $O(N \lg N)$ bits. Applying $S(n) = O(n \lg n)$ for the base case, we see that the first term is $O((\lg N/\lg\lg N) \cdot N \cdot \lg\lg N) = O(N \lg N)$ bits, and the space used by the header information is indeed negligible.

3.2   The Slab-Rank and Slab-Select Problems

The input to each recursive problem of size n is given in local coordinates (i.e. from $[n] \times [n]$). Upon decomposing the query to this problem, we need to solve the following *slab-rank* problem (with a symmetric variant for vertical slabs):

Given a point $p = (i^*, j^*)$ in top-level coordinates, which is mapped to (i, j) in a recursive problem of size n, such that (i, j) that lies in a horizontal slab of size $n \times k$, map (i, j) to the appropriate position (i', j') in the size k problem represented by this slab.

We formalize the "inverse" *slab-select* problem as follows:

Given a rectangle R in the coordinate system of a recursive problem, return the top-level coordinates of all points that lie within R.

The following lemma assumes and builds upon the four "global" data structures mentioned after the statement to Theorem 1 (proof omitted):

Lemma 4. *The slab-rank problem can be solved in $O(\lg N / \lg \lg N)$ time, and the slab-select problem in $O(\lg N / \lg \lg N)$ time as well, provided that R contains at most $O(\sqrt{\lg N})$ points.*

3.3 Encoding 2-Sided Queries

We now show that to answer 2-sided range maxima queries at point q ($RMQ(q)$ hereafter), a linear number of bits suffice to encode priority information:

Lemma 5. *Given a set S of n points from $\mathbb{R}^2$ and relative priorities given as a permutation π on $[n]$, the query $RMQ(q)$ can be reduced to point location of q in a collection of at most n horizontal semi-open line segments, whose left endpoints are points from S, and whose right endpoints have x-coordinate equal to the x-coordinate of some point from S. Further, given at most $3n$ bits of extra information, the collection of line segments can be reconstructed from S.*

Proof. Assume the points are in general position and that the 2-sided query is open to the top and left. Associate each point $p = (x(p), y(p)) \in S$ with a horizontal semi-open *line of influence*, possibly of length zero, whose left endpoint (included in the line) is p itself, and is denoted by $Inf(p)$, and contains all points q such that $y(q) = y(p)$, $x(q) \geq x(p)$ and $RMQ(q) = p$. It can be seen that (see e.g. [16]) the answer to $RMQ(q)$ for any $q \in \mathbb{R}^2$ can be obtained by shooting a vertical ray upward from q until the first line $Inf(p)$ is encountered; the answer to $RMQ(q)$ is then p (if no line is encountered then there is no point in the 2-sided region specified by q). See Fig. 3 for an example.

The set $Inf(S) = \{Inf(p) | p \in S\}$ can be computed by sweeping a vertical line from left to right. At any given position $x = t$ of the sweep line, the sweep line will intersect $Inf(S')$ for some set S' (initially $S' = \emptyset$). If $S' = p_{i_1}, \ldots, p_{i_r}$ such that $y(p_{i_1}) > \ldots > y(p_{i_r})$ then it follows that $\pi(i_1) < \ldots < \pi(i_r)$ (the current lines of influence taken from top to bottom represent points with increasing priorities). Upon reaching the next point p_s such that $y(p_{i_j}) < y(p_s) < y(p_{i_{j+1}})$, either (i) $\pi(s) < \pi(i_j)$—in this case $Inf(p_s)$ is empty—or (ii) $\pi(k) > \pi(i_j)$. In the latter case, it may be that $\pi(k) > \pi(i_{j+1}), \ldots \pi(i_{j+k})$ for some $k \geq 0$, which would mean that $Inf(p_{i_{j+1}}), \ldots, Inf(p_{i_{j+k}})$ are terminated, with their right endpoints being $x(p_s)$. To construct $Inf(S)$, therefore, only $O(n)$ bits of information are

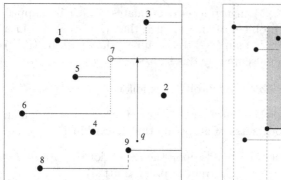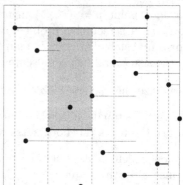

Fig. 3. (Left) Example for Lemma 5. The horizontal lines are the lines of influence. Vertical dotted lines show where a point has terminated the line of influence of another point. The arrow shows how point location in the lines of influence answers the 2-sided query with lower right hand corner at q, returning the point with priority 7. (Right) We select the thick lines of influence, and create a vertical decomposition of all lines of influence shown by vertical dashed lines. One of the regions obtained is highlighted.

needed: for each point, one bit is needed to indicate whether case (i) or (ii) holds, and in the latter case, the value of k needs to be stored. However, k can be stored in unary using $k + 1$ bits, and the total value of k, over the course of the entire sweep, is at most n, giving a total of at most $3n$ bits[2]. $\Box$

3.4   Data Structures for 2-Sided Queries

In this section we show the following lemma. The given 2-sided problem of size n is viewed as a point location problem among $O(n)$ horizontal line segments as in Lemma 5. As the space available is only $O(n\sqrt{\lg n})$ bits, we use an implicit representation of the problem, using and building upon the four "global" data structures mentioned after the statement to Theorem 1.

Lemma 6. *Given a recursive sub-problem of size n, we can answer 2-sided queries on this problem in $O(\lg N)$ time using $O(n\sqrt{\lg n})$ bits of space.*

We begin with an overview. We first create $Inf(T)$, and choose a parameter $\lambda = \sqrt{\lg n}$. We then select a set of points $T' \subseteq T$ with the following properties: (a) $|T'| = O(n/\lambda)$; (b) the *vertical decomposition*, whereby we shoot vertical rays upward and downward from each endpoint of each segment in $Inf(T')$ until they hit another segment, of the plane induced by $Inf(T')$[3] decomposes the plane into $O(n/\lambda)$ rectangular regions each of which has at most $O(\lambda)$ points from T and parts of line segments from $Inf(T)$ in it (see Fig. 3(R)). T' always

---

[2] A tight bound of $n \lg 5 + o(n) \sim 2.33n$ bits can be shown [11].

[3] Note that the extent of a line segment in $Inf(T')$ is defined, as originally, wrt points in T, and not wrt the points in T'.

exists and can be found by plane sweep [3, Section 3],[9, Section 4.3]. We store a standard point location data structure, called the *skeleton*, on T': this requires $O((n/\lambda)\lg n) = O(n\sqrt{\lg n})$ bits. We also store $O(\lambda)$ bits of information with each region (including the $O(\lambda)$-bit encoding of priority information from Lemma 5).

Given a query point q, we first perform a point location query on the skeleton to determine the region R in which q lies. We now need to solve the original point location problem within R, and perform a slab-select to determine the points of T that lie within this region. This, together with the priority information, allows us to partially—but not fully, since lines of influence may originate from outside R—reconstruct the point location problem within R. To handle lines of influence starting outside R, we do a binary search with $O(\lg \lambda)$ steps, each step including a a slab-select, giving the claimed bound. The details are as follows.

Preprocessing. Let R be any region, and let $Left(R)$ $(Right(R))$ be the set of line segments from $Inf(T)$ that intersect the left (right) boundaries of R, and let $P(R)$ be the set of points from T in R. We store the following data for R:

1. For each line segment $\ell \in Left(R)$, ordered top-to down by y-axis, a bit that indicates whether the right endpoint of ℓ is in R or not; similarly for $\ell \in Right(R)$, a bit indicating whether ℓ begins in R or not.
2. If the left boundary of R is adjacent to other regions $R_1, R_2, \ldots$ (taken from top to bottom) and $l_i \geq 0$ represents the number of line segments from $Left(R)$ that also intersect R_i, then we store a bit-string $0^{l_1}10^{l_2}1\ldots$. A similar bit-string is stored for the right boundary of R.
3. For each point in $P(R)$ and each line segment in $Left(R)$, a bit-string of length $|P(R)| + |Left(R)|$ whose i-th bit indicates whether the i-th largest y-coordinate in $P(R) \cup Left(R)$ is from $P(R)$ or $L(R)$.
4. Suppose that a line segment $\ell = Inf(p)$ for some $p \in T$ crosses $m \geq \lambda$ regions. Then, in every λth region that ℓ crosses, we explicitly store the region containing p, and p's local coordinates.
5. Finally, for each point $p \in P(R)$, we store the sequence of bits from Lemma 5, which indicates whether p has a non-empty $Inf(p)$ and if so, for how many lines from $Left(R) \cup Inf(P(R))$, p is a right endpoint (p cannot be a right endpoint of any other line in $Inf(T)$, by the construction of the skeleton).

The purpose of (1) and (2) is to trace a line segment ℓ as it crosses multiple regions: if ℓ crosses from a region R' to a region R'' on its right, then given its position in $Right(R')$, we can deduce its position in $Left(R'')$. Using (4), after tracing $\ell = Inf(p)$ through $\leq \lambda$ regions, we will discover (at least) the region containing p. The skeleton takes $O(n\sqrt{\lg n})$ bits, so we now add up the space used by (1)-(5). By construction, $|Left(R)|$, $|Right(R)|$ and $|P(R)|$, summed over all regions R, is $O(n)$. The space bound for (1) and (3) is therefore $O(n)$ bits. The number of 1s in the bit string of (3), summed over all regions, is $O(n/\lambda)$, as there are $O(n/\lambda)$ regions and the graph which indicates adjacency of regions is planar; the number of 0s is $O(n)$ as before. The space used by (4) is $O(n\sqrt{\lg n})$ bits again, as for every $O(\sqrt{\lg n})$ portions of line segments in the regions we store $O(\lg n)$ bits. Finally, the space used for (5) is $O(n)$ bits by Lemma 5.

Query algorithm. Given a query point q in a sub-problem of size n (assume that we have q's local and top-level coordinates), we answer $RMQ(q)$ as follows:

(a) Do a planar point location in the skeleton, and find a region R in which the point q lies. Perform slab-select on R to get $P(R)$.

(b) As we know how many segments from $Left(R)$ lie vertically between any pair of points in $P(R)$, when we are given the data in (5) above, we are able to determine whether the x-coordinate of a given point p in $P(R)$ is the right endpoint of a line from either $Left(R)$ or $Inf(P(R))$. Thus, we have enough information to determine $Inf(p)$ for all $p \in P(R)$ (at least until the right boundary of R). Furthermore, for each line in $Left(R)$ that terminates in R, we also know (the top-level coordinates of) its right endpoint.

(c) Using the top-level coordinates of q, we determine the nearest segment from $Inf(P(R))$ that is above q.

(d) Using the top-level coordinates of q we also find the set of segments from $Left(R)$ whose right endpoints are not to the left of q. Let this set be $Left^*(R)$. We now determine the nearest segment from $Left^*(R)$ that is above q. Unfortunately, although $|Left^*(R)| = O(\lambda)$, since the segments in $Left^*(R)$ originate in points outside R, we do *not* have their y-coordinates. Hence, we need to perform the following binary search on $Left^*(R)$:

(d1) Take the line segment $\ell \in Left^*(R)$ with median y-coordinate, and suppose that $\ell = Inf(p)$. The first task is to find the region R_p containing p, as follows. Use (2) to determine which of the adjacent regions of R ℓ intersects, say this is R'. If ℓ ends in R', or $R' = R_p$ and we are done. Otherwise, use (1) to locate ℓ in $Left(R')$ and continue.

(d2) Once we have found R_p, we perform a slab-select on R' to determine $P(R_p)$, and sort $P(R_p)$ by y-axis. Then we perform (c) above on $P(R_p)$, thus determining which points of $P(R_p)$ have lines of influence that reach the right boundary of R_p. Using this we can now determine the (top-level) coordinates of p.

(d3) We compare the top-level y-coordinates of p and q and recurse.

(e) We take the lower of the lines found in (d) and (e) and use it to return a candidate. Observe that we have the top-level coordinates of this candidate.

We now derive the time complexity of a 2-sided query. Step (a) takes $O(\lg n)$ for the point location, and $O(\lg N / \lg \lg N)$ for the slab-select. Step (b) can be done in $O(\lg n) = O(\lg N)$ time by running the plane sweep algorithm of Lemma 5 (recall that $|P(R)| = O(\sqrt{\lg n})$—a quadratic algorithm will suffice). Step (c) likewise can be done by a simple plane sweep in $O(\lg n)$ time. Step (d1) is iterated at most $O(\sqrt{\lg n})$ times before R_p is found since every λ-th region intersected by ℓ contains information about p. Each iteration of (d1) takes $O(1)$ time: operations on the bit-strings are done either by table lookup if the bit-string is short ($O(\lambda)$ bits), or else using rank and select operations [19], if the bit string is long (as e.g. the bit-string in (2) may be) – these entirely standard tricks are not described in detail. Step (d2) takes $O(\lg N / \lg \lg N)$ time as before. Steps (d1)-(d3) are performed $O(\lg \lambda) = O(\lg \lg N)$ times, so this takes $O(\lg N)$ time overall. Step (e) is trivial. We have thus shown Lemma 6.

3.5 Putting Things Together

Section 3.1 shows that our data structure occupies $O(N)$ words. The dominant term in the running time is due to solving $O(\lg \lg N)$ 2-sided queries using Lemma 6, taking $O(\lg N \lg \lg N)$ time. The $O(\lg \lg N)$ square-aligned queries are solved in $O(1)$ time each. The $O(1)$ problems at the bottom of the recursion are solved in $O((\lg \lg N)^2)$ time. We scan all $O(\lg \lg N)$ candidates to find the answer (any candidates given in local coordinates are converted to top-level coordinates in $O(\lg N / \lg \lg N)$ time each, or $O(\lg N)$ time overall). This proves Theorem 1.

4 Succinct Indices for 2-Sided Queries

We now consider succinct indices for 2-sided range maxima queries over N points in rank space. Our results are stand-alone variants of Lemma 6 and reuse its structure (proofs can be found in [11]). The indices encode the priority information, but not the point coordinates, which are assumed to be accessible in one of two ways. First, we consider the case where points are reported through an orthogonal range reporting query, such that a query that results in k points being reported takes $T(N, k)$ time (assume that $T(N, O(k)) = O(T(N, k))$). Then:

Lemma 7. *Let* $\lambda \geq 2$ *be some parameter. There is a succinct index of size* $O(N + (N \lg N)/\lambda)$ *bits such that 2-sided range maxima queries can be answered in* $O(\lg N + \lg \lambda(\lambda + T(N, \lambda)))$ *time.*

In the *permuted-point* model of [4], the point coordinates are stored in read-only memory, and the i-th point (according to an ordering specified by the data structure) can be accessed in $O(1)$ time. We can show:

Lemma 8. *There is a succinct index of* $N \lg 5 + o(N) = 2.33N + o(N)$ *bits such that 2-sided range maxima queries can be answered in* $O(\lg \lg N)$ *time in the permuted-point model.*

5 Conclusions

We have introduced a new approach to producing space-efficient data structures for orthogonal range queries, and have applied our approach to give the first linear-space data structure for 2D range maxima that improves upon Chazelle's 1985 linear-space data structure. It would be interesting to try to obtain (say) $O(\lg \lg N)$ running time as in [8] in linear space, or to apply these ideas to related problems such as top-k queries.

References

1. Afshani, P., Arge, L., Larsen, K.D.: Orthogonal range reporting: query lower bounds, optimal structures in 3-D, and higher-dimensional improvements. In: Snoeyink, J., de Berg, M., Mitchell, J.S.B., Rote, G., Teillaud, M. (eds.) Symposium on Computational Geometry, pp. 240–246. ACM (2010)
2. Barbay, J., He, M., Munro, J.I., Rao, S.S.: Succinct indexes for strings, binary relations and multi-labeled trees. In: Bansal, N., Pruhs, K., Stein, C. (eds.) SODA, pp. 680–689. SIAM, Philadelphia (2007)

3. Bender, M.A., Cole, R., Raman, R.: Exponential Structures for Efficient Cache-Oblivious Algorithms. In: Widmayer, P., Triguero, F., Morales, R., Hennessy, M., Eidenbenz, S., Conejo, R. (eds.) ICALP 2002. LNCS, vol. 2380, pp. 195–207. Springer, Heidelberg (2002)
4. Bose, P., Chen, E.Y., He, M., Maheshwari, A., Morin, P.: Succinct geometric indexes supporting point location queries. In: Mathieu, C. (ed.) SODA, pp. 635–644. SIAM, Philadelphia (2009)
5. Bose, P., He, M., Maheshwari, A., Morin, P.: Succinct Orthogonal Range Search Structures on a Grid with Applications to Text Indexing. In: Dehne, F., Gavrilova, M., Sack, J.-R., Tóth, C.D. (eds.) WADS 2009. LNCS, vol. 5664, pp. 98–109. Springer, Heidelberg (2009)
6. Brodal, G.S., Davoodi, P., Rao, S.S.: On Space Efficient Two Dimensional Range Minimum Data Structures. In: de Berg, M., Meyer, U. (eds.) ESA 2010, Part II. LNCS, vol. 6347, pp. 171–182. Springer, Heidelberg (2010)
7. Brodal, G.S., Jørgensen, A.G.: Data Structures for Range Median Queries. In: Dong, Y., Du, D.-Z., Ibarra, O. (eds.) ISAAC 2009. LNCS, vol. 5878, pp. 822–831. Springer, Heidelberg (2009)
8. Chan, T.M., Larsen, K.G., Pătraşcu, M.: Orthogonal range searching on the RAM, revisited. In: Proceedings of the 27th Annual ACM Symposium on Computational Geometry, SoCG 2011, pp. 1–10. ACM, New York (2011), http://doi.acm.org/10.1145/1998196.1998198
9. Chan, T.M., Patrascu, M.: Transdichotomous results in computational geometry, I: Point location in sublogarithmic time. SIAM J. Comput. 39(2), 703–729 (2009)
10. Chazelle, B.: A functional approach to data structures and its use in multidimensional searching. SIAM J. Comput. 17(3), 427–462 (1988), prel. vers. FOCS 1985 (1985)
11. Farzan, A., Munro, J.I., Raman, R.: Succinct indices for range queries with applications to orthogonal range maxima. Tech. Rep. CS-TR-12-001, U. Leicester (April 2012), http://arxiv.org/abs/1204.4835
12. Gabow, H.N., Bentley, J.L., Tarjan, R.E.: Scaling and related techniques for geometry problems. In: Proc. 16th Annual ACM Symposium on Theory of Computing, pp. 135–143. ACM (1984)
13. Golin, M.J., Iacono, J., Krizanc, D., Raman, R., Rao, S.S.: Encoding 2D Range Maximum Queries. In: Asano, T., Nakano, S.-I., Okamoto, Y., Watanabe, O. (eds.) ISAAC 2011. LNCS, vol. 7074, pp. 180–189. Springer, Heidelberg (2011)
14. JáJá, J., Mortensen, C.W., Shi, Q.: Space-Efficient and Fast Algorithms for Multidimensional Dominance Reporting and Counting. In: Fleischer, R., Trippen, G. (eds.) ISAAC 2004. LNCS, vol. 3341, pp. 558–568. Springer, Heidelberg (2004)
15. Karpinski, M., Nekrich, Y.: Space Efficient Multi-Dimensional Range Reporting. In: Ngo, H.Q. (ed.) COCOON 2009. LNCS, vol. 5609, pp. 215–224. Springer, Heidelberg (2009)
16. Makris, C., Tsakalidis, A.K.: Algorithms for three-dimensional dominance searching in linear space. Inf. Process. Lett. 66(6), 277–283 (1998)
17. Mehta, D.P., Sahni, S. (eds.): Handbook of Data Structures and Applications. Chapman & Hall/CRC (2009)
18. Nekrich, Y.: Orthogonal range searching in linear and almost-linear space. Comput. Geom. 42(4), 342–351 (2009)
19. Rahman, N., Raman, R.: Rank and select operations on binary strings. In: Kao, M.Y. (ed.) Encyclopedia of Algorithms, Springer (2008)
20. Yuan, H., Atallah, M.J.: Data structures for range minimum queries in multidimensional arrays. In: Charikar, M. (ed.) SODA 2010, pp. 150–160. SIAM, Philadelphia (2010)

Universal Factor Graphs[*]

Uriel Feige and Shlomo Jozeph

Department of Computer Science and Applied Mathematics,
The Weizmann Institute of Science, Rehovot, Israel
{uriel.feige,shlomo.jozeph}@weizmann.ac.il

Abstract. The factor graph of an instance of a symmetric constraint satisfaction problem on n Boolean variables and m constraints (CSPs such as k-SAT, k-AND, k-LIN) is a bipartite graph describing which variables appear in which constraints. The factor graph describes the instance up to the polarity of the variables, and hence there are up to 2^{km} instances of the CSP that share the same factor graph. It is well known that factor graphs with certain structural properties make the underlying CSP easier to either solve exactly (e.g., for tree structures) or approximately (e.g., for planar structures). We are interested in the following question: is there a factor graph for which if one can solve every instance of the CSP with this particular factor graph, then one can solve every instance of the CSP regardless of the factor graph (and similarly, for approximation)? We call such a factor graph *universal*. As one needs different factor graphs for different values of n and m, this gives rise to the notion of a family of universal factor graphs.

We initiate a systematic study of universal factor graphs, and present some results for max-kSAT. Our work has connections with the notion of preprocessing as previously studied for closest codeword and closest lattice-vector problems, with proofs for the PCP theorem, and with tests for the long code. Many questions remain open.

1 Introduction

A constraint satisfaction problem (CSP) has a set of n variables and a set of m constraints (also referred to as clauses, or factors). Every constraint involves a subset of the variables, and is satisfied by some assignments to the variables and not satisfied by others. An instance of a CSP is satisfiable if there is an assignment to the variables that satisfies all constraints. When variables are Boolean and constraints are symmetric a constraint is fully specified by the set of literals that it contains (where a literal is either a variable or its negation), and is satisfied if and only if the appropriate number of literals is set to true (e.g., at least one for SAT, an odd number for XOR, all for AND, the majority for MAJ, and at least one but not all for NAE). To simplify the presentation, we shall consider in this paper CSPs that are Boolean and symmetric, though we remark that much of what we discuss can be extended to non-Boolean and non-symmetric CSPs.

---

[*] An extended version of the paper appears at http://arxiv.org/abs/1204.6484

A. Czumaj et al. (Eds.): ICALP 2012, Part I, LNCS 7391, pp. 339–350, 2012.
© Springer-Verlag Berlin Heidelberg 2012

The *factor graph* of an instance of a CSP is a bipartite graph. Vertices on one side represent the variables, vertices on the other side represent the constraints (also known as *factors*), and edges connect constraints to the variables that they contain. For Boolean symmetric CSPs, a factor graph together with a labeling of the edges with ±1 (indicating whether the corresponding variable has positive or negative polarity in the underlying clause) completely specifies an instance of the CSP. Without the edge labels, there are many instances of the CSP that share the same factor graph and differ only in the polarity of the variables.

As is well known, deciding satisfiability for CSPs is NP-hard for a large class of predicates (including, SAT, MAJ and NAE). See [21] for a complete classification. Here we shall consider NP-hard CSPs. The research question that motivates our current paper is to understand what are the obstacles for obtaining efficient algorithms for solving CSPs. Specifically, are algorithms having trouble in "understanding" the structure of the factor graph, and this translates to difficulties in solving the underlying CSP? Alternatively, are the computational difficulties a result of the combinatorial richness of the polarities?

The structure of the factor graph may cause the underlying CSP instance to be easy. For example, if the factor graph is a tree (or more generally, of bounded treewidth), then the underlying CSP instance can be solved in polynomial time (by dynamic programming). Our research question (once properly formalized) can be viewed as asking whether in other cases, the structure of the factor graph might be the major contributing factor to making a CSP hard.

The playing field of our research agenda is greatly enriched once optimization versions of CSPs are considered, namely max-CSP: find an assignment to the variables that satisfies as many constraints as possible. As is well known, even some polynomial time solvable CSPs (such as XOR, or 2SAT) become NP-hard when their optimization version is considered. See [7] for a classification. A standard way of dealing with NP-hard max-CSP instances is via approximation algorithms that in polynomial time find an assignment that is guaranteed to satisfy a number of constraints that is at least ρ times the maximum number of constraints that can be satisfied, for some $0 < \rho < 1$. For many CSPs, the best possible ρ is known, in the sense that the approximation ratios provided by known approximation algorithms are matched by hardness of approximation results that show that better approximation ratios would imply that P=NP. For example, $\rho = 7/8$ is a tight approximation threshold for max-3SAT [12]. Moreover, for all CSPs, an algorithm (based on semidefinite programming) with the optimal approximation ratio is given by Raghavendra [20], assuming the *Unique Games Conjecture* of Khot [14]. However, despite the optimality of this algorithm, it is difficult to figure out which approximation ratio it guarantees, and consequently there are CSPs for which the value of this threshold is not known. (And of course, if the Unique Games Conjecture is false then the approximation ratio implied by this algorithm need not be tight.)

Our research agenda naturally extends to max-CSP. One may ask whether approximation algorithms are having trouble in "understanding" the structure of the factor graph, and whether this translates to difficulties in approximating the underlying CSP. Moreover, now the question acquires also a quantitative aspect, and one may ask to what extent does the factor graph contribute to the approximation difficulty. For example, if algorithms had no difficulty in "understanding" factor graphs, could the approximation ratio for max-3SAT be improved from 7/8 to 8/9?

As in the case of tree factor graphs for decision versions, there are known families of factor graphs (such as planar graphs, or more generally, families of graphs excluding a fixed minor) on which the underlying CSP instance has improved approximation ratios, or even a PTAS ($\rho > 1 - \epsilon$ for every $\epsilon > 0$). On the other hand, it appears that for some CSPs, almost every factor graph is difficult. For example, there is no known approximation algorithm that runs in polynomial time on random 3CNF formulas (with say $m = n \log n$ constraints) and approximates max-3SAT within a ratio better than 7/8. This suggests (though does not prove) that there is no need for clever design of the factor graph in order to make the underlying CSP instance difficult – almost any factor graph would do.

In contrast, for unique games (which is a special family of CSPs with two non-Boolean variables per constraint), the approximation ratios achievable on random factor graphs [4] are much better than those currently known to be achievable on arbitrary factor graphs. (Technically, the graphs considered by Arora et al. [4] have variables as vertices and constraints as edges, but there is a one-to-one correspondence between such graphs and factor graphs.) The same holds for some other classes of graphs [22,16]. Can we (and should we) identify more factor graphs on which unique games are easy? Is there a "universal" graph (e.g., a generalized Kneser constraint graph?) such that if unique games are easy on it, then the Unique Games Conjecture is false? Such questions lead naturally to the notion that we call here *universal factor graphs*.

1.1 Preprocessing

How can we provide evidence that algorithms for max-3SAT should be spending substantial time in analyzing the factor graph? Here is a possible formal approach. Reveal the input instance in two stages. In the first stage, only the factor graph is revealed. At this point the algorithm is allowed to run for arbitrary time and record (in polynomial space) whatever information about the factor graph that it may hope to find useful (e.g., an optimal tree decomposition of the factor graph, or a minimum dominating set in the factor graph, both of which are pieces of information that take exponential time to compute). Thereafter the polarities of the variables are revealed. At this stage the algorithm has only polynomial time, and it needs to find an optimal solution to the max-3SAT instance. If there is a combination of algorithms (unbounded time for stage 1, polynomial time for stage 2) that can do this on every instance, this establishes that a good understanding of the factor graph suffices for solving 3SAT instances. If this cannot be done, this establishes that at least some substantial portion of the running time is a result of the combinatorial richness of space of possibilities for polarities of the variables. Refined versions of the preprocessing approach either require less of the stage 2 algorithm (finding nearly optimal solutions rather than optimal ones) or give it extra power (allow subexponential time), and may lead to a more quantitative understanding of the value of preprocessing.

To derive positive results in this model, it suffices to provide the respective algorithms and their analysis. But how does one provide negative results? This is where the notion of *universal factor graphs* comes in. Informally, these are factor graphs on which preprocessing is unlikely to help, because if it does, then all instances (regardless of their factor graph) can be solved even without preprocessing.

1.2   Universal Factor Graphs

We consider infinite families of factor graphs. Basically, for every value of $N, M > 0$, a family includes at most one factor graph with N variables and M constraints. However, for convenience in intended future uses, members of the family are indexed by two auxiliary indices that are called n and m. Definition 1 does not exclude the possibility that several factor graphs in the family share the same values of N and M, but their number is upper bounded by some polynomial in $N + M$.

Definition 1. *Consider an arbitrary CSP with k variables per-constraint. For integers $n > 0$ and $0 < m \leq 2^k \binom{n}{k}$, let $N(n, m)$ and $M(n, m)$ be two functions, each lower bounded by n and upper bounded by a polynomial in $n + m$. A family of factor graphs associates with each pair of values of n and m a factor graph with $N(n, m)$ variables and $M(n, m)$ constraints. The family is uniform if there is an algorithm running in time polynomial in $n + m$ that given n, m produces the associated factor graph.*

Every member of a family of factor graphs for a k-CSP can give rise to 2^{kM} instances of the CSP, depending on how one sets the polarities of the variables in the constraints. Given any such instance as input, we shall consider computational tasks such as *satisfiability* (find a satisfying assignment if one exists), *optimization* (find an assignment satisfying as many clauses as possible) and *approximation* (get close to optimal).

The algorithms that perform the above tasks will be limited in their running times. In this work, we shall be interested in two classes of running times. One is the standard *polynomial time* (P) notion, which in our case will mean polynomial in $(N + M)$. The other is *subexponential time*, (SUBEXP) which in this paper is taken to mean time time $2^{O(N^{1-\epsilon})}$ for some $\epsilon > 0$.

Recall that in computational complexity theory, one distinguishes between uniform models of computation (such as Turing machines) and non-uniform models (such as families of circuits). This distinction is relevant in our context. The notion of preprocessing the factor graph can be captured by allowing for nonuniform algorithms. Hence we shall be dealing with the complexity classes P/poly, SUBEXP/poly and SUBEXP/subexp (the parameters /poly and /subexp correspond to the length of advice that the preprocessing stage is allowed to record). For simplicity in our presentation, in each of our definitions below we shall specify one particular complexity class (either P/poly or SUBEXP/poly), but we note that our results extend to other complexity classes as well (such as P instead of P/poly, or SUBEXP/subexp instead of SUBEXP/poly).

In this work we will show that for some uniform families of factor graphs solving satisfiability or approximation tasks are hard. These families of factor graphs will be referred to as *universal*, and with slight abuse of terminology, individual factor graphs within these families will be referred to as *universal factor graphs*. The hardness results will be proved under some complexity assumption. If the complexity assumption is widely believed, such as that NP is not contained in P/poly, then the universal factor graphs support the view that the complexity of the underlying CSP cannot be attributed entirely to the factor graph and is at least partly due to the polarities of the variables, because the nonuniform algorithms could preprocess the factor graph for arbitrary time prior to receiving the polarities of the variables. If the complexity assumption is not as widely believed (such as the Unique Games Conjecture), the interpretation of these hardness result can be that if one wishes to refute the complexity assumption, it would suffice to design algorithms that are specifically tailored to work on instances with factor graphs as in the universal family.

We now present formal definitions that are tailored to match those results that we can prove in this paper. It is straightforward to adapt these definitions to other variations as well.

Definition 2. *For a given CSP, a uniform family of factor graphs is P-universal if there is no P/poly algorithm for instances of the CSP with factor graphs from this family, unless NP is contained in P/poly.*

Definition 3. *For a given CSP, a uniform family of factor graphs is subexp-universal if there is no SUBEXP/poly algorithm for instances of the CSP with factor graphs from this family, unless there is a SUBEXP/poly algorithm for all instances of the CSP.*

Definition 4. *For a given CSP and $0 < \rho < 1$, a uniform family of factor graphs is ρ-universal if there is no P/poly approximation algorithm with approximation ratio better than ρ on the instances of the CSP with factor graphs from this family, unless NP is contained in P/poly. This notion is referred to as threshold-universal. If ρ is equal to the best approximation ratio known for the underlying CSP, we will refer to this as a* tight *threshold. When we do not wish to specify a particular value for ρ, we call the family APX-universal. A variation on ρ-universality is (c, s)-universality with $0 < s < c \leq 1$, where instead of approximation within a ratio of ρ, one considers distinguishing between instances with at least a c-fraction of the clauses being satisfiable, and instances with at most s-fraction being satisfiable. For a CSP for which the decision variant is NP-hard (e.g. 3SAT), ρ-universality will be taken to mean $(1, \rho)$-universal.*

More generally, for optimization versions we shall allow vertices (representing constraints) of universal factor graphs to have nonnegative weights, thus representing instances in which one wishes to find an assignment that maximizes the weight (rather than the number) of satisfied constraints. As the weights will be fixed (independently of the subsequent polarities given to variables), this is in essence a condensed representation of an unweighted universal factor graph (which can be obtained by duplicating each vertex a number of times proportional to its weight, rounded to the nearest integer – details omitted).

1.3   Some Research Goals

The notion of universal factor graphs opens up many research directions that we find interesting. In our current work we attempt to answer questions such as: Does 3SAT have P-universal factor graphs? Subexp-universal factor graphs? Does max-3SAT have APX-universal factor graphs? Does max-3SAT have 7/8-universal factor graphs? These questions are part of a wider research agenda that concerns questions such as: Do all CSPs have tight threshold-universal factor graphs? Which CSPs do not have tight threshold-universal factor graphs? Other questions of interest include: How do universal factor graphs look like? Can knowledge of their structure help us either in designing new algorithms, or in reductions that prove new hardness results?

1.4   Related Work

There has been work showing that CSPs on particular factor graphs are NP-hard, and using such results to help in reductions establishing further NP-hardness results. For example, it is known that 3SAT is NP-hard even when the factor graph is planar [17], and this was used (for example) in showing that *minimum-length rectangular partitioning of a rectilinear polygon* (with holes) is NP-hard [18]. Our notion of universal factor graphs is stronger as it requires at most one particular factor graph for each instance size, rather than a whole family of factor graphs (e.g., the n by n grid, rather than all planar graphs).

A line of work that closely relates to our research agenda is that of preprocessing for NP-hard problems. As the universal factor graph is fixed, one may consider preprocessing it for arbitrary (exponential) time in order to produce a polynomial size "advice", prior to getting the polarities of the variables. Preprocessing was extensively studied for some NP-hard problems, and hardness results in the context of preprocessing amount to designing instances that are universal (in our terminology). Naor and Bruck [6] show that the nearest code word problem remains NP-hard even when the code can be preprocessed. Nearest lattice vector (CVP) when the lattice can be preprocessed was shown to be NP-hard and APX-hard by Feige and Micciancio [10]. The tightest hardness results for lattice problems with preprocessing currently known are by Khot et al. [15]. An earlier work by Alekhnovich et al. [2] has some partial overlap with our current work, because it uses PCP theory and in the process gives hardness of approximation results with preprocessing for additional problems. See more details in Section 2.2.

The above results on coding and lattice problems with preprocessing are motivated by the fact that in these problems, it is indeed often the case that part of the input is fixed in advance (the code, or a basis for the lattice), and part of the input (a noisy word that one wishes to decode, or a vector for which one wishes to find the closest lattice point) is a query that is received only later. Moreover, multiple queries are expected to be received on the same fixed input. In these cases it really makes sense to invest much time in preprocessing the fixed part of the input, if this later helps answering the multiple queries more quickly. In contrast, our notion of universal factor graphs is independent of such practical concerns. Our motivation is to understand the source of difficulties in solving NP-hard problems. In particular, it is irrelevant to us whether there really is any real life situation in which one receives the factor graph of a 3CNF formula in advance, and then is asked a sequence of queries about it, each time with different polarities of the variables.

Is it at all plausible that preprocessing can help? For lattice problems, this indeed appears to be the case. There are no known approximation algorithms with subexponential ratios for CVP, but if preprocessing is allowed, than polynomial approximation ratios are known (by using an exponential time preprocessing procedure that derives a so called *reduced basis* of the lattice). For CSPs, the authors are aware of only much weaker evidence that preprocessing may help. This relates to the case that polarities of variables are random rather than arbitrary.

There is a refutation algorithm that is poly-time on random 3CNF formulas with more than $n^{1.5}$ clauses. The obstacle to extending this to lower density of $n^{1.4}$ is graph-theoretic: if one knew how to efficiently find certain substructures in the factor graphs (that almost surely exist), this would suffice [11]. Preprocessing the factor graph would allow finding these structures. Hence at these densities, random factor graphs are not expected to be universal (with respect to *random* polarities).

In the current paper we consider arbitrary polarities for the variables rather than random polarities. Nevertheless, we remark that the case of random polarities is also well motivated, and related to possible cryptographic application. See [3] as an example showing how results from [11] can be used in a proposal of new public key cryptographic primitives.

More generally, cryptography offers many examples where preprocessing is believed to help (it will lead to the discovery of a so called *trapdoor* that would make solving future instances easy), but as this typically relates to computational problems that are believed not to be NP-hard, further discussion of this is omitted from the current manuscript.

1.5   Our Results

The first theorem is based on a straightforward reduction and we have no doubt that it was previously known (perhaps using different terminology).

Theorem 1. *There are P-universal factor graphs for 3SAT.*

For the P-universal factor graphs constructed by our proof for Theorem 1, an algorithm running in time $2^{N^{1-\epsilon}}$ on instances of the universal family would correspond to time $2^{n^{3-3\epsilon}}$ on general instances. Hence they are not subexp-universal. The next theorem addresses this issue.

Theorem 2. *There are subexp-universal factor graphs for 3SAT.*

We would have liked to prove that there are 7/8-universal factor graphs for max-3SAT, matching the tight threshold of approximability for max-3SAT. However, we only managed to prove weaker bounds.

Theorem 3. *There are 77/80-universal factor graphs for max-3SAT.*

Is there any CSP for which we can obtain tight threshold-universal families? We do not know, but we do have almost tight results.

Theorem 4. *For every $\epsilon > 0$ there is an integer k for which there is a family of factor graphs that are $\left(1 - (1-\epsilon)\,2^{-k}\right)$-universal for max-EkSAT.*

Theorem 4 in nearly tight because every instance of max-EkSAT is $(1 - 2^{-k})$-satisfiable, and consequently there are several algorithms with a $(1 - 2^{-k})$ approximation ratio. To actually get tight results we would need to switch the order of quantifiers in Theorem 4 (show that for some k the result holds for every ϵ), but doing so remains an open question.

Using the techniques developed in our work and known reductions among CSPs one can obtain APX-universal factor graphs for additional CSPs. In particular, we derive APX-universal factor graphs for max-2LIN, thus illustrating that for approximating unique games (max-2LIN is a unique game) at least part of the difficulty comes from the polarities of variables rather than from the structure of the factor graph.

2   Overview of Proofs

Because of space limitations, most of our proofs are omitted from this manuscript. They can be found in [9].

At a high level, to show that a factor graph is universal, one shows that any other factor graph (of the appropriate size) can be reduced to it. The details of how this is done depend on the context.

The proof of Theorem 1 is elementary and is only sketched here. Consider the 3CNF formula on n variables that contains all possible $2^3\binom{n}{3}$ clauses. Add one auxiliary variable x_0 to every clause, giving a 4CNF formula F, and add to F a few extra clauses that force x_0 to be set to true in every satisfying assignment. Every 3CNF formula f can be embedded in this 4CNF formula by negating x_0 only in these clauses of F that appear in f. To change F into a 3CNF formula, break every 4-clause into two 3-clauses using a fresh auxiliary variable and its negation. The factor graph of the resulting 3CNF formula is P-universal for 3SAT. For more details, see [9].

2.1   Subexp-Universal Families

Our proof of Theorem 2 combines two ingredients. One is a variation on a result of Impagliazzo et al. [13]. It can be leveraged to show that for the purpose of constructing subexp-universal factor graphs it suffices to consider 3CNF instances with a linear number of clauses.

The other ingredient is a reduction with a tighter connection between $n + m$ and N compared to the one used in our proof of Theorem 1.

Lemma 1. *There is a factor graph with $N = O(m \log m \log n)$ variables that is P-universal with respect to 3SAT instances with n variables and m clauses.*

Our proof of Lemma 1 makes use of oblivious sorting networks (specifically, the one of Ajtai et al. [1]).

More details on those two ingredients and how they are combined to prove Theorem 2 appear in [9].

2.2   Threshold-Universal Families

For our proof of Theorem 3 we use a notion that we call a *factor graph preserving reduction* (FGPR). It is an algorithm that transforms a *source* 3CNF instance f_s to a *target* 3CNF instance f_t. The transformation has the following properties:

1. Polynomiality. The transformation algorithm runs in polynomial time (in the size of f_s). Consequently, the size of f_t is polynomial in the size of f_s.
2. Faithfulness. If f_s is satisfiable, so is f_t, and vice versa.
3. Factor graph preserving. Any two instances f_s and f'_s with the same factor graph are reduced to two instances f_t and f'_t that have the same factor graph.

To be useful for our purposes, we would like the FGPR to also have a *gap amplification* aspect. Namely, if f_s is not satisfiable, then the fraction of clauses satisfiable in f_t is smaller than the fraction of clauses satisfiable in f_s.

Theorem 3 will be broken into two sub-theorems, each of which is proved using FGPRs.

Theorem 5. *There are APX-universal factor graphs for max-3SAT.*

Theorem 6. *There is a reduction from APX-universal factor graphs for max-3SAT to 77/80-universal ones.*

The proof of Theorem 5 strongly relates to the work of Alekhnovich et al. [2]. As explained in Section 1.4, in that work various APX-hardness results with preprocessing were obtained. Among them, there were APX-hardness results with preprocessing for certain CSPs (satisfying quadratic equations). It is not difficult to use these results in order to obtain APX-universal factor graphs for max-3SAT. However, we present an alternative proof because [2] claims the relevant theorem without providing a proof[1]. Our proof is patterned after a proof of the PCP theorem due to Dinur [8].

Recall that Dinur's proof is based on a sequence of gap amplification steps. However, some of these transformations are not factor graph preserving. Our proof performs a sequence of gap amplifying FGPRs, starting with the outcome of Theorem 1, and eventually proving Theorem 5. Every FGPR is based on modifying Dinur's proof (or more exactly, on modifying a variation on Dinur's proof that is given in [19]). The modifications are related to those discussed below for the long code (though our proof for Theorem 5 uses a quadratic code rather than the long code).

The proof of Theorem 6 involves an FGPR from APX-universal factor graphs for max-3SAT to 77/80-universal ones. Our proof is based on a modification of the proof of Bellare et al. [5], and consequently obtains the same hardness ratio of 77/80. The main difficulty we encounter is the following. Tight or nearly tight hardness of approximation results use the so called *long code*. A major reason why it is used is that its high redundancy allows one to replace explicit queries that check whether an underlying predicate is satisfied by an implicit operation (referred to as *folding*) that allows one to avoid making these queries. The only queries that need to be made are those that check whether the encoding is really (close to) a long code. The saving in queries translates to stronger hardness of approximation results. The problem with folding is that it is sensitive to the predicate that needs to be checked, and a change in the predicate (e.g., changing the polarity of a single variable in a 3SAT clause) changes the folding. As a result, query locations change, and the resulting reduction is not an FGPR. To overcome this problem we introduce a notion of *oblivious folding* of the long code, which does allow us to eventually obtain an FGPR. We remark that it was not a-priori obvious that a construct such as oblivious folding should exist at all. In particular, tight hardness of approximation results for 3SAT by Hastad [12] use a notion related to folding but somewhat stronger, that is called *conditioning* of the long code. We were unable to find an "oblivious" version of conditioning that can replace the conditioning used by Hastad, and consequently we do not know if 7/8-universal factor graphs for 3SAT exist.

For the full proofs of Theorems 5 and 6, see [9].

2.3   Threshold-Universal Families with Nearly Tight Bounds

Recall that the prefix E (for *exact*) in EkSAT indicates that every clause in the CNF formula contains exactly k literals (rather than at most) and no two literals in a clause correspond to the same variable. It is not difficult to see that the proof of Theorem 3 in fact gives E3CNF formulas, and not just 3CNF formulas (and even if not, there are simple FPGRs from max-3SAT to max-E3SAT, with only

---

[1] Quoting from [2]: "The proof of this theorem, which is a laborious and an almost exact mimic of the proof of the PCP Theorem, is beyond the scope of this version of the paper." A subsequent paper [15] that extends [2] no longer uses this theorem, and hence does not contain the proof either.

a bounded loss in the approximation ratio). Our proof of Theorem 4 is based on a direct reduction from instances of max-E3SAT to instances of max-EkSAT. This reduction has the property that mere APX-hardness of max-E3SAT suffices in order to get nearly tight hardness of approximation ratios for the resulting max-EkSAT instances, if k is sufficiently large.

Proof. Theorem 3 implies that there is a $(1 - \gamma)$-universal family of factor graphs for E3-CNF formulas, for some $0 < \gamma < \frac{1}{8}$. We shall use this in an FGPR to prove Theorem 4. For simplicity of the presentation we shall describe our reduction as a reduction from a single E3-CNF formula ϕ_3 to a single Ek-CNF formula ϕ_k. As the factor graph resulting for ϕ_k will be independent of polarities of variables in ϕ_3, this will be an FGPR.

Let ϕ_3 be an E3-CNF formula with n variables and m clauses for which one wants to distinguish between the case that it is satisfiable and the case that it is at most $(1 - \gamma)$-satisfiable. Formula ϕ_k will be obtained from a combination of 2^q auxiliary Ek-CNF formulas called ψ_i, for $0 \le i \le 2^q - 1$. Let $q = k - 3$. Introduce q fresh variables $y_1, \ldots, y_q$, and 3 fresh variables z_1, z_2, z_3. Formula ψ_0 is obtained from ϕ_3 by adding the y variables (all in negative polarity) to each clause of ϕ_3. As to the other formulas indexed by $i \ge 1$, each such formula ψ_i has eight clauses, where each clause contains the variables $y_1, \ldots y_q, z_1, z_2, z_3$. Excluding the all negative polarity combination, there are $2^q - 1$ remaining combinations of polarities for the q variables of type y. Each such combination of polarities will be associated with the clauses of one ψ_i for $i \ge 1$. One may think of the binary representation of i as specifying the polarity of the y variables in clauses of ψ_i, where if the j'th bit of i is 0 then y_j is negative, and if the j'th bit of i is 1 then y_j is positive. As to the z variables, there are 8 possible combinations of polarities. Within a formula ψ_i there are 8 clauses, and each of them has a different combination of polarities for the z variables.

The formula ϕ_k will be a weighted mixture of the ψ_i (see [9] regarding an unweighted version). Formula ψ_0 is taken with weight $\frac{1}{8\gamma}$ (which is larger than 1 because $\gamma < \frac{1}{8}$), spreading this weight equally among its m clauses. Each of the other ψ_i is taken with weight 1, spreading the weight equally among its 8 clauses. The total weight of ϕ_k is $2^q - 1 + \frac{1}{8\gamma}$.

If ϕ_3 is satisfiable, so is ϕ_k: an assignment to the original variables of ϕ_3 that satisfies ϕ_3 also satisfies ψ_0, and assigning *true* to all y variables satisfies all ψ_i for $i \ge 1$. If ϕ_3 is only $1 - \gamma$ satisfiable then the weight of unsatisfied clauses in ϕ_k is at least $\frac{1}{8}$: if all variables y are assigned *true*, this results from ψ_0, and in all other cases, this results from one of the other ψ_i.

The total weight of ϕ_k is $W = 2^q - 1 + \frac{1}{8\gamma}$, and for q satisfying $2^q \ge \frac{1-\epsilon}{\epsilon}(\frac{1}{8\gamma} - 1)$ we have that $W \le \frac{2^q}{(1-\epsilon)}$ which implies that $\frac{1}{8} \ge \frac{W(1-\epsilon)}{2^k}$. Hence ϕ_k is at most $\left(1 - \frac{(1-\epsilon)}{2^k}\right)$-satisfiable, as desired. $\square$

Acknowledgments. Work supported in part by The Israel Science Foundation (grant No. 873/08).

References

1. Ajtai, M., Komlós, J., Szemerédi, E.: An O(n log n) sorting network. In: STOC 1983, pp. 1–9. ACM (1983)
2. Alekhnovich, M., Khot, S.A., Kindler, G., Vishnoi, N.K.: Hardness of approximating the closest vector problem with pre-processing. In: FOCS 2005, pp. 216–225 (2005)
3. Applebaum, B., Barak, B., Wigderson, A.: Public-key cryptography from different assumptions. In: STOC 2010, pp. 171–180 (2010)
4. Arora, S., Khot, S.A., Kolla, A., Steurer, D., Tulsiani, M., Vishnoi, N.K.: Unique games on expanding constraint graphs are easy. In: STOC 2008, pp. 21–28 (2008)
5. Bellare, M., Goldreich, O., Sudan, M.: Free bits, pcps, and nonapproximability—towards tight results. SIAM Journal on Computing 27(3), 804–915 (1998)
6. Bruck, J., Naor, M.: The hardness of decoding linear codes with preprocessing. IEEE Transactions on Information Theory 36(2), 381–385 (1990)
7. Creignou, N., Khanna, S., Sudan, M.: Complexity classifications of boolean constraint satisfaction problems. Society for Industrial and Applied Mathematics (2001)
8. Dinur, I.: The PCP theorem by gap amplification. J. ACM 54(3), 12 (2007)
9. Feige, U., Jozeph, S.: Universal Factor Graphs (2012), http://arxiv.org/abs/1204.6484
10. Feige, U., Micciancio, D.: The inapproximability of lattice and coding problems with preprocessing. In: CCC 2002, pp. 32–40 (2002)
11. Feige, U., Kim, J.H., Ofek, E.: Witnesses for non-satisfiability of dense random 3CNF formulas. In: FOCS 2006, pp. 497–508 (2006)
12. Håstad, J.: Some optimal inapproximability results. J. ACM 48, 798–859 (2001)
13. Impagliazzo, R., Paturi, R., Zane, F.: Which problems have strongly exponential complexity? J. Comput. Syst. Sci. 63, 512–530 (2001)
14. Khot, S.: On the power of unique 2-prover 1-round games. In: STOC 2002, pp. 767–775 (2002)
15. Khot, S., Popat, P., Vishnoi, N.: $2^{\log^{1-\epsilon} n}$ Hardness for Closest Vector Problem with Preprocessing. ECCC Report, 119 (2011), (To appear in STOC 2012)
16. Kolla, A.: Spectral algorithms for unique games. In: CCC 2010, pp. 122–130 (2010)
17. Lichtenstein, D.: Planar formulae and their uses. SIAM Journal on Computing 11(2), 329–343 (1982)
18. Lingas, A., Pinter, R., Rivest, R., Shamir, A.: Minimum edge length decompositions of rectilinear figure. In: Proceedings of 12th Annual Allerton Conference on Communication, Control, and Computing (1982)
19. Radhakrishnan, J., Sudan, M.: On Dinur's proof of the PCP theorem. B. AMS 44(1), 19–61 (2007)
20. Raghavendra, P.: Optimal algorithms and inapproximability results for every CSP? In: STOC 2008, pp. 245–254 (2008)
21. Schaefer, T.J.: The complexity of satisfiability problems. In: STOC 1978, pp. 216–226 (1978)
22. Trevisan, L.: Approximation algorithms for unique games. In: FOCS 2005, pp. 197–205 (2005)

Parameterized Approximation via Fidelity Preserving Transformations

Michael R. Fellows[1], Ariel Kulik[2], Frances Rosamond[1], and Hadas Shachnai[3],[*]

[1] School of Engineering and IT, Charles Darwin Univ., Darwin, NT Australia 0909
{michael.fellows,frances.rosamond}@cdu.edu.au
[2] Computer Science Department, Technion, Haifa 32000, Israel
ariel.kulik@gmail.com
[3] Computer Science Department, Technion, Haifa 32000, Israel
hadas@cs.technion.ac.il

Abstract. We motivate and describe a new parameterized approximation paradigm which studies the interaction between performance ratio
and running time for any parametrization of a given optimization problem. As a key tool, we introduce the concept of α-*shrinking transformation*, for $\alpha \geq 1$. Applying such transformation to a parameterized
problem instance decreases the parameter value, while preserving approximation ratio of α (or α-*fidelity*).

For example, it is well-known that *Vertex Cover* cannot be approximated within any constant factor better than 2 [24] (under usual assumptions). Our parameterized α-approximation algorithm for k-*Vertex
Cover*, parameterized by the solution size, has a running time of
$1.273^{(2-\alpha)k}$, where the running time of the best FPT algorithm is 1.273^k
[10]. Our algorithms define a continuous tradeoff between running times
and approximation ratios, allowing practitioners to appropriately allocate computational resources.

Moving even beyond the performance ratio, we call for a new type of
approximative kernelization race. Our α-shrinking transformations can
be used to obtain kernels which are smaller than the best known for a
given problem. For the *Vertex Cover* problem we obtain a kernel size of
$2(2-\alpha)k$. The smaller "α-fidelity" kernels allow us to solve exactly problem instances more efficiently, while obtaining an approximate solution
for the original instance.

We show that such transformations exist for several fundamental problems, including *Vertex Cover*, *d-Hitting Set*, *Connected Vertex Cover* and
Steiner Tree. We note that most of our algorithms are easy to implement
and are therefore practical in use.

1 Introduction

Given the common belief that most NP-hard problems cannot be solved, or
even well-approximated, in polynomial time, it is natural for us to turn to a

* Work partially supported by the Technion V.P.R. Fund, by Smoler Research Fund,
and by the Ministry of Trade and Industry MAGNET program through the NEGEV
Consortium (www.negev-initiative.org).

A. Czumaj et al. (Eds.): ICALP 2012, Part I, LNCS 7391, pp. 351–362, 2012.
© Springer-Verlag Berlin Heidelberg 2012

generalization of polynomial time, *fixed-parameter tractability*, to develop a paradigm of *parameterized approximation*.

Parameterized complexity approaches hard computational problems through a multivariate analysis of the running time. Instead of expressing the running time as a function of the input size n only, the running time is expressed as a function of n and k, where k is a well-defined parameter of the input instance. We say that a problem (with a particular parameter k) is *fixed-parameter tractable (FPT)* if it can be solved in time $f(k) \cdot p(n)$, where f is an arbitrary function depending only on k. Thus we relax polynomial time by committing the exponential explosion to the parameter k. For further background on parameterized complexity we refer the reader to the textbooks [13,21,27], and the recent surveys in [14,16].

Extensive research since the beginning of the 70's has led to results exhibiting limits to the approximability of NP-hard problems. Comprehensive surveys of works on classical approximation algorithms can be found, e.g., in [23,31,33]. Formally, given a maximization (minimization) problem Π, we say that $\mathcal{A}$ is an r-approximation algorithm for some $r \geq 1$, if for any instance I of Π $\mathcal{A}$ yields a solution that satisfies $OPT(I)/\mathcal{A}(I) \leq r$ $(\mathcal{A}(I)/OPT(I) \leq r)$, where $OPT(I)$ is the value of an optimal solution for I. Thus, for instance, *Maximum Independent Set* on a graph $G = (V, E)$, with $|V| = n$, is inapproximable within ratio better than $n^{1-\epsilon}$, for some $\epsilon > 0$, unless P = NP [34]. Assuming the Unique Games Conjecture (UGC), the *Vertex Cover* problem cannot be approximated within any constant factor better than 2, and the best constant-factor approximation for *d-Hitting Set* is d [24]. These results lead to the question that is at the heart of our study.

> "Given an optimization problem, Π, that is hard to approximate within factor ρ, for some $\rho > 1$: can we devise a family of α-approximation algorithms, A_α, such that A_ρ is polynomial, A_1 has the running time of the best FPT algorithm for Π, and A_α defines a continuous tradeoff between approximation ratios and running times"?

We will see later that our parameterized approximation algorithms have performance ratios better than the best possible polynomial-time approximation algorithms (under the common assumption that $P \neq NP$, and assuming that UGC holds). Our technique enables us to obtain any ratio $\alpha \in [1, \rho(\Pi)]$ for a given problem Π, where $\rho(\Pi)$ is the best known polynomial-time approximation ratio for the problem, and α is the approximation ratio achieved, depending on the desired running-time of the algorithm. In developing a general paradigm for parameterized approximation, we combine tools used in approximation algorithms with the framework of parameterized complexity. We move now to an overview of our results, after which will follow an in-depth presentation of α-shrinking transformations.

1.1   Our Results

In this paper, we describe a new parameterized approximation paradigm which relates parameterized complexity and polynomial-time approximation. While many earlier studies refer to parametrization by solution size, or, more generally, by the value of the objective function, our approximation approach can be applied for *any parametrization* of a given problem. We demonstrate our techniques with several fundamental problems, including *Vertex Cover*, *d-Hitting Set*, *Connected Vertex Cover*, and *Steiner Tree*.

We summarize our results in Table 1. For each of the studied problems, we specify the kernel size obtained by our algorithms (when applicable), as well as the running time of the algorithm as function of the approximation ratio, $\alpha \geq 1$, and the best known running time of an exact FPT algorithm for the problem.

Table 1. Approximations via α-fidelity Shrinking: Four Examples

Problem	Parameter	Kernel size	Running time	Best FPT algorithm
Vertex cover	solution size	$2(2-\alpha)k$	$1.273^{(2-\alpha)k}$	1.273^k [9]
Connected vertex cover	solution size	No $k^{O(1)}$	$2^{k(2-\alpha)}$	2^k [12]
3-Hitting set	solution size	$\frac{5(3-\alpha)^2}{4}k^2 + \frac{3-\alpha}{2}k$	$2.076^{k(3-\alpha)/2}$	2.076^k [32]
Steiner tree	size of terminal set	No $k^{O(1)}$	$2^{(3-\alpha)k/2}$	2^k [4]

One of the most important practical techniques in parameterized complexity is *kernelization*. Here one takes a problem specified by $(x, k) \in \Sigma^* \times N$ and produces, typically in polynomial time, a small version of the problem: (x', k') such that (x, k) is a yes instance iff (x', k') is a yes instance, and moreover $|x'| \leq g(k)$ and usually $k' \leq k$. This technique is widely used in practice as it usually relies on a number of easily implementable reduction rules.

There are two types of *races* in parameterized complexity research: the race for the smallest possible function $f(k)$ in the running time of an exact algorithm, and the race for the smallest possible function $g(k)$ to bound the size of a kernel. These races are well-established, and the current leader boards are exhibited on the FPT community wiki [28]. Our parameterized approximation paradigm gives rise to a new kind of race, *approximative kernelization*.

As a key tool in our study, we introduce (in Section 2) the concept of α-*shrinking transformation*, for $\alpha \geq 1$. We shall see that applying such transformation to a parameterized problem instance decreases the parameter value, while preserving α-fidelity in the approximation ratio. We show that α-shrinking transformations can be used also as a tool for *approximative kernelization*, to obtain kernels which are smaller than the best known for a given problem. Thus, we define the notion of α-*fidelity kernel*, for $\alpha \geq 1$, where the special case of $\alpha = 1$ is a standard kernel. Such smaller α-fidelity kernels will allow us to solve exactly problem instances more efficiently, while obtaining an α-approximate solution for the original instance.

Our technique yields a continuous tradeoff between the approximation ratios achieved by an algorithm and the running times. This positive feature will allow practitioners to obtain as much accuracy as they can computationally afford. We note that most of our algorithms are easy to implement and are therefore practical in use.

As shown in [18], our approximation technique utilizes α-shrinking transformations to their full power in solving *Vertex Cover, Connected Vertex Cover*, and *d-Hitting Set*, as long as the transformations are *linear*. Specifically, the running times of our α-approximation algorithms (as well as the sizes of the α-fidelity kernels), are in fact the smallest possible.

In developing our approximation algorithm for Steiner Tree (in Section 4), we make non-standard use of a result of Björklund et al. [4] for solving efficiently the Steiner Tree problem for a subset of the terminals in a given instance.

1.2   Related Work

Recently, it has been proposed that the notion of approximability can be investigated in the framework of fixed-parameter tractability, and various models have been suggested (see, e.g., [8,10,15]. These models seek, for example, an FPT-algorithm which on input k either delivers "a no size k dominating set" or produces one of size $2k$. Marx and Razgon [26] follow this approach and present an algorithm with running time $f(k)n^{O(1)}$ that, given an instance of the Edge Multicut problem and an integer $k \geq 1$, either finds a solution of size $2k$ or correctly concludes that no solution of size k exists. The general subject of parameterized complexity and approximation is well-surveyed by Marx in [25].

A different but very interesting kind of trade-off between exact computation and polynomial approximation has been studied by [30], which proposes to cope with hardness through the usage of *hybrid* algorithms.

Other research has studied the FPT approximability of W-hard problems. The algorithms developed for such problems yield a solution of value $g(k)$ for a problem parameterized by k, where k is the solution size (see, e.g., [22,15,17]).

Parameterized approximations for NP-hard problems by "moderately exponential time" algorithms has been studied with the goal of devising algorithms with exponential running time $O(2^{n/r})$ and r large enough. For *Vertex Coloring* the first $O^*(2^{n/0.77})$-time algorithm by Lawler was then improved in a series of papers culminating in a breakthrough $O^*(2^n)$ bound by Björklund et al. [3]. Bourgeois et al. [5] used such algorithms to improve the best known approximation ratios for subgraph maximization and minimum covering problems. The paper [5] also gives results similar to ours for Vertex Cover, however, the technique used seems to be specialized for Vertex Cover and cannot be applied to other problems. A similar approach was developed by Cygan et al. [11]. Fernau [20] applied a related approach in deriving parameterized approximation schemes for a class of graph minimization problems. Moderately exponential approximation has been investigated by [8,10,15], though with objectives oriented towards development of fixed-parameter algorithms.

Recent works by Brankovic and Fernau [6] and by Fernau [7] present parameterized β-approximation algorithms for Vertex Cover and 3-Hitting Set, for certain values of β, through accelerated branching. The technique was used also in [19], to obtain parameterized approximation algorithms for Total Vertex Cover. The algorithms in [6,7] outperform our algorithms for Vertex Cover and 3-Hitting Set in terms of running times, however, they can be used to obtain only a restricted set of approximation ratios and are significantly more complicated. We note that our approach for obtaining parameterized α-approximations for these problems can be combined with the techniques used in [6] and [7] to obtain improved running times for some values of $\alpha > 1$.

To our knowledge, there have been few studies that link approximation and kernelization. However, in a method for kernelizing vertex deletion problems whose goal graphs can be characterized by forbidden induced subgraphs, van Bevern et al. [29] show how polynomial time approximation results can be exploited in kernelization.

Due to space constraints, some of the proofs are omitted. The detailed results appear in [18].

2   Main Technique: Fidelity Preserving Transformations

We consider languages that consist of words in $U = \{0, 1\}^* \times \mathbb{N}$. Define a language to be $\mathcal{L} \subseteq U$, such that $(x, k) \in \mathcal{L}$ implies that $(x, k + 1) \in \mathcal{L}$. Such a language can represent any minimization problem in which $k \geq 0$ is the objective value.

For some $\alpha \geq 1$, we say that an algorithm $\mathcal{A}$ is α-approximation for a language $\mathcal{L}$ if the following conditions hold. For any $(x, k) \in U$: (i) if $(x, k) \in \mathcal{L}$ then $\mathcal{A}(x, k)$ returns $true$, and (ii) if $\mathcal{A}(x, k)$ returns $true$ then $(x, \alpha k) \in \mathcal{L}$.

Note that this is the standard definition of an approximation algorithm, with the problem described as a language. We consider problems which also have a $parametrization$, that is a function $\kappa : U \to N$. Often, the parametrization of the problem is $\kappa(x, k) = k$. For $\alpha > 1$, our objective is to find α-approximation algorithm (or, a family of algorithms with varying α values) for a given problem $\mathcal{L}$, whose running time is of the form $f(\kappa(x, k)) \cdot |x|^{O(1)}$. Such an algorithm is called $fixed\ parameter\ \alpha$-$approximation$.[1] When $\alpha = 1$, we get a fixed-parameter algorithm for the problem. If there exists such an algorithm for a language $\mathcal{L}$, we say that $\mathcal{L}$ is $fixed$-$parameter\ tractable$ ($\mathcal{L} \in FPT$).

To obtain such an algorithm, we first define the notion of fidelity preserving transformations.

Definition 1. *Given a language $\mathcal{L}$, a transformation $t : U \to U$ is α-fidelity preserving, for a given $\alpha \geq 1$, if the following hold: For any $(x, k) \in U$, (i) if $(x, k) \in \mathcal{L}$ then $t(x, k) \in \mathcal{L}$, and (ii) if $t(x, k) \in \mathcal{L}$ then $(x, \alpha k) \in \mathcal{L}$.*

Indeed, a kernelization of a problem is a 1-fidelity preserving transformation which guarantees that, for any $(x', k') = t(x, k)$, $|x'| \leq f(\kappa(x, k))$ for some function f and $\kappa(x', k') \leq \kappa(x, k)$.

---

[1] We may view a family of such algorithms, which yield α-approximation for any $\alpha > 1$, as the parameterized analog of an *efficient polynomial time approximation scheme (EPTAS)*, since the running times are polynomial in $|x|$, but may depend arbitrarily on $\kappa(x, k)$.

We now introduce the notion of α-shrinking transformation, an α-fidelity transformation which reduces the magnitude of the parameter κ.

Definition 2. *Given a language $\mathcal{L}$ with parametrization κ, a transformation $t : U \to U$ is α-shrinking of order f if*

(i) t is α-fidelity preserving transformation with respect to $\mathcal{L}$.
(ii) For any $(x, k) \in U$ and $(x', k') = t(x, k)$ it holds that $\kappa(x', k') \leq f(\kappa(x, k))$.

If the transformation t can also be evaluated in polynomial time in $|(x, k)|$, we refer to t as a polynomial α-shrinking *of order f.*

2.1   Approximation via Shrinking

We now show that with α-shrinking transformations, we can significantly improve the running time, if we are wiling to settle for an approximation. Given an α-shrinking transformation t of order f, and a parameterized algorithm $\mathcal{A}$ for a problem $\mathcal{L}$, a parameterized approximation algorithm for $\mathcal{L}$ can be obtained as follows. For any $(x, k) \in U$ we simply run $\mathcal{A}(t(x, k))$. If the output of the algorithm is *true* then $t(x, k) \in \mathcal{L}$, and since t is α-fidelity preserving, we have that $(x, \alpha k) \in \mathcal{L}$. Also, if $(x, k) \in \mathcal{L}$ we get that $t(x, k) \in \mathcal{L}$, therefore $\mathcal{A}(t(x, k))$ returns *true*. It follows, that $\mathcal{A}$ is an α-approximation algorithm for $\mathcal{L}$. The running time of $\mathcal{A}$ is of the form $g(\kappa(x, k)) \cdot poly(|x|)$, and therefore the running time of $\mathcal{A}(t(x, k))$ is $g(f(\kappa(x, k))) \cdot poly(|x|)$ plus the time for applying the transformation. When the transformation is polynomial, we get a parameterized α-approximation algorithm. We note that the function g is often exponential in κ, thus any reduction of the value of $f(\kappa(x, k))$ yields a significant improvement in the running time of the algorithm.

For example, in Section 3.2, we present an α-shrinking transformation for *Vertex Cover (VC)* of order $(2-\alpha)k$, for any $1 \leq \alpha \leq 2$. The best known running time of an FPT algorithm for VC is 1.273^k (ignoring polynomial factors), due to [9]. By combining the two, we obtain a parameterized α-approximation for VC, whose running time is $1.273^{(2-\alpha)k}$. For $k = 160$, if we are willing to settle for a 1.25-approximation, we get running time of about 2^{41} as contrasted with 2^{55} for an exact algorithm.

2.2 α-Fidelity Kernels

We can also use α-shrinking to generate α-fidelity kernels, defined as follows.

Definition 3. *Given a language $\mathcal{L}$ with parametrization κ, a transformation $t : U \to U$ is an α-fidelity kernel of size f if*

(i) t is an α-fidelity preserving transformation with respect to $\mathcal{L}$.
(ii) There is a function f such that, for any $(x, k) \in U$ and $(x', k') = t(x, k)$, it holds that $\kappa(x', k') \leq \kappa(x, k)$, and $|(x', k')| \leq f(\kappa(x, k))$.
(iii) t can be evaluated in polynomial time in $|(x, k)|$.

We see that α-fidelity kernels generalize the standard notion of kernels. As often enumeration over a kernel turns out to be faster than branch and bound algorithms (either in running time, or the time required to implement them), it makes sense to find an α-fidelity kernel for a problem (whose size is smaller than the 1-fidelity kernel) and then use enumeration to find an approximate solution.

Given a kernelization algorithm, which yields a kernel of size $g(k)$ for a problem $\mathcal{L}$, and a *polynomial α-shrinking* t of order f for the problem, we can generate an α-fidelity kernel similar to the way we used shrinking to obtain approximation algorithm. For any $(x, k) \in U$, we run the kernelization algorithm over $t(x, k)$. We see that the resulting transformation is an α-fidelity kernel of size $g(f(k))$. For *Vertex Cover*, using the α-shrinking of Section 3.2, this leads to an α-fidelity kernel of size $2(2 - \alpha)k$, for any $1 \le \alpha \le 2$.

3   Parametrization by Problem Objective

The reduction steps we use to obtain the α-shrinking are quite simple. We describe them here, and in the following section show how they are applied. Throughout this section, we consider problems for which the parametrization is $\kappa(x, k) = k$. For simplicity, we ignore the κ notation and simply use k.

3.1   Obtaining α-Shrinking by Simple Reduction Steps

To efficiently obtain polynomial α-shrinking, we use as a key building block the following reduction step.

Definition 4. *Given a language $\mathcal{L}$, a transformation $r : U \to U$ is an (a, b)-reduction step if, for any $(x, k) \in U$ and $(x', k') = r(x, k)$,*

(i) $k' = k - a$
(ii) If $(x, k) \in \mathcal{L}$ then $(x', k') \in \mathcal{L}$.
(iii) For any integer $n \ge 0$, if $(x', k' + n) \in \mathcal{L}$ then $(x, k + b + n) \in \mathcal{L}$.

This reduction step is useful due to the next lemma.

Lemma 1. *Given a language $\mathcal{L}$, an (a, b)-reduction step r and $\alpha \le \frac{a+b}{a}$ such that r can be evaluated in polynomial time, there is polynomial α-shrinking of order $\left(k \cdot \frac{b+a-\alpha a}{b}\right)$ for $\mathcal{L}$.[2]*

Proof. We note that if r is an (a, b)-reduction step, then r^ℓ is $(a\ell, b\ell)$-reduction step. We use this property as follows. Given $(x, k) \in U$, we select $\ell = k \cdot \frac{(\alpha-1)}{b}$ and apply r^ℓ on (x, k). Let t denote the resulting transformation. Now notice, if $t(x, k) \in \mathcal{L}$ then $(x, k + b\ell) = (x, \alpha k) \in \mathcal{L}$. Also, if $(x, k) \in \mathcal{L}$ then $t(x, k) \in \mathcal{L}$, and as r can be evaluated in polynomial time, t can be evaluated in polynomial time as well. This means that t is α-shrinking, and its order is $k' = k - a\ell = \left(k \cdot \frac{b+a-\alpha a}{b}\right)$.   ∎

For many problems, finding such a reduction step is easy, as described in Section 3.2. In all cases, we rely heavily on ideas used in local-ratio algorithms for the problems. For more details on the local ratio technique, see, e.g., [2] .

---

[2] For $\alpha = \frac{a+b}{b}$ an α-approximation for the problem can be obtained by iteratively applying the reduction step.

3.2 Applications of the Technique

In this section we apply our α-shrinking technique to obtain parameterized approximations for Vertex Cover and d-Hitting Set. In [18] we apply the technique to Connected Vertex Cover and Steiner Tree, parameterized by solution size.

Vertex Cover: The *Vertex Cover (VC)* problem is defined as follows. Given a graph $G = (V, E)$, a subset of vertices $S \subseteq V$ is a cover of G if, for any edge $(v, u) \in E$, either $v \in S$ or $u \in S$. The VC problem is to find a cover of G of minimum cardinality. As a language, Vertex Cover can be defined by

$$VC = \{(G, k) \mid \text{there is a cover of } G \text{ of size at most } k\}.$$

Given an instance (G, k), we use the following reduction step. For an arbitrarily selected edge (u, v), let $G' = G \setminus \{u, v\}$. We take $r(G, k) = (G', k - 1)$. Let $(G, k) \in U$, denote $(G', k') = r(G, k)$, and let (u, v) be the edge selected by the transformation r. Then, if $(G, k) \in \mathcal{L}$, there is a vertex cover C of G with $|C| \leq k$; either $u \in C$ or $v \in C$. Therefore, $C' = C \setminus \{v, u\}$ is of size at most $k - 1$, and C' is a cover of G'. Hence, $(G', k') \in \mathcal{L}$. We also note that if $(G', k' + n) \in \mathcal{L}$, then there is a cover C' of G' of size at most $k' + n = k - 1 + n$. Let $C = C' \cup \{u, v\}$, then we see that C is a vertex cover of G of size at most $k' + n + 2 \leq k + 1 + n$.

This implies that r is a $(1, 1)$-reduction for VC. The reduction r can be evaluated in polynomial time and therefore, by Lemma 1, there is a polynomial α-shrinking for VC of order $k \cdot (2 - \alpha)$, for any $1 \leq \alpha \leq 2$. As mentioned above, such shrinking can be used to obtain a parameterized α-approximation algorithm for VC, with running time $1.273^{(2-\alpha)k}$ (ignoring polynomial factors) and an α-fidelity kernel of size $2(2 - \alpha)k$, for any $1 \leq \alpha \leq 2$.

d-Hitting Set: The *d-Hitting Set (d-HS)* problem is the following extension of Vertex Cover to hypergraphs. Given a hypergraph $G = (V, E)$ with edge sizes bounded by d, a set $S \subseteq V$ is a cover of G if, for any $e \in E$, it holds that $e \cap S \neq \emptyset$. The d-HS problem is to find a cover of G of minimum cardinality. As a language, d-hitting-set can be defined by d-HS $= \{(G, k) \mid \text{there is a cover of } G \text{ of size } k\}$.

For any fixed $d \geq 2$, we show a $(1, d-1)$-reduction step for d-HS, which extends the reduction used for VC. Given an instance (G, k), arbitrarily select an edge $e \in E$ and let $G' = (V, E')$, where $E' = E \setminus \{e\}$. Consider $r(G, k) = (G', k - 1)$. It can be easily shown that r is indeed a $(1, d - 1)$-reduction step for d-HS, which can be evaluated in polynomial time. By Lemma 1, there is a polynomial α-shrinking for d-HS of order $k \cdot \frac{d-\alpha}{d-1}$, for any $1 \leq \alpha \leq d$.

The best known parameterized algorithm for 3-HS, due to Wahlström [32], has running time 2.076^k. Combining α-shrinking with this algorithm, we obtain a parameterized α-approximation algorithm with running time $2.076^{k \cdot \frac{3-\alpha}{2}}$, for any $1 \leq \alpha \leq 3$. Abu-Khzam showed a kernelization for d-HS of size $(2d - 1)k^{d-1} + k$ [1]. Combining this kernelization with our α-shrinking, we have an α-fidelity kernel for d-HS of size $(2d - 1) \left(k \cdot \frac{d-\alpha}{d-1} \right)^{d-1} + k \cdot \frac{d-\alpha}{d-1}$, for any $1 \leq \alpha \leq d$.

4 The Parametrized Steiner Tree Problem

The *Steiner Tree (ST)* problem is defined as follows. Given are an undirected graph $G = (V, E)$, a set of terminals $T \subseteq V$, and a value $k \geq 1$. We say that a

subset of edges E' is a *Steiner tree*, if E' forms a tree, and for any $v \in T$ there
is $(u, v) \in E'$. Our objective is to determine if T has a Steiner tree in G of size
k or less. Formally, denote by $ST_G(T)$ a Steiner tree of T of minimum size in G,
and let $ST = \{(G, T, k) \mid |ST_G(T)| \le k\}$.

We consider ST with its standard parametrization, by the number of termi-
nals, that is, $\kappa(G, T, k) = |T|$. We define below an α-shrinking transformation.
While the running time of our shrinking procedure is non-polynomial, it still
yields a significant improvement over the running time of the exact algorithm.

4.1   The Shrinking Technique for Parameterized Steiner Tree

Overview: Our shrinking technique is based on the following observations.

(1) Given a subset $S \subseteq T$, the graph G and the set of terminals T can be
    reduced to G' and T', respectively, such that (*i*) $|T'| = |T| - |S| + 1$, (*ii*) if
    $(G', T', k) \in ST$ then $(G, T, k + |ST_G(S)|) \in ST$, and (*iii*) if $(G, T, k) \in ST$
    then $(G', T', k) \in ST$.
(2) For any $\ell \ge 1$, there is $S \subseteq T$ of size ℓ, such that $ST_G(S) \le ST_G(T) \cdot \frac{2 \cdot \ell}{|T|}$.
(3) For any $\ell \ge 1$, a subset $S \subseteq T$ of size ℓ for which $|ST_G(S)|$ is minimal can be
    found in time $h(|T|, \ell)$ (ignoring polynomial factors), where $h(|T|, \ell)$ is the
    number of subsets of T of size at most ℓ.

Using the above observations, we define our shrinking procedure as follows. We
select $\ell = \frac{(\alpha-1)}{2}|T|$ and find a subset $S \subseteq T$ of size ℓ for which $|ST_G(S)|$ is
minimal. By (2), we have that $|ST_G(S)| \le (\alpha-1)|ST_G(T)|$; therefore, by (1), the
graph G' and the set T' are α-shrinking of (G, T, k) of order $f(|T|) = \frac{3-\alpha}{2}|T|+1$.

Reducing the Graph: For any $S \subseteq T$, we define $G_S = (V_S, E_S), T_S$, which is
basically the graph G after merging all vertices in S to a single vertex, as follows.
The set of vertices is $V_S = V \cup \{s\} \setminus S$, where s is a new vertex. The set of edges is
$E_S = (E \cap (V_S \times V_S)) \cup \{(v, s) | \text{there is } u \in S \text{ such that } (v, u) \in E\}$, and the set
of terminals is $T_S = T \cup \{s\} \setminus S$. Notice that $|T_S| = |T| - |S| + 1$. Given a Steiner
tree H of T in G, let its projection on G_S be $H_S = (H \cap E_S) \cup \{(u, s) | (u, v) \in
H, v \in S\}$. We note that H_S is a connected component in G_S, which spans the
vertices in T_S. Thus, $|ST_{G_S}(T_S)| \le |H_S| \le |H|$.

Now, given a Steiner tree H_S of T_S in G, let H consist of all edges of H_S which
are in G, an edge (u, v) for each $(u, s) \in H$, where $v \in S$ is arbitrarily chosen,
and also $ST_G(S)$. It is not difficult to see that H is a connected component in G
which spans T; therefore, we have that $|ST_G(T)| \le |H_S| + |ST_G(S)|$. The next
lemma shows the existence of a good subset of size ℓ.

Lemma 2.   *For any $\ell \ge 1$ satisfying $|T| \bmod \ell = 0$, there is $S \subseteq T$ of size ℓ,
such that $ST_G(S) \le ST_G(T) \cdot \frac{2\ell}{|T|}$.*

Finding a Good Subset: To find a subset $S \subseteq T$ of size ℓ, such that
$|ST_G(S)|$ is minimal, we use a slight adaptation of the algorithm of [4] for the
(parametrized) Steiner Tree problem. The algorithm uses the following recursive
formula. For any $q \in V$ and $X \subseteq T \setminus \{q\}$,

$$|ST_G(\{q\} \cup X)| = \min_{p \in V} \{|ST_G(\{p,q\})| + g_p(X)\},$$

where

$$g_p(X) = \min_{\emptyset \subset D \subset X} \{|ST_G(\{p\} \cup D)| + |ST_G(\{p\} \cup (X \setminus D))|\}.$$

While a simple bottom up evaluation of the formula has running time $3^{|T|}$, the algorithm of [4] is based on evaluating $g_p(X), |ST_G(\{q\} \cup X)|$ for sets $X \subseteq T$ of increasing size, by using a subset convolution algorithm, with running time $2^{|T|}$. This results in a total running time of $2^{|T|}$ (ignoring polynomial factors).

To find the desired set S, we need to evaluate $g_p(X), |ST_G(\{q\} \cup X)|$ for all sets $X \subseteq T$ satisfying that $|X| \leq \ell$. While not explicitly mentioned in [4], we note that, given the values of $|ST_G(\{q\} \cup X)|$ for any $X \subseteq T$ of size at most r, we can evaluate $g_p(X)$, for any $X \subseteq T$ of size at most $r+1$, in time $h(|T|, r+1)$ (ignoring polynomial factor). This is done by using the convolution algorithm only over sets of size at most $r+1$. Therefore, we can evaluate $|ST_G(\{q\} \cup X)|$, for all sets $X \subseteq T$ such that $|X| \leq \ell$, in time $h(|T|, \ell)$ (ignoring polynomial factors). Now, we can find the set S for which $|ST_G(S)|$ is minimal, by going over all the subsets. Thus, we have

Lemma 3. *For any given ℓ, a subset $S \subseteq T$ of size ℓ for which $|ST_G(S)|$ is minimal can be found in time $h(|T|, \ell)$ (ignoring polynomial factors), where $h(|T|, \ell)$ is the number of subsets of T of size at most ℓ.*

Combining the previous results, and using Stirling's approximation to evaluate the running time, we summarize in the following theorem.

Theorem 1. *For any $1 \leq \alpha \leq 3/2$, there is an α-shrinking of order $f_\alpha(|T|) = \frac{3-\alpha}{2}|T|+1$ for parameterized Steiner Tree. The shrinking can be evaluated in time $\left(\left(\frac{1}{\beta}\right)^\beta \cdot \left(\frac{1}{1-\beta}\right)^{(1-\beta)}\right)^n$, where $\beta = (1-\alpha)/2$ (ignoring polynomial factors).*

4.2   Applicability

Define $g(\beta) = \left(\frac{1}{\beta}\right)^{\left(\frac{\beta}{1-\beta}\right)} \cdot \frac{1}{1-\beta}$, and note that the running time of the shrinking procedure can be written as $g(\beta)^{f_\alpha(|T|)} = g\left(\frac{3-\alpha}{2}\right)^{f_\alpha(|T|)}$ (f_α and β are defined as in Thm 1). Apply the α-shrinking for the given input, and run the algorithm of [4] on the reduced instance. We obtain an α-approximation algorithm for the Steiner Tree problem. The running time of the algorithm is $g\left(\frac{3-\alpha}{2}\right)^{f_\alpha(|T|)} + 2^{f_\alpha(|T|)}$ (ignoring polynomial factors). For any $1 \leq \alpha \leq 1.4$, we have $g\left(\frac{3-\alpha}{2}\right) \leq 2$; therefore, the running time of the algorithm is $2^{f_\alpha(|T|)} = 2^{\left(\frac{3-\alpha}{2}|T|\right)}$.

Theorem 2. *There is a parametrized α approximation algorithm for the Steiner Tree problem, parametrized by the number of terminals, whose running time is $2^{\left(\frac{3-\alpha}{2}\kappa\right)}$ (ignoring polynomial factors), for any $1 \leq \alpha \leq 1.4$.*

5 Discussion

We introduced a new parameterized approximation paradigm with important and general features. Our algorithms, which obtain any approximation ratio between 1 and the best known P-time ratio for a given problem, yield a continuous trade-off between approximation and running times.

We showed how our key tool of α-shrinking transformations can be applied to obtain parameterized approximation algorithms for several fundamental problems. We further showed that, even when the running time of our shrinking procedure is non-polynomial (as in the Steiner Tree problem), it can still yield significant improvement over the running time of an exact algorithm. Finally, we note that in applying our technique, problem parameter is not restricted to be the solution size.

We point to a few of the many avenues for future work.

- Further explore the generic approach of approximations based on α-fidelity shrinking and seek efficient application for other problems, such as Feedback Vertex Set, Edge Dominating Set, and others.
- Further explore approximative kernelization. For example, can non-linear reduction (kernelization) rules be used to obtain decreased running time?
- Extend the approach to problems with no FPT algorithm.

References

1. Abu-Khzam, F.N.: Kernelization Algorithms for D-Hitting Set Problems. In: Dehne, F., Sack, J.-R., Zeh, N. (eds.) WADS 2007. LNCS, vol. 4619, pp. 434–445. Springer, Heidelberg (2007)
2. Bar-Yehuda, R.: One for the price of two: a unified approach for approximating covering problems. Algorithmica 27(2), 131–144 (2000)
3. Björklund, A., Husfeldt, T.: Inclusion–exclusion algorithms for counting set partitions. In: FOCS, pp. 575–582 (2006)
4. Björklund, A., Husfeldt, T., Kaski, P., Koivisto, M.: Fourier meets Möbius: fast subset convolution. In: STOC, pp. 67–74 (2007)
5. Bourgeois, N., Escoffier, B., Paschos, V.T.: Efficient Approximation of Combinatorial Problems by Moderately Exponential Algorithms. In: Dehne, F., Gavrilova, M., Sack, J.-R., Tóth, C.D. (eds.) WADS 2009. LNCS, vol. 5664, pp. 507–518. Springer, Heidelberg (2009)
6. Brankovic, L., Fernau, H.: Combining Two Worlds: Parameterised Approximation for Vertex Cover. In: Cheong, O., Chwa, K.-Y., Park, K. (eds.) ISAAC 2010. LNCS, vol. 6506, pp. 390–402. Springer, Heidelberg (2010)
7. Brankovic, L., Fernau, H.: Parameterized Approximation Algorithms for HITTING SET. In: Solis-Oba, R., Persiano, G. (eds.) WAOA 2011. LNCS, vol. 7164, pp. 63–76. Springer, Heidelberg (2012)
8. Cai, L., Huang, X.: Fixed-parameter approximation: Conceptual framework and approximability results. Algorithmica 57(2), 398–412 (2010)
9. Chen, J., Kanj, I.A., Xia, G.: Improved Parameterized Upper Bounds for Vertex Cover. In: Královič, R., Urzyczyn, P. (eds.) MFCS 2006. LNCS, vol. 4162, pp. 238–249. Springer, Heidelberg (2006)
10. Chen, Y.-J., Grohe, M., Grüber, M.: On Parameterized Approximability. In: Bodlaender, H.L., Langston, M.A. (eds.) IWPEC 2006. LNCS, vol. 4169, pp. 109–120. Springer, Heidelberg (2006)

11. Cygan, M., Kowalik, L., Pilipczuk, M., Wykurz, M.: Exponential-time approximation of hard problems. CoRR abs/0810.4934 (2008)
12. Cygan, M., Nederlof, J., Pilipczuk, M., Pilipczuk, M., van Rooij, J.M.M., Wojtaszczyk, J.O.: Solving connectivity problems parameterized by treewidth in single exponential time. In: FOCS 2011 (2011)
13. Downey, R.G., Fellows, M.R.: Parameterized Complexity. Springer (1999)
14. Downey, R.G., Fellows, M.R., Langston, M.A.: The computer journal special issue on parameterized complexity: Foreword by the guest editors. Comput. J. 51(1), 1–6 (2008)
15. Downey, R.G., Fellows, M.R., McCartin, C., Rosamond, F.A.: Parameterized approximation of dominating set problems. Inf. Process. Lett. 109(1), 68–70 (2008)
16. Downey, R.G., Thilikos, D.M.: Confronting intractability via parameters. Computer Science Review 5(4), 279–317 (2011)
17. Drescher, M., Vetta, A.: An approximation algorithm for the maximum leaf spanning arborescence problem. ACM Transactions on Algorithms 6(3) (2010)
18. Fellows, M.R., Kulik, A., Rosamond, F., Shachnai, H.: Parameterized approximation via fidelity preserving transformations. full version,
    http://www.cs.technion.ac.il/~hadas/PUB/FKRS_approx_param.pdf/
19. Fernau, H.: Saving on phases: Parametrized approximation for total vertex cover. In: IWOCA 2012 (2012)
20. Fernau, H.: A systematic approach to moderately exponential-time approximation schemes. Manusctript (2012)
21. Flum, J., Grohe, M.: Parameterized Complexity Theory. An EATCS Series: Texts in Theoretical computer Science. Springer (1998)
22. Grohe, M., Grüber, M.: Parameterized Approximability of the Disjoint Cycle Problem. In: Arge, L., Cachin, C., Jurdziński, T., Tarlecki, A. (eds.) ICALP 2007. LNCS, vol. 4596, pp. 363–374. Springer, Heidelberg (2007)
23. Hochbaum, D.S.: Approximation Algorithms for NP-Hard Problems. PWS Publishing Company (1997)
24. Khot, S., Regev, O.: Vertex cover might be hard to approximate to within 2-epsilon. J. Comput. Syst. Sci. 74(3), 335–349 (2008)
25. Marx, D.: Parameterized complexity and approximation algorithms. Comput. J. 51(1), 60–78 (2008)
26. Marx, D., Razgon, I.: Constant ratio fixed-parameter approximation of the edge multicut problem. Information Processing Letters 109(20), 1161–1166 (2009)
27. Niedermeier, R.: Invitation to Fixed-Parameter Algorithms. Oxford Lecture Series in Mathematics and Its Applications. Oxford Univerity Press (2006)
28. Parameterized Complexity community Wiki., http://fpt.wikidot.com/
29. van Bevern, R., Moser, H., Niedermeier, R.: Kernelization Through Tidying. In: López-Ortiz, A. (ed.) LATIN 2010. LNCS, vol. 6034, pp. 527–538. Springer, Heidelberg (2010)
30. Vassilevska, V., Williams, R., Woo, S.L.M.: Confronting hardness using a hybrid approach. In: SODA 2006, pp. 1–10 (2006)
31. Vazirani, V.V.: Approximation Algorithms. Springer (2001)
32. Wahlström, M.: Algorithms, Measures and Upper Bounds for Satisfiability and Related Problems. PhD thesis, Department of Computer and Information Science. Linkopings University, Sweden (2007)
33. Williamson, D.P., Shmoys, D.B.: The Design of Approximation Algorithms. Cambridge University Press (2011)
34. Zuckerman, D.: Linear degree extractors and the inapproximability of max clique and chromatic number. In: STOC 2006, pp. 681–690 (2006)

Backdoors to Acyclic SAT[*]

Serge Gaspers[1,2] and Stefan Szeider[2]

[1] School of Computer Science and Engineering, The University of New South Wales,
Sydney, Australia
gaspers@kr.tuwien.ac.at
[2] Institute of Information Systems, Vienna University of Technology,
Vienna, Austria
stefan@szeider.net

Abstract. Backdoor sets contain certain key variables of a CNF formula F that make it easy to solve the formula. More specifically, a *weak backdoor set* of F is a set X of variables such that there exits a truth assignment τ to X that reduces F to a satisfiable formula $F[\tau]$ that belongs to a polynomial-time decidable base class C. A *strong backdoor set* is a set X of variables such that for all assignments τ to X, the reduced formula $F[\tau]$ belongs to C.

We study the problem of finding backdoor sets of size at most k with respect to the base class of CNF formulas with acyclic incidence graphs, taking k as the parameter. We show that

1. the detection of weak backdoor sets is W[2]-hard in general but fixed-parameter tractable for r-CNF formulas, for any fixed $r \geq 3$, and
2. the detection of strong backdoor sets is fixed-parameter approximable.

Result 1 is the the first positive one for a base class that does not have a characterization with obstructions of bounded size. Result 2 is the first positive one for a base class for which strong backdoor sets are more powerful than deletion backdoor sets.

Not only SAT, but also #SAT can be solved in polynomial time for CNF formulas with acyclic incidence graphs. Hence Result 2 establishes a new structural parameter that makes #SAT fixed-parameter tractable and that is incomparable with known parameters such as treewidth and clique-width. We obtain the algorithms by a combination of an algorithmic version of the Erdős-Pósa Theorem, Courcelle's model checking for monadic second order logic, and new combinatorial results on how disjoint cycles can interact with the backdoor set.

1 Introduction

Since the advent of computational complexity in the 1970s it quickly became apparent that a large number of important problems are intractable [16]. This predicament motivated significant efforts to identify tractable special cases. For the propositional satisfiability problem (SAT), dozens of such "islands of tractability" have been identified [14]. Whereas it may seem unlikely that a real-world

---

[*] The full version of the paper is available on arXiv [17].

A. Czumaj et al. (Eds.): ICALP 2012, Part I, LNCS 7391, pp. 363–374, 2012.
© Springer-Verlag Berlin Heidelberg 2012

instance belongs to a known island of tractability, it may be "close" to one. In this paper we study the question of whether we can exploit the proximity of a SAT instance to the island of acyclic formulas algorithmically.

For SAT, the distance to an island of tractability (or *base class*) C is most naturally measured in terms of the number of variables that need to be instantiated to put the formula into C. Williams *et al.* [32] introduced the term "*backdoor set*" for sets of such variables, and distinguished between weak and strong backdoor sets. A set B of variables is a *weak C-backdoor set* of a CNF formula F if for at least one partial truth assignment $\tau : B \to \{0,1\}$, the restriction $F[\tau]$ is satisfiable and belongs to C. The set B is a *strong C-backdoor set* of F if for every partial truth assignment $\tau : B \to \{0,1\}$ the restriction $F[\tau]$ belongs to C.

1.1   Weak Backdoor Sets

If we are given a weak C-backdoor set of F of size k, we know that F is satisfiable, and we can verify the satisfiability of F by checking whether at least one of the 2^k assignments to the backdoor variables leads to a satisfiable formula that belongs to C. If the base class allows to find an actual satisfying assignment in polynomial time, as is usually the case, we can find a satisfying assignment of F in $2^k n^{O(1)}$ time. Can we find such a backdoor set quickly if it exists? For all reasonable base classes C it is NP-hard to decide, given a CNF formula F and an integer k, whether F has a strong or weak C-backdoor set of size at most k. On the other hand, the problem is clearly solvable in time $n^{k+O(1)}$. The question is whether we can get k out of the exponent, and find a backdoor set in time $f(k)n^{O(1)}$, i.e., is weak backdoor set detection *fixed-parameter tractable (FPT)* in k? Over the last couple of years, this question has been answered for various base classes C; Table 1 gives an overview of some of the known results. See [18] for a survey.

For general CNF, the detection of weak C-backdoor sets is W[2]-hard for all reasonable base classes C. For some base classes the problem becomes FPT if clause lengths are bounded. All FPT results for weak backdoor set detection in Table 1 are due to the fact that for r-CNF formulas, where $r \geq 3$ is a fixed constant, membership in the considered base class can be characterized by certain obstructions of bounded size. Formally, say that a base class C has the *small obstruction property* if there is a family $\mathcal{F}$ of CNF formulas, each with a finite number of clauses, such that for every CNF formula F, $F \in C$ if and only if F contains no subset of clauses isomorphic to a formula in $\mathcal{F}$. Hence, if a base class C has this property, fixed-parameter tractability for weak C-backdoor set detection for r-CNF formulas can be established by a bounded search tree algorithm.

The base class FOREST is another class for which the detection of weak backdoor sets is W[2]-hard for general CNF formulas (Theorem 3). For r-CNF formulas the above argument does not apply because FOREST does not have the small obstruction property. Nevertheless, we can still show that the weak FOREST backdoor set detection problem is FPT for r-CNF formulas, for every fixed $r \geq 3$ (Theorem 4). This is our first main result.

Table 1. The parameterized complexity of finding weak and strong backdoor sets of CNF formulas and r-CNF formulas, where $r \geq 3$ is a fixed integer

Base Class	Weak		Strong	
	CNF	r-CNF	CNF	r-CNF
HORN	W[2]-h [22]	FPT	FPT [22]	FPT [22]
2-CNF	W[2]-h [22]	FPT	FPT [22]	FPT [22]
UP	W[P]-c [30]	W[P]-c [30]	W[P]-c [30]	W[P]-c [30]
RHORN	W[2]-h [18]	W[2]-h [18]	W[2]-h [18]	open
CLU	W[2]-h [23]	FPT	W[2]-h [23]	FPT [23]

1.2   Strong Backdoor Sets

Given a strong $\mathcal{C}$-backdoor set of size k of a formula F, one can decide whether F is satisfiable by 2^k polynomial checks. In Table 1, HORN and 2-CNF are the only base classes for which strong backdoor set detection is FPT in general. A possible reason for the special status of these two classes is the fact that they have the *deletion property*: for $\mathcal{C} \in \{\text{HORN}, 2\text{-CNF}\}$ a set X of variables is a strong $\mathcal{C}$-backdoor set of F if and only if X is a *deletion $\mathcal{C}$-backdoor set* of F, i.e., the formula obtained from F by deleting all positive and negative occurrences of the variables in X, is in $\mathcal{C}$. The advantage of the deletion property is that it simplifies the search for a strong backdoor set. Its disadvantage is that the backdoor set cannot "repair" a formula differently for different truth assignments of the backdoor variables, and thus it does not use the full power of all the partial assignments. Indeed, for other base classes one can construct formulas with small strong backdoor sets whose smallest deletion backdoor sets are arbitrarily large. In view of these results, one wonders whether a small strong backdoor set can be found efficiently for a base class that does not have the deletion property. Our second main result provides a positive answer. Namely we exhibit an FPT algorithm, which, for a CNF formula F and a positive integer parameter k, either concludes that F has no strong FOREST-backdoor set of size k or concludes that F has a strong FOREST-backdoor set of size at most 2^k (Theorem 5).

This FPT-approximation result is interesting for several reasons. First, it implies that SAT and #SAT are FPT, parameterized by the size of a smallest strong FOREST-backdoor set. Second, (unlike the size of a smallest deletion FOREST-backdoor set) the size of a smallest strong FOREST-backdoor set is incomparable to the treewidth of the incidence graph. Hence the result applies to formulas that cannot be solved efficiently by other known methods. Finally, it exemplifies a base class that does not satisfy the deletion property, for which strong backdoor sets are FPT-approximable.

1.3   #SAT and Implied Cycle Cutsets

Our second main result, Theorem 5, has applications to the model counting problem #SAT, a problem that occurs, for instance, in the context of Bayesian

Reasoning [2,26]. #SAT is #P-complete [31] and remains #P-hard even for monotone 2-CNF formulas and Horn 2-CNF formulas, and it is NP-hard to approximate the number of models of a formula with n variables within $2^{n^{1-\epsilon}}$ for $\epsilon > 0$, even for monotone 2-CNF formulas and Horn 2-CNF formulas [26]. A common approach to solve #SAT is to find a small *cycle cutset* (or feedback vertex set) of variables of the given CNF formula, and by summing up the number of satisfying assignments of all the acyclic instances one gets by setting the cutset variables in all possible ways [7]. Such a cycle cutset is nothing but a *deletion* FOREST-backdoor set. By considering *strong* FOREST-backdoor sets instead, one can get super-exponentially smaller sets of variables, and hence a more powerful method. A strong FOREST-backdoor set can be considered as a an *implied cycle cutset* as it can cut cycles by removing clauses that are satisfied by certain truth assignments to the backdoor variables. Theorem 5 states that we can find a small implied cycle cutset efficiently if one exists.

2   Preliminaries

We refer to standard textbooks for background in parameterized complexity [9,12] and graph theory [8].

Backdoors. A *literal* is a propositional variable x or its negation $\neg x$. A *clause* is a disjunction of literals that does not contain a complementary pair x and $\neg x$. A *propositional formula* in *conjunctive normal form* (CNF formula) is a conjunction of clauses. An r-CNF formula is a CNF formula where each clause contains at most r literals. For a clause c, we write $\mathsf{lit}(c)$ and $\mathsf{var}(c)$ for the sets of literals and variables occurring in c, respectively. For a CNF formula F we write $\mathsf{cla}(F)$ for its set of clauses, $\mathsf{lit}(F) = \bigcup_{c \in \mathsf{cla}(F)} \mathsf{lit}(c)$ for its set of literals, and $\mathsf{var}(F) = \bigcup_{c \in \mathsf{cla}(F)} \mathsf{var}(c)$ for its set of variables.

Let F be a CNF formula and $X \subseteq \mathsf{var}(F)$. We denote by 2^X the set of all mappings $\tau : X \to \{0,1\}$, the *truth assignments* on X. A truth assignment on X can be extended to the literals over X by setting $\tau(\neg x) = 1 - \tau(x)$ for all $x \in X$. Given a $\tau \in 2^X$, $F[\tau]$ denotes the formula obtained from F by removing all clauses c such that τ sets a literal of c to 1, and removing the literals set to 0 from all remaining clauses. F is *satisfiable* if there is some $\tau \in 2^{\mathsf{var}(F)}$ with $\mathsf{cla}(F[\tau]) = \emptyset$. SAT is the NP-complete problem of deciding whether a given CNF formula is satisfiable [5,21]. #SAT is the #P-complete problem of determining the number of distinct $\tau \in 2^{\mathsf{var}(F)}$ with $\mathsf{cla}(F[\tau]) = \emptyset$ [31].

Backdoor Sets (BDSs) are defined with respect to a fixed class $\mathcal{C}$ of CNF formulas, the *base class*. Let $B \subseteq \mathsf{var}(F)$. B is a *strong $\mathcal{C}$-BDS* of F if $F[\tau] \in \mathcal{C}$ for each $\tau \in 2^B$. B is a *weak $\mathcal{C}$-BDS* of F if there is an assignment $\tau \in 2^B$ such that $F[\tau]$ is satisfiable and $F[\tau] \in \mathcal{C}$. B is a *deletion $\mathcal{C}$-BDS* of F if $F - B \in \mathcal{C}$, where $\mathsf{cla}(F - B) = \{c \setminus \{x, \neg x : x \in B\} : c \in \mathsf{cla}(F)\}$.

The challenging problem is to find a strong, weak, or deletion $\mathcal{C}$-BDS of size at most k if it exists. This leads to the following backdoor detection problems for any base class $\mathcal{C}$.

STRONG $\mathcal{C}$-BDS DETECTION

Input: A CNF formula F and an integer $k \geq 0$.
Parameter: The integer k.
Question: Does F have a strong $\mathcal{C}$-backdoor set of size at most k?

The problems WEAK $\mathcal{C}$-BDS DETECTION and DELETION $\mathcal{C}$-BDS DETECTION are defined similarly.

Acyclic Formulas. The *incidence graph* of a CNF formula F is the bipartite graph $\mathsf{inc}(F) = (V, E)$ with $V = \mathsf{var}(F) \cup \mathsf{cla}(F)$ and for a variable $x \in \mathsf{var}(F)$ and a clause $c \in \mathsf{cla}(F)$ we have $xc \in E$ if $x \in \mathsf{var}(c)$. The edges of G may be annotated by a function $\mathsf{sign} : E \to \{+, -\}$. The *sign* of an edge xc is

$$\mathsf{sign}(xc) = \begin{cases} + & \text{if } x \in \mathsf{lit}(c), \text{ and} \\ - & \text{if } \neg x \in \mathsf{lit}(c) . \end{cases}$$

A *cycle* in F is a cycle in $\mathsf{inc}(F)$. The formula F is *acyclic* if $\mathsf{inc}(F)$ is acyclic. We denote by FOREST the set of all acyclic CNF formulas.

The satisfiability of formulas from FOREST can be decided in polynomial time, and even the number of satisfying assignments of formulas from FOREST can be determined in polynomial time [11,28].

The *strong clause-literal graph* of F is the graph $\mathsf{slit}(F) = (V, E)$ with $V = \mathsf{lit}(F) \cup \mathsf{cla}(F)$. There is an edge $uc \in E$ with $u \in \mathsf{lit}(F)$ and $c \in \mathsf{cla}(F)$ if $u \in \mathsf{lit}(c)$, and there is an edge $uv \in E$ with $u, v \in \mathsf{lit}(F)$ if $u = \neg v$ or $\neg u = v$.

Lemma 1. *Let F be a CNF formula, τ be an assignment to $B \subseteq \mathsf{var}(F)$. The formula $F[\tau]$ is acyclic if and only if $\mathsf{slit}(F) - N[\mathsf{true}(\tau)]$ is acyclic.*

It follows that there is a bijection between assignments τ such that $F[\tau]$ is acyclic and independent sets $Y \subseteq \mathsf{lit}(F)$ in $\mathsf{slit}(F)$ such that $\mathsf{slit}(F) - N[Y]$ is acyclic.

3 Background and Methods

The simplest type of FOREST-BDSs are deletion FOREST-BDSs. In the incidence graph, they correspond to feedback vertex sets that are subsets of $\mathsf{var}(F)$. Therefore, algorithms solving slight generalizations of FEEDBACK VERTEX SET can be used to solve the DELETION FOREST-BDS DETECTION problem. By results from [4] and [13], DELETION FOREST-BDS DETECTION is FPT and can be solved in time $5^k \cdot \|F\|^{O(1)}$ and in time $1.7548^n \cdot \|F\|^{O(1)}$, where n is the number of variables of F and $\|F\| = \sum_{c \in \mathsf{cla}(F)} |\mathsf{lit}(c)|$ denotes the formula length.

Any deletion FOREST-BDS B of a CNF formula F is also a strong FOREST-BDS of F, and if F is satisfiable, then B is also a weak FOREST-BDS. In recent years SAT has been studied with respect to several width parameters of graphs and hypergraphs associated with formulas [1,11,15,24,28,29]. Several parameters, such as the treewidth of incidence graphs and the clique-width of signed incidence graphs, are more general than the size of a smallest deletion FOREST-BDS.

The size of a smallest weak or strong FOREST-BDS is incomparable to treewidth and clique-width. On one hand, one can construct formulas with arbitrary large FOREST-BDSs by taking the disjoint union of formulas with bounded width. On the other hand, consider an $r \times r$ grid of variables and subdivide each edge by a clause. Now, add a variable x that is contained positively in all clauses subdividing horizontal edges and negatively in all other clauses. The set $\{x\}$ is a weak and strong FOREST-BDS of this formula, but the treewidth and clique-width of the formula depend on r. Therefore, weak and strong FOREST-BDSs have the potential of augmenting the tractable fragments of SAT formulas.

In the remainder of this section we outline our algorithms. To find a weak or strong FOREST-BDS, consider the incidence graph $G = \mathrm{inc}(F)$ of the input formula F. By Robertson and Seymour's Grid Minor Theorem [25] there is a function $f : \mathbb{N} \to \mathbb{N}$ such that for every integer r, either $\mathrm{tw}(G) \leq f(r)$ or G has an $r \times r$ grid minor. Here, $\mathrm{tw}(G)$ denotes the treewidth of G. Choosing r to be a function of the parameter k, it suffices to solve the problems for incidence graphs whose treewidth is upper bounded by a function of k, and for incidence graphs that contain an $r \times r$ grid minor, where r is lower bounded by a function of k. The former case can be solved by invoking Courcelle's theorem [6], as the FOREST-BDS DETECTION problems can be defined in Monadic Second Order Logic. In the latter case we can make use of the fact that G contains many vertex-disjoint cycles and we consider several cases how these cycles might disappear from $\mathrm{inc}(F)$ by assigning values to variables.

In order to obtain slightly better bounds, instead of relying on the Grid Minor Theorem, we use the Erdős-Pósa Theorem [10] and an algorithmization by Bodlaender [3] to distinguish between the cases where G has small treewidth (in fact, a small feedback vertex set) or many vertex-disjoint cycles.

Theorem 1 ([10]). *Let $k \geq 0$ be an integer. There exists a function $f(k) = O(k \log k)$ such that every graph either contains k vertex-disjoint cycles or has a feedback vertex set of size $f(k)$.*

Theorem 2 ([3]). *Let $k \geq 2$ be an integer. There exists an $O(n)$ time algorithm, taking as input a graph G on n vertices, that either finds k vertex-disjoint cycles in G or finds a feedback vertex set of G of size at most $12k^2 - 27k + 15$.*

We will use Theorem 2 to distinguish between the case where G has a feedback vertex set of size $\mathsf{fvs}(k)$ and the case where G has $\mathsf{cycles}(k)$ vertex-disjoint cycles, for some function $\mathsf{cycles} : \mathbb{N} \to \mathbb{N}$, where $\mathsf{fvs}(k) = 12(\mathsf{cycles}(k))^2 - 27\mathsf{cycles}(k) + 15$.

Suppose G has a feedback vertex set W of size $\mathsf{fvs}(k)$. By adding W to every bag of an optimal tree decomposition of $G - W$, we obtain a tree decomposition of G of width at most $\mathsf{fvs}(k) + 1$. We then define the WEAK and STRONG FOREST-BDS DETECTION problems in Monadic Second Order Logic (MSO) and use Courcelle's theorem [6] to conclude.

Our main arguments come into play when Bodlaender's algorithm returns a set $\mathcal{C}$ of $\mathsf{cycles}(k)$ vertex-disjoint cycles of G. The algorithms will then compute a set $S^* \subseteq \mathrm{var}(F)$ whose size is upper bounded by a function of k such that every weak/strong FOREST-BDS of size at most k contains a variable from S^*.

A standard branching argument will then be used to recurse. In the case of WEAK FOREST-BDS DETECTION, F has a weak FOREST-BDS of size at most k if and only if there is a variable $x \in S^*$, such that $F[x = 0]$ or $F[x = 1]$ has a weak FOREST-BDS of size at most $k - 1$. In the case of STRONG FOREST-BDS DETECTION, F has no strong FOREST-BDS of size at most k if for every variable $x \in S^*$, $F[x = 0]$ or $F[x = 1]$ has no strong FOREST-BDS of size at most $k - 1$; and if $F[x = 0]$ and $F[x = 1]$ have strong FOREST-BDSs B and B' of size at most $2^{k-1} - 1$, then $B \cup B' \cup \{x\}$ is a strong FOREST-BDS of F of size at most $2^k - 1$. This leads to a factor $2^k/k$ approximation.

In order to compute the set S^*, the algorithms consider how the cycles in $\mathcal{C}$ can interact with a BDS. Let x be a variable and C a cycle in G. In the case of weak FOREST-BDSs, we say that x kills[1] C if either inc($F[x = 1]$) or inc($F[x = 0]$) does not contain C. In the case of strong FOREST-BDSs, we say that x kills C if neither inc($F[x = 1]$) nor inc($F[x = 0]$) contain C. We say that x kills C internally if $x \in C$, and that x kills C externally if x kills C but does not kill it internally. In any FOREST-BDS of size at most k, at most k cycles from $\mathcal{C}$ can be killed internally, since all cycles from $\mathcal{C}$ are vertex-disjoint. The algorithms go over all possible choices of selecting k cycles from $\mathcal{C}$ that may be killed internally. All other cycles $\mathcal{C}'$ need to be killed externally. The algorithms now aim at computing a set S such that every weak/strong FOREST-BDS of size at most k which is a subset of var(F) $\setminus \bigcup_{C \in \mathcal{C}'}$ var(C) contains a variable from S.

Computing the set S is the most challenging part of this work. In the algorithm for weak FOREST-BDSs there is an intricate interplay between several cases, making use of bounded clause lengths. In the algorithm for strong FOREST-BDSs a further argument is needed to obtain a more structured interaction between the considered cycles and their external killers.

4   Weak Forest-BDSs

By a parameterized reduction from HITTING SET, WEAK FOREST-BDS DETECTION is easily shown to be W[2]-hard.

Theorem 3. WEAK FOREST-BDS DETECTION *is* W[2]-*hard*.

In the remainder of this section, we consider the WEAK FOREST-BDS DETECTION problem for r-CNF formulas, for any fixed integer $r \geq 3$. Let F be an r-CNF formula, and consider its incidence graph $G = \text{inc}(F)$. We use Theorem 2 to distinguish between the case where G has many vertex-disjoint cycles and the case where G has a small feedback vertex set. In the latter case the problem is expressed in MSO and solved by Courcelle's theorem.

Lemma 2. *Given a feedback vertex set of* inc(F) *of size* fvs(k), WEAK FOREST-BDS DETECTION *is fixed-parameter tractable.*

---

[1] We apologize for the violent language.

Let $\mathcal{C} = \{C_1, \ldots, C_{\mathsf{cycles}(k)}\}$ denote vertex-disjoint cycles in G, with $\mathsf{cycles}(k) = 2k + 1$. We describe an algorithm that finds a set S^* of $O(r4^k k^6)$ variables from $\mathsf{var}(F)$ such that every weak FOREST-BDS of F of size at most k contains a variable from S^*. We will use several functions of k in our arguments. Let

$$\mathsf{ext\text{-}cycles}(k) := \mathsf{cycles}(k) - k, \qquad \mathsf{supp}(k) := (r - 3) \cdot (k^3 + 9) + 4k^2 + k,$$

$$\mathsf{multi}(k) := 4k, \text{ and} \qquad \mathsf{overlap}(k) := (r - 2) \cdot (k \cdot \mathsf{multi}(k))^2 + k.$$

Let C be a cycle in G and $x \in \mathsf{var}(F)$. Recall that x *kills C internally* if $x \in C$. In this case, x is an *internal killer* for C. We say that x *kills C externally* if $x \notin C$ and there is a clause $u \in \mathsf{cla}(F) \cap C$ such that $xu \in E$. In this case, x is an *external killer* for C. We first dispense with cycles that are killed internally. Our algorithm goes through all $\binom{\mathsf{cycles}(k)}{k}$ ways to choose k cycles from $\mathcal{C}$ that may be killed internally. W.l.o.g., let $C_{\mathsf{ext\text{-}cycles}(k)+1}, \ldots, C_{\mathsf{cycles}(k)}$ denote the cycles that may be killed internally. All other cycles $\mathcal{C}' = \{C_1, \ldots, C_{\mathsf{ext\text{-}cycles}(k)}\}$ need to be killed externally. Let $\mathsf{var}'(F) = \mathsf{var}(F) \setminus \bigcup_{i=1}^{\mathsf{ext\text{-}cycles}(k)} \mathsf{var}(C_i)$ denote the variables that may be selected in a weak FOREST-BDS killing no cycle from $\mathcal{C}'$ internally. From now on, consider only external killers from $\mathsf{var}'(F)$. The algorithm will find a set S of $O(rk^6)$ variables such that S contains a variable from every weak FOREST-BDS $B \subseteq \mathsf{var}'(F)$ of F with $|B| \leq k$. The algorithm first computes the set of external killers (from $\mathsf{var}'(F)$) for each of these cycles. Then the algorithm applies the first applicable from the following rules.

Rule 1 (No External Killer). *If there is a $C_i \in \mathcal{C}'$ that has no external killer, then set $S := \emptyset$.*

For each $i \in \{1, \ldots, \mathsf{ext\text{-}cycles}(k)\}$, let x_i be an external killer of C_i that has a maximum number of neighbors in C_i.

Rule 2 (Multi-Killer Unsupported). *If there is a $C_i \in \mathcal{C}'$ such that x_i has $\ell \geq \mathsf{multi}(k)$ neighbors in C_i and at most $\mathsf{supp}(k)$ external killers of C_i have at least $\ell/(2k)$ neighbors in C_i, then include all these external killers in S.*

Rule 3 (Multi-Killer Supported). *If there is a $C_i \in \mathcal{C}'$ such that x_i has $\ell \geq \mathsf{multi}(k)$ neighbors in C_i and more than $\mathsf{supp}(k)$ external killers of C_i have at least $\ell/(2k)$ neighbors in C_i, then set $S := \{x_i\}$.*

Rule 4 (Large Overlap). *If there are two cycles $C_i, C_j \in \mathcal{C}'$, with at least $\mathsf{overlap}(k)$ common external killers, then set $S := \emptyset$.*

Rule 5 (Small Overlap). *Include in S all vertices that are common external killers of at least two cycles from $\mathcal{C}'$.*

Lemma 3. *Rules 1–5 are sound.*

Lemma 4. *There is an FPT algorithm that, given an r-CNF formula F, a positive integer parameter k, and $\mathsf{cycles}(k)$ vertex-disjoint cycles in $\mathsf{inc}(F)$, finds a set S^* of $O(r4^k k^6)$ variables in F such that every weak FOREST-BDS of F of size at most k contains a variable from S^*.*

Our FPT algorithm for WEAK FOREST-BDS DETECTION, restricted to r-CNF formulas, $r \geq 3$, is now easily obtained.

Theorem 4. *For every fixed* $r \geq 3$, WEAK FOREST-BDS DETECTION *is fixed-parameter tractable for* r-*CNF formulas.*

5   Strong Forest-BDSs

In this section we design an algorithm that, given a CNF formula F and an integer k, either concludes that F has no strong FOREST-BDS of size at most k or concludes that F has a strong FOREST-BDS of size at most 2^k.

Let $G = (V, E) = \mathsf{inc}(F)$. Again, we consider the cases where G has a small feedback vertex set or a large number of vertex-disjoint cycles separately. Let

$$\mathsf{cycles}(k) = k^2 2^{k-1} + k + 1,$$

$$\mathsf{ext\text{-}cycles}(k) = \mathsf{cycles}(k) - k, \text{ and}$$

$$\mathsf{fvs}(k) = 12(\mathsf{cycles}(k))^2 - 27\mathsf{cycles}(k) + 15.$$

The case where G has a small feedback vertex set is again solved by formulating the problem in MSO and using Courcelle's theorem.

Lemma 5. *Given a feedback vertex set of* $\mathsf{inc}(F)$ *of size* $\mathsf{fvs}(k)$, STRONG FOREST-BDS DETECTION *is fixed-parameter tractable.*

Let $\mathcal{C} = \{C_1, \ldots, C_{\mathsf{cycles}(k)}\}$ denote vertex-disjoint cycles in G. We refer to these cycles as $\mathcal{C}$-cycles. The aim is to compute a set $S^* \subseteq \mathsf{var}(F)$ of size $O(k^{2k}2^{k^2-k})$ such that no strong FOREST-BDS of F of size at most k is disjoint from S^*.

Let C be a cycle in G and $x \in \mathsf{var}(F)$. Recall that x *kills* C *internally* if $x \in C$. In this case, x is an *internal killer* for C. We say that x *kills* C *externally* if $x \notin C$ and there are two clauses $u, v \in \mathsf{cla}(F) \cap C$ such that $x \in \mathsf{lit}(u)$ and $\neg x \in \mathsf{lit}(v)$. In this case, x is an *external killer* for C and x kills C externally *in* u and v. As described earlier, our algorithm goes through all $\binom{\mathsf{cycles}(k)}{k}$ ways to choose k $\mathcal{C}$-cycles that may be killed internally. W.l.o.g., let $C_{\mathsf{ext\text{-}cycles}(k)+1}, \ldots, C_{\mathsf{cycles}}(k)$ denote the cycles that may be killed internally. All other cycles $\mathcal{C}' = \{C_1, \ldots, C_{\mathsf{ext\text{-}cycles}(k)}\}$ need to be killed externally. We refer to these cycles as $\mathcal{C}'$-cycles. Let $\mathsf{var}'(F) = \mathsf{var}(F) \setminus \bigcup_{i=1}^{\mathsf{ext\text{-}cycles}(k)} \mathsf{var}(C_i)$ denote the set of variables that may be selected in a strong FOREST-BDS killing no $\mathcal{C}'$-cycle internally. From now on, consider only external killers from $\mathsf{var}'(F)$. The algorithm will find a set S of at most 2 variables such that S contains a variable from every strong FOREST-BDS $B \subseteq \mathsf{var}'(F)$ of F with $|B| \leq k$. External killers and $\mathcal{C}'$-cycles might be adjacent in many different ways. The following procedure defines Cx-cycles that have a much more structured interaction with their external killers.

For each cycle $C_i \in \mathcal{C}'$ consider vertices x_i, u_i, v_i such that (i) $x_i \in \mathsf{var}'(F)$ kills C_i externally in u_i and v_i, and (ii) there is a path P_i from u_i to v_i along the cycle C_i such that if any variable from $\mathsf{var}'(F)$ kills C_i externally in two clauses u_i' and v_i' such that $u_i', v_i' \in P_i$, then $\{u_i, v_i\} = \{u_i', v_i'\}$. Let Cx_i denote the cycle $P_i \cup x_i$. We refer to the cycles in $Cx = \{Cx_1, \ldots, Cx_{\mathsf{ext\text{-}cycles}(k)}\}$ as Cx-cycles.

Observation 1. *Every external killer y of a Cx-cycle Cx_i is incident to u_i and v_i and* $\mathsf{sign}(yu_i) \neq \mathsf{sign}(yv_i)$.

Indeed, an external killer of C_i that is adjacent to two vertices from P_i with distinct signs is adjacent to u_i and v_i. Moreover, any external killer of Cx_i is a killer of C_i that is adjacent to two vertices from P_i with different signs. Thus, any external killer of Cx_i is adjacent to u_i and v_i.

We will be interested in external killers of $\mathcal{C}'$-cycles that also kill the corresponding Cx-cycles. That is, we are going to restrict our attention to vertices in $\mathsf{var}'(F)$ that kill Cx_i. An external killer of a $\mathcal{C}'$-cycle C_i is *interesting* if it is in $\mathsf{var}'(F)$ and it kills Cx_i. As each variable that kills a Cx-cycle Cx_i also kills C_i, and each Cx cycle needs to be killed by a variable from any strong FOREST-BDS, we may indeed restrict our attention to interesting external killers of $\mathcal{C}'$-cycles.

We are now ready to formulate the rules to construct the set S containing at least one variable from any strong FOREST-BDS $B \subseteq \mathsf{var}'(F)$ of F of size at most k. These rules are applied in the order of their appearance.

Rule 6 (No External Killer). *If there is a $Cx_i \in Cx$ such that Cx_i has no external killer, then set $S := \{x_i\}$.*

Rule 7 (Killing Same Cycles). *If there are vertices y and z and at least $2^{k-1}+1$ $\mathcal{C}'$-cycles such that both y and z are interesting external killers of each of these $\mathcal{C}'$-cycles, then set $S := \{y, z\}$.*

Rule 8 (Killing Many Cycles). *If there is a $y \in \mathsf{var}'(F)$ such that y is an interesting external killer of at least $k \cdot 2^{k-1}+1$ $\mathcal{C}'$-cycles, then set $S := \{y\}$.*

Rule 9 (Too Many Cycles). *Set $S := \emptyset$.*

Lemma 6. *Rules 6–9 are sound.*

Lemma 7. *There is an FPT algorithm that, given a CNF formula F, a positive integer parameter k, and $\mathsf{cycles}(k)$ vertex-disjoint cycles of $\mathsf{inc}(G)$, computes a set S^* of $O(k^{2k}2^{k^2-k})$ variables from $\mathsf{var}(F)$ such that every strong FOREST-BDS of F of size at most k includes a variable from S^*.*

This can now be used in an FPT-approximation algorithm for STRONG FOREST-BDS DETECTION. From this algorithm, it follows that SAT and #SAT, parameterized by the size of a smallest strong FOREST-BDS, are FPT.

Theorem 5. *There is an FPT algorithm that, given a CNF formula F and a positive integer parameter k, either concludes that F has no strong FOREST-BDS of size at most k or computes a strong FOREST-BDS of F of size at most 2^k.*

6   Conclusion

Our methods offer various ways of generalization. For instance, instead of strong backdoor sets of smallest size one could consider backdoor trees with minimum height or a minimum number of leaves [27]. Another possibility is to consider base classes that properly entail FOREST. Indeed, a similar approach has very recently been used to design FPT-approximation algorithms for the detection of strong backdoor sets to the base classes of nested CNF formulas [20] and CNF formulas with incidence graphs of bounded treewidth [19]. Finally, it might be possible to use elements from our algorithms and proofs for other problems that are tractable for instances with small feedback vertex sets, and where instantiations of a smaller number of variables could already lead to acyclic instances.

Acknowledgments. The authors acknowledge support from the European Research Council (COMPLEX REASON, 239962). Serge Gaspers acknowledges partial support from the Australian Research Council (DE120101761).

References

1. Alekhnovich, M., Razborov, A.A.: Satisfiability, branch-width and Tseitin tautologies. In: FOCS 2002, pp. 593–603 (2002)
2. Bacchus, F., Dalmao, S., Pitassi, T.: Algorithms and complexity results for #SAT and Bayesian inference. In: FOCS 2003, pp. 340–351 (2003)
3. Bodlaender, H.L.: On disjoint cycles. Int. J. Found. Comput. Sci. 5(1), 59–68 (1994)
4. Chen, J., Fomin, F.V., Liu, Y., Lu, S., Villanger, Y.: Improved algorithms for feedback vertex set problems. J. Comput. Syst. Sci. 74(7), 1188–1198 (2008)
5. Cook, S.A.: The complexity of theorem-proving procedures. In: STOC 1971, pp. 151–158 (1971)
6. Courcelle, B.: Graph rewriting: an algebraic and logic approach. In: Handbook of Theoretical Computer Science, vol. B, pp. 193–242. Elsevier (1990)
7. Dechter, R.: Constraint Processing. Morgan Kaufmann (2003)
8. Diestel, R.: Graph Theory, 4th edn. Graduate Texts in Mathematics. Springer (2010)
9. Downey, R.G., Fellows, M.R.: Parameterized Complexity. Monographs in Computer Science. Springer, New York (1999)
10. Erdős, P., Pósa, L.: On independent circuits contained in a graph. Canadian Journal of Mathematics 17, 347–352 (1965)
11. Fischer, E., Makowsky, J.A., Ravve, E.R.: Counting truth assignments of formulas of bounded tree-width or clique-width. Discr. Appl. Math. 156(4), 511–529 (2008)
12. Flum, J., Grohe, M.: Parameterized Complexity Theory. Texts in Theoretical Computer Science. An EATCS Series, vol. XIV. Springer, Berlin (2006)
13. Fomin, F.V., Gaspers, S., Pyatkin, A.V., Razgon, I.: On the minimum feedback vertex set problem: exact and enumeration algorithms. Algorithmica 52(2), 293–307 (2008)
14. Franco, J., Martin, J.: A history of satisfiabilty. In: Biere, A., Heule, M., van Maaren, H., Walsh, T. (eds.) Handbook of Satisfiability, ch. 1, pp. 3–97. IOS Press (2009)

15. Ganian, R., Hlinený, P., Obdrzálek, J.: Better algorithms for satisfiability problems for formulas of bounded rank-width. In: FSTTCS 2010. LIPIcs, vol. 8, pp. 73–83. Schloss Dagstuhl - Leibniz-Zentrum fuer Informatik (2010)

16. Garey, M.R., Johnson, D.R.: Computers and Intractability. W. H. Freeman and Company, New York (1979)

17. Gaspers, S., Szeider, S.: Backdoors to acyclic SAT. Technical Report 1110.6384, arXiv (2011)

18. Gaspers, S., Szeider, S.: Backdoors to Satisfaction. In: Bodlaender, H.L., Downey, R.G., Fomin, F.V., Marx, D. (eds.) Fellows Festschrift. LNCS, vol. 7370, pp. 287–317. Springer, Heidelberg (2012)

19. Gaspers, S., Szeider, S.: Strong backdoors to bounded treewidth SAT. Technical Report 1204.6233, arXiv (2012)

20. Gaspers, S., Szeider, S.: Strong backdoors to nested satisfiability. In: SAT 2012. LNCS, vol. 7317. Springer (to appear, 2012)

21. Levin, L.: Universal sequential search problems. Problems of Information Transmission 9(3), 265–266 (1973)

22. Nishimura, N., Ragde, P., Szeider, S.: Detecting backdoor sets with respect to Horn and binary clauses. In: SAT 2004, pp. 96–103 (2004)

23. Nishimura, N., Ragde, P., Szeider, S.: Solving #SAT using vertex covers. Acta Informatica 44(7-8), 509–523 (2007)

24. Ordyniak, S., Paulusma, D., Szeider, S.: Satisfiability of acyclic and almost acyclic CNF formulas. In: FSTTCS 2010. LIPIcs, vol. 8, pp. 84–95. Schloss Dagstuhl - Leibniz-Zentrum fuer Informatik (2010)

25. Robertson, N., Seymour, P.D.: Graph minors. V. Excluding a planar graph. J. Combin. Theory Ser. B 41(1), 92–114 (1986)

26. Roth, D.: On the hardness of approximate reasoning. Artif. Intell. 82(1-2), 273–302 (1996)

27. Samer, M., Szeider, S.: Backdoor trees. In: AAAI 2008, pp. 363–368. AAAI Press (2008)

28. Samer, M., Szeider, S.: Algorithms for propositional model counting. J. Discrete Algorithms 8(1), 50–64 (2010)

29. Szeider, S.: On Fixed-Parameter Tractable Parameterizations of SAT. In: Giunchiglia, E., Tacchella, A. (eds.) SAT 2003. LNCS, vol. 2919, pp. 188–202. Springer, Heidelberg (2004)

30. Szeider, S.: Backdoor sets for DLL subsolvers. Journal of Automated Reasoning 35(1-3), 73–88 (2005)

31. Valiant, L.G.: The complexity of computing the permanent. Theoretical Computer Science 8(2), 189–201 (1979)

32. Williams, R., Gomes, C., Selman, B.: Backdoors to typical case complexity. In: IJCAI 2003, pp. 1173–1178. Morgan Kaufmann (2003)

Dominators, Directed Bipolar Orders, and Independent Spanning Trees*

Loukas Georgiadis[1] and Robert E. Tarjan[2]

[1] Department of Computer Science, University of Ioannina, Greece
loukas@cs.uoi.gr
[2] Department of Computer Science, Princeton University,
Princeton, NJ, 08540, and Hewlett-Packard Laboratories
ret@cs.princeton.edu

Abstract. We consider problems related to dominators and independent spanning trees in flowgraphs and provide linear-time algorithms for their solutions. We introduce the notion of a *directed bipolar order*, generalizing a previous notion of Plein and Cheriyan and Reif. We show how to construct such an order from information computed by several known algorithms for finding dominators. We show how to concurrently verify the correctness of a dominator tree D and a directed bipolar order O very simply, and how to construct from D and O two spanning trees whose paths are disjoint except for common dominators. Finally, we describe alternative ways to verify dominators without using a directed bipolar order.

1 Introduction

A *flowgraph* is a directed graph with a distinguished root vertex r such that every vertex is reachable from r. Throughout this paper $G = (V, A, r)$ is a flowgraph with vertex set V, arc set A, distinguished vertex r, and no arc entering r: arcs entering r can be deleted without affecting any of the concepts we study. We denote the number of vertices by n and the number of arcs by m. To simplify bounds we assume $n > 1$. Since $m \geq n - 1$, this implies $m = \Omega(n)$.

A fundamental concept in flowgraphs is that of *dominators*. A vertex u is a *dominator* of a vertex v (u *dominates* v) if every path from r to v contains u; u is a *proper dominator* of v if u dominates v and $u \neq v$. The dominator relation is reflexive and transitive. Its transitive reduction is a rooted tree, the *dominator tree* D: u dominates v if and only if u is an ancestor of v in D. Tree D has root r and vertex set V; it is not in general a spanning tree of G since

---

* Research at the University of Ioannina partially funded by the John S. Latsis Public Benefit Foundation. The sole responsibility for the content of this paper lies with its authors. Research at Princeton University partially supported by NSF grants CCF-0830676 and CCF-0832797. Research while visiting Stanford University partially supported by an AFOSR MURI grant. The information contained herein does not necessarily reflect the opinion or policy of the federal government and no official endorsement should be inferred.

A. Czumaj et al. (Eds.): ICALP 2012, Part I, LNCS 7391, pp. 375–386, 2012.
© Springer-Verlag Berlin Heidelberg 2012

its arcs need not be in A. If $v \neq r$, the parent of v in D, denoted by $d(v)$, is the *immediate dominator* of v: it is the unique proper dominator of v that is dominated by all proper dominators of v. Dominators have applications in diverse areas including program optimization and code generation [6], constraint programming [19], circuit testing [3], theoretical biology [1], memory profiling [15], and connectivity and path-determination problems [7,8,13]. Lengauer and Tarjan [14] gave two quasi-linear-time algorithms for computing D that run fast in practice and have been used in many of these applications. The simpler of these runs in $O(m \log_{(m/n+2)} n)$ time, the other runs in $O(m\alpha(m,n))$ time, where α is a functional inverse of Ackermann's function [22]. Subsequently, even more-complicated but truly linear-time algorithms were discovered [2,4,9].

The genesis of the work we report here was a question asked to the second author by Steve Weeks in 1999: how does one know that the output produced by these fast but complicated algorithms for finding dominators is correct? That is, is there a simple way to verify that a given tree is the dominator tree of a given graph? Here what "simple" means is subjective: since linear time is necessary and sufficient for verification, running time cannot be the measure of simplicity. Nevertheless, one might hope for a verification method that avoids at least some of the technical complications of the fast algorithms for finding dominators.

In 2005 [10] we gave one answer to this question by proposing a linear-time verification algorithm that avoids some but not all of the complications of finding dominators. Here we give what we think is a much more satisfying answer. As a beneficial side effect, we develop new theory about dominators and related concepts, theory that has other applications. The key to verifying dominators is that verification becomes easy, indeed entirely straightforward, given some additional information that serves as a *certificate of correctness* [16]. We want the certificate C to have the property that the correctness of D and C can be verified in linear time by a simple computation. We obtain a suitable certificate by generalizing some results of Whitty [24], Plein [18], and Cheriyan and Reif [5] on disjoint spanning trees. The key notion is that of a *directed bipolar order O*. Given a spanning tree T rooted at r, a directed bipolar order O of T is a preorder of T such that for all $v \neq r$, $p_T(v) <_O v$, where $p_T(v)$ is the parent of v in T, and if $(p_T(v), v) \notin A$, then there are arcs (u,v) and (w,v) with $u <_O v <_O w$ and w is not a descendant of v. Such an order need not be unique. We prove (1) D has a directed bipolar order O; (2) D and O can be verified very simply in linear time, (3) a directed bipolar order O of D can be constructed in linear time from D and additional information computed by several of the fast dominators algorithms; and (4) given D and O, two spanning trees of G can be constructed in linear time whose paths meet only at common dominators. The constructions in (3) and (4) are of independent interest as they have applications to other graph problems. They were previously considered only for the special case of flowgraphs where $d(v) = r$ for all $v \neq r$. (See Section 2.) We also give two other algorithms for verifying D that are simpler than the one in our previous paper, although not as simple as the algorithm that verifies D and O together.

The remainder of our paper consists of three sections. Section 2 contains additional terminology and discusses the results of Whitty, Plein, and Cheriyan and Reif and how our results generalize and improve theirs. Section 3 gives our algorithms for (2), (3), and (4); our algorithm for (3) proves (1) by construction. Section 4 presents alternative algorithms to verify D, which simplify the verification algorithm from [10].

2   Terminology and Related Work

Let T be a rooted tree whose vertices are in V. If v is an ancestor of w, $T[v, w]$ is the path in T from v to w. Tree T is *flat* if the root is the parent of every other vertex. The dominator tree D of G is flat if and only if r is the only proper dominator. Tree T is *valid* if v an ancestor of w implies v dominates w, and *co-valid* if v and w in T and v not an ancestor of w implies v does not dominate w. The dominator tree is the unique tree on vertex set V that is both valid and co-valid. Thus to verify $T = D$ it suffices to verify that T is a tree (easy in linear time), that the vertex set of T is V (also easy in linear time), and that T is both valid and co-valid. The following two lemmas show that validity and co-validity need only hold for certain pairs of vertices.

Lemma 1. *Tree T is valid if and only if for each $w \neq r$, $p_T(w)$ dominates w.*

Proof. If there are vertices w and $v = p_T(w)$ such that v does not dominate w, then T is not valid by definition. Suppose the condition in the lemma holds. Let v be an ancestor of w. The path $T[v, w]$ consists of a sequence of vertices $v = v_1, v_2, \ldots, v_k = w$ such that $v_i = p_T(v_{i+1})$. By the condition in the lemma, v_i dominates v_{i+1}; by induction and the transitivity of domination, v dominates w. Thus T is valid.   □

Lemma 2. *If T is valid, T is co-valid if and only if T has the* sibling property: *if v and w are siblings, then v does not dominate w.*

Proof. If T does not have the sibling property, there are siblings v and w such that v dominates w, which means that T is not co-valid. Suppose T is valid and has the sibling property. Let v and w be unrelated vertices in T. Let x and y be the siblings that are ancestors of v and w in T, respectively. Since T is valid, x dominates v and y dominates w. If v dominates w, then both x and y must be on $D[r, w]$, which implies that x dominates y or y dominates x, but neither can be the case since T has the sibling property. Thus v does not dominate w. It follows that T is co-valid.   □

In Section 3 we give an easy-to-test sufficient condition for validity; the hard part of dominator verification is verifying co-validity; that is, verifying the sibling property. As we shall see, a directed bipolar order makes this task much easier.

Suppose D is flat; that is, each vertex $v \neq r$ has only one proper dominator, r. Then for any vertex $v \neq r$, there is no vertex other than r and v common to all paths from r to v. By Menger's Theorem [17], there are two paths from r to v containing no common vertex other than r and v. Whitty [24] proved that such paths can be realized for all v by a pair of trees: there are spanning trees B and R rooted at r such that for any vertex v, $B[r,v]$ and $R[r,v]$ contain only r and v in common. We call such a pair of trees *disjoint*. Whitty actually proved something stronger: there are disjoint spanning trees B and R rooted at r such that, for any pair of distinct vertices v and w, either $B[r,v]$ and $R[r,w]$ contain only r in common, or $B[r,w]$ and $R[r,v]$ do. We call such a pair of trees *strongly disjoint*. Plehn [18] and independently Cheriyan and Reif [5] gave simpler proofs of Whitty's result using what Cheriyan and Reif called a *directed st-numbering* as an intermediary. This is a numbering of the vertices from 1 to n such that r is numbered 1 and each vertex $v \neq r$ has an entering arc (r,v) or two entering arcs (u,v) and (w,v) such that u is numbered less than v and w is numbered greater than v. The proofs of Whitty, of Plehn, and of Cheriyan and Reif give polynomial-time constructions of pairs of strongly disjoint spanning trees and of directed st-numberings, but their constructions seem to require $\Omega(nm)$ time in the worst case. Huck [12] later gave an $O(nm)$-time algorithm to find two disjoint spanning trees.

We generalize these definitions and results to arbitrary flowgraphs and give linear-time algorithms. The definition of a directed bipolar order given in Section 1 generalizes the notion of a directed st-numbering to a non-flat dominator tree. A pair of spanning trees B and R rooted at r is *independent* if for all v, $B[r,v]$ and $R[r,v]$ contain only dominators of v in common; an independent pair of trees B and R is *strongly independent* if for every pair of distinct vertices v and w, either $B[r,v]$ and $R[r,w]$ contain only common dominators of v and w in common, or $B[r,w]$ and $R[r,v]$ do. In the special case of a flat dominator tree, these two definitions specialize exactly to those of a disjoint pair of trees or a strongly disjoint pair of trees, respectively. In our previous paper [10] we showed by a linear-time construction that every flowgraph has a pair of independent spanning trees. In Section 3 we extend this result to show by linear-time constructions that every flowgraph has a directed bipolar order of its dominator tree as well as a pair of strongly independent spanning trees.

3   Linear-Time Verification and Construction Algorithms

Our verification algorithm uses sufficient conditions for validity and co-validity, conditions that hold for D. The sufficient condition for validity is simple to test; the sufficient condition for co-validity is the existence of a directed bipolar order.

Lemma 3. *If T is a tree with vertices in V, T is valid if it has the* parent property: *for all arcs (u,v), $p_T(v)$ is an ancestor of u.*

Proof. Suppose T is not valid. By Lemma 1 there is a vertex y with parent x such that x does not dominate y. Thus there is a path from r to y that avoids

x. Let (u, v) be the first arc on this path such that v is a descendant of x. Then $v \neq x$, so the parent of v is a descendant of x. But u is not a descendant of x, so the parent of v is not an ancestor of u, violating the condition in the lemma. $\square$

Lemma 4. *The dominator tree D has the parent property.*

Proof. Let (u, v) be an arc, and let x be the parent of v. (By assumption, no arc enters r, so v has a parent.) Suppose x is not an ancestor of u. Then x does not dominate u, so there is a path from r to u that avoids x, and hence a path to v that avoids x. But x dominates v, a contradiction. Thus x is an ancestor of u. $\square$

Lemma 5. *Let T be a valid tree rooted at r whose vertex set is V. If T has a directed bipolar order, then T is co-valid.*

Proof. Let T be a valid tree with a directed bipolar order O. Construct two subgraphs B and R as follows: for every vertex $w \neq r$, add to B some arc (x, w) such that $x <_O w$; add to R either $(p_T(w), w)$ if this is an arc or if not some arc (y, w) such that $w <_O y$. Each vertex in B is reachable from r by a path of vertices in increasing directed bipolar order, so B is a spanning tree; we show that R is also a spanning tree. To that end, we first note that for each $v \neq r$, R contains a path from $p_T(v)$ to v containing only $p_T(v)$, v, and vertices ordered higher than v. This fact implies (by induction on v in increasing depth in T) that there is a path from r to v in R, thus R is also a spanning tree.

Now we show that B and R are strongly independent. Let $v \neq r$. By Lemma 1, $p_T(v)$ dominates v. By construction $B[p_T(v), v]$ and $R[p_T(v), v]$ contain only $p_T(v)$ and v in common. By Lemma 3, both of these paths contain only descendants of $p_T(v)$ in T that are not proper descendants of v in T. By induction on v in increasing directed bipolar order, $B[r, v]$ and $R[r, v]$ contain only ancestors of v in T in common, each of which dominates v. Thus B and R are independent. Let $v \neq w$, and let u be their nearest common ancestor in T. Without loss of generality, suppose $v <_O w$. If v is an ancestor of w in T then $u = v$. Suppose now that v is not an ancestor of w in T. Path $B[u, v]$ contains only vertices ordered no greater than v. Suppose $B[u, v]$ and $R[u, w]$ contain a vertex $z \neq u$ in common. Choose v so that it has minimum depth in T. Then $z \leq_O v <_O w$, which implies that z dominates w and u dominates z. If z dominates v then $z = u$, a contradiction. Otherwise, $B[u, p_T(v)]$ and $R[u, w]$ have z in common. This violates the choice of v, a contradiction. Thus $B[u, v]$ and $R[u, w]$ have only u in common. Furthermore by Lemma 3 both of these paths contain only descendants of u. Since B and R are independent, $B[r, u]$ and $R[r, u]$ contain only dominators of u in common, and contain only vertices that are not proper descendants of u. It follows that $B[r, v]$ and $R[r, w]$ contain only common dominators of v and w in common.

Let v and w be siblings in T such that $v <_O w$. Then $B[r, v]$ avoids w and $R[r, w]$ avoids v. Hence neither v nor w dominates the other. Therefore T satisfies the sibling property, so it is co-valid. $\square$

Note that the proof of Lemma 5 shows not only how to construct a pair of strongly independent trees but also how to choose, for a given pair of vertices v and w, a pair of paths that meet only at the common dominators of v and w.

Lemma 6. *The dominator tree D has a directed bipolar order.*

Proof. We provide a simple construction which simplifies and generalizes the construction of a directed st-numbering given by Cheriyan and Reif [5]. Let B and R be two independent spanning trees of G. We substitute each tree arc (x, y) with (z, y) where z is the sibling of y that is an ancestor of x. (See Section 4.1). The construction runs in two phases, each consisting of $n - 1$ rounds. To avoid special cases we consider that $p_B(v) = p_R(v)$ for all vertices v such that $(d(v), v)$ is an arc of G. It is straightforward to modify any pair of independent spanning trees so that they satisfy this requirement. We begin with the graph G^n formed by the arcs in $B^n = B$ and $R^n = R$. (G^n contains a duplicate arc (x, y) when $x = p_B(y) = p_R(y)$.)

Phase 1. During the i-th round of the first phase we may remove or replace some arcs in G^{n-i+1} to form a graph G^{n-i} with $n - i$ vertices. We also perform the corresponding changes to B^{n-i+1} and R^{n-i+1}, forming B^{n-i} and R^{n-i}. This process stops when we reach $G^1 = (\{r\}, \emptyset, r)$, for which a valid directed bipolar order O^1 is immediately obtained by setting $r =_{O^1} 1$. At each round of the first phase we maintain the following invariants: (1a) G^{n-i} has $n - i$ vertices and $2(n - i) - 2$ arcs. (1b) G^{n-i} has at least one vertex w with out-degree at most one such that $p_{B^{n-1}}(w) \neq p_{R^{n-1}}(w)$, or at least one vertex with out-degree zero. (1c) B^{n-i} and R^{n-i} are independent spanning trees of G^{n-i}. (1d) For all vertices v such that $(d(v), v)$ is an arc of G^{n-i}, $p_{B^{n-i}}(v) = p_{R^{n-i}}(v)$. Consider the i-th round of this phase. At the beginning of this round we have a graph G^{n-i+1} and two independent spanning trees B^{n-i+1} and R^{n-i+1}. Let w be a vertex that satisfies (1b). Then w must be a leaf in at least one of the two spanning trees. Assume that w is a leaf in B^{n-i+1}. (If w is a leaf in R^{n-i+1} we apply the symmetric steps.) First we remove w and its adjacent arcs (entering or leaving w). If w is also a leaf in R^{n-i+1} then we are done. Otherwise, let y be the child of w in R^{n-i+1}. Also, let v be the parent of y in R^{n-i+1}. We form G^{n-i} by inserting the arc (v, y). We form R^{n-i} by making v the new parent of y. This completes the description of the i-th round.

Phase 2. During the second phase we perform the reverse sequence of operations and extend a directed bipolar order O^i of G^i to a directed bipolar order O^{i+1} of G^{i+1}. That is, we maintain the following invariant: (2a) O^i is a directed bipolar order of G^i. Consider the i-th round. At the beginning of this round we have a graph G^i, two independent spanning trees B^i and R^i of G^i, and a directed bipolar order O^i of G^i. Let w be the vertex we removed during the $(n - i)$-th round of the first phase. Our goal is to assign an appropriate number for w. Suppose w was a leaf in B^{i+1}. (We apply the symmetric steps if y was a leaf in R^{i+1}.) Let u be the parent of w in B^{i+1} and v be the parent of w in R^{i+1}. If $v = u$ then it suffices to set $w >_{O^{i+1}} v$. Now suppose $v \neq u$. If w is also a leaf in R^{i+1} then we can assign w any number between u and v. Otherwise, w has a child y in R^{i+1}. Let $x = p_{B^{i+1}}(y)$. Suppose $v <_{O^{i+1}} u$. If $y >_{O^{i+1}} x$ then we can assign w any number between u and v. Otherwise, if $y <_{O^{i+1}} x$ then we can assign w a number immediately larger than v. Finally suppose $v >_{O^{i+1}} u$. If $y >_{O^{i+1}} x$ we can assign w a number immediately smaller than v. Otherwise, if $y <_{O^{i+1}} x$ then we can assign w any number between u and v.

We prove by induction on i that the invariants are maintained. Consider the first phase. For $i = 1$ we have the graph G^n and its spanning trees $B^n = B$ and $R^n = R$. Since B and R are independent, (1a) and (1c) hold. Also, (1d) holds by the construction of B and R. It remains to show (1b). To that end let $X = \{x \in V \mid p_B(x) = p_R(x)\} \cup \{r\}$. Then, for any $x \in X - r$, $d(x) = p_B(x) = p_R(x)$. If $X = V$ then the definition of X implies that there is at least one vertex with out-degree zero. If $X \subset V$ then $|X| \geq 2$ and there are at least $2|X| - 1$ arcs (v, w) with $v \in X$. Then, the number of arcs (v, w) with $v \notin X$ is at most $2(n - |X|) - 1$. Therefore, there is at least one vertex in $V \setminus X$ with out-degree less than or equal to one. For the induction step, suppose that the invariants hold for G^{n-i+1}, B^{n-i+1} and R^{n-i+1}. Consider the graph G^{n-i} and the trees B^{n-i} and R^{n-i} obtained by removing vertex w, as described above. Then B^{n-i} and R^{n-i} are independent spanning trees of G^{n-i}, which implies (1a), (1b) and (1c) as in the base case. It remains to show that (1d) also holds. Suppose to the contrary that (1d) is violated after removing w. Then w must have a child y in one of the two trees, since y is the only vertex that is assigned a new parent v in one of the two trees. Without loss of generality, assume that (u, w) is in B^{n-i+1} and (v, w) and (w, y) are in R^{n-i+1}. Since (1d) is violated for G^{n-i}, B^{n-i} and R^{n-i}, we must have $d(y) = v$ in G^{n-i}. But then $d(y) = v$ also in G^{n-i+1}. Then v dominates w in G^{n-i+1}. Since (v, w) is an arc in G^{n-i+1}, the induction hypothesis implies $u = v$. But then w must have out-degree zero, a contradiction.

Now consider the second phase. Invariant (2a) trivially holds for $i = 1$. For the induction step assume that O^i is a valid numbering for G^i. For O^{i+1} we only need to consider w and y (if it exists), since these are the only vertices that change parents. First we consider the case where w is a leaf in both B^{i+1} and R^{i+1}. If $v = u$ then (v, w) is the only arc entering w in G^{i+1}, so O^{i+1} is valid if we assign w any number greater than v. If $v \neq u$, then O^{i+1} is valid if we assign w any number between v and w. Now suppose that y exists. By the choice of w and invariant (1d), $u \neq v$ and $v \neq x$. Suppose $v <_{O^{i+1}} u$. If $y >_{O^{i+1}} x$ then $y <_{O^{i+1}} v$. Then $x <_{O^{i+1}} y <_{O^{i+1}} v <_{O^{i+1}} u$, so we can assign w any number between u and v. Otherwise, if $y <_{O^{i+1}} x$ then $y >_{O^{i+1}} v$. We have $v <_{O^{i+1}} y <_{O^{i+1}} x <_{O^{i+1}} u$, or $v <_{O^{i+1}} y <_{O^{i+1}} u <_{O^{i+1}} x$, or $v <_{O^{i+1}} u <_{O^{i+1}} y <_{O^{i+1}} x$. In all cases, we can assign w a number immediately larger than v. Finally suppose $v >_{O^{i+1}} u$. If $y >_{O^{i+1}} x$ then $y <_{O^{i+1}} v$. We have $x <_{O^{i+1}} y <_{O^{i+1}} u <_{O^{i+1}} v$, or $x <_{O^{i+1}} u <_{O^{i+1}} y <_{O^{i+1}} v$, or $u <_{O^{i+1}} x <_{O^{i+1}} y <_{O^{i+1}} v$. In all cases, we can assign w a number immediately smaller than v. Otherwise, if $y <_{O^{i+1}} x$ then $y >_{O^{i+1}} v$. Then $u <_{O^{i+1}} v <_{O^{i+1}} y <_{O^{i+1}} x$, so we can assign w any number between u and v. $\square$

Lemma 5 and Lemma 6 imply the following result.

Theorem 1. *A valid tree is co-valid if and only if it has a directed bipolar order.*

Given a tree T and a directed bipolar order O of T, we verify that $T = D$ and O is indeed a directed bipolar order of T as follows. Assume that O is given by a numbering of the vertices from 1 to n such that $v <_O w$ if and only if v is numbered less than w. We execute the following steps:

1. Verify that T is a tree rooted at r with vertex set V. This is straightforward in $O(m)$ time.
2. Do a depth-first traversal of T to number the vertices from 1 to n in both preorder and postorder. This gives an $O(1)$-time ancestor test: v is an ancestor of w if and only if v precedes w in preorder and follows w in postorder [20]. This is straightforward in $O(n)$ time.
3. Verify that T has the parent property. This takes $O(1)$ time per arc, given an $O(1)$-time ancestor test, for a total of $O(m)$ time.
4. Verify that O is a directed bipolar order of T. This requires an examination of the arcs entering each vertex to verify that the one or two needed arcs are present, and takes $O(1)$ time per arc given an $O(1)$-time ancestor-descendant test, for a total of $O(m)$ time.

Our algorithm for finding a directed bipolar order is an efficient implementation of the construction given in the proof of Lemma 6. We present an extended version of the algorithm, which during the vertex-deletion process of the first phase verifies that the B and R are indeed trees. During this process, we can also verify that the alleged dominator tree T corresponds to the tree implied by B and R, and compute the number of descendants of every vertex in T. In the second phase, as above, we reinsert the vertices, computing a directed bipolar order. If nothing has gone wrong so far, the alleged tree is co-valid. We then use the computed preorder numbers and numbers of descendants to test validity. Therefore, we can construct the directed bipolar order and at the same time verify the dominator tree. Now we provide the details of our algorithm. During the first phase every vertex z has a size $size(z)$, initially one, a number of descendants $numdes(z)$, initially one, and a list of vertices $list(z)$, initially containing just z. The list contains vertices whose parent in T the alleged dominator tree needs checking. While there is more than one vertex, find a vertex $w \neq r$ with out-degree at most one and $p_B(w) \neq p_R(w)$ or with out-degree zero, and do the following. Let (u, w) and (v, w) be the arcs into w, (w, y) the arc out of w if there is one, (x, y) is the other arc into y if y exists. Choose v so that (v, w) and (w, y) are in the same tree (R or B). Arcs (u, w) and (x, y) are in the other tree. If y exists and equals v, stop with failure: one of the "trees" contains a cycle. Otherwise, continue. Add $size(w)$ to $size(v)$. If $u = v$, verify that each vertex z on $list(w)$ has $u = v$ as its parent in T, and add $numdes(z)$ to $numdes(u)$. If $u \neq v$, add the vertices in $list(w)$ to $list(v)$ (or equivalently to $list(u)$). We can do this by list catenation, so it only takes $O(1)$ time. Delete w and its incident arcs. If y exists, add arc (v, y) to the tree that previously contained (v, w) and (w, y). If the vertex deletion process finishes, r will have size n. Now process the vertices in the opposite order to their deletion. Each vertex z will get assigned an interval in $[1, n]$ such that, if any integer in the interval is the number of z, the numbering is directed bipolar. The interval for z is $[num(z), num(z) + size(z) - 1]$. Initially $num(r) = 1$. From the first phase, $size(r) = n$. To process the next vertex w, let u and v be the corresponding vertices as defined above. Set $size(v) = size(v) - size(w)$. If $num(v) \leq num(u)$, set $num(w) = num(v) + size(v)$; otherwise, set $num(w) = num(v)$ and then set $num(v) = num(w) + size(w)$. Once the second phase is complete, the size of every vertex is back to one, the num function is a directed bipolar order, and the $numdes$ function gives the number of descendants of each vertex in T, which can be used to test validity. Note that if we are only interested in computing a directed bipolar order then we only need to maintain the $size$ and num functions in this algorithm.

The running time of the above algorithm is $O(n)$ if two independent spanning trees of G are given. In [10] we showed that such two independent spanning trees can be computed in $O(n)$ time if for all $v \neq r$ we store the following information: the semidominator $s(v)$ of v, the parent of v in the corresponding depth-first search tree, and a vertex w such that (w, v) is an arc of a high path from $s(v)$ to v. (See Section 4.2.) Since, by [2,4], semidominators can be computed in $O(m)$ time, our algorithm gives the following generalization of Plehn's and Cheriyan and Reif's construction:

Theorem 2. *Any flowgraph has a directed bipolar order, constructible in linear time.*

The construction combined with the proof of Lemma 5 gives the following generalization of Whitty's theorem:

Theorem 3. *Any flowgraph has a pair of strongly independent spanning trees, constructible in linear time.*

4 Alternative Verification Algorithms

Here we present alternative linear-time algorithms to verify a dominator tree. The algorithm of Section 4.1 uses the concept of *headers* which are computed in [4], while the algorithm of Section 4.2 uses the concept of *semidominators* which are computed in [2,4,14]. Due to limited space we omit the proofs of the results stated in this section. The proofs will appear in the full version of the paper.

4.1 Using Headers

To motivate this algorithm we first consider the simple case where G is acyclic and T is flat.

Lemma 7. *Suppose G is acyclic. The root vertex r is the only proper dominator in G if and only if each vertex other than r has an entering arc from r or at least two entering arcs.*

To extend Lemma 7 to general acyclic graphs, we introduce the notion of the *support* $sp(v, w)$ of an arc (v, w) with respect to a valid tree T, defined as follows: if $v = p_T(w)$, $sp(v, w) = v$; otherwise, $sp(v, w)$ is the child of $p_T(w)$ that is an ancestor of v.

Theorem 4. *Suppose G is acyclic and T is valid. Then T is co-valid if and only if, for every vertex $w \neq r$, there is an arc (v, w) with $sp(v, w) = p_T(w)$, or two arcs (x, w) and (y, w) with $sp(x, w) \neq w$, $sp(y, w) \neq w$, and $sp(x, w) \neq sp(y, w)$.*

Theorem 4 allows us to test if T is co-valid when G is acyclic by computing the supports of all the arcs. This we can do by a radix sort. Number the vertices from 1 to n in preorder with respect to a fixed sibling order on T. For each i, $2 \leq i \leq n$, initialize a set $B(i)$ to empty. Process the arcs (v, w) one at a time. To process (v, w), if $v = p_T(w)$, set $sp(v, w) = v$; otherwise, add (v, w) to $B(pre(v))$. Once all the arcs are processed, empty the sets $B(i)$ into a collection of stacks $C(u)$, one for each vertex u, as follows: Initialize each stack $C(u)$ to empty. Empty the sets $B(i)$ in decreasing order on i. To empty $B(i)$, remove each arc (v, w) and push it onto the front of $C(p_T(w))$. Now each $C(u)$ contains the arcs (v, w) such that $u = p_T(w)$ and $v \neq u$, in non-decreasing order on $pre(v)$. The final step is to empty each stack $C(u)$, computing the support of each arc it contains. This amounts to a merge of the children of u with the arcs on $C(u)$. Begin with the first two children x and y of u. Pop arcs (v, w) from $C(u)$, setting $sp(v, w) = x$, up to but not including the first arc (v, w) such that $pre(v) \leq pre(y)$. Then replace x by y and y by the sibling after y, and continue. Once x is the last sibling, set $sp(v, w) = x$ for all arcs (v, w) remaining on $C(u)$. Computing the supports in this way takes $O(m)$ time. Once all the supports are computed, applying the test in Theorem 4 takes $O(m)$ time.

To extend the above test to general graphs, we need to deal with cycles, which we do using *headers* [23]. Headers are defined with respect to a depth-first spanning tree. Let F be a spanning tree generated by a depth-first search of G starting from r, with vertices numbered in reverse postorder with respect to the search. The header $h(v)$ of a vertex v is the maximum-numbered proper ancestor u of v such that there is a path from v to u containing only descendants of u; if there is no such u, $h(v) = null$. The headers define the *header forest* H by $p_H(v) = h(v)$. Graph G is acyclic if and only if all headers are null. Tarjan [23] gave an algorithm that computes headers in $O(m\alpha(m, n))$ time using disjoint set union and graph search; Buchsbaum et al. [4] gave an $O(m)$-time algorithm. Computing headers requires less machinery than computing dominators (see [4]). We define the *support set* $S(w)$ of a vertex w by $S(w) = \{ sp(x, y) \mid y \text{ is a descendant of } w \text{ in } H \}$.

Theorem 5. *A valid tree T is co-valid if and only if, for every vertex $w \neq r$, $S(w)$ contains either $p_T(w)$ or at least two vertices numbered less than w.*

To implement the test of Theorem 5 do a depth-first search of G, generating a depth-first spanning tree F and number the vertices in reverse postorder. Compute the corresponding headers and the support of every arc. Then compute, for each vertex $w \neq r$, the set $S_2(w)$ containing the two smallest-numbered vertices in $S(w)$ that are numbered less than w. (If there is only one such vertex, $S(w)$ contains this single vertex.) To do this, start from the leaves of H and proceed bottom-up, setting $S_2(w)$ to contain the two smallest-numbered vertices numbered less than w in $\{sp(x, w)\} \cup \{S_2(z) \mid h(z) = w\}$. Computing the headers is the most complicated part of the algorithm, but it can be done in almost-linear [23] or even linear [4] time, and it is simpler than computing dominators from scratch. Computing the arc supports and the sets $S_2(w)$ takes $O(m)$ time. Once the sets $S_2(w)$ are computed, the test in Theorem 5 takes $O(n)$ time.

For an important special case, the Theorem 5 test reduces to the Theorem 4 test. A graph is *reducible* if every strongly connected subgraph G' (which does not contain r) contains a single vertex v that dominates all vertices in G' [21]. Structured programs have reducible control flow graphs; reducibility simplifies various global code optimizations [11]. Reducibility has a characterization in terms of headers: a graph is reducible if and only if, for every vertex $w \neq r$ and every arc (x, y) such that y is a descendant of w in H, either $y = w$ or x is a descendant of w in H [21]. It follows that if G is reducible, each vertex in $S_2(w)$ is the support of an arc entering w.

4.2   Using Semidominators

Let F be a depth-first spanning tree of G. Number the vertices in preorder with respect to F. A path from v to w is *high* if every vertex on the path except v is numbered no less than w. The *semidominator* $s(w)$ of a vertex w is the smallest numbered vertex v such that there is a high path from v to w.

Theorem 6. *A valid tree T is co-valid if and only if, for every arc (x, w) of F, either $x = p_T(w)$ or $s(w)$ is numbered less than $sp(x, w)$.*

We use Theorem 6 to verify $T = D$ in $O(m)$ time as follows: As well as T, store F and the corresponding vertex preorder numbering, and the semidominators computed by the dominator algorithm. Verify that T is valid. Then compute the supports of the arcs in F as in Section 4.1, and apply the test in Theorem 6.

References

1. Allesina, S., Bodini, A.: Who dominates whom in the ecosystem? Energy flow bottlenecks and cascading extinctions. Journal of Theoretical Biology 230(3), 351–358 (2004)
2. Alstrup, S., Harel, D., Lauridsen, P.W., Thorup, M.: Dominators in linear time. SIAM Journal on Computing 28(6), 2117–2132 (1999)
3. Amyeen, M.E., Fuchs, W.K., Pomeranz, I., Boppana, V.: Fault equivalence identification using redundancy information and static and dynamic extraction. In: Proceedings of the 19th IEEE VLSI Test Symposium (March 2001)
4. Buchsbaum, A.L., Georgiadis, L., Kaplan, H., Rogers, A., Tarjan, R.E., Westbrook, J.R.: Linear-time algorithms for dominators and other path-evaluation problems. SIAM Journal on Computing 38(4), 1533–1573 (2008)

 5. Cheriyan, J., Reif, J.H.: Directed s-t numberings, rubber bands, and testing digraph k-vertex connectivity. Combinatorica, 435–451 (1994); also in SODA 1992
 6. Cytron, R., Ferrante, J., Rosen, B.K., Wegman, M.N., Zadeck, F.K.: Efficiently computing static single assignment form and the control dependence graph. ACM Transactions on Programming Languages and Systems 13(4), 451–490 (1991)
 7. Georgiadis, L.: Testing 2-Vertex Connectivity and Computing Pairs of Vertex-Disjoint s-t Paths in Digraphs. In: Abramsky, S., Gavoille, C., Kirchner, C., Meyer auf der Heide, F., Spirakis, P.G. (eds.) ICALP 2010. LNCS, vol. 6198, pp. 738–749. Springer, Heidelberg (2010)
 8. Georgiadis, L.: Approximating the Smallest 2-Vertex Connected Spanning Subgraph of a Directed Graph. In: Demetrescu, C., Halldórsson, M.M. (eds.) ESA 2011. LNCS, vol. 6942, pp. 13–24. Springer, Heidelberg (2011)
 9. Georgiadis, L., Tarjan, R.E.: Finding dominators revisited. In: Proc. 15th ACM-SIAM Symp. on Discrete Algorithms, pp. 862–871 (2004)
10. Georgiadis, L., Tarjan, R.E.: Dominator tree verification and vertex-disjoint paths. In: Proc. 16th ACM-SIAM Symp. on Discrete Algorithms, pp. 433–442 (2005)
11. Hecht, M.S., Ullman, J.D.: Flow graph reducibility. In: Proceedings of the Fourth Annual ACM Symposium on Theory of Computing, STOC 1972, pp. 238–250 (1972)
12. Huck, A.: Independent trees in graphs. Graphs and Combinatorics 10, 29–45 (1994)
13. Italiano, G.F., Laura, L., Santaroni, F.: Finding strong bridges and strong articulation points in linear time. Theoretical Computer Science (in press, 2012)
14. Lengauer, T., Tarjan, R.E.: A fast algorithm for finding dominators in a flowgraph. ACM Transactions on Programming Languages and Systems 1(1), 121–141 (1979)
15. Maxwell, E.K., Back, G., Ramakrishnan, N.: Diagnosing memory leaks using graph mining on heap dumps. In: Proceedings of the 16th ACM SIGKDD International Conference on Knowledge Discovery and Data Mining, KDD 2010, pp. 115–124 (2010)
16. McConnell, R.M., Mehlhorn, K., Näher, S., Schweitzer, P.: Certifying algorithms. Computer Science Review 5(2), 119–161 (2011)
17. Menger, K.: Zur allgemeinen kurventheorie. Fund. Math. 10, 96–115 (1927)
18. Plehn, J.: Über die Existenz und das Finden von Subgraphen. PhD thesis. University of Bonn, Germany (May 1991)
19. Quesada, L., Van Roy, P., Deville, Y., Collet, R.: Using Dominators for Solving Constrained Path Problems. In: Van Hentenryck, P. (ed.) PADL 2006. LNCS, vol. 3819, pp. 73–87. Springer, Heidelberg (2005)
20. Tarjan, R.E.: Depth-first search and linear graph algorithms. SIAM Journal on Computing 1(2), 146–159 (1972)
21. Tarjan, R.E.: Testing flow graph reducibility. In: Proceedings of the Fifth Annual ACM Symposium on Theory of Computing, pp. 96–107 (1973)
22. Tarjan, R.E.: Efficiency of a good but not linear set union algorithm. Journal of the ACM 22(2), 215–225 (1975)
23. Tarjan, R.E.: Edge-disjoint spanning trees and depth-first search. Acta Informatica 6(2), 171–185 (1976)
24. Whitty, R.W.: Vertex-disjoint paths and edge-disjoint branchings in directed graphs. Journal of Graph Theory 11, 349–358 (1987)

Hardness of Approximation for Quantum Problems

Sevag Gharibian[1] and Julia Kempe[2,3]

[1] David R. Cheriton School of Computer Science and Institute for Quantum Computing,
University of Waterloo, Waterloo, Canada
[2] CNRS & LIAFA, University Paris Diderot - Paris 7, Paris, France
[3] Blavatnik School of Computer Science, Tel Aviv University, Tel Aviv, Israel

Abstract. The polynomial hierarchy plays a central role in classical complexity theory. Here, we define a quantum generalization of the polynomial hierarchy, and initiate its study. We show that not only are there natural complete problems for the second level of this quantum hierarchy, but that these problems are in fact hard to approximate. Our work thus yields the first known hardness of approximation results for a quantum complexity class. Using these techniques, we also obtain hardness of approximation for the class QCMA. Our approach is based on the use of dispersers, and is inspired by the classical results of Umans regarding hardness of approximation for the second level of the classical polynomial hierarchy (Umans 1999). We close by showing that a variant of the local Hamiltonian problem with hybrid classical-quantum ground states is complete for the second level of our quantum hierarchy.

1 Introduction and Results

Over the last decades, the Polynomial Hierarchy (PH), a natural generalization of the class NP, has been the focus of much study in classical computational complexity. Of particular interest is the second level of PH, denoted Σ_2^p. Here, we say a problem is in Σ_2^p if it has an efficient verifier with the property that for any YES instance $x \in \{0,1\}^n$ of the problem, *there exists* a polynomial length proof y such that *for all* polynomial length proofs z, the verifier accepts x, y and z. Note that the *alternation* from an existential quantifier over y to a for-all quantifier over z is crucial here — keeping only the existential quantifier reduces us to NP.

It turns out that introducing such alternating quantifiers makes Σ_2^p a powerful class believed to be *beyond* NP. For example, there exist natural and important problems known to be in Σ_2^p but not in NP. Such problems range from "does the optimal assignment to a 3SAT instance satisfy *exactly k* clauses?" to practically relevant problems related to circuit minimization, such as "given a boolean formula C in Disjunctive Normal Form (DNF), what is the smallest DNF formula C' equivalent to C?" (see, e.g. [Uma99]). The study of Σ_2^p has also led to a host of other fundamental theoretical results, such as the Karp-Lipton theorem, which states that NP $\not\subseteq$ P$_{/\text{poly}}$ unless PH collapses to Σ_2^p. Σ_2^p has even been used to prove that SAT cannot be solved simultaneously in linear time and logarithmic space [For00,FLvMV05]. For these reasons, Σ_2^p and more generally PH have occupied a central role in complexity theoretic research.

A. Czumaj et al. (Eds.): ICALP 2012, Part I, LNCS 7391, pp. 387–398, 2012.
© Springer-Verlag Berlin Heidelberg 2012

Moving to the quantum setting, the study of quantum proof systems and a natural quantum generalization of NP, the class Quantum Merlin Arthur (QMA) [KSV02], has been a very active area of research over the last decade. Roughly, a problem is in QMA if for any YES instance of the problem, there exists a polynomial size *quantum* proof convincing a quantum verifier of this fact with high probability. With the notion of quantum proofs in mind, we thus ask the natural question: *Can a quantum generalization of Σ_2^p be defined, and what types of problems might it contain and characterize?* Perhaps surprisingly, to date there are almost no known results in this direction.

Our results: In this work, we introduce a quantum generalization of Σ_2^p, which we call cq-Σ_2, and initiate its study. Our results include cq-Σ_2-completeness and cq-Σ_2-hardness of approximation for a number of new problems we define. Our techniques also yield hardness of approximation for the complexity class known as QCMA. We now describe these results in further detail.

1. Hardness of approximation for cq-Σ_2. To begin, we informally define cq-Σ_2 (see Section 2 for formal definitions).

Definition 1 (cq-Σ_2 **(informal)**). *A problem Π is in cq-Σ_2 if there exists an efficient quantum verifier V satisfying the following property for any input $x \in \{0,1\}^*$:*

- *If x is a YES instance of Π, then there exists a poly-size classical proof y such that for all poly-size quantum proofs $|z\rangle$, V accepts $(x, y, |z\rangle)$ with high probability.*
- *If x is a NO instance of Π, then for all poly-size classical proofs y, there exists a poly-size quantum proof $|z\rangle$ such that V rejects $(x, y, |z\rangle)$ with high probability.*

We believe this is a natural quantum generalization of Σ_2^p. Here, the prefix *cq* in cq-Σ_2 follows since the existential proof is classical, while the for-all proof is quantum. One can also consider variations of this scheme such as qq-Σ_2, qc-Σ_2, or cc-Σ_2 (with a quantum verifier), defined analogously. In this paper, however, our focus is on cq-Σ_2, as it is the natural setting for the computational problems for which we wish to prove hardness of approximation. Note also that unlike for Σ_2^p, the definition of cq-Σ_2 is bounded error — this is due to the use of a quantum verifier for cq-Σ_2. This implies, for instance, that the quantum analogue of the classically non-trivial result BPP $\subseteq \Sigma_2^p$ [Sip83,Lau83], i.e. BQP $\subseteq$ cq-Σ_2, holds trivially. Finally, one can extend the definition of cq-Σ_2 to an entire hierarchy of quantum classes analogous to PH by adding further levels of alternating quantifiers, attaining presumably different classes depending on whether the quantifier at any particular level runs over classical or quantum proofs.

To next discuss hardness of approximation for cq-Σ_2, we recall two classical problems crucial to our work here. First, in the NP-complete problem SET COVER, one is given a set of subsets $\{S_i\}$ whose union covers a ground set U, and we are asked for the smallest number of the S_i whose union still covers U. If, however, the S_i are represented *succinctly* as the on-set[1] of a 3-DNF formula ϕ_i, we obtain a more difficult problem known as SUCCINCT SET COVER (SSC). SSC, along with a related problem IRREDUNDANT (IRR), are not just NP-hard, but are Σ_2^p-complete (indeed, they are even Σ_2^p-hard to approximate [Uma99]). SSC and IRR are defined as:

---

[1] By *on-set*, we mean the set of assignments which cause ϕ_i to be true.

Definition 2 (SUCCINCT SET COVER (SSC) [Uma99]). *Given a set* $S = \{\phi_i\}$ *of* 3-*DNF formulae such that* $\bigvee_{i \in S} \phi_i$ *is a tautology, what is the size of the smallest* $S' \subseteq S$ *such that* $\bigvee_{i \in S'} \phi_i$ *a tautology?*

Definition 3 (IRREDUNDANT (IRR) [Uma99]). *Given a DNF formula* $\phi = t_1 \vee t_2 \vee \cdots \vee t_n$, *what is the size of the smallest* $S \subseteq \{t_i\}_{i=1}^n$ *such that* $\phi = \bigvee_{i \in S} t_i$?

Our work introduces and studies quantum generalizations of SSC and IRR. In particular, analogous to the classically important task of circuit minimization, the quantum generalizations we define are arguably natural and related to what one might call "Hamiltonian minimization" — given a sum of Hermitian operators $H = \sum_i H_i$, what is the smallest subset of terms $\{H_i\}$ whose sum approximately preserves certain spectral properties of H? We hope that such questions may be useful to physicists in a lab who wish to simulate the simplest Hamiltonian possible while retaining the desired characteristics of a complex Hamiltonian involving many interactions. We remark that at a high level, the connection to cq-Σ_2 for the task of Hamiltonian minimization is as follows: The classical existential proof encodes the subset of terms $\{H_i\}$, while the quantum for-all proof encodes complex unit vectors which achieve certain energies against H. We now define the problem QUANTUM SUCCINCT SET COVER.

Definition 4. QUANTUM SUCCINCT SET COVER *(QSSC) (informal) Given a set of local Hamiltonians* $\{H_i\}$ *such that* $\sum_i H_i$ *has smallest eigenvalue at least* α, *what is the size of the smallest subset* S *of the* H_i *such that* $\sum_{H_i \in S} H_i$ *has smallest eigenvalue at least* α? *Any subset satisfying this property is called a* cover.

Here, a *local Hamiltonian* is a sum of Hermitian operators, each of which acts non-trivially on at most $k \in \Theta(1)$ qubits (hence the name k-local Hamiltonian). Intuitively, the goal in QSSC is to cover the entire Hilbert space using as few interaction terms H_i as possible. Hence, we associate the notion of a "cover" with obtaining large eigenvalues, as opposed to small ones, making QSSC a direct quantum analogue of SSC. We remark that since SSC is a classical constraint satisfaction problem, we believe the language of *quantum* constraint satisfaction, i.e. Hamiltonian constraints, is a natural avenue for defining QSSC. Our first result concerns QSSC, and is as follows.

Theorem 1. *QSSC is* cq-Σ_2-*complete, and moreover is* cq-Σ_2-*hard to approximate within* $N^{1-\epsilon}$ *for all* $\epsilon > 0$, *where* N *is the encoding size of the QSSC instance.*

By *hard to approximate*, we mean that any problem in cq-Σ_2 can be reduced to an instance of QSSC via a polynomial time mapping or Karp reduction such that the gap between the sizes of the optimal cover in the YES and NO cases scales as $N^{1-\epsilon}$. In other words, it is cq-Σ_2-hard to determine whether the smallest cover size of an arbitrary instance of QSSC is at most g or at least g' for $g'/g \in \Omega(N^{1-\epsilon})$ (where $g' \geq g$). We next define the problem QUANTUM IRREDUNDANT (QIRR).

Definition 5. QUANTUM IRREDUNDANT *(QIRR) (informal) Given a set of succinctly described orthogonal projection operators* $\{H_i\}$ *acting on* N *qubits, and a set* $\{c_i \geq 0\} \subseteq \mathbb{R}$, *define* $H := \sum_i c_i H_i$. *Then, what is the size of the smallest subset* $S \subseteq \{H_i\}$ *such that for* $H' = \sum_{H_i \in S} c_i H_i$, *vectors achieving high and low energies against* H *continue to obtain high and low energies against* H', *respectively?*

Here, by a *succinctly* described projector, we mean a possibly non-local operator which is the tensor product of k-local projectors for some $k \in \Theta(1)$. This non-local structure naturally generalizes IRR, where the DNF formula is allowed to be non-local. Our next result is the following.

Theorem 2. *QIRR is cq-Σ_2-hard to approximate within $N^{\frac{1}{2}-\epsilon}$ for all $\epsilon > 0$, where N is the encoding size of the QIRR instance.*

2. Hardness of Approximation for QCMA. The techniques from above can also be used to show hardness of approximation for QCMA. Here, the class QCMA [AN02] is defined as cq-Σ_2 with the second (quantum) proof omitted, and can hence be thought of as the first level of our "cq-hierarchy". By defining the problem QUANTUM MONO-TONE MINIMUM SATISFYING ASSIGNMENT (QMSA) (see Section 3), we show:

Theorem 3. *QMSA is QCMA-complete, and moreover is QCMA-hard to approximate within $N^{1-\epsilon}$ for all $\epsilon > 0$, where N is the encoding size of the QMSA instance.*

3. A Canonical cq-Σ_2-Complete Problem. Our last result concerns a canonical Σ_2^p-complete problem, Σ_iSAT, and its generalization to the quantum setting. Specifically, given a boolean formula ϕ, Σ_iSAT asks whether:

$$\exists x_1 \forall x_2 \exists x_3 \cdots \forall x_i \text{ such that } \phi(x_1, x_2, x_3, \ldots, x_i) = 1 . \qquad (1)$$

Here, we assumed i is even; for odd i, the last quantifier is a $\exists$. The terms x_j are vectors of boolean variables. For $i = 2$, one can define a natural quantum generalization of this problem, denoted cq-Σ_2LH and defined in Section 4, using local Hamiltonians whose ground states are tensor products of a classical string and a quantum state. We show:

Theorem 4. *cq-Σ_2LH is cq-Σ_2-complete.*

Proof ideas: Our proofs are inspired by the classical work of Umans [Uma99,Hem02], and are achieved in a few steps. First, we show a *gap-introducing* reduction from an arbitrary cq-Σ_2 problem to a problem we call QUANTUM MONOTONE MINI-MUM WEIGHT WORD (QMW) using *dispersers* (see e.g., [SZ94,TSUZ07]). We then show the following *gap-preserving* reductions, where $\leq_K$ denotes a mapping or Karp reduction:

$$\text{QMW} \leq_K \text{QSSC} \leq_K \text{QIRR} . \qquad (2)$$

This yields hardness ratios of N^ϵ for some $\epsilon > 0$. To obtain the stronger results claimed in Section 1, we finally apply the gap amplification of Umans [Uma99] and improved disperser construction of Ta-Shma, Umans, and Zuckerman [TSUZ07].

In the classical setting, Umans [Uma99,Hem02] used dispersers to attain hardness of approximation results relative to Σ_2^p for the classical problems MMWW (the classical version of QMW), SSC and IRR. To extend his techniques to the quantum setting, the most involved aspects of our work are the gap-preserving reductions from QMW to QSSC to QIRR. Here, an intricate balancing act involving carefully defined local Hamiltonian terms is needed to construct operators with the spectral properties required for our reductions. To analyze the resulting sums of non-commuting Hamiltonians, we require heavier machinery, such as the specific structure of Kitaev's local Hamiltonian construction [KSV02], the Projection Lemma of Kempe, Kitaev, and Regev [KKR06], and the Geometric Lemma of Kitaev [KSV02].

Finally, to show cq-Σ_2-completeness of cq-Σ_2LH, we study the interplay between classical-quantum proofs and Kempe and Regev's [KR03] 3-local Hamiltonian construction. A careful analysis reveals that any cq-Σ_2 verification circuit can be modified in such a way that fixing the value c of its classical proof register leads to an *effective* Hamiltonian H_c. We then study the spectrum of H_c to achieve the desired result.

Previous and related work: To the best of our knowledge, our work is the first to obtain hardness of approximation results for a quantum complexity class. The related question of whether a *quantum* PCP theorem holds is currently one of the biggest open problems in quantum complexity theory (see, e.g., [Aar06,AALV09,Ara10,Has12]). Regarding quantum generalizations of PH, we remark that Yamakami has proposed an approach [Yam02] differing from ours as follows: It is based on quantum Turing machines (we work with quantum circuits), allows *quantum* inputs (our inputs are classical strings, as in QMA), and considers quantum quantifiers at each level of the hierarchy (our scheme allows alternating classical and quantum quantifiers between levels).

Significance and open questions: The classical polynomial hierarchy plays an important role in classical complexity theory, both as a generalization of NP and as a proof tool in itself. It is hoped that the scheme we propose here for generalizing PH to the quantum setting will find similar applications in quantum complexity theory. Second, the problems we show to be cq-Σ_2-complete here are arguably natural, and in embodying a generalization of classical circuit minimization or optimization, may hopefully be related to practical scenarios in a lab. Further, although the alternation between classical and quantum quantifiers in cq-Σ_2 may a priori seem odd, the notion of relating a classical proof to, say, subsets of local Hamiltonian terms, and the quantum proof to quantum states achieving certain energies is in itself quite natural, and in our opinion justifies the study of such a combination of quantifiers. Third, with respect to hardness of approximation, since whether a quantum PCP theorem holds remains a challenging open question, it is all the more interesting that one is able to prove hardness of approximation in a quantum setting here using an entirely different tool, namely that of dispersers. We remark that dispersers and their two-sided analogues, extractors, have been used classically to amplify existing PCP inapproximability results [SZ94,Zuc96]. However, as far as we are aware, neither are known to directly yield PCP constructions.

We leave a number of questions open: How do the different classes cq-Σ_2, qc-Σ_2, qq-Σ_2, and cc-Σ_2 relate to each other? What results about Σ_2^p carry over to cq-Σ_2? Where do the different variants of our quantum hierarchy sit with respect to known complexity classes? Can an appropriately defined approximation version of cq-Σ_2LH be shown cq-Σ_2-hard to approximate? We hope the answers to such questions will help establish classes like cq-Σ_2 as fundamental concepts in the study of quantum computational complexity.

Organization of this paper: Section 2 introduces formal definitions and states useful lemmas. In Section 3, we prove that QSSC and QIRR are hard to approximate for cq-Σ_2, and similarly for QMSA relative to QCMA. Section 4 shows cq-Σ_2-completeness of cq-Σ_2LH.

2   Definitions

We now set our notation, define relevant classes and problems, and state lemmas which prove useful in our analysis.

Beginning with notation, the term $A \succeq B$ means operator $A - B$ is positive semidefinite. The spectral norm of A is $\| A \|_\infty := \max\{\| A|v\rangle \|_2 : \| |v\rangle \|_2 = 1\}$. The projector onto space S is Π_S. The set of natural numbers is $\mathbb{N}$. For convenience, we define $\mathcal{B} := \mathbb{C}^2$, and for a set S of matrices over $\mathbb{C}$, let $H_S := \sum_{H_i \in S} H_i$.

We next give a formal definition of cq-Σ_2. Here, a promise problem is a pair $A = (A_{\text{yes}}, A_{\text{no}})$ such that $A_{\text{yes}}, A_{\text{no}} \subseteq \{0,1\}^*$ and $A_{\text{yes}} \cap A_{\text{no}} = \emptyset$.

Definition 6 (cq-Σ_2). *Let $A = (A_{\text{yes}}, A_{\text{no}})$ be a promise problem. We say that $A \in$ cq-Σ_2 if there exist polynomially bounded functions $t, c, q : \mathbb{N} \mapsto \mathbb{N}$, and a deterministic Turing machine M acting as follows. For every n-bit input x, M outputs in time $t(n)$ a description of a quantum circuit V_x such that V_x takes in a $c(n)$-bit proof $|c\rangle$, a $q(n)$-qubit proof $|q\rangle$, and outputs a single qubit. We say V_x accepts $|c\rangle|q\rangle$ if measuring its output qubit in the computational basis yields 1. Then:*

- *Completeness: If $x \in A_{\text{yes}}$, then $\exists |c\rangle$ such that $\forall |q\rangle$, V_x accepts $|c\rangle|q\rangle$ with probability $\geq 2/3$.*
- *Soundness: If $x \in A_{\text{no}}$, then $\forall |c\rangle$, $\exists |q\rangle$ such that V_x rejects $|c\rangle|q\rangle$ with probability $\geq 2/3$.*

Using the standard approach of repeating V_x polynomially many times in parallel (see, e.g. [AN02]), the completeness and soundness parameters can be amplified to values exponentially close to 1. Throughout this paper, we refer to this as *error reduction*.

We next define the terms cQMA circuit, monotone set, QMW, and QSSC.

Definition 7 (cQMA **circuit**). *Let $n, m \in \mathbb{N}^+$. A cQMA circuit V is a quantum circuit receiving n bits in an INPUT register and m qubits in a CHOICE register, and outputting a single qubit $|a\rangle$. We say that:*

- *V accepts $x \in \{0,1\}^n$ in INPUT if for all $|y\rangle \in \mathcal{B}^{\otimes m}$ in CHOICE, measuring $|a\rangle$ in the computational basis yields 1 with probability at least $2/3$.*
- *V rejects $x \in \{0,1\}^n$ in INPUT if there exists a $|y\rangle \in \mathcal{B}^{\otimes m}$ in CHOICE such that measuring $|a\rangle$ in the computational basis yields 0 with probability at least $2/3$.*

Definition 8 (**Monotone set**). *A set $S \subseteq \{0,1\}^n$ is called* monotone *if for any $x \in S$, any string obtained from x by flipping one or more zeroes in x to one is also in S.*

Definition 9 (QUANTUM MONOTONE MINIMUM WEIGHT WORD (**QMW**)). *Given a cQMA circuit V accepting exactly a non-empty monotone set $S \subseteq \{0,1\}^n$, and integer thresholds $0 \leq g \leq g' \leq n$, output:*

- *YES if there exists an $x \in \{0,1\}^n$ of Hamming weight at most g accepted by V.*
- *NO if all $x \in \{0,1\}^n$ of Hamming weight at most g' are rejected by V.*

Note that clearly QMW $\in$ cq-Σ_2.

Definition 10 (QUANTUM SUCCINCT SET COVER (**QSSC**)). *Let* $S := \{H_i\}$ *be a set of 5-local Hamiltonians* H_i *acting on* N *qubits such that* $\sum_{H_i \in S} H_i \succeq \alpha I$ *for* $\alpha > 0$. *Then, given* $\beta \in \mathbb{R}$ *such that* $\alpha - \beta \geq 1$ *and integer thresholds* $0 \leq g \leq g'$, *output:*

- *YES if there exists* $S' \subseteq S$ *of cardinality at most* g *such that* $\sum_{H_i \in S'} H_i \succeq \alpha I$.
- *NO if for all* $S' \subseteq S$ *of size at most* g', $\sum_{H_i \in S'} H_i$ *has an eigenvalue at most* β.

Any S' *satisfying the YES case is called a* cover.

Roughly, QSSC asks how many local interaction terms in a local Hamiltonian one can discard while maintaining the value of the worst assignment. This is intended to mimic the idea of maintaining a tautology for a 3-DNF formula in SSC classically. Note also that requiring $\alpha - \beta \in \Omega(1)$ above is without loss of generality, as any instance of QSSC with gap $1/p(N)$ for p a polynomially bounded function can be modified to obtain an equivalent instance with constant gap by multiplying each H_i by $p(N)$ [Wat09].

Next, the key tool enabling the creation of a gap in our reductions is a *disperser*.

Definition 11 (Disperser). *Let* $G = (L, R, E)$ *be a bipartite graph with* $|L| = 2^n$, $|R| = 2^m$ *and left-degree* 2^d. *Then,* G *is called a* (k, ϵ)-*disperser if, for any subset* $L' \subseteq L$ *of size* $|L'| \geq 2^k$, L' *has at least* $(1 - \epsilon)|R|$ *neighbors in* R. *Moreover, if for any pair* $(v \in L, i)$, *one can compute the ith neighbor of* v *in time polynomial in* n, *then the disperser is called* explicit.

Finally, we recall useful known facts from Hamiltonian complexity theory. We first state a lemma used to bound the eigenvalues of a pair of non-commuting operators.

Lemma 1 (Kempe, Kitaev, Regev [KKR06], Projection Lemma). *Let* $Y = Y_1 + Y_2$ *act on Hilbert space* $\mathcal{H} = \mathcal{S} + \mathcal{S}^{\perp}$ *for Hamiltonians* Y_1 *and* Y_2. *Denote the zero eigenspace of* Y_2 *as* $\mathcal{S}$, *and assume the* Y_2 *eigenvectors in* $\mathcal{S}^{\perp}$ *have eigenvalue at least* $J > 2 \|Y_1\|_{\infty}$. *Then, for* $\lambda(Y)$ *the smallest eigenvalue of* Y *and* $Y|_{\mathcal{S}} := \Pi_{\mathcal{S}} Y \Pi_{\mathcal{S}}$,

$$\lambda(Y_1|_{\mathcal{S}}) - \frac{\|Y_1\|_{\infty}^2}{J - 2\|Y_1\|_{\infty}} \leq \lambda(Y) \leq \lambda(Y_1|_{\mathcal{S}}) \; . \tag{3}$$

We next define Kitaev's 5-local circuit-to-Hamiltonian construction [KSV02]. Given a cq-Σ_2 verification circuit $V = V_L \cdots V_1$ (where without loss of generality, each V_i is a one- or two-qubit unitary) acting on n proof bits (register A), m proof qubits (register B), and p ancilla qubits (register C), this construction outputs a 5-local Hamiltonian H acting on $A \otimes B \otimes C \otimes D$, where D is a clock register of dimension L. We then have $H := H_{\text{in}} + H_{\text{out}} + H_{\text{prop}} + H_{\text{stab}}$, for *penalty* terms as defined below:

$$H_{\text{in}} := I_{A,B} \otimes \left(\sum_{i=1}^{p} |1\rangle\langle 1|_{C_i} \right) \otimes |0\rangle\langle 0|_D \; , \tag{4}$$

$$H_{\text{out}} := I_A \otimes |0\rangle\langle 0|_{B_1} \otimes I_C \otimes |L\rangle\langle L|_D \; , \tag{5}$$

$$H_{\text{prop}} := \sum_{j=1}^{L} H_j, \text{ where } H_j \text{ is defined as} \tag{6}$$

$$-\frac{1}{2}V_j \otimes |j\rangle\langle j-1|_D - \frac{1}{2}V_j^\dagger \otimes |j-1\rangle\langle j|_D + \frac{1}{2}I \otimes (|j\rangle\langle j| + |j-1\rangle\langle j-1|)_D ,$$

$$H_{\text{stab}} := I_{A,B,C} \otimes \sum_{i=1}^{L-1} |01\rangle\langle 01|_D . \tag{7}$$

Note that time t in clock register D is implicitly encoded in unary as $|1^t 0^{L-t}\rangle$. We use two important properties of this construction. First, the null space of $H_{\text{in}} + H_{\text{prop}} + H_{\text{stab}}$ is the space of *history states*, which for arbitrary $|\psi\rangle_{A,B}$ are defined as

$$|\psi\rangle_{\text{hist}} := \frac{1}{\sqrt{L+1}} \sum_{i=0}^{L} V_i \cdots V_1 |\psi\rangle_{A,B} \otimes |0\rangle_C \otimes |i\rangle_D . \tag{8}$$

For cq-Σ_2 circuits V, it is further convenient to define for $c \in \{0,1\}^n$ and $|q\rangle \in \mathcal{B}^m$ the shorthand $|c,q\rangle_{\text{hist}} := |\psi\rangle_{\text{hist}}$ for $|\psi\rangle = |c\rangle|q\rangle$. The second important property of H we use is that its spectrum is related to V as follows.

Lemma 2 ([KSV02]). *The construction above maps V to (H, a, b) satisfying:*

- *If there exists a proof $|\psi\rangle$ accepted by V with probability at least $1 - \epsilon$, then $|\psi\rangle_{\text{hist}}$ achieves $\text{Tr}(H|\psi\rangle\langle\psi|_{\text{hist}}) \leq a$ for $a := \epsilon/(L+1)$.*
- *If V rejects all proofs $|\psi\rangle$, then $H \succeq bI$ for $b := \Theta\left(\frac{1-\epsilon}{L^3}\right)$.*

3   Hardness of Approximation

We now show hardness of approximation for the problems QMW, QSSC, QIRR, and QMSA. We begin with a gap-introducing reduction to QMW.

Theorem 5. *There exists a poly-time reduction which, given an instance of an arbitrary cq-Σ_2 problem, outputs an instance of QMW with thresholds g and g' satisfying $g'/g \in \Theta(N^\epsilon)$ for some $\epsilon > 0$, where N is the encoding size of the QMW instance.*

Proof (Sketch). The reduction follows the proof of Theorem 1 of Umans [Uma99] closely; we explicitly note the main differences in the quantum setting. To map an instance Π of an arbitrary cq-Σ_2 problem with verifier V to a cQMA circuit W for QMW, we first construct an explicit disperser $G = (L, R, E)$, which W uses as follows. The vertices in L and R roughly correspond to possible assignments to V's and W's classical registers, respectively. Given $R_y \subseteq R$ and quantum proof $|z\rangle$, if $|R_y|$ is "small", W decodes R_y to obtain a set of assignments $L_y \subseteq L$ for V, where the encoding scheme is carefully constructed so that by the disperser's expansion property, the decoding can be done in polynomial time. W then calls V as a subroutine to check if there exists an $x \in L_y$ causing V to accept on proof $|z\rangle$. If so, W accepts. With this approach, if Π is a YES instance, one need only choose a "small" subset R_y to encode an accepting $x \in L$ for V. Note that in the quantum setting, we must be careful in constructing a different "bootstrapping" procedure for V above than that of Umans to accommodate for multiple copies of (possibly entangled) proofs $|z\rangle$.

If, however, Π is a NO instance, then no encoded $x \in L$ causes W to accept (with high probability in the quantum case) as above. Thus, we design W to have a "default" option for acceptance — if $|R_y|$ is "large", W accepts immediately. A close analysis thus yields that $|R_y|$ differs by a large gap for YES and NO instances Π, as desired. □

We next show a gap-preserving reduction from QMW to QSSC. Its proof requires Lemmas 3 and 4, which are stated subsequently.

Theorem 6. *QSSC is in* cq-Σ_2. *Further, there exists a poly-time reduction which, given an instance of QMW with thresholds f and f', outputs an instance of QSSC with thresholds $g = f + 2$ and $g' = f' + 2$, respectively.*

Proof. That QSSC is in cq-Σ_2 follows using Kitaev's verifier [KSV02] for putting k-local Hamiltonian in QMA. As for cq-Σ_2-hardness of QSSC, suppose we are given a cQMA circuit $V = V_L \cdots V_1$ accepting exactly a non-empty monotone set $T \subseteq \{0,1\}^n$ with soundness and completeness error $\epsilon := 1 - 2^{-4(n+m)}$, and threshold parameters f and f'. By applying Kitaev's circuit-to-Hamiltonian construction from Section 2 to V, we obtain 3-tuple $(H = H_{in} + H_{out} + H_{prop} + H_{stab}, a, b)$. Now, set $\alpha := 1 - (\zeta + 1)\epsilon$ for $\zeta := 2(1 + 2^{2(n+m)})/(L + 1)$, $\beta := 1 - b$, $g := f + 2$, $g' := f' + 2$, and let S consist of the elements (intuition to follow)

$$G_1 := (L + 1)|0\rangle\langle 0|_{A_1} \otimes I_{B,C} \otimes |0\rangle\langle 0|_D \,, \tag{9}$$

$$\vdots \tag{10}$$

$$G_n := (L + 1)|0\rangle\langle 0|_{A_n} \otimes I_{B,C} \otimes |0\rangle\langle 0|_D \,, \tag{11}$$

$$G_{n+1} := (\Delta + 1)(H_{in} + H_{prop} + H_{stab}) \,, \tag{12}$$

$$G_{n+2} := I - (H_{in} + H_{prop} + H_{stab} + H_{out}) \,, \tag{13}$$

for $\Delta \in \Omega(n^2 L^5/\epsilon)$. Intuitively, the terms in S play the following roles: G_{n+1} penalizes assignments which are not valid history states. G_{n+2} penalizes valid history states accepted by V. Finally, the G_i for $i \in [n]$ penalize valid history states rejected by V (recall that V accepts a monotone set, and so flipping a one to a zero in register A may lead V to reject). Thus, we cover the entire space. We now make this rigorous.

As required by Definition 10, we begin by showing that S itself is a cover, i.e. that $G_S \succeq \alpha I_{A,B,C,D}$. To attain this, it suffices to prove that

$$\Delta(H_{in} + H_{prop} + H_{stab}) + \left(\sum_{i=1}^{n} G_i\right) - H_{out} \succeq -(\zeta + 1)\epsilon I \,. \tag{14}$$

We show this using Lemma 1, the Projection Lemma, by setting $Y_1 := (\sum_{i=1}^{n} G_i) - H_{out}$ and $Y_2 := \Delta(H_{in} + H_{prop} + H_{stab})$. The Projection Lemma yields that we can focus our attention on the smallest eigenvalue of Y_1 restricted to the space of all *valid* history states, $\mathcal{S}_{hist}$, i.e. states of the form of (8). By calling Lemma 4 to bound $-Y_1|_{\mathcal{S}_{hist}} \preceq \zeta \epsilon I$ and Lemma 3 to bound the smallest non-zero eigenvalue of Y_2 as $\Omega(\Delta/L^3)$, the claim that S is a cover follows.

We now show the desired reduction. The forward direction, i.e. if V accepts a string x of Hamming weight k, follows similarly to above. As for the converse, suppose V rejects any string x of Hamming weight at most k. For any $S' \subseteq S$ with $|S'| \leq k+2$, we claim that $G_{S'}$ has an eigenvalue at most β. To see this, we first argue that $G_{n+1}, G_{n+2} \in S'$, as otherwise the state $|1^n, y\rangle_{\text{hist}}$, for example, attains $\text{Tr}(G_{S'}|1^n, y\rangle\langle 1^n, y|_{\text{hist}}) = 0$. This implies that S' contains at most k terms G_i for $i \in [n]$. Then, consider the string x which has ones precisely at these at most k positions $i \in [n]$ corresponding to $G_i \in S'$. Since V rejects all strings of Hamming weight at most k, a $|y\rangle \in \mathcal{B}^{\otimes m}$ such that

$$\text{Tr}\left(G_{n+2}|x, y\rangle\langle x, y|_{\text{hist}}\right) = 1 - \text{Tr}\left(H|x, y\rangle\langle x, y|_{\text{hist}}\right) \leq 1 - b = \beta \quad (15)$$

exists by the definition of a cQMA circuit and Lemma 2.   □

As mentioned earlier, we now state two lemmas which were used in the proof of Theorem 6. Their statements assume the notation of Theorem 6, and their proofs require, among other tools, the Geometric Lemma of Reference [KSV02].

Lemma 3. *The smallest non-zero eigenvalue of* $Y_2 = \Delta(H_{\text{in}} + H_{\text{prop}} + H_{\text{stab}})$ *scales as* $\Omega(\Delta/L^3)$.

Lemma 4. *Let* Π_{hist} *project onto* $\mathcal{S}_{\text{hist}}$, $\zeta := 2(1 + 2^{2(n+m)})/(L+1)$, *and consider* $T \subseteq [n]$. *Then, if* V *outputs one with probability at least* $1 - \epsilon$ *for inputs* $(x, |y\rangle)$ *with* $x \in \{0, 1\}^n$ *such that* $x_i = 1$ *for all* $i \in T$ *and for all m-qubit* $|y\rangle$, *one has* $\Pi_{\text{hist}}[H_{\text{out}} - \sum_{i \in T} G_i]\Pi_{\text{hist}} \preceq \zeta \epsilon I$.

We next show that QIRR is cq-Σ_2-hard to approximate. The intuition behind the proof is as follows. QIRR is stated in terms of projectors F_j (up to scalar multiplication), whereas QSSC is stated in terms of Hermitian operators G_i. Hence, beginning with the reduction of Theorem 6, a natural idea is to treat each local Hamiltonian term in the sums comprising G_{n+1} and G_{n+2} as distinct terms F_j. However, in order to rigorously argue that the gap between thresholds g and g' for QSSC is preserved when defining thresholds h and h' for QIRR, we would like, for example, that *all* terms F_j making up G_{n+1} are chosen together. This is attained by introducing *chaperone* qubits, which force all F_j coming from G_{n+1} and G_{n+2} to be chosen in any candidate cover.

Theorem 7. *There exists a poly-time reduction which, given an instance of an arbitrary* cq-Σ_2 *problem, outputs an instance of QIRR with thresholds h and h' satisfying $h'/h \in \Theta(N^\epsilon)$ for some $\epsilon > 0$, where N is the encoding size of the QIRR instance.*

Finally, the hardness gaps of Theorems 5, 6, and 7 can be improved to those claimed in Section 1 as follows. Specifically, the gap in Theorem 5 can be amplified by first following a classical idea of Umans to compose the cQMA circuit W with itself. Note that this composition W' must be defined more carefully in the quantum setting due to bounded-error quantum circuits and possible entanglement between copies of proofs. Using W' and the improved disperser construction of Reference [TSUZ07] (see Theorem 7.2 therein) in the proof of Theorem 5 then yields the result below. This, in turn, yields the improved hardness ratios for QSSC and QIRR via our previous reductions.

Theorem 8. *QMW is cq-Σ_2-hard to approximate with gap $N^{1-\epsilon}$ for any $\epsilon > 0$, for N the encoding size of the QMW instance.*

Finally, by straightforwardly extending Umans' classical result [Uma99] showing NP-hardness of approximation for the problem MONOTONE MINIMUM SATISFYING ASSIGNMENT, one can show hardness of approximation for QCMA. Specifically, define QUANTUM MONOTONE MINIMUM SATISFYING ASSIGNMENT (QMSA) analogously to QMW, except with the definition of a cQMA circuit V modified to drop the second (quantum) proof, i.e. V now only takes one input register comprised of n classical bits. Then, it is straightforward to re-run the proofs of Theorems 5 and 8 without the existence of a second quantum proof register, leading to Theorem 3.

4   A Canonical cq-Σ_2-Complete Problem

We now show the following quantum generalization of the canonical Σ_2^p-complete problem Σ_2SAT is cq-Σ_2-complete.

Definition 12 (cq-Σ_2LH). *Given a 3-local Hamiltonian H acting on $N = n + m$ qubits, and $a, b \in \mathbb{R}$ such that $a \leq b$ for $b - a \geq 1$, output:*

- *YES if $\exists\, x \in \{0,1\}^n$ such that $\forall\, |y\rangle \in \mathcal{B}^{\otimes m}$, $\mathrm{Tr}(H|x\rangle\langle x| \otimes |y\rangle\langle y|) \geq b$.*
- *NO if $\forall\, x \in \{0,1\}^n$, $\exists\, |y\rangle \in \mathcal{B}^{\otimes m}$ such that $\mathrm{Tr}(H|x\rangle\langle x| \otimes |y\rangle\langle y|) \leq a$.*

Proof (Sketch of Theorem 4). That cq-Σ_2LH $\in$ cq-Σ_2 follows from Kitaev's verifier for placing k-local Hamiltonian in QMA [KSV02]. To reduce an instance of a problem in cq-Σ_2 with verification circuit V'' to cq-Σ_2LH, we first modify V'' to ensure the contents of its classical proof register remain unchanged throughout the verification (this ensures history states have a tensor product structure across the classical-quantum cut), and we negate the output of V'' (in the YES case we wish to have large energy *for all* proofs $|y\rangle$, including proofs normally accepted by the verification circuit); call the new circuit V. We then apply Kitaev's construction from Section 2 on V to obtain a 5-local Hamiltonian H. (We can also use the 3-local construction of [KR03].)

Now suppose we have a YES instance of Π, i.e. there exists $|c\rangle$ such that for all $|q\rangle$, the circuit V'' accepts proof $|c\rangle \otimes |q\rangle$ with probability at least $1 - \epsilon$. Letting $\Pi_c :=$ $(|c\rangle\langle c|_A \otimes I_{B,C,D})$ for the accepted $|c\rangle$ above, we must show that for all $|\psi\rangle_{B,C,D}$,

$$\langle c| \otimes \langle\psi|H|c\rangle \otimes |\psi\rangle = \langle c| \otimes \langle\psi|\Pi_c H \Pi_c|c\rangle \otimes |\psi\rangle \geq b . \tag{16}$$

By carefully analyzing the terms $\Pi_c H_{\mathrm{in}} \Pi_c$, $\Pi_c H_{\mathrm{out}} \Pi_c$, $\Pi_c H_{\mathrm{prop}} \Pi_c$, and $\Pi_c H_{\mathrm{stab}} \Pi_c$, we find that $\langle c| \otimes \langle\psi|H|c\rangle \otimes |\psi\rangle = \langle\psi|H_c|\psi\rangle$ for some effective Hamiltonian H_c dependant on c. We are thus reduced to showing $H_c \succeq bI$. If we now think of c not as an input to V, but rather as indexing a set of circuits V_c, where V_c is V with c hardwired into its classical register, it turns out that Kitaev's construction applied to V_c yields H_c. By Lemma 2, the claim thus follows. The converse direction is similar. $\qquad\square$

Acknowledgements. We thank Richard Cleve, Ashwin Nayak, Sarvagya Upadhyay, and John Watrous for interesting discussions, and especially Oded Regev for many helpful insights, including the suggestion to think about a quantum version of PH. SG is supported by the NSERC CGS, NSERC CGS-MSFSS, and EU-Canada Transatlantic Exchange Partnership programs, and the David R. Cheriton School of Computer Science. JK is supported by an Individual Research Grant of the Israeli Science Foundation, by European Research Council (ERC) Starting Grant QUCO and by the Wolfson Family Charitable Trust.

References

AALV09.   Aharonov, D., Arad, I., Landau, Z., Vazirani, U.: The detectibility lemma and quantum gap amplification. In: 41st ACM Syposium on Theory of Computing, vol. 287, pp. 417–426 (2009)

Aar06.   Aaronson, S.: The quantum PCP manifesto (2006),
          http://scottaaronson.com/blog/?p=139

AN02.   Aharonov, D., Naveh, T.: Quantum NP - A survey. Preprint at arXiv:quant-ph/0210077v1 (2002)

Ara10.   Arad, I.: A note about a partial no-go theorem for quantum PCP. Preprint at arXiv:quant-ph/1012.3319 (2010)

FLvMV05. Fortnow, L., Lipton, R., van Melkebeek, D., Viglas, A.: Time-space lower bounds for satisfiability. Journal of the ACM 52, 835–865 (2005)

For00.   Fortnow, L.: Time-space tradeoffs for satisfiability. Journal of Computer and System Sciences 60(2), 337–353 (2000)

Has12.   Hastings, M.B.: Trivial low energy states for commuting hamiltonians, and the quantum PCP conjecture. Preprint at arXiv:quant-ph/1201.3387 (2012)

Hem02.   Hemaspaandra, L.: SIGACT news complexity theory column 38. ACM SIGACT News 33(4) (2002); Guest column by Schaefer, M., Umans, C.

KKR06.   Kempe, J., Kitaev, A., Regev, O.: The complexity of the local Hamiltonian problem. SIAM Journal on Computing 35(5), 1070–1097 (2006)

KR03.   Kempe, J., Regev, O.: 3-local Hamiltonian is QMA-complete. Quantum Information & Computation 3(3), 258–264 (2003)

KSV02.   Kitaev, A., Shen, A., Vyalyi, M.: Classical and Quantum Computation. American Mathematical Society (2002)

Lau83.   Lautemann, C.: BPP and the polynomial time hierarchy. Information Processing Letters 17, 215–218 (1983)

Sip83.   Sipser, M.: A complexity theoretic approach to randomness. In: 15th Symposium on Theory of Computing, pp. 330–335. ACM Press (1983)

SZ94.   Srinivasan, A., Zuckerman, D.: Computing with very weak random sources. In: 35th Symposium on Foundations of Computer Science, pp. 264–275 (1994)

TSUZ07.   Ta-Shma, A., Umans, C., Zuckerman, D.: Lossless condensers, unbalanced expanders, and extractors. Combinatorica 27(2), 213–240 (2007)

Uma99.   Umans, C.: Hardness of approximating Σ_2^p minimization problems. In: 40th Symposium on Foundations of Computer Science, pp. 465–474 (1999)

Wat09.   Watrous, J.: Quantum computational complexity. In: Meyers, R. (ed.) Encyclopedia of Complexity and Systems Science, ch. 17, pp. 7174–7201. Springer (2009)

Yam02.   Yamakami, T.: Quantum NP and a quantum hierarchy. In: 2nd IFIP International Conference on Theoretical Computer Science, pp. 323–336. Kluwer Academic Publishers (2002)

Zuc96.   Zuckerman, D.: On unapproximable versions of NP-complete problems. SIAM Journal on Computing 25(6), 1293–1304 (1996)

The Complexity of Computing the Sign of the Tutte Polynomial (and Consequent #P-hardness of Approximation)[*]

Leslie Ann Goldberg[1] and Mark Jerrum[2]

[1] Department of Computer Science, University of Liverpool, Ashton Building,
Liverpool L69 3BX, United Kingdom
[2] School of Mathematical Sciences Queen Mary, University of London, Mile End
Road, London E1 4NS, United Kingdom

Abstract. We study the complexity of computing the sign of the Tutte
polynomial of a graph. As there are only three possible outcomes (positive, negative, and zero), this seems at first sight more like a decision
problem than a counting problem. Surprisingly, however, there are large
regions of the parameter space for which computing the sign of the Tutte
polynomial is actually #P-hard. As a trivial consequence, approximating
the polynomial is also #P-hard in this case. Thus, *approximately* evaluating the Tutte polynomial in these regions is as hard as *exactly* counting
the satisfying assignments to a CNF Boolean formula. For most other
points in the parameter space, we show that computing the sign of the
polynomial is in FP, whereas approximating the polynomial can be done
in polynomial time with an NP oracle. As a special case, we completely
resolve the complexity of computing the sign of the chromatic polynomial — this is easily computable at $q = 2$ and when $q \leq 32/27$, and is
NP-hard to compute for all other values of the parameter q.

1 Introduction

The Tutte polynomial of a graph is two-variable polynomial that captures many
interesting properties of the graph such as (by making appropriate choices of the
two variables) the number of q-colourings, the number of nowhere-zero q-flows,
the number of acyclic orientations, and the probability that the graph remains
connected when edges are deleted at random.

Much work [1,2,3,4,9,12], has studied the difficulty of evaluating the polynomial (exactly or approximately) when the values of the variables are fixed, and
a graph is given as input.

Our early paper [3] identified a large region of points where the approximate
evaluation of the polynomial is NP-hard and a short hyperbola segment along
which approximate evaluation is even #P-hard. Thus, an approximation of the
polynomial at a point on this short hyperbola segment would enable one to

---

[*] This work was partially supported by the EPSRC grant *Computational Counting*.
A full version of this paper is available at http://arxiv.org/abs/1202.0313.

A. Czumaj et al. (Eds.): ICALP 2012, Part I, LNCS 7391, pp. 399–410, 2012.
© Springer-Verlag Berlin Heidelberg 2012

exactly solve a problem in #P. In this paper, we show that, in fact, for most of these NP-hard points (and more), approximation is #P-hard. Moreover, it is #P-hard for a very simple reason: determining the sign of the polynomial — i.e., whether the evaluation of the polynomial is positive, negative or zero — is #P-hard. This seems surprising since determining the sign of the polynomial is nearly a decision problem (there are only three possible outcomes) but it is #P-hard nearly everwhere (at all of the red points in the plane in Figure 1).

Past work [7] has studied the sign of the Tutte polynomial — in particular, Jackson and Sokal sought to determine for which choices of the two variables the sign is "trivial" in the sense that it does not depend on the input graph (or it depends only very weakly on the input graph, for example when it depends only on the number of vertices in the graph).

To illustrate how our work fits in with the work of Jackson and Sokal, we start with an important univariate case. The *chromatic polynomial* $P(G; q)$ of an n-vertex graph G is the unique degree-n polynomial in the variable q such that $P(G; q)$ is the number of proper q-colourings of G. Jackson [6, Theorem 5] showed that for $q \in (1, 32/27]$ the sign of $P(G; q)$ depends upon G in an essentially trivial way. In particular, for every connected simple graph with $n \geq 2$ vertices and b blocks, $P(G; q)$ is non-zero with sign $(-1)^{n+b-1}$. The sign of $P(G; q)$ is also known to be a trivial function of G for $q \leq 1$. (See, for example, [7, Theorem 1.1].) Jackson [6, Theorem 12] demonstrated the significance of the value $32/27$ by constructing an infinite family of graphs such that $P(G; q) = 0$ at a value of q which is arbitrarily close to $32/27$. In fact, Jackson and Sokal conjectured [7, Conjecture 10.3(e)] that the value $32/27$ is a phase transition in the sense that, for every q above this critical value, the sign of $P(G; q)$ is a non-trivial function of G. In particular, they conjectured that for any fixed $q > 32/27$, and all sufficiently large n and m, there are 2-connected graphs G with n vertices and m edges that make $P(G; q)$ non-zero with either sign.

It turns out that this intuition is correct (see Corollary 1) and that $q = 32/27$ is, in some sense, a phase transition for the complexity of computing the sign of $P(G; q)$:

- As was known, for $q \leq 32/27$, the sign of $P(G; q)$ is a trivial function of G, which is easily computed.
- At $q = 2$, $P(G; q)$ is the number of 2-colourings of G. The sign of $P(G; q)$ is positive if G is bipartite, and is 0 otherwise. Thus, the sign of $P(G; q)$ is not a trivial function of G, but $P(G; q)$ is still easily computed in polynomial time.
- However, for *every* other fixed $q > 32/27$, computing the sign of $P(G; q)$ is NP-hard.

However the full version of Jackson and Sokal's conjecture turns out to be incorrect. See Observations 38 and 40 of our full paper for counter-examples.

While computing the sign of $P(G; q)$ is NP-hard for every $q \neq 2$ which is greater than $32/27$, the precise complexity of compufting the sign does actually depend upon q. We show (see Corollary 1) that for each fixed *non-integer* $q > 32/27$, the complexity of computing the sign of $P(G; q)$ is #P-hard. This means that a polynomial-time algorithm for computing the sign of $P(G; q)$, given G, would give a polynomial-time algorithm for exactly solving every problem in #P.

On the other hand, for integers $q > 2$, the problem of computing the sign of $P(G; q)$ is merely NP-complete.

As one would expect, both of these results have ramifications for the complexity of approximating $P(G; q)$. A fully polynomial approximation scheme (FPRAS) for evaluating $P(G; q)$, given G, can be used as a polynomial-time randomised algorithm for computing the sign of $P(G; q)$. Thus, we can deduce that if q is a non-integer which is greater than $32/27$, then there is no FPRAS for $P(G; q)$ unless there is a randomised polynomial-time algorithm for exactly solving every problem in #P. See the full version for a more thorough discussion of this claim.

On the other hand, for integer values $q > 32/27$, we show that the problem of evaluating $P(G; q)$ is in the complexity class #P$_\mathbb{Q}$, which is defined as follows.

Definition 1. FP *is the class of functions computable by polynomial-time algorithms. We say that a function* $f : \Sigma^* \to \mathbb{Q}$ *is in the class* #P$_\mathbb{Q}$ *if* $f(x) = a(x)/b(x)$, *where* $a, b : \Sigma^* \to \mathbb{N}$, *and* $a \in$ #P *and* $b \in$ FP.

If f is in #P$_\mathbb{Q}$ then there is an approximation scheme for f that runs in polynomial time, using an oracle for an NP predicate (for a more detailed discussion, see [3, Section 2.2]). Thus, it is presumably much easier to approximate $P(G; q)$ when q is an integer greater than $32/27$, as compared to a non-integer.

All of these considerations generalise smoothly to the Tutte polynomial, which we now define. Since we will later need the multivariate generalisation [11] of the polynomial, we use the "random cluster" formulation of the Tutte polynomial, which for a graph $G = (V, E)$, is defined as a polynomial in inderminates q and γ as follows,

$$Z(G; q, \gamma) = \sum_{A \subseteq E} q^{\kappa(V, A)} \gamma^{|A|}, \tag{1}$$

where $\kappa(V, A)$ denotes the number of connected components in the graph (V, A). The chromatic polynomial studied earlier is related to the Tutte polynomial via the identity [7, (2.15)] $P(G; q) = Z(G; q, -1)$.

In fact, Tutte defined the Tutte polynomial using a different, two-variable parameterisation, in terms of variables x and y. This polynomial is defined for a graph $G = (V, E)$ by

$$T(G; x, y) = \sum_{A \subseteq E} (x - 1)^{\kappa(V, A) - \kappa(V, E)} (y - 1)^{|A| - |V| + \kappa(V, A)}. \tag{2}$$

It is well known (see, for example, [11, (2.26)]) that when $q = (x - 1)(y - 1)$ and $\gamma = y - 1$ we have

$$T(G; x, y) = (y - 1)^{-|V|} (x - 1)^{-\kappa(V, E)} Z(G; q, \gamma). \tag{3}$$

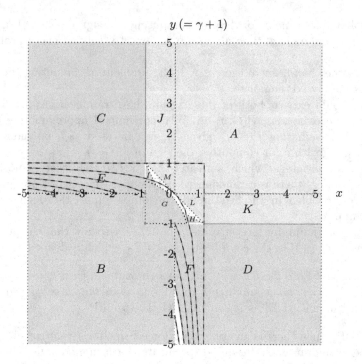

Fig. 1. The regions. Computing the sign of the Tutte polynomial is #P-hard at red points, and is in FP at green points. It is NP-complete at blue points. We have not resolved the complexity at white points. At red points, approximating the Tutte polynomial is also #P-hard. At blue and green points, it can be done in polynomial time with an NP oracle. Guide for the greyscale version: The red points appear as a darker grey in regions B, C, D, E, F, G, H and I. The green points appear as a lighter grey in regions A, J, K, L and M and also as dashed hyperbola segments and at the points $(-1, 0)$, $(-1, -1)$, $(0, -1)$ and $(0, -5)$. The blue points are $(-2, 0)$, $(-3, 0)$, $(-4, 0)$, $(-5, 0)$ and $(0, -2)$.

This paper studies the complexity of computing the sign of the (random cluster) Tutte polynomial. Figure 1 gives a map of the (x, y) plane, illustrating our results.[1] The colours depict the complexity of computing the sign of the polynomial for a fixed point (x, y). If the point (x, y) is coloured red, then the problem

---

[1] For convenience, our proofs use the random cluster formulation of the Tutte polynomial (1). However, in order to make our results easily comparable to other results in the literature such as [3] and [9], we classify points using the (x, y)-coordinatisation of (2). This is without loss of generality, since it is easy to go from one coordinate system to the other using (3). However, the reader should note that if $y = 1$ then $\gamma = 0$ and $q = (x - 1)(y - 1) = 0$ so computing $Z(G; q, \gamma)$ is trivial, whereas the complexity of computing $T(G; x, y)$ is unclear. In general, any two-parameter version of the Tutte polynomial will omit some points. This issue is discussed further in [5, Section 1].

of computing the sign is #P-hard. If the point (x, y) is coloured green, then the problem of computing the sign is in FP. Finally, if the point (x, y) is coloured blue, the the problem of computing the sign is NP-complete. (There are still some points for which we have not resolved the complexity — these are coloured white.)

Once again, there are ramifications for the complexity of approximating the Tutte polynomial. Since an FPRAS for $Z(G; q, \gamma)$ gives a randomised algorithm for computing its sign, we can again deduce that there is no FPRAS for points that are coloured red (unless there is a randomised polynomial-time algorithm for exactly solving every problem in #P). By contrast, for all of the points that are coloured green or blue, we also show that the problem of computing $Z(G; q, \gamma)$ is in the complexity class #P$_\mathbb{Q}$. Thus, the polynomial can be approximated in polynomial-time using an NP oracle.

In order to reach into some of the regions, for example F, it has been necessary to use gadgets that go beyond the series-parallel graphs that have so-far proved adequate in this area. For example, in exploring region F has necessitated the use of a gadget based on the Petersen graph.

Before giving a few details, and briefly describing proof techniques, we summarise our conclusions and contributions.

- Figure 1 gives a nearly-complete classification of the complexity of computing the sign of the Tutte polynomial — there are not too many unresolved points (coloured white in the figure).
- It is interesting that the hardness results are actually #P-hardness, even though the results apply to the problem of computing the sign, which has only three possible outcomes (so it feels more like a decision problem than a counting problem).
- It is interesting that whether q is an integer or not makes an actual difference to the complexity in Regions E and F (contrary to the conjecture of Jackson and Sokal) but it makes no difference in other regions (such as regions B and G).
- Though it is not apparent from the figure, we have completely resolved the complexity of computing the sign of the chromatic polynomial, which corresponds to the x axis (see Corollary 1).
- Even though the results are much stronger than previous work (#P-hardness, rather than NP-hardness), the proofs do not become technically overwhelming. Somehow, it helps technically to know that the hardness actually comes from the difficulty of computing the sign, rather than from issues of numerical approximation.
- Resolving the unresolved points on the y-axis (region F) would require progress on a long-standing open problem related to nowhere-zero q-flows (see Section 5.4 of the full version and the comments below).

The proofs of the theorems described in Figure 1 are given in the full version of the paper. Here, in Section 3, we give a brief glimpse of the hardness results — proving hardness for some of the points in Region B, and briefly discussing the other hardness proofs (without presenting details). In Section 4, we briefly discuss the tractability results. Details are given in the full paper. Finally, Section 5 contains a corollary which collects the relevant results for $q \geq 32/27$ and $|y| < 1$. The proof is given in the full version.

2   Preliminaries

2.1   The Tutte Polynomial

It will be helpful to define the multivariate version of the random cluster formulation of the Tutte polynomial. Let γ be a function that assigns a (rational) weight γ_e to every edge $e \in E$. We refer to γ as a "weight function". We define

$$Z(G; q, \gamma) = \sum_{A \subseteq E} q^{\kappa(V,A)} \prod_{e \in A} \gamma_e.$$

Given a graph $G = (V, E)$ with distinguished nodes s and t, $Z_{st}(G; q, \gamma)$ denotes the contribution to $Z(G; q, \gamma)$ arising from edge-sets A in which s and t are in the same component of (V, A). That is,

$$Z_{st}(G; q, \gamma) = \sum_{A \subseteq E: s \text{ and } t \text{ in same component}} q^{\kappa(V,A)} \prod_{e \in A} \gamma_e.$$

Similarly, $Z_{s|t}$ denotes the contribution arising from edge-sets A in which s and t are in different components, so $Z(G; q, \gamma) = Z_{st}(G; q, \gamma) + Z_{s|t}(G; q, \gamma)$.

2.2   Implementing New Edge Weights, Series Compositions and Parallel Compositions

Implementations and compositions are described in detail in [4, Section 2.1] and in the full version. We give a very brief account here. Let W be a set of (rational) edge weights and fix a value q. Let w^* be a weight (which may not be in W) which we want to "implement". Suppose that there is a graph Υ, with distinguished vertices s and t and a weight function $\widehat{\gamma} : E(\Upsilon) \to W$ such that

$$w^* = qZ_{st}(\Upsilon; q, \widehat{\gamma})/Z_{s|t}(\Upsilon; q, \widehat{\gamma}). \tag{4}$$

In this case, we say that Υ and $\widehat{\gamma}$ implement w^* (or even that W implements w^*).

Let G be a graph with weight function γ. Let f be some edge of G with weight $\gamma_f = w^*$. Suppose that W implements w^*. Let Υ be a graph with distinguished vertices s and t with a weight function $\widehat{\gamma}$ satisfying (4). Construct the weighted graph G' by replacing edge f with a copy of Υ as follows: Identify s with either endpoint of f (it doesn't matter which one) and identify t with the other endpoint of f and remove edge f. Define the weight function γ' of G' as follows: $\gamma'_e = \widehat{\gamma}_e$ if $e \in E(\Upsilon)$ and $\gamma'_e = \gamma_e$ otherwise. Then the definition of the multivariate Tutte polynomial gives

$$Z(G'; q, \gamma') = \frac{Z_{s|t}(\Upsilon; q, \widehat{\gamma})}{q^2} Z(G; q, \gamma). \tag{5}$$

Two especially useful implementations are series and parallel compositions (see [7, Section 2.3]). Parallel composition is the case in which Υ consists of two parallel edges with endpoints s and t. Series composition is the case in which Υ is a length-2 path from s to t. The k-thickening of [9] is the parallel composition of k edges of weight α. It implements $\alpha' = (1 + \alpha)^k - 1$ Similarly, the k-stretch is the series composition of k edges of weight α. It implements an α' satisfying

$$1 + \frac{q}{\alpha'} = \left(1 + \frac{q}{\alpha}\right)^k.$$

Since it is useful to switch freely between (q, α) coordinates and (x, y) coordinates we also refer to the implementation in Equation (4) as an implementation of the point $(x, y) = (q/w^* + 1, w^* + 1)$ using the points

$$\{(x, y) = (q/w + 1, w + 1) \mid w \in W\}.$$

2.3 Computational Problems

For fixed rational numbers q, γ and $\gamma_1, \ldots, \gamma_k$, we consider the following computational problems in which the goal is to compute the sign of the Tutte polynomial.

Name: SIGNTUTTE(q, γ).
Instance: A graph $G = (V, E)$.
Output: Determine whether the sign of $Z(G; q, \gamma)$ is positive, negative, or 0.

Name: SIGNTUTTE$(q; \gamma_1, \ldots, \gamma_k)$.
Instance: A graph $G = (V, E)$ and a weight function $\gamma : E \to \{\gamma_1, \ldots, \gamma_k\}$.
Output: Determine whether the sign of $Z(G; q, \gamma)$ is positive, negative, or 0.

We say that a point (x, y) is #P-hard, NP-complete, or in FP, if, for $\gamma = y - 1$ and $q = (x - 1)(y - 1)$, the corresponding problem SIGNTUTTE$(q; \gamma)$ is #P-hard, NP-complete, or in FP, respectively. #P-hardness is defined with respect to randomised polynomial-time Turing reductions. NP-hardness is defined by a many-one reduction from an NP-complete decision problem, whose instance is a "yes instance" if the corresponding instance of SIGNTUTTE$(q; \gamma)$ has a positive sign, and a "no instance" otherwise.

3   A Glimpse at the Hardness Result

The main lemma used in the #P-hardness proofs is Lemma 1 below. Informally, it shows the following. Suppose that we can implement an edge weight $\gamma_1 \in (-2, -1)$ and that we can also implement an edge weight $\gamma_2 \notin [-2, 0]$. then we can use an oracle for SIGNTUTTE$(q; \gamma_1, \gamma_2)$ to exactly count the minimum-cardinality (s, t)-cuts of a graph (which was shown to be #P-complete by Provan and Ball [10]).

Lemma 1. *Suppose $q > 1$ and that $\gamma_1 \in (-2, -1)$ and $\gamma_2 \notin [-2, 0]$. Then* SIGNTUTTE$(q; \gamma_1, \gamma_2)$ *is #P-hard.*

Proof. A complete proof is given in the full version — we just give a sketch here. We will give a polynomial-time Turing reduction from #MINIMUM CARDINAL-ITY (s, t)-CUT to SIGNTUTTE$(q; \gamma_1, \gamma_2)$.

Let G be a graph (an instance of #MINIMUM CARDINALITY (s, t)-CUT) with distinguished vertices s and t. Let $n = |V(G)|$ and $m = |E(G)|$. Assume without loss of generality that G has no edge from s to t and that it is connected and that $m \geq n$ is sufficiently large. Let k be the size of a minimum cardinality (s, t)-cut in G and let C be the number of size-k (s, t)-cuts.

We start by letting h be a sufficiently large polynomial in m (details are given in the full version). By an h-thickening using γ_2, we can implement the (exponentially large) quantity $M = (\gamma_2 + 1)^h - 1$. Let δ be the small quantity $\delta = (2q)^m / M$. Let $\boldsymbol{M}$ be the constant weight function which gives every edge weight M. We will use the following facts.

$$(1 - \delta) M^m q \leq Z_{st}(G; q, \boldsymbol{M}) \leq (1 + \delta) M^m q. \tag{6}$$

$$C M^{m-k} q^2 (1 - \delta) \leq Z_{s|t}(G; q, \boldsymbol{M}) \leq C M^{m-k} q^2 (1 + \delta). \tag{7}$$

Fact (6) follows from the fact that each of the (at most 2^m) terms in $Z_{st}(G; q, \boldsymbol{M})$ other than the term with all edges in A has value at most $M^{m-1} q^n$ and $2^m M^{m-1} q^n \leq \delta M^m q$. (We don't actually need the $1 - \delta$ in the lower bound, but it is easier, in the full version, to have symmetric bounds on the error.) Fact (7) follows from the fact that all terms in $Z_{s|t}(G; q, \boldsymbol{M})$ are (s, t)-cuts. Each term that is not a size-k (s, t)-cut has value at most $M^{m-k-1} q^{n+2}$ and

$$2^m M^{m-k-1} q^{n+2} \leq \delta C M^{m-k} q^2.$$

For a parameter ε in the open interval $(0, 1)$ which we will tune below, let $\gamma' = -1 - \varepsilon \in (-2, -1)$. We will discuss the implementation of γ' below. Let G' be the graph formed from G by adding an edge from s to t. Let $\boldsymbol{\gamma}$ be the edge-weight function for G' that assigns weight M to every edge of G and assigns weight γ' to the new edge. Then, using the definition of the Tutte polynomial,

$$Z(G'; q, \boldsymbol{\gamma}) = Z_{st}(G; q, \boldsymbol{M})(1 + \gamma') + Z_{s|t}(G; q, \boldsymbol{M})\left(1 + \frac{\gamma'}{q}\right)$$

$$= -\varepsilon Z_{st}(G; q, \boldsymbol{M}) + Z_{s|t}(G; q, \boldsymbol{M})\left(1 - \frac{1 + \varepsilon}{q}\right). \tag{8}$$

Now suppose $\varepsilon = 0$. Then $Z(G'; q, \boldsymbol{\gamma}) = Z_{s|t}(G; q, \boldsymbol{M})\left(1 - \frac{1}{q}\right)$, which is positive. On the other hand, using the definition of M and Facts (6) and (7) above, we can confirm that, when $\varepsilon = 1$, $Z(G'; q, \boldsymbol{\gamma})$ is negative. The idea is to perform binary search on the range $(0, 1)$ to find an ε where $Z(G'; q, \boldsymbol{\gamma}) = 0$. For this value of ε, we have $\varepsilon Z_{st}(G; q, \boldsymbol{M}) = Z_{s|t}(G; q, \boldsymbol{M})\left(1 - \frac{1 + \varepsilon}{q}\right)$. It turns out that the approximations (6) and (7) above will give us enough information that we'll be able to calculate C exactly from this equality.

As one would expect, there are small technical complications. Since we are somewhat constrained in what values ε we can implement, we won't be able to discover the exact value of ε that we need, but we will be able to approximate it sufficiently closely to compute C exactly from this equality. For technical reasons to do with this approximation, it helps to note when $\varepsilon = M^{-2m}$, $Z(G'; q, \gamma)$ is still positive (this is proved in the full version) — thus, we can start the binary search at value around M^{-2m}, which is bounded away from 0. Also, when $\varepsilon = q - 1$ we have $Z(G'; q, \gamma) = -(q-1)Z_{st}(G; q, M) < 0$.

Thus we have a range from $\varepsilon = M^{-2m}$ to $\varepsilon = \min(1, q - 1)$ of length at most 1 which contains some value ε for which $Z(G'; q, \gamma) = 0$. We'll perform binary search on this interval. Suppose for a moment that we are able, for a given $\varepsilon \in (M^{-2m}, \min(1, q - 1))$, to compute the sign of $Z(G'; q, \gamma)$. Our basic strategy will be binary search, sub-dividing the initial interval $\lceil m^2 \lg M \rceil$ times, so eventually we'll get an interval of width at most M^{-m^2} which contains an ε where $Z(G'; q, \gamma) = 0$.

To do this, we need to address the issue of computing the sign of $Z(G'; q, \gamma)$ using an oracle for $\text{SignTutte}(q; \gamma_1, \gamma_2)$. We have already seen above that is easy to implement the weight M using γ_2 (and that the implementation has polynomial size) — we now need to consider the implementation of $\gamma' = -1 - \varepsilon$ (where $\varepsilon \in (M^{-2m}, \min(1, q - 1))$ is the particular value that is being queried). In the full version we show that, while we may not be able to query the exact value of ε that we want to (because we can't quite implement the corresponding γ') we can query a value that is between $\varepsilon - \pi$ and ε, where π is any given positive quantity. The size of the graph used in the implementation is at most a polynomial in $\log(\pi^{-1})$.

In the full version, we set $\pi = M^{-m^2}/3$ and we tune the binary search appropriately. The overall result is that we can find a subinterval of width at most M^{-m^2} which contains an ε where $Z(G'; q, \gamma) = 0$. Now let ε be an endpoint of this subinterval. Let $\rho = 2^m q^m M^m M^{-m^2}$. In the full version, we use (6), (7), and (8) to show

$$\frac{(1 - 2 \cdot 4^{-m})\varepsilon M^m q}{\left(1 - \frac{1+\varepsilon}{q}\right) M^{m-k} q^2 (1 + 4^{-m})} \leq C \leq \frac{\varepsilon M^m q(1 + 2 \cdot 4^{-m})}{\left(1 - \frac{1+\varepsilon}{q}\right) M^{m-k} q^2 (1 - 4^{-m})}. \tag{9}$$

Now the point is that C is an integer between 1 and 2^m. Even though the value of k is not known, the fact that $M > 4^m$ means that there can only be one integer k such that the above interval contains an integer between 1 and 2^m (so k can easily be deduced). All of the other quantities in the lower and upper bounds in (9) are known. Now let $R = \frac{\varepsilon M^k}{(1 - \frac{1+\varepsilon}{q})q}$, so (9) becomes

$$\left(\frac{1 - 2 \cdot 4^{-m}}{1 + 4^{-m}}\right) R \leq C \leq R \left(\frac{1 + 2 \cdot 4^{-m}}{1 - 4^{-m}}\right). \tag{10}$$

Now, $R < 2^{m+1}$, since otherwise the left-hand-side of (10) is greater than 2^n. Also, multiplying through by $(1 + 4^{-m})(1 - 4^{-m})$, the width of the interval is at most $6 \cdot 4^{-m} R < 1$ so the width of the interval in (10) is less than 1, so the (integral) value of C can be calculated exactly.

Many of the hardness results in the paper follow directly from Lemma 1 by implementing the relevant points γ_1 and γ_2. Here is an easy example which shows hardness for some of the points in Region B.

Lemma 2. *Suppose* (x, y) *is a point with* $x < -1$ *and* $y < -1$. *Then* (x, y) *is* #P-*hard.*

Proof. Let $q = (x - 1)(y - 1)$. First, we will show that we can use (x, y) to implement a point (x_1, y_1) with $y_1 \in (-1, 0)$ so $\gamma_1 = y - 1 \in (-2, -1)$, as required by Lemma 1. Let j be an odd positive integer which is sufficiently large that $|x|^j + 1 > q$. Implement $(x', y') = (x^j, q/(x^j - 1) + 1)$ from (x, y) with a j-stretch. Note that $y' \in (0, 1)$. Now, for a sufficiently large positive integer k, implement (x_1, y_1) using the parallel composition of (x, y) with k copies of (x', y') so $y_1 = y'^k y \in (-1, 0)$. The point $(x_2, y_2) = (x, y)$ satisfies $y_2 \notin [-1, 1]$. Since $q > 1$, the result follows from Lemma 1.

Of course, many of the hardness proofs in the paper require more difficult implementations than Lemma 2. Some of this is essentially "technical work" (for example, in Region G). More interesting issues arise in Regions E and F. In these regions, we only have hardness when q is a non-integer (unless all problems in #P can be exactly solved by randomised polynomial-time algorithms!). In Region E, the most significant challenge is implementing a point (x', y') with $y' < 0$ when $q > 2$. We implement the desired point using the graph K_n minus an edge, where $n = \lfloor q \rfloor + 2$ with edge weights that are very close to -1. The analysis of the implementation (Lemma 17 in the full version) proceeds by studying the chromatic polynomial of K_n and K_n minus an edge.

Dually, in Region F, the most significant challenge is implementing a point (x', y') with $x' < 0$ when $q > 2$ is not an integer. Our implementations use edge weights that are very close to $-q$. When the edge weights are all $-q$, the Tutte polynomial specialises to the so-called flow polynomial. A q-*flow* of an undirected graph $G = (V, E)$ is defined as follows [11, Section 2.4]. Choose an arbitrary direction for each edge. Let H be any Abelian group of order q. A q-flow is a mapping $\psi : E \to H$ such that the flow into each vertex is equal to the flow out (doing arithmetic in H). A q-flow ψ of a graph $G = (V, E)$ is said to be *nowhere-zero* if, for every $e \in E$, $\psi(e) \neq 0$. If q is a positive integer and all edge weights are $-q$, then the Tutte polynomial counts the nowhere-zero q-flows of a graph. (It is a non-trivial fact that the number of nowhere-zero q-flows only depends upon q, the size of H, and not on H itself.)

Our construction for $q \in (3, 4)$ proceeds by analysing the flow polynomial of the Petersen graph. This is zero at $q = 3$ and $q = 4$, since this graph has no nowhere-zero 3-flow or 4-flow but it is positive for $q > 4$ (hence negative between 3 and 4). On the other hand, the graph obtained by removing an edge has a positive flow polynomial for $q > 3$. The fact that the signs of these polynomials are different is key to the construction (see the proof of Lemma 22 in the full version for details). A similar construction works for q between 2 and 3. The construction breaks down for $q > 4$ because both graphs have positive flow polynomial. It is conceivable that the lemma could be proved for non-integer q in the range $4 < q < 6$ by using a generalised Petersen graph rather than a Petersen graph in the construction. Indeed, Jacobsen and Salas have shown [8] that there are generalised Petersen graphs whose flow polynomials have roots between 5 and 6. Given the current state of knowledge, we are pessimistic about the prospects of proving the lemma for all $q > 4$. Currently, it is an open question [8] whether there is a some uniform upper bound Q for real zeros of arbitrary bridgeless graphs (so that every bridgeless graph G would have a positive flow polynomial for all $q > Q$). If so, then computing the sign of the flow polynomial will be trivial for $q > Q$, so computing the sign of the Tutte polynomial will also be trivial for $y < -Q + 1$ along the y-axis. If not, then it seems likely that the hardness construction can be extended. (Thus, it doesn't seem to be possible to resolve all of the unresolved points in Region F without solving the open problem about flow polynomials.)

4   A Very Brief Glimpse at the Tractability Results

For the green points in Figure 1, the sign of the Tutte polynomial can be computed in polynomial time and the evaluation of the polynomial is in $\#P_{\mathbb{Q}}$. This is most interesting in regions J, K, L and M. It is easier to prove these positive results for binary matroids, rather than for graphs. A graph can be viewed as a special case of a binary matroid. The advantage of working more generally is that the results in regions K and M follow by duality from the results in Regions J and L. The algorithms are recursive. For example, in Lemma 45 of the full paper we show that if $q < 0$ and $\mathcal{M}$ is a loopless matroid in which all edge weights are between -2 and 0 then the sign of the Tutte polynomial of $\mathcal{M}$ is positive, and the polynomial can be evaluted in $\#P_{\mathbb{Q}}$. The algorithm and its correctness proof are recursive. If $\mathcal{M}$ has full rank then the result is easy. Otherwise, we apply contraction and deletion, and recurse on the minors of $\mathcal{M}$. See the full version for details.

5   Putting Things Together for Points with $|y| < 1$

In this section we state a corollary of our work which collects the results depicted in Figure 1 for regions with $q \geq 32/27$ and $|y| < 1$. This is Corollary 55 in the full version. Together with the work of Jackson [6], it completely resolves the complexity of computing the sign of the chromatic polynomial of a graph.

Corollary 1. *Suppose (x, y) is a point satisfying $|y| < 1$ such that $q = (x - 1)(y - 1) \geq 32/27$. Let $\gamma = y - 1$.*

- *If $(x, y) = (-1, 0)$ then* SIGNTUTTE(q, γ) *and* TUTTE(q, γ) *are in* FP.
- *If $(x, y) = (x, 0)$ for any integer $x < -1$ then* SIGNTUTTE(q, γ) *is NP-complete.* TUTTE(q, γ) *is in $\#P_{\mathbb{Q}}$.*

– *If $x \leq -1$ and $0 < y < 1$ and q is an integer then $Z(G; q, \gamma) > 0$ so* SignTutte(q, γ) *is in* FP. *Also,* Tutte(q, γ) *is in* #P$_{\mathbb{Q}}$.
– *Otherwise,* SignTutte(q, γ) *is #P-hard.*

References

1. Dell, H., Husfeldt, T., Wahlén, M.: Exponential Time Complexity of the Permanent and the Tutte Polynomial. In: Abramsky, S., Gavoille, C., Kirchner, C., Meyer auf der Heide, F., Spirakis, P.G. (eds.) ICALP 2010. LNCS, vol. 6198, pp. 426–437. Springer, Heidelberg (2010)
2. Goldberg, L.A., Jerrum, M.: Approximating the Partition Function of the Ferromagnetic Potts Model. In: Abramsky, S., Gavoille, C., Kirchner, C., Meyer auf der Heide, F., Spirakis, P.G. (eds.) ICALP 2010. LNCS, vol. 6198, pp. 396–407. Springer, Heidelberg (2010)
3. Goldberg, L.A., Jerrum, M.: Inapproximability of the Tutte polynomial. Inform. and Comput. 206(7), 908–929 (2008)
4. Goldberg, L.A., Jerrum, M.: Inapproximability of the Tutte polynomial of a planar graph. CoRR, abs/0907.1724 (2009); To appear "Computational Complexity"
5. Goldberg, L.A., Jerrum, M.: Approximating the Tutte polynomial of a binary matroid and other related combinatorial polynomials. CoRR, abs/1006.5234 (2010)
6. Jackson, B.: A zero-free interval for chromatic polynomials of graphs. Combinatorics, Probability & Computing 2, 325–336 (1993)
7. Jackson, B., Sokal, A.D.: Zero-free regions for multivariate Tutte polynomials (alias potts-model partition functions) of graphs and matroids. J. Comb. Theory, Ser. B 99(6), 869–903 (2009)
8. Jacobsen, J.L., Salas, J.: Is the five-flow conjecture almost false? ArXiv e-prints (September 2010)
9. Jaeger, F., Vertigan, D.L., Welsh, D.J.A.: On the computational complexity of the Jones and Tutte polynomials. Math. Proc. Cambridge Philos. Soc. 108(1), 35–53 (1990)
10. Provan, J.S., Ball, M.O.: The complexity of counting cuts and of computing the probability that a graph is connected. SIAM J. Comput. 12(4), 777–788 (1983)
11. Sokal, A.: The multivariate Tutte polynomial. In: Surveys in Combinatorics, Cambridge University Press (2005)
12. Vertigan, D.: The computational complexity of Tutte invariants for planar graphs. SIAM J. Comput. 35(3), 690–712 (electronic) (2005)

Stochastic Vehicle Routing with Recourse[*]

Inge Li Gørtz[1,**], Viswanath Nagarajan[2], and Rishi Saket[2]

[1] Technical University of Denmark, DTU Informatics
[2] IBM T.J. Watson Research Center

Abstract. We study the classic *Vehicle Routing Problem* in the setting of stochastic optimization with recourse. StochVRP is a two-stage problem, where demand is satisfied using two routes: fixed and recourse. The fixed route is computed using only a demand distribution. Then after observing the demand instantiations, a recourse route is computed – but costs here become more expensive by a factor λ.

We present an $O(\log^2 n \cdot \log(n\lambda))$-approximation algorithm for this stochastic routing problem, under arbitrary distributions. The main idea in this result is relating StochVRP to a special case of *submodular orienteering*, called *knapsack rank-function orienteering*. We also give a better approximation ratio for *knapsack rank-function orienteering* than what follows from prior work. Finally, we provide a Unique Games Conjecture based $\omega(1)$ hardness of approximation for StochVRP, even on star-like metrics on which our algorithm achieves a logarithmic approximation.

1 Introduction

Consider a distribution problem involving a depot location and a set of customer locations. There is a vehicle of capacity Q that is used to distribute items. The demand at customer locations is random with a known (joint) distribution $\mathcal{D}$. The distributor wants to plan a *fixed route* for this capacitated vehicle, that will be employed on a daily basis. However due to the stochastic nature of demands, the fixed route might be insufficient to meet all demands. Therefore the distributor also plans a secondary *recourse strategy*, that satisfies all unmet demands after the fixed route. Each morning the distributor receives the precise demand quantities from all customers (drawn from $\mathcal{D}$). Based on this he/she decides which subset of customers will be satisfied along the fixed route, and then plans a recourse route to satisfy the remaining customers. The goal is to minimize the cost of the fixed route plus the expected cost of the recourse route. Examples of real-world applications are local deposit collection from bank branches, garbage collection, home heating oil delivery, and forklift routing [1,4].

A solution based on fixed routes is desirable for several reasons, and is commonly used in practice; see [28,15] for more detailed discussions on this. In our context, there are at least two advantages. First, the driver can get familiar with the road/traffic conditions which results in time savings. Moreover, having fixed

---

[*] A full version of this extended abstract appears as [17].
[**] Supported by the Danish Council for Independent Research | Natural Sciences.

A. Czumaj et al. (Eds.): ICALP 2012, Part I, LNCS 7391, pp. 411–423, 2012.
© Springer-Verlag Berlin Heidelberg 2012

routes simplifies the everyday route planning process: the incremental recourse step will typically contain fewer demands.

Fixed-route problems are often modeled in the framework of two-stage stochastic optimization. *A priori* optimization handles some natural but simple recourse strategies: eg., short-cutting over customers without demand in TSP [5,30], and refill-visits from the depot in the Vehicle Routing Problem (VRP) [4,19]. Recently, more complex recourse actions have been considered: adding penalty terms in deadline TSP [8], and using backup vehicles in VRP [1].

In this paper, we penalize the cost of the recourse route by an inflation factor $\lambda \geq 1$. This is also a common approach for two-stage stochastic optimization with recourse. Furthermore, in the stochastic VRP we consider, recourse strategies are non-trivial since it also involves choosing the subset of realized demands served by the fixed route. In this respect it is unlike most previously studied 2-stage stochastic problems (eg. [27,32,20]) where the recourse step is just a deterministic instance of the same problem. Before describing the results of this paper, we define the deterministic and two-stage stochastic VRP below.

Vehicle Routing Problem (VRP). There is a vehicle of capacity Q, metric (V, d) with root/depot $r \in V$ and demands $\{q_v \leq Q\}_{v \in V}$. The goal is to find a minimum cost tour of the vehicle that delivers q_v units to each $v \in V$. The demands are "unsplittable", i.e. the demand at any vertex must be satisfied in a single visit. Any VRP solution corresponds to a sequence of round-trips from the depot, where at most Q units of demands are served during each round-trip. It is well-known [2] that an α-approximation ratio for TSP implies an $(\alpha + 2)$-approximation algorithm for VRP.

Two-Stage Stochastic VRP (StochVRP). The setting is same as above, with a capacity Q vehicle, metric (V, d) and depot $r \in V$. Here the demands $\{q_v\}_{v \in V}$ are random variables given by a joint demand distribution $\mathcal{D}$ on $\{0, 1, \ldots, Q\}^V$, available as a black-box that can be sampled from. We are also given an inflation parameter $\lambda \geq 1$. The goal is to compute a fixed route solution with a recourse strategy.

- In the first stage the algorithm computes a *fixed tour* τ, without knowledge of the actual demand. The tour τ consists of several round-trips from the depot: each round-trip is a cycle containing r (henceforth called r-tour). We represent τ as a concatenation $\{\tau_1, \ldots, \tau_F\}$ of r-tours. It is important to note that τ only represents the vehicle route, and does not specify demand deliveries (this will be decided after demand instantiations). In particular, a vertex v may appear in multiple r-tours of τ; and even if v appears in τ the instantiated demand at v may not eventually be satisfied by τ.

- In the second stage, the demands $\overline{q}$ are instantiated from $\mathcal{D}$. Knowing this, an algorithm chooses to satisfy subset $\overline{q}_A \subseteq \overline{q}$ of demands using the fixed tour τ, subject to the vehicle capacity of Q. That is, for each r-tour $\{\tau_i\}_{i=1}^F$ the algorithm chooses a subset $S_i \subseteq \tau_i$ of vertices to serve, where $\sum_{v \in S_i} \overline{q}_v \leq Q$; and sets $\overline{q}_A \equiv \{\overline{q}_v : v \in \cup_{i=1}^F S_i\}$. Then the algorithm computes a *recourse tour* σ meeting all residual demands $\overline{q}_B = \overline{q} \setminus \overline{q}_A$. That is, σ is a solution to the deterministic VRP instance with demands $\{\overline{q}_v : v \in V \setminus \cup_{i=1}^F S_i\}$.

Note that the demands $\overline{q}_A$ satisfied by the fixed tour τ differs based on the instantiation $\overline{q}$; however the route taken by the vehicle stays fixed. So the first stage cost is just the length $d(\tau)$ of the fixed tour. The recourse tour σ clearly depends on the demand instantiation. The second stage cost under demand $\overline{q}$ is $\lambda \cdot d(\sigma(\overline{q}))$, the length of the recourse tour inflated by a parameter λ. The objective in StochVRP is to minimize the expected total cost:

$$d(\tau) \quad + \quad \lambda \cdot \mathbb{E}_{\overline{q} \leftarrow \mathcal{D}}\left[d(\sigma(\overline{q})) \right]$$

For any integer $I \geq 0$, we let $[I] := \{1, \ldots, I\}$. For a given StochVRP instance, opt will denote its optimal value. We let $n = |V|$ denote the number of vertices in the metric and $D = \max_{u,v} d(u,v)$ the diameter of the metric.

Our Results, Techniques and Outline. In this paper we show:

Theorem 1. *There is a randomized $O(\log^2 n \cdot \log(n\lambda))$-approximation algorithm for StochVRP under arbitrary distributions.*

Using a sampling-based reduction [10] we show that the objective *value* under any black-box distribution can be well-approximated by another demand distribution having support size $m = poly(n, \lambda)$. Then, in Section 2 we present an $O(\log^2 n \cdot \log(nm))$-approximation algorithm for StochVRP where m is the support size of the distribution. This is a set-cover type algorithm that uses the *submodular orienteering problem* [12,7] as a subroutine. In the submodular orienteering problem there is a metric (V, d) with root r, length bound B and monotone submodular function $f : 2^V \rightarrow \mathbb{R}_+$; and the goal is to find an r-tour of length at most B visiting some subset $S \subseteq V$ of vertices so as to maximize $f(S)$. Direct use of algorithms from [11,7] yields an approximation ratio worse than Theorem 1 by a factor of $\log^\epsilon n$. Instead we give a better result for submodular orienteering on objective functions of the type encountered in StochVRP, called *knapsack rank-function orienteering* (KnapRankOrient). In particular, we consider the *ratio* KnapRankOrient problem where instead of the length-bound, the objective is to maximize the ratio of function value to the length.

Theorem 2. *There is a deterministic $O(\log^2 n)$-approximation algorithm for ratio knapsack rank-function orienteering.*

The main idea here is to use LP rounding techniques for the related group Steiner problem [16,25], augmented with an *alteration* step (for the analysis). While alteration has been widely used with LP-rounding, eg. [33], we are not aware of an application in context of the group Steiner tree problem. This step only bounds the function-value and length in expectation (separately). In order to bound their ratio, we adapt the group Steiner derandomization from Charikar et al. [9] to our context. We defer further discussion and details on KnapRankOrient to the full version [17].

Combined with the sampling-based reduction this suffices to approximate the objective *value* of StochVRP under black-box distributions. However, more work is required in order to provide an approximate *solution*. This is because the recourse step in StochVRP is quite non-trivial, and a solution must specify an algorithm to construct the recourse tour for *any* possible demand (not merely the m sampled points). It turns out that the recourse step corresponds to solving an "outlier" version of VRP. Although this problem does not admit any true approximation ratio (by a relation to generalized assignment [26]), in Section 3 we give an LP-based $O(1)$ *bicriteria* approximation: this suffices for Theorem 1.

Our second main result is a UGC-based hardness of approximation:

Theorem 3. *Assuming the Unique Games Conjecture, it is NP-hard to approximate* StochVRP *to within a constant factor, even on star-like metrics.*

This is proved in Section 4 and involves a reduction from the vertex cover problem on k-uniform hypergraphs: we use a result by Bansal and Khot [3] which says that it is UGC-hard to distinguish between the (yes) case when the hypergraph is almost k-partite and the (no) case when any vertex cover is almost the entire vertex-set. We remark that this super-constant hardness holds in star-like metrics, where our algorithm achieves an $O(\log(n\lambda))$-approximation. Our algorithm loses additional log-factors in going from (i) stars to trees, and then (ii) trees to general metrics: these overheads are similar to the best known results for the related *group Steiner tree* problem [16].

Finally, we consider the special case when demands are independent across vertices. Using a different algorithm we obtain a better ratio (see full version).

Theorem 4. *There is a randomized* $O(\frac{\log(n\lambda)}{\log\log(n\lambda)})$-*approximation algorithm for* StochVRP *under independent demand distributions.*

Related Work. The VRP [35] is an extensively studied routing problem that combines aspects of both TSP and bin-packing. Several stochastic variants of the basic problem have received attention, eg. [34,4,14,1,15]. Approximation algorithms for VRP with independent stochastic demands (in the a priori model) were given in [4,19]. This paper takes a different approach, that of two-stage stochastic optimization with recourse (along the lines of [22,27,32,20] etc). To the best of our knowledge no prior approximation results are known for vehicle routing problems in this model.

Stochastic optimization [6] is a broad area dealing with probabilistic input. Approximation algorithms for two-stage stochastic problems were introduced by Immorlica et al. [22] and Ravi and Sinha [27]. Gupta et al. [20] and Shmoys and Swamy [32] gave general frameworks for approximating a number of stochastic optimization problems; the former result is combinatorial using certain cost-sharing properties, whereas the latter is LP-based. However, these approaches do not seem directly applicable to StochVRP. The results in [20,32] hold in the most general distribution model, where an algorithm only receives independent samples from a black-box. Charikar et al. [10] showed that any arbitrary distribution can be reduced to one having polynomial support (under certain conditions). We also make use of this result in proving Theorem 1. For most other combinatorial optimization problems that have been considered in the two-stage stochastic model (with proportional cost inflation), it has been observed that approximation ratios are the same order of magnitude as the underlying deterministic problem [22,27,20,32,31]. A notable exception is minimum cost max-matching [23], for which an $\Omega(\log n)$-hardness of approximation was shown. In the case of VRP Theorem 3 shows (under UGC) that the stochastic approximation ratio is necessarily worse than its deterministic counterpart, even in very special metrics.

2 Algorithm for Polynomial Scenarios

Here we consider the case when the demand distribution $\mathcal{D}$ is specified as a list of possible outcomes. Later on we show how the general case of a black-box distribution can be reduced to this case. Formally $\mathcal{D}$ is a multiset $\{\overline{q}^1, \ldots, \overline{q}^m\}$ where the actual demand $\overline{q} = \overline{q}^i$ (for some $i \in [m]$) with probability $1/m$.

The main idea of our algorithm is to recast the problem as an instance of set-cover with an exponential number of sets. Then we show that the greedy subproblem is an instance of *submodular orienteering* (SOP) for which a polylogarithmic approximation is known [7,12]. In fact, for the type of SOP instances obtained from StochVRP we give a better approximation ratio in the full version. Altogether, this implies Theorem 1 for polynomial scenarios.

Set Cover Instance $\mathcal{I}$. The groundset U consists of tuples $\langle i, v \rangle$ for all scenarios $i \in [m]$ and vertices $v \in V$, which denotes $\overline{q}^i(v)$ demand units at v under scenario i. For any $\langle i, v \rangle \in U$ we use $q(\langle i, v \rangle) := \overline{q}^i(v)$, and for any subset $S \subseteq U$, $q(S) := \sum_{t \in S} q(t)$. Instance $\mathcal{I}$ has the following two types of sets:

1. $S := \cup_{i=1}^m S_i$ is a **first-stage set** iff $S_i \subseteq \{\langle i, v \rangle : v \in V\}$ and $q(S_i) \le Q$ for all $i \in [m]$. The cost of this set S is the minimum length of an r-tour that contains all the vertices represented in S.
2. For any scenario $i \in [m]$, $T \subseteq \{\langle i, v \rangle : v \in V\}$ is a **second-stage set** iff $q(T) \le Q$. The cost of set T is λ/m times the minimum length of an r-tour containing all vertices of T.

Lemma 1. *The set cover instance $\mathcal{I}$ is equivalent to* StochVRP.

Proof. Recall that any feasible StochVRP solution is specified by:

- *The fixed tour τ.* It will be convenient to view this as a collection $\{\tau_1, \ldots, \tau_F\}$ of r-tours, each of which is a round-trip from the depot.
- *For each scenario $i \in [m]$, the demands $\overline{q}^i_A \subseteq \overline{q}^i$ satisfied by the fixed tour.* Again this is viewed as follows: for each r-tour $\{\tau_j\}_{j=1}^F$, $S_{i,j} \subseteq \{\langle i, v \rangle : v \in V\}$ denotes the demands satisfied in τ_j. Note that by definition, $\bigcup_{j \in [F]} S_{i,j} \equiv \overline{q}^i_A$. Also due to the capacity constraint, $q(S_{i,j}) \le Q$ for each $j \in [F]$.
- *For each scenario $i \in [m]$, the recourse tour σ_i which satisfies residual demands $\overline{q}^i \setminus \overline{q}^i_A$.* Again we view this as a collection $\{\sigma_{i,1}, \ldots, \sigma_{i,L_i}\}$ of r-tours. For $k \in [L_i]$ let $T_{i,k} \subseteq \{\langle i, v \rangle : v \in V\}$ denote the demands satisfied in $\sigma_{i,k}$. Clearly, $\bigcup_{k \in [L_i]} T_{i,k} \equiv \overline{q}^i \setminus \overline{q}^i_A$. Again $q(T_{i,k}) \le Q$ for all $k \in [L_i]$.

Note that corresponding to each first-stage r-tour τ_j, the set $\bigcup_{i=1}^{m} S_{i,j}$ is a valid *first-stage set* in $\mathcal{I}$ since for all $i \in [m]$ (a) $S_{i,j} \subseteq \{\langle i, v \rangle : v \in V\}$ and (b) $q(S_{i,j}) \leq Q$. Moreover the cost of this set in $\mathcal{I}$ is at most $d(\tau_j)$.

Similarly, for each scenario $i \in [m]$ and second-stage r-tour $\sigma_{i,k}$ ($k \in [L_i]$), set $T_{i,k}$ is a valid *second-stage* set. The cost of this set in $\mathcal{I}$ is at most $\frac{\lambda}{m} \cdot d(\sigma_{i,k})$.

Finally, these sets cover U in $\mathcal{I}$ since for each scenario $i \in [m]$, we have:

$$\left(\cup_{j=1}^{F} S_{i,j} \right) \bigcup \left(\cup_{k \in [L_i]} T_{i,k} \right) \;\; = \;\; \{\langle i, v \rangle : v \in V\}$$

The total cost of this solution to $\mathcal{I}$ is at most:

$$\sum_{j=1}^{F} d(\tau_j) \; + \; \frac{\lambda}{m} \cdot \sum_{i=1}^{m} \sum_{k=1}^{L_i} d(\sigma_{i,k}) \;\; = \;\; d(\tau) \; + \; \lambda \cdot \mathbb{E}_{\bar{q} \leftarrow \mathcal{D}} \left[d(\sigma(\bar{q})) \right],$$

which is just the StochVRP objective value. The reverse relation (from $\mathcal{I}$ to StochVRP) can be shown in a similar manner, and the lemma follows. $\square$

Thus it suffices to solve the set cover instance $\mathcal{I}$. We use the greedy algorithm for set cover which requires solving the following *max-coverage subproblem*: given $U' \subseteq U$ find a set (of either first/second type) that maximizes the ratio of the number of U'-elements it covers to its cost. We give separate algorithms for this problem, under the two types of sets.

Max-Coverage for Second-Stage Sets. We give a constant approximation in this case. Assume that the algorithm knows by enumeration (i) the cost B of the best ratio set (up to a factor two), and (ii) the scenario $i \in [m]$ corresponding to it. Then it suffices to find a set $T \subseteq U' \bigcap \{\langle i, v \rangle : v \in V\}$ maximizing $|T|$ such that $q(T) \leq Q$ and $\text{cost}(T) \leq B$. By the definition of second-stage sets, this reduces to finding an r-tour visiting the maximum vertices $W \subseteq \{u \in V : \langle i, u \rangle \in U'\}$, having length at most $\frac{m}{\lambda} \cdot B$ *and* with $\sum_{u \in W} \bar{q}^i(u) \leq Q$. This is just an instance of the *knapsack-orienteering* problem, for which a constant-factor approximation is known [18].

Max-Coverage for First-Stage Sets. In this case, we obtain a poly-logarithmic approximation. Again, we assume that the algorithm knows the cost B of the best ratio set (up to a factor two). Recall that unlike the previous case, one first-stage set can cover elements from several scenarios. By definition, each first-stage set S corresponds to an r-tour visiting vertices $W \subseteq V$ *and* subsets $S_i \subseteq \{\langle i, v \rangle : v \in W\}$ for each $i \in [m]$ such that $\{q(S_i) \leq Q\}_{i=1}^{m}$ and $S = \bigcup_{i=1}^{m} S_i$. Among all first-stage sets visiting a *fixed vertex-set* $W \subseteq V$, the maximum coverage of U' equals:

$$f(W) \;\; := \;\; \sum_{i=1}^{m} \max \left\{ |S_i| \; : \; S_i \subseteq \{u \in W : \langle i, u \rangle \in U'\}, \sum_{v \in S_i} \bar{q}_v^i \leq Q \right\}$$

For each $i \in [m]$ let $f_i(W)$ denote the term inside the above summation. Recall that the cost of all first-stage sets visiting vertices W is the same, namely the minimum TSP on $\{r\} \cup W$. Thus the subproblem we wish to solve is:

$$\max \; f(W) \; : \; \text{there is an } r\text{-tour visiting } W \subseteq V \text{ of length } \leq B. \tag{1}$$

Recall the submodular orienteering problem (SOP) where given metric (V, d) with root r, bound B and submodular function $g : 2^V \to \mathbb{R}_+$, the goal is to find an r-tour visiting some subset $W \subseteq V$ of vertices, having length at most B that maximizes $g(W)$. If f were submodular then we can use the algorithm [7,12] to solve this. But f is not submodular: eg. if $f(W) = \max\{|S| : S \subseteq W, \sum_{v \in S} q_v \leq Q\}$ on groundset $U = \{a, b, c\}$ with $q_a = q_b = 1$, $q_c = 3$ and $Q = 3$, then $f(\{a, c\}) + f(\{b, c\}) = 2 < 3 = f(\{a, b, c\}) + f(\{c\})$. Still, we show below that f can be well approximated by a submodular function g.

We approximate each f_i (point-wise) by a submodular function g_i. Let $V_i := \{u \in V : \langle i, u \rangle \in U'\}$ denote the vertices appearing with scenario i in U'. Define:

$$g_i(W) := \max \left\{ \sum_{v \in V_i \cap W} x_v : \sum_{v \in W} \overline{q}_v^i \cdot x_v \leq Q, \ 0 \leq x_v \leq 1, \ \forall v \in W \right\}$$

Observe that $g_i(W)$ is just an LP relaxation for a maximization $\{0, 1\}$-knapsack problem. So its value is given by the greedy algorithm that increases x_v (up to 1) in increasing order of $\{\overline{q}_v^i : v \in V_i \cap W\}$. On the other hand, $f_i(W)$ is the value of the same *integral* knapsack problem. Now, function g_i can be rewritten as the rank function of a *polymatroid* [29] which is submodular; see eg. [13]. Moreover, the integrality gap of the natural LP for max-knapsack is two. Thus,

Claim 1. g_i *is monotone submodular and* $\frac{g_i(W)}{2} \leq f_i(W) \leq g_i(W), \forall W \subseteq V$.

So if we define $g(W) := \sum_{i=1}^m g_i(W)$ then it is submodular and maximizing g in (1) is equivalent to maximizing f (up to factor two). Hence, assuming a ρ-approximation algorithm for SOP, we obtain a 2ρ-approximation algorithm for (1). This suffices to give an $O(\rho)$-approximation for the max-coverage subproblem. We have $\rho = O(\log^{2+\epsilon} n)$ in polynomial time using the bicriteria approximation in Calinescu-Zelikovsky [7], and $\rho = O(\log n)$ in *quasi-polynomial* time using the true approximation in Chekuri-Pal [12]. In the full version [17] we directly consider the *ratio* objective corresponding to (1), called *ratio knapsack rank-function orienteering*, i.e.

$$\max \left\{ \frac{f(V(\tau))}{d(\tau)} : \tau \text{ is an } r\text{-tour visiting vertices } V(\tau) \right\},$$

and obtain an improved polynomial time $O(\log^2 n)$-approximation algorithm.

Finally, we lose an additional $\log |U| = O(\log(mn))$ factor to solve the set cover instance $\mathcal{I}$ (which is equivalent to StochVRP). Thus we obtain:

Theorem 5. *There is a polynomial time $O(\log^2 n \cdot \log(mn))$-approximation algorithm for* StochVRP *for a polynomial number m of scenarios and n vertices. This ratio improves to $O(\log n \cdot \log(mn))$ in quasi-polynomial time.*

3   Algorithm for General Distributions

In this section we prove Theorem 1 under an arbitrary distribution $\mathcal{D}$ that is accessed by sampling. We denote the input StochVRP instance by $\mathcal{J}$. In the full version [17] we apply a sampling-based reduction from [10] to obtain an equivalent StochVRP instance $\mathcal{J}'$ with $m = poly(n, \lambda)$ scenarios. This allows us to apply the algorithm from the previous section to approximate the *optimal value* of instance $\mathcal{J}$. However a solution to $\mathcal{J}$ must also specify a valid recourse strategy for *every* outcome $\bar{q} \in \mathcal{D}$, and not just for the m outcomes in instance $\mathcal{J}'$. It turns out that the recourse step is captured by an "outlier" version of VRP (defined below), and we give an LP-based constant-factor bicriteria approximation for it.

The recourse strategy involves the *outlier VRP* problem: given a fixed tour τ (as collection $\{\tau_1, \ldots, \tau_F\}$ of r-tours) and outcome $\bar{q} \in \{0, \ldots, Q\}^V$, find

- a subset of vertices whose demands $\bar{q}_A \subseteq \bar{q}$ can be served by the existing route τ, subject to the capacity constraint of Q on its r-tours; *and*
- a minimum cost VRP solution to the residual demands $\bar{q} - \bar{q}_A$.

A special case of outlier VRP is the *restricted assignment* problem [26]; see [17] for details. So it is NP-hard to obtain any true approximation ratio for outlier VRP. Instead we give an $(O(1), O(1))$ *bicriteria* approximation algorithm, which suffices to obtain an approximation algorithm for StochVRP, proving Theorem 1.

The algorithm is based on a natural LP relaxation to outlier VRP. Consider a solution with $S \subseteq V$ as the vertices chosen to be served by τ. Then:

- There is an assignment $\phi : S \to [F]$ such that (1) $v \in \tau_{\phi(v)}$ for all $v \in S$; and (2) for each r-tour $j \in [F]$, the total demand assigned to it $\sum_{v \in \phi^{-1}(j)} \bar{q}_v \leq Q$.
- The objective value is the optimum VRP on metric (V, d), depot r, capacity Q and demands $\{\bar{q}_v : v \in V \setminus S\}$. Using known lower-bounds for VRP [21,2], at the loss of a constant factor, this is just $\mathsf{MST}(V \setminus S) + \mathsf{Flow}(V \setminus S)$ where for any $T \subseteq V$, $\mathsf{MST}(T) =$ length of minimum spanning tree on $\{r\} \bigcup T$, and $\mathsf{Flow}(T) := \frac{1}{Q} \sum_{v \in T} \bar{q}_v \cdot d(r, v)$.

Thus we can write the following integer programming formulation for outlier VRP, at the loss of an $O(1)$-factor.

$$\min \quad \sum_{e \in E} d_e \cdot z_e \; + \; \frac{1}{Q} \sum_{v \in V} d(r, v) \cdot \bar{q}_v \cdot (1 - x_v) \tag{2}$$

$$\text{s.t.} \quad \sum_{v \in \tau_j} \bar{q}_v \cdot y_{v,j} \leq Q \qquad \forall j \in [F], \tag{3}$$

$$\sum_{j \in [F] : v \in \tau_j} y_{v,j} = x_v \qquad \forall v \in V, \tag{4}$$

$$\sum_{e \in \delta(U)} z_e \geq 1 - x_v \qquad \forall U \not\ni r, \; \forall v \in U, \tag{5}$$

$$x_v, y_{v,j} \in \{0, 1\} \qquad \forall v \in V, \; \forall j \in [F],$$

$$z_e \geq 0 \qquad \forall e \in E.$$

Above x_v is one iff $v \in S$, i.e. served by τ. Variables $y_{v,j}$ denote the assignment $\phi : S \to [F]$. Constraint (4) ensures that each $v \in S$ is assigned to some $\phi(v)$ such that $v \in \tau_{\phi(v)}$. Constraint (3) enforces the total assignment to each r-tour is at most Q. Also $E = \binom{V}{2}$ denotes the edge-set of the metric, and for any

$U \subseteq V$, $\delta(U)$ denotes the edges with exactly one vertex in U. Constraint (5) says that $\{z_e : e \in E\}$ is a fractional spanning tree connecting the vertices $\{v : x_v = 0\} = V \setminus S$ to r. In the objective (2), the first term is the length of the fractional spanning tree (corresponding to $\mathsf{MST}(V \setminus S)$), and the second term is $\mathsf{Flow}(V \setminus S)$. Dropping the integrality gives us an LP relaxation $\mathsf{LP}(\tau, \bar{q})$ which can be solved in polynomial time. Then we round the resulting fractional solution (details in [17]) to obtain:

Theorem 6. *There is an $(O(1), 5)$-bicriteria approximation algorithm for outlier VRP, that uses the fixed tour at most five times.*

4 UGC Hardness of Approximation

In this section we prove a $\omega(1)$ UGC-hardness of approximation for StochVRP even for a very simple star-like metric with a setting of λ that renders the recourse tour trivial. Our hardness result is based on the Unique Games Conjecture (UGC) of Khot [24]. Based on UGC, Bansal and Khot [3] proved the following hardness of approximation result for minimum vertex cover on almost k-partite k-uniform hypergraphs, which shall be the starting point of our reduction.

Theorem 7. *[3] Assuming the Unique Games Conjecture, for any $\varepsilon > 0$ and positive integer $k \geq 2$, given a k-uniform hypergraph G with vertex set U and hyperedge set E, it is NP-hard to distinguish between the following two cases:*
YES CASE: *There is a partition of U into $k + 1$ disjoint subsets $X, U_1, \ldots, U_k$ such that $|X| \leq \varepsilon|U|$ and the hypergraph induced by $U \setminus X$ is k-partite with $U_1, \ldots, U_k$ as the k-partition. That is, any hyperedge e has at most one vertex from any U_i. This implies that $X \cup U_i$ is a vertex cover in G for each $i = 1, \ldots, k$, and that the minimum vertex cover in G has size at most $(1/k + \varepsilon)|U|$.*
NO CASE: *The size of the maximum independent set in G is at most $\varepsilon|U|$, and therefore the size of the minimum vertex cover in G is at least $(1 - \varepsilon)|U|$.*

In the rest of this section we shall give a hardness reduction from the problem of distinguishing between k-uniform hypergraphs which are almost k-partite (as in the YES case of Theorem 7) from those that have a very small maximum independent set (as in the NO case of Theorem 7).

Hardness Reduction. Fix any positive integer $k \geq 2$. Suppose we are given a k-uniform hypergraph G on vertex set U and with hyperedge set E as a hard instance from Theorem 7, where we shall fix the parameter ε in Theorem 7 later. We transform $G(U, E)$ into an instance of StochVRP as follows. For clarity, in this section the nomenclature of "vertices" shall be in context of the hypergraph, while "points" shall be used for corresponding elements in the metric.
Metric (V, d). The set of points V in the metric is $U \cup \{r\}$, where r is the *root*. The distances d are defined as follows. Let $d(r, u) = L$, where $L = (|U|/2k + 1/2)$, for all $u \in U$. Further, for each pair $u, u' \in U$, $u \neq u'$, let $d(u, u') = 1$. It is easy to see that d is a metric. This simple metric can be realized by the shortest paths in a star-like tree of distances as illustrated in Figure 1.

Capacity and Demands. The capacity $Q = 1$ and demands will be $\{0, 1\}$.
Demand Distribution $\mathcal{D}$. There are polynomially many scenarios $m = |E|$, each
having uniform probability. Every hyperedge $e \in E$ is a scenario having demand
of one at all points in e, and zero demand elsewhere.
Parameter λ. We set $\lambda = 2m|U|(k + 1)$.

Before we proceed to the analysis of this reduction, we note that the cost of
the minimum cost r-tour covering points $S \subseteq V \setminus \{r\}$, is simply $|U|/k + |S|$.
Also, the optimal value is at most λ/m. Consider the fixed tour consisting of k
identical r-tours each covering U: since each scenario has at most k demands,
this solution never uses a recourse tour, and has cost $k \cdot (|U|/k + |U|) < \lambda/m$. So
we may assume that *the optimal solution has no recourse tour*: if the recourse
tour is non-empty in any scenario then its cost is at least λ/m.

YES Case. Suppose that $G(U, E)$ is
a YES instance of Theorem 7 with
$X, U_1, \ldots, U_k$ as the partition of U
with the properties as stated in the
theorem. Consider the r-tours $\tau_1, \ldots,$
τ_k, where τ_i is an r-tour that covers points $X \cup U_i$ (in addition to r).
Since every scenario in our instance
of StochVRP corresponds to a hyperedge in G, using the property in the
YES case that each hyperedge has at
most one vertex from each U_i, we see
that the r-tours $\tau_1, \ldots, \tau_k$ satisfy all
the scenarios. As noted earlier the cost
of each r-tour that covers $S \subseteq V \setminus \{r\}$
is $|U|/k + |S|$. Therefore the total cost
of the k r-tours $\tau_1, \ldots, \tau_k$ is,

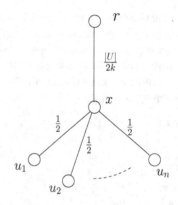

Fig. 1. Tree of distances realizing metric
d, with intermediate point x and $V = \{r, u_1, \ldots, u_n\}$

$$k \cdot (|U|/k) + \sum_{i=1}^{k} |X \cup U_i| \;\leq\; |U| + (1 + k\varepsilon)|U| \;=\; (2 + k\varepsilon)|U|,$$

by the properties of the partition $X, U_1, \ldots, U_k$ of U.

NO Case. Suppose that $G(U, E)$ is a NO instance of Theorem 7, so that the
maximum independent set in G is of size at most $\varepsilon|U|$. In this case we shall
prove that the total cost of any set of r-tours that satisfy all scenarios is at least
$k(1 - f_k(\varepsilon))|U|$, where $f_k(\varepsilon) \to 0$ as $\varepsilon \to 0$ for any fixed positive integer $k \geq 2$.
We may assume that the number of r-tours in the optimal solution is at most
k^2, otherwise the total cost will be at least $k^2(|U|/k) = k|U|$ and we shall be
done. Therefore, let $\gamma_1, \ldots, \gamma_T$ be the r-tours in an optimal fixed tour, where
$T \leq k^2$. We shall estimate the number of points in U which occur in at most
$k - 1$ of these r-tours. For any subset $I \subseteq [T]$, let $U(I) \subseteq U$ be the points which
do not occur in $\{\gamma_i : i \in [T] \setminus I\}$. We have the following simple lemma.

Lemma 2. *For any $I \subseteq [T]$ with $|I| = k - 1$, $U(I)$ is an independent set in G.*

Proof. For a contradiction, suppose that e is a hyperedge induced by $U(I)$. Since $|e| = k$, the scenario corresponding to e will not be satisfied by our solution as the k vertices of e appear (as points) in at most $k - 1$ of the r-tours, namely those given by $I \subseteq [T]$. Recall that each r-tour can serve only one demand. $\square$

The total number of points in U that appear in at most $k - 1$ of the r-tours is upper bounded by,

$$\sum_{I \subseteq [T], |I| = k-1} |U(I)|.$$

There are $\binom{T}{k-1} \leq 2^T \leq 2^{k^2}$ choices for the subsets I in the above expression. Using the fact that any independent set in G has size at most $\varepsilon|U|$, the fraction of points in U that occur in at most $k - 1$ of the r-tours is at most $\varepsilon 2^{k^2} =: f_k(\varepsilon)$. Each of the remaining $(1 - f_k(\varepsilon))|U|$ points appears in at least k of the r-tours; so the total cost of the fixed tour is $k(1 - f_k(\varepsilon))|U|$.

Hardness Factor. In the YES case there is a solution of cost at most $(2+k\varepsilon)|U|$, whereas in the NO case any solution has cost at least $k(1 - f_k(\varepsilon))|U|$. For any positive integer $k \geq 2$ and arbitrarily small $\delta > 0$, choosing $\varepsilon > 0$ to be small enough in Theorem 7, we obtain a hardness factor of $k/2 - \delta$.

References

1. Ak, A., Erera, A.L.: A paired-vehicle recourse strategy for the vehicle-routing problem with stochastic demands. Transportation Science 41(2), 222–237 (2007)
2. Altinkemer, K., Gavish, B.: Heuristics for unequal weight delivery problems with a fixed error guarantee. Operations Research Letters 6, 149–158 (1987)
3. Bansal, N., Khot, S.: Inapproximability of Hypergraph Vertex Cover and Applications to Scheduling Problems. In: Abramsky, S., Gavoille, C., Kirchner, C., Meyer auf der Heide, F., Spirakis, P.G. (eds.) ICALP 2010, Part I. LNCS, vol. 6198, pp. 250–261. Springer, Heidelberg (2010)
4. Bertsimas, D.J.: A vehicle routing problem with stochastic demand. Operations Research 40(3), 574–585 (1992)
5. Bertsimas, D.J., Jaillet, P., Odoni, A.R.: A priori optimization. Operations Research 38(6), 1019–1033 (1990)
6. Birge, J.R., Louveaux, F.: Introduction to Stochastic Programming. Springer, New York (1997)
7. Calinescu, G., Zelikovsky, A.: The polymatroid steiner problems. Journal of Combinatorial Optimization 9(3), 281–294 (2005)
8. Campbell, A.M., Thomas, B.W.: Probabilistic traveling salesman problem with deadlines. Transportation Science 42(1), 1–21 (2008)
9. Charikar, M., Chekuri, C., Goel, A., Guha, S.: Rounding via trees: Deterministic approximation algorithms for group steiner trees and k-median. In: Proc. STOC 1998, pp. 114–123 (1998)
10. Charikar, M., Chekuri, C., Pál, M.: Sampling Bounds for Stochastic Optimization. In: Chekuri, C., Jansen, K., Rolim, J.D.P., Trevisan, L. (eds.) APPROX 2005 and RANDOM 2005. LNCS, vol. 3624, pp. 257–269. Springer, Heidelberg (2005)

11. Chekuri, C., Even, G., Kortsarz, G.: A greedy approximation algorithm for the group steiner problem. Discrete Applied Mathematics 154(1), 15–34 (2006)
12. Chekuri, C., Pál, M.: A recursive greedy algorithm for walks in directed graphs. In: Proc. FOCS, pp. 245–253 (2005)
13. Dean, B.C., Goemans, M.X., Vondrák, J.: Approximating the stochastic knapsack problem: The benefit of adaptivity. Math. Oper. Res. 33(4), 945–964 (2008)
14. Dror, M.: Vehicle Routing with Stochastic Demands: Models & Computational Methods. In: Modeling Uncertainty. International Series In Operations Research & Management Science, vol. 46(8), pp. 625–649. Springer, Heidelberg (2005)
15. Erera, A.L., Savelsbergh, M.W.P., Uyar, E.: Fixed routes with backup vehicles for stochastic vehicle routing problems with time constraints. Networks 54(4), 270–283 (2009)
16. Garg, N., Konjevod, G., Ravi, R.: A Polylogarithmic Approximation Algorithm for the Group Steiner Tree Problem. Journal of Algorithms 37(1), 66–84 (2000)
17. Gørtz, I.L., Nagarajan, V., Saket, R.: Stochastic vehicle routing with recourse. CoRR, abs/1202.5797 (2012)
18. Gupta, A., Krishnaswamy, R., Nagarajan, V., Ravi, R.: Approximation Algorithms for Stochastic Orienteering. In: Proc. SODA 2012, pp. 245–253 (2012)
19. Gupta, A., Nagarajan, V., Ravi, R.: Approximation Algorithms for VRP with Stochastic Demands. Operations Research 60(1), 123–127 (2012)
20. Gupta, A., Pál, M., Ravi, R., Sinha, A.: Sampling and cost-sharing: Approximation algorithms for stochastic optimization problems. SIAM J. Comput. 40(5), 1361–1401 (2011)
21. Haimovich, M., Rinnooy Kan, A.H.G.: Bounds and heuristics for capacitated routing problems. Mathematics of Operations Research 10(4), 527–542 (1985)
22. Immorlica, N., Karger, D.R., Minkoff, M., Mirrokni, V.S.: On the costs and benefits of procrastination: approximation algorithms for stochastic combinatorial optimization problems. In: Proc. SODA 2004, pp. 691–700 (2004)
23. Katriel, I., Mathieu, C.K., Upfal, E.: Commitment under uncertainty: Two-stage stochastic matching problems. Theor. Comput. Sci. 408(2-3), 213–223 (2008)
24. Khot, S.: On the power of unique 2-prover 1-round games. In: Proc. STOC 2002, pp. 767–775 (2002)
25. Konjevod, G., Ravi, R., Srinivasan, A.: Approximation algorithms for the covering steiner problem. Random Struct. Algorithms 20(3), 465–482 (2002)
26. Lenstra, J.K., Shmoys, D.B., Tardos, E.: Approximation algorithms for scheduling unrelated parallel machines. Mathematical Programming 46, 259–271 (1990)
27. Ravi, R., Sinha, A.: Hedging uncertainty: Approximation algorithms for stochastic optimization problems. Math. Program. 108(1), 97–114 (2006)
28. Savelsbergh, M.W.P., Goetschalkx, M.: A comparison of the efficiency of fixed versus variable vehicle routes. J. Business Logistics 16, 163–187 (1995)
29. Schrijver, A.: Combinatorial optimization: polyhedra and efficiency. Springer, Berlin (2003)
30. Shmoys, D.B., Talwar, K.: A Constant Approximation Algorithm for the a priori Traveling Salesman Problem. In: Lodi, A., Panconesi, A., Rinaldi, G. (eds.) IPCO 2008. LNCS, vol. 5035, pp. 331–343. Springer, Heidelberg (2008)
31. Shmoys, D.B., Sozio, M.: Approximation Algorithms for 2-Stage Stochastic Scheduling Problems. In: Fischetti, M., Williamson, D.P. (eds.) IPCO 2007. LNCS, vol. 4513, pp. 145–157. Springer, Heidelberg (2007)

32. Shmoys, D.B., Swamy, C.: An approximation scheme for stochastic linear programming and its application to stochastic integer programs. J. ACM 53(6), 978–1012 (2006)
33. Srinivasan, A.: New approaches to covering and packing problems. In: Proc. SODA 2001, pp. 567–576 (2001)
34. Stewart, W., Golden, B.: Stochastic vehicle routing: A comprehensive approach. Eur. Jour. Oper. Res. 14, 371–385 (1983)
35. Toth, P., Vigo, D.: The vehicle routing problem. Society for Industrial and Applied Mathematics (2001)

The Online Metric Matching Problem for Doubling Metrics

Anupam Gupta[1,*] and Kevin Lewi[2,**]

[1] Computer Science Department, Carnegie Mellon University
[2] Computer Science Department, Stanford University

Abstract. In the online minimum-cost metric matching problem, we are given an instance of a metric space with k servers, and must match arriving requests to as-yet-unmatched servers to minimize the total distance from the requests to their assigned servers. We study this problem for the line metric and for doubling metrics in general. We give $O(\log k)$-competitive randomized algorithms, which reduces the gap between the current $O(\log^2 k)$-competitive randomized algorithms and the constant-competitive lower bounds known for these settings.

We first analyze the "harmonic" algorithm for the line, that for each request chooses one of its two closest servers with probability inversely proportional to the distance to that server; this is $O(\log k)$-competitive, with suitable guess-and-double steps to ensure that the metric has aspect ratio polynomial in k. The second algorithm embeds the metric into a random HST, and picks a server randomly from among the closest available servers in the HST, with the selection based upon how the servers are distributed within the tree. This algorithm is $O(1)$-competitive for HSTs obtained from embedding doubling metrics, and hence gives a randomized $O(\log k)$-competitive algorithm for doubling metrics.

1 Introduction

In the online minimum-cost metric matching problem, the input is a metric space (V, d) with k pre-specified *servers* $S \subseteq V$. The requests $R = r_1, r_2, \ldots, r_k$ (with each $r_i \in V$) arrive online one-by-one; upon arrival each request must be immediately and irrevocably matched to an as-yet-unmatched server. The cost of matching request r to server $f(r) \in S$ is the distance $d(r, f(r))$ in the underlying metric space. The goal is to find a matching f that approximately minimizes the total cost $\sum_i d(r_i, f(r_i))$. We study the problem in the framework of competitive analysis, comparing the cost of our algorithm's matching to the cost of the best offline matching from R to S. (This minimum cost bipartite perfect matching problem can be easily solved offline.)

The online problem was introduced in the early 1990's by Kalyanasundaram and Pruhs [4], and by Khuller, Mitchell, and Vazirani [6]. Both papers gave

---

* Supported in part by NSF awards CCF-0964474 and CCF-1016799.
** Work done when at the Computer Science Department, Carnegie Mellon University, Pittsburgh PA 15217.

A. Czumaj et al. (Eds.): ICALP 2012, Part I, LNCS 7391, pp. 424–435, 2012.

deterministic $2k - 1$-competitive algorithms, which is the best possible even when the metric is a star with k leaves with the servers at the leaves. For the star example, any *randomized* algorithm must be $\Omega(H_k)$-competitive, and the natural randomized greedy algorithm is indeed $O(H_k)$-competitive, where H_k is the kth harmonic number. In 2006, Meyerson et al. [9] showed that the randomized greedy algorithm, which assigns to a uniformly random closest server, is $O(\log k)$-competitive when the metric is a α-hierarchically well-separated tree (α-HST) with suitably large separation α between levels, namely with $\alpha = \Omega(\log k)$. This implies an $O(\log^3 k)$-competitive randomized algorithm for general metrics using randomized embeddings into HSTs [2]. Bansal et al. [1] gave a different algorithm which is $O(\log k)$-competitive on 2-HSTs, resulting in an $O(\log^2 k)$ competitive algorithm on general metrics. It remains an open problem to close the gap between $O(\log^2 k)$ and $\Omega(\log k)$ for general metrics.

The gap is even worse when we consider natural special classes of metrics such as the line, or grids, or doubling metrics. For points on the line, the best *deterministic* lower bound is only 9.001 [3] (with the randomized lower bound being even weaker), and no algorithms better than those that apply to general metrics are known for neither the line nor doubling metrics, both in the deterministic and randomized settings.

Results and Techniques. In this paper, we give randomized algorithms for restricted classes of metrics. In particular, we show $O(\log k)$-competitive randomized algorithms for the online metric matching problem on the line metric and on doubling metrics. (For the rest of this section, we assume that the *aspect ratio* of the metric, namely the maximum-to-minimum distance ratio, is $O(k^3)$—this can be achieved with only a constant factor loss using the guess-and-double framework, details in the full version.)

Our first algorithm is the natural randomized *Harmonic* algorithm: letting s_L and s_R be the closest available left and right servers to the current request r, we assign r to s_L with probability

$$\frac{1/d(r,s_L)}{1/d(r,s_L)+1/d(r,s_R)} = \frac{d(r,s_R)}{d(s_L,s_R)},$$

and to s_R with the remaining probability. If $d_{\max}$ and $d_{\min}$ are the largest and smallest distances between any two servers, we show:

Theorem 1. *The* Harmonic *algorithm is* $O(\log \frac{d_{\max}}{d_{\min}})$-*competitive for the line. Hence, with using guess-and-double to ensure* $d_{\max}/d_{\min} = O(k^3)$, *we get an* $O(\log k)$-*competitive algorithm for the line.*

Our proof uses a coupling argument: we consider two runs of the *Harmonic* algorithm, the first starting with some set S of servers and the second with the set $(S \cup \{s_1\} \setminus \{s_2\})$—i.e., differing from S in exactly one server. We show that the expected difference in cost between these two runs of *Harmonic* is $O(\log \frac{d_{\max}}{d_{\min}}) \cdot d(s_1, s_2)$. Now, if we construct a sequence of hybrid algorithms (each of which first follows the optimal algorithm, and at some point switches to the *Harmonic* algorithm), we can use the coupling argument to compare the runs of adjacent pairs on these sequences to bound the difference between our algorithm and Opt. This idea is similar to the path-coupling idea of Bubley and Dyer, used to show mixing of Markov chains.

Our second algorithm *Random-Subtree* generalizes to the broader class of doubling metrics. It first embeds the metric into a random distance-preserving Δ-degree α-HST, and then runs a certain randomized greedy algorithm on this new instance, where Δ and α are constants that depend on the doubling dimension. At first glance, using a $O(1)$-HST seems bad, since Meyerson et al. showed that their randomized greedy algorithm requires a large separation α in the HST. However, we avoid the lower bound (a) by using the fact that the bad examples require large degrees, whereas HSTs obtained from the line and doubling metrics have small degree, and (b) by altering the randomized greedy algorithm slightly (in a way we will soon describe). At a high level, we show that if a metric can be embedded into an α-HST where each vertex has at most Δ children, our randomized algorithm is $O(H_\Delta/\epsilon)$-competitive on such an HST, so long as $\alpha \geq (1 + \epsilon)H_\Delta$. (See Theorem 5 for a precise statement.) Since all doubling metrics admit such embeddings (for values of Δ, α depending only on the doubling dimension) with $O(\log k)$ expected stretch, we get:

Theorem 2. *The randomized algorithm* Random-Subtree *is* $O(\log k)$-*competitive for online metric matching on doubling metrics, and hence also for the line.*

The improvement from $O(\log^3 k)$ in [9] (for general metrics) to $O(\log k)$ (for doubling metrics) is due to both the nature of doubling metrics and the HSTs arising from them, and also due to our algorithm *Random-Subtree* differing from that of [9]. Instead of picking a *uniformly* random available server closest to the request in the HST, we use the following procedure: starting off at the lowest ancestor of the request that contains an available server, our algorithm repeatedly moves us to a uniformly random subtree of this node that has an available server until we reach a leaf/server. Note that our process does not pick a random closest server, but biases towards available servers in subtrees with few available servers. This results in such subtrees being empty earlier, which in turn results in fewer choices higher up in the tree for future requests. Our potential function-based analysis refines the one from [9] by using this property.

The rest of the paper is as follows. We present some notation and preliminaries in Section 2. The *Harmonic* algorithm is analyzed in Section 3, and the *Random-Subtree* algorithm presented and analyzed in Section 4. Due to lack of space, many proofs are deferred to the full version. Also in the full version, we present a third algorithm that is $O(\log k)$ competitive for matching on the line. This algorithm also embeds the line into a random HST, but then runs deterministically on the resulting HST to give this guarantee.

Other Related Work. The paper [5, Section 2.2] gave a lower bound of 9 for deterministic algorithms on the line via a reduction from the so-called cowpath problem; they conjectured this lower bound was tight for the line, which was disproved in [3]. [5] also conjectured that the work function algorithm (see, e.g., [8]) obtains an $O(1)$-competitive ratio on the line; this was disproved by Koutsoupias and Nanavati [7], who showed an $\Omega(\log k)$ lower bound (and an $O(k)$ upper bound) for the work function algorithm. There is no algorithm, either randomized or deterministic, currently conjectured to be $O(1)$-competitive.

2 Notation and Preliminaries

An instance of the problem is given by a metric (V, d) with servers at $S \subseteq V$, where $|S| = k$. As mentioned in [9], we can assume without loss of generality that all requests also arrive at vertices in S (with only a constant factor loss in the competitive ratio). Hence, in the rest of the paper, we assume that $S = V$, and hence $|V| = |S| = k$. Moreover, we assume there is only one server at each node, as this is only for ease of exposition and the algorithms easily extend to multiple servers at nodes.

An α-HST (Hierarchically well-Separated Tree) is defined as a rooted tree where all edges at depth i have weight c/α^i for some fixed constant c. Here, the edges at depth 0 are those incident to the root, etc. An HST is Δ-ary if each node has at most Δ children. For the case of the line metric, we assume that the aspect ratio of the points containing the servers (which, recall, is defined as $\frac{\max_{x,y \in S} d(x,y)}{\min_{x,y \in S} d(x,y)}$) is $O(k^3)$; in the full version, we show that this loses only a constant factor in the competitiveness. This allows us to embed these points into distributions of dominating *binary* 2-HSTs with expected stretch $O(\log k)$. Furthermore, for HSTs that are constructed from the line, we refer to the *width* of a tree as the maximum line-distance between any two points within the tree. For doubling metrics we cannot make such a general assumption on the aspect ratio; however, by suitably guessing the value of Opt and running the HST construction algorithms only for the top $O(\log k)$ levels, one can still give a reduction to the problem on bounded-degree HSTs with only an $O(\log k)$-expected loss. (Details in the full version.)

For a node a of a tree, let $T(a)$ represent the subtree rooted at a. Also, define the *level of a* to be the maximum number of edges on a path from a to a leaf of $T(a)$. When referring to servers to be assigned by requests, we will refer to servers that have not yet been assigned to as "available", "free", or "unassigned". We will use Opt to denote both the optimum matching as well as its cost.

3 The Harmonic Algorithm for the Line

To prove the performance guarantee for the *Harmonic* algorithm given by Theorem 1, we first give a lemma which analyzes the expected difference in cost between running *Harmonic* on all the requests and running the optimal algorithm *for just the first step*, and *Harmonic* thenceforth. Then, we show that using this bound in a "hybrid argument" proves Theorem 1. (This is essentially the path-coupling idea of Bubley and Dyer.) For a request sequence $\sigma = r_1, \ldots, r_k$, let g_σ be the matching obtained by assigning $r_1, \ldots, r_k$ using *Harmonic*. Let

$N(r_t)$ be the set of available neighboring servers to r_t—those which are closest to r_t on the left or right and available at time t^-. Define h_σ to be a matching obtained by first matching r_1 to an given server $s_1 \in N(r_1)$, and then using *Harmonic* to assign $r_2, \ldots, r_k$. We will use G for the algorithm producing g_σ and H for h_σ.

Lemma 3 (Hybrid Lemma). *If distances between servers are in* $\mathbb{Z} \cap [0, \Gamma]$,

$$E_G\left[\sum_{i=1}^{k} d(r_i, g_\sigma(r_i))\right] - E_H\left[\sum_{i=1}^{k} d(r_i, h_\sigma(r_i))\right] \leq O(\log \Gamma) \cdot d(r_1, s_1).$$

In other words, the expected cost of G for any request sequence is at most the expected cost of H on the same request sequence *plus* $O(\log \Gamma) \cdot d(r_1, s_1)$—the difference is proportional to the length of this forced initial assignment. This immediately gives us Theorem 1—let us show this fact before we prove Lemma 3.

Proof of Theorem 1. Given any request sequence σ and an optimal matching f_σ for this sequence such that $f_\sigma(r_1) \in N(r_1)$, we can define a sequence of hybrid matchings $\{h_\sigma^t\}_{t=0}^{k}$, where h_σ^t is obtained by matching the first t requests $r_1, \ldots, r_t$ in σ to $f_\sigma(r_1), \ldots, f_\sigma(r_t)$ and the remaining requests $r_{t+1}, \ldots, r_k$ to $g_\sigma(r_{t+1}), \ldots, g_\sigma(r_k)$. Note that h_σ^0 is just the *Harmonic* matching g_σ, and h_σ^k produces the optimal matching f_σ. Moreover, by ignoring the servers in $\{f_\sigma(r_i) \mid i \leq t\}$ and just considering $r_{t+1}, \ldots, r_k$ as the request sequence, Lemma 3 implies

$$E[\textstyle\sum_{i=t+1}^{k} d(r_i, h_\sigma^t(r_i))] \leq E[\textstyle\sum_{i=t+1}^{k} d(r_i, h_\sigma^{t+1}(r_i))] + O(\log \Gamma) \cdot d(r_{t+1}, f_\sigma(r_{t+1})),$$

since we can regard the assignment $r_{t+1} \to f_\sigma(r_{t+1}) \in N(r_{t+1})$ as the assignment $r_1 \to s_1$ used in Lemma 3. (It can be checked that the optimal assignment indeed assigns r_1 to a server in $N(r_1)$.) Now, by adding $\sum_{i=1}^{t} d(r_i, f_\sigma(r_i))$ to both sides,

$$E[\textstyle\sum_{i=1}^{k} d(r_i, h_\sigma^t(r_i))] \leq E[\textstyle\sum_{i=1}^{k} d(r_i, h_\sigma^{t+1}(r_i))] + O(\log \Gamma) \cdot d(r_{t+1}, f_\sigma(r_{t+1})).$$

Summing this over all values of $t \leq k - 1$, and using that $h_\sigma^0 = g_\sigma$ and $h_\sigma^k = f_\sigma$,

$$E[\textstyle\sum_{i=1}^{k} d(r_i, g_\sigma(r_i))] \leq E[\textstyle\sum_{i=1}^{k} d(r_i, f_\sigma(r_i))] + O(\log \Gamma) \cdot \textstyle\sum_{i=1}^{k} d(r_i, f_\sigma(r_i)).$$

The left side is the expected cost of *Harmonic*, and the right side is the cost of the optimal matching, which proves Theorem 1.   ∎

3.1   Proof of the Hybrid Lemma: A Coupling Argument

We now prove Lemma 3. Here is the high-level idea: recall that G is just a run of *Harmonic*, whereas H first forces $r_1 \to s_1$ (a server adjacent to r_1 on the line) and then runs *Harmonic*. So, just after r_1 has been assigned, either both G and H have the same set of free servers, or their symmetric difference is a pair of servers with no other free servers between them. (Think of the location of the free servers in one run as being obtainable from the free servers in the other run by moving a single server without jumping over any free servers, and let δ_1 be this random distance.) Now we will couple the random runs of G and H so that this property will continue to hold (or the set of servers will become the same, after which they will proceed in lock-step). We prove that the expected difference in the costs of the two algorithms will be $O(\log \Gamma) \cdot E[\delta_1]$. Since $E[\delta_1] = d(r_1, s_1)$ by the probabilities in the *Harmonic* algorithm, Lemma 3 follows.

For convenience, we will say that the request r_t is assigned at time t, and we refer to the situation just before this assignment as being at time t^-, and the situation just after as time t^+; note that $(t-1)^+ = t^-$. Let $\mathsf{A}_G(t)$ be the set of free servers at time t^+ when running algorithm G, and $\mathsf{A}_H(t)$ be similarly defined for algorithm H. Note that if at time t^+, $\mathsf{A}_G(t) = \mathsf{A}_H(t)$, then the expected difference in costs between algorithms G and H on requests $r_{t+1}, \ldots, r_k$ is 0. Thus, we can without loss of generality only consider the time instants where $\mathsf{A}_G(t) \neq \mathsf{A}_H(t)$.

Let g_1 be the only element of $\mathsf{A}_G(1) \setminus \mathsf{A}_H(1)$ and h_1 the only element of $\mathsf{A}_H(1) \setminus \mathsf{A}_G(1)$. Let $\delta_t = d(g_t, h_t)$ be the distance between these two servers. We now give a coupling π between the executions of G and H—equivalently a coupling between the evolutions of sets $\mathsf{A}_G(t)$ and $\mathsf{A}_H(t)$—from the two different starting configurations. For a valid coupling, the marginals should give us a faithful execution of *Harmonic* on $\mathsf{A}_G(1)$ and $\mathsf{A}_H(1)$ respectively. We define a coupling π maintaining the invariant that $|\mathsf{A}_G(t) \setminus \mathsf{A}_H(t)| = 1 = |\mathsf{A}_H(t) \setminus \mathsf{A}_G(t)|$, so we need to also define the coupling only on such pairs of states. By symmetry, assume that g_1 is to the left of h_1; we will maintain the invariant that g_t lies to the left of h_t. Also, when we start, there are no available servers between g_1 and h_1, and we will also maintain the invariant that there are no available servers between g_t and h_t, so we need only define the coupling over such pairs of states.

For the coupling π, we will write $\Pr_\pi[\mathcal{E}]$ to denote the probability of an event $\mathcal{E}$. This coupling also induces marginals on G and H, which we indicate by $\Pr_G[\mathcal{E}]$. We write $r \to_G s$ if G assigns r to s, and $N_G(r_t)$ will be the (at most two) neighboring free servers to the request r_t in G. (Analogous definitions hold for H.) If $r \to_G s_g$ and $r \to_H s_h$, then $\Delta c(r) := d(r, s_g) - d(r, s_h)$. Note that $\Delta c(r)$ can be negative.

Now, for the coupling π, there are four cases to consider when the request r_t arrives:

- **Case 0:** r_t's neighboring servers are identical in both G and H,
- **Case 1:** r_t lies to the left of both g_t and h_t but $g_t \in N_G(r_t)$,
- **Case 2:** r_t lies between g_t and h_t (so $g_t \in N_G(r_t)$ and $h_t \in N_H(r_t)$), and
- **Case 3:** r_t lies to the right of both g_t and h_t but $h_t \in N_H(r_t)$.

The fact that these are the only four cases follows from the invaraints we maintain in the coupling. Let us now define the coupling for these cases. (For lack of space, we defer the first and last cases to the full version.)

– For Case 1, we have the following situation:

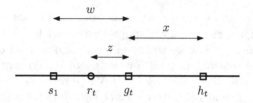

Set $p = \frac{w-z}{w}$ and $q = \frac{w-z}{w+x}$. We define the coupling π for this case:

Event	Assignments	Pr_π	$\delta_{t+1} - \delta_t$	$\Delta c(r_t)$
1	$r_t \to_G s_1, r_t \to_H s_1$	$1-p$	0	0
2	$r_t \to_G g_t, r_t \to_H s_1$	$p-q$	w	$2z-w$
3	$r_t \to_G g_t, r_t \to_H h_t$	q	$-x$	$-x$

Note that r_t goes to s_1 with probability $1-p$ and to g_t with probability p, hence the coupling is faithful run of G. And r_t goes to s_1 with probability $1-q$ and to h_t with probability q, as it should in H.

- In Case 2, we have the following situation:

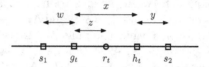

Set $p = \frac{x+y-z}{x+y}$ and $q = \frac{w+z}{w+x}$, and define the coupling as follows:

Event	Assignments	Pr_π	$\delta_{t+1} - \delta_t$	$\Delta c(r_t)$
1	$r_t \to_G s_2, r_t \to_H h$	$1-p$	y	y
2	$r_t \to_G g, r_t \to_H h$	$p+q-1$	$-x$	$2z-x$
3	$r_t \to_G g, r_t \to_H s_1$	$1-q$	w	$-w$

Note that r_t goes to g_t with probability p and to s_2 with probability p, hence the coupling is faithful run of G. And r_t goes to s_1 with probability $1-q$ and to h_t with probability q, as it should in H.

Define $Q_{i,n}$ to be the worst-case probability of the distance between $g_{t'}$ and $h_{t'}$ eventually going above n (for some future time t'), conditioned on $d(g_t, h_t) = i$ at time t—here the worst-case is taken over all possible future request sequences, and all feasible arrangements of any number of common servers in $\mathsf{A}_G(t) \cap \mathsf{A}_H(t)$, subject to the constraint that the distance between g_t and h_t is equal to i (and where g_t is to the left of h_t and no free servers between them).

Lemma 4. *The coupling maintains the following properties:*
(i) At each step t, if $\delta_{t+1} \neq 0$, then $\Delta c(r_t) \leq \delta_{t+1} - \delta_t$. If $\delta_{t+1} = 0$, then $\Delta c(r_t) \leq \delta_t$.
(ii) $Q_{i,n} \leq i/n$.

Proof. Property (i) follows by inspection of the above tables. For the proof of Property (ii), clearly $Q_{n,n} = 1$ for all n. Now, fix some n, and suppose we know that for all $j > \delta_t$, we have $Q_{j,n} \leq j/n$. Note that for the above four cases, each of these requests at time t either makes $\delta_{t+1} = 0$ (after which it can never reach n), keeps $\delta_{t+1} = \delta_t$, or makes distance δ_{t+1} more than δ_t (upon which we can apply induction to get a bound on $Q_{\delta_t,n}$). We thus enumerate over all four possibilities:

- For case 0, the distance does not change, and so there is nothing to show.
- For case 1, we get $Q_{x,n} = (p-q)Q_{x+w,n} + (1-p)Q_{x,n}$. This gives us $Q_{x,n} = \frac{p-q}{p}Q_{x+w,n}$, and using the inductive hypothesis for $Q_{x+q,n}$, we get $Q_{x,n} \leq (1-q/p)(x+w)/n = x/n$.
- For case 2, we have $Q_{x,n} = (1-p)Q_{x+y,n} + (1-q)Q_{x+w,n} \leq \frac{z}{x+y}(x+y)/n + \frac{x-z}{x+w}(x+w)/n = x/n$.
- For case 3, we get $Q_{x,n} = (p-q)Q_{x+w,n} + (1-p)Q_{x,n}$. This gives us $Q_{x,n} = \frac{p-q}{p}Q_{x+w,n}$, and so $Q_{x,n} \leq (1-q/p)(x+w)/n = x/n$.

In all cases, assuming that $Q_{j,n} \leq j/n$ for all $j > x$, we see that $Q_{x,n} \leq x/n$. This completes the proof of the lemma. ∎

We can now prove Lemma 3. We want to bound $E_G[\sum_{i=1}^{k} d(r_i, g_\sigma(r_i))] - E_H[\sum_{i=1}^{k} d(r_i, h_\sigma(r_i))]$, but since π's marginals are faithfully running G and H, we can use linearity of expectatitions to bound

$$E_\pi \left[\sum_{i=1}^{k} d(r_i, g_\sigma(r_i)) - \sum_{i=1}^{k} d(r_i, h_\sigma(r_i)) \right] = E_\pi \left[\sum_{i=1}^{k} \Delta c(r_k) \right].$$

But by Lemma 4(i), we know that $\Delta c(r_k) \leq (\delta_2 - \delta_1) + (\delta_3 - \delta_2) + \cdots + (\delta_q - \delta_{q-1}) + \delta_q$, where $\delta_{q+1} = 0$ for the first time. This is at most $2\delta_q = 2\delta_{\max}$. So it remains to bound $E_\pi[\delta_{\max}]$. We see that

$$E_\pi[\delta_{\max} \mid \delta_1] = \sum_{l=1}^{\Gamma} \Pr_\pi[\delta_{\max} \geq l \mid \delta_1] \leq \sum_l Q_{\delta_1,l} \leq \sum_{l=1}^{\Gamma} \delta_1/l = O(\log \Gamma) \cdot \delta_1$$

by Lemma 4(ii) and the definition of $Q_{j,n}$. So $E_\pi[\delta_{\max}] = O(\log k) \cdot E_\pi[\delta_1]$. Now if the two servers adjacent to r_1 were s_1 and h_1, then we have $E[\delta_1] = \frac{d(r_1,h_1)}{d(h_1,s_1)} \cdot 0 + \frac{d(r_1,s_1)}{d(h_1,s_1)} \cdot d(h_1, s_1) = d(r_1, s_1)$. This proves the hybrid lemma (Lemma 3).

4   The Random-Subtree Algorithm

We now turn to showing that a different randomized algorithm gives an $O(\log k)$ competitive ratio for the line; the proof generalizes to doubling metrics too. To start off, we use the fact that *binary* 2-HSTs approximate the line metric with $O(\log k)$ expected stretch. It is not difficult to show that the (deterministic) greedy algorithm on a binary 2-HST is $O(\log k)$-competitive compared to the optimal solution on the tree, which implies an $O(\log^2 k)$-competitive ratio in all. In this section, we show that randomization helps: a certain randomized greedy algorithm is $O(1)$-competitive on the binary 2-HST, giving us a different $O(\log k)$-competitive algorithm for the line. In fact, the proof extends to HSTs obtained from doubling metrics, and hence proves Theorem 2.

The Algorithm. Let us define the algorithm *Random-Subtree* for online metric matching on an arbitrary HST as follows: when a request r comes in, consider its lowest ancestor node a whose subtree $T(a)$ also contains a free server. Now we choose a random free server in the subtree rooted at a as follows: from among those of a's children whose subtrees contain a free server under them, we choose such a child of a uniformly at random, and repeat this process until we reach a leaf/server s—we then map r to server s. Observe that ours is a different randomized greedy algorithm from that in [9], which would have chosen a server uniformly at random from among all of the servers in $T(a)$. Our main theorem is the following.

Theorem 5. *The algorithm* Random-Subtree *is* $2(1 + 1/\epsilon)H_\Delta$-*competitive on* Δ-*ary* α-*HSTs, as long as* $\alpha \geq \max((1 + \epsilon)H_\Delta, 2)$.

Since the line embeds into binary 2-HSTs with expected stretch $O(\log k)$, we get an $O(\log k)$-competitive randomized algorithm for the line. Moreover, in the full version, we show that an algorithm for Δ-ary α-HSTs satisfying the property above (with $\Delta = O(1)$) implies an algorithm for doubling metrics with an additional loss of $O(\log k)$; this proves Theorem 2.

The proof of the theorem goes thus: we first just consider the edges incident to the root (which we call root-edges) of an Δ-ary α-HST, and count the number of times these edges are used. Specifically, we show that for any sequence of requests, the number of requests that use the root-edges in our algorithm is at most H_Δ times the minimum number of requests that must use these root-edges. This "root-edges lemma" is the technical heart of our analysis; getting H_Δ instead of H_k (obtained in [9]) requires defining the right potential function, and carefully accounting for the gain we get from using the *Random-Subtree* algorithm rather than the randomized greedy algorithm of [9].

Having proved the root-edges lemma, notice that for any fixed vertex v in an HST, the subtree rooted at v is another HST on which we can apply the root-edges lemma to bound the cost incurred on the edges incident to v. Consequently, applying this for every internal vertex in the HST and summing up the results shows that the total cost remains at most $O(H_\Delta) \cdot \mathsf{Opt}$, as long as the parameter α for the HST is larger than H_Δ.   •

The Analysis. Consider a Δ-ary α-HST T with a set of requests $R \cup R'$ such that the requests in R originate at the leaves of T, and those in R' originate at the root. We assume that the number of servers in T is at least $|R \cup R'|$. Let $T_1, T_2, \ldots, T_\Delta$ denote the Δ child subtrees of T. Without loss of generality, we assume that T has exactly Δ child subtrees. We will use $R(T_i)$ to denote the set of requests that originate in subtree T_i. Let n_i be the number of servers in T_i, and let $M^* = \sum_{i=1}^{\Delta} \max(|R \cap R(T_i)| - n_i, 0)$. The following fact gives a lower bound for Opt.

Fact 6. *In any assignment of requests in $R \cup R'$ to servers, at least $M^* + |R'|$ requests use root-edges.*

The following crucial lemma upper-bounds the expected cost incurred by the algorithm on just the root edges.

Lemma 7 (Root-Edges Lemma). *Let the random variable M count the number of requests in $R \cup R'$ that use a root-edge when assigned by the algorithm* Random-Subtree.

$$E[M] \leq H_\Delta \cdot (M^* + |R'|).$$

Proof. Let the k requests $R \cup R'$ be labeled $r_1, r_2, \ldots, r_k$, where r_1 is the earliest request and r_k is the latest request. The request r_t is assigned at time t, and we refer to the situation just before this assignment as being at time t^-, and the situation just after as time t^+. Note that t^- for $t = 1$ (denoted as 1^-) represents the time before any request assignments have been made, and t^+ for $t = k$ (denoted as k^+) represents the time after all request assignments have been made. Let $R_t = \{r_t, r_{t+1}, \ldots, r_k\}$, the set of requests at time t^- that have yet to arrive. At time t^-, let $n_{i,t}$ be the number of available servers in tree T_i. A subtree T_i is said to be *open* at time t^- if $n_{i,t} > 0$ (there are available servers at time t^- in T_i). Let η_t be the number of open subtrees of T at time t^-.

Define the first $\min(n_{i,t}, |R_t \cap R(T_i)|)$ requests of T_i to be the *local requests of T_i at time t^-* (these are the ones in $R(T_i)$ that have the lowest numbered indices), and the remaining requests in T_i to be the *global requests of T_i at time t^-*.[1] Let $\mathcal{L}_{i,t}$ and $\mathcal{G}_{i,t}$ be the set of local and global requests in T_i at time t^-, and let $\mathcal{L}_t := \cup_i \mathcal{L}_{i,t}$ and $\mathcal{G}_t := \cup_i \mathcal{G}_{i,t}$. For convenience, we say that a request r_j *becomes global* at time t if r_j is local at time t^-, but r_j is global at time t^+. Let requests in $\mathcal{R}_t := R_t \cap R'$ be called *root requests of T at time t^-*.

As a sanity check, note that at the beginning (at time 1^-), the set of pending requests $R_1 = R \cup R'$, the number of pending requests in subtree T_i is $n_{i,1} = n_i$, the number of global requests in T_i is $|\mathcal{G}_{i,1}| = \max(|R \cap R(T_i)| - n_i, 0)$ (so the total number of global requests at time 1^- is M^*), and the number of root requests is $|\mathcal{R}_1| = |R'|$.

Recall that global requests of T_i must assign to servers outside of T_i: while an optimal offline algorithm can identify where to assign these global requests, an online algorithm may assign a global request from T_i to some subtree T_j that only has as many servers as future requests, which causes some local request in T_j to become global. Hence we want to upper-bound the number of future requests in R_{t+1} that become global due to our assignment for r_t. We associate with each request in R_t a "cost" at time t^- which represents this upper bound. Later, we will use the cost function to define the potential function. The cost function at time t^- is $F_t : R_t \to \mathbb{Z}_{\geq 0}$; we say it is *well-formed* if it satisfies two properties:

- $F_t(r_j) = 0$ if and only if $r_j \in \mathcal{L}_t$ (i.e., it is a local request at time t^-), and

---

[1] The idea behind calling requests local/global is this: assuming no servers in T_i are used up by requests from other subtrees, the local servers will be assigned within T_i by our algorithm, whereas the global ones will be assigned to other subtrees (and hence use a root-edge). Of course, as servers within T_i are used by requests in other subtrees, some local requests become global.

- for all global and root requests $r_j \in \mathcal{G}_t \cup \mathcal{R}_t$, $F_t(r_j)$ is an upper bound on the random variable η_j, the number of open subtrees at time j^-.

Constructing the Well-Formed Cost Functions. We set $F_1(r_j) = \Delta$ (the degree of the tree) for all $r_j \in \mathcal{G}_1 \cup \mathcal{R}_1$ (global and root requests at time 1^-), and $F_1(r_j) = 0$ for all $r_j \in \mathcal{L}_1$ (local requests at time 1^-). It is immediate that the map F_1 is well-formed.

Now at each time t^+, we will define the next function F_{t+1} using F_t. For this, first consider time t^-, and suppose that the map F_t is well-formed. The easy case first: If $r_t \in \mathcal{L}_t$, then define $F_{t+1}(r) = F_t(r)$ for all $r \in R_t$. In this case if a request in R_t is a local/global/root request at time t^-, it remains a local/global/root request at time t^+, so F_{t+1} is still well-formed.

On the other hand, suppose $r_t \in \mathcal{G}_t \cup \mathcal{R}_t$, i.e., it is a global or root request. Recall there are η_t open subtrees at time t^-. Each open subtree T_i contains $|R_t \cap R(T_i)|$ requests and $n_{i,t}$ free servers, so if $|R_t \cap R(T_i)| \geq n_{i,t}$ then assigning r_t to a server in this subtree would cause some request r_j in $R_t \cap R(T_i)$ to become global at time t (because $n_{i,t+1}$ would become $n_{i,t} - 1$). In this case, define $a_t(T_i) := j$, the index of the request r_j that turns global in subtree T_i. Else, if no request in $R_t \cap R(T_i)$ would become global, set $a_t(T_i) := k + i$ (which cannot be the index of any request, since there are only k requests). Let $A_t = \{a_t(T_i) \mid T_i \text{ open at time } t^-\}$; note that $|A_t| = \eta_t$. Now denote the elements of A_t by $\{p_j\}_{j=1}^{\eta_t}$ such that $p_1 < p_2 < \cdots < p_{\eta_t}$.

(Another sanity check: we claim that the last entry $p_{\eta_t} > k$; indeed, if r_t is a global or root request, there must be some open subtree T_i which has more available servers than requests.) Now, let T_i be the subtree that r_t assigns to, chosen by picking out of the open subtrees uniformly at random. We now define the map F_{t+1} at time t^+. There are two cases to consider:

- If $a_t(T_i) > k$ (i.e., none of the requests in $R(T_i) \cap R_{t+1}$ become global due to assigning r_t), then we set $F_{t+1}(r) = F_t(r)$ for all requests $r \in R_{t+1}$.
- If $a_t(T_i) \leq k$, then say $a_t(T_i) = p_{\eta_t - q + 1}$ in the ordering given above (i.e., $a_t(T_i)$ was the qth largest value in A_t). Now assign $F_{t+1}(r) = F_t(r)$ for all $r \in R_{t+1} \setminus \{r_{a_t(T_i)}\}$, and $F_{t+1}(r_{a_t(T_i)}) = q - 1$.

Showing that this map F_{t+1} is well-formed is deferred to the full version. Note that maps F_t and F_{t+1} are either the same or differ on at most one request r_j that becomes global at time t, in which case $F_{t+1}(r_j)$ becomes positive. Moreover, $F_{t'}(r_j) = F_{t+1}(r_j)$ for all times $t' \in [t+1, j]$.

The Potential Function Analysis. We are now in a position to define the potential function,

$$\Phi_t = \sum_{r \in R_t} H_{F_t(r)}, \tag{4.1}$$

where we consider $H_0 = 0$. Also, define ρ_t to be the number of requests that our algorithm has already matched outside of their subtrees at time t^-. The root-edges lemma follows immediately from the following claim, proved using induction.

Lemma 8. *For all* $t \in [1, k+1]$, $E[\Phi_t + \rho_t] \leq H_\Delta \cdot (M^* + |R'|)$.

(Proof given in the full version.) Since $\rho_{k+1} = M$ and $\Phi_{k+1} = 0$, using Lemma 8 with $t = k + 1$ finishes the proof of the root-edges lemma. ∎

The next lemma bounds the total cost, not just the cost on the root edges, by considering every subtree in the HST and applying the root-edges lemma to each subtree.

Lemma 9. *Consider a* Δ*-ary* α*-HST* T*, any set* R *of requests at the leaves of* T*, and requests* R' *at the root of* T*, such that* $|R \cup R'|$ *is at most the number of servers in* T*. If* $\mathsf{Alg}(R \cup R', T)$ *denotes the cost of* Random-Subtree *on requests* $R \cup R'$ *on tree* T*, and* $\mathsf{Opt}(R \cup R', T)$ *the cost of the optimal solution, we have*

$$E[\mathsf{Alg}(R \cup R', T)] \leq c \cdot H_\Delta \cdot \mathsf{Opt}(R \cup R', T)$$

for $c = 2(1 + 1/\epsilon)$ *as long as* $\alpha \geq \max\{2, (1 + \epsilon)H_\Delta\}$*.*

The lemma above directly proves Theorem 5. As an aside, note that 2-HSTs that have large degree, or binary HST's that have $\alpha \approx 1$ (say $\alpha = 1 + 1/\log k$), can both simulate star metrics, on which we have an $\Omega(\log k)$ lower bound—hence we do need some relationship between α and Δ.

References

1. Bansal, N., Buchbinder, N., Gupta, A., Naor, J.S.: An $o(\log^2 k)$-competitive algorithm for metric bipartite matching. In: Proceedings of the 15th Annual European Symposium on Algorithms, pp. 522–533 (2007)
2. Fakcharoenphol, J., Rao, S., Talwar, K.: A tight bound on approximating arbitrary metrics by tree metrics. In: STOC 2003: Proceedings of the Thirty-Fifth Annual ACM Symposium on Theory of Computing, pp. 448–455 (2003)
3. Fuchs, B., Hochstattler, W., Kern, W.: Online matching on a line. Theoretical Computer Science 332, 251–264 (2005)
4. Kalyanasundaram, B., Pruhs, K.: Online weighted matching. J. Algorithms 14(3), 478–488 (1993)
5. Kalyanasundaram, B., Pruhs, K.: Online Network Optimization Problems. In: Fiat, A. (ed.) Online Algorithms 1996. LNCS, vol. 1442, pp. 268–280. Springer, Heidelberg (1998)
6. Khuller, S., Mitchell, S.G., Vazirani, V.V.: On-line algorithms for weighted bipartite matching and stable marriages. Theor. Comput. Sci. 127(2), 255–267 (1994)
7. Koutsoupias, E., Nanavati, A.: The Online Matching Problem on a Line. In: Solis-Oba, R., Jansen, K. (eds.) WAOA 2003. LNCS, vol. 2909, pp. 179–191. Springer, Heidelberg (2004)
8. Koutsoupias, E., Papadimitriou, C.H.: On the k-server conjecture. J. ACM 42, 971–983 (1995)
9. Meyerson, A., Nanavati, A., Poplawski, L.: SODA 2006: Proceedings of the Seventeenth Annual ACM-SIAM Symposium on Discrete Algorithm. In: SODA 2006: Proceedings of the Seventeenth Annual ACM-SIAM Symposium on Discrete Algorithm, pp. 954–959 (2006)

Approximating Sparse Covering Integer Programs Online*

Anupam Gupta[1,**] and Viswanath Nagarajan[2]

[1] Computer Science Department, Carnegie Mellon University
[2] IBM T.J. Watson Research Center

Abstract. A *covering integer program* (CIP) is a mathematical program of the form:

$$\min\{c^\top x \mid Ax \geq 1, \ 0 \leq x \leq u, \ x \in \mathbb{Z}^n\},$$

where $A \in R_{\geq 0}^{m \times n}, c, u \in \mathbb{R}_{\geq 0}^n$. In the online setting, the constraints (i.e., the rows of the constraint matrix A) arrive over time, and the algorithm can only increase the coordinates of x to maintain feasibility. As an intermediate step, we consider solving the *covering linear program* (CLP) online, where the requirement $x \in \mathbb{Z}^n$ is replaced by $x \in \mathbb{R}^n$.

Our main results are (a) an $O(\log k)$-competitive online algorithm for solving the CLP, and (b) an $O(\log k \cdot \log \ell)$-competitive randomized online algorithm for solving the CIP. Here $k \leq n$ and $\ell \leq m$ respectively denote the maximum number of non-zero entries in any row and column of the constraint matrix A. By a result of Feige and Korman, this is the best possible for polynomial-time online algorithms, even in the special case of set cover (where $A \in \{0, 1\}^{m \times n}$ and $c, u \in \{0, 1\}^n$).

The novel ingredient of our approach is to allow the dual variables to increase and decrease throughout the course of the algorithm. We show that the previous approaches, which either only raise dual variables, or lower duals only within a guess-and-double framework, cannot give a performance better than $O(\log n)$, even when each constraint only has a single variable (i.e., $k = 1$).

1 Introduction

Covering Integer Programs (CIPs) have long been studied, giving a very general framework which captures a wide variety of natural problems. CIPs are mathematical programs of the following form:

$$\min \quad \sum_{i=1}^{n} c_i x_i \tag{IP1}$$

$$\text{subject to:} \quad \sum_{i=1}^{n} a_{ij} x_i \geq 1 \qquad \forall j \in [m], \tag{1.1}$$

$$0 \leq x_i \leq u_i \qquad \forall i \in [n], \tag{1.2}$$

$$x \in \mathbb{Z}^n. \tag{1.3}$$

---

* A full version of this extended abstract appears as [11].
** Supported in part by NSF awards CCF-0964474 and CCF-1016799.

A. Czumaj et al. (Eds.): ICALP 2012, Part I, LNCS 7391, pp. 436–448, 2012.
© Springer-Verlag Berlin Heidelberg 2012

Above, all the entries a_{ij}, c_i, and u_i are non-negative. The *constraint matrix* is denoted $A = (a_{ij})_{i \in [n], j \in [m]}$. We define k to be the *row sparsity* of A, i.e., the maximum number of non-zeroes in any constraint $j \in [m]$. For each row $j \in [m]$ let $T_j \subseteq [n]$ denote its non-zero columns; we say that the variables indexed by T_j "appear in" constraint j. Let ℓ denote the *column sparsity* of A, i.e., the maximum number of constraints that any variable $i \in [n]$ appears in. Dropping the integrality constraint (1.3) gives us a covering linear program (CLP).

In this paper we study the *online* version of these problems, where the constraints $j \in [m]$ arrive over time, and we are required to maintain a monotone (i.e., non-decreasing) feasible solution $\mathbf{x}$ at each point in time. Our main results are (a) an $O(\log k)$-competitive algorithm for solving CLPs online, and (b) an $O(\log k \cdot \log \ell)$-competitive randomized online algorithm for CIPs. In settings where $k \ll n$ or $\ell \ll m$ our results give a significant improvement over the previous best bounds of $O(\log n)$ for CLPs [8] and $O(\log n \cdot \log m)$ for CIPs that can be inferred from rounding of these fractional solutions. Analyzing performance guarantees for covering/packing integer programs in terms of row (k) and column (ℓ) sparsity has received much attention in the offline setting, e.g. [16,18,12,15,6]. This paper obtains tight bounds in terms of these parameters for *online* covering integer programs.

Our Techniques. Our algorithms use online primal-dual framework of Buchbinder and Naor [7]. To solve the covering LP, we give an algorithm that monotonically raises the primal. However, we *both raise and lower the dual variables* over the course of the algorithm; this is unlike typical applications of the online primal-dual approach, where both primal and dual variables are only increased (except possibly within a "guess and double" framework—see the discussion in the related work section). This approach of lowering duals is crucial for our bound of $O(\log k)$, since we show a primal-dual gap of $\Omega(\log n)$ for algorithms that lower duals only within the guess-and-double framework, even when $k = 1$.

The algorithm for covering IP solves the LP relaxation and then rounds it. It is well-known that the natural LP relaxation is too weak: so we extend our online CLP algorithm to also handle Knapsack Cover (KC) inequalities from [9]. This step has an $O(\log k)$-competitive ratio. Then, to obtain an integer solution, we adapt the method of *randomized rounding with alterations* to the online setting. Direct randomized rounding as in [1] results in a worse $O(\log m)$ overhead, so to get the $O(\log \ell)$ loss we use this different approach.

Related Work. The powerful online primal-dual framework has been used to give algorithms for set cover [1], graph connectivity and cut problems [2], caching [19,4,5], packing/covering IPs [8], and many more problems. This framework usually consists of two steps: obtaining a fractional solution (to an LP relaxation) online, and rounding the fractional solution online to an integral solution. (See the monograph of Buchbinder and Naor [7] for a lucid survey.)

In most applications of this framework, the fractional online algorithm raises both primal and dual variables monotonically, and the competitive ratio is given by the primal to dual ratio. For CLPs, Buchbinder and Naor [8] showed that if we increase dual variables monotonically, the primal-dual gap can be $\Omega(\log \frac{a_{max}}{a_{min}})$. In order to obtain an $O(\log n)$-competitive ratio, they used a *guess-and-double* framework [8, Theorem 4.1] that changes duals in a partly non-monotone manner as follows: *The algorithm proceeds in phases, where each phase r corresponds to the primal value being roughly 2^r. Within a phase the primal and dual are raised monotonically. But the algorithm resets duals to zero at the beginning of each phase—this is the only form of dual reduction.*

For the special case of fractional set cover (where $A \in \{0, 1\}^{m \times n}$), they get an improved $O(\log k)$-competitive ratio using this guess-and-double framework [8, Section 5.1]. However, we show in the full version [11] that such dual update processes do not extend to obtain an $o(\log n)$ ratio for general CLPs. So our algorithm reduces the dual variables more continuously throughout the algorithm, giving an $O(\log k)$-competitive ratio for general CLPs.

Other online algorithms: Koufogiannakis and Young [14] gave a k-competitive deterministic online algorithm for CIPs based on a greedy approach; their result holds for a more general class of constraints and for submodular objectives. Our $O(\log k \log \ell)$ approximation is incomparable to this result. Feige and Korman [13] show that no randomized *polynomial-time* online algorithm can achieve a competitive ratio better than $O(\log k \log \ell)$.

Offline algorithms. CLPs can be solved optimally offline in polynomial time. For CIPs in the absence of variable upper bounds, randomized rounding gives an $O(\log m)$-approximation ratio. Srinivasan [16] gave an improved algorithm using the FKG inequality (where the approximation ratio depends on the optimal LP value). Srinivasan [17] also used the method of alterations in context of CIPs and gave an RNC algorithm achieving the bounds of [16]. An $O(\log \ell)$-approximation algorithm for CIPs (no upper bounds) was obtained in [18] using the Lovász Local Lemma. Using KC-inequalities and the algorithm from [18], Kolliopoulos and Young [12] gave an $O(\log \ell)$-approximation algorithm for CIPs with variable upper bounds. Our algorithm matches this $O(\log \ell)$ loss in the online setting. Finally, the knapsack-cover (KC) inequalities were introduced by Carr et al. [9] to reduce the integrality gap for CIPs. These were used in [12,10], and also in an online context by [5] for the generalized caching problem.

2   An Algorithm for a Special Class for Covering LPs

In this section, we consider CLPs without upper bounds on the variables:

$$\min \left\{ \sum_{i=1}^n c_i x_i \ \middle| \ \sum_i a_{ij} x_i \geq 1 \ \forall j \in [m], \ x \geq \mathbf{0} \right\} \tag{2.4}$$

and give an $O(\log k)$-competitive deterministic online algorithm for solving such LPs, where k is an (upper bound) on the row-sparsity of $A = (a_{ij})$. The dual is the packing linear program:

$$\max \left\{ \sum_{j=1}^m y_j \ \middle| \ \sum_{j:i \in T_j} a_{ij} y_j \leq c_i \ \forall i \in [n], \ y \geq \mathbf{0} \right\} \tag{2.5}$$

We assume that c_i's are strictly positive for all i, else we can drop all constraints containing variable i.

Algorithm I. In the online algorithm, we want a solution pair (x, y), where we monotonically increase the value of x, but the dual variables can move up or down as needed. We want a feasible primal, and an approximately feasible dual. The primal update step is the following:

When constraint h (i.e., $\sum_i a_{ih}x_i \geq 1$) arrives,
(a) define $d_{ih} = \frac{c_i}{a_{ih}}$ for all $i \in [n]$, and $d_{m(h)} = \min_i d_{ih} = \min_{i \in T_h} d_{ih}$.
(b) while $\sum_i a_{ih}x_i < 1$, update the x's by

$$x_i^{new} \leftarrow \left(1 + \frac{d_{m(h)}}{d_{ih}}\right) x_i^{old} \ + \ \frac{1}{k \cdot a_{ih}} \frac{d_{m(h)}}{d_{ih}}, \qquad \forall i \in T_h.$$

Let t_h be the number of times this update step is performed for constraint h.

As stated, the algorithm assumes we know k, but this is not required. We can start with the estimate $k = 2$ and increase it any time we see a constraint with more variables than our current estimate. Since this estimate for k only increases over time, the analysis below will go through unchanged. (We can assume that k is a power of 2—which makes $\log k$ an integer; we will need that $k \geq 2$.)

Lemma 1. *For any constraint h, the number of primal updates $t_h \leq 2 \log k$.*

Proof. Fix some h, and consider the value i^* for which $d_{i^*h} = d_{m(h)}$. In each round the variable $x_{i^*} \leftarrow 2x_{i^*} + 1/(k \cdot a_{i^*h})$; hence after t rounds its value will be at least $(2^t - 1)/(k \cdot a_{i^*h})$. So if we do $2 \log k$ updates, this variable alone will satisfy the hth constraint. ∎

Lemma 2. *The total increase in the value of the primal is at most $2 t_h d_{m(h)}$.*

Proof. Consider a single update step that modifies primal variables from x^{old} to x^{new}. In this step, the increase in each variable $i \in T_h$ is $\frac{d_{m(h)}}{d_{ih}} \cdot x_i^{old} + \frac{1}{k \cdot a_{ih}} \frac{d_{m(h)}}{d_{ih}}$. So the increase in the primal objective is:

$$\sum_{i \in T_h} c_i \cdot \left[\frac{d_{m(h)}}{d_{ih}} \cdot x_i^{old} + \frac{1}{k \cdot a_{ih}} \frac{d_{m(h)}}{d_{ih}}\right] = d_{m(h)} \sum_{i \in T_h} a_{ih} \cdot x_i^{old} + d_{m(h)} \cdot \frac{|T_h|}{k} \leq 2 \cdot d_{m(h)}$$

The inequality uses $|T_h| \leq k$ and $\sum_{i \in T_h} a_{ih} \cdot x_i^{old} \leq 1$ which is the reason an update was performed. The lemma now follows since t_h is the number of update steps. ∎

To show approximate optimality, we want to change the dual variables so that the dual increase is (approximately) the primal increase, and so that the dual remains (approximately) feasible. To achieve the first goal, we raise the newly arriving dual variable, and to achieve the second we also decrease the "first few" dual variables in each dual constraint where the new dual variable appears.

For the hth primal constraint, let $d_{ih}, d_{m(h)}, t_h$ be given by the primal update process.

(a) Set $y_h \leftarrow d_{m(h)} \cdot t_h$.
(b) For each $i \in T_h$, do the following for dual constraint $\sum_j a_{ij} y_j \leq c_i$:
  (i) If $\sum_{j<h} a_{ij} y_j \leq (10 \log k) c_i$, do nothing; else
  (ii) Let $k_i < h$ be the largest index such that $\sum_{j \leq k_i} a_{ij} y_j \leq (5 \log k) c_i$; let $P_i = \{j \leq k_i \mid i \in T_j\}$ be the indices of these first few dual variables that are active in the ith dual constraint. For all $j \in P_i$,

$$y_j^{new} \leftarrow \left(1 - \frac{d_{m(h)}}{d_{ih}}\right) \cdot y_j^{old}.$$

Observe that the dual update process starts each dual variable y_j off at some value $d_{m(j)} t_j$ and subsequently *only decreases* this dual variable, and that the dual variables remain non-negative.

Lemma 3. *When primal constraint h arrives, the left-hand-side of each dual constraint i increases due to the variable y_h by $a_{ih} \cdot d_{m(h)} \cdot t_h \leq (2 \log k) c_i$.*

Proof. We set the initial value of the dual variable y_h to $d_{m(h)} \cdot t_h$. By Lemma 1, $t_h \leq 2 \log k$. By definition, $d_{m(h)} \leq c_i / a_{ih}$. Hence, for any $i \in T_h$, the increase in the left-hand-side of dual constraint i is at most $a_{ih} \cdot (2 \log k)(c_i / a_{ih}) = (2 \log k) c_i$. This proves the lemma. ∎

Lemma 4. *When primal constraint h arrives, if the dual update reaches step b (ii) for some $i \in T_h$, then k_i is well-defined and the set P_i is non-empty; moreover, $\frac{\sum_{j \in P_i} a_{ij} y_j}{c_i \log k} \in [3, 5]$.*

Proof. For each $j < h$ we have $y_j \leq 2 \log k \cdot d_{m(j)}$, since dual variable y_j was initialized to $t_j d_{m(j)} \leq 2 \log k \cdot d_{m(j)}$ (by Lemma 1) and subsequently never increased—so $a_{ij} \cdot y_j \leq 2 \log k \cdot d_{m(j)} \cdot a_{ij} \leq 2 \log k \cdot c_i$, using $d_{m(j)} \leq d_{ij} = c_i / a_{ij}$. If the dual update reaches step b(ii) then we have $\sum_{j<h} a_{ij} y_j > (10 \log k) c_i$, but each $j < h$ contributes at most $2 \log k \cdot c_i$, so k_i is well-defined, and P_i is non-empty. Moreover, by the choice of k_i, we have $\sum_{j \leq k_i+1} a_{ij} y_j > (5 \log k) c_i$, so $\sum_{j \leq k_i} a_{ij} y_j > (5 \log k) c_i - a_{i,k_i+1} \cdot y_{k_i+1} \geq (3 \log k) \cdot c_i$, as claimed. ∎

Lemma 5. *After each dual update step, each dual constraint i satisfies $\sum_j a_{ij} y_j \leq (12 \log k) c_i$. Hence the dual is $(12 \log k)$-feasible.*

Proof. Consider the dual update process when the primal constraint h arrives, and look at any dual constraint $i \in T_h$ (the other dual constraints are unaffected). If case b(i) happens, then by Lemma 3 the left-hand-side of the constraint will be at most $(12 \log k) c_i$. Else, case b(ii) happens. Each y_j for $j \in P_i$ decreases by $y_j \cdot d_{m(h)} / d_{ih}$, and so the decrease in $\sum_{j \in P_i} a_{ij} y_j$ is at least $\sum_{j \in P_i} a_{ij} y_j \cdot (d_{m(h)} / d_{ih})$. Using Lemma 4, this is at least

$$\frac{d_{m(h)}}{d_{ih}} \cdot c_i \, (3 \log k) = \frac{d_{m(h)}}{c_i / a_{ih}} \cdot c_i \, (3 \log k) = d_{m(h)} \cdot a_{ih} \cdot (3 \log k).$$

But since the increase due to y_h is at most $a_{ih} \cdot d_{m(h)} \, t_h \leq a_{ih} \cdot d_{m(h)} \cdot (2 \log k)$, there is no net increase in the LHS, so it remains at most $(12 \log k) \, c_i$. ∎

Lemma 6. *The net increase in the dual value due to handling primal constraint h is at least $\frac{1}{2} d_{m(h)} \cdot t_h$.*

Proof. The increase in the dual value due to y_h itself is $d_{m(h)} \cdot t_h$. What about the decrease in the other y_j's? These decreases could happen due to any of the k dual constraints $i \in T_h$, so let us focus on one such dual constraint i, which reads $\sum_{j:i \in T_j} a_{ij} y_j \leq c_i$. Now for $j < h$, define $\gamma_{ij} := \frac{y_j}{t_j \, d_{ij}}$. Since y_j was initially set to $t_j \, d_{m(j)} \leq t_j \, d_{ij}$ and subsequently never increased, we know that at this point in time,

$$\gamma_{ij} \quad \leq \quad \frac{d_{m(j)}}{d_{ij}} \quad \leq \quad 1. \tag{2.6}$$

The following claim, whose proof appears after this lemma, helps us bound the total dual decrease.

Claim 1. *If we are in case b(ii) of the dual update, then $\sum_{j \in P_i} \frac{\gamma_{ij} t_j}{a_{ij}} \leq \frac{1}{2k} \cdot \frac{1}{a_{ih}}$.*

Using this claim, we bound the loss in dual value caused by dual constraint i:

$$\sum_{j \in P_i} \frac{d_{m(h)}}{d_{ih}} \cdot y_j = \frac{d_{m(h)}}{d_{ih}} \cdot \sum_{j \in P_i} \gamma_{ij} \cdot t_j \, d_{ij} = \frac{d_{m(h)}}{c_i / a_{ih}} \cdot \sum_{j \in P_i} \gamma_{ij} \cdot t_j \, (c_i / a_{ij})$$

$$= d_{m(h)} \, a_{ih} \cdot \sum_{j \in P_i} \gamma_{ij} \cdot \frac{t_j}{a_{ij}} \leq_{\text{(Claim 1)}} d_{m(h)} \, a_{ih} \cdot \frac{1}{2k} \cdot \frac{1}{a_{ih}} = \frac{d_{m(h)}}{2k}.$$

Summing over the $|T_j| \leq k$ dual constraints affected, the total decrease is at most $\frac{1}{2} d_{m(h)} \leq \frac{1}{2} d_{m(h)} t_h$ (since there is no decrease when $t_h = 0$). Subtracting from the increase of $d_{m(h)} \cdot t_h$ gives the lemma. ∎

Proof of Claim 1. Consider the primal constraints j such that $i \in T_j$: when they arrived, the value of primal variable x_i may have increased. (In fact, if some primal constraint j does not cause the primal variables to increase, y_j is set to 0 and never plays a role in the subsequent algorithm, so we will assume that for each primal constraint j there is some increase and hence $t_j > 0$.)

The first few among the constraints j such that $i \in T_j$ lie in the set P_i: when $j \in P_i$ arrived, we added at least $\frac{1}{k \cdot a_{ij}} \frac{d_{m(j)}}{d_{ij}}$ to x_i's value[1], and did so t_j times. Hence the value of x_i after seeing the constraints in P_i is at least $\sum_{j \in P_i} \frac{d_{m(j)} t_j}{k \cdot a_{ij} \cdot d_{ij}} \geq \sum_{j \in P_i} \frac{\gamma_{ij} t_j}{k \cdot a_{ij}}$, using (2.6).

---

[1] More precisely, x_i increased by at least $\frac{1}{k_j \cdot a_{ij}} \frac{d_{m(j)}}{d_{ij}}$ where $k_j \leq k$ was the estimate of the row-sparsity at the arrival of constraint j, and k is the current row-sparsity estimate.

If χ_i is the value of x_i after seeing the constraints in P_i, and χ_i' is its value after seeing the rest of the constraints in $Q_i := (\{j < h \mid i \in T_j\} \setminus P_i)$. Then

$$\frac{\chi_i'}{\chi_i} \geq \prod_{j \in Q_i} \left(1 + \frac{d_{m(j)}}{d_{ij}}\right)^{t_j} \geq_{(2.6)} \prod_{j \in Q_i} (1 + \gamma_{ij})^{t_j} \geq_{(\gamma_{ij} \leq 1)} e^{\frac{1}{2} \sum_{j \in Q_i} \gamma_{ij} t_j} \geq 2k^2.$$

$$(2.7)$$

The last inequality uses the fact that $k \geq 2$, and that:

$$\sum_{j \in Q_i} \gamma_{ij} t_j = \sum_{j \in Q_i} y_j / d_{ij} = \sum_{j \in Q_i} \frac{y_j \cdot a_{ij}}{c_i} = \frac{1}{c_i} \left(\sum_{j < h} a_{ij} y_j - \sum_{j \in P_i} a_{ij} y_j \right) > 5 \log k,$$

where the inequality is because we are in case b(ii) and $\sum_{j \in P_i} a_{ij} y_j \leq (5 \log k) \cdot c_i$ by Lemma 4.

Finally, when doing the primal/dual update steps for constraint h, the value of x_i just before this must have been $\chi_i' < 1/a_{ih}$ (otherwise constraint h would have already been satisfied just by variable x_i). And χ_i is at least $\sum_{j \in P_i} \frac{\gamma_{ij} t_j}{k \cdot a_{ij}}$, by the first calculations. And $\chi_i'/\chi_i \geq 2k^2$ by (2.7). Putting these together gives

$$\sum_{j \in P_i} \frac{\gamma_{ij} t_j}{k \cdot a_{ij}} \leq \frac{1}{2k^2} \cdot \frac{1}{a_{ih}},$$

and hence the claim.   ∎

Lemma 6 and Lemma 2 imply that the dual increase is at least $1/4$ the primal increase, and Lemma 5 implies we have an $O(\log k)$-feasible dual, implying the following theorem:

Theorem 1. *Algorithm I is an $O(\log k)$-competitive online algorithm for covering linear programs without upper-bound constraints, where k is the row-sparsity of the constraint matrix.*

3   The Online Algorithm for CIPs

We now want to solve CLPs with variable upper bounds, en route to solving general CIPs of the form (IP1). However, it is well-known that when we have variable upper-bounds, the natural relaxation has a large integrality gap even with a single constraint.[2] Hence, Carr et al. [9] suggested adding the *knapsack cover* (KC) inequalities—defined below—to reduce the integrality gap significantly. In this section, we first show how to extend Algorithm I to get an $O(\log k)$-competitive algorithm for the natural CLP relaxation (with upper bounds) where we also

---

[2] The trivial CIP $\min\{x_1 \mid M x_1 \geq 1\}$ has integrality gap M, no upper bounds needed. However, if we truncate the a_{ij}s to be at most 1 (which is the right-hand-side value), and we have no upper bound constraints, this gap disappears. Introducing upper bounds brings back large integrality gaps, as the example $\min\{x_1 \mid x_1 + (1 - \epsilon)x_2 \geq 1, x_2 \leq 1\}$ shows, which has an integrality gap of $1/\epsilon$.

satisfy some suitable KC inequalities. Next, we round (in an online fashion) such a fractional solution to get a randomized $O(\log \ell \cdot \log k)$-competitive online algorithm for general k-row-sparse and ℓ-column-sparse CIPs.

Knapsack Cover Inequalities. Given a CIP of the form (IP1), the KC-inequalities for a particular covering constraint $\sum_{i \in [n]} a_{ij} x_i \geq 1$ are defined as follows: for *any* subset $H \subseteq [n]$ of variables, the maximum possible contribution of the variables in H to the constraint is $a_j(H) := \sum_{i \in H} a_{ij} u_i$, and if $a_j(H) < 1$ then at least a contribution of $1 - a_j(H)$ must come from variables $[n] \setminus H$. Moreover, in any *integral solution* $\mathbf{x}$, since each positive variable x_i is at least one, we get the inequality:

$$\sum_{i \in [n] \setminus H} \min\{a_{ij}, 1 - a_j(H)\} \cdot x_i \quad \geq \quad 1 - a_j(H) \qquad (3.8)$$

Since (3.8) is not be true for an arbitrary *fractional* solution satisfying $\sum_{i \in [n]} a_{ij} x_i \geq 1$, we add this additional constraint to the LP, for each original constraint j and $H \subseteq [n]$ where $a_j(H) < 1$. There are exponentially many such KC-inequalities, and it is not known how to separate exactly over these in poly-time[3]. But as in previous works [9,12,5], the randomized rounding algorithm just needs us to enforce one specific KC-inequality for each constraint j—namely for the set $H_j := \{i \in [n] \mid x_i \geq \tau \cdot u_i\}$ with some suitable threshold $\tau > 0$. We call this the "special" KC-inequality for constraint j.

3.1   Fractional Solution with Upper Bounds and KC-inequalities

In extending Algorithm I from the previous section to also handle "box constraints" (those of the form $0 \leq x_i \leq u_i$), and the associated KC-inequalities, the high-level idea is to create a "wrapper" procedure around Algorithm I which ensures these new inequalities: when a constraint $\sum_{i \in T_j} a_{ij} x_i \geq 1$ arrives, we start to apply the primal update step from Algorithm I. Now if some variable x_p gets "close" to its upper bound u_p, we could then consider setting $x_p = u_p$, and feeding the new inequality $\sum_{i \in T_j \setminus p} a_{ij} x_i \geq 1 - a_{pj} u_p$ (or rather, a knapsack cover version of it) to Algorithm I, and continuing. Implementing this idea needs a little more work. For the rest of the discussion, $\tau \in (0, \frac{1}{2})$ is a threshold fixed later.

Suppose we want a solution to:

$$(IP) \quad \min \left\{ \sum_i c_i x_i \mid \sum_{i \in S_j} a_{ij} x_i \geq 1 \ \forall j \in [m], \ 0 \leq x_i \leq u_i, x_i \in \mathbb{Z} \ \forall i \in [n], \right\}$$

where constraint j has $|S_j| \leq k$ non-zero entries. The natural LP relaxation is:

$$(P) \quad \min \left\{ \sum_i c_i x_i \mid \sum_{i \in S_j} a_{ij} x_i \geq 1 \ \forall j \in [m], \ 0 \leq x_i \leq u_i \ \forall i \in [n] \right\}$$

---

[3] KC-inequalities can be separated in pseudo-polynomial time via a dynamic program for the knapsack problem.

We obtain an online algorithm to find a feasible fractional solution to this LP relaxation (P), along with some additional KC-inequalities. This algorithm maintains a vector $\mathbf{x} \in \mathbb{R}^n$ that need not be feasible for the covering constraints in (P). However $\mathbf{x}$ implicitly defines the "real solution" $\overline{\mathbf{x}} \in \mathbb{R}^n$ as follows:

$$\overline{x}_i = \begin{cases} x_i & \text{if } x_i < \tau u_i \\ u_i & \text{otherwise} \end{cases}, \qquad \forall i \in [n]$$

Let $\mathbf{x}^{(j)}$ and $\overline{\mathbf{x}}^{(j)}$ denote the vectors immediately after the jth constraint to (IP) has been satisfied. Due to lack of space, the following theorem is proved in [11].

Theorem 2. *Given the constraints of the CIP (IP) online, there is an algorithm that produces $\mathbf{x}$ (and hence $\overline{\mathbf{x}}$) satisfying the following:*

(i) The solution $\overline{\mathbf{x}}$ is feasible for (P).
(ii) The cost $\sum_{i=1}^{n} c_i \cdot x_i = O(\log k) \cdot \mathrm{opt}_{IP}$.
(iii) For each $j \in [m]$ let $H_j = \{i \in [n] \mid x_i^{(j)} \geq \tau \cdot u_i\}$ and $a_j(H_j) = \sum_{r \in H_j} a_{rj} u_r$. Then the solution $\mathbf{x}^{(j)}$ satisfies the KC-inequality corresponding to constraint j with the set H_j, i.e., if $a_j(H_j) < 1$ then:

$$\sum_{i \in S_j \setminus H_j} \min\{a_{ij}, 1 - a_j(H_j)\} \cdot x_i^{(j)} \geq 1 - a_j(H_j).$$

Furthermore, the vectors $\mathbf{x}$ and $\overline{\mathbf{x}}$ are non-decreasing over time.

Again, the value of row-sparsity k is not required in advance—the algorithm just uses the current estimate as before.

3.2   Online Rounding

We now complete the algorithm for CIPs by showing how to round the online fractional solution generated by Theorem 2 also in an online fashion. This rounding algorithm also does randomized rounding on the incremental change like in [1], but to get a loss of $O(\log \ell)$ instead $O(\log m)$, we use the method of randomized rounding with alterations [3,17]. Recall $\ell \leq m$ is the column-sparsity of the constraint matrix A—the maximum number of constraints any variable x_i participates in. (The $O(\log \ell)$ bound for offline CIPs given by [18,12] uses a derandomization of the Lovász Local Lemma via pessimistic estimators, and is not applicable in the online setting.)

Given that the constraints of a CIP arrive online, we run the algorithm from Theorem 2 to maintain vectors $\mathbf{x}$ and $\overline{\mathbf{x}}$. For this section, we set the threshold τ to $\frac{1}{8} \cdot \frac{1}{\log \ell}$. Before any constraints arrive, pick a uniformly random value $\rho_i \in [0,1]$ for each variable $i \in [n]$—this is the only randomness used by the algorithm. We will maintain an integer solution $\mathbf{X} \in \mathbb{Z}_{\geq 0}^n$; again let $\mathbf{X}^{(j)}$ denote this solution right after primal constraint j has been satisfied. We start off with $\mathbf{X}^{(0)} = \mathbf{0}$. When the jth constraint arrives and the (fractional) x_i values have been increased in response to this constraint, we do the following.

1. Define the "rounded unaltered" solution:

$$Z_i = \begin{cases} 0 & \text{if } x_i < \tau\rho_i \\ \lceil x_i/\tau \rceil & \text{if } \tau\rho_i \le x_i < \tau u_i \ , \qquad \forall i \in [n]. \\ u_i & \text{if } x_i \ge \tau u_i \end{cases}$$

2. *Maintain monotonicity.* Define:

$$X_i^{new} = \max\{X_i^{(j-1)},\ Z_i\}, \qquad \forall i \in [n].$$

Observe that this rounding ensures that $X_i \in \{0, 1, \ldots, u_i\}$ for all $i \in [n]$.

3. *Perform potential alterations.* If we are unlucky and the arriving constraint j is not satisfied by $\mathbf{X}^{new}$, we increase $\mathbf{X}^{new}$ to cover this constraint j as follows. Let $H_j := \{i \in [n] \mid x_i^{(j)} \ge \tau \cdot u_i\}$ be the frozen variables in the fractional solution; note that $Z_i = u_i$ for all $i \in H_j$, so these variables cannot be increased. Recall that $a_j(H_j) := \sum_{r \in H_j} a_{rj} \cdot u_r$. Since constraint j is not satisfied, $a_j(H_j) < 1$ and the algorithm performs the following alteration for constraint j. Consider the residual constraint on variables $[n] \setminus H_j$ after applying the KC-inequality on H_j, i.e.

$$\sum_{i \in [n] \setminus H_j} \min\{a_{ij},\, 1 - a_j(H_j)\} \cdot w_i \ \ge \ 1 - a_j(H_j).$$

Set $\bar{a}_{ij} = \min\left\{1, \frac{a_{ij}}{1-a_j(H_j)}\right\}$ for all $i \in [n] \setminus H_j$. Consider the following covering knapsack problem:

$$\min\ \sum_{i \in [n] \setminus H_j} c_i \cdot w_i \qquad\qquad (IP_K)$$

$$\text{subject to:}\ \ \sum_{i \in [n] \setminus H_j} \bar{a}_{ij} \cdot w_i \ge 1$$

$$0 \le w_i \le u_i, \qquad \forall i \in [n] \setminus H_j$$

$$w_i \in \mathbb{Z}, \qquad \forall i \in [n] \setminus H_j$$

Note that there is only one covering constraint in this problem. Let W denote an approximately optimal integral solution obtained by the natural greedy algorithm. It is clear that W satisfies the residual constraint j on variables $[n] \setminus H_j$. Define $\mathbf{X}^{(j)}$ as follows.

$$X_i^{(j)} = \begin{cases} X_i^{new} & \text{for } i \in H_i \\ \max\{X_i^{new}, W_i\} & \text{for } i \in [n] \setminus H_j \end{cases}$$

This completes the description of the algorithm. By construction, it outputs a feasible integral solution to the constraints so far, so it remains to bound its expected cost.

Remark: This algorithm does not require knowledge of the final column-sparsity ℓ in advance. At each step, we use the current value of ℓ. Notice that this only affects τ and the definition of $\mathbf{Z}$. However, for fixed values of x_i and ρ_i (any

$i \in [n]$) the value of Z_i is non-decreasing with ℓ: so vector $\mathbf{Z}$ is monotone over time (since ℓ is non-decreasing). We also require a slightly more general version of Theorem 2 where we have multiple thresholds $\tau_1 \leq \tau_2 \leq \cdots \leq \tau_m$ and replace τ by τ_j in condition (iii). This extension is straightforward and details are omitted.

Cost of Z. Consider the rounding algorithm immediately after all m constraints have been satisfied. If $x_i/\tau \in [0,1]$, then $\mathbb{E}[Z_i] = \Pr[\rho_i \leq x_i/\tau] = x_i/\tau$; if $x_i/\tau \geq 1$, then $Z_i \leq \lceil x_i/\tau \rceil \leq 2x_i/\tau$ with probability 1. Hence:

$$\mathbb{E}\left[\sum_{i=1}^n c_i \cdot Z_i\right] \leq (2/\tau)\sum c_i x_i = O(\log k \cdot \log \ell) \cdot \mathsf{opt}_{IP},$$

where we use $1/\tau = O(\log \ell)$, and Theorem 2(ii) to bound $\sum_i c_i x_i$.

Cost of X − Z. To account for $\mathbf{X} - \mathbf{Z}$, we need to bound the expected cost of any alterations. In the sequel, let ℓ_j, k_j and τ_j denote the respective values of ℓ, k and τ at the arrival of constraint j. When j is clear from context we will drop the subscript.

Recall that $H_j := \{i \in [n] \mid x_i^{(j)} \geq \tau_j \cdot u_i\}$ are the frozen variables in the fractional solution after handling constraint j, and note $Z_i = u_i$ for $i \in H_j$. Define $A_j := \{i \in [n] \mid x_i^{(j)} < \tau_j\}$. Note that the randomness only plays a role in the values of $\{Z_i \mid i \in A_j\}$, since all variables in $[n] \setminus A_j$ deterministically are set to $Z_i = \min\left\{\lceil x_i^{(j)}/\tau_j \rceil, u_i\right\}$. Let $\mathcal{E}_j$ denote the event that an alteration was performed for constraint j. The event $\mathcal{E}_j$ occurs exactly when $\sum_{i \in [n]} a_{ij} \cdot X_i^{new} < 1$. Since variables $r \in H_j$ have $X_r^{new} = Z_r = u_r$ with probability 1, event $\mathcal{E}_j$ is the same as $a_j(H_j) < 1$ (which is a deterministic condition) and $\sum_{i \in [n] \setminus H_j} a_{ij} \cdot X_i^{new} < 1 - a_j(H_j)$.

Lemma 7. *The probability of an alteration for constraint j is* $\Pr[\mathcal{E}_j] \leq \frac{1}{\ell_j^2}$.

Proof. Let $b = 1 - a_j(H_j)$, for $\mathcal{E}_j$ to occur we have $b > 0$. Set $\bar{a}_{ij} = \min\{a_{ij}/b, 1\}$ for $i \in [n] \setminus H_j$. Now since $\mathbf{Z} \leq \mathbf{X}$ and both are integer-valued, $\Pr[\mathcal{E}_j]$

$$= \Pr\left[\sum_{i \in [n] \setminus H_j} a_{ij} \cdot X_i^{new} < b\right] \leq \Pr\left[\sum_{i \in [n] \setminus H_j} a_{ij} \cdot Z_i < b\right] = \Pr\left[\sum_{i \in [n] \setminus H_j} \bar{a}_{ij} \cdot Z_i < 1\right].$$

Theorem 2(iii) guarantees that $\sum_{i \in [n] \setminus H_j} \bar{a}_{ij} \cdot x_i^{(j)} \geq 1$. Among $i \in [n] \setminus H_j$,

- $Z_i = \lceil x_i^{(j)}/\tau \rceil$ deterministically for $i \in [n] \setminus (H_j \cup A_j)$, and
- $Z_i \in \{0,1\}$ with $\mathbb{E}[Z_i] = x_i^{(j)}/\tau$ independently for $i \in A_j$.

So $\mathbb{E}\left[\sum_{i \in [n] \setminus H_j} \bar{a}_{ij} \cdot Z_i\right] \geq \frac{1}{\tau}$. Now Chernoff bound implies for a collection of $[0,1]$-valued independent random variables, that the probability of their sum being less than $\tau = 1/(8 \log \ell_j)$ times their expectation is at most $1/\ell_j^2$. ∎

Lemma 8. *Conditioned on $\mathcal{E}_j$, the cost of incrementing $\mathbf{X}^{new}$ to $\mathbf{X}^{(j)}$ is at most* $36 \sum_{i \in S_j} c_i \cdot x_i^{(j)}$; *here $S_j \subseteq [n]$ are the non-zero columns in constraint j.*

Proof. The fractional solution $\mathbf{x}^{(j)}$ satisfies the KC inequality for set H_j, by Theorem 2(iv). In particular, setting $w_i' = x_i^{(j)}$ for $i \in S_j \setminus H_j$ (and zero otherwise) gives a feasible *fractional* solution to the LP relaxation of the covering knapsack subproblem (IP_K). It suffices to show that the greedy integral solution W to (IP_K) costs $36 \sum_{i \in S_j} c_i \cdot w_i'$. It is crucial that $w_i' \leq \tau \cdot u_i < u_i/2$ for all $i \in [n] \setminus H_j$, as in general the integrality gap due to relaxing (IP_K) is unbounded.

The greedy algorithm orders columns $i \in [n] \setminus H_j$ in non-decreasing $c_i/\bar{a}_{ij}$ order, and increases W_i variables integrally (up to their u_is) until $\sum_i \bar{a}_{ij} \cdot W_i \geq 1$. Since all $\bar{a}_{ij} \leq 1$, it is easy to show that this algorithm achieves a 2-approximation for covering knapsack (IP_K).

To complete the proof, we show the optimal integral solution to (IP_K) costs at most $18 \sum_{i \in S_j} c_i \cdot w_i'$: we give a rounding algorithm to obtain an integral solution W' from w' with only a factor 18 increase in cost. Set $W_i' \sim \text{Binom}(u_i, 2w_i'/u_i)$ for all $i \in [n] \setminus H_j$—this definition is valid since $w_i' \leq u_i/2$. Clearly W' always satisfies the upper bounds u_i and has expected cost $2\mathbf{c} \cdot \mathbf{w'}$. Moreover, each W_i' is a binomial r.v. and $\bar{a}_{ij} \leq 1$, so $\sum_i \bar{a}_{ij} \cdot W_i'$ can be viewed as a sum of independent $[0,1]$-valued random variables. The expectation $\mathbb{E}\left[\sum_i \bar{a}_{ij} \cdot W_i'\right] \geq 2$, so a Chernoff bound gives $\Pr\left[\sum_i \bar{a}_{ij} \cdot W_i' < 1\right] \leq 8/9$. Using Markov's inequality, $\Pr\left[\mathbf{c} \cdot \mathbf{W'} > 18\mathbf{c} \cdot \mathbf{w'}\right] < 1/9$. So with positive probability, W' satisfies (IP_K) and costs at most $18\mathbf{c} \cdot \mathbf{w'}$, showing that $\text{Opt}(IP_K)$ is at most this cost. ∎

Thus the total expected cost of alterations after m constraints is:

$$\sum_{j=1}^{m} \Pr[\mathcal{E}_j] \cdot 36 \sum_{i \in S_j} c_i \cdot x_i^{(j)} \leq 36 \sum_{j=1}^{m} \frac{1}{\ell_j^2} \cdot \sum_{i \in S_j} c_i \cdot x_i^{(j)} \leq 36 \sum_{i=1}^{n} c_i \cdot x_i^{(m)} \left(\sum_{j : i \in S_j} \frac{1}{\ell_j^2} \right)$$

$$\leq 36 \sum_{i=1}^{n} c_i \cdot x_i^{(m)} \left(\frac{1}{1^2} + \frac{1}{2^2} + \cdots + \frac{1}{\ell_j^2} \right) \leq 9\pi^2 \sum_{i=1}^{n} c_i \cdot x_i^{(m)}.$$

The second inequality uses the monotonicity of the fractional solution $\mathbf{x}$, and the third inequality uses that for any $i \in [n]$, the value ℓ_j is at least q upon arrival of the qth constraint containing variable i.

Combining the expected cost of $O(\mathbf{c} \cdot \mathbf{x})$ for the alterations with the expected cost of $O(\log \ell) \cdot (\mathbf{c} \cdot \mathbf{x})$ for the initial rounding, and Theorem 2(ii), we get the main result for this section:

Theorem 3. *There is an $O(\log k \cdot \log \ell)$-competitive randomized online algorithm for covering integer programs with row-sparsity k and column-sparsity ℓ.*

Again, we note that the algorithm does not assume knowledge of the eventual k or ℓ values; it works with the current values after each constraint. Furthermore, the algorithm clearly does not need the entire cost function in advance: it suffices to know the cost coefficient c_i of each variable i at the arrival time of the first constraint that contains i.

References

1. Alon, N., Awerbuch, B., Azar, Y., Buchbinder, N., Naor(Seffi), J.: The online set cover problem. In: STOC 2003, pp. 100–105 (2003)
2. Alon, N., Awerbuch, B., Azar, Y., Buchbinder, N., Naor(Seffi), J.: A general approach to online network optimization problems. ACM Trans. Algorithms 2(4), 640–660 (2006)
3. Alon, N., Spencer, J.: The Probabilistic Method. Wiley-Interscience, New York (2008)
4. Bansal, N., Buchbinder, N., Naor(Seffi), J.: A primal-dual randomized algorithm for weighted paging. In: FOCS 2007, pp. 507–517 (2007)
5. Bansal, N., Buchbinder, N., Naor(Seffi)., J.: Randomized competitive algorithms for generalized caching. In: STOC 2008, pp. 235–244. ACM, New York (2008)
6. Bansal, N., Korula, N., Nagarajan, V., Srinivasan, A.: On k-Column Sparse Packing Programs. In: Eisenbrand, F., Shepherd, F.B. (eds.) IPCO 2010. LNCS, vol. 6080, pp. 369–382. Springer, Heidelberg (2010)
7. Buchbinder, N., Naor(Seffi)., J.: The design of competitive online algorithms via a primal-dual approach. Found. Trends Theor. Comput. Sci. 3(2-3), 93–263 (2007)
8. Buchbinder, N., Naor(Seffi)., J.: Online primal-dual algorithms for covering and packing. Math. Oper. Res. 34(2), 270–286 (2009)
9. Carr, R.D., Fleischer, L.K., Leung, V.J., Phillips, C.A.: Strengthening integrality gaps for capacitated network design and covering problems. In: SODA 2000, pp. 106–115 (2000)
10. Chakrabarty, D., Grant, E., Könemann, J.: On Column-Restricted and Priority Covering Integer Programs. In: Eisenbrand, F., Shepherd, F.B. (eds.) IPCO 2010. LNCS, vol. 6080, pp. 355–368. Springer, Heidelberg (2010)
11. Gupta, A., Nagarajan, V.: Approximating sparse covering integer programs online. CoRR, abs/1205.0175 (2012)
12. Kolliopoulos, S.G., Young, N.E.: Approximation algorithms for covering/packing integer programs. J. Comput. Syst. Sci. 71(4), 495–505 (2005)
13. Korman, S.: On the use of randomness in the online set cover problem. M.Sc. thesis, Weizmann Institute of Science (2005)
14. Koufogiannakis, C., Young, N.E.: Greedy Δ-Approximation Algorithm for Covering with Arbitrary Constraints and Submodular Cost. In: Albers, S., Marchetti-Spaccamela, A., Matias, Y., Nikoletseas, S., Thomas, W. (eds.) ICALP 2009, Part I. LNCS, vol. 5555, pp. 634–652. Springer, Heidelberg (2009)
15. Pritchard, D., Chakrabarty, D.: Approximability of sparse integer programs. Algorithmica 61(1), 75–93 (2011)
16. Srinivasan, A.: Improved approximation guarantees for packing and covering integer programs. SIAM J. Comput. 29(2), 648–670 (1999)
17. Srinivasan, A.: New approaches to covering and packing problems. In: SODA 2001, pp. 567–576 (2001)
18. Srinivasan, A.: An extension of the Lovász Local Lemma, and its applications to integer programming. SIAM J. Comput. 36(3), 609–634 (2006)
19. Young, N.E.: The k-server dual and loose competitiveness for paging. Algorithmica 11(6), 525–541 (1994)

Streaming and Communication Complexity of Clique Approximation

Magnús M. Halldórsson[1,*], Xiaoming Sun[2,**],
Mario Szegedy[3,***], and Chengu Wang[4,†]

[1] ICE-TCS, School of Computer Science, Reykjavik University, Iceland
[2] Institute of Computing Technology, Chinese Academy of Sciences
[3] Department of Computer Science, Rutgers University, New Jersey
[4] Institute for Interdisciplinary Information Sciences, Tsinghua University, Beijing

Abstract. We consider the classic clique (or, equivalently, the independent set) problem in two settings. In the streaming model, edges are given one by one in an adversarial order, and the algorithm aims to output a good approximation under space restrictions. In the communication complexity setting, two players, each holds a graph on n vertices, and they wish to use a limited amount of communication to distinguish between the cases when the union of the two graphs has a low or a high clique number. The settings are related in that the communication complexity gives a lower bound on the space complexity of streaming algorithms.

We give several results that illustrate different tradeoffs between clique separability and the required communication/space complexity under randomization. The main result is a lower bound of $\Omega(\frac{n^2}{r^2 \log^2 n})$-space for any r-approximate randomized streaming algorithm for maximum clique. A simple random sampling argument shows that this is tight up to a logarithmic factor. For the case when $r = o(\log n)$, we present another lower bound of $\Omega(\frac{n^2}{r^4})$. In particular, it implies that any constant approximation randomized streaming algorithm requires $\Omega(n^2)$ space, even if the algorithm runs in exponential time. Finally, we give a third lower bound that holds for the extremal case of $s - 1$ vs. $\mathcal{R}(s) - 1$, where $\mathcal{R}(s)$ is the s-th Ramsey number. This is the extremal setting of clique numbers that can be separated. The proofs involve some novel combinatorial structures and sophisticated combinatorial constructions.

1 Introduction

Streaming for cliques. In the *streaming model* for graph problems, edges are presented sequentially in the form of a data stream, and the objective is to

---

* Research partially supported by Icelandic Research Fund grant 90032021.
** Research partially supported by the National Natural Science Foundation of China Grant 61170062, 61061130540, and the National Basic Research Program of China Grant 2011CBA00300, 2011CBA00301.
*** Research partially supported by NSF grant CCF-0832787.
† Research partially supported by the National Basic Research Program of China Grant 2011CBA00300, 2011CBA00301, the National Natural Science Foundation of China Grant 61033001, 61061130540, 61073174.

A. Czumaj et al. (Eds.): ICALP 2012, Part I, LNCS 7391, pp. 449–460, 2012.
© Springer-Verlag Berlin Heidelberg 2012

compute a good near-optimal solution using working space significantly less than the size of the data stream. The motivation for streaming comes from practical applications of managing massive data sets such as, e.g., real-time network traffic, on-line auctions, and telephone call records. These data sets are huge and arrive at a very high rate, making it impossible to store more than a small part of the input.

We consider the space requirements of finding or approximating the maximum clique in a graph, or equivalently the maximum independent set. We assume that the graph is given as a stream of edges, where the algorithm can view the stream several times. In this paper, we assume the algorithm only views the stream a constant number of times. More generally, we treat the following *gap* problem: given a graph G and numbers U and L with $L \leq U$, decide whether G contains a U-clique, or contains no $(L + 1)$-clique. When the clique number is greater than L or less than U, the algorithm can answer arbitrarily. Here, U and L can be functions of the order n of the input graph.

Several graph problems have been considered in the streaming setting, including bipartite matching (weighted and unweighted cases) [14], diameter and shortest paths [14,15], min-cut [1], and graph spanners [15]. Except for certain counting problems, such as counting triangles [5], cycles [24], $K_{3,3}$ bipartite cliques [9] and small graph minors [8], these use $n \cdot polylog(n)$ space.

Limited attention has been given to streaming algorithms for NP-hard problems; exceptions include Max-Cut [1,27] and certain clustering problems (e.g., [19]). In [17], the independent set problem in graphs and hypergraphs was considered, but with the primary focus on the fine-grained space requirements of matching the Turán bound on sparse (hyper)graphs. Some additional upper bounds are given in [23], but with a focus on general hypergraphs. We are not aware of any lower bounds for the space complexity of computing any classical NP-hard graph parameter like clique number (except for max-cut [27]).

The MAX-CLIQUE problem, and its sister the independent set problem, is one of the central problems in optimization, and graph theory. For instance, the algorithm textbook of Kleinberg and Tardos uses variations of the independent set problem as a common theme for the whole book. It has long been one of the cornerstones of complexity theory, including monotone circuit complexity [3], decision tree complexity [7], fixed-parameter intractability [10], and interactive proofs and approximation hardness [18]. The current best intractability bound for MAX-CLIQUE is $n/2^{(\log n)^{3/4}+\epsilon}$ [21], and the best approximation result is $O(n(\log \log n)^2 / \log^3 n)$ [13].

Communication Complexity. Communication complexity, introduced by Yao [26], is a powerful tool to solve a variety of problems in areas as disparate as VLSI design, decision trees, data structures, and circuit complexity [22]. It is a game between two parties, Alice and Bob, with unlimited computing power, that want to compute the value of a function $f : X \times Y \mapsto \{0, 1\}$. Alice only knows $x \in X$, while Bob only knows $y \in Y$. To perform the computation, they are allowed to send messages to each other in order to converge on a shared output $P(x, y)$. In a randomized protocol, Alice and Bob toss coins, and the messages can depend on the coin flips. We say a randomized (deterministic) protocol P computes f if $\Pr[P(x,y) = f(x,y)] \geq 2/3$ $(P(x,y) = f(x,y))$ for any input x, y, and define the randomized (deterministic) *communication complexity* $R_{1/3}(f)$ $(D(f))$ to be the number of bits communicated for the worst input under the best randomized (deterministic) protocol computing f, respectively. Here, $1/3$ refers to the error rate. Since a deterministic protocol is a randomized protocol, $D(f) \geq R_{1/3}(f)$.

Our Results. We give several constructions that imply communication lower bounds for clique separation, resulting in equivalent lower bounds for the space complexity of streaming algorithms. The constructions differ in their range of parameters U and L, as well as the strengths of the lower bounds.

The results are summarized in the table. $R_{1/3}(\text{CLIQUE-GAP}(U, L))$ denotes the randomized communication complexity to determine whether the clique number of the union of two graphs is at least U or at most L, and $\mathcal{R}(s)$ refers the s-th diagonal Ramsey number (see Sec. 2 for formal definitions).

Table 1. A summary of our results

U	L	$R_{1/3}(\text{CLIQUE-GAP}(U, L))$
r	$10 \cdot 2^{1/\epsilon} \log n$	$\Omega(n^2/r^2)$
r	s	$O(n^2/(r/s)^2)$
r	$2\sqrt{r} - 1$	$\Omega(n^2/r^2)$
$r = \mathcal{R}(s) - 1$	$s - 1$	$\Omega\left(\max\left(n/r, \dfrac{n^2}{r^3 \exp(10\sqrt{\log r \log \frac{n}{2r^2}})}\right)\right)$
$r = \mathcal{R}(s)$	$s - 1$	$O(1)$

The first two results in the table match up to a logarithmic factor. Thus, except for the case of very small or very large cliques, this gives a fairly precise characterization of what cliques can be separated. For smaller cliques, the bounds are still open to a large extent. The third result shows that any constant approximation requires quadratic space, which is a supplement to the first result when the clique number is a constant. Finally, the last two bounds give a sharp threshold within which we can separate cliques: constant space suffices below the threshold, while non-trivial and even superlinear space is necessary above the threshold.

We note that our results hold equally for the Max Independent Set problem. While the optimization and approximation of cliques and independent sets are equivalent in general graphs, the streaming problems are not identical since the stream is formed by edges and not non-edges. This distinction disappears in the communication problem, as well as in the sampling-based upper bounds.

The clique problem appears at first to be strongly related to the previously studied problem of counting triangles [5,6], and in fact, the known hardness of detecting triangles and short cycles in stream [5,15] yields a starting point for proving hardness of clique computation. Nevertheless, while a large clique implies many triangles, the converse is not true (viz. complete 3-partite graphs). Indeed, different arguments are needed for the clique problem.

While our hardness results involve reductions to the prototypical problem of set disjointness, our proofs involve some novel connections between Ramsey theory and additive combinatorics. Obtaining superlinear constant-pass lower bounds on graph problems via disjointness is often hampered by dependencies between edges. The use of designs and random partition to get around this here may be useful for proving such lower bounds for other graph problems in the semi-streaming model.

Outline of the Paper. We define the problems and notation formally in Section 2, and introduce our methodology in Section 3. The bulk of the paper is in Section 4, where we give several different space-approximation tradeoffs for the clique problem. Some upper bounds are given in Section 5. Some proofs of lemmas have been deferred to the full version.

2   Problem Definitions

A clique in a graph is a subset of mutually adjacent vertices. The MAX-CLIQUE problem is that of finding a clique of approximately maximum size. Let $\omega(G)$ denote the clique number of graph G. Let n denote the number of vertices of the graph input to MAX-CLIQUE. A t-subgraph refers to a subgraph induced by t vertices. The Ramsey number $\mathcal{R}(r)$ is the smallest n so that for any graph G of size n, either G or its complement, $\overline{G}$, has a r-clique. By the classic results of [11,12], $\mathcal{R}(r) = 2^{\theta(r)}$, and in particular $\sqrt{2}^r < \mathcal{R}(r) < 4^r$.

Let $[n] = \{1, 2, \ldots, n\}$. An edge stream is formally defined to be a sequence $\langle a_1, a_2, \ldots, a_m \rangle$, where $a_j \in \binom{[n]}{2}$, inducing the undirected graph $G = (V, E)$ on n vertices with $V = [n]$ and $E = \{a_j : j \in [m]\}$. Each edge may appear more than once. Only in Sec. 4.1 do we need to allow edges to appear more than once (specifically, twice), and only when $r > \sqrt{n}$.

Set disjointness, denoted DISJ, is a communication complexity problem where Alice and Bob hold two subsets, x and y, of $[N]$, respectively, and they want to determine whether the intersection of their subsets is empty. Improving a result in [4], Kalyanasundaram and Schnitger [20] proved that $R_{1/3}(\text{DISJ}) = \Omega(N)$.

The clique gap problem is the communication complexity problem for clique approximation, where Alice and Bob hold two subgraphs $G_A = \langle V_n, E_A \rangle$ and $G_B = \langle V_n, E_B \rangle$ and they want to approximately determine the clique number of the combined graph $G_A \cup G_B = \langle V_n, E_A \cup E_B \rangle$. We define the value of the function CLIQUE-GAP(U, L) to be 1 if $\omega(G) \geq U$, 0 if $\omega(G) \leq L$, and arbitrary (0 or 1) otherwise.

The communication complexity of a decision problem is closely related to the space complexity of the problem, in that the former gives a lower bound for the latter. Namely, for any decision problem Π, it holds that space$_{1/3}(\Pi) \geq R_{1/3}(\Pi)$, where space$_{1/3}(\Pi)$ denotes the space complexity of a randomized streaming algorithm that answers correctly with at least 2/3-probability on any instance of Π. This holds, up to constant factors, even if we allow the streaming algorithm passes through the input constant times.

3   Our Methodology

Reduction from the set disjointness problem is generally the method of choice for proving communication complexity lower bounds for graph problems. Yet, to come up with reductions with near-optimal parameters to CLIQUE-GAP(U, L) involves a number of combinatorial challenges.

Our starting point was the following reduction from the set disjointness problem with parameter $N = (n/4)^2$ to CLIQUE-GAP$(4, 2)$:

For any input of set disjointness problem, where Alice holds $x \in \{0, 1\}^{(n/4)^2}$ and Bob holds $y \in \{0, 1\}^{(n/4)^2}$, we construct an input for the clique problem as follows. We denote the vertices by $\{v_{i,j} | i = 1, 2, 3, 4; j = 1, 2, 3, \cdots, n/4\}$. Alice has edges $(v_{1,j}, v_{3,j'})$ and $(v_{2,j}, v_{4,j'})$ if $x[j, j'] = 1$. Bob has edges $(v_{1,j}, v_{4,j'})$ and $(v_{2,j}, v_{3,j'})$ if $y[j, j'] = 1$. Finally, both of them have the edges $(v_{1,j}, v_{2,j})$ and $(v_{3,j}, v_{4,j})$, for $j = 1, 2, ..., n/4$. In this construction, the graph has a 4-clique if x intersects with y, and the clique number is only 2 if x doesn't intersect with y.

The above construction can be viewed as an extension of constructions from [5,15] on detecting triangles in streams. This argument can, however, not be extended further: proving an $\Omega(n^2)$ lower bound for CLIQUE-GAP$(5, 2)$ is impossible because of the counting version of the Szemerédi's Regularity Lemma [25]. We will detail the reason and give a weaker lower bound for CLIQUE-GAP$(5, 2)$ in Sec. 4.2. This obstacle shows that some non-trivial combinatorics lies beneath our problem. We overcome this and other obstacles for different U, L pairs by applying different arguments, and by exploiting properties of the worst case distribution for the set disjointness problem. Along the way, we create some interesting combinatorial structures, such as the one in Lemma 1, which we could not find elsewhere in the literature.

4   Lower Bounds

We reduce the set disjointness problem to the approximate clique determination problem, thereby obtaining lower bounds on space for streaming algorithms approximating cliques. We give several constructions that apply to different combinations of the parameters U and L.

The structure of the arguments is as follows. Given an instance (x, y) of DISJ, we form a graph $\tilde{G}$ that is a packing of "gadgets", or clique subgraphs, each corresponding to a single bitpair of the vectors x and y. Some of the edges of each gadget are reserved for Alice, and the remaining edges for Bob. The actual graphs G_A and G_B handed to Alice and Bob are subgraphs of $\tilde{G}$, where Alice (Bob) receives her (his) edges of gadget i only if the corresponding bit x_i (y_i) is set, respectively. This ensures that if $x_i = y_i = 1$ – the case of a positive set intersection instance – then the corresponding gadget is a clique, yielding a positive answer to the clique separation problem. The main issue is to ensure that for negative instances, the clique size of the whole graph $G_A \cup G_B$ remains small.

We present three constructions. The first gives optimal space lower bounds, up to logarithmic factor, for all but very large clique numbers. The second yields weaker lower bounds, but holds for sub-logarithmic values of L. The third one gives optimal $\Omega(n^2)$-space lower bound for the case of constant clique sizes.

4.1 r vs. $\log n$

Theorem 1. *For* $0 < \epsilon < 1$, $r = n^{1-\epsilon}$ *and* $s = 100 \cdot 2^{2/\epsilon} \log n$, *it holds that* $R_{1/3}(\text{CLIQUE-GAP}(r, s)) = \Omega(n^{2\epsilon})$. *Thus, for some constant c, any randomized streaming algorithm for* MAX-CLIQUE *with approximation ratio* $\frac{c \cdot r}{\log n}$ *requires* $\Omega(n^2/r^2)$ *space (when* $r = O(n^{1-\epsilon})$).

We reduce DISJ to CLIQUE-GAP(r, s) in such a way that positive instances will have clique-size r, while negative instances will be like the Erdös-Renyi random graphs $G_{n,p}$, and thus have clique-size $s = O(\log_{1/p} n)$ (we shall specify p later).

We construct optimal reductions (up to a factor of $\log n$) from the set-disjointness when $r = O(n^{1-\epsilon})$. At the heart of the reduction, there is a combinatorial lemma:

Lemma 1. *For every* $n > 2^{2/\epsilon}$ *and every* $r < n/2$, *there is a set system* $\mathcal{C}$ *on* $[n]$ *with* n^2/r^2 *sets of size* r *each, such that each pair of distinct points is covered by at most d sets from* $\mathcal{C}$, *where* $d = \lceil 2/\epsilon \rceil - 2$.

Proof. Let P be the largest prime with the property that $rP \leq n$. Then, $rP > n/2$, by Bertrand's postulate. We identify $[P]$ with GF_P performing all arithmetic modulo P. We also identify $[r]$ with an arbitrary subset of GF_P^d, and assume that there is an injective mapping $f : [r] \mapsto GF_P^d$ because $P^d \geq (\frac{n}{2r})^d > r$. For $(x, y) \in GF_P^2$ we define the set

$$C_{x,y} = \{(a_1, a_2, \ldots, a_d, a) \mid a = a_d x^d + \ldots + a_1 x - y \text{ and } (a_1, a_2, \ldots, a_d) \in f([r])\}.$$

Notice that $C_{x,y}$ has size exactly r, since given x and y the values of $a_1, a_2, \ldots, a_d$ determine the value of a. In particular, this implies that for two distinct points that $C_{x,y}$ covers, the first d coordinates are always different. Consider now two distinct points $(a_1, a_2, \ldots, a_d, a)$ and $(b_1, b_2, \ldots, b_d, b)$. If they are covered by the same $C_{x,y}$, we get that $a_d x^d + \ldots + a_1 x - y = a$ and $b_d x^d + \ldots + b_1 x - y = b$, implying that

$$(a_d - b_d)x^d + \ldots + (a_1 - b_1)x = a - b. \tag{1}$$

Notice that $C_{x,y}$ and $C_{x,y'}$ are disjoint whenever $y \neq y'$. Thus, if $C_{x,y}$ and $C_{x',y'}$ intersect in a point, and $(x, y) \neq (x', y')$, then it is necessary that $x \neq x'$. Thus, in particular, if there are $(x_1, y_1), \ldots, (x_{d+1}, y_{d+1})$ such that C_{x_i, y_i} cover the same two points, then $x_1, \ldots, x_{d+1}$ are all distinct, and Eqn. 1 holds for all $x_1, \ldots, x_{d+1}$. Since by our earlier remark $(a_1, a_2, \ldots, a_d) \neq (b_1, b_2, \ldots, b_d)$, we get a contradiction by discovering that a degree d polynomial (namely $(a_d - b_d)x^d + \ldots + (a_1 - b_1)x - a + b$) has $d + 1$ roots.

Our reduction from the set disjointness problem of size $N = n^2/r^2$ will be the following. First, we define N cliques (gadgets) of size r on n nodes, as shown in the above lemma. Let us denote the ith clique by C_i ($1 \le i \le N$). We then associate each edge of C_i to Alice or Bob with probability $1/2$ independently. We call the set of edges associated this way to Alice and Bob C_i^A and C_i^B, respectively. Note that it is possible that the same edge of the graph is associated to both Alice and Bob, since an edge may occur in up to d different C_is.

The graph G_A given to Alice consists of the edges in the union of those C_i^As, for which the bit x_i in the set disjointness problem is set to 1. Similarly, we give to Bob the graph G_B, which is the union of those C_i^Bs, for which the bit y_i is set to 1. Clearly, if $x_i = y_i = 1$ then the combined graph $G_A \cup G_B$ will contain all of C_i, and thus have clique size at least r.

We argue now that in the negative case, we can embed the resulting graph in an Erdös-Renyi random graph $G_{n,p}$ with edge probability $p = 1 - 1/2^d$.

Lemma 2. *For any negative instance (x, y) of* DISJ *on s bits ($x \cap y = \emptyset$), let q_1 be the probability that the graph $G_A \cup G_B$, generated by the above described randomized map of (x, y), contains an s-clique. Let q_2 be the probability that an Erdös-Renyi random graph, where each edge is drawn with probability $1 - 1/2^d$, contains an s-clique. Then $q_1 < q_2$.*

Proof. For an edge e in the graph $G_A \cup G_B$ generated by (x, y), we consider the set of cliques $\mathcal{C}_e = \{C_i | e \in C_i \text{ and } i \in x \cup y\}$. In the method we described above, we choose e in each clique in $\mathcal{C}_e$ with probability $1/2$ independently, and e appears in $G_A \cup G_B$ if e is chosen in any clique in $\mathcal{C}_e$. Thus, the probability that $G_A \cup G_B$ contains e is $1 - 1/2^{|\mathcal{C}_e|} \le 1 - 1/2^d$, because $|\mathcal{C}_e| \le |\{C_i | e \in C_i\}| \le d$ by Lemma 1. However, in the Erdös-Renyi random graph, each edge is chosen with probability $1 - 1/2^d$. Therefore, $G_A \cup G_B$ has sparser edges, and it has an s-clique with less probability. $\square$

*Proof (**Proof of Theorem 1**).* Given instance x, y to DISJ, we form and hand the graphs G_A and G_B to Alice and Bob, as expressed above. On positive instance, when $x_i = y_i = 1$, for some bit i, the corresponding subgraph in $G_A \cup G_B$ is an r-clique. On negative instances, $G_A \cup G_B$ is sparser than the Erdös-Renyi random graph $G_{n,p}$, with $p = 1 - 1/2^d$. As shown by Grimmett and McDiarmid [16], $\omega(G_{n,p}) \le 2 \log n / \log(1/p) + o(\log n) \le 2^{d+1} \log n + o(\log n)$, with high probability. The theorem now follows. $\square$

4.2 $\mathcal{R}(s) - 1$ vs. $s - 1$

When proving an $\Omega(n^2)$ lower bound for CLIQUE-GAP(5,2), the $s = 3$ case of CLIQUE-GAP($\mathcal{R}(s) - 1, s - 1$), we run into obstacles if we use the approach for CLIQUE-GAP(4, 2). To do so, we must pack $\Theta(n^2)$ 5-clique gadgets in a graph on n vertices. We then need to partition the $\binom{5}{2}$ edges into two parts, one for Alice and the other for Bob, such that each part has no triangles. In fact, the partition is unique up to a permutation, and it does not contain "hard-wired" edges like the gadget in the proof of CLIQUE-GAP(4, 2) does. Furthermore, we require more properties of the packing: all the gadgets are edge-disjoint and each triangle must lie fully within one gadget. The Triangle Removal Lemma, which can be proven from Szemerédi's Regularity Lemma [25], states that we can remove $o(n^2)$ edges from a graph containing $o(n^3)$ triangles to make it triangle-free. If we take one triangle from each gadget, these $\Theta(n^2)$ triangles are edge-disjoint and $o(n^2)$ edges do not suffice to destroy them all. Therefore, we cannot pack $\Theta(n^2)$ gadgets in a graph of size n.

Instead, we can prove the following result, using a different packing requirement.

Theorem 2. *For any* r, $R_{1/3}(\text{CLIQUE-GAP}(r, s - 1)) = \Omega\left(\frac{n^2}{r^3 \exp(10\sqrt{\log r \log \frac{n}{2r^2}})}\right)$, *where* $r = \mathcal{R}(s) - 1$.

For instance, this gives a $n^2 / \exp(O(\sqrt{\log n})) = n^{2-o(1)}$ lower bound for CLIQUE-GAP$(5, 2)$. That is the best we can hope for in the sense that CLIQUE-GAP$(6, 2)$ has a trivial upper bound, as we shall see in Section 5.

We shall use the following combinatorial structure and theorem of Alon and Shapira.

Definition 1 (h-Sum-Free). *[2] A set* $X \subseteq [n]$ *is called h-sum-free if for every three positive integers* $a, b, c \leq h$ *such that* $a + b = c$, *if* $x, y, z \in X$ *satisfy the equation* $ax + by = cz$, *then* $x = y = z$. *That is, whenever* $a + b = c$, *and* $a, b, c \leq h$, *the only solution to the equation that uses values from* X, *is one of the* $|X|$ *trivial solutions.*

Theorem 3. *[2] For every positive integer n, there exists an h-sum-free subset* $X \subseteq [n]$ *of size at least* $|X| \geq \frac{n}{e^{10\sqrt{\log h \log n}}} \doteq g(n, h)$.

We say that a set system $\mathcal{C} = \{C_i\}_i$ is *edge-disjoint* if any pair of points is contained in at most a single set, and that it is *triangle-free* if whenever $u, v \in C_i$, $v, w \in C_j$ and $w, u \in C_k$, for some $C_i, C_j, C_k \in \mathcal{C}$, then $C_i = C_j = C_k$.

Lemma 3. *For any n, there is an edge-disjoint triangle-free set system on $[n]$ with* $g(n/(2r^2), r) \cdot n/r = \Omega(n^2/(r^3 \exp(10\sqrt{\log r \log n/(2r^2)})))$ *sets of size r each.*

Proof. We first pick an r-sum-free set $Z \subseteq [\frac{n}{2r^2}]$ such that

$$|Z| = m \geq g(n/(2r^2), r) = \frac{n/(2r^2)}{\exp(10\sqrt{\log r \log \frac{n}{2r^2}})}.$$

Suppose $Z = \{z_1, \ldots, z_m\}$. For $i \in [m]$, let $S_i = (z_i r + 1) \cdot [r] = \{(z_i r + 1)a : a \in [r]\}$. We denote the set shift j from S_i by $S_i^{(j)}$, namely we define $S_i^{(j)} = S_i + jr$, for $i \in [m]$, $j \in [n/(2r)]$. Finally, we define the set family $\mathcal{C} = \{S_i^{(j)} | i \in [m], j \in [n/(2r)]\}$, and let $\tilde{G} = ([n], E)$, where $E = \{(u, v) | \exists S \in \mathcal{C}, u, v \in S\}$. It is clear that for each $S \in \mathcal{C}$, the subgraph on S induces an r-clique in $\tilde{G}$.

The lemma follows from the following two claims.

Claim. $\mathcal{C}$ is edge-disjoint, i.e., any $S_{i_1}^{(j_1)}$ and $S_{i_2}^{(j_2)}$ intersect in at most one element if $(i_1, j_1) \neq (i_2, j_2)$.

Proof. Suppose they have two common elements u and v. From $u, v \in S_{i_1}^{(j_1)}$, by definition we have $u = (z_{i_1}r + 1)b_1 + j_1 r$ and $v = (z_{i_1}r + 1)c_1 + j_1 r$, for some $b_1, c_1 \in [r]$. Similarly, there are $b_2, c_2 \in [r]$, such that $u = (z_{i_2}r + 1)b_2 + j_2 r$, and $v = (z_{i_2}r + 1)c_2 + j_2 r$. So we have

$$u = (z_{i_1}r+1)b_1+j_1 r = (z_{i_2}r+1)b_2+j_2 r \ , \text{ and } v = (z_{i_1}r+1)c_1+j_1 r = (z_{i_2}r+1)c_2+j_2 r. \tag{2}$$

Modulo r, we have $b_1 = b_2$ and $c_1 = c_2$ (because $|b_i|, |c_i| < r$). We denote $b = b_1 = b_2$ and $c = c_1 = c_2$. By computing $u - v$, $(z_{i_1}r+1)(b-c) = (z_{i_2}r+1)(b-c)$. Now, $b \neq c$ because $u \neq v$. So, $z_{i_1} = z_{i_2}$, then we have $i_1 = i_2$. By (2) then, $j_1 = j_2$. Therefore, $(i_1, j_1) = (i_2, j_2)$, which is a contradiction.

Claim. $\mathcal{C}$ is triangle-free, i.e., for any distinct u, v, w, if $v, w \in S_{i_1}^{(j_1)}$, $w, u \in S_{i_2}^{(j_2)}$, and $u, v \in S_{i_3}^{(j_3)}$, then $(i_1, j_1) = (i_2, j_2) = (i_3, j_3)$.

*Proof (**Proof of Theorem 2**).* We reduce the set disjointness problem with $N = t \cdot q$ bits to

$\text{CLIQUE-GAP}(r, s - 1)$, where $t = \dfrac{n/(2r^2)}{\exp(10\sqrt{\log r \log(n/(2r^2))})} = g(n/(2r^2), r)$ and $q = n/(2r)$.

By the definition of Ramsey number, for each $S_i^{(j)}$, there exists a subgraph $Q_i^{(j)}$ of the clique on $S_i^{(j)}$, such that neither $Q_i^{(j)}$ nor $\overline{Q_i^{(j)}}$ has a clique of size s.

Given a DISJ instance $x, y \subseteq [t] \times [q]$, we consider each $S_i^{(j)}$ as a gadget and construct a clique separation instance, in which we give Alice $G_A = \bigcup_{(i,j) \in x} Q_i^{(j)}$, and give Bob $G_B = \bigcup_{(i,j) \in y} \overline{Q_i^{(j)}}$. We are going to prove that $G_A \cup G_B$ has an r-clique if $x \cap y \neq \emptyset$, and has no s-clique if $x \cap y = \emptyset$.

On positive DISJ instances, when $x_{i,j} = y_{i,j} = 1$, the corresponding gadget $S_i^{(j)}$ induces an r-clique in $G_A \cup G_B$. On negative DISJ instance, for each (i, j), each subgraph induced by $S_i^{(j)}$ in $G_A \cup G_B$ is one of three possibilities: $Q_i^{(j)}$, $\overline{Q_i^{(j)}}$ or empty. By construction, none of these contain a $(2 \log r)$-clique, so if $G_A \cup G_B$ contains one, there exists a triangle (u, v, w) which is not in any $S \in \mathcal{C}$. This contradicts the triangle-freeness property of $\mathcal{C}$.

Therefore, if we have a protocol of the $\text{CLIQUE-GAP}(r, s-1)$ problem, we have a protocol of set disjointness problem with the same communication complexity. Hence, $\text{CLIQUE-GAP}(r, s - 1)$ problem has communication complexity lower of $\Omega(N) = \Omega(t \cdot q)$.

For larger values of r (e.g., $r = n/\text{polylog}(n)$), a naive packing gives better bounds: Simply combine $\lfloor n/r \rfloor$ *vertex*-disjoint r-cliques. This yields an edge-disjoint triangle-free set system with n/r sets of size r each.

Theorem 4. *For any n and any s, $R_{1/3}(\text{CLIQUE-GAP}(r, s - 1)) = \Omega(n/r)$, where $r = \mathcal{R}(s) - 1$.*

4.3 r^2 vs. $2r - 1$

We now focus on graphs of constant clique number, for which we obtain optimal quadratic space lower bounds.

Theorem 5. *For any number $r \geq 18$, $R_{1/3}(\text{CLIQUE-GAP}(r^2, 2r - 1)) = \Omega(n^2/r^4)$. Thus, any randomized ρ-approximation streaming algorithm for* MAX-CLIQUE *requires $\Omega(n^2/\rho^4)$ space.*

We construct a gadget $H = \langle V_H, E_H \rangle$ on r^2 vertices corresponding to a single bit in DISJ. We shall ensure that H is clique if the corresponding bits of both Alice and Bob are both 1, and that H contains no $2r$-clique otherwise. The vertex set V_H consists of r groups, r vertices each: $V_H = \{u_{i,j} | i, j \in [r]\}$. We color all the $\binom{r^2}{2}$ edges with three colors: A (Alice), B (Bob), and C (Common).

We say that a triplet $u, v, w \in V_H$ is a *colorful triangle* if all three mutual edges are differently colored. The proof of the following lemma is based on the probabilistic method.

Lemma 4. *For large r, there exists a coloring of E_H satisfying*

1. *Edge $\{u_{i,j}, u_{i',j'}\}$ is with color C if and only if $i = i'$ and $j \neq j'$.*
2. *Any $2r$-subgraph of H contains a colorful triangle.*

Let P be a prime in the range $[n/(2r^2), n/r^2]$. We reduce the CLIQUE-GAP problem of size n from DISJ problem of size $N = P^2$ by packing P^2 gadgets in a graph of size n, where each gadget is of size r^2. We isolate the remaining $n - r^2P$ vertices, and focus on the r^2P vertices $\{v_{i,j,k} | i, j \in [r], k \in [P]\}$. On these vertices, the edges are given by $E_G^C = \{\{v_{i,j,k}, v_{i,j',k}\} | i, j, j' \in [r], k \in [P], j \neq j'\}$, $E_G^A = \{\{v_{i,j,(s+ti) \bmod P}, v_{i',j',(s+ti') \bmod P}\} | i, i', j, j' \in [r], i \neq i'$ and $\{u_{i,j}, u_{i',j'}\}$ is with color A and $x_{s,t} = 1\}$, and $E_G^B = \{\{v_{i,j,(s+ti) \bmod P}, v_{i',j',(s+ti') \bmod P}\} | i, i', j, j' \in [r], i \neq i'$ and $\{u_{i,j}, u_{i',j'}\}$ is with color B and $y_{s,t} = 1\}$.

Alice is given the edges in $E_G^A \cup E_G^C$, and Bob given the edges in $E_G^B \cup E_G^C$.

Lemma 5. *If* DISJ$(x, y) = 1$, *then $\omega(G) = r^2$.*

Lemma 6. *If* DISJ$(x, y) = 0$, *then $\omega(G) < 2r$.*

Proof (Proof of Theorem 5). We reduce DISJ problem of size $N = P^2$ to the CLIQUE-GAP$(r^2, 2r - 1)$ problem. Since $\mathcal{R}(\text{DISJ}) = \Omega(P^2)$, any randomized protocol to separate graphs with r^2-cliques from those with only $(2r - 1)$-cliques requires $\Omega(n^2/r^4)$ communication.

5   Upper Bounds

The following simple random sampling argument shows that the lower bound of Thm. 1 is within a logarithmic factor of optimal.

Theorem 6. *There   is   a   randomized   streaming   algorithm   for* CLIQUE-GAP$(r, r/\rho)$ *that   uses* $O((n/\rho)^2)$ *space   (for* $\rho = O(n/\sqrt{\log n})$*). Thus,*
$$R_{1/3}(\text{CLIQUE-GAP}(r, r/\rho)) \leq \text{space}_{1/3}(\text{CLIQUE-GAP}(r, r/\rho) = O((n/\rho)^2).$$

Proof. Assuming that the vertices are numbered $0, 1, \ldots, n-1$, we initially choose a random number h from $[n]$. This specifies a set S consisting of the n/ρ vertices numbered h through $h+n/\rho-1$ (mod n). In processing the stream, we only store edges between pairs of vertices in S and afterwards output the maximum clique within S. The probability that any given vertex falls within S is $(n/\rho)/n = 1/\rho$. Thus, by linearity of expectation, the expected number of vertices within any r-clique that fall inside S is r/ρ.

Finally, we cannot expect to get a non-trivial lower bound on the separation of $(s-1)$-cliques vs. $\mathcal{R}(s)$-cliques using communication complexity. Namely, by the definition of Ramsey numbers, any 2-coloring of an $\mathcal{R}(s)$-clique – or a splitting of the clique edges between Alice and Bob – leaves a monochromatic s-clique. Thus, at least one of Alice and Bob can detect a s-clique, without any communication. The gap in Thm. 2 is therefore best possible, even though the space lower bound is not optimal. In fact, we get a sharp transition for CLIQUE-GAP(U, L) in terms of the values of U and L for which non-trivial communication is needed.

Theorem 7. *There   is   a   deterministic   communication   protocol   for* CLIQUE-GAP$(\mathcal{R}(s), s-1)$ *that uses* $O(1)$-*bits. That is,* $D(\text{CLIQUE-GAP}(\mathcal{R}(s), s-1)) = O(1)$.

References

1. Ahn, K.J., Guha, S.: Graph Sparsification in the Semi-streaming Model. In: Albers, S., Marchetti-Spaccamela, A., Matias, Y., Nikoletseas, S., Thomas, W. (eds.) ICALP 2009. LNCS, vol. 5556, pp. 328–338. Springer, Heidelberg (2009)
2. Alon, N., Shapira, A.: A characterization of easily testable induced subgraphs. In: SODA 2004, pp. 942–951. SIAM (2004)
3. Alon, N., Boppana, R.B.: The monotone circuit complexity of boolean functions. Combinatorica 7(1), 1–22 (1987)
4. Babai, L., Frankl, P., Simon, J.: Complexity classes in communication complexity theory. In: FOCS 1986, pp. 337–347. IEEE Computer Society (1986)
5. Bar-Yossef, Z., Kumar, R., Sivakumar, D.: Reductions in streaming algorithms, with an application to counting triangles in graphs. In: SODA 2002, pp. 623–632 (2002)
6. Becchetti, L., Boldi, P., Castillo, C., Gionis, A.: Efficient semi-streaming algorithms for local triangle counting in massive graphs. In: KDD 2008, pp. 16–24 (2008)

7. Bollobás, B.: Complete subgraphs are elusive. Journal of Combinatorial Theory, Series B 21(1), 1–7 (1976)
8. Bordino, I., Donato, D., Gionis, A., Leonardi, S.: Mining Large Networks with Subgraph Counting. In: 8th IEEE International Conference on Data Mining, pp. 737–742. IEEE Computer Society (2008)
9. Buriol, L.S., Frahling, G., Leonardi, S., Sohler, C.: Estimating Clustering Indexes in Data Streams. In: Arge, L., Hoffmann, M., Welzl, E. (eds.) ESA 2007. LNCS, vol. 4698, pp. 618–632. Springer, Heidelberg (2007)
10. Downey, R.G., Fellows, M.R.: Fixed-parameter tractability and completeness. II. On completeness for W[1]. Theoretical Computer Science 141(1-2), 109–131 (1995)
11. Erdős, P.: Some remarks on the theory of graphs. Bull. Amer. Math. Soc. 53, 292–294 (1947)
12. Erdős, P., Szekeres, G.: A combinatorial problem in geometry. Compositio. Math. 2, 463–470 (1935)
13. Feige, U.: Approximating maximum clique by removing subgraphs. SIAM J. Discrete Math. 18(2), 219–225 (2004)
14. Feigenbaum, J., Kannan, S., McGregor, A., Suri, S., Zhang, J.: On graph problems in a semi-streaming model. Theoretical Computer Science 348(2), 207–216 (2005)
15. Feigenbaum, J., Kannan, S., McGregor, A., Suri, S., Zhang, J.: Graph distances in the data-stream model. SIAM J. Comput. 38(5), 1709–1727 (2008)
16. Grimmett, G.R., McDiarmid, C.J.H.: On colouring random graphs. Mathematical Proceedings of the Cambridge Philosophical Society 77, 313–324 (1975)
17. Halldórsson, B.V., Halldórsson, M.M., Losievskaja, E., Szegedy, M.: Streaming Algorithms for Independent Sets. In: Abramsky, S., Gavoille, C., Kirchner, C., Meyer auf der Heide, F., Spirakis, P.G. (eds.) ICALP 2010. LNCS, vol. 6198, pp. 641–652. Springer, Heidelberg (2010)
18. Håstad, J.: Clique is hard to approximate within $n^{1-\epsilon}$. Acta Mathematica 182, 105–142 (1999)
19. Indyk, P., Price, E.: K-median clustering, model-based compressive sensing, and sparse recovery for earth mover distance. In: STOC 2011, pp. 627–636 (2011)
20. Kalyanasundaram, B., Schnitger, G.: The probabilistic communication complexity of set intersection. SIAM J. Discrete Math. 5(4), 545–557 (1992)
21. Khot, S., Ponnuswami, A.K.: Better Inapproximability Results for maxClique, Chromatic Number and Min-3Lin-Deletion. In: Bugliesi, M., Preneel, B., Sassone, V., Wegener, I. (eds.) ICALP 2006. LNCS, vol. 4051, pp. 226–237. Springer, Heidelberg (2006)
22. Kushilevitz, E., Nisan, N.: Communication Complexity. Cambridge Univ. Pr. (1997)
23. Losievskaja, E.: Approximation Algorithms for Independent Set Problems on Hypergraphs. PhD thesis. Reykjavik University (January 2010)
24. Manjunath, M., Mehlhorn, K., Panagiotou, K., Sun, H.: Approximate Counting of Cycles in Streams. In: Demetrescu, C., Halldórsson, M.M. (eds.) ESA 2011. LNCS, vol. 6942, pp. 677–688. Springer, Heidelberg (2011)
25. Ruzsa, I., Szemerédi, E.: Triple systems with no six points carrying three triangles. Combinatorics (Keszthely, 1976), Coll. Math. Soc. J. Bolyai 18, 939–945 (1976)
26. Yao, A.C.C.: Some complexity questions related to distributive computing (preliminary report). In: STOC 1979, pp. 209–213. ACM (1979)
27. Zelke, M.: Intractability of min- and max-cut in streaming graphs. Inf. Process. Lett. 111(3), 145–150 (2011)

Distributed Private Heavy Hitters

Justin Hsu, Sanjeev Khanna*, and Aaron Roth**

University of Pennsylvania, Philadelphia PA 19104, USA
{justhsu,sanjeev,aaroth}@cis.upenn.edu

Abstract. In this paper, we give efficient algorithms and lower bounds for solving the *heavy hitters* problem while preserving *differential privacy* in the fully distributed *local* model. In this model, there are n parties, each of which possesses a single element from a universe of size N. The heavy hitters problem is to find the identity of the most common element shared amongst the n parties. In the local model, there is no trusted database administrator, and so the algorithm must interact with each of the n parties separately, using a differentially private protocol. We give tight information-theoretic upper and lower bounds on the accuracy to which this problem can be solved in the local model (giving a separation between the local model and the more common centralized model of privacy), as well as computationally efficient algorithms even in the case where the data universe N may be exponentially large.

1 Introduction

Consider the problem of a website administrator who wishes to know what his most common traffic sources are. Each of n visitors arrives with a single *referring site*: the name of the last website that she visited, which is drawn from a vast universe N of possible referring sites (N here is the set of all websites on the internet). There is value in identifying the most popular referring site (the *heavy hitter*): the site administrator may be able to better tailor the content of his webpage, or better focus his marketing resources. On the other hand, the identity of each individual's referring site might be embarrassing or otherwise revealing, and is therefore private information. We can therefore imagine a world in which this information must be treated "privately." In this situation, visitors are communicating directly with the servers of the websites that they visit: i.e. there is no third party who might be trusted to aggregate all of the referring website data and provide privacy preserving statistics to the website administrator. In this setting, how well can the website administrator estimate the heavy hitter while being able to provide formal privacy guarantees to his visitors?

This situation can more generally be modeled as the *heavy hitters* problem under the constraint of *differential privacy*. There are n individuals $i \in [n]$ each of whom is associated with an element $v_i \in N$ of some large data universe N. The *heavy hitter* is the most frequently occurring element $x \in N$ among the

---

* Supported in part by NSF Awards CCF-0635084 and IIS-0904314.

** Supported in part by NSF Awards CCF-1101389 and CNS-1065060.

A. Czumaj et al. (Eds.): ICALP 2012, Part I, LNCS 7391, pp. 461–472, 2012.

set $\{v_1, \ldots, v_n\}$, and we would like to be able to identify that element, or one that occurs almost as frequently as the heavy hitter. Moreover, we wish to solve this problem while preserving *differential privacy* in the fully distributed (local) model. We define this formally in section 2, but roughly speaking, an algorithm is differentially private if changes to the data of single individuals only result in small changes in the output distribution of the algorithm. Moreover, in the fully distributed setting, each individual (who can be viewed as a database of size 1) must interact with the algorithm independently of all of the other individuals, using a differentially private algorithm. This is in contrast to the more commonly studied centralized model, in which a trusted database administrator may have (exact) access to all of the data, and coordinate a private computation.

We study this problem both from an information theoretic point of view, and from the point of view of efficient algorithms. We say that an algorithm for the private heavy hitters problem is *efficient* if it runs in time poly($n, \log N$): i.e. polynomial in the database size, but only polylogarithmic in the universe size (i.e. in what we view as the most interesting range of parameters, the universe may be exponentially larger than the size of the database). We give tight information theoretic upper and lower bounds on the accuracy to which the heavy hitter can be found in the private distributed setting (separating this model from the private centralized setting), and give several efficient algorithms which achieve good, although information-theoretically sub-optimal accuracy. We leave open the question of whether *efficient* algorithms can match the information theoretic bounds we prove for the private heavy hitters problem in the this setting.

1.1   Our Results

In this section, we summarize our results. The bounds we discuss here are informal and hide many of the parameters which we have not yet defined. The formal bounds are given in the main body of the paper.

First, we provide an information theoretic characterization of the accuracy to which any algorithm (independent of computational constraints) can solve the heavy hitters problem in the private distributed setting. We say that an algorithm is α-accurate if it returns a universe element which occurs with frequency at most an additive α smaller than the true heavy hitter. In the centralized setting, an application of the exponential mechanism [1] gives an α-accurate mechanism for the heavy-hitters problem where $\alpha = O(\log |N|)$, which is independent of the number of individuals n. In contrast, we show that in the fully distributed setting, no algorithm can be α-accurate for $\alpha = \Omega(\sqrt{n})$ even in the case of $|N| = 2$. Conversely, we give an almost matching upper bound (and an algorithm with run-time linear in N) which is α-accurate for $\alpha = O(\sqrt{n \log N})$.

Next, we consider *efficient* algorithms which run in time only polylogarithmic in the universe size $|N|$. Here, we give two algorithms. One is an application of a compressed sensing algorithm of Gilbert et al. [2], which is α-accurate for $\alpha = \tilde{O}(n^{5/6} \log \log N)$. Then, we give an algorithm based on group-testing using pairwise independent hash functions, which has an incomparable bound. Roughly speaking, it guarantees to return the exact heavy hitter (i.e. $\alpha = 0$) whenever the frequency of the heavy hitter is larger than the ℓ_2-norm of the frequencies of the remaining elements. Depending on how these frequencies are distributed, this can correspond to a bound of α-accuracy for α ranging anywhere between the optimal $\alpha = O(\sqrt{n})$ to $\alpha = O(n)$.

1.2 Our Techniques

Our upper bounds, both information theoretic, and those with efficient algorithms, are based on the general technique of *random projection* and *concentration of measure*. To prove our information theoretic upper bound, we observe that to find the heavy hitter, we may view the private database as a histogram v in N dimensional space. Then, it is enough to find the index $i \in [N]$ of the universe element which maximizes $\langle v, e_i \rangle$, where e_i is the i'th standard basis vector. Both v and each e_i have small ℓ_1-norm, and so each of these inner products can be approximately preserved by taking a random projection into $\tilde{O}(\log N)$ dimensional space. Moreover, we can project each individual's data into this space independently in the fully distributed setting, incurring a loss of only $O(\sqrt{n})$ in accuracy. This mechanism, however, is not efficient, because to find the heavy hitter, we must enumerate through all $|N|$ basis vectors e_i in order to find the one that maximizes the inner product with the projected database. Similar ideas lead to our efficient algorithms, albeit with worse accuracy guarantees. For example, in our first algorithm, we apply techniques from compressed sensing to the projected database to recover (approximately) the heavy hitter, rather than checking basis vectors directly. In our second algorithm, we take a projection using a particular family of pairwise-independent hash functions, which are linear functions of the data universe elements. Because of this linearity, we are able to efficiently "invert" the projection matrix in order to find the heavy hitter.

Our lower bound separates the distributed setting from the centralized setting by applying an anti-concentration argument. Roughly speaking, we observe that in the fully distributed setting, if individual data elements were selected uniformly *i.i.d.* from the data universe N, then even after conditioning on the messages exchanged with any differentially private algorithm, they remain independently distributed, and approximately uniform. Thus, by the Berry-Esseen theorem, after any algorithm computes its estimate of the heavy hitter, the true distribution over counts remains approximately normally distributed. Since the Gaussian distribution exhibits strong anti-concentration properties, we arrive at an $\Omega(\sqrt{n})$ lower bound for any algorithm in the fully distributed setting.

1.3 Related Work

Differential privacy was introduced in a sequence of papers culminating in [3], and has since become the standard "solution concept" for privacy in the theoretical computer science literature. There is by now a very large literature on this topic, which is too large to summarize here. Instead, we focus only on the most closely related work, and refer the curious reader to a survey of Dwork [4].

Most of the literature on differential privacy focuses on the *centralized* model, in which there is a trusted database administrator. In this paper, we focus on the *local* or *fully distributed* model, introduced by [5] and studied also by [6], in which each individual holds their own data (i.e. there are n databases, each of size 1), and the algorithm must interact with each one in a differentially private manner. There has been little work in this more restrictive model–the problems of *learning* [5] and *query release* [7] in the local model are well understood [1], but only up to polynomial factors that do not imply tight bounds for the heavy hitters problem. The *two-party* setting (which is intermediate between the centralized and fully distributed setting), in which the data is divided between two databases without a trusted central administrator, was considered by [8]. They proved a separation between the two-party setting and the centralized setting for the problem of computing the Hamming distance between two strings. In this work, we prove a separation between the fully distributed setting and the centralized setting for the problem of estimating the heavy hitter.

A variant of the private heavy hitters problem has been considered in the setting of *pan-private streaming algorithms* [9,10]. This work considers a different (although related) problem in a different (although related) setting. [9,10] consider a setting in which a stream of elements is presented to the algorithm, and the algorithm must estimate the *approximate count* of frequently occurring elements (i.e. the number of "heavy hitters"). In this setting, the universe elements themselves are the individuals appearing in the stream, and so it is not possible to reveal the identity of the heavy hitter. In contrast, in our work, individuals are distinct from universe elements, which merely label the individuals. Moreover, our goal here is to actually identify a specific universe element which is the heavy hitter, or which occurs almost as frequently. Also, [9,10] work in the centralized setting, but demand *pan-privacy*, which roughly requires that the internal state of the algorithm itself remain differentially private. In contrast, we work in the *local privacy* setting which gives a guarantee which is strictly stronger than pan-privacy. Because algorithms in the local privacy setting only interact with individuals in a differentially private way, and never have any other access to the private data, any algorithm in the local privacy model can never have its state depend on data in a non-private way, and such algorithms therefore also preserve pan-privacy. Therefore, our upper bounds hold also in the setting of pan-privacy, whereas our lower bounds do not necessarily apply to algorithms which only satisfy the weaker guarantee of pan-privacy.

Finally, we note that many of the upper bound techniques we employ have been previously put to use in the centralized model of data privacy i.e. random projections [11,12] and compressed sensing (both for lower bounds [13] and

---

[1] Roughly, the set of concepts that can be *learned* in the local model given polynomial sample complexity is equal to the set of concepts that can be learned in the SQ model given polynomial query complexity [5], and the set of queries that can be *released* in the local model given polynomial sample complexity is equal to the set of concepts that can be agnostically learned in the SQ model given polynomial query complexity [7], but the polynomials are not equal.

algorithms [14]). As algorithmic techniques, these are rarely optimal in the centralized privacy setting. We remark that they are particularly well suited to the fully distributed setting which we study here, because in a formal sense, algorithms in the local model of privacy are constrained to only access the private data using noisy linear queries, which is exactly the form of access used by random linear projections and compressed sensing measurements.

2 Preliminaries

A database v consists of n records from a data universe N, one corresponding to each of n individuals: for $i \in [n]$, $v^i \in N$ and $v = \{v^1, \ldots, v^n\}$ which may be a *multiset*. Without loss of generality, we will index the elements of the data universe from 1 to $|N|$. It will be convenient for us to represent databases as *histograms*. In this representation, $v \in \mathbb{N}^{|N|}$, where v_i represents the number of occurrences of the i'th universe element in the database. Further, we write $v^i \in \mathbb{N}^{|N|}$ for each individual $i \in [n]$, where $v^i_j = 1$ if individual i is associated with the j'th universe element, and $v^i_{j'} = 0$ for all other $j' \neq j$. Note that in this histogram notation, we have: $v = \sum_{i=1}^{n} v^i$. In the following, we will usually use the histogram notation for mathematical convenience, with the understanding that we can in fact more concisely represent the database as a multiset.

Given a database v, the *heavy hitter* is the universe element that occurs most frequently in the database: $hh(v) = \arg\max_{i \in N} v_i$. We refer to the frequency with which the heavy hitter occurs as $fhh(v) = v_{hh(v)}$. We want to design algorithms which return universe elements that occur *almost as frequently as the heavy hitter*.

Definition 1. *An algorithm A is (α, β)-accurate for the heavy hitters problem if for every database $v \in \mathbb{N}^{|N|}$, with probability at least $1 - \beta$: $A(v) = i^*$ such that $v_{i^*} \geq fhh(v) - \alpha$.*

2.1 Differential Privacy

Differential privacy constrains the sensitivity of a randomized algorithm to individual changes in its input.

Definition 2. *An algorithm $A : \mathbb{N}^{|N|} \to R$ is (ϵ, δ)-differentially private if for all $v, v' \in \mathbb{N}^{|N|}$ such that $||v - v'||_1 \leq 1$, and for all events $S \subseteq R$: $\Pr[A(v) \in S] \leq \exp(\epsilon) \Pr[A(v') \in S] + \delta$*

Typically, we will want δ to be negligibly small, whereas we think of ϵ as being a small constant (and never smaller than $\epsilon = O(1/n)$).

Additional preliminaries (including the formal definition of the local privacy model) can be found in the full version [15]. Informally speaking, an algorithm A operates in the local privacy model if for each individual i, the only access that A has to v^i is through the output of $A_i(v^i)$, where A_i is itself an (ϵ, δ)-differentially private algorithm that operates on a database of size $||v^i||_1 = 1$.

2.2   Probabilistic Tools

We will make use of several useful probabilistic tools which can be found in the full version. In particular, we use the well-known Johnson-Lindenstrauss lemma. Informally, it says that any set of q points in a high dimensional space can be *obliviously* embedded into a space of dimension $m = O(\log q)$ such that w.h.p. this embedding essentially preserves pairwise distances. Moreover, the embedding is linear and can be accomplished with a random projection matrix with entries taken to be independently uniformly drawn from $\{-1/\sqrt{m}, 1/\sqrt{m}\}$.

3   Information Theoretic Upper and Lower Bounds.

In this section we present upper and lower bounds on the accuracy to which any algorithm in the fully distributed model can privately approximate heavy hitters. Our upper bound can be viewed as an algorithm, albeit one that runs in time linear in $|N|$ and so is not what we consider to be efficient.

3.1   An Upper Bound via Johnson-Lindenstrauss Projections

We present here our first algorithm, referred to as *JL-HH*, that solves the heavy hitters problem in the local model using the Johnson-Lindenstrauss lemma. The algorithm JL-HH is outlined in Algorithm 1. We write e_i to refer to the i'th standard basis vector in $\mathbb{R}^N$, and write RandomProjection$(m, N+1)$ for a sub-routine which returns a linear embedding of $N + 1$ points into m dimensions using a random $\pm 1/\sqrt{m}$ valued projection matrix, as specified by the Johnson-Lindenstrauss lemma. By this lemma, for any set of $N + 1$ elements, this map approximately preserves pairwise distances with high probability.

JL-HH is based on the following straightforward idea: if v is a private histogram, we will estimate the count of the i'th element ($\langle v, e_i \rangle$), by estimating $\langle Av, Ae_i \rangle$, and return the largest count. By the Johnson-Lindenstrauss lemma, since we are using the random projections matrix, we have that with high probability, inner products between points in the set $V = \{e_1 \cdots e_N, v\}$ are approximately preserved under A. However, we cannot access Av directly since v is private data. To preserve differential privacy, our mechanism must add noise z to Av, and work only with the noisy samples. Our analysis will thus focus on bounding the error introduced by this noise term. First, though, we show that JL-HH is differentially private; the proof is simple and can be found in the full version.

Lemma 1. *JL-HH operates in the local privacy model and is (ϵ, δ)-differentially private.*

We next prove an accuracy bound for the mechanism:

Theorem 1. *For any $\beta > 0$, JL-HH mechanism is (α, β)-accurate for the heavy hitters problem, with $\alpha = O\left(\frac{\sqrt{n \log(N/\beta)} \log(1/\delta)}{\epsilon} \right)$.*

Algorithm 1. JL-HH Mechanism

Input: Private histograms $v^i \in \mathbb{N}^N, i \in [n]$. Privacy parameters $\epsilon, \delta > 0$. Failure probability $\beta > 0$.

Output: p^*, index of the heavy hitter.

$\gamma \leftarrow 1/n^2$

$m \leftarrow \frac{\log(N+1)\log(2/\beta)}{\gamma^2}$

$A \leftarrow \text{RandomProjection}(m, N+1)$

for $p = 1$ to N indices **do**

    for $i = 1$ to n users **do**

        $z^i \sim \left\{ \text{Lap}\left(\frac{\sqrt{8\log(1/\delta)}}{\epsilon}\right) \right\}^m$

        $q^i = Av^i + z^i$

        $r_{ip} = \langle Ae_p, q^i \rangle$

    end for

    $c_p \leftarrow \sum_{i=1}^n r_{ip}$

end for

$p^* \leftarrow \text{argmax}_p\, c_p$

return p^*

The proof, which proceeds by bounding two sources of error (one from the random projection, and one from the added noise), can be found in the full version.

It is worthwhile to compare JL-HH with a more naive approaches. A simpler differentially private algorithm to solve the distributed heavy hitters problem is to have each user simply add noise $\text{Lap}(1/\epsilon)$ to each entry in the user's private histogram, and report this vector to the central party, which sums the noisy vectors and estimates the most frequently occurring item. This is differentially private, as any neighboring histogram will change exactly one entry in a user's histogram. However, this method requires having each user transmit $O(N)$ bits of information to the central party. JL-HH achieves similar accuracy compared to this naive approach, but since the clients compress the histogram first, only $O(\log N)$ information must be communicated. Even though JL-HH runs in time linear in N, there are natural situations where long running time can be tolerated, but large communication complexity cannot, for instance, if the central party is a server farm with considerable computational resources, but the communication with users must happen over standard network links.

3.2 A Lower Bound via Anti-concentration

We next show that our upper bound is essentially optimal: for any $\epsilon < 1/2$ and any δ bounded away from 1 by a constant, no (ϵ, δ)-private mechanism in the fully distributed setting can be α-accurate for the heavy hitters problem for $\alpha = \Omega(\sqrt{n})$, even in the case in which $|N| = 2$. We remark that our technique (while specific to the local privacy model) holds for (ϵ, δ)-differential privacy, even when $\delta > 0$. This is similar to lower bounds based on reconstruction arguments [16], and in contrast to other techniques for proving lower bounds in the centralized model, such as the elegant packing arguments used in [17,18], which are specific to $(\epsilon, 0)$-differential privacy. We use an independence argument also used by [8] to prove a lower bound in the two-party setting, and by [6] to prove a lower bound in the fully distributed setting.

Theorem 2. *For any $\epsilon \leq 1/2$ and $\delta < 1$ bounded away from 1, there exists an $\alpha = \Omega(\sqrt{n})$ and a $\beta = \Omega(1)$ such that no (ϵ, δ)-private mechanism in the local model is (α, β)-accurate for the heavy hitters problem.*

Proof (Sketch). The proof can be found in the full version – here we include just a sketch. We construct a lower bound instance over a universe of size 2: $N = \{0, 1\}$, in which the universe element s_i for each individual i is selected independently and uniformly at random. We observe that even after conditioning on the private interaction with the mechanism, because we are in the local model, the random variables s_i remain independent. Moreover, because our mechanism is (ϵ, δ)-differentially private, for each i with probability $1 - \delta$ (over the choices of the mechanism) the conditional probability that s_i is 1 remains in the range $[\exp(-\epsilon)/2, \exp(\epsilon/2)]$ and in particular, bounded away from 0 and 1 by a constant. Therefore, by the Berry-Esseen theorem, conditioned on the outcome of the mechanism, the distribution over the number of individuals with $s_i = 1$ converges to a Gaussian distribution with standard deviation $\sigma = \Omega(\sqrt{n})$. But the Gaussian distribution exhibits strong anti-concentration properties: with constant probability it deviates from its expectation by an additive $\Omega(\sqrt{n})$. Therefore, no private local mechanism can be accurate to within this factor.

4   Efficient Algorithms

In the last section, we saw the Johnson-Lindenstrauss algorithm which gave almost optimal accuracy guarantees, but had running time linear in $|N|$. In this section, we consider efficient algorithms with running time poly$(n, \log |N|)$. The first is an application of a sublinear time algorithm from the compressed sensing literature, and the second is a group-testing approach made efficient by the use of a particular family of pairwise-independent hash functions.

4.1   GLPS Sparse Recovery

In this section we adapt a sophisticated algorithm from compressed sensing. Gilbert, et al. [2] present a sparse recovery algorithm (we refer to it as the GLPS algorithm) that takes linear measurements from a sparse vector, and reconstructs the original vector to high accuracy. Importantly, the algorithm runs in time polylogarithmic in $|N|$, and polynomial in the sparsity parameter of the vector. We remark that our database v is n-sparse: it has at most n non-zero components. In the rest of this section, we will write v_s to denote the vector v truncated to contain only its s largest components.

Let s be a sparsity parameter, and let γ be a tunable approximation level. The GLPS algorithm runs in time $O((s/\gamma) \log^c N)$, and makes $m = O(s \log(N/s)/\gamma)$

measurements from a specially constructed (randomized) $\{-1, 0, 1\}$ valued matrix, which we will denote Φ. Given measurements $u = \Phi v + z$ (where z is arbitrary noise), the algorithm returns an approximation $\hat{v}$, with error

$$\|v - \hat{v}\|_2 \le (1 + \gamma)\|v - v_s\|_2 + \gamma \log(s)\frac{\|z\|_2}{\kappa} \tag{1}$$

with probability at least $3/4$, where $\kappa = O(\log^2(s) \log(N/s))$.

Though the GLPS bound only occurs with probability $3/4$, the success probability can be made arbitrarily close to 1 by running this algorithm several times. In particular, using the amplification lemma from [19], the failure probability can be driven down to β at a cost of only a factor of $\log(1/\beta)$ in the accuracy. In what follows, we analyze a single run of the algorithm, with $\gamma = 1$.

Algorithm 2. GLPS-HH Mechanism

Input: Private histograms $v^i \in \mathbb{N}^N, i \in [n]$. GLPS matrix Φ. Privacy parameters $\epsilon, \delta > 0$.
Output: p^*, estimated index of heavy hitter.
   $m \leftarrow s \log(N/s)$
   $b \leftarrow \sqrt{8m \log(1/\delta)}/\epsilon$
   for $i = 1$ to n users **do**
      $z^i \sim \{\mathrm{Lap}(b)\}^m$
      $q^i \leftarrow \Phi v^i + z^i$
   end for
   $c \leftarrow \sum_{i=1}^n q^i$
   $\hat{v} \leftarrow GLPS(c, \Phi)$
   $p^* \leftarrow \mathrm{argmax}_p \hat{v}_p$
   return p^*

First, we will show that GLPS-HH is (ϵ, δ)-differentially private (proof in the full version).

Theorem 3. *GLPS-HH operates in the local privacy model and is (ϵ, δ)-differentially private.*

Next, we will bound the error that we introduce by adding noise for differential privacy.

Theorem 4. *Let $\beta > 0$ be given. GLPS-HH is $(\alpha, 3/4 - \beta)$-accurate for the heavy hitters problem, with $\alpha = O\left(\dfrac{n^{5/6} \log^{1/3}(1/\beta) \log\log N \log^{1/6}(1/\delta)}{\epsilon^{1/3}}\right)$*

The proof, which proceeds by bounding the total error added and applying the GLPS bound, can be found in the full version.

Algorithm 3. The Bucket Mechanism

Input: Private labels $v^i \in [N], i \in [n]$. Failure probability $\beta > 0$. Privacy parameters $\epsilon, \delta > 0$.

Output: p^*, the index of the heavy hitter.

$F \leftarrow \{0,1\}^{\log N} \setminus 0$

for $i = 1$ to $8\log(1/\beta)$ trials **do**

$\quad H \in \{0,1\}^{\log(12N) \times \log N} \leftarrow$ Draw $\log(12N)$ rows from F, uniformly at random.

$\quad u \in \mathbb{R}^{\log(12N)} \leftarrow 0$

$\quad$ **for** $j = 1$ to n users **do**

$\quad\quad b \in \{0,1\}^{\log N} \leftarrow$ binary expansion of v^j.

$\quad\quad s \leftarrow Hb \pmod 2$

$\quad\quad z \sim \left\{ \mathrm{Lap}\left(\dfrac{8\sqrt{\log(12N)\log(1/\beta)\log(1/\delta)}}{\epsilon} \right) \right\}^{\log(12N)}$

$\quad\quad u \leftarrow u + s + z$

$\quad$ **end for**

$\quad$ **for** $k = 1$ to $\log(12N)$ hash functions **do**

$\quad\quad b_k \leftarrow \begin{cases} 1 : u_k > n/2 \\ 0 : \text{otherwise} \end{cases}$

$\quad$ **end for**

$\quad w_i \leftarrow \begin{cases} x_0 : Hx_0 = b \pmod 2 \\ \bot \ : Hx = b \pmod 2 \text{ infeasible} \end{cases}$

end for

$w^* \leftarrow$ most frequent w_i, ignoring $\bot$

return $p^* \leftarrow w^*$ converted from binary

4.2   The Bucket Mechanism

In this section we present a second computationally efficient algorithm, based on group-testing and a specific family of pairwise independent hash functions.

At a high level, our algorithm, referred to as the *Bucket mechanism*, runs $O(\log(1/\beta))$ *trials* consisting of $O(\log N)$ 0/1 valued hash functions in each trial. For a given trial, the mechanism hashes each universe element into one of two buckets for each hash function. Then, the mechanism tries to find an element that hashes into the bucket with more weight (the *majority bucket*) for all the hash functions. If there is such an element, it is a candidate for the heavy hitter for that trial. Finally, the mechanism takes a majority vote over the candidates from each trial to output a final heavy hitter.

For efficiency purposes we do not use truly random hash functions, but instead rely on a particular family of pairwise-independent hash functions which can be expressed as linear functions on the bits of a universe element. Specifically, each function h in the family maps $[N]$ to $\{0,1\}$, and is parameterized by a bit-string $r \in \{0,1\}^{\log|N|}$. In particular, given any bit-string $r \in \{0,1\}^{\log|N|}$, we define $h_r(x) = \langle r, b(x) \rangle$, where $b(x)$ denotes the binary representation of x. r is chosen uniformly at random from the set of all strings $r \in \{0,1\}^{\log|N|} \setminus 0^{\log|N|}$. Given hash functions of this form, and a list of target buckets, the problem of finding an element that hashing to all of the target buckets is equivalent to solving a linear system mod 2, which can be done efficiently. Our family of hash functions operates on the element label in binary, hence the conversions to and from binary in the algorithm.

We will now show that the bucket mechanism is (ϵ, δ)-differentially private, runs in time $\mathrm{poly}(n, \log |N|)$, and assuming a certain condition on the distribution over universe elements, returns the exact heavy hitter. The accuracy analysis proceeds in two steps: first, we argue that with constant probability $> 1/2$, the heavy hitter is the unique element hashed into the larger bucket by every hash function in a given trial. Then, we argue that with high probability, the proceeding event indeed occurs in the majority of trials, and so the majority vote among all trials returns the true heavy hitter.

Theorem 5. *The Bucket mechanism operates in the local model and is (ϵ, δ)-differentially private.*

We now consider the accuracy of the mechanism. The analysis proceeds by a series of lemmas, and can be found in the full version.

Theorem 6. *Without loss of generality, suppose that the elements are labeled in decreasing order of count, with counts $v_1 \geq v_2 \geq \cdots \geq v_N$. The Bucket mechanism is $(0, \beta)$-accurate whenever $v_1 \geq \tilde{\Omega}\left(\dfrac{\sqrt{\log |N|}\left(\sqrt{\sum_{i=2}^{N} v_i^2} + \sqrt{n \log \frac{1}{\beta} \log \frac{1}{\delta}}\right)}{\epsilon} \right)$.*

We note that the accuracy guarantee of the bucket mechanism is incomparable to those of our other mechanisms. While the other mechanisms guarantee (without conditions) to return an element which occurs within some additive factor α as frequently as the true heavy hitter, the bucket mechanism always returns the true heavy hitter, so long as a certain condition on v is satisfied. We remark that this condition is not unreasonable: if universe elements follow a power law distribution, such as a Zipf distribution, the condition will be satisfied with overwhelming probability.

5   Discussion and Open Questions

We have initiated the study of the *private heavy hitters* problem in the fully distributed (local) privacy model. Our information theoretic understanding of this problem is almost tight, but we leave open the question of whether there exist *efficient algorithms* in the local model which can match this information theoretically optimal bound.

Acknowledgments. We would like to thank Martin Strauss for providing valuable insights about [2], and the anonymous reviewers for their helpful suggestions. We would also like to thank Andreas Haeberlen for suggesting that we study the heavy hitters problem in the fully distributed setting, and Andreas, Marco Gaboardi, Benjamin Pierce, and Arjun Narayan for valuable discussions.

References

1. McSherry, F., Talwar, K.: Mechanism design via differential privacy. In: Proceedings of the 48th Annual Symposium on Foundations of Computer Science (2007)

2. Gilbert, A., Li, Y., Porat, E., Strauss, M.: Approximate sparse recovery: optimizing time and measurements. In: Proceedings of the 42nd ACM Symposium on Theory of Computing, pp. 475–484. ACM (2010)
3. Dwork, C., McSherry, F., Nissim, K., Smith, A.: Calibrating Noise to Sensitivity in Private Data Analysis. In: Halevi, S., Rabin, T. (eds.) TCC 2006. LNCS, vol. 3876, pp. 265–284. Springer, Heidelberg (2006)
4. Dwork, C.: Differential Privacy: A Survey of Results. In: Agrawal, M., Du, D.-Z., Duan, Z., Li, A. (eds.) TAMC 2008. LNCS, vol. 4978, pp. 1–19. Springer, Heidelberg (2008)
5. Kasiviswanathan, S., Lee, H., Nissim, K., Raskhodnikova, S., Smith, A.: What Can We Learn Privately? In: IEEE 49th Annual IEEE Symposium on Foundations of Computer Science, FOCS 2008, pp. 531–540 (2008)
6. Beimel, A., Nissim, K., Omri, E.: Distributed Private Data Analysis: Simultaneously Solving How and What. In: Wagner, D. (ed.) CRYPTO 2008. LNCS, vol. 5157, pp. 451–468. Springer, Heidelberg (2008)
7. Gupta, A., Hardt, M., Roth, A., Ullman, J.: Privately Releasing Conjunctions and the Statistical Query Barrier. In: Proceedings of the 43rd annual ACM Symposium on the Theory of Computing. ACM, New York (2011)
8. McGregor, A., Mironov, I., Pitassi, T., Reingold, O., Talwar, K., Vadhan, S.: The limits of two-party differential privacy. In: Proceedings of the 51st Annual IEEE Symposium on Foundations of Computer Science (FOCS), pp. 81–90. IEEE (2010)
9. Dwork, C., Naor, M., Pitassi, T., Rothblum, G., Yekhanin, S.: Pan-private streaming algorithms. In: Proceedings of ICS (2010)
10. Mir, D., Muthukrishnan, S., Nikolov, A., Wright, R.: Pan-private algorithms via statistics on sketches. In: Proceedings of the 30th Symposium on Principles of Database Systems of Data, pp. 37–48. ACM (2011)
11. Blum, A., Ligett, K., Roth, A.: A learning theory approach to non-interactive database privacy. In: Proceedings of the 40th Annual ACM Symposium on Theory of Computing, pp. 609–618. ACM (2008)
12. Blum, A., Roth, A.: Fast private data release algorithms for sparse queries. CoRR, abs/1111.6842 (2011)
13. Dwork, C., McSherry, F., Talwar, K.: The price of privacy and the limits of LP decoding. In: Proceedings of the Thirty-Ninth Annual ACM Symposium on Theory of Computing, p. 94. ACM (2007)
14. Li, Y., Zhang, Z., Winslett, M., Yang, Y.: Compressive mechanism: Utilizing sparse representation in differential privacy. In: Proceedings of the 10th Annual ACM Workshop on Privacy in the Electronic Society, pp. 177–182. ACM (2011)
15. Hsu, J., Khanna, S., Roth, A.: Distributed private heavy hitters. Arxiv preprint arXiv:1202.4910 (2012)
16. Dinur, I., Nissim, K.: Revealing information while preserving privacy. In: 22nd ACM SIGACT-SIGMOD-SIGART Symposium on Principles of Database Systems (PODS 2003), pp. 202–210 (2003)
17. Beimel, A., Kasiviswanathan, S., Nissim, K.: Bounds on the sample complexity for private learning and private data release. Theory of Cryptography, 437–454 (2010)
18. Hardt, M., Talwar, K.: On the Geometry of Differential Privacy. In: The 42nd ACM Symposium on the Theory of Computing, STOC 2010 (2010)
19. Gupta, A., Ligett, K., McSherry, F., Roth, A., Talwar, K.: Differentially Private Combinatorial Optimization. In: Proceedings of the ACM-SIAM Symposium on Discrete Algorithms (2010)

A Thirty Year Old Conjecture
about Promise Problems

Andrew Hughes[1], A. Pavan[2,*], Nathan Russell[1], and Alan Selman[1]

[1] Department of Computer Science and Engineering, University at Buffalo
{ahughes6,nrussell,selman}@buffalo.edu
[2] Department of Computer Science. Iowa State University
pavan@cs.iastate.edu

Abstract. Even, Selman, and Yacobi [ESY84, SY82] formulated a conjecture that in current terminology asserts that there do not exist disjoint NP-pairs all of whose separators are NP-hard viaTuring reductions. In this paper we consider a variant of this conjecture—there do not exist disjoint NP-pairs all of whose separators are NP-hard via bounded-truth-table reductions. We provide evidence for this conjecture. We also observe that if the original conjecture holds, then some of the known probabilistic public-key cryptosystems are not NP-hard to crack.

1 Introduction

Even, Selman and Yacobi [ESY84, SY82] conjectured that there do not exist certain promise problems all of whose solutions are NP-hard. Specifically, there do not exist disjoint NP-pairs all of whose separators are NP-hard. This conjecture has fascinating (and largely believable) consequences, including that NP differs from co-NP and NP is not equal to UP. Even though this conjecture is 30 years old, we do not know of concrete evidence in support of the conjecture. (We don't know hypotheses that imply all of its consequences.) In this paper, we report some exciting progress on this conjecture. We consider variants of the conjecture and show that under some reasonable hypotheses these variants of the conjecture hold.

A promise problem can be thought of as a disjoint pair—a pair of disjoint sets (Π_y, Π_n), Π_y is called the set of "yes" instances and Π_n is the set of "no" instances. Their union $\Pi_y \cup \Pi_n$ is called the *promise*. The motivation to study disjoint pairs/promise problems stems from their connections to a wide range of question from disperate areas such as public-key cryptosystems, propositional proof systems, study of complete problems for semantic classes, and approximation algorithms. For a recent survey on promise problems, we refer the reader to a survey by Goldreich [Gol06].

For a promise problem (Π_y, Π_n), one is interested in the following computational question: Is there an efficient algorithm that tells whether an instance x lies in Π_y or not, under the promise that x is in $\Pi_y \cup \Pi_n$. The algorithm may give an arbitrary answer if the promise does not hold, i.e., $x \notin \Pi_y \cup \Pi_n$. More

* Research supported in part by CCF: 0916797.

A. Czumaj et al. (Eds.): ICALP 2012, Part I, LNCS 7391, pp. 473–484, 2012.
© Springer-Verlag Berlin Heidelberg 2012

formally a *solution/separator* of a promise problem is any set S that includes Π_y and is disjoint from Π_n. A promise problem is considered easy if it admits a solution in P and is hard if every solution is computationally difficult. The ESY conjecture concerns the computational difficulty of disjoint NP-pairs.

The ESY conjecture has some interesting implication regarding the hardness of public key cryptosystems. Even, Selman, and Yacobi [ESY84] observed that the problem of cracking a PKCS may not formalize as a straightforward decision problem, and it is more natural to formulate it as a promise problem. They associated a promise problem (Π_y, Π_n) to a model of public-key cryptosystems such that both Π_y and Π_n are in NP. A PKCS that fits the model cannot be deemed secure, if the underlying promise problem admits at least one efficient solution. On the other hand if every solution is NP-hard then the system is NP-hard to crack. Thus the ESY conjecture implies that PKCS that fit the model are not NP-hard to crack. We will discuss this further in a later section.

The ESY conjecture is also related to the study of propositional proof systems [Raz94, Pud01]. Razborov observed that every propositional proof system f can be identified with a canonical disjoint NP-pair (SAT^*, REF_f) where REF_f is the set of all formulas that have short proofs of unsatisfiability with respect to f, and SAT^* is a padded version of SAT. Conversely, Glaßer, Selman, and Zhang [GSZ07] showed that for every disjoint NP-pair (A, B) there is a proof system f such that (A, B) is many-one equivalent to (SAT^*, REF_f). Because of this equivalence between propositional proof systems and disjoint NP-pairs, several interesting questions regarding propositional proof systems are related to the structure of disjoint NP-pairs. One of the open questions on propositional proof systems is whether optimal proof systems exist and the belief is that they do not exist. This question is related to the ESY conjecture. It is known that if optimal proof systems do not exist, then a variant of the ESY conjecture holds [GSSZ04].

In addition to connections with PKCS and propositional proof systems, the ESY conjecture has several believable consequences in complexity theory. It is known that this conjecture implies NP differs from co-NP, NP differs from UP, and satisfying assignments of boolean formulas cannot be computed by single-valued NP-machines [ESY84, GS88].

Given its relation to public-key cryptosystems, propositional proof systems, and complexity theory, it is important to understand the power of the ESY conjecture. Is there a reasonable hypothesis that implies the conjecture? To date we do not know reasonable hypotheses that imply ESY, although the analogue of ESY to the c.e. sets is a known theorem [Sch60]. It seems to be difficult to formulate reasonable hypotheses that imply the ESY conjecture, because of its wide range of consequences. Any hypothesis that implies the ESY conjecture immediately implies that NP $\neq$ co-NP, NP $\neq$ UP, and SAT is not in NPSV. None of the standard hypotheses used in complexity theory, such as PH is infinite, E has high circuit complexity, the measure of NP is not zero, and so on, are known to imply all of the above mentioned consequences. This seems to be the root difficulty.

In this paper we make progress toward this question. We consider variants of the ESY conjecture and show that under some reasonable hypotheses these variants follow. Note that ESY states that every disjoint NP-pair has a solution that is not NP-hard via *adaptive reductions*. We can obtain variants of the conjecture by replacing adaptive reductions with more restrictive reductions. Given a reduction type r, the ESY-r conjecture states that every disjoint NP-pair has a solution that is not NP-hard via r-reductions. We know already that if we take r to be many-one reductions, then the ESY-m conjecture is equivalent to NP $\neq$ co-NP [GSSZ04]. What if we take r to be truth-table reductions or bounded truth table reductions?

We first observe that the ESY conjecture for truth-table reductions also has the same set of complexity-theoretic consequences such as NP $\neq$ co-NP, NP $\neq$ UP and SAT is not in NPSV. This suggests that obtaining evidence for the ESY-tt conjecture could be as hard as obtaining evidence for the original conjecture. In this paper we consider bounded-truth-table reductions, these are nonadaptive reductions that make a fixed number of queries,

The first main result of the paper is that if NP $\neq$ co-NP, then every disjoint NP-pair has a solution that is not NP-hard via length-increasing bounded-truth-table reductions (i.e., the ESY conjecture for btt length-increasing reductions hold). By using a stronger hypothesis, we remove the length-increasing restriction. We show that if NP contains certain type of generic sets, then every disjoint NP-pair has a solution that is not NP-hard via bounded-truth-table reductions.

As noted earlier, one of the motivations for introducing the ESY conjecture was its relation to NP-hardness of public key cryptosystems. The analysis of Even, Selman, and Yacobi pertained to the deterministic public-key cryptosystems of that time. In the final section we observe that the ESY conjecture remains relevant to some of the current probabilistic PKCS. If the cracking problem of a current public-key cryptosystem also can be formulated as a disjoint NP pair, and the ESY conjecture holds, then these cryptosystems are not NP-hard to crack.

2   Preliminaries

We assume the standard lexicographic order on strings. We use $x-1$ to denote the immediate predecessor of x in this order. Given a language L and a string x, $L(x)$ is 1 if $x \in L$ otherwise $L(x) = 0$, and $L|x$ is defined as $L(\lambda)L(0)L(1)\cdots L(x-1)$.

A language A is k-truth table reducible to a language B ($A \leq^P_{ktt} B$) if there exist two polynomial-time computable functions f and t such that for every x, $f(x) = \langle q_1, \cdots, q_k \rangle$ and $t(x, B(q_1), \cdots, B(q_k)) = A(x)$. We say that A is bounded truth-table reducible to B if there is a exists a constant $k > 0$ such that $A \leq^P_{ktt} B$.

Definition 2.1. *A function $f : \Sigma^* \to \Sigma^*$ is SNP computable if there is a nondeterministic polynomial-time bounded Turing machine M that for every x at least one path of $M(x)$ outputs $f(x)$ and no path outputs a wrong answer. Some paths may output $\perp$.*

We will also consider strong nondeterministic reductions. These reductions are originally defined by Adleman and Manders [AM77]. We slightly modify their definition to suit our purposes.

Definition 2.2. *Let A and B be two languages. We say that A is* strong nondeterministic k-truth table reducible *to B (denoted $A \leq_{ktt}^{SNP} B$), if there is a polynomial-time computable function f and a SNP computable function t such that every $f(x) = \langle q_1, \cdots q_k \rangle$, and $t(x, B(q_1), \cdots, B(q_k)) = A(x)$.*

Remark. Note that in this definition, the reduction does not use nondeterminism to produce the queries. The original definition of Adleman and Manders allows the query generator f also to be SNP computable.

We say that A is *strong nondeterministic bounded truth table reducible* to B ($A \leq_{btt}^{SNP} B$) if there exists a $k > 0$ such that $A \leq_{ktt}^{SNP} B$.

Definition 2.3. *We say that A is reducible to B via* length-increasing, strong nondeteministic, k-truth table reductions (denoted $\leq_{ktt,li}^{SNP}$) *if $A \leq_{ktt}^{SNP} B$ and the length of every query is larger than the length of the input.*

Notions of length-increasing are defined similarly for $\leq_{ktt}^{P}$, $\leq_{btt}^{P}$, and $\leq_{btt}^{SNP}$ reductions.

2.1   ESY Conjecture

Let (A, B) be a disjoint pair of sets. We say that a set S is a *separator* for (A, B) if $B \subseteq S$ and $A \subseteq \overline{S}$. We now state the original conjecture of Even, Selman, and Yacobi [ESY84].

ESY Conjecture. For every pair of disjoint sets in NP, there is a separator that is not Turing hard for NP.

Although the original conjecture talks about Turing hardness, we can generalize it to arbitrary reductions. Let r be a reduction.

ESY-r Conjecture. For every pair of disjoint sets in NP, there is a separator that is not r-hard for NP.

Although the ESY conjecture stipulates a condition about arbitrary pairs of sets in NP, we can always take one of the sets to be SAT. We state the following observation whose proof appears in the full paper.

Observation 1. *The ESY-r conjecture is equivalent to the following statement: For every set B in NP that is disjoint from SAT, there is a separator that is not r-hard for NP.*

We observe that the ESY-tt conjecture also has the same set of consequences as the original conjecture.

Observation 2. *The ESY-tt conjecture implies that NP $\neq$ UP, NP $\neq$ co-NP, and satisfying assignments of boolean formulas cannot be computed by single-valued NP-machines.*

This observations suggest that providing evidence for the ESY-tt conjecture could be as difficult as providing evidence for the original conjecture. Thus we consider the ESY-btt conjecture.

2.2   Unpredictability

Our results make use of the notion of *unpredictability*, which is similar to the notion of genericity.

Definition 2.4. *We say that a nondeterministic machine M is strong if for every input x, exactly one of the following conditions hold: 1) at least one path of M accepts x and no path rejects, 2) at least one path of M rejects x and no path accepts. Some paths of the machine may output $\bot$.*

Definition 2.5. *Let M be strong nondeterministic machine and L be a language. We say that M is a predictor for L, if for every $x \in L$, M accepts $\langle x, L|x \rangle$ and for every $x \notin L$, M rejects $\langle x, L|x \rangle$.*

Definition 2.6. *Let $t(n)$ be any time bound. We say that a language L is SNTIME($t(n)$)-unpredictable if for every strong nondeterministic machine M that predicts L, M runs for more than $t(n)$ time for all but finitely many inputs of form $\langle x, L|x \rangle$.*

Remark. The running time $t(n)$ of the predictor is in terms of the length of the input which is $\langle x, L|x \rangle$. Measured in terms of the length of x, this time is roughly $t(|x| + 2^{|x|})$. The notion of unpredictability is very similar to the notion of *genericity* [ASFH87, ASNT96]. In fact it is known that for deterministic computations, these two notions are equivalent [BM95].

Definition 2.7. *Let A and L be two languages. We say that L is SNTIME($t(n)$)-unpredictable within A if $L \subseteq A$ and for every strong nondeterministic machine M that predicts L, for all but finitely many x in A, M runs for more than $t(n)$ time on inputs of form $\langle x, L|x \rangle$.*

The following theorem can be shown using strong diagonalization techniques. We omit the proof in the conference version.

Theorem 2.8. *For every $k > 0$, there is a set R such that R is SNTIME($2^{\log^k n}$)-unpredictable within $\overline{\text{SAT}}$.*

3   ESY Conjecture for Bounded Truth-Table Reductions

In this section we provide evidence for the ESY-$\leq^P_{btt}$ conjecture. Before we present our results, we describe the ideas and intuition behind our proofs. Let (SAT, B) be a disjoint NP-pair. Our goal is to exhibit a separator S that is not NP-hard. One trivial way to achieve this is by chosing S to be an easy set—a set in P. However, this approach is not feasible because if NP differs from UP, or

P does not equal NP ∩ co-NP, then (SAT, B) does not have separators in P (for some $B \in$ NP) [GS88]. Our first observation is that there exist "computationally difficult" sets that are not NP-hard, thus we can achieve our goal by taking S to be a difficult set.

Results that concern separating NP-completeness notions [LM96, ASB00, PS04] show that if H is an unpredictable set, then H does not reduce to $H \cup B$. This suggests that we can take $H \cup B$ as our separator and claim that it is not NP-hard. However, we run into at least two major problems. The set $H \cup B$ may not be disjoint from SAT and thus cannot be a separator. In fact, one can show that an unpredictable set H must have an infinite intersection with SAT. We get around this problem by taking H as an unpredictable set within $\overline{SAT}$. This ensures that S is a separator.

The second and the more serious problem is that showing H does not reduce to $H \cup B$ does not imply that $H \cup B$ is not NP-hard as the set H may not be in NP. Instead of working with H, we will argue that SAT does not reduce to S. We will show that if it does, then either we get a predictor for H or making use of nondeterminism, we can reduce the number of queries.

Our first observation is that any reduction from SAT to S must infinitely often produce *relevant queries*—these are queries whose answers, given answers to all other queries, uniquely determine the output of the reduction. We then show that these relevant queries must lie outside of SAT $\cup$ B, if not we can reduce the number of queries by making use of *strong nondeterminsim*. Next we argue that if a query q is relevant, then knowing answers to all other queries help us to determine the membership of $q \in S$, and if q lies outside of SAT $\cup$ B, then this contradicts the unpredictability of the set H.

3.1   Length-Increasing Reductions

In this section we prove that the if NP does not equal co-NP, then the ESY conjecture holds for length-increasing bounded-truth table reductions. In fact, we will show that the conjecture even holds for reductions that use nondeterminism.

Theorem 3.1. *If* NP $\neq$ co-NP, *then the ESY-$\leq_{btt,li}^{SNP}$ conjecture is true.*

Proof. Let (SAT, B) be a disjoint NP-pair. Let Q_1 and Q_2 be two polynomial-time computable relations for SAT and B respectively. Assume that the length of witnesses (for positive instances in SAT and in B) is bounded by n^r, $r > 0$. By Theorem 2.8, there is a set R that is SNTIME($2^{\log^{2r} n}$)-unpredictable with in $\overline{SAT}$. Consider the separator $S = R \cup B$. Suppose that S is $\leq_{ktt,li}^{SNP}$ hard for some $k \geq 0$. We achieve a contradiction to our hypothesis NP $\neq$ co-NP.

We prove this by induction. The base case is when the number of queries is zero. This means that there is an SNP computable function t such that $t(x) = SAT(x)$. This implies that NP = co-NP, a contradiction.

As inductive hypothesis, assume that S is not $\leq_{(\ell-1)tt,li}^{SNP}$-hard. Now assume that SAT $\leq_{\ell tt,li}^{SNP}$-reduces to S via $\langle f, t \rangle$. Given x, let $f(x) = \langle q_1, \cdots, q_\ell \rangle$. We assume that q_ℓ is the largest query and denote it with b_x. We say that a query q_i is *relevant* if the following holds

$$t(x, S(q_1), \cdots S(q_i), \cdots S(q_\ell)) \neq t(x, S(q_1), \cdots \overline{S(q_i)}, \cdots S(q_\ell)).$$

In other words, if q_i is relevant then knowing answers to the all other queries still does not help us determine $\text{SAT}(x)$.

Observation 3. *There exist infinitely many x such that b_x is relevant.*

Proof. Suppose not. Then we have a $\leq^{\text{SNP}}_{(\ell-1)tt,li}$ reduction from SAT to S and this contradicts the induction hypothesis.

Let

$$T = \{x \mid b_x \text{ is relevant}\}.$$

Lemma 3.2. *There exist infinitely many $x \in T$ such that $b_x \notin \text{SAT} \cup B$.*

Proof. Suppose not. For all but finitely many $x \in T$, the query b_x is relevant and belongs to $\text{SAT} \cup B$. Now consider the following reduction $\langle f', t' \rangle$ from SAT to S: On input x, f' will first compute $f(x) = \langle q_1, q_2, \cdots, q_{\ell-1}, b_x \rangle$ and outputs the queries $\langle q_1, \cdots, q_{\ell-1} \rangle$. We now describe t':

1. Let $c_1 = S(q_1), \cdots, c_{\ell-1} = S(q_{\ell-1})$.
2. Determine whether b_x is relevant or not by comparing $t(x, c_1, \cdots, c_{\ell-1}, 0)$ with $t(x, c_1, \cdots, c_{\ell-1}, 1)$. If b_x is not relevant, then output $t(x, c_1, \cdots, c_{\ell-1}, 0)$.

3. Guess a witness $w \in \Sigma^{n^r}$. If $Q_1(b_x, w)$ holds, then output $t(x, c_1, \cdots, c_{\ell-1}, 0)$.
4. If $Q_1(b_x, w)$ does not hold, then guess a witness $u \in \Sigma^{n^r}$. If $Q_2(b_x, u)$ holds then output $t(x, c_1, \cdots, c_{\ell-1}, 1)$, else output $\perp$.

We claim that the above is a $\leq^{\text{SNP}}_{(\ell-1)tt,li}$ reduction from SAT to S. Clearly f' produces only $\ell - 1$ queries. If b_x is not relevant, then the reduction is correct. Suppose that b_x is relevant. By our assumption $b_x \in \text{SAT} \cup B$. If $b_x \in \text{SAT}$, then $b_x \notin S$. Thus $t(x, c_1, \cdots, c_{k-1}, 0) = \text{SAT}(x)$. If $b_x \in B$, then $b_x \in S$. Thus $t(x, c_1, \cdots, c_{k-1}, 1) = \text{SAT}(x)$. Thus the reduction is always correct.

It remains to show that this is an SNP reduction. Clearly all queries are produced by a deterministic polynomial-time process. Step 2 computes the function t. However t is SNP-computable. So this step can be done via an SNP-machine. Suppose $b_x \in \text{SAT}$. Then there is a $w \in \Sigma^{n^r}$ such that $Q_1(b_x, w)$ holds, and thus this path outputs the correct answer. Since SAT is disjoint from B, for every $u \in \Sigma^{n^r}$ $Q_2(b_x, u)$ does not hold. Thus no path outputs the wrong answer. A similar argument shows that when $b_x \in B$, at least one path outputs the correct answer and no path outputs the wrong answer.

Thus SAT $\leq^{\text{SNP}}_{(\ell-1)tt,li}$ reduces to S. This contradicts our induction hypothesis. This complete the proof of the lemma.

Now, we return to the proof of the theorem. Lemma 3.2 gives the following corollary.

Corollary 3.3. *There exist infinitely many $y \notin \text{SAT} \cup B$ with the following property: There exists an x, $|x| < |y|$ such that $y = b_x$ and y is relevant.*

This enables us to build the following predictor for R. Let M be a strong nondeterministic algorithm that decides R.

1. Input $\langle y, R|y \rangle$.
2. If $y \in \text{SAT} \cup B$, then run $M(y)$ and stop.
3. Search for an x such that $|x| < |y|$ and $b_x = y$. If no such x is found run $M(y)$ and stop.
4. Let $f(x) = \langle q_1, \cdots, q_{\ell-1}, y \rangle$. Compute $c_i = S(q_i)$, $1 \leq i \leq \ell - 1$ by
   (a) Decide the membership of $q_i \in B$, by running a brute force algorithm for B
   (b) Decide the membership of $q_i \in R$, by looking at $R|y$.
5. Check if y is relevant or not by comparing $t(x, c_1, \cdots, c_{\ell-1}, 0)$ and $t(x, c_1, \cdots, c_{\ell-1}, 1)$. If y is not relevant, then run $M(y)$ and stop.
6. Now we know that y is relevant. Compute $\text{SAT}(x)$. Find the unique bit b such that $\text{SAT}(x) = t(x, c_1, \cdots, c_{\ell-1}, b)$.
7. Accept if and only if b equals 1.

Claim. The above predictor correctly decides R and for infinitely many strings from $\overline{\text{SAT}}$ runs in time $2^{\log^{2^r} n}$.

Proof. Let I be the set of all y for which the conditions of Corollary 3.3 holds. The above predictor runs $M(y)$ on any y that is not in I and thus is correct on all such y. Let $y \in I$. We know that $\text{SAT}(x) = t(x, c_1, \cdots, c_{\ell-1}, S(y))$. Since y is relevant $\text{SAT}(x) \neq t(x, c_1, \cdots, c_{\ell-1}, \overline{S(y)})$. Thus $b = S(y)$. Since $y \notin \text{SAT} \cup B$, $y \in S$ if and only if $y \in R$. Thus the above predictor correctly decides every y in I.

Now we will show that for every $y \in I$, the above predictor halts in quasi-polynomial time. Let $|y| = m$, note that the length of x found in step 3 is at most m. Checking for membership of y in $\text{SAT} \cup B$ takes $O(2^{m^r})$ time. Since $y = b_x$ is the largest query produced, $|q_i| \leq m$, $1 \leq i \leq \ell - 1$. Since B can be decided in time 2^{n^r}, Step 4a takes $O(2^{m^r})$ time. Since $y > q_i$, $1 \leq i \leq \ell - 1$, Step 4b takes polynomial time. Computing $\text{SAT}(x)$ takes $O(2^m)$ time. The predictor computes the function t. However t is SNP computable. Thus the total time taken is $O(2^{m^{r+1}})$. Note that the run time of the predictor is measured in terms of length of $\langle y, R|y \rangle$ which is at least 2^m. Thus for every $y \in I$, the predictor runs in time $2^{\log^{2^r} n}$ time. Since I is an infinite set and by definition is a subset of $\overline{\text{SAT}}$, the claim follows.

We have shown that S is not $\leq_{\ell tt,li}^{\text{SNP}}$ hard for NP. This completes the induction step. Thus S is not $\leq_{btt,li}^{\text{SNP}}$ hard for NP. This completes the proof of the Theorem.

Since every length increasing bounded truth-table reduction is trivially a $\leq_{btt,li}^{\text{SNP}}$ reduction, our main result of this section is a corollary of the above theorem.

Theorem 3.4. *If $\text{NP} \neq \text{co-NP}$, then the ESY-$\leq_{btt,li}^{\text{P}}$ conjecture holds.*

3.2 General Reductions

In this section we show that if NP contains unpredictable sets, then the ESY-$\leq^P_{btt}$ conjecture holds without the length increasing restriction.

Theorem 3.5. *If* NP *has an* SNTIME(n^2) *unpredictable set, then the ESY-$\leq^P_{btt}$ conjecture holds.*

Due to lack of space we present the proof of this theorem in the full paper.

Power of the Hypothesis. We will now make a few remarks about the hypothesis in the above theorem and connect it to earlier used hypotheses. We will first make a few informal observations. The notion of unpredictability attempts to capture the difficulty of a language given some auxiliary information: For a language L how easy/difficult is to determine membership of $x \in L$ *given $L|x$ as auxiliary information*? Many natural problems (for example SAT) turn out to be very easy in this model. Do there exist languages that are difficult to compute even when the partial characteristic sequence is given as auxiliary input? It turns out that EXP, somewhat surprisingly, contains such languages. Our hypothesis asserts that NP also contains such languages.

We now connect our hypothesis to a few hypotheses regarding bi-immunity and genericity . Say that a language is NP ∩ co-NP bi-immune if every strong nondeterministic machine that decides L takes more than polynomial-time on all but finitely many inputs. It is easy to see that if our hypothesis holds, then NP contains NP ∩ co-NP bi-immune sets.

Our hypothesis is similar to, but stronger than, the genericity hypothesis of Ambos-Spies et al. The genericity hypothesis asserts that NP contains n^2-generic languages. This hypothesis is shown to have several interesting and believable consequences [ASFH87] [ASNT96]. In the definition of unpredictability, if we replace strong nondeterministic machines by deterministic machines, then it coincides with genericity [BM95]. That is, the statements "L is DTIME($t(n)$) unpredictable" and "L is $t(n)$-generic" are equivalent. Since our hypothesis concerns with strong nondeterministic predictors, our hypothesis can be taken as "NP contains SNTIME(n^2) (or simply NP ∩ co-NP) generic sets".

4 Application to Probabilistic Encryption

Although the ESY conjecture was originally formulated to capture the difficulty of cracking deterministic PKCS, we observe that if it holds, then certain "probabilistic" encryption schemes including the Goldwasser-Micali [GM84], Gentry [Gen09] and Ajtai-Dwork [AD97] systems also cannot be NP-hard to crack.

A probabilistic public-key cryptosystem consists of three publicly known, polynomial-time computable functions, encryption function E, decryption function D and a key generator G. For a randomly generated string X, $G(X)$ generates the pair (k_1, k_2), where k_1 is the public key and k_2 is the private private key. Given a plain text m, the encryption function randomly picks a string r, and generates cipher text $E(m, r, k_1) = c$. The decryption function D has the property that $D(c, k_2) = m$, if c is a valid cipher text for m. We say that the cryptosystem is *error-free* if whenever m and m' are two distinct messages, then for every r and public key k_1, $E(m, r, k_1) \neq E(m', r, k_1)$.

We will now observe that the cracking problem of every error-free public-key cryptosystem can be formulated as a disjoint NP pair. Given such a cryptosystem, let $\Pi_n = \{\langle c, k_1, m' \rangle \mid \exists m, r, X \text{ and } k_2 \text{ such that } E(m, r, k_1) = c \text{ and } G(X) = \langle k_1, k_2 \rangle \text{ and } m < m'\}$, and let $\Pi_y = \{\langle c, k_1, m' \rangle \mid \exists m, r, X \text{ and } k_2 \text{ such that } E(m, r, k_1) = c \text{ and } G(X) = \langle k_1, k_2 \rangle \text{ and } m \geq m'\}$. Since the cryptosystem is error free we have that $\Pi_y \cap \Pi_n = \emptyset$ and both Π_y and Π_n are in NP. Thus (Π_y, Π_n) is a disjoint NP pair. Clearly a separator for this pair can be used to crack the cryptosystem. Thus if the ESY conjecture holds, then this problem has a separator that is not NP-hard. Since the cryptosystem of Goldwasser-Micali and Gentry's [Gen09] homomorphism cryptosystem are error free, we have the following result.

Theorem 4.1. *If the ESY conjecture holds, then the Goldwasser-Micali cryptosystem, as well as Gentry's homomorphic cryptosystem, cannot be NP-hard to crack.*

Now we consider the Ajtai-Dwork [AD97] cryptosystem. This cryptosystem has the property that 0 is encrypted as a lattice point near one of the hidden hyperplanes that constitute the private key. The bit 1 is encrypted as a random lattice point so that, with low probability, 1 might be encrypted as a cyphertext that is also a valid encryption of 0, and thus the system is not error-free. We now formulate the cracking problem in terms of detecting encryptions of 0. We let $\Pi_y = \{\langle k_1, c \rangle \mid \exists X \text{ such that } G(X) = \langle k_1, k_2 \rangle \text{ for some } k_2 \text{ and } D(c, k_2) = 0\}$ and $\Pi_n = \{\langle k_1, c \rangle \mid \exists X \text{ such that } G(X) = \langle k_1, k_2 \rangle \text{ for some } k_2 \text{ and } D(c, k_2) = 1\}$. Clearly, both Π_y and Π_n are in NP. The pair is disjoint since no message decrypts to boith 0 and 1. Thus we have the following result.

Theorem 4.2. *If the ESY conjecture holds, then in the Ajtai-Dwork cryptosystem, it is not NP-hard to determine whether a given cypher text is a valid encryption of 0.*

We note that Nguyen and Stern [NS98] showed that if the polynomial-time hierarchy is infinite, then the Ajtai-Dwork cryptosystem is not NP-hard to crack.

5   Discussion

In this paper we provide evidence that the ESY conjecture holds when we restrict the reduction types. We note that we can relax the length-increasing restriction in Theorem 3.1 to "all queries are lexicographically larger than the input". Theorem 3.5 removes the restriction but requires a much stronger hypothesis. Can we weaken this hypothesis? One way to proceed is to show that every $\leq_{btt}^{P}$-hard set for NP is hard via length-increasing $\leq_{btt}^{P}$ reductions. We note that there are several results that indicate that all many-one complete sets for NP are complete via length-increasing reductions [Agr02, HP07, BHHT10, GHP10]. Perhaps one can use similar ideas to show that $\leq^{P}_{btt}$-hard sets are hard via length-increasing reductions. Another question is whether we can replace btt reductions with reductions that make $O(\log n)$ (or n^ϵ) nonadaptive queries.

As noted in the preliminaries, the ESY conjecture and ESY-tt conjecture both imply that NP differs from UP, and we believe that the ESY-btt conjecture does not imply NP $\neq$ UP. Is there an oracle relative to which the ESY-$\leq^{P}_{btt}$ conjecture holds and NP = UP? As noted in the introduction, the ESY-m conjecture is equivalent to NP $\neq$ co-NP, and there is an oracle relative to which the ESY-m conjecture holds and NP = UP [For]. Can we show that the ESY-btt conjecture is also equivalent to NP $\neq$ co-NP or is there an oracle against it? We mention that there exists an oracle relative to which the original ESY conjecture holds [GSSZ04].

References

[AD97]   Ajtai, M., Dwork, C.: A public-key cryptosystem with worst-case/average-case equivalence. In: Proceedings of the Twenty-Ninth Annual ACM Symposium on Theory of Computing, STOC 1997, pp. 284–293. ACM, New York (1997)

[Agr02]   Agrawal, M.: Pseudo-random generators and structure of complete degrees. In: 17th Annual IEEE Conference on Computational Complexity, pp. 139–145 (2002)

[AM77]   Adleman, L., Manders, K.: Reducibility, randomness, and intractability. In: Proc. 9th ACM Symp. Theory of Computing, pp. 151–163 (1977)

[ASB00]   Ambos-Spies, K., Bentzien, L.: Separating NP-completeness under strong hypotheses. Journal of Computer and System Sciences 61(3), 335–361 (2000)

[ASFH87]   Ambos-Spies, K., Fleischhack, H., Huwig, H.: Diagonalizations over polynomial time computable sets. Theoretical Computer Science 51, 177–204 (1987)

[ASNT96]   Ambos-Spies, K., Neis, H., Terwijn, A.: Genericity and measure for exponential time. Theoretical Computer Science 168(1), 3–19 (1996)

[ASTZ97]   Ambos-Spies, K., Terwijn, A., Zheng, X.: Resource bounded randomness and weakly complete problems. Theoretical Computer Science 172(1), 195–207 (1997)

[BHHT10]   Buhrman, H., Hescott, B., Homer, S., Torenvliet, L.: Non-uniform reductions. Theory of Computing Systems 47(2), 317–241 (2010)

[BM95]   Balcazar, J., Mayordomo, E.: A note on genericty and bi-immunity. In: Proceedings of the Tenth Annual IEEE Conference on Computational Complexity, pp. 193–196 (1995)

[ESY84]   Even, S., Selman, A., Yacobi, Y.: The complexity of promise problems with applications to public-key cryptography. Information and Control 61(2), 159–173 (1984)

[For]   Fortnow, L.: Personal Communication

[Gen09]   Gentry, C.: Fully homomorphic encryption using ideal lattices. In: STOC 2009, pp. 169–178 (2009)

[GHP10]   Gu, X., Hitchcock, J., Pavan, A.: Collapsing and separating completeness notions under average-case and worst-case hypotheses. In: STACS 2010. LIPIcs, vol. 5, pp. 429–440. Schloss Dagstuhl - Leibniz-Zentrum fuer Informatik (2010)

[GM84]   Goldwasser, S., Micali, S.: Probabilistic encryption. J. Comp. System Sci. 28, 270–299 (1984)

[Gol06]   Goldreich, O.: On Promise Problems: A Survey. In: Goldreich, O., Rosenberg, A.L., Selman, A.L. (eds.) Shimon Even Festchrift. LNCS, vol. 3895, pp. 254–290. Springer, Heidelberg (2006)

[GS88]   Grollmann, J., Selman, A.: Complexity measures for public-key cryptosystems. SIAM Journal on Computing 17(2), 309–355 (1988)

[GSSZ04]   Glaßer, C., Selman, A., Sengupta, S., Zhang, L.: Disjoint NP-pairs. SIAM J. Comput. 33(6), 1369–1416 (2004)

[GSZ07]   Glaßer, C., Selman, A., Zhang, L.: Canonical disjoint NP-pairs of propositional proof systems. Theoretical Computer Science 370(1), 60–73 (2007)

[HP07]   Hitchcock, J., Pavan, A.: Comparing reductions to NP-complete sets. Information and Computation 205(5), 694–706 (2007); In: Bugliesi, M., Preneel, B., Sassone, V., Wegener, I. (eds.) ICALP 2006. LNCS, vol. 4051, pp. 465–476. Springer, Heidelberg (2006)

[LM96]   Lutz, J.H., Mayordomo, E.: Cook versus Karp-Levin: Separating completeness notions if NP is not small. Theoretical Computer Science 164, 141–163 (1996)

[NS98]   Nguyên, P.Q., Stern, J.: Cryptanalysis of the Ajtai-Dwork Cryptosystem. In: Krawczyk, H. (ed.) CRYPTO 1998. LNCS, vol. 1462, pp. 223–242. Springer, Heidelberg (1998)

[PS02]   Pavan, A., Selman, A.: Separation of NP-completeness notions. SIAM Journal on Computing 31(3), 906–918 (2002)

[PS04]   Pavan, A., Selman, A.: Bi-immunity separates strong NP-completeness notions. Information and Computation 188, 116–126 (2004)

[Pud01]   Pudlak, P.: On reducibility and symmetry of disjoint NP-pairs. In: Electronic Colloquium on Computational Complexity, technical reports (2001)

[Raz94]   Razborov, A.: On provably disjoint NP pairs. Technical Report 94-006, ECCC (1994)

[Sch60]   Schoenfield, J.: Degrees of models. Journal of Symbolic Logic 25, 233–237 (1960)

[SY82]   Selman, A., Yacobi, Y.: The Complexity of Promise Problems. In: Nielsen, M., Schmidt, E.M. (eds.) ICALP 1982. LNCS, vol. 140, pp. 502–509. Springer, Heidelberg (1982)

Minimum Latency Submodular Cover[*]

Sungjin Im[1],[**], Viswanath Nagarajan[2], and Ruben van der Zwaan[3]

[1] Department of Computer Science, University of Illinois, USA
im3@illinois.edu
[2] IBM T. J. Watson Research Center, USA
viswanath@us.ibm.com
[3] Maastricht University, The Netherlands
r.vanderzwaan@maastrichtuniversity.nl

Abstract. We study the submodular ranking problem in the presence of metric costs. The input to the *minimum latency submodular cover* (MLSC) problem consists of a metric (V, d) with source $r \in V$ and m monotone submodular functions $f_1, f_2, ..., f_m : 2^V \rightarrow [0, 1]$. The goal is to find a path originating at r that minimizes the total cover time of all functions; the cover time of function f_i is the smallest value t such that f_i has value one for the vertices visited within distance t along the path. This generalizes many previously studied problems, such as *submodular ranking* [1] when the metric is uniform, and *group Steiner tree* [14] when $m = 1$ and f_1 is a coverage function. We give a polynomial time $O(\log \frac{1}{\epsilon} \cdot \log^{2+\delta} |V|)$-approximation algorithm for MLSC, where $\epsilon > 0$ is the smallest non-zero marginal increase of any $\{f_i\}_{i=1}^m$ and $\delta > 0$ is any constant. This result is enabled by a simpler analysis of the submodular ranking algorithm from [1].

We also consider the *stochastic submodular ranking* problem where elements V have random instantiations, and obtain an adaptive algorithm with an $O(\log 1/\epsilon)$ approximation ratio, which is best possible. This result also generalizes several previously studied stochastic problems, eg. *adaptive set cover* [15] and *shared filter evaluation* [24,23].

1 Introduction

Ordering a set of elements so as to be simultaneously good for several valuations is an important issue in web-search ranking and broadcast scheduling. A formal model for this was introduced by Azar et al. [2] where they studied the *multiple intents re-ranking* problem (a.k.a.*generalized min-sum set cover* [3]). Subsequently, Azar and Gamzu [1] studied the *submodular ranking problem* where the valuations can be arbitrary monotone submodular functions.

In this paper, we extend these models to the setting of general metric switching costs. This allows us to handle additional issues such as: *Data locality:* the time taken to read/transmit data j after data i is $d(i, j)$; and *Context switching:* it takes $d(i, j)$ time for a user to parse data j when scheduled after data i.

---

[*] A full version of this extended abstract appears as [21].
[**] This work was partially supported by NSF grant CCF-1016684.

A. Czumaj et al. (Eds.): ICALP 2012, Part I, LNCS 7391, pp. 485–497, 2012.
© Springer-Verlag Berlin Heidelberg 2012

We study the *minimum latency submodular cover* problem (MLSC), which is the metric version of submodular ranking [1], and its interesting special case, the *latency covering Steiner tree* problem (LCST), which extends *generalized min-sum set cover* [2,3]. The formal definitions follow shortly. We obtain polylogarithmic approximation guarantees for both problems. We remark that due to a relation to the well-known *group Steiner tree* [14] problem, any significant improvement on our results would lead to a similar improvement for group Steiner tree. The MLSC problem is a common generalization of several previously studied problems [14,22,13,18,8,2,1]; see also Figure 1.

In a somewhat different direction, we also study the *stochastic submodular ranking* problem, where the goal is to *adaptively* schedule stochastic elements so as to minimize the expected total cover time. This problem models situations where web pages can be adaptively presented using user feedback. We obtain an $O(\log \frac{1}{\epsilon})$-approximation algorithm for this problem, which is known to be best possible even in the deterministic setting [1]. Moreover, this result generalizes and gives a unified analysis of many previous results [15,24,23,1].

Problem Definition. We let V denote the ground set of elements/vertices, and $d : \binom{V}{2} \to \mathbb{R}_+$ a distance function that is symmetric and satisfies triangle inequality. Recall that a function $f : 2^V \to \mathbb{R}_+$ is *submodular* if, for any $A, B \subseteq V$, $f(A) + f(B) \geq f(A \cup B) + f(A \cap B)$; and it is *monotone* if for any $A \subseteq B$, $f(A) \leq f(B)$. We assume some familiarity with submodular functions [26].

The input to the *minimum latency submodular cover problem* (MLSC) consists of a metric (V, d) with a specified root vertex $r \in V$, and m monotone submodular functions $f_1, \ldots, f_m : 2^V \to \mathbb{R}_+$ representing the valuations of different users. We assume, without loss of generality by truncation, that $f_i(V) = 1$ for all $i \in [m]$. Function f_i is said to be *covered* (or satisfied) by set $S \subseteq V$ if $f_i(S) = 1 = f_i(V)$. The *cover time* of function f_i in a path π is the length of the shortest prefix of π that has f_i value one, i.e.

$$\min \, t \, : \, f_i(\{v \in V : v \text{ appears within distance } t \text{ on } \pi\}) = 1.$$

The objective in MLSC is to compute a path originating at r and visiting all vertices that minimizes the sum of cover times of all functions. We recover the *submodular ranking* problem [1] as the special case when metric d is uniform (i.e. all pairwise distances are one). A technical parameter defined in [1] (also implicit in [27]) that we use to measure performance is ϵ which is the smallest non-zero marginal increase of any function $\{f_i\}_{i=1}^m$.

As shown in [1], the *submodular ranking* (SR) problem contains *set-cover* as a special case (even when $m = 1$). Similarly, MLSC generalizes the *group Steiner tree* problem [14], where given a metric (V, d) and N groups of vertices $\{g_i \subseteq V\}_{i=1}^N$, the goal is to find a minimum length tree that contains at least one vertex from each group $\{g_i\}_{i=1}^N$.

The *latency covering Steiner tree* problem (LCST) is a natural special case of MLSC, where each function f_i is associated with a group $g_i \subseteq V$ and requirement $k_i \leq |g_i|$ where $f_i(S) = \min\{|g_i \cap S|/k_i, 1\}$. Here $\epsilon = 1/\max_{i=1}^m k_i$. The uniform metric special case of LCST reduces to *generalized min-sum set cover* [2,3]. When $\max_{i=1}^m k_i = 1$ in LCST, we obtain *latency group Steiner tree* [18,8].

Figure 1 shows the relationship between previously studied special cases of MLSC. Due to lack of space we defer further discussion to the full version [21].

In *stochastic submodular ranking* we are given a set V of stochastic elements, each having an independent distribution over certain domain Δ. The submodular functions are also defined on ground set Δ, i.e. $f_1, ..., f_m : 2^\Delta \rightarrow [0, 1]$. In addition, each element $i \in V$ has a deterministic cost/time ℓ_i to be scheduled. The realization (from Δ) of any element is known immediately after scheduling it. The goal is to find an adaptive ordering of V that minimizes the total *expected* cover time. Since elements are stochastic, it is possible that a function is never covered: in such cases we just fix the cover time to be $\sum_{i \in V} \ell_i$ (which is the total duration of any schedule). We are concerned with *adaptive* algorithms. Such an algorithm is allowed to decide the next element to schedule based on the instantiations of the previously scheduled elements.

Remark: Our approach does not seem to extend directly to the *stochastic* MLSC (i.e. on general metrics). We leave this as an open question.

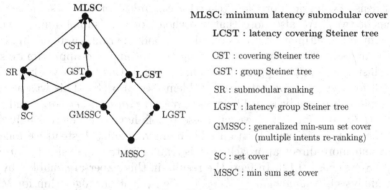

MLSC: minimum latency submodular cover

LCST : latency covering Steiner tree

CST : covering Steiner tree

GST : group Steiner tree

SR : submodular ranking

LGST : latency group Steiner tree

GMSSC : generalized min-sum set cover
(multiple intents re-ranking)

SC : set cover

MSSC : min sum set cover

Fig. 1. An arrow from X to Y means X is a special case of Y

Results, Techniques and Outline. We start with the *minimum latency submodular cover problem* (MLSC) problem, for which we obtain:

Theorem 1. *For any constant $\delta > 0$, there is an $O(\log \frac{1}{\epsilon} \cdot \log^{2+\delta} |V|)$-approximation algorithm for the minimum latency submodular cover problem.*

Note that in the special case of *group Steiner tree*, this result is larger only by a factor of $O(\log^\delta |V|)$ than its best known approximation ratio of $O(\log N \log^2 |V|)$, due to Garg et al. [14]. Our algorithm uses the framework of [1] and the *submodular orienteering problem* (SOP) [11] as a sub-routine. The input to SOP consists of metric (V, d), root r, monotone submodular function $f : 2^V \rightarrow \mathbb{R}_+$ and length bound B. The goal is to find a path originating at r having length at most B that maximizes $f(S)$, where $S \subseteq V$ is the set of vertices visited in the path. Specifically, we show that an (ρ, σ)-bicriteria approximation algorithm for SOP can be used to obtain an $O(\rho \sigma \cdot \log \frac{1}{\epsilon})$-approximation algorithm for MLSC; here the algorithm is allowed to return a path of length at most σB. To obtain Theorem 1 we use an $\left(O(1), O(\log^{2+\delta}|V|)\right)$-bicriteria approximation for SOP that follows from [6,10].

Our algorithm for MLSC is an extension of the elegant "adaptive residual updates scheme" of Azar and Gamzu [1] for *submodular ranking* (i.e. uniform metric MLSC). As shown in [1], an interesting aspect of this problem is that the natural greedy algorithm, based on absolute contribution of elements, performs very poorly. Instead they used a modified greedy algorithm that selects one element at a time according to residual coverage. In the MLSC setting of general metrics, our algorithm uses a similar residual coverage *function* to repeatedly augment the solution. However our augmentations are paths of geometrically increasing lengths, instead of just one element. A crucial point in our algorithm is that the residual coverage functions are always submodular, and hence we can use *submodular orienteering* (SOP) in the augmentation step.

We remark that the approach of covering the maximum number of functions within geometrically increasing lengths fails because the residual coverage function here is non-submodular; in fact as noted in [3] this subproblem contains the difficult dense-k-subgraph problem (even for *generalized min-sum set cover* with requirement two). We also note that the choice of our (submodular) residual coverage function ultimately draws on the submodular ranking algorithm [1].

The analysis in [1] was based on viewing the optimal and approximate solutions as histograms. This approach was first used in this line of work by Feige et al. [13] for the *min-sum set cover* problem (see also [4]). This was also the main framework of analysis in [2] for *generalized min-sum set cover* and then for *submodular ranking* [1]. However, these proofs have been increasingly difficult as the problem in consideration adds more generality. Instead we follow a different and more direct approach that is similar to the analysis of *minimum latency TSP*, see eg. [9,12]. In fact, the results in this paper are enabled by our simpler analysis of *submodular ranking* [1]. We present our algorithm for MLSC in Section 2, which yields Theorem 1.

Our second main result is for the *latency covering Steiner tree* (LCST) problem. Note that Theorem 1 implies directly an $O(\log k_{max} \cdot \log^{2+\delta}|V|)$-approximation algorithm for LCST. In the full version [21] we show:

Theorem 2. *There is a polynomial-time $O(\log^2 |V|)$-approximation algorithm for latency covering Steiner tree.*

The main idea in this result is a new LP relaxation for covering Steiner tree (using *Knapsack Cover* type inequalities [7]) having a poly-logarithmic integrality gap. All previous algorithms [22,19] for covering Steiner tree were based on iteratively solving an LP with large integrality gap– this approach does not seem suitable to the *latency* version we consider. As shown in [25], any improvement over Theorem 2 even in the $k_{max} = 1$ special case (i.e. *latency group Steiner tree*) would yield an improved approximation ratio for *group Steiner tree*, which is a long-standing open question.

Our final result is for the *stochastic submodular ranking* problem. As shown in [15,16], even special cases of this problem have polynomially large adaptivity gap (ratio between the optimal non-adaptive and adaptive solutions).

This motivates adaptive algorithms, and we obtain the following result in Section 3.

Theorem 3. *There is an adaptive $O(\log \frac{1}{\epsilon})$-approximation algorithm for the stochastic submodular ranking problem.*

In particular, we show that the most natural stochastic extension of the algorithm from [1] achieves this approximation factor. We remark that the analysis in [1] of deterministic submodular ranking assumed unit costs, whereas Theorem 3 holds for the stochastic setting even with non-uniform costs $\{\ell_i\}$.

As mentioned before, our results generalize the results in [15,24,23] which study (some variants of) *stochastic set cover*. Our analysis is arguably simpler and more transparent than [23], which gave the first tight analysis of these problems. We note that [23] used an intricate charging scheme with "dual prices" and it does not seem directly applicable to general submodular functions.

Previous Work. The *submodular ranking* problem was introduced by Azar and Gamzu [1] where they obtained a tight $O(\log \frac{1}{\epsilon})$-approximation algorithm.

The first poly-logarithmic approximation for group Steiner tree was $O(\log N \log^2 |V|)$ due to Garg et al. [14] which is still the best known bound. Calinescu and Zelikovsky [6] building on [10], gave an algorithm for covering any submodular function in a metric. We use this algorithm in the submodular orienteering (SOP) subroutine for our MLSC result. We note that an $\Omega(\log^{2-\delta} |V|)$ hardness of approximation is known for *group Steiner tree* (even on tree metrics) due to Halperin and Krauthgamer [20].

The *adaptive set cover* problem introduced in [15] is clearly a special case of stochastic SR that we consider; the authors showed a large adaptivity gap for set cover, and a logarithmic approximation for a variant with multiplicities. A related problem in context of fast query evaluation was studied in [24], where the authors gave a triple logarithmic approximation. This bound was improved to the optimal logarithmic ratio by [23]; this result was also applicable to adaptive set cover. Another related paper is [16], where they defined a general property "adaptive submodularity" and showed nearly optimal approximation guarantees for several objectives. The result in [16] most relevant to stochastic SR problem is the 4-approximation for stochastic *min sum set cover*. This required a *fixed* submodular function f such that the objective is $\mathbb{E}\left[\sum_{t\geq 0} f(\overline{V}) - f(\overline{\pi}_t)\right]$ where $\overline{\pi}_t$ is the instantiation of elements scheduled within time t and $\overline{V}$ denotes the instantiation of all elements. However, as mentioned earlier this is not the case even for *generalized min-sum set cover* with requirements two. Recently [17] studied the *submodular ranking* problem in an online regret setting, which is different from the adaptive model we consider.

Preliminaries. When dealing with a submodular function $f : 2^V \to \mathbb{R}_+$, we use the standard convention that a value oracle for f is available. The oracle takes as input a subset $S \subseteq V$ and returns the function value $f(S)$ in constant time. We use the following claim from [1].

Claim 1 (Claim 2.3 in [1]). *Given a monotone function* $f : 2^{[n]} \to [0,1]$ *and sets* $\emptyset = S_0 \subseteq S_1 \subseteq \cdots \subseteq S_\ell \subseteq [n]$, *we have (using the convention* $0/0 = 0$)

$$\sum_{k=1}^{\ell} \frac{f(S_k) - f(S_{k-1})}{1 - f(S_{k-1})} \leq 1 + \ln \frac{1}{\epsilon}.$$

Here $\epsilon > 0$ *is such that for any* $A \subseteq B$, *if* $f(B) - f(A) > 0$ *then* $f(B) - f(A) \geq \epsilon$.

We use ALG to denote the cost of the algorithm we consider. For notational simplicity, we let OPT denote the optimal solution itself or the optimal cost depending on the context. For any integer $k \geq 1$ we let $[k] := \{1, 2, \ldots, k\}$.

2   Algorithm for Minimum Latency Submodular Cover

As mentioned earlier, our algorithm for MLSC uses as sub-routine an algorithm for submodular orienteering (SOP). We assume a (ρ, σ)-bicriteria approximation algorithm, i.e., on any SOP instance with objective function f and length bound B, it returns a path P of length at most $\sigma \cdot B$ and $f(V(P)) \geq \text{OPT}/\rho$. We recall the following known result on SOP.

Theorem 4 ([6]). *For any constant* $\delta > 0$ *there is a polynomial time* $(O(1), O(\log^{2+\delta} |V|))$ *bicriteria approximation algorithm for SOP.*

Algorithm ALG-MLSC below uses the (ρ, σ) bicriteria approximation algorithm ALG-SOP. Here $\alpha = 1 + \ln \frac{1}{\epsilon}$. Note the difference from the submodular ranking algorithm [1]: here each augmentation is a path possibly covering several vertices. Despite the similarity of ALG-MLSC to the min-latency TSP type algorithms [9,12] an important difference is that we *do not* try to directly maximize the number of covered functions in each augmentation: as noted before this subproblem is at least as hard as dense-k-subgraph, for which the best approximation ratio known is only polynomial [5]. Instead we maximize in each step some proxy residual coverage function f^S that suffices to eventually cover all functions quickly. This function is a natural extension of the single-element coverage values used in ALG-AG [1]. It is important to note that in Line (4), $f^S(\cdot)$ is defined based on the current set S of visited vertices in each iteration. Moreover, since each function f_i is monotone submodular, so is f^S for any $S \subseteq V$. In Line (6), $\pi \cdot P$ implies the concatenation of π and P.

We prove the following theorem, which implies Theorem 1.

Theorem 5. *ALG-MLSC is an* $O(\alpha \rho \sigma)$-*approximation algorithm for MLSC.*

We now analyze ALG-MLSC. We say that the algorithm is in the j-th *phase*, when the variable k of the for loop in Line (2) has value j. Observe that the final solution visits all vertices that are added in the j-th phase within time $16\alpha\rho 2^j$. This can be easily shown as follows: the final solution is a concatenation of the paths that were found in Line (5). Since all these paths are stitched at the root r, the length of π at the end of phase j is at most $\sum_{k=1}^{j} 2 \cdot 4\alpha\rho \cdot \sigma 2^k \leq 16\alpha\rho\sigma \cdot 2^j$. The following proposition easily follows.

Algorithm 1. ALG-MLSC

INPUT: $(V, d), r \in V; \{f_i : 2^V \to [0,1]\}_{i=1}^m$.

1: $S \leftarrow \emptyset, \pi \leftarrow \emptyset$.
2: **for** $k = 0, 1, 2, \dots$ **do**
3: **Repeat** the following $4\alpha\rho$ times.
4:   Define submodular function $f^S(T) := \sum_{i \in [m], f_i(S) < 1} \frac{f_i(S \cup T) - f_i(S)}{1 - f_i(S)}, \forall T \subseteq V$.
5:   Use ALG-SOP to find a path P of length at most $\sigma \cdot 2^k$ starting from r that maximizes $f^S(V(P))$ (up to factor ρ); $V(P)$ is the set of vertices visited by P.
6:   $S \leftarrow S \cup V(P)$ and $\pi \leftarrow \pi \cdot P$.
7: **end for**

OUTPUT: Output solution π.

Proposition 1. *Any vertex v added to S in the j-th phase is visited by π within distance $16\alpha\rho\sigma \cdot 2^j$.*

Let $R(t) \subseteq [m]$ denote the set of (indices of) functions that are not covered by ALG-MLSC earlier than time t; $R(t)$ includes the functions that are covered exactly at time t as well. We interchangeably use $i \in R(t)$ and $f_i \in R(t)$ for notational simplicity. Let $R_j := R(16\alpha\rho\sigma\, 2^j)$. Similarly, we let $R^*(t)$ denote the set of functions that are not covered by OPT earlier than time t and let $R_j^* = R^*(2^j)$. Note that R_j and R_j^* refer to different times. For notational convenience, we let $R_{-1} := \emptyset$.

We show the following key lemma. It shows that the number of uncovered functions by ALG-MLSC must decrease fast as j grows, unless the corresponding number by the optimal solution is comparable.

Lemma 1. *Consider any $j \geq 0$. Then we have $|R_j| \leq \frac{1}{4}|R_{j-1}| + |R_j^*|$.*

Proof. The lemma trivially holds when $j = 0$, hence consider any fixed phase $j \geq 1$. Let S_0 denote the set of vertices that were added to S up to the end of phase $j - 1$. Let $H = 4\alpha\rho$ and $T_1, T_2, \dots, T_H$ be the sets of vertices that were added in Line (6) in the j-th phase. Let $S_h = S_0 \cup T_1 \cup T_2 \cup \dots \cup T_h, \forall 1 \leq h \leq H$. We prove Lemma 1 by lower and upper bounding the quantity

$$\Delta_j := \sum_{h=1}^H f^{S_{h-1}}(T_h) = \sum_{h=1}^H \sum_{i \in [m]: f_i(S_{h-1}) < 1} \frac{f_i(S_h) - f_i(S_{h-1})}{1 - f_i(S_{h-1})}$$

We first *lower bound* Δ_j. Let T^* denote the set of vertices that OPT visited within time 2^j. Observe that in Line (5), ALG-MLSC could have visited all nodes in T^* by choosing P as the prefix of length 2^j of OPT. Via the approximation guarantee of ALG-SOP, we obtain

Proposition 2. *For any $h \in [H]$ we have $f^{S_{h-1}}(T_h) \geq \frac{1}{\rho} \cdot f^{S_{h-1}}(T^*)$.*

Observe that by definition of sets R_js, for any $h \in [H]$ and $i \in R_j$, $f_i(S_{h-1}) < 1$. Moreover, by definition of R_j^*s, for each $i \notin R_j^*$, $f_i(T^*) = 1$. So

$$f^{S_{h-1}}(T^*) \geq \sum_{i \in R_j \backslash R_j^*} \frac{f_i(S_{h-1} \cup T^*) - f_i(S_{h-1})}{1 - f_i(S_{h-1})} \geq |R_j \backslash R_j^*|, \quad \forall h \in [H].$$

Using this in the above proposition and summing over $h \in [H]$,

$$\Delta_j \geq \frac{1}{\rho} \sum_{h=1}^{H} f^{S_{h-1}}(T^*) \geq \frac{H}{\rho}(|R_j| - |R_j^*|) = 4\alpha(|R_j| - |R_j^*|) \qquad (1)$$

We now *upper bound* Δ_j. Note that for any $i \notin R_{j-1}$, $f_i(S_0) = 1$ (it is already covered before phase j) and therefore f_i does not contribute to Δ_j. So,

$$\Delta_j = \sum_{i \in R_{j-1}} \sum_{h \in [H], f_i(S_{h-1}) < 1} \frac{f_i(S_h) - f_i(S_{h-1})}{1 - f_i(S_{h-1})} \leq \alpha|R_{j-1}|.$$

The inequality is by Claim 1, which implies that each function $f_i \in R_{j-1}$ contributes at most α. Combining this with (1) completes the proof of Lemma 1.

Proof (Theorem 5). Given Lemma 1, we proceed as follows:

$$\mathsf{ALG} = \sum_{j \geq 0} \sum_{16\alpha\rho\sigma 2^j \leq t < 16\alpha\rho\sigma 2^{j+1}} |R(t)| + \sum_{0 \leq t < 16\alpha\rho\sigma} |R(t)|$$

$$\leq \sum_{j \geq 0} 16\alpha\rho\sigma(2^{j+1} - 2^j)|R_j| + 16\alpha\rho\sigma\mathsf{OPT}$$

$$[\text{Since } |R(t)| \text{ is non-increasing, and for any } t \geq 0, |R(t)| \leq m \leq \mathsf{OPT}]$$

$$= 16\alpha\rho\sigma \sum_{j \geq 0} 2^{j+1} \left(|R_j| - \frac{1}{4}|R_{j-1}|\right) + 16\alpha\rho\sigma\mathsf{OPT} \quad [\text{Using } R_{-1} = \emptyset]$$

$$\leq 16\alpha\rho\sigma \sum_{j \geq 0} 2^{j+1}|R_j^*| + 16\alpha\rho\sigma\mathsf{OPT} \qquad [\text{By Lemma 1}]$$

$$\leq 64\alpha\rho\sigma \sum_{j \geq 1} \left(\sum_{2^{j-1} \leq t < 2^j} |R^*(t)| \right) + 32\alpha\rho\sigma|R_0^*| + 16\alpha\rho\sigma\mathsf{OPT} \leq 112\alpha\rho\sigma\mathsf{OPT}.$$

3   Stochastic Submodular Ranking

In this section, we study the *stochastic submodular ranking* problem. Here we are given a set $\mathcal{A} = \{X_1, ..., X_n\}$ of n independent random variables (called elements), each of which takes values from some domain Δ. The distribution of each $\{X_i\}_{i=1}^{n}$ is known to the algorithm, but the realization $x_i \in \Delta$ of X_i is only known after scheduling X_i. Each element X_i (for $i \in [n]$) also has a deterministic integer length ℓ_i, which denotes the amount of time taken to schedule X_i. We are also given a set of m monotone submodular functions $f_1, ..., f_m : 2^\Delta \to [0, 1]$ on groundset Δ.

A feasible solution (or policy) is an adaptive ordering of $\mathcal{A}$, represented naturally by a decision tree with nodes corresponding to scheduled elements and branches corresponding to their realizations. We use $\langle \pi(1), \ldots, \pi(n) \rangle$ to denote this ordering, where each $\pi(l)$ is a random variable denoting the index of the l-th scheduled element. The element $X_{\pi(l)} \in \mathcal{A} \setminus \{X_{\pi(1)}, X_{\pi(2)}, \ldots, X_{\pi(l-1)}\}$ is chosen at time $\ell_{\pi(1)} + \ell_{\pi(2)} + \ldots + \ell_{\pi(l-1)}$, after observing the realizations $x_{\pi(1)}, \ldots, x_{\pi(k-1)}$.

Given any policy as above, the cover time $\mathsf{cov}(f_i)$ of function f_i is defined as the earliest time t such that f_i has value one on the realization of elements that are completely scheduled within time t. More formally, $\mathsf{cov}(f_i)$ is the earliest time t such that $f_i(\{x_{\pi(1)}, \ldots, x_{\pi(k_t)}\}) = 1$ where k_t is the maximum integer such that $\ell_{\pi(1)} + \ell_{\pi(2)} + \ldots + \ell_{\pi(k_t)} \le t$. If the function value never reaches one (due to the stochastic nature of elements) then $\mathsf{cov}(f_i) = \ell_1 + \ell_2 + \ldots + \ell_n$ the total length of any ordering. Note that the cover time is a random value. The goal in stochastic submodular ranking is to find a policy minimizing $\mathbb{E}\left[\sum_{i \in [m]} \mathsf{cov}(f_i)\right]$.

We prove Theorem 3, by obtaining an $O(\log \frac{1}{\epsilon})$-approximate adaptive policy. This result has many applications, that are described in the full version [21].

To formally describe our algorithm, we first define the probability spaces we are concerned with. We use $\Omega = \Delta^n$ to denote the outcome space of $\mathcal{A}$. We use the same notation Ω to denote the probability space induced by the outcomes. For any $S \subseteq \mathcal{A}$ and its realization s, let $\Omega(s)$ denote the outcome subspace that conforms to s. We can naturally define the probability space $\Omega(s)$ as follows: the probability that $w \in \Omega(s)$ occurs is $\Pr_\Omega[w]/\Pr_\Omega[\Omega(s)]$. We also use $\Omega(s)$ to denote this probability space. The algorithm ALG-AG-STO is a natural extension of the deterministic case [1]. At any point with $S \subseteq \mathcal{A}$ being the previously scheduled elements and s their instantiations, choose:

$$X_e \;=\; \arg \max_{X_e \in \mathcal{A} \setminus S} \frac{1}{\ell_e} \cdot \mathbb{E}_{\,\Omega(s)} \left[\sum_{i \in [m], f_i(s) < 1} \frac{f_i(s \cup \{X_e\}) - f_i(s)}{1 - f_i(s)} \right] \qquad (2)$$

Observe that taking expectation over $\Omega(s)$ is the same as expectation over the distribution of X_e since $X_e \notin S$ and the elements are independent. Also note that this algorithm implicitly defines a decision tree.

We now show that this algorithm implies Theorem 3. Let $\alpha = 1 + \ln \frac{1}{\epsilon}$. Let $R(t)$ denote the (random) set of functions that are not satisfied by ALG-AG-STO before time t. Note that the set $R(t)$ includes the functions that are satisfied exactly at time t. Analogously, the set $R^*(t)$ is defined for the optimal policy. We use $i \in R(t)$ interchangeably with $f_i \in R(t)$. Let $C(t) := \{f_1, \ldots, f_m\} \setminus R(t)$ and $C^*(t) := \{f_1, \ldots, f_m\} \setminus R^*(t)$. Note that all the sets $C(\cdot), C^*(\cdot), R(\cdot), R^*(\cdot)$ are stochastic. We set $\mathsf{ALG} := \sum_{t \in [n]} |R(t)|$ and $\mathsf{OPT} := \sum_{t \in [n]} |R^*(t)|$ that are also stochastic quantities.

We will be interested in the number of unsatisfied functions at times $\{8\alpha 2^j : j \in \mathbb{Z}_+\}$ by ALG-AG-STO and the number of unsatisfied functions at times $\{2^j : j \in \mathbb{Z}_+\}$ by the optimal policy. Let $R_j := R(8\alpha 2^j)$ and $R_j^* = R^*(2^j)$. It is important to note that R_j and R_j^* are concerned with different times, and they are stochastic. For notational simplicity, we let $R_{-1} := \emptyset$.

We show the following key lemma. Using this lemma we can show that $\mathbb{E}\text{ALG} = O(\alpha) \cdot \mathbb{E}\text{OPT}$ (as in the proof of Theorem 5 from Lemma 1). This suffices to prove Theorem 3.

Lemma 2. *For any $j \geq 0$, we have $\mathbb{E}[|R_j|] \leq \frac{1}{4}\mathbb{E}[|R_{j-1}|] + \mathbb{E}[|R_j^*|]$.*

Proof. The lemma trivially holds for $j = 0$, so we consider any $j \geq 1$. For any $t \geq 1$, we use s_{t-1} to denote the set of elements *completely* scheduled by ALG-AG-STO by time $t - 1$ along with their instantiations. Also, for $t \geq 1$ let $\sigma(t) \in [n]$ denote the (random) index of the element being scheduled during time slot $(t - 1, t]$. Note that s_{t-1} determines $\sigma(t)$ precisely, but not the instantiation of $X_{\sigma(t)}$.

Let $E_j^* \subseteq \mathcal{A}$ be the (stochastic) set of elements that is completely scheduled by the optimal policy within time 2^j. For a stochastic set (or element) S, we denote its realization under an outcome $w \in \Omega$ as $S(w)$. For example, $X_i(w) \in \Delta$ is the realization of X_i under w; and $E_j^*(w)$ is the set of elements completely scheduled by OPT by time 2^j (under w) along with their realizations.

For any time t and corresponding outcome $s_{t-1} \subseteq \Delta$, define a set function:

$$f^{s_{t-1}}(D) := \sum_{i \in [m], f_i(s_{t-1}) < 1} \frac{f_i(s_{t-1} \cup D) - f_i(s_{t-1})}{1 - f_i(s_{t-1})}, \qquad \forall D \subseteq \Delta.$$

We also use $f_i^{s_{t-1}}(D)$ to denote the term inside the above summation. It is easy to see that the function $f^{s_{t-1}} : 2^\Delta \to \mathbb{R}_+$ is submodular. Also define:

$$F^{s_{t-1}}(X_e) := \mathbb{E}_{w \leftarrow \Omega(s_{t-1})}[f^{s_{t-1}}(X_e(w))], \qquad \forall X_e \in \mathcal{A}. \tag{3}$$

Observe that this is zero for elements $X_e \in s_{t-1}$. By the choice (2),

Proposition 3. *Consider any time $t \in [n]$ and outcome s_{t-1}. Note that s_{t-1} determines $\sigma(t)$. Then $\frac{1}{\ell_{\sigma(t)}} \cdot F^{s_{t-1}}(X_{\sigma(t)}) \geq \frac{1}{\ell_i} \cdot F^{s_{t-1}}(X_i)$, for all $X_i \in \mathcal{A}$.*

Define *expected gain* in step t: $\quad G_t := \mathbb{E}_{s_{t-1}}\left[\frac{1}{\ell_{\sigma(t)}} F^{s_{t-1}}(X_{\sigma(t)})\right]. \tag{4}$

Define expected total gain: $\quad \Delta_j := \sum_{t=8\alpha 2^{j-1}+1}^{8\alpha 2^j} G_t \ . \tag{5}$

We complete the proof of Lemma 2 by upper and lower bounding Δ_j.

Lower bound for Δ_j. Consider any $8\alpha 2^{j-1} < t \leq 8\alpha 2^j$. We lower bound G_t. Condition on s_{t-1}; this determines $\sigma(t)$ (but not $x_{\sigma(t)}$). Note that $\sum_{i=1}^n \ell_i \cdot \Pr[X_i \in E_j^* | s_{t-1}] \leq 2^j$ by definition of E_j^* being the elements that are completely scheduled by time 2^j in OPT. So $\sum_{X_i \in \mathcal{A}} \frac{\ell_i}{2^j} \cdot \Pr[X_i \in E_j^* | s_{t-1}] \leq 1$.

Applying Proposition 3 with the convex multipliers (over i) given above,

$$\frac{1}{\ell_{\sigma(t)}} F^{s_{t-1}}(X_{\sigma(t)}) \geq \sum_{X_i \in \mathcal{A}} \frac{\ell_i}{2^j} \Pr[X_i \in E_j^* | s_{t-1}] \cdot \frac{1}{\ell_i} F^{s_{t-1}}(X_i)$$

$$= \frac{1}{2^j} \sum_{X_i \in \mathcal{A}} \Pr[X_i \in E_j^* | s_{t-1}] \sum_{x_i \in \Delta} \Pr[X_i = x_i | s_{t-1}] \cdot f^{s_{t-1}}(x_i)$$

$$= \frac{1}{2^j} \sum_{X_i \in \mathcal{A}} \sum_{x_i \in \Delta} \Pr[X_i \in E_j^* \wedge X_i = x_i | s_{t-1}] \cdot f^{s_{t-1}}(x_i)$$

$$= \frac{1}{2^j} \sum_{w \in \Omega(s_{t-1})} \Pr[w | s_{t-1}] \sum_{X_i \in E_j^*(w)} f^{s_{t-1}}(X_i(w)) \qquad (6)$$

The first equality is by definition of $F^{s_{t-1}}(\cdot)$ from (3). The second equality holds since the optimal policy must decide whether to schedule X_i (by time 2^j) without knowing the realization of X_i: i.e. events $X_i = x_i$ and $X_i \in E_j^*$ are independent (even conditioned on s_{t-1}). Now for each $w \in \Omega(s_{t-1})$, due to submodularity of the function $f^{s_{t-1}}(\cdot)$, we get

$$\sum_{X_i \in E_j^*(w)} f^{s_{t-1}}(X_i(w)) \geq f^{s_{t-1}}(E_j^*(w)) = \sum_{i \in [m], f_i(s_{t-1}) < 1} \frac{f_i(E_j^*(w)) - f_i(s_{t-1})}{1 - f_i(s_{t-1})}$$

$$\geq |C_j^*(w)| - |C(t, w)| \qquad (7)$$

Recall that $E_j^*(w)$ denotes the set of elements scheduled by time 2^j in OPT(conditional on w), as well as the realizations of these elements. The equality comes from the definition of $f^{s_{t-1}}$. The last inequality holds because $C(t, w) = \{i \in [m] : f_i(s_{t-1}) = 1\}$ and set $E_j^*(w)$ covers functions $C_j^*(w)$. Combining (6) and (7) gives: $\frac{1}{\ell_{\sigma(t)}} F^{s_{t-1}}(X_{\sigma(t)}) \geq \frac{1}{2^j}\left(\mathbb{E}\left[|C_j^*| \mid s_{t-1} \right] - \mathbb{E}[|C(t)| \mid s_{t-1}] \right)$. Deconditioning this inequality (taking expectation over s_{t-1}) and by (4),

$$G_t \geq \frac{1}{2^j} \cdot \left(\mathbb{E}[|C_j^*|] - \mathbb{E}[|C(t)|] \right) \geq \frac{1}{2^j} \cdot \left(\mathbb{E}[|C_j^*|] - \mathbb{E}[|C_j|] \right),$$

where the last inequality uses $\mathbb{E}[C(t)]$ is non-decreasing and $t \leq 8\alpha 2^j$.
   Now summing over all $t \in (8\alpha 2^{j-1}, 8\alpha 2^j]$ yields:

$$\Delta_j = \sum_{t=8\alpha 2^{j-1}}^{8\alpha 2^j} G_t \geq 4\alpha\left(\mathbb{E}[|C_j^*|] - \mathbb{E}[|C_j|] \right) = 4\alpha\left(\mathbb{E}[|R_j|] - \mathbb{E}[|R_j^*|] \right) \qquad (8)$$

The following *upper bound* for Δ_j (which uses Claim 1) is proved in [21].

$$\Delta_j \leq \alpha \mathbb{E}[|R_{j-1}|] \qquad (9)$$

Combining (8) and (9), we obtain Lemma 2.   □

References

1. Azar, Y., Gamzu, I.: Ranking with submodular valuations. In: SODA 2011, pp. 1070–1079 (2011)
2. Azar, Y., Gamzu, I., Yin, X.: Multiple intents re-ranking. In: STOC 2009, pp. 669–678 (2009)
3. Bansal, N., Gupta, A., Krishnaswamy, R.: A constant factor approximation algorithm for generalized min-sum set cover. In: SODA 2010, pp. 1539–1545 (2010)
4. Bar-Noy, A., Bellare, M., Halldórsson, M.M., Shachnai, H., Tamir, T.: On chromatic sums and distributed resource allocation. Inf. Comput. 140(2), 183–202 (1998)
5. Bhaskara, A., Charikar, M., Chlamtac, E., Feige, U., Vijayaraghavan, A.: Detecting high log-densities: an $n^{1/4}$ approximation for densest k-subgraph. In: STOC 2010, pp. 201–210 (2010)
6. Calinescu, G., Zelikovsky, A.: The polymatroid steiner problems. Journal of Combinatorial Optimization 9(3), 281–294 (2005)
7. Carr, R.D., Fleischer, L., Leung, V.J., Phillips, C.A.: Strengthening integrality gaps for capacitated network design and covering problems. In: SODA 2000, pp. 106–115 (2000)
8. Chakrabarty, D., Swamy, C.: Facility Location with Client Latencies: Linear Programming Based Techniques for Minimum Latency Problems. In: Günlük, O., Woeginger, G.J. (eds.) IPCO 2011. LNCS, vol. 6655, pp. 92–103. Springer, Heidelberg (2011)
9. Chaudhuri, K., Godfrey, B., Rao, S., Talwar, K.: Paths, trees, and minimum latency tours. In: FOCS 2003, pp. 36–45 (2003)
10. Chekuri, C., Even, G., Kortsarz, G.: A greedy approximation algorithm for the group steiner problem. Discrete Applied Mathematics 154(1), 15–34 (2006)
11. Chekuri, C., Pál, M.: A recursive greedy algorithm for walks in directed graphs. In: FOCS 2005, pp. 245–253 (2005)
12. Fakcharoenphol, J., Harrelson, C., Rao, S.: The k-traveling repairmen problem. ACM Transactions on Algorithms 3(4) (2007)
13. Feige, U., Lovász, L., Tetali, P.: Approximating min sum set cover. Algorithmica 40(4), 219–234 (2004)
14. Garg, N., Konjevod, G., Ravi, R.: A polylogarithmic approximation algorithm for the group steiner tree problem. J. Algorithms 37(1), 66–84 (2000)
15. Goemans, M.X., Vondrák, J.: Stochastic Covering and Adaptivity. In: Correa, J.R., Hevia, A., Kiwi, M. (eds.) LATIN 2006. LNCS, vol. 3887, pp. 532–543. Springer, Heidelberg (2006)
16. Golovin, D., Krause, A.: Adaptive submodularity: A new approach to active learning and stochastic optimization. In: COLT 2010, pp. 333–345 (2010)
17. Guillory, A., Bilmes, J.A.: Online submodular set cover, ranking, and repeated active learning. In: NIPS 2011 (2011)
18. Gupta, A., Nagarajan, V., Ravi, R.: Approximation Algorithms for Optimal Decision Trees and Adaptive TSP Problems. In: Abramsky, S., Gavoille, C., Kirchner, C., Meyer auf der Heide, F., Spirakis, P.G. (eds.) ICALP 2010, Part I. LNCS, vol. 6198, pp. 690–701. Springer, Heidelberg (2010)
19. Gupta, A., Srinivasan, A.: An improved approximation ratio for the covering steiner problem. Theory of Computing 2(1), 53–64 (2006)
20. Halperin, E., Krauthgamer, R.: Polylogarithmic inapproximability. In: STOC 2003, pp. 585–594 (2003)

21. Im, S., Nagarajan, V., van der Zwaan, R.: Minimum latency submodular cover. CoRR, abs/1110.2207 (2011)
22. Konjevod, G., Ravi, R., Srinivasan, A.: Approximation algorithms for the covering steiner problem. Random Struct. Algorithms 20(3), 465–482 (2002)
23. Liu, Z., Parthasarathy, S., Ranganathan, A., Yang, H.: Near-optimal algorithms for shared filter evaluation in data stream systems. In: SIGMOD 2008, pp. 133–146 (2008)
24. Munagala, K., Srivastava, U., Widom, J.: Optimization of continuous queries with shared expensive filters. In: PODS 2007, pp. 215–224 (2007)
25. Nagarajan, V.: Approximation Algorithms for Sequencing Problems. PhD thesis. Tepper School of Business, Carnegie Mellon University (2009)
26. Schrijver, A.: Combinatorial optimization: polyhedra and efficiency. Springer, Berlin (2003)
27. Wolsey, L.A.: An analysis of the greedy algorithm for the submodular set covering problem. Combinatorica 2(4), 385–393 (1982)

Constant-Time Algorithms for Sparsity Matroids

Hiro Ito[1], Shin-Ichi Tanigawa[2], and Yuichi Yoshida[3]

[1] School of Informatics, Kyoto University
itohiro@kuis.kyoto-u.ac.jp
[2] Research Institute for Mathematical Sciences, Kyoto University
tanigawa@kurims.kyoto-u.ac.jp
[3] School of Informatics, Kyoto University, and Preferred Infrastructure, Inc.
yyoshida@kuis.kyoto-u.ac.jp

Abstract. A graph $G = (V, E)$ is called (k, ℓ)-sparse if $|F| \leq k|V(F)| - \ell$ for any $F \subseteq E$ with $F \neq \emptyset$. Here, $V(F)$ denotes the set of vertices incident to F. A graph $G = (V, E)$ is called (k, ℓ)-full if G contains a (k, ℓ)-sparse subgraph with $|V|$ vertices and $k|V| - \ell$ edges. The family of edge sets of (k, ℓ)-sparse subgraphs forms a family of independent sets of a matroid on E, known as the sparsity matroid of G. In this paper, we give a constant-time algorithm that approximates the rank of the sparsity matroid associated with a degree-bounded undirected graph. This algorithm leads to a constant-time tester for (k, ℓ)-fullness in the bounded-degree model, (i.e., we can decide with high probability whether the input graph satisfies a property or far from it). Depending on the values of k and ℓ, our algorithm can test various properties of graphs such as connectivity, rigidity, and how many spanning trees can be packed in a unified manner.

Based on this result, we also propose a constant-time tester for (k, ℓ)-edge-connected-orientability in the bounded-degree model, where an undirected graph G is called (k, ℓ)-edge-connected-orientable if there exists an orientation $\boldsymbol{G}$ of G with a vertex $r \in V$ such that $\boldsymbol{G}$ contains k arc-disjoint dipaths from r to each vertex $v \in V$ and ℓ arc-disjoint dipaths from each vertex $v \in V$ to r.

A tester is called a one-sided error tester for P if it always accepts a graph satisfying P. We show, for any $k \geq 2$ and (proper) $\ell \geq 0$, every one-sided error tester for (k, ℓ)-fullness and (k, ℓ)-edge-connected-orientability requires $\Omega(n)$ queries.

1 Introduction

In *property testing*, given an instance I, we are supposed to distinguish the case that I satisfies a predetermined property P from the case that I is "far" from satisfying P. The definition of farness varies depending on model. The main objective of property testing is designing efficient algorithms that run even in constant time, independent of the input size.

In this paper, we study about testing algorithms for two strongly related properties of undirected graphs, (k, ℓ)-sparsity and (k, ℓ)-edge-connected-orientability. A graph $G = (V, E)$ is called (k, ℓ)-*sparse* if $|F| \leq k|V(F)| - \ell$ for any $F \subseteq E$

A. Czumaj et al. (Eds.): ICALP 2012, Part I, LNCS 7391, pp. 498–509, 2012.

with $F \neq \emptyset$, where $V(F)$ denotes the set of vertices incident to edges in F. We note that (k, ℓ)-sparsity becomes meaningful only when $2k - \ell \geq 1$. If otherwise, any non-empty graph cannot be (k, ℓ)-sparse since just an edge violates the condition. Hence, we assume $k \geq 1, \ell \geq 0$ and $2k - \ell \geq 1$ throughout the paper. A graph G is called (k, ℓ)-*tight* if G is (k, ℓ)-sparse and $|E| = kn - \ell$, where n is the number of vertices in G. A graph G is called (k, ℓ)-*full* if G contains a (k, ℓ)-tight subgraph with n vertices. Checking (k, ℓ)-fullness of a graph is one of main topics in this paper.

Another topic studied in this paper is orientability of undirected graphs. A (di)graph is called k-*edge-connected* (resp., k-*vertex-connected*) if deletion of any $k-1$ edges (resp., vertices) leaves the graph connected. By Menger's theorem, this is equivalent to asking k edge-disjoint (resp., k internally-disjoint) paths between any pair of vertices. A digraph $D = (V, A)$ is called (k, ℓ)-*edge-connected* with a root $r \in V$ if, for each $v \in V \setminus \{r\}$, D has k arc-disjoint dipaths from r to v and ℓ arc-disjoint dipaths from v to r. An undirected graph $G = (V, E)$ is called (k, ℓ)-*edge-connected-orientable* ((k, ℓ)-ec-orientable, in short) if one can assign an orientation to each edge so that the resulting digraph is (k, ℓ)-edge-connected with some root $r \in V$. We note that the choice of r is actually not important, and we may specify any vertex as r.

Nash-Williams' graph-orientation theorem [15] implies that a graph G admits an orientation such that the resulting digraph is k-edge-connected if and only if G is $2k$-edge-connected. This implies that (k, k)-ec-orientability of a graph is equivalent to $2k$-edge-connectivity. Another famous result of Nash-Williams [17] for the forest-partition problem shows that an undirected graph G contains k edge-disjoint spanning trees if and only if G is (k, k)-full. This theorem, combined with Edmonds' arc-disjoint branching theorem [3], implies that G is $(k, 0)$-ec-orientable if and only if G is (k, k)-full. In this sense, (k, ℓ)-ec-orientability can be considered as a unified concept of the sparsity and the conventional edge-connectivity.

In this paper, we give constant-time testers for (k, ℓ)-fullness and (k, ℓ)-ec-orientability in the bounded-degree model.

Definition 1 (Bounded-degree model [9]). *In the bounded-degree model with a degree bound d, we consider graphs with maximum degree at most d. A graph $G = (V, E)$ of n vertices is represented by an oracle $\mathcal{O}_G$ satisfying the followings:*

- *For each vertex $v \in V$, there exists an injection $\pi_v : E_G(v) \rightarrow [d]$ where $E_G(v)$ is a set of edges incident to v.*
- *The oracle $\mathcal{O}_G$, on two values $u \in V, i \in \mathbb{N}$, returns v such that $uv \in E$ and $\pi_u(uv) = i$. If no such vertex v exists, it returns a special character $\bot$.*

Algorithms are given V, n, d, and the access to $\mathcal{O}_G$. For an error parameter $\epsilon > 0$, a graph is called ϵ-far from a property P, if we must add or remove at least $\frac{\epsilon dn}{2}$ edges to make G satisfy P.[1]

---

[1] Sometimes it is required that the resulting graph must satisfy the degree bound. We use the present model in order not to make the argument unnecessarily involved.

An edge $e = uv$ is called the i-th edge of u if $\pi_u(e) = i$. The *query complexity* of an algorithm is the number of accesses to $\mathcal{O}_G$. An algorithm is called a *tester* for a property P if it accepts graphs satisfying P with probability at least $\frac{2}{3}$ and rejects graphs ϵ-far from P with probability at least $\frac{2}{3}$.

Our main results are summarized as follows.

Theorem 1. *In the bounded-degree model with a degree bound d, there is a tester for (k, ℓ)-fullness with query complexity $(k + d)^{O(1/\epsilon'^2)}(\frac{1}{\epsilon'})^{O(1/\epsilon')}$, where $\epsilon' = \frac{\epsilon}{k+d\ell}$.*

Theorem 2. *In the bounded-degree model with a degree bound d, there is a tester for (k, ℓ)-ec-orientability with query complexity $(k + d)^{O(1/\epsilon'^2)}(\frac{1}{\epsilon'})^{O(1/\epsilon')}$, where $\epsilon' = \max(\frac{\epsilon}{dk}, \frac{d\epsilon}{\ell})$.*

The second result resolves an open problem raised by Orenstein [20], which asks the existence of a constant-time tester for (k, ℓ)-ec-orientability. As mentioned later, the first result has numerous applications to both theoretical and practical problems.

An algorithm is called a $(1, \beta)$-*approximation algorithm* for a value x^* if, with probability $\frac{2}{3}$, it outputs x such that $x^* - \beta \leq x \leq x^*$. For a graph $G = (V, E)$, it is known that the family of edge sets of (k, ℓ)-sparse subgraphs forms a family of independent sets of a matroid on E. This matroid is called the (k, ℓ)-*sparsity matroid* of G, denoted by $\mathcal{M}_{k,\ell}(G)$, and the rank function by $\rho_{k,\ell} : 2^E \to \mathbb{Z}$. Although detailed properties will be discussed in the next section, we should note that G is (k, ℓ)-full if and only if $\rho_{k,\ell}(E) = kn - \ell$. To test (k, ℓ)-fullness, we actually develop a constant-time $(1, \epsilon n)$-approximation algorithm for $\rho_{k,\ell}(E)$.

Theorem 3. *Let $G = (V, E)$ be a graph with n vertices. In the bounded-degree model with a degree bound d, there exists a $(1, \epsilon n)$-approximation algorithm for the rank of $\mathcal{M}_{k,\ell}(G)$ with query complexity $(k + d)^{O(1/\epsilon'^2)}(\frac{1}{\epsilon'})^{O(1/\epsilon')}$ where $\epsilon' = \frac{\epsilon}{k+d\ell}$.*

A tester is called a *one-sided error tester* for a property P if it always accepts graphs satisfying P. A general tester is sometimes called a *two-sided error tester* for comparison. Our testers for (k, ℓ)-fullness and (k, ℓ)-ec-orientability are two-sided error testers. On the contrary, we give the following lower bounds for one-sided error testers.

Theorem 4. *Let $k \geq 2$. In the bounded-degree model, any one-sided error tester for (k, ℓ)-fullness requires $\Omega(n)$ queries where n is the number of vertices in the input graph.*

The corresponding linear lower bound also holds for testing (k, ℓ)-ec-orientability for $k \geq 2$ and $k > \ell$ in the bounded-degree model. It is not hard to show that there are one-sided error testers for $(1, \ell)$-fullness and $(1, \ell)$-ec-orientability. Also, we have one-sided error testers for (k, ℓ)-ec-orientability when $k \leq \ell$.

We briefly mention why we use the bounded-degree model. Another famous model for graphs is the *adjacency matrix model*, in which a graph is represented by an oracle $\mathcal{O}_G$ such that, given two vertices u and v, $\mathcal{O}_G$ answers whether there is an edge between u and v. A graph G is called ϵ-far from P in this model if we must modify $\frac{\epsilon n^2}{2}$ edges to make G satisfy P. We see that testing (k, ℓ)-fullness is trivial in this model. Note that we can make any graph (k, ℓ)-full by adding $kn - \ell$ edges. Thus, any graph is at most $O(\frac{1}{n})$-far. Thus, for any $\epsilon > 0$, when $n = \Omega(\frac{1}{\epsilon})$, we can safely accept graphs without any computation. When $n = O(\frac{1}{\epsilon})$, we can test (k, ℓ)-fullness using a standard polynomial-time algorithm. We have the same issue also for (k, ℓ)-ec-orientability.

It may be interesting to consider our theorems can be generalized to the *general graph model* [12], in which ϵ-farness is measured with respect to the number of edges in the original graph.

Related Works. In the bounded-degree model, many testers are known for several fundamental graph properties (see e.g., [8]). The most relevant works are testers for connectivity. For undirected graphs, k-edge-connectivity [9] and k-vertex-connectivity [24] are known to be testable in constant time for any $k \geq 1$. Those results are extended to digraphs and simplified [20,25]. We stress that the idea behind all the algorithms above is to detect a small evidence that a graph does not satisfy the property we are concerned with. However, as we discuss later, for (k, ℓ)-sparsity and (k, ℓ)-ec-orientability, there may not be any such small evidence. This fact makes our testers more involved.

One of final goals in property testing is arguably characterizing testable properties. Planarity, or more generally H-minor-freeness, are known to be testable in the bounded-degree model, and testability of them can be described from their *hyperfiniteness*. Roughly speaking, a property is called hyperfinite if any graph with the property can be decomposed into constant-size components by removing small fraction of edges. In contrast, as noted in [19], we do not know any "general" reason so far why k-edge-connectivity is testable. Our result suggests that the matroid theory and the edge-augmentation theory might be a key tool to characterize non-hyperfinite testable properties.

As for exact and deterministic algorithms for checking (k, ℓ)-fullness of graphs, Imai [11] proposed an algorithm for computing the rank of $\mathcal{M}_{k,k}(G)$ in $O(n^2)$ time and that of $\mathcal{M}_{k,\ell}(G)$ in $O(nm)$ time for general ℓ, where n is the number of vertices and m is the number of edges. Improved algorithms were proposed by Gabow and Westermann [7], which run in $O(n\sqrt{m + n \log n})$ time for $k = \ell$ and in $O(n^2)$ time for $k = 2$ and $\ell = 3$. An efficient algorithm for computing the rank of $\mathcal{M}_{k,\ell}(G)$ for general k and ℓ is the so-called pebble algorithm by Lee and Streinu [14], which runs in $O(n^2)$ time.

As (k, ℓ)-sparsity has a wide range of applications in rigidity theory and scene analysis (see e.g., [23]), it is recognized as an important open problem to improve the $O(n^2)$ upper-bound for computing ranks of (k, ℓ)-sparsity matroids (see e.g., [2, Open Problem 4.1]). To the best of our knowledge, our result is the first sub-quadratic algorithm that approximates ranks of (k, ℓ)-sparsity matroids.

Applications. It is elementary to see that a graph is a forest if and only if it is $(1, 1)$-sparse, and the concept of $(1, 1)$-fullness coincides with the connectivity of graphs. As a variant of forests, a graph is called a *pseudoforest* if each connected component contains at most one cycle [7]. It is known that a graph is a pseudo-forest if and only if it is $(1,0)$-sparse. As we mentioned above, Nash-Williams [17] proved that a graph contains k edge-disjoint spanning trees if and only if it is (k,k)-full. Motivated by an application to rigidity theory, Whiteley [23] and Haas [10] proved a generalization of Nash-Williams' theorem to (k, ℓ)-sparse graphs by mixing trees and pseudoforests. Our result leads to constant-time testers for these properties.

Another important application of (k, ℓ)-sparse graphs is the rigidity of graphs. A classical theorem by Laman [13] implies that a $(2,3)$-full graph has a special property of being a generically rigid bar-joint framework on the plane, by re-garding each vertex as a joint and each edge as a bar. Whiteley [22] further showed that some other (k, ℓ)-sparsity matroid characterize generic rigidity of graphs embedded on surfaces.

We note that the (k, k)-fullness of a graph can be decided by checking the rank of the union of k graphic matroids. This problem is usually solved via a matroid intersection problem. This leaves us several unsolved questions: for which matroids can we approximate the rank of their union, and for which matroids $\mathcal{M}_1, \mathcal{M}_2$ can we approximate the size of the largest common independent set in $\mathcal{M}_1$ and $\mathcal{M}_2$ with a constant number of queries?

Organization and Proof Overview. In Section 2, we review properties of $\mathcal{M}_{k,\ell}(G)$. Then, in Sections 3.1 and 3.2, we give a tester for (k, ℓ)-fullness. We first develop a $(1, \epsilon n)$-approximation algorithm for $\rho_{k,\ell}(E)$ running in constant time (Theorem 3). Then, we can test (k, ℓ)-fullness since a (k, ℓ)-full graph has rank $k|V| - \ell$, and a graph ϵ-far from (k, ℓ)-fullness has rank at most $k|V| - \ell - \epsilon n$.

A natural way to estimate the rank of $\mathcal{M}_{k,\ell}(G)$ is locally simulating the greedy algorithm, i.e., we add edges one by one, and if a newly added edge forms a circuit w.r.t. $\mathcal{M}_{k,\ell}(G)$, we discard it. The main obstacle to simulate this algorithm is that, in general, we cannot detect any circuit in constant time. For example, a circuit in $\mathcal{M}_{1,1}(G)$ corresponds to a cycle in G. However, there is a d-regular graph in which any cycle has length $\Omega(\log_d n)$. Thus, we need to estimate the rank without seeing any circuit. We note that $\rho_{1,1}(E) = n - c$ holds for the matroid $\mathcal{M}_{1,1}(G)$, where c is the number of connected components. Using this fact, constant-time approximation algorithms for $\rho_{1,1}(E)$ are given in [1]. However, there is no such formula for general k and ℓ.

Our strategy to overcome this issue is as follows: First, we remove constant-size circuits w.r.t. $\mathcal{M}_{k,\ell}(G)$, and let $G' = (V, E')$ be the resulting graph. We can show that $\rho_{k,\ell}(E) = \rho_{k,\ell}(E')$. A crucial fact is that $\rho_{k,\ell}(E')$ is close to $\rho_{k,0}(E')$. Thus, it amount to estimate $\rho_{k,0}(E')$ efficiently. It is known that $\rho_{k,0}(E')$ equals the size of the maximum matching of an auxiliary graph, and we can compute it using a constant-time approximation algorithm for the maximum matching given by [18,26].

In Section 4, we provide a constant-time tester for (k, ℓ)-ec-orientability. Our algorithm is based on the characterization of the number of edges we need to add to make a graph (k, ℓ)-ec-orientable by Frank and Király [6]. Although this characterization is not so simple as the case of the edge-connectivity augmentation problem, we are able to show that, if G is ϵ-far, either there are many small evidences or G is globally sparse which can be measured by (k, k)-fullness (Theorem 8). As mentioned before, the (k, ℓ)-ec-orientability has strong relations to sparsity as well as to edge-connectivity. Indeed, our algorithm can be seen as a combination of the idea used to test k-edge-connectivity [9,24,20] and the idea used here to test (k, k)-sparsity in Section 3.2.

Due to space limit, linear lower bounds of one-sided error testers are defer to the full version. In [20], Orenstein proved linear lower bounds of one-sided error testers for $(k, 0)$-ec-orientability (or equivalently, (k, k)-fullness). Orenstein's proof made use of Tutte-and-Nash-Williams' tree packing theorem, which is a special property of (k, k)-fullness. However, we show that Orenstein's approach can be applied to the case for general ℓ with some graph operation that preserves (k, ℓ)-fullness.

Due to space limit, proofs of most claims are omitted. Instead, we attach the full paper in appendix for completeness.

2   Preliminaries

For an integer n, we denote by $[n]$ the set $\{1, \ldots, n\}$. Let $G = (V, E)$ be a graph. For a vertex set $S \subseteq V$, $G[S]$ denotes the subgraph of G induced by S. For an edge set $F \subseteq E$, we define $V_G(F)$ as the set of vertices incident to F.

For a graph $G = (V, E)$ and integers $k \geq 1, \ell \geq 0$, we define a function $f_{k,\ell} : 2^E \to \mathbb{Z}$ by $f_{k,\ell}(F) = k|V(F)| - \ell$ for $F \subseteq E$. In the (k, ℓ)-*sparsity matroid* $\mathcal{M}_{k,\ell}(G)$, $F \subseteq E$ is independent if and only if $|I| \leq f_{k,\ell}(I)$ holds for any nonempty $I \subseteq F$. This matroid is also called the (k, ℓ)-*count matroid* of G and is known as the matroid induced by the non-decreasing submodular function $f_{k,\ell}$ (see e.g.,[5]). The rank function and the closure operator are denoted by $\rho_{k,\ell}$ and $\mathrm{cl}_{k,\ell}$, respectively. We note that $\rho_{k,\ell}(F)$ equals the size of the largest (k, ℓ)-sparse edge set contained in F. This implies that G is (k, ℓ)-tight iff the rank of $\mathcal{M}_{k,\ell}(G)$ is $kn - \ell$.

A set $F \subseteq E$ is called a (k, ℓ)-*connected set* if, for any pair $e, e' \in F$, F has a circuit of $\mathcal{M}_{k,\ell}(G)$ that contains e and e'. For simplicity of exposition, a singleton $\{e\}$ is also considered as a (k, ℓ)-connected set. A maximal (k, ℓ)-connected set w.r.t. edge inclusion is called a (k, ℓ)-*connected component*. The following property of (k, ℓ)-connected sets is just a restatement of a general fact on matroid-connectivity for our purpose.

Proposition 1. $\mathcal{M}_{k,\ell}(G)$ *has the following properties: (i) For two (k, ℓ)-connected sets F_1 and F_2 with $F_1 \cap F_2 \neq \emptyset$, $F_1 \cup F_2$ is (k, ℓ)-connected. (ii) We can uniquely partition E into (k, ℓ)-connected components $\{C_1, \ldots, C_t\}$, and $\rho_{k,\ell}(E) = \sum_{i=1}^{t} \rho_{k,\ell}(C_i)$.*

A (k, ℓ)-connected set (or component) is called *trivial* if it is singleton, otherwise *non-trivial*. We remark that $\{e\}$ is a trivial (k, ℓ)-connected component if and only if e is a *coloop* in $\mathcal{M}_{k,\ell}(G)$ (i.e., every base contains e) since $\mathcal{M}_{k,\ell}(G)$ has no loop (in the matroid sense) if $2k - \ell \geq 1$. Hence, if we denote the family of non-trivial (k, ℓ)-connected components in $\mathcal{M}_{k,\ell}(G)$ by $\{C_1, \ldots, C_s\}$, then Proposition 1(ii) implies

$$\rho_{k,\ell}(E) = |E \setminus \bigcup_{i=1}^{s} C_i| + \sum_{i=1}^{s} \rho_{k,\ell}(C_i). \tag{1}$$

We need the following known properties of $\mathcal{M}_{k,\ell}(G)$.

Lemma 1. $\mathcal{M}_{k,\ell}(G)$ has the following properties: (i) For any circuit C of $\mathcal{M}_{k,\ell}(G)$, $\rho_{k,\ell}(C) = f_{k,\ell}(C)$. (ii) For any non-trivial (k, ℓ)-connected set $F \subseteq E$, $\rho_{k,\ell}(F) = f_{k,\ell}(F)$. Namely, F is (k, ℓ)-full.

We also note the following relation between $\mathcal{M}_{k,\ell}(G)$ and $\mathcal{M}_{k,\ell'}(G)$ for distinct ℓ and ℓ', which trivially follows from $|F| \leq k|V(F)| - \ell \leq k|V(F)| - \ell'$.

Lemma 2. Any (k, ℓ)-sparse set $F \subseteq E$ is (k, ℓ')-sparse for every $\ell' \leq \ell$.

3   Testing (k, ℓ)-Fullness

3.1   Approximating the Rank of $\mathcal{M}_{k,0}(G)$

In this section, we present a constant-time approximation algorithm for the rank $\rho_{k,0}(E)$ of $\mathcal{M}_{k,0}(G)$ for a graph $G = (V, E)$. A crucial fact is that computing $\rho_{k,0}(E)$ can be reduced to computing the size of the maximum matching in an auxiliary bipartite graph G_k obtained from G. The vertex set of G_k is $E \cup (V \times [k])$ where E and $V \times [k]$ form the partition, and G_k has an edge between $e \in E$ and $(v, i) \in V \times [k]$ iff e is incident to v in the original graph G. From the celebrated Hall's marriage theorem, the following result easily follows (see e.g., [11] for more details):

Proposition 2. Let $G = (V, E)$ be a graph and k be an integer. Then, G_k contains a matching covering $F \subseteq E$ if and only if F is $(k, 0)$-sparse.

Proposition 2 implies that the rank of $\mathcal{M}_{k,0}(G)$ is equal to the size of the maximum matching in G_k. We use the following algorithm.

Lemma 3 ([26]). In the bounded-degree model with a degree bound d, there exists a $(1, \epsilon n)$-approximation algorithm for the size of the maximum matching of a graph with query complexity $d^{O(1/\epsilon^2)} (\frac{1}{\epsilon})^{O(1/\epsilon)}$.

To run the algorithm given in Lemma 3 on G_k, we want to make an oracle access $\mathcal{O}_{G_k}$ to G_k using the oracle access $\mathcal{O}_G$ to G. However, since we do not have a method to access E directly, the vertex set $E \cup (V \times [k])$ is inconvenient to design $\mathcal{O}_{G_k}$.

Although the detailed description is omitted in this extended abstract, we invent a slightly different auxiliary graph G'_k, which is nearly identical to G_k, to deal with this issue. Then, we show that an oracle access $\mathcal{O}_{G'_k}$ to G'_k can be realized by asking the original oracle $\mathcal{O}_G$ at most d times. We thus obtain a constant-time approximation algorithm for $\rho_{k,0}$. (The detail is deferred to the full version.)

Lemma 4. *In the bounded-degree model with a degree bound d, there exists a* $(1, \epsilon n)$*-approximation algorithm for the rank of* $\mathcal{M}_{k,0}(G)$ *with query complexity* $(k + d)^{O(1/\epsilon'^2)}(\frac{1}{\epsilon'})^{O(1/\epsilon')}$ *where* $\epsilon' = \frac{\epsilon}{k+d}$.

3.2   Approximating the Rank of $\mathcal{M}_{k,\ell}(G)$

In this section, we describe a constant-time approximation algorithm for the rank of $\mathcal{M}_{k,\ell}(G)$ for a graph $G = (V, E)$. Let t be a parameter determined later by using the error parameter ϵ. We say that a subset $S \subseteq E$ is *large* if $|S| \geq t$; otherwise called *small*.

For an edge $e = uv$ and an integer $r > 0$, let $G_r(e)$ be the graph induced by the set of vertices whose distance to u or v is at most r. Also, let $E_r(e)$ be the set of edges in $G_r(e)$. The core of our approximation algorithm is an efficient implementation of an algorithm Component(e), which (approximately) decides whether a given edge $e \in E$ is in a large (k, ℓ)-connected set or not. As a subroutine, we first prepare an algorithm called SmallCircuits(e) in Algorithm 1 and then show Component(e) in Algorithm 2.

Algorithm 1. SmallCircuits(e): returns the union of small circuits containing an edge e

---

1: Set $t = \frac{\ell d}{\epsilon}$ and $S = \{e\}$.
2: **while** there is an unchecked small circuit $C \subseteq E_t(e)$ containing e **do**
3:     Check C.
4:     $S = S \cup C$.
5:     **if** $|S| \geq t$ **then**
6:         **return Large** (a special symbol).
7: **return** S.

Algorithm 2. Component(e): decides whether e is contained in a large (k, ℓ)-connected set

---

1: Set $t = \frac{\ell d}{\epsilon}$ and $S = \{e\}$.
2: **while** there is an unchecked element f in S **do**
3:     Check f.
4:     **if** SmallCircuits(f) = **Large then**
5:         **return Large.**
6:     $S = S \cup$ SmallCircuits(f)
7:     **if** $|S| \geq t$ **then**
8:         **return Large.**
9: **return** S.

The following lemma shows structural properties of outputs of Component(e). (The detail is deferred to the full version.)

Lemma 5. *For any* $e \in E$*, Component(e) is a small (k, ℓ)-connected set unless it returns* ***Large***. *Moreover, if Component(e) is not* ***Large***, *then Component(e) = Component(f) for any* $f \in$ *Component(e).*

Let $L = \{e \in E \mid \mathsf{Component}(e) = \mathbf{Large}\}$, and let $\{S_1, S_2, \ldots, S_m\}$ be the set of subsets of E such that $S_i = \mathsf{Component}(e)$ for some $e \in E$. Then, by Lemma 5, $\{L, S_1, \ldots, S_m\}$ forms a partition of E. The following lemma states that $\rho_{k,\ell}$ is well approximated by $\rho_{k,0}$ after replacing each connected component (w.r.t. $\mathcal{M}_{k,\ell}(G)$) by its base.

Theorem 5. *Let $\{L, S_1, \ldots, S_m\}$ be the partition of E defined as above. For each i with $1 \leq i \leq m$, let B_i be a base of S_i in $\mathcal{M}_{k,\ell}(G)$, and let $E' = L \cup \bigcup_{i=1}^{m} B_i$. Then, $\rho_{k,0}(E') - \frac{\ell dn}{t} \leq \rho_{k,\ell}(E) \leq \rho_{k,0}(E')$.*

Proof. Since B_i is a base of S_i in $\mathcal{M}_{k,\ell}(G)$, we have $S_i \subseteq \mathrm{cl}_{k,\ell}(B_i) \subseteq \mathrm{cl}_{k,\ell}(E')$ for each i. This implies $\rho_{k,\ell}(E') = \rho_{k,\ell}(E)$. Also, by Lemma 2, we have $\rho_{k,\ell}(E') \leq \rho_{k,0}(E')$.

To see $\rho_{k,0}(E') - \frac{\ell dn}{t} \leq \rho_{k,\ell}(E')$, recall that (k, ℓ)-connected components of $\mathcal{M}_{k,\ell}(G)|E'$ partitions E' by Proposition 1(ii) (where $\mathcal{M}_{k,\ell}(G)|E'$ denotes the restriction of $\mathcal{M}_{k,\ell}(G)$ to E'). We have the following properties of these connected sets. (The detail is deferred to the full version.)

Claim. Any $e \in L$ is contained in a large (k, ℓ)-connected component in $\mathcal{M}_{k,\ell}(G)|E'$.

Claim. Every non-trivial (k, ℓ)-connected component in $\mathcal{M}_{k,\ell}(G)|E'$ is large.

Let $\{C_1, C_2, \ldots, C_s\}$ be the family of non-trivial (k, ℓ)-connected components in $\mathcal{M}_{k,\ell}(G)|E'$. Note that $s \leq \frac{dn}{t}$ holds by the second Claim given above. Therefore, $\rho_{k,\ell}(E') = |E' \setminus \bigcup_{i=1}^{s} C_i| + \sum_{i=1}^{s} \rho_{k,\ell}(C_i) = |E' \setminus \bigcup_{i=1}^{s} C_i| + \sum_{i=1}^{s}(k|V(C_i)| - \ell) \geq |E' \setminus \bigcup_{i=1}^{s} C_i| + \sum_{i=1}^{s} k|V(C_i)| - \frac{\ell dn}{t}$, where the first equality follows from Equation (1), the second follows from Lemma 1(ii), and the third inequality follows from $s \leq \frac{dn}{t}$. On the other hand, from submodularity of $\rho_{k,0}$, we also have $\rho_{k,0}(E') \leq |E' \setminus \bigcup_{i=1}^{s} C_i| + \sum_{i=1}^{s} \rho_{k,0}(C_i) \leq |E' \setminus \bigcup_{i=1}^{s} C_i| + \sum_{i=1}^{s} k|V(C_i)|$. Comparing these two inequalities we obtain the desired result.

Proof (of Theorem 3). Let $G' = (V, E')$ where E' is as in Theorem 5. Set $t = \frac{\ell d}{\epsilon}$. Our algorithm computes $\rho_{k,0}(E')$ based on the algorithm given in Lemma 4 for the error threshold ϵ and just returns this value. By Lemma 4 and Theorem 5, this value approximates $\rho_{k,\ell}(E)$ with additive error ϵn. Therefore, if we can make an oracle access $\mathcal{O}_{G'}$ to the graph G', we are done.

For a query $\mathcal{O}_{G'}(v, i)$, we decide the output as follows. If $\mathcal{O}_G(v, i) = \bot$, we return $\bot$. Suppose that $\mathcal{O}_G(v, i) = e$. Then, we invoke $\mathsf{Component}(e)$. If $\mathsf{Component}(e)$ returns $\mathbf{Large}$, we return e. Otherwise, we take any base B of the returned set of $\mathsf{Component}(e)$ by an existing algorithm. We return e if $e \in B$ and return $\bot$ if otherwise. Note that for other edges $f \in S$, we use the same base B.

To analyze the query complexity, note that, during $\mathsf{Component}(e)$, we perform queries $\mathcal{O}_G(v, i)$ only for vertices v in $G_{3t}(e)$. So, to perform $\mathsf{Component}(e)$, we need $d^{3t} = d^{3\ell d/\epsilon}$ queries to $\mathcal{O}_G$. In total, we need $d^{3\ell d/\epsilon}(k+d)^{O(1/\epsilon'^2)}(\frac{1}{\epsilon'})^{O(1/\epsilon')} = (k+d)^{O(1/\epsilon'^2)}(\frac{1}{\epsilon'})^{O(1/\epsilon')}$, where $\epsilon' = \frac{\epsilon}{k+d\ell}$.

Theorem 1 directly follows from Theorem 3.

4   Testing (k, ℓ)-Edge-Connected-Orientability

In this section, we present a tester for the (k, ℓ)-edge-connected-orientability of a graph $G = (V, E)$.

A multiset $\mathcal{F} = \{V_1, \ldots, V_s\}$ of subsets of V is said to be *regular* if each element of V belongs to the same number of subsets in $\mathcal{F}$. For a regular multiset $\mathcal{F} = \{V_1, \ldots, V_s\}$ of subsets of V, let $d_G(\mathcal{F}) = \sum_{i=1}^{s} \frac{d_G(V_i)}{2}$. If $\mathcal{F}$ is a partition of V, $d_G(\mathcal{F})$ amounts to the number of edges connecting distinct subsets of $\mathcal{F}$.

In [4], Frank gave a characterization of orientability of graphs, called the supermodular covering condition. This theorem includes the following characterization of (k, ℓ)-ec-orientability as a special case (see e.g., [6]).

Theorem 6 (Frank [4]). *Let $G = (V, E)$ be a graph. Then, G admits a (k, ℓ)-edge-connected-orientation if and only if $d_G(\mathcal{F}) \geq k(|\mathcal{F}| - 1) + \ell$ for any partition $\mathcal{F}$ of V into non-empty subsets with $|\mathcal{F}| \geq 2$.*

This theorem motivates us to look at the following deficiency function: for $\ell > 0$

$$\eta_{k,\ell}(G) = \max\{0, \max\{k(|\mathcal{F}| - 1) + \ell - d_G(\mathcal{F}) \mid \text{ a partition } \mathcal{F} \text{ of } V \text{ with } |\mathcal{F}| \geq 2\}\},$$

and for $\ell = 0$

$$\eta_{k,0}(G) = \max\{k(|\mathcal{F}| - 1) - d_G(\mathcal{F}) \mid \text{ a partition } \mathcal{F} \text{ of } V\}.$$

Notice $\eta_{k,0}(G) \geq 0$ (consider $\mathcal{F} = \{V\}$). Notice also $\eta_{k,\ell}(G) \leq \eta_{k,0}(G) + \ell$, where the equality holds if $\eta_{k,0}(G) > 0$. Hence, we also have $\eta_{k,0}(G) \leq \eta_{k,\ell}(G)$. Namely,

$$\eta_{k,0}(G) \leq \eta_{k,\ell}(G) \leq \eta_{k,0}(G) + \ell. \tag{2}$$

The celebrated Tutte-and-Nash-Williams tree packing theorem [21,16] asserts $\rho_{k,k}(E) = k(n-1) - \eta_{k,0}(G)$, and hence $\eta_{k,0}(G)$ can be computed from $\rho_{k,k}(G)$. Therefore, the approximation algorithm for $\rho_{k,k}(G)$ proposed in Theorem 3 can be modified to compute $\eta_{k,\ell}(G)$.

Corollary 1. *Let G be a graph with n vertices, and $k \geq 1, \ell \geq 0$ be integers with $2k - \ell \geq 1$. In the bounded-degree model with a degree bound d, there exists a $(1, \ell + \epsilon n)$-approximation algorithm for $\eta_{k,\ell}(G)$ with query complexity $(k + d)^{O(1/\epsilon'^2)}(\frac{1}{\epsilon'})^{O(1/\epsilon')}$ where $\epsilon' = \frac{\epsilon}{dk}$.*

For testing (k, ℓ)-ec-orientability, we need a certificate to decide whether G is ϵ-far from (k, ℓ)-ec-orientable. This part relies on a structural property of the connectivity argumentation problem proved by Frank and Király [6]. A family $\{X_1, \ldots, X_s\}$ of subsets of $X \subseteq V$ is called a *co-partition* of X if $\{V \setminus X_1, \ldots, V \setminus X_s\}$ forms a partition of $V \setminus X$. Also, for two multisets $\mathcal{F}_1$ and $\mathcal{F}_2$, $\mathcal{F}_1 + \mathcal{F}_2$ denotes their union as a multiset.

Theorem 7 (Frank and Király [6]). *A graph G can be made (k, ℓ)-ec-orientable by adding γ edges iff the following two conditions hold:*

(A) *$\gamma \geq k(|\mathcal{F}| - 1) + \ell - d_G(\mathcal{F})$ for every partition $\mathcal{F}$ of V with $|\mathcal{F}| \geq 2$.*

(B) $2\gamma \geq |\mathcal{F}_1|k + |\mathcal{F}_2|\ell - d_G(\mathcal{F})$ *for every multiset* $\mathcal{F} = \mathcal{F}_1 + \mathcal{F}_2$ *such that* $\mathcal{F}_1$ *is a partition of some* $X \subset V$, $\mathcal{F}_2$ *is a co-partition of* $V \setminus X$, *and every member of* $\mathcal{F}_2$ *is the complement of the union of some members of* $\mathcal{F}_1$.

By Corollary 1, the condition (A) is efficiently checkable. The non-trivial part is an algorithm for checking the second condition. Let

$$\xi_{k,\ell}(G) = \max_{\mathcal{F} = \mathcal{F}_1 + \mathcal{F}_2} \{|\mathcal{F}_1|k + |\mathcal{F}_2|\ell - d_G(\mathcal{F})\} \tag{3}$$

where the maximum is taken over all multisets $\mathcal{F} = \mathcal{F}_1 + \mathcal{F}_2$ satisfying the property specified in Theorem 7(B). By carefully counting the number of edges appeared in the right hand side, $\xi_{k,\ell}$ can be simplified as follows. (See Theorem 5.4 for the proof.)

Lemma 6. *Let* $g_{k,\ell}(X) = k + \ell - d(X) + \eta_{k,0}(G[X])$ *for* $X \subseteq V$. *Then,*

$$\xi_{k,\ell}(G) = \max\left\{\sum_{i=1}^{s} g_{k,\ell}(X_i) \mid \text{a sub-partition } \mathcal{P} = \{X_1, \ldots, X_s\} \text{ of } V\right\}. \tag{4}$$

We say that $X \subseteq V$ is *deficient* if $g_{k,\ell}(X) > 0$. By Theorem 7 and (4), $g_{k,\ell}(X) \leq 0$ holds for every X with $\emptyset \neq X \subsetneq V$ if G is (k, ℓ)-ec-orientable. The following theorem is a key result to develop a constant-time tester. (The detail is deferred to the full version.)

Theorem 8. *For a given* ϵ, *let* $c = \frac{\epsilon^2 d^2}{16k\ell}$ *and* $t = \frac{4\ell}{\epsilon d}$. *Suppose that* $\xi_{k,\ell}(G) \geq \epsilon dn$. *Then, at least one of the followings holds:*

(i) *There are at least* cn *disjoint small deficient sets, where a set is called small if the cardinality is less than* t;

(ii) $\eta_{k,0}(G) \geq \frac{1}{4}\epsilon dn$. *Namely,* G *is* $\frac{\epsilon}{2}$-*far from* (k, k)-*fullness.*

A tester for the (k, ℓ)-ec-orientability of a graph $G = (V, E)$ is given in Algorithm 3. In Line 7, $V_t(v)$ denotes the set of vertices whose distances to $v \in V$ are at most t.

Algorithm 3. Testing the (k, ℓ)-ec-orientability of a bounded-degree graph G

1: Take any ϵ'' such that $\epsilon'' < \epsilon$.
2: Run a $(1, \frac{\epsilon'' dn}{4})$-approximation algorithm for $\eta_{k,0}(G)$.
3: **if** the obtained value x^* satisfies $x^* > 0$ **then**
4:   **reject** G.
5: Choose a set S of $\frac{8k\ell}{\epsilon^2 d^2}$ vertices uniformly at random from G.
6: **for** $v \in S$ **do**
7:   Compute $X_v = \text{argmax}\{g_{k,\ell}(X) : X \subseteq V_t(v), X \neq \emptyset\}$ with $t = \frac{4\ell}{\epsilon d}$.
8:   **if** $g_{k,\ell}(X_v) > 0$ **then**
9:     **reject** G.
10: **accept** G.

The query complexity of Algorithm 3 can be easily bounded as claimed in Theorem 2. The correctness follows from Corollary 1 and Theorem 8 (the detailed description is omitted in this extended abstract), and thus we obtain Theorem 2.

References

1. Chazelle, B., Rubinfeld, R., Trevisan, L.: Approximating the Minimum Spanning Tree Weight in Sublinear Time. SIAM Comp. 34(6), 1370–1379 (2005)
2. Demaine, E., O'Rourke, J.: Geometric Folding Algorithms: Linkages, Origami, Polyhedra, Reprint edition. Cambridge University Press, New York (2008)
3. Edmonds, J.: Edge disjoint branchings. In: Rustin, B. (ed.) Combinatorial Algorithms, pp. 91–96. Algorithmics Press (1973)
4. Frank, A.: On the orientation of graphs. J. Comb. Theory, B 28(3), 251–261 (1980)
5. Frank, A.: Connections in Combinatorial Optimization. Oxford University Press (2011)
6. Frank, A., Király, T.: Combined connectivity augmentation and orientation problems. Discrete Appl. Math. 131, 401–419 (2003)
7. Gabow, H., Westermann, H.: Forests, frames, and games: algorithms for matroid sums and applications. Algorithmica 7(1), 465–497 (1992)
8. Goldreich, O.: Intriduction to testing graph properties. Technical report. Electronic Colloquium on Computational Complexity, ECCC (2010)
9. Goldreich, O., Ron, D.: Property testing in bounded degree graphs. Algorithmica 32(2), 302–343 (2002)
10. Haas, R.: Characterizations of arboricity of graphs. Ars Comb. 63, 129–138 (2002)
11. Imai, H.: Network flow algorithms for lower truncated transversal polymatroids. Journal of the Operations Research Society of Japan 26(3), 186–210 (1983)
12. Kaufman, T., Krivelevich, M., Ron, D.: Tight bounds for testing bipartiteness in general graphs. SIAM Journal on Computing 33(6), 1441–1483 (2004)
13. Laman, G.: On graphs and rigidity of plane skeletal structures. Journal of Engineering Mathematics 4(4), 331–340 (1970)
14. Lee, A., Streinu, I.: Pebble game algorithms and sparse graphs. Discrete Mathematics 308(8), 1425–1437 (2008)
15. Nash-Williams, C.: On orientations, connectivity and odd vertex pairings in finite graphs. Canad. J. Math. 12, 555–567 (1960)
16. Nash-Williams, C.: Edge-disjoint spanning trees of finite graphs. Journal of the London Mathematical Society 1(1), 445–450 (1961)
17. Nash-Williams, C.: Decomposition of finite graphs into forests. Journal of the London Mathematical Society 1(1), 12 (1964)
18. Nguyen, H.N., Onak, K.: Constant-time approximation algorithms via local improvements. In: Proc. of FOCS 2008, pp. 327–336 (2008)
19. Newman, I., Sohler, C.: Every property of hyperfinite graphs is testable. In: Proc. of STOC 2011, pp. 675–684 (2011)
20. Orenstein, Y.: Property testing in directed graphs. Master's thesis, Tel-Aviv University (2010)
21. Tutte, W.T.: On the problem of decomposing a graph into n connected factors. Journal of the London Mathematical Society 36, 221–230 (1961)
22. Whiteley, W.: The union of matroids and the rigidity of frameworks. SIAM Journal on Discrete Mathematics 1(2), 237–255 (1988)
23. Whiteley, W.: Some matroids from discrete applied geometry. Contemporary Mathematics 197, 171–312 (1996)
24. Yoshida, Y., Ito, H.: Property testing on k-vertex-connectivity of graphs. Algorithmica 62(3), 701–712 (2012)
25. Yoshida, Y., Ito, H.: Testing k-edge-connectivity of digraphs. Journal of System Science and Complexity 23(1), 91–101 (2010)
26. Yoshida, Y., Yamamoto, M., Ito, H.: An improved constant-time approximation algorithm for maximum matchings. In: Proc. of STOC 2009, pp. 225–234 (2009)

CRAM: Compressed Random Access Memory

Jesper Jansson[1], Kunihiko Sadakane[2], and Wing-Kin Sung[3]

[1] Laboratory of Mathematical Bioinformatics, Institute for Chemical Research,
Kyoto University, Gokasho, Uji, Kyoto 611-0011, Japan
jj@kuicr.kyoto-u.ac.jp
[2] National Institute of Informatics, 2-1-2 Hitotsubashi, Chiyoda-ku,
Tokyo 101-8430, Japan
sada@nii.ac.jp
[3] National University of Singapore, 13 Computing Drive, Singapore 117417
ksung@comp.nus.edu.sg

Abstract. We present a new data structure called the *Compressed Random Access Memory* (CRAM) that can store a dynamic string T of characters, e.g., representing the memory of a computer, in compressed form while achieving asymptotically almost-optimal bounds (in terms of empirical entropy) on the compression ratio. It allows short substrings of T to be decompressed and retrieved efficiently and, significantly, characters at arbitrary positions of T to be modified quickly during execution *without decompressing the entire string*. This can be regarded as a new type of data compression that can update a compressed file directly. Moreover, at the cost of slightly increasing the time spent per operation, the CRAM can be extended to also support insertions and deletions. Our key observation that the empirical entropy of a string does not change much after a small change to the string, as well as our simple yet efficient method for maintaining an array of variable-length blocks under length modifications, may be useful for many other applications as well.

1 Introduction

Certain modern-day information technology-based applications require random access to very large data structures. For example, to do genome assembly in bioinformatics, one needs to maintain a huge graph [18]. Other examples include dynamic programming-based problems, such as optimal sequence alignment or finding maximum bipartite matchings, which need to create large tables (often containing a lot of redundancy). Yet another example is in image processing, where one sometimes needs to edit a high-resolution image which is too big to load into the main memory of a computer all at once. Additionally, a current trend in the mass consumer electronics market is cheap mobile devices with limited processing power and relatively small memories; although these are not designed to process massive amounts of data, it could be economical to store non-permanent data and software on them more compactly, if possible.

The standard solution to the above problem is to employ secondary memory (disk storage, etc.) as an extension of the main memory of a computer. This

A. Czumaj et al. (Eds.): ICALP 2012, Part I, LNCS 7391, pp. 510–521, 2012.
© Springer-Verlag Berlin Heidelberg 2012

technique is called *virtual memory*. The drawback of virtual memory is that the processing time will be slowed down since accessing the secondary memory is an order of magnitude slower than accessing the main memory. An alternative approach is to compress the data T and store it in the main memory. By using existing data compression methods, T can be stored in $nH_k + o(n \log \sigma)$-bits space [2,8] for every $0 \le k < \log_\sigma n$, where n is the length of T, σ is the size of the alphabet, and $H_k(T)$ denotes the k-th order empirical entropy of T. Although greatly reducing the amount of storage needed, it does not work well because it becomes computationally expensive to access and update T.

Motivated by applications that would benefit from having a large virtual memory that supports fast access- and update-operations, we consider the following task: Given a memory/text $T[1..n]$ over an alphabet of size σ, maintain a data structure that stores T compactly while supporting the following operations. (We assume that $\ell = \Theta(\log_\sigma n)$ is the length of one machine word.)

- access(T, i): Return the substring $T[i..(i + \ell - 1)]$.
- replace(T, i, c): Replace $T[i]$ by a character $c \in [\sigma]$. [1]
- delete(T, i): Delete $T[i]$, i.e., make T one character shorter.
- insert(T, i, c): Insert a character c into T between positions $i - 1$ and i, i.e., make T one character longer.

Compressed Read Only Memory: When only the access operation is supported, we call the data structure *Compressed Read Only Memory*. Sadakane and Grossi [17], González and Navarro [6], and Ferragina and Venturini [4] developed storage schemes for storing a text succinctly that allow constant-time access to any word in the text. More precisely, these schemes store $T[1..n]$ in $nH_k + \mathcal{O}\left(n \log \sigma \left(\frac{k \log \sigma + \log \log n}{\log n}\right)\right)$ bits[2] and access(T, i) takes $\mathcal{O}(1)$ time, and both the space and access time are optimal for this task. Note, however, that none of these schemes allow T to be modified.

Compressed Random Access Memory (CRAM): When the operations access and replace are supported, we call the data structure *Compressed Random Access Memory* (CRAM). As far as we know, it has not been considered previously in the literature, even though it appears to be a fundamental and important data structure.

Extended CRAM: When all four operations are supported, we call the data structure *extended CRAM*. It is equivalent to *the dynamic array* [16] and also solves *the list representation problem* [5]. Fredman and Saks [5] proved a cell probe lower bound of $\Omega(\log n / \log \log n)$ time for the latter, and also showed that $n^{\Omega(1)}$ update time is needed to support constant-time access. Raman *et al.* [16] presented an $n \log \sigma + o(n \log \sigma)$-bit data structure which supports access, replace, delete, and insert in $\mathcal{O}(\log n / \log \log n)$ time. Navarro and Sadakane [15] recently gave a data structure using $nH_0(T) + \mathcal{O}(n \log \sigma / \log^\epsilon n + \sigma \log^\epsilon n)$ bits that supports access, delete, and insert in $\mathcal{O}(\frac{\log n}{\log \log n}(1 + \frac{\log \sigma}{\log \log n}))$ time.

---

[1] The notation $[\sigma]$ stands for the set $\{1, 2, \ldots, \sigma\}$.
[2] Reference [17] has a slightly worse space complexity.

1.1   Our Contributions

This paper studies the complexity of maintaining the CRAM and extended CRAM data structures. We assume the uniform-cost word RAM model with word size $w = \Theta(\log n)$ bits, i.e., standard arithmetic and bitwise boolean operations on w-bit word-sized operands can be performed in constant time [9]. Also, we assume the memory consists of a sequence of bits, and each bit is identified with an address in $0, \ldots, 2^w - 1$. Furthermore, any consecutive w bits can be accessed in constant time. (Note that this memory model is equivalent under the word RAM model to a standard memory model consisting of a sequence of words of some fixed length.) At any time, if the highest address of the memory used by the algorithm is s, the space used by the algorithm is said to be $s + 1$ bits [10].

Our main results for the CRAM are summarized in:

Theorem 1. *Given a text $T[1..n]$ over an alphabet of size σ and any $\epsilon > 0$, after $\mathcal{O}(n \log \sigma / \log n)$ time preprocessing, the CRAM data structure for $T[1..n]$ can be stored in $nH_k(T) + \mathcal{O}\left(n \log \sigma \left((k+1)\epsilon + \frac{k \log \sigma + \log \log n}{\log n}\right)\right)$ bits for every $0 \leq k < \log_\sigma n$ simultaneously, where $H_k(T)$ denotes the k-th order empirical entropy of T, while supporting* access(T, i) *in $\mathcal{O}(1)$ time and* replace(T, i, c) *for any character c in $\mathcal{O}(1/\epsilon)$ time.*

Theorem 1 is proved in Section 5 below.

Next, by setting $\epsilon = \max\{\frac{\log \sigma}{\log n}, \frac{\log \log n}{(k+1) \log n}\}$, we obtain:

Corollary 1. *Given a text $T[1..n]$ over an alphabet of size σ and any $k = o(\log_\sigma n)$, after $\mathcal{O}(n \log \sigma / \log n)$ time preprocessing, the CRAM data structure for $T[1..n]$ can be stored in $nH_k(T) + \mathcal{O}\left(n \log \sigma \cdot \frac{k \log \sigma + \log \log n}{\log n}\right)$ bits while supporting* access(T, i) *in $\mathcal{O}(1)$ time and* replace(T, i, c) *for any character c in $\mathcal{O}(\min\{\log_\sigma n, (k+1) \log n / \log \log n\})$ time.*

For the extended CRAM, we have:

Theorem 2. *Given a text $T[1..n]$ over an alphabet of size σ, after spending $\mathcal{O}(n \log \sigma / \log n)$ time on preprocessing, the extended CRAM data structure for $T[1..n]$ can be stored in $nH_k(T) + \mathcal{O}\left(n \log \sigma \cdot \frac{k \log \sigma + (k+1) \log \log n}{\log n}\right)$ bits for every $0 \leq k < \log_\sigma n$ simultaneously, where $H_k(T)$ denotes the k-th order empirical entropy of T, while supporting all four operations in $\mathcal{O}(\log n / \log \log n)$ time.*

(Due to space limitations, the proof of Theorem 2 has been omitted from the conference version of our paper.)

Table 1 shows a comparison with existing data structures. Many existing dynamic data structures for storing compressed strings [7,11,13,15] use the fact $nH_0(S) = \log \binom{n}{n_1, \ldots, n_\sigma}$ where n_c is the number of occurrences of character c in the string S. However, this approach is helpful for small alphabets only because of the size of the auxiliary data. For large alphabets, generalized wavelet trees [3] can be used to decompose a large alphabet into smaller ones, but this slows down

Table 1. Comparison between previously existing data structures and the new ones in this paper. For simplicity, we assume $\sigma = o(n)$. The upper table lists results for the Compressed Read Only Memory (the first line) and the CRAM (the second and third lines), and the lower table lists results for the extended CRAM.

access	replace	Space (bits)	Ref.
$\mathcal{O}(1)$	—	$nH_k(T) + \mathcal{O}\left(n \log \sigma \cdot \frac{k \log \sigma + \log \log n}{\log n}\right)$	[4,6]
$\mathcal{O}(1)$	$\mathcal{O}(\min\{\log_\sigma n, \frac{(k+1)\log n}{\log \log n}\})$	$nH_k(T) + \mathcal{O}\left(n \log \sigma \cdot \frac{k \log \sigma + \log \log n}{\log n}\right)$	New
$\mathcal{O}(1)$	$\mathcal{O}(\frac{1}{\epsilon})$	$nH_k(T) +$ $\mathcal{O}\left(n \log \sigma \left(\frac{k \log \sigma + \log \log n}{\log n} + (k+1)\epsilon\right)\right)$	New

access/replace/insert/delete	Space (bits)	Ref.
$\mathcal{O}(\frac{\log^2 n}{\log \sigma})$	$nH_k(T) + o(n \log \sigma)$	[15]
$\mathcal{O}(\frac{\log \sigma \log n}{(\log \log n)^2})$	$nH_0(T) + \mathcal{O}\left(n \log \sigma \cdot \frac{1}{\log^\epsilon n}\right)$	[15]
$\mathcal{O}(\frac{\log n}{\log \log n})$	$nH_0(T) + \mathcal{O}\left(n \log \sigma \cdot \frac{\log \log n}{\log n}\right)$	New
$\mathcal{O}(\frac{\log n}{\log \log n})$	$nH_k(T) + \mathcal{O}\left(n \log \sigma \cdot \frac{k \log \sigma + (k+1) \log \log n}{\log n}\right)$	New

the access and update times. For example, if $\sigma = \sqrt{n}$, the time complexity of those data structures is $\mathcal{O}((\log n / \log \log n)^2)$, while ours is $\mathcal{O}(\log n / \log \log n)$, or even constant. Also, a technical issue when using large alphabets is how to update the code tables for encoding characters to achieve the entropy bound. Code tables that achieve the entropy bound will change when the string changes, and updating the entire data structure with the new code table is time-consuming.

Our results depend on a new analysis of the empirical entropies of *similar* strings in Section 3. We prove that *the empirical entropy of a string does not change a lot after a small change to the string* (Theorem 4). By using this fact, we can delay updating the entire code table. Thus, after each update operation to the string, we just change a part of the data structure according to the new code table. In Section 5, we show that the redundancy in space usage by this method is negligible, and we obtain Theorem 1.

Looking at Table 1, we observe that Theorem 1 can be interpreted as saying that for arbitrarily small, fixed $\epsilon > 0$, by spending $\mathcal{O}(n \log \sigma \cdot \epsilon(k+1))$ bits space more than the best existing data structures for Compressed Read Only Memory, we can also get $\mathcal{O}(1/\epsilon)$ (i.e., constant) time **replace** operations.

1.2 Organization of the Paper

Section 2 reviews the definition of the empirical entropy of a string and the data structure of Ferragina and Venturini [4]. In Section 3, we prove an important result on the empirical entropies of similar strings. In Section 4, we describe a technique for maintaining an array of variable-length blocks. Section 5 explains how to implement the CRAM to achieve the bounds stated in Theorems 1 above. Finally, Section 6 gives some concluding remarks.

2   Preliminaries

2.1   Empirical Entropy

The compression ratio of a data compression method is often expressed in terms of the *empirical entropy* of the input strings [12]. We first recall the definition of this concept. Let T be a string of length n over an alphabet $\mathcal{A} = [\sigma]$. Let n_c be the number of occurrences of $c \in \mathcal{A}$ in T. Let $\{P_c = n_c/n\}_{c=1}^{\sigma}$ be the empirical probability distribution for the string T. The 0-th order empirical entropy of T is defined as $H_0(T) = -\sum_{c=1}^{\sigma} P_c \log P_c$. We also use $H_0(p)$ to denote the 0-th order empirical entropy of a string whose empirical probability distribution is p.

Next, let k be any non-negative integer. If a string $s \in \mathcal{A}^k$ precedes a symbol c in T, s is called the *context of* c. We denote by $T^{(s)}$ the string that is the concatenation of all symbols, each of whose context in T is s. The k-th order empirical entropy of T is defined as $H_k(T) = \frac{1}{n}\sum_{s \in \mathcal{A}^k} |T^{(s)}|H_0(T^{(s)})$. It was shown in [14] that for any $k \geq 0$, $H_k(T) \geq H_{k+1}(T)$ holds, and $nH_k(T)$ is a lower bound for the output size of any compressor that encodes each symbol of T with a code that only depends on the symbol and its context of length k.

To prove our new results, we shall use the following theorem in Section 3:

Theorem 3 ([1, Theorem 16.3.2]). *Let p and q be two probability mass functions on $\mathcal{A}$ such that $||p - q||_1 \equiv \sum_{c \in \mathcal{A}} |p(c) - q(c)| \leq \frac{1}{2}$. Then $|H_0(p) - H_0(q)| \leq -||p - q||_1 \log \frac{||p-q||_1}{|\mathcal{A}|}$.*

The technique of *blocking*, i.e., to conceptually merge consecutive symbols to form new symbols over a larger alphabet, is used to reduce the redundancy of Huffman encoding for compressing a string. A string T of length n is partitioned into $\frac{n}{\ell}$ blocks of length ℓ each, then Huffman or other entropy codings are applied to compress a new string T_ℓ of those blocks. We call this operation *blocking of length ℓ*.

2.2   Review of Ferragina and Venturini's Data Structure

Here, we briefly review the data structure of Ferragina and Venturini from [4]. It uses the same basic idea as Huffman coding: replace every fixed-length block of symbols by a variable-length code in such a way that frequently occurring blocks get shorter codes than rarely occurring blocks.

To be more precise, consider a text $T[1..n]$ over an alphabet $\mathcal{A}$ where $|\mathcal{A}| = \sigma$ and $\sigma < n$. Let $\ell = \frac{1}{2}\log_\sigma n$ and $\tau = \log n$. Partition $T[1..n]$ into $\frac{n}{\tau\ell}$ super-blocks, each contains $\tau\ell$ characters. Each super-block is further partitioned into τ blocks, each contains ℓ characters. Denote the $\frac{n}{\ell}$ blocks by $T_i = T[(i-1)\ell + 1..i\ell]$ for $i = 1, 2, \ldots, n/\ell$.

Since each block is of length ℓ, there are at most $\sigma^\ell = \sqrt{n}$ distinct blocks. For each block $P \in \mathcal{A}^\ell$, let $f(P)$ be the frequency of P in $\{T_1, \ldots, T_{n/\ell}\}$. Let $r(P)$ be the rank of P according to the decreasing frequency, i.e., the number of distinct blocks P' such that $f(P') \geq f(P)$, and $r^{-1}(j)$ be its inverse function. Let $enc(j)$ be the rank j-th binary string in $[\epsilon, 0, 1, 00, 01, 10, 11, 000, \ldots]$.

The data structure of Ferragina and Venturini consists of four arrays:

- $V = enc(r(T_1)) \ldots enc(r(T_{n/\ell}))$.
- $r^{-1}(j)$ for $j = 1, \ldots, \sqrt{n}$.
- Table $T_{Sblk}[1..\frac{n}{\ell_\tau}]$ stores the starting position in V of the encoding of every super-block.
- Table $T_{blk}[1..\frac{n}{\ell}]$ stores the starting position in V of the encoding of every block relative to the beginning of its enclosing super-block.

The algorithm for access(T, i) is simple: Given i, compute the address where the block for $T[i]$ is encoded by using T_{Sblk} and T_{blk} and obtain the code which encodes the rank of the block. Then, from r^{-1}, obtain the substring. In total, this takes $\mathcal{O}(1)$ time. This yields:

Lemma 1 ([4]). *Any substring $T[i..j]$ can be retrieved in $\mathcal{O}(1+(j-i+1)/\log_\sigma n)$ time.*

Using the data structure of Ferragina and Venturini, $T[1..n]$ can be encoded using $nH_k + \mathcal{O}(\frac{n}{\log_\sigma n}(k \log \sigma + \log \log n))$ bits according to the next lemma.

Lemma 2 ([4]). *The space needed by V, r^{-1}, T_{Sblk}, and T_{blk} is as follows:*

- *V is of length $nH_k + 2 + \mathcal{O}(k \log n) + \mathcal{O}(nk \log \sigma/\ell)$ bits, simultaneously for all $0 \le k < \log_\sigma n$.*
- *$r^{-1}(j)$ for $j = 1, \ldots, \sqrt{n}$ can be stored in $\sqrt{n} \log n$ bits.*
- *$T_{Sblk}[1..\frac{n}{\ell_\tau}]$ can be stored in $\mathcal{O}(\frac{n}{\ell})$ bits.*
- *$T_{blk}[1..\frac{n}{\ell}]$ can be stored in $\mathcal{O}(\frac{n}{\ell} \log \log n)$ bits.*

3   Entropies of Similar Strings

In this section, we prove that the empirical entropy of a string does not change much after a small change to it. This result will be used to bound the space complexity of our main data structure in Section 5.4. Consider two strings T and T' of length n and n', respectively, such that the edit distance between T and T' is one. That is, T' can be obtained from T by replacement, insertion, or deletion of one character. We show that the empirical entropies of the two strings do not differ so much.

Theorem 4. *For two strings T and T' of length n and n', respectively, over an alphabet $\mathcal{A}$ such that the edit distance between T and T' is one, it holds for any integer $k \ge 0$ that $|nH_k(T) - n'H_k(T')| = \mathcal{O}((k+1)(\log n + \log |\mathcal{A}|))$.*

To prove Theorem 4, we first prove the following:

Lemma 3. *Let T be a string of length n over an alphabet $\mathcal{A}$, T^- be a string made by deleting a character from T at any position, T^+ be a string made by inserting a character into T at any position, and T' be a string by replacing a character of T into another one at any position. Then the following holds:*

$$|nH_0(T) - (n-1)H_0(T^-)| \le 4 \log n + 3 \log |\mathcal{A}| \qquad (if\ n \ge 1) \qquad (1)$$

$$|nH_0(T) - (n+1)H_0(T^+)| \le 4 \log(n+1) + 4 \log |\mathcal{A}| \quad (if\ n \ge 0) \qquad (2)$$

$$|nH_0(T) - nH_0(T')| \le 4 \log(n+1) + 3 \log |\mathcal{A}| \qquad (if\ n \ge 0) \qquad (3)$$

Proof. Let $P(x)$, $P^-(x)$, $P^+(x)$, and $P'(x)$ denote the empirical probability of a character $x \in \mathcal{A}$ in T, T^-, T^+, and T', respectively, and let n_x denote the number of occurrences of $x \in \mathcal{A}$ in T. It holds that $P(x) = \frac{n_x}{n}$ for any $x \in \mathcal{A}$.

If a character c is removed from T, then $P^-(c) = \frac{n_c-1}{n-1}$, and $P^-(x) = \frac{n_x}{n-1}$ for any other $x \in \mathcal{A}$. Then $||P - P^-||_1 = \frac{n-n_c}{n(n-1)} + \sum_{x\in\mathcal{A},x\neq c}\frac{n_x}{n(n-1)} = \frac{2(n-n_c)}{n(n-1)}$. If $n = 1$, it holds $H_0(T) = 0$, and therefore $nH_0(T) - (n-1)H_0(T^-) = 0$ and the claim holds. If $n = n_c$, which means that all characters in T are c, it holds $H_0(T) = H_0(T^-) = 0$ and the claim holds. Otherwise, $\frac{2}{n(n-1)} \leq ||P - P^-||_1 \leq \frac{2}{n}$ holds. If $||P - P^-||_1 \leq \frac{1}{2}$, from Theorem 3, $|H_0(P) - H_0(P^-)| \leq -||P - P^-||_1 \log\frac{||P-P^-||_1}{|\mathcal{A}|} \leq \frac{2}{n}\log\frac{|\mathcal{A}|n(n-1)}{2}$. Then $|nH_0(T) - (n-1)H_0(T^-)| \leq n|H_0(P) - H_0(P^-)| + H_0(P^-) \leq 4\log n + 3\log|\mathcal{A}|$. If $||P - P^-||_1 > \frac{1}{2}$, which implies $n < 4$, $|nH_0(T) - (n-1)H_0(T^-)| \leq 3\log|\mathcal{A}|$. This proves the claim for T^-.

If a character c is inserted into T, then $P^+(c) = \frac{n_c+1}{n+1}$, and $P^+(x) = \frac{n_x}{n+1}$ for any other $x \in \mathcal{A}$. Then $||P - P^+||_1 = \frac{2(n-n_c)}{n(n+1)}$. If $n = 0$, $H_0(T) = H_0(T^+) = 0$ and the claim holds. If $n = n_c$, which means that T^+ consists of only the character c, $H_0(T) = H_0(T^+) = 0$ and the claim holds. Otherwise, $\frac{2}{n(n-1)} \leq ||P-P^+||_1 \leq \frac{2}{n}$ holds. If $||P-P^+||_1 \leq \frac{1}{2}$, $|nH_0(T)-(n+1)H_0(T^+)| \leq n|H_0(P)-H_0(P^+)| + H_0(P^-) \leq 4\log n + 3\log|\mathcal{A}|$. If $||P - P^+||_1 > \frac{1}{2}$, which implies $n < 4$, $|nH_0(T) - (n+1)H_0(T^+)| \leq 4\log|\mathcal{A}|$. This proves the claim for T^+.

If a character c of T is replaced with another character $c' \in \mathcal{A}$ $(c' \neq c)$, then $||P - P'||_1 = \sum_{\alpha\in\mathcal{A}}|P(\alpha) - P'(\alpha)| = |\frac{n_c}{n} - \frac{n_c-1}{n}| + |\frac{n_{c'}}{n} - \frac{n_{c'}+1}{n}| = \frac{2}{n}$. If $||P - P'||_1 \leq \frac{1}{2}$, $|nH_0(T) - nH_0(T')| \leq n|H_0(P) - H_0(P')| \leq 4\log n + 2\log|\mathcal{A}|$. If $||P - P'||_1 > \frac{1}{2}$, which implies $n < 4$, $|nH_0(T) - nH_0(T')| \leq 3\log|\mathcal{A}|$. If $c' = c$, $T' = T$ and $|nH_0(T) - nH_0(T')| = 0$. This completes the proof. $\square$

Proof. (of Theorem 4) From the definition of the empirical entropy, $nH_k(T) = \sum_{s\in\mathcal{A}^k}|T^{(s)}|H_0(T^{(s)})$. Therefore, for each context $s \in \mathcal{A}^k$, we estimate the change of 0-th order entropy. Because the edit distance between T and T' is one, we can write $T = T_1cT_2$ and $T' = T_1c'T_2$ using two (possibly empty) strings T_1, T_2 and two (possibly empty) characters c, c'. For the context $T_1[n_1-k+1..n_1]$ $(n_1 = |T_1|)$, denoted by s_0, the character c in the string $T^{(s_0)}$ will change to c'. The character $T_2[i]$ $(i = 1, 2, \ldots, k)$ has the context $T_1[n_1 - k + 1 + i..n_1]cT_2[1..i-1]$, denoted by s_i, in T, but the context will change to $s_i' = T_1[n_1 - k + 1 + i..n_1]c'T_2[1..i-1]$ in T'. Thus, a character $T_2[i]$ is removed from the string $T^{(s_i)}$ and inserted into $T'^{(s_i)}$. Therefore, the entropies will change in at most $2k + 1$ strings $(T^{(s_0)}, T^{(s_1)}, \ldots, T^{(s_k)}, T'^{(s_1)}, \ldots, T'^{(s_k)})$. By Lemma 3, each one will change only $\mathcal{O}(\log n + \log|\mathcal{A}|)$. This proves the claim. $\square$

4   Memory Management

This section presents a data structure for storing a set B of m variable-length strings over the alphabet $\{0,1\}$, which is an extension of the one in [15]. The data structure allows the contents of the strings and their lengths to change, but the value of m must remain constant. We assume a unit-cost word RAM model with word size w bits. The memory consists of consecutively ordered bits, and any consecutive w bits can be accessed in constant time, as stated above. A string over $\{0,1\}$ of length at most b is called a $(\leq b)$-block. Our data structure stores a set B of m such $(\leq b)$-blocks, while supporting the following operations:

- address(i): Return a pointer to where in the memory the i-th $(\leq b)$-block is stored $(1 \leq i \leq m)$.
- realloc(i, b'): Change the length of the i-th $(\leq b)$-block to b' bits $(0 \leq i \leq m)$. The physical address for storing the block (address(i)) may change.

Theorem 5. *Given that $b \leq m$ and $\log m \leq w$, consider the unit-cost word RAM model with word size w. Let $B = \{B[1], B[2], \ldots, B[m]\}$ be a set of $(\leq b)$-blocks and let s be the total number of bits of all $(\leq b)$-blocks in B. We can store B in $s + \mathcal{O}(m \log m + b^2)$ bits while supporting address in $\mathcal{O}(1)$ time and realloc in $\mathcal{O}(b/w)$ time.*

Theorem 6. *Given a parameter $b = \mathcal{O}(w)$, consider the unit-cost word RAM model with word size w. Let $B = \{B[1], B[2], \ldots, B[m]\}$ be a set of $(\leq b)$-blocks, and let s be the total number of bits of all $(\leq b)$-blocks in B. We can store B in $s + \mathcal{O}(w^4 + m \log w)$ bits while supporting address and realloc in $\mathcal{O}(1)$ time.*

(Due to lack of space, the proofs of Theorems 5 and 6 have been omitted from the conference version of our paper.)

From here on, we say that the data structure has parameters (b, m).

5   A Data Structure for Maintaining the CRAM

This section is devoted to proving Theorem 1. Our aim is to dynamize Ferragina and Venturini's data structure [4] by allowing replace operations. Ferragina and Venturini's data structure uses a code table for encoding the string, while our data structure uses two code tables, which will change during update operations.

Given a string $T[1..n]$ defined over an alphabet $\mathcal{A}$ ($|\mathcal{A}| = \sigma$), we support two operations. (1) access(T, i): which returns $T[i..i + \frac{1}{2}\log_\sigma n - 1]$; and (2) replace($T, i, c$): which replaces $T[i]$ with a character $c \in \mathcal{A}$.

We use blocking of length $\ell = \frac{1}{2}\log_\sigma n$ of T. Let $T'[1..n']$ be a string of length $n' = \frac{n}{\ell}$ on an alphabet $\mathcal{A}^\ell$ made by blocking of T. The alphabet size is $\sigma^\ell = \sqrt{n}$. Each character $T'[i]$ corresponds to the string $T[((i-1)\ell+1)..i\ell]$. A super-block consists of $1/\epsilon$ consecutive blocks in T' (ℓ/ϵ consecutive characters in T), where ϵ is a predefined constant.

Our algorithm runs in phases. Let $n'' = \epsilon n'$. For every $j \geq 1$, we refer to the sequence of the $(n''(j-1)+1)$-th to $(n''j)$-th replacements as *phase j*. The preprocessing stage corresponds to phase 0. Let $T^{(j)}$ denote the string just before phase j. (Hence, $T^{(1)}$ is the input string T.) Let $F^{(j)}$ denote the frequency table of blocks $b \in \mathcal{A}^\ell$ in $T^{(j)}$, and $C^{(j)}$ and $D^{(j)}$ a code table and a decode table defined below. The algorithm also uses a bit-vector $R^{(j-1)}[1..n'']$, where $R^{(j-1)}[i] = 1$ means that the i-th super-block in T is encoded by code table $C^{(j-1)}$; otherwise, it is encoded by code table $C^{(j-2)}$.

During the execution of the algorithm, we maintain the following invariant:

- At the beginning of phase j, the string $T^{(j)}$ is encoded with code table $C^{(j-2)}$ (we assume $C^{(-1)} = C^{(0)} = C^{(1)}$), and the table $F^{(j)}$ stores the frequencies of blocks in $T^{(j)}$.
- During phase j, the i-th super-block is encoded with code table $C^{(j-2)}$ if $R^{(j-1)}[i] = 0$, or $C^{(j-1)}$ if $R^{(j-1)}[i] = 1$. The code tables $C^{(j-2)}$ and $C^{(j-1)}$ do not change.
- During phase j, $F^{(j+1)}$ stores the correct frequency of blocks of the current T.

5.1   Phase 0: Preprocessing

First, for each block $b \in \mathcal{A}^\ell$, we count the numbers of its occurrences in T' and store it in an array $F^{(1)}[b]$. Then we sort the blocks $b \in \mathcal{A}^\ell$ in decreasing order of the frequencies $F^{(1)}[b]$, and assign a code $C^{(1)}[b]$ to encode them. The code for a block b is defined as follows. If the length of the code $enc(b)$, defined in Section 2.2, is at most $\frac{1}{2} \log n$ bits, then $C^{(1)}[b]$ consists of a bit '0', followed by $enc(b)$. Otherwise, it consists of a bit '1', followed by the binary encoding of b, that is, the block is stored without compression. The code length for any block b is upper bounded by $1 + \frac{1}{2} \log n$ bits. Then we construct a table $D^{(1)}$ for decoding a block. The table has $2^{1+\frac{1}{2}\log n} = \mathcal{O}(\sqrt{n})$ entries and $D^{(1)}[x] = b$ for all binary patterns x of length $1 + \frac{1}{2} \log n$ such that a prefix of x is equal to $C^{(1)}[b]$. Note that this decode table is similar to r^{-1} defined in Section 2.2.

Next, for each block $T'[i]$ ($i = 1, \ldots, n'$), compute its length using $C^{(1)}[T'[i]]$, allocate space for storing it using the data structure of Theorem 6 with parameters $(1 + \ell \log \sigma, \frac{n}{\ell}) = (1 + \frac{1}{2} \log n, \frac{2n \log \sigma}{\log n})$, and $w = \log n$. From Lemma 2 and Theorem 6, if follows that the size of the initial data structure is $nH_k(T) + \mathcal{O}\left(\frac{n \log \sigma}{\log n}(k \log \sigma + \log \log n)\right)$ bits. Finally, for later use, copy the contents of $F^{(1)}$ to $F^{(2)}$, and initialize $R^{(0)}$ by 0. By sorting the blocks by a radix sort, the preprocessing time becomes $\mathcal{O}(n \log \sigma / \log n)$.

5.2   Algorithm for Access

The algorithm for $\mathtt{access}(T, i)$ is: Given the index i, compute the block number $x = \lfloor (i - 1)/\ell \rfloor + 1$ and the super-block number y containing $T[i]$. Obtain the pointer to the block and the length of the code by $\mathtt{address}(x)$. Decode the block using the decode table $D^{(j-2)}$ if $R^{(j-1)}[x] = 0$, or $D^{(j-1)}$ if $R^{(j-1)}[x] = 1$. This takes constant time.

5.3   Algorithm for Replace

We first explain a naive, inefficient algorithm. If $b = T'[i]$ is replaced with b', we change the frequency table $F^{(1)}$ so that $F^{(1)}[b]$ is decremented by one and

$F^{(1)}[b']$ is incremented by one. Then new code table $C^{(1)}$ and decode table $D^{(1)}$ are computed from updated $F^{(1)}$, and all blocks $T'[j]$ $(j = 1, \ldots, n')$ are re-encoded by using the new code table. Obviously, this algorithm is too slow.

To get a faster algorithm, we can delay updating code tables for the blocks and re-writing the blocks using new code tables because of Theorem 4. Because the amount of change in entropy is small after a small change in the string, we can show that the redundancy of using code tables defined according to an old string can be negligible. For each single character change in T, we re-encode a super-block (ℓ/ϵ characters in T). After $\epsilon n'$ changes, the whole string will be re-encoded. To specify which super-block to be re-encoded, we use an integer array $G^{(j-1)}[1..n'']$. It stores a permutation of $(1, \ldots, n'')$ and indicates that at the x-th replace operation in phase j we rewrite the $G^{(j-1)}[x]$-th super-block. The bit $R^{(j-1)}[x]$ indicates if the super-block has been already rewritten or not. The array $G^{(j-1)}$ is defined by sorting super-blocks in increasing order of lengths of codes for encoding super-blocks.

We implement $\texttt{replace}(T, i, S)$ as follows. In the x-th update in phase j,

1. If $R^{(j-1)}[G^{(j-1)}[x]] = 0$, i.e., if the $G^{(j-1)}[x]$-th super-block is encoded with $C^{(j-2)}$, decode it and re-encode it with $C^{(j-1)}$, and set $R^{(j-1)}[G^{(j-1)}[x]] = 1$.
2. Let y be the super-block number containing $T[i]$, that is, $y = \lfloor \epsilon(i - 1)/\ell \rfloor$.
3. Decode the y-th super-block, which is encoded with $C^{(j-2)}$ or $C^{(j-1)}$ depending on $R^{(j-1)}[y]$. Let S' denote the block containing $T[i]$. Make a new block S from S' by applying the $\texttt{replace}$ operation.
4. Decrement the frequency $F^{(j+1)}[S']$ and increment the frequency $F^{(j+1)}[S]$.
5. Compute the code for encoding S using $C^{(j-1)}$ if the y-th super-block is already re-encoded ($R^{(j-1)}[y] = 1$), or $C^{(j-2)}$ otherwise ($R^{(j-1)}[y] = 0$).
6. Compute the lengths of the blocks in y-th super-block and apply $\texttt{realloc}$ for those blocks.
7. Rewrite the blocks in the y-th super-block.
8. Construct a part of tables $C^{(j)}$, $D^{(j)}$, $G^{(j)}$, and $R^{(j)}$ (see below).

To prove that the algorithm above maintains the invariant, we need only to prove that the tables $C^{(j-1)}$, $F^{(j)}$, and $G^{(j-1)}$ are ready at the beginning of phase j. In phase j, we create $C^{(j)}$ based on $F^{(j)}$. This is done by just radix-sorting the frequencies of blocks, and therefore the total time complexity is $\mathcal{O}(\sigma^l) = \mathcal{O}(\sqrt{n})$. Because phase j consists of n'' $\texttt{replace}$ operations, the work for creating $C^{(j)}$ can be distributed in the phase. We represent the array $G^{(j-1)}$ implicitly by $(1/\epsilon)(1 + \frac{1}{2}\log n)$ doubly-linked lists L_d; L_d stores super-blocks of length d. By retrieving the lists in decreasing order of d we can enumerate the elements of $G^{(j)}$. If all the elements of a list have been retrieved, we move to the next non-empty list. This can be done in $\mathcal{O}(1/\epsilon)$ time if we use a bit-vector of $(1/\epsilon)(1 + \frac{1}{2}\log n)$ bits indicating which lists are non-empty. We copy $F^{(j)}$ to $F^{(j+1)}$ in constant time by changing pointers to $F^{(j)}$ and $F^{(j+1)}$. For each $\texttt{replace}$ in phase j, we re-encode a super-block, which consists of $1/\epsilon$ blocks. This takes $\mathcal{O}(1/\epsilon)$ time. Therefore the time complexity for $\texttt{replace}$ is $\mathcal{O}(1/\epsilon)$ time.

Note that during phase j, only the tables $F^{(j)}$, $F^{(j+1)}$, $C^{(j-2)}$, $C^{(j-1)}$, $C^{(j)}$, $D^{(j-2)}$, $D^{(j-1)}$, $D^{(j)}$, $G^{(j-1)}$, $G^{(j)}$, $R^{(j-1)}$, and $R^{(j)}$ are stored. The other tables are discarded.

5.4 Space Analysis

Let $s(T)$ denote the size of the encoding of T by our dynamic data structure. At the beginning of phase j, the string $T^{(j)}$ is encoded with code table $C^{(j-2)}$, which is based on the string $T^{(j-2)}$. Let $L^{(j)} = nH_k(T^{(j)})$ and $L^{(j-2)} = nH_k(T^{(j-2)})$.

After the preprocessing, $s(T^{(1)}) \leq L^{(1)} + \mathcal{O}\left(\frac{n \log \sigma}{\log n}(k \log \sigma + \log \log n)\right)$. If we do not re-encode the string, for each `replace` operation we write at most $1 + \frac{1}{2} \log n$ bits. Therefore $s(T^{(j)}) \leq s(T^{(j-2)}) + \mathcal{O}(n'' \log n)$ holds. Because $T^{(j)}$ is made by $2(n'' + \sqrt{n})$ character changes to $T^{(j-2)}$, from Theorem 4, we have $|L^{(j)} - L^{(j-2)}| = \mathcal{O}(n''(k+1)(\log n + \log \sigma))$. Therefore we obtain $s(T^{(j)}) \leq L^{(j)} + \mathcal{O}(\epsilon(k+1)n \log \sigma)$. The space for storing the tables $F^{(j)}$, $C^{(j)}$, $D^{(j)}$, $G^{(j)}$, $H^{(j)}$, and $R^{(j)}$ is $\mathcal{O}(\sqrt{n} \log n)$, $\mathcal{O}(\sqrt{n} \log n)$, $\mathcal{O}(\sqrt{n} \log n)$, $\mathcal{O}(n'' \log n) = \mathcal{O}(\epsilon n \log \sigma)$, $\mathcal{O}(n'' \log n)$, $\mathcal{O}(n'')$ bits, respectively.

Next we analyze the space redundancy caused by the re-encoding of super-blocks. We re-encode the super-blocks with the new code table in increasing order of their lengths, that is, the shortest one is re-encoded first. This guarantees that at any time, the space does not exceed $\max\{s(T^{(j)}), s(T^{(j-2)})\}$. This completes the proof of Theorem 1.

6 Concluding Remarks

We have presented a data structure called Compressed Random Access Memory (CRAM), which compresses a string T of length n into its k-th order empirical entropy in such a way that any consecutive $\log_\sigma n$ bits can be obtained in constant time (the `access` operation), and replacing a character (the `replace` operation) takes $\mathcal{O}(\min\{\log_\sigma n, (k+1) \log n / \log \log n\})$ time. The time for `replace` can be reduced to constant ($\mathcal{O}(1/\epsilon)$) time by allowing an additional $\mathcal{O}(\epsilon(k+1)n \log \sigma)$ bits redundancy. The extended CRAM data structure also supports the `insert` and `delete` operations, at the cost of increasing the time for `access` to $\mathcal{O}(\log n / \log \log n)$ time, which is optimal under this stronger requirement, and the time for each update operation also becomes $\mathcal{O}(\log n / \log \log n)$.

Preliminary experimental results indicate that our CRAM data structure supports faster reads/writes of short segments (from 16 to 256 bytes) than when using `gzip`. These experimental results will be reported in another paper.

An open problem is how to improve the running time of `replace` for the CRAM data structure to $\mathcal{O}(1)$ without using the $\mathcal{O}(\epsilon(k+1)n \log \sigma)$ extra bits.

Acknowledgments. JJ was funded by The Hakubi Project at Kyoto University. KS was supported in part by Funding Program for World-Leading Innovative R&D on Science and Technology (FIRST Program). WKS was supported in part by the MOE's AcRF Tier 2 funding R-252-000-444-112.

References

1. Cover, T.M., Thomas, J.A.: Elements of Information Theory. Wiley Interscience (1991)
2. Ferragina, P., Manzini, G.: Indexing compressed text. Journal of the ACM 52(4), 552–581 (2005)
3. Ferragina, P., Manzini, G., Mäkinen, V., Navarro, G.: Compressed representations of sequences and full-text indexes. ACM Transactions on Algorithms 3(2), article No. 20 (2007)
4. Ferragina, P., Venturini, R.: A simple storage scheme for strings achieving entropy bounds. Theoretical Computer Science 372(1), 115–121 (2007)
5. Fredman, M.L., Saks, M.E.: The cell probe complexity of dynamic data structures. In: Proceedings of ACM STOC, pp. 345–354 (1989)
6. González, R., Navarro, G.: Statistical Encoding of Succinct Data Structures. In: Lewenstein, M., Valiente, G. (eds.) CPM 2006. LNCS, vol. 4009, pp. 294–305. Springer, Heidelberg (2006)
7. González, R., Navarro, G.: Rank/select on dynamic compressed sequences and applications. Theoretical Computer Science 410(43), 4414–4422 (2009)
8. Grossi, R., Gupta, A., Vitter, J.S.: High-order entropy-compressed text indexes. In: Proceedings of ACM-SIAM SODA, pp. 841–850 (2003)
9. Hagerup, T.: Sorting and searching on the word RAM. In: Proceedings of Symposium on Theory Aspects of Computer Science (STACS 1998), pp. 366–398 (1998)
10. Hagerup, T., Raman, R.: An Efficient Quasidictionary. In: Penttonen, M., Schmidt, E.M. (eds.) SWAT 2002. LNCS, vol. 2368, pp. 1–18. Springer, Heidelberg (2002)
11. He, M., Munro, J.I.: Succinct Representations of Dynamic Strings. In: Chavez, E., Lonardi, S. (eds.) SPIRE 2010. LNCS, vol. 6393, pp. 334–346. Springer, Heidelberg (2010)
12. Kosaraju, S.R., Manzini, G.: Compression of low entropy strings with Lempel-Ziv algorithms. SIAM Journal on Computing 29(3), 893–911 (1999)
13. Mäkinen, V., Navarro, G.: Dynamic entropy-compressed sequences and full-text indexes. ACM Transactions on Algorithms 4(3), article No. 32 (2008)
14. Manzini, G.: An analysis of the Burrows-Wheeler transform. Journal of the ACM 48(3), 407–430 (2001)
15. Navarro, G., Sadakane, K.: Fully-functional static and dynamic succinct trees. Submitted for Journal Publication (2010), http://arxiv.org/abs/0905.0768; A preliminary version appeared in Proc. ACM-SIAM SODA, pp. 134–149 (2010)
16. Raman, R., Raman, V., Rao, S.S.: Succinct Dynamic Data Structures. In: Dehne, F., Sack, J.-R., Tamassia, R. (eds.) WADS 2001. LNCS, vol. 2125, pp. 426–437. Springer, Heidelberg (2001)
17. Sadakane, K., Grossi, R.: Squeezing succinct data structures into entropy bounds. In: Proceedings of ACM-SIAM SODA, pp. 1230–1239 (2006)
18. Simpson, J.T., Wong, K., Jackman, S.D., Schein, J.E., Jones, S.J.M., Birol, İ.: ABySS: A parallel assembler for short read sequence data. Genome Research 19(6), 1117–1123 (2009), http://dx.doi.org/10.1101/gr.089532.108

Improving Quantum Query Complexity of Boolean Matrix Multiplication Using Graph Collision

Stacey Jeffery[1,2], Robin Kothari[1,2], and Frédéric Magniez[3]

[1] David R. Cheriton School of Computer Science, University of Waterloo, Canada
[2] Institute for Quantum Computing, University of Waterloo, Canada
[3] LIAFA, Univ. Paris Diderot, CNRS, Paris, France

Abstract. The quantum query complexity of Boolean matrix multiplication is typically studied as a function of the matrix dimension, n, as well as the number of 1s in the output, ℓ. We prove an upper bound of $\tilde{O}(n\sqrt{\ell})$ for all values of ℓ. This is an improvement over previous algorithms for all values of ℓ. On the other hand, we show that for any $\varepsilon < 1$ and any $\ell \leq \varepsilon n^2$, there is an $\Omega(n\sqrt{\ell})$ lower bound for this problem, showing that our algorithm is essentially tight.

We first reduce Boolean matrix multiplication to several instances of graph collision. We then provide an algorithm that takes advantage of the fact that the underlying graph in all of our instances is very dense to find all graph collisions efficiently.

1 Introduction

Quantum query complexity has been of fundamental interest since the inception of the field of quantum algorithms [BBBV97, Gro96, Sho97]. The quantum query complexity of Boolean matrix multiplication was first studied by Buhrman and Špalek [BŠ06]. In the Boolean matrix multiplication problem, we want to multiply two $n \times n$ matrices A and B over the Boolean semiring, which consists of the set $\{0, 1\}$ with logical OR ($\vee$) as the addition operation and logical AND ($\wedge$) as the multiplication operation.

For this problem it is standard to consider an additional parameter in the complexity: the number of 1s in the product $C := AB$, which we denote by ℓ. We study the query complexity as a function of both n and ℓ, and obtain improvements for all values of ℓ.

The problem of Boolean matrix multiplication is of fundamental interest, in part due to its relationship to a variety of graph problems, such as the triangle finding problem and the all-pairs shortest path problem.

Classically, it was shown by Vassilevska Williams and Williams that "practical advances in triangle detection would imply practical [Boolean matrix multiplication] algorithms" [VW10]. The previous best quantum algorithm for Boolean matrix multiplication, by Le Gall, is based on a subroutine for finding triangles in graphs *with a known tripartition* [Gal12a], already suggesting that the relationship between Boolean matrix multiplication and triangle finding might be

A. Czumaj et al. (Eds.): ICALP 2012, Part I, LNCS 7391, pp. 522–532, 2012.

more complex for quantum query complexity. We give further evidence for this by bypassing the triangle finding subroutine entirely.

Despite its fundamental importance, much has remained unknown about the quantum query complexity of Boolean matrix multiplication and its relationship with other query problems in the quantum regime. Even for the simpler decision problem of Boolean matrix product verification, where we are given oracle access to three $n \times n$ Boolean matrices, A, B and C, and must decide whether or not $AB = C$, the quantum query complexity is unknown. The best upper bound is $O(n^{3/2})$ [BŠ06], whereas the lower bound was recently improved from the trivial $\Omega(n)$ to $\Omega(n^{1.055})$ by Childs, Kimmel, and Kothari [CKK11].

A better understanding of these problems may lead to an improved understanding of quantum query complexity in general. We contribute to this by closing the gap (up to logarithmic factors) between the best known upper and lower bounds for Boolean matrix multiplication for all $\ell \leq \varepsilon n^2$ for any constant $\varepsilon < 1$.

Previous Work. We are interested in the query complexity of Boolean matrix multiplication, where we count the number of accesses (or *queries*) to the input matrices A and B. Buhrman and Špalek [BŠ06, Section 6.2] describe how to perform Boolean matrix multiplication using $\widetilde{O}(n^{3/2}\sqrt{\ell})$ queries, by simply quantum searching for a pair $(i, j) \in [n] \times [n]$ such that there is some $k \in [n]$ for which $A[i, k] = B[k, j] = 1$, where $[n] = \{1, \ldots, n\}$. By means of a classical reduction relating Boolean matrix multiplication and triangle finding, Vassilevska Williams and Williams [VW10] were able to combine the quantum triangle finding algorithm of Magniez, Santha and Szegedy [MSS07] with a classical strategy of Lingas [Lin09] to get a quantum algorithm for Boolean matrix multiplication with query complexity $\widetilde{O}(\min\{n^{1.3}\ell^{17/30}, n^2 + n^{13/15}\ell^{47/60}\})$.

Recently, Le Gall [Gal12a] improved on their work by noticing that the triangle finding needed for Boolean matrix multiplication involves a tripartite graph with a known tripartition. He then recast the known quantum triangle finding algorithm of [MSS07] for this special case and improved the query complexity of Boolean matrix multiplication. He then further improved the algorithm for large ℓ by adapting the strategy of Lingas to the quantum setting.

Our Contributions. Since previous quantum algorithms for Boolean matrix multiplication are based on a triangle finding subroutine, a natural question to ask is whether triangle finding is a bottleneck for this problem. We show that this is not the case by bypassing the triangle finding problem completely to obtain a nearly tight result for Boolean matrix multiplication.

A key ingredient of the best known quantum algorithm for triangle finding is an efficient algorithm for the graph collision problem. Our main contribution is to build an algorithm directly on graph collision instead, bypassing the use of a triangle finding algorithm. Surprisingly, we do not use the graph collision algorithm that is used as a subroutine in the best known quantum algorithm for triangle finding. That algorithm is based on Ambainis' quantum walk for the element distinctness problem [Amb04]. Our algorithm, on the other hand, does not have any quantum walks.

We would like to emphasize two main ideas. First, we can reduce the Boolean matrix multiplication problem to several instances of the graph collision problem. Second, the instances of graph collision that arise depend on ℓ; in particular, they have at most ℓ non-edges. Moreover, we need to find all graph collisions, not just one. We present an algorithm to find a graph collision in query complexity $\widetilde{O}(\sqrt{\ell} + \sqrt{n})$, or to find all graph collisions in time $\widetilde{O}(\sqrt{\ell} + \sqrt{n\lambda})$, where $\lambda \geq 1$ is the number of graph collisions. Combining these ideas yields the aforementioned $\widetilde{O}(n\sqrt{\ell})$ upper bound. In addition, Le Gall [Gal12b] notes that these ideas yield an algorithm with time complexity $\widetilde{O}(n\sqrt{\ell} + \ell\sqrt{n})$.

A lower bound of $\Omega(n\sqrt{\ell})$ for all values of $\ell \leq \varepsilon n^2$ for any constant $\varepsilon < 1$ follows from a simple reduction to ℓ-THRESHOLD, which we state in Theorem 5.

This paper is organized as follows. After presenting some preliminaries in Section 2, we describe in Section 3, the graph collision problem, its relationship to Boolean matrix multiplication, and a subroutine for finding all graph collisions when there are at most ℓ non-edges. In Section 4, we apply our graph collision subroutine to get the stated upper bound for Boolean matrix multiplication, and then describe a tight lower bound that applies to all values of $\ell \leq \varepsilon n^2$ for $\varepsilon < 1$.

2   Preliminaries

2.1   Quantum Query Framework

For a more thorough introduction to the quantum query model, see [BBC+01]. For Boolean matrix multiplication, we assume access to two query operators that act as follows on a Hilbert space spanned by $\{|i, j, b\rangle : i, j \in [n], b \in \{0, 1\}\}$:

$$\mathcal{O}_A : |i, j, b\rangle \mapsto |i, j, b \oplus A[i, j]\rangle \quad \mathcal{O}_B : |i, j, b\rangle \mapsto |i, j, b \oplus B[i, j]\rangle$$

In the quantum query model, we count the uses of $\mathcal{O}_A$ and $\mathcal{O}_B$, and ignore the cost of implementing other unitaries that are independent of A and B. We call $\mathcal{O}_A$ and $\mathcal{O}_B$ the *oracles*, and each access a *query*. The query complexity of an algorithm is the maximum number of oracle accesses used by the algorithm, taken over all inputs.

A search problem P is a map $\mathcal{X} \to 2^{\mathcal{Y}}$, where $P(x) \subseteq \mathcal{Y}$ denotes the set of valid outputs on input x. We say a quantum algorithm A solves a problem P: $\mathcal{X} \to 2^{\mathcal{Y}}$ with bounded error $\delta(|x|)$ if for all $x \in \mathcal{X}$, $\Pr[A(x) \in P(x)] \geq 1 - \delta(|x|)$, where $|x|$ is the size of the input. The quantum query complexity of P is the minimum query complexity of any quantum algorithm that solves P with bounded error $\delta(|x|) \leq 1/3$.

We will use the phrase *with high probability* to mean probability at least $1 - \frac{1}{\text{poly}}$ for some super-linear polynomial. We ensure that all of our subroutines succeed with high probability, to finally achieve a bounded-error algorithm. Consequently, we will necessarily incur polylog factors in the query complexity. We will use the notation $\widetilde{O}$ to indicate that we are suppressing polylog factors. More precisely, $f(n) \in \widetilde{O}(g(n))$ means $f(n) \in O(g(n) \log^k n)$ for some constant k.

Boolean Matrices. We let $\mathbb{B}$ denote the *Boolean semiring*, which is the set $\{0, 1\}$ under the operations $\vee, \wedge$. The problem we will be considering is formally defined as the following:

BOOLEAN MATRIX MULTIPLICATION
Oracle Input: Two Boolean matrices $A, B \in \mathbb{B}^{n \times n}$.
Output: $C \in \mathbb{B}^{n \times n}$ such that $C = AB$.

In $\mathbb{B}^{n \times n}$, we say that $C = AB$ if for all $i, j \in [n]$, $C[i, j] = \bigvee_{k=1}^{n} A[i, k] \wedge B[k, j]$. We will use the notation $A + B$ to denote the entry-wise $\vee$ of two Boolean matrices.

2.2   Quantum Search Algorithms

In this section we examine some well-known variations of the search problem that we require. The reader familiar with quantum search algorithms may skip to Section 3.

Any search problem can be recast as searching for a marked element among a given collection, U. In order to formalize this, let $f : U \rightarrow \{0, 1\}$ be a function whose purpose is to identify marked elements. An element is marked if and only if $f(x) = 1$. Define $t_f = |f^{-1}(1)|$. In Grover's search algorithm, the algorithm can directly access f, and the overall complexity can be stated as the number of queries to f. In the following $t \geq 1$ is an integer parameter.

Theorem 1 ([Gro96]). *There is a quantum algorithm,* GroverSearch(t), *with query complexity* $\widetilde{O}(\sqrt{|U|/t})$ *to* f, *such that, if* $t/2 \leq t_f \leq t$, *then* GroverSearch(t) *finds a marked element with probability at least* $1 - 1/\text{poly}(|U|)$.
Moreover, if $t_f = 0$, *then* GroverSearch(t) *declares with probability* 1 *that there is no marked element.*

There are several ways to generalize the above statement when no approximation of t is known. Most of the generalizations in the literature are stated in terms of expected query complexity, such as in [BBHT98]. Nonetheless, one can derive from [BBHT98, Lemma 2] an algorithm in terms of worst case complexity, when only a lower bound t on t_f is known. The algorithm consists of T iterations of one step of the original Grover algorithm where T is chosen uniformly at random from $[0, \sqrt{|U|/t}]$. This procedure is iterated $O(\log |U|)$ times in order to get bounded error $1/\text{poly}(|U|)$.

Corollary 1. *There is a quantum algorithm* Search(t) *with query complexity* $\widetilde{O}(\sqrt{|U|/t})$ *to* f, *such that, if* $t_f \geq t$, *then* Search(t) *finds a marked element with probability at least* $1 - 1/\text{poly}(|U|)$.
Moreover, if $t_f = 0$, *then* Search(t) *declares with probability* 1 *that there is no marked element.*

One consequence of Corollary 1 is that we can always apply Search(t) with $t = 1$, when no lower bound on t_f is given. In that case, we simply refer to the resulting algorithm as Search. Its query complexity to f is then $\widetilde{O}(\sqrt{|U|})$.

Another simple generalization is for finding all marked elements. This generalization is stated in the literature in various ways for expected and worst case complexity. For the sake of clarity we explicitly describe one version of this procedure using GroverSearch as a subroutine. This version is robust in the sense that it works even when the number of marked elements may decrease arbitrarily. This may occur, for example, when the finding of one marked element may cause several others to become unmarked. This situation will naturally occur in the context of Boolean matrix multiplication. Then the complexity will only depend on the number of elements that are actually in the output, as opposed to the number of elements that were marked at the beginning of the algorithm.

SearchAll

1. Let $t = |U|$, and $V = U$
2. While $t \geq 1$
   (a) Apply GroverSearch(t) to V
   (b) If a marked element x is found: Output x; Set $V \leftarrow V - \{x\}$ and $t \leftarrow t-1$
       Else: $t \leftarrow t/2$
3. If no marked element has been found, declare 'no marked element'

Corollary 2. SearchAll *has query complexity* $\widetilde{O}(\sqrt{|U|(t_f + 1)})$ *to* f, *and finds all marked elements with probability at least* $1 - 1/\text{poly}(|U|)$.
Moreover, if $t_f = 0$, *then* SearchAll *declares with probability* 1 *that there is no marked element.*

We end this section with an improvement of GroverSearch when we are looking for an optimal solution for some notion of maximization.

Theorem 2 ([DH96, DHHM06]). *Given a function* $g : U \to \mathbb{R}$, *there is a quantum algorithm,* FindMax(g), *with query complexity* $\widetilde{O}(\sqrt{|U|})$ *to* f, *such that* FindMax(g) *returns* $x \in f^{-1}(1)$ *such that* $g(x) = \max_{x' \in f^{-1}(1)} g(x')$ *with probability at least* $1 - 1/\text{poly}(|U|)$. *Moreover, if* $t_f = 0$, *then* FindMax(g) *declares with probability* 1 *that there is no marked element.*

3   Graph Collision

In this section we describe the graph collision problem, and its relation to Boolean matrix multiplication. We then describe a method for solving the special case of graph collision in which we are interested.

3.1   Problem Description

Graph collision is the following problem. Let $G = (\mathcal{A}, \mathcal{B}, E)$ be a balanced bipartite graph on $2n$ vertices. We will suppose $\mathcal{A} = [n]$ and $\mathcal{B} = [n]$, though we note that in the bipartite graph, the vertex labelled by i in $\mathcal{A}$ is distinct from the vertex labelled by i in $\mathcal{B}$.

GRAPH COLLISION(G)
Oracle Input: A pair of Boolean functions $f_A : \mathcal{A} \to \{0,1\}$ and $f_B : \mathcal{B} \to \{0,1\}$.
Output: $(i,j) \in \mathcal{A} \times \mathcal{B}$ such that $f_A(i) = f_B(j) = 1$ and $(i,j) \in E$, if such a pair exists, otherwise reject.

The graph collision problem was introduced by Magniez, Santha and Szegedy as a subproblem in triangle finding [MSS07]. The subroutine used to solve an instance of graph collision is based on Ambainis' quantum walk algorithm for element distinctness [Amb04], and has query complexity $O(n^{2/3})$. The same subroutine is used in the current best triangle finding algorithm of Belovs [Bel11]. However, the best known lower bound for this problem is $\Omega(\sqrt{n})$. It is an important open problem to close this gap.

To obtain our upper bound, we do not use the quantum walk algorithm for graph collision, but rather, a new algorithm that takes advantage of two special features of our problem. The first is that we always know an upper bound, ℓ, on the number of non-edges. When $\ell \leq n$, we can find a graph collision in $O(\sqrt{n})$ queries. The second salient feature of our problem is that we need to find all graph collisions.

3.2   Relation to Boolean Matrix Multiplication

Recall that the Boolean matrix product of A and B, can be viewed as the sum (entry-wise $\vee$) of n outer products: $C = \sum_{k=1}^{n} A[\cdot,k]B[k,\cdot]$, where $A[\cdot,k]$ denotes the kth column of A and $B[k,\cdot]$ denotes the kth row of B.

For a fixed k, if there exists some $i \in [n]$ and some $j \in [n]$ such that $A[i,k] = 1$ and $B[k,j] = 1$, then we know that $C[i,j] = 1$, and we say that k is a *witness* for (i,j). We are interested in finding all such pairs (i,j). For each index k, we could search for all pairs (i,j) with $A[i,k] = B[k,j] = 1$; however, this could be very inefficient, since a pair (i,j) may have up to n witnesses. Instead, we will keep a matrix $\widetilde{C}$ such that $\widetilde{C}[i,j] = 1$ if we have already found a one at position (i,j). Thus, we want to find a pair (i,j) such that $A[i,k] = B[k,j] = 1$ and $\widetilde{C}[i,j] = 0$. That is, we want to find a graph collision in the graph with bi-adjacency matrix $\overline{\widetilde{C}}$, the entry-wise complement of $\widetilde{C}$, and $f_A = A[\cdot,k]$, $f_B = B[k,\cdot]$.

This gives the following natural algorithm for Boolean matrix multiplication, whose details and full analysis can be found in Section 4.1:

First, let $\widetilde{C} = 0$.

Search for an index k such that the graph collision problem on k with $\overline{\widetilde{C}}$ as the underlying graph has a collision.

If no such k is found then we are done, and $\widetilde{C}$ is the product of A and B.

Otherwise, find all the graph collisions on the graph defined by $\overline{\widetilde{C}}$ with oracles $A[\cdot,k]$ and $B[k,\cdot]$ and record them in $\widetilde{C}$.

Eliminate this k from future searches and search for another index k again.

3.3   Algorithm for Graph Collision

When G is a complete bipartite graph, then the relation between $\mathcal{A}$ and $\mathcal{B}$ defined by G is trivial. In that case, there is a very simple algorithm to find a graph collision: Search for some $i \in [n]$ such that $f_A(i) = 1$. Then search for some $j \in [n]$ such that $f_B(j) = 1$. Then (i, j) is a graph collision pair. The query complexity of this is $O(\sqrt{n} + \sqrt{n})$. However, when G is not a complete bipartite graph, there is a nontrivial relation between $\mathcal{A}$ and $\mathcal{B}$. The best known algorithm solves this problem using a quantum walk.

In our case, we can take advantage of the fact that the graph we are working with always has at most ℓ non-edges — it is never more than distance ℓ from the complete bipartite graph, which we know is easy to deal with. We are therefore interested in the query complexity of finding a graph collision in some graph with m non-edges, which we denote $\mathfrak{GC}(n, m)$. In our case, ℓ will always be an upper bound on m.

For larger values of ℓ, we will also make use of the fact that for some k, we will have multiple graph collisions to find. We let $\mathfrak{GC}_{\text{all}}(n, m, \lambda)$ denote the query complexity of finding all graph collisions in a graph with m non-edges, where λ is the number of graph collisions. It is not necessary to know λ a priori.

Again we note that if G is a complete bipartite graph, then we can accomplish the task of finding all graph collisions using SearchAll to search for all marked elements on each of f_A and f_B, and output $f_A^{-1}(1) \times f_B^{-1}(1)$. Letting $t_A = |f_A^{-1}(1)|$ and $t_B = |f_B^{-1}(1)|$, so the total number of graph collision pairs is $\lambda = t_A t_B$, the query complexity of this method is $O(\sqrt{nt_A} + \sqrt{nt_B}) \in O(\sqrt{n\lambda})$. So if G is close to being a complete bipartite graph, we would like to argue that we can do nearly as well. This motivates the following algorithm.

$\text{AllGC}_G(f_A, f_B)$

Let d_i be the degree of the ith vertex in $\mathcal{A}$, and let $c_i := n - d_i$. Let the vertices in $\mathcal{A}$ be arranged in decreasing order of degree, so that $d_1 \geq d_2 \geq \ldots \geq d_n$.

1. Find the highest degree marked vertex in $\mathcal{A}$ using FindMax. Let r denote the index of this vertex. $\tilde{O}(\sqrt{n})$
2. Case 1: If $c_r \leq \sqrt{m}$
   (a) Find all marked neighbors of r by SearchAll. Output any graph collisions found. $\tilde{O}(\sqrt{n\lambda})$
   (b) Delete all unmarked neighbors of r. Read the values of all non-neighbors of r. $O(\sqrt{m})$
   (c) Let $\mathcal{A}'$ denote the subset of $\mathcal{A}$ consisting of all $i \in \mathcal{A}$ with a marked neighbour in $\mathcal{B}$. Find all marked vertices in $\mathcal{A}'$ by SearchAll. $\tilde{O}(\sqrt{n\lambda})$
3. Case 2: If $c_r \geq \sqrt{m}$
   (a) Delete the first $r - 1$ vertices in $\mathcal{A}$ since they are unmarked.
   (b) Read the values of all remaining vertices in $\mathcal{A}$. $O(\sqrt{m})$
   (c) Let $\mathcal{B}'$ denote the subset of $\mathcal{B}$ consisting of all $j \in \mathcal{B}$ with a marked neighbour in $\mathcal{A}$. Find all marked vertices in $\mathcal{B}'$ by SearchAll. $\tilde{O}(\sqrt{n\lambda})$

Theorem 3. *For all* $\lambda \geq 1$, $\mathfrak{GC}_{\mathrm{all}}(n, m, \lambda) \in \widetilde{O}(\sqrt{n\lambda} + \sqrt{m})$ *and* $\mathfrak{GC}(n, m) \in \widetilde{O}(\sqrt{n} + \sqrt{m})$.

Proof. We will analyze the complexity of $\mathsf{AllGC}_G(f_A, f_B)$ step by step.

Step 1 has query complexity $\widetilde{O}(\sqrt{n})$ by Theorem 2. Steps 2a, 2c and 3c have query complexity $\widetilde{O}(\sqrt{n\lambda})$ by Corollary 2. In Case 1, r has $c_r \leq \sqrt{m}$ non-neighbours, so we can certainly query them all in step 2b with $O(\sqrt{c_r}) \in O(\sqrt{m})$ queries.

Consider Case 2, when $c_r \geq \sqrt{m}$. We can ignore the first $r - 1$ vertices, since they are unmarked. Since the remaining $n - r + 1$ vertices all have $c_i \geq c_r \geq \sqrt{m}$, and the total number of non-edges is m, we have $(n - r + 1) \times \sqrt{m} \leq m \Rightarrow (n - r + 1) \leq \sqrt{m}$. Thus, there are at most $\sqrt{m}$ remaining vertices and querying them all costs at most $O(\sqrt{m})$ queries.

The query complexity of this algorithm is therefore $\widetilde{O}(\sqrt{n\lambda} + \sqrt{m})$, and it outputs all graph collisions. To check if there is at least one graph collision, instead of finding them all, we can replace finding all marked vertices using SearchAll in steps 2a, 2c and 3c, with a procedure to check if there is any marked vertex, Search, and this only requires $\widetilde{O}(\sqrt{n})$ queries by Corollary 1, rather than $\widetilde{O}(\sqrt{n\lambda})$.

4   Boolean Matrix Multiplication

In this section we show how the graph collision algorithm from the previous section can be used to obtain an efficient algorithm for Boolean matrix multiplication, and then prove a lower bound.

4.1   Algorithm

What follows is a more precise statement of the high level procedure described in Section 3.2.

BMM(A, B)

---

1. Let $\widetilde{C} = 0$, $t = n$, and $V = [n]$.
2. While $t \geq 1$:
    (a) GroverSearch(t) for an index $k \in V$ such that the graph collision problem on k with $\widetilde{\widetilde{C}}$ as the underlying graph has a collision.
    (b) If such a k is found:
        Compute AllGC on the graph defined by $\widetilde{\widetilde{C}}$ with oracles $A[\cdot, k]$ and $B[k, \cdot]$ and record all output graph collisions in $\widetilde{C}$.
        Set $V \leftarrow V - \{k\}$ and $t \leftarrow t - 1$.
    (c) Else: $t \leftarrow t/2$.
3. Output $\widetilde{C}$.

Theorem 4. *The query complexity of* BOOLEAN MATRIX MULTIPLICATION *is* $\widetilde{O}(n\sqrt{\ell})$.

Proof. We will analyze the complexity of the algorithm BMM(A, B). We begin by analyzing the cost of all the iterations in which we don't find a marked k. We have by Theorem 1 that GroverSearch(t) costs $\widetilde{O}(\sqrt{n/t})$ queries to a procedure that checks if there is a collision in the graph defined by $\widetilde{\overline{C}}$ with respect to $A[\cdot, k]$ and $B[k, \cdot]$, each of which costs $\mathfrak{GC}(n, m_i)$, where $m_i \leq \ell$ is the number of 1s in $\widetilde{C}$ at the beginning of the ith iteration. The cost of these steps is at most the following:

$$\widetilde{O}\left(\sum_{i=0}^{\log n} \sqrt{\frac{n}{2^i}}\mathfrak{GC}(n, m_i)\right) \in \widetilde{O}\left(\sum_{i=0}^{\log n} \sqrt{\frac{n}{2^i}}(\sqrt{n} + \sqrt{m_i})\right)$$

$$\in \widetilde{O}\left((n + \sqrt{n\ell})\sum_{i=0}^{\log n}\left(\frac{1}{\sqrt{2}}\right)^i\right) \in \widetilde{O}\left(n + \sqrt{n\ell}\right)$$

We now analyze the cost of all the iterations in which we do find a marked witness k. Let T be the number of witnesses found by BMM, that is, the number of times we execute step **2(b)**. Of course, T is a random variable that depends on which witnesses k are found, and in which order. We always have $T \leq \min\{n, \ell\}$.

Let $i_1, \ldots, i_T$ be the indices of rounds where we find a witness. Let t_j be the value of t in round j. Since there must be at least 1 marked element in the last round in which we find a marked element, we have $t_{i_T} \geq 1$. Since we find and eliminate at least 1 marked element in each round, we also have $t_{i_{(T-j-1)}} \geq t_{i_{(T-j)}} + 1$, which yields $t_{i_{(T-j)}} \geq j + 1 \Rightarrow t_{i_j} \geq T - j + 1$.

Let λ_j be the number of graph collisions found on the jth successful iteration, that is, the number of pairs witnessed by the jth witness, k_j, that have not been recorded in $\widetilde{C}$ *at the time we find* k_j. Then λ_j is also a random variable depending on which other witnesses k have been found already, but we always have $\sum_{j=1}^{T} \lambda_j = \ell$.

Then we can upper bound the cost of all the iterations in which we do find a witness by the following:

$$\widetilde{O}\left(\sum_{j=1}^{T}\left(\sqrt{\frac{n}{t_{i_j}}}\mathfrak{GC}(n, m_{i_j}) + \mathfrak{GC}_{\text{all}}(n, m_{i_j}, \lambda_j)\right)\right) \tag{1}$$

$$\in \widetilde{O}\left(\sum_{j=1}^{T}\left(\sqrt{\frac{n}{T-j+1}}\mathfrak{GC}(n, m_{i_j}) + \mathfrak{GC}_{\text{all}}(n, m_{i_j}, \lambda_j)\right)\right) \tag{2}$$

$$\in \tilde{O}\left(\sqrt{nT}\mathfrak{GC}(n,\ell) + \sum_{j=1}^{T} \mathfrak{GC}_{\mathrm{all}}(n, m_{i_j}, \lambda_j)\right) \tag{3}$$

$$\in \tilde{O}\left(\sqrt{nT}\left(\sqrt{\ell} + \sqrt{n}\right) + \sum_{j=1}^{T}\left(\sqrt{n\lambda_j} + \sqrt{\ell}\right)\right) \tag{4}$$

$$\in \tilde{O}\left(\sqrt{nT\ell} + n\sqrt{T} + \sqrt{n\ell T} + T\sqrt{\ell}\right) \tag{5}$$

$$\in \tilde{O}\left(\sqrt{n\min\{n,\ell\}\ell} + n\sqrt{\min\{n,\ell\}} + \sqrt{\min\{n,\ell\}n\ell} + \min\{n,\ell\}\sqrt{\ell}\right) \tag{6}$$

$$\in \tilde{O}\left(n\sqrt{\ell}\right) \tag{7}$$

In (3), we use the fact that $m_{i_j} \leq \ell$, and in (5), we use $\sum_{j=1}^{T}\sqrt{\lambda_j} \leq \sqrt{T}\sqrt{\sum_j \lambda_j} = \sqrt{\ell T}$, which follows from the Cauchy–Schwarz inequality.

4.2 Lower Bound

Theorem 5. *The query complexity of* BOOLEAN MATRIX MULTIPLICATION *is* $\Omega(n\sqrt{\min\{\ell, n^2 - \ell\}})$.

Proof. We will reduce the problem of ℓ-THRESHOLD, in which we must determine whether an input oracle f has $\geq \ell$ or $< \ell$ marked elements, to BOOLEAN MATRIX MULTIPLICATION.

Consider an instance of ℓ-THRESHOLD of size n^2, $f : [n^2] \to \{0,1\}$. We can construct an instance of BOOLEAN MATRIX MULTIPLICATION as follows. Set A to the identity, and let B encode f. Finding AB then gives the solution to the ℓ-THRESHOLD instance. By [BBC+01], ℓ-THRESHOLD (with inputs of size n^2) requires $\Omega(\sqrt{n^2 \min\{\ell, n^2 - \ell\}})$ queries to solve with bounded error.

This lower bound implies that our algorithm is tight for any $\ell \leq \varepsilon n^2$ for any constant $\varepsilon < 1$. However, it is not tight for $\ell = n^2 - o(n)$. We can search for pairs (i,j) such that there is no $k \in [n]$ that witnesses (i,j) in cost $n^{3/2}$. If there are m 0s, we can find them all in $\tilde{O}(n^{3/2}\sqrt{m})$, which is $o(n\sqrt{\ell})$ when $m \in o(n)$. It remains open to close the gap between $\tilde{O}(n^{3/2}\sqrt{m})$ and $\Omega(n\sqrt{m})$ when $m \in o(n^2)$.

Acknowledgements. This work was partially supported by NSERC, MITACS, QuantumWorks, the French ANR Defis project ANR-08-EMER-012 (QRAC), and the European Commission IST STREP project 25596 (QCS).

References

[Amb04]   Ambainis, A.: Quantum walk algorithm for element distinctness. In: Proceedings of the 45th IEEE Symposium on Foundations of Computer Science, pp. 22–31 (2004)

[BBBV97] Bennett, C.H., Bernstein, E., Brassard, G., Vazirani, U.: Strengths and weaknesses of quantum computing. SIAM Journal on Computing (Special Issue on Quantum Computing) 26, 1510–1523 (1997), arXiv:quant-ph/9701001v1

[BBC+01] Beals, R., Buhrman, H., Cleve, R., Mosca, M., de Wolf, R.: Quantum lower bounds by polynomials. Journal of the ACM 48 (2001)

[BBHT98] Boyer, M., Brassard, G., Høyer, P., Tapp, A.: Tight bounds on quantum searching. Fortschritte der Physik 46(4-5), 493–505 (1998)

[Bel11] Belovs, A.: Span programs for functions with constant-sized 1-certificates. Technical Report arXiv:1105.4024, arXiv (2011)

[BŠ06] Buhrman, H., Špalek, R.: Quantum verification of matrix products. In: Proceedings of the 17th ACM-SIAM Symposium On Discrete Algorithms, pp. 880–889 (2006)

[CKK11] Childs, A., Kimmel, S., Kothari, R.: The quantum query complexity of read-many formulas. Technical Report arXiv:1112.0548v1, arXiv (2011)

[DH96] Dürr, C., Høyer, P.: A quantum algorithm for finding the minimum. Technical Report arXiv:quant-ph/9607014v2, arXiv (1996)

[DHHM06] Dürr, C., Heiligman, M., Høyer, P., Mhalla, M.: Quantum query complexity of some graph problems. SIAM Journal on Computing 35(6), 1310–1328 (2006)

[Gal12a] Le Gall, F.: Improved output-sensitive quantum algorithms for Boolean matrix multiplication. In: Proceedings of the 23rd ACM-SIAM Symposium on Discrete Algorithms, pp. 1464–1476 (2012)

[Gal12b] Le Gall, F.: Improved time-efficient output-sensitive quantum algorithms for Boolean matrix multiplication, arXiv:1201.6174 (2012)

[Gro96] Grover, L.K.: A fast quantum mechanical algorithm for database search. In: Proceedings of the 28th ACM Symposium on Theory of Computing, pp. 212–219 (1996)

[Lin09] Lingas, A.: A Fast Output-Sensitive Algorithm for Boolean Matrix Multiplication. In: Fiat, A., Sanders, P. (eds.) ESA 2009. LNCS, vol. 5757, pp. 408–419. Springer, Heidelberg (2009)

[MSS07] Magniez, F., Santha, M., Szegedy, M.: Quantum algorithms for the triangle problem. SIAM Journal on Computing 37(2), 413–424 (2007)

[Sho97] Shor, P.W.: Polynomial-time algorithms for prime factorization and discrete logarithms on a quantum computer. SIAM Journal on Computing 26, 1484–1509 (1997)

[VW10] Vassilevska Williams, V., Williams, R.: Sub-cubic equivalences between path, matrix and triangle problems. In: Proceedings of the 51st IEEE Symposium on Foundations of Computer Science, pp. 645–654 (2010)

Faster Fully Compressed Pattern Matching by Recompression[*]

Artur Jeż[**]

Institute of Computer Science, University of Wrocław, Poland
aje@cs.uni.wroc.pl

Abstract. In this paper, a fully compressed pattern matching problem is studied. The compression is represented by straight-line programs (SLPs), i.e. a context-free grammar generating exactly one string; the term fully means that *both* the pattern *and* the text are given in the compressed form. The problem is approached using a recently developed technique of local recompression: the SLPs are refactored, so that substrings of the pattern and text are encoded in both SLPs in the same way. To this end, the SLPs are *locally decompressed* and then *recompressed* in a uniform way.

This technique yields an $\mathcal{O}((n + m) \log M \log(n + m))$ algorithm for compressed pattern matching, where n (m) is the size of the compressed representation of the text (pattern, respectively), while M is the size of the decompressed pattern. Since $M \leq 2^m$, this substantially improves the previously best $\mathcal{O}(m^2 n)$ algorithm.

Since LZ compression standard reduces to SLP with $\log(N/n)$ overhead and in $\mathcal{O}(n \log(N/n))$ time, the presented algorithm can be applied also to the fully LZ-compressed pattern matching problem, yielding an $\mathcal{O}(s \log s \log M)$ running time, where $s = n \log(N/n) + m \log(M/m)$.

Keywords: Pattern matching, Compressed pattern matching, Straight-line programms, Lempel-Ziv compression, Algorithms for compressed data.

1 Introduction

Compression and Straight-Line Programms. Due to ever-increasing amount of data, compression methods are widely applied in order to decrease the data's size. Still, the stored data is accessed and processed. Decompressing it on each such an occasion basically wastes the gain of reduced storage size; especially that we do not even know in advance, which data is relevant to our queries and we decompress many completely irrelevant files. Thus there is a large demand for algorithms dealing directly with the compressed data, without the explicit decompression. The commonly investigated problem is the *compressed pattern*

---

[*] The full version of this paper is available at http://arxiv.org/abs/1111.3244
[**] Supported by NCN grant number DEC-2011/01/D/ST6/07164, 2011–2014.

A. Czumaj et al. (Eds.): ICALP 2012, Part I, LNCS 7391, pp. 533–544, 2012.

matching i.e. a pattern matching in which the text is supplied in a compressed form.

Processing compressed data is not as hopeless, as it may seem: it is a popular outlook, that compression basically extracts the hidden structure of the text and if the compression rate is high, the data has a lot of internal structure. And it is natural to assume, that such a structure will help devising methods dealing directly with the compressed representation. Indeed, efficient algorithms for fundamental text operations (pattern matching, equality testing, etc.) are known for various practically used compression methods (LZ, LZW, their variants, etc.) [3,4,5,6,7,18].

The compression standards differ in the main idea as well as in details. Thus when devising algorithms for compressed data, quite early one needs to focus on the exact compression method, to which the algorithm is applied. The most practical (and challenging) choice is one of the widely used standards, like LZW or LZ. However, a different approach is also pursued: for some applications (and most of theory-oriented considerations) it would be useful to *model* one of the practical compression standard by a more mathematically well-founded method. This idea, among other, lays at the foundations of the notion of *Straight-Line Programms* (SLP), which are simply context-free grammars generating exactly one string.

SLPs are the most popular theoretical model of compression. This is on one hand motivated by a simple, 'clean' and appealing definition, on the other hand, they model the LZ compression standard: each LZ compressed text can be converted into an equivalent SLP with only $\log(N/n)$ overhead and in $\mathcal{O}(n \log(N/n))$ time (where N is the size of the decompressed text [19]), while each SLP can be converted to an equivalent LZ with just a constant overhead (and in linear time).

The approach of modelling LZ by SLPs in order to develop efficient algorithms turned out to be fruitful: the recent state-of-the-art (and is some sense optimal) algorithm for pattern matching in LZ-compressed texts changes the LZ-compression into the SLP-one as its first step [4]. To author's best knowledge, there are no algorithms for FCPM specific for LZ, instead, the translation to SLP is used in such a case. On the other hand, algorithmic problems for SLP-compressed input strings were considered and successfully solved [13,14,18].

Problem Statement. The problem considered in this paper is the *fully compressed membership problem* (FCPM), i.e. we are given a text of length N and pattern of length M, represented by SLPs of size n and m, respectively. We are to answer, whether the pattern appears in the text and give a compact representation of all such appearances in the text.

Previous and Related Results. The first algorithmic result dealing with the SLPs is for the compressed equality testing, i.e. the question whether two SLPs represent the same text. This was solved by Plandowski in 1994 [18], with $\mathcal{O}(n^4)$ running time. The first solution for FCPM by Karpiński et al. followed a year later [12], its main drawback was that the proposed algorithm did not return positions of all pattern appearances in the text. Next, a polynomial algorithm for computing various combinatorial properties of SLP-generated texts, in particular pattern matching, was given by Gąsieniec et al. [7], the same authors presented also a faster randomised algorithm for FCPM [8]; both these algorithms returned compact representation of all pattern appearances. In 1997 Miyazaki et al. [17] constructed new $\mathcal{O}(n^2m^2)$ algorithm for FCPM. A faster $\mathcal{O}(mn)$ algorithm for a special sub-case (restricting the form of SLPs) was given in 2000 by Hirao et al. [9]. Finally, in 2007, a state of the art $\mathcal{O}(nm^2)$ algorithm was given by Lifshits [13].

Concerning related problems, pattern matching in which the text is compressed using LZW method and the pattern is supplied uncompressed was proposed and recently a linear-time algorithm was given [3]. A variant in which the pattern is also compressed using LZW was also considered and a linear-time algorithm was recently developed [6]. Pattern matching for multiple patterns in LZW compressed text was also studied [5].

Similar work was carried also for the LZ-compressed text, for which the problem becomes substantially harder than in LZW case. In 2011, an $\mathcal{O}(n \log(N/n) + m)$ algorithm, which is in some sense optimal, was proposed [4].

The paradigm employed in all mentioned work and constructed algorithms, was to consider the combinatorial properties of strings described by appropriate compression methods; our method uses a new paradigm.

Our Results and Techniques. We give an $\mathcal{O}((n + m) \log M \log(n + m))$ algorithm for FCPM, i.e. pattern matching problem in which both the text and the pattern are supplied as SLPs. This outperforms the previously-best $\mathcal{O}(m^2n)$ algorithm [13].

Theorem 1. *Algorithm* FCPM *returns an* $\mathcal{O}((n+m) \log(n+m))$ *representation of all pattern appearances, where* n *(m) is the size of the SLP-compressed text (pattern, respectively) and* M *is the size of the decompressed pattern. It runs in* $\mathcal{O}((n + m) \log M \log(n + m))$ *time. The space consumption is* $\mathcal{O}((n + m) \log(n + m))$.

This representation allows calculation of the number of pattern appearances, and if N *fits in* $\mathcal{O}(1)$ *codewords, also the position of the first, last etc. pattern; in other case the space consumption increases to* $\mathcal{O}((n+m) \log(N+M) \log(n+m))$.

Our approach to the problem is essentially different than all previously applied for compressed pattern matching. We do *not* consider any combinatorial properties of the encoded strings. Instead, we analyse and change the way strings are described by the SLPs in the instance. That is, we focus on the SLPs alone, ignoring any properties of the encoded strings. Roughly speaking, our algorithm aims at having all the strings in the instance compressed 'in the same way'. To achieve this goal, we decompress the SLPs. Since the compressed text can be exponentially long, we do this *locally*: we introduce explicit strings into the right-hand sides of the productions. Then, we recompress these explicit strings uniformly: roughly, a fixed pair of letters ab is replaced by a new letter c in both the string and the pattern; such a procedure is applied for every possible pair of letters. Since such pieces of text are compressed in the same way, we can 'forget' about the original substrings of the input and treat the introduced nonterminals as atomic letters. Such recompression shortens the pattern (and the text) significantly: roughly one 'round' of recompression in which every pair of letters that was present at the beginning of the 'round' is compressed shortens the encoded strings by a constant factor. Thus, there are $\mathcal{O}(\log M)$ rounds.

Although it is not so hard to believe that this high level idea can work, it is much less believable that this can be turned into a fast, efficient and simple algorithm. However, by choosing wisely the parts of the text to be recompressed and keeping the overall size of the instance low, we manage to achieve the goal.

Similar techniques. While application the idea of recompression to pattern matching is new, related approaches were previously employed: most notably the idea of replacing short strings by a fresh letter and iterating this procedure was used by Mehlhorn et al. [16], in their work on data structure for equality testing for dynamic strings (cf. also an improved implementation of a similar data structure by Brodal et al. [1]). They viewed this process as 'hashing'. In particular their method can be straightforwardly applied to equality testing for SLPs, yielding a nearly quadratic algorithm (as observed by Gawrychowski [2]). However, the inside technical details of the construction makes extension to FCPM problematic: while this method can be used to build 'canonical' SLPs for the text and the pattern, there is no apparent way to control how these SLPs actually look like and how do they encode the strings.

In the area of compressed membership problems, from which the presented method emerge, recent work of Mathissen and Lohrey [15] already implemented the idea of replacing strings with fresh letters as well as modifications of the instance so that such replacement is possible. However, the replacement was not iterated, and the newly introduced letters could not be further compressed.

Other applications of the technique. A more crude variant of recompression technique has been used in order to solve an old open problem regarding fully compressed membership problem for NFAs [10]. Furthermore, a variant of this method can also be applied in the area of word equations. While not claiming any essentially new results, the recompression approach yielded much simpler proofs and faster algorithms of many classical results in the area, like PSPACE algorithm for solving word equations, double exponential bound on the size of the solution, exponential bound on the exponent of periodicity, *etc.* [11].

Computational Model and Positions in Text. Our algorithm uses Radix-Sort and we assume that the codeword is of size $\Omega(\log(n + m))$. However, we do not make such assumptions on N and M. Changing the model into pointer machine introduces a $\log(n + m)$ factor to the running time.

The position of the first appearance of the pattern in the text might be exponential in n, and so it is infeasible to output it within the given bounds. However, if we assume that N fits in a constant amount of codewords, our algorithm can also output the position of the first, last etc. position of the pattern.

2 Basic Notions, Outline of the Algorithm

Straight Line Programmes. Formally, a *Straight-Line Programme* (SLP) is a context free grammar G over the alphabet Σ with a set of nonterminals $\mathcal{X}$, generating a one-word language. For normalisation reasons, it is assumed that G is in a *Chomsky normal form*, i.e. each production is either of the form $X \to YZ$ or $X \to a$. We denote the string defined by nonterminal A by val(A), like *value*; this notion extends to val(α) for $\alpha \in (\mathcal{X} \cup \Sigma)^*$ in the usual way. We also use first$[X_i]$ (last$[X_i]$) to denote the first (last, respectively) letter of val(X_i). The tables first$[]$ and last$[]$ are stored by the algorithm FCPM .

Without loss of generality we may assume that Σ consists of consecutive natural numbers (starting from 1): it is enough to sort the input letters and number them $1, \ldots, |\Sigma|$. During our algorithm, the alphabet Σ is increased many times and whenever this happens, the new letter is assigned number $|\Sigma|+1$ (and Σ's size increases by 1 as well). The $|\Sigma|$ does not become large in this way: it remains of size $\mathcal{O}((n + m) \log(n + m) \log M)$, see Lemma 8.

For our purposes it is more convenient to treat the two SLPs as a single context free grammar G with a set of nonterminals $\mathcal{X} = \{X_1, \ldots, X_{n+m}\}$, the text being given by X_{n+m} and the pattern by X_m. We assume, however, that X_m is not referenced by any other nonterminal. Furthermore, in our constructions, it is essential to *relax* the usual assumption that G is in a Chomsky normal form, instead we only require that G satisfies the conditions:

each X_i has exactly one production, which has at most 2 noterminals,   (1a)

if X_j appears in the rule for X_i then $j < i$,   (1b)

if val$(X_i) = \epsilon$ then X_i is not on the right-hand side of any production,.   (1c)

We refer to these conditions collectively as (1). Let $X_i \to \alpha_i$, then a substring $u \in \Sigma^+$ of α_i appears *explicitly* in the rule; this notion is introduced to distinguish them from the substrings of val(X_i). Note that (1) does not exclude the case, when $X_i \to \epsilon$ and allowing such a possibility streamlines the analysis.

The size $|G|$ is the sum of length of the right-hand sides of G's rules. The size of G kept by the algorithm will be small: $\mathcal{O}((n + m) \log(n + m))$, see Lemma 8.

There may be exponentially many appearances of the pattern in the text (consider text a^{2^n} and pattern a^{2^m}), and so naive outputting all of them is infeasible. Instead, our algorithm provides an $\mathcal{O}((n + m) \log(m + n))$ SLP, in which appearances of a designated letter correspond to the pattern appearance in the original instance.

(Non) Crossing Appearance. The main part of the presented FCPM consists of recompression, i.e. replacing strings appearing in val(X_m) by shorter ones throughout val(X_1), $\ldots$, val(X_{n+m}). In some cases, such replacing is harder, in

other easier. It is intuitively clear, that this depends on the position of the pair
with regard to the nonterminals: suppose that we are to compress a pair ab. If b is
a first letter of some $\mathrm{val}(X_i)$ and aX_i appears explicitly in the grammar, then the
compression seems hard, as it requires modification of G. On the other hand, if
none such, nor symmetrical, situation appears then replacing all explicit abs in G
should do the job. Thus, before stating the algorithm, we introduce classification
of pairs into 'easy' and 'hard'.

We first formalise the notion, that a nonterminal generates some substring of
$\mathrm{val}(X_i)$. We say that X_i generates $\mathrm{val}(X_i)$ starting at position 1; furthermore, if
X_j generates $\mathrm{val}(X_i)$ starting at position p and $X_j \to \alpha X_k \alpha'$, then X_k generates
$\mathrm{val}(X_i)$ starting at position $p + |\mathrm{val}(\alpha)|$. Symmetrically, X_j generates $\mathrm{val}(X_i)$
ending at some position. We use this notions only to say that X_i generates
pattern $(\mathrm{val}(X_m))$ or text $(\mathrm{val}(X_{n+m}))$ at some position.

We say that a letter $a \in \Sigma$ is *to the left* of X_i, if, for some position p, a is
p-th letter of $\mathrm{val}(X_{n+m})$ (or $\mathrm{val}(X_m)$) and X_i generates the text (or pattern,
respectively) from position $p + 1$; in such a case we say that X_i is to the right of
a. In the symmetric situation, we say that a is *to the right* of X_i.

A pair of letters ab is a *crossing pair* if there is a nonterminal X_i such that a
is to the left of X_i and $first[X_i] = b$ or, symmetrically, b is to the right of X_i and
$last[X_i] = a$; otherwise ab is *non-crossing*. Intuitively ab 'crosses' the symbols
in some production $u\,\mathrm{val}(X_j)v\,\mathrm{val}(X_k)w$. Unless explicitly written, we use this
notion only to pairs of *different* letters.

The notions of (non-) crossing pairs is usually not applied to pairs of the form
aa, instead, for a letter $a \in \Sigma$ we say that a^ℓ is a a's *maximal block* of length ℓ,
if there exist two letters $x, y \in \Sigma$, where $x \neq a \neq y$ such that $xa^\ell y$ is a substring
of $\mathrm{val}(X_{n+m})$ (or $\mathrm{val}(X_m)$). We say that a letter $a \in \Sigma$ *has a crossing block*, if
the pair aa is crossing. The crossing pairs and letters with crossing blocks are
intuitively hard to compress.

The definition of the crossing pairs (and letters with crossing blocks) is very
'global' in the sense that is uses $\mathrm{val}(X_{n+m})$ and $\mathrm{val}(X_m)$. However, it turns
out that the set of crossing (non-crossing) pairs, letter with (without) crossing
blocks can be easily established by reading G. The number of such pairs (blocks)
is linear.

Lemma 1. *There are at most $2(n+m)$ different letters with crossing blocks and
at most $4(n+m)$ different crossing-pairs and at most $|G|$ noncrossing pairs. For
a letter a there are at most $|G| + 4(n+m)$ different lengths of a's maximal blocks
in $\mathrm{val}(X_1), \ldots, \mathrm{val}(X_{n+m})$.*

The set of crossing (non-crossing) pairs can be calculated in $\mathcal{O}(|G|)$ time.

Outline of the Algorithm. The main operations of our algorithm are two
types of compressions performed on strings encoded by G:

Pair Compression of ab. For two different letters ab appearing in $\mathrm{val}(X_m)$
replace each of ab in $\mathrm{val}(X_1), \ldots, \mathrm{val}(X_{n+m})$ by a fresh letter c.

a's **Block Compression.** For each maximal block a^ℓ, with $\ell > 1$, that appears in $\mathrm{val}(X_m)$, replace all a^ℓs in $\mathrm{val}(X_1), \ldots, \mathrm{val}(X_{n+m})$ by a fresh letter a_ℓ.

We adopt the following notational convention throughout rest of the paper: whenever we refer to a letter a_ℓ, it means that the last blocks compression was done for a and a_ℓ is the letter that replaced a^ℓ.

We call the ℓth iteration of the main loop, i.e. the one in line 1, the ℓth *phase*. Ideally, each phase of FCPM compresses each consecutive letters into one letter, this gives $\log M$ iterations of this loop. This is true, up to a constant factor.

Lemma 2. *There are $\mathcal{O}(\log M)$ executions of the main loop of* FCPM .

Remark. Notice, that pair compression of ab to c is in fact introducing a new nonterminal with a production $c \to ab$, similarly, block compression for a introduces new nonterminals with rules $a_\ell \to a^\ell$. Hence, FCPM creates new SLPs for text and pattern. This justifies the name 'recompression' used for the whole process.

Still, these new nonterminals are never expanded by FCPM and are always treated as individual symbols; thus it is better to think of them as letters. In particular, the running time analysis of FCPM use the fact that no new nonterminals are ever introduced to G.

Algorithm 1. FCPM : outline

1: **while** $|\mathrm{val}(X_m)| > 1$ **do**
2:     $P \leftarrow$ list of non-crossing pairs
3:     $P' \leftarrow$ list of crossing pairs
4:     $L \leftarrow$ list of letters
5:     fix the beginning and end
6:     **for** each $ab \in P$ **do**
7:         compress pair ab
8:     **for** $ab \in P'$ **do**
9:         compress pair ab
10:     **for** each $a \in L$ **do**
11:         compress blocks of a
12: Output the answer.

Major Challenges. Before we proceed to describing the details of FCPM , we would like to point out, what are the main problems we are dealing with. The non-crossing pair (and blocks) compression are easy to implement and are not an issue: it is enough to read G and replace the appropriate explicit strings. When it comes to a crossing pair compression, a simple transformation of the instance changes the crossing pair ab into a non-crossing one: whenever a is to the left of X_i and $\mathrm{val}(X_i) = bw$ we modify the productions for X_i, so that $\mathrm{val}(X_i) = w$ and replace X_i by bX_i in every rule; similar transformation are applied to the nonterminals X_j to the left of b such that $\mathrm{val}(X_j) = w'a$. This makes ab a noncrossing pair.

Similar approach works for crossing blocks compression, this time though we need to remove a-prefix (a-suffix) from each nonterminal to the right (left, respectively) of a. This removes all crossing blocks of a so that it blocks can be compressed. Notice, that this is all easy to perform, except that we may introduce explicit blocks of a that have exponential length to G. These can be conveniently represented: a^ℓ is simply denoted as (a, ℓ), with ℓ encoded in binary.

The ends of $\mathrm{val}(X_m)$ have to be treated somehow special: consider pattern abc and text $aabccb$. When aa is replaced by a_2, and cc by c_2 the obtained text

a_2bc_2b no longer contains the pattern, which is still abc. This is fixed by enforcing that the leading pair of the pattern (ab in this case) is compressed as first. The situation complicates, when the val(X_m) begins with an ℓ-block of a, in this case we tune the block compression a little.

Simplifications. There are some simplifications and additional assumptions made in the extended abstract, done in order to increase the readability; the full version of this paper has no such simplifications nor additional assumptions. *Simplified statements*: The lemmas are stated in a simplified way, omitting some of the technical details, but highlighting the intuitively important properties. *Size of code-word*: We assume that N and M fit in $\mathcal{O}(1)$ code words. This allows the explicit calculations of the *lengths* of val(X_m) and val(X_{n+m}). In the full version of this paper the same results are shown under the weaker assumption that n and m fit in $\mathcal{O}(1)$ code words.

3   Details

Grammar. The grammar kept by FCPM is closely related to the input one:

SLP   The set of used nonterminals is a subset of $\mathcal{X} = \{X_1, \ldots, X_{n+m}\}$ and the productions are of the form described in (1).

FCPM preserves (SLP), in particular, we always assume, that the input of the subroutines satisfies (SLP). We assume more for the input instance: we want it to obey the Chomsky normal form, instead of the relaxed conditions (1).

Compression of Non-crossing Pair. We start by describing the compression of a non-crossing pair ab, as it is the easiest to explain. Intuitively, whenever ab appears in string encoded by G, the letters a and b cannot be split between nonterminals. Thus, it should be enough to replace their explicit appearances.

Lemma 3. *The non-crossing pairs compression can be performed in $\mathcal{O}(|G|)$ time.*

We read G and list all pairs' appearances and flag them, depending on whether these appearances are crossing or not. We then group these appearances by the pair, i.e. for a fixed pair we have a list of all appearances of this pair. For a fixed non-crossing pair ab, we go through the corresponding list of appearances and replace each explicit ab in G by a fresh letter c.

Compression of Crossing Pairs. Let ab be a crossing pair because a is to the left of nonterminal X_i such that first$[X_i] = b$. To remedy this we 'pop' the leading b from X_i: if val(X_i) = bw

---

Algorithm 2. LeftPop (X_i)

1: let $X_i \to \alpha$ and b the α's first symbol
2: remove leading b from α
3: replace each X_i in the rules by bX_i

---

we modify G so that $\mathrm{val}(X_i) = w$. This is implemented in LeftPop . Such a procedure is applied to each non-terminal that is to the right of a. Symmetric procedure is applied for a letter b and nonterminals X_i such that b is to the right of X_i and $last[X_i] = a$.

When the pair ab is no longer crossing, it can be compressed in the way described above.

Lemma 4. *The* PairComp *properly compresses a crossing pair ab.*

Changing all crossing pairs to noncrossing ones can be done in parallel, similarly as in the case of the compression of the noncrossing pairs. However, this can be done under the assumption that these pairs do not over-

---
Algorithm 3. PairComp (ab)

1: **for** $i \leftarrow 1 \ldots m + n$ **do**
2:     **if** a is to the left of X_i and $first[X_i] = b$ **then**
3:         LeftPop (X_i)
4:     **if** b is to the right of X_i and $last[X_i] = a$ **then**
5:         RightPop (X_i)
6:     compress the pair ab

---

lap, where ab and $a'b'$ overlap if $a = b'$ or $b = a'$. The general case is obtained by partitioning all crossing pairs into $2\log(n+m)$ groups, such that within each of the groups the pairs are not overlapping. The partition is found by a simple greedy method, similar to approximation of a vertex cover.

Lemma 5. *Pairs from P' can be partitioned into $2\log(n+m)$ groups, such that performing the* PairComp *(with appropriate implementation) for pairs in one group takes $\mathcal{O}(|G|)$ time.*

Blocks Compression. Now, we turn our attention to the block compression. Suppose first that G has no letters with a crossing block. Then a procedure similar to the one compressing non-crossing pairs can be performed: when reading G, we establish all maximal blocks of letters. We group these appearances according to the letter, i.e. for each letter a we create a list of a's maximal blocks in G and we sort this list according to the lengths of the blocks. We go through such list and we replace each appearance of a^ℓ by a fresh letter a_ℓ.

However, usually there are letters with crossing blocks. We deal with this similarly as in the case of crossing pairs: a letter a has a crossing block if and only if aa is a crossing pair. So suppose that a is to the left of X_i and $first[X_i] = a$, in such a case we left-pop a letter from X_i. In general, this does not solve the problem as it may happen that still $first[X_i] = a$. So we keep on left-popping until $first[X_i] \neq a$. In other words, we remove the a-prefix of $\mathrm{val}(X_i)$. Symmetric procedure is applied to X_j such $last[X_j] = a$ and X_j is to the left of a.

It turns out that even a simplified approach works: for each nonterminal X_i, where $first[X_i] = a$ and $last[X_i] = b$, it is enough to 'pop' its a-prefix and b-suffix, see RemCrBlocks .

Observe that during the procedure, long blocks of a (up to 2^{n+m}) may be explicitly written in the rules. This is conveniently represented: a^ℓ is simply denoted as (a, ℓ), with ℓ encoded in binary. When ℓ fits in one code word, the a^ℓ representation is still of constant size and everything works smoothly.

After RemCrBlocks , every letter a has no crossing blocks and we may compress maximal blocks using the already described method.

Lemma 6. *After application of* RemCrBlocks *there are no crossing blocks. The time consumption of* RemCrBlocks *and following block compression is* $\mathcal{O}(|G|)$.

Algorithm 4. RemCrBlocks :

1: **for** $i \leftarrow 1..m+n$, except n and $n+m$ **do**
2:     let $X_i \rightarrow \alpha_i$ be the production for X_i
3:     let $a = first[X_i]$
4:     calculate and remove the a-prefix a^{ℓ_i} of α_i
5:     let $b = last[X_i]$
6:     calculate and remove the b-suffix b^{r_i} of α_i
7:     replace each X_i in rule's bodies by $a^{\ell_1} X_i b^{r_i}$
8:     **if** $val(X_i) = \epsilon$ **then**
9:         remove X_i from the rules' bodies

First and Last Letter of Pattern. We have to treat the 'ends' of the pattern in a careful way: consider a text $ababa$ and a pattern bab. Then compression of ab into c results in a text cca and pattern bc, which no longer appears in the text. The other problem appears during the block compression: consider pattern aab and text $aaab$. Then after the block compression the pattern is replaced with $a_2 b$ and text with $a_3 b$.

In general, the problems arise because the compression applied by FCPM is done partially on the pattern appearance and partially outside it, so it cannot be reflected in the compression of the pattern. We say, that the compression *spoils pattern's beginning* (*end*) when such partial compression appears on pattern appearance beginning (end, respectively). In the working example, spoiling of the pattern beginning can be circumvented by enforcing a compression of the pair ab in the first place: when two first letters of the pattern are replaced by a fresh letter c, then the beginning of the pattern no longer can be spoiled in this phase (as c will not be compressed in this phase). We say, that pattern's beginning (end) is *fixed* by a pair or block compression, if after this compression a first (last, respectively) letter of the pattern is a fresh letter. Notice, that the same compression can at the same time fix the beginning and spoil the end: for instance, compressing ba into c does so in the working example. Our goal is to fix both the beginning and the end, without spoiling any of them.

If the first two letters of the pattern are ab for $a \neq b$, then we can fix the beginning by compressing the pair ab, before any other pairs (or blocks) are compressed. This cannot be applied if $val(X_m)$ has a leading ℓ-block of letters a, for $\ell > 1$. The problem is that each m-block for $m \geq \ell$ can begin an appearance of the pattern in the text. This is circumvented by applying a tuned version of block compression: each m-block of length $m \leq \ell$ is replaced by a fresh letter a_m and each m-block of length $m > \ell$ is replaced by a pair of letters $a_m a_\ell$. For instance, in the example of $aaaba$ and $aaba$ above we obtain $a_3 a_2 b a_1$ as new text and $a_2 b a_1$ as a new pattern; clearly the pattern has an appearance in the text. In this way we fix the pattern beginning.

When $first[X_m] \neq last[X_m]$ then fixing the beginning does not spoil the end and afterwards we simply fix the end in a symmetrical way. When $first[X_m] = last[X_m]$ we need to apply a mixture of these two techniques, but still both the beginning and the end can be fixed, without prior spoiling. Roughly, when a^ℓ and a^r are the a-prefix and a-suffix (where, without loss of generality, $\ell \geq r \geq 1$) we first make the block compression of a, in which m-blocks are replaced with: a_m for $m < r$; $a_r a_m$ for $r \leq m < \ell$; $a_r a_\ell$ for $m = \ell$; $a_r a_m a_\ell$ for $m > \ell$. Unfortunately, for some values of ℓ, r (for instance, take $\ell = r = 1$), this might actually enlarge the text (in the example, a is replaced by $a_r a_\ell$). However, by enforcing compression of pairs of the form $a_\ell b$ and $b a_r$ for $b \in \Sigma$ and some simple tricks, the compression can be achieved.

Lemma 7. *In $\mathcal{O}(|G|)$ time we can fix both the beginning and end without prior spoiling them.*

Grammar and Alphabet Sizes. The subroutines of FCPM run in time dependant on $|G|$ and $|\Sigma|$, we bound these sizes.

Lemma 8. *During* FCPM *, $|G| = \mathcal{O}((n + m)\log(n + m))$ and $|\Sigma| = \mathcal{O}((n + m)\log(n + m)\log|M|)$.*

The proof is straightforward: using an argument similar to Lemma 2 we show that the size of each rule shortens by a constant factor in each phase. On the other hand, only LeftPop , RightPop and RemCrBlocks introduce new letters to the rules and it can be estimated, that in total they introduces $\mathcal{O}(\log(n + m))$ letters to a rule in each phase. Thus, bound $\mathcal{O}(\log(n + m))$ on each rules' length holds. Concerning $|\Sigma|$, new letters appear as a result of a compression. Since each compression decreases the size of $|G|$ by at least 1, there are no more than $|G|$ of them in a phase, which yields the bound.

Memory Consumption. FCPM uses $\mathcal{O}((n + m)\log(n + m))$ space, the same holds if we want to retrieve first/last positions etc. of the pattern, under the assumption that N and M fit in $\mathcal{O}(1)$ codewords. If only n and m fit in $\mathcal{O}(1)$ codewords, the space consumption increases by a factor representing the length of text and pattern, i.e. $\log(N + M)$.

Sketch of the Main Proof. The cost of one phase of FCPM is $\mathcal{O}(|G| + (n + m) + (m + n)\log(n + m))$, by Lemmas 3, 5–7 while Lemma 8 shows that $|G| = \mathcal{O}((n+m)\log(n+m))$ and Lemma 2 shows that there are $\mathcal{O}(\log M)$ phases. So the total running time is $\mathcal{O}((n + m)\log M \log(n + m))$.

Acknowledgements. I would like to thank Paweł Gawrychowski for introducing me to the topic, for pointing out the relevant literature [13,15,16] and discussions [2].

References

1. Alstrup, S., Brodal, G.S., Rauhe, T.: Pattern matching in dynamic texts. In: Proc. 11th Annual ACM-SIAM Symposium on Discrete Algorithms, pp. 819–828 (2000)
2. Gawrychowski, P.: personal communication (2011)
3. Gawrychowski, P.: Optimal pattern matching in LZW compressed strings. In: Randall, D. (ed.) SODA, pp. 362–372. SIAM (2011)
4. Gawrychowski, P.: Pattern Matching in Lempel-Ziv Compressed Strings: Fast, Simple, and Deterministic. In: Demetrescu, C., Halldórsson, M.M. (eds.) ESA 2011. LNCS, vol. 6942, pp. 421–432. Springer, Heidelberg (2011)
5. Gawrychowski, P.: Simple and Efficient LZW-Compressed Multiple Pattern Matching. In: Kärkkäinen, J. (ed.) CPM 2012. LNCS, vol. 7354, pp. 232–242. Springer, Heidelberg (2012)
6. Gawrychowski, P.: Tying up the loose ends in fully LZW-compressed pattern matching. In: Dürr, C., Wilke, T. (eds.) STACS 2012. LIPIcs, vol. 14, pp. 624–635. Schloss Dagstuhl — Leibniz-Zentrum fuer Informatik (2012)
7. Gąsieniec, L., Karpiński, M., Plandowski, W., Rytter, W.: Efficient Algorithms for Lempel-Ziv Encoding. In: Karlsson, R., Lingas, A. (eds.) SWAT 1996. LNCS, vol. 1097, pp. 392–403. Springer, Heidelberg (1996)
8. Gąsieniec, L., Karpiński, M., Plandowski, W., Rytter, W.: Randomized Efficient Algorithms for Compressed Strings: The Finger-Print Approach (Extended Abstract). In: Hirschberg, D.S., Meyers, G. (eds.) CPM 1996. LNCS, vol. 1075, pp. 39–49. Springer, Heidelberg (1996)
9. Hirao, M., Shinohara, A., Takeda, M., Arikawa, S.: Fully compressed pattern matching algorithm for balanced straight-line programs. In: SPIRE 2000, pp. 132–138 (2000)
10. Jeż, A.: Compressed membership for NFA (DFA) with compressed labels is in NP (P). In: Dürr, C., Wilke, T. (eds.) STACS 2012. LIPIcs, vol. 14, pp. 136–147. Schloss Dagstuhl - Leibniz-Zentrum fuer Informatik (2012)
11. Jeż, A.: Recompression: a simple and powerful technique for word equations. In: CoRR 1203.3705 (submitted, 2012)
12. Shibata, Y., Takeda, M., Shinohara, A., Arikawa, S.: Pattern Matching in Text Compressed by Using Antidictionaries. In: Crochemore, M., Paterson, M. (eds.) CPM 1999. LNCS, vol. 1645, pp. 37–49. Springer, Heidelberg (1999)
13. Lifshits, Y.: Processing Compressed Texts: A Tractability Border. In: Ma, B., Zhang, K. (eds.) CPM 2007. LNCS, vol. 4580, pp. 228–240. Springer, Heidelberg (2007)
14. Lifshits, Y., Lohrey, M.: Querying and Embedding Compressed Texts. In: Královič, R., Urzyczyn, P. (eds.) MFCS 2006. LNCS, vol. 4162, pp. 681–692. Springer, Heidelberg (2006)
15. Lohrey, M., Mathissen, C.: Compressed Membership in Automata with Compressed Labels. In: Kulikov, A., Vereshchagin, N. (eds.) CSR 2011. LNCS, vol. 6651, pp. 275–288. Springer, Heidelberg (2011)
16. Mehlhorn, K., Sundar, R., Uhrig, C.: Maintaining dynamic sequences under equality tests in polylogarithmic time. Algorithmica 17(2), 183–198 (1997)
17. Miyazaki, M., Shinohara, A., Takeda, M.: An Improved Pattern Matching Algorithm for Strings in Terms of Straight-Line Programs. In: Hein, J., Apostolico, A. (eds.) CPM 1997. LNCS, vol. 1264, pp. 1–11. Springer, Heidelberg (1997)
18. Plandowski, W.: Testing Equivalence of Morphisms on Context-Free Languages. In: van Leeuwen, J. (ed.) ESA 1994. LNCS, vol. 855, pp. 460–470. Springer, Heidelberg (1994)
19. Rytter, W.: Application of Lempel-Ziv factorization to the approximation of grammar-based compression. Theor. Comput. Sci. 302(1-3), 211–222 (2003)

NNS Lower Bounds via Metric Expansion
for l_∞ and *EMD*

Michael Kapralov[1,*] and Rina Panigrahy[2]

[1] Stanford iCME, Stanford, CA
kapralov@stanford.edu
[2] MSR Silicon Valley, Mountain View, CA
rina@microsoft.com

Abstract. We give new lower bounds for randomized NNS data structures in the cell probe model based on *robust* metric expansion for two metric spaces: l_∞ and Earth Mover Distance (EMD) in high dimensions. In particular, our results imply stronger non-embedability for these metric spaces into l_1. The main components of our approach are a strengthening of the isoperimetric inequality for the distribution on l_∞ introduced by Andoni et al [FOCS'08] and a robust isoperimetric inequality for EMD on quotients of the boolean hypercube.

1 Introduction

In the Nearest Neighbor Problem we are given a data set of n points $x_1, ..., x_n$ lying in a metric space V. The goal is to preprocess the data set into a data structure such that when given a query point $y \in V$, it is possible to recover the data set point which is closest to y by querying the data structure at most t times. The goal is to keep both the querying time t and the data structure space m as small as possible. Nearest Neighbor Search is a fundamental problem in data structures with numerous applications to web algorithms, computational biology, information retrieval, machine learning, etc. As such it has been researched extensively.

Natural metric spaces include the spaces $\Re^d$ equipped with the ℓ_1 or ℓ_2 distance that have been extensively studied in terms of upper and lower bounds. But other metrics such as ℓ_∞, edit distance and earth mover distance may be more appropriate in some settings [3,9]. Naturally, the time space tradeoff of known solutions crucially depend upon the underlying metric space. The known upper bounds exhibit the 'curse of dimensionality': for d dimensional spaces either the space or time complexity is exponential in d – thus encouraging research on approximate solutions. In the c-approximate nearest neighbor version, one returns a neighbor that is at most distance c times that to the nearest neighbor [10], [12], [9], [2] –for example there is an algorithm to obtain a c-approximate near neighbor in time $\tilde{O}(1)$ and space $n^{1+O(1/c)}$ using locality sensitive hashing in the l_1 metric; for the l_2 metric the space drops to $n^{1+O(1/c^2)}$ [2]. For the ℓ_∞ metric Indyk [9] shows how to compute a $O(\log_{1/\epsilon} \log d)$-approximate NNS using space $n^{\Omega(1/\epsilon)}$; most of these algorithms are *randomized*, while the algorithm of

---

* Research supported by NSF grant 0904325. Part of the work was done while the author was an intern at Microsoft Research Silicon Valley.

A. Czumaj et al. (Eds.): ICALP 2012, Part I, LNCS 7391, pp. 545–556, 2012.

Indyk [9] is deterministic. Our lower bounds for ℓ_∞ show that the space/approximation tradeoff in [9] is essentially optimal even if randomization is allowed.

There is a substantial body of work on lower bounds covering various metric spaces and parameter settings many of which assume the algorithm to be deterministic. Most previous papers are concerned with the Hamming distance over the d-dimensional hypercube. The cases of exact or deterministic algorithms were handled in a series of papers[7], [6],[13], [5]. These lower bounds hold for any polynomial space. In contrast the known upper bounds are both approximate and randomized, and with polynomial space can retrieve the output with one query. Chakrabarti and Regev [8] allow for both randomization and approximation, with polynomial space and show a tight bound for the *nearest* neighbor problem. Patrascu and Thorup[18] showed lower bounds on the query time of near neighbor problems with a stronger space restriction (near linear space), although their bound holds for deterministic or exact algorithms. Traditionally, cell probe lower bounds for data structures have been shown using communication complexity arguments [15]. Patrascu and Thorup [18] use a direct sum theorem along with the richness technique to obtain lower bounds for deterministic algorithms. Andoni, Indyk and Patrascu [4] showed randomized lower bounds using communication complexity lower bounds for Lopsided Set Disjointness. In [16,17], a more direct geometric argument was used to show lower bounds for randomized algorithms based on different variants of expansion of the underlying metric space.

The metric ℓ_∞ is considered in an intriguing paper by Andoni *et al.*[1] who prove a lower bound for deterministic algorithms. The paper uses the richness lemma though the crux of the proof is an interesting isoperimetric bound on ℓ_∞ for a carefully chosen measure. The lower bound they provide is tight for constant query time and matches the upper bound from [9]. In this work we obtain lower bounds for randomized algorithms for two new metric spaces: ℓ_∞ and Earth Movers Distance (EMD). For the ℓ_∞ metric we extend the tight lower bounds of [1] from deterministic to randomized algorithms by computing the notion of 'robust expansion' introduced in [17]. Our's is the first work that looks at the hardness of NNS in EMD metric. Inspired from the Fourier based techniques in the non-embeddability results from [11] we show hardness of NNS in the EMD metric over point sets in the d-dimensional hamming cube. We prove the following hardness guarantees for the case when cell size is $n^{o(1)}$, where n is the number of points in the database. For a given distribution of points and query, a randomized algorithm for (approximate) NNS is one that produces an (approximate) near neighbor with probability at least $2/3$.

Theorem 1. *1. For a $O(\log_{1/\epsilon} \log d)$ approximate NNS in ℓ_∞, any (randomized) t-probe data structure needs space at least $n^{\Omega(1/(\epsilon t))}$*

2. There is distribution of sets from the Hamming cube $\{0,1\}^d$ so that any (randomized) t-probe data structure for an α approximate NNS the EMD metric on this set needs space at least $e^{\Omega(d/(\alpha t))}$ (each set in the distribution can be specified explicitly using $O(d)$ bits).

It is interesting to note that approximate NNS for EMD under this distribution takes exponential space for approximation $O(d^{1-\epsilon})$ for all constant $\epsilon > 0$. Note that lower bounds on NNS on a metric space are stronger than non-embeddability results as once a metric space can be embedded into a well-studied metric space, the algorithms for NNS from the latter will carry over with the appropriate distortion. Thus our results automatically imply robust non-embeddibility results for these metric spaces. While it was known that these metric spaces do not embed into l_1 or l_2 with constant distortion, we now know that they are also not gap embeddable. In particular for the EMD metric on point sets from a d-dimensional hypercube Khot et al[11] showed that it doesn't embed into the ℓ_1 metric with distortion less than d. Our bound generalizes this to gap-inembeddibility:

Theorem 2. *There is no embedding M from the EMD metric space induced by the hamming metric on point sets over $\{0,1\}^d$ to the l_1 metric that satisfies the following gap distortion guarantees.*

$$EMD(u,v) \leq \omega(1) \implies |M(u) - M(v)|_1 \leq 1$$

$$EMD(u,v) = \Omega(d) \implies |M(u) - M(v)|_1 \geq 2$$

We will now review the different notions of metric-expansion from [17] that produce lower bounds for different classes of algorithms, deterministic and randomized. The bounds hold even for even in the average case when the points are chosen uniformly from a certain distribution.

1.1 Expansion and Its Relation to Complexity of NNS

The results in [17] show a relation between the expansion of the metric space and the complexity of NNS. It works with the version of the *Near Neighbor* version of the problem that is parameterized by a search radius r. As in the Nearest Neighbor Search Problem given a query point y the goal is to determine whether the data set contains a point of distance at most r from y. Expansion can be used to consider the case when points are chosen randomly from a distrubution and the query point is a random point from a ball of radius r around one of the database points. Intuitively expansion is the amount by which a set of points expands when we include points in their r neighborhood. If distribution of points is such that the distance between any pair of database points is at least cr then this lower bound also implies hardness for c-approximate NNS.

To compute the expansion we construct an undirected bipartite graph $G = (U, V, E)$ where U and V are all the points in the metric space and and edge is placed between a pair of nodes from U and V if they are at most distance r apart. The data set comes by choosing n points randomly from U and query is a random neighbor from V of a random database point from U (these distrubutions may be non-uniform which we specify in detail later).

Definition 1 (Vertex expansion). *The δ-vertex expansion of the graph is defined as*

$$\Phi_v(\delta) := \min_{A \subset V, |A| \leq \delta |U|} \frac{|N(A)|}{|A|}.$$

Here $N(A)$ denotes the neighborhood of the set A in G. For $A \subset V$, $B \subset U$, let $E(A, B)$ denote the set of edges between A and B in the bipartite graph G. Assume that $|A| = \delta|U|$. Observe that if $E(A, B) = E(A, U)$ then $|B| \geq \Phi_v(\delta)|A|$. In other words $\Phi_v(\delta)$ bounds the measure of the sets that cover all the edges incident on a set of measure δ. The notion of *robust expansion* relaxes this by requiring B to cover at least a γ-fraction of the edges incident on A. This idea is captured in the definition below. For simplicity we assume that $V = U$ and that G is regular. A more subtle definition which takes into account other non-regular graphs is presented later.

Definition 2 (Robust expansion). *G has robust-expansion $\Phi_r(\delta, \gamma)$ if $\forall A, B \subseteq V$ satisfying $|A| \leq \delta|V|, |B| \leq \Phi(\delta, \gamma)|A|$, it is the case that $\frac{|E(A,B)|}{|E(A,V)|} \leq \gamma$. Note that $\Phi_r(\delta, 1) = \Phi_v(\delta)$.*

Lower bounds for NNS based on the above notions of expansion were proven in [17]; the deterministic lower bounds use expansion and the randomized lower bounds make use of robust-expansion We now state the bounds for randomized algorithms. For technical reasons, it also assumes that the metric space satisfies a property called *weak-independence* which simply means that two balls of radius r centered at randomly chosen points are sufficiently disjoint with high probability $1 - o(1/n^2)$. Here m denotes the number of cells used by the algorithm where each cell can hold a word of size w bits.

Theorem 3. *[17] There exists an absolute constant γ such that the following holds. Any randomized algorithm for a* weakly-independent *instance of Near Neighbor problem which is correct with probability at least half (where the probability is taken over the sampling of the input and the algorithm), satisfies the following inequalities:*

$$\frac{m^t w}{n} \geq \Phi_r\left(\frac{1}{m^t}, \frac{\gamma}{t}\right) \tag{1}$$

These theorems, combined with known isoperimetric inequalities yield most known cell probe lower bounds for near neighbor problems. There is also some evidence that the connection between expansion and hardness of NNS is tight for constant t – this has been shown to hold for cases when the graph G is symmetric [17].

The bipartite graph $G = (U, V, E)$ may be weighted by a a probability distribution e over the edges E. Let $\mu(u) = e(u, V) = \sum_{v \in V} e(u, v)$ be the induced distribution on U, and let $\nu(v) = e(U, v)$ be the induced distribution on V. For $x \in U$, we denote by ν_x the conditional distribution of the endpoints in V of edges incident on u, i.e. $\nu_x(y) = e(x, y)/e(x, V)$. Thus ν_y is a distribution over (or concentrated over) the r-neighborhood of y. In this case we select n points $x_1, \ldots, x_n$ independently from the distribution μ uniformly at random. This defines the database distribution. To generate the query, we pick an $i \in [n]$ uniformly at random, and sample y independently from ν_{x_i}. The tuple (G, e) satisfies γ-*weak independence (WI)* if $\Pr_{x,z \sim \mu, y \sim \nu_x}[(y, z) \in E] \leq \frac{\gamma}{n}$. Thus, weak independence ensures that with probability $(1 - \gamma)$, for the instance generated as above, x is indeed the unique neighbor in G of y in $\{x_1, \ldots, x_n\}$. The following definition generalizes the notion of robust-expansion to weighted bipartite graphs.

Definition 3. *[17] [Robust Expansion] The γ-robust expansion of a set $A \subseteq V$ is*

$$\phi_r(A, \gamma) \overset{def}{=} \min_{B \subseteq U : e(B,A) \geq \gamma e(U,A)} \mu(B)/\nu(A).$$

2   Robust Expansion of l_∞

In this section we prove a bound on the robust expansion of l_∞ under a variant of the distribution introduced in [1]. Let $G_d = (U, V, E)$ be the l_∞ graph on $U = V = [1, \ldots, m]^d$, i.e. $u \in U$ is connected to $v \in V$ iff $||u - v||_\infty \leq 1$. We now define a distribution τ on the edges of G_d.

We start by defining the distribution for G_1, the one-dimensional l_1 graph (see Fig. 1). The distribution on G_d for general d will be the product of distributions on G_1. We let

$$\tau_{i,j}^1 = 2^{-(1/\epsilon)^i} \text{ if } j = i + 1 \text{ and } i \text{ is odd}$$

$$\tau_{i,j}^1 = 2^{-(1/\epsilon)^j} \text{ if } j = i - 1 \text{ and } i \text{ is odd}$$

$$\tau_{1,0}^1 = 1 - \sum_{i \geq 1} 2^{-(1/\epsilon)^i}, \text{ and } \tau_{i,j}^1 = 0 \text{ o.w.}$$

We denote the induced one-dimensional distributions by

$$\mu_u^1 = \sum_{v \in N^1(u)} \tau_{(u,v)}^1, \nu_v^1 = \sum_{u \in N^1(v)} \tau_{(u,v)}^1.$$

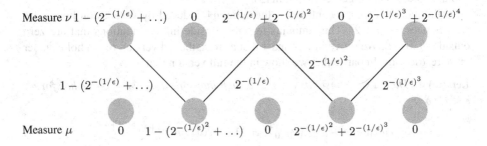

Fig. 1. Distribution on G_1

The d-dimensional distribution τ^d over edges is defined by $\tau_{(u,v)}^d = \prod_{i=1}^d \tau_{u_i, v_i}^1$. We fist note that this induces a product distribution on the vertices $u \in U$, where $\mu^d(u) = \prod_{i=1}^d \mu_1(u_i)$. In what follows we will use the notation $e_d(A, B) = \sum_{e \in E \cap (A \times B)} \tau_e^d$. We also omit superscripts in μ^d, ν^d, e_d and τ_e^d whenever this does not cause confusion.

The main component of our lower bound is a strengthened isoperimetric inequality for l_∞ under the distribution that we just defined. The main technical lemma will be

Lemma 1. *Let $G_d = (U, V, E)$ denote the l_∞ graph. For any $A \subseteq U, B \subseteq V$ one has $e(A, B) \leq (\mu(A)\nu(B))^{1/(1+\delta)}$ for some $\delta = \Theta(\epsilon)$ and all sufficiently small ϵ.*

A bound on robust expansion follows from Lemma 1 (details are deferred to the full version):

Lemma 2. *Let $G_d = (U, V, E)$ denote the l_∞ graph. For any $A \subseteq U, B \subseteq V$ such that $e(B, A) \geq \gamma e(A, V)$ one has $\nu(B) \geq \gamma^{1+\delta}(\mu(A))^\delta$ for some $\delta = \Theta(\epsilon)$ and sufficiently small ϵ.*

The proof Lemma 1 is by induction on the dimension, and we start by outlining the proof strategy for the base case, i.e. $d = 1$. For $d = 1$, Lemma 1 turns into

Lemma 3. *Let G_1 denote the l_∞ graph in dimension 1 with the measure τ defined as above. There exist constants $\gamma, \epsilon^* > 0$ such that for every $x, y \in \mathbb{R}_+^V$ for $\epsilon < \epsilon^*$ and $\delta = \gamma\epsilon$ one has*

$$\sum_{(i,j) \in E(G_1)} x_i \tau_{i,j} y_j \leq \left(\sum_i \mu_i x_i^{1+\delta} \right)^{1/(1+\delta)} \left(\sum_i \nu_i y_i^{1+\delta} \right)^{1/(1+\delta)} \tag{2}$$

It will be convenient to make a substitution to ensure that the rhs is the product of unweighted $(1 + \delta)$-norms. Set $u_i = \mu_i^{1/(1+\delta)} x_i, v_i = \nu_i^{1/(1+\delta)} y_i$, so that (2) becomes

$$\sum_{(i,j) \in E} u_i \mu_i^{-1/(1+\delta)} \tau_{ij} \nu_j^{-1/(1+\delta)} v_j \leq ||u||_{1+\delta} ||v||_{1+\delta}. \tag{3}$$

We prove the bound (3) in two steps. In particular, we break the graph G_1 into two pieces that overlap by one vertex, prove stronger versions of (3) for both subproblems, and then piece them together to obtain (3).

In the first step we concentrate on the subgraph induced by vertices on both sides with indices in $[0 : 2]$. This amounts to only considering distributions that are zero outside of $[0 : 2]$. We prove in Lemma 4 that a strengthened version of (3) holds under these restrictions. In particular, we show in the full version that

Lemma 4. *There exist constants $\epsilon^*, \gamma > 0$ such that for all $v_0, v_2 \geq 0$ one has for all $\epsilon < \epsilon^*, \delta = \gamma\epsilon$*

$$\tau_{10} v_0 + \tau_{12} v_2 \leq \left(\nu_0 v_0^{1+\delta} + (1 - \Omega(\delta^5)) \nu_1 v_1^{1+\delta} \right)^{1/(1+\delta)} \tag{4}$$

It should be noted that while (3) depends on both u and v, the inequality in (4) only depends on u. This is because only the single vertex v_1 has a nonzero weight among vertices in $[0 : 2]$, and hence can be cancelled from both sides. The $1 + O(\delta^5)$ term multiplying u_2 on the lhs represents the main strengthening, and will be crucially important for combining the inequalities for different parts of the graph later.

In the second step we consider the subgraph of G_1 induced by vertices with indices in $[2 : +\infty]$. This amounts to considering distributions that are zero on the the first two vertices on each side of the graph. For this case we prove

Lemma 5. *Let G_1 denote the l_∞ graph in dimension 1 with the measure τ defined as above. There exist constants $\gamma, \epsilon^* > 0$ such that for every $x, y \in \mathbb{R}_+^V$ for $\epsilon < \epsilon^*$ and $\delta = \gamma\epsilon$ one has*

$$\sum_{(i,j) \in E(G_1), i>1} x_i \tau_{i,j} y_j \leq 2^{-1/\epsilon} \left(\sum_i \mu_i x_i^{1+\delta} \right)^{1/(1+\delta)} \left(\sum_i \nu_i y_i^{1+\delta} \right)^{1/(1+\delta)} \tag{5}$$

The $2^{-1/\epsilon}$ term represents the strengthening with respect to (3) and will be crucial for combining (4) and (5). Combining (4) and (5), we then get the result (essentially) by an application of Cauchy-Schwarz and norm inequalities. One complication will be the fact that (4) and (5) overlap by v_2, but we will be able to handle this since the strengthened inequalities ensure that v_2 appears in (4) and (5) with weights that sum up to at most 1. We now give

Proof of Lemma 5: We need to bound $\sum_{(i,j)\in E, i\geq 2} u_i \mu_i^{-1/(1+\delta)} \tau_{ij} \nu_j^{-1/(1+\delta)} v_j$. In order to do that, we decompose the edges of G_1 restricted to $[2 : +\infty]$ into two edge disjoint matchings M_1 and M_2: $M_1 = \{(i,j) \in E(G_1) : j = i - 1, i,j \geq 2\}$, $M_2 = \{(i,j) \in E(G_1) : j = i+1, i,j \geq 2\}$.

First, suppose that $(i,j) \in M_1$, i.e. $j = i - 1$ and $i = 2k + 1$, where $k \geq 1$ since we are considering distributions restricted to $[2 : +\infty]$. We have

$$\mu_i^{-1/(1+\delta)} \tau_{ij} \nu_j^{-1/(1+\delta)} \leq 2^{(1/\epsilon)^{(k+1)}/(1+\delta)} \cdot 2^{-(1/\epsilon)^{k+1}} \cdot 2^{(1/\epsilon)^k/(1+\delta)} = 2^{(1/\epsilon)^k(1-\delta/\epsilon)/(1+\delta)}.$$

For $\delta \geq 4\epsilon$ and sufficiently small constant ϵ $\mu_i^{-1/(1+\delta)} \tau_{ij} \nu_j^{-(1-2\epsilon)} \leq 2^{-2(1/\epsilon)^k} \leq 2^{-2/\epsilon}$, where we used the fact that $k \geq 1$. A similar argument shows that the same holds for all $(i,j) \in M_2$. Thus, for $r = 1, 2$

$$\sum_{(i,j)\in M_r} u_i \mu_i^{-1/(1+\delta)} \tau_{ij} \nu_j^{-1/(1+\delta)} v_j \leq 2^{-2/\epsilon} \sum_{(i,j)\in E, i\geq 2} u_i v_j \leq 2^{-2/\epsilon} \sqrt{\sum_{i\geq 2} u_i^2} \sqrt{\sum_{j\geq 2} v_j^2}$$

by Cauchy-Schwarz. Since for all x one has $||x||_p \geq ||x||_q$ when $p \leq q$, we conclude that for $r = 1, 2$

$$\sum_{(i,j)\in M_r} u_i \mu_i^{-1/(1+\delta)} \tau_{ij} \nu_j^{-1/(1+\delta)} v_j \leq 2^{-2/\epsilon} \left(\sum_{i\geq 2} u_i^{1+\delta}\right)^{1/(1+\delta)} \left(\sum_{j\geq 2} v_j^{1+\delta}\right)^{1/(1+\delta)},$$

as required. Putting the estimates for M_1 and M_2 together, we get

$$\sum_{(i,j)\in E(G_1), i\geq 2} x_i \tau_{i,j} y_j \leq 2^{-1/\epsilon} \left(\sum_i \mu_i x_i^{1+\delta}\right)^{1/(1+\delta)} \left(\sum_i \nu_i y_i^{1+\delta}\right)^{1/(1+\delta)}.$$

$\square$

We now prove Lemma 3, and then use it as the base case for induction on dimension.

Proof of Lemma 3: By Lemma 4 we have

$$\sum_{(i,j)\in E(G_1), i,j\leq 2} x_i \tau_{i,j} y_j \leq \left(\mu_1 x_1^{1+\delta}\right)^{1/(1+\delta)} \left(\nu_0 y_0^{1+\delta} + (1 - \Omega(\delta^5))\nu_2 y_2^{1+\delta}\right)^{1/(1+\delta)},$$

(6)

For convenience, let $A := \left(\mu_1 x_1^{1+\delta}\right)^{1/(1+\delta)}$, $B := \left(\nu_1 y_1^{1+\delta} + (1 - \Omega(\delta^5))\nu_2 y_2^{1+\delta}\right)^{1/(1+\delta)}$. Furthermore, by Lemma 5

$$\sum_{(i,j)\in E(G_1), i\geq 2, j\geq 2} x_i \tau_{i,j} y_j \leq 2^{-1/\epsilon} \left(\sum_i \mu_i x_i^{1+\delta}\right)^{1/(1+\delta)} \left(\sum_i \nu_i y_i^{1+\delta}\right)^{1/(1+\delta)},$$

(7)

and we define for convenience $C := \left(\sum_i \mu_i x_i^{1+\delta}\right)^{1/(1+\delta)}$ and $D := 2^{-1/\epsilon} \left(\sum_i \nu_i y_i^{1+\delta}\right)^{1/(1+\delta)}$.

First, we get by combining (6) and (7) that

$$\sum_{(i,j)\in E(G_1)} x_i \tau_{i,j} y_j \le A \cdot B + C \cdot D \tag{8}$$

Applying Cauchy-Schwarz and norm inequalities to the rhs of (8), we get

$$A \cdot B + C \cdot D \le \sqrt{A^2 + C^2} \sqrt{B^2 + D^2} \le \left(A^{1+\delta} + C^{1+\delta}\right)^{1/(1+\delta)} \left(B^{1+\delta} + D^{1+\delta}\right)^{1/(1+\delta)}. \tag{9}$$

Combining (8) and (9), we obtain

$$\sum_{(i,j)\in E(G_1)} x_i \tau_{i,j} y_j \le \left(\nu_0 y_0^{1+\delta} + \nu_2(1 - \Omega(\delta^5) + 2^{-(1+\delta)/\epsilon}) y_2^{1+\delta} + \sum_{j>2} \nu_j y_j^{1+\delta}\right)^{1/(1+\delta)}$$

$$\cdot \left(\mu_1 x_1^{1+\delta} + \sum_{i>1} \mu_i x_i^{1+\delta}\right)^{1/(1+\delta)} \le \left(\sum_{i\ge0} \mu_i x_i^{1+\delta}\right)^{\frac{1}{1+\delta}} \left(\sum_{j\ge0} \nu_j y_j^{1+\delta}\right)^{\frac{1}{1+\delta}}$$

$\square$

Proof of Lemma 1: We use induction on d. The base case $d = 1$ is given by Lemma 3. We now describe the inductive step $d - 1 \to d$.

Let $A \subseteq U, B \subseteq V$. For each i let $A_i = \{u \in A : u_i = i\}, B_i = \{u \in A : u_i = i\}$. Then by our definition of edge weights $e_d(A, B) = \sum_{(i,j)\in E(G_1)} \tau_{ij} e_{d-1}(A_i, B_j)$. By the inductive hypothesis we have $e_{d-1}(A_i, B_j) \le (\mu_{d-1}(A_i)\mu_{d-1}(B_j))^{1/(1+\delta)}$, and hence

$$e_d(A, B) \le \sum_{(i,j)\in E(G_1)} \tau_{ij} (\mu_{d-1}(A_i)\mu_{d-1}(B_j))^{1/(1+\delta)}.$$

Now by Lemma 3 we have

$$\sum_{(i,j)\in E(G_1)} \tau_{ij} (\mu_{d-1}(A_i)\mu_{d-1}(B_j))^{1/(1+\delta)} \le \left(\sum_i \mu_i^1 \mu_{d-1}(A_i) \sum_j \mu_j^1 \mu_{d-1}(B_j)\right)^{1/(1+\delta)}$$

$$= (\mu_d(A)\mu_d(B))^{1/(1+\delta)}.$$

$\square$

Theorem 4. $O(\log_{1/\epsilon} \log d)$-*approximate NNS for* l_∞ *requires space* $n^{\Omega(1/(\epsilon t))}$ *even with randomization.*

Proof. The proof follows by first showing that the distance between a pair of points drawn from our distribution is $\Omega(\log_{1/\epsilon} \log d)$ and applying Theorem 3 together with Lemma 2. The details are deferred to the full version.

3   Earth Mover Distance

In this section we derive lower bounds on the cell probe complexity of nearest neighbor search for Earth mover distance (also known, as transportation cost metric) over $\mathbb{F}_2^d$. Our approach is based on lower bounding the robust expansion of EMD over quotients of $\mathbb{F}_2^d$ with respect to the dual of a random linear code. Quotients of $\mathbb{F}_2^d$ with respect to random linear codes have been used in [11] to derive non-embeddability results for EMD over $\mathbb{F}_2^d$ into l_1. Here we extend these non-embeddability results to hardness of nearest neighbor search. As a by-product of our approach, we also prove that EMD over $\mathbb{F}_2^d$ is not gap-embeddable into l_1 with distortion less than $\Omega(d)$.

Let (X, d) be a metric space. The earth mover distance between two sets $A, B \subseteq X$, such that $|A| = |B|$ is defined by

$$EMD(A, B) = \min_{\pi: A \to B} \sum_{x \in A} d(x, \pi(x)), \qquad (10)$$

where the minimum is taken over all bijective mappings π from A to B. For the purposes of our lower bounds, the metric space (X, d) will be the binary hypercube $(\mathbb{F}_2^d, || \cdot ||_1)$ with Hamming distance as the metric, and A, B will be subsets of $\mathbb{F}_2^d$ of a special form. In particular, A and B will be cosets of $\mathbb{F}_2^d$ with respect to the action of a carefully chosen group (in fact, a linear code with large minimum distance).

Let C denote a linear code, i.e. a linear subspace of $\mathbb{F}_2^d$ of dimension $\Omega(d)$ and minimum distance $\Omega(d)$. Such codes are known to exist [14]. In particular, it can be seen that a random linear code of dimension $\Omega(d)$ satisfies this conditions with high probability. We will use the notation for the dual code

$$C^\perp = \{y \in \mathbb{F}_2^d : (y, x) \equiv 0 \mod 2, \forall x \in C\},$$

where $(x, y) = \sum_{i=1}^d x_i y_i$. For a vector $u \in \mathbb{F}_2^d$ we denote the coset of u with respect to the dual code $C^\perp$ by $\mathbf{u} = \{w \in \mathbb{F}_2^d : w - u \in C^\perp\}$. Thus, $\mathbf{u}$ is the set of vectors in $\mathbb{F}_2^d$ that can be obtained from u by translating it by an element of $C^\perp$. In what follows we consider EMD on such subsets $\mathbf{u}$ of the hypercube. The following simple property of EMD restricted to cosets of $\mathbb{F}_2^d$ with respect to $C^\perp$ will be very useful. Recall that by (10) $EMD(\mathbf{u}, \mathbf{v})$ is the cost of the bijective mapping π from A to B that minimizes total movement $\sum_{x \in A} ||x - \pi(x)||_1$. We now show that when EMD is restricted to cosets of $C^\perp$, i.e. $A = \mathbf{u}, B = \mathbf{v}$ for some $u, v \in \mathbb{F}_2^d$, the minimum over mappings π is achieved for a mapping that simply translates each element of a coset $\mathbf{u}$ by a fixed vector w to get $\mathbf{v}$ (the proof is deferred to the full version.):

Fact 5. *For* $\mathbf{u}, \mathbf{v} \in \mathbb{F}_2^d/C^\perp$ *one has* $EMD(\mathbf{u}, \mathbf{v}) = |C^\perp| \cdot \min_{a \in \mathbf{u}, b \in \mathbf{v}} ||a - b||_1$.

Our estimates of robust expansion of EMD on $\mathbb{F}_2^d/C^\perp$ will use Fourier analysis on the hypercube, so we give the necessary definitions now. The Fourier basis is given by Walsh functions $W_A : \mathbb{F}_2^d \to \mathbb{R}, A \subseteq \{1, \ldots, d\}$ is denoted by

$$W_A(x) = (-1)^{\sum_{j \in A} x_j}, x = (x_1, \ldots, x_d) \in \mathbb{F}_2^d.$$

Thus, $\{W_A : A \subseteq \{1, \ldots, d\}\}$ is an orthonormal basis of $L_2(\mathbb{F}_2^d, \sigma)$, where $\sigma(x) = 2^{-d}, x \in \mathbb{F}_2^d$ is the uniform measure on $\mathbb{F}_2^d$. For each $f : \mathbb{F}_2^d \to \mathbb{R}$ one has $f =$

$\sum_{A \subseteq \{1,...,d\}} \hat{f}(A) W_A$, where $\hat{f}(A) = \int_{\mathbb{F}_2^d} f(x) W_A(x) d\sigma(x)$. Parseval's indentity states that

$$\int_{\mathbb{F}_2^d} f(x) g(x) d\sigma(x) = \sum_{A \subseteq \{1,...,d\}} \hat{f}(A) \hat{g}(A)$$

for all $f, g \in L_2(\mathbb{F}_2^d, \sigma)$. We will often use the notation $(f, g) = \int_{\mathbb{F}_2^d} f(x) g(x) d\sigma(x)$. We will also use the non-uniform measure $\sigma_\epsilon(x) = \epsilon^{\sum_{i=1}^d x_i} (1 - \epsilon)^{d - \sum_{i=1}^d x_i}$.

We now define the distribution on inputs that we will use for our lower bounds. For $r \in (0, d)$ let $G = (U, V, E)$, where $U = V = \mathbb{F}_2^d / C^\perp$ denote the complete bipartite graph. We now define distributions on U, V and the edges of G. Let μ and ν denote the uniform distribution on U and V respectively. The distribution on pairs is given first sampling $\mathbf{u} \in U$ uniformly, and then letting

$$\mathbf{v} = \mathbf{u} + Z, \tag{11}$$

where $\mathbf{Pr}[Z = z] = \sigma_{r/d}(z)$, i.e. Z is a point in $\mathbb{F}_2^d$ obtained by setting each coordinate independently to 1 with probability r/d and 0 with probability $1 - r/d$. Here for a coset $\mathbf{u}$ and a point $z \in \mathbb{F}_2^d$ we write $\mathbf{u} + z$ to denote the coset obtained from $\mathbf{u}$ by adding z to each $u \in \mathbf{u}$. We note that this is equivalent to sampling a uniformly random $\mathbf{u}$, then sampling a uniformly random point $u \in \mathbf{u}$, letting $v = u + Z$ and declaring $\mathbf{v}$ to be the resulting coset. In particular, this yield the following distribution on edges;

$$\tau_{\mathbf{u},\mathbf{v}} = \frac{1}{2^d} \sum_{u \in \mathbf{u}, v \in \tau} \sigma_{r/d}(u - v). \tag{12}$$

The distance between $\mathbf{u}$ and $\mathbf{v}$ sampled according to this distribution is $O(r)$ with high probability: $\mathbf{Pr}_{(\mathbf{u},\mathbf{v}) \in E}[EMD(\mathbf{u}, \mathbf{v}) > \gamma r] \leq e^{-\Omega((\gamma-1)r)}$, i.e. pairs sampled from our distribution are nearby with high probability. On the other hand, two uniformly random cosets are at distance $\Omega(d)$ with high probability:

Lemma 6. *Let* $\mathbf{u}, \mathbf{v}$ *denote uniformly random points in* $\mathbb{F}_2^d / C^\perp$. *Then* $\mathbf{Pr}[EMD(\mathbf{u}, \mathbf{v}) > c'd] \geq 1 - 2^{-\Omega(d)}$ *for a constant* $c' > 0$.

We now turn to lower bounding the robust expansion. It will be convenient to use the following notation. For $A \in \mathbb{F}_2^d / C^\perp$ we will write $\mathbf{1}_A$ to denote the indicator function of A lifted to $\mathbb{F}_2^d$, i.e. $\mathbf{1}_A(x)$ equals 1 if $x \mod C^\perp = A$ and 0 otherwise. Our main lemma relies on the following crucial property of functions that are constant on cosets of $C^\perp$, proved in [11]. In particular, any such function necessarily has zero Fourier coefficients corresponding to non-empty sets of small size:

Lemma 7. *[11] Assume that* $f : \mathbb{F}_2^d \to \mathbb{R}$ *satisfies for every* $x \in \mathbb{F}_2^d$ *and for all* $y \in C^\perp$, $f(x + y) = f(x)$. *Suppose that the minimum distance of* C *is* d_0. *Then* $\hat{f}(S) = 0$ *for all* $|S| < d_0, S \neq \emptyset$.

The function $\mathbf{1}_A(x)$ satisfies the preconditions of Lemma 7 for $A \in \mathbb{F}_2^d / C^\perp$, and hence we have $\hat{\mathbf{1}}_A(S) = 0$ for $|S| \leq c'd, S \neq \emptyset$.

We now bound the robust expansion of EMD under our distribution. Similarly to section 2, we first bound the weight of edges going between a pair of sets A, B. As before, we use the notation $e(A, B) = \sum_{u \in A, v \in B} \tau_{u,v}$. It will be convenient to express $e(A, B)$ in terms of the Bonami-Beckner operator $T_\rho : L_2(\mathbb{F}_2^d, \sigma) \to L_2(\mathbb{F}_2^d, \sigma)$. For a function $f \in L_2(\mathbb{F}_2^d, \sigma)$ one has $T_\rho f(x) = \mathbf{E}_{z \sim \sigma_{1-2\rho}}[f(x + z)]$, where we will use $\rho = 1 - 2r/d$. The proof of the following claim is given in the full version:

Claim 6. *For any* $A, B \in \mathbb{F}_2^d/C^\perp$ *one has* $e(A, B) = (T_\rho 1_A, 1_B)$, *where* $(f, g) = \int_{\mathbb{F}_2^d} f(x)g(x)\sigma(x)$.

Our main lemma, which bounds the weight of edges going between a pair $A, B \in V$ is

Lemma 8. *Let* C *be a linear code of dimension* $\Omega(d)$ *and minimum distance* $\Omega(d)$. *Let* $\mathbb{F}_2^d/C^\perp$ *denote the quotient of* $\mathbb{F}_2^d$ *with respect to the dual code* $C^\perp$, *and consider the distribution over edges given by the noise operator with parameter* $\rho = 1 - 2r/d$ *as in* (11). *Then for any* $r < d/4$ *one has* $e(A, B) \leq \mu(A)\mu(B) + e^{-\Omega(r)}\sqrt{\mu(A)\mu(B)}$.

Proof. Consider any two sets $A, B \subseteq \mathbb{F}_2^d/C^\perp$. By Claim 6, we have $e(A, B) = (T_\rho 1_A, 1_B)$. We now use the fact that 1_A is constant on quotients of $C^\perp$, and hence by Lemma 7 one has $\hat{1}_A(S) = 0$ for all $S \subseteq \{0, 1\}^n$, $S \neq \emptyset$, with $|S| \leq cd$. Since

$$T_\rho 1_A = \sum_{S \subseteq \{0,1\}^d} (1 - 2\rho)^{|S|} \hat{1}_A(S) W_S, \tag{13}$$

we have $\|T_\rho f\| \leq e^{-cr}\|f\|$ for all $f \in L_2(\mathbb{F}_2^d, \sigma)$, such that $(f, 1_1) = 0$. Here we denote the constant function equal to 1 by 1. We also use the fact that if $(f, 1) = 0$, then $(T_\rho f, 1) = 0$, as can be seen directly from (13). For $A \subset \mathbb{F}_2^d/C^\perp$ we will write $|1_A|$ to denote l_1-norm of 1_A (in particular, $|1_A| = |C^\perp| \cdot |A|$), where $|A|$ is the number of elements in A. We now have

$$(T_\rho 1_A, 1_B) = \left(\frac{|1_A|}{2^d} 1 + T_\rho(1_A - \frac{|1_A|}{2^d} 1), \frac{|1_B|}{2^d} 1 + (1_B - \frac{|1_B|}{2^d} 1) \right)$$

$$= \left(\frac{|1_A|}{2^d} 1, \frac{|1_B|}{2^d} 1 \right) + \left(T_\rho(1_A - \frac{|1_A|}{2^d} 1), 1_B - \frac{|1_B|}{2^d} 1 \right)$$

since the cross terms cancel due to orthogonality. Thus,

$$(T_\rho 1_A, 1_B) \leq 2^{-2d}|1_A||1_B| + e^{-2\rho cd}\|1_A - \frac{|1_A|}{2^d} 1\|\|1_B - \frac{|1_B|}{2^d} 1\|,$$

and since $\rho d = r$, we get

$$e(A, B) \leq \frac{|1_A|}{2^d} \cdot \frac{|1_B|}{2^d} + e^{-2\rho cd}\sqrt{\frac{|1_A|}{2^d} \cdot \frac{|1_B|}{2^d}} \leq \mu(A)\mu(B) + e^{-\Omega(r)}\sqrt{\mu(A)\mu(B)}.$$

Using Lemma 8 we can now bound the robust expansion of EMD over $\mathbb{F}_2^d/C^\perp$:

Lemma 9. *Let* C *be a linear code of dimension* $d/4$ *such that the distance of* $C^\perp$ *is at least* $c'd$ *for some constant* $c > 0$. *Then the* γ-*robust expansion of EMD over* $\mathbb{F}_2^d/C^\perp$ *at distance* r *is at least* $(\gamma/2)^2 e^{\Omega(r)}$.

Theorem 7. α-*approximate NNS with t probes for d-dimensional EMD requires $e^{\Omega(d/(\alpha t))}$ space, even with randomization.*

Proof. Set $r = \Theta(d/\alpha)$. By Lemma 6 the distance between points is $\Omega(d)$ whenever $d \geq c \log n$ for a sufficiently large $c > 0$, which gives the weak independence property. The distance to the near point is $\Theta(r)$ with probability $1 - n^{-\Omega(1)}$. The robust expansion is at least $(\gamma/2)^2 e^{\Omega(r)}$ by Lemma 9, so the result follows by Theorem 3.

Proof of Theorem 2: Suppose that such an embedding exists. Then one can build a NNS data structure of size $n^{O(1)}$ to solve $3/2$-approximate NNS in l_1, implying a $o(d)$-approximate NNS for EMD. However, this would contradict Theorem 7 when $d = \Omega(\log n)$. □

References

1. Andoni, A., Croitoru, D., Patrascu, M.: Hardness of nearest neighbor under l-infinity. In: FOCS 2008 (2008)
2. Andoni, A., Indyk, P.: Near-optimal hashing algorithms for approximate nearest neighbor in high dimensions. Commun. ACM 51, 117–122 (2008)
3. Andoni, A., Indyk, P., Krauthgamer, R.: Earth mover distance over high-dimensional spaces. In: SODA 2008, pp. 343–352 (2008)
4. Andoni, A., Indyk, P., Patrascu, M.: On the optimality of the dimensionality reduction method. In: FOCS 2006, pp. 449–458 (2006)
5. Barkol, O., Rabani, Y.: Tighter bounds for nearest neighbor search and related problems in the cell probe model. In: STOC 2000, pp. 388–396 (2000)
6. Borodin, A., Ostrovsky, R., Rabani, Y.: Lower bounds for high dimensional nearest neighbor search and related problems. In: STOC 1999, pp. 312–321 (1999)
7. Chakrabarti, A., Chazelle, B., Gum, B., Lvov, A.: A lower bound on the complexity of approximate nearest-neighbor searching on the hamming cube. In: STOC 1999, pp. 305–311 (1999)
8. Chakrabarti, A., Regev, O.: An optimal randomised cell probe lower bound for approximate nearest neighbour searching. In: FOCS 2004, pp. 473–482 (2004)
9. Indyk, P.: On approximate nearest neighbors under l_∞ norm. J. Comput. Syst. Sci. 63 (2001)
10. Indyk, P., Motwani, R.: Approximate nearest neighbors: Towards removing the curse of dimensionality. In: STOC 1998, pp. 604–613 (1998)
11. Khot, S., Naor, A.: Nonembeddability theorems via fourier analysis. In: FOCS 2005 (2005)
12. Kushilevitz, E., Ostrovsky, R., Rabani, Y.: Efficient search for approximate nearest neighbor in high dimensional spaces. In: STOC 1998, pp. 614–623 (1998)
13. Liu, D.: A strong lower bound for approximate nearest neighbor searching. Inf. Process. Lett. 92(1), 23–29 (2004)
14. MacWilliams, F.J., Sloane, N.J.A.: The Theory of Error-Correcting Codes. North-Holland, New York (1977)
15. Miltersen, P.B., Nisan, N., Safra, S., Wigderson, A.: On data structures and asymmetric communication complexity. J. Comput. Syst. Sci. 57(1), 37–49 (1998)
16. Panigrahy, R., Talwar, K., Wieder, U.: A geometric approach to lower bounds for approximate near-neighbor search and partial match. In: FOCS 2008, pp. 414–423 (2008)
17. Panigrahy, R., Talwar, K., Wieder, U.: Lower bounds on near neighbor search via metric expansion. In: FOCS 2010 (2010)
18. Patrascu, M., Thorup, M.: Higher lower bounds for near-neighbor and further rich problems. In: FOCS 2006, pp. 646–654 (2006)

Quantum Adversary (Upper) Bound*

Shelby Kimmel

Center for Theoretical Physics, Massachusetts Institute of Technology,
Cambridge, USA
skimmel@mit.edu

Abstract. We describe a method for upper bounding the quantum query complexity of certain boolean formula evaluation problems, using fundamental theorems about the general adversary bound. This non-constructive method gives an upper bound on query complexity without producing an algorithm. For example, we describe an oracle problem that we prove (non-constructively) can be solved in $O(1)$ queries, where the previous best quantum algorithm uses a polynomial number of queries. We then give an explicit $O(1)$ query algorithm based on span programs, and show that for a special case of this problem, there exists a $O(1)$ query algorithm that uses the quantum Haar transform. This special case is a potentially interesting problem in its own right, which we call the HAAR PROBLEM.

1 Introduction

The general adversary bound has proven to be a powerful concept in quantum computing. Originally formulated as a lower bound on quantum query complexity [1], it has been shown to be tight with respect to the quantum query complexity of evaluating any function, and in fact is tight with respect to the more general problem of state conversion [2]. The general adversary bound is in some sense the culmination of a series of adversary methods [3,4]. While the adversary method in its various forms has been useful in finding lower bounds on quantum query complexity [5,6,7], the general adversary bound itself can be difficult to apply, as the quantity for even simple, few-bit functions must usually be calculated numerically [1,7].

One of the nicest properties of the general adversary bound is that it behaves well under composition [2]. This fact has been used to lower bound the query complexity of composed total functions, and to create optimal algorithms for composed total functions [7]. In this work we extend one of the composition results to partial boolean functions, and use it to obtain an *upper* bound on query complexity by upper bounding the general adversary bound.

Generally, finding an upper bound on the general adversary bound is just as difficult as finding an algorithm, as they are dual problems [2]. However, using the composition property of the general adversary bound, given an algorithm for

---

* Full version can be found at http://arxiv.org/abs/1101.0797

A. Czumaj et al. (Eds.): ICALP 2012, Part I, LNCS 7391, pp. 557–568, 2012.

a boolean function f composed d times, we upper bound the general adversary bound of f. Due to the tightness of the general adversary bound and query complexity, this procedure gives an upper bound on the query complexity of f, but because it is nonconstructive, it doesn't give any hint as to what the corresponding algorithm for f might look like. The procedure a bit counter-intuitive: we obtain information about an algorithm for a simpler function by creating an algorithm for a more complicated function. This is similar in spirit to the tensor-product trick, where an inequality between two terms is proved by considering tensor powers of those terms[1].

We describe a class of oracle problems called CONSTANT-FAULT DIRECT TREES (introduced by Zhan et al. [8]), for which this method proves the existence of a $O(1)$ query algorithm, where the previous best known query complexity is poly-nomial in the size of the problem. While this method does not give an explicit algorithm, we show that a span program algorithm achieves this bound.

We show that a special case of CONSTANT-FAULT DIRECT TREES can be solved in a single query using an algorithm based on the quantum Haar trans-form. The quantum Haar transform has appeared as a subroutine in other al-gorithms [9,10], and a 3-dimensional wavelet transform is the workhorse of an algorithm due to Liu [11]. We describe a new problem, the HAAR PROBLEM, that also can be solved with the quantum Haar transform. While the HAAR PROBLEM requires only 1 quantum query, it requires $\Omega(\log n)$ classical queries (where the oracle is an n-bit function). The HAAR PROBLEM is somewhat like period finding and may have interesting applications.

2   A Nonconstructive Upper Bound on Query Complexity

Our theorem for creating a nonconstructive upper bound on query complexity relies on the tightness of the general adversary bound with respect to query com-plexity, and the properties of the general adversary bound under composition. The actual definition of the general adversary bound is not necessary for our purposes, but can be found in [6].

Our theorem applies to boolean functions. f is boolean if $f : S \to \{0, 1\}$ with $S \subseteq \{0, 1\}^n$. Given a boolean function f and a natural number d, we define f^d, "f composed d times," recursively as $f^d = f \circ (f^{d-1}, \ldots, f^{d-1})$, where $f^1 = f$.

Now we can state the main theorem:

Theorem 1. *Suppose we have a (possibly partial) boolean function f that is composed d times, f^d, and a quantum algorithm for f^d that requires $O(J^d)$ queries. Then $Q(f) = O(J)$, where $Q(f)$ is the bounded-error quantum query complexity of f.*

(For background on bounded-error quantum query complexity and quantum al-gorithms, see [3].) There are seemingly similar results in the literature; for exam-ple, Reichardt proves in [12] that the query complexity of a function composed

---

[1] See Terence Tao's blog, *What's New* "Tricks Wiki article: The tensor power trick," http://terrytao.wordpress.com/2008/08/25/tricks-wiki-article-the-tensor-product-trick/

d times, when raised to the $1/d^{th}$ power, is equal to the adversary bound of the function, in the limit that d goes to infinity. This result is meant to give understanding of the exact query complexity of a function, whereas our result is a tool for upper bounding query complexity, possibly without gaining any knowledge of the exact query complexity of the function.

One might think that Theorem 1 is useless because an algorithm for f^d usually comes from composing an algorithm for f, and one expects the query complexity of the algorithm for f^d to be at least J^d if J is the query complexity of the algorithm for f.

Luckily for us, this is not always correct. If there is a quantum algorithm for f that uses J queries, where J is not optimal (i.e. is larger than the true bounded error quantum query complexity of f), then the number of queries used when the algorithm is composed d times can be much less than J^d. If this is the case, and if the non-optimal algorithm for f is the best known, Theorem 1 promises the existence of an algorithm for f that uses fewer queries than the best known algorithm, but, as Theorem 1 is nonconstructive, it gives no hint as to what the algorithm looks like.

We need two lemmas to prove Theorem 1:

Lemma 1. (Based on Lee et al. [2]) *For any boolean function* $f : S \to \{0,1\}$ *with* $S \subseteq \{0,1\}^n$ *and natural number* d,

$$\mathrm{ADV}^\pm(f^d) \geq (\mathrm{ADV}^\pm(f))^d. \tag{1}$$

The proof of this lemma can be found in the full version. Høyer et al. [1] prove Lemma 1 for *total* boolean functions[2], and the result is extended to more general total functions in [2]. Our contribution is to extend the result to partial boolean functions. While Theorem 1 still holds for total functions, the example we will describe later in the paper requires it to hold for partial functions.

Lemma 2. (Lee, et al. [2]) *For any function* $f : S \to E$, *with* $S \in D^n$, *and* E, D *finite sets, the bounded-error quantum query complexity of* f, $Q(f)$, *satisfies*

$$Q(f) = \Theta(\mathrm{ADV}^\pm(f)). \tag{2}$$

We now prove Theorem 1:

Proof. Given an algorithm for f^d that requires $O(J^d)$ queries, by Lemma 2,

$$\mathrm{ADV}^\pm(f^d) = O(J^d). \tag{3}$$

Combining Eq. (3) and Lemma 1, we have

$$(\mathrm{ADV}^\pm(f))^d = O(J^d). \tag{4}$$

---

[2] While the statement of Theorem 11 in [1] seems to apply to partial functions, it is mis-stated; their proof actually assumes total functions.

Raising both sides to the $1/d^{th}$ power, we obtain

$$\text{ADV}^{\pm}(f) = O(J). \tag{5}$$

At this point, we have the critical upper bound on the general adversary bound of f. Finally, using Lemma 2 again, we have

$$Q(f) = O(J). \tag{6}$$

3   Example Where the General Adversary Upper Bound Is Useful

In this section we will describe a function, called the 1-FAULT NAND TREE, for which Theorem 1 gives a better upper bound on query complexity than any known quantum algorithm. The 1-FAULT NAND TREE was proposed by Zhan et al. [8] to obtain a superpolynomial speed-up for a boolean formula with a promise on the inputs, and is a specific type of CONSTANT-FAULT DIRECT TREE, which is mentioned in Section 1. We will first define a NAND TREE, and then explain the promise of the 1-FAULT NAND TREE.

The NAND TREE is a complete, binary tree of depth n, where each node is assigned a bit value. The leaves are assigned arbitrary values, and any internal node v is given the value NAND($val(v_1), val(v_2)$), where v_1 and v_2 are v's children, and $val(v_i)$ denotes the value of that node.

To evaluate the NAND TREE, one must find the value of the root given an oracle for the values of the leaves. (The NAND TREE is equivalent to solving NAND^n, although the composition we will use for Theorem 1 is not the composition of the NAND function, but of the NAND TREE as a whole.) For arbitrary inputs, Farhi et al. showed that there exists an optimal algorithm in the Hamiltonian model to solve the NAND TREE in $O(2^{0.5n})$ time [13], and this was subsequently extended to a standard discrete algorithm with quantum query complexity $O(2^{0.5n})$ [14,15]. Classically, the best algorithm requires $O(2^{0.753n})$ queries [16]. Here, we will consider the 1-FAULT NAND TREE, for which there is a promise on the values of the inputs.

Definition 1. (1-FAULT NAND TREE [8]) *Consider a* NAND TREE *of depth* n, *(as described above). Then to each node* v, *with child nodes* v_1 *and* v_2, *we assign an integer* $\kappa(v)$ *such that:*

- $\kappa(v) = 0$ *for leaf nodes.*
- $\kappa(v) = \max_i \kappa(v_i)$, *if* $val(v_1) = val(v_2)$
- *Otherwise* $val(v_1) \neq val(v_2)$. *Let* v_i *be the node such that* $val(v_i) = 0$. *Then* $\kappa(v) = 1 + \kappa(v_i)$.

A tree satisfies the 1-fault condition if $\kappa(v) \leq 1$ *for any node* v *in the tree.*

Notation: When $val(v_1) \neq val(v_2)$ we call the node v a **fault**. (Since NAND(0, 1) = 1, fault nodes must have value 1, although not all 1-valued nodes are faults.)

The 1-fault condition is a limit on the amount and location of faults within the tree. In a 1-FAULT NAND TREE, if a path moving from a root to a leaf encounters any fault node and then passes through the 0-valued child of the fault node, there can be no further fault nodes on the path. An example of a 1-FAULT NAND TREE is given in Figure 1.

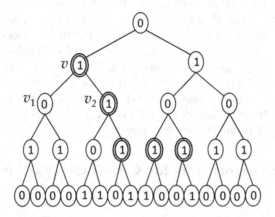

Fig. 1. An example of a 1-FAULT NAND TREE of depth 4. Fault nodes are highlighted by a double circle. The node v is a fault since one of its children (v_1) has value 0, and one (v_2) has value 1. Among v_1 and its children, there are no further faults, as required by the 1-fault condition. At v_2, we can have faults below the 1-valued child of v_2, but there can be no faults below the 0-valued child.

Zhan et al. [8] propose a quantum algorithm for an n level 1-FAULT NAND TREE that requires $O(n^2)$ queries to an oracle for the leaves. However, when the 1-FAULT NAND TREE is composed $\log n$ times, they apply their algorithm and find it only requires $O(n^3)$ queries. (Here we see an example where the number of queries required by an algorithm composed d times does not scale exponentially in d, which is critical for applying Theorem 1.) By applying Theorem 1 to the algorithm for the 1-FAULT NAND TREE composed $\log n$ times, we find that an upper bound on the query complexity of the 1-FAULT NAND TREE is $O(1)$. This is a large improvement over $O(n^2)$ queries. Zhan et al. prove $\Omega(\text{poly} \log n)$ is a lower bound on the classical query complexity of 1-FAULT NAND TREES. An identical argument can be used to show that CONSTANT-FAULT NAND TREES (from Definition 1, trees satisfying $\kappa(v) \leq c$ with c a constant) have query complexity $O(1)$.

In fact, Zhan et al. find algorithms for a broad range of trees, where instead of NAND, the evaluation tree is made up of a type of boolean function they call a *direct* function. A direct function is a generalization of a monotonic boolean function, and includes functions like majority, threshold, and their negations. For the exact definition, which involves span programs, see [8]. Applying Theorem 1 to their algorithm for trees made of direct functions proves the existence of

$O(1)$ query algorithms for CONSTANT-FAULT DIRECT TREES (a generalization of CONSTANT-FAULT NAND TREES to trees composed of direct functions rather than NAND). The best quantum algorithm of Zhan et. al requires $O(n^2)$ queries, and again they prove $\Omega(\text{poly} \log n)$ is a lower bound on the classical query complexity of CONSTANT-FAULT DIRECT TREES.

The structure of CONSTANT-FAULT DIRECT TREES can be quite complex, and it is not obvious that there should be a $O(1)$ query algorithm. Inspired by the knowledge of the algorithm's existence, thanks to Theorem 1, we found a span program algorithm for CONSTANT-FAULT DIRECT TREES that requires $O(1)$ queries. In the next section we will briefly describe this algorithm. However, as with many span program algorithms, it is hard to gain intuition about the algorithm. Thus in later sections we will describe a quantum algorithm based on the Haar transform that solves the 1-FAULT NAND TREE in 1 query in the special case that there is exactly one fault on every path from the root to a leaf, and those faults all occur at the same level.

4 Quantum Algorithms for CONSTANT-FAULT DIRECT TREES

4.1 Span Program Algorithm

Span programs are linear algebraic representations of boolean functions, which have an intimate relationship with quantum algorithms. In particular, Reichardt proves [12] that given a span program P for a function f, there is a function of the span program, called the witness size, such that one can create a quantum algorithm for f with query complexity $Q(f)$ such that

$$Q(f) = O(\text{WITNESS SIZE}(P)) \qquad (7)$$

Thus, creating a span program for a function is equivalent to creating a quantum query algorithm.

There have been many iterations of span program quantum algorithms, due to Reichardt and others [2,12,7]. In [8], Zhan et al. create span programs for direct boolean functions [8] using the span program formulation described in Definition 2.1 in [12], one of the earliest versions (we will not go into the details of span programs in this paper). Using the more recent advancements in span program technology, we show here:

Theorem 2. *Given an evaluation tree composed of the direct boolean function f, with the promise that the tree satisfies the k-fault condition, (k a natural number), there is a quantum algorithm that evaluates the tree using $O(w^k)$ queries, where w is a constant that depends on f. In particular, for a CONSTANT-FAULT DIRECT TREE, (k a constant), the algorithm requires $O(1)$ queries.*

Properties of direct boolean functions and precise definitions for the k-fault condition can be found in the full version, as well as the proof of Theorem 2. The proof combines the properties of the witness size of direct boolean functions with a more current version of span program algorithms, due to Reichardt [12]. (For more details on direct boolean functions, see [8].)

Thus, while Theorem 1 promises the existence of $O(1)$ query quantum algorithms for CONSTANT-FAULT DIRECT TREES, Theorem 2 gives an explicit $O(1)$ query quantum algorithm for these problems.

4.2   Quantum Haar Transform Algorithm

In this section we will describe a quantum algorithm for solving the 1-FAULT NAND TREE in a single query when there is exactly one fault node in each path from the root to a leaf, and all those faults occur at the same level, as in Figure 2b. We call this problem the HAAR TREE.

Let's consider the values of the leaves on a HAAR TREE. When there are no faults in a NAND TREE, as in Figure 2a, then all even depth nodes have the same value as the root, and all odd depth nodes have the opposite value. Since faults can only occur at nodes with value 1 (since NAND$(0, 1) = 1$), the level of the tree containing faults must occur at even depth if the root has value 1 or at odd depth if the root has value 0. Thus if all the faults are at height h (so their depth is $n - h$), then the value of the root is PARITY$(n - h + 1)$.

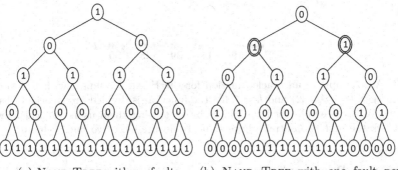

(a) NAND TREE with no faults   (b) NAND TREE with one fault per path

Fig. 2. Figure (a) shows a NAND TREE with no faults, and Figure (b) shows a HAAR TREE. In Figure (a), at each depth, all nodes have the same value, depending on the parity of the level. In Figure (b), since the root is 0, the level of faults occurs at odd depth. (Faults are double circled.) The first half of the leaves descending from a fault node have one value, and the next half have the opposite value.

Now consider the leaves descending from a fault node v when there are no further faults at any nodes descending from v (as in Figure 2b). If v is at height h, then it has 2^h leaves descending from it. Because one of $v's$ children has value 0, and one has value 1, the 2^{h-1} leaves descending from one child will all have the same value, b, and the 2^{h-1} leaves descending from the other child will have the value ¬b. For a HAAR TREE, since we are promised all faults are at the same height h, the values of the leaves will come in blocks of 2^h, where within each block, the first 2^{h-1} leaves will have one value, and the next 2^{h-1} leaves will have the negation of the value in the first set of leaves.

We can now reformulate the HAAR TREE outside of the context of boolean evaluation trees. We define a new problem, the HAAR PROBLEM, to which the HAAR TREE reduces. For the HAAR PROBLEM, one is given access to an oracle for a function $x : \{0, \ldots, 2^n - 1\} \rightarrow \{0, 1\}$. We call the ith output of the oracle x_i. The function x is promised to have a certain form: there exists an integer $h^* \in \{1, \ldots, n\}$ and boolean variables b_l for $l \in \{1, \ldots, 2^{n-h^*}\}$ such that

$$x_i = \begin{cases} b_l, & \text{if } 2^{h^*}(l - 1) \le i < 2^{h^*}(l - \frac{1}{2}) \\ \neg b_l, & \text{if } 2^{h^*}(l - \frac{1}{2}) \le i < 2^{h^*}l. \end{cases} \tag{8}$$

See Figure 3 for an example of a HAAR PROBLEM oracle.

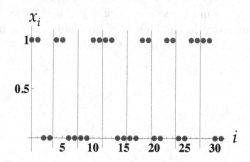

Fig. 3. An example of an oracle function for the HAAR PROBLEM with $n = 5$ (so i is an integer, $0 \le i < 32$) and $h = 2$ (so the function is divided into blocks of length $2^2 = 4$). We have emphasized the blocks by separating them using vertical lines. In each block the first two outputs have value 1 and the next two have value 0, or vice versa.

The HAAR PROBLEM is almost like period finding. We are promised that the function is divided into blocks of length 2^{h^*}, and we need to find the length of these blocks. But instead of the output being the same in each block, each block has one degree of freedom: within the lth block, there is a choice of $b_l = 0$ or $b_l = 1$, where the first half of the outputs have value b_l, and second half have value ¬b_l.

Note that any oracle for the HAAR PROBLEM is also an oracle for the HAAR TREE; to solve the HAAR TREE, simply solve the HAAR PROBLEM, and then calculate PARITY($n - h^* + 1$).

The quantum algorithm for solving the HAAR PROBLEM requires making a measurement in the Haar wavelet basis [17,18]. The Haar basis is based on the following step-like function:

$$\psi(t) = \begin{cases} 1 & \text{if } 0 \le t < 1/2 \\ -1 & \text{if } 1/2 \le t < 1 \\ 0 & \text{otherwise.} \end{cases} \tag{9}$$

On the 2^n dimensional Hilbert space, with standard basis states $\{|i\rangle\}$, $i \in \{0, \ldots, 2^n-1\}$, the (un-normalized) Haar basis consists of the states $\{|\phi_0\rangle, |\psi_{h,l}\rangle\}$:

$$|\phi_0\rangle = \sum_{j=0}^{2^n-1} |i\rangle, \qquad |\psi_{h,l}\rangle = \sum_{i=0}^{2^n-1} \psi(2^{-h}i - (l-1))|i\rangle \tag{10}$$

where $h \in \{1, \ldots, n\}$ and $l \in \{1, \ldots, 2^{n-h}\}$. Several Haar basis states for $n = 3$ are shown in Figure 4a.

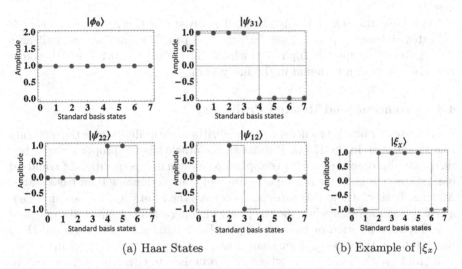

(a) Haar States   (b) Example of $|\xi_x\rangle$

Fig. 4. Figure (a) shows four of the eight un-normalized Haar basis states for $n = 3$. The x-axis depicts the standard basis states $\{|0\rangle, |1\rangle, \ldots, |7\rangle\}$, while the y-axis shows the un-normalized amplitude corresponding to each basis state. The line graphs represent the underlying functions $\psi(2^{-h}i - (l-1))$ that give the states their form, while the amplitudes themselves are represented by dots. Figure (b) shows $|\xi_x\rangle$ for x with $n = 3$, $h^* = 2$, $b_0 = 1$, and $b_1 = 0$, plotted as a function of the non-normalized amplitude of each standard basis state.

We suppose that we have access to a phase-flip oracle O_x such that $O_x|i\rangle = (-1)^{x_i}|i\rangle$ where $\{x_i\}$ satisfy the promise of the HAAR PROBLEM oracle. Then the following algorithm solves the HAAR PROBLEM in one query:

(1) Create an equal superposition of standard basis states: $|\xi\rangle = \dfrac{1}{\sqrt{2^n}} \displaystyle\sum_{j=0}^{2^n-1} |i\rangle$

(2) Apply the phase flip oracle, giving $|\xi_x\rangle = \dfrac{1}{\sqrt{2^n}} \displaystyle\sum_{j=0}^{2^n-1} (-1)^{x_i} |i\rangle$

(3) Measure $|\xi_x\rangle$ in the Haar basis. If the state $|\psi_{h,l}\rangle$ is measured, return h.

It is especially easy to see why the algorithm works graphically. Suppose we are given an oracle x with $n = 3$ and $h^* = 2$. Then $|\xi_x\rangle$ (the state in step (2) of the algorithm) is a superposition of all standard basis states, with amplitudes as shown, for example, in Figure 4b. One can see by comparing the graphs in Figure 4a and Figure 4b that the amplitudes completely destructively interfere for the inner product of $|\xi_x\rangle$ and any Haar basis states except $\{|\psi_{2,l}\rangle\}$ (since here $h^* = 2$).

Classically, the HAAR PROBLEM can be solved in $\tilde{\Theta}(\log n)$ queries, where $\tilde{\Theta}$ indicates tightness up to $\log\log$ factors. The proof of this fact, as well as a description of a subset of inputs on which the 1-FAULT NAND TREE becomes classically easy, can be found in the full version.

4.3   Extensions and Related Problems

There are other oracle problems whose algorithms naturally involve the quantum Haar transform. In the HAAR PROBLEM, the oracle has the property that when the phase flip oracle operation is applied to an equal superposition of standard basis states, the outcome is a superposition of non-overlapping Haar basis states. All Haar basis states in this superposition have the form $|\psi_{h^*,l}\rangle$. One can design a new oracle such that when the the phase flip operation is applied, the outcome is still a superposition of non-overlapping Haar basis states, but now all Haar basis states in the superposition share a new common feature. For example, they could all have the form $|\psi_{h_j,l}\rangle$ where h_j is promised to either be even or odd. In this case, the goal would be to determine whether $\{h_j\}$ are even or odd, and a single quantum query in the Haar basis will give the answer.

This new promise problem (determining whether $\{h_j\}$ are even or odd) is equivalent to solving a 1-FAULT NAND TREE where each path from the root to the leaves contains exactly one fault, but those faults are now not all on the same level.

The HAAR PROBLEM is closely related to the PARITY PROBLEM introduced by Bernstein and Vazirani [19]. Let $x : \{0,1\}^n \to \{0,1\}$ such that $x_i = i \cdot k$ where $k \in \{0,1\}^n$. Then the PARITY PROBLEM is: given an oracle for x, find k. The PARITY PROBLEM can also be solved in a single quantum query.

Notice that any oracle that satisfies the promise required by the PARITY PROBLEM also satisfies the promise required by the HAAR PROBLEM (although the converse is not true). The algorithm for the PARITY PROBLEM is similar to the quantum Haar transform algorithm described in Section 4.2, except in step (3), one measures in the Hadamard basis rather than the Haar basis, and obtains the output k. It is not hard to show that the Bernstein and Vazirani algorithm can also be used to solve the HAAR PROBLEM; the value of h^* is the location of the first non-zero bit of the outcome of the PARITY PROBLEM, counting from least significant to most significant bits. While both the HAAR and PARITY PROBLEMS are similar, the HAAR PROBLEM has a less stringent promise, and is slightly more natural, when viewed as finding the period of a function with some freedom within each period.

5   Conclusions and Future Work

We describe a method to upper bound the quantum query complexity of boolean functions using the general adversary bound. Using this method, we show that CONSTANT-FAULT DIRECT TREES can always be solved in $O(1)$ queries. Furthermore, we create an algorithm with a matching upper bound using improved span program technology. For the more restricted case of the HAAR TREE we give a single query algorithm using a reduction to the HAAR PROBLEM. The HAAR PROBLEM is a new oracle problem that can be solved in a single quantum query using the quantum Haar transform, but which requires $\Omega(\log n)$ classical queries to solve. This problem seems to fall somewhere in between the PARITY PROBLEM of Bernstein and Vazirani [19] and period finding. Period finding has been shown to have useful applications, most notably in factoring [20]. Thus we hope that a new application for the HAAR PROBLEM or the quantum Haar transform can be found. In particular, the fact that the quantum Haar transform can be used find the length of blocks in the HAAR PROBLEM, while ignoring the extra degree of freedom in each block, seems like a useful property.

We would like to find other examples where Theorem 1 is useful, although we suspect that CONSTANT-FAULT DIRECT TREES are a somewhat unique case. Our span program algorithm suggests that Theorem 1 will not be useful for composed functions where the base function is created using span programs. However, there could be other types of quantum walk algorithms, for example, to which Theorem 1 might be applied. In any case, this work suggests that new ways of upper bounding the general adversary bound could give us a second window into quantum query complexity beyond algorithms.

Acknowledgements. Thanks to Rajat Mittal for generously explaining the details of the composition theorem for the general adversary bound. Thanks to the anonymous FOCS reviewer for pointing out problems with the previous version, and also for encouraging me to find a constant query span program algorithm. Thanks to Bohua Zhan, Avinatan Hassidim, Eddie Farhi, Andy Lutomirski, Paul Hess, and Scott Aaronson for helpful discussions. This work was supported by NSF Grant No. DGE-0801525, *IGERT: Interdisciplinary Quantum Information Science and Engineering* and by the U.S. Department of Energy under cooperative research agreement Contract Number DE-FG02-05ER41360.

References

1. Hoyer, P., Lee, T., Spalek, R.: Negative weights make adversaries stronger. In: Proc. 39th ACM STOC 2007, pp. 526–535 (2007)
2. Lee, T., Mittal, R., Reichardt, B., Spalek, R., Szegedy, M.: Quantum query complexity of state conversion. In: Proc. 52nd IEEE FOCS 2011, pp. 344–353 (2011)
3. Ambainis, A.: Quantum lower bounds by quantum arguments. In: Proc. 32nd ACM STOC 2000, pp. 636–643 (2000)
4. Ambainis, A.: Polynomial degree vs. quantum query complexity. J. Comput. Syst. Sci. 72(2), 220–238 (2006)
5. Ambainis, A., Magnin, L., Roetteler, M., Roland, J.: Symmetry-assisted adversaries for quantum state generation. In: Proc. 24th IEEE CCC 2011, pp. 167–177 (2011)
6. Hoyer, Neerbek, Shi: Quantum complexities of ordered searching, sorting, and element distinctness. Algorithmica 34, 429–448 (2008)
7. Reichardt, B.W., Spalek, R.: Span-program-based quantum algorithm for evaluating formulas. In: Proc. 40th ACM STOC 2008, pp. 103–112 (2008)
8. Zhan, B., Kimmel, S., Hassidim, A.: Super-polynomial quantum speed-ups for boolean evaluation trees with hidden structure. In: Proc. 3rd ACM ITCS 2012, pp. 249–265 (2012)
9. Park, S., Bae, J., Kwon, Y.: Wavelet quantum search algorithm with partial information. Chaos, Solitons and Fractals 32(4), 1371–1374 (2007)
10. Hoyer, P.: Quantum ordered searching. Abstract from Talk at QIP (2001) (unpublished)
11. Liu, Y.K.: Quantum algorithms using the curvelet transform. In: Proc. 41st ACM STOC 2009, pp. 391–400 (2009)
12. Reichardt, B.W.: Span programs and quantum query complexity: The general adversary bound is nearly tight for every boolean function. In: Proc. 50th IEEE FOCS 2009, pp. 544–551 (2009)
13. Farhi, E., Goldstone, J., Gutmann, S.: A quantum algorithm for the hamiltonian nand tree. Theory of Computing 4(1), 169–190 (2008)
14. Childs, A.M., Cleve, R., Jordan, S.P., Yeung, D.: Discrete-query quantum algorithm for NAND trees. Theory of Computing 5(1), 119–123 (2009)
15. Reichardt, B.W.: Reflections for quantum query algorithms. In: Proc. 22nd ACM-SIAM SODA 2011, pp. 560–569 (2011)
16. Saks, M., Wigderson, A.: Probabilistic boolean decision trees and the complexity of evaluating game trees. In: Proc. 27th IEEE FOCS 1986, pp. 29–38 (1986)
17. Haar, A.: On the Theory of Orthogonal Function Systems. Mathematische Annalen 69(3), 331–371 (1910)
18. Nievergelt, Y.: Wavelets Made Easy. Birkhäuser, Washington (1999)
19. Bernstein, E., Vazirani, U.: Quantum complexity theory. In: Proc. 25th ACM STOC 1993, pp. 11–20 (1993)
20. Shor, P.W.: Algorithms for quantum computation: discrete logarithms and factoring. In: Proc. 35th IEEE FOCS 1994, pp. 124–134. IEEE Computer Society (1994)

Solving PLANAR k-TERMINAL CUT in $O(n^{c\sqrt{k}})$ Time

Philip N. Klein[1,*] and Dániel Marx[2,**]

[1] Computer Science Department,Brown University,Providence, RI
klein@brown.edu
[2] Computer and Automation Research Institute, Hungarian Academy of Sciences (MTA
SZTAKI), Budapest, Hungary
dmarx@cs.bme.hu

Abstract. The problem PLANAR k-TERMINAL CUT is as follows: given an undirected planar graph with edge-costs and with k vertices designated as terminals, find a minimum-cost set of edges whose removal pairwise separates the terminals. It was known that the complexity of this problem is $O(n^{2k-4}\log n)$. We show that there is a constant c such that the complexity is $O(n^{c\sqrt{k}})$. This matches a recent lower bound of Marx showing that the $c\sqrt{k}$ term in the exponent is best possible up to the constant c (assuming the Exponential Time Hypothesis).

1 Introduction

MULTIWAY CUT (also called MULTITERMINAL CUT) is a generalization of the classical minimum $s - t$ cut problem: given a undirected graph G with edge-costs and given a subset T of k vertices specified as terminals, the task is to find a minimum-cost set of edges whose deletion pairwise separates the k terminal vertices from each other. The study of the computational complexity of this problem was initiated almost thirty years ago in a widely circulated paper by Dahlhaus, Johnson, Papadimitriou, Seymour, and Yannakakis (eventually published [4,5]). They showed the problem is NP-hard even for $k = 3$, and they gave a 2-approximation algorithm, which has since been improved [1,3,8].

They showed that if k can be arbitrarily large, even the restriction to planar graphs is NP-hard. Therefore, for each positive integer k, they consider the problem PLANAR k-TERMINAL CUT and give an algorithm with a running time of $O((4k)^k n^{2k-1}\log n)$. This bound was since improved by roughly a factor of n^3, to $O(k4^k n^{2k-4}\log n)$, by Hartvigsen [6].[1]

We show that the dependence on k of the exponent of n can be improved from $2k - 4$ to $c\sqrt{k}$ for a constant c. In particular, we give an algorithm with running time $d^k \cdot n^{c\sqrt{k}}$ for constants c, d. This shows that the complexity of PLANAR k-TERMINAL CUT is $O(n^{c\sqrt{k}})$. A companion paper [9] shows that this is best possible (up to the particular constant c), assuming the Exponential Time Hypothesis [7].

* Supported in part by National Science Foundation Grant CCF-0964037.
** Research supported by the European Research Council (ERC) grant "PARAMTIGHT: Parameterized complexity and the search for tight complexity results," reference 280152.

[1] The much simpler algorithm of [10] is incorrect; see [2].

A. Czumaj et al. (Eds.): ICALP 2012, Part I, LNCS 7391, pp. 569–580, 2012.

Dahlhaus et al. observed that a solution of PLANAR MULTIWAY CUT in the *dual* graph is a planar graph with $O(k)$ branch vertices connected by paths. Thus an algorithm can guess the branch vertices of this planar graph in the dual in time $n^{O(k)}$ and then find min-cost paths between them, subject to constraints about enclosing terminals–constraints that are not readily incorporated into shortest-path computation. Dahlhaus et al. achieve their result by exploiting some structural properties of these paths. Our approach is very different: Our algorithm computes a min Steiner tree on the terminals in the dual graph and cuts the plane open along this tree, thereby forming a cycle on which all the terminals lie, and adds zero-cost edges inside the cycle We prove that there is an optimum solution that uses $O(k)$ zero-cost edges; thus the solution after cutting the tree open can still be described by a planar graph having $O(k)$ vertices and therefore treewidth $O(\sqrt{k})$. Since all the terminals lie on a cycle, the topological constraint that certain paths enclose certain terminals can be completely expressed by requiring that the paths cross the cycle in a certain order. Therefore dynamic programming on a tree decomposition suffices to find the solution in the cut-open graph.

2   Preliminaries

Let G be an undirected graph. For a set X of vertices, $\delta_G(X)$ denotes the set of edges uv such that $u \in X, v \notin X$. Such a set is called a *cut*. A cut is *simple* if both X and $V(G) - X$ induce connected components. For nodes u, v in G, a set S of edges *separates* u and v in G if every u-to-v path includes an edge of S.

Fact 2.1 *S separates u, v iff there is a cut $\delta_G(X)$ such that $u \in X, v \notin X$ and $\delta_G(X) \subseteq S$; moreover, the cut can be chosen to be simple.*

We assume basic knowledge of the definitions of *planar embedded graph*, *faces*, and the planar dual. Let G be a connected planar embedded graph, and let G^* be its dual.

Fact 2.2 *Edge-set S forms a simple cut in G iff S forms a simple cycle in G^*.*

Definition 2.3. *For nodes v_1, v_2 of G, edge-set S dual-separates v_1 and v_2 in G if S does not include any edge incident to v_1 or v_2, and, for a face f_1 incident to v_1 and a face f_2 incident to v_2, S separates f_1 and f_2 in the planar dual G^*.*

Lemma 2.4. *If S dual-separates v_1 and v_2 in G then G contains a simple cycle of edges in S that dual-separates v_1 and v_2.*

Proof. For $i = 1, 2$, let f_i be a face of G incident to v_i. By Fact 2.1, S contains the edges of a simple cut in the planar dual G^* that separates f_1 and f_2 in G^*. By Fact 2.2, the edges of this simple cut form a simple cycle in G.   □

Definition 2.5. *For edge-set S, let H^* be the subgraph of G^* consisting of S. Each face f of H^* corresponds to a collection X_f of faces of G^* (those embedded in f). We say f encloses the faces in X_f. For x a vertex or edge of H^*, we say f encloses x if f encloses all the faces that have x on their boundary. If f is not the infinite face, we consider the faces and vertices enclosed by f to be also enclosed by H^*.*

3   Reducing the Problem to the Biconnected Case

For a pair (G,T) where G is an undirected graph and T is a subset of vertices (the *terminals*), an T-*mcut* (a multiway cut with respect to terminal set T) is a set S of edges such that $G - S$ contains no path between distinct terminals. For disjoint subsets $X, Y \subset T$, we define an (X,Y)-*mcut* to be a set S of edges such that $G - S$ contains no path between vertices of X and no path from X to Y.

For a planar embedded graph G, we say a pair (X,Y) of sets of vertices is *biconnectivity-inducing* in G if every minimum-cost (X,Y)-mcut forms a biconnected subgraph of G^*.

Fix a planar embedded graph G_{in} with positive edge-costs and n vertices. We define two problems:

- *Problem A:* given a set T of k vertices, find a minimum-cost T-mcut.
- *Problem B:* given a pair (X,Y) of vertex-sets where $k = |X| + |Y|$, find an (X,Y)-mcut S such that if (X,Y) is a biconnectivity-inducing pair, then S is guaranteed to be a minimum-cost (X,Y)-mcut.

We show that Problem A can be solved by 2^k calls to an algorithm for Problem B, plus additional $O(3^k)$ time. Let $a(T)$ be the minimum cost of a multiway cut for terminal set T. Let $b(X,Y)$ be a function such that

- if (X,Y) is 2-connectivity-inducing, then $b(X,Y)$ is the minimum cost of an (X,Y)-mcut, and
- otherwise, $b(X,Y)$ is the cost of *some* (X,Y)-mcut.

We use a dynamic program based on the recurrence relation

Lemma 3.1. $a(T) = \min_{\emptyset \neq X \subseteq T} b(X, T - X) + a(T - X)$

Proof. It is trivial that the left-hand side is at most the right hand side: the $(X, T - X)$-mcut and the multiway cut of $T - X$ together gives a multiway cut for T. Our goal is to show that the left-hand side is at least the right-hand side.

We generalize the notion of a multiway cut as follows. Let $X_1, \dots, X_p$ be a partition of T (p is arbitrary). An $(X_1, \dots, X_p)$-mcut is a tuple $(S_1, \dots, S_{p-1})$ of mutually disjoint edge-sets of G such that, for $i = 1, \dots, k-1$, $G - S_i$ contains no path between distinct nodes of X_i and no path from a node in X_i to a node in $X_{i+1} \cup X_{i+2} \cup \dots \cup X_p$. If X_p is singleton then $S_1 \cup \dots \cup S_{p-1}$ is a multiway cut separating all terminals in T.

The cost of a tuple $(S_1, \dots, S_p)$ is the sum of costs of the edges. We say a partition $X_1, \dots, X_p$ is *perfect* if $|X_p| = 1$ and the minimum-cost of an $(X_1, \dots, X_p)$-mcut equals $a(T)$. Observe that a perfect partition always exist: in particular, $(T - \{t\}, \{t\})$ is a perfect partition for every $t \in T$.

Among all perfect partitions of T, let $\hat{X}_1, \dots, \hat{X}_p$ be the finest, and let $(\hat{S}_1, \dots, \hat{S}_{p-1})$ be a minimum $(\hat{X}_1, \dots, \hat{X}_p)$-mcut. We claim that $(\hat{X}_1, \hat{X}_2 \cup \dots \cup \hat{X}_p)$ is 2-connectivity-inducing. Indeed, if $(\hat{X}_1, \hat{X}_2 \cup \dots \cup \hat{X}_p)$ were not 2-connectivity-inducing—if there were a minimum-cost solution S that was not 2-connected in the dual—the partition $\hat{X}_1, \dots, \hat{X}_p$ could be refined by breaking $\hat{X}_i$ into two parts according to the 2-connected components of S in the dual.

As $(\hat{S}_1, \dots, \hat{S}_{p-1})$ has cost $a(T)$, we have that $a(T)$ is at least the sum of the cost of an $(\hat{X}_1, \hat{X}_2 \cup \dots \cup \hat{X}_p)$-mcut and the cost of a multiway cut for $\hat{X}_2 \cup \dots \cup \hat{X}_p$. By the claim

Fig. 1. Illustrates the reduction. The lines are the edges in the planar dual of
a minimum-cost $(\hat{X}, \hat{Y})$-mcut. The disks represent terminals. The thin lines
represent $\hat{S}$, and the small disks are the terminals enclosed by $\hat{S}$.

Fig. 2. Each terminal is replaced by a cycle. The size of the cycle is
the original degree of the terminal, and the the edges forming the cycle
all have cost M.

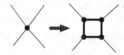

in the previous paragraph, $(\hat{X}_1, \hat{X}_2 \cup \cdots \cup \hat{X}_p)$ is 2-connectivity-inducing, thus the first
term is at least $b(\hat{X}_1, \hat{X}_2 \cup \cdots \cup \hat{X}_p)$. The second term is at least $a(\hat{X}_2 \cup \cdots \cup \hat{X}_p)$. Thus
with the choice $X = \hat{X}_1$ shows that the left-hand side is at least the right-hand side.   □

4   Algorithm for Problem B

Here is pseudocode for the algorithm for Problem B.

Procedure BSOLVE(G_{in}, X, Y):
input: planar graph G_{in}, pair of disjoint terminal sets (X, Y)
output: (X, Y)-mcut that is min-cost if (X, Y) is biconnectivity-inducing.

Let M be a number greater than the sum of all costs
0 For each terminal t,
1   replace t by a size-degree(t) cycle of edges of cost M
      let t^* (called the *rep* of t) denote the face thus formed
Let $\hat{G}_{in}$ be the resulting graph and let $\hat{G}_{in}^*$ denote its planar dual
2 In $\hat{G}_{in}^*$, find min-cost Steiner tree T^* connecting all terminal reps
3 Replace each edge of T^* with two copies, and replace each node v on T^*
      with degree(v) copies connected by a star of $d - 1$ zero-cost edges
4 Let G_1 denote the planar embedded graph derived in this way from $\hat{G}_{in}^*$
5 Let $C(G_1)$ be the cycle in G_1 formed by copies of edges of T^*
   · Label terminal reps by $1, 2, \ldots, k$ in clockwise order about $C(G_1)$
6 return the minimum-cost set in
      $\{\text{RE}(H, M, G_1) : (H, M) \text{ an } X\text{-valid representative topology}, |M| \leq \beta k\}$

Line 1 is illustrated in Fig. 2. Line 3 is illustrated in Figures 3 and 4. In Line 6, β is a
constant to be determined.

Line 6 uses the notion of *topology* and the procedure $\text{RE}(H, M, G_1)$. We will presently
define this notion. The basic idea underlying the procedure BSOLVE is to enumerate
topologies and, for each, find the minimum-cost solution "consistent" with that topology.

• Of course, the procedure cannot enumerate *all* topologies. We will define what it
means for topologies to be isomorphic; the procedure will enumerate representatives of
distinct isomorphism classes.

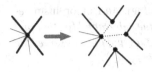

Fig. 3. Figure shows part of graph before and after duplicating tree edges (thick edges). Node v on tree is replaced by degree(v) copies connected by a zero-cost star. Graph edges not in tree remain incident to copies of v so as to preserve the embedding.

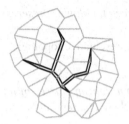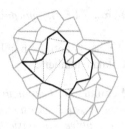

Fig. 4. Cutting along T^* and adding new (dotted) zero-cost edges between copies of the vertices

- Furthermore, we will show it suffices that the procedure consider only representative topologies of *small size*, and that there are not too many such topologies.
- We describe a property, X-*validity*, that captures what a topology must do in order to correspond to an (X,Y)-mcut. The procedure considers only valid representative topologies.
- In Line 6, the procedure RE is invoked on each valid small representative topology. We would like to say that RE finds a minimum-cost topology in G_1 that is isomorphic to the valid representative topology. This is not necessarily true; instead, the procedure finds a valid solution in G_1 whose cost is no greater than the minimum cost of a topology in G_1 isomorphic to the representative.

Definition 4.1. A label structure *is a planar embedded graph H containing*
- *a simple cycle $C(H)$ that strictly encloses no nodes, and*
- *a subset of nodes of $C(H)$ labeled $1, 2, \ldots, k$ in clockwise order along the   cycle* (the *terminal reps, short for representatives*).

Note: *The graph G_1 with the cycle $C(G_1)$ in Lines 4-5 is a label structure.*

Let H be a label structure and let M be a subset of edges. We say M is a feasible solution *for H if no edges of M are incident to labeled nodes. For a subset $X \subset \{1, \ldots, k\}$, we say M is X-valid for H if M dual-separates every element of X from every other labeled node in H.*

Let $M_1 =$ edges strictly enclosed by $C(H)$ and $M_2 = M - M_1$. We say (H, M) is a topology in H *if, for $i = 1, 2$, the edges of M_i form a forest with leaves on $C(H)$. The size of (H, M) is $|V(H)|$.*

Definition 4.2. *For a topology (G_1, M_1), where G_1 is the graph obtained in Line 4, the* solution induced in G_{in} *is the set of edges of M_1 that are in G_{in} (including edges of T^* with copies in M_1).*

The definition of dual-separates implies the following lemma.

Lemma 4.3. *An X-valid topology induces an $(X, \{1, \ldots, k\} - X)$-mcut.*

Definition 4.4. *Suppose that, for $i = 1, 2$, (G_i, M_i) is a topology. An* isomorphism *between (H_1, M_1) and (H_2, M_2) is a homeomorphism between the subgraph M_1 of H_1 and the subgraph M_2 of H_2 that maps interior edges to interior edges and that preserves the order on the cycle of $\{endpoints\ of\ interior\ edges\} \cup \{labeled\ nodes\}$.*

Lemma 4.5. *Isomorphism between topologies preserves X-validity.*

We can bound the number of representative topologies considered in Line 6 by using Catalan numbers:

Lemma 4.6. *The number of isomorphism classes of topologies of size at most s is at most α^s, and representatives of these classes can be enumerated in $O(\alpha^s)$ time, where α is a universal constant.*

This bound depends on the size of the topologies considered; the following theorem, proved in Section 5, states that only small ones need be considered.

Theorem 4.7. *If (X, Y) is biconnectivity-inducing then there is an X-valid topology (G, M) of size at most βk that is isomorphic to a topology in G_1 whose cost is at most that of an optimal (X, Y)-mcut in G_{in}, where β is a universal constant.*

The following theorem is proved in Section 6.

Theorem 4.8. *There is a procedure $\mathrm{RE}(H, M, G_1)$ that returns a feasible solution M_1 with the following properties:*
1) If M is X-valid for H then M_1 is X-valid for G_1.
2) If there is a topology (G_1, M_1') isomorphic to (H, M) then M_1 is no more costly than M_1'.
3) The time required is at most $n^{\gamma \sqrt{|V(H)|}}$ for a constants γ.

Finally, putting these results together, we obtain

Theorem 4.9. $\mathrm{BSOLVE}(G_{\mathrm{in}}, X, Y)$ *finds an (X, Y)-mcut in G_{in} that is optimal if (X, Y) is biconnectivity-inducing, and the procedure takes at most $\alpha^{\beta k} n^{c \sqrt{\beta k}}$ time.*

Proof. By Property 1 of Theorem 4.8, BSOLVE returns an X-valid topology of G_1, which by Lemma 4.3 induces an (X, Y)-mcut. We choose the constant β in Line 6 according to Theorem 4.7. Therefore, there exists some small X-valid topology (G, M), among those considered in Line 6, that is isomorphic to a topology (G_1, M_1') in G_1 that induces an optimal (X, Y)-mcut. Therefore, by Property 2 of Theorem 4.8, $\mathrm{RE}(G, M, G_1)$ returns a feasible solution M_1 for G_1 whose cost is at most that of M_1' and therefore at most the optimal cost of an (X, Y)-mcut. The running time is dominated by having to call RE at most $\alpha^{\beta k}$ times (Lemma 4.6), each taking time $n^{\gamma \sqrt{\beta k}}$. □

This theorem plus the reduction to the biconnected case yields our main result, an algorithm for planar k-terminal cut that requires $O(d^k n^{c \sqrt{k}})$ time.

5   Proof of Theorem 4.7

Suppose (X,Y) is biconnectivity-inducing in G_{in}, and let $S \subset E(G_{\text{in}})$ be a minimum-cost (X,Y)-cut in G_{in}, breaking ties by minimizing the number of edges not in T^*. Because of the transformation of Line 3 of BSOLVE, the edges of S alone do not dual-separate terminals in G_1, so S is not X-valid for G_1: some zero-cost edges are needed. For a set A of external edges of G_1, define $\text{cr}(A)$ as follows: if A contains edges incident to different copies of the same node of G_{in}^*, include in $\text{cr}(A)$ the internal edges forming a simple path between the different copies. We refer to the edges in $\text{cr}(A)$ as *crossings*.

Lemma 5.1. *For any set A of external edges of G_1, if A induces the solution S in G_{in} then $A \cup \text{cr}(A)$ is X-valid for G_1.*

Among all sets A that induce S, let A_S be one that minimizes $|\text{cr}(A)|$. Without loss of generality, we assume that A_S does not include more than one copy of an edge of S. If G_1 contained a cycle consisting of edges of A_S then G_{in}^* would contain a cycle C consisting of edges of S such that C did not enclose any terminal, so S would not be minimum. Thus A_S is a forest in G_1. A similar argument shows that all the leaves of A_S are endpoints of $\text{cr}(A_S)$. Thus $(G_1, A_S \cup \text{cr}(A_S))$ is a topology in G_1, and it is X-valid by Lemma 5.1. Moreover, since the number of leaves is $\leq 2|\text{cr}(A_S)|$, at most $2|\text{cr}(A_S)|$ nodes have three or more incident edges in A_S. This implies that there is a topology (H,M) isomorphic to $(G_1, A_S \cup \text{cr}(A_S))$ of size at most $3|\text{cr}(A_S)|$. We next show $|\text{cr}(A_S)| \leq 24k$, which implies Theorem 4.7.

Define a *branchpoint* of a graph to be a node of degree three or greater. We refer to the edges of S as *red* edges, and to the subgraph of $\hat{G}_{\text{in}}^*$ they form as the *red graph*. We refer to its faces as *red faces*. The *red degree* of a node of $\hat{G}_{\text{in}}^*$ is the number of incident red edges. We use *spliced red graph* to refer to the graph obtained from the red graph by splicing out degree-two vertices. By minimality of S, each face of the red graph encloses at least one terminal. Euler's formula then implies $e \leq 3(k-2)$, so the sum of degrees of branchpoints of the red graph is at most $6(k-2)$.

Recall that T^* is a minimum Steiner tree in $\hat{G}_{\text{in}}^*$, which we call the *blue graph*. (The red and the blue graphs can share edges.) Each leaf is a terminal rep, so there are k leaves, so the spliced blue graph has at most $2k-3$ edges, so the sum of degrees of branchpoints in the unspliced blue graph is at most $2(2k-3)$.

For a singular red face R, define a *blue ear of R* to be a path B of blue edges such that B connects two nodes on the boundary of a singular red face and each internal node of B is strictly enclosed by R and has blue degree two.

We prove the bound on the number of crossings by a charging scheme, where we charge the crossings to the red branch nodes, blue branch nodes, terminals, and blue ears. We already have a bound of $O(k)$ on the total degree of the branch nodes. The following lemma gives a similar bound on the blue ears.

Lemma 5.2. *The number of blue ears of singular red faces is at most $14k$.*

The proof is illustrated in Figure 5. Let R be a red face, and let R' be the graph obtained from R by including the blue ears of R. Let R'' be the graph obtained from R' by splicing out nodes of blue degree two that are strictly enclosed by R. Consider the planar dual

 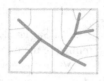 

Fig. 5. Proof of Theorem 5.2. On the left is a singular red face (the box) enclosing some blue edges. In the middle is the subgraph of the dual induced by the enclosed faces; it is a tree. As illustrated by the figure on the right, every tree node of degree zero or two is a face that either encloses a terminal or has a red branchpoint on its boundary.

of R'', and let G_R denote the subgraph of the planar dual consisting of the edges of blue ears. Because every edge of G_R is a cut-edge, we infer that G_R is a tree.

A face of R'' is a *red-blue* face if its boundary consists of a red path and a blue path, and is a *red-blue-red-blue* face if it consists of two red and two blue paths (alternating). The leaves of G_R are red-blue faces in R'', and the degree-two nodes of G_R are red-blue-red-blue faces.

Proposition 5.3. *Every red-blue face either encloses a terminal or has a red branchpoint on its boundary.*

Proof. Suppose PQ is the boundary of a red-blue face, where P is red and Q is blue. If $\text{len}(P) < \text{len}(Q)$ then Q could be replaced in the Steiner tree by P, reducing the length, a contradiction. Therefore $\text{len}(Q) \leq \text{len}(P)$. If PQ does not enclose a terminal and P does not have a branchpoint, replacing P by Q in the optimal solution yields an optimal solution with fewer non-blue edges, a contradiction.   □

Proposition 5.4. *The only red-blue-red-blue faces are those that enclose terminals and those that have red branchpoints on their boundary.*

Fig. 6. Illustrates the proof of Lemma 5.4. The horizontal line segments are red, as is the dashed curve, and the vertical line segments are blue. The solid circle represents a terminal in the same red face as the red-blue-red-blue face.

Proof. Suppose F is a red-blue-red-blue face of R'' that does not enclose a terminal and does not have a red branchpoint on its boundary. See Figure 6. The boundary of F is $pqrs$ where p and r are blue and q and s are red, and p divides R'' into a part enclosing F and a part enclosing a terminal.

If $\text{len}(p) \leq \text{len}(q)$, then replacing q with p in the red path yields a solution that is no more expensive but has fewer non-blue edges, a contradiction. Thus $\text{len}(p) > \text{len}(q)$. Similarly $\text{len}(p) > \text{len}(s)$. Removing the path p from the blue graph yields two disconnected components. If the one not containing r contains the intersection of p with s, the graph obtained from the blue graph by replacing p with s is a cheaper solution, a contradiction. The other case is similar.   □

Fig. 7. There are two crossings, u_1u_2 and e. P_1 includes no nodes of multiplicity greater than two and no nodes of red degree greater than two. Therefore the path P_1 can be replaced by the dotted line and the two crossings eliminated.

The proof of Lemma 5.2 now follows from the fact that G_R is a tree and from Prop. 5.4 and Prop. 5.3, which bound the number of leaves and degree-two nodes in terms of terminals and red branchpoints.

To complete the proof of the theorem, we now bound the crossings by charging to branchnodes, blue ears, and terminals.

Recall that G_1^* is obtained from $\hat{G}_{in}^*$ by cutting along the edges of T^*, so every edge of T^* is represented in G_1^* by two copies, and every node u of T^* is represented by a number of copies equal to the degree of u in T^*. The *multiplicity* of one such copy is the number of copies, i.e. the degree of u in T^*. If a copy has multiplicity greater than two then u is a branchpoint of the blue graph. The *red degree* of one such copy is defined to be u's red degree in $\hat{G}_{in}^*$ (so here we may count red edges incident to u that are no longer incident to a given copy of u). Let $u_1u_2 \in \mathrm{cr}(A_S)$. In the following, for each case, we assume the previous cases do not hold. By definition of $\mathrm{cr}(A_S)$, there are red edges incident to u_1 and u_2. For $i = 1, 2$, let P_i be a maximal path, starting with u_i, of edges in G_1^* that are both red and blue, such that every node of P_i except possibly the last has red degree two and multiplicity two.

Case 1: P_1 or P_2 ends at a branchpoint of the red graph. In this case we charge the crossing to the red branchpoint. The number of crossings charged to such a branchpoint is at most the degree of the branchpoint, so at most $6k$ crossings are charged in this way.

Case 2: P_1 or P_2 ends at a node of multiplicity greater than two. In this case, we charge the crossing to the branchpoint of the blue graph. The number of crossings charged to a branchpoint w by this rule is at most the degree of w in the blue graph. Thus the total number of such crossings is at most $4k$.

Case 3: P_1 or P_2 ends at a node with no incident red edge in G_1^*. Since the red edges form a two-connected subgraph of $\hat{G}_{in}^*$, the last node of P_i has red degree two or more. It follows that in G_1^* some $e \in \mathrm{cr}(A_S)$ is incident to the last node of P_i. However, since every node in P_1 and in P_2 has multiplicity at most two, the configuration is as shown in in Figure 7, and the two crossings can be eliminated, a contradiction.

Case 4: For $i = 1$ and $i = 2$, P_i ends at a node v of red degree two and multiplicity two, but the second red edge incident to v and the second blue edge incident to v differ. Let u be the node of $\hat{G}_{in}^*$ whose copies are u_1 and u_2. Since the red edges form a two-connected subgraph of $\hat{G}_{in}^*$, the neighbors of u in this subgraph are connected in the subgraph by a path Q that avoids u. Let Q' be the cycle obtained from Q by adding the red edges incident to u. (See Figure 8.) For $i = 1, 2$, let P_i' be the path obtained from P_i by appending the second red edge incident to the end of P_i.

Because all the nodes of $P_1 \cup P_2$ have red degree two, Q' includes all the edges corresponding to those in $P_1' \cup P_2'$. Let b_i be the blue edge of G_1^* incident to the end of P_i, and let b_i' be the corresponding edge of $\hat{G}_{in}^*$. The cycle Q' shows that b_1' and b_2' are in different faces f_1 and f_2 of the red graph. Because the nodes of $P_1 \cup P_2$ have red

Fig. 8. Case 4. On left, at end of P_i, red path and blue path diverge. Red edge incident to the end of P_i differs from blue edge b_i incident to the end of P_i. Right figure shows $\hat{G}_{in}^*$: a path Q joining the red neighbors of u, forming a cycle Q'. Edges b_1' and b_2' are in different but neighboring red faces.

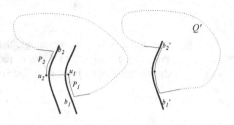

degree two, the edges of $P_1' \cup P_2'$ belong to the boundaries of f_1 and f_2. The faces f_1 and f_2 cannot both be plural faces, else the edges between them could be removed while maintaining feasibility. Assume without loss of generality that f_2 is a singular face. Let B be a maximal path of blue edges starting with b_2' such that every node except the last has blue degree two and is strictly internal to f_2.

Subcase a: The last node of B has blue degree one. That last node is a terminal. We charge the crossing to the terminal. There are at most k crossings thus charged.

Subcase b: The last node of B has blue degree greater than two. We charge the crossing to this blue branchpoint. The number of crossings charged to this branchpoint is at most its degree, so the total number of crossings thus charged is at most $4k$.

Subcase c: B forms a path between two nodes on the boundary of f_2. In this case, we charge the crossing to the blue ear. The number of such ears is bounded by Lemma 5.2.

6   Realization

In this section, we prove Theorem 4.8. Given a topology, we try to find a realization of minimum cost in a label structure:

Definition 6.1. *Let (H, M) be a topology of some label structure H, and let G be another label structure. A realization of (H, M) in G consists of a mapping $\phi_v : V(M) \to V(G)$ and a mapping $\phi_e : E(M) \to 2^{E(G)}$ such that*

- *ϕ_v preserves the order among {endpoints of interior edges} $\cup$ {labeled nodes} on the cycles $C(H)$ and $C(G)$.*
- *For every interior edge $xy \in E(M)$, $\phi_e(xy)$ is an interior edge between $\phi_v(x)$ and $\phi_v(y)$.*
- *For every exterior edge $xy \in E(M)$, $\phi_e(xy)$ is a path of exterior edges between $\phi_v(x)$ and $\phi_v(y)$.*

The cost of a realization is $\sum_{xy \in E(M)} cost(\phi_e(xy))$.

Lemma 6.2. *If (H_1, M_1) and (H_2, M_2) are isomorphic topologies and (H_1, M_1) has a realization of cost R in label structure G, then so does (H_2, M_2).*

The following lemma shows that a realization of a valid topology is indeed a solution:

Lemma 6.3. *Let (H, M) be an X-valid topology. Let (ϕ_v, ϕ_e) be a realization of (H, M) in a label structure G having cost R. Then there is an X-valid set $S \subseteq E(G)$ of weight at most R that is X-valid in G.*

Proof. Let S be the union of the edge sets of the path $\phi_e(uv)$ for every edge $uv \in M$. It is clear that the total cost of the edge set S is at most R, the cost of the realization. We claim that S is X-valid in G. Let $i \in X$ be a labeled node and let $j \neq i$ be some other labeled node. By the definition of valid topology and Lemma 2.4, there is a cycle C_1 in M dual-separating i and j. Replacing each interior edge $uv \in C_1$ with the edge $\phi_e(uv)$ and each exterior edge $uv \in C_1$ with the path $\phi_e(uv)$, we can obtain a closed walk C_2 of G. We claim that C_2 dual-separates i and j in G^*.

Let R_{ij}^1 and R_{ij}^2 be the segment of $C(H)$ (resp., $C(G)$) between labeled nodes i and j in clockwise direction. Let I_1 be the interior edges I_1 with exactly one endpoint on R_{ij}^1. We claim that $|I_1|$ is odd. As C_1 is a simple cycle that dual-separates i and j, there is a dual path Q_1 (i.e., a sequence of faces and edges) in the exterior of H from a face of i to a face of j such that Q_1 contains exactly one edge of C_1. Let R_1 be the set of vertices that can be reached from $R_{ij}^1 - \{i, j\}$ on exterior edges without using an edge of Q_1 or going through i or j. By planarity, R_1 does not contain any vertex of the cycle of H outside R_{ij}^1. A simple parity argument shows that the number of edges in the cycle C_1 with exactly one endpoint in R_1 is even. As C_1 does not go through i and j (by the definition of topology), every such edge is either in Q_1 (there is exactly one such edge) or it is an interior edge with exactly one endpoint in R_{ij}^1. Thus there are exactly $|I_1| + 1$ such edges and hence $|I_1|$ is odd.

Let $I_2 \subseteq E(G)$ contain those edges of S used by C_2 that have exactly one endpoint in R_{ij}^2. Observe that $|I_1| = |I_2|$: by the definition of realization, the order on the cycle is preserved and hence each edge of I_1 is mapped to a distinct edge of I_2. It also follows that C_2 uses each edge of I_2 only once. Suppose that C_2 does not dual-separate i and j: there is a dual path Q_2 in the exterior of G from a face of i to a face of j. Let R_2 be the set of vertices that can be reached from $R_{ij}^2 - \{i, j\}$ on exterior edges of G without using an edge of Q_2 or going through i or j. As C_2 does not go through i and j (by definition of dual-separate) and disjoint from Q_2, only the edges in I_2 have exactly one endpoint in R_2. We have observed that C_2 uses each such edge exactly once and $|I_2| = |I_1|$ is odd, a contradiction.   □

In light of Lemma 6.3, all we need is to find minimum-cost realizations of valid topologies. We will use the following embedding result, whose proof uses standard dynamic programming techniques on tree decompositions.

Theorem 6.4. *Let D be a directed graph, U a set of elements, and functions $cv: V(D) \times U \to \mathbb{Z}^+ \cup \{\infty\}$, $ce: V(D) \times V(D) \times U \times U \to \mathbb{Z}^+ \cup \{\infty\}$. In time $|U|^{O(\text{tw}(D))}$, we can find a mapping $\phi: V(D) \to U$ that minimizes*

$$\sum_{v \in V(D)} cv(v, \phi(v)) + \sum_{(u,v) \in E(D)} ce(u, v, \phi(u), \phi(v)).$$

Lemma 6.5. *Given a topology (H, M) and another label structure G, a minimum-cost realization of (H, M) in G can be found in time $|V(G)|^{O(\sqrt{|V(M)|})}$.*

Proof. Let D be the directed graph obtained as an arbitrary orientation of the subgraph of H spanned by M. For every edge $\vec{xy}$ of D arising from an interior edge of H, we define $ce(x, y, x', y')$ to be 0 if $x'y'$ is an interior edge of G and ∞ otherwise. If $\vec{xy}$ arises from an exterior edge, then $ce(x,y,x',y')$ is the cost of the shortest path from x' to y' in G containing only exterior edges. We introduce some further directed edges as follows. If x,y are two vertices that are endpoints of interior edges of H such that x is between terminal vertices i and $i+1$ on the cycle and y is the next vertex (in clockwise direction) with this property, then we introduce a directed edge $\overrightarrow{xy}$ and define $ce(x,y,x',y')$ to be 0 if terminal i, vertex x', vertex y', terminal $i+1$ follow each other in this order (in clockwise direction) and ∞ otherwise.

If $x \in V(H)$ (resp., $x' \in V(G)$) is an endpoint of an interior edge of H (resp., G) and it is between i and $i+1$ (resp., i' and $i'+1$) on the cycle in clockwise direction, then we define $cv(x,x') = 0$ if $i = i'$ and $cv(x,x') = \infty$ otherwise. If $x \in V(H)$ is not an endpoint of an interior vertex, then we set $cv(x,x') = 0$ for every $x' \in V(G)$.

Let us use the algorithm of Theorem 6.4 to find a mapping ϕ_v. As D is planar, its treewidth is $O(\sqrt{|V(M)|})$. Therefore, the running time of this step is $|V(G)|^{O(\sqrt{|V(M)|})}$. For every interior edge $xy \in E(M)$, we define $\phi_e(xy)$ to be the interior edge $\phi_v(x)\phi_v(y)$, while if $xy \in E(M)$ is exterior, then we define it to be a shortest path between $\phi_v(x)$ and $\phi_v(y)$ using only the exterior edges of G. It is easy to verify that (ϕ_v, ϕ_e) is a realization of (H,M) in G and its cost is the cost of the mapping ϕ. Furthermore, every realization can be transformed into a mapping with the same cost. Thus the realization obtained this way is indeed a minimum-cost realization.   □

To prove Theorem 4.8, the procedure $\mathrm{RE}(G,M,G_1)$ uses the algorithm of Lemma 6.5 to find a minimum-cost realization of (G,M) in G_1. By Lemma 6.3, the result is X-valid. The second statement of Theorem 4.8 follows from Lemma 6.2. The running time follows from the statement of Lemma 6.5.

References

1. Calinescu, G., Karloff, H., Rabani, Y.: An improved approximation algorithm for multiway cut. In: STOC 1998, pp. 48–52 (1998)
2. Cheung, K.K., Harvey, K.: Revisiting a simple algorithm for the planar multiterminal cut problem. Operations Research Letters 38(4), 334–336 (2010)
3. Cunningham, W., Tang, L.: Optimal 3-Terminal Cuts and Linear Programming. In: Cornuéjols, G., Burkard, R.E., Woeginger, G.J. (eds.) IPCO 1999. LNCS, vol. 1610, pp. 114–125. Springer, Heidelberg (1999)
4. Dahlhaus, E., Johnson, D., Papadimitriou, C., Seymour, P., Yannakakis, M.: The complexity of multiway cuts. In: STOC 1992, pp. 241–251. ACM (1992)
5. Dahlhaus, E., Johnson, D.S., Papadimitriou, C.H., Seymour, P.D., Yannakakis, M.: The complexity of multiterminal cuts. SIAM J. Comput. 23(4), 864–894 (1994)
6. Hartvigsen, D.: The planar multiterminal cut problem. Discrete Applied Mathematics 85(3), 203–222 (1998)
7. Impagliazzo, R., Paturi, R., Zane, F.: Which problems have strongly exponential complexity? J. Comput. System Sci. 63(4), 512–530 (2001)
8. Karger, D.R., Klein, P.N., Stein, C., Thorup, M., Young, N.E.: Rounding algorithms for a geometric embedding of minimum multiway cut. In: STOC 1999, pp. 668–678 (1999)
9. Marx, D.: A tight lower bound for planar multiway cut with fixed number of terminals. In: Czumaj, A., et al. (eds.) ICALP 2012, Part I. LNCS, vol. 7391, pp. 677–688. Springer, Heidelberg (2012)
10. Yeh, W.-C.: A simple algorithm for the planar multiway cut problem. J. Algorithms 39(1), 68–77 (2001)

Fixed-Parameter Tractability of Multicut in Directed Acyclic Graphs*

Stefan Kratsch[1,**], Marcin Pilipczuk[2,***], Michał Pilipczuk[3,†],
and Magnus Wahlström[4]

[1] Utrecht University, Utrecht, the Netherlands
s.kratsch@uu.nl
[2] University of Warsaw, Warsaw, Poland
malcin@mimuw.edu.pl
[3] University of Bergen, Bergen, Norway
michal.pilipczuk@ii.uib.no
[4] Max-Planck-Institute for Informatics, Saarbrücken, Germany
wahl@mpi-inf.mpg.de

Abstract. The MULTICUT problem, given a graph G, a set of terminal pairs $\mathcal{T} = \{(s_i, t_i) \mid 1 \le i \le r\}$ and an integer p, asks whether one can find a *cutset* consisting of at most p non-terminal vertices that separates all the terminal pairs, i.e., after removing the cutset, t_i is not reachable from s_i for each $1 \le i \le r$. The fixed-parameter tractability of MULTICUT in undirected graphs, parameterized by the size of the cutset only, has been recently proven by Marx and Razgon [2] and, independently, by Bousquet et al. [3], after resisting attacks as a long-standing open problem. In this paper we prove that MULTICUT is fixed-parameter tractable on directed acyclic graphs, when parameterized both by the size of the cutset and the number of terminal pairs. We complement this result by showing that this is implausible for parameterization by the size of the cutset only, as this version of the problem remains $W[1]$-hard.

1 Introduction

Parameterized complexity is an approach for tackling hard problems by designing algorithms that perform robustly, when the input instance is in some sense simple; its difficulty is measured by an integer that is additionally appended to the input, called the *parameter*. Formally, we say that a problem is *fixed-parameter tractable* (FPT), if it can be solved by an algorithm that runs in time $f(k)n^c$ for n being the length of the input and k being the parameter, where f is some computable function and c is a constant independent of the parameter.

* The full version of this paper is available online [1].
** Supported by the Netherlands Organization for Scientific Research (N.W.O.), project "KERNELS: Combinatorial Analysis of Data Reduction".
*** Partially supported by NCN grant N206567140 and Foundation for Polish Science.
† Partially supported by European Research Council (ERC) grant "Rigorous Theory of Preprocessing", reference 267959.

A. Czumaj et al. (Eds.): ICALP 2012, Part I, LNCS 7391, pp. 581–593, 2012.
© Springer-Verlag Berlin Heidelberg 2012

The search for fixed-parameter algorithms resulted in the introduction of a number of new algorithmic techniques, and gave fresh insight into the structure of many classes of problems. One family that received a lot of attention recently is the so-called *graph cut* problems, where the goal is to make the graph satisfy a global separation requirement by deleting as few edges or vertices as possible (depending on the variant). Graph cut problems in the context of fixed-parameter tractability were to our knowledge first introduced explicitly in the seminal work of Marx [4], where it was proved that (i) MULTIWAY CUT (separate all terminals from each other by a cutset of size at most p) in undirected graphs is FPT when parameterized by the size of the cutset; (ii) MULTICUT in undirected graphs is FPT when parameterized by both the size of the cutset and the number of terminal pairs. Fixed-parameter tractability of MULTICUT parameterized by the size of the cutset only was left open by Marx [4]; resolved much later (see below).

The probably most fruitful contribution of the work of Marx [4] is the concept of *important separators*, which proved to be a tool almost perfectly suited to capturing the bounded-in-parameter character of sensible cutsets. The technique proved to be extremely robust and serves as the key ingredient in a number of FPT algorithms [2,4,5,6,7,8,9,10,11]. In particular, the fixed-parameter tractability of SKEW MULTICUT in directed acyclic graphs, obtained via a simple application of important separators, enabled the first FPT algorithm for DIRECTED FEEDBACK VERTEX SET [5], resolving another long-standing open problem.

However, important separators have a drawback in that not all graph cut problems admit solutions with "sensible" cutsets in the required sense. This is particularly true in directed graphs, where, with the exception of the aforementioned SKEW MULTICUT problem in DAGs, for a long time few fixed-parameter tractable graph cut problems were known; in fact, up until very recently it was open whether MULTIWAY CUT in directed graphs admits an FPT algorithm even in the restricted case of two terminals. The same complication arises in the undirected MULTICUT problem parameterized by the size of the cutset.

After a long struggle, MULTICUT was shown to be FPT by Marx and Razgon [2] and, independently, by Bousquet et al. [3]. The key component in the algorithm of Marx and Razgon [2] is the technique of *shadow removal*, which, in some sense, serves to make the solutions to cut problems more well-behaved. This was adapted to the directed case by Chitnis et al. [11], who proved that MULTIWAY CUT, parameterized by the size of the cutset, is fixed-parameter tractable for an arbitrary number of terminals, by a simple and elegant application of the shadow removal technique. This gives hope that, in general, shadow removal may be helpful for the application of important separators to the directed world.

As for the directed MULTICUT problem, it was shown by Marx and Razgon [2] to be $W[1]$-hard when parameterized only by the size of the cutset, but otherwise had unknown status, even for a constant number of terminals in a DAG. We note that this case is known to be NP-hard and APX-hard [12].

Our Results. The main result of this paper is the proof of fixed-parameter tractability of the MULTICUT IN DAGs problem, formally defined as follows:

MULTICUT IN DAGS   **Parameter:** $p + r$
Input: Directed acyclic graph G, set of terminal pairs $\mathcal{T} = \{(s_i, t_i) \mid 1 \leq i \leq r\}$, $s_i, t_i \in V(G)$ for $1 \leq i \leq r$, and an integer p.
Question: Does there exist a set Z of at most p non-terminal vertices of G, such that for any $1 \leq i \leq r$ the terminal t_i is not reachable from s_i in $G \setminus Z$?

Theorem 1. MULTICUT IN DAGS *can be solved in* $O^*(2^{O(r^2 p + r 2^{O(p)})})$ *time.*

Note, that throughout the paper we use O^*-notation to suppress polynomial factors. Note also that we focus on vertex cuts; it is well known that in the directed acyclic setting the arc- and vertex-deletion variants are equivalent (cf. [11]).

Our algorithm makes use of the shadow removal technique introduced by Marx and Razgon [2], adjusted to the directed setting by Chitnis et al. [11], as well as the basic important separators toolbox that can be found in [11]. We remark that the shadow removal is but one of a number of ingredients of our approach: in essence, the algorithm combines the shadow removal technique with a degree reduction for the sources in order to carefully prepare the structure of the instance for a simplifying branching step.

We complement the main result with two lower bounds. First, we show that the dependency on r in the exponent is probably unavoidable.[1]

Theorem 2 (♣). MULTICUT IN DAGS, *parameterized by the size of the cutset only, is* $W[1]$-*hard.*

Thus, we complete the picture of parameterized complexity of MULTICUT IN DAGS. We hope that it is a step towards fully understanding the parameterized complexity of MULTICUT in general directed graphs.

Second, we establish NP-completeness of SKEW MULTICUT, a special case of MULTICUT IN DAGS where we are given d sources $(s_i)_{i=1}^d$ and d sinks $(t_i)_{i=1}^d$, and the set of terminal pairs is defined as $\mathcal{T} = \{(s_i, t_j) : 1 \leq i \leq j \leq d\}$. Recall that the FPT algorithm for SKEW MULTICUT is the core subroutine of the algorithm for DIRECTED FEEDBACK VERTEX SET of Chen et al. [5].

Theorem 3 (♣). SKEW MULTICUT *is NP-complete even in the restricted case of two sinks and two sources.*

2   Preliminaries

For a directed graph G, by $V(G)$ and $E(G)$ we denote its vertex- and arc-set, respectively. For a vertex $v \in V(G)$, we define its *in-neighbourhood* $N_G^-(v) = \{u : (u, v) \in E(G)\}$ and *out-neighbourhood* $N_G^+(v) = \{u : (v, u) \in E(G)\}$; these definitions are extended to sets $X \subseteq V(G)$ by $N_G^-(X) = (\bigcup_{v \in X} N_G^-(v)) \setminus X$ and $N_G^+(X) = (\bigcup_{v \in X} N_G^+(v)) \setminus X$. The *in-degree* and *out-degree* of v are defined as $|N_G^-(v)|$ and $|N_G^+(v)|$, respectively. In this paper we consider simple directed

---

[1] Proofs of statements marked with ♣ are deferred to the full version of the paper [1].

graphs only; if at any point a modification of the graph results in a multiple arc, we delete all copies of the arc except for one. By G^{rev} we denote the graph G with all the arcs reversed, i.e., $G^{\mathrm{rev}} = (V(G), \{(v, u) : (u, v) \in E(G)\})$.

A *path* in G is a sequence of pairwise different vertices $P = (v_1, v_2, \ldots, v_d)$ such that $(v_i, v_{i+1}) \in E(G)$ for any $1 \leq i < d$. If v_1 is the first vertex of the path P and v_d is the last vertex, we say that P is a $v_1 v_d$-*path*. We extend this notion to sets of vertices: if $v_1 \in X$ and $v_d \in Y$ for some $X, Y \subseteq V(G)$, then P is a XY-path as well. For a path $P = (v_1, v_2, v_3, \ldots, v_d)$ the vertices $v_2, v_3, \ldots, v_{d-1}$ are the *internal vertices* of P. The set of internal vertices of a path P is the *interior* of P. We say that a vertex v is *reachable* from a vertex u in G if there exists a uv-path in G. As the considered digraphs are simple, each path $P = (v_1, v_2, \ldots, v_d)$ has a unique *first arc* (v_1, v_2) and a unique *last arc* (v_{d-1}, v_d).

Let $(G, \mathcal{T}, p)$ be a MULTICUT IN DAGs instance with a set of r terminal pairs $\mathcal{T} = \{(s_i, t_i) : 1 \leq i \leq r\}$. We call the terminals s_i *source terminals* and the terminals t_i *sink terminals*. We let $T^s = \{s_i : 1 \leq i \leq r\}$, $T^t = \{t_i : 1 \leq i \leq r\}$ and $T = T^s \cup T^t$. By an easy reduction, we may assume that all terminals are pairwise distinct and that $N_G^-(s_i) = N_G^+(t_i) = \emptyset$ for all $1 \leq i \leq r$. In our algorithm, the set of terminal pairs $\mathcal{T}$ is never modified, and neither in-neighbors of a source nor out-neighbors of a sink are added.

Fix a topological order $\leq_\tau$ of G. For any sets $X, Y \subseteq V(G)$, we may order the vertices of X and Y with respect to $\leq_\tau$, and compare X and Y lexicographically; we refer to this order on subsets of $V(G)$ as the *lexicographical order*.

A set $Z \subseteq V(G)$ is called a *multicut* in $(G, \mathcal{T}, p)$, if Z contains no terminals, but for each $1 \leq i \leq r$, t_i is not reachable from s_i in $G \setminus Z$. Given a MULTICUT IN DAGs instance $\mathcal{I} = (G, \mathcal{T}, p)$ a multicut Z is called a *solution* if $|Z| \leq p$. A solution Z is called a *lex-min solution* if Z is lexicographically minimum solution in $\mathcal{I}$ among solutions of minimum possible size.

For $v \in V(G)$, by $S(G, v)$ we denote set of source terminals s_i for which there exists a $s_i v$-path in G. For a set $S \subseteq T^s$ by $V(G, S)$ we denote the set of nonterminal vertices v for which $S(G, v) = S$.

2.1   Important Separators and Shadows

In the rest of this section we recall the notion of important separators by Marx [4], adjusted to the directed case by Chitnis et al. [11], as well as the shadow removal technique of Marx and Razgon [2] and Chitnis et al. [11].

Definition 4 (separator, [11], Definition 2.2). *Let G be a directed graph with terminals $T \subseteq V(G)$. Given two disjoint non-empty sets $X, Y \subseteq V(G)$, we call a set $Z \subseteq V(G)$ an $X - Y$ separator if (i) $Z \cap T = \emptyset$, (ii) $Z \cap (X \cup Y) = \emptyset$, (iii) there is no path from X to Y in $G \setminus Z$. An $X - Y$ separator Z is called minimal if no proper subset of Z is a $X - Y$ separator.*

By $\mathrm{cut}_G(X, Y)$ we denote the size of a minimum $X - Y$ separator in G; $\mathrm{cut}_G(X, Y) = \infty$ if G contains an arc going directly from X to Y. By Menger's theorem, $\mathrm{cut}_G(X, Y)$ equals the maximum possible size of a family of XY-paths with pairwise disjoint interiors.

Definition 5 (important separator, [11], Definition 4.1). *Let G be a directed graph with terminals $T \subseteq V(G)$ and let $X, Y \subseteq V(G)$ be two disjoint non-empty sets. Let Z and Z' be two $X - Y$ separators. We say that Z' is behind Z if any vertex reachable from X in $G \setminus Z$ is also reachable from X in $G \setminus Z'$. A minimal $X - Y$ separator is an important separator if no other $X - Y$ separator Z' satisfies $|Z'| \leq |Z|$ while being also behind Z.*

Lemma 6 ([11], Lemma B.4). *Let G be a directed graph with terminals $T \subseteq V(G)$. For two disjoint non-empty sets $X, Y \subseteq V(G)$, there exists exactly one minimum size important $X - Y$ separator.*

Definition 7 (closest mincut). *Let G be a directed graph with terminals $T \subseteq V(G)$. For two disjoint non-empty sets $X, Y \subseteq V(G)$, the unique minimum size important $X - Y$ separator is called the $X - Y$ mincut closest to Y. The $X - Y$ mincut closest to X is the $Y - X$ mincut closest to X in G^{rev}.*

Lemma 8 (♣). *Let G be a directed graph with terminals $T \subseteq V(G)$ and let $X, Y \subseteq V(G)$ be two disjoint non-empty sets. Let B be the unique minimum size important $X - Y$ separator, that is, the $X - Y$ mincut closest to Y, and let $v \in B$ be an arbitrary vertex. Construct a graph G' from G as follows: delete v from G and add an arc (x, w) for each $x \in X$ and $w \in N_G^+(v) \setminus X'$, where X' is the set of vertices reachable from X in $G \setminus B$. Then the size of any $X - Y$ separator in G' is strictly larger than $|B|$.*

We use the technique of shadows [2,11] to identify vertices separated from all sources in a given MULTICUT IN DAGs instance. We note that we do not use the full power of the shadow removal technique in directed graphs: the delicate part of the result of Chitnis et al. [11] is to remove forward and backward shadows at once; in our work we need to remove only one type of the shadows.

Definition 9 (source shadow). *Let $(G, \mathcal{T}, p)$ be a MULTICUT IN DAGs instance and $Z \subseteq V(G)$ be a subset of nonterminals in G. We say that $v \in V(G)$ is in source shadow of Z if Z is a $T^s - v$ separator.*

Lemma 10 (derandomized random sampling for source shadows, ♣). *There is an algorithm that, given a MULTICUT IN DAGs instance $(G, \mathcal{T}, p)$, produces in time $O^*(2^{2^{O(p)}})$ a family $\mathcal{A}$ of size $2^{2^{O(p)}} \log |V(G)|$ of subsets of nonterminals of G such that if $(G, \mathcal{T}, p)$ is a YES instance and Z is the lex-min solution to $(G, \mathcal{T}, p)$, then there exists $A \in \mathcal{A}$ such that $A \cap Z = \emptyset$ and all vertices of source shadows of Z in G are contained in A.*

3   The Algorithm

3.1   Potential Function and Simple Operations

Our algorithm consists of a number of branching steps. To measure the progress of the algorithm, we introduce the following potential function.

Definition 11 (potential). *Given a* MULTICUT IN DAGs *instance* $\mathcal{I} = (G, \mathcal{T}, p)$, *we define its potential* $\phi(\mathcal{I})$ *as* $\phi(\mathcal{I}) = (r+1)p - \sum_{i=1}^{r} cut_G(s_i, t_i)$.

Observe, that if $\mathcal{I} = (G, \mathcal{T}, p)$ is a MULTICUT IN DAGs instance, in which $cut(s_i, t_i) > p$ for some $(s_i, t_i) \in \mathcal{T}$, then we can immediately conclude that $\mathcal{I}$ is a NO instance. Therefore, w.l.o.g. we can henceforth assume that $cut(s_i, t_i) \leq p$ for all $(s_i, t_i) \in \mathcal{T}$ in all the appearing instances of MULTICUT IN DAGs.

In many places we perform the following simple operations on MULTICUT IN DAGs instances $(G, \mathcal{T}, p)$. We formalize their properties in subsequent lemmata.

Definition 12 (killing a vertex). *For a* MULTICUT IN DAGs *instance* $(G, \mathcal{T}, p)$ *and a nonterminal vertex* v *of* G, *by killing the vertex* v *we mean the following operation: we delete the vertex* v *and decrease* p *by one.*

Definition 13 (bypassing a vertex). *For a* MULTICUT IN DAGs *instance* $(G, \mathcal{T}, p)$ *and a nonterminal vertex* v *of* G, *by bypassing the vertex* v *we mean the following operation: we delete the vertex* v *and for each in-neighbour* v^- *of* v *and each out-neighbour* v^+ *of* v *we add an arc* (v^-, v^+).

Lemma 14 (♣). *Let* $\mathcal{I}' = (G', \mathcal{T}, p-1)$ *be obtained from* MULTICUT IN DAGs *instance* $\mathcal{I} = (G, \mathcal{T}, p)$ *by killing a vertex* v. *Then* $\mathcal{I}'$ *is a YES instance if and only if* $\mathcal{I}$ *is a YES instance that admits a solution that contains* v. *Moreover,* $\phi(\mathcal{I}') < \phi(\mathcal{I})$.

Lemma 15 (♣). *Let* $\mathcal{I}' = (G', \mathcal{T}, p)$ *be obtained from* MULTICUT IN DAGs *instance* $\mathcal{I} = (G, \mathcal{T}, p)$ *by bypassing a vertex* v. *Then:*

1. *any multicut in* $\mathcal{I}'$ *is a multicut in* $\mathcal{I}$ *as well;*
2. *any multicut in* $\mathcal{I}$ *that does not contain* v *is a multicut in* $\mathcal{I}'$ *as well;*
3. $S(G, u) = S(G', u)$ *for any* $u \in V(G') = V(G) \setminus \{v\}$;
4. $\phi(\mathcal{I}') \leq \phi(\mathcal{I})$.

We note that bypassing a vertex corresponds to the torso operation of Chitnis et al. [11] and, if we perform a series of bypass operations, the result does not depend on their order.

3.2 Degree Reduction

In this section we introduce the second — apart from the source shadow reduction in Lemma 10 — main tool used in the algorithm. In an instance $(G, \mathcal{T}, p)$, let B_i be the $s_i - t_i$ mincut closest to s_i and let Z be a solution. If we know that a vt_i-path survives in $G \setminus Z$ for some $v \in B_i$, we may add an arc (v, t_i) and then bypass the vertex v, strictly increasing the value $cut_G(s_i, t_i)$ (and thus decreasing the potential) by Lemma 8. Therefore, we can branch: we either guess the pair (i, v), or guess that none such exist; in the latter branch we do not decrease potential but instead we may modify the set of arcs incident to the sources to get some structure, as formalized in the following definition.

Definition 16 (degree-reduced graph). *For a* MULTICUT IN DAGS *instance* $(G, \mathcal{T}, p)$ *the* degree-reduced *graph* G^* *is a graph constructed as follows. For* $1 \le i \le r$, *let* B_i *be the* $s_i - t_i$ *mincut closest to* s_i. *We start with* $V(G^*) = V(G)$, $E(G^*) = E(G \setminus T^s)$ *and then, for each* $1 \le i \le r$, *we add an arc* (s_i, v) *for all* $v \in B_i$ *and for all* $v \in \bigcup_{1 \le i' \le r} B_{i'}$ *for which* $s_i \in S(G, v)$ *but* v *is not reachable from* B_i *in* G.

The following two lemmata formalize the properties of the degree-reduced graph and the aforementioned branching step. Recall that we assume that each vertex s_i (t_i) has in- (out-) degree zero.

Lemma 17 (properties of the degree-reduced graph,♣). *For any* MUL-TICUT IN DAGS *instance* $\mathcal{I} = (G, \mathcal{T}, p)$ *and the degree-reduced graph* G^* *of* $\mathcal{I}$, *the following holds:*

1. $|N_{G^*}^+(T^s)| \le rp$.
2. *for each* $1 \le i \le r$, B_i *is the* $s_i - t_i$ *mincut closest to* s_i *in* G^*.
3. $\phi(\mathcal{I}') = \phi(\mathcal{I})$, *where* $\mathcal{I}' = (G^*, \mathcal{T}, p)$.
4. $Z \subseteq V(G)$ *is a multicut in* $(G^*, \mathcal{T})$ *if and only if* Z *is a multicut in* $(G, \mathcal{T})$ *satisfying the following property: for each* $1 \le i \le r$, *for each* $v \in B_i$, *the vertex* v *is either in* Z *or* Z *is an* $v - t_i$ *separator; in particular,* $\mathcal{I}'$ *is a YES instance if and only if* $\mathcal{I}$ *is a YES instance that admits a solution satisfying the above property.*
5. *for each* $v \in V(G)$ *we have* $S(G^*, v) \subseteq S(G, v)$; *moreover, if* (s_i, v) *is an arc in* G^* *for some* $1 \le i \le r$ *then* $S(G^*, v) = S(G, v)$.

Lemma 18 (♣). *There exists an algorithm that, given a* MULTICUT IN DAGS *instance* $\mathcal{I} = (G, \mathcal{T}, p)$, *in polynomial time generates a sequence of instances* $(\mathcal{I}_j = (G_j, \mathcal{T}_j, p_j))_{j=1}^d$ *satisfying the following properties. Let* $\mathcal{I}_0 = (G^*, \mathcal{T}, p)$;

1. *if* Z *is a multicut* Z *in* $\mathcal{I}_j$ *for some* $0 \le j \le d$, *then* $Z \subseteq V(G)$ *and* Z *is a multicut in* $\mathcal{I}$ *too;*
2. *for any multicut* Z *in* $\mathcal{I}$, *there exists* $0 \le j \le d$ *such that* Z *is a multicut in* $\mathcal{I}_j$ *too;*
3. *for each* $1 \le j \le d$, $p_j = p$, $\mathcal{T}_j = \mathcal{T}$ *and* $\phi(\mathcal{I}_j) < \phi(\mathcal{I})$;
4. $d \le rp$.

3.3 Overview on the Branching Step

In order to prove Theorem 1, we show the following lemma that encapsulates a single branching step of the algorithm.

Lemma 19. *There exists an algorithm that, given a* MULTICUT IN DAGS *instance* $\mathcal{I} = (G, \mathcal{T}, p)$ *with* $|\mathcal{T}| = r$, *in time* $O^*(2^{r+2^{O(p)}})$ *either correctly concludes that* $\mathcal{I}$ *is a NO instance, or computes a sequence of instances* $(\mathcal{I}_j = (G_j, \mathcal{T}_j, p_j))_{j=1}^d$ *such that:*

1. $\mathcal{I}$ *is a YES instance if and only if at least one instance* $\mathcal{I}_j$ *is a YES instance;*

2. *for each* $1 \leq j \leq d$, $V(G_j) \subseteq V(G)$, $p_j \leq p$, $\mathcal{T}_j = \mathcal{T}$ *and* $\phi(\mathcal{I}_j) < \phi(\mathcal{I})$;

3. $d \leq 4 \cdot 2^{r+2^{O(p)}} rp \log |V(G)|$.

The algorithm of Theorem 1 applies Lemma 19 and solves the output instances recursively; the time bound follows from inequality $\log^k n \leq 2^{k^{3/2}} n^{o(1)}$.

In rough overview of the proof of Lemma 19, we describe a sequence of steps where in each step, either the potential of the instance is decreased or more structure is forced onto the instance. For example, consider Lemma 18. The result is a branching into polynomially many branches, where in every branch but one the potential strictly decreases, and in the remaining branch, the degrees of the source terminals are bounded. Thus we may treat this step as "creating" a degree-reduced instance.

In somewhat more detail, let Z be the lex-min solution to $\mathcal{I}$. We guess a set $S \subseteq T^s$ such that there is some $v \in Z$ with $S(G, v) = S$, but no $v' \in Z$ with $S(G, v') \subsetneq S$; bypass any vertex u with $S(G, u) \subsetneq S$. By appropriately combining degree reduction with shadow removal, we may further assume that no vertex in $V(G, S)$ is in source-shadow of Z, and that the sources S have bounded degree. Consider now the first vertex $v \in V(G, S)$ under $\leq_\tau$ (if any) which has its set of seen sources modified by Z, i.e., $v \in V(G, S) \setminus Z$, $S(G \setminus Z, v) \subsetneq S$, and v is $\leq_\tau$-minimal among all such vertices. Let w be an in-neighbour of v. The important observation is that since $S(G, w)$ is by assumption not modified by Z, every such vertex w must be either a source or deleted. Since v is not in source shadow of Z, there is an arc (s, v) in G for some $s \in S$, and by the degree reduction, there is only a bounded number of such vertices. Thus, if any vertex is modified by Z in this sense, then we may find one by branching on the out-neighbours of S, decreasing the potential.

Otherwise, we know that if $v \in V(G, S)$, then either $v \in Z$ or $S(G \setminus Z, v) = S$. Thus, we may "flatten" the graph, by making every $v \in V(G, S)$ a direct out-neighbour of every $s \in S$. By further degree reduction, we can now identify a polynomially sized set out of which at least one vertex must be deleted.

3.4   Branchings and Reductions

We now proceed with the formal proof of Lemma 19. The proof contains a sequence of *branching rules* (when we generate a number of subcases, some of them already ready to output as one instance $\mathcal{I}_j$), or *reduction rules* (when we reduce the graph without changing the answer).

If the input instance $\mathcal{I}$ is YES instance, by Z we denote its lex-min solution. Whenever we perform a branching or reduction step, in the new instance we consider the topological order that is induced by the old one; all the reductions and branchings add arcs only directed from vertices smaller in $\leq_\tau$ to bigger. This also ensures that during the course of the algorithm all the directed graphs in the instances are acyclic.

We start with the obvious rule that was already mentioned in Section 3. Then, we roughly localize one vertex of Z.

Reduction rule 1. *If $cut_G(s_i, t_i) > p$ (in particular, if $(s_i, t_i) \in E(G)$) for some $1 \leq i \leq r$, conclude that $\mathcal{I}$ is a NO instance.*

Branching rule 2. *Branch into $2^r - 1$ subcases, labeled by nonempty sets $S \subseteq T^s$. In the case labeled S we assume that Z contains a vertex v with $S(G, v) = S$, but no vertex v' with $S(G, v')$ being a proper subset of S.*

As Z is a lex-min solution (in case of $\mathcal{I}$ being a YES instance), Z cannot contain any vertex v with $S(G, v) = \emptyset$. In each branch we can bypass some vertices.

Reduction rule 3. *In each subcase, as long as there exists a nonterminal vertex $u \in V(G)$ with $S(G, u) \subsetneq S$ bypass u. Let $(G^1, \mathcal{T}, p)$ be the reduced instance.*

By Lemma 15, an application of the above rule cannot turn a NO instance into a YES instance. Moreover, in the branch where S is guessed correctly, Z remains the lex-min solution to $(G^1, \mathcal{T}, p)$. By Lemma 15, $\phi((G^1, \mathcal{T}, p)) \leq \phi(\mathcal{I})$.

We now apply the reduction of source degrees.

Branching rule 4. *In each subcase, let S be its label and $(G^1, \mathcal{T}, p)$ be the instance. Invoke Lemma 18 on the instance $(G^1, \mathcal{T}, p)$. Output all instances $\mathcal{I}_j$ for $1 \leq j \leq d$ as part of the output instances in Lemma 19. Keep the instance $\mathcal{I}_0$ for further analysis in this subcase and denote $\mathcal{I}_0 = (G^2, \mathcal{T}, p)$; G^2 is the degree-reduced graph of G^1.*

Let us summarize what Lemma 18 implies on the outcome of Branching 4. We output at most $2^r r p$ instances, and keep one instance for further analysis in each branch. Each output instance has strictly decreased potential, while $\phi((G^2, \mathcal{T}, p)) \leq \phi(G^1, \mathcal{T}, p)$. If $\mathcal{I}$ is a NO instance, all the generated instances — both the output and kept ones — are NO instances. If $\mathcal{I}$ is a YES instance, then it is possible that all the output instances are NO instances only if in the branch where the set S is guessed correctly, the solution Z is a solution to $(G^2, \mathcal{T}, p)$ as well. Moreover, as any solution to $(G^2, \mathcal{T}, p)$ is a solution to $\mathcal{I}$ as well by Lemma 17, in this case Z is the lex-min solution to $(G^2, \mathcal{T}, p)$.

Let us now investigate more deeply the structure of the kept instances.

Lemma 20 (♣). *In a branch, let S be its label, $(G^1, \mathcal{T}, p)$ the instance on which Lemma 18 is invoked and $(G^2, \mathcal{T}, p)$ the kept instance. For any $v \in V(G^1) = V(G^2)$ with $S(G, v) = S$, we have $S(G^1, v) = S$ and $S(G^2, v) \in \{\emptyset, S\}$.*

Recall that if $\mathcal{I}$ is a YES instance and all instances output so far are NO instances, then in some subcase S the set Z is the lex-min solution to $(G^2, \mathcal{T}, p)$. In this case Z does not contain any vertex from $V(G^2, \emptyset)$ and we can remove these vertices, as they are not contained in any $s_i t_i$-path for any $1 \leq i \leq r$.

Reduction rule 5. *In each branch, let S be its label and $(G^2, \mathcal{T}, p)$ the kept instance. As long as there exists a nonterminal vertex $v \in V(G^2)$ with $S(G^2, v) = \emptyset$, delete v. Denote the output instance by $(G^3, \mathcal{T}, p)$.*

Reduction 5 does not interfere with any $s_i t_i$-paths, thus $\phi((G^3, \mathcal{T}, p)) = \phi((G^2, \mathcal{T}, p))$. Again, if $\mathcal{I}$ is a NO instance, all instances $(G^3, \mathcal{T}, p)$ are NO instances as well, and if $\mathcal{I}$ is a YES instance, but all output instances produced so far are NO instances, Z is the lex-min solution to $(G^3, \mathcal{T}, p)$ in some branch S. Moreover, in G^3 each source has out-degree at most rp and there is no vertex v with $S(G^3, v) \subsetneq S$ (note that Reduction 5 does not change reachability of a vertex from a fixed source). We apply the source shadow reduction to $(G^3, \mathcal{T}, p)$.

Branching rule 6. *In each branch, let S be its label, and $(G^3, \mathcal{T}, p)$ be the remaining instance. Invoke Lemma 10 on $(G^3, \mathcal{T}, p)$, obtaining a family $\mathcal{A}_S$. Branch into $|\mathcal{A}_S|$ subcases, labeled by pairs (S, A) for $A \in \mathcal{A}_S$. In each subcase, obtain a graph $(G^4, \mathcal{T}, p)$ by bypassing (in arbitrary order) all vertices of $A \setminus N_{G^3}^+(T^s)$.*

Note that the graph G^4 does not depend on the order in which we bypass vertices of $A \setminus N_{G^3}^+(T^s)$. By Lemma 15, bypassing some vertices cannot turn a NO instance into a YES instance. Moreover, by Lemma 10, if $(G^3, \mathcal{T}, p)$ is a YES instance and Z is the lex-min solution to $(G^3, \mathcal{T}, p)$, then there exists $A \in \mathcal{A}_S$ that contains all vertices of source shadows of Z, but no vertex of Z. Note that no out-neighbour of a source may be contained in a source shadow; therefore, $A \setminus N_{G^3}^+(T^s)$ contains all vertices of source shadows of Z as well. We infer that in the branch (S, A), $(G^4, \mathcal{T}, p)$ is a YES instance and, as bypassing a vertex only shrinks the set of solutions, Z is still the lex-min solution to $(G^4, \mathcal{T}, p)$. Moreover, there are no source shadows of Z in $(G^4, \mathcal{T}, p)$.

At this point we have at most $2^{r+2^{O(p)}} \log |V(G)|$ subcases and at most $2^r rp$ already output instances. In each subcase, we have $\phi((G^4, \mathcal{T}, p)) \leq \phi((G^3, \mathcal{T}, p))$ by Lemma 15. The following observation is crucial for further branching.

Lemma 21 (♣). *Take an instance $(G^4, \mathcal{T}, p)$ obtained in a branch labeled with (S, A). Assume that $(G^4, \mathcal{T}, p)$ is a YES instance and let Z be its lex-min solution. Moreover, assume that there are no source shadows of Z in $(G^4, \mathcal{T}, p)$. Then the following holds: if there exists a vertex $v' \in (V(G, S) \cap V(G^4)) \setminus Z$ with $S(G^4 \setminus Z, v') \neq S$, then the first such vertex in the topological order $\leq_\tau$ (denoted v) belongs to $N_{G^4}^+(T^s)$. Moreover, v has at least one in-neighbour in G^4 that is not in T^s, and all such in-neighbours belong to Z.*

Branching rule 7. *In each branch, let (S, A) be its label and $(G^4, \mathcal{T}, p)$ the remaining instance. Output at most rp instances $\mathcal{I}_v$, labeled by vertices $v \in N_{G^4}^+(T^s) \cap V(G, S)$ for which $N_{G^4}^-(v) \not\subseteq T^s$: the instance $\mathcal{I}_v$ is created from $(G^4, \mathcal{T}, p)$ by killing all non-terminal in-neighbours of v and bypassing v. Moreover, create one remaining instance $(G^5, \mathcal{T}, p)$ as follows: delete from G^4 all arcs that have their ending vertices in $V(G, S) \cap V(G^4)$ and for each $v \in V(G, S) \cap V(G^4)$ and $s_i \in S$ add an arc (s_i, v).*

By Lemmata 14 and 15, the output instances have strictly smaller potential than $(G^4, \mathcal{T}, p)$ and are NO instances if $(G^4, \mathcal{T}, p)$ is a NO instance. On the other hand, assume that $(G^4, \mathcal{T}, p)$ is a YES instance with lex-min solution Z such that there are no source shadows of Z. If there exist vertices v' and v as in the statement of Lemma 21, then the instance $\mathcal{I}_v$ is computed and $Z \setminus N_{G^4}^-(v)$ (i.e., Z without the killed vertices) is a solution to $\mathcal{I}_v$. Otherwise, we claim that $(G^5, \mathcal{T}, p)$ represents the remaining case:

Lemma 22 (♣). *Let $(G^4, \mathcal{T}, p)$ be an instance obtained in the branch (S, A).*

1. $\phi((G^5, \mathcal{T}, p)) \leq \phi((G^4, \mathcal{T}, p))$.
2. *Any multicut Z in $(G^5, \mathcal{T}, p)$ is a multicut in $(G^4, \mathcal{T}, p)$ as well.*
3. *Assume additionally that $(G^4, \mathcal{T}, p)$ is a YES instance whose lex-min solution Z satisfies the following properties: there are no source shadows of Z and for each $v \in V(G, S) \cap V(G^4)$, either $v \in Z$ or $S(G^4 \setminus Z, v) = S$. Then $(G^5, \mathcal{T}, p)$ is a YES instance and Z is its lex-min solution.*

The structure of $V(G, S) \cap V(G^5)$ is quite simple in $(G^5, \mathcal{T}, p)$. Recall that, if $\mathcal{I}$ is a YES instance, but no instance output so far is a YES instance, then in at least one branch (S, A) we have that the lex-min solution Z to $\mathcal{I}$ is the lex-min solution to $(G^5, \mathcal{T}, p)$ and $Z \cap V(G, S) \cap V(G^5) \neq \emptyset$. We would like to guess one vertex of $Z \cap V(G, S) \cap V(G^5)$. Although, $V(G, S) \cap V(G^5)$ may still be large, each vertex $v \in V(G, S) \cap V(G^5)$ has $N_{G^5}^-(v) = S$. Therefore we may limit the size of $V(G, S) \cap V(G^5)$ by applying once again the degree reduction branching.

Branching rule 8. *In each branch, let (S, A) be its label and $(G^5, \mathcal{T}, p)$ the remaining instance. Apply Lemma 18 on $(G^5, \mathcal{T}, p)$, obtaining a sequence of instances $(\mathcal{I}_j)_{j=1}^d$ and the remaining instance $(G^6, \mathcal{T}, p)$, where G^6 is the degree-reduced graph G^5. Output all instances $\mathcal{I}_j$ for $1 \leq j \leq d$ and keep $(G^6, \mathcal{T}, p)$ for further analysis.*

By Lemma 18, if $(G^5, \mathcal{T}, p)$ is a NO instance, all the output instances as well as $(G^6, \mathcal{T}, p)$ are NO instances. Otherwise, if $(G^5, \mathcal{T}, p)$ is a YES instance with the lex-min solution Z, but the instances $\mathcal{I}_j$ are all NO instances, then Z is the lex-min solution to $(G^6, \mathcal{T}, p)$.

Note that, by Lemma 18, all output instances have potential strictly smaller than $\phi((G^5, \mathcal{T}, p))$, whereas $\phi((G^6, \mathcal{T}, p)) = \phi((G^5, \mathcal{T}, p))$. Moreover, applications of Branching 8 in all subcases output at most $2^{r+2^{O(p)}} rp \log |V(G)|$ instances in total.

We are left with the final observation.

Lemma 23. *In each subcase, let (S, A) be its label and $(G^6, \mathcal{T}, p)$ the remaining instance. Then at most rp vertices $v \in V(G, S) \cap V(G^6)$ have $S(G^6, v) \neq \emptyset$.*

Proof. Note that $V(G^4) = V(G^5) = V(G^6)$. Take $v \in V(G, S) \cap V(G^6)$. Recall that $N_{G^5}^-(v) = S$ and G^6 differs from G^5 only on the set of arcs incident to the sources, so $S(G^6, v) = N_{G^6}^-(v)$. The lemma follows from Lemma 17, Claim 1. □

Reduction rule 9. *In each branch, let (S, A) be its label and $(G^6, \mathcal{T}, p)$ be the remaining instance. As long as there exists a nonterminal vertex $v \in V(G^6)$ with $S(G^6, v) = \emptyset$, delete v. Denote the output instance by $(G^7, \mathcal{T}, p)$.*

As in the case of Reduction 5, Z is the lex-min solution to $(G^6, \mathcal{T}, p)$ if and only if Z is the lex-min solution to $(G^7, \mathcal{T}, p)$. Moreover, $\phi((G^6, \mathcal{T}, p)) = \phi((G^7, \mathcal{T}, p))$. By Lemma 23, $|V(G, S) \cap V(G^7)| \leq rp$. Now we can perform final branching.

Branching rule 10. *In each subcase, let (S, A) be its label and $(G^7, \mathcal{T}, p)$ the remaining instance. For each $v \in V(G, S) \cap V(G^7)$ output an instance $\mathcal{I}_v$ created from $(G^7, \mathcal{T}, p)$ by killing the vertex v.*

Note that if $V(G, S) \cap V(G^7) = \emptyset$, then this rule results in no branches created.

By Lemma 14, if $(G^7, \mathcal{T}, p)$ is a NO instance, so are the output instances $\mathcal{I}_v$. On the other hand, assume that $\mathcal{I}$ is a YES instance with the lex-min solution Z. Then in at least one subcase (S, A), if no previously output instance is a YES instance, then the instance $(G^7, \mathcal{T}, p)$ is a YES instance, Z is its lex-min solution, and $Z \cap V(G, S) \cap V(G^7) \neq \emptyset$. Then the instance $\mathcal{I}_v$ for any $v \in Z \cap V(G, S) \cap V(G^7)$ is a YES instance; in particular, $V(G, S) \cap V(G^7)$ is nonempty. To conclude the proof of Lemma 19 note that $\phi(\mathcal{I}_v) < \phi((G^7, \mathcal{T}, p))$ for each output instance $\mathcal{I}_v$.

4   Conclusions

The results of this paper unravel the full picture of the parameterized complexity of MULTICUT IN DAGS. A natural follow-up question is the complexity of MULTICUT in general directed graphs, where we so far know only that the case of two terminal pairs is FPT [11] and the cutset parameterization is W[1]-hard [2]. The assumption of acyclicity seems to be crucial for our approach in Lemma 21 and subsequent Branching 7. We also note that, although an existence of a polynomial kernelization algorithm for most graph separation problems in directed graphs was recently refuted [13], the question of a polynomial kernel for MULTICUT IN DAGS remains open.

References

1. Kratsch, S., Pilipczuk, M., Pilipczuk, M., Wahlström, M.: Fixed-parameter tractability of multicut in directed acyclic graphs. CoRR, abs/1202.5749 (2012)
2. Marx, D., Razgon, I.: Fixed-parameter tractability of multicut parameterized by the size of the cutset. In: Proc. of STOC 2011, pp. 469–478 (2011)
3. Bousquet, N., Daligault, J., Thomassé, S.: Multicut is FPT. In: Proc. of STOC 2011, pp. 459–468 (2011)
4. Marx, D.: Parameterized graph separation problems. Theor. Comput. Sci. 351(3), 394–406 (2006)
5. Chen, J., Liu, Y., Lu, S., O'Sullivan, B., Razgon, I.: A fixed-parameter algorithm for the directed feedback vertex set problem. J. ACM 55(5) (2008)
6. Cygan, M., Pilipczuk, M., Pilipczuk, M., Wojtaszczyk, J.O.: Subset Feedback Vertex Set is Fixed-Parameter Tractable. In: Aceto, L., Henzinger, M., Sgall, J. (eds.) ICALP 2011. LNCS, vol. 6755, pp. 449–461. Springer, Heidelberg (2011)
7. Razgon, I., O'Sullivan, B.: Almost 2-SAT is fixed-parameter tractable. J. Comput. Syst. Sci. 75(8), 435–450 (2009)
8. Chen, J., Liu, Y., Lu, S.: An improved parameterized algorithm for the minimum node multiway cut problem. Algorithmica 55(1), 1–13 (2009)
9. Guillemot, S.: FPT algorithms for path-transversal and cycle-transversal problems. Discrete Optimization 8(1), 61–71 (2011)

10. Lokshtanov, D., Marx, D.: Clustering with Local Restrictions. In: Aceto, L., Henzinger, M., Sgall, J. (eds.) ICALP 2011. LNCS, vol. 6755, pp. 785–797. Springer, Heidelberg (2011)
11. Chitnis, R.H., Hajiaghayi, M., Marx, D.: Fixed-parameter tractability of directed multiway cut parameterized by the size of the cutset. In: Proc. of SODA 2012, pp. 1713–1725 (2012)
12. Bentz, C.: On the hardness of finding near-optimal multicuts in directed acyclic graphs. Theor. Comput. Sci. 412(39), 5325–5332 (2011)
13. Cygan, M., Kratsch, S., Pilipczuk, M., Pilipczuk, M., Wahlström, M.: Clique cover and graph separation: New incompressibility results. CoRR abs/1111.0570 (2011)

Preserving Terminal Distances Using Minors*

Robert Krauthgamer and Tamar Zondiner

Weizmann Institute of Science, Rehovot, Israel
{robert.krauthgamer,tamar.zondiner}@weizmann.ac.il

Abstract. We introduce the following notion of compressing an undirected graph G with (nonnegative) edge-lengths and terminal vertices $R \subseteq V(G)$. A *distance-preserving minor* is a minor G' (of G) with possibly different edge-lengths, such that $R \subseteq V(G')$ and the shortest-path distance between every pair of terminals is exactly the same in G and in G'. We ask: what is the smallest $f^*(k)$ such that every graph G with $k = |R|$ terminals admits a distance-preserving minor G' with at most $f^*(k)$ vertices?

Simple analysis shows that $f^*(k) \leq O(k^4)$. Our main result proves that $f^*(k) \geq \Omega(k^2)$, significantly improving over the trivial $f^*(k) \geq k$. Our lower bound holds even for planar graphs G, in contrast to graphs G of constant treewidth, for which we prove that $O(k)$ vertices suffice.

1 Introduction

A *graph compression* of a graph G is a small graph G^* that preserves certain features (quantities) of G, such as distances or cut values. This basic concept was introduced by Feder and Motwani [FM95], although their definition was slightly different technically. (They require that G^* has fewer edges than G, and that each graph can be quickly computed from the other one.) Our paper is concerned with preserving the selected features of G *exactly* (i.e., lossless compression), but in general we may also allow the features to be preserved approximately.

The algorithmic utility of graph compression is readily apparent – the compressed graph G^* may be computed as a preprocessing step, and then further processing is performed on it (instead of on G) with lower runtime and/or memory requirement. This approach is clearly beneficial when the compression can be computed very efficiently, say in linear time, in which case it may be performed on the fly, but it is useful also when some computations are to be performed (repeatedly) on a machine with limited resources such as a smartphone, while the preprocessing can be executed in advance on much more powerful machines.

For many features, graph compression was already studied and many results are known. For instance, a *k-spanner* of G is a subgraph G^* in which all pairwise distances approximate those in G within a factor of k [PS89]. Another example, closer in spirit to our own, is a *sourcewise distance preserver* of G with respect

---

* A full version appears at http://arxiv.org/abs/1202.5675. This work was supported in part by The Israel Science Foundation (grant #452/08), by a US-Israel BSF grant #2010418, and by the Citi Foundation.

A. Czumaj et al. (Eds.): ICALP 2012, Part I, LNCS 7391, pp. 594–605, 2012.

to a set of vertices $R \subseteq V(G)$; this is a subgraph G^* of G that preserves (exactly) the distances in G for all pairs of vertices in R [CE06]. We defer the discussion of further examples and related notions to Section 1.2, and here point out only two phenomena: First, it is common to require G^* to be structurally similar to G (e.g., a spanner is a subgraph of G), and second, sometimes only the features of a subset R need to be preserved (e.g., distances between vertices of R).

We consider the problem of compressing a graph so as to maintain the shortest-path distances among a set R of required vertices. From now on, the required vertices will be called *terminals*.

Definition 1. *Let G be a graph with edge lengths $\ell : E(G) \to \mathbb{R}_+$ and a set of terminals $R \subseteq V(G)$. A distance-preserving minor (of G with respect to R) is a graph G' with edge lengths $\ell' : E(G') \to \mathbb{R}_+$ satisfying:*

1. *G' is a minor of G; and*
2. *$d_{G'}(u, v) = d_G(u, v)$ for all $u, v \in R$.*

Here and throughout, d_H denotes the shortest-path distance in a graph H. It also goes without saying that the terminals R must survive the minor operations (they are not removed, but might be merged with non-terminals, due to edge contractions), and thus $d_{G'}(u, v)$ is well-defined; in particular, $R \subseteq V(G')$. For illustration, suppose G is a path of n unit-length edges and the terminals are the path's endpoints; then by contracting all the edges, we can obtain G' that is a single edge of length n.

The above definition basically asks for a minor G' that preserves all terminal distances exactly. The minor requirement is a common method to induce structural similarity between G' and G, and in general excludes the trivial solution of a complete graph on the vertex set R (with appropriate edge lengths).

This definition may be viewed as a conceptual contribution of our paper. Indeed, our main motivation is its mathematical elegance, but let us mention one potential algorithmic application. Suppose we need to solve multiple TSP instances involving altogether relatively few vertices in a large (perhaps planar) graph; then it makes sense to reduce the graph (to a minor of it).

We raise the following question, which to the best of our knowledge was not studied before. Its main point is to bound the size of G' independently of the size of G.

Question 1. *What is the smallest $f^*(k)$, such that for every graph G with k terminals, there is a distance-preserving minor G' with at most $f^*(k)$ vertices?*

Before describing our results, let us provide a few initial observations, which may well be folklore or appear implicitly in literature. Consider the naive method depicted in Algorithm 1. It is straightforward to see that these steps reduce the number of non-terminals without affecting terminal distances, and a simple analysis proves that this algorithm always produces a minor with $O(k^4)$ vertices and edges and runs in polynomial time (details omitted from this version). It follows that $f^*(k)$ exists, and furthermore

$$f^*(k) \leq O(k^4).$$

Algorithm 1. REDUCEGRAPHNAIVE (graph G, required vertices R)

---

(1) Remove all vertices and edges in G that do not participate in any shortest-path between terminals.

(2) Repeat while the graph contains a non-terminal v of degree two: merge v with one of its neighbors (by contracting the appropriate edge), thereby replacing the 2-path $w_1 - v - w_2$ with a single edge (w_1, w_2) of the same length as the 2-path.

Moreover, if G is a tree then G' has at most $2k - 2$ vertices, and this last bound is in fact tight (attained by a complete binary tree) whenever k is a power of 2.

1.1 Our Results

Our first and main result directly addresses Question 1, by providing the lower bound $f^*(k) \geq \Omega(k^2)$. The proof uses only simple planar graphs, leading us to study the restriction of $f^*(k)$ to specific graph families, defined as follows.[1]

Definition 2. *For a family $\mathcal{F}$ of graphs, define $f^*(k, \mathcal{F})$ as the minimum value such that every graph $G = (V, E, \ell) \in \mathcal{F}$ with k terminals admits a distance-preserving minor G' with at most $f^*(k, \mathcal{F})$ vertices.*

Theorem 1. *Let* Planar *be the family of all planar graphs. Then*

$$f^*(k) \geq f^*(k, \text{Planar}) \geq \Omega(k^2).$$

Our proof of this lower bound uses a two-dimensional grid graph, which has super-constant treewidth. This stands in contrast to graphs of treewidth 1, because we already mentioned that

$$f^*(k, \text{Trees}) \leq 2k - 2,$$

where Trees is the family of a all tree graphs. It is thus natural to ask whether bounded-treewidth graphs behave like trees, for which $f^* \leq O(k)$, or like planar graphs, for which $f^* \geq \Omega(k^2)$. We answer this question as follows.

Theorem 2. *Let* Treewidth(p) *be the family of all graphs with treewidth at most p. Then for all $k \geq p$,*

$$\Omega(pk) \leq f^*(k, \text{Treewidth}(p)) \leq O(p^3 k).$$

We summarize our results together with some initial observations in the table below.

---

[1] We use (V, E, ℓ) to denote a graph with vertex set V, edge set E, and edge lengths $\ell : E \to \mathbb{R}_+$. As usual, the definition of a family $\mathcal{F}$ of graphs refers only to the vertices and edges, and is irrespective of the edge lengths.

Graph Family $\mathcal{F}$	Bounds on $f^*(k, \mathcal{F})$		
Trees	$= 2k - 2$		omitted
Treewidth p	$\Omega(pk)$	$O(p^3k)$	Theorem 2
Planar Graphs	$\Omega(k^2)$	$O(k^4)$	Theorem 1
All Graphs	$\Omega(k^2)$	$O(k^4)$	Theorem 1

All our upper bounds are algorithmic and run in polynomial time. In fact, they can be achieved using the naive algorithm described above.

1.2 Related Work

Coppersmith and Elkin [CE06] studied a problem similar to ours, except that they seek subgraphs with few edges (rather than minors). Among other things, they prove that for every weighted graph $G = (V, E)$ and every set of $k = O(|V|^{\frac{1}{4}})$ terminals, there exists a weighted subgraph $G' = (V, E')$ with $|E'| \le O(|V|)$, that preserves terminal distances exactly. They also show a nearly-matching lower bound on $|E'|$.

Some compressions preserve cuts and flows in a given graph G rather than distances. A Gomory-Hu tree [GH61] is a weighted tree that preserves all st-cuts in G (or just between terminal pairs). A so-called mimicking network preserves all flows and cuts between subsets of the terminals in G [HKNR98].

Terminal distances can also be approximated instead of preserved exactly. In fact, allowing a constant factor approximation may be sufficient to obtain a compression G^* without any non-terminals. Gupta [Gup01] introduced this problem and proved that for every weighted tree T and set of terminals, there exists a weighted tree T' without the non-terminals that approximates all terminal distances within a factor of 8. It was later observed that this T' is in fact a minor of T [CGN+06], and that the factor 8 is tight [CXKR06]. Basu and Gupta [BG08] claimed that a constant approximation factor exists for weighted outerplanar graphs as well. It remains an open problem whether the constant factor approximation extends also to planar graphs (or excluded-minor graphs in general). Englert et al. [EGK+10] proved a randomized version of this problem for all excluded-minor graph families, with an expected approximation factor depending only on the size of the excluded minor.

The relevant information (features) in a graph can also be maintained by a data structure that is not necessarily graphs. A notable example is Distance Oracles – low-space data structures that can answer distance queries (often approximately) in constant time [TZ05]. These structures adhere to our main requirement of "compression" and are designed to answer queries very quickly. However, they might lose properties that are natural in graphs, such as the triangle inequality or the similarity of a minor to the given graph, which may be useful for further processing of the graph.

2   A Lower Bound of $\Omega(k^2)$

In this section we prove Theorem 1 using an even stronger assertion: there exist planar graphs G such that every distance-preserving *planar graph* H (a planar graph with $R \subseteq V(H)$ that preserves terminal distances) has $|V(H)| \geq \Omega(k^2)$. Since any minor G' of G is planar, Theorem 1 follows.

Our proof uses a $k \times k$ grid graph with k terminals, whose edge-lengths are chosen so that terminal distances are essentially "linearly independent" of one another. We use this independence to prove that no distance-preserving minor G' can have a small vertex-separator. Since G' is planar, we can apply the planar separator theorem [LT79], and obtain the desired lower bound.

Theorem 3. *For every $k \in \mathbb{N}$ there exists a planar graph $G = (V, E, \ell)$ (in particular, the $k \times k$ grid) and k terminals $R \subseteq V$, such that every distance-preserving planar graph $G' = (V', E', \ell')$ has $\Omega(k^2)$ vertices. In particular, $f^*(k, \text{Planar}) \geq \Omega(k^2)$.*

Proof. For simplicity we shall assume that k is even. Consider a grid graph G of size $k \times k$ with vertices (x, y) for $x, y \in [0, k-1]$. Let the length function ℓ be such that the length of all horizontal edges $((x, y), (x+1, y))$ is 1, and the length of each vertical edge $((x, y), (x, y+1))$ is $1 + \frac{1}{2^{x^2} \cdot k}$. Let $R_1 = \{(0, y) : y \in [0, \frac{k}{2} - 1]\}$, and $R_2 = \{(x, x) : x \in [\frac{k}{2}, k-1]\}$. Let the terminals in the graph be $R = R_1 \cup R_2$, so $|R| = k$. See Figure 1 for illustration.

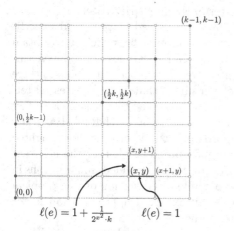

Fig. 1. A grid graph G and terminals R

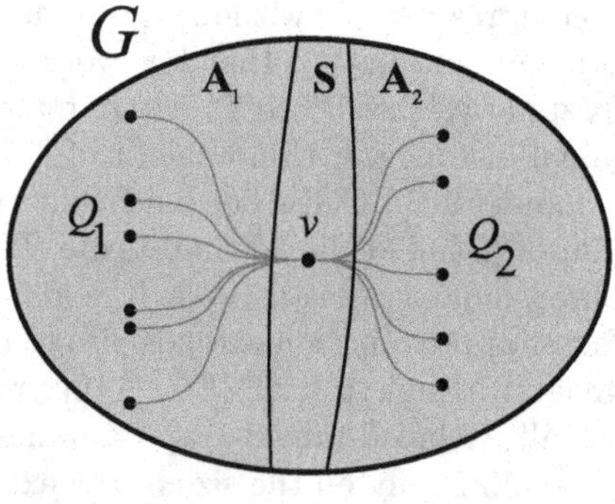

Fig. 2. Terminals on different sides connected by paths going through $v \in S$

It is easy to see that the shortest-path between a vertex $(0, y) \in R_1$ and a vertex $(x, x) \in R_2$ includes exactly x horizontal edges and $x - y$ vertical edges. Indeed, such paths have length smaller than $x + (x - y)(1 + \frac{1}{k}) \leq 2x - y + 1$. Any other path between these vertices will have length greater than $2x - y + 2$. Furthermore, the shortest path with x horizontal edges and $x - y$ vertical edges starting at vertex $(0, y)$ makes horizontal steps before vertical steps, since the vertical edge-lengths decrease as x increases, hence

$$d_G((0, y), (x, x)) = 2x - y + \frac{x - y}{2^{x^2} \cdot k}. \tag{1}$$

Assume towards contradiction that there exists a planar graph G' with less than $\frac{k^2}{1600}$ vertices that preserves terminal distances exactly. Since G' is planar, by the weighted version of the planar separator theorem by Lipton and Tarjan [LT79] with vertex-weight 1 on terminals and 0 on non-terminals, there exists a partitioning of V' into three sets A_1, S, and A_2 such that $w(S) \leq |S| \leq 2.5 \cdot \sqrt{\frac{k^2}{1600}} < \frac{3k}{40}$, each of A_1 and A_2 has at most $\frac{2k}{3}$ terminals, and there are no edges going between A_1 and A_2. Hence, for $i \in \{1, 2\}$ it holds that $w(A_i \cup S) \geq k/3$ and $w(A_i) \geq \frac{k}{3} - \frac{3k}{40} > \frac{k}{4}$.

Without loss of generality, we claim that $A_1 \cap R_1$ and $A_2 \cap R_2$ each have $\Theta(k)$ terminals. To see this, suppose without loss of generality that A_1 is the heavier of the two sets (i.e. $w(A_1) \geq \frac{k}{2} - \frac{3k}{40}$ and $\frac{k}{4} \leq w(A_2) \leq \frac{k}{2}$). Suppose also that $w(A_2 \cap R_2) \geq w(A_2 \cap R_1)$. Then $w(A_2 \cap R_2) \geq \frac{k}{8}$, and $w(A_2 \cap R_1) \leq \frac{1}{2} \cdot w(A_2) \leq \frac{k}{4}$, implying that $w(A_1 \cap R_1) \geq w(R_1) - (w(R_1 \cap A_2) + w(R_1 \cap S)) \geq \frac{k}{2} - (\frac{k}{4} + \frac{3k}{40}) = \frac{k}{5}$. In conclusion, without loss of generality it holds that $w(A_1 \cap R_1) \geq \frac{k}{5}$ and $w(A_2 \cap R_2) \geq \frac{k}{8}$. Let $Q_1 \subseteq A_1 \cap R_1$ and $Q_2 \subseteq A_2 \cap R_2$ be two sets with the exact sizes $\frac{k}{5}$ and $\frac{k}{8}$.

Every path between a terminal in Q_1 and a terminal in Q_2 goes through at least one vertex of the separator S. Overall, the vertices in the separator participate in $\frac{k}{8} \times \frac{k}{5}$ paths between Q_1 and Q_2. See Figure 2 for illustration. We will need the following lemma, which is proved below.

Lemma 1. *Let G', S, Q_1 and Q_2 be as described above. Then every vertex $v \in S$ participates in at most $|Q_1| + |Q_2| = \frac{k}{5} + \frac{k}{8}$ shortest paths between Q_1 and Q_2.*

Applying Lemma 1 to every vertex in S, at most $\frac{3k}{40} \cdot \frac{13k}{40} = \frac{39k^2}{1600} < \frac{k^2}{40}$ shortest paths between Q_1 and Q_2 go through S, which contradicts the fact that all $\frac{k}{8} \cdot \frac{k}{5} = \frac{k^2}{40}$ shortest-paths between Q_1 and Q_2 in G' go through the separator, and proves Theorem 3.   □

Proof (of Lemma 1). Define a bipartite graph H on the sets Q_1 and Q_2, with an edge between $(0, y) \in Q_1$ and $(x, x) \in Q_2$ whenever a shortest path in G' between $(0, y)$ and (x, x) uses the vertex v. We shall show that H does not contain an even-length cycle. Since H is bipartite, it contains no odd-length cycles either, making H a forest with $|E(H)| < |Q_1| + |Q_2| = \frac{k}{5} + \frac{k}{8}$, thereby proving the lemma.

Let us consider a potential $2s$-length (simple) cycle in H on the vertices $(0, y_1)$, (x_1, x_1), $(0, y_2)$, (x_2, x_2), ..., $(0, y_s)$, (x_s, x_s) (in that order), for particular $(0, y_i) \in Q_1$ and $(x_i, x_i) \in Q_2$. Every edge $((0, y), (x, x)) \in E(H)$ represents a shortest path in G' that uses v, thus

$$d_G((0, y), (x, x)) = d_{G'}((0, y), v) + d_{G'}(v, (x, x)). \tag{2}$$

If the above cycle exists in H, then the following equalities hold (by convention, let $y_{s+1} = y_1$). Essentially, we get that the sum of distances corresponding to "odd-numbered" edges in the cycle equals the one corresponding to "even-numbered" edges in the cycle.

$$\sum_{i=1}^{s} d_G((0, y_i), (x_i, x_i)) \overset{(2)}{=} \sum_{i=1}^{s} d_{G'}((0, y_i), v) + \sum_{i=1}^{s} d_{G'}(v, (x_i, x_i))$$

$$= \sum_{i=1}^{s} d_{G'}(v, (0, y_{i+1})) + \sum_{i=1}^{s} d_{G'}((x_i, x_i), v)$$

$$\overset{(2)}{=} \sum_{i=1}^{s} d_G((x_i, x_i), (0, y_{i+1})).$$

Plugging in the distances as described in (1) and simplifying, we obtain

$$\sum_{i=1}^{s} (2x_i - y_i + (x_i - y_i) \cdot \frac{1}{2^{x_i^2} \cdot k}) = \sum_{i=1}^{s} (2x_i - y_{i+1} + (x_i - y_{i+1}) \cdot \frac{1}{2^{x_i^2} \cdot k}),$$

or equivalently,

$$\sum_{i=1}^{s} \frac{y_i}{2^{x_i^2}} = \sum_{i=1}^{s} \frac{y_{i+1}}{2^{x_i^2}}$$

Suppose without loss of generality that $x_1 = \min\{x_i : i \in [1, s]\}$ (otherwise we can rotate the notations along the cycle), and that $y_1 > y_2$ (otherwise we can change the orientation of the cycle). Then we obtain

$$\frac{y_1 - y_2}{2^{x_1^2}} = \sum_{i=2}^{s} \frac{y_{i+1} - y_i}{2^{x_i^2}}.$$

However, since $y_1 > y_2$, the lefthand side is at least $\frac{1}{2^{x_1^2}}$, whereas the righthand side is $\sum_{i=2}^{s} \frac{y_{i+1}-y_i}{2^{x_i^2}} \leq (s-1) \cdot \frac{k}{2^{(x_1+1)^2}} \leq \frac{k^2}{2^{(x_1+1)^2}}$. Therefore it must hold that $2^{2x_1+1} \leq k^2$. Since $x_1 \geq \frac{k}{2}$, this inequality does not hold. Hence, for all s, no cycle of size $2s$ exists in H, completing the proof of Lemma 1.   □

3 $\Theta(k)$ Bounds for Constant Treewidth Graphs

In this section we prove Theorem 2, which bounds $f^*(k, \mathsf{Treewidth}(p))$. The upper and the lower bound are proved separately in Theorems 4 and 5 below.

3.1   An Upper Bound of $O(p^3 k)$

Theorem 4. *Every graph* $G = (V, E, \ell)$ *with treewidth* p *and a set* $R \subseteq V$ *of* k *terminals admits a distance-preserving minor* $G' = (V', E', \ell')$ *with* $|V'| \leq O(p^3 k)$. *In other words,* $f^*(k, \mathsf{Treewidth}(p)) \leq O(p^3 k)$.

The graph G' can in fact be computed in time polynomial in $|V|$ (see Remark 1).

Without loss of generality, we may assume that $k \geq p$, since otherwise the $O(k^4)$ bound mentioned in the introduction applies. To prove Theorem 4 we introduce the algorithm REDUCEGRAPHTW (depicted in Algorithm 2 below), which follows a divide-and-conquer approach. We use the small separators guaranteed by the treewidth p, to break the graph recursively until we have small, almost-disjoint subgraphs. We execute REDUCEGRAPHNAIVE (Algorithm 1) on each of these subgraphs with an altered set of terminals — the original terminals in the subgraph, plus the separator (*boundary*) vertices which disconnect these terminals from the rest of the graph — and we get many small distance-preserving minors; these are then combined into a distance-preserving minor G' of the original graph G.

Proof (of Theorem 4). The divide-and-conquer technique works as follows. Given a partitioning of V into the sets A_1, S and A_2, such that removing S disconnects A_1 from A_2, the graph G is divided into the two subgraphs $G[A_i \cup S]$ (the subgraph of G induced on $A_i \cup S$) for $i \in \{1,2\}$. For each $G[A_i \cup S]$, we compute a distance-preserving minor with respect to terminals set $(R \cap A_i) \cup S$, and denote it $\hat{G}_i = (\hat{V}_i, \hat{E}_i, \hat{\ell}_i)$. The two minors are then combined into a distance-preserving minor of G with respect to R, according to the following definition.

We define the *union* $H_1 \cup H_2$ of two (not necessarily disjoint) graphs $H_1 = (V_1, E_1, \ell_1)$ and $H_2 = (V_2, E_2, \ell_2)$ to be the graph $H = (V_1 \cup V_2, E_1 \cup E_2, \ell)$ where the edge lengths are $\ell(e) = \min\{\ell_1(e), \ell_2(e)\}$ (assuming infinite length when $\ell_i(e)$ is undefined). A crucial point here is that H_1, H_2 need not be disjoint – overlapping vertices are merged into one vertex in H, and overlapping edges are merged into a single edge in H.

Lemma 2. *The graph $\hat{G} = \hat{G}_1 \cup \hat{G}_2$ is a distance-preserving minor of G with respect to R.*

Proof (of Lemma 2). Note that since the *boundary vertices* in S exist in both $\hat{G}_1$ and $\hat{G}_2$, they are never contracted into other vertices. In fact, the only minor-operation allowed on vertices in S is the removal of edges (s_1, s_2) for two vertices $s_1, s_2 \in S$, when shorter paths in $G[A_1 \cup S]$ or $G[A_2 \cup S]$ are found. It is thus possible to perform both sequences of minor-operations independently, making $\hat{G}$ a minor of G.

A path between two vertices $t_1, t_2 \in R$ can be split into subpaths at every visit to a vertex in $R \cup S$, so that each subpath between $v, u \in R \cup S$ does not contain any other vertices in $R \cup S$. Since there are no edges between A_1 and A_2, each of these subpaths exists completely inside $G[A_1 \cup S]$ or $G[A_2 \cup S]$. Hence, for every subpath between $v, u \in R \cup S$ it holds that $d_G(v, u) = d_{G[A_i \cup S]}(v, u) = d_{\hat{G}_i}(v, u)$ for some $i \in \{1, 2\}$. Altogether, the shortest path in G is preserved in $\hat{G}$. It is easy to see that shorter paths will never be created, as these too can be split into subpaths such that the length of each subpath is preserved. Hence, $\hat{G}$ is a distance-preserving minor of G. $\square$

The graph G has bounded treewidth p, hence for every nonnegative vertex-weights $w(\cdot)$, there exists a set $S \subseteq V$ of at most $p + 1$ vertices (to simplify the analysis, we assume this number is p) whose removal separates the graph into two parts A_1 and A_2, each with $w(A_i) \leq \frac{2}{3}w(V)$. It is then natural to compute a distance-preserving minor for each part A_i by recursion, and then combine the two solutions using Lemma 2. We can use the weights $w(\cdot)$ to obtain a balanced split of the terminals, and thus $|R \cap A_i|$ is a constant factor smaller than $|R|$. However, when solving each part A_i, the boundary vertices S must be counted as "additional" terminals, and to prevent those from accumulating too rapidly, we compute (à la [Bod89]) a second separator S^i with different weights $w(\cdot)$ to obtain a balanced split of the boundary vertices accumulated so far.

Algorithm REDUCEGRAPHTW receives, in addition to a graph H and a set of terminals $R \subseteq V(H)$, a set of boundary vertices $B \subseteq V(H)$. Note that a terminal that is also on the boundary is counted only in B and not in R, so that $R \cap B = \emptyset$.

The procedure SEPARATOR(H, U) returns the triple $\langle A_1, S, A_2 \rangle$ of a separator S and two sets A_1 and A_2 such that $|S| \leq p$, no edges between A_1 and A_2 exist in G, and $|A_1 \cap U|, |A_2 \cap U| \leq \frac{2}{3}|U|$, i.e., using $w(\cdot)$ that is unit-weight inside U and 0 otherwise.

Algorithm 2. REDUCEGRAPHTW (graph H, required vertices R, boundary vertices B)

1: **if** $|R \cup B| \leq 18p$ **then**
2:     **return** REDUCEGRAPHNAIVE$(H, R \cup B)$ (see Algorithm 1)
3: $\langle A_1, S, A_2 \rangle \leftarrow$ SEPARATOR(H, R)
4: **for** $i = 1, 2$ **do**
5:     $\langle A_i^1, S^i, A_i^2 \rangle \leftarrow$ SEPARATOR$(H[A_i \cup S], (B \cap A_i) \cup S)$
6:     $R^i \leftarrow R \setminus (S \cup S^i)$
7:     $B^i \leftarrow B \cup S \cup S^i$
8:     **for** $j = 1, 2$ **do**
9:         $\hat{G}_i^j \leftarrow$ REDUCEGRAPHTW$(H[A_i^j \cup S^i], R^i \cap A_i^j, B^i \cap (A_i^j \cup S^i))$
10: **return** $(\hat{G}_1^1 \cup \hat{G}_1^2) \cup (\hat{G}_2^1 \cup \hat{G}_2^2)$.

See Figure 3 for an illustration of a single execution. Consider the recursion tree T on this process, starting with the invocation of REDUCEGRAPHTW$(G, R, \emptyset)$. A node $a \in V(T)$ corresponds to an invocation REDUCEGRAPHTW(H_a, R_a, B_a). The execution either terminates at line 2 (the stop condition), or performs 4 additional invocations b_i for $i \in [1, 4]$, each with $|R_{b_i}| \leq \frac{2}{3}|R_a|$. As the process continues, the number of terminals in R_a decreases, whereas the number of boundary vertices may increase. We show the following upper bound on the number of boundary vertices B_a.

Lemma 3. *For every $a \in V(T)$, the number of boundary vertices $|B_a| < 6p$.*

Proof (of Lemma 3). Proceed by induction on the depth of the node in the recursion tree. The lemma clearly holds for the root of the recursion-tree, since initially $B = \emptyset$. Suppose it holds for an execution with values H_a, R_a, B_a. When

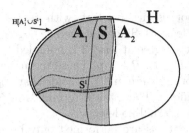

Fig. 3. The separators S (from line 3) and S^1 (from line 7), and the subgraph $H[A_1^1 \cup S^1]$ to be processed recursively (in line 11)

partitioning $V(H_a)$ into A_1, S, and A_2, the separator S has at most p vertices. From the induction hypothesis, $|B_a| < 6p$, making $|B_a \cup S| < 7p$.

The algorithm constructs another separator, this time separating the boundary vertices $B_a \cup S$. For $i = 1, 2$ and $j = 1, 2$ it holds that, $|S^i| \leq p$, $|A_i^j| \leq \frac{2}{3} \cdot |B_a \cup S| \leq \frac{2}{3} \cdot 7p = \frac{14}{3}p$, and so $|A_i^j \cup S^i| \leq \frac{14}{3}p + p < 6p$. The execution corresponding to the node a either terminates in line 2, or invokes executions with the values $A_i^j \cup S^i$ for $i, j = 1, 2$, hence all new invocations have less than $6p$ boundary vertices. □

We also prove the following lower bound on the number of terminals R_a.

Lemma 4. *Every $a \in V(T)$ is either a leaf of the tree T, or it has at least two children, denoted b_1, b_2, such that $|R_{b_1}|, |R_{b_2}| \geq p$.*

Proof (of Lemma 4). Consider a node $a \in V(T)$. If this execution terminates at line 2, a is a leaf and the lemma is true. Otherwise it holds that $|R_a \cup B_a| \geq 18p$. Since Lemma 3 states that $|B_a| \leq 6p$ it must holds that $|R_a| \geq 12p$.

When performing the separation of $V(H_a)$ into A_1, S, and A_2, the vertices R_a are distributed between A_1, S, and A_2, such that $|R_a \cap (A_i \cup S)| \geq \frac{1}{3}|R_a| = 4p$ for $i = 1, 2$. Since $|S| \leq p$ it must holds that $|(R_a \setminus S) \cap A_i| = |(R_a \cap (A_i \cup S)) \setminus S| \geq 3p$. When the next separation is performed, at most p of these $3p$ terminals belong to S^i, while the remaining terminals belong to R^i and are distributed between A_i^1 and A_i^2. At least one of these sets, without loss of generality A_i^1, gets $|R^i \cap A_i^1| \geq \frac{1}{2}2p = p$. This is a value of R_b for a child b of a in the recursion tree. Since this holds for both A_1 and A_2, at least two invocations b_1, b_2 with $|R_{b_i}| \geq p$ are made. □

The following observation is immediate from Lemma 3.

Observation 1. *Every node $a \in V(T)$ such that $|R_a| < p$ has $|R_a \cup B_a| \leq 7p$, thus is a leaf in T.*

To bound the size of the overall combined graph G' returned by the first call to REDUCEGRAPHTW, we must bound the number of leaves in T. To do that, we first consider the recursion tree T' created by removing those nodes a with $|R_a| < p$; these are leaves from Observation 1. From Lemma 4 every node in this tree (except the root) is either a leaf (with degree 1) or has at least two children (with degree at least 3). Since the average degree in a tree is less than 2, the number of nodes with degree at least 3 is bounded by the number of leaves. Every leaf b in the tree T' has $|R_b| \geq p$. These terminals do not belong to any boundary, so for every other leaf b' in T' it holds that $R_b \cap (R_{b'} \cup B_{b'}) = \emptyset$ and these p terminals are unique. There are k terminals in G, so there are $O(k/p)$ such leaves, and $O(k/p)$ internal nodes.

From Lemma 4, invocations are performed only by internal vertices in T'. Each internal vertex has 4 children, hence there are $O(k/p)$ invocations overall. Each leaf in T has $|R_a \cup B_a| \leq O(p)$, hence the graph returned from REDUCEGRAPHNAIVE(H_a) is a distance-preserving minor with $O(p^4)$ vertices. Using Lemma 2, the combination of these graphs is a distance-preserving minor $\hat{G}$ of G with respect to R. The minor $\hat{G}$ has $O(k/p \cdot p^4) = O(k \cdot p^3)$ vertices, proving Theorem 4.   □

Remark 1. Every action (edge or vertex removals, as well as edge contractions) taken by REDUCEGRAPHTW, is actually performed during a call to REDUCEGRAPHNAIVE, and an equivalent action to it would have been taken in executing the naive algorithm directly on G with respect to terminals R. Therefore, the naive algorithm returns distance-preserving minors of size $O(k \cdot p^3)$ to any graph with treewidth p. (When $p > k$ this statement holds by the $O(k^4)$ bound.)

3.2   A Lower Bound of $\Omega(pk)$

Theorem 5. *For every p and $k \geq p$ there is a graph $G = (V, E, \ell)$ with treewidth p and k terminals $R \subseteq V$, such that every distance-preserving minor G' of G with respect to R has $|V'| \geq \Omega(k \cdot p)$. In other words, $f^*(k, \mathsf{Treewidth}(p)) \geq \Omega(pk)$.*

Proof. Consider the bound shown in Theorem 3. The graph used to obtain this bound is a $k \times k$ grid, and has treewidth k. The following corollary holds.

Corollary 1. *For every $p \in \mathbb{N}$ there exists a graph G with treewidth p and p terminals $R \subseteq V$, such that every distance-preserving minor G' of G with respect to R has $|V'| \geq \Omega(p^2)$.*

Let the graph G consist of $\frac{k}{p}$ disjoint graphs G_i with p terminals, treewidth p, and distance-preserving minors with $|V'| \geq \Omega(p^2)$ as guaranteed by Corollary 1. Any distance-preserving minor of the graph G must preserve (in disjoint components) the distances between the terminals in each G_i. The graph G has k terminals, treewidth p, and any distance-preserving minor of it has $|V'| \geq \Omega(k \cdot p)$, thus proving Theorem 5.   □

4   Concluding Remarks

The algorithms mentioned in this paper (including the naive one) actually satisfy a stronger property: They output a minor $G' = (V', E', \ell')$ where $V' \subset V$ (meaning that every vertex in G' can be mapped back to a vertex in G) and

$$d_{G'}(u, v) \geq d_G(u, v) \qquad \forall u, v \in V'. \tag{3}$$

However, it is not hard to construct instances G (say, using Euclidean distances between random points in the plane, which yields in particular a planar graph), for which every distance-preserving minor G' satisfying the stronger property (3) must have $\Omega(k^4)$ vertices. Therefore, narrowing the gap between the current bounds $\Omega(k^2) \leq f^*(k) \leq O(k^4)$, might require, even for planar graphs, breaking away from the above paradigm.

References

[BG08]   Basu, A., Gupta, A.: Steiner point removal in graph metrics (2008),
           http://www.math.ucdavis.edu/~abasu/papers/SPR.pdf
           (unpublished manuscript)

[Bod89]   Bodlaender, H.L.: NC-algorithms for graphs with small treewidth. In: 14th
           International Workshop on Graph-Theoretic Concepts in Computer Sci-
           ence, pp. 1–10. Springer (1989)

[CE06]   Coppersmith, D., Elkin, M.: Sparse sourcewise and pairwise distance pre-
           servers. SIAM J. Discrete Math. 20, 463–501 (2006)

[CGN+06]   Chekuri, C., Gupta, A., Newman, I., Rabinovich, Y., Sinclair, A.: Embed-
           ding k-outerplanar graphs into ℓ_1. SIAM J. Discret. Math. 20(1), 119–136
           (2006)

[CXKR06]   Chan, T.-H.H., Xia, D., Konjevod, G., Richa, A.W.: A Tight Lower Bound
           for the Steiner Point Removal Problem on Trees. In: Díaz, J., Jansen, K.,
           Rolim, J.D.P., Zwick, U. (eds.) APPROX 2006 and RANDOM 2006. LNCS,
           vol. 4110, pp. 70–81. Springer, Heidelberg (2006)

[EGK+10]   Englert, M., Gupta, A., Krauthgamer, R., Räcke, H., Talgam-Cohen, I.,
           Talwar, K.: Vertex Sparsifiers: New Results from old Techniques. In: Serna,
           M., Shaltiel, R., Jansen, K., Rolim, J. (eds.) APPROX and RANDOM 2010,
           LNCS, vol. 6302, pp. 152–165. Springer, Heidelberg (2010)

[FM95]   Feder, T., Motwani, R.: Clique partitions, graph compression and speeding-
           up algorithms. J. Comput. Syst. Sci. 51(2), 261–272 (1995)

[GH61]   Gomory, R.E., Hu, T.C.: Multi-terminal network flows. Journal of the So-
           ciety for Industrial and Applied Mathematics 9, 551–570 (1961)

[Gup01]   Gupta, A.: Steiner points in tree metrics don't (really) help. In: 12th Annual
           ACM-SIAM Symposium on Discrete Algorithms, pp. 220–227. SIAM (2001)

[HKNR98]   Hagerup, T., Katajainen, J., Nishimura, N., Ragde, P.: Characterizing
           multiterminal flow networks and computing flows in networks of small
           treewidth. J. Comput. Syst. Sci. 57, 366–375 (1998)

[LT79]   Lipton, R.J., Tarjan, R.E.: A separator theorem for planar graphs. SIAM
           J. Appl. Math. 36(2), 177–189 (1979)

[PS89]   Peleg, D., Schäffer, A.A.: Graph spanners. J. Graph Theory 13(1), 99–116
           (1989)

[TZ05]   Thorup, M., Zwick, U.: Approximate distance oracles. J. ACM 52(1), 1–24
           (2005)

A Rounding by Sampling Approach
to the Minimum Size k-Arc Connected
Subgraph Problem

Bundit Laekhanukit[1], Shayan Oveis Gharan[2], and Mohit Singh[3]

[1] School of Computer Science, McGill University
[2] Department of Management Science and Engineering, Stanford University
[3] McGill University and Microsoft Research, Redmond

Abstract. In the k-arc connected subgraph problem, we are given a directed graph G and an integer k and the goal is the find a subgraph of minimum cost such that there are at least k-arc disjoint paths between any pair of vertices. We give a simple $(1 + 1/k)$-approximation to the unweighted variant of the problem, where all arcs of G have the same cost. This improves on the $1 + 2/k$ approximation of Gabow et al. [GGTW09].

Similar to the 2-approximation algorithm for this problem [FJ81], our algorithm simply takes the union of a k in-arborescence and a k out-arborescence. The main difference is in the selection of the two arborescences. Here, inspired by the recent applications of the rounding by sampling method (see e.g. [AGM+10, MOS11, OSS11, AKS12]), we select the arborescences randomly by sampling from a distribution on unions of k arborescences that is defined based on an *extreme point solution* of the linear programming relaxation of the problem. In the analysis, we crucially utilize the sparsity property of the extreme point solution to upper-bound the size of the union of the sampled arborescences.

To complement the algorithm, we also show that the integrality gap of the minimum cost strongly connected subgraph problem (i.e., when $k = 1$) is at least $3/2 - \epsilon$, for any $\epsilon > 0$. Our integrality gap instance is inspired by the integrality gap example of the asymmetric traveling salesman problem [CGK06], hence providing further evidence of connections between the approximability of the two problems.

1 Introduction

In the *minimum cost k-arc connected spanning subgraph* (min-cost k-ACSS) problem, we are given a directed graph $G = (V, A)$ with cost $c : A \to R$ on the arcs and a connectivity requirement k. The goal is to find a spanning subgraph $G' = (V, A')$ of G of minimum total cost which is k-arc connected, i.e., every pair of vertices have at least k-arc disjoint paths between them. The special case of $k = 1$, 1-ACSS problem, is called the *minimum cost strongly connected subgraph* problem. In the unweighted variant of k-ACSS, the *minimum size k-arc connected spanning subgraph* (min-size k-ACSS) problem, where all arcs of G have the same cost, we want to minimize the number of arcs that we choose.

A. Czumaj et al. (Eds.): ICALP 2012, Part I, LNCS 7391, pp. 606–616, 2012.
© Springer-Verlag Berlin Heidelberg 2012

The min-cost k-ACSS problem has a 2-approximation algorithm [FJ81], and it has been a long standing open problem to improve this bound. Significant attention has been given to the unweighted variant of the problem. In particular, the minimum size strongly connected subgraph problem is very well studied [FJ81, KRY94, KRY96, Vet01, ZNI03], and the current best approximation ratio is 3/2, which is due to Vetta [Vet01]. The min-size k-ACSS problem has been shown to be easier as k increases [CT00, Gab04, GGTW09], and the best approximation ratio is $1 + 2/k$ that is given in the work of Gabow et al. [GGTW09]. This approximation ratio is almost tight as the min-size k-ACSS problem does not admit $(1 + \epsilon/k)$-approximation, for some fixed $\epsilon > 0$, unless P=NP [GGTW09]. Similar to the directed case, the *minimum size k-edge connected subgraph spanning* problem, an undirected variant of the min-size k-ACSS problem, is known to be easier as k increases, and the best known approximation ratio for this problem is $1 + 1/(2k) + O(1/k^2)$ due to Gabow and Gallagher [GG08].

1.1   Our Results

In this paper, we give improved upper and lower bounds for the k-ACSS problem. We first show the following improved algorithms for the min-size k-ACSS problem.

Theorem 1. *For any $k \geq 1$, there is a $\min\{7/4, 1 + 1/k\}$-approximation algorithm for the min-size k-ACSS problem.*

Similar to the simple 2-approximation algorithm for the minimum-cost k-ACSS problem, our algorithm takes the union of a k in-arborescence and a k out-arborescence. The main difference is in the selection of the two arborescences. Here, we select the arborescences randomly by sampling from a distribution on unions of k arborescences that is defined by the linear programming relaxation of the problem. In particular, we write a convex combination of the unions of k-arborescences such that the marginal probability of each arc is bounded above by its fraction in the solution of LP relaxation.

The algorithm essentially employs the rounding by sampling method that recently has been applied to various problems in the algorithm design and on-line optimization literature (c.f. [AGM+10, MOS11, OSS11, AKS12]), while the analysis is much simpler in our setting. Here, the main technical difference is a crucial use of the extreme point solutions of LP relaxation. In particular, because of the sparsity of the extreme point solutions, we can argue that the union of k in-arborescences and k out-arborescences is not much larger than the size of the support of the LP extreme point solution and thus the size of the optimum.

Our result improves on the $(1 + \frac{2}{k})$-approximation of Gabow et al. [GGTW09] for the min-size k-ACSS problem, for any $k > 0$. Furthermore, for the minimum size strongly connected subgraph problem, while we do not improve the approximation factor of $\frac{3}{2}$ [Vet01], our algorithm is much simpler and gives a possible direction for weighted version of the problem.

To complement the positive results, we prove that the integrality gap of the natural linear programming relaxation of the strongly connected subgraph problem is bounded below by $3/2 - \epsilon$ for any $\epsilon > 0$.

Theorem 2. *For any $\epsilon > 0$, the integrality gap of the standard linear programming relaxation for the minimum cost strongly connected subgraph problem is at least $\frac{3}{2} - \epsilon$.*

To the best of our knowledge, there is no explicit construction that gives a lower bound on the integrality gap of the minimum cost strongly connected subgraph problem. Our integrality gap example builds on a similar construction for the asymmetric traveling salesman problem [CGK06] and shows stronger connections between the two problems.

1.2 Notations

Let $\delta_G^+(U) := \{(u,v) \in E : u \in U, v \in V \setminus U\}$ denote the set of arcs leaving U in a graph G; if G is clear in the context, we will skip the subscript.

A graph G is k-*arc connected* if and only if every (proper) subset of vertices $U \subset V$ have at least k leaving arcs, i.e., $|\delta_G^+(U)| \geq k$, and G is *strongly connected* if it is 1-arc connected. We may drop the subscript if G is clear in the context. We use the following Linear Programming relaxation for k-ACSS.

$$\text{(LP-ACSS)} \quad \text{minimize} \quad \sum_{a \in A} c_a x_a$$

$$\text{subject to} \quad x(\delta^+(U)) \geq k \quad \forall U \neq \emptyset, U \subsetneq V$$

$$0 \leq x_a \leq 1 \quad \forall e \in E,$$

where $x(\delta^+(U)) = \sum_{a \in \delta^+(U)} x_a$. Throughout the paper x will always be an optimum solution of the (LP-ACSS).

For any vector $y : A \to \mathbb{R}$, and a set $F \subset A$ of arcs, $y(F) := \sum_{a \in F} y_a$, is the sum of the values of the arcs in F, and $c(F) := \sum_{a \in F} c_a$ is the sum of the cost of the arcs in F. Also, $\chi(F)$ denotes the characteristic vector of the set F, i.e., $\chi(F)_a = 1$ if $a \in F$ and $\chi(F)_a = 0$ otherwise.

2 An Approximation Algorithm for Min-Size k-ACSS

In this section, we prove Theorem 1: given a graph G, we give a polynomial time algorithm that finds a k-arc connected subgraph of G such that it has no more than $\min\{1+1/k, 7/4\}$ of the arcs of the optimum solution. Before describing the algorithm, we need to recall some of the properties of arborescences in directed graphs.

Given a directed graph G and a (root) vertex $r \in V$, an r-*out arborescence* T of G is a directed tree rooted at r that contains a path from r to every other vertex of G. An *r-out k-arborescence* is a subgraph T of G that is the union of k arc-disjoint r-out arborescences. An *r-in arborescence* and an *r-in k-arborescence* are defined analogously. The following polyhedron plays an important role in the design and analysis of our algorithm.

$$P^{out} = \left\{ y : y(\delta^+(U)) \geq k, \quad \forall \emptyset \neq U \subsetneq V \setminus \{r\}, 0 \leq y \leq 1 \right\}$$

Frank [Fra79] showed that P^{out} is the up hull of the convex hull of r-out k-arborescences (see Corollary 53.6a [Sch03]), and it can be seen that every feasible solution of (LP-ACSS) is a point in P^{out}. Vempala and Carr [CV02] gave a polynomial-time algorithm that allows us to write a point $x \in P^{out}$ as a convex combination of k arc-disjoint arborescences. Their algorithm requires a polynomial-time algorithm for finding an r-out k-arborescences [Edm73, Gab91].

Lemma 1. *[Fra79, CV02, Edm73, Gab91] P^{out} is the convex hull of subsets of A containing r-out k-arborescences. Moreover, given any fractional solution $y \in P^{out}$, there is a polynomial time algorithm that finds a convex combination of r-out k-arborescences, $T_1, \ldots, T_l$, such that*

$$y \geq \sum_{i=1}^{l} \lambda_i \chi(T_i).$$

The above lemma holds analogously for the r-in arborescences. Now, since $x \in P^{out}$, we can write a distribution of r-out(in) k-arborescences such that probability of each arc $a \in A$ chosen in a random k-arborescence is bounded above by x_a:

Corollary 1. *There are distributions $\mathcal{D}_{in}(r)$ and $\mathcal{D}_{out}(r)$ of r-in k-arborescences and r-out k-arborescences, such that the marginal value of each arc $a \in A$ is bounded above by x_a, i.e., for all arcs $a \in A$,*

$$\mathbf{P}_{T \sim \mathcal{D}_{in}(r)} [a \in T] \leq x_a,$$
$$\mathbf{P}_{T \sim \mathcal{D}_{out}(r)} [a \in T] \leq x_a.$$

Moreover, these distributions can be computed in polynomial time.

Now, we are ready to describe our algorithm. We sample k-arborescences T_{in} and T_{out} independently from $\mathcal{D}_{in}$ and $\mathcal{D}_{out}$, respectively, and we then return $T_{in} \cup T_{out}$ as an output. The details are described in Algorithm 1.

Next, we show that the approximation ratio of the above algorithm is no more than $1 + 1/k$.

Theorem 3. *For any directed graph G, Algorithm 1 always produces a k-arc connected subgraph of G such that the expected size of the solution is no more than $\min\{7/4, 1 + 1/k\}$ of the optimum.*

Proof. First, we show that the union of any pair of r-in and r-out k-arborescences is k-arc connected. Let $T_{in}(T_{out})$ be a r-in (r-out) k-arborescence, and $H =$

Algorithm 1. Approximation Algorithm for Min-Size k-ACSS

1: Solve (LP-ACSS) to get an optimum *extreme point* solution x.
2: Find distributions $\mathcal{D}_{in}(r)$ and $\mathcal{D}_{out}(r)$ on r-in and r-out k-arborescences, respectively, such that the marginal value of each arc $a \in A$ is bounded above by x_a.
3: Sample an r-in k-arborescence T_{in} from $\mathcal{D}_{in}(r)$ and an r-out k-arborescence T_{out}, *independently*, from $\mathcal{D}_{out}(r)$.
4: **return** $T_{in} \cup T_{out}$.

$T_{in} \cup T_{out}$. Since both T_{in} and T_{out} are unions of k arc-disjoint arborescences, there are k arc-disjoint paths from each of the vertices to r and k arc-disjoint paths from r to each of the vertices. Therefore, H remains strongly connected after removing any set of $k - 1$ arcs. Hence, H is k-arc connected.

It remains to show that the expected size of the solution is no more than $\min\{1 + 1/k, 7/4\}$ of the optimum, i.e.,

$$\frac{\mathbf{E}_{T_{in} \sim \mathcal{D}_{in}(r), T_{out} \sim \mathcal{D}_{out}(r)} \left[|T_{in} \cup T_{out}|\right]}{|\mathrm{OPT}|} \le \min\left\{\frac{7}{4}, 1 + \frac{1}{k}\right\}.$$

To simplify the notation, we will skip the subscript and write $\mathbf{E}\left[|T_{in} \cup T_{out}|\right]$ to mean $\mathbf{E}_{T_{in} \sim \mathcal{D}_{in}(r), T_{out} \sim \mathcal{D}_{out}(r)} \left[|T_{in} \cup T_{out}|\right]$. Similarly, we will skip the subscripts for $\mathbf{P}_{T_{in} \sim \mathcal{D}_{in}(r)}\left[a \in T_{in}\right]$ and $\mathbf{P}_{T_{out} \sim \mathcal{D}_{out}(r)}\left[a \in T_{out}\right]$.

Since T_{in} and T_{out} are chosen independently,

$$\mathbf{E}\left[|T_{in} \cup T_{out}|\right] = \sum_{a \in A} \left\{\mathbf{P}\left[a \in T_{in}\right] + \mathbf{P}\left[a \in T_{out}\right] - \mathbf{P}\left[a \in T_{in}\right] \cdot \mathbf{P}\left[a \in T_{out}\right]\right\}$$

$$\le \sum_{a \in A} 2x_a - \sum_{a \in A} x_a^2.$$

The last inequality follows from Corollary 1 and the fact that $x_a \le 1$ for all $a \in A$. Let $F := \{a : 0 < x_a < 1\}$ be the set of the fractional arcs (i.e., set of arcs with non-integer values in the solution of (LP-ACSS)). Since x is an optimal solution of (LP-ACSS), $|\mathrm{OPT}| \ge \sum_{a \in A} x_a$. Therefore,

$$\frac{\mathbf{E}\left[|T_{in} \cup T_{out}|\right]}{|\mathrm{OPT}|} \le 1 + \frac{\sum_{a \in A} x_a - \sum_{a \in A} x_a^2}{\sum_{a \in A} x_a}$$

$$= 1 + \frac{x(F) - \sum_{a \in F} x_a^2}{x(A)}$$

$$\le 1 + \frac{x(F) - x(F)^2/|F|}{x(A)}, \tag{1}$$

where the last inequality follows from Jenson's inequality and the fact that $f(t) = -t^2$ is a concave function.

Since x is an extreme point solution of (LP-ACSS), x is a sparse vector. It follows from the work of Melkonian and Tardos [MT04] (see also [GGTW09]), that the number of fractional arcs, $|F|$, is no more than $4n$. Hence,

$$\frac{x(F) - x(F)^2/|F|}{x(A)} \le \frac{x(F) - x(F)^2/4n}{x(A)} \le \frac{n}{x(A)} \le \frac{1}{k}, \tag{2}$$

where the second inequality follows since $x(F) - x(F)^2/4n$ attains its maximum at $x(F) = 2n$, and the last inequality follows from the fact that $x(A) = \sum_{v \in V} x(\delta^+(v)) \geq nk$. On the other hand, since $x(F) \leq x(A)$, we get

$$\frac{x(F) - x(F)^2/|F|}{x(A)} \leq \frac{1}{2} + \frac{x(F) - x(F)^2/2n}{2x(A)} \leq \frac{1}{2} + \frac{n}{4x(A)} \leq \frac{3}{4}. \tag{3}$$

The theorem simply follows by putting equations (1),(2),(3) together.   □

Remark 1. Since the distributions $\mathcal{D}_{in}(r)$ and $\mathcal{D}_{out}(r)$ can be constructed such that the support of each distribution has size only polynomially large in n, the algorithm can be derandomized simply by choosing a pair of k-arborescences that have the minimum number of arcs in their union.

3   A Lower Bound on the Integrality Gap

In this section, we prove Theorem 2: we show a lower-bound of $1.5 - \epsilon$, for any arbitrary small $\epsilon > 0$, on the integrality gap of (LP-ACSS) for $k = 1$. Our construction is based on the LP-gap construction of the asymmetric traveling saleman problem by Charikar, Goemans and Karloff [CGK06].

3.1   Construction

Let $r > 0$ be an integral parameter that will be defined later. We start by defining the integrality gap example, $G(d, s, t)$, by a recursive construction of depth d. In any graph $G(d, s, t)$, d is the *depth*, r is the number of *columns*, s, t are the *source, sink* vertices, respectively. We allow s and t to be the same vertex. We will construct $G(d, s, t)$ inductively such that it contains exactly r copies of $G(d - 1, ., .)$.

We start by describing $G(1, s, t)$. The graph consists of s, t and r distinct vertices $v_1, \ldots, v_r$. Let $v_0 = s$ and $v_{r+1} = t$; note that v_0 and v_{r+1} may be the same depending on the given parameters s and t. For any $1 \leq i \leq r + 1$, we include arcs (v_i, v_{i-1}) and (v_{i-1}, v_i) in $G(1, s, t)$. Therefore,

$$A(G(1, s, t)) := \{(v_{i-1}, v_i), (v_i, v_{i-1}), 1 \leq i \leq r + 1\}.$$

Next, we define $G(d, s, t)$. The graph consists of s, t and r distinct copies of $G(d - 1, ., .)$. In particular, let $v_1, \ldots, v_r, u_1, \ldots, u_r$ be $2r$ distinct vertices, and $v_0 = u_{r+1} = s$ and $v_{r+1} = u_0 = t$. For any $1 \leq i \leq r$, include a distinct copy of $G(d - 1, ., .)$ with source u_i and sink v_i. Also, for any $1 \leq i \leq r + 1$, include the arcs (v_i, v_{i-1}) and (u_{i-1}, u_i). Therefore,

$$A(G(d, s, t)) := \{(u_{i-1}, u_i), (v_i, v_{i-1}), 1 \leq i \leq r+1\} \cup \left\{ \bigcup_{i=1}^{r} A(G(d - 1, u_i, v_i)) \right\}.$$

Figure 3.1 illustrates the graph $G(3, s, s)$ for $r = 3$.

Our integrality gap example is $G_d := G(d, s, s)$, where the source and the sink are unified. The ith column of G_d is defined to be the ith copy of the $G(d-1, ., .)$, i.e., $G_d^{(i)} := G(d-1, u_i, v_i)$. The set of arcs that connect the r columns with s and t, i.e., $A(G_d) \setminus \bigcup_{i=1}^{r} A(G_d^{(i)})$, are denoted by d^{th} level arcs. Similarly, the lth level arcs of G_d are defined to be set of arcs included at the lth level of induction. For example, the $(d-1)^{th}$ level arcs of G_d are $\bigcup_{i=1}^{r} \left(A(G_d^{(i)}) \setminus \bigcup_{j=1}^{r} A(G_d^{(i;j)}) \right)$, where $G_d^{(i;j)}$ is the jth column of $G_d^{(i)}$.

We define the costs of the arcs of G_d such that, for any $1 \le l \le d$, the total cost of the arcs at level l is equal to 1. In other words, the cost of each arc at level l, $c_d(l)$, is the reciprocal of the number of arcs at level l. By the construction of G_d, we have

$$c_d(l) := \frac{1}{2(r+1)r^{d-l}}. \qquad (4)$$

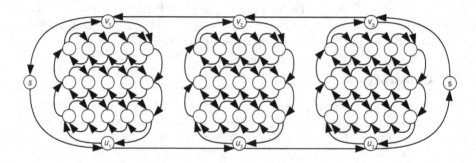

Fig. 1. An illustration of the graph $G(3, s, s)$, for $r = 3$. Note that the vertices labeled "s" on the left and on the right are the same.

3.2   Lower Bounding the Integrality Gap

We show that for any $d > 0$, and for a sufficiently large r, the integrality gap of the instance $G(d, s, s)$ is at least $3/2 - O(1/d)$.

Theorem 4. *For any $d > 0$ and $r \ge d$, the integrality gap of the instance $G(d, s, s)$ is at least $3/2 - 8/d$.*

First, we show that the optimal value of the LP is at most $d/2$. Define $x_a^* := 1/2$ for all arcs $a \in A(G_d)$. Charikar et al. [CGK06] show that x^* belongs to the Held-Karp relaxation polytope [HK70]. Since any solution of the Held-Karp relaxation polytope is a feasible solution to (LP-ACSS) for $k = 1$, x^* is also a feasible solution to (LP-ACSS). Furthermore, since the sum of the cost of the arcs of G_d is d, i.e., $c(A(G_d)) = d$, we have $\sum_a c(a)x_a^* = d/2$. Hence, the optimal value of LP is at most $d/2$.

Lemma 2 (Charikar et al. [CGK06]). *For $k = 1$, the optimum value of* (LP-ACSS) *for the graph G_d is at most $d/2$.*

For any $d > 0$, let H_d be the minimum cost strongly connected subgraph of G_d, and $T(d) := c(A(H_d))$ be the cost of H_d. In the rest of the section, we prove the following lemma:

Lemma 3. *For all $d > 0$,*

$$T(d) \geq \frac{3d-1}{4} - \frac{3d}{r}. \tag{5}$$

Let $H_d^{(i)} := H_d \cap G_d^{(i)}$ be the ith column of H_d. Observe that $H_d^{(i)}$ can be incident to (at most) four arcs of the dth level arcs of H_d. Let

$$A_d(i) := \{(v_i, v_{i-1}), (v_{i+1}, v_i), (u_{i-1}, u_i), (u_i, u_{i+1})\} \cap A(H_d),$$

be the set of those arcs. We can lower-bound $c(A(H_d^{(i)}))$ based on the number of arcs that is incident to $H_d^{(i)}$ (note that since H_d is strongly connected, $|A_d(i)| \geq 2$):

Case 1: $|A_d(i)| \geq 3$
In this case, we must have

$$c(A(H_d^{(i)})) \geq T(d-1)/r. \tag{6}$$

The inequality essentially follows from the fact that $H_d^{(i)}$ is a strongly connected subgraph of G_{d-1}. This is because the remaining arcs of the graph, $H_d \setminus H_d^{(i)}$, can only connect (or unify) the source and sink of $H_d^{(i)}$, i.e., u_i and v_i. The $1/r$ factor follows from the fact that the cost of each arc of G_{d-1} is r times the corresponding arc in $G_d^{(i)}$.

Case 2: $|A_d(i)| = 2$, and each of u_i and v_i is incident to exactly one arc of $A_d(i)$
Similar to the previous case, here we have

$$c(A(H_d^{(i)})) \geq T(d-1)/r. \tag{7}$$

As we will see in Lemma 4, at most two columns of H_d may satisfy this case. Therefore, although we have the worse lower-bound on $c(H_d^{(i)})$ in this case, it has an insignificant effect on the final lower-bound.

Case 3: $|A_d(i)| = 2$, and one of u_i or v_i is incident to none of the arcs of $A_d(i)$
Here we obtain a better lower-bound. For $1 \leq j \leq r$, let $H_d^{(i;j)}$ be the jth column of $H_d^{(i)}$ with source $u_{i,j}$ and sink $v_{i,j}$. It follows that the only u_i, v_i (or v_i, u_i) path in H_d is the one that is made by the $d-1$ level arcs connecting the columns of $H_d^{(i)}$, i.e., $u_i, u_{i,1}, u_{i,2}, \ldots, u_{i,r}, v_i$ (resp. $v_i, v_{i,r}, v_{i,r-1}, \ldots, v_{i,1}, u_i$). Therefore, $H_d^{(i)}$ must contain all of the $(d-1)^{th}$ level arcs of $G_d^{(i)}$. Since each

column of $H_d^{(i)}$ is incident to 4 arcs of level $(d-1)^{th}$, by repeated application of case 1, we obtain

$$c(A(H_d^{(i)})) \geq 2(r+1)c_d(d-1) + \sum_{j=1}^{r} c(A(H_d^{(i;j)}))$$

$$= 2(r+1)c_d(d-1) + \frac{T(d-2)}{r}. \tag{8}$$

Next, we show that there are at most 2 columns satisfying the second case.

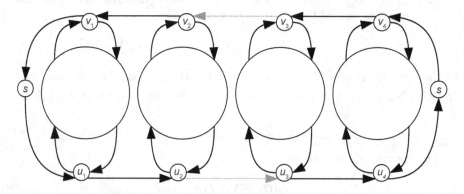

Fig. 2. An illustration of H_d where the second column satisfies Case 2. The black arcs represent the arcs of H_d, and grey arcs represent the removed arcs. Observe that every arc at level d is a min-cut of H_d.

Lemma 4. *At most two columns of H_d satisfy the second case.*

Proof. The proof is a simple case analysis argument. First, observe that there exists a column satisfying the second case in H_d if and only if $(v_i, v_{i-1}), (u_{i-1}, u_i) \notin H_d$ for some $1 \leq i \leq r+1$. Now, suppose this is the case. It then follows that H_d must contain all arcs at level d except these two arcs because each of the other arcs is a min-cut of H_d. See Figure 3.2. Therefore, all except (at most) two of the columns of H_d are adjacent to exactly 4 arcs at level d.   □

Now we are ready to prove Lemma 3.

Proof of Lemma 3. We prove by induction. First observe that $T(0) = 0$ and $T(1) = 1/2$ satisfying (5). Let $N_1, N_2, (r - N_1 - N_2)$ be the number of columns satisfying case 1, 2, 3, respectively. We divide the cost of each arc at level d equally between the columns incident to it. This incurs a cost of $3c_d(d)/2$ to the columns satisfying case 1, $c_d(d)$ to the rest of the columns and at least $c_d(d)$ to the source vertex s (note that s is adjacent to at least two arcs at level d). Using equations (6), (7), (8) we get:

$$T(d) \geq c_d(d) + \min_{0 \leq N_1, N_2 \leq r} \left\{ N_1 \left(\frac{3c_d(d)}{2} + \frac{T(d-1)}{r} \right) + N_2 \left(c_d(d) + \frac{T(d-1)}{r} \right) \right.$$

$$\left. + (r - N_1 - N_2) \left(c_d(d) + 2(r+1)c_d(d-1) + \frac{T(d-2)}{r} \right) \right\}$$

$$\geq \min_{0 \leq N \leq r} \left\{ N \left(\frac{3c_d(d)}{2} + \frac{T(d-1)}{r} \right) \right.$$

$$\left. + (r - N) \left(c_d(d) + 2(r+1)c_d(d-1) + \frac{T(d-2)}{r} \right) \right\}$$

$$\geq \min_{0 \leq \alpha \leq 1} \left\{ \alpha \left(\frac{3r}{4(r+1)} + T(d-1) \right) + (1 - \alpha) \left(\frac{3r}{2(r+1)} + T(d-2) \right) \right\}$$

$$\geq \min \{ 3/4 + T(d-1), 3/2 + T(d-2) \} - 3/r.$$

The second inequality follows from the fact that $N_2 \leq 2$. The third inequality follows from equation (4), and the last one follows from a simple algebra.

Now, we may apply the induction hypothesis to $T(d-1)$ and $T(d-2)$. We get

$$T(d) \geq \min \left\{ \frac{3}{4} + \frac{3(d-1)-1}{4} - \frac{3(d-1)}{r}, \frac{3}{2} + \frac{3(d-2)-1}{4} - \frac{3(d-2)}{r} \right\} - \frac{3}{r}$$

$$\geq \frac{3d-1}{4} - \frac{3d}{r},$$

which completes the proof. □

This completes the proof of Theorem 4.

4 Conclusion

We presented a simple $(1 + 1/k)$-approximation algorithm based on the rounding by sampling method for the minimum size k-arc connected subgraph problem. Unlike recent applications of the rounding by sampling method [AGM+10, OSS11], our algorithm has a flavor of the *iterated rounding method* [Jai01] in its particular use of the extreme point solutions. The main open problem is to find a better than factor 2-approximation for the minimum cost strongly connected subgraph problem.

We also showed that the integrality gap of the minimum cost strongly connected subgraph problem is at least $1.5 - \epsilon$, for any $\epsilon > 0$. This leaves an interesting open question whether the lower bound of $1 + \Omega(1/k)$ is achievable for the minimum size k-arc connected subgraph problem as well.

Acknowledgments. We thank Joseph Cheriyan for useful discussions on the preliminary construction of the integrality-gap instance.

References

[AGM+10] Asadpour, A., Goemans, M.X., Madry, A., Gharan, S.O., Saberi, A.: An $O(\log n / \log \log n)$-approximation algorithm for the asymmetric traveling salesman problem. In: SODA, pp. 379–389 (2010)

[AKS12]   An, H.-C., Kleinberg, R., Shmoys, D.B.: Improving Christofides' algorithm for the s-t path tsp. In: STOC (to appear, 2012)

[CGK06]   Charikar, M., Goemans, M.X., Karloff, H.J.: On the integrality ratio for the asymmetric traveling salesman problem. Math. Oper. Res. 31(2), 245–252 (2006); Preliminary version in FOCS 2004

[CT00]   Cheriyan, J., Thurimella, R.: Approximating minimum-size k-connected spanning subgraphs via matching. SIAM J. Comput. 30(2), 528–560 (2000); Preliminary version in FOCS 1996

[CV02]   Carr, R.D., Vempala, S.: Randomized metarounding. Random Struct. Algorithms 20(3), 343–352 (2002)

[Edm73]   Edmonds, J.: Edge-disjoint branchings. In: Combinatorial algorithms (Courant Comput. Sci. Sympos. 9, New York Univ., New York, 1972), pp. 91–96. (1973)

[FJ81]   Frederickson, G.N., JáJá, J.: Approximation algorithms for several graph augmentation problems. SIAM J. Comput. 10(2), 270–283 (1981)

[Fra79]   Frank, A.: Covering branchings. Acta Scientiarum Mathematicarum (Szeged) 41, 77–81 (1979)

[Gab91]   Gabow, H.N.: A matroid approach to finding edge connectivity and packing arborescences. In: STOC, pp. 112–122 (1991)

[Gab04]   Gabow, H.N.: Special edges, and approximating the smallest directed k-edge connected spanning subgraph. In: SODA, pp. 234–243 (2004)

[GG08]   Gabow, H.N., Gallagher, S.: Iterated rounding algorithms for the smallest k-edge connected spanning subgraph. In: SODA, pp. 550–559 (2008)

[GGTW09]   Gabow, H.N., Goemans, M.X., Tardos, É., Williamson, D.P.: Approximating the smallest k-edge connected spanning subgraph by LP-rounding. Networks 53(4), 345–357 (2009); Preliminary version in SODA 2005

[HK70]   Held, M., Karp, R.: The traveling salesman problem and minimum spanning trees. Operations Research 18, 1138–1162 (1970)

[Jai01]   Jain, K.: A factor 2 approximation algorithm for the generalized steiner network problem. Combinatorica 21(1), 39–60 (2001); Preliminary version in FOCS 1998

[KRY94]   Khuller, S., Raghavachari, B., Young, N.E.: Approximating the minimum equivalent digraph. In: Proceedings of the Fifth Annual ACM-SIAM Symposium on Discrete Algorithms, SODA 1994, pp. 177–186. Society for Industrial and Applied Mathematics, Philadelphia (1994)

[KRY96]   Khuller, S., Raghavachari, B., Young, N.: On strongly connected digraphs with bounded cycle length. Discrete Applied Mathematics 69(3), 281–289 (1996)

[MOS11]   Manshadi, V.H., Gharan, S.O., Saberi, A.: Online stochastic matching: Online actions based on offline statistics. In: SODA, pp. 1285–1294 (2011)

[MT04]   Melkonian, V., Tardos, E.: Algorithms for a network design problem with crossing supermodular demands. Networks 43(4), 256–265 (2004)

[OSS11]   Gharan, S.O., Saberi, A., Singh, M.: A randomized rounding approach to the traveling salesman problem. In: FOCS, pp. 550–559 (2011)

[Sch03]   Schrijver, A.: Combinatorial Optimization. Springer (2003)

[Vet01]   Vetta, A.: Approximating the minimum strongly connected subgraph via a matching lower bound. In: SODA, pp. 417–426 (2001)

[ZNI03]   Zhao, L., Nagamochi, H., Ibaraki, T.: A linear time 5/3-approximation for the minimum strongly-connected spanning subgraph problem. Inf. Process. Lett. 86, 63–70 (2003)

Classical and Quantum Partition Bound and Detector Inefficiency*

Sophie Laplante[1], Virginie Lerays[1], and Jérémie Roland[2]

[1] LRI, Université Paris-Sud 11
[2] ULB, QuIC, Ecole Polytechnique de Bruxelles

Abstract. We study randomized and quantum efficiency lower bounds in communication complexity. These arise from the study of zero-communication protocols in which players are allowed to abort. Our scenario is inspired by the physics setup of Bell experiments, where two players share a predefined entangled state but are not allowed to communicate. Each is given a measurement as input, which they perform on their share of the system. The outcomes of the measurements should follow a distribution predicted by quantum mechanics; however, in practice, the detectors may fail to produce an output in some of the runs. The efficiency of the experiment is the probability that neither of the detectors fails.

When the players share a quantum state, this leads to a new bound on quantum communication complexity (eff*) that subsumes the factorization norm. When players share randomness instead of a quantum state, the efficiency bound (eff), coincides with the partition bound of Jain and Klauck. This is one of the strongest lower bounds known for randomized communication complexity, which subsumes all the known combinatorial and algebraic methods including the rectangle (corruption) bound, the factorization norm, and discrepancy. The lower bound is formulated as a convex optimization problem. In practice, the dual form is more feasible to use, and we show that it amounts to constructing an explicit Bell inequality (for eff) or Tsirelson inequality (for eff*). For one-way communication, we show that the quantum one-way partition bound is tight for classical communication with shared entanglement up to arbitrarily small error.

1 Introduction

1.1 Communication Complexity and the Partition Bound

Recently, Jain and Klauck [1] proposed a new lower bound on randomized communication complexity which subsumes two families of methods: the algebraic methods, including the nuclear norm and factorization norm, and combinatorial methods, including discrepancy and the rectangle or corruption bound.

A longstanding open problem is whether there are total functions for which there is an exponential gap between classical and quantum communication complexities. Many partial results have been given [2,3,4,5], most recently [6]. These

---

* Full version available as arXiv:1203.4155 and ECCC TR12-023.

A. Czumaj et al. (Eds.): ICALP 2012, Part I, LNCS 7391, pp. 617–628, 2012.
© Springer-Verlag Berlin Heidelberg 2012

strong randomized lower bounds all use the distributional model, in which the randomness of the protocol is replaced by randomness in the choice of inputs, which are sampled according to some hard distribution. The equivalence of the randomized and distributional models, due to Yao's minmax theorem [7], comes from strong duality of linear programming. This appears to be non-applicable to quantum communication complexity (see for instance [8] which considers a similar question in the setting of query complexity), and the rectangle bound, as a result, is understood to be an inherently classical method for lower bounds.

1.2   Bell Experiments

Quantum nonlocality gives us a different viewpoint from which to consider lower bounds for communication complexity. A fundamental question of quantum mechanics is to establish experimentally whether nature is truly *nonlocal*, as predicted by quantum mechanics, or whether there is a purely classical (i.e., *local*) explanation to the phenomena that have been observed in the lab. In an experimental setting, two players share an entangled state and each player is given a measurement to perform. The outcomes of the measurements are predicted by quantum mechanics and follow some probability distribution $p(a, b|x, y)$, where a is the outcome of Alice's measurement x, and b is the outcome of Bob's measurement y. (We write $\mathbf{p}$ for the distribution, and $p(a, b|x, y)$ for the individual probabilities.) A Bell test [9] consists of estimating all the probabilities $p(a, b|x, y)$ and computing a Bell functional, or linear function, on these values. The Bell functional $B(\mathbf{p})$ is chosen together with a threshold τ so that any local classical distribution $\mathbf{p}'$ verifies $B(\mathbf{p}') \leq \tau$, but the chosen distribution $\mathbf{p}$ violates this inequality: $B(\mathbf{p}) > \tau$.

Although there have been numerous experiments that have validated the predictions of quantum mechanics, none so far has been totally "loophole-free". A loophole can be introduced, for instance, when the state preparation and the measurements are imperfect, or when the detectors are partially inefficient so that no measurement is registered in some runs of the experiment, or if the entangled particles are so close that communication may have taken place in the course of a run of the experiment. In such cases, there are classical explanations for the results of the experiment. For instance, if the detectors were somehow coordinating their behavior, they may choose to discard a run, and though the conditional probability (conditioned on the run not having been discarded) may look quantum, the unconditional probability may very well be classical. This is called the detection loophole. When an experiment aborts with probability at most $1 - \eta$, we say that the efficiency is η. (Here we assume that individual runs are independent of one another.) To close the detection loophole, the efficiency has to be high enough so that the classical explanations are ruled out.

What can Bell tests tell us about communication complexity? Both are measures of how far a distribution is from the set of local distributions (those requiring no communication), and one would expect that if a Bell test shows a large violation for a distribution, simulating this distribution should require a lot of communication, and vice versa. Degorre et al. showed that the factorization norm amounted to finding large Bell inequality violations for a particular class of Bell inequalities [10]. Here, we show that the partition bound also corresponds to a class of Bell inequalities.

1.3 Summary of Results

If we assume there is a c-bit classical communication protocol where Alice and Bob output a, b with distribution $p(a, b|x, y)$ when Alice's input is x and Bob's input is y, then there is a protocol without communication that outputs according to $\mathbf{p}$ (conditioned on the run not being discarded) that uses shared randomness and whose efficiency is 2^{-c}: both players guess a transcript, and if they disagree with the transcript, they abort. Otherwise they follow the protocol using the transcript. As others have noticed [11,12], one can immediately derive a lower bound: let η be the maximum efficiency of a protocol without communication that successfully simulates $\mathbf{p}$ with shared randomness. We define $\mathrm{eff}(\mathbf{p}) = 1/\eta$, and $\log(\mathrm{eff}(\mathbf{p}))$ is a lower bound on the communication complexity of simulating $\mathbf{p}$. Though this may sound naïve, this gives a surprisingly strong bound which coincides with the partition bound (in the special case of computing functions).

When we turn to the dual formulation, we get a natural physical interpretation, that of Bell inequalities. To prove a lower bound amounts to finding a good Bell inequality and proving a large violation. This is similar to finding a hard distribution and proving a lower bound in the distributional model of communication; but it is much stronger since the Bell functional is not required to have positive coefficients that sum to one.

Our approach leads naturally to a "quantum partition bound" which gives a lower bound on quantum communication complexity. Let $\mathrm{eff}^*(\mathbf{p}) = 1/\eta^*$, where η^* is the maximum efficiency of a protocol without communication that successfully simulates $\mathbf{p}$ with shared entanglement. In the one-way setting, our quantum partition bound is tight up to arbitrarily small error.

Simulating distributions while allowing for runs to be discarded with some probability is a stronger requirement than allowing a probability of error since the errors are flagged. Lee and Shraibman give a proof of the factorization norm (γ_2) lower bound on (quantum) communication complexity based on the best bias one can achieve with no communication [13, Theorem 60] (attributed to Buhrman; see also Degorre et al. [10]). In light of our formulation of the (quantum) partition bound, it is an easy consequence that the (quantum) partition bound is an upper bound on γ_2 (see e.g. [14] for definitions of the factorization norm γ_2 and the related nuclear norm ν, as well as [10] for their extensions to the communication complexity of distributions).

The following gives a summary of our results. Full definitions and statements are given in the main text. Let $\mathrm{prt}(\mathbf{p})$ be the partition bound for a distribution $\mathbf{p}$ (defined in Sect. 3.1). $R_0(\mathbf{p})$ denotes the communication complexity of simulating $\mathbf{p}$ exactly using shared randomness and classical communication, and $Q_0^*(\mathbf{p})$ denote the communication complexity of simulating $\mathbf{p}$ exactly using shared entanglement and quantum communication. One-way communication, where only Alice sends a message to Bob, is denoted $\rightarrow$. In the simultaneous messages model, denoted $\|$, each player sends a message to the referee, who does not know the inputs of either player, and has to produce the output. Shared entanglement is indicated by the superscript $*$. For any distribution $\mathbf{p}$,

- Theorem 2: $\mathrm{prt}(\mathbf{p}) = \mathrm{eff}(\mathbf{p})$,
- Theorem 3: $Q_0^*(\mathbf{p}) \geq \frac{1}{2}\log(\mathrm{eff}^*(\mathbf{p}))$,
- Theorem 4: $\gamma_2(\mathbf{p}) \leq 2\,\mathrm{eff}^*(\mathbf{p})$ and $\nu(\mathbf{p}) \leq 2\,\mathrm{eff}(\mathbf{p})$ (for nonsignaling $\mathbf{p}$),
- Theorem 5: $R_\varepsilon^{*,\|}(\mathbf{p}) \leq O(\mathrm{eff}^*(\mathbf{p}))$ and $R_\varepsilon^*(\mathbf{p}) \leq O(\sqrt{\mathrm{eff}^*(\mathbf{p})})$.

In the case of one-way communication, the upper bounds are much tighter. The one-sided efficiency measure, which we denote $\mathrm{eff}^{\rightarrow}$ is given in Definition 5.

- Theorem 6: $\frac{1}{2}\log(\mathrm{eff}^{*,\rightarrow}(\mathbf{p})) \leq Q_0^{*,\rightarrow}(\mathbf{p})$; $Q_\varepsilon^{*,\rightarrow}(\mathbf{p}) \leq \log(\mathrm{eff}^{*,\rightarrow}(\mathbf{p})) + O(1)$.

We can use smoothing to handle ϵ error, and demonstrate in the full paper how this is done in practice for some specific examples. For simplicity we have omitted these details in this summary. In the case of boolean functions, this is equivalent to relaxing the exactness constraints in the linear programs.

1.4   Related Work

The question of simulating quantum distributions in the presence of inefficient detectors has long been the object of study, since the reality of the experimental setups is that whenever the detectors can be placed far apart enough to prevent the communication loophole (typically in optics setups), the efficiency is extremely small (on the order of 10%). Gisin and Gisin show that the EPR correlations can be reproduced classically using only 75% detector efficiency [15].

Massar exhibits a Bell inequality that is more robust against detector inefficiency based on the distributed Deutsch Josza game [11]. The Bell inequality is derived from the lower bound on communication complexity for this promise problem [16,17]. Massar shows an upper bound of $\mathrm{eff}(\mathbf{p})$ on expected communication complexity of simulating $\mathbf{p}$. He also states, but does not claim to prove, that a lower bound can be obtained as the logarithm of the efficiency. Buhrman et al. [12,18] show how to get Bell inequalities with better resistance to detector inefficiency by considering multipartite scenarios where players share GHZ type entangled states. Their technique is based on the rectangle bound and they derive a general tradeoff between monochromatic rectangle size, efficiency, and communication. They show a general lower bound on multiparty communication complexity which is exactly as we describe above.

Buhrman et al. [19] show gaps between quantum and classical winning probability for games where the players are each given inputs and attempt, without communication, to produce outputs that satisfy some predicate. In the classical case they use shared randomness and entanglement in the quantum case. Winning probabilities are linear so these translate to large Bell inequality violations.

Lower bounds for communication complexity of simulating distributions were first studied in a systematic way by Degorre et al. [10]. These bounds are shown to be closely related to the nuclear norm and factorization norm [14], and the dual expressions are interpreted as Bell inequality violations. Jain and Klauck define a Las Vegas partition bound where protocols are allowed to abort [1].

2   Preliminaries

2.1   Classical Partition Bound

The partition bound of Jain and Klauck [1] is given as a linear program, following the approach of Lovász [20] and studied in more depth by Karchmer et al. [21].

Definition 1 (Partition bound [1]). *Let $f : \mathcal{X} \times \mathcal{Y} \to \mathcal{Z}$ be any partial function whose domain we write f^{-1}. Then $\mathrm{prt}_\epsilon(f)$ is defined to be the optimal value of the following linear program, where R ranges over the rectangles from $\mathcal{X} \times \mathcal{Y}$ and z ranges over the set $\mathcal{Z}$:*

$$\mathrm{prt}_\epsilon(f) = \min_{w_{R,z} \geq 0} \sum_{R,z} w_{R,z} \text{ subject to } \sum_{R:(x,y)\in R} w_{R,f(x,y)} \geq 1 - \epsilon \quad \forall x,y \in f^{-1}$$

$$\sum_z \sum_{R:(x,y)\in R} w_{R,z} = 1 \quad \forall x,y \in \mathcal{X} \times \mathcal{Y} .$$

Jain and Klauck [1] show that $R_\epsilon(f) \geq \log(\mathrm{prt}_\epsilon(f))$. The partition bound subsumes almost all previously known techniques [1], in particular the factorization norm [14], rectangle or corruption bound [7], and discrepancy [22,23].

2.2   Local and Quantum Distributions

Given a distribution $\mathbf{p}$, how much communication is required if Alice is given $x \in \mathcal{X}$, Bob is given $y \in \mathcal{Y}$, and their goal is to output $a,b \in \mathcal{A} \times \mathcal{B}$ with probability $p(a,b|x,y)$?

Some classes of distributions are of interest and have been widely studied in quantum information theory since the seminal paper of Bell [9]. The local deterministic distributions, denoted $\ell \in \mathcal{L}_{\mathrm{det}}$, are the ones where Alice outputs according to a deterministic strategy, i.e., a (deterministic) function of x, and Bob independently outputs as a function of y. The local distributions $\mathcal{L}$ are any distribution over the local deterministic strategies. Mathematically this corresponds to taking convex combinations of the local deterministic distributions, and operationally to zero-communication protocols with shared randomness.

We will also consider local strategies that are allowed to output $\perp$ when the players abort the protocol. We will use the notation $\mathcal{L}_{\mathrm{det}}^\perp$ and $\mathcal{L}^\perp$ to denote these strategies, where $\perp$ is added to the possible outputs for both players, and $\perp \notin \mathcal{A} \cup \mathcal{B}$. Therefore, when $\ell \in \mathcal{L}_{\mathrm{det}}^\perp$ or $\mathcal{L}^\perp$, $l(a,b|x,y)$ is not conditioned on $a,b \neq \perp$ since $\perp$ is a valid output for such distributions.

The quantum distributions, denoted $\mathbf{q} \in \mathcal{Q}$, result from applying measurements to a shared quantum state. Each player outputs the measurement outcome. In communication complexity, these are zero-communication protocols with shared entanglement. When players are allowed to abort, the corresponding set of distributions is denoted $\mathcal{Q}^\perp$.

Consider a boolean function $f : \mathcal{X} \times \mathcal{Y} \to \{0,1\}$ whose communication complexity we wish to study. First, we split the output so that if $f(x,y) = 0$, Alice and Bob are required to output the same bit, and if $f(x,y) = 1$, they output different bits. Let us further require Alice's marginal distribution to be uniform, likewise for Bob. Call the resulting distribution $\mathbf{p}_f$. Computing f reduces to computing $\mathbf{p}_f$ and Alice sending her outcome to Bob. For any f, p_f is nonsignaling, that is, the marginals $p(a|x,y) = p(a|xy')$ for any a, x, y, y', and $p(b|x,y) = p(b|x',y)$ for any b, x, x', y.

2.3   Communication Complexity Measures

$R_\epsilon(\mathbf{p})$ is the minimum amount of communication necessary to reproduce the distribution $\mathbf{p}$ in the worst case, up to ϵ in total variation distance for all x, y. We write $|\mathbf{p} - \mathbf{p}'|_1 \leq \epsilon$ to mean that for any x, y, $\sum_{a,b} |p(a,b|x,y) - p'(a,b|x,y)| \leq \epsilon$. We use Q to denote quantum communication, and the superscript $*$ denotes the presence of shared entanglement. We use superscripts $\rightarrow$ for one-way communication (i.e, when only Alice can send a message to Bob), and $\|$ for simultaneous messages (i.e., when Alice and Bob cannot communicate to each other, but are only allowed to send a message to a third party who should produce the final output of the protocol). The usual relation $Q_\epsilon^*(\mathbf{p}) \leq R_\epsilon(\mathbf{p})$ holds for any $\epsilon, \mathbf{p}$.

For all the models of randomized communication, we assume shared randomness between the players. Except in the case of simultaneous messages, this is the same as private randomness up to a logarithmic additive term [24]. (Ref. [10] sketches how to adapt Newman's theorem to the case of distributions.)

3   Partition Bound and Detector Inefficiency

3.1   The Partition Bound for Distributions

We extend the partition bound to the more general setting of simulating a distribution $p(a,b|x,y)$ instead of computing a function.

Definition 2. *For any distribution* $\mathbf{p} = p(a,b|x,y)$, *over inputs* $x \in \mathcal{X}, y \in \mathcal{Y}$ *and outputs* $a \in \mathcal{A}, b \in \mathcal{B}$, *define* $\mathrm{prt}(\mathbf{p})$ *to be the optimal value of the following linear program. The variables of the program are* $w_{R,\ell}$, *where* R *ranges over the rectangles from* $\mathcal{X} \times \mathcal{Y}$ *and* ℓ *ranges over the local deterministic distributions with outputs in* $\mathcal{A} \times \mathcal{B}$.

$$\mathrm{prt}(\mathbf{p}) = \min_{w_{R,\ell} \geq 0} \sum_{R,\ell} w_{R,\ell}$$

subject to $\displaystyle\sum_{R,\ell:x,y \in R} w_{R,\ell} \cdot l(a,b|x,y) = p(a,b|x,y) \quad \forall x,y,a,b \in \mathcal{X} \times \mathcal{Y} \times \mathcal{A} \times \mathcal{B}$.

For randomized communication with error, $\mathrm{prt}_\epsilon(\mathbf{p}) = \min_{|p'-p|_1 \leq \epsilon} \mathrm{prt}(\mathbf{p}')$.

In the special case of a distribution $\mathbf{p}_f$ arising from a function f, we have as expected $\mathrm{prt}_\epsilon(\mathbf{p}_f) = \mathrm{prt}_\epsilon(f)$. For the general case of a distribution $\mathbf{p}$, we can show that $R_\epsilon(\mathbf{p}) \geq \log \mathrm{prt}_\epsilon(\mathbf{p})$. Rather than proving this directly, we will first show that this partition bound is equivalent to another bound based on the notion of efficiency.

3.2 The Efficiency Bound

For any distribution $\mathbf{p}$, $\mathrm{eff}(\mathbf{p})$ is the inverse of the maximum efficiency sufficient to simulate it classically with shared randomness, without communication.

Definition 3. *For any distribution* $\mathbf{p}$ *with inputs* $\mathcal{X} \times \mathcal{Y}$ *and outputs in* $\mathcal{A} \times \mathcal{B}$, $\mathrm{eff}(\mathbf{p}) = 1/\zeta_{\mathrm{opt}}$, *where* ζ_{opt} *is the optimal value of the following linear program. The variables are* ζ *and* q_ℓ, *where* ℓ *ranges over local deterministic protocols with inputs taken from* $\mathcal{X} \times \mathcal{Y}$ *and outputs in* $\mathcal{A} \cup \{\bot\} \times \mathcal{B} \cup \{\bot\}$.

$$\zeta_{\mathrm{opt}} = \max_{\zeta, q_\ell \geq 0} \zeta$$

$$\text{subject to} \sum_{\ell \in \mathcal{L}_{\mathrm{det}}^{\bot}} q_\ell l(a, b | x, y) = \zeta p(a, b | x, y) \quad \forall x, y, a, b \in \mathcal{X} \times \mathcal{Y} \times \mathcal{A} \times \mathcal{B}$$

$$\sum_{\ell \in \mathcal{L}_{\mathrm{det}}^{\bot}} q_\ell = 1 .$$

For randomized communication with error, define $\mathrm{eff}_\epsilon(\mathbf{p}) = \min_{|p'-p|_1 \leq \epsilon} \mathrm{eff}(\mathbf{p}')$.

The first constraint says that the local distribution, conditioned on both outputs differing from $\bot$, equals $\mathbf{p}$, and the second is a normalization constraint. Note that the efficiency ζ is the same for every input x, y. This is surprisingly important and the relaxation $\zeta_{x,y} \geq \zeta$ does not appear to coincide with the partition bound. Other more realistic variants (for the Bell setting), such as players aborting independently of one another, could be considered as well. We note that this would not result in a linear program.

Theorem 1 ([11,12]). $R_\epsilon(\mathbf{p}) \geq \log \mathrm{eff}_\epsilon(\mathbf{p})$.

Proof (sketch). Let P be a randomized protocol for a distribution $\mathbf{p}'$ with $|\mathbf{p} - \mathbf{p}'|_1 \leq \epsilon$, using t bits of communication. We assume that the total number of bits exchanged is independent of the execution of the protocol, introducing dummy bits at the end of the protocol if necessary. Let q_l be the following distribution over local deterministic protocols ℓ: Alice and Bob pick a transcript $T \in \{0, 1\}^t$ using shared randomness. If T is consistent with P, Alice outputs according to P, otherwise she outputs $\bot$; similarly for Bob. Only one transcript is valid for Alice and Bob simultaneously, so the probability that neither player outputs $\bot$ is exactly 2^{-t}. This satisfies the constraints of $\mathrm{eff}(\mathbf{p}')$ with $\zeta = 2^{-t}$. □

Theorem 2. *For any distribution* $\mathbf{p}$, $\mathrm{eff}(\mathbf{p}) = \mathrm{prt}(\mathbf{p})$.

Proof. In the partition bound, a pair (ℓ, R), where ℓ is a local distribution with outputs in $\mathcal{A} \times \mathcal{B}$ and R is a rectangle, defines a local distribution ℓ_R with outputs in $(\mathcal{A} \cup \{\bot\}) \times (\mathcal{B} \cup \{\bot\})$, where Alice outputs as in ℓ if $x \in R$, and outputs $\bot$ otherwise (similarly for Bob). Let $(a_0, b_0) \in \mathcal{A} \times \mathcal{B}$ be an arbitrary pair of outputs. In the efficiency bound, a distribution $\ell \in \mathcal{L}_{\mathrm{det}}^{\bot}$ defines both a rectangle being the set of inputs where neither Alice nor Bob abort, and a local distribution $\ell' \in \mathcal{L}_{\mathrm{det}}$ where Alice outputs as ℓ if the output is different from

$\perp$ and a_0 otherwise (similarly for Bob with b_0). We can transform the linear program for $\mathrm{prt}(\mathbf{p})$ into the linear program for $\mathrm{eff}(\mathbf{p})$ by making the change of variables: $\zeta = \left(\sum_{R,\ell} w_{R,\ell}\right)^{-1}$ and $q_{\ell_R} = \zeta\, w_{R,\ell}$. $\qquad\qquad\square$

3.3 Lower Bound for Quantum Communication Complexity

By replacing local distributions by quantum distributions we get a strong new lower bound on quantum communication that subsumes the factorization norm.

Definition 4. *For any distribution* $\mathbf{p}$ *with inputs* $\mathcal{X} \times \mathcal{Y}$ *and outputs* $\mathcal{A} \times \mathcal{B}$, $\mathrm{eff}^*(\mathbf{p}) = 1/\eta^*$, *with* η^* *the optimal value of the following (non-linear) program.*

$$\max_{\zeta,\mathbf{q}\in\mathcal{Q}^\perp} \zeta \quad subject\ to \quad q(a,b|x,y) = \zeta p(a,b|x,y) \quad \forall x,y,a,b \in \mathcal{X}\times\mathcal{Y}\times\mathcal{A}\times\mathcal{B} \ .$$

As before, we let $\mathrm{eff}^*_\epsilon(\mathbf{p}) = \min_{|p'-p|_1 \le \epsilon} \mathrm{eff}^*(\mathbf{p}')$.

Theorem 3. $Q^*_\epsilon(\mathbf{p}) \ge \frac{1}{2}\log\mathrm{eff}^*_\epsilon(\mathbf{p})$.

The proof follows the lines of the proof for eff, except that we first use teleportation to replace quantum communication by entanglement-assisted classical communication.

Since the local distributions form a subset of the quantum distributions, $\mathrm{eff}^*(\mathbf{p}) \le \mathrm{eff}(\mathbf{p})$ for any $\mathbf{p}$. We can show that the efficiency is bounded below by the factorization norm.

Theorem 4. *For any nonsignaling* $\mathbf{p}$, $\nu(\mathbf{p}) \le 2\,\mathrm{eff}(\mathbf{p})$ *and* $\gamma_2(\mathbf{p}) \le 2\,\mathrm{eff}^*(\mathbf{p})$.

The proof is provided in the long version of the article, and is based on the fact that a reject outcome can be replaced by a random outcome.

3.4 Proving Concrete Lower Bounds Using the Dual

To prove lower bounds, we use the dual formulation, and give a feasible solution.

Lemma 1 (Dual formulation). *For any distribution* $\mathbf{p}$,

$$\mathrm{eff}(\mathbf{p}) = \max_{B_{abxy}} \sum_{a,b,x,y\in\mathcal{A}\times\mathcal{B}\times\mathcal{X}\times\mathcal{Y}} B_{abxy}\, p(a,b|x,y)$$

$$subject\ to \quad \sum_{a,b,x,y\in\mathcal{A}\times\mathcal{B}\times\mathcal{X}\times\mathcal{Y}} B_{abxy}\, l(ab|xy) \le 1 \quad \forall \ell \in \mathcal{L}^\perp_{\det} \ .$$

For $\mathrm{eff}^*(\mathbf{p})$ *the expression is identical save for replacing* ℓ *by* $\mathbf{q} \in \mathcal{Q}^\perp$.

The first equality (for eff) uses linear programming duality and the second (for eff^*) can be shown using Lagrange multipliers.

Concretely, how does one go about finding a feasible solution to the dual? Consider a distribution $\mathbf{p}$ for which we would like to find a lower bound. We construct a Bell inequality $B(\mathbf{p}) = \sum_{a,b,x,y} B_{abxy} p(a,b|x,y)$ so that $B(\mathbf{p})$ is large, and $B(\ell)$ is small for every $\ell \in \mathcal{L}^\perp$. The goal is to balance the coefficients so that they correlate well with the distribution $\mathbf{p}$ and badly with local strategies.

In the full paper, we give an example for a distribution based on the Hidden Matching problem [4,5,19]; we also study the Khot Vishnoi game for which there is a large Bell inequality violation [25,19]. We reformulate it as a quantum distribution $\mathbf{p} \in \mathcal{Q}$ (that is, $Q_0(\mathbf{p}) = 0$) and prove a lower bound $R_0(\mathbf{p}) = \Omega(\log(n))$. The proofs use many of the techniques Burhman et al. used to establish large Bell inequality violations [19].

4   Upper Bounds for One- and Two-Way Communication

The efficiency bound subsumes most known lower bound techniques for randomized communication complexity. How close is it to being tight? An upper bound on randomized communication is proven by Massar [11]. We give a similar bound for quantum communication complexity in terms of eff^*.

Theorem 5. *For any distribution* $\mathbf{p}$ *with outputs in* $\mathcal{A}, \mathcal{B}$,

1. $R_\epsilon^\parallel(\mathbf{p}) \leq \mathrm{eff}(\mathbf{p}) \log(\frac{1}{\epsilon}) \log(\#(\mathcal{A} \times \mathcal{B}))$ *[11]*,
2. $R_\epsilon^{*,\parallel}(\mathbf{p}) \leq \mathrm{eff}^*(\mathbf{p}) \log(\frac{1}{\epsilon}) \log(\#(\mathcal{A} \times \mathcal{B}))$,
3. $R_\epsilon^*(\mathbf{p}) \leq O\left(\sqrt{\mathrm{eff}^*(\mathbf{p}) \log(\frac{1}{\epsilon})}\right)$.

Proof (Sketch). For the first two items, the players simulate a zero-communication protocol $\lceil \log(\frac{1}{\epsilon}) \mathrm{eff}(\mathbf{p}) \rceil$ times and send the outcomes to the referee, who outputs a non-aborting run. For the third item, a quadratic speedup is possible by using a quantum protocol for disjointness [16,26,27] on the input u, v, where u_i is 0 if Alice aborts in the ith run and 0 otherwise, similarly for v with Bob.   □

The partition and efficiency bounds can easily be tailored to the case of one-way communication protocols. For the partition bound, we consider only rectangles of the form $X \times Y$ with $Y = \mathcal{Y}$. For the efficiency bound, this amounts to only letting Alice abort the protocol. The set of local (resp. quantum) distributions where only Alice can abort is denoted $\mathcal{L}_{\mathrm{det}}^{\perp A}$ (resp. $\mathcal{Q}^{\perp A}$).

Definition 5. *Define* $\mathrm{eff}^{\rightarrow}$ *in the same way as* eff, *replacing* $\mathcal{L}_{\mathrm{det}}^\perp$ *with* $\mathcal{L}_{\mathrm{det}}^{\perp A}$; *and* $\mathrm{eff}^{*,\rightarrow}$, *by replacing* $\mathcal{Q}^\perp$ *with* $\mathcal{Q}^{\perp A}$ *in the definition of* eff^*.

The dual can also be interpreted as violations of Bell inequalities.

Lemma 2 (Dual formulation for one-way efficiency). *The dual for the one-way efficiency is as in the dual for* eff, *replacing* $\mathcal{L}_{\mathrm{det}}^\perp$ *with* $\mathcal{L}_{\mathrm{det}}^{\perp A}$.

Theorem 6. $R_0^{\rightarrow}(\mathbf{p}) \geq \log \mathrm{eff}^{\rightarrow}(\mathbf{p})$ *and* $Q_0^{*,\rightarrow}(\mathbf{p}) \geq \frac{1}{2} \log \mathrm{eff}^{*,\rightarrow}(\mathbf{p})$.

The proof is similar to the two-way case. Here we show that the one-way partition bound is tight, up to arbitrarily small error. We give the results for quantum communication since the rectangle bound is already known to be tight for randomized communication complexity [28].

Theorem 7. $Q_\epsilon^{*,\rightarrow}(\mathbf{p}) \leq \log(\mathrm{eff}^{*,\rightarrow}(\mathbf{p})) + \log\log(1/(\epsilon))$.

Proof. Let $(\zeta, \mathbf{q})$ be an optimal solution for $\mathrm{eff}^{*,\rightarrow}(\mathbf{p})$. For any x, y, if we sample a, b according to $\mathbf{q}$, $\mathrm{Pr}_\mathbf{q}[a \neq \perp|x] = \zeta$ and $\mathrm{Pr}_\mathbf{q}[a, b|x, y] = \zeta p(a, b|x, y)$ for all $a, b \neq \perp$ and all x, y. Let Alice and Bob simulate this quantum distribution $N = \lceil\log(\frac{1}{\epsilon})\frac{1}{\zeta}\rceil$ times, keeping a record of the outputs (a_i, b_i) for $i \in [N]$. Since this distribution is quantum, this requires no communication (only shared entanglement). Alice then communicates an index $i \in [N]$ such that $a_i \neq \perp$, if such an index exists, or just a random index if $a_i = \perp$ for all $i \in [N]$. Alice and Bob output (a_i, b_i) corresponding to this index.

The correctness of the protocol follows from the fact that $\mathrm{Pr}_\mathbf{q}[a_i = \perp(\forall i)] = (1 - \zeta)^N \leq e^{-\zeta N} \leq \epsilon$. The protocol then requires $\log N$ bits of communication. $\qquad\square$

5   Conclusion and Open Problems

There are many questions to explore. In experimental setups, in particular with optics, one is faced with the very real problem that in most runs of an experiment, no outcome is recorded. The frequency with which apparatus don't yield an outcome is called detector inefficiency. Can we find explicit Bell inequalities for quantum distributions that are very resistant to detector inefficiency? For experimental purposes, it is also important for the distribution to be feasible to implement. One way to achieve this could be to prove stronger bounds for the inequalities based on the GHZ paradox given by Buhrman et al. [18]. Their analysis is based on a tradeoff derived from the rectangle bound. It may be possible to give sharper bounds with our techniques. Another is to consider asymmetric Bell inequalities and dimension witnesses [29,30]. Here, Alice prepares a state and Bob makes a measurement. The goal is to have a Bell inequality demonstrating that Alice's system has to be large. The dimension is exponential in the size of Alice's message to Bob, so proving a lower bound on one-way communication complexity gives a lower bound on the dimension. In order to close the detection loophole, one can also consider more realistic models of inefficiency, where the failure to produce a measurement outcome is the result of either the entangled state not being produced, or the detector of each player failing independently. This could be exploited by defining a stronger version of the partition/efficiency bound that also takes into account the probabilities of events where only one of the players produces a valid outcome. While such a variation of the efficiency bound is meaningful for Bell tests, we have not considered it here as it might not be a lower bound on communication complexity.

We would like to see more applications. For the Khot Vishnoi distribution, we are not aware of any nontrivial upper bound so there is a gap to be improved.

A family of lower bound techniques still not subsumed by the partition bound are the information theoretic bounds such as information complexity [31]. It was recently shown that information complexity is an upper bound on discrepancy [32], and this upper bound was subsequently extended to a relaxation of the partition bound [33]. This *relaxed* partition bound also subsumes most algebraic and combinatorial lower bound techniques, with the notable exception of the partition bound itself, and we would therefore like to see connections one way or the other between information complexity and the partition bound.

Finally, the quantum partition bound is of particular interest. It is hard to apply since it is not linear, and it amounts to finding a Tsirelson inequality, a harder task than finding a good Bell inequality, that can nevertheless be approached via semidefinite programming [34,35]. On the other hand, it is a very strong bound and one can hope to get a better upper bound on quantum communication complexity. Finding tight bounds complexity would be an important step to proving the existence, or not, of exponential gaps for total functions.

Acknowledgements. We wish to particularly thank Raghav Kulkarni and Iordanis Kerenidis for many fruitful discussions. Research funded in part by the EU grant QCS, ANR Jeune Chercheur CRYQ, ANR Blanc QRAC and EU ANR Chist-ERA DIQIP.

References

1. Jain, R., Klauck, H.: The partition bound for classical complexity and query complexity. In: Proc. 25th CCC 2010, pp. 247–258 (2010)
2. Newman, I., Szegedy, M.: Public vs. private coin flips in one round communication games. In: Proc. 28th STOC 1996, pp. 561–570 (1996)
3. Buhrman, H., Cleve, R., Watrous, J., de Wolf, R.: Quantum fingerprinting. Phys. Rev. Lett. 87(16), 167902 (2001)
4. Bar-Yossef, Z., Jayram, T.S., Kerenidis, I.: Exponential separation of quantum and classical one-way communication complexity. SIAM J. Comput. 38(1), 366–384 (2008)
5. Gavinsky, D., Kempe, J., Kerenidis, I., Raz, R., de Wolf, R.: Exponential separation for one-way quantum communication complexity, with applications to cryptography. SIAM J. Comput. 38(5), 1695–1708 (2008)
6. Klartag, B., Regev, O.: Quantum one-way communication can be exponentially stronger than classical communication. In: Proc. 43rd STOC 2011, pp. 31–40 (2011)
7. Yao, A.C.: Lower bounds by probabilistic arguments. In: Proc. 24th FOCS 1983, pp. 420–428 (1983)
8. de Graaf, M., de Wolf, R.: On Quantum Versions of the Yao Principle. In: Alt, H., Ferreira, A. (eds.) STACS 2002. LNCS, vol. 2285, pp. 347–358. Springer, Heidelberg (2002)
9. Bell, J.S.: On the Einstein Podolsky Rosen paradox. Physics 1, 195 (1964)
10. Degorre, J., Kaplan, M., Laplante, S., Roland, J.: The communication complexity of non-signaling distributions. Quantum Information and Computation 11(7-8), 649–676 (2011)
11. Massar, S.: Non locality, closing the detection loophole and communication complexity. Phys. Rev. A 65, 032121 (2002)
12. Buhrman, H., Høyer, P., Massar, S., Röhrig, H.: Combinatorics and quantum non-locality. Phys. Rev. Lett. 91, 048301 (2003)
13. Lee, T., Shraibman, A.: Lower bounds in communication complexity. Foundations and Trends in Theoretical Computer Science 3(4), 263–399 (2009)

14. Linial, N., Shraibman, A.: Lower bounds in communication complexity based on factorization norms. Random Structures and Algorithms 34(3), 368–394 (2009)
15. Gisin, B., Gisin, N.: A local hidden variable model of quantum correlation exploiting the detection loophole. Phys. Lett. A 260, 323–327 (1999)
16. Buhrman, H., Cleve, R., Wigderson, A.: Quantum vs classical communication and computation. In: Proc. 30th STOC 1998, pp. 63–68 (1998)
17. Brassard, G., Cleve, R., Tapp, A.: Cost of exactly simulating quantum entanglement with classical communication. Phys. Rev. Lett. 83, 1874–1877 (1999)
18. Buhrman, H., Høyer, P., Massar, S., Röhrig, H.: Multipartite nonlocal quantum correlations resistant to imperfections. Phys. Rev. A 73, 012321 (2006)
19. Buhrman, H., Regev, O., Scarpa, G., de Wolf, R.: Near-optimal and explicit Bell inequality violations. In: Proc. 26th CCC 2011, pp. 157–166 (2011)
20. Lovász, L.: Communication Complexity: a Survey. In: Paths, Flows, and VLSI Layout, B.H. Korte edition. Springer (1990)
21. Karchmer, M., Kushilevitz, E., Nisan, N.: Fractional covers and communication complexity. SIAM J. Discrete Math. 8(1), 76–92 (1995)
22. Chor, B., Goldreich, O.: Unbiased bits from sources of weak randomness and probabilistic communication complexity. In: Proc. 26th FOCS 1985, pp. 429–442 (1985)
23. Babai, L., Nisan, N., Szegedy, M.: Multiparty protocols and logspace-hard pseudorandom sequences. In: Proc. 21st STOC 1989, pp. 1–11 (1989)
24. Newman, I.: Private vs. common random bits in communication complexity. Information Processing Letters 39(2), 61–71 (1991)
25. Khot, S., Vishnoi, N.: The unique games conjecture, integrality gap for cut problems and embeddability of negative type metrics into l_1. In: Proc. 46th FOCS 2005, pp. 53–62 (2005)
26. Høyer, P., de Wolf, R.: Improved Quantum Communication Complexity Bounds for Disjointness and Equality. In: Alt, H., Ferreira, A. (eds.) STACS 2002. LNCS, vol. 2285, pp. 299–310. Springer, Heidelberg (2002)
27. Aaronson, S., Ambainis, A.: Quantum search of spatial regions. Theory of Computing 1, 47–79 (2005)
28. Jain, R., Klauck, H., Nayak, A.: Direct product theorems for communication complexity via subdistribution bounds. In: Proc. 40th STOC 2008, pp. 599–608 (2008)
29. Brunner, N., Pironio, S., Acín, A., Gisin, N., Méthot, A., Scarani, V.: Testing the dimension of Hilbert spaces. Phys. Rev. Lett. 100, 210503 (2008)
30. Vértesi, T., Pironio, S., Brunner, N.: Closing the detection loophole in Bell experiments using qudits. Phys. Rev. Lett. 104, 060401 (2010)
31. Chakrabarti, A., Shi, Y., Wirth, A., Yao, A.: Informational complexity and the direct sum problem for simultaneous message complexity. In: Proc. 42nd FOCS 2001, pp. 270–278 (2001)
32. Braverman, M., Weinstein, O.: A discrepancy lower bound for information complexity. Technical Report 12-164, ECCC (2011)
33. Kerenidis, I., Laplante, S., Lerays, V., Roland, J., Xiao, D.: Lower bounds on information complexity via zero-communication protocols and applications. Technical Report 12-038, ECCC (2012)
34. Navascués, M., Pironio, S., Acín, A.: A convergent hierarchy of semidefinite programs characterizing the set of quantum correlations. New Journal of Physics 10(7), 073013 (2008)
35. Doherty, A.C., Liang, Y.-C., Toner, B., Wehner, S.: The quantum moment problem and bounds on entangled multi-prover games. In: Proc. 23rd CCC 2008, pp. 199–210 (2008)

Testing Similar Means

Reut Levi[1], Dana Ron[1,*], and Ronitt Rubinfeld[1,2,**]

[1] Tel Aviv University
[2] MIT

Abstract. We consider the problem of testing a basic property of collections of distributions: having similar means. Namely, the algorithm should accept collections of distributions in which all distributions have means that do not differ by more than some given parameter, and should reject collections that are relatively far from having this property. By 'far' we mean that it is necessary to modify the distributions in a relatively significant manner (measured according to the ℓ_1 distance averaged over the distributions) so as to obtain the property. We study this problem in two models. In the first model (the *query model*) the algorithm may ask for samples from any distribution of its choice, and in the second model (the *sampling model*) the distributions from which it gets samples are selected randomly. We provide upper and lower bounds in both models. In particular, in the query model, the complexity of the problem is polynomial in $1/\epsilon$ (where ϵ is the given distance parameter). While in the sampling model, the complexity grows roughly as $m^{1-\mathrm{poly}(\epsilon)}$, where m is the number of distributions.

1   Introduction

We consider testing a basic property of collections of distributions: having similar means. Namely, given a collection $\mathcal{D} = (D_1, \ldots, D_m)$ of distributions over $\{0, \ldots, n\}$, and parameters γ and ϵ, we would like to determine whether the means of all distributions reside in an interval of size γn (in which case they have the property of γ-*similar means*), or whether the collection is ϵ-far from having this property. By "ϵ-far" we mean that for every collection $\mathcal{D}^* = (D_1^*, \ldots, D_m^*)$ that has the property, $\frac{1}{m} \sum_{i=1}^{m} d(D_i, D_i^*) > \epsilon$, where $d(\cdot, \cdot)$ is some predetermined distance measure between distributions.

The problem of determining whether a collection of distributions consists of distributions that have similar means arises in many contexts: Suppose one is given a collection of coins and would like to determine whether they have the same (or very similar) bias. Alternatively, suppose one would like to compare mean behavior of multiple groups in a scientific experiment. As we discuss in some more detail in Subsection 1.2, related questions have been studied in the Statistics literature, resulting in particular in the commonly used family of procedures ANOVA (Analysis of Variance), used for deciding whether a collection of normal distributions all have the same mean. As stated above, we consider distributions over a discrete domain but other than that we do not

* Research supported by the ISF grant number 246/08.
** Research supported by NSF grants CCF-1065125 and CCF-0728645, Marie Curie Reintegration grant PIRG03-GA-2008-231077 and the ISF grant nos. 1147/09 and 1675/09.

A. Czumaj et al. (Eds.): ICALP 2012, Part I, LNCS 7391, pp. 629–640, 2012.

make any assumptions regarding the distributions. Our formulation of the problem falls within the framework of property testing [11,4,2], so that in particular it allows for a small fraction of "outlier" distributions.

1.1   Our Contributions

We consider two models, proposed in previous work [8], that describe possible access patterns to multiple distributions $D_1, \ldots, D_m$ over the same domain $\{0, \ldots, n\}$. In the *query model* the algorithm is allowed to specify $i \in \{1, \ldots, m\}$ and receives j that is distributed according to D_i. We refer to each such request for a sample from D_i as a *query*. In the *(uniform) sampling model*, the algorithm receives pairs of the form (i, j) where i is selected uniformly in $\{1, \ldots, m\}$ and j is distributed according to D_i.

The ℓ_1 distance between two probability distributions, $d(D_1, D_2) = \sum_{j=0}^{n} |D_1(j) - D_2(j)|$, is perhaps the most standard measure of distance between distributions, as it measures the maximum difference between the probability of *any event* (i.e., set $S \subseteq \{0, \ldots, n\}$) occurring according to one distribution as compared to the other distribution. In other words, if the distance is small, then the distributions are essentially indistinguishable in terms of their behavior. Hence, we take it as our default distance measure when testing properties of distributions. However, for specific properties one may consider other distance measures that are appropriate. In this study, since the property is related to the means of the distributions and thus the numerical values of the domain elements are meaningful (as opposed to symmetric properties of distributions), we also consider the Earth Mover's Distance (EMD).[1] We prove our upper and lower bounds for the case where the underlying distance measure is the ℓ_1 distance and show by a simple observation that all our results hold for the case which the underlying distance measure is EMD. Hence, unless stated explicitly otherwise, in all that follows the underlying distance measure is the ℓ_1 distance.

RESULTS IN THE QUERY MODEL. We give an algorithm whose query complexity is $\tilde{O}(1/\epsilon^2)$. which is almost tight as there is a simple lower bound of $\Omega(1/\epsilon^2)$ (even for the $\{0, 1\}$ case).

Consider first a basic algorithm that works by obtaining very good estimates of the means of a sufficient number of randomly selected distributions. If the collection is ϵ-far from having γ-similar means, then (with high probability) after performing $\tilde{\Theta}(1/\epsilon^3)$ queries, the algorithm will obtain two distributions whose estimated means are sufficiently far from each other. Thus, this algorithm essentially uses estimates of means as estimates of the distance to having a certain mean.

We design and analyze an improved (almost optimal) algorithm that, roughly speaking, tries to directly estimate the distance to having a certain mean. The more direct estimate is done by estimating means as well, albeit these are means of "mutations" of the original distribution in which the weight of the distribution is either shifted higher or lower. By obtaining such estimates we can apply a "bucketing" technique that allows us to save a factor of $\tilde{\Theta}(1/\epsilon)$ in the query complexity.

---

[1] Informally, if the distributions are interpreted as two different ways of piling up a certain amount of earth over the region D, the EMD is the minimum cost of turning one pile into the other, where the cost is assumed to be amount of earth moved times the distance by which it is moved. A formal definition appears in the full version of the paper [9].

RESULTS IN THE SAMPLING MODEL. While in the query model the complexity of the problem of testing similar means has no dependence on the number of distributions, m, this is no longer the case in the sampling model. We prove that the number of samples required is lower bounded by $(1 - \gamma)m^{1-\tilde{O}((\epsilon/\gamma)^{1/2})}$. Thus, for any fixed γ (bounded away from 0 and 1), the sample complexity approaches a linear dependence on m as ϵ is decreased. On the positive side, we can show the following. First, by emulating the algorithm designed for the query model, we get an algorithm in the sampling model whose sample complexity is $\tilde{O}(1/\epsilon^2)m^{1-\tilde{\Omega}(\epsilon^2)}$. If we restrict our attention to the case where the domain is $\{0, 1\}$, then we can get a better dependence on ϵ in the exponent (at a cost of a higher dependence in the factor that depends only on ϵ). We also observe that (for the $\{0, 1\}$ case), if $\gamma < \epsilon/c$ for some sufficiently large constant c, then we can use an algorithm from [7] whose sample complexity is $\text{poly}(1/\epsilon)\sqrt{m}$ (we note that it is not possible to go below $\sqrt{m}$).

In order to prove the abovementioned lower bound we construct a pair of collections of distributions, one that has the property of γ-similar means, and one that is ϵ-far from having this property. We prove that when taking $(1 - \gamma)m^{1-\tilde{O}((\epsilon/\gamma)^{1/2})}$ samples, these two collections are indistinguishable. The heart of the proof is the construction of two random variables that on one hand have the same first t moments (for $t = \tilde{O}((\gamma/\epsilon)^{1/2})$) and on the other hand differ in the maximal distance between pairs of elements in the support. These random variables can then be transformed into collections of distributions that cannot be distinguished (with the abovementioned number of samples) but differ in the distance between the maximal and minimal means in the collection. The construction of the random variables is based on Chebyshev polynomials [3], whose roots, and their derivatives at the roots, have useful properties that we exploit.

1.2   Related Work

The work that is most closely related to the work presented in this paper appears in [8]. The testing models used here were introduced in [8], where the main focus was on the property of equivalence of a collection of distributions. Namely, the goal is to distinguish between the case that all distributions in the collection are the same (or possibly very similar), and the case in which the collection is far from having this property. When the domain of the distributions is $\{0, 1\}$, then the problem of testing similar means for $\gamma = 0$ is the same as testing whether the distributions are equivalent. Therefore, an algorithm with sampling complexity $\text{poly}(1/\epsilon)\sqrt{m}$ that is given in [8] for testing equivalence in the sampling model, carries over directly to our problem, when $\gamma = 0$ and the domain is $\{0, 1\}$). In fact, a *tolerant* version of this algorithm [7] implies the same complexity for $\gamma \leq \epsilon/c$ for a sufficiently large constant c. However, these results do not have any implications for larger γ, and the problems are very different when the domain is larger.

Testing and approximating properties of single and pairs of distributions has been studied quite extensively in the past (see e.g. [2,1,14]).

Statistical techniques for determining whether sets of populations have the same mean are in wide use. Paired difference tests, and in particular the Student's and Welch's t-tests, are commonly used to study whether the mean of two normally distributed populations are equal [10,12,15]. The family of procedures ANOVA (Analysis of Variance), applies when there are more than two normally distributed populations (see [6, Chapter 12]), where the difficulty is that the pairwise comparison of all the populations increases the chance of incorrectly failing collections of populations that do in fact all have the same mean. In all of these procedures, the problem solved is more stringent than in our property testing setting, but the assumptions made in all settings are quite strong, e.g., assuming the normality of the distributions and assuming that all distributions have the same variance, and thus the sample complexity bounds are incomparable to those in our setting.

ORGANIZATION. In Section 2 we give our results for the query model and in Section 3 we give our results for the sampling model. All missing proofs as well as the EMD extension and suggestions for further research can be found in the full version of the paper [9].

2   Results for the Query Model

In this section we provide an algorithm for testing γ-similar means in the query model, and give an almost matching simple lower bound.

For a distribution D over $\{0, \ldots, n\}$, we shall use the notation $\mu(D) \overset{\text{def}}{=} \sum_{i=1}^{n} i \cdot D(i)$ for the mean of D. for a value $0 \leq z \leq n$ let $d_1(D, z) \overset{\text{def}}{=} \min_{D' : \mu(D')=z} \{\|D - D'\|_1\}$ denote the minimum ℓ_1 distance between D and a distribution that has mean z.

A BASIC ALGORITHM. Consider first a basic algorithm that works by randomly selecting $\Theta(1/\epsilon)$ distributions, and estimating each of their means to within $O(\epsilon n)$ additive error. This can be done by querying each selected distribution so as to obtain $\tilde{O}(1/\epsilon^2)$ samples from each. The resulting query complexity is $\tilde{O}(1/\epsilon^3)$. The correctness of this algorithm is based on Lemma 1 (stated below), which gives an upper bound on the "cost", in terms of the ℓ_1-distance, for modifying the mean of a distribution by ϵn. Note that in general this cost is not necessarily linear in ϵ. For example, consider the case in which ϵn is an integer and D has all its weight on $n(1 - \epsilon)$, so that $\mu(D) = n(1 - \epsilon)$. Suppose we want to *increase* D's mean by ϵn. The only distribution whose mean is n is the distribution whose weight is all on n, and the ℓ_1 distance between D and this distribution is 1. On the other hand, if we wanted to *decrease* the mean of D by ϵn, then this can easily be done with a cost linear in ϵ, by moving $\epsilon/(1 - \epsilon)$ weight from $n(1 - \epsilon)$ to 0.

Lemma 1. *Let D be a distribution over $\{0, \ldots, n\}$, let $\mu = \mu(D)$, and let $\epsilon \leq 1/16$. If $\mu \geq n/2$, then for every $\mu' \in [\mu - \epsilon n, \mu]$ there exists D' such that $\mu(D') = \mu'$ and $\|D - D'\|_1 \leq 4\epsilon$. If $\mu \leq n/2$, then for every $\mu' \in [\mu, \mu + \epsilon n]$ there exists D' such that $\mu(D') = \mu'$ and $\|D - D'\|_1 \leq 4\epsilon$.*

Lemma 2 can be shown to follow from Lemma 1.

Lemma 2. *Let $\mathcal{D}$ be a collection of distributions. If $\mathcal{D}$ is ϵ-far from having γ-similar means, then there exists an interval $[x, y] \subseteq [n]$ where $y - x \geq \gamma n + \epsilon n/8$ such that*

$\sum_{i:\mu(D_i)>y} d_1(D_i, y) > (\epsilon/4)m$ and $\sum_{i:\mu(D_i)<x} d_1(D_i, x) > (\epsilon/4)m$. In particular, there are more than $(\epsilon/4)m$ distributions whose mean is at most x and more than $(\epsilon/4)m$ distributions whose mean is at least y.

The correctness of the basic algorithm follows from Lemma 2: If $\mathcal{D}$ is ϵ-far from having γ-similar means, then by selecting $\Theta(1/\epsilon)$ distributions and estimating the means of each to within $O(\epsilon n)$, with high constant probability the algorithm finds evidence for a pair of distributions with means outside both sides of the interval defined in Lemma 2. On the other hand, if $\mathcal{D}$ has γ-similar means, then the probability of such an event is small.

AN IMPROVED ALGORITHM. We can modify the basic algorithm so as to obtain a lower complexity (which we later show is almost optimal). One ingredient of the modification (similar to that applied for example in [5]) is roughly the following. Consider the following two (extreme) cases where the collection is ϵ-far from having γ-similar means. In the first case, there is an interval $[x, y]$ of size $\gamma n + 2\epsilon n$, such that half of the distributions have mean x and half of the distributions have mean y. If we select just a constant number of distributions, and for each we estimate its mean to within $\epsilon n/2$, then we shall have sufficient evidence for rejection. In the second case, all but $2\epsilon m$ of the distributions have a mean that resides in an interval of size γn, say, $[0, \gamma n]$ and the remaining $2\epsilon m$ distributions have a mean of n. In this case we need to sample $\Theta(1/\epsilon)$ distributions so as to "hit" one of the high-mean distributions, but then it suffices to take a constant size sample so as to detect that it has a high mean.

If the distributions were over $\{0, 1\}$, then by generalizing the above discussion we can get a certain trade-off between the number of selected distributions and the required quality of the estimate of their means. When dealing with general domains, estimating the means might not suffice. As noted previously, a distribution might have a mean that is very close to a certain value, while the distribution is very far, in terms of the ℓ_1 distance, from any distribution that has this mean. Therefore, rather than estimating means as a "proxy" for estimating the ℓ_1 distance to having a certain mean, we estimate the latter directly.

To make the above notion of estimation more precise, we introduce some notation. For $0 \leq \beta \leq 1$ and D such that $d_1(D, n) \geq \beta$ (where $d_1(\cdot, \cdot)$ is as defined at the beginning of this section), let $\mu_\beta^>(D)$ equal $\mu > \mu(D)$ such that $d_1(D, \mu) = \beta$ and for D such that $d_1(D, 0) \geq \beta$, let $\mu_\beta^<(D)$ equal $\mu < \mu(D)$ such that $d_1(D, \mu) = \beta$. If $d_1(D, n) < \beta$, then $\mu_\beta^>(D) \stackrel{\text{def}}{=} n$ and if $d_1(D, 0) < \beta$, then $\mu_\beta^<(D) \stackrel{\text{def}}{=} 0$. Observe that if the domain is $\{0, 1\}$, then $\mu_\beta^>(D) = \min\{\mu(D) + \beta, 1\}$ and $\mu_\beta^<(D) = \max\{\mu(D) - \beta, 0\}$ (while if the domain is larger, then $\mu_\beta^>(D) - \mu(D)$ and $\mu(D) - \mu_\beta^<(D)$ might be much smaller than βn).

We first describe a procedure that given sampling access to a distribution D and a parameter β, outputs a pair of estimates such that with high probability one is between $\mu(D)$ and $\mu_\beta^>(D)$ and the other is between $\mu_\beta^<(D)$ and $\mu(D)$. The number of samples that the procedure takes is quadratic in $1/\beta$. We later show how to apply this procedure so as to obtain a testing algorithm with query complexity $\tilde{O}(1/\epsilon^2)$.

The idea behind the procedure is the following. Consider a distribution D. For any $0 \leq a \leq a' \leq n$, $\sum_{i>a} i \cdot D(i) + \sum_{i \leq a} a' \cdot D(i) \geq \mu(D)$. On the other hand, by the definition of $\mu_\beta^>(D)$, if a is such that $\Pr_D[i \leq a] \leq \beta$, then $\sum_{i>a} i \cdot D(i) + \sum_{i\leq a} n \cdot D(i) \leq \mu_\beta^>(D)$. Let a indeed be a value that satisfies $\Pr_D[i \leq a] \leq \beta$, let $a' = a + (n-a)/2 = (a+n)/2$ and let $\mu^a(D) \overset{\text{def}}{=} \sum_{i>a} i \cdot D(i) + a' \cdot \Pr_D[i \leq a]$. Then on one hand $\mu^a(D) \geq \mu(D) + ((n-a)/2) \cdot \Pr_D[i \leq a]$ and on the other hand $\mu^a(D) \leq \mu_\beta^>(D) - ((n-a)/2) \cdot \Pr_D[i \leq a]$. If $\Pr_D[i \leq a] \geq \beta/c$ for some constant c, then by estimating $\mu^a(D)$ to within an additive error of $(n-a)\beta/4c$, we get a value between $\mu(D)$ and $\mu_\beta^>(D)$. Since $\mu^a(D)$ is the mean of a distribution (we describe this distribution formally in the proof of Lemma 3) whose support is in the interval $[a, n]$, this can be done by taking a sample of size $\Theta(1/\beta^2)$. A technical issue that arises is that it is possible that no such value a exists because $\Pr_D[i = a]$ is relatively large. But then we can slightly modify the definition of $\mu^a(D)$ and still obtain the desired estimate. A similar argument can give us $\mu_\beta^<(D) \leq y \leq \mu(D)$ (with high probability).

Procedure. `GetBounds` (D, β, δ)

1. Take a sample of size $s_1 = \Theta(\log(1/\delta)/\beta)$ from D and let $i_1 \leq \cdots \leq i_{s_1}$ be the selected points (ordered from small to large).
2. Set $a = i_{(\beta/4)s_1}, b = i_{(1-\beta/4)s_1}, a' = (a+n)/2$ and $b' = b/2$.
3. Take a sample of size $s_2 = \Theta(\log(1/\delta)/\beta^2)$ from D. Let $\hat{\alpha}(a)$ be the fraction of sampled points $i = a$ and let $\hat{\alpha}(b)$ be the fraction of sampled points $i = b$.
4. If $\hat{\alpha}(a) \leq \beta/4$, then let $a'' = a'$, else let $a'' = \frac{\beta}{4\hat{\alpha}(a)} \cdot a' + \left(1 - \frac{\beta}{4\hat{\alpha}(a)}\right) \cdot a$. Similarly, if $\hat{\alpha}(b) \leq \beta/4$ then let $b'' = b'$, else let $b'' = \frac{\beta}{4\hat{\alpha}(b)} \cdot b' + \left(1 - \frac{\beta}{4\hat{\alpha}(b)}\right) \cdot b$.
5. Take a sample of size $s_3 = \Theta(\log(1/\delta)/\beta^2)$ from D, and denote the sampled points by $i_1, \ldots, i_{s_3}$. Let $u = \frac{1}{s_3}\left(\sum_{i_j>a} i_j + \sum_{i_j<a} a' + \sum_{i_j=a} a''\right)$ and $\ell = \frac{1}{s_3}\left(\sum_{i_j<b} i_j + \sum_{i_j>b} b' + \sum_{i_j=b} b''\right)$.
6. Return (u, ℓ).

Lemma 3. *The procedure* GetBounds(D, β, δ) *returns* u *and* ℓ *such that with probability at least* $1 - \delta$ *(over its internal coin flips),* $\mu(D) \leq u \leq \mu_\beta^>(D)$ *and* $\mu_\beta^<(D) \leq \ell \leq \mu(D)$.

Proof: We prove the claim for u, and an analogous (symmetric) analysis holds for ℓ. Let a, a', a'' be as determined in Procedure GetBounds, and let

$$\tilde{D}(i) \overset{\text{def}}{=} \begin{cases} D(i) & \text{if } i > a, i \neq a', i \neq a'' \\ D(i) + \Pr_D[i < a] & \text{if } i = a' \\ D(i) + D(a) & \text{if } i = a'' \\ 0 & \text{o.w.} \end{cases} \tag{1}$$

By the definition of the distribution $\tilde{D}$ we have that $\mu(\tilde{D}) = \sum_{i>a} i \cdot D(i) + a' \cdot \Pr_D[i < a] + a'' \cdot \Pr_D[i = a]$. Observe that in Step 5, the procedure takes s_3 independent samples from $\tilde{D}$ and that $\mathrm{E}[u] = \mu(\tilde{D})$.

By a multiplicative Chernoff bound, with probability at least $1 - \delta/4$ (for a sufficiently large constant in the Θ notation for s_1) we have that $\Pr_D[i < a] \leq \beta/3$ and $\Pr_D[i \leq a] \geq \beta/8$. Next, by an additive Chernoff bound, with probability at least $1 - \delta/4$ (for a sufficiently large constant in the Θ notation for s_2) we have that $\Pr_D[i = a] - \beta/4 \leq \hat{\alpha}(a) \leq \Pr_D[i = a] + \beta/4$. From this point on assume that the above inequalities indeed hold. If $\hat{\alpha}(a) \leq \beta/4$ (so that $\Pr_D[i \leq a] \leq \beta/3 + \beta/4 + \beta/4 < \beta$), then (as explained in the discussion preceding the algorithm), on one hand, $\mu(\widetilde{D}) \geq \mu(D) + \frac{n-a}{2} \cdot \Pr_D[i \leq a] \geq \mu(D) + (n - a) \cdot (\beta/16)$, and on the other hand $\mu(\widetilde{D}) \leq \mu_\beta^\geq(D) - \frac{n-a}{2} \cdot \Pr_D[i \leq a] \leq \mu_\beta^\geq(D) - (n - a) \cdot (\beta/16)$. If $\hat{\alpha}(a) > \beta/4$ (so that $\Pr_D[i < a] + \min\{1, \beta/4\hat{\alpha}(a)\} \cdot \Pr_D[i = a] \leq \beta/3 + \beta/2 < \beta$), then

$$\mu(\widetilde{D}) \geq \mu(D) + \frac{n-a}{2} \cdot \left(\Pr_D[i < a] + \frac{\beta}{4\hat{\alpha}(a)} \cdot \Pr_D[i = a] \right) \geq \mu(D) + \frac{n-a}{2} \cdot \frac{\beta}{32}$$

and similarly $\mu(\widetilde{D}) < \mu_\beta^\geq(D) - (n - a) \cdot (\beta/32)$. By the definition of u and an additive Chernoff bound, with probability at least $1 - \delta/4$ (for a sufficiently large constant in the Θ notation for s_3), we have that $|u - \mu(\widetilde{D})| \leq (n - a) \cdot (\beta/32)$ implying that $\mu(D) \leq u \leq \mu_\beta^\geq(D)$. ∎

Algorithm 1. Testing γ-similar means

---

1. For $q = 1$ to r, where $r = \lceil \log(8/\epsilon) \rceil$ do:
   - Select $t(q) = \Theta(2^q \log(1/\epsilon))$ distributions from the collection, and denote them by $D_1^q, \ldots, D_{t(q)}^q$.
   - For each D_j^q selected let $(u_j^q, \ell_j^q) = \text{GetBounds}\left(D_j^q, (\epsilon/8)2^{q-1}, \frac{1}{(6rt(q))} \right)$
2. Let $\hat{x} = \max_{q,j}\{u_j^q\}$ and $\hat{y} = \min_{q,j}\{\ell_j^q\}$. If $\hat{y} - \hat{x} > \gamma n$, then REJECT, otherwise, ACCEPT.

Theorem 1. *Algorithm 1 tests γ-similar means in the query model. The algorithm's query complexity is $O(\log^2(1/\epsilon)/\epsilon^2)$.*

Proof: Let E_g denote the event that all pairs (u_j^q, ℓ_j^q) returned by the procedure GetBounds are as specified in Lemma 3. Since each call to GetBounds in iteration q is done with $\delta = 1/(6rt(q))$, by Lemma 3 the probability that E_g holds is at least $5/6$. If $\mathcal{D}$ has γ-similar means, then, conditioned on E_g, the algorithm accepts.

We now turn to the case that $\mathcal{D}$ is ϵ-far from having γ-similar means. Let $[x, y]$ be an interval as described in Lemma 2. We partition the distributions D_i such that $\mu(D_i) < x$ into *buckets* B_q^L, for $1 \leq q \leq r$, where $B_q^L = \{i : (\epsilon/8)2^{q-1} < d_1(D_i, x) \leq (\epsilon/8)2^q\}$, and similarly we partition the distributions D_i such that $\mu(D_i) > y$ into buckets B_q^R, where $B_q^R = \{i : (\epsilon/8)2^{q-1} < d_1(D_i, y) \leq (\epsilon/8)2^q\}$. Since $\sum_{i:\mu(D_i)<x} d_1(D_i, x) > (\epsilon/4)m$ and $\sum_{i:d_1(D_i,x)\leq\epsilon/8} d_1(D_i, x) \leq (\epsilon/8)m$, we have that there exists an index q^L such that $|B_{q^L}^L| > \left((\epsilon/8)m/r \right)/\left((\epsilon/8)2^{q^L} \right) = \Omega\left(m/(\log(1/\epsilon)2^{q^L}) \right)$, and similarly there exists an index q^R such that $|B_{q^R}^R| = \Omega\left(m/(\log(1/\epsilon)2^{q^R})\right)$. But in such a case, with high constant probability, the algorithm will select a distribution D_i such that $i \in B_{q^L}^L$ in iteration q^L, and a distribution D_j such that $j \in B_{q^R}^R$ in iteration q^R, and conditioned on the event E_g, will reject, as required.

Let $s(q)$ denote the number of queries performed in iteration q by the procedure Get-Bounds for each distribution it is called on. The query complexity of the algorithm is $\sum_{q=1}^r t(q) \cdot s(q) = O\left(\sum_{q=1}^r 2^q \log(1/\epsilon) \cdot \frac{\log(1/\epsilon)}{2^{2q}\epsilon^2}\right) = O(\log^2(1/\epsilon)/\epsilon^2)$ and the theorem follows. ∎

A LOWER BOUND. We end this section with a lower bound (almost matching our upper bound) by reducing the testing problem to the problem of distinguishing two coins.

Fact 4. *Distinguishing an unbiased coin from a coin with bias ϵ with constant success probability requires $\Omega(1/\epsilon^2)$ samples.*

Corollary 2. *Testing γ-similar means in the query model requires $\Omega(1/\epsilon^2)$ samples.*

3   Results for the Sampling Model

As opposed to the query model, where the algorithms had no dependence on the number of distributions, m, we show that in the sampling model there is a strong dependence on m. We start by giving a lower bound for the sampling complexity of this problem, and continue with several upper bounds.

3.1   A Lower Bound

In this section we prove the following theorem.

Theorem 3. *For every $n \geq 1$, testing γ-similar means in the uniform sampling model requires $(1 - \gamma) \cdot m^{1-\tilde{O}((\epsilon/\gamma)^{1/2})}$ samples.*

In particular, when γ is a constant we get a lower bound of $m^{1-\tilde{O}(\epsilon^{1/2})}$. We also note that we may assume without loss of generality that $1 - \gamma = \Omega(\epsilon)$, or else the algorithm can accept automatically.

In order to prove Theorem 3 we construct a pair of collections of distributions, one that has the property of γ-similar means, the YES instance, and one that is ϵ-far from having this property, the NO instance. We prove that when taking $m^{1-\tilde{O}((\epsilon/\gamma)^{1/2})}$ samples, these pair of collections are indistinguishable and thus prove a lower bound on the sample complexity of the problem. The main part of this proof is the construction of two random variables that on one hand have the same first t moments (where t will be defined later) and on the other hand differ in the maximal distance between pairs of elements in the support. These random variables can then be transformed into collections of distributions that cannot be distinguished (with the abovementioned number of samples) but differ in the distance between the maximal and minimal means in the collection. The next lemma is central to the proof of Theorem 3. In the lemma and what follows we shall use the notation $[k] \stackrel{\text{def}}{=} \{1, \ldots, k\}$.

Lemma 5. *Given sequences $\{d_i\}_{i=1}^{t}$ and $\{\alpha_i\}_{i=1}^{t}$ that satisfy $0 \le |d_i|, \alpha_i \le 1$ for every $i \in [t]$ and $\sum_{i=1}^{t} \alpha_i = 1$, we define a random variable $X = X(\{d_i\}, \{\alpha_i\})$ over $[0, 1]$ as follows: $\Pr[X = d_i] = \alpha_i$. For every even integer t, there exist sequences $\{d_i^{+}\}_{i=1}^{t}$, $\{\alpha_i^{+}\}_{i=1}^{t}$, $\{d_i^{-}\}_{i=1}^{t+1}$ and $\{\alpha_i^{-}\}_{i=1}^{t+1}$ that obey the aforementioned constraints and for which the following holds:*

1. *For the random variables $X^{+} = X(\{d_i^{+}\}, \{\alpha_i^{+}\})$ and $X^{-} = X(\{d_i^{-}\}, \{\alpha_i^{-}\})$ we have*

$$\mathrm{E}\left[\left(X^{+}\right)^{i}\right] = \mathrm{E}\left[\left(X^{-}\right)^{i}\right] \quad \forall i \in [t]. \tag{2}$$

2. *The sequences are symmetric around zero. Namely, $d_{t/2+1}^{-} = 0$, and for every $1 \le i \le t/2$, we have that $d_i^{+} = -d_{t+1-i}^{+}$ and $\alpha_i^{+} = \alpha_{t+1-i}^{+}$ as well as $d_i^{-} = -d_{t+2-i}^{-}$ and $\alpha_i^{-} = \alpha_{t+2-i}^{-}$.*

3. *If we denote by $d_{\max}^{+}$ ($d_{\min}^{+}$) the maximal (minimal) non-negative element in the support of X^{+} (so that $d_{\max}^{+} = d_1^{+}$ and $d_{\min}^{+} = d_{t/2}^{+}$) and by $\alpha_{\max}^{+}$ ($\alpha_{\min}^{+}$) the corresponding probability, and let $d_{\max}^{-}, d_{\min}^{-}, \alpha_{\max}^{-}, \alpha_{\min}^{-}$ be defined analogously, then $\alpha_{\max}^{-}(d_{\max}^{-} - d_{\max}^{+}) = \tilde{\Theta}\left(\frac{1}{t^2}\right)$ and $d_{\max}^{+} - d_{\min}^{+} = \Theta(1)$.*

We prove Lemma 5 subsequently, and first show how Theorem 3 follows from it. Let $\{\alpha_i^{+}\}_{i=1}^{t}$, $\{d_i^{+}\}_{i=1}^{t}$, $\{\alpha_i^{-}\}_{i=1}^{t+1}$ and $\{d_i^{-}\}_{i=1}^{t+1}$ be as defined in Lemma 5. We assume for simplicity that $\{\alpha_i^{+}m\}$ and $\{\alpha_i^{-}m\}$ are integers. We deal with the issue of rounding in [9]. For a parameter δ, we define the collection of distributions $\mathcal{D}_t^{+}$ (the YES instance) as follows. For every $1 \le i \le t/2$ there are $\alpha_i^{+}m$ distributions $D \in \mathcal{D}_t^{+}$ of the following form:

$$D(j) \stackrel{\text{def}}{=} \begin{cases} \frac{1}{2} \cdot \left(1 + d_i^{+}\delta\right) & \text{if } j = 0 \\ \frac{1}{2} \cdot \left(1 - d_i^{+}\delta\right) & \text{if } j = n \\ 0 & \text{o.w.} \end{cases} \tag{3}$$

and another $\alpha_i^{+}m$ of the distributions $D \in \mathcal{D}_t^{+}$ are of the following form:

$$D(j) \stackrel{\text{def}}{=} \begin{cases} \frac{1}{2} \cdot \left(1 - d_i^{+}\delta\right) & \text{if } j = 0 \\ \frac{1}{2} \cdot \left(1 + d_i^{+}\delta\right) & \text{if } j = n \\ 0 & \text{o.w.} \end{cases} \tag{4}$$

The collection $\mathcal{D}_t^{-}$ is defined analogously based on $\{\alpha_i^{-}\}_{i=1}^{t+1}$ and $\{d_i^{-}\}_{i=1}^{t+1}$, where for $i = t/2 + 1$ there are $\alpha_{-t/2+1}m$ distributions $D \in \mathcal{D}_t^{-}$ such that $D(0) = D(n) = 1/2$ (recall that $d_{t/2+1}^{-} = 0$). The proof of the next lemma appears in [9].

Lemma 6. *For every even integer $t \le m^{1/2}$, in order to distinguish between $\mathcal{D}_t^{+}$ and $\mathcal{D}_t^{-}$ in the uniform sampling model (with success probability at least $2/3$), it is necessary to take $\Omega\left(m^{1-1/t}(1 - d_{\max}^{+}\delta)\right)$ samples.*

Proof of Theorem 3: Define γ such that $\mathcal{D}_t^{+}$ has the property of γ-similar means, i.e. $\gamma = \frac{1}{2} \cdot (1 + d_{\max}^{+}\delta) - \frac{1}{2} \cdot (1 - d_{\max}^{+}\delta) = d_{\max}^{+}\delta$. To change $\mathcal{D}_t^{-}$ into a γ-similar means instance, we have to either change the means of $\alpha_{\max}^{-}$ fraction of the distributions from $\frac{1}{2} \cdot (1 + d_{\max}^{-}) \cdot n$ to $\frac{1}{2} \cdot (1 + d_{\max}^{-}) \cdot n$ or change the means of $\alpha_{\max}^{-}$ distributions from

$\frac{1}{2} \cdot (1 - d_{\max}^-) \cdot n$ to $\frac{1}{2} \cdot (1 - d_{\max}^+) \cdot n$. Letting $\epsilon = \alpha_{\max}^+ \cdot (d_{\max}^+ - d_{\max}^-) \delta$, we get that $\mathcal{D}^-$ is at least ϵ-far from γ-similar means. By Lemma 5 we have that $\frac{\epsilon}{\gamma} = \frac{\alpha_{\max}^+ \cdot (d_{\max}^+ - d_{\max}^-)}{d_{\max}^+} = \tilde{\Theta}\left(\frac{1}{t^2}\right)$. We note that for every $\epsilon/\gamma \leq 1/\log^2 m$ we get that $m^{1 - \tilde{O}((\epsilon/\gamma)^{1/2})} = \Omega(m)$. Hence we can assume without loss of generality that $\epsilon/\gamma = \tilde{\Omega}(m^{-1/2})$ and thus by setting $1/t = \tilde{\Theta}(\epsilon^{1/2}/\gamma^{1/2})$, the theorem follows from Lemma 6. ∎

The random variables described in Lemma 5 are constructed via a polynomial f: the support of X^+ (respectively, X^-) is the set of roots of f with a negative (respectively, positive) derivative. If f has an odd number of roots then the sign of the derivative at the largest root is the same as the sign at the smallest root. If it is positive, then the support of X^- resides in an interval which contains the support of X^+. To prove a lower bound, X^- needs to be far from similar means (more precisely, the collection of coins that corresponds to X^-) and indistinguishable from X^+. To make X^- far from having similar means, f should maximize the size of X^-'s interval (compared to X^+'s interval) and the weight on the extreme roots.

As suggested by Lemma 7 (stated below) X^- and X^+ have matching moments if the probability to take the value x_i, where x_i is a root of f, is $1/|f'(x_i)|$, up to normalization. In this case, a small derivative on the extreme roots would result with X^- which is far from having similar means. When the roots of f is taken to be the value of the Sine function at equal distances, the derivative at the extreme roots, that is at -1 and 1, is small. As we see next, these roots are the roots of the Chebyshev polynomials.

The proof of Lemma 5 requires some preliminaries concerning Chebyshev polynomials, which we provide next. Let T_ℓ be the ℓ-th *Chebyshev polynomial of the first kind*, which is defined by the recurrence relation:

$$T_0(x) = 1, \quad T_1(x) = x, \quad \text{and} \quad T_{\ell+1}(x) = 2x T_\ell(x) - T_{\ell-1}(x) . \tag{5}$$

Let U_ℓ be the ℓ-th *Chebyshev polynomial of the second kind*, which is defined by the recurrence relation:

$$U_0(x) = 1. \quad U_1(x) = 2x, \quad \text{and} \quad U_{\ell+1}(x) = 2x U_\ell(x) - U_{\ell-1}(x) . \tag{6}$$

Then we have that

$$\frac{dT_\ell(x)}{dx} = \ell \cdot U_{\ell-1} , \tag{7}$$

and that

$$U_{\ell-1}(\cos(x)) = \frac{\sin(\ell x)}{\sin x} . \tag{8}$$

T_ℓ has ℓ different simple roots:

$$x_i = \cos\left(\frac{\pi}{2} \cdot \frac{2i - 1}{\ell}\right) \quad i = 1, \ldots, \ell \tag{9}$$

and the following equalities hold: $T_\ell(1) = 1$ and $T_\ell(-1) = (-1)^\ell$.

We shall also use the next lemma concerning properties of (derivatives of) polynomials.

Lemma 7 ([13]). *Let $f(x)$ be a polynomial of degree ℓ whose roots $\{x_i\}$ are real and distinct. Letting f' denote the derivative of f, for every $j \leq \ell - 2$ we have that $\sum_{i=1}^{\ell} \frac{x_i^j}{f'(x_i)} = 0$.*

We are now ready to prove Lemma 5.

Proof of Lemma 5:   Consider the following polynomial: $f(x) \stackrel{\text{def}}{=} (x - 1)(x + 1)T_\ell(x)$, where $T_\ell(\cdot)$ is the ℓ-th Chebyshev polynomial of the first kind and $\ell = 2t - 1$. The polynomial $f(\cdot)$ has $\ell + 2$ roots, which, by decreasing order, are: $1, \cos\left(\frac{\pi}{2} \cdot \frac{1}{\ell}\right), \cos\left(\frac{\pi}{2} \cdot \frac{3}{\ell}\right), \ldots, -1$. The derivative of $f(\cdot)$ is $f'(x) = 2x \cdot T_\ell(x) + (x^2 - 1) \cdot T_\ell'(x)$ and thus $\frac{1}{|f'(1)|} = \frac{1}{2T_\ell(1)} = \frac{1}{2}$ and $\frac{1}{|f'(-1)|} = \frac{1}{|2T_\ell(-1)|} = \frac{1}{2}$. While for x_i which is a root of T_ℓ we have: $\frac{1}{|f'(x_i)|} = \frac{1}{|(x_i^2-1)\cdot T_\ell'(x_i)|}$. By Equations (7) and (8): $\frac{1}{|T_\ell'(x_i)|} = \left| \frac{\sin\left(\frac{\pi}{2}\cdot\frac{2i-1}{\ell}\right)}{\ell\cdot\sin\left(\frac{\pi}{2}\cdot(2i-1)\right)} \right| = \frac{1}{\ell} \cdot \left|\sin\left(\frac{\pi}{2}\cdot\frac{2i-1}{\ell}\right)\right|$ where we used the fact that $\left|\sin\left(\frac{\pi}{2}\cdot(2i-1)\right)\right| = 1$. Therefore by Equation (9) and the identity $1 - \cos^2 x = \sin^2 x$ we obtain:

$$\frac{1}{|f'(x_i)|} = \frac{1}{\ell} \cdot \frac{1}{\left|\sin\left(\frac{\pi}{2}\cdot\frac{2i-1}{\ell}\right)\right|} . \tag{10}$$

Since $g(x) = \sin x / x$ is monotone decreasing for $0 < x \leq \pi/2$, from the fact that $g(\pi/2) = 2/\pi$ we get that $\sin x > (2/\pi)x$ for $0 < x \leq \pi/2$. Thus for $i \leq \ell/2$, $\frac{1}{|f'(x_i)|} \leq \frac{1}{\ell} \cdot \frac{\pi}{2} \cdot \frac{1}{\left(\frac{\pi}{2}\cdot\frac{2i-1}{\ell}\right)} = \frac{1}{2i-1}$. Therefore, for $\{x_i\}$, the roots of $T_\ell(\cdot)$,

$$\sum_{i=1}^{\ell} \frac{1}{|f'(x_i)|} \leq 2 \sum_{x_i \geq 0} \frac{1}{|f'(x_i)|} = O(\log \ell) . \tag{11}$$

We take $\{d_i^-\}_{i=1}^{t+1}$ to be those roots x_j of $f(\cdot)$ for which $f'(x_j) > 0$ and set $\alpha_i^- = \frac{1}{|f'(d_i^-)|} \cdot \beta^-$, where $\beta^- = 1 / \left(\sum_{i=1}^{t+1} \frac{1}{|f'(d_i^-)|}\right)$ is a normalization factor. Similarly we take $\{d_i^+\}_{i=1}^{t}$ to be the roots with the negative derivative. Then $d_{\max}^- = 1$ and by Equation (11), $\alpha_{\max}^- = \Omega(1/\log \ell)$. On the other hand, $d_{\max}^+ = \cos\left(\frac{\pi}{2}\cdot\frac{1}{\ell}\right)$. Due to the identity $1 - \cos x = \sin x \cdot \tan(x/2)$, we get that: $\lim_{x\to 0} \frac{1-\cos x}{x^2} = \lim_{x\to 0} \frac{\sin x}{x} \cdot \frac{\sin(x/2)}{x} \cdot \frac{1}{\cos(x/2)} = \frac{1}{2}$, and so $d_{\max}^- - d_{\max}^+ = \Theta(1/\ell^2)$. Since ℓ is odd and the sign of the derivative alternates between roots we get that $d_{\min}^- = \cos\left(\frac{\pi}{2}\right) = 0$ while $d_{\min}^+ = \cos\left(\frac{\pi}{2}\cdot\frac{\ell-2}{\ell}\right) = \sin\left(\frac{\pi}{2}\cdot\frac{2}{\ell}\right)$. Thus $d_{\min}^- - d_{\min}^+ = \Theta\left(1/\ell\right)$. By Equations (10) and (11) we get $\alpha_{\min}^+ = \tilde{\Theta}(1/\ell)$. Therefore the requirements in Item 3 of Lemma 5 are satisfied. Equation (2) follows from Lemma 7. Since the roots of the Chebyshev polynomials are symmetric around zero, we get that the roots of $f(\cdot)$ are also symmetric. For an odd ℓ we get that zero is one of the roots and thus each one of the sequences $\{d_i^+\}_{i=1}^{t}$, $\{d_i^-\}_{i=1}^{t+1}$ is symmetric around zero, as desired. ∎

3.2 Upper Bounds

The lower bound stated in Theorem 3 does not leave much room for an algorithm in the sampling model with sample complexity that is sublinear in m. In particular note that for a constant γ, if $\epsilon = o(1/\log^2 m \log\log m)$, then we get a linear dependence on m. However, for ϵ that is not too small, we may still ask whether we can get an upper bound that is sublinear in m. This observation immediately provides a test for γ-similar means in the sampling model that has $m^{1-\tilde{\Omega}(\epsilon^2)}$ sample complexity (conditioned on $\epsilon > c \log\log m/\log m$ for some sufficiently large constant c.)

When the domain of the distributions is $\{0, 1\}$ and m is sufficiently larger than $1/\epsilon$, we can get an improved upper bound. We note that the more efficient algorithm is not obtained by reducing to the query model. The complexity of this algorithm is $\exp(\tilde{O}(1/\epsilon))m^{1-\tilde{\Omega}(\epsilon)}$. If $\gamma \leq \epsilon/c$ for a sufficiently large constant c (and the domain is $\{0, 1\}$) then it is possible to significantly reduce the complexity to $\tilde{O}(\sqrt{m} \cdot \text{poly}(1/\epsilon))$ by using an algorithm presented in [7]. Furthermore, it is not possible to reduce the dependence on m below $\sqrt{m}$.

References

1. Batu, T., Fortnow, L., Fischer, E., Kumar, R., Rubinfeld, R., White, P.: Testing random variables for independence and identity. In: Proceedings of FOCS, pp. 442–451 (2001)
2. Batu, T., Fortnow, L., Rubinfeld, R., Smith, W.D., White, P.: Testing that distributions are close. In: Proceedings of FOCS, pp. 259–269 (2000)
3. Chebyshev, P.L.: Théorie des mécanismes connus sous le nom de parallélogrammes. Mémoires des Savants étrangers présentés â l'Académie de Saint-Pétersbourg 7, 539–586 (1854)
4. Goldreich, O., Goldwasser, S., Ron, D.: Property testing and its connection to learning and approximation. Journal of the ACM 45(4), 653–750 (1998)
5. Goldreich, O., Ron, D.: Property testing in bounded degree graphs. Algorithmica, 302–343 (2002)
6. Larsen, R.J., Marx, M.L.: An introduction to mathematical statistics and its applications, vol. 1. Pearson Prentice Hall (2006)
7. Levi, R., Ron, D., Rubinfeld, R.: Testing properties of collections of distributions. Technical Report TR10-157, Electronic Colloquium on Computational Complexity (ECCC) (2010)
8. Levi, R., Ron, D., Rubinfeld, R.: Testing properties of collections of distributions. In: Proceedings of ICS, pp. 179–194 (2011); See also ECCC TR10-157
9. Levi, R., Ron, D., Rubinfeld, R.: Testing similar means. Technical Report TR12-055, Electronic Colloquium on Computational Complexity (ECCC) (2012)
10. Mendenhall, W., Beaver, R.J., Beaver, B.M.: Introduction to probability and statistics. Brooks/Cole, Cengage Learning (2009)
11. Rubinfeld, R., Sudan, M.: Robust characterization of polynomials with applications to program testing. SIAM Journal on Computing 25(2), 252–271 (1996)
12. Student. The probable error of a mean. Biometrika 6, 1–25 (1908)
13. Valiant, G., Valiant, P.: Estimating the unseen: an $n/\log(n)$-sample estimator for entropy and support size, shown optimal via new CLTs. In: Proceedings of STOC, pp. 685–694 (2011)
14. Valiant, P.: Testing symmetric properties of distributions. In: Proceedings of STOC, pp. 383–392 (2008)
15. Welch, B.L.: The generalization of 'student's' problem when several different population variances are involved. Biometrika 34, 28–35 (1947)

The Parameterized Complexity of k-Edge Induced Subgraphs*

Bingkai Lin and Yijia Chen

Shanghai Jiao Tong University
bing314159@sjtu.edu.cn, yijia.chen@cs.sjtu.edu.cn

Abstract. We prove that finding a k-edge induced subgraph is fixed-parameter tractable, thereby answering an open problem of Leizhen Cai [2]. Our algorithm is based on several combinatorial observations, Gauss' famous *Eureka* theorem [1], and a generalization of the well-known fpt-algorithm for the model-checking problem for first-order logic on graphs with locally bounded tree-width due to Frick and Grohe [13]. On the other hand, we show that two natural counting versions of the problem are hard. Hence, the k-edge induced subgraph problem is one of the very few known examples in parameterized complexity that are easy for decision while hard for counting.

1 Introduction

Induced subgraphs are one of the most natural substructures in graphs. They capture many different combinatorial objects, e.g., clique, independent set, chordless path. Thus, a great number of algorithmic problems are about finding certain induced subgraphs, and their complexity is among the mostly extensively studied in algorithmic graph theory [3,7,14]. In this paper, we are mainly interested in the problem of finding an induced subgraph which contains exactly k edges, i.e., a k-edge induced subgraph. This problem is equivalent to solving a special 0-1 quadratic Diophantine equation $x^T A x = k$, where A is the adjacent matrix of G, $x \in \{0,1\}^n, n = |V(G)|$.

It is not difficult to prove that the k-edge induced subgraph problem is NP-hard by a reduction from the clique problem. So we approach the problem via parameterized complexity [9,12] and treat k as the parameter:

---

p-EDGE-INDUCED-SUBGRAPH
 Instance: A graph G and $k \in \mathbb{N}$.
 Parameter: k.
 Problem: Decide whether G contains a k-edge induced subgraph.

As the main result of our paper, we show that p-EDGE-INDUCED-SUBGRAPH is fixed-parameter tractable. In fact, there are special cases of p-EDGE-INDUCED-SUBGRAPH whose fixed-parameter tractability has been known for a while. Since

---

* Full version available at http://arxiv.org/abs/1105.0477

A. Czumaj et al. (Eds.): ICALP 2012, Part I, LNCS 7391, pp. 641–652, 2012.

we can define a k-edge induced subgraph by a first-order sentence, using logic machinery, it can be shown that p-EDGE-INDUCED-SUBGRAPH is fixed-parameter tractable if the graph G has bounded tree-width [8], bounded local tree-width [13], etc., or most generally locally bounded expansion [10]. Unfortunately, the class of all graphs containing a k-edge induced subgraph does not possess any of these bounded measures. As another previously known case, using his *Random Separation* method [5] and Ramsey's Theorem, Cai [4] gave a very nice combinatorial algorithm that solves p-EDGE-INDUCED-SUBGRAPH when the parameter k is a *triangular number*, i.e., $k = \binom{m}{2}$ for some $m \in \mathbb{N}$. However, it looks very difficult to adapt Cai's algorithm to handle arbitrary k. Therefore neither logic nor combinatorial approach so far seems to be sufficient to settle the complexity of p-EDGE-INDUCED-SUBGRAPH by its own. So our fpt-algorithm is a rather tricky combination of these two methods.

1.1   Our Approach

As just mentioned, our starting pointing is that the existence of a k-edge induced subgraph can be characterized by a sentence of first-order logic (FO) which depends on k only. It is a well-known result of Frick and Grohe [13] that the model-checking problem for FO on graphs of bounded *local tree-width* is fixed-parameter tractable. The local tree-width for a graph is a function bounding the tree-width of the induced subgraphs on the neighborhoods within a certain radius of every vertex. For instance, bounded-degree graphs have bounded local tree-width. These give immediately the fixed-parameter tractability of p-EDGE-INDUCED-SUBGRAPH on graphs with bounded degree[1].

With some more efforts, the above result can be extended to graphs G with degree bounded by a function of the parameter k. In that case, we can say the degree $\deg(v)$ of each vertex v is sufficiently small. The corresponding fpt-algorithm generalizes Frick and Grohe's Theorem to graphs with local tree-width bounded by a function of both the radius of the neighborhoods and an additional parameter. As a dual, if $\deg(v)$ of each vertex v in G is sufficiently large, or more precisely, the complement of G has degree bounded by a function of k, then we can decide p-EDGE-INDUCED-SUBGRAPH in fpt time, too.

Moving one step further, we consider graphs in which each $\deg(v)$ is either sufficiently small or sufficiently large, e.g., an n-star. We call such graphs *degree-extreme*. Using the same logic machinery as above, we are able to show the fixed-parameter tractability of p-EDGE-INDUCED-SUBGRAPH on degree-extreme graphs.

Assume that the graph G is not degree-extreme, i.e., there exists a vertex v_0 whose degree is neither sufficiently small nor sufficiently large. We partition the vertex set of G into two sets V_1 and V_2, where V_1 contains all vertices adjacent to v_0 and V_2 the remaining vertices. Then both V_1 and V_2 are relatively large. Note

---

[1] This is also a direct consequence of Seese's result that the model-checking problem for FO on bounded-degree graphs is fixed-parameter tractable [15]. But we find it more natural to work with bounded local tree-with in the following generalization.

possibly there are many edges between V_1 and V_2. Nevertheless, we can compute a vertex set B in G such that every edge between V_1 and V_2 has one vertex in B; and if B is large enough, we can show that G contains a k-edge induced subgraph. Otherwise, the graph G consists of two induced subgraphs $G[V_1]$ and $G[V_2]$, plus the edges between V_1 and V_2 adjacent to the set B of bounded size. In case $G[V_1]$ and $G[V_2]$ are both degree-extreme, we call such a graph G a *bridge* (of two degree-extreme graphs). By the logic method again, we prove that p-EDGE-INDUCED-SUBGRAPH is fixed-parameter tractable on bridges.

Now we are left with the case that at least one of $G[V_1]$ and $G[V_2]$ is not degree-extreme, say $G[V_1]$. Then we repeat the above procedure on $G[V_1]$ to get a partition $V_{11} \mid V_{12}$ of V_1. And again, both V_{11} and V_{12} are sufficiently large. Arguing as before, either we already know $G[V_1]$, and hence G, contains a k-edge induced subgraph, or there is a set B_1 of bounded size such that every edge between V_{11} and V_{22} intersects B_1.

Finally we remove the vertex set $B_0 := B \cup B_1$ from G. Then $G[V \setminus B_0]$ is the disjoint union of $G[V_{11} \setminus B_0]$, $G[V_{12} \setminus B_0]$ and $G[V_2 \setminus B_0]$. Moreover, all the three induced subgraphs are so large that, by Ramsey's Theorem, either one of them contains a large independent set, or we have three large disjoint cliques which are not adjacent to each other. For both cases, we show that $G[V \setminus B_0]$, and hence G, contains a k-edge induced subgraph. As a matter of fact, the second case is an easy consequence of a famous number-theoretic result of Gauss which states that *every natural number is the sum of three triangular numbers*.

We should mention that the running time of our algorithm in terms of the parameter k is *triple exponential* at least. On the other hand, it is linear in the size of the graph. We leave the detailed analysis in the full version of the paper.

1.2   Counting k-Edge Induced Subgraphs

We also study the parameterized complexity of computing the number of k-edge induced subgraphs. For most natural problems, if the decision version is easy, then so is the counting problem. However, it turns out that two natural counting versions of p-EDGE-INDUCED-SUBGRAPH are both hard. To the best of our knowledge, there are only few natural problems which exhibit such a phenomenon [11,6].

1.3   Organization of Our Paper

In Section 2 we introduce necessary background and fix our notations. We prove all required combinatorial results in Section 3. In particular, we present several simple structures in a graph which, if exist, guarantee the existence of a k-edge induced subgraph. Then in Section 4 we establish the fixed-parameter tractability of p-EDGE-INDUCED-SUBGRAPH on the degree-extreme graphs and the bridges using model-checking problems for FO. We present our fpt-algorithm for p-EDGE-INDUCED-SUBGRAPH by putting all the pieces together in Section 5. Finally in Section 6 we prove the hardness of the counting problems. Due to the space limitations, for some proofs we refer to the full version of this paper.

2 Preliminaries

$\mathbb{N}$ and $\mathbb{N}^+$ denote the sets of natural numbers (that is, nonnegative integers) and positive integers, respectively. For a natural number n let $[n] := \{1, \ldots, n\}$. We denote the alphabet $\{0, 1\}$ by Σ and identify problems with subsets Q of Σ^*. Clearly, as done mostly, we present concrete problems in a verbal, hence uncodified form over Σ. For every set S we use $|S|$ to denote its size. Moreover we let $\binom{S}{2}$ be the set of all two-element subsets of S, i.e., $\{\{a, b\} \mid a, b \in S \text{ and } a \neq b\}$. A triangular number is $\binom{k}{2} := \left|\binom{[k]}{2}\right|$ for some $k \in \mathbb{N}$. In particular, $\binom{0}{2} = \binom{1}{2} = 0$.

2.1 Parameterized Complexity

A *parameterized problem* is a pair (Q, κ) consisting of a classical problem $Q \subseteq \Sigma^*$ and a polynomial time computable *parameterization* $\kappa : \Sigma^* \to \mathbb{N}$.

An algorithm $\mathbb{A}$ is an *fpt-algorithm with respect to a parameterization* κ if for every $x \in \Sigma^*$ the running time of $\mathbb{A}$ on x is bounded by $f(\kappa(x)) \cdot |x|^{O(1)}$ for a computable function $f : \mathbb{N} \to \mathbb{N}$. Or equivalently, we say that the algorithm $\mathbb{A}$ runs in fpt time. A parameterized problem (Q, κ) is *fixed-parameter tractable* if there is an fpt-algorithm with respect to κ that decides Q.

2.2 Graphs

We only consider *simple* graphs, that is, finite nonempty undirected graphs without loops and parallel edges. Every graph $G = (V, E)$ is thus determined by a nonempty vertex set V and an edge set $E \subseteq \binom{V}{2}$. For an edge $\{u, v\} \in E$ we say that u is *adjacent* to v, and vice versa. Often we also use $V(G)$ and $E(G)$ to denote the vertex set and the edge set of G, respectively.

Let $G = (V, E)$ be a graph. For every vertex $v \in V$ the set $N^G(v)$ contains all vertices in G that are adjacent to v, i.e., $N^G(v) := \{u \mid \{u, v\} \in E\}$. Moreover, for every $S \subseteq V$ we let $N^G(S) := \bigcup_{v \in S} N^G(v)$. Note the degree of v, written $\deg^G(v)$, is $|N^G(v)|$. If $\deg^G(v) = 0$, then v is an *isolated* vertex. The distance $d^G(u, v)$ between two vertices $u, v \in V$ is the length of a shortest path from u to v in the graph G. If it is clear from the context, we omit the superscript G in the above notations and write $N(v)$, $\deg(v)$, etc., instead.

Every nonempty subset $S \subseteq V(G)$ induces a subgraph $G[S]$ with the vertex set S and the edge set $E(G[S]) := \binom{S}{2} \cap E(G)$. Consequently, a graph H is an *induced subgraph* of G if $H = G[V(H)]$. Recall that H is a k-*edge* induced subgraph of G for $k := |E(H)|$.

Again, let S be a set of vertices in G. Then S is a *clique*, if for every $u, v \in S$ we have either $u = v$ or $\{u, v\} \in E(G)$. On the other hand, the set S is an *independent set* in G, if $\{u, v\} \notin E(G)$ for all $u, v \in S$. For every $k \in \mathbb{N}$, there exists a constant $\mathscr{R}_k$, known as the *Ramsey number*, such that every graph G with $|V(G)| \geq \mathscr{R}_k$ has either a clique of size k or an independent set of size k. It is well-known that $\mathscr{R}_k < 2^{2 \cdot k}$ for every $k \in \mathbb{N}$.

2.3 Relational Structures and First-Order Logic

A *vocabulary* τ is a finite set of relation symbols. Each relation symbol has an *arity*. A *structure* $\mathcal{A}$ of vocabulary τ, or simply structure, consists of a nonempty set A called the *universe*, and an interpretation $R^{\mathcal{A}} \subseteq A^r$ of each r-ary relation symbol $R \in \tau$. For example, a graph G can be identified with a structure $\mathcal{A}(G)$ of vocabulary $\tau_{\text{graph}} := \{E\}$ with the binary relation symbol E such that $A(G) := V(G)$ and $E^{\mathcal{A}(G)} := \{(u, v) \mid \{u, v\} \in E(G)\}$.

The *disjoint union* of two τ-structures $\mathcal{A}_1$ and $\mathcal{A}_2$ is again a τ-structure, denoted by $\mathcal{A}_1 \,\dot\cup\, \mathcal{A}_2$, whose universe is $A_1 \,\dot\cup\, A_2$, and where for each relation symbol $R \in \tau$ we let $R^{\mathcal{A}_1 \,\dot\cup\, \mathcal{A}_2} := R^{\mathcal{A}_1} \,\dot\cup\, R^{\mathcal{A}_2}$.

Let $\mathcal{A}$ be a structure of a vocabulary τ. Then the *Gaifman graph* of $\mathcal{A}$ is $G(\mathcal{A}) := (V, E)$ with $V := A$ and $E := \{\{a, b\} \mid a, b \in A, a \neq b, \text{and for some} R \in \tau, \text{and some tuple}(a_1, \ldots, a_r) \in R^{\mathcal{A}}, \{a, b\} \subseteq \{a_1, \ldots, a_r\}\}$. Note any unary relation in $\mathcal{A}$ has no influence on E.

Let $r \in \mathbb{N}$ and $a \in A$. Then the *r-neighborhood* of a is $N_r^{\mathcal{A}}(a) := \{b \in A \mid d^{G(\mathcal{A})}(a, b) \leq r\}$. Moreover, the structure $\mathcal{N}_r^{\mathcal{A}}(a)$ induced by the r-neighborhood of a has universe $N_r^{\mathcal{A}}(a)$, and for each r-ary relation symbol $R \in \tau$ the interpretation $\{(a_1, \ldots, a_r) \in R^{\mathcal{A}} \mid a_1, \ldots, a_r \in N_r^{\mathcal{A}}(a)\}$.

Formulas of first-order logic of vocabulary τ are built up from atomic formulas $x = y$ and $Rx_1 \ldots x_r$ where $x, y, x_1, \ldots, x_r$ are variables and $R \in \tau$ is of arity r, using the boolean connectives and existential and universal quantification.

2.4 Tree-Width and Local Tree-Width

We assume that the reader is familiar with the notion of *tree-width* $\text{tw}(G)$ of a graph G. Recall that the tree-width $\text{tw}(\mathcal{A})$ of a structure $\mathcal{A}$ is simply $\text{tw}(G(\mathcal{A}))$, that is, the tree-width of the Gaifman graph of $\mathcal{A}$. In fact, to understand most parts of our proofs and algorithms, it is sufficient to know that for every structure $\mathcal{A}$ we have $\text{tw}(\mathcal{A}) < |A|$.

Now we are ready to define the *local tree-width* of a structure $\mathcal{A}$. For every $r \in \mathbb{N}$ let $\text{ltw}(\mathcal{A}, r) := \max\{\text{tw}(\mathcal{N}_r^{\mathcal{A}}(a)) \mid a \in A\}$. Let $g : \mathbb{N} \times \mathbb{N} \to \mathbb{N}$ be a function and $p \in \mathbb{N}$. We say a structure $\mathcal{A}$ has *local tree-width bounded by g with respect to p* if $\text{ltw}(\mathcal{A}, r) \leq g(r, p)$ for every $r \in \mathbb{N}$. This slightly generalizes the usual notion of local tree-width bounded by a *unary* function [13].

3 Some Easy Positive Instances

In this section, let $k \in \mathbb{N}$ and $G = (V, E)$ be a graph.

Definition 1. *G contains a k-independent-set-matching structure on vertices $u_1, \ldots, u_k, v_1, \ldots, v_k$ if $u_1, \ldots, u_k, v_1, \ldots, v_k$ are pairwise distinct; for every $i, j \in [k]$ we have $\{u_i, v_j\} \in E$ if and only if $i = j$; and $\{u_1, \ldots, u_k\}$ is an independent set in G.*

Lemma 1. *Every graph containing a k-independent-set-matching structure has a k-edge induced subgraph.*

Proof: The case for $k = 0$ is trivially true. So assume $k \geq 1$ and G contains a k-independent-set-matching structure on the vertices $u_1, \ldots, u_k, v_1, \ldots, v_k$. We choose the maximum $k' \leq k$ such that $\ell := \left| E(G[\{v_1, \ldots, v_{k'}\}]) \right| \leq k$. If $k' = k$, then $G[V']$ with $V' := \{u_1, \ldots, u_{k-\ell}\} \cup \{v_1, \ldots, v_k\}$ is a k-edge induced subgraph of G. Otherwise, $k' < k$. In particular, $\left| E(G[\{v_1, \ldots, v_{k'}, v_{k'+1}\}]) \right| > k$. As $v_{k'+1}$ can contribute at most k' many new edges, we have $\ell + k' > k$, i.e., $k - \ell < k'$. Then $G[V']$ with $V' := \{u_1, \ldots, u_{k-\ell}\} \cup \{v_1, \ldots, v_{k'}\}$ is a k-edge induced subgraph of G. □

Definition 2. *G contains a k-clique-matching structure on vertices $u_1, \ldots, u_k$, $v_1, \ldots, v_k$ if $u_1, \ldots, u_k, v_1, \ldots, v_k$ are pairwise distinct; for every $i, j \in [k]$ we have $\{u_i, v_j\} \in E$ if and only if $i = j$; and $\{u_1, \ldots, u_k\}$ is a clique in G.*

Lemma 2. *If G contains a k-clique-matching structure, then G has a k-edge induced subgraph.*

Proof: The cases for $k \leq 2$ are trivial. So we consider $k \geq 3$. Let k_0 be maximum with $\binom{k_0}{2} \leq k$ and set $r := k - \binom{k_0}{2}$. It is easy to verify that $k \geq k_0 + r$ by $k \geq 3$ and $k_0 > r$. Now assume G contains a k-clique-matching-structure on the vertices $u_1, \ldots, u_k, v_1, \ldots, v_k$. Then, we choose the maximum $r' \leq r$ such that $\ell := \left| E(G[\{v_1, \ldots, v_{r'}\}]) \right| \leq r$. If $r' = r$, then $G[V']$ with $V' := \{v_1, \ldots, v_r\} \cup \{u_1, \ldots, u_{r-\ell}, u_{r+1}, \ldots, u_{k_0+\ell}\}$ is a k-edge induced subgraph of G. Otherwise, $r' < r$ and by the maximality of r' we have $\left| E(G[\{v_1, \ldots, v_{r'}, v_{r'+1}\}]) \right| > r$. As $v_{r'+1}$ can add at most r' many new edges, we have $\ell + r' > r$, or equivalently $r - \ell < r'$. It follows that $G[V']$ with $V' := \{v_1, \ldots, v_{r'}\} \cup \{u_1, \ldots, u_{r-\ell}, u_{r'+1}, \ldots, u_{r'+k_0-r+\ell}\}$ has exactly k edges. □

Definition 3. *We say that G contains a k-apex structure on v_0, A and B if*

(A1) $A, B \subseteq V$ are disjoint with $|A| \geq k$ and $|B| \geq \mathscr{R}_k$, $v_0 \in V$;
(A2) A is a clique in G;
(A3) $\{u, v_0\} \in E$ for every $u \in A$ and $\{v, v_0\} \notin E$ for every $v \in B$;
(A4) $\{u, v\} \in E$ for every $u \in A$ and $v \in B$.

Lemma 3. *If G contains a k-apex structure, then it has a k-edge induced subgraph.*

Proof: The case for $k \leq 1$ is trivially true. So let $k \geq 2$. Moreover, let v_0, A, B be as stated in Definition 3. Since $|B| \geq \mathscr{R}_k$, $G[B]$ contains either a clique of size k or an independent set of size k. If $G[B]$ contains an independent set $B' \subseteq B$ with $|B'| = k$. Then for every $u \in A$ the induced subgraph $G[B' \cup \{u\}]$ has exactly k edges by (A4). Now assume that there is a clique B' in $G[B]$ of size k. Observe by (A3) and $k \geq 2$, we have $v_0 \notin (A \cup B')$. Furthermore, it is easy to see that we can write $k = \binom{k_0}{2} + r$ for some appropriate $k \geq k_0 \geq r$. We select arbitrary subsets $A' \subseteq A$ and $B'' \subseteq B'$ with $|A'| = r$ and $|B''| = k_0 - r$. Then it is straightforward to check that $G[A' \cup B'' \cup \{v_0\}]$ has exactly k edges. □

Lemma 4. *Assume there exists three disjoint cliques S_1, S_2, S_3 in G, all of size k; and there are no edges between any distinct S_i and S_j. Then G has a k-edge induced subgraph.*

It is easy to that Lemma 4 is a direct consequence of Gauss' famous Eureka Theorem [1].

Theorem 1. *For every $k \in \mathbb{N}$ there exist $k_0, k_1, k_2 \in \mathbb{N}$ such that $k = \binom{k_0}{2} + \binom{k_1}{2} + \binom{k_2}{2}$.*

Lemma 5. *Let $k \in \mathbb{N}^+$ and $G = (V, E)$ be a graph without isolated vertices. If G contains an independent set of size $(k-1)^2 + 1$, then it has a k-edge induced subgraph.*

To prove the above lemma, we need some further preparation.

Lemma 6. *Let $m, n \in \mathbb{N}^+$ and $A, B \subseteq V$ be disjoint. If for every $u \in A$ we have $|N(u) \cap B| \geq 1$ and $|A| > (m-1)(n-1)$, then*

(i) either there are m vertices $u_1, \ldots, u_m$ in A, and a vertex v in B with $\{u_i, v\} \in E$ for every $i \in [m]$,

(ii) or there are n vertices $u_1, \ldots, u_n$ in A and n vertices $v_1, \ldots, v_n$ in B such that for all $i, j \in [n]$ we have $\{u_i, v_j\} \in E$ if and only if $i = j$.

Proof: [of Lemma 5] Let $S \subseteq V$ be an independent set in G with $|S| > (k-1)^2$. Since G has no isolated vertex, $|N(u) \cap N(S)| \geq 1$ for every $u \in S$. So we can apply Lemma 6 on $A \leftarrow S$, $B \leftarrow N(S)$, $m \leftarrow k$, and $n \leftarrow k$. If (i) holds, then we have an induced k-star of exactly k edges. Otherwise, we have (ii). Hence, there exist vertices $u_1, \ldots, u_k \in S$ and $v_1, \ldots, v_k \in N(S)$ such that G contains a k-independent-set-matching structure on those vertices. The result follows from Lemma 1. □

Definition 4. *Let $d \in \mathbb{N}$. We define*

$$V_{[1,d]}^G := \{v \in V \mid 1 \leq \deg(v) \leq d\}.$$

It is well-known that if a graph contains many small-degree vertices, then it has a large independent set. As a result, the following is an easy consequence of Lemma 5.

Lemma 7. *Let $d, k \in \mathbb{N}^+$. If $\left|V_{[1,d]}^G\right| > (d+1) \cdot (k-1)^2$, then G contains a k-edge induced subgraph.*

3.1 A Further Combinatorial Lemma

For later purpose, we need a generalization of Lemma 6.

Lemma 8. *Let $m, n, p \in \mathbb{N}^+$ and $A, B \subseteq V$ be disjoint in the graph G. If for every $u \in A$, $|N(u) \cap B| \geq p$ and $|A| > (m-1)(n-1)^p$, then*

(i) either there are m vertices $u_1, \ldots, u_m$ in A and p vertices $v_1, \ldots, v_p$ in B with $\{u_i, v_j\} \in E$ for every $i \in [m]$ and $j \in [p]$,

(ii) or there are n vertices $u_1, \ldots, u_n$ in A and n vertices $v_1, \ldots, v_n$ in B such that for all $i, j \in [n]$ we have $\{u_i, v_j\} \in E$ if and only if $i = j$.

4   Easy Instances by Model-Checking

In this section we show the fixed-parameter tractability of p-EDGE-INDUCED-SUBGRAPH on some restricted classes of graphs via the model-checking problem for first-order logic. The following is a generalization of a well-known result due to Frick and Grohe [13].

Theorem 2. *For every computable function* $g : \mathbb{N} \times \mathbb{N} \to \mathbb{N}$ *the problem*

> p-MC-LTW$_g$-FO
>   *Instance: A structure* $\mathcal{A}$, $p \in \mathbb{N}$ *and an FO-sentence* φ *such that*
> $\mathcal{A}$ *has local tree-width bounded by* g *with respect to* p.
> *Parameter:* $p + |\varphi|$.
> *Problem: Decide whether* $\mathcal{A} \models \varphi$.

is fixed-parameter tractable.

Definition 5. *Let* $d \in \mathbb{N}$ *and* $G = (V, E)$ *be a graph. If* $\deg(v) \leq d$ *or* $\deg(v) \geq |V| - 1 - d$ *for every* $v \in V$, *then the graph* G *is* d-degree-extreme.

Proposition 1. *Let* $D : \mathbb{N} \to \mathbb{N}$ *be a computable function. Then the problem*

> *Instance:* A graph G and $k \in \mathbb{N}$ such that G is $D(k)$-degree-
>   extreme.
> *Parameter:* k.
> *Problem:* Decide whether G contains a k-edge induced subgraph.

is fixed-parameter tractable.

Definition 6. *Let* $d, b \in \mathbb{N}$. *Then* (G, V_1, V_2, B) *is a* (d, b)-bridge *(of the two degree-extreme graphs) if :*

(B1) $V = V_1 \cup V_2$ *for some disjoint* V_1 *and* V_2. *(B2)* $G[V_1]$ *and* $G[V_2]$ *are both* d-degree-extreme.
(B3) B *is a subset of* V *with* $|B| = b$ *such that for every edge* $\{u, v\}$ *with* $u \in V_1$ *and* $v \in V_2$ *we have either* $u \in B$ *or* $v \in B$.

Similar to Proposition 1, we can prove:

Proposition 2. *Let* $D : \mathbb{N} \to \mathbb{N}$ *be a computable function. Then the problem*

> *Instance:* A graph $G = (V, E)$, $V_1, V_2, B \subseteq V$ and $k \in \mathbb{N}$ such that
> (G, V_1, V_2, B) is a $(D(k), |B|)$-bridge.
> *Parameter:* $k + |B|$.
> *Problem:* Decide whether G contains a k-edge induced subgraph.

is fixed-parameter tractable.

5   The Algorithm

The main component of our fpt-algorithm for p-EDGE-INDUCED-SUBGRAPH is the following procedure that either already solves the problem or decomposes the given graph into potentially a bridge of two large degree-extreme graphs (cf. Definition 6).

For every $k \in \mathbb{N}$ we let $p_k := 2^{2 \cdot k} (> \mathscr{R}_k)$.

Lemma 9. *For every computable function $D : \mathbb{N} \to \mathbb{N}$ there is an fpt-algorithm $\mathbb{A}_D$ such that for every graph $G = (V, E)$ and every $k \in \mathbb{N}$ exactly one of following conditions is satisfied.*

(S1) G is $D(k)$-degree-extreme and $\mathbb{A}_D$ correctly decides whether G contains a k-edge induced subgraph.

(S2) G is not $D(k)$-degree-extreme and $\mathbb{A}_D$ correctly outputs that G contains a k-edge induced subgraph.

(S3) G is not $D(k)$-degree-extreme and $\mathbb{A}_D$ outputs three subsets $V_1, V_2, B \subseteq V$ such that

(S3.1) $V = V_1 \cup V_2$ with $|V_1| > D(k)$ and $|V_2| > D(k) + 1$;

(S3.2) every edge between V_1 and V_2 in G has one vertex in B and $|B| \leq (p_k - 1)^{p_k + 1} + (p_k - 1)^2$.

Proof: Let $G = (V, E)$ be a graph and $k \in \mathbb{N}$. If G is $D(k)$-degree-extreme, then we apply Proposition 1 to achieve (S1). Otherwise let $v_0 \in V$ be a vertex with

$$D(k) < \deg(v_0) < |V| - 1 - D(k). \tag{1}$$

Then we set $V_1 := N(v_0)$ and $V_2 := V \setminus V_1$. By (1) it holds that $|V_1| > D(k)$ and $|V_2| = |V| - |V_1| = |V| - \deg(v_0) > D(k) + 1$, i.e., (S3.1). Let $W_1 := \Big\{ u \in V_1 \ \Big| \ |N(u) \cap V_2| \geq p_k \Big\}$ and $W_2 := V_1 \setminus W_1$.

Claim 1. If $|W_1| > (p_k - 1)^{p_k + 1}$, then G contains a k-edge induced subgraph.

Proof of the claim. We apply Lemma 8 on $A \leftarrow W_1$, $B \leftarrow V_2$, $m \leftarrow p_k$, $n \leftarrow p_k$, and $p \leftarrow p_k$. So there are p_k vertices $u_1, \ldots, u_{q_k}$ in W_1 and p_k vertices $v_1, \ldots, v_{p_k}$ in V_2 such that

(i) either $\{u_i, v_j\} \in E$ for every $i, j \in [p_k]$,

(ii) or for all $i, j \in [p_k]$ we have $\{u_i, v_j\} \in E$ if and only if $i = j$.

Recall $p_k > \mathscr{R}_k$, so there is a subset $S \subseteq \{u_1, \ldots, u_{p_k}\}$ such that S is either an independent set or a clique. If S is an independent set, then $G[S \cup \{v_0\}]$ has exactly k edges. So suppose S is a clique. Assume that (i) is true, then G contains a k-apex structure on $v_0, S, \{v_1, \ldots, v_{p_k}\}$. Hence, Lemma 3 implies the claim. Otherwise (ii) holds. And say $S = \{u_{i_1}, \ldots, u_{i_k}\}$. Then the graph G contains a k-clique-matching structure on $u_{i_1}, \ldots, u_{i_k}, v_1, \ldots, v_k$. The result follows from Lemma 2.   ⊣

Claim 2. If $|N(W_2) \cap V_2| > (p_k - 1)^2$, then G contains a k-edge induced subgraph.

Proof of the claim. It is easy to verify that we can apply Lemma 6 on $A \leftarrow N(W_2) \cap V_2$, $B \leftarrow W_2$, $m \leftarrow p_k$, and $n \leftarrow p_k$. So,

(i) either there are p_k vertices $u_1, \ldots, u_{q_k}$ in $N(W_2) \cap V_2$ and a vertex v in W_2 such that $\{u_i, v\} \in E$ for every $i \in [p_k]$,

(ii) or there are p_k vertices $u_1, \ldots, u_{p_k}$ in $N(W_2) \cap V_2$ and p_k vertices $v_1, \ldots, v_{p_k}$ in W_2 such that for all $i, j \in [p_k]$ we have $\{u_i, v_j\} \in E$ if and only if $i = j$.

But (i) contradicts our definition of W_2, i.e., for every $u \in W_2$ we have $|N(u) \cap V_2| < p_k$, therefore (ii) must hold. Recall $p_k > \mathscr{R}_k$, hence $G[\{v_1, \ldots, v_{p_k}\}]$ contains either a clique of size of k or an independent set of size k. Without loss of generality, let $\{v_1, \ldots, v_k\} \subseteq W_2 \subseteq V_1$ be a clique or an independent set.

For the independent set case, as $v_0 \notin V_1$, then $G[\{v_0, v_1, \ldots, v_k\}]$ is a k-induced subgraph. For the clique case, G contains a k-clique-matching structure on $u_1, \ldots, u_k, v_1, \ldots, v_k$. We are done by Lemma 2.   ⊣

Let $B := W_1 \cup (N(W_2) \cap V_2)$. If $|B| > (p_k - 1)^{p_k + 1} + (p_k - 1)^2$, then, by Claim 1 and Claim 2, the graph G contains a k-edge induced subgraph, and (S2) follows. Otherwise $|B| \leq (p_k - 1)^{p_k + 1} + (p_k - 1)^2$. Observe that every edge between V_1 and V_2 has at least one vertex in B. Thus, we achieve (S3) by outputting (V_1, V_2, B).   □

Finally we are ready to present our fpt-algorithm for p-EDGE-INDUCED-SUBGRAPH.

Theorem 3. p-EDGE-INDUCED-SUBGRAPH *is fixed-parameter tractable.*

Proof: We define a computable function $D_0 : \mathbb{N} \to \mathbb{N}$ by

$$D_0(k) := 2 \cdot \left((p_k - 1)^{p_k + 1} + (p_k - 1)^2\right) + 2^{2 \cdot ((k-1)^2 + 1)}. \tag{2}$$

Then let $\mathbb{A}_{D_0}$ be the algorithm as stated in Lemma 9 for the function D_0.

Let (G, k) with $G = (V, E)$ be an instance of p-EDGE-INDUCED-SUBGRAPH. First, we remove all the isolated vertices in G. For simplicity, the resulting graph is denoted by G again. Then, we simulate the algorithm $\mathbb{A}_{D_0}$ on (G, k). If the result is either (S1) or (S2) in Lemma 9, we already get the correct answer. Otherwise, $\mathbb{A}_{D_0}$ outputs three subsets $V_1, V_2, B \subseteq V$ satisfying (S3.1) and (S3.2).

If $G[V_1]$ and $G[V_2]$ are both $D_0(k)$-degree-extreme, then (G, V_1, V_2, B) is a $(D_0(k), |B|)$-bridge with $|B|$ bounded by an appropriate computable function of k. The fixed-parameter tractability of whether G contains a k-edge induced subgraph follows from Proposition 2. Otherwise, either $G[V_1]$ or $G[V_2]$ is not $D_0(k)$-degree-extreme.

We assume that $G[V_1]$ is not $D_0(k)$-degree-extreme. (The case for $G[V_2]$ is symmetric.) Then we simulate the algorithm $\mathbb{A}_{D_0}$ on $(G[V_1], k)$. Observe that the result cannot be (S1). If the output is (S2), since $G[V_1]$ is an induced subgraph of G, we conclude that G has an induced subgraph of exactly k edges.

Now we are left with case (S3). In particular, there are subsets $V_{11}, V_{12}, B_1 \subseteq V_1$ such that the corresponding properties of (S3.1) and (S3.2) are satisfied. Let $U_1 := V_{11} \setminus (B \cup B_1)$, $U_2 := V_{12} \setminus (B \cup B_1)$, and $U_3 := V_2 \setminus (B \cup B_1)$. Observe that in G if we remove the vertex set B, then there is no edge left between V_1 and V_2. Similarly, if we remove the vertex set B_1, every edge between V_{11} and V_{12}

is destroyed. Thus, by (S3.2), in the original graph G, there is no edge between each pair of U_1, U_2 and U_3. Moreover by (S3.1) and (S3.2) for every $i \in [3]$

$$|U_i| > D_0(k) - 2 \cdot \left((p_k - 1)^{p_k+1} + (p_k - 1)^2\right) = 2^{2 \cdot ((k-1)^2+1)} > \mathscr{R}_{(k-1)^2+1},$$

where the equality is by (2).

We use Ramsey's Theorem again. If there is an independent set of size $(k - 1)^2 + 1$ in one of the U_1, U_2 and U_3, as G has no isolated vertex, then G contains a k-edge induced subgraph by Lemma 5. Otherwise every U_i contains a clique of size $(k - 1)^2 + 1 \geq k$. As we have seen that there is no edge between U_1, U_2 and U_3 in G, Lemma 4 implies that G contains an induced subgraph of exactly k edges. □

6 Counting k-Edge Induced Subgraphs

The most natural counting version of p-EDGE-INDUCED-SUBGRAPH is:

---

p-#EDGE-INDUCED-SUBGRAPH
    Instance: A graph G and $k \in \mathbb{N}$.
    Parameter: k.
    Problem: Compute the number of k-edge induced subgraphs in G.

In fact, the hardness of p-#EDGE-INDUCED-SUBGRAPH is rather easy to show. We observe that the vertex set of every induced subgraph *without any edge* is an independent set, and vice versa. Hence the *first slice* of p-#EDGE-INDUCED-SUBGRAPH, i.e., counting the number of 0-edge induced subgraphs is exactly the classical problem #INDEPENDENT-SET of counting the number of independent sets in a given graph. Recall that #INDEPENDENT-SET is #P-hard [16]. Hence:

Theorem 4. *Assume* #P $\neq$ P. *Then* p-#EDGE-INDUCED-SUBGRAPH *is not fixed-parameter tractable.*

One might attribute the above hardness result to the fact that we allow induced subgraphs to have isolated vertices. Note these isolated vertices play no role in the decision problem p-EDGE-INDUCED-SUBGRAPH. Therefore, it also makes sense to consider:

---

p-#EDGE-INDUCED-SUBGRAPH*
    Instance: A graph G and $k \in \mathbb{N}$.
    Parameter: k.
    Problem: Compute the number of k-edge induced subgraphs *without isolated vertices* in G.

Then we can show:

Theorem 5. p-#EDGE-INDUCED-SUBGRAPH* *is hard for* #W[1].

Acknowledgement. We thank Leizhen Cai for bringing the problem p-EDGE-INDUCED-SUBGRAPH to our attention, and Jörg Flum for comments on earlier versions of this paper. This research has been partly supported by the National Nature Science Foundation of China (60970011, 61033002).

References

1. Andrews, G.: Eureka! num $= \Delta + \Delta + \Delta$. Journal of Number Theory 23(3), 285–293 (1986)
2. Bodlaender, H.L., Cai, L., Chen, J., Fellows, M.R., Telle, J.A., Marx, D.: Open problems in parameterized and exact computation - IWPEC 2006. Technical Report UU-CS-2006-052, Department of Information and Computing Sciences, Utrecht University (2006)
3. Cai, L.: Fixed-parameter tractability of graph modification problems for hereditary properties. Information Processing Letters 58(4), 171–176 (1996)
4. Cai, L.: Private communication (2008)
5. Cai, L., Chan, S.M., Chan, S.O.: Random Separation: A New Method for Solving Fixed-Cardinality Optimization Problems. In: Bodlaender, H.L., Langston, M.A. (eds.) IWPEC 2006. LNCS, vol. 4169, pp. 239–250. Springer, Heidelberg (2006)
6. Chen, Y., Flum, J.: On parameterized path and chordless path problems. In: Proceedings of 22nd Annual IEEE Conference on Computational Complexity (CCC 2007), pp. 250–263. IEEE Computer Society Press (2007)
7. Chen, Y.-J., Thurley, M., Weyer, M.: Understanding the Complexity of Induced Subgraph Isomorphisms. In: Aceto, L., Damgård, I., Goldberg, L.A., Halldórsson, M.M., Ingólfsdóttir, A., Walukiewicz, I. (eds.) ICALP 2008, Part I. LNCS, vol. 5125, pp. 587–596. Springer, Heidelberg (2008)
8. Courcelle, B.: Graph rewriting: An algebraic and logic approach. In: Van Leeuwen, J. (ed.) Handbook of Theoretical Computer Science, pp. 192–242. Elsevier Science Publishers, Amsterdam (1990)
9. Downey, R.G., Fellows, M.R.: Parameterized Complexity. Springer (1999)
10. Dvorak, Z., Král, D., Thomas, R.: Deciding first-order properties for sparse graphs. In: Proceedins of the 51th Annual IEEE Symposium on Foundations of Computer Science (FOCS 2010), pp. 133–142. IEEE Computer Society (2010)
11. Flum, J., Grohe, M.: The parameterized complexity of counting problems. SIAM Journal on Computing 33(4), 892–922 (2004)
12. Flum, J., Grohe, M.: Parameterized Complexity Theory. Springer (2006)
13. Frick, M., Grohe, M.: Deciding first-order properties of locally tree-decomposable structures. Journal of ACM 48(6), 1184–1206 (2001)
14. Khot, S., Raman, V.: Parameterized complexity of finding subgraphs with hereditary properties. Theoretical Computer Science 289(2), 997–1008 (2002)
15. Seese, D.: Linear time computable problems and first-order descriptions. Mathematical Structures in Computer Science 6(6), 505–526 (1996)
16. Valiant, L.G.: The complexity of enumeration and reliability problems. SIAM Journal on Computing 8(3), 410–421 (1979)

Converting Online Algorithms to Local Computation Algorithms*

Yishay Mansour[1,**], Aviad Rubinstein[1], Shai Vardi[1,***], and Ning Xie[2,†]

[1] School of Computer Science, Tel Aviv University, Israel
[2] CSAIL, MIT, Cambridge MA 02139, USA
{mansour,shaivarl}@post.tau.ac.il
aviadrub@mail.tau.ac.il,
ningxie@csail.mit.edu

Abstract. We propose a general method for converting online algorithms to local computation algorithms,[1] by selecting a random permutation of the input, and simulating running the online algorithm. We bound the number of steps of the algorithm using a *query tree*, which models the dependencies between queries. We improve previous analyses of query trees on graphs of bounded degree, and extend this improved analysis to the cases where the degrees are distributed binomially, and to a special case of bipartite graphs.

Using this method, we give a local computation algorithm for maximal matching in graphs of bounded degree, which runs in time and space $O(\log^3 n)$.

We also show how to convert a large family of load balancing algorithms (related to balls and bins problems) to local computation algorithms. This gives several local load balancing algorithms which achieve the same approximation ratios as the online algorithms, but run in $O(\log n)$ time and space.

Finally, we modify existing local computation algorithms for hypergraph 2-coloring and k-CNF and use our improved analysis to obtain better time and space bounds, of $O(\log^4 n)$, removing the dependency on the maximal degree of the graph from the exponent.

1 Introduction

1.1 Background

The classical computation model has a single processor which has access to a given input, and using an internal memory, computes the output. This is essentially the von

---

* The full version of the paper can be found at http://arxiv.org/abs/1205.1312
** Supported in part by the Google Inter-university center for Electronic Markets and Auctions, by a grant from the Israel Science Foundation, by a grant from United States-Israel Binational Science Foundation (BSF), by a grant from Israeli Centers of Research Excellence (ICORE), and by a grant from the Israeli Ministry of Science (MoS).
*** Supported in part by the Google Inter-university center for Electronic Markets and Auctions.
† Supported by NSF grants CCF-0728645, CCF-0729011 and CCF-1065125.
[1] For a given input x, *local computation algorithms* support queries by a user to values of specified locations y_i in a legal output $y \in F(x)$.

A. Czumaj et al. (Eds.): ICALP 2012, Part I, LNCS 7391, pp. 653–664, 2012.
© Springer-Verlag Berlin Heidelberg 2012

Newmann architecture, which has been the driving force since the early days of computation. The class of polynomial time algorithms is widely accepted as the definition of *efficiently computable* problems. Over the years many interesting variations of this basic model have been studied, focusing on different issues.

Online algorithms (see, e.g., [6]) introduce limitations in the time domain. An online algorithm needs to select actions based only on the history it observed, without access to future inputs that might influence its performance. Sublinear algorithms (e.g. [9, 12]) limit the space domain, by limiting the ability of an algorithm to observe the entire input, and still strive to derive global properties of it.

Local computation algorithms (LCAs) [13] are a variant of sublinear algorithms. The LCA model considers a computation problem which might have multiple admissible solutions, each consisting of multiple bits. The LCA can return queries regarding parts of the output, in a consistent way, and in poly-logarithmic time. For example, the input for an LCA for a job scheduling problem consists of the description of n jobs and m machines. The admissible solutions might be the allocations of jobs to machines such that the makespan is at most twice the optimal makespan. On any query of a job, the LCA answers quickly the job's machine. The correctness property of the LCA guarantees that different query replies will be consistent with some admissible solution.

1.2 Our Results

1.2.1 Bounds on Query Trees

Suppose that we have an online algorithm where the reply to a query depends on the replies to a small number of previous queries. The reply to each of those previous queries depends on the replies to a small number of other queries and so on. These dependencies can be used to model certain problems using *query trees* – trees which model the dependency of the replies to a given query on the replies to other queries.

Bounding the size of a query tree is central to the analyses of our algorithms. We show that the size of the query tree is $O(\log n)$ w.h.p., where n is the number of vertices. d, the degree bound of the dependency graph, appears only in the constant. [2] This answers in the affirmative the conjecture of [1]. Previously, Alon et al. [1] show that the expected size of the query tree is constant, and $O(\log^{d+1} n)$ w.h.p.[3] Our improvement is significant in removing the dependence on d from the exponent of the logarithm. We also show that when the degrees of the graph are distributed binomially, we can achieve the same bound on the size of the query tree. In addition, in the full version of this paper, we show a trivial lower bound of $\Omega(\log n / \log \log n)$.

We use these results on query trees to obtain LCAs for several online problems – maximal matching in graphs of bounded degree and several load balancing problems. We also use the results to improve the previous algorithms for hypergraph 2-coloring and k-CNF.

---

[2] Note that, however, the hidden constant is exponentially dependent on d. Whether or not this bound can be improved to have a polynomial dependency on d is an interesting open question.

[3] Notice that bounding the expected size of the query tree is not enough for our applications, since in LCAs we need to bound the probability that *any* query fails.

1.2.2 Hypergraph 2-Coloring

We modify the algorithm of [1] for an LCA for hypergraph 2-coloring, and coupled with our improved analysis of query tree size, obtain an LCA which runs in time and space $O(\log^4 n)$, improving the previous result, an LCA which runs $O(\log^{d+1} n)$ time and space.

1.2.3 k-CNF

Building on the similarity between hypergraph 2-coloring and k-CNF, we apply our results on hypergraph 2-coloring to give an an LCA for k-CNF which runs in time and space $O(\log^4 n)$.

We use the query tree to transform online algorithms to LCAs. We simulate online algorithms as follows: first a random permutation of the items is generated on the fly. Then, for each query, we simulate the online algorithm on a stream of input items arriving according to the order of the random permutation. Fortunately, because of the nature of our graphs (the fact that the degree is bounded or distributed binomially), we show that in expectation, we will only need to query a constant number of nodes, and only $O(\log n)$ nodes w.h.p. We now state our results:

1.2.4 Maximal Matching

We simulate the greedy online algorithm for maximal matching, to derive an LCA for maximal matching which runs in time and space $O(\log^3 n)$.

1.2.5 Load Balancing

We give several LCAs to load balancing problems which run in $O(\log n)$ time and space. Our techniques include extending the analysis of the query tree size to the case where the degrees are selected from a binomial distribution with expectation d, and further extending it to bipartite graphs which exhibit the characteristics of many balls and bins problems, specifically ones where each ball chooses d bins at random. We show how to convert a large class of the "power of d choices" online algorithms (see, e.g., [2, 5, 14]) to efficient LCAs.

1.3 Related Work

Nguyen and Onak [11] focus on transforming classical approximation algorithms into constant-time algorithms that approximate the size of the optimal solution of problems such as vertex cover and maximum matching. They generate a random number $r \in [0, 1]$, called the rank, for each node. These ranks are used to bound the query tree size.

Rubinfeld et al. [13] show how to construct polylogarithmic time local computation algorithms to maximal independent set computations, scheduling radio network broadcasts, hypergraph coloring and satisfying k-SAT formulas. Their proof technique uses Beck's analysis in his algorithmic approach to the Lovász Local Lemma [3], and a reduction from distributed algorithms. Alon et al. [1], building on the technique of [11], show how to extend several of the algorithms of [13] to perform in polylogarithmic space as well as time. They further observe that we do not actually need to assign each query a rank, we only need a random permutation of the queries. Furthermore, assuming the query tree is bounded by some k, the query to any node depends on at most k queries to other nodes, and so a k-wise independent random ordering suffices. They show how to construct a $1/n^2$-almost k-wise independent random ordering[4] from a seed of length $O(k \log^2 n)$.

Recent developments in sublinear time algorithms for sparse graph and combinatorial optimization problems have led to new constant time algorithms for approximating the size of a minimum vertex cover, maximal matching, maximum matching, minimum dominating set, and other problems (cf. [12, 9, 11, 16]), by randomly querying a constant number of vertices. A major difference between these algorithms and LCAs is that LCAs require that w.h.p., the output will be correct on any input, while optimization problems usually require a correct output only on *most* inputs. More importantly, LCAs reuire a consistent output for each query, rather than only approximating a given global property.

There is a vast literature on the topic of balls and bins and the power of d choices. (e.g. [2, 5, 8, 14]). For a survey on the power of d choices, we refer the reader to [10].

2 Preliminaries

Let $G = (V, E)$ be an undirected graph. We denote by $N_G(v) = \{u \in V(G) : (u, v) \in E(G)\}$ the neighbors of vertex v, and by $deg_G(v)$ we denote the degree of v. When it is clear from the context, we omit the G in the subscript. Unless stated otherwise, all logarithms in this paper are to the base 2. We use $[n]$ to denote the set $\{1, \ldots, n\}$, where $n \geq 1$ is a natural number.

We present our model of local computation algorithms (LCAs): Let F be a computational problem and x be an input to F. Let $F(x) = \{y \mid y$ is a valid solution for input $x\}$. The *search problem* for F is to find any $y \in F(x)$.

A $(t(n), s(n), \delta(n))$-*local computation algorithm* $\mathcal{A}$ is a (randomized) algorithm which solves a search problem for F for an input x of size n. However, the LCA $\mathcal{A}$ does not output a solution $y \in F(x)$, but rather implements query access to $y \in F(x)$. $\mathcal{A}$ receives a sequence of queries $i_1, \ldots, i_q$ and for any $q > 0$ satisfies the following: (1) after each query i_j it produces an output y_{i_j}, (2) With probability at least $1 - \delta(n)$ $\mathcal{A}$ is *consistent*, that is, the outputs $y_{i_1}, \ldots, y_{i_q}$ are substrings of some $y \in F(x)$. (3) $\mathcal{A}$ has access to a random tape and local computation memory on which it can perform current computations as well as store and retrieve information from previous computations.

We assume that the input x, the local computation tape and any random bits used are all presented in the RAM word model, i.e., $\mathcal{A}$ is given the ability to access a word of any of these in one step. The running time of $\mathcal{A}$ on any query is at most $t(n)$, which is sublinear in n, and the size of the local computation memory of $\mathcal{A}$ is at most $s(n)$. Unless stated otherwise, we always assume that the error parameter $\delta(n)$ is at most some constant, say, $1/3$. We say that $\mathcal{A}$ is a *strongly local computation algorithm* if both $t(n)$ and $s(n)$ are upper bounded by $O(\log^c n)$ for some constant c.

---

[4] A random ordering D_r is said to be ϵ-*almost* k-*wise independent* if the statistical distance between D_r and some k-wise independent random ordering by at most ϵ.

Two important properties of LCAs are as follows. We say an LCA $\mathcal{A}$ is *query order oblivious* (*query oblivious* for short) if the outputs of $\mathcal{A}$ do not depend on the order of the queries but depend only on the input and the random bits generated on the random tape of $\mathcal{A}$. We say an LCA $\mathcal{A}$ is *parallelizable* if $\mathcal{A}$ supports parallel queries, that is $\mathcal{A}$ is able to answer multiple queries simultaneously so that all the answers are consistent.

3   Bounding the Size of a Random Query Tree

3.1   The Problem and Our Main Results

In online algorithms, queries arrive in some unknown order, and the reply to each query depends only on previous queries (but not on any future events). The simplest way to transform online algorithms to LCAs is to process the queries in the order in which they arrive. This, however, means that we have to store the replies to all previous queries, so that even if the time to compute each query is polylogarithmic, the overall space is linear in the number of queries. Furthermore, this means that the resulting LCA is not query-oblivious. The following solution can be applied to this problem ([11] and [1]): Each query v is assigned a random number, $r(v) \in [0, 1]$, called its *rank*, and the queries are performed in ascending order of rank. Then, for each query x, a query tree can be constructed, to represent the queries on which x depends. If we can show that the query tree is small, we can conclude that each query does not depend on many other queries, and therefore a small number of queries need to be processed in order to reply to query x. We formalize this as follows:

Let $G = (V, E)$ be an undirected graph. The vertices of the graph represent queries, and the edges represent the dependencies between the queries. A real number $r(v) \in [0, 1]$ is assigned independently and uniformly at random to every vertex $v \in V$; we call $r(v)$ the *rank* of v. This models the random permutation of the vertices. Each vertex $v \in V$ holds an input $x(v) \in R$, where the range R is some finite set. The input is the content of the query associated with v. A randomized function F is defined inductively on the vertices of G such that $F(v)$ is a (deterministic) function of $x(v)$ as well as the values of F at the neighbors w of v for which $r(w) < r(v)$. F models the output of the online algorithm. We would like to upper bound the number of queries to vertices in the graph needed in order to compute $F(v_0)$ for any vertex $v_0 \in G$, namely, the time to simulate the output of query v_0 using the online algorithm.

To upper bound the number of queries to the graph, we turn to a simpler task of bounding the size of a certain d-regular tree, which is an upper bound on the number of queries. Consider an infinite d-regular tree $\mathcal{T}$ rooted at v_0. Each node w in $\mathcal{T}$ is assigned independently and uniformly at random a real number $r(w) \in [0, 1]$. For every node w other than v_0 in $\mathcal{T}$, let parent(w) denote the parent node of w. We grow a (possibly infinite) subtree T of $\mathcal{T}$ rooted at v as follows: a node w is in the subtree T if and only if parent(w) is in T and $r(w) < r(\text{parent}(w))$ (for simplicity we assume all the ranks are distinct real numbers). That is, we start from the root v_0, add all the children of v_0 whose ranks are smaller than that of v_0 to T. We keep growing T in this manner where a node $w' \in T$ is a leaf node in T if the ranks of its d children are all larger than $r(w')$. We call the random tree T constructed in this way a *query tree* and we denote by $|T|$ the random variable that corresponds to the size of T. Note that $|T|$ is an upper bound on the number of queries since each node in T has at least as many neighbors as that in G and if a node is connected to some previously queried nodes, this can only decrease the number of queries. Therefore the number of queries is bounded by the size of T. Our goal is to find an upper bound on $|T|$ which holds with high probability.

We improve the upper bound on the query tree of $O(\log^{d+1} N)$ given in [1] for the case when the degrees are bounded by a constant d and extend our new bound to the case that the degrees of G are binomially distributed, independently and identically with expectation d, i.e., $deg(v) \sim B(n, d/n)$.

Our main result in this section is bounding, with high probability, the size of the query tree T as follows.

Lemma 1. *Let G be a graph whose vertex degrees are bounded by d or distributed independently and identically from the binomial distribution: $deg(v) \sim B(n, d/n)$. Then there exists a constant $C(d)$ which depends only on d, such that*

$$\Pr[|T| > C(d) \log n] < 1/n^2,$$

where the probability is taken over all the possible permutations $\pi \in \Pi$ of the vertices of G, and T is a random query tree in G under π.

3.2   Overview of the Proof

Our proof of Lemma 1 consists of two parts. Following [1], we partition the query tree into $L = 3d$ levels. The first part of the proof is an upper bound on the size of a single (sub)tree on any level. For the bounded degree case, this was already proved in [1]. We extend their proof to the binomial case; that is, we prove the following, where $T_i^{(j)}$ is the j-th subtree on level i of the tree.

Proposition 1. *Let $\mathcal{T}$ be a tree with vertex degree distributed i.i.d. binomially with $deg(v) \sim B(n, d/n)$. For any $1 \le i \le L$ and any $1 \le j \le t_i$, $\Pr[|T_i^{(j)}| \ge m] \le \sum_{i=m}^{\infty} 2^{-ci} \le 2^{-\Omega(m)}$, for $n \ge \beta$, for some constant $\beta > 0$.*

The proof can be found in the full version of this paper.

The second part, which is a new ingredient of our proof, inductively upper bounds the number of vertices on each level, as the levels increase. For this to hold, it crucially depends on the fact that all subtrees are generated independently and that the probability of any subtree being large is exponentially small. The main idea is to show that although each subtree, in isolation, can reach a logarithmic size, their combination is not likely to be much larger. We use the distribution of the sizes of the subtrees, in order to bound the aggregate of multiple subtrees.

3.3   Bounding the Increase in Subtree Size as We Go Up Levels

From [1] and Proposition 1 we know that the size of any subtree, in particular $|T_1|$, is bounded by $O(\log n)$ with probability at least $1 - 1/n^3$ in both the bounded degree and the binomial degree cases (see the full version for a more complete discussion). Our next step in proving Lemma 1 is to show that, as we increase the levels, the size of the tree does not increase by more than a constant factor for each level. That is, there exists an absolute constant η depending on d only such that if the number of vertices on level k is at most $|T_k|$, then the number of vertices on level $k + 1$, $|T_{k+1}|$ satisfies $|T_{k+1}| \le \eta \sum_{i=1}^{k} |T_i| + O(\log n) \le 2\eta |T_k| + O(\log n)$. Since there are L levels in total, this implies that the number of vertices on all L levels is at most $O((2\eta)^L \log n) = O(\log n)$.

The following Proposition establishes our inductive step.

Proposition 2. *For any infinite query tree $\mathcal{T}$ with constant bounded degree d (or degrees i.i.d. $\sim B(n, d/n)$), for any $1 \le i < L$, there exist constants $\eta_1 > 0$ and $\eta_2 > 0$ s.t. if $\sum_{j=1}^{t_i} |T_i^{(j)}| \le \eta_1 \log n$ then $\Pr[\sum_{j=1}^{t_{i+1}} |T_{i+1}^{(j)}| \ge \eta_1 \eta_2 \log n] < 1/n^2$ for all $n > \beta$, for some $\beta > 0$.*

Proof. Denote the number of vertices on level k by Z_k and let $Y_k = \sum_{i=1}^{k} Z_i$. Assume that each vertex i on level $\le k$ is the root of a tree of size z_i on level $k + 1$. Notice that $Z_{k+1} = \sum_{i=1}^{Y_k} z_i$.

From [1] and Proposition 1, there are absolute constants c_0 and β depending on d only such that for any subtree $T_k^{(i)}$ on level k and any $n > \beta$, $\Pr[|T_k^{(i)}| = n] \le e^{-c_0 n}$. Therefore, given $(z_1, \ldots, z_{Y_k})$, the probability of the forest on level $k + 1$ consisting of exactly trees of size $(z_1, \ldots, z_{Y_k})$ is at most $\prod_{i=1}^{Y_k} e^{-c_0(z_i - \beta)} = e^{-c_0(Z_{k+1} - \beta Y_k)}$.

Notice that, given Y_k (the number of nodes up to level k), there are at most $\binom{Z_{k+1} + Y_k - 1}{Y_k - 1} < \binom{Z_{k+1} + Y_k}{Y_k}$ vectors $(z_1, \ldots, z_{Y_k})$ that can realize Z_{k+1}.

We want to bound the probability that $Z_{k+1} = \eta Y_k$ for some (large enough) constant $\eta > 0$. We can bound this as follows:

$$
\begin{aligned}
\Pr[|T_{k+1}| = Z_{k+1}] &< \binom{Z_{k+1} + Y_k}{Y_k} e^{-c_0(Z_{k+1} - \beta Y_k)} \\
&< \left(\frac{e \cdot (Z_{k+1} + Y_k)}{Y_k} \right)^{Y_k} e^{-c_0(Z_{k+1} - \beta Y_k)} \\
&= (e(1 + \eta))^{Y_k} e^{-c_0(\eta - \beta)Y_k} \\
&= e^{Y_k(-c_0(\eta - \beta) + \ln(\eta + 1) + 1)} \\
&\le e^{-c_0 \eta Y_k / 2},
\end{aligned}
$$

It follows that there is some absolute constant c' which depends on d only such that $\Pr[|T_{k+1}| \ge \eta Y_k] \le e^{-c' \eta Y_k}$. That is, if $\eta Y_k = \Omega(\log n)$, the probability that $|T_{k+1}| \ge \eta Y_k$ is at most $1/n^3$. Adding the vertices on all L levels and applying the union bound, we conclude that with probability at most $1/n^2$, the size of T is at most $O(\log n)$. $\quad\square$

4 Hypergraph 2-Coloring and k-CNF

We use the bound on the size of the query tree of graphs of bounded degree to improve the analysis of [1] for hypergraph 2-coloring. We also modify their algorithm slightly to further improve the algorithm's complexity. Due to space limitations, we only state our main theorems for hypergraph 2-coloring and k-CNF; the proofs can be found in the full version of this paper.

Theorem 1. *Let H be a k-uniform hypergraph s.t. each hyperedge intersects at most d other hyperedges. Suppose that $k \geq 16 \log d + 19$.*

Then there exists an $(O(\log^4 n),\ O(\log^4 n), 1/n)$-local computation algorithm which, given H and any sequence of queries to the colors of vertices $(x_1, x_2, \ldots, x_s)$, with probability at least $1 - 1/n^2$, returns a consistent coloring for all x_i's which agrees with a 2-coloring of H. Moreover, the algorithm is query oblivious and parallelizable.

Theorem 2. *Let H be a k-CNF formula with $k \geq 2$. Suppose that each clause intersects no more than d other clauses, and furthermore suppose that $k \geq 16 \log d + 19$.*

Then there exists a $(O(\log^4 n), O(\log^4 n), 1/n)$-local computation algorithm which, given a formula H and any sequence of queries to the truth assignments of variables $(x_1, x_2, \ldots, x_s)$, with probability at least $1 - 1/n^2$, returns a consistent truth assignment for all x_i's which agrees with some satisfying assignment of the k-CNF formula H. Moreover, the algorithm is query oblivious and parallelizable.

5   Maximal Matching

We consider the problem of *maximal matching in a bounded-degree graph*. We are given a graph $G = (V, E)$, where the maximal degree is bounded by some constant d, and we need to find a maximal matching. A matching is a set of edges with the property that no two edges share a common vertex. The matching is maximal if no other edge can be added to it without violating the matching property.

Assume the online scenario in which the edges arrive in some unknown order. The following greedy online algorithm can be used to calculate a maximal matching: When an edge e arrives, we check whether e is already in the matching. If it is not, we check if any of the neighboring edges are in the matching. If none of them is, we add e to the matching. Otherwise, e is not in the matching.

We turn to the local computation variation of this problem. We would like to query, for some edge $e \in E$, whether e is part of some maximal matching. (Recall that all replies must be consistent with some maximal matching).

We use the technique of [1] to produce an almost $O(\log n)$-wise independent random ordering on the edges, using a seed length of $O(\log^3 n)$.[5] When an edge e is queried, we use a BFS (on the edges) to build a DAG rooted at e. We then use the greedy online algorithm on the edges of the DAG (examining the edges with respect to the ordering), and see whether e can be added to the matching.

As the query tree is an upper-bound on the size of the DAG, we derive the following theorem from Lemma 1.

Theorem 3. *Let $G = (V, E)$ be an undirected graph with n vertices and maximum degree d. Then there is an $(O(\log^3 n), O(\log^3 n), 1/n)$ - local computation algorithm which, on input an edge e, decides if e is in a maximal matching. Moreover, the algorithm gives a consistent maximal matching for every edge in G.*

---

[5] Since the query tree is of size $O(\log n)$ w.h.p., we don't need a complete ordering on the vertices; an almost $O(\log n)$-wise independent ordering suffices.

6   The Bipartite Case and Local Load Balancing

We consider a general "power of d choices" online algorithm for load balancing. In this setting there are n balls that arrive in an online manner, and m bins. Each ball selects a random subset of d bins, and queries these bins. (Usually the query is simply the current load of the bin.) Given this information, the ball is assigned to one of the d bins (usually to the least loaded bin). We denote by $\mathcal{LB}$ such a generic algorithm (with a decision rule which can depend in an arbitrary way on the d bins that the ball is assigned to). Our main goal is to simulate such a generic algorithm.

The load balancing problem can be represented by a bipartite graph $G = (\{V, U\}, E)$, where the balls are represented by the vertices V and the bins by the vertices U. The random selection of a bin $u \in U$ by a ball $v \in V$ is represented by an edge. By definition, each ball $v \in V$ has degree d. Since there are random choices in the algorithm $\mathcal{LB}$ we need to specify what we mean by a simulation. For this reason we define the input to be the following: a graph $G = (\{V, U\}, E)$, where $|V| = n$, $|U| = m$, and $n = cm$ for some constant $c \geq 1$. We also allocate a rank $r(u) \in [0, 1]$ to every $u \in U$. This rank represents the ball's arrival time: if $r(v) < r(u)$ then vertex v arrived before vertex u. Furthermore, all vertices can have an input value $x(w)$. (This value represents some information about the node, e.g., the weight of a ball.) Given this input, the algorithm $\mathcal{LB}$ is deterministic, since the arrival sequence is determined by the ranks, and the random choices of the balls appear as edges in the graph. Therefore by a simulation we will mean that given the above input, we generate the same allocation as $\mathcal{LB}$.

We consider the following stochastic process: Every vertex $v \in V$ uniformly and independently at random chooses d vertices in U. Notice that from the point of view of the bins, the number of balls which chose them is distributed binomially with $X \sim B(n, d/m)$. Let X_v and X_u be the random variables for the number of neighbors of vertices $v \in V$ and $u \in U$ respectively. By definition, $X_v = d$, since all balls have d neighbors, and hence each X_u is independent of all X_v's. However, there is a dependence between the X_u's (the number of balls connected to different bins). Fortunately this is a classical example where the random variables are negatively dependent (see e.g. [8]). [6]

6.1   The Bipartite Case

Recall that in Section 3, we assumed that the degrees of the vertices in the graph were independent. We would like to prove an $O(\log n)$ upper bound on the query tree T for our bipartite graph. As we cannot use the theorems of Section 3 directly, we show that the query tree is smaller than another query tree which meets the conditions of our theorems.

The query tree for the binomial graph is constructed as follows: a root $v_0 \in V$ is selected for the tree. (v_0 is the ball whose bin assignment we are interested in determining.) Label the vertices at depth j in the tree by W_j. Clearly, $W_0 = \{v_0\}$. At each depth j, we add vertices one at a time to the tree, from left to right, until the depth is

---

[6] We remind the reader that two random variables X_1 and X_2 are negatively dependent if $\Pr[X_1 > x | X_2 = a] < \Pr[X_1 > x | X_2 = b]$, for $a > b$ and vice-versa.

"full" and then we move to the next depth. Note that at odd depths $(2j + 1)$ we add bin vertices and at even depths $(2j)$ we add ball vertices.

Specifically, at odd depths $(2j + 1)$ we add, for each $v \in W_{2j}$ its d neighbors $u \in N(v)$ as children, and mark each by u.[7] At even depths $(2j)$ we add for each node marked by $u \in W_{2j-1}$ all its (ball) neighbors $v \in N(u)$ such that $r(v) < r(parent(u))$, if they have not already been added to the tree. Namely, all the balls that are assigned to u by time

A leaf is a node marked by a bin u_ℓ for whom all neighboring balls $v \in N(u_\ell) - \{parent(u_\ell)\}$ have a rank larger than its parent, i.e., $r(v) > r(parent(u_\ell))$. Namely, $parent(u_\ell)$ is the first ball to be assigned to bin u_ℓ. This construction defines a stochastic process $F = \{F_t\}$, where F_t is (a random variable for) the size of T at time t. (We start at $t = 0$ and t increases by 1 for every vertex we add to the tree).

We now present our main lemma for bipartite graphs. The proof can be found in the full version of the paper.

Lemma 2. *Let* $G = (\{V, U\}, E)$ *be a bipartite graph,* $|V| = n$ *and* $|U| = m$ *and* $n = cm$ *for some constant* $c \geq 1$, *such that for each vertex* $v \in V$ *there are* d *edges chosen independently and at random between* v *and* U. *Then there is a constant* $C(d)$ *which depends only on* d *such that*

$$\Pr[|T| < C(d) \log n] > 1 - 1/n^2,$$

where the probability is taken over all of the possible permutations $\pi \in \Pi$ *of the vertices of* G, *and* T *is a random query tree in* G *under* π.

6.2   Local Load Balancing

The following theorem states our basic simulation result.

Theorem 4. *Consider a generic online algorithm* $\mathcal{LB}$ *which requires constant time per query, for* n *balls and* m *bins, where* $n = cm$ *for some constant* $c > 0$. *There exists an* $(O(\log n), O(\log n), 1/n)$-*local computation algorithm which, on query of a (ball) vertex* $v \in V$, *allocates* v *a (bin) vertex* $u \in U$, *such that the resulting allocation is identical to that of* $\mathcal{LB}$ *with probability at least* $1 - 1/n$.

Proof. Let $K = C(d) \log |U|$ for some constant $C(d)$ depending only on d. K is the upper bound given in Lemma 2. (In the following we make no attempt to provide the exact values for $C(d)$ or K.)

We now describe our $(O(\log n), O(\log n), 1/n)$-local computation algorithm for $\mathcal{LB}$. A query to the algorithm is a (ball) vertex $v_0 \in V$ and the algorithm will chose a (bin) vertex from the d (bin) vertices connected to v_0.

We first build a query tree as follows: Let v_0 be the root of the tree. For every $u \in N(u_0)$, add to the tree the neighbors of u, $v \in V$ such that $r(v) < r(v_0)$. Continue inductively until either K nodes have been added to the random query tree or no more

---

[7] A bin can appear several times in the tree. It appears as different nodes, but they are all marked so that we know it is the same bin. Recall that we assume that all nodes are unique, as this assumption can only increase the size of the tree.

nodes can be added to it. If K nodes have been added to the query tree, this is a failure event, and assign to v_0 a random bin in $N(v_0)$. From Lemma 2, this happens with probability at most $1/n^2$, and so the probability that some failure event will occur is at most $1/n$. Otherwise, perform $\mathcal{LB}$ on all of the vertices in the tree, in order of addition to the tree, and output the bin to which ball v_0 is assigned to by $\mathcal{LB}$. $\square$

A reduction from various load balancing algorithms gives us the following corollaries to Theorem 4.

Corollary 1. *(Using [5]) Suppose we wish to allocate m balls into n bins of uniform capacity, $m \geq n$, where each ball chooses d bins independently and uniformly at random. There exists a $(\log n, \log n, 1/n)$ LCA which allocates the balls in such a way that the load of the most loaded bin is $m/n + O(\log \log n / \log d)$ w.h.p.*

Corollary 2. *(Using [15]) Suppose we wish to allocate n balls into n bins of uniform capacity, where each ball chooses d bins independently at random, one from each of d groups of almost equal size $\theta(\frac{n}{d})$. There exists a $(\log n, \log n, 1/n)$ LCA, which allocates the balls in such a way that the load of the most loaded bin is $\ln \ln n / (d-1) \ln 2 + O(1)$ w.h.p. [8]*

Corollary 3. *(Using [4]) Suppose we wish to allocate m balls into $n \leq m$ bins, where each bin i has a capacity c_i, and $\sum_i c_i = m$. Each ball chooses d bins at random with probability proportional to their capacities. There exists a $(\log n, \log n, 1/n)$ LCA which allocates the balls in such a way that the load of the most loaded bin is $2 \log \log n + O(1)$ w.h.p.*

Corollary 4. *(Using [4]) Suppose we wish to allocate m balls into $n \leq m$ bins, where each bin i has a capacity c_i, and $\sum_i c_i = m$. Assume that the size of a large bin is at least $rn \log n$, for large enough r. Suppose we have s small bins with total capacity m_s, and that $m_s = O((n \log n)^{2/3})$. There exists a $(\log n, \log n, 1/n)$ LCA which allocates the balls in such a way that the expected maximum load is less than 5.*

Corollary 5. *(Using [7]) Suppose we have n bins, each represented by one point on a circle, and n balls are to be allocated to the bins. Assume each ball needs to choose $d \geq 2$ points on the circle, and is associated with the bins closest to these points. There exists a $(\log n, \log n, 1/n)$ LCA which allocates the balls in such a way that the load of the most loaded bin is $\ln \ln n / \ln d + O(1)$ w.h.p.*

6.3 Random Ordering

In the above we assume that we are given a random ranking for each ball. If we are not given such random rankings (in fact, a random permutation of the vertices in U will also suffice), we can generate a random ordering of the balls. Specifically, since w.h.p. the

---

[8] In fact, in this setting the tighter bound is $\frac{\ln \ln n}{d \ln \phi_d} + O(1)$, where ϕ_d is the ratio of the d-step Fibonacci sequence, i.e. $\phi_d = \lim_{k \to \infty} \sqrt[k]{F_d(k)}$, where for $k < 0$, $F_d(k) = 0$, $F_d(1) = 1$, and for $k \geq 1$ $F_d(k) = \sum_{i=1}^{d} F_d(k-i)$.

size of the random query is $O(\log n)$, an $O(\log n)$-*wise independent random ordering*[9] suffices for our local computation purpose. Using the construction in [1] of $1/n^2$-almost $O(\log n)$-wise independent random ordering over the vertices in U which uses space $O(\log^3 n)$, we obtain $(O(\log^3 n), O(\log^3 n), 1/n)$-local computation algorithms for balls and bins.

References

[1] N. Alon, R. Rubinfeld, S. Vardi, and N. Xie. Space-efficient local computation algorithms. In *Proc. 23rd ACM-SIAM Symposium on Discrete Algorithms*, pages 1132–1139, 2012.

[2] Y. Azar, A. Z. Broder, A. R. Karlin, and E. Upfal. Balanced allocations. *SIAM Journal on Computing*, 29(1):180–200, 1999.

[3] J. Beck. An algorithmic approach to the Lovász Local Lemma. *Random Structures and Algorithms*, 2:343–365, 1991.

[4] P. Berenbrink, A. Brinkmann, T. Friedetzky, and L. Nagel. Balls into non-uniform bins. In *Proceedings of the 24th IEEE International Parallel and Distributed Processing Symposium (IPDPS)*, pages 1–10. IEEE, 2010.

[5] P. Berenbrink, A. Czumaj, A. Steger, and B. Vöcking. Balanced allocations: The heavily loaded case. *SIAM J. Comput.*, 35(6):1350–1385, 2006.

[6] A. Borodin and Ran El-Yaniv. *Online Computation and Competitive Analysis*. Cambridge University Press, 1998.

[7] John W. Byers, Jeffrey Considine, and Michael Mitzenmacher. Simple load balancing for distributed hash tables. In *Proc. of Intl. Workshop on Peer-to-Peer Systems(IPTPS)*, pages 80–87, 2003.

[8] D. Dubhashi and D. Ranjan. Balls and bins: A study in negative dependence. *Random Structures and Algorithms*, 13:99–124, 1996.

[9] S. Marko and D. Ron. Distance approximation in bounded-degree and general sparse graphs. In *APPROX-RANDOM'06*, pages 475–486, 2006.

[10] M. Mitzenmacher, A. Richa, and R. Sitaraman. The power of two random choices: A survey of techniques and results. In *Handbook of Randomized Computing, Vol. I, edited by P. Pardalos, S. Rajasekaran, J. Reif, and J. Rolim*, pages 255–312. Norwell, MA: Kluwer Academic Publishers, 2001.

[11] H. N. Nguyen and K. Onak. Constant-time approximation algorithms via local improvements. In *Proc. 49th Annual IEEE Symposium on Foundations of Computer Science*, pages 327–336, 2008.

[12] M. Parnas and D. Ron. Approximating the minimum vertex cover in sublinear time and a connection to distributed algorithms. *Theoretical Computer Science*, 381(1–3), 2007.

[13] R. Rubinfeld, G. Tamir, S. Vardi, and N. Xie. Fast local computation algorithms. In *Proc. 2nd Symposium on Innovations in Computer Science*, pages 223–238, 2011.

[14] K. Talwar and U. Wieder. Balanced allocations: the weighted case. In *Proc. 39th Annual ACM Symposium on the Theory of Computing*, pages 256–265, 2007.

[15] Berthold Vöcking. How asymmetry helps load balancing. *J. ACM*, 50:568–589, July 2003.

[16] Y. Yoshida, Y. Yamamoto, and H. Ito. An improved constant-time approximation algorithm for maximum matchings. In *Proc. 41st Annual ACM Symposium on the Theory of Computing*, pages 225–234, 2009.

---

[9] See [1] for the formal definitions of k-wise independent random ordering and almost k-wise independent random ordering.

Assigning Sporadic Tasks to Unrelated Parallel Machines

Alberto Marchetti-Spaccamela[1], Cyriel Rutten[2,*],
Suzanne van der Ster[3], and Andreas Wiese[1,**]

[1] Sapienza University of Rome, Rome, Italy
{alberto,wiese}@dis.uniroma1.it
[2] Maastricht University, Maastricht, The Netherlands
c.rutten@maastrichtuniversity.nl
[3] Vrije Universiteit, Amsterdam, The Netherlands
suzanne.vander.ster@vu.nl

Abstract. We study the problem of assigning sporadic tasks to unrelated machines such that the tasks on each machine can be feasibly scheduled. Despite its importance for modern real-time systems, this problem has not been studied before. We present a polynomial-time algorithm which approximates the problem with a constant speedup factor of $11 + 4\sqrt{3} \approx 17.9$ and show that any polynomial-time algorithm needs a speedup factor of at least 2, unless $P = NP$. In the case of a constant number of machines we give a polynomial-time approximation scheme. Key to these results are two new relaxations of the demand bound function which yields a sufficient and necessary condition for a task system on a single machine to be feasible.

1 Introduction

The *sporadic task model* is a model of recurrent processes in hard real-time systems that has received great attention in the last years; see for example [4], [8], and references therein. A sporadic task $\tau = (c_\tau, d_\tau, t_\tau)$ is characterized by a worst-case execution time c_τ, a relative deadline d_τ, and a minimum interarrival separation t_τ. Such a sporadic task generates a potentially infinite sequence of jobs with successive job arrivals separated by at least t_τ time units, it has an execution requirement less than or equal to c_τ and a deadline that occurs d_τ time units after its arrival time. A sporadic task system is comprised of several such sporadic tasks.

A sporadic task system is said to be *feasible* upon a specified platform if it is possible to schedule the system on the platform such that all jobs of all tasks will meet their deadlines, under all permissible combinations of job arrival sequences by the different tasks comprising the system.

---

* Supported by the METEOR International Travel Grant.
** Supported by the German Academic Exchange Service (DAAD).

A. Czumaj et al. (Eds.): ICALP 2012, Part I, LNCS 7391, pp. 665–676, 2012.

The feasibility analysis of sporadic task systems on single processors has been extensively studied [4]. It is known that the Earliest Deadline First (EDF) algorithm, that schedules at any time the job with the earliest absolute deadline, is optimal in the sense that for any sequence of jobs it produces a valid schedule whenever a valid schedule exists [22]. However it is co-NP hard to decide whether a task system is feasible on a single machine [15].

On multiprocessor systems, there are two main paradigms for scheduling: *global* and *partitioned* scheduling. In the former, all tasks can use all machines, and jobs can even be migrated from one machine to another. In the partitioned scheduling approach each task has to be assigned to one of the machines such that all its jobs have to be executed on this specific machine. Since the process of partitioning tasks among processors reduces a multiprocessor scheduling problem to a series of single processor problems, the optimality of EDF for preemptive single processor scheduling makes EDF a reasonable algorithm to use as the run-time scheduling algorithm on each machine.

In recent years, hardware design has seen a highly visible trend towards heterogeneous processors. In particular, modern hardware architectures often contain specialized processors for certain tasks (e.g. graphical processors, floating-point units). To model such settings we assume that the given machines are *unrelated*; i.e., we assume that the processing time of each task depends on the machine where it is executed.

Related Work. The hardness result for the feasibility problem on a single machine motivated the study of approximate feasibility tests that run efficiently but introduce an error in the decision process, controlled by an accuracy parameter α (the speedup). If an α-*approximation test* returns "feasible", then the task set is guaranteed to be feasible on an α-speed processor(s); if the test returns "infeasible", the task set is guaranteed to be infeasible on a unit-speed processor(s). For the case of a single processor, a FPTAS feasibility test for EDF has been proposed [11] (i.e. for any $\epsilon > 0$, there exists a $(1 + \epsilon)$-approximation test with running time polynomial in the number of tasks and in $1/\epsilon$).

In the case of m identical processors assuming the global paradigm, the natural EDF-policy is no longer optimal, but it is known that any feasible collection of jobs on m machines of unit speed is schedulable using EDF on m machines of speed $2 - \frac{1}{m}$ [23]. Also, a corresponding test for task sets is known [7] [9]. Recently, Anand et al. [2] presented an online algorithm needing only a speedup factor of $e/(e - 1) \approx 1.58$.

In the case of the partitioned paradigm, in the special case of implicit deadline systems in which all tasks have their deadlines equal to their period parameters (i.e., $d_\tau = t_\tau$ for all τ) it is known that a set of tasks on a machine is feasible if and only if the sum of the utilizations c_τ/t_τ are at most one; therefore the problem reduces to BIN-PACKING. Recall that for bin packing an asymptotic FPTAS exists [19]. In the general case, Baruah and Fisher [5] propose an algorithm which can partition any set of tasks that is feasible on m machines such that the assignment is feasible if the machines run $(4 - \frac{2}{m})$ times faster; a similar result is given if the tasks are scheduled according to static priorities, rather than with the more powerful EDF-policy [16]. Chen and Chakraborty [12] improved upon these results by showing that a deadline-monotonic policy with approximate demand bound functions leads to $\frac{3e-1}{e} - \frac{1}{m} \approx 2.6322 - \frac{1}{m}$ approximation test in case of implicit deadlines and a $3 - \frac{1}{m}$ approximation test for the general case.

When taking the number of needed machines as objective function, a PTAS has been proposed if tasks are scheduled according to fixed priorities using resource augmentation [14]. Also, the existence of an asymptotic FPTAS has been ruled out, thus showing that the problem is indeed harder than BIN-PACKING.

In [3] a 3- approximation test is presented for partitioning a set of implicit deadline tasks on a platform of related processors.

As mentioned above, we assume the machines to be unrelated, meaning that the processing times of the tasks can differ on the different machines and we observe that results are known only for job scheduling. Lenstra, Shmoys and Tardos [21] showed a 2-approximation algorithm for the problem of minimizing the makespan of a set of jobs and that it is NP-hard to achieve a performance ratio of at most 1.5. Despite a lot of effort, the only improvements in the setting of an arbitrary number of machines are a 1.75-approximation algorithm for the graph balancing case [13] and a $33/17 \approx 1.94$-estimation algorithm for the restricted assignment case [24].

For a constant number of machines polynomial-time approximation schemes are known [18], [21]; recently a PTAS has been proposed for the case that each machine belongs to one of a fixed number of types, and processing time of each job depends only on the job and the type of the machine it is assigned to [10].

Our Contribution. To the best of our knowledge, no non-trivial algorithm is known for assigning a set of sporadic tasks to a set of unrelated machines. We first present $11 + 4\sqrt{3} \approx 17.9$ approximation test. Also, we show that no polynomial-time algorithm can compute a task assignment needing a speedup factor of $2 - \epsilon$ for $\epsilon > 0$, unless $P = NP$. Note that this bound is stronger than the best known $(3/2 - \epsilon)$-hardness result for the contained problem of minimizing the makespan when scheduling jobs on unrelated machines [21].

If the number of machines is fixed, we present a polynomial-time algorithm that either finds a feasible task assignment on m machines that are $(1 + \epsilon)$ times as fast or guarantees that no solution exists on unit-speed processors.

In order to be able to achieve these results, we need deep understanding of the *demand bound function* (dbf) which yields a necessary and sufficient condition for a task system to be feasible on one machine. In particular, we present two new relaxations for handling this well-studied function. For our result for an arbitrary number of machines, we give a set of sparse linear constraints which approximate the dbf up to a constant factor. Due to the sparsity and using other techniques we are able to design an efficient iterative rounding procedure.

For the case of a constant number of machines we cannot exploit the technique of partitioning the task set into "big" and "small" tasks as in the job scheduling problem. A task having small execution times or small utilization might still be very tight in the sense that its relative deadline is fairly small. Therefore, assigning small tasks is tricky, i.e., they cannot be simply poured onto machines which still have still some capacity left after scheduling the large tasks. An important feature of our dbf-relaxation is that the feasibility test of assigning a task with deadline D to a machine having tasks with deadlines $< D$ already assigned to it, requires only limited information of the previously assigned tasks. Namely, the approximated dbf for each task only needs to be evaluated at a constant number of points. Afterwards, we just approximate the dbf by the task's utilization. We exploit this feature and other tricks to polynomially bound the running time of a dynamic programming algorithm.

2   Preliminaries

Given is a set M of m parallel unrelated machines and a sporadic task system T, with $|T| = n$. Each task $\tau \in T$ is characterized by a set of values $(\{c_{i,\tau}\}_{i=1,\ldots,m}, d_\tau, t_\tau)$, where $c_{i,\tau}$ is its execution time on machine i, d_τ is its deadline, relative to its arrival time, and t_τ denotes the minimum interarrival time between two jobs of τ and is called the period. We assume all parameters to be integer and strictly positive. We study the problem of finding a task assignment $\mathcal{T} = \{\mathcal{T}_i\}_{i\in M}$ such that $\cup_i \mathcal{T}_i = T$ and $\mathcal{T}_i \cap \mathcal{T}_{i'} = \emptyset$ for any two machines $i \neq i'$. A task assignment is *feasible* if for any machine i, any job arrival sequence of the tasks in $\mathcal{T}_i$ can be feasibly scheduled on i, allowing jobs to be preempted. Once tasks are partitioned among the machines, EDF will be our scheduling algorithm of choice.

An α-*approximation test* for the problem of assigning tasks to unrelated machines is an algorithm that runs in polynomial time and which either guarantees that there is no feasible integral assignment of the tasks to the given machines (running at unit speed), or finds an integral assignment which is feasible if the machines run at speed α.

By $u_{i,\tau}$ we denote the *utilization* of task τ on machine i and we define it as $u_{i,\tau} = c_{i,\tau}/t_\tau$. Given a task assignment $\mathcal{T}$, we define the utilization of each machine i by $u_i = \sum_{\tau \in \mathcal{T}_i} u_{i,\tau}$. A necessary (but not sufficient) condition for feasibility of an assignment then is that $u_i \leq 1$ for all $i \in M$ [22]. It follows that a task τ for which $u_{i,\tau} > 1$ will never be assigned to machine i. Similarly, in case $c_{i,\tau} > d_\tau$ task τ will not be assigned to machine i.

Feasibility Test. The synchronous arrival sequence for task system T is defined to be the collection of job arrivals in which each task in T generates a job at time-instant zero, and subsequent jobs arrive as soon as legally permitted (i.e., task τ generates a job at each time-instant $kt_\tau, k = 0, 1, 2, \ldots$).

It is known [6] that a set of sporadic tasks $\mathcal{T}_i$ is EDF-schedulable on machine i if and only if the following conditions are satisfied:

1. the utilization of the task system does not exceed 1, i.e. $u_i = \sum_{\tau \in \mathcal{T}_i} u_{i,\tau} \leq 1$,
2. all jobs with deadlines $[0, lcm_{\tau \in \mathcal{T}_i}(t_\tau)]$ in the synchronous arrival sequence of $\mathcal{T}_i$ meet their deadlines (here lcm denotes the least common multiple).

This immediately yields an exponential-time test to check whether $\mathcal{T}_i$ is EDF-schedulable; however we recall that the problem is co-NP hard [15] and that it is not known whether it can be determined in pseudo-polynomial time.

A necessary and sufficient condition for a task system T to be schedulable is based on the *demand bound function*. We refer to [1,4,6,7,9,12] and references therein for the study of the properties and the complexity of the demand bound function.

In the case of unrelated machines we have the following.

Proposition 1. *[6] An assignment* $\mathcal{T} = \{T_i\}_{i \in M}$ *is feasible for task system T if and only if for all $i \in M$*

$$dbf_{\mathcal{T},i}(s) := \sum_{\tau \in T_i : d_\tau \le s} \left\lfloor \frac{s + t_\tau - d_\tau}{t_\tau} \right\rfloor c_{i,\tau} \le s \qquad \forall s \ge 0,$$

We write dbf_i instead of $dbf_{\mathcal{T},i}$ whenever the assignment $\mathcal{T}$ is clear from the context. Further, we define $dbf_i(\tau, s) := \lfloor (s + t_\tau - d_\tau)/t_\tau \rfloor c_{i,\tau}$; $dbf_i(\tau, s)$ denotes the contribution of task τ to $dbf_{\mathcal{T},i}(s)$.

3 Arbitrary Number of Machines

In this section we present an $\alpha = 11 + 4\sqrt{3} \approx 17.9$-approximation test for assigning tasks to unrelated machines and we show that the problem is NP-hard to approximate with a ratio of $2 - \epsilon$ for any $\epsilon > 0$.

We will formulate the problem of assigning tasks to unrelated machines as a linear program, such that the tasks on each machine can be feasibly scheduled, using the EDF-scheduler. First, we derive a set of linear inequalities which are

- necessary, meaning that they are fulfilled by any feasible assignment,
- approximately sufficient, meaning that any assignment which (approximately) fulfills the constraints is feasible if the speed of the machine is increased by some constant factor, and
- sparse, meaning that in our constraints every variable occurs only twice.

We introduce a variable $y_{i,\tau}$ for each pair of a machine i and a task τ, modeling to assign τ to machine i. The first constraints are utilization bounds on all tasks assigned to the same machine i. Formally, we demand that $\sum_{\tau \in T} u_{i,\tau} y_{i,\tau} \le 1$ for each machine i. Secondly, we require that for all tasks with deadline in the interval $(2^{k-1}, 2^k]$, the sum of their execution time is at most 2^k. Formally, we require $\sum_{\tau \in T : d_\tau \in (2^{k-1}, 2^k]} c_{i,\tau} y_{i,\tau} \le 2^k$ for each machine i and each $k \in \mathbb{N}$. We call these conditions the *relaxed dbf-constraints*. It is clear that these constraints have to be fulfilled by any feasible task assignment. Since they are linear, they can be used in an LP-relaxation for the problem. Their sparsity gives the potential to derive efficient rounding schemes which result in integral solutions, violating the relaxed *dbf*-constraints only by constant factors. Below we will present such an algorithm; to this end, the following lemma shows that—even when violated up to constant factors—they are approximately sufficient.

Lemma 1. *Let $\mathcal{T}$ be an assignment for the task system T such that, for all machines i, $\sum_{\tau \in T_i} u_{i,\tau} \le \beta$ and $\sum_{\tau \in T_i : d_\tau \in (2^{k-1}, 2^k]} c_{i,\tau} \le \beta \cdot 2^k$. Then $dbf_{\mathcal{T},i}(s) \le 6\beta s$ for all $s \ge 0$ and $\mathcal{T}$ is a feasible assignment under a speedup factor of 6β.*

Let $\rho > 1$. Assume we are given an instance of our problem. We define the function $r(x) := \rho^{\lceil \log_\rho x \rceil}$. Let $d_{\max} := \max_{\tau \in T} d_\tau$ and define the set $\mathcal{D}_\rho := \{\rho^0, \rho^1, \ldots, r(d_{\max})\}$. We formulate the problem with the following linear program, denoted by ASS-LP.

$$\sum_{i \in M} y_{i,\tau} = 1 \qquad \forall \tau \in T \tag{1a}$$

$$\sum_{\tau \in T} u_{i,\tau} y_{i,\tau} \le 1 \qquad \forall i \in M \tag{1b}$$

$$\sum_{\tau \in T: r(d_\tau) \le D} c_{i,\tau} y_{i,\tau} \le D \qquad \forall D \in \mathcal{D}_\rho, \forall i \in M \tag{1c}$$

$$y_{i,\tau} \ge 0 \qquad \forall \tau \in T, \forall i \in M : u_{i,\tau} \le 1 \wedge c_{i,\tau} \le d_\tau \tag{1d}$$

If ASS-LP is infeasible, then there can be no feasible (integral) task assignment. Now assume that it is feasible and we have computed a feasible solution $\mathbf{y}^*$. For each machine i and deadline $D \in \mathcal{D}_\rho$ we extract a value $U_{i,D} := \sum_{\tau \in T: r(d_\tau)=D} c_{i,\tau} y_{i,\tau}^*$. Based on these values, we define a strengthened variation of ASS-LP, denoted by ASS2-LP in the sequel. We obtain the latter by replacing the constraints (1c) by the following set of constraints:

$$\sum_{\tau \in T: r(d_\tau)=D} c_{i,\tau} y_{i,\tau} \le U_{i,D} \qquad \forall D \in \mathcal{D}_\rho, \forall i \in M \tag{1c'}$$

Clearly if $\mathbf{y}^*$ is a feasible solution for ASS-LP it is also a feasible solution for ASS2-LP and if ASS2-LP is infeasible then no feasible task assignment exists. We now round $\mathbf{y}^*$ to an integral vector which *approximately* satisfies ASS2-LP. Namely, we follow an iterative rounding approach, similar to [17], which derives an integer solution $\hat{\mathbf{y}}$ that satisfies constraints (1a) and (1d) and the following two inequalities

$$\sum_{\tau \in T} u_{i,\tau} \hat{y}_{i,\tau} \le 4 \qquad \forall i \in M \tag{2}$$

$$\sum_{\tau \in T: r(d_\tau)=D} c_{i,\tau} \hat{y}_{i,\tau} \le U_{i,D} + 3D \qquad \forall D \in \mathcal{D}_\rho, \forall i \in M \tag{3}$$

The idea of our iterative rounding procedure is the following. In each iteration k, we first compute an extreme point solution $\mathbf{y}^k$ of a linear program LP^k where LP^0 equals ASS2-LP and each LP^k is obtained by fixing some variables and removing some constraints of LP^{k-1}.

Given a feasible fractional solution $\mathbf{y}^k$ to define LP^{k+1}, we first fix all variables which are integral in $\mathbf{y}^k$, i.e., those variables are not allowed to be changed anymore in the remainder of the procedure. Then, we check whether there exists a constraint of either type (1b) or type (1c'), with at most three fractional variables. We obtain LP^{k+1} by dropping this constraint. If such a constraint does not exist, then we obtain an integral solution by rounding all variables of $\mathbf{y}^k$ in a suitable way. The key lemma in our procedure is the following.

Lemma 2. *Let $\mathbf{y}^k$ be an extreme point solution to LP^k. Then either,*
(i) there is a machine i for which there is a constraint of type (1b) in LP^k such that there are at most three tasks τ with $y_{i,\tau} \in (0,1)$ (i.e., $y_{i,\tau}$ is fractional), or
(ii) there is a machine i and a deadline $D \in \mathcal{D}_\rho$ for which there is a constraint of type (1c') in LP^k and there are at most three tasks τ with $r(d_\tau) = D$ and $y_{i,\tau} \in (0,1)$, or
(iii) for each machine i there are exactly four tasks τ with $y_{i,\tau} \in (0,1)$.

Proof (sketch). Let n, w and z be the number of constraints of types (1a), (1b) and (1c'), respectively, which are still in LP^k. With $I(\mathbf{y}^k)$ and $F(\mathbf{y}^k)$ being the number of entries in $\mathbf{y}^k$ equal to one and the number of fractional entries, respectively, it holds that $n \le I(\mathbf{y}^k) + F(\mathbf{y}^k)/2$. Using also that $\mathbf{y}^k$ is an extreme point solution, we have that $I(\mathbf{y}^k) + F(\mathbf{y}^k) \le n + w + z$ and we get that $F(\mathbf{y}^k) \le 2w + 2z$. If $z > w$ then there are less than $4z$ fractional variables and by the pigeonhole-principle in one constraint of type (1c') at most three of them appear (as each variable appears in only one constraint of this type). Similarly, if $z < w$ there must be a constraint of type (1b) with this property. Finally, if $z = w$ there are at most $4w$ fractionals and either Case (i) or Case (iii) applies. $\square$

If Case (i) (Case (ii)) of Lemma 2 applies for a machine i (a machine i and a deadline D), then we obtain LP^{k+1} by dropping the corresponding constraint $\sum_{\tau \in T} u_{i,\tau} y_{i,\tau} \le 1$ ($\sum_{\tau \in T: r(d_\tau)=D} c_{i,\tau} y_{i,\tau} \le U_{i,D}$). Since we fixed the integer variables, for any solution for the remaining LP it holds that $\sum_{\tau \in T} u_{i,\tau} y_{i,\tau} \le 4$ ($\sum_{\tau \in T: r(d_\tau)=D} c_{i,\tau} y_{i,\tau} \le U_{i,D} + 3D$). Therefore, disregarding this constraint in the sequel, ensures that the right-hand side of this constraint is violated by at most an amount of 3 (or $3D$) no matter how the involved variables are rounded in the remaining iterations.

If Case (iii) of Lemma 2 applies, using the following lemma we assign all remaining tasks at once. The claim of the lemma can be shown by taking a fractional matching (representing the fractionally assigned tasks) and transforming it to an integral matching.

Lemma 3. *Assume that in $\mathbf{y}^k$ for each machine i there are exactly four tasks τ such that $y^k_{i,\tau} \in (0,1)$. Then in polynomial time we can compute an assignment of all these tasks to the machines such that at most two such tasks are assigned to a single machine.*

If either all constraints have been removed or if Lemma 3 has been applied we obtain a task assignment $\hat{\mathbf{y}}$ which satisfies $\sum_{\tau \in T} u_{i,\tau} \hat{y}_{i,\tau} \le 4$ for each machine i and $\sum_{\tau \in T: r(d_\tau)=D} c_{i,\tau} \hat{y}_{i,\tau} \le U_{i,D} + 3D$ for all machines i and all deadlines $D \in \mathcal{D}_\rho$. Observe that $U_{i,D} \le D$ for all $D \in \mathcal{D}_\rho$ and all machines i, and hence the vector $\hat{\mathbf{y}}$ satisfies the relaxed *dbf*-constraints up to a factor 4. Also, Lemma 1 directly implies that the task assignment given by the vector $\hat{\mathbf{y}}$ is feasible with a speedup of 24 if we choose $\rho = 2$. However, using the definition of $U_{i,D}$ and a more careful calculation, we can bound the needed speedup even further.

Theorem 1. *There is a $(11 + 4\sqrt{3})$-approximation test for the problem of assigning tasks to unrelated machines.*

Our rounding scheme hinges on the sparsity of the coefficient matrix as the one proposed in [20]. We remark that our rounding scheme *cannot* be derived via the result of [20] since here we need to ensure that the constraints (1a) must be *exactly* satisfied by the computed integral solution.

Finally, we show that it is NP-hard to decide whether a task system T has an assignment which is feasible on m unrelated machines, even with a speedup factor of 2; the proof follows the lines of the $(\frac{3}{2} - \epsilon)$-hardness result for makespan minimization in [21].

Theorem 2. *Let $\epsilon > 0$. There is no $(2 - \epsilon)$-approximation test for the problem of assigning tasks to unrelated machines, unless $P = NP$.*

Note that our hardness result is different from the one by Andersson and Tovar [3]. They show that, given a task system with implicit deadlines that is feasible on a platform of m related parallel machines when migration is allowed, then any partitioned algorithm needs a speedup factor of at least $2 - \epsilon$ for finding a feasible partition of the tasks.

4   Constant Number of Machines

Assuming that the number of machines is bounded by a constant, we present a dynamic programming algorithm (DP) that gives a $(1+\epsilon)$-approximation test for any $\epsilon > 0$. For having a DP-table of bounded size we introduce an approximation of the demand bound function such that the contribution of each task can be derived by using only a constant number of values.

Let $\epsilon > 0$, and we assume w.l.o.g. that $\epsilon < 1/2$. Let L be the minimum integer which satisfies $1 \leq (1 + \epsilon)^{L-1}\epsilon^2$. We define the function dbf^*:

$$dbf_i^*(\tau, s) := \begin{cases} \left\lfloor \frac{s+t_\tau-d_\tau}{t_\tau} \right\rfloor c_{i,\tau} & \text{if } s < (1+\epsilon)^L \cdot d_\tau \\ \frac{c_{i,\tau}}{t_\tau}s & \text{otherwise.} \end{cases}$$

Given a task assignment $\mathcal{T}$ of tasks in T to the machines, we define $dbf_{\mathcal{T},i}^*(s) := \sum_{\tau \in \mathcal{T}_i} dbf_i^*(\tau, s)$, for all $s > 0$. Further, for having clean notation, we write $dbf_i^*(s)$ instead of $dbf_{\mathcal{T},i}^*(s)$ in case the assignment $\mathcal{T}$ is clear from the context.

The key observation is that for computing the function $dbf_i^*(\tau, \cdot)$ for a fixed task τ, it suffices to know the utilization of the task τ and the values the demand bound function $dbf_i(\tau, s)$ for $s \in [d_\tau, (1+\epsilon)^L \cdot d_\tau)$. Exploiting the properties of the functions $dbf_{\mathcal{T},i}(s)$ and $dbf_{\mathcal{T},i}^*(s)$, we have that $dbf_{\mathcal{T},i}^*$ is a $(1 + \epsilon)$-approximation of the "real" demand bound function.

Lemma 4. *Given an assignment $\mathcal{T}$ and let $\epsilon < 1/2$. Then, for all machines i,*
(i) if $dbf_{\mathcal{T},i}^*(r) \leq \alpha \cdot r$ for all $r \geq 0$, then $dbf_{\mathcal{T},i}(s) \leq (1 + \epsilon) \cdot \alpha \cdot s$ for all $s \geq 0$;
(ii) if $dbf_{\mathcal{T},i}(r) \leq r$ for all $r \geq 0$, then $dbf_{\mathcal{T},i}^*(s) \leq (1 + \epsilon) \cdot s$ for all $s \geq 0$.

Note that in contrast to other approximations of the demand bound function considered in the literature (e.g. [1]), in Lemma 4 we do not use an analysis task by task, and we do not bound the ratio $dbf(\tau, s)/dbf^*(\tau, s)$. In fact, the latter can be unbounded.

The following proposition shows that at the cost of a $(1+\epsilon)$-speedup it suffices to check whether the condition $dbf^*_{T,i}(s) \leq s$ is (approximately) satisfied at powers of $1 + \epsilon$. It is useful for our DP as it suffices to characterize each task τ only by its utilization and the constantly many values $dbf^*_i(\tau, (1+\epsilon)^k)$ for integers k such that $d_\tau \leq (1 + \epsilon)^k < (1 + \epsilon)^L \cdot d_\tau$ (on each machine i).

Proposition 2. *Consider a task assignment T and a machine i. If for all $k \in \mathbb{N}$ $dbf^*_{T,i}((1 + \epsilon)^k) \leq \alpha \cdot (1 + \epsilon)^k$ then $dbf^*_{T,i}(s) \leq \alpha \cdot s \cdot (1 + \epsilon)$ for all $s \geq 0$.*

For each task τ, each machine i and $\ell \in \mathbb{N}_0$, we introduce a vector $v(i, \tau)$ by defining position $v(i, \tau)_\ell := dbf^*_i(\tau, (1 + \epsilon)^\ell)/(1 + \epsilon)^\ell$.

Proposition 3. *Consider an assignment T. For all machines $i \in M$, we have $\left\|\sum_{\tau \in T_i} v(i, \tau)\right\|_\infty \leq \alpha$ iff $dbf^*_{T,i}(s) \leq \alpha s$, for each $s = (1 + \epsilon)^k$, $k \in \mathbb{N}$.*

We present a dynamic programming algorithm which either (i) asserts that there is no feasible assignment of the tasks to the machines by showing that there is no assignment T of tasks to machines such that $\left\|\sum_{\tau \in T_i} v(i, \tau)\right\|_\infty \leq 1$ for each machine i, or (ii) finds an assignment T such that $\left\|\sum_{\tau \in T_i} v(i, \tau)\right\|_\infty \leq 1 + O(\epsilon)$ for each machine i. In the latter case, Lemma 4 and the above proposition imply an approximation test for the problem of assigning tasks to a constant number of unrelated machines.

Assume w.l.o.g. that the tasks $\tau_1, ..., \tau_n$ are ordered such that $d_{\tau_p} \leq d_{\tau_{p+1}}$ for each p. We partition the tasks into groups $G_k := \{\tau | (1+\epsilon)^k \leq d_\tau < (1+\epsilon)^{k+1}\}$ for each $k \in \mathbb{N}$. Our DP works in phases; one phase for each task. The key idea is that when trying to assign task $\tau \in G_k$, we need only a constant number of values from the assignment of the previously considered tasks. With $L^{(k)} := \min\{k, L\}$ (s.t. $k - L^{(k)} \geq 0$), for each machine i we need the sum $\sum_{\tau \in T_i \cap (\cup_{k'=0}^{k-L^{(k)}} G_{k'})} u_{i,\tau}$, the sum $\sum_{\tau \in T_i \cap G_{k'}} u_{i,\tau}$, for all $k' : k - L^{(k)} < k' \leq k$, and the sum $\sum_{\tau \in T_i \cap G_{k'}} v(i, \tau)_\ell$ for all $\ell : k \leq \ell \leq k + L$ and all $k' : \ell - L^{(\ell)} < k' \leq k$.

Ideally, we would like to store all possible combinations of the above quantities that can result from assigning the tasks of previous iterations. Then we could compute the values for the next iteration by taking each combination of values from the last iteration and compute the values we get by additionally assigning τ to one of the machines. Unfortunately, the number of possible combinations of the above values is not polynomially bounded. In order to bound them, we round entries of the vectors $v(i, \tau)$. We perform the described procedure with the rounded vectors. This will result in a polynomial time procedure.

We now formally present the dynamic programming algorithm. Consider a task $\tau \in G_k$, $k \in \mathbb{N}$. For each i we define $v'(i, \tau)_\ell := \frac{\epsilon}{n} \left\lfloor \frac{n}{\epsilon} \cdot v(i, \tau)_\ell \right\rfloor$ for each $\ell < k + L$, and $v'(i, \tau)_{\ell'} := u'_{i,\tau} := \frac{\epsilon}{n} \left\lfloor \frac{n}{\epsilon} \cdot u_{i,\tau} \right\rfloor$ for each $\ell' \geq k + L$. The following lemma bounds our rounding error.

Lemma 5. *Let i be a machine and T_i be a set of tasks. For all ℓ, it holds that* $\sum_{\tau \in T_i} v'(i,\tau)_\ell \le \sum_{\tau \in T_i} v(i,\tau)_\ell \le \epsilon + \sum_{\tau \in T_i} v'(i,\tau)_\ell.$

Note that we can also describe each rounded vector $v'(i,\tau)$ with only constantly many pieces of information. When working with the rounded vectors, for the quantities mentioned above there are only a polynomial number of combinations (assuming that m is a constant). In particular, we obtain a dynamic programming table of polynomial size. Formally, our dynamic programming table consists of entries of the form $(p, \mathbf{z}, \mathbf{w}, \mathbf{c})$ where

- $p \in \{0, ..., n\}$ denotes the phase of the DP. In phase p, task τ_p is being assigned to a machine. Let k be an integer such that $\tau_p \in G_k$;
- for each machine i, the value z_i is of the form $\ell \cdot \frac{\epsilon}{n}$ for some integer ℓ, denoting the rounded aggregated utilization of machine i due to the tasks having a deadline at least a factor of $(1 + \epsilon)^L$ smaller with respect to the deadline of task τ_p;
- for each machine i and each k' with $k - L^{(k)} < k' \le k$, the value $w_{i,k'}$ is of the form $\ell \cdot \frac{\epsilon}{n}$ for some integer ℓ, denoting the rounded utilization of tasks in $G_{k'} \cap T_i$.
- for each triple $(i, k', k'') \in C_p$ with $C_p = \{(i, k', k'') : 1 \le i \le m; k \le k'' < k + L\}$ and $k'' - L^{(k'')} < k' \le k$, the value $c_{i,k',k''}$ is of the form $\ell \cdot \frac{\epsilon}{n}$ for some integer ℓ, denoting the quantity $\sum_{\tau \in T_i \cap G_{k'}} v'(i,\tau)_{k''}$. Intuitively, it expresses how much the vectors of the tasks in $G_{k'}$ on machine i contribute towards dimension k''.

We require the following set of conditions to be satisfied for a DP-cell $(p, \mathbf{z}, \mathbf{w}, \mathbf{c})$ to exist; for each machine $i \in M$ and all $k'' \in \{k, \ldots, k + L\}$

$$z_i + \sum_{k'=k-L^{(k)}+1}^{k''-L^{(k'')}} w_{i,k'} + \sum_{k'=k''-L^{(k'')}+1}^{k} c_{i,k',k''} \le 1 + \epsilon \tag{4}$$

This condition implies that, for all parameters, z_i, $w_{i,k'}$, $c_{i,k',k''} \le 1 + \epsilon$.

Proposition 4. *The number of DP-entries is bounded by $n \cdot ((1 + \epsilon)n/\epsilon)^{m^3 \cdot L^3}$.*

In each entry $(p, \mathbf{z}, \mathbf{w}, \mathbf{c})$ of the DP-table we store either "YES" or "NO", which represents whether or not there is an assignment of the tasks $\tau_1, ..., \tau_p$ to the machines which yields the quantities given by the vectors $\mathbf{z}, \mathbf{w}, \mathbf{c}$.

We now describe how to fill the DP-table. We initialize the table by assigning a "YES"-entry to $(0, \mathbf{0}, \mathbf{0}, \mathbf{0})$ and a "NO"-entry to any other entry with $p = 0$. Assume that for some phase p, all entries of the form $(p - 1, \mathbf{z}^{(p-1)}, \mathbf{w}^{(p-1)}, \mathbf{c}^{(p-1)})$ have been computed. We extend the DP-table in phase p by considering each combination of assigning task τ_p to some machine i and each DP-cell $(p - 1, \mathbf{z}^{(p-1)}, \mathbf{w}^{(p-1)}, \mathbf{c}^{(p-1)})$ with a "YES"-entry. Intuitively, we compute what values for $\mathbf{z}^{(p)}$, $\mathbf{w}^{(p)}$, and $\mathbf{c}^{(p)}$ we get if we take the task assignment encoded in $(p - 1, \mathbf{z}^{(p-1)}, \mathbf{w}^{(p-1)}, \mathbf{c}^{(p-1)})$ and additionally add τ_p to i.

Formally, let tasks τ_{p-1} and τ_p be in group G_h and G_k, respectively. Almost all entries of the vectors are equal and hence we only list the values which differ.

If $h = k$, then we have: $w_{i,k}^{(p)} = w_{i,k}^{(p-1)} + u'_{i,\tau}$, and $c_{i,k,k''}^{(p)} = c_{i,k,k''}^{(p-1)} + v'(i,\tau)_{k''}$ for all $k'' \in \{k, \ldots, k+L\}$. If $h \neq k$ we may assume w.l.o.g. that $h = k-1$ by creating dummy tasks of zero processing requirement. Then, $z_g^{(p)} = z_g^{(p-1)} + w_{g,k-L^{(k)}}^{(p-1)}$ for all machines $g \in M$; $w_{g,k'}^{(p)} = w_{g,k'}^{(p-1)}$ for machines $g \in M$ and all $k' : k - L^{(k)} < k' < k$; $w_{i,k}^{(p)} = u'_{i,\tau_p}$ and $w_{g,k}^{(p)} = 0$ for all machines $g \neq i$; $c_{g,k',k''}^{(p)} = c_{g,k',k''}^{(p-1)}$ for all machines g, all $k'' : k \leq k'' \leq k + L$ and all $k' : k'' - L^{(k'')} < k' < k$; $c_{i,k,k''}^{(p)} = v'(i,\tau)_{(k'')}$ for all $k'' : k \leq k'' \leq k + L$; and $c_{g,k,k''}^{(p)} = 0$ for all machines $g \neq i$ and all $k'' : k \leq k'' \leq k + L$.

Finally, we check whether the computed values $\mathbf{z}^{(p)}, \mathbf{w}^{(p)}$ and $\mathbf{c}^{(p)}$ satisfy the condition given in (4). If this is the case, then we store a "YES"-entry in the corresponding DP-cell $(p, \mathbf{z}^{(p)}, \mathbf{w}^{(p)}, \mathbf{c}^{(p)})$ and we say that it *extends* the DP-cell $(p - 1, \mathbf{z}^{(p-1)}, \mathbf{w}^{(p-1)}, \mathbf{c}^{(p-1)})$. In case there does not exist a DP-cell $(p - 1, \mathbf{z}^{(p-1)}, \mathbf{w}^{(p-1)}, \mathbf{c}^{(p-1)})$ which can be extended to the DP-cell $(p, \mathbf{z}^{(p)}, \mathbf{w}^{(p)}, \mathbf{c}^{(p)})$, the latter DP-cell is filled with a "NO"-entry.

We fill the DP-table inductively, phase by phase, until each cell in the DP-table is filled.

Lemma 6. *For phase p, there exists a DP-cell of the form $(p, \mathbf{z}^{(p)}, \mathbf{w}^{(p)}, \mathbf{c}^{(p)})$ with a "YES"-entry if and only if there exists task assignment $\mathcal{T}$ of the first p tasks to the machines, such that for each $i \in M$ it holds that $\left\|\sum_{\tau \in \mathcal{T}_i} v'(i,\tau)\right\|_\infty \leq 1+\epsilon$.*

Combining Proposition 2 and the Lemmas 4, 5 and 6 yields a $(1 + 8\epsilon)$-approximation test, for any ϵ and a constant number of machines. The claim on the running time follows from Proposition 4. Redefining ϵ yields our main theorem.

Theorem 3. *For any $\epsilon > 0$ there exists a $(1 + \epsilon)$-approximation test if the number of machines is constant, that runs in time polynomial in the number of tasks.*

References

1. Albers, K., Slomka, F.: An event stream driven approximation for the analysis of real-time systems. In: Proceedings of the 16th Euromicro Conference on Real-Time Systems (ECRTS 2004), pp. 187–195 (2004)
2. Anand, S., Garg, N., Megow, N.: Meeting Deadlines: How Much Speed Suffices? In: Aceto, L., Henzinger, M., Sgall, J. (eds.) ICALP 2011. LNCS, vol. 6755, pp. 232–243. Springer, Heidelberg (2011)
3. Andersson, B., Tovar, E.: Competitive analysis of partitioned scheduling on uniform multiprocessors. In: Proceedings of Parallel and Distributed Processing Symposium (IPDPS), pp. 1–8 (2007)
4. Baker, T.P., Baruah, S.K.: Schedulability analysis of multiprocessor sporadic task systems. In: Handbook of Real-Time and Embedded Systems, ch. 3. CRC Press (2007)
5. Baruah, S., Fisher, N.: The partitioned multiprocessor scheduling of sporadic task systems. In: Proc. 26th IEEE Real-Time Systems Symposium, pp. 321–329. IEEE (2005)

6. Baruah, S., Mok, A., Rosier, L.: Preemptively scheduling hard-real-time sporadic tasks on one processor. In: Proc. 11th IEEE Real-Time Systems Symposium, pp. 182–190. IEEE (1990)

7. Baruah, S.K., Bonifaci, V., Marchetti-Spaccamela, A., Stiller, S.: Improved multi-processor global schedulability analysis. Real-Time Systems 46, 3–24 (2010)

8. Baruah, S.K., Pruhs, K.: Open problems in real-time scheduling. Journal of Scheduling 13, 577–582 (2010)

9. Bonifaci, V., Marchetti-Spaccamela, A., Stiller, S.: A constant-approximate feasibility test for multiprocessor real-time scheduling. Algorithmica 62, 1034–1049 (2012)

10. Bonifaci, V., Wiese, A.: Scheduling unrelated machines of few different types (unpublished manuscript)

11. Chakraborty, S., Künzli, S., Thiele, L.: Approximate schedulability analysis. In: Proc. 23rd IEEE Real-Time Systems Symposium, pp. 159–168. IEEE (2002)

12. Chen, J.-J., Chakraborty, S.: Resource augmentation bounds for approximate demand bound functions. In: Proceedings of 32nd IEEE Real-Time Systems Symposium, pp. 272–281. IEEE (2011)

13. Ebenlendr, T., Krcal, M., Sgall, J.: Graph balancing: A special case of scheduling unrelated parallel machines. In: Proc. 19th Symp. on Discrete Algorithms, pp. 483–490 (2008)

14. Eisenbrand, F., Rothvoß, T.: A PTAS for Static Priority Real-Time Scheduling with Resource Augmentation. In: Aceto, L., Damgård, I., Goldberg, L.A., Halldórsson, M.M., Ingólfsdóttir, A., Walukiewicz, I. (eds.) ICALP 2008, Part I. LNCS, vol. 5125, pp. 246–257. Springer, Heidelberg (2008)

15. Eisenbrand, F., Rothvoß, T.: EDF-schedulability of synchronous periodic task systems is coNP-hard. In: Proc. 21st Symp. on Discrete Algorithms, pp. 1029–1034 (2010)

16. Fisher, N., Baruah, S., Baker, T.P.: The partitioned scheduling of sporadic tasks according to static-priorities. In: Proc. 18th Euromicro Conf. on Real-Time Systems, pp. 118–127 (2006)

17. Jain, K.: A factor 2 approximation algorithm for the generalized Steiner network problem. Combinatorica 21, 39–60 (2001)

18. Jansen, K., Porkolab, L.: Improved approximation schemes for scheduling unrelated parallel machines. In: Proc. 31st Symp. on Theory of Computing, pp. 408–417 (1999)

19. Karmarkar, N., Karp, R.M.: An efficient approximation scheme for the one-dimensional bin-packing problem. In: Proc. of the 23rd Annual Symposium on Foundations of Computer Science, pp. 312–320 (1982)

20. Karp, R.M., Leighton, F.T., Rivest, R.L., Thompson, C.D., Vazirani, U.V., Vazirani, V.V.: Global wire routing in two-dimensional arrays. Algorithmica 2, 113–129 (1987)

21. Lenstra, J.K., Shmoys, D.B., Tardos, E.: Approximation algorithms for scheduling unrelated parallel machines. Mathematical Programming 46, 259–271 (1990)

22. Liu, C., Layland, J.: Scheduling algorithms for multiprogramming in a hard real-time environment. Journal of the ACM 20, 46–61 (1973)

23. Phillips, C.A., Stein, C., Torng, E., Wein, J.: Optimal time-critical scheduling via resource augmentation. Algorithmica 32, 163–200 (2002)

24. Svensson, O.: Santa claus schedules jobs on unrelated machines. In: Proc. 43rd Symp. on Theory of Computing, pp. 617–626. ACM Press (2011)

A Tight Lower Bound for Planar Multiway Cut with Fixed Number of Terminals

Dániel Marx[*]

Computer and Automation Research Institute,
Hungarian Academy of Sciences (MTA SZTAKI), Budapest, Hungary
dmarx@cs.bme.hu

Abstract. Given a planar graph with k terminal vertices, the PLANAR MULTIWAY CUT problem asks for a minimum set of edges whose removal pairwise separates the terminals from each other. A classical algorithm of Dahlhaus et al. [2] solves the problem in time $n^{O(k)}$, which was very recently improved to $2^{O(k)} \cdot n^{O(\sqrt{k})}$ time by Klein and Marx [6]. Here we show the optimality of the latter algorithm: assuming the Exponential Time Hypothesis (ETH), there is no $f(k) \cdot n^{o(\sqrt{k})}$ time algorithm for PLANAR MULTIWAY CUT. It also follows that the problem is W[1]-hard, answering an open question of Downey and Fellows [3].

1 Introduction

MULTIWAY CUT (also called MULTITERMINAL CUT) is a generalization of the classical minimum $s - t$ cut problem: given an undirected graph G with subset T of k vertices specified as terminals, the task is to find a set of edges having minimum total weight whose deletion pairwise separates the k terminal vertices from each other. While the problem is polynomial-time solvable for $k = 2$, it becomes NP-hard for $k = 3$ on general graphs. The special case of the problem on planar graphs, PLANAR MULTIWAY CUT, is also NP-hard if k can be arbitrarily large, but can be solved in time $O((4k)^k n^{2k-1} \log n)$ [2] or in time $O(k4^k n^{2k-4} \log n)$ [4]. That is, perhaps somewhat unexpectedly, the problem is polynomial-time solvable on planar graphs for every fixed k. In the companion paper [6], the dependence of the running time on the number of terminals was significantly improved: an algorithm with running time $2^{O(k)} \cdot n^{O(\sqrt{k})}$ was given for PLANAR MULTIWAY CUT.

How much further the dependence on k can be improved? Dahlhaus et al. [2] asked if PLANAR MULTIWAY CUT can be solved in time $c^k \cdot n^{O(1)}$, which would be a significant improvement over all known algorithms. More generally, Downey and Fellows asked in the open problem list of their classical 1999 monograph [3] if the problem parameterized by the number of terminals is *fixed-parameter tractable*, that is, can be solved in time $f(k) \cdot n^{O(1)}$ for some computable function

[*] Research supported by the European Research Council (ERC) grant "PARAMTIGHT: Parameterized complexity and the search for tight complexity results," reference 280152.

f depending only on k. The main result of the paper is a negative answer to this question: the problem is W[1]-hard parameterized by the number of terminals, making it unlikely to be fixed-parameter tractable. Moreover, our reduction shows that there is no $f(k) \cdot n^{o(\sqrt{k})}$ time algorithm for any computable function f, unless the Exponential Time Hypothesis (ETH) fails. ETH is the assumption that n-variable m-clause 3SAT cannot be solved in time $2^{o(n)} \cdot m^{O(1)}$, see [5,7]. Therefore, the $2^{O(k)} \cdot n^{O(\sqrt{k})}$ time algorithm of [6] is optimal in the sense that the exponent of n cannot be better than $O(\sqrt{k})$. We present the hardness proof for the version of the problem where weights are allowed on the edges. However, the weights are polynomially large in the reductions, thus the results can be easily transferred to the case where each edge has unit weight by replacing an edge of weight c by c parallel edges (or parallel paths if one wishes to state the hardness result for simple graphs).

2   The Reduction

It will be convenient to present the reduction to PLANAR MULTIWAY CUT from the following W[1]-hard problem:

GRID TILING

Input: Integers k, n, and k^2 nonempty sets $S_{i,j} \subseteq [n] \times [n]$ ($1 \leq i, j \leq k$).

Find: For each $1 \leq i, j \leq k$, a value $s_{i,j} \in S_{i,j}$ such that
- If $s_{i,j} = (x, y)$ and $s_{i,j+1} = (x', y')$, then $x = x'$.
- If $s_{i,j} = (x, y)$ and $s_{i+1,j} = (x', y')$, then $y = y'$.

The W[1]-hardness of GRID TILING essentially follows from [8]. Note that the reduction transforms the problem of finding a k-clique into a $k \times k$ GRID TILING instance (we will need this fact for the tight lower bound in Corollary 5).

Lemma 1. GRID TILING *is* W[1]-*hard parameterized by* k.

To prove the W[1]-hardness of PLANAR MULTIWAY CUT, we construct gadgets of the following form. An $n \times n$ *gadget* is an embedded planar graph G_n with a set of $4n+8$ distinguished vertices (see Figure 1). These distinguished vertices all appear on the boundary of the graph (i.e, on the infinite face) in the clockwise order $UL, u_1, \ldots, u_{n+1}, UR, r_1, \ldots, r_{n+1}, DR, d_{n+1}, \ldots, d_1, DL, \ell_{n+1}, \ldots, \ell_1$. Note that these distinguished vertices are a subset of the vertices on the boundary of the gadget, thus e.g., u_i and u_{i+1} are not necessarily adjacent. The four vertices UL, UR, DR, DL are the only terminal vertices in the gadget. We say that a multiway cut M of the gadget *represents* the pair $(x, y) \in [n]^2$ if $G_n \setminus M$ has four components that partition the distinguished vertices into the following classes:

$$\{UL, u_1, \ldots, u_y, \ell_1, \ldots, \ell_x\} \qquad \{UR, u_{y+1}, \ldots, u_{n+1}, r_1, \ldots, r_x\}$$
$$\{DL, d_1, \ldots, d_y, \ell_{x+1}, \ldots, \ell_{n+1}\} \, \{DR, d_{y+1}, \ldots, d_{n+1}, r_{x+1}, \ldots, r_{n+1}\}$$

The main part of the hardness proof is to show that certain gadgets exist:

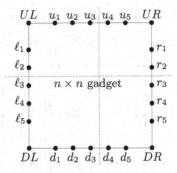

Fig. 1. The distinguished vertices of a $n \times n$ gadget for $n = 4$. The dashed lines indicate a multiway cut of UL, UR, DR, DL that represents the pair $(2, 3)$.

Lemma 2. *Given a subset $S \subseteq [n]^2$, we can construct in polynomial time a gadget G_S and an integer D such that the following properties hold:*

1. *For every $(x, y) \in S$, the gadget G_S has a multiway cut of weight D representing (x, y).*
2. *If a multiway cut of G_S has weight D, then it represents some $(x, y) \in S$.*
3. *Every multiway cut of G_S has weight at least D.*

The proof of Lemma 2 appears in Section 3. Assuming that such gadgets can be constructed, we can prove that PLANAR MULTIWAY CUT is W[1]-hard.

Theorem 3. PLANAR MULTIWAY CUT *is* W[1]*-hard parameterized by the number of terminal vertices.*

Proof. We reduce GRID TILING to PLANAR MULTIWAY CUT. Let $S_{i,j} \subseteq [n]^2$ $(1 \le i, j \le k)$ be the subsets in a GRID TILING instance. For every $1 \le i, j \le k$, we use Lemma 2 to construct the $n \times n$ gadget $G_{i,j}$ and compute the integer $D_{i,j}$ corresponding to the set $S_{i,j}$. Let $D = \sum_{1 \le i, j \le k} D_{i,j}$. We construct a planar graph G by attaching the gadgets the following way:

- for every $1 \le i \le k$, $1 \le j < k$, we identify vertices UR, r_1, $\dots$, r_{n+1}, DR of $G_{i,j}$ with vertices of UL, ℓ_1, $\dots$, ℓ_{n+1}, DL of $G_{i,j+1}$, respectively, and
- for every $1 \le i < k$, $1 \le j \le k$, we identify vertices DL, d_1, $\dots$, d_{n+1}, DR of $G_{i,j}$ with vertices of UL, u_1, $\dots$, u_{n+1}, UR of $G_{i+1,j}$, respectively.

Note that we glue together the gadgets only at the distinguished vertices, not along the whole boundary. It is easy to see that G is planar. With these identifications, the $4k^2$ terminal vertices of the k^2 gadgets are identified into a set T of exactly $(k + 1)^2$ terminal vertices. For the sake of analysis, if there are two gadgets that have edges between two vertices v and u, then we keep both edges as parallel edges in the graph G. This way, we can say the the edge set of G is the disjoint union of the edge sets of all the gadgets.

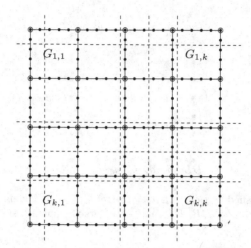

Fig. 2. Constructing the instance by identifying the distinguished vertices of k^2 gadgets. The circled vertices are terminals, the dashed lines indicate a multiway cut that corresponds to a clique.

We claim that there is a multiway cut of weight D separating the terminals in T if and only if the GRID TILING instance has a solution. Suppose first that $s_{i,j} \in S_{i,j}$ ($1 \le i, j \le k$) is a solution of the GRID TILING instance. By property 1 of Lemma 2, every gadget $G_{i,j}$ has a multiway cut $M_{i,j}$ of weight $D_{i,j}$ that represents $s_{i,j}$. We claim that the union of these $M_{i,j}$'s is a multiway cut separating T. This follows from the fact that the multiway cuts are consistent in the following sense: if distinguished vertices v_1 and v_2 of gadget $G_{i,j}$ are identified with vertices v'_1 and v'_2, respectively, of gadget $G_{i,j+1}$, then v_1 and v_2 are in the same component of $G_{i,j} \setminus M_{i,j}$ if and only if v'_1 and v'_2 are in the same component of $G_{i,j+1} \setminus M_{i,j+1}$. For example, consider vertices r_{s_1} and r_{s_2} of $G_{i,j}$, which are identified with vertices ℓ_{s_1} and ℓ_{s_2} of $G_{i,j+1}$. Suppose that $s_{i,j} = (x, y)$ and $s_{i,j+1} = (x', y')$ with $x = x'$. Then from the fact that $M_{i,j}$ represents (x, y), we have that r_{s_1} and r_{s_2} are in different components of $G_{i,j} \setminus M_{i,j}$ if and only if $s_1 \le x < s_2$. Similarly, ℓ_{s_1} and ℓ_{s_2} are in different components of $G_{i,j+1} \setminus M_{i,j+1}$ if and only if $s_1 \le x' < s_2$. The consistency of the multiway cuts implies that if we look at a terminal vertex, say DR of $G_{i,j}$, then its component in $G \setminus M$ is exactly the union of the component of DR in $G_{i,j} \setminus M_{i,j}$, the component of DL in $G_{i,j+1} \setminus M_{i,j+1}$, the component of UR in $G_{i+1,j} \setminus M_{i,j}$, and the component of UL in $G_{i+1,j+1} \setminus M_{i+1,j+1}$. Therefore, the terminals in T are indeed separated from each other in $G \setminus M$.

For the other direction of the proof, suppose that M is a multiway cut of T. In particular, this means that M is a multiway cut of the four terminals of each gadget. Since the gadgets are edge disjoint, M can be partitioned into disjoint sets $M_{i,j}$ ($1 \le i, j \le k$) such that $M_{i,j}$ is a multiway cut of gadget $G_{i,j}$. As every multiway cut of gadget $G_{i,j}$ has weight at least $D_{i,j}$ (Property 3 of Lemma 2) and M has weight at most D, it follows that $M_{i,j}$ has weight exactly

$D_{i,j}$. Therefore, by Property 2 of Lemma 2, $M_{i,j}$ represents some set $s_{i,j} \in S_{i,j}$. We claim the the pairs $s_{i,j}$ form a solution for GRID TILING. We verify that if $s_{i,j} = (x,y)$ and $s_{i,j+1} = (x',y')$, then $x = x'$. Suppose first that $x < x'$. Then r_{x+1} of $G_{i,j}$ is in the same component as DR of $G_{i,j}$, while ℓ_{x+1} of $G_{i+1,j}$ (which is actually identified with r_{x+1} of $G_{i,j}$) is in the same component as UL of $G_{i,j+1}$ (as $x + 1 \leq x'$). Therefore, two terminal vertices are in the same component of $G \setminus M$, a contradiction. The case $x > x'$, as well as the proof that $s_{i,j}$ and $s_{i+1,j}$ agree in the second component, is analogous. $\qquad\square$

To obtain a lower bound on the exponent of n in the running time of algorithms for PLANAR MULTIWAY CUT, we can use the following lower bound on CLIQUE:

Theorem 4 ([1]). *An $f(k)n^{o(k)}$ algorithm for* CLIQUE *implies that ETH fails.*

Observe that, given an instance of CLIQUE, the two reductions in Lemma 1 and Theorem 3 create an instance of PLANAR MULTIWAY CUT with $(k+1)^2$ terminals. Thus by Theorem 4, we have the following lower bound:

Corollary 5. *If there is an $f(k)n^{o(\sqrt{k})}$ algorithm for* PLANAR MULTIWAY CUT, *then ETH fails.*

3   Gadget Construction

The goal of this section is to construct a gadget that satisfies the requirements of Lemma 2. Section 3.1 describes the construction of the gadget, Section 3.2 proves Property 1 of Lemma 2 by showing how a pair $(x,y) \in S$ defines a cheap multiway cut, while Section 3.3 proves Properties 2 and 3 by showing how a cheap multiway cut defines a pair in S.

3.1   Construction

Let $N := n^2 + 2n + 1$. We start the construction of the gadget G_S with an $(N+1) \times (N+1)$ grid: let us introduce vertices $g[i,j]$ $(0 \leq i, j \leq N)$ such that vertices $g[i,j]$ and $g[i',j']$ are adjacent if and only if $|i - i'| + |j - j'| = 1$. The grid is pictured as $g[0,0]$ being the upper left corner, $g[0,N]$ the upper right corner, etc. We call the edges on the horizontal path from $g[i,0]$ to $g[i,N]$ the *row* R_i, while the edges on the vertical path from $g[0,j]$ to $g[N,j]$ is the *column* C_j. We also say that the horizontal edge $\{g[i,j], g[i,j+1]\}$ of row R_i has *column number* j and the vertical edge $\{g[i,j], g[i+1,j]\}$ of column C_j has *row number* i.

For ease of notation, we define the functions $\alpha(s) = N - n - 2 + s$ and $\beta(x,y) = x + yn$ (observe that $\beta(n+1,y) = \beta(1,y+1)$ and $n+1 \leq \beta(x,y) \leq n^2 + n = N - n - 1 = \alpha(1)$ for every $1 \leq x, y \leq n$). The distinguished vertices of the gadget are defined as follows (see Figure 3): for every $1 \leq s \leq n + 1$, we set

$$UL = g[0,0] \quad UR = g[0,N] \quad u_s = g[0, \beta(1,s)] \quad d_s = g[N, \beta(1,s)]$$
$$DL = g[N,0] \quad DR = g[N,N] \quad \ell_s = g[\alpha(s),0] \quad r_s = g[s,N].$$

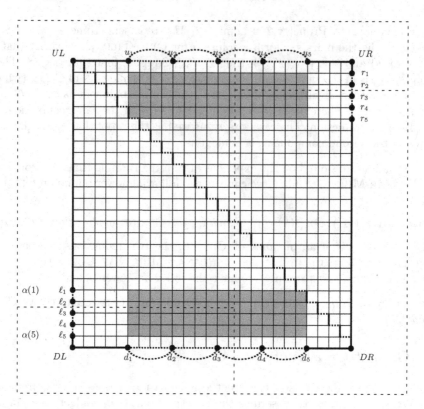

Fig. 3. A $n \times n$ gadget with $n = 4$ (and hence $N = n^2 + 2n + 1 = 25$). The black dots represent the distinguished vertices, strong edges have weight ∞, dotted strong edges have weight at least W^3, normal edges have weight between W^2 and $W^2 + NW$. Some of the shaded cells are marked special (the cell edges are omitted from the figure). The dashed lines show the 4 components created by the multiway cut corresponding to the pair $(2, 3)$.

We refer to the edges of the grid as *grid edges*. We define now the weight of the grid edges. Let $W := 100N^2$. Let us set the weights first as follows:

– For $0 \le i \le N$, $0 \le j \le N - 1$, vertical edge $\{g[i, j], g[i + 1, j]\}$ has weight

$$
\begin{cases}
\infty & \text{if } i = j - 1, \\
\infty & \text{if } i = 0 \text{ and } j \notin [\alpha(1), \alpha(n)], \\
W^3 + W^2 & \text{if } i = 0 \text{ and } j \in [\alpha(1), \alpha(n)], \\
\infty & \text{if } i = N \text{ and } j \notin [1, n], \\
W^3 + W^2 & \text{if } i = N \text{ and } j \in [1, n], \\
W^2 & \text{otherwise.}
\end{cases}
$$

- For $0 \leq i \leq N - 1$, $0 \leq j < N - 1$, horizontal edge $\{g[i,j], g[i, j+1]\}$ has weight

$$
\begin{cases}
\infty & \text{if } i = 0 \text{ and } j \notin [\beta(1,1), \beta(n,n)], \\
W^3 + W^2 + jW & \text{if } i = 0 \text{ and } j \in [\beta(1,1), \beta(n,n)], \\
W^2 + jW & \text{if } 0 < i < j, \\
W^3 + W^2 - i^2W - (N-i)^2W & \text{if } i = j, \\
W^2 + (N-j)W & \text{if } j < i < N, \\
W^3 + W^2 + (N-j)W & \text{if } j = N \text{ and } j \in [\beta(1,1), \beta(n,n)], \\
\infty & \text{if } j = N \text{ and } j \notin [\beta(1,1), \beta(n,n)].
\end{cases}
$$

Note that the only part of the boundary with finite edges are the horizontal path between u_1 and u_{n+1}, the horizontal path between d_1 and d_{n+1}, the vertical path between ℓ_1 and ℓ_{n+1}, and the vertical path between r_1 and r_{n+1}.

Let us consider a column number $\beta(1,1) \leq z \leq \beta(n,n)$. The horizontal edges with column number z have weight either W^2 or $W^3 + W^2$ plus or minus some lower-order terms. If we sum the weights of these horizontal edges, then the extra terms jW in rows less than z and the extra terms $(N-j)W$ for rows greater than z are canceled by the negative terms in the weight of the edge in row R_z. Thus the total weight of these edges is the same for every column number z.

Claim 6. *For every $\beta(1,1) \leq z \leq \beta(n,n)$, the total weight of all the horizontal edges with column number z is exactly $3W^3 + (N+1) \cdot W^2$.*

For every $1 \leq s \leq n$, we add the *upper ear edge* $\{u_s, u_{s+1}\}$ and the *lower ear edge* $\{d_s, d_{s+1}\}$, both having weight W^3.

The *cell* $C[i,j]$ is the face of the grid with the *corners* $g[i,j]$, $g[i+1,j]$, $g[i+1, j+1]$, $g[i, j+1]$ on its boundary. We mark each cell either as *normal* or *special* and add edges to the cell accordingly (we will call these new edges the *cell edges*). If the cell $C[i,j]$ is normal, then we add new (parallel) edges $\{g[i,j], g[i+1,j]\}$, $\{g[i+1,j], g[i+1, j+1]\}$, $\{g[i+1, j+1], g[i, j+1]\}$, $\{g[i, j+1], g[i,j]\}$, all with weight 2 (see Figure 4). If the cell $C[i,j]$ is special, then we add the edges $\{g[i,j], g[i+1,j]\}$ and $\{g[i, j+1], g[i+1, j+1]\}$ having weight 1, as well as the edges $\{g[i,j], g[i, j+1]\}$, $\{g[i+1,j], g[i, j+1]\}$ having weight 2.

The crucial properties of the cell edges are the following. If the two upper corners $g[i,j]$, $g[i, j+1]$ are separated from the two lower corners $g[i+1,j]$, $g[i+1, j+1]$, then the cell edges connecting these vertices have to be in the multiway cut. Observe that the total weight of these cell edges is exactly 4 both in a normal cell and in a special cell. Similarly, if the two corners $g[i,j]$, $g[i+1,j]$ on the left are separated from the two corners $g[i, j+1]$, $g[i+1, j+1]$ on the right, then the weight of the edges that need to be in the multiway cut is exactly 4 for both type of cells. However, there is a difference if we want to partition the four corners of the cell $C[i,j]$ into three components $\{g[i,j]\}$, $\{g[i+1,j]\}$, $\{g[i, j+1], g[i+1, j+1]\}$. For normal cells, the edges $\{g[i,j], g[i+1,j]\}$, $\{g[i+1,j], g[i+1, j+1]\}$, $\{g[i, j+1], g[i,j]\}$, having total weight 6, need to be in the multiway cut. On the other hand, for special cells, the edges $\{g[i,j], g[i+1,j]\}$,

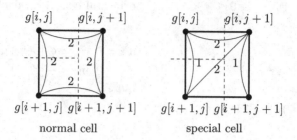

Fig. 4. The cell edges within a normal and a special cell. The strong edges are the 4 grid edges forming the boundary of the cell. The dashed lines show a 3-way partition of the corners that costs 6 in a normal cell and only 5 in a special cell.

$\{g[i,j], g[i,j+1]\}$, $\{g[i+1,j], g[i,j+1]\}$ that need to be in the multiway cut have total weight 5. Similarly, a multiway cut with partition $\{g[i,j], g[i+1,j]\}$, $\{g[i,j+1]\}$, $\{g[i+1,j+1]\}$ need to contain cell edges of total weight 6 in a normal cell, but weight 5 is sufficient in a special cell.

We complete the construction of the gadget by encoding the set S: for every $(x,y) \in S$, we mark the two cells $C[x, \beta(x,y)]$ and $C[\alpha(x), \beta(x,y)]$ special. Finally, we set $D := 7W^3 + (2N+2)W^2 + 4(2N-3) + 10 < 8W^3$.

3.2   Pair $(x,y) \Rightarrow$ Multiway Cut

Given a pair $(x,y) \in S$, we construct a multiway cut M representing (x,y) the following way. We partition the vertices of gadget the following way: vertex $g[i,j]$ is in the same class as

- UL if $i \leq \alpha(x)$ and $j \leq \beta(x,y)$,
- UR if $i \leq x$ and $j > \beta(x,y)$,
- DL if $i > \alpha(x)$ and $j \leq \beta(x,y)$,
- DR if $i > x$ and $j > \beta(x,y)$.

Observe that this partition indeed represents the pair (x,y). The multiway cut M contains all the edges that connect edges between two different classes:

1. ear edges $\{u_y, u_{y+1}\}$, $\{d_y d_{y+1}\}$,
2. $\{g[s, \beta(x,y)], g[s, \beta(x,y)+1]\}$ for every $0 \leq s \leq N$,
3. $\{g[\alpha(x), s], g[\alpha(x)+1, s]\}$ for every $0 \leq s \leq \beta(x,y)$,
4. $\{g[x, s], g[x+1, s]\}$ for every $\beta(x,y) < s \leq N$,
5. and some number of cell edges.

The total weight of the two edges in the first group is $2W^3$. The second group contains all the horizontal edges with column number $\beta(x,y)$, hence their total weight is exactly $3W^3 + (N+1) \cdot W^2$ by Claim 6. Groups 3 and 4 contain $N+1$ vertical edges in total, two of them having weight $W^3 + W^2$ and the rest having weight W^2. Note that none of these edges has weight ∞: it is not possible that $\alpha(x) = s - 1$ for some $s \leq \beta(x,y)$ (as $\alpha(x) \geq N - n - 1$ and $\beta(x,y) \leq N - n - 1$)

or that $x = s - 1$ for some $s > \beta(x,y)$ (as $x \leq n$ and $\beta(x,y) \geq n+1$). Thus the total weight of the edges in Groups 3 and 4 is $2W^3 + (N+1)W^2$. We can conclude that the total weight of the grid edges and ear edges in the multiway cut is $7W^3 + (2N+2)W^2$.

What remains to be shown is that the weight of the cell edges in M is at most $4(2N-3) + 10$ in the multiway cut. Let us analyze how the corners of the cell $C[i,j]$ are partitioned by the multiway cut.

1. Horizontal cut: $\{g[i,j], g[i,j+1]\}, \{g[i+1,j], g[i+1,j+1]\}$ if $i = \alpha(x)$ and $j < \beta(x,y)$ holds, or if $i = x$ and $j > \beta(x,y)$ holds.
2. Vertical cut: $\{g[i,j], g[i+1,j]\}, \{g[i,j+1], g[i+1,j+1]\}$ if $j = \beta(x,y)$ and $i \notin \{x, \alpha(x)\}$.
3. 3-way $\dashv$ cut: $\{g[i,j]\}, \{g[i+1,j]\}, \{g[i,j+1], g[i+1,j+1]\}$ if $i = \alpha(x)$ and $j = \beta(x,y)$.
4. 3-way $\vdash$ cut: $\{g[i,j], g[i+1,j]\}, \{g[i,j+1]\}, \{g[i+1,j+1]\}$ if $i = x$ and $j = \beta(x,y)$.
5. All corners are in the same class otherwise.

For each cell in the first two groups, the weight of the cell edges in the multiway cut is exactly 4 (regardless if the cell is normal or special). As $(x,y) \in S$, the cells $C[\alpha(x), \beta(x,y)]$, $C[x, \beta(x,y)]$ were marked as special. Therefore, each of the two cells in groups 3 and 4 contribute weight 5 to the multiway cut. It follows that the total weight of the cell edges in M is exactly $4(2N-3)+10$, as required. This proves Property 1 of Lemma 2.

3.3   Multiway Cut $\Rightarrow$ pair (x,y)

To prove Properties 2 and 3 of Lemma 2, consider a multiway cut M of weight at most D. We prove that M has to be of the form shown in Figure 3: it consists of a "vertical cut" with two "horizontal cuts" on its two sides. Moreover, the the two cells where these cuts meet should be special. Taking into account the way in which the special cells are located, we can conclude that the two horizontal cuts have the same "vertical position" and that the pair (x,y) is indeed in S.

Let us denote by K_{UL} the component of $G \setminus M$ containing the vertex UL (and we define K_{UR} etc. similarly). We call the path $UL = g[0,0], g[0,1], g[1,1], g[1,2], \ldots, g[N-1, N-1], g[N-1, N], DR = g[N,N]$ the *diagonal path*.

Claim 7. *Multiway cut M contains exactly one upper ear edge, exactly one lower ear edge, exactly one edge of the diagonal path, and exactly edge from each of C_0, C_N, R_0, and R_N.*

Proof. The vertex u_1 is in K_{UL} (as horizontal edges of weight ∞ connect it to UL) and u_{n+1} is in K_{UR}. This means that at least one of the upper ear edges is in M. Similarly, at least one of the lower ear edges have to be in M. It is also clear that M has to contain at least one edge from the diagonal path and each of C_0, C_N, R_0, and R_N, as each one of these 5 edge sets connects two distinct terminals. Every edge shared by these 5 sets has weight ∞ (note that edge $\{g[0,0], g[0,1]\}$

appears both on the diagonal path and R_0, while $\{g[N-1,N], g[N,N]\}$ appears both on the diagonal path and C_N). Therefore, M contains at least 5 edges from these 5 sets. As every ear edge and every edge in these 5 sets have weight either ∞ or at least W^3 and the weight of M is at most $D < 8W^3$, the multiway cut contains exactly one edge from each of these sets. □

By Claim 7, the multiway cut M contains

- $\{u_{y_1}, u_{y_1+1}\}$ for some $1 \le y_1 \le n$,
- $\{d_{y_2}, d_{y_2+1}\}$ for some $1 \le y_2 \le n$,
- $\{\ell_{x_1}, \ell_{x_1+1}\}$ for some $1 \le x_1 \le n$,
- $\{r_{x_2}, r_{x_2+1}\}$ for some $1 \le x_2 \le n$,
- $\{g[0, z_1], g[0, z_1 + 1]\}$ for some $\beta(1,1) \le z_1 \le \beta(n,n)$,
- $\{g[N, z_2], g[N, z_2 + 1]\}$ for some $\beta(1,1) \le z_2 \le \beta(n,n)$, and
- $\{g[z, z], g[z, z+1]\}$ for some $0 < z < N$.

As C_0 contains exactly one edge of the multiway cut, every vertex on C_0 is in $K_{UL} \cup K_{DL}$. We can argue similarly for the other three sides.

Claim 8. $V(C_0) \subseteq K_{UL} \cup K_{DL}$, $V(C_N) \subseteq K_{UR} \cup K_{DR}$, $V(R_0) \subseteq K_{UL} \cup K_{UR}$, and $V(R_N) \subseteq K_{DL} \cup K_{DR}$.

Observe that every horizontal grid edge on the diagonal path has weight at least $W^3 + W^2/2$. Therefore, the 7 edges given by Claim 7 (4 on the boundary of the grid, 1 on the diagonal path, and 2 ear edges) have total weight at least $7W^3 + 4W^2 + W^2/2$. This implies that the remaining edges have total weight less than $(2N - 2)W^2$, otherwise the weight would be at least $7W^3 + (2N + 2)W^2 + W^2/2 > D$. In particular, M can contain at most $2N - 3$ further grid edges, that is, the total number of grid edges in M is at most $2N + 2$.

Claim 9. M contains exactly one edge from each of R_i and C_i $(0 \le i, j \le N)$.

Proof. As $V(C_0) \subseteq K_{UL} \cup K_{DL}$ and $V(C_N) \subseteq K_{UR} \cup K_{DR}$ (by Claim 8), the multiway cut M has to contain at least one edge of R_i for every $1 \le i \le N-1$. In a similar way, M contains at least one edge C_j for every $1 \le j \le N-1$. As M contains at most $2N + 2$ grid edges, it immediately follows that M contains exactly one edge of each row and column. □

Observe that every vertical grid edge inside the grid has weight exactly W^2 if its weight is finite. Therefore, we know the exact weight of the vertical grid edges in M and hence can bound the total weight of the horizontal grid edges.

Claim 10. *The total weight of vertical grid edges in M is exactly $2W^3 + (N + 1)W^2$ and therefore the total weight of horizontal grid edges in M is at most $3W^3 + (N + 1)W^2 + 4(2N - 3) + 10 < 3W^3 + (N + 1)W^2 + W$.*

Claim 11. *M contains $\{g[i, z], g[i, z + 1]\}$ for every $0 \le i \le N$ and the total weight of the horizontal grid edges in M is exactly $3W^3 + (N + 1)W^2$.*

Proof. Since $\{g[z, z], g[z, z+1]\}$ is the unique edge of the multiway cut that is on the diagonal path from UL to DR, we have $g[z, z] \in K_{UL}$ and $g[z, z+1] \in K_{DR}$. Therefore, the multiway cut has to contain an edge of the vertical path from $g[z, z] \in K_{UL}$ to $g[N, z] \in R_N \subseteq K_{DL} \cup K_{DR}$ (by Claim 8). Since we know that the multiway cut M contains exactly one edge of C_z (Claim 9), it follows that M does not contain any edge of the vertical path from $g[z, z]$ to $g[0, z]$, that is, every vertex on this path is in K_{UL}. The multiway cut has to separate the vertices on this vertical path from the vertices of $C_N \subseteq K_{UR} \cup K_{DR}$ (Claim 8), thus for every $0 \leq i < z$ the unique edge in $M \cap R_i$ has column number at least z. Therefore, the total weight of these z edges is at least $W^3 + zW^2 + z^2W$, with equality only if every edge has column number exactly z (and $\beta(1, 1) \leq z \leq \beta(n, n)$ to ensure that $\{g[0, z], g[0, z+1]\}$ has finite weight). A similar argument shows that every edge in $M \cap R_i$ for $i > z$ has to have column number at most z, and hence the total weight of these $N - z$ edges is at least $W^3 + (N-z)W^2 + (N-z)^2W$ with equality only if all these edges have the same column number $\beta(1, 1) \leq z \leq \beta(n, n)$. Taking into account also the edge $\{g[z, z], g[z, z+1]\}$, we get that the total weight of the horizontal grid edges is at least $3W^3 + (N+1) \cdot W^2$, with equality only if all of them have column number z. Furthermore, as the weight of every grid edge is a multiple of W, if not every horizontal edge has column number z, then the weight is at least $3W^2 + (N+1)W^2 + W$, contradicting Claim 10. Thus every horizontal edge has the same column number $\beta(1, 1) \leq z \leq \beta(n, n)$. □

For $i = 0$ and $i = N$, Claim 11 implies $z = z_1 = z_2$ and hence $\beta(1, 1) \leq z \leq \beta(n, n)$ (to avoid the selection of edges with weight ∞ on the boundary).

Claim 12. *The unique edge of $M \cap C_j$ is $\{g[x_1, j], g[x_1 + 1, j]\}$ if $j \leq z$ and $\{g[x_2, j], g[x_2 + 1, j]\}$ if $j > z$.*

Proof. Observe that for every $i \leq x_1$ and $j \leq z$, vertex $g[i, j]$ is in K_{UL}: the multiway cut does not contain any of the edges on the vertical path from UL to $g[i, 0]$ and on the horizontal path from $g[i, 0]$ to $g[i, j]$. Similarly, vertex $g[i, j]$ is in K_{DL} if $i > x_1$ and $j \leq z$. These two statements together imply that the edge $\{g[x_1, j], g[x_1 + 1, j]\}$ has to be in the multiway cut for every $0 \leq j \leq z$. The argument for $j > z$ is analogous. □

By Claim 12, the $2N + 2$ grid edges in the multiway cut are arranged as in Figure 3. It follows that the multiway cut contains cell edges from exactly $2N - 1$ cells. As the weight of M is at most D, the total weight of these cell edges is at most $4(2N - 3) + 10$.

Claim 13. *The cells edges in M have total weight exactly $4(2N - 3) + 10$ and the cells $C[x_1, z]$ and $C[x_2, z]$ are special.*

Proof. For every $0 \leq j < z$, the two upper corners of the cell $C[x_1, j]$ are separated from the two lower corners. Therefore, the weight of the cell edges from $C[x_1, j]$ in the multiway cut is 4 (no matter whether the cell is special or not). The same is true for every cell $C[x_2, j]$ with $j > z$. For the cells $C[i, z]$ with $i \notin \{x_1, x_2\}$, the two corners on the left are separated from the two corners on the right, which again means that the weight contributed by the cell edges of $C[i, z]$ is 4. Thus the weight contributed by the cell edges of these $2N - 3$ cells is at least $4(2N - 3)$. As the total weight of the cell edges is at most $4(2N - 3) + 10$, the contribution of the two cells $C[x_1, z]$ and $C[x_2, z]$ is at most 10. The corners of $C[x_1, z]$ are partitioned as $\{g[x_1, z]\}$, $\{g[x_1 + 1, z]\}$, $\{g[x_1, z + 1], g[x_1 + 1, z + 1]\}$ by the multiway cut, while the corners of $C[x_2, z]$ are partitioned as $\{g[x_2, z], g[x_2 + 1, z]\}$, $\{g[x_2, z+1]\}$, $\{g[x_1 + 1, z+1]\}$. In both cases, the weight of the edges contained in the multiway cut is 5 if the cell is special and 6 if it is normal. Therefore, both of these cells have to be special.   □

Suppose that $z = \beta(x, y)$ for some $1 \leq x, y \leq n$ (recall that $\beta(1, 1) \leq z \leq \beta(n, n)$). There are at most two cells with column number z that are special. By construction, $C[x_1, z]$ and $C[x_2, z]$ are special only if $(x, y) \in S$ and we have $x_1 = \alpha(x)$, $x_2 = x$. That is, the multiway cut M contains the edge $\{g[\alpha(x), 0], g[\alpha(x) + 1, 0]\} = \{\ell_x, \ell_x + 1\}$ and the edge $\{g[x, N], g[x + 1, N]\} = \{r_x, r_x + 1\}$. Finally, as the multiway cut contains $\{g[0, z], g[0, z + 1]\}$ and $\{g[N, z], g[N, z + 1]\}$, the two ear edges contained in the multiway cut should be $\{u_y, u_y + 1\}$ and $\{d_y, d_y + 1\}$. This proves that the multiway cut represents the pair $(x, y) \in S$, what we had to show. Claim 13 also shows that there is no multiway cut with weight $< D$.

References

1. Chen, J., Huang, X., Kanj, I.A., Xia, G.: Strong computational lower bounds via parameterized complexity. Journal of Computer and System Sciences 72(8), 1346–1367 (2006)
2. Dahlhaus, E., Johnson, D.S., Papadimitriou, C.H., Seymour, P.D., Yannakakis, M.: The complexity of multiterminal cuts. SIAM J. Comput. 23(4), 864–894 (1994)
3. Downey, R.G., Fellows, M.R.: Parameterized Complexity. Monographs in Computer Science. Springer, New York (1999)
4. Hartvigsen, D.: The planar multiterminal cut problem. Discrete Applied Mathematics 85(3), 203–222 (1998)
5. Impagliazzo, R., Paturi, R., Zane, F.: Which problems have strongly exponential complexity? J. Comput. System Sci. 63(4), 512–530 (2001)
6. Klein, P.N., Marx, D.: Solving planar k-terminal cut in time $O(n^{c\sqrt{k}})$. In: Czumaj, A., et al. (eds.) ICALP 2012, Part I. LNCS, vol. 7391, pp. 569–580. Springer, Heidelberg (2012)
7. Lokshtanov, D., Marx, D., Saurabh, S.: Lower bounds based on the Exponential Time Hypothesis. Bulletin of the EATCS 84, 41–71 (2011)
8. Marx, D.: On the optimality of planar and geometric approximation schemes. In: FOCS 2007, pp. 338–348 (2007)

The Power of Recourse for Online MST and TSP

Nicole Megow[1], Martin Skutella[1,*], José Verschae[2,**], and Andreas Wiese[3,***]

[1] Department of Mathematics, Technische Universität Berlin, Germany
{nmegow,skutella}@math.tu-berlin.de
[2] Departamento de Ingeniería Industrial, Universidad de Chile, Chile
jverscha@ing.uchile.cl
[3] Department of Computer and System Sciences, Sapienza University of Rome, Italy
wiese@dis.uniroma1.it

Abstract. We consider the online MST and TSP problems with recourse. The nodes of an unknown graph with metric edge cost appear one by one and must be connected in such a way that the resulting tree or tour has low cost. In the standard online setting, with irrevocable decisions, no algorithm can guarantee a constant competitive ratio. In our model we allow recourse actions by giving a limited budget of edge rearrangements per iteration. It has been an open question for more than 20 years if an online algorithm equipped with a constant (amortized) budget can guarantee constant-approximate solutions [7].

As our main result, we answer this question affirmatively in an amortized setting. We introduce an algorithm that maintains a nearly optimal tree when given constant amortized budget. In the non-amortized setting, we specify a promising proof technique and conjecture a structural property of optimal solutions that would prove a constant competitive ratio with a single recourse action. It might seem rather optimistic that such a small budget should be sufficient for a significant cost improvement. However, we can prove such power of recourse in the offline setting in which the sequence of node arrivals is known. Even this problem prohibits constant approximations if there is no recourse allowed. Surprisingly, already a smallest recourse budget significantly improves the performance guarantee from non-constant to constant.

Unlike in classical TSP variants, the standard double-tree and short-cutting approach does not give constant guarantees in the online setting. However, a non-trivial robust shortcutting technique allows to translate online MST results into TSP results at the loss of small factors.

1 Introduction

In the *online Minimum Spanning Tree* (MST) problem and *online Traveling Salesman Problem* (TSP) we aim at constructing low-cost spanning trees, resp.

---

* Supported by the DFG Research center MATHEON in Berlin.
** Supported by the Berlin Mathematical School (BMS) and Nucleo Milenio Información y Coordinación en Redes ICM/FIC P10-024F.
*** Supported by the German Academic Exchange Service (DAAD).

A. Czumaj et al. (Eds.): ICALP 2012, Part I, LNCS 7391, pp. 689–700, 2012.
© Springer-Verlag Berlin Heidelberg 2012

tours, for an unknown graph that is revealed online. In each iteration a new node becomes known, together with all connections to previously revealed nodes, and an algorithm must make an irrevocable decision on how to connect it. When a tree must be maintained, then the decision concerns inserting one edge, whereas for maintaining a tour, an edge must be removed and two new ones inserted. Such problems appear naturally in applications related to multicast routing in multimedia distribution systems, video-conferencing, software delivery, and groupware [12,13]. They have been studied extensively, in particular online MST and Steiner tree variants. Here, constant competitive ratios are not achievable; the best possible performance ratio is $\Theta(\log t)$, where t is the number of iterations [7].

However, in many of the above-mentioned applications it is possible to adapt solutions in some limited way when the node set changes [14,16]. Clearly, such *recourse actions* allow for improved solutions. However, an increasing number of adaptations—in particular, a complete reconstruction of solutions—might not be feasible or may cause unacceptable additional cost, and is therefore not desirable. Our goal is to understand the tradeoff between the amount of adaptivity and the quality of solutions. As a main problem, we want to determine the amount of recourse that is necessary to allow for provably near-optimal solutions. Looking from a different perspective, we construct solutions that satisfy some adequate concept of *robustness*, where we measure robustness by the (amortized) *recourse budget* that is necessary to guarantee solutions of a particular quality.

More precisely, we consider the *online MST* and *online TSP with recourse*: An undirected complete metric graph is revealed online node by node. In each iteration a new node becomes known and all edges (with corresponding costs) to previously arrived nodes are revealed. The objective is to construct in each iteration a low-cost spanning tree, resp. tour, of the revealed vertices, without any assumption on the vertices that might arrive in future. We measure the quality of the solution sequence with standard competitive analysis by comparing online solutions to the offline optimum, i.e., the MST, resp. TSP tour, on the currently known subgraph. We control the amount of recourse, i.e., how much the solution changes along iterations, by limiting the number *rearrangements*. More precisely, we give a budget for the number of edges that can be inserted in each iteration. We say that an algorithm needs budget k if the number of inserted edges in each iteration is bounded by k. Similarly, the algorithm uses an *amortized budget* k if up to iteration t the total number of inserted edges is at most $t \cdot k$. Notice that by this definition, the standard online MST problem equips algorithms with a budget of 1 whereas the online TSP without recourse requires a minimum budget of 2.

It has been a longstanding open question if constant budget suffices to maintain constant-approximate solutions [7,2]. In this paper we not only answer this question affirmatively, but we also show the surprising fact that already a smallest amount of recourse suffices to significantly improve the solution.

Related Work. The online MST and Steiner Tree problems have been studied intensively. The best possible competitive ratio for online algorithms is known to be $\Theta(\log t)$, where t is the number of iterations [7]. A simple greedy algorithm that connects a new node to the current tree through a shortest edge achieves this bound. Even in the special case of Euclidean distances, there is a lower bound of $\Omega(\frac{\log t}{\log\log t})$ on the competitive guarantee of any online algorithm [1].

Unlike in stochastic programming [3], where recourse actions are an important concept when optimizing under limited information, the literature on recourse models for online optimization seems rather sparse. Regarding our model, we are aware only of the work by Imase and Waxman [7] that deals with the online Minimum Steiner Tree problem with recourse (or *Dynamic Steiner Tree*). The model they introduce is slightly more general as nodes do not only arrive but may also *depart* from the terminal set. For this setting, they give an algorithm that is 8-competitive and performs in t iterations at most $O(t^{3/2})$ rearrangements. This translates by our definitions to an algorithm that requires an amortized recourse budget of $O(t^{1/2})$. In the more restricted setting considered in this paper, with no node leaving the terminal set, their algorithm achieves a competitive guarantee of 4. Furthermore, for the online MST problem, their algorithm is even 2-competitive when given the mentioned non-constant amortized budget. The question if there is a constant-competitive algorithm that uses only constant (amortized) budget has been left open, but was conjectured to be true.

Interestingly, the maximization variant of our problem, i.e., finding a sequence of spanning trees of *maximum* cost, does not admit constant-competitive algorithms with a low budget [15]. However, this changes when comparing the solution to an optimal solution under the same limited recourse budget, instead of optimal MSTs. For this case, there is a 2-competitive algorithm, even in the general context of matroids and a constant competitive ratio for the intersection of a constant number of matroids [15].

A related online MST variant has been studied in [6]. Here, in each iteration the cost of some edge increases or decreases by one. The task is to maintain a sequence of *optimal* MSTs with the goal to minimize the number of rearrangements. They give a best possible deterministic algorithm that is $O(t^2)$-competitive and a randomized algorithm with expected competitive ratio $O(t \cdot \log t)$.

The TSP is one of the most prominent problems in combinatorial optimization.Despite a remarkable recent progress, the best known approximation algorithm for the offline metric TSP is still the classic 3/2-approximation by Christofides [4]. In the online setting, when the nodes appear over time as the salesman traverses its tour, constant-competitive algorithms are known [8]. In a different online TSP model, related to graph exploration, where nodes are revealed when the salesman moves to one of its neighbors, constant-competitive algorithms are known even without any assumption on the cost function [9,11]. Both online models clearly differ from our setting.

Our Contribution.[1] Our main contribution, presented in Sect. 3, is an online algorithm for the online MST problem with recourse that is $(1 + \varepsilon)$-competitive, for any $\varepsilon > 0$, when given an amortized budget of $O(\frac{1}{\varepsilon} \log \frac{1}{\varepsilon})$. This is the first and significant improvement on a 2-competitive algorithm with non-constant budget

---

[1] Due to space limitations for many proofs and details we refer to the full version of this paper (see also [15]).

by Imase and Waxman [7]. We complement our result by showing that any $(1+\varepsilon)$-competitive algorithm for the online MST problem needs an amortized budget of $\Omega(\frac{1}{\varepsilon})$. Thus, our algorithm is best possible up to logarithmic factors. Using a standard argument, we immediately obtain a $(2 + \varepsilon)$-competitive algorithm for the online Steiner Tree problem with the same amortized budget.

Our algorithm is simple and easy to implement, but it captures subtleties in the structure of the problem that allow the improved analysis. Similarly to the algorithm proposed in [7], we implement the following natural idea: when a new node appears, we (i) connect it to its closest neighbor, and (ii) iteratively perform edge swaps if they yield a sufficient improvement of the solution. The key difficulty when implementing this idea is to balance the number of swaps and the cost of the solution. As our crucial refinement of this approach, we introduce two *freezing rules* that effectively avoid performing unnecessary swaps. The first rule prevents from removing edges whose cost is very small. The second rule is more subtle and prohibits an edge swap if the removed edge can be traced back to a subgraph whose MST has negligible cost compared to the current MST.

Our results imply that amortized budget is much more powerful than its non-amortized counterpart. Indeed, we contrast our findings for constant amortized budget with a simple example showing that no online algorithm can be $(2 - \varepsilon)$-competitive, for any $\varepsilon > 0$, if it uses non-amortized constant budget. It is, however, an important and longstanding open question [2,7] whether there exists a constant-competitive algorithm with constant budget. In Sect. 4, we contribute towards an affirmative answer of this question as follows: We consider a simple online greedy algorithm with budget 2 that is similar to one given in [7]. We state a structural condition on the behavior of optimal solutions that guarantees this algorithm to be constant competitive. We conjecture that this condition is satisfied for every input sequence. We believe that this conjecture is an important step towards understanding the problem structure and will foster further research.

It might seem rather optimistic that a single extra rearrangement should be sufficient to yield a factor-$(\log n)$ improvement in the competitive ratio. To support our conjecture, we study the problem in the case of full information, i. e., the offline variant in which the input sequence and the cost function are known in advance. Even in this case, any algorithm with unit budget has a competitive factor in $\Omega(\log t)$. However, allowing just one additional edge insertion leads to a significant improvement. We give a polynomial time 14-competitive algorithm with budget 2. Furthermore, we show how to obtain a constant competitive ratio with "almost" unit budget, i. e., we allow one additional swap every k iterations, for some constant k. These two offline results prove that only a smallest relaxation of the unit-budget restriction may lead to significantly better solutions.

A very natural approach to solve the online TSP with recourse is to combine the algorithms proposed for the online MST problem with the folkloric classical double-tree and shortcutting technique [10]. Indeed, with this technique most offline variants of TSP are equivalent to MST from an approximation point of view and performance guarantees differ only in a factor of 2. Hence, one might be tempted to assume that the same conversion technique applies directly to the online model with recourse. However, we observe that this is not true. We give examples in which two trees differ in just a single edge, but the standard short-cutting technique leads to completely different tours, no matter which Eulerian walk on the doubled tree edges is chosen. In Sect. 5 we overcome this difficulty by introducing a *robust* variant of the shortcutting technique: we choose the Eulerian tour in a specific way and keep track of which copy of a node in the Eulerian tour is visited by the TSP tour. With this robust shortcutting technique we show that any algorithm for the online MST problem with recourse can be converted to an algorithm for the online TSP by increasing the competitive ratio by a factor 2 and the budget by a factor 4.

2   Problem Definitions

An instance of the online MST problem with recourse is defined as follows. A sequence of nodes $v_0, v_1, \ldots$ arrives online one by one. In iteration $t \geq 0$, node v_t appears together with all edges $v_t v_s$ for $s \in \{0, \ldots, t-1\}$. The cost $c(e) \geq 0$ of an edge e is revealed when the edge appears. We assume that the edges are undirected and that the costs satisfy the triangular inequality, that is, $c(vw) \leq c(vz) + c(zw)$ for all nodes v, w, z. For each iteration t, the current graph is denoted by $G_t = (V_t, E_t)$ where $V_t = \{v_0, \ldots, v_t\}$ and $E_t = V_t \times V_t$, that is, G_t is a complete graph. We are interested in constructing an online sequence $T_0, T_1, T_2, \ldots$ where $T_0 = \emptyset$ and for each $t \geq 1$ tree T_t is a spanning tree of G_t. We say that the sequence needs budget k if $|T_t \setminus T_{t-1}| \leq k$ for all $t \geq 1$. A re-laxed version of this concept is obtained by considering the average or *amortized* budget k, $\sum_{s=1}^{t} |T_s \setminus T_{s-1}| \leq k \cdot t$.

In online TSP with recourse, the nodes of a complete metric graph arrive in the same online fashion as described above and yield a sequence of graphs $G_0, \ldots, G_t, \ldots$. The objective, now, is to construct a sequence of TSP tours $Q_2, Q_3, \ldots$ for graphs $G_2, G_3, \ldots$ with minimum cost for each tour. We apply the same budget constraints as for trees.

We measure the performance of our online algorithms using classic competitive analysis. Let OPT_t be the cost of an MST, resp. TSP tour, of G_t, and for a given set of edges E denote $c(E) := \sum_{e \in E} c(e)$. We say that an algorithm is α-competitive for some $\alpha \geq 1$, if for any input sequence the algorithm computes a solution sequence $X_0, X_1, \ldots$ such that $c(X_t) \leq \alpha \cdot \text{OPT}_t$ for each t.

3   An Online PTAS with Amortized Constant Budget

In this section we give a $(1+\varepsilon)$-competitive algorithm for the online MST problem with constant amortized budget for any $\varepsilon > 0$. This improves on a previous 2-competitive algorithm that requires non-constant amortized budget [7]. We also show that our budget bound is best possible up to logarithmic factors.

A natural approach to solve our problem is as follows. Let T_{t-1} be the tree solution in iteration $t-1$. To construct T_t, we first find the closest connection between the new node v_t and V_{t-1}, edge g_t, and initialize T_t as $T_{t-1} \cup \{g_t\}$. We can diminish the cost of T_t by subsequently inserting a low cost edge f to T_t and removing the largest edge h in the formed cycle. Indeed, performing this swapping operation often enough will eventually turn T_t into the optimal solution, i.e., the MST. The difficulty lies in balancing the number of swaps that increases the budget, on the one hand, and the closeness of the tree to the MST, and thus, the total cost, on the other hand. We cope with this challenge by introducing two *freezing rules* that effectively avoid performing unnecessary swaps.

The intuition behind these freezing rules is as follows. Note that if at iteration t the optimal value OPT_t is much higher than OPT_s for some $s < t$, e.g., $OPT_s \in O(\varepsilon OPT_t)$ then the edges in T_s—whose total cost is approximately OPT_s—are already very cheap. Thus, replacing these edges by cheaper ones would only waste rearrangements. To avoid technical difficulties we use $OPT_t^{\max} := \max\{OPT_s : 1 \le s \le t\}$ instead of OPT_t to determine whether $OPT_s \in O(\varepsilon OPT_t)$; note that since we assume the triangle inequality for the costs of the edges, $OPT_t \le OPT_t^{\max} \le 2OPT_t$ holds.

With this in mind, we define $\ell(t)$ as the largest iteration with ignorable edges with respect to $OPT_t^{\max}$, i.e., $\ell(t) \le t - 1$ is the largest non-negative integer such that $OPT_{\ell(t)}^{\max} \le \varepsilon OPT_t^{\max}$.

For our first freezing rule we consider sequences of edges $(g_s^0, \dots, g_s^{i(s)})$, where g_s^0 corresponds to the greedy edge added at iteration s (that is, an edge connecting v_s to one of its closest neighbors in V_{s-1}). At the moment when edge g_s^0 is removed from our solution we define g_s^1 as the element that replaces g_s^0. In general g_s^i is the edge that was swapped in for edge g_s^{i-1}. In this way, the only edge in the sequence that belongs to the current solution is $g_s^{i(s)}$. Note that $i(s)$ changes through the iterations. Notationally, $i(s)$ will refer to the value at the iteration in consideration in the current context (unless it is stated otherwise).

With this construction, we *freeze* a sequence $(g_s^0, \dots, g_s^{i(s)})$ in iteration t if $s \le \ell(t)$. Note that since $\ell(\cdot)$ is non-decreasing, once the sequence is frozen $g_s^{i(s)}$ will stay indefinitely in the solution. Our second freezing rule is somewhat simpler. We skip swaps that remove edges that are too small, namely, smaller than $\varepsilon OPT_t^{\max}/(t - \ell(t))$. Combining these ideas we propose the following algorithm.

Algorithm Sequence-Freeze

Define $T_0 = \emptyset$. For each iteration $t \ge 1$ do as follows.

1. Let g_t^0 be any minimum cost edge in $\{v_t v_s : 0 \le s \le t - 1\}$.
2. Initialize $T_t := T_{t-1} \cup \{g_t^0\}$ and $i(t) := 0$.
3. While there exists a pair of edges $(f, h) \in (E_t \setminus T_t) \times T_t$ such that $(T_t \cup \{f\}) \setminus h$ is a tree, and the following three conditions are satisfied
   (C1) $c(h) > (1 + \varepsilon) \cdot c(f)$,
   (C2) $h = g_s^{i(s)}$ for some $s \ge \ell(t) + 1$, and
   (C3) $c(h) > \varepsilon \frac{OPT_t^{\max}}{t - \ell(t)}$,
   then set $T_t := (T_t \cup \{f\}) \setminus \{h\}$, $i(s) := i(s) + 1$ and $g_s^{i(s)} := f$.
4. Return T_t.

Conditions (C2) and (C3) correspond to the two freezing rules described above. In the following we show that this algorithm is $(1 + \varepsilon)$-competitive and uses amortized budget $O\left(\frac{1}{\varepsilon} \log \frac{1}{\varepsilon}\right)$.

Competitive Analysis. To prove that our algorithm is $(1 + O(\varepsilon))$-competitive, we first show that Conditions (C1) and (C3) imply a cost increase of at most a factor $(1 + 3\varepsilon)$. Then we show that skipping swaps because of Condition (C2) can increase the cost of the solution by at most $O(\varepsilon \mathrm{OPT}_t)$.

Consider an iteration t and let $\ell := \ell(t)$. We partition the tree T_t into two disjoint subsets, $T_t = T_t^{\mathrm{old}} \cup T_t^{\mathrm{new}}$ where $T_t^{\mathrm{old}} := \{g_1^{i(1)}, \ldots, g_\ell^{i(\ell)}\}$ and $T_t^{\mathrm{new}} := \{g_{\ell+1}^{i(\ell+1)}, \ldots, g_t^{i(t)}\}$.

Lemma 1. *For each iteration t it holds that* $c(T_t^{\mathrm{new}}) \leq (1 + 3\varepsilon)\mathrm{OPT}_t$.

For bounding the cost of T_t^{old}, we use induction over the iterations. The inductive step is given in the following lemma.

Lemma 2. *Let $\varepsilon < \frac{1}{7}$. Consider an iteration t and suppose that $c(T_{\ell(t)}) \leq (1 + 7\varepsilon)\mathrm{OPT}_{\ell(t)}$. Then it holds that $c(T_t^{\mathrm{old}}) \leq 4\varepsilon \, \mathrm{OPT}_t$.*

The above reasoning implies the following lemma.

Lemma 3. *Algorithm* SEQUENCE-FREEZE *is $(1 + 7\varepsilon)$-competitive for any $\varepsilon < \frac{1}{7}$.*

Amortized Budget Bound. To show the constant amortized budget bound, we define $k_q := |T_q \setminus T_{q-1}|$ and prove that for every $t \geq 1$ it holds that $\sum_{q=1}^{t} k_q \leq D_\varepsilon \cdot t$, where $D_\varepsilon \in O\left(\frac{1}{\varepsilon} \log \frac{1}{\varepsilon}\right)$.

Lemma 4. *Assume that $\sum_{q=\ell(t)+1}^{t} k_q \leq C_\varepsilon \cdot (t - \ell(\ell(t) + 1))$ for every $t \geq 1$ with $C_\varepsilon \in O\left(\frac{1}{\varepsilon} \log \frac{1}{\varepsilon}\right)$. Then for every $t \geq 1$ it holds that $\sum_{q=1}^{t} k_q \leq 2C_\varepsilon \cdot t$.*

It remains to prove that the assumption of Lemma 4 holds. The two freezing rules, Conditions (C2) and (C3), are crucial for this purpose. Indeed, we will bound the length of the sequences $(g_s^0, \ldots, g_s^{i(s)})$, which will give a direct bound on $\sum_{q=\ell(t)+1}^{t} k_q \leq C_\varepsilon \cdot (t - \ell(\ell(t) + 1))$. This can be done since by Condition (C1), we only swap edges when the cost is decreased by a $(1 + \varepsilon)$ factor, that is, $c(g_s^j) \leq c(g_s^{j-1})/(1 + \varepsilon)$ for each j. Thus, the length of this sequence is bounded by $\log_{1+\varepsilon} c(g_s^0) - \log_{1+\varepsilon} c(g_s^{i(s)-1}) + 1$. We can bound this quantity further by lower bounding the cost $g_s^{i(s)-1}$ with our freezing rules and by exploiting a particular cost structure of greedy edges g_s^0.

More precisely, consider the values $i(s)$ at the end of iteration t, and let $i'(s)$ be the value of $i(s)$ at the beginning of iteration $\ell(t) + 1$ (and $i'(s) := 0$ for $s \geq \ell(t) + 1$). By Condition (C2) in the algorithm, in iterations $\ell(t) + 1$ to t we only touch edges belonging to $\{g_s^{i'(s)}, g_s^{i'(s)+1}, \ldots, g_s^{i(s)}\}$ for some $s \in \{\ell(\ell(t) + 1) + 1, \ldots, t\}$. Let us denote $r := \ell(\ell(t) + 1)$. Then,

$$\sum_{q=\ell(t)+1}^{t} k_q \leq \sum_{s=r+1}^{t} (i(s) - i'(s) + 1) = 2(t - r) + \sum_{s=r+1}^{t} (i(s) - 1 - i'(s)). \quad (1)$$

We now upper bound each term $i(s) - 1 - i'(s)$ for $s \in \{r+1, \ldots, t\}$, which corresponds to the length of the sequence $(g_s^{i'(s)+1}, g_s^{i'(s)+2}, \ldots, g_s^{i(s)-1})$.

Lemma 5. *For each $s \in \{r, r+1 \ldots, t\}$ it holds that*

$$i(s) - 1 - i'(s) \leq \frac{1}{\ln(1+\varepsilon)} \cdot \left(\ln c(g_s^0) - \ln c(g_s^{i(s)-1}) \right). \tag{2}$$

Proof. The lemma follows due to Condition (C1), since whenever we add an edge g_s^j and remove g_s^{j-1}, then $c(g_s^j) < c(g_s^{j-1})/(1+\varepsilon)$. □

In the next claims, we lower bound $c(g_s^{i(s)-1})$ and upper bound $\sum_{s=r+1}^t \ln c(g_s^0)$. These bounds applied to Equation (2) will lead with (1) to the desired bound on $\sum_{q=\ell(t)+1}^t k_q$.

Proposition 1. *Due to Condition (C3), it holds that either $c(g_s^{i(s)-1}) \geq \varepsilon^2 \frac{\mathrm{OPT}_t^{\max}}{(t-r)}$ or $i(s) - 1 - i'(s) \leq 0$.*

Recall that for any s, g_s^0 is a closest connection between v_s and any element in $\{v_0, \ldots, v_{s-1}\}$. Such greedy edges are known to have a special cost structure [1].

Lemma 6 (Alon and Azar [1]). *Let $e_1, \ldots, e_t$ be greedy edges reindexed such that $c(e_1) \geq c(e_2) \geq \ldots \geq c(e_t)$. Then, $c(e_j) \leq \frac{2\mathrm{OPT}_t}{j}$ for all $j \in \{1, \ldots, t\}$.*

Lemma 7. $\sum_{s=r+1}^t \ln c(g_s^0) \leq (t-r) \cdot (\ln(2 \cdot \mathrm{OPT}_t^{\max}) - \ln(t-r) + 1)$.

Proof. We rename edges $\{g_{r+1}^0, \ldots, g_t^0\} = \{e_1, \ldots, e_{t-r}\}$ such that $c(e_1) \geq \ldots \geq c(e_{t-r})$. Lemma 6 implies that $c(e_j) \leq 2\frac{\mathrm{OPT}_t}{j} \leq 2\frac{\mathrm{OPT}_t^{\max}}{j}$ for all j. We conclude

$$\sum_{s=r+1}^t \ln c(g_s^0) = \sum_{j=1}^{t-r} \ln c(e_j) \leq (t-r)\ln(2 \cdot \mathrm{OPT}_t^{\max}) - \sum_{j=1}^{t-r} \ln j.$$

The lemma follows since for any $n \in \mathbb{N}_{>0}$ it holds that $\sum_{j=1}^n \ln j \geq n \ln(n) - n$. □

The above statement and basic arithmetics imply the desired bound.

Lemma 8. *For each $t \geq 1$ it holds that $\sum_{q=1}^t k_q \leq D_\varepsilon \cdot t$, where $D_\varepsilon \in O\left(\frac{1}{\varepsilon} \log \frac{1}{\varepsilon}\right)$.*

Our main result follows from Lemmas 3 and 8.

Theorem 2. *There exists a $(1+\varepsilon)$-competitive algorithm for the online MST problem with amortized recourse budget $O\left(\frac{1}{\varepsilon} \log \frac{1}{\varepsilon}\right)$.*

Finally, we show that the amortized budget of our algorithm is best possible up to logarithmic factors. This result also implies that 1-competitive solutions need non-constant amortized budget.

Theorem 3. *Any $(1+\varepsilon)$-competitive algorithm for the online MST problem requires an amortized recourse budget of $\Omega(\frac{1}{\varepsilon})$.*

4   The Non-amortized Scenario

For the amortized setting we have seen that with sufficient (but constant) budget we can obtain a competitive ratio of $1 + \epsilon$. In the non-amortized setting however, there can be no $(2 - \epsilon)$-competitive algorithm with constant budget. Nevertheless, it is a long standing open question [7] whether in the non-amortized setting one can obtain constant-competitive online algorithms with constant budget. We contribute two pieces of evidence for an affirmative answer to this question: we state a conjecture about a structural property of optimal solutions and show that a natural greedy algorithm with budget 2 is constant-competitive if the conjecture holds true. Furthermore, we show that in the full information scenario even one extra recourse action every $O(1)$ iterations is enough to obtain a constant-competitive algorithm.

4.1   A Greedy Algorithm with Budget 2

A natural online algorithm with budget 2 works as follows: In each iteration t find the shortest connection g_t from v_t to V_{t-1} and find a pair of edges f_t, h_t maximizing $c(h_t) - c(f_t)$ among all edges such that $T_t := (T_{t-1} \cup \{g_t, f_t\}) \setminus \{h_t\}$ is a spanning tree for V_t. Then T_t is the tree for iteration t. Imase and Waxman [7] suggested to require that $c(f_t) \leq c(h_t)/2$ and f_t is adjacent to v_t. If no such edges f_t, h_t exist they define $T_t := T_{t-1} \cup \{g_t\}$. With this modification, we prove a relation between the competitive factor of this algorithm and a structural property of optimal solutions that we conjecture to be true.

To state our conjecture we need the following definition: given a complete graph $G = (V, E)$ and a non-negative cost function c on the edges, we say that the graph is 2-*metric* if for every cycle $C \subseteq E$ it holds that $c(e) \leq 2 \cdot c(C \setminus \{e\})$ for all $e \in C$. Moreover, for a given real number x we define $x_+ := \max\{x, 0\}$ and $x_- := \max\{-x, 0\}$. Note that $x = x_+ - x_-$. Also, denote $\Delta\mathrm{OPT}_t := \mathrm{OPT}_t - \mathrm{OPT}_{t-1}$.

Conjecture 4. *There exists a constant $\alpha \geq 1$ satisfying the following. Consider any input sequence $G_0, G_1, \ldots, G_n$ of the online MST problem with recourse, with a cost function c' on the edges such that G_t is 2-metric for all $t \geq 0$. If OPT_t denotes the optimal cost of the tree in iteration t for cost function c', then for all $t \geq 1$ it holds that $\sum_{s=1}^{t} (\Delta\mathrm{OPT}_s)_- \leq \alpha \cdot \mathrm{OPT}_t$.*

We do not know how to show that the conjecture holds. However, it is easy to see that it holds if OPT_t, as a function of t, is *unimodal*, i.e., there exists a t^* such that OPT_t is non-decreasing for $t < t^*$, and non-increasing for $t \geq t^*$.

Theorem 5. *If Conjecture 4 holds then the greedy algorithm with budget 2 is $(2 \cdot (\alpha + 1))$-competitive.*

To show the theorem let $I \subseteq \mathbb{N}_0$ be the subset of iterations t such that $T_t = T_{t-1} \cup \{g_t\}$. In particular, for all iterations $t \notin I$ we have that $c(T_t) \leq c(T_{t-1})$. Setting $\Delta T_t := c(T_t) - c(T_{t-1})$ for each t, we can decompose the cost of each tree

T_t by $c(T_t) = \sum_{s=1}^{t} \Delta T_t$, and by our previous observation $c(T_t) \le \sum_{s \in I, s \le t} \Delta T_s$. It remains to bound the latter value. To this end, we use the fact that in the iterations in I the algorithm did not find any pair of edges f_t, h_t to swap such that $c(f_t) \le c(h_t)/2$ with f_t being adjacent to v_t. As we will see this implies the same for the optimal solution. In particular, if we double the cost of all edges of the form $v_t v_s$ with $t \in I, s \le t - 1$ and $v_t v_s \ne g_t$, then the optimal solution would not see any gain in inserting any other edge but g_t in this iteration. More precisely, consider the following set $E_I := \{v_t v_s : t \in I, s \le t - 1\} \setminus \{g_t : t \in I\}$. We define a new cost function c' by setting $c'(e) := 2 \cdot c(e)$ for all $e \in E_I$ and $c'(e) := c(e)$ for all $e \notin E_I$.

Let us denote by OPT'_t the cost of an MST in iteration t with cost function c' and note that $\text{OPT}'_t \le 2 \cdot \text{OPT}_t$. Also, since c is metric, graph G_t with cost function c' is 2-metric. The next lemma implies that with the new cost function $\text{OPT}'_t - \text{OPT}'_{t-1} = c'(g_t) = c(g_t) = c(T_t) - c(T_{t-1}) = \Delta T_t$.

Lemma 9. *For all t there exists a minimum spanning tree T_t^{**} for graph $G_t = (V_t, E_t)$ with cost function c' such that $T_t^{**} \cap E_I = \emptyset$. In particular $T_t^{**} = T_{t-1}^{**} \cup \{g_t\}$ if $t \in I$, where g_t is a shortest connection between v_t and any vertex in V_{t-1}.*

We can show Theorem 5 using the bound $c(T_t) \le \sum_{s \in I, s \le t} \Delta T_s$, the last lemma, our conjecture, and a short computation.

4.2   The Full Information Scenario

To support the conjecture that there is a constant-competitive online algorithm with budget 2, we show that there always exists a sequence of spanning trees with constant competitive ratio, even when allowing one unit of budget in each iteration and one additional edge swap exactly every k iterations. More formally, we say that a sequence of trees $T_0, T_1, \dots$ needs budget $1 + 1/k$ when: $|T_t \setminus T_{t-1}| \le 2$ if $t \equiv 0 \bmod k$, and $|T_t \setminus T_{t-1}| = 1$ if $t \not\equiv 0 \bmod k$.

First, we present a constant-competitive algorithm with budget 2. Given the cost function c and graphs $G_0, \dots, G_n$ in advance, we compute a 2-approximate Hamiltonian path X_n (which is a tree) for graph G_n by computing an MST and using the folkloric shortcutting technique[2]. Taking shortcuts of this path to vertices in G_t we obtain a Hamiltonian path X_t for G_t with cost at most 2OPT_n. It is not hard to see that the budget used by the sequence $X_1, \dots, X_n$ is at most 2. However, it is not necessarily constant-competitive. By carefully embedding this idea into the classic *doubling framework* [5] we obtain the following result.

Theorem 6. *Under full information there exists a 14-competitive algorithm with budget 2.*

Finally, skipping most of the edge swaps done by the above sequence still yields constant-competitive solutions. The intuitive idea is that we group the edges

---

[2] The *shortcutting* technique takes a tree, obtains a Eulerian graph by doubling the edges, finds a Eulerian walk, and visits the first copy of each node in the walk [10].

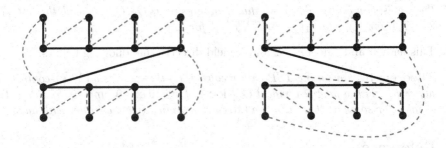

Fig. 1. Two trees (solid) differing in one edge such that standard shortcutting of a(ny) walk in the double-tree yields arbitrarily different tours (dotted)

into sets $\{g_s^0, ..., g_s^{i(s)}\}$ where g_s^0 is the greedy edge added in iteration s and each edge g_s^i is at some point swapped out for the edge g_s^{i+1} with $c(g_s^i) \leq 2 \cdot c(g_s^{i+1})$. Then, we can construct a new sequence with budget $1 + \frac{1}{k}$ which performs at least one out of $O(k)$ swaps of each sequence.

Theorem 7. *Let $k \in \mathbb{N}$. In the full information scenario, there exists a $2^{O(k)}$-competitive algorithm with budget $1 + \frac{1}{k}$.*

5 Applications to TSP

In this section we consider the online TSP with recourse. In a natural approach we aim for combining our algorithms for the online MST problem with the classic *shortcutting* technique [10], which yields a sequence of tours, each with cost at most twice the cost of the tree. This implies a $(2 + \varepsilon)$-competitive algorithm. However, bounding the budget is intricate. Obviously, MSTs that differ in few edges might have quite different Eulerian walks and thus TSP tours. However, even when adapting the Eulerian walks as much as possible, the standard short-cutting might lead to very different TSP tours. In fact, there are examples (see Figure 1) in which two trees T, T' differ in just one edge and shortcutting *any* Eulerian walk on (doubled) T' in the standard way yields a tour Q' which differs from Q for T in an unbounded number of edges.

Our key ingredients for solving this problem are as follows: In case of an edge swap, we decompose the Eulerian walk W corresponding to the tour Q before the swap into 4 sub-walks defined by the swap edges, which we concatenate then in an appropriate way. Furthermore, we find a *robust* variant of the shortcutting technique: instead of shortcutting the new Eulerian walk by visiting the *first* appearance of a node, we remember the copy of each node that we visit in W to construct Q, and then we visit the same copy when constructing Q'. In general we cannot expect to obtain the same tour, but we prove that Q and Q' differ in at most 4 edges which may be necessary when concatenating the sub-walks.

Lemma 10. *Given an online algorithm computing a sequence of trees $T_0, T_1, \ldots,$ there exists an online algorithm that computes tours $Q_2, Q_3, \ldots$ such that $c(Q_t) \leq 2 \cdot c(T_t)$ and $|Q_t \setminus Q_{t-1}| \leq 4 \cdot |T_t \setminus T_{t-1}|$ for any $t \geq 1$.*

This lemma and Theorems 2 and 5 yield directly the following results.

Theorem 8. *For online TSP with recourse, (i) there exists a $(2+\varepsilon)$-competitive algorithm with amortized budget $O(\frac{1}{\varepsilon} \log \frac{1}{\varepsilon})$ for any $\varepsilon > 0$, and (ii) if Conjecture 4 holds for some α, there exists a $(4(\alpha+1))$-competitive algorithm with budget 8.*

References

1. Alon, N., Azar, Y.: On-line Steiner trees in the euclidean plane. Discrete Comp. Geom. 10, 113–121 (1993)
2. Bafna, V., Kalyanasundaram, B., Pruhs, K.: Not all insertion methods yield constant approximate tours in the euclidean plane. Theor. Comput. Sci. 125, 345–360 (1994)
3. Birge, J.R., Louveaux, F.: Introduction to Stochastic Programming. Springer Series in Operations Research. Springer, New York (1997)
4. Christofides, N.: Worst-case analysis of a new heuristic for the travelling salesman problem. Report 388, Graduate School of Industrial Administration, CMU (1976)
5. Chrobak, M., Kenyon-Mathieu, C.: Competitiveness via doubling. SIGACT NEWS 37, 115–126 (2006)
6. Dynia, M., Korzeniowski, M., Kutylowski, J.: Competitive Maintenance of Minimum Spanning Trees in Dynamic Graphs. In: van Leeuwen, J., Italiano, G.F., van der Hoek, W., Meinel, C., Sack, H., Plášil, F. (eds.) SOFSEM 2007. LNCS, vol. 4362, pp. 260–271. Springer, Heidelberg (2007)
7. Imase, M., Waxman, B.M.: Dynamic Steiner tree problem. SIAM J. Discrete Math. 4, 369–384 (1991)
8. Jaillet, P., Wagner, M.R.: Online vehicle routing problems: A survey. In: The Vehicle Routing Problem: Latest Advances and New Challenges. Operations Research/Computer Science Interfaces, vol. 43, pp. 221–237. Springer (2008)
9. Kalyanasundaram, B., Pruhs, K.: Constructing competitive tours from local information. Theor. Comput. Sci. 130, 125–138 (1994)
10. Lawler, E.L., Lenstra, J.K., Rinnooy Kan, A.H.G., Shmoys, D.B.: The Traveling Salesman Problem. John Wiley and Sons, Chichester (1985)
11. Megow, N., Mehlhorn, K., Schweitzer, P.: Online Graph Exploration: New Results on Old and New Algorithms. In: Aceto, L., Henzinger, M., Sgall, J. (eds.) ICALP 2011, Part II. LNCS, vol. 6756, pp. 478–489. Springer, Heidelberg (2011)
12. Oliveira, C.A.S., Pardalos, P.M.: A survey of combinatorial optimization problems in multicast routing. Comput. & Oper. Res. 32, 1953–1981 (2005)
13. Pansiot, J.-J., Grad, D.: On routes and multicast trees in the internet. ACM SIGCOMM Comp. Comm. Review 28, 41–50 (1998)
14. Subramanian, N., Liu, S.: Centralized multi-point routing in wide area networks. In: SAC 1991, pp. 46–52 (1991)
15. Verschae, J.: The Power of Recourse in Online Optimization. PhD thesis, Technische Universität Berlin, Germany (2012)
16. Waxman, B.M.: Routing of multipoint connections. IEEE J. Sel. Area Comm. 6, 1617–1622 (1988)

Geometry of Online Packing Linear Programs*

Marco Molinaro and R. Ravi

Carnegie Mellon University

Abstract. We consider packing LP's with m rows where all constraint coefficients are normalized to be in the unit interval. The n columns arrive in random order and the goal is to set the corresponding decision variables irrevocably when they arrive to obtain a feasible solution maximizing the expected reward. Previous $(1 - \epsilon)$-competitive algorithms require the right-hand side of the LP to be $\Omega(\frac{m}{\epsilon^2} \log \frac{n}{\epsilon})$, a bound that worsens with the number of columns and rows. However, the dependence on the number of columns is not required in the single-row case and known lower bounds for the general case are also independent of n.

Our goal is to understand whether the dependence on n is required in the multi-row case, making it fundamentally harder than the single-row version. We refute this by exhibiting an algorithm which is $(1 - \epsilon)$-competitive as long as the right-hand sides are $\Omega(\frac{m^2}{\epsilon^2} \log \frac{m}{\epsilon})$. Our techniques refine previous PAC-learning based approaches which interpret the online decisions as linear classifications of the columns based on sampled dual prices. The key ingredient of our improvement comes from a non-standard covering argument together with the realization that only when the columns of the LP belong to few 1-d subspaces we can obtain small such covers; bounding the size of the cover constructed also relies on the geometry of linear classifiers. General packing LP's are handled by perturbing the input columns, which can be seen as making the learning problem more robust.

1 Introduction

Traditional optimization models usually assume that the input is known a priori. However, in most applications the data is either revealed over time or only coarse information about the input is known, often modeled in terms of a probability distribution. Consequently, much effort has been directed towards understanding the quality of solutions that can be obtained without full knowledge of the input, which led to the development of online and stochastic optimization [6,7]. Emerging problems such as allocating advertisement slots to advertisers and yield management in the internet are of inherent online nature and have further accelerated this development [1].

Linear programming is arguably the most important and thus well-studied optimization problem. Therefore, understanding the limitations of solving linear programs when complete data is not available is a fundamental theoretical problem with a slew of applications, including the ad allocation and yield management problems above. Indeed, a simple linear program with one uniform knapsack constraint, the Secretary Problem,

---

* Full version available at http://arxiv.org/abs/1204.5810. The first author is supported by NSF grant CMMI1024554 and the second author is supported in part by NSF award CCF-1143998.

was one of the first online problems to be considered and an optimal solution was already obtained by the early 60's [13,15]. Although the single knapsack case is currently well-understood under different models of how information is revealed [4], much less is known about problems with multiple knapsacks and only recently algorithms with solution guarantees have been developed [1,10,14].

The Model. We study online packing LP's in the *random permutation model*. Consider a fixed but unknown LP with n columns $a^1, a^2, \ldots, a^n \in [0,1]^m$, whose associated variables are constrained to be in $[0,1]$, and m packing constraints:

$$\text{OPT} = \max \sum_{t=1}^{n} \pi_t x_t$$

$$\sum_{t=1}^{n} a^t x_t \leq B \qquad \text{(LP)}$$

$$x_t \in [0,1].$$

We know B in advance but columns and their associated π_t's are presented in uniformly random order, and when a column is presented we are required to irrevocably choose the value of its corresponding variable. We assume that the number of columns n is known.[1] The goal is to obtain a feasible solution while maximizing its value. We use OPT to denote the optimum value of the (offline) LP.

By scaling down rows as necessary, we assume without loss of generality that all entries of B are the same, which we also denote with some overload of notation by B. Due to the packing nature of the problem, we also assume without loss of generality that all the π_t's are non-negative and all the a^t's are non-zero: we can simply ignore columns which do not satisfy the first property and always set to 1 the variables associated to the remaining columns which do not satisfy the second property. Finally, we assume that the columns a^t's are in *general position*: for all $p \in \mathbb{R}^m$, there are at most m different $t \in [n]$ such that $\pi_t = pa^t$. Notice that perturbing the input randomly by a tiny amount achieves this property with probability one, while the effect of the perturbation is absorbed in our approximation guarantees [1,11].

Related work. The random permutation model has grown in popularity [4,11,16] since it avoids strong lower bounds of the pessimistic adversarial-order model [8] while still capturing the lack of total information about the input. Different online problems have already been studied in this model, including bin-packing [19], matchings [16,18], the AdWords Problem [11] and different generalizations of the Secretary Problem [2,4,5,17,23]. Closest to our work are packing problems with a single knapsack constraint. In [20], Kleinberg considered the B-Choice Secretary Problem, where the goal is to select at most B items coming online in random order to maximize profit. The author presented an algorithm with competitive ratio $1 - O(1/\sqrt{B})$ and showed that $1 - \Omega(1/\sqrt{B})$ is best possible. Generalizing the B-Choice Secretary Problem, Babaioff et al. [3] considered the Secretary Knapsack Problem and presented a $(1/10e)$-competitive algorithm. Notice that in both cases the competitive ratio does not depend on n.

---

[1] Knowing n up to $1 \pm \epsilon$ factor is enough; this is required for non-trivial competitive ratios [11].

Despite all these works, results for the more general online packing LP's considered here were only recently obtained by Feldman et al. [14] and Agrawal et al. [1]. The first paper presents an algorithm that obtains with high probability a solution of value at least $(1 - \epsilon)$OPT whenever $B \geq \Omega(\frac{m \log n}{\epsilon^3})$ and OPT $\geq \Omega(\frac{\pi_{\max} m \log n}{\epsilon})$, where $\pi_{\max}$ is the largest profit. In the second paper, the authors present an algorithm which obtains a solution of expected value at least $(1 - \epsilon)$OPT under the weaker assumptions $B \geq \Omega(\frac{m}{\epsilon^2} \log \frac{n}{\epsilon})$ or OPT $\geq \Omega(\frac{\pi_{\max} m^2}{\epsilon^2} \log \frac{n}{\epsilon})$. One other way of stating this result is that the algorithm has competitive ratio $1 - O(\sqrt{m \log(n) \log B}/\sqrt{B})$; this guarantee degrades as n increases. The current lower bound on B to allow $(1 - \epsilon)$-competitive algorithms is $B \geq \frac{\log m}{\epsilon^2}$, also presented in [1]. We remark that these algorithms actually work for more general allocation problems, where a set of columns representing various options arrive at each step and the solution may choose at most one of the options.

Both of the above algorithms use a connection between solving the online LP and PAC-learning [9] a linear classification of its columns, which was initiated by Devanur and Hayes [11] in the context of the AdWords problem. Here we further explore this connection and our improved bounds can be seen as a consequence of making the learning algorithm more robust by suitably changing the input LP. Robustness is a topic well-studied in learning theory [12,21], although existing results do not seem to apply directly to our problem. We remark that a component of robustness more closely related to the standard PAC-learning literature was also used by Devanur and Hayes [11].

In recent work, Devanur et al. [10] consider the weaker *i.i.d. model* for the general allocation problem. While in the random permutation model one assumes that columns are sampled without replacement, in the i.i.d. model they are sampled with replacement. Making use of the independence between samples, Devanur et al. substantially improve requirement on B to $\Omega(\frac{\log(m/\epsilon)}{\epsilon^2})$ while showing that the lower bound $\Omega(\frac{\log m}{\epsilon^2})$ still holds in this model. We remark, however, that these models can present very different behaviors: as a simple example, consider an LP with n columns, $m = 1$ constraints and budget $B = 1$, where only one of the columns has $\pi_1 = a^1 = 1$ and all others have $\pi_t = a^t = 0$; in the random permutation model the expected value of the optimal solution is 1, while in the i.i.d. model this value is $1 - (1 - 1/n)^n \to 1 - 1/e$. The competitiveness of the algorithm of [10] under the random permutation model is still unknown and was left as an open problem by the authors.

Our results. Our focus is to understand how large B is required to be in order to allow $(1 - \epsilon)$-competitive algorithms. In particular, the requirements for B in the above algorithms degrade as the number of columns in the LP increases, while the the lower bound does not. With the trend of handling LP's with larger number of columns (e.g. columns correspond to the keywords in the ad allocation problem, which in turn correspond to visits of a search engine's webpage), this gap is very unsatisfactory from a practical point of view. Furthermore, given that guarantees for the single knapsack case do not depend on the number of columns, it is important to understand if the multi-knapsack case is fundamentally more difficult. In this work, we give a precise indication of why the latter problem was resistant to arguments used in the single knapsack case, and overcome this difficulty to exhibit an algorithm with dimension-independent guarantee.

We show that a modification of the DPA algorithm from [1] that we call *Robust DPA* obtains a $(1 - \epsilon)$-competitive solution for online packing LP's with m constraints in the random permutation model whenever $B \geq \Omega(\frac{m^2}{\epsilon^2} \log \frac{m}{\epsilon})$. Another way of stating this result is that the algorithm has competitive ratio $1 - O(m\sqrt{\log B}/\sqrt{B})$. Contrasting to previous results, our guarantee does not depend on n and in the case $m = 1$ matches the bounds for the B-Choice Secretary Problem (up to lower order terms) and improves [3] for large B. We remark that we can replace the requirement $B \geq \Omega(\frac{m^2}{\epsilon^2} \log \frac{m}{\epsilon})$ by OPT $\geq \Omega(\frac{\pi_{\max} m^3}{\epsilon^2} \log \frac{m}{\epsilon})$ exactly as done in Section 5.1 of [1].

High-level outline. As mentioned before, we use the connection between solving an on-line LP and PAC-learning a good linear classification of its columns; in order to obtain the improved guarantee, we focus on tightening the bounds for the generalization error of the learning problem. More precisely, solving the LP can be seen as classifying the columns into 0/1, which corresponds to setting their associated variable to 0/1. Consider a family $\mathcal{X} \subseteq \{0, 1\}^n$ of linear classifications of the columns. Our algorithms sample a set S of columns and learn a classification $x^S \in \mathcal{X}$ which is 'good' for the columns in S (i.e., obtains large proportional revenue while not filling up the proportionally scaled budget too much). The goal is to upper bound the probability that x^S is not good for the whole LP; this is typically done via a union bound over the classifications in $\mathcal{X}$ [1,11].

To obtain improved guarantees, we refine this bound using an argument akin to covering: we consider *witnesses* (Section 2.2), which are representatives of groups of 'similar' bad classifications that can be used to bound the probability that *any* classification in the group is learned; for that we need to use a non-standard measure of similarity between classifications which is based on the budget of the LP. The problem is that, when the columns (π_t, a^t)'s do not lie in a two-dimensional subspace of $\mathbb{R}^m$, the set $\mathcal{X}$ may contain a large number of mutually dissimilar bad classifications; this is a roadblock for obtaining a small set of witnesses. In stark contrast, when these columns do lie in a two-dimensional subspace (e.g., $m = 1$), these classifications have a much nicer structure which admits a small set of witnesses. This indicates that the latter learning problem is intrinsically more robust than the former, which seem to precisely capture the increased difficulty in obtained good bounds for the multi-row case.

Motivated by this discussion, we first consider LP's whose columns a^t's lie in *few* one-dimensional subspaces (Section 2). For each of these subspaces, we are able to approximate the classifications induced in the columns lying in the subspace by considering a small subset of the induced classifications; patching together these partial classifications gives us a witness set for $\mathcal{X}$. However, this strategy as stated does not make use of the fact that the subspaces are embedded in an m-dimensional space, and hence leads to large witness sets. By establishing a connection between the 'useful' patching possibilities with faces of a hyperplane arrangement in $\mathbb{R}^m$ (Lemma 7), we are able to make use of the dimension of the host space and exhibit witness sets of much smaller sizes, which leads to improved bounds.

For a general packing LP, we perturb the columns a^t's to make them lie in few one-dimensional subspaces that form an 'ϵ-net' of the space, while not altering the feasibility and optimality of the LP by more than a $(1 \pm \epsilon)$ factor (Section 3). Finally, we tighten the bound by using the idea of periodically recomputing the classification, following [1] (Section 4). We remark that omitted proofs are presented in the full version [22].

2 OTP for Almost 1-dim Columns

In this section we describe and analyze the algorithm OTP (One-Time Pricing) over LP's whose columns are contained in few 1-dimensional subspaces of $\mathbb{R}^m$. The overall goal is to find an appropriate dual (perhaps infeasible) solution p for (LP) and use it to classify the columns of the LP. More precisely, given $p \in \mathbb{R}^m$, we define $x(p)_t = 1$ if $\pi_t > pa^t$ and $x(p)_t = 0$ otherwise. Thus, $x(p)$ is the result of classifying the columns (π_t, a^t)'s with the homogeneous hyperplane in $\mathbb{R}^{m+1}$ with normal $(-1, p)$. The motivation behind this classification is that it selects the columns which have positive reduced cost with respect to the dual solution p, or alternatively, it solves to optimality the Lagrangian relaxation that uses p as multipliers.

Sampling LP's. In order to obtain a good dual solution p we use the (random) LP consisting on the first s columns of (LP) with appropriately scaled right-hand side.

$$\max \sum_{t=1}^{s} \pi_{\sigma(t)} x_{\sigma(t)} \qquad ((s,\delta)\text{-LP})$$

$$\sum_{t=1}^{s} a^{\sigma(t)} x_{\sigma(t)} \leq \frac{s}{n} \delta B$$

$$x_{\sigma(t)} \in [0,1] \quad t = 1, \ldots, s.$$

$$\min \frac{s}{n} \delta B \sum_{i=1}^{m} p_i + \sum_{t=1}^{s} \alpha_{\sigma(t)}$$
$$((s,\delta)\text{-Dual})$$

$$pa^{\sigma(t)} + \alpha_{\sigma(t)} \geq \pi_{\sigma(t)} \quad t = 1, \ldots, s$$

$$p \geq 0$$

$$\alpha \geq 0.$$

Here σ denotes the random permutation of the columns of the LP. We use $\mathrm{OPT}(s, \delta)$ to denote the optimal value of (s, δ)-LP and $\mathrm{OPT}(s)$ to denote the optimal value of $(s, 1)$-LP.

The static pricing algorithm OTP of [1] can then be described as follows.[2]

1. Wait for the first ϵn columns of the LP (indexed by $\sigma(1), \sigma(2), \ldots, \sigma(\epsilon n)$) and solve $(\epsilon n, 1 - \epsilon)$-Dual. Let (p, α) be the obtained dual optimal solution.
2. Use the classification given by p as above by setting $x_{\sigma(t)} = x(p)_{\sigma(t)}$ for $t = \epsilon n + 1, \epsilon n + 2, \ldots$ for as long as the solution obtained remains valid. From this point on set all further variables to zero.

Note that by definition this algorithm outputs a feasible solution with probability one. Our goal is then to analyze the quality of the solution produced, ultimately leading to the following theorem.

Theorem 1. *Fix $\epsilon \in (0, 1]$. Suppose that there are $K \geq m$ 1-dim subspaces of $\mathbb{R}^m$ containing the columns a^t's and that $B \geq \Omega\left(\frac{m}{\epsilon^3} \log \frac{K}{\epsilon}\right)$. Then algorithm OTP returns a feasible solution with expected value at least $(1 - 5\epsilon)OPT$.*

Let $S = \{\sigma(1), \ldots, \sigma(\epsilon n)\}$ be the (random) index set of the columns sampled by OTP. We use p^S to denote the optimal dual solution obtained by OTP; notice that p^S is completely determined by S. To simplify the notation, we also use x^S to denote $x(p^S)$.

---

[2] To simplify the exposition, we assume that ϵn is an integer.

Notice that, for all the scenarios where x^S is feasible, the solution returned by OTP is identical to x^S with its components $x^S_{\sigma(1)}, \ldots, x^S_{\sigma(\epsilon n)}$ set to zero. Given this observation and the fact that $\mathbb{E}[\sum_{t \le \epsilon n} \pi_{\sigma(t)} x^S_{\sigma(t)}] \le \epsilon \mathrm{OPT}$, one can prove that the following proposition implies Theorem 1.

Proposition 1. *Fix* $\epsilon \in (0, 1]$. *Suppose that there are* $K \ge m$ *1-dim subspaces of* $\mathbb{R}^m$ *containing the columns* a^t's *and that* $B \ge \Omega\left(\frac{m}{\epsilon^3} \log \frac{K}{\epsilon}\right)$. *Then with probability at least* $(1 - \epsilon)$, x^S *is a feasible solution for* (LP) *with value at least* $(1 - 3\epsilon)\mathrm{OPT}$.

2.1   Connection to PAC Learning

We assume from now on that $B \ge \Omega(\frac{m}{\epsilon^3} \log \frac{K}{\epsilon})$. Let $\mathcal{X} = \{x(p) : p \in \mathbb{R}^m_+\} \subseteq \{0, 1\}^n$ denote the set of all possible linear classifications of the LP columns which can be generated by OTP. With slight overload in the notation, we identify a vector $x \in \{0, 1\}^n$ with the subset of $[n]$ corresponding to its support.

Definition 1 (Bad solution). *Given a scenario, we say that* x^S *is bad if it does not satisfy the properties of Proposition 1, namely* x^S *is either infeasible or has value less than* $(1 - 3\epsilon)\mathrm{OPT}$. *We say that* x^S *is good otherwise.*

As noted in previous work, since our decisions are made based on reduced costs it suffices to analyze the *budget occupation* (or complementary slackness) of the solution in order to understand its *value*. To make this precise, given $x \in \{0, 1\}^n$ let $a_i(x) = \sum_{t \in x} a^t_i$ be its occupation of the ith budget and let $a^S_i(x) = \frac{1}{\epsilon} \sum_{t \in x \cap S} a^t_i$ be its appropriately scaled occupation of ith budget in the sampled LP (recall $|S| = \epsilon n$).

Lemma 1. *Consider a scenario where* x^S *satisfies: (i) for all* $i \in [m]$, $a_i(x^S) \le B$ *and (ii) for all* $i \in [m]$ *with* $p^S_i > 0$, $a_i(x^S) \ge (1 - 3\epsilon)B$. *Then* x^S *is good.*

Moreover, since we are making decisions based on the *optimal* reduced cost for the sampled LP, our solution satisfies the above properties for the sampled LP.

Lemma 2. *In every scenario,* x^S *satisfies the following: (i) for all* $i \in [m]$, $a^S_i(x^S) \le (1 - \epsilon)B$ *and (ii) for every* $i \in [m]$ *with* $p^S_i > 0$, $a^S_i(x^S) \ge (1 - 2\epsilon)B$.

Given that $a_i(x) = \mathbb{E}[a^S_i(x)]$ for all x, the idea is to use concentration inequalities to argue that the conditions in Lemma 1 hold with good probability. Although concentration of $a^S_i(x)$ for *fixed* x can be achieved via Chernoff-type bounds, the quantity $a^S_i(x^S)$ has undesired correlations; obtaining an effective bound is the main technical contribution of this paper.

Definition 2 (Badly learnable). *For a given scenario, we say that* $x \in \mathcal{X}$ *can be badly learned for budget* i *if either (i)* $a^S_i(x) \le (1 - \epsilon)B$ *and* $a_i(x) > B$ *or (ii)* $a^S_i(x) \ge (1 - 2\epsilon)B$ *and* $a_i(x) < (1 - 3\epsilon)B$.

Essentially these are the classifications which look good for the sampled $(\epsilon n, 1 - \epsilon)$-LP but are actually bad for (LP). Putting Lemmas 1 and 2 together and unraveling the definitions gives that

$$\Pr\left(x^S \text{ is bad}\right) \leq \Pr\left(\bigvee_{i\in[m],x\in\mathcal{X}} x \text{ can be badly learned for budget } i\right).$$

Notice that the right-hand side of this inequality does not depend on x^S, it is only a function of how skewed $a_i^S(x)$ is as compared to its expectation $a_i(x)$ (over all $x \in \mathcal{X}$).

Usually the right-hand side in the previous equation is upper bounded by taking a union bound over all its terms [1]. Unfortunately this is too wasteful: when x and x' are 'similar' there is a large overlap between the scenarios where $a_i^S(x)$ is skewed and those where $a_i^S(x')$ is skewed. In order to obtain improved guarantees, we introduce in the next section a new way of bounding the right-hand side of the above expression.

2.2 Similarity via Witnesses

First, we partition the classifications which can be badly learned for budget i into two sets, depending on why they are bad: for $i \in [m]$, let $\mathcal{X}_i^+ = \{x \in \mathcal{X} : a_i(x) > B\}$ and $\mathcal{X}_i^- = \{x \in \mathcal{X} : a_i(x) < (1 - 3\epsilon)B\}$. In order to simplify the notation, given a set x we define $\text{skewm}_i(\epsilon, x)$ to be the event that $a_i^S(x) \leq (1 - \epsilon)B$ and $\text{skewp}_i(\epsilon, x)$ to be the event that $a_i^S(x) \geq (1 - 2\epsilon)B$. Notice that if $x \in \mathcal{X}_i^+$, then $\text{skewm}_i(\epsilon, x)$ is the event that $a_i^S(x)$ is significantly smaller than its expectation (skewed in the minus direction), while for $x \in \mathcal{X}_i^-$ $\text{skewp}_i(\epsilon, x)$ is the event that $a_i^S(x)$ is significantly larger than its expectation (skewed in the plus direction). These definitions directly give the equivalence

$$\Pr\left(\bigvee_{i,x\in\mathcal{X}} x \text{ can be badly learned for budget } i\right) = \Pr\left(\bigvee_{i,x\in\mathcal{X}_i^+} \text{skewm}_i(\epsilon, x) \vee \bigvee_{i,x\in\mathcal{X}_i^-} \text{skewp}_i(\epsilon, x)\right).$$

In order to introduce the concept of witnesses, consider two sets x, x', say, in $\mathcal{X}_i^+$. Take a subset $w \subseteq x \cap x'$; the main observation is that, since $a^t \geq 0$ for all t, for all scenarios we have $a_i^S(w) \leq a_i^S(x)$ and $a_i^S(w) \leq a_i^S(x')$. In particular, the event $\text{skewm}_i(\epsilon, x) \vee \text{skewm}_i(\epsilon, x')$ is contained in $\text{skewm}_i(\epsilon, w)$. The set w serves as a witness for scenarios which are skewed for either x or x'; if additionally $a_i(w)$ reasonably larger than $(1 - \epsilon)B$, we can then use concentration inequalities over $\text{skewm}_i(\epsilon, w)$ in order to bound probability of $\text{skewm}(\epsilon, x) \vee \text{skewm}(\epsilon, x')$. This ability of bounding multiple terms of the right-hand side of (2.2) simultaneously is what gives an improvement over the naive union bound.

Definition 3 (Witness). *We say that $\mathcal{W}_i^+$ is a witness set for $\mathcal{X}_i^+$ if: (i) for all $w \in \mathcal{W}_i^+$, $a_i(w) \geq (1-\epsilon/2)B$ and (ii) for all $x \in \mathcal{X}_i^+$ there is $w \in \mathcal{W}_i^+$ contained in x. Similarly, we say that $\mathcal{W}_i^-$ is a witness set for $\mathcal{X}_i^-$ if: (i) for all $w \in \mathcal{W}_i^-$, $a_i(w) \leq (1 - 3\epsilon/2)B$ and (ii) for all $x \in \mathcal{X}_i^-$ there is $w \in \mathcal{W}_i^-$ containing x.*

As indicated by the previous discussion, given witness sets $\mathcal{W}_i^+$ and $\mathcal{W}_i^-$ for $\mathcal{X}_i^+$ and $\mathcal{X}_i^-$, we directly get the bound

$$\Pr\left(\bigvee_{i,x\in\mathcal{X}_i^+} \text{skewm}(\epsilon, x) \vee \bigvee_{i,x\in\mathcal{X}_i^-} \text{skewp}(\epsilon, x)\right) \leq \Pr\left(\bigvee_{i,w\in\mathcal{W}_i^+} \text{skewm}(\epsilon, w) \vee \bigvee_{i,w\in\mathcal{W}_i^-} \text{skewp}(\epsilon, w)\right).$$

$$(2.1)$$

Putting together the last three displayed equations and using Chernoff-type bounds, we can get an upper estimate on the probability that x^S is bad in terms of the size of witnesses sets.

Lemma 3. *Suppose that, for all $i \in [m]$, there are witness sets for $\mathcal{X}_i^+$ and $\mathcal{X}_i^-$ of size at most M. Then $\Pr(x^S$ is bad $) \leq 8mM \exp\left(-\frac{\epsilon^3 B}{33}\right)$.*

One natural choice of a witness set for, say, $\mathcal{X}_i^+$ is the collection of all of its minimal sets; unfortunately this may not give a witness set of small enough size. But notice that a witness set need not be a subset of $\mathcal{X}_i^+$ (or even $\mathcal{X}$). Allowing elements outside $\mathcal{X}_i^+$ gives the flexibility of obtaining witnesses which are associated to multiple "similar" minimal elements of $\mathcal{X}_i^+$, which is effective in reducing the size of witness sets.

2.3 Small Witness Sets for Almost 1-dim Columns

Given the previous lemma, our task is to find small witness sets. Unfortunately, when the (π_t, a^t)'s lie in a space of dimension at least 3, $\mathcal{X}_i^+$ and $\mathcal{X}_i^-$ may contain many ($\Omega(n)$) disjoint sets [22], which shows that in general we cannot find small witness sets directly. This sharply contrasts with the case where the (π_t, a^t)'s lie in a 2-dimensional subspace of $\mathbb{R}^{m+1}$, where one can show that $\mathcal{X}$ is a union of 2 chains with respect to inclusion. In the special case where the a^t's lie in a 1-dimensional subspace of $\mathbb{R}^m$, we show that $\mathcal{X}$ is actually a single chain (Lemma 5) and therefore we can take $\mathcal{W}_i^+$ as *the* minimal set of $\mathcal{X}_i^+$ and $\mathcal{W}_i^-$ as *the* maximal set of $\mathcal{X}_i^-$.

Due to the above observations, we focus on LP's whose a^t's lie in *few* 1-dimensional subspaces. In this case, $\mathcal{X}_i^+$ and $\mathcal{X}_i^-$ are sufficiently well-behaved so that we can find small (independent of n) witness sets.

Lemma 4. *Suppose that there are $K \geq m$ 1-dimensional subspaces of $\mathbb{R}^m$ which contain the a^t's. Then there are witness sets for $\mathcal{X}_i^+$ and $\mathcal{X}_i^-$ of size at most $(O(\frac{K}{\epsilon} \log \frac{K}{\epsilon}))^m$.*

To prove this lemma, assume its hypothesis and partition the index set $[n]$ into $C_1, C_2, \ldots, C_K$ such that for all $j \in [K]$ the columns $\{a^t\}_{t \in C_j}$ belong to the same 1-dimensional subspace. Equivalently, for each $j \in [K]$ there is a vector c^j of ℓ_∞-norm 1 such that for all $t \in C_j$ we have $a^t = \|a^t\|_\infty c^j$. An important observation is that now we can order the columns (locally) by the ratio of profit over budget occupation: without loss of generality assume that for all $j \in [K]$ and $t, t' \in C_j$ with $t < t'$, we have $\frac{\pi_t}{\|a^t\|_\infty} \geq \frac{\pi_{t'}}{\|a^{t'}\|_\infty}$.[3]

Given a classification x, we use $x|_{C_j}$ to denote its projection onto the coordinates in C_j; so $x|_{C_j}$ is the induced classification on columns with indices in C_j. Similarly, we define $\mathcal{X}|_{C_j} = \{x|_{C_j} : x \in \mathcal{X}\}$ as the set of all classifications induced in the columns in C_j. The most important structure that we get from working with 1-d subspaces, which is implied by the local order of the columns, is the following.

Lemma 5. *For each $j \in [K]$, the sets in $\mathcal{X}|_{C_j}$ are prefixes of C_j.*

---

[3] Notice that this ratio is well-defined since by assumption $a^t \neq 0$ for all $t \in [n]$.

To simplify the notation fix $i \in [m]$ for the rest of this section, so we aim at providing witness sets for $\mathcal{X}_i^+$ and $\mathcal{X}_i^-$. The idea is to group the classifications according to their budget occupation caused by the different column classes C_j's. To make this formal, start by covering the interval $[0, B+m]$ with intervals $\{I_\ell\}_{\ell \in L}$, where $I_0 = [0, \frac{\epsilon B}{4K})$ and $I_\ell = [\frac{\epsilon B}{4K}(1 + \frac{\epsilon}{4})^{\ell-1}, \frac{\epsilon B}{4K}(1 + \frac{\epsilon}{4})^\ell)$ for $\ell > 0$ and $L = \{0, \ldots, \lceil \log_{1+\epsilon/4} \frac{8K}{\epsilon} \rceil\}$ (note that since $B \geq m$, we have $B + m \leq 2B$). Define $\mathcal{B}_{i,j}^\ell$ as the set of partial classifications $y \in \mathcal{X}|_{C_j}$ whose budget occupation $a_i(y)$ lie in the interval I_ℓ. For $v \in L^K$ define the family of classifications $\mathcal{B}_i^v = \{(y^1, y^2, \ldots, y^K) : y^j \in \mathcal{B}_{i,j}^{v_j}\}$. The $\mathcal{B}_i^v$'s then provide the desired grouping of the classifications. Note that the $\mathcal{B}_i^v$'s may include classifications not in $\mathcal{X}$ and may not include classifications in $\mathcal{X}$ which have occupation $a_i(.)$ greater than $B + m$.

Now consider a non-empty $\mathcal{B}_i^v$. Let $\underline{w}_i^v$ be the inclusion-wise smallest element in $\mathcal{B}_i^v$. Notice that such unique smallest element exists: since $\mathcal{X}|_{C_j}$ is a chain, so is $\mathcal{B}_{i,j}^{v_j}$, and hence $\underline{w}_i^v$ is the product (over j) of the smallest elements in the sets $\{\mathcal{B}_{i,j}^{v_j}\}_j$. Similarly, let $\overline{w}_i^v$ denote the largest element in $\mathcal{B}_i^v$. Intuitively, $\underline{w}_i^v$ and $\overline{w}_i^v$ will serve as witnesses for all the sets in $\mathcal{B}_i^v$.

Finally, define the witness sets by adding the $\underline{w}_i^v$ and $\overline{w}_i^v$'s of appropriate size corresponding to meaningful $\mathcal{B}_i^v$'s: set $\mathcal{W}_i^+ = \{\underline{w}_i^v : v \in L^K, \mathcal{B}_i^v \cap \mathcal{X} \neq \emptyset, a_i(\underline{w}_i^v) \geq (1 - \epsilon/2)B\}$ and $\mathcal{W}_i^- = \{\overline{w}_i^v : v \in L^K, \mathcal{B}_i^v \cap \mathcal{X} \neq \emptyset, a_i(\overline{w}_i^v) \leq (1 - 3\epsilon/2)B\}$.

It is not too difficult to see that, say, $\mathcal{W}_i^+$ is a witness set for $\mathcal{X}_i^+$: If $x \in \mathcal{X}_i^+$ belongs to some $\mathcal{B}_i^v$, then $\underline{w}_i^v$ belongs to $\mathcal{W}_i^+$ and is easily shown to be a witness for x. However, if x does not belong to any $\mathcal{B}_i^v$, by having too large $a_i(x)$, the idea is to find $x' \subseteq x$ which belongs to some $\mathcal{B}_i^v$ and to $\mathcal{X}$, and then use $\underline{w}_i^v$ as witness for x. We note that ignoring induced classifications with occupation larger than $B + m$ and ignoring $\mathcal{B}_i^v$'s which do not intersect $\mathcal{X}$ is very important for guaranteeing that $\mathcal{W}_i^+$ and $\mathcal{W}_i^-$ are small.

Lemma 6. *The sets $\mathcal{W}_i^+$ and $\mathcal{W}_i^-$ are witness sets for $\mathcal{X}_i^+$ and $\mathcal{X}_i^-$.*

Bounding the size of witness sets. Clearly the witness sets $\mathcal{W}_i^+$ and $\mathcal{W}_i^-$ have size at most $|L|^K$. Although this size is independent of n, it is still unnecessarily large since it only uses locally (for each C_j) the fact that $\mathcal{X}$ consists of linear classifications; in particular, it does not use the dimension of the ambient space $\mathbb{R}^m$. Now we sketch the argument for an improved bound, and details are provided in the full version.

First notice that the partial classification $x(p)|_{C_j}$ is completely defined by the value pc^j. Thus, if $J \subseteq [K]$ is such that the directions $\{c^j\}_{j \in J}$ form a basis of $\mathbb{R}^m$ then knowing pc^j for all $j \in J$ completely determines the whole classification $x(p)$. Similarly, if we know that $x(p)|_{C_j} \in \mathcal{B}_i^{v_j}$ for all $j \in J$, then for each $j \notin J$ we should have fewer possible $\mathcal{B}_i^{v_j}$'s where the partial classification $x(p)|_{C_j}$ can belong to; this indicates that some of the sets $\{\mathcal{B}_i^v\}_{v \in L^K}$ do not contain any element from $\mathcal{X}$, which implies a reduced size for the witness sets.

In order to capture this idea, we focus on the space of dual vectors p and define the sets $P_j^\ell = \{p \in \mathbb{R}_+^m : x(p)|_{C_j} \in \mathcal{B}_{i,j}^\ell\}$ and $P^v = \{p \in \mathbb{R}_+^m : x(p) \in \mathcal{B}_i^v\}$. Notice that $P^v = \bigcap_j P_j^{v_j}$ and that $\mathcal{B}_i^v$ is empty iff P^v is. The main step is to show that each P_j^ℓ is a polyhedron with 'few' facets, which uses the definition of $x(p)$ and Lemma 5. We then consider the arrangement of the hyperplanes which are facet-defining for the P_j^ℓ's and conclude that the P^v's are given by unions of the cells in this arrangement;

classical bounds on the number of cells in a hyperplane arrangement in $\mathbb{R}^m$ then allow us to upper bound the number of nonempty P^v's. This gives the following.

Lemma 7. *At most* $(O(\frac{K}{\epsilon} \log \frac{K}{\epsilon}))^m$ *of the* $\mathcal{B}_i^v$*'s contain an element from* $\mathcal{X}$.

This implies that both $\mathcal{W}_i^+$ and $\mathcal{W}_i^-$ have size at most $(O(\frac{K}{\epsilon} \log \frac{K}{\epsilon}))^m$, which then proves Lemma 4. Finally, applying Lemma 3 we conclude the proof of Proposition 1.

3   Robust OTP

In this section we consider (LP) with columns that may not belong to few 1-dimensional subspaces. Given the results of the previous section we would like to perturb the columns of this LP so that it belongs to few 1-dim subspaces, and such that an approximate solution for this perturbed LP is also an approximate solution for the original one. More precisely, we obtain a set of vectors $Q \subseteq \mathbb{R}^m$ and transform each column a^t into a column $\tilde{a}^t$ which is a scaling of a vector in Q, and we let the rewards π_t remain unchanged. The crucial observation is that the solutions of an LP are robust to slight changes in the the constraint matrix.

Lemma 8. *Consider real numbers* $\pi_1, \dots, \pi_n$ *and vectors* $a^1, \dots, a^n$ *and* $\tilde{a}^1, \dots, \tilde{a}^n$ *in* $\mathbb{R}_+^m$ *such that* $\|\tilde{a}^t - a^t\|_\infty \le \frac{\epsilon}{m+1}\|a^t\|_\infty$. *If* x *is an* ϵ-*approximate solution for* (LP) *with columns* $(\pi_t, \tilde{a}^t)$ *and right-hand side* $(1 - \epsilon)B$, *then* x *is a* $(1 - 2\epsilon)$-*approximate solution for* (LP).

Perturbing the columns. To simplify the notation, set $\delta = \frac{\epsilon}{m+1}$; for simplicity of exposition we assume that $1/\delta$ is integral. When constructing Q we want the rays spanned by the each of its vectors to be "uniform" over $\mathbb{R}_+^m$. Using ℓ_∞ as normalization, let Q be a δ-net of the unit ℓ_∞ sphere, namely let Q be the vectors in $\{0, \delta, 2\delta, 3\delta, \dots, 1\}^m$ which have ℓ_∞ norm 1. Note that $|Q| = (O(\frac{m}{\epsilon}))^m$.

Given a vector $a^t \in \mathbb{R}^m$ we let $\tilde{a}^t = \|a^t\|_\infty q^t$, where q^t is the vector in Q closest (in ℓ_∞) to $\frac{a^t}{\|a^t\|_\infty}$. By definition of Q, for every vector $v \in \mathbb{R}^m$ with $\|v\|_\infty = 1$ there is a vector $q \in Q$ with $\|v - q\|_\infty \le \delta$. It then follows from positive homogeneity of norms that the $\tilde{a}^t$'s satisfy the property required in Lemma 8: $\|a^t - \tilde{a}^t\|_\infty \le \delta\|a^t\|_\infty$.

Algorithm Robust OTP. One way to think of the algorithm Robust OTP is that it works in two phases. First, it transforms the vectors a^t into $\tilde{a}^t$ as described above. Then it returns the solution obtained by running the algorithm OTP over the LP with columns $(\pi_t, \tilde{a}^t)$ and right-hand side $(1 - \epsilon)B$. Notice that this algorithm can indeed be implemented to run in an online fashion.

Putting together the discussion in the previous paragraphs and the guarantee of OTP for almost 1-dim columns given by Theorem 1 with $K = |Q| = (O(\frac{m}{\epsilon}))^m$, we obtain the following theorem.

Theorem 2. *Fix* $\epsilon \in (0, 1]$ *and suppose* $B \ge \Omega\left(\frac{m^2}{\epsilon^3} \log \frac{m}{\epsilon}\right)$. *Then algorithm Robust OTP returns a solution to the online* (LP) *with expected value at least* $(1 - 10\epsilon)OPT$.

4   Robust DPA

In this section we describe our final algorithm, which has an improved dependence on $1/\epsilon$. Following [1], the idea is to update the dual vector used in the classification as new columns arrive: we use the first $2^i\epsilon n$ columns to classify columns $2^i\epsilon n+1,\ldots,2^{i+1}\epsilon n$. This leads to improved generalization bounds, which in turn give the reduced dependence on $1/\epsilon$. The algorithm Robust DPA (as the algorithm DPA) can be seen as a combination of solutions to multiple sampled LP's, obtained via a modification of OTP denoted by (s,δ)-OTP.

Algorithm (s,δ)-OTP. This algorithm aims at solving the program $(2s,1)$-LP and can be described as follows: it finds an optimal dual solution (p,α) for $(s,(1-\delta))$-LP and sets $x_{\sigma(t)} = x(p)_{\sigma(t)}$ for $t = s+1, s+2,\ldots,t' \le 2s$ such that t' is the maximum one guaranteeing $\sum_{t=s+1}^{2s} a^{\sigma(t)}x_{\sigma(t)} \le \frac{s}{n}B$ (for all other t's it sets $x_{\sigma(t)} = 0$).
   The analysis of (s,δ)-OTP is similar to the one employed for OTP. The main difference is that this algorithm tries to approximate the value of the *random* LP $(2s,1)$-LP. This requires a partition of the bad classifications which is more refined than simply splitting into $\mathcal{X}_i^+$ and $\mathcal{X}_i^-$, and witness sets need to be redefined appropriately. Nonetheless, using these ideas we can prove the following guarantee for (s,δ)-OTP. Again let $S = \{\sigma(1),\sigma(2),\ldots,\sigma(s)\}$ be the random index set of the first s columns of the LP, let $T = \{\sigma(s+1),\sigma(s+2),\ldots,\sigma(2s)\}$ and $U = S \cup T$.

Proposition 2. *Suppose that there are $K \ge m$ 1-dim subspaces of $\mathbb{R}^m$ containing the columns a^t's. Fix an integer s and a real number $\delta \in (0,1/10)$ such that $\frac{\delta^2 sB}{n} \ge \Omega(m\ln\frac{K}{\delta})$. Then algorithm (s,δ)-OTP returns a solution x satisfying $a_i^T(x) \le B$ for all $i \in [m]$ with probability 1 and with expected value $\mathbb{E}[\sum_{\tau \in U} \pi_\tau x_\tau] \ge (1 - 3\delta)\mathbb{E}[OPT(2s)] - \mathbb{E}[OPT(s)] - \delta^2 OPT$.*

Algorithm Robust DPA. In order to simplify the description of the algorithm, we assume in this section that $\log(1/\epsilon)$ is an integer.
   Again the algorithm Robust DPA can be thought as acting in two phases. In the first phase it converts the vectors a^t into $\tilde{a}^t$, just as in the first phase of Robust OTP. In the second phase, for $i = 0,\ldots,\log(1/\epsilon) - 1$, it runs $(\epsilon 2^i n, \sqrt{\epsilon/2^i})$-OTP over (LP) with columns $(\pi_t, \tilde{a}^t)$ and right-hand side $(1 - \epsilon)B$ to obtain the solution x^i. The algorithm finally returns the solution x consisting of the 'union' of x^i's: $x = \sum_i x^i$.
   Note that the second phase corresponds exactly to using the first $\epsilon 2^i n$ columns to classify the columns $\epsilon 2^i n + 1,\ldots,\epsilon 2^{i+1}n$. This relative increase in the size of the training data for each learning problem allow us to reduce the dependence of B on ϵ in each of the iterations, while the error from all the iterations telescope and are still bounded as before. Furthermore, notice that Robust DPA can be implemented to run online.
   The analysis of Robust DPA reduces to that of (s,δ)-OTP. That is, using the definition of the parameters of (s,δ)-OTP used in Robust DPA and Proposition 2, it is routine to check that the algorithm produces a feasible solution which has expected value $(1 - \epsilon)OPT$. This is formally stated in the following theorem.

Theorem 3. *Fix $\epsilon \in (0, 1/100)$ and suppose that $B \geq \Omega(\frac{m^2}{\epsilon^2} \ln \frac{m}{\epsilon})$. Then the algorithm Robust* DPA *returns a solution to the online LP* (LP) *with expected value at least* $(1 - 50\epsilon)OPT$.

5   Open Problems

A very interesting open question is whether the techniques introduced in this work can be used to obtain improved algorithms for generalized allocation problems [14]. The difficulty in these problems is that the classifications of the columns are not linear anymore; they essentially come from a conjunction of linear classifiers. Given this additional flexibility, having the columns in few 1-dimensional subspaces does not seem to impose strong enough properties in the classifications. It would be interesting to find the appropriate geometric structure of the columns in this case.

Of course a direct open question is to improve the lower or upper bound on the dependence on the right-hand side B to obtain $(1 - \epsilon)$-competitive algorithms. One possibility is to investigate how much the techniques presented here can be pushed and what are their limitations. Another possibility is to analyze the performance of the algorithm from [10] under the random permutation model.

References

1. Agrawal, S., Wang, Z., Ye, Y.: A dynamic near-optimal algorithm for online linear programming, http://arxiv.org/abs/0911.2974
2. Babaioff, M., Dinitz, M., Gupta, A., Immorlica, N., Talwar, K.: Secretary problems: weights and discounts. In: SODA (2009)
3. Babaioff, M., Immorlica, N., Kempe, D., Kleinberg, R.D.: A Knapsack Secretary Problem with Applications. In: Charikar, M., Jansen, K., Reingold, O., Rolim, J.D.P. (eds.) RANDOM 2007 and APPROX 2007. LNCS, vol. 4627, pp. 16–28. Springer, Heidelberg (2007)
4. Babaioff, M., Immorlica, N., Kempe, D., Kleinberg, R.: Online auctions and generalized secretary problems. SIGecom Exchanges 7(2) (2008)
5. Bateni, M., Hajiaghayi, M., Zadimoghaddam, M.: Submodular Secretary Problem and Extensions. In: Serna, M., Shaltiel, R., Jansen, K., Rolim, J. (eds.) APPROX and RANDOM 2010, LNCS, vol. 6302, pp. 39–52. Springer, Heidelberg (2010)
6. Birge, J.R., Louveaux, F.: Introduction to Stochastic Programming. Springer Series in Operations Research and Financial Engineering. Springer (1997)
7. Borodin, A., El-Yaniv, R.: Online computation and competitive analysis. Cambridge University Press (1998)
8. Buchbinder, N., Naor, J.S.: Online primal-dual algorithms for covering and packing. Mathematics of Operations Research 34, 270–286 (2009)
9. Cucker, F., Zhou, D.X.: Learning Theory: An Approximation Theory Viewpoint. Cambridge University Press (2007)
10. Devanur, N.R., Jain, K., Sivan, B., Wilkens, C.A.: Near optimal online algorithms and fast approximation algorithms for resource allocation problems. In: EC (2011)
11. Devenur, N.R., Hayes, T.P.: The adwords problem: online keyword matching with budgeted bidders under random permutations. In: EC (2009)
12. Devroye, L., Wagner, T.: Distribution-free performance bounds for potential function rules. IEEE Transactions on Information Theory 25, 601–604 (1979)

13. Dynkin, E.B.: The optimum choice of the instant for stopping a Markov process. Soviet Mathematics Doklady 4 (1963)

14. Feldman, J., Henzinger, M., Korula, N., Mirrokni, V.S., Stein, C.: Online Stochastic Packing Applied to Display Ad Allocation. In: de Berg, M., Meyer, U. (eds.) ESA 2010. LNCS, vol. 6346, pp. 182–194. Springer, Heidelberg (2010)

15. Gilbert, J.P., Mosteller, F.: Recognizing the Maximum of a Sequence. Journal of the American Statistical Association 61(313), 35–73 (1966)

16. Goel, G., Mehta, A.: Online budgeted matching in random input models with applications to adwords. In: SODA (2008)

17. Im, S., Wang, Y.: Secretary problems: Laminar matroid and interval scheduling. In: SODA (2011)

18. Karp, R.M., Vazirani, U.V., Vazirani, V.V.: An optimal algorithm for on-line bipartite matching. In: STOC (1990)

19. Kenyon, C.: Best-fit bin-packing with random order. In: SODA (1996)

20. Kleinberg, R.: A multiple-choice secretary algorithm with applications to online auctions. In: SODA (2005)

21. Kutin, S., Niyogi, P.: Almost-everywhere algorithmic stability and generalization error. In: Uncertainty in Artificial Intelligence, pp. 275–282 (2002)

22. Molinaro, M., Ravi, R.: Geometry of online packing linear programs, http://arxiv.org/abs/1204.5810

23. Soto, J.A.: Matroid secretary problem in the random assignment model. In: SODA (2011)

Self-assembly with Geometric Tiles*

Bin Fu[1],[**], Matthew J. Patitz[2],[* * *], Robert T. Schweller[3],[***],
and Robert Sheline[4],[***]

Department of Computer Science, University of Texas - Pan American
{binfu,mpatitz,schwellerr}@cs.panam.edu,
b.sheline@gmail.com

Abstract. In this work we propose a generalization of Winfree's abstract Tile Assembly Model (aTAM) in which tile types are assigned rigid shapes, or geometries, along each tile face. We examine the number of distinct tile types needed to assemble shapes within this model, the temperature required for efficient assembly, and the problem of designing compact geometric faces to meet given compatibility specifications. We pose the following question: can complex geometric tile faces arbitrarily reduce the number of distinct tile types to assemble shapes? Within the most basic generalization of the aTAM, we show that the answer is no. For almost all n at least $\Omega(\sqrt{\log n})$ tile types are required to uniquely assemble an $n \times n$ square, regardless of how much complexity is pumped into the face of each tile type. However, we show for all n we can achieve a matching $O(\sqrt{\log n})$ tile types, beating the known lower bound of $\Theta(\log n / \log \log n)$ that holds for almost all n within the aTAM. Further, our result holds at temperature $\tau = 1$. Our next result considers a geometric tile model that is a generalization of the 2-handed abstract tile assembly model in which tile aggregates must move together through obstacle free paths within the plane. Within this model we present a novel construction that harnesses the collision free path requirement to allow for the unique assembly of any $n \times n$ square with a sleek $O(\log \log n)$ distinct tile types. This construction is of interest in that it is the first tile self-assembly result to harness collision free planar translation to increase efficiency, whereas previous work has simply used the planarity restriction as a desireable quality that could be achieved at reduced efficiency. This surprisingly low tile type result further emphasizes a fundamental open question: Is it possible to assemble $n \times n$ squares with $O(1)$ distinct tile types? Essentially, how far can the trade off between the number of distinct tile types required for an assembly and the complexity of each tile type itself be taken?

---

* A full version of this paper can be found at [13].
** This author's research was supported in part by National Science Foundation Early Career Award 0845376.
* * * This author's research was supported in part by National Science Foundation Grant CCF-1117672.

A. Czumaj et al. (Eds.): ICALP 2012, Part I, LNCS 7391, pp. 714–725, 2012.
© Springer-Verlag Berlin Heidelberg 2012

1   Introduction

The stunning diversity of biological tissues and structures found in nature, including examples such as signaling axons stretching from neurons, powerfully contracting muscle tissue, and specifically tailored coats protecting viral payloads, are composed of basic molecular building blocks called proteins. These proteins, in turn, are assembled from an amazingly small set of only around 20 amino acids. So how is it that so much structural and functional variety can be derived from so few unique components? The simplified answer is "geometry". Essentially, a protein's function is determined by its 3-dimensional shape, or geometry. The exact sequence of amino acids which compose a protein (along with environmental influences such as temperature and pH levels) determine how that particular string of amino acids will fold into a protein's characteristic 3-dimensional structure. However, as simple as it may sound, the resulting geometries are often extremely complex, and predicting them has proven to be computationally intractable. It is from such geometrically intricate structure that nearly all of the complexity of life as we know it arises.

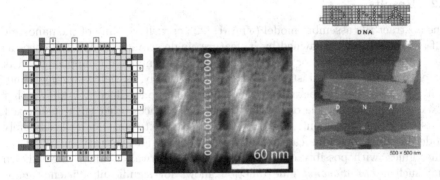

Fig. 1. The use of jigsaw faced macro tiles for self-assembly is emerging in both theoretical and experimental work. This figure contains three separate recent examples. The first figure depicts a macro tile assembled using staged assembly from smaller tile types [9]. The second figure depicts the experimental work of [24] to create geometric tiles made from DNA origami which encode a binary number on their edges. Finally, the third figure depicts the experimental work of [12] in which a jigsaw geometry on the face of tiles is created with the DNA origami technique.

Scientists and inventors have always recognized nature as providing invaluable examples and inspiration, and as for many other fields, this is also true for the study of artificial self-assembling systems. Self-assembling systems are systems in which sets of relatively simple components begin in disconnected and disorganized initial states, and then spontaneously and autonomously combine to form more complex structures. Self-assembling systems are pervasive in nature, and their power for creating intricate structures at even the nano-scale have inspired researchers to design artificial systems which self-assemble. One such productive line of research has followed from the introduction of the Tile Assembly Model (TAM) by Winfree in [22]. As a basic model, the TAM has proven powerful, providing a basis for laboratory implementations [4, 6, 14, 16, 17, 20] as well as copious amounts of theoretical work [5, 8, 10, 11, 15, 21]. However, in this work, we've once again looked to nature's guidance, this time in terms of the power and importance of the geometric complexity of the components of self-assembling systems, to extend the TAM in an attempt to harness that power.

1.1   Overview

We introduce a generalization of the abstract Tile Assembly Model (aTAM) in which tile types are assigned rigid shapes, or geometries, along each tile face. This model is motivated by the plausibility of implementing novel sophisticated nanoscale shapes with technology such as DNA origami [18]. We show that this model permits substantially greater efficiency in terms of tile type complexity when compared to assembling shapes in the basic temperature 2 aTAM. Furthermore, these efficiency improvements hold even at temperature 1.

1.2   Results

The abstract tile assembly model (aTAM) [22], as well as many of the nanoscale self-assembly models spawned by it, feature single stranded DNA sequences as the primary mechanism for decision making. This commonality applies to weak systems such as deterministic temperature-1 assembly, as well as stronger ones that rely on higher temperatures or stochastic methods. Since it is known that DNA strands are capable of hybridizing with sequences other than their exact Watson-Crick complements, it is therefore reasonable to consider a tile assembly model in which one glue can potentially bond with an arbitrary subset of the other glues, with possibly differing strengths. Aggarwal et. al. [7] have shown that such a *non-diagonal* glue function allows for significant efficiency gains in terms of the numbers of unique tile types used to assemble a target shape. Despite this potentially promising result, it is also true that designing non-specific hybridization pairs, while possible, is severely limited in a practical sense, and would likely introduce a potential for error in a much greater sense than is already present in laboratory experiments.

Table 1. Summary of our Results. σ denotes the set of distinct glues of a tile system to be simulated, with σ_n and σ_w denoting only the north/south and west/east glue types respectively.

$n \times n$ square	Tile Types	Temperature	Geometry Size
aTAM (previous work) [1, 19]	$\Theta(\log n / \log\log n)$	2	-
GTAM (Thms. 1,2)	$\Theta(\sqrt{\log n})$	1	$O(\sqrt{\log n})$
2GAM (Thm. 5)	$O(\log\log n)$	2	$O(\log n \log\log n)$

Zig-zag simulation	Tile Type Scale	Glues	Temperature	Geometry Size
Theorem 3	$O(1)$	$O(|\sigma_w|)$	1	$\log|\sigma_n| + \log\log|\sigma_n| + O(1)$
Theorem 4	$O(1)$	1	1	$\log|\sigma| + \log\log|\sigma| + O(1)$

Fig. 2. Examples of geometric tiles. Note that only the black portions on the corners are binding surfaces with glues, while the "teeth" in between provide potential geometric hindrance. Left: Compatible tiles. Right: Incompatible tiles (colliding teeth, which prevent the glue pads from coming together, are circled).

If non-specific binding is impractical or impossible to implement, but powerful in theory, the question remains: are there any other mechanisms by which this power can be realized? One possible answer to this question is motivated by advances in DNA origami [12,18] in which DNA strands can be folded into blocks with semi-rigid jig-saw faces (see the rightmost image in Figure 1). In this work we introduce a generalization of the aTAM in which tile faces are given some rigid shape (which we hereon refer to as geometry). As suggested in Figure 2, the *geometric hindrance* which can be provided by this geometry is capable of simulating non-diagonal glue functions by creating a set of compatible and non-compatible faces. We show that this new model realizes much of the power of non-specific hybridization. Among our results, we show that $n \times n$ squares can be assembled in $\Theta(\sqrt{\log n})$ distinct tile types, which meets an information theoretic lower bound for the model and improves what is possible without geometric tiles from $\Theta(\log n / \log \log n)$ (see [19]). In addition, this tile efficient construction requires only a temperature threshold of 1, thus showing this model can mimic both non-specific glue functions and temperature 2 self-assembly simultaneously.

Next, we show that temperature-1 systems utilizing geometry can efficiently simulate a powerful class of temperature-2 aTAM systems. This class of systems, called *zig-zag* systems, is capable of simulating arbitrary Turing machines and therefore universal computation. Furthermore, the simulation performed using geometric tiles is efficient in that it requires no asymptotic increase in tile complexity (i.e. the number of unique tile types required) or in the size of the assembly. This is especially notable due to the fact that it is conjectured (although currently unproven) that temperature-1 systems in the aTAM are not computationally universal.

While tile geometries provide a method for greatly reducing the tile complexity required to build squares in a seeded model like the aTAM (i.e. one in which tiles can only combine with a growing assembly one at a time), our next result holds for geometric tiles considered within the 2-handed assembly model (sometimes referred to by other names [2,7,9,15,23]). We show that, in this model, the tile complexity required to build a square is reduced to only $O(\log \log n)$ tile types, while the complexity of the geometries increases to $O(\log n \log \log n)$.

In the full version of this paper [13] we additionally conduct a detailed analysis of problems related to computing necessary patterns for tile geometries given specifications of the desired compatibility matrices (i.e. the listings of which tile sides should be compatible and incompatible with each other), with the goal being to minimize the size of the necessary geometries (as well as the running time of the computations). While these results are omitted in this version, we apply some of these results in our constructions for this paper.

2   Model

In this section we define the basic *geometric tile assembly model* (GTAM) and the *two-handed planar geometric tile assembly model* (2GAM). See [13] for a more detailed technical definition of these models.

2.1   Basics

A tile type is a unit square with four sides, each having a glue consisting of a label (a finite string) and strength (0, 1, or 2). We assume a finite set T of tile types, but an infinite number of copies of each tile type, each copy referred to as a tile. A supertile (a.k.a., assembly) is a positioning of tiles on the integer lattice $\mathbb{Z}^2$. Two adjacent tiles in a supertile interact if the glues on their abutting sides are equal. Each supertile induces a binding graph, a grid graph whose vertices are tiles, with an edge between two tiles if they interact. The supertile is τ-stable if every cut of its binding graph has strength at least τ, where the weight of an edge is the strength of the glue it represents. That is, the supertile is stable if at least energy τ is required to separate the supertile into two parts. A seeded tile assembly system (TAS) is a triple $T = (T, \tau, s)$, where T is a finite tile set, τ is the temperature, usually 1 or 2, and $s \in T$ is a special tile type denoted as the *seed*. Given a TAS $T = (T, \tau, s)$, a supertile is producible if either it is the seed tile, or it is the τ-stable result of attaching a single tile $r \in T$ to a producible supertile. A supertile α is terminal if for every tile type $r \in T$, r cannot be τ-stably attached to α. A TAS is directed (a.k.a., deterministic or confluent) if it has only one terminal, producible supertile. Given a connected shape $X \subset \mathbb{Z}^2$, a TAS T produces X uniquely if every producible, terminal supertile places tiles only on positions in X (appropriately translated if necessary).

2.2   Geometric Tiles and the Basic (GTAM)

In this paper we generalize the basic aTAM by assigning a geometric pattern to each side of a tile type along with its glue. For each tile set in the GTAM, fix two values $w, \ell \in \mathbb{N}$. While at a high-level we still consider tiles as occupying unit squares within the plane, in order to determine whether or not adjacent tiles are *geometrically compatible* with each other, we define a *tile body* to be an $\ell \times \ell$ square (see Figure 3), and we define a *(tile face) geometry* to be a subset of $\mathbb{Z}_w \times \mathbb{Z}_\ell$. A *geometric* tile type consists of a tile body which has both a glue and a geometry assigned to each side. For a tile type t, let $northGeometry(t)$ denote the geometry assigned to the north side of t. Define $eastGeometry(t)$, $southGeometry(t)$, and $westGeometry(t)$ analogously. Intuitively, the geometry of a tile type face represents the positions of inflexible bumps, or "filled-in" locations of the $w \times \ell$ rectangle, that can prevent two tiles from lining up adjacently to one another so that the rectangles of their adjacent geometries completely overlap. Only if the $w \times \ell$ geometries on adjacent sides of two combining tiles can completely overlap so that no location contains a filled-in portion of both, can any glues on those adjacent sides interact. Formally, we say a tile type t is *east incompatible* with tile type r if $eastGeometry(t) \cap westGeometry(r) \neq \emptyset$. We define *north*, *south*, and *west* incompatibility analogously. Seeded Geometric Tile Assembly takes place in the same manner as in the aTAM, with the added requirement that a tile type cannot be attached to a supertile at a position in which the tile type is either east, west, north, or south incompatible with another adjacent tile type in the supertile at a position west, east, south, or north, respectively, of the attachment position. As in the original aTAM, tiles are not allowed to rotate and must always maintain their pre-specified orientation, even while moving into position to attach to an assembly.

2.3   Two-Handed Geometric Tile Assembly Model

The Two-Handed Geometric Tile Assembly Model (2GAM) extends the GTAM by allowing large assembled supertiles to attach to one another. As in the GTAM, tiles are composed of tile bodies and tile face geometries as shown in Figure 3. Within the 2GAM, two tiles may attach if 1) there exists a collision free path within the 2D plane to shift the tiles into an adjacent position in which the east (or south) geometry box of one tile exactly overlaps the west (or north) geometry box of the second tile, and 2) the east (north) and west (south) glues of each tile are equal and have strength at least τ.

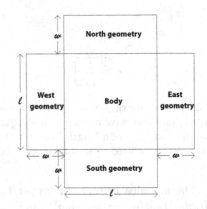

Fig. 3. Definition of a geometric tile

More generally, preassembled multiple tile supertiles may come together if there is a collision free path in which the supertiles line up to create a τ-stable assembly. The set of *producible* supertiles within the 2GAM is defined recursively: As a base case, all singleton supertiles consisting of a single tile are producible. Recursively, for any two producible supertiles α and β such that there exists a collision free path within the plane to shift α and β into a τ-stable configuration γ, then the supertile γ is also producible. The subset of producible assemblies of a 2GAM system to which no producible assembly can attach defines the *terminally produced* supertiles. Intuitively, this set represents the set of assemblies we expect to see from a system if it is given enough time to assemble, and we refer to this as the output of the system. A 2GAM is directed (e.g., deterministic, confluent) if it has only one terminal, producible supertile. Given a connected shape $X \subseteq \mathbb{Z}^2$, a 2GAM Γ produces X uniquely if every producible, terminal supertile places tiles only on positions in X (appropriately translated if necessary).

3   GTAM Complexities: Squares and $\tau = 1$ Assembly

In this section we examine the power of the GTAM in the context of efficiently building squares and simulating temperature $\tau = 2$ aTAM systems at $\tau = 1$. Our first result shows that the tile complexity of $n \times n$ squares in the GTAM is $O(\sqrt{\log n})$ for all n by providing a $O(\sqrt{\log n})$ tile complexity upper bound construction for all n. This result is notable in that it beats a lower bound of $\Omega(\log n / \log \log n)$ for almost all n for the aTAM [19], and further, does so at $\tau = 1$ (all known sublinear tile complexities in the aTAM use at least $\tau = 2$).

Theorem 1. *The minimum tile complexity required to assemble an $n \times n$ square in the GTAM is $O(\sqrt{\log n})$. Further, this complexity can be achieved by a temperature $\tau = 1$ system with $O(\sqrt{\log n})$ size geometry.*

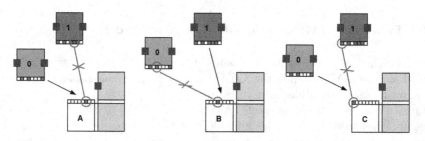

Fig. 4. Geometric tile compatibility can mimic *flexible* glue functions in which non-equal glues may have non-zero bonding strength. This flexible glue compatibility allows for the assembly of a length $\log n$ bit string using only order $\sqrt{\log n}$ tile types, which in turn can be used to seed a counter for the assembly of an $n \times n$ square.

The key idea behind this upper bound is the fact that geometry can mimic *flexible* (a.k.a. *non-diagonal*) glue functions in which non-equal glues are permitted to have bonding strength. Figure 4 provides a simple example in which the southern geometry of a pair of tiles interacts with a large set of different tile types A, B, and C. In this example, the 0 tile is compatible with A and C, and the 1 tile is compatible with tile B. As shown in [3], flexible glue functions permit the assembly of squares in $O(\sqrt{\log n})$ tile types. By modifying this construction to utilize geometry instead of flexible glues, we obtain the same result, with the added bonus of only requiring temperature $\tau = 1$ and only a single glue type.

Our next result shows an asymptotically tight lower bound for almost all n.

Theorem 2. *For almost all integers n, the minimum tile complexity required to assemble an $n \times n$ square in the GTAM is $\Omega(\sqrt{\log n})$.*

The possibility of efficient assembly at $\tau = 1$ extends beyond just squares to a computationally powerful class of assembly systems called *zig-zag* systems in which growth proceeds upward row by row, each row alternating left and right.

Definition 1. Zig-Zag System. *A system $\Gamma = (T, \tau, s)$ is a zig-zag system if:*

1. *The location and type of the ith tile to attach is the same for all assembly sequences.*
2. *The ith tile attachment occurs to the north, west, or east (not south) of the previously placed tile attachment in all assembly sequences.*
3. *For finite assemblies, the final tile type placed does not occur anywhere else in the assembly.*

The first theorem shows how to simulate zig-zag temperature $\tau = 2$ systems at temperature $\tau = 1$ by utilizing geometry to mimic the cooperative binding effect of temperature $\tau = 2$.

Theorem 3. *Any $\tau = 2$ zig-zag aTAM system $\Gamma = (T, 2, s)$ can be simulated by a $\tau = 1$ GTAM system $\Upsilon = (R, 1, q)$ with tile type scale $|R|/|T| = O(1)$. The simulation utilizes geometry size at most $\log |\sigma_n| + \log \log |\sigma_n| + O(1)$ where σ_n is the set of distinct north/south glue types represented in T.*

The next result extends the temperature $\tau = 1$ result by further showing that the number of distinct glues in the simulating system can be reduced to use just a single glue type.

Theorem 4. *Any $\tau = 2$ zig-zag aTAM system $\Gamma = (T, 2, s)$ can be simulated by a $\tau = 1$ GTAM system $\Upsilon = (R, 1, q)$ using only 1 glue type and tile type scale $|R|/|T| = O(1)$. The geometry size of Υ is at most $\log |\sigma| + \log \log |\sigma| + O(1)$ where σ is the set of distinct glue types represented in T.*

4 $n \times n$ 2GAM Squares with $O(\log \log n)$ Tile Types

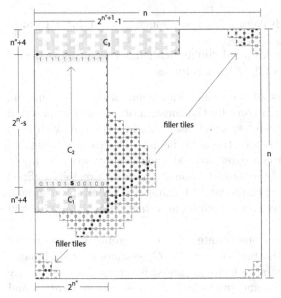

Fig. 5. Construction building a square in the 2GAM

In this section, we explore the theoretical limits achievable when utilizing geometric tiles by designing tiles whose edges contain highly complex geometries. Furthermore, we move to the 2-handed variant of the GTAM, the 2GAM, to allow for the geometric hindrances experienced by individual tiles to be grouped and combined to provide more complex interactions between larger supertiles. The goal, rather than providing a realistic and potentially experimentally realizable set of constructions, is to gain further understanding into the interplay between geometry and the types of computations which can be carried out via algorithmic self-assembly. Our construction reduces the tile complexity required to self-assemble an $n \times n$ square to a mere $O(\log \log n)$ tile types, while requiring a geometry size of $O(\log n \log \log n)$. It requires the constraint of planarity, in which components are not allowed to float into position from above or below the assembly, but must always be able to slide into position with a series of translations along only the x and y axes. However, the intricate geometric designs and complex series of movements require that individual tile geometries are composed of disconnected components. (Note that in [13] we show how to extend the tiles into the third dimension, utilizing a total of 4 planes, in a manner which results in connected tiles and also implicitly enforces the restriction that only tile translations along the x and y axes must be sufficient to allow for tile attachments.)

Theorem 5. *For every* $n \in \mathbb{N}$, *there exists a 2GAM tile system* $\Gamma = (T, 2)$ *which uniquely produces an* $n \times n$ *square, where* $|T| = O(\log \log n)$, *and with* $O(\log n \log \log n)$ *size geometry.*

To prove Theorem 5, we present the following construction.

These values are based on the particular dimensions of the square to be formed and are used throughout the following discussion:

- n: dimensions of the square to self-assemble
- n': $\lceil \log n \rceil$
- n'': $\lceil \log n' \rceil$
- s: $2^{n'} + 2^{n''} + 2n'' + 8 - n$
- h: $2^{n''-1} - 1$
- C_1: 2-handed counter which counts from 0 through $2^{n''} - 1$ for a total of $2^{n''}$ columns
- C_2: standard counter which counts from s through $2^{n'} - 1$ for a total of $2^{n'} - s$ columns
- C_3: 2-handed counter with additional "buffer" columns which counts from 0 through $2^{n''} - 1$ for a total of $2^{n''+1} - 1$ columns

Figure 5 shows a high-level view of the main components of this construction. Without loss of generality, we can consider the construction to be composed of a series of sub-assemblies, or modules, which assemble in sequence, with each module completely assembling before the next begins. The careful design of all modules ensures that none can grow so that they occupy space required by another, and that each will be able to terminally grow to precisely defined dimensions that result in the final combination forming exactly an $n \times n$ square. For the rest of this discussion, we will describe the formation of the modules in such a sequence.

This construction makes use of one counter, C_1, to assemble an encoding of a number which in turn seeds another counter, C_2. C_1 assembles in a 2-handed manner, meaning that each number which is counted is represented by exactly one one-tile-wide column of tiles, and individual columns form separately and then combine to form the full counter of length $2^{n''}$ (similar in design to counters found in [11]). Each column of the counter, besides representing a counter value, is used to represent (on the north face of the northernmost tile) one bit of the seed value s for C_2. Each column can form in one of two versions: one that represents a 0, and one that represents a 1. The east and west sides of the tiles forming the columns contain geometries which force the columns, in order to combine, to "wiggle" up and down in patterns based on the counter values of those columns. See Figure 6 for an example pair of compatible columns. The topmost tiles of the columns consist of tiles with geometries which "read" those patterns of wiggling and allow columns to combine with each other if and only if they are the correct versions of the counter columns, namely those with the bit values of s which correctly correspond to their location in the counter. It is the tiles of this component as well as those of the counter C_3 in which the intricate $O(\log n \log \log n)$ geometries are contained.

C_2 is a standard binary counter (i.e. one that would also assemble correctly in the aTAM) which utilizes 16 tile types and grows to complete the majority of the western side of the square. Next, a small set

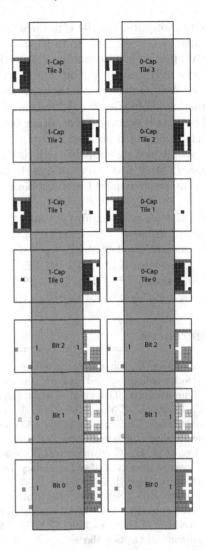

Fig. 6. Example columns for the counter C_1. Note that all colored areas are filled-in, and areas colored white are empty, although they may be outlined for reference.

of 7 "filler" tile types fill in the majority of the square, and once they have filled in a sufficient portion of the northern portion they provide a platform to which C_3 can attach (as long as C_3 is fully formed). In order to provide a directed system with only one terminal assembly, the "incorrect" columns (those which couldn't become part of C_1 due to their nondeterministic selection of 0 or 1 cap tiles) are able to combine into the 2-handed counter structure C_3 via some extra buffer columns. Finally, the filler tiles are able to complete the formation of the square. Note that the tile types which make up C_2 and the filler tiles require no geometries but only standard glues.

By utilizing the assembly of supertiles (i.e. sub-assemblies of grouped tiles) and carefully designing geometries which force the supertiles forming C_1 to move in well-defined patterns as they attach, we are able to essentially "transmit" information about tiles in one location of a supertile to the interfaces where potential binding is occurring with other tiles in the same supertile. By concatenating this information from such a group of distant tiles, the binding "decision" can be made based on an arbitrarily large amount of information (as long as the geometry sizes scale appropriately). This results in a dramatic lowering of the tile complexity required to assemble an $n \times n$ square, with the tradeoff being an increase in the complexity of the tiles themselves.

C_1 and C_3 each require a constant number of tile types for each of the n'' bit positions, and thus $O(\log \log n)$ tile types, and C_2 and the filler tiles consist of a constant number of tile types, for an overall tile complexity of $O(\log \log n)$. The geometries defined for all tiles in this construction consist of rectangles of dimensions $(2^{n''} + h + 4) \times (n'' + 2) = (2^{n''} + \lceil 2^{n''}/2 \rceil + 4) \times (n'' + 2) = O(\log n \times \log \log n)$, and therefore the geometry size is $O(\log n \log \log n)$.

References

1. Adleman, L., Cheng, Q., Goel, A., Huang, M.-D.: Running time and program size for self-assembled squares. In: Proceedings of the Thirty-third Annual ACM Symposium on Theory of Computing, pp. 740–748. ACM, New York (2001)
2. Adleman, L., Cheng, Q., Goel, A., Huang, M.-D., Wasserman, H.: Linear self-assemblies: Equilibria, entropy and convergence rates. In: Sixth International Conference on Difference Equations and Applications. Taylor and Francis (2001)
3. Aggarwal, G., Goldwasser, M.H., Kao, M.-Y., Schweller, R.T.: Complexities for generalized models of self-assembly. In: Proceedings of ACM-SIAM Symposium on Discrete Algorithms (2004)
4. Barish, R.D., Schulman, R., Rothemund, P.W., Winfree, E.: An information-bearing seed for nucleating algorithmic self-assembly. Proceedings of the National Academy of Sciences 106(15), 6054–6059 (2009)
5. Chandran, H., Gopalkrishnan, N., Reif, J.: The Tile Complexity of Linear Assemblies. In: Albers, S., Marchetti-Spaccamela, A., Matias, Y., Nikoletseas, S., Thomas, W. (eds.) ICALP 2009. LNCS, vol. 5555, pp. 235–253. Springer, Heidelberg (2009)
6. Chen, H.-L., Schulman, R., Goel, A., Winfree, E.: Reducing facet nucleation during algorithmic self-assembly. Nano Letters 7(9), 2913–2919 (2007)
7. Cheng, Q., Aggarwal, G., Goldwasser, M.H., Kao, M.-Y., Schweller, R.T., de Espanés, P.M.: Complexities for generalized models of self-assembly. SIAM Journal on Computing 34, 1493–1515 (2005)
8. Cook, M., Fu, Y., Schweller, R.T.: Temperature 1 self-assembly: Deterministic assembly in 3d and probabilistic assembly in 2d. In: Randall, D. (ed.) Proceedings of the Twenty-Second Annual ACM-SIAM Symposium on Discrete Algorithms, SODA 2011, pp. 570–589. SIAM (2011)
9. Demaine, E.D., Demaine, M.L., Fekete, S.P., Ishaque, M., Rafalin, E., Schweller, R.T., Souvaine, D.L.: Staged self-assembly: nanomanufacture of arbitrary shapes with $O(1)$ glues. Natural Computing 7(3), 347–370 (2008)

10. Doty, D., Lutz, J.H., Patitz, M.J., Summers, S.M., Woods, D.: Intrinsic universality in self-assembly. In: Proceedings of the 27th International Symposium on Theoretical Aspects of Computer Science, pp. 275–286 (2009)
11. Doty, D., Patitz, M.J., Reishus, D., Schweller, R.T., Summers, S.M.: Strong fault-tolerance for self-assembly with fuzzy temperature. In: Proceedings of the 51st Annual IEEE Symposium on Foundations of Computer Science (FOCS 2010), pp. 417–426 (2010)
12. Endo, M., Sugita, T., Katsuda, Y., Hidaka, K., Sugiyama, H.: Programmed-assembly system using DNA jigsaw pieces. Chemistry: A European Journal, 5362–5368 (2010)
13. Fu, B., Patitz, M.J., Schweller, R., Sheline, R.: Self-assembly with geometric tiles. Arxiv preprint arXiv:1104.2809 (2012)
14. LaBean, T.H., Winfree, E., Reif, J.H.: Experimental progress in computation by self-assembly of DNA tilings. DNA Based Computers 5, 123–140 (1999)
15. Luhrs, C.: Polyomino-Safe DNA Self-assembly via Block Replacement. In: Goel, A., Simmel, F.C., Sosík, P. (eds.) DNA. LNCS, vol. 5347, pp. 112–126. Springer, Heidelberg (2009)
16. Mao, C., LaBean, T.H., Relf, J.H., Seeman, N.C.: Logical computation using algorithmic self-assembly of DNA triple-crossover molecules. Nature 407(6803), 493–496 (2000)
17. Reif, J.H., Sahu, S., Yin, P.: Compact Error-Resilient Computational DNA Tiling Assemblies. In: Ferretti, C., Mauri, G., Zandron, C. (eds.) DNA 2004. LNCS, vol. 3384, pp. 293–307. Springer, Heidelberg (2005)
18. Rothemund, P.W.K.: Folding DNA to create nanoscale shapes and patterns. Nature 440(7082), 297–302 (2006)
19. Rothemund, P.W.K., Winfree, E.: The program-size complexity of self-assembled squares (extended abstract). In: STOC 2000: Proceedings of the Thirty-Second Annual ACM Symposium on Theory of Computing, Portland, Oregon, United States, pp. 459–468. ACM Press (2000)
20. Schulman, R., Winfree, E.: Synthesis of crystals with a programmable kinetic barrier to nucleation. Proceedings of the National Academy of Sciences 104(39), 15236–15241 (2007)
21. Soloveichik, D., Winfree, E.: Complexity of self-assembled shapes. SIAM Journal on Computing 36(6), 1544–1569 (2007)
22. Winfree, E.: Algorithmic self-assembly of DNA. Ph.D. thesis, California Institute of Technology (June 1998)
23. Winfree, E.: Self-healing tile sets. In: Chen, J., Jonoska, N., Rozenberg, G. (eds.) Nanotechnology: Science and Computation. Natural Computing Series, pp. 55–78. Springer (2006)
24. Woo, S., Rothemund, P.W.K.: Stacking bonds: Programming molecular recognition based on the geometry of dna nanostructures. Nature Chemistry 3, 620–627 (2011)

Quasi-polynomial Local Search for Restricted Max-Min Fair Allocation*

Lukas Polacek[1] and Ola Svensson[2]

[1] KTH Royal Institute of Technology, Sweden
polacek@csc.kth.se
[2] EPFL, Switzerland
ola.svensson@epfl.ch

Abstract. The restricted max-min fair allocation problem (also known as the restricted Santa Claus problem) is one of few problems that enjoys the intriguing status of having a better estimation algorithm than approximation algorithm. Indeed, Asadpour et al. [1] proved that a certain configuration LP can be used to estimate the optimal value within a factor $1/(4 + \epsilon)$, for any $\epsilon > 0$, but at the same time it is not known how to efficiently find a solution with a comparable performance guarantee.

A natural question that arises from their work is if the difference between these guarantees is inherent or because of a lack of suitable techniques. We address this problem by giving a quasi-polynomial approximation algorithm with the mentioned performance guarantee. More specifically, we modify the local search of [1] and provide a novel analysis that lets us significantly improve the bound on its running time: from $2^{O(n)}$ to $n^{O(\log n)}$. Our techniques also have the interesting property that although we use the rather complex configuration LP in the analysis, we never actually solve it and therefore the resulting algorithm is purely combinatorial.

1 Introduction

We consider the problem of indivisible resource allocation in the following classical setting: a set $\mathcal{R}$ of available resources shall be allocated to a set $\mathcal{P}$ of players where the value of a set of resources for player i is given by the function $f_i : 2^{\mathcal{R}} \mapsto \mathbb{R}$. This is a very general setting and dependent on the specific goals of the allocator several different objective functions have been studied.

One natural objective, recently studied in [7,8,11,15], is to maximize the social welfare, i.e., to find an allocation $\pi : \mathcal{R} \mapsto \mathcal{P}$ of resources to players so as to maximize $\sum_{i \in \mathcal{P}} f_i(\pi^{-1}(i))$. However, this approach is not suitable in settings where the property of "fairness" is desired. Indeed, it is easy to come up with examples where an allocation that maximizes the social welfare assigns all resources to even a single player. In this paper we address this issue by studying algorithms for finding "fair" allocations. More specifically, fairness is modeled by evaluating an allocation with respect to the satisfaction of the least happy player, i.e., we wish to find an allocation π that maximizes $\min_{i \in \mathcal{P}} f_i(\pi^{-1}(i))$.

---

* A full version of this paper is available at http://arxiv.org/abs/1205.1373. This research was supported by ERC Advanced investigator grants 228021 and 226203.

A. Czumaj et al. (Eds.): ICALP 2012, Part I, LNCS 7391, pp. 726–737, 2012.
© Springer-Verlag Berlin Heidelberg 2012

In contrast to maximizing the social welfare, the problem of maximizing fairness is already NP-hard when players have linear value functions. In order to simplify notation for such functions we denote $f_i(j)$ by $v_{i,j}$ and hence we have that $f_i(\pi^{-1}(i)) = \sum_{j \in \pi^{-1}(i)} v_{i,j}$. This problem has recently received considerable attention in the literature and is often referred to as the *max-min fair allocation* or the *Santa Claus* problem.

One can observe that the max-min fair allocation problem is similar to the classic problem of scheduling jobs on unrelated machines to minimize the makespan, where we are given the same input but wish to find an allocation that minimizes the maximum instead of one that maximizes the minimum. In a classic paper [13], Lenstra, Shmoys & Tardos gave a 2-approximation algorithm for the scheduling problem and proved that it is NP-hard to approximate the problem within a factor less than 1.5. The key step of their 2-approximation algorithm is to show that a certain linear program, often referred to as the assignment LP, yields an additive approximation of $v_{\max} = \max_{i,j} v_{i,j}$. Bezáková and Dani [5] later used these ideas for max-min fair allocation to obtain an algorithm that always finds a solution of value at least $OPT - v_{\max}$, where OPT denotes the value of an optimal solution. However, in contrast to the scheduling problem, this algorithm and more generally the assignment LP gives no approximation guarantee for max-min fair allocation in the challenging cases when $v_{\max} \geq OPT$.

In order to overcome this obstacle, Bansal & Sviridenko [3] proposed a stronger linear program relaxation, known as the configuration LP, for the max-min fair allocation problem. The configuration LP that we describe in detail in Section 2 has been vital to the recent progress on better approximation guarantees. Asadpour & Saberi [2] used it to obtain a $\Omega(1/\sqrt{|\mathcal{P}|}(\log |\mathcal{P}|)^3)$-approximation algorithm which was later improved by Bateni et al. [4] and Chakrabarty et al. [6] to algorithms that return a solution of value at least $\Omega(OPT/|\mathcal{P}|^\epsilon)$ in time $O(|\mathcal{P}|^{1/\epsilon})$.

The mentioned guarantee $\Omega(OPT/|\mathcal{P}|^\epsilon)$ is rather surprising because the integrality gap of the configuration LP is no better than $O(OPT/\sqrt{|\mathcal{P}|})$ [3]. However, in contrast to the general case, the configuration LP is significantly stronger for the prominent special case where values are of the form $v_{i,j} \in \{v_j, 0\}$. This case is known as the *restricted* max-min fair allocation or the restricted Santa Claus problem and is the focus of our paper. The worst known integrality gap for the restricted case is $1/2$ and it is known [5] that it is NP-hard to beat this factor (which is also the best known hardness result for the general case). Bansal & Sviridenko [3] first used the configuration LP to obtain an $O(\log \log \log |\mathcal{P}| / \log \log |\mathcal{P}|)$-approximation algorithm for the restricted max-min fair allocation problem. They also proved several structural properties that were later used by Feige [9] to prove that the integrality gap of the configuration LP is in fact constant in the restricted case. The proof is based on repeated use of Lovász local lemma and was turned into a polynomial time algorithm [12].

The approximation guarantee obtained by combining [9] and [12] is a large constant and is far away from the best known analysis of the configuration LP by Asadpour et al. [1]. More specifically, they proved in [1] that the integrality gap is lower bounded by $1/4$ by designing a beautiful local search algorithm that eventually finds a solution with the mentioned approximation guarantee, but is only known to converge in exponential time. As the configuration LP can be solved up to any precision in polynomial time, this means that we can approximate the value of an optimal solution within a factor $1/(4 + \epsilon)$ for any $\epsilon > 0$ but it is not known how to efficiently find a solution with a comparable performance guarantee. Few other problems enjoy this intriguing status (see e.g. the overview article by Feige [10]). One of them is the restricted assignment problem[1], for which the second author in [14] developed the techniques from [1] to show that the configuration LP can be used to approximate the optimal makespan within a factor $33/17 + \epsilon$ improving upon the 2-approximation by Lenstra, Shmoys & Tardos [13]. Again it is not known how to efficiently find a schedule of the mentioned approximation guarantee. However, these results indicate that an improved understanding of the configuration LP is likely to lead to improved approximation algorithms for these fundamental allocation problems.

In this paper we make progress that further substantiates this point. We modify the local search of [1] and present a novel analysis that allows us to significantly improve the bound on the running time from an exponential guarantee to a quasi-polynomial guarantee.

Theorem 1. *For any $\epsilon \in (0, 1]$, we can find a $\frac{1}{4+\epsilon}$-approximate solution to restricted max-min fair allocation in time $n^{O\left(\frac{1}{\epsilon} \log n\right)}$, where $n = |\mathcal{P}| + |\mathcal{R}|$.*

In Section 3.1, we give an overview of the local search of [1] together with our modifications. The main modification is that at each point of the local search, we carefully select which step to take in the case of several options, whereas in the original description [1] an arbitrary choice was made. We then use this more stringent description with a novel analysis (Section 3.3) that uses the dual of the configuration LP as in [14]. The main advantage of our analysis (of the modified local search) is that it allows us to obtain a better upper bound on the search space of the local search and therefore also a better bound on the run-time. Furthermore, our techniques have the interesting property that although we use the rather complex configuration LP in the analysis, we never actually solve it. This gives hope to the interesting possibility of a polynomial time algorithm that is purely combinatorial and efficient to implement (in contrast to solving the configuration LP) with a good approximation ratio.

Finally, we note that our approach currently has a similar dependence on ϵ as in the case of solving the configuration LP since, as mentioned above, the linear program itself can only be solved approximately. However, our hidden constants are small and for a moderate ϵ we expect that our combinatorial approach is already more attractive than solving the configuration LP.

2 The Configuration LP

The intuition of the configuration linear program (LP) is that any allocation of value T needs to allocate a bundle or configuration C of resources to each player

---

[1] Also here the restricted version of the problem is the special case where $v_{ij} \in \{v_j, \infty\}$ (∞ instead of 0 since we are minimizing).

i so that $f_i(C) \geq T$. Let $\mathcal{C}(i,T)$ be the set of those configurations that have value at least T for player i. In other words, $\mathcal{C}(i,T)$ contains all those subsets of resources that are feasible to allocate to player i in an allocation of value T. For a guessed value of T, the configuration LP therefore has a decision variable $x_{i,C}$ for each player $i \in \mathcal{P}$ and configuration $C \in \mathcal{C}(i,T)$ with the intuition that this variable should take value one if and only if the corresponding set of resources is allocated to that player. The configuration LP $CLP(T)$ is a feasibility program and it is defined as follows:

$$
\sum_{C \in \mathcal{C}(i,T)} x_{i,C} \geq 1 \qquad i \in \mathcal{P}
$$

$$
\sum_{i,C:j\in C, C\in\mathcal{C}(i,T)} x_{i,C} \leq 1 \qquad j \in \mathcal{R}
$$

$$
x \geq 0
$$

The first set of constraints ensures that each player should receive at least one bundle and the second set of constraints ensures that a resource is assigned to at most one player.

If $CLP(T_0)$ for some T_0 is feasible, then $CLP(T)$ is also feasible for all $T \leq T_0$, because $\mathcal{C}(i,T_0) \subseteq \mathcal{C}(i,T)$ and thus a solution to $CLP(T_0)$ is a solution to $CLT(T)$ as well. Let T_{OPT} be the maximum of all such values. Every feasible allocation is a feasible solution of configuration LP, hence T_{OPT} is an upper bound on the value of the optimal allocation.

We note that the LP has exponentially many constraints; however, it is known that one can approximately solve it to any desired accuracy by designing a polynomial time (approximate) separation algorithm for the dual [3]. Although our approach does not require us to solve the linear program, the dual shall play an important role in our analysis. By associating a variable y_i with each constraint in the first set of constraints, a variable z_j with each constraint in the second set of constraints, and letting the primal have the objective function of minimizing the zero function, we obtain the dual program:

$$
\max \sum_{i\in\mathcal{P}} y_i - \sum_{j\in\mathcal{R}} z_j
$$

$$
y_i \leq \sum_{j\in C} z_j \qquad i \in \mathcal{P}, C \in \mathcal{C}(i,T)
$$

$$
y, z \geq 0
$$

3 Local Search with Better Run-Time Analysis

In this section we modify the algorithm by Asadpour et al. [1] in order to significantly improve the run-time analysis: we obtain a $1/(4 + \epsilon)$-approximate solution in run-time bounded by $n^{O(1/\epsilon \log n)}$ whereas the original local search is only known to converge in time $2^{O(n)}$. For better comparison, we can write $n^{O(1/\epsilon \log n)} = 2^{O(1/\epsilon \log^2 n)}$. Moreover, our modification has the nice side effect that we actually never solve the complex configuration LP — we only use it in the analysis.

3.1 Description of Algorithm

Throughout this section we assume that T — the guessed optimal value — is such that $CLP(T)$ is feasible. We shall find an $1/\alpha$ approximation where α is a parameter such that $\alpha > 4$. As we will see, the selection of α has the following trade-off: the closer α is to 4 the worse bound on the run-time we get.

We note that if $CLP(T)$ is not feasible and thus T is more than T_{OPT}, our algorithm makes no guarantees. It might fail to find an allocation, which means that $T > T_{OPT}$. We can use this for a standard binary search on the interval $[0, \frac{1}{|\mathcal{P}|} \sum_i v_i]$ so that in the end we find an allocation with a value at least T_{OPT}/α.

Max-min Fair Allocation Is a Bipartite Hypergraph Problem. Similar to [1], we view the max-min fair allocation problem as a matching problem in the bipartite hypergraph $G = (\mathcal{P}, \mathcal{R}, E)$. Graph G has an hyperedge $\{i\} \cup C$ for each player $i \in \mathcal{P}$ and configuration $C \subseteq \mathcal{R}$ that is feasible with respect to the desired approximation ratio $1/\alpha$, i.e., $f_i(C) \geq T/\alpha$, and minimal in the sense that $f_i(C') < T/\alpha$ for all $C' \subset C$. Note that the graph might have exponentially many edges and the algorithm therefore never keeps an explicit representation of all edges.

From the construction of the graph it is clear that a matching covering all players corresponds to a solution with value at least T/α. Indeed, given such a matching M in this graph, we can assign matched resources to the players and everyone gets resources with total value of at least T/α.

Alternating Tree of "Add" and "Block" Edges. The algorithm of Asadpour et al. [1] can be viewed as follows. In the beginning we start with an empty matching and then we increase its size in every iteration by one, until all players are matched. In every iteration we build an alternating tree rooted in a currently unmatched player p_0 in the attempt to find an alternating path to extend our current matching M. The alternating tree has two types of edges: edges in the set A that we wish to *add* to the matching and edges in the set B that are currently in the matching but intersect edges in A and therefore *block* them from being added to the matching. While we are building the alternating tree to find an alternating path, it is important to be careful in the selection of edges, so as to guarantee eventual termination. As in [1], we therefore define the concept of addable and blocking edges.

Before giving these definitions, it will be convenient to introduce the following notation. For a set of edges F, we denote by $F_\mathcal{R}$ all resources contained in edges in F and similarly $F_\mathcal{P}$ denotes all players contained in edges in F. We also write $e_\mathcal{R}$ instead of $\{e\}_\mathcal{R}$ for an edge e and use $e_\mathcal{P}$ to denote the player in e.

Definition 1. *We call an edge e addable if $e_\mathcal{R} \cap (A_\mathcal{R} \cup B_\mathcal{R}) = \emptyset$ and $e_\mathcal{P} \in \{p_0\} \cup A_\mathcal{P} \cup B_\mathcal{P}$.*

Definition 2. *An edge b in the matching M is blocking e if $e_\mathcal{R} \cap b_\mathcal{R} \neq \emptyset$.*

Note that an addable edge matches a player in the tree with resources that currently do not belong to any edge in the tree and that the edges blocking an edge e are exactly those in the matching that prevent us from adding e. For a more intuitive understanding of these concepts see Figure 1 in Section 3.2.

The idea of building an alternating tree is similar to standard matching algorithms using augmenting paths. However, one key difference is that the matching can be extended once an alternating path is found in the graph case, whereas the situation is more complex in the considered hypergraph case, since a single hyperedge might overlap several hyperedges in the matching. It is due to this complexity that it is more difficult to bound the running time of the hypergraph matching algorithm of [1] and our improved running time is obtained by analyzing a modified version where we carefully select in which order the edges should be added to the alternating tree and drop edges from the tree beyond certain distance.

We divide resources into 2 groups. *Fat resources* have value at least T/α and *thin resources* have less than T/α. Thus any edge containing a fat resource contains only one resource and is called *fat edge*. Edges containing thin resources are called *thin edges*. Our algorithm always selects an addable edge of minimum distance to the root p_0 according to the following convention. The length of a thin edge in the tree is one and the length of a fat edge in the tree is zero. Edges not in the tree have infinite length. Hence, the *distance of a vertex* from the root is the number of thin edges between the vertex and the root and, similarly, the *distance of an edge e* is the number of thin edges on the way to e from p_0 including e itself. We also need to refer to distance of an addable edge that is not yet in the tree. In that case we take the distance as if it was in the tree. Finally, by the *height of the alternating tree* we refer to the maximum distance of a resource from the root.

Algorithm for Extending a Partial Matching. Algorithm 1 summarizes the modified procedure for increasing the size of a given matching by also matching a previously unmatched player p_0. For better understanding of the algorithm, we included an example of an algorithm execution in Figure 1 in Section 3.2.

Input : A partial matching M
Output: A matching of increased size assuming that T is at most T_{OPT}

Find an unmatched player $p_0 \in \mathcal{P}$, make it a root of the alternating tree
while *there is an addable edge within distance* $2\log_{(\alpha-1)/3}(|\mathcal{P}|) + 1$ **do**
    Find an addable edge e of minimum distance from the root
    $A \leftarrow A \cup \{e\}$
    if *e has blocking edges* $b_1, \ldots, b_k$ **then**
        $B \leftarrow B \cup \{b_1, \ldots, b_k\}$
    else// collapse procedure
        while *e has no blocking edges* **do**
            if *there is an edge* $e' \in B$ *such that* $e'_{\mathcal{P}} = e_{\mathcal{P}}$ **then**
                $M \leftarrow M \setminus \{e'\} \cup \{e\}$
                $A \leftarrow A \setminus \{e\}$
                $B \leftarrow B \setminus \{e'\}$
                Let $e'' \in A$ be the edge that e' was blocking
                $e \leftarrow e''$
            else
                $M \leftarrow M \cup \{e\}$
                return M
            end if
        end while
        Drop from A and B all edges of greater or the same distance as e
    end if
end while
return T_{OPT} is less than T

Algorithm 1. Increase the size of the matching

Note that the algorithm iteratively tries to find addable edges of minimum distance to the root. On the one hand, if the picked edge e has blocking edges that prevents it from being added to the matching, then the blocking edges are added to the alternating tree and the algorithm repeatedly tries to find addable edges so as to make progress by removing the blocking edges.

On the other hand, if edge e has no blocking edges, then this means that the set of resources $e_{\mathcal{R}}$ is free, so we make progress by adding e to the matching M. If the player was not previously matched, it is the root p_0 and we increased the size of the matching. Otherwise the player $e_{\mathcal{P}}$ was previously matched by an edge $e' \in B$ such that $e'_{\mathcal{P}} = e_{\mathcal{P}}$, so we remove e' from M and thus it is not a blocker anymore and can be removed from B. This removal has decreased the number of blockers for an edge $e'' \in A$. If e'' has 0 blockers, we recurse and repeat the same procedure as with e. Note that this situation can be seen on Figure 1(b) and 1(c) in Section 3.2.

3.2   Example of Algorithm Execution

Figure 1 is a visualization of an execution of Algorithm 1. The right part of every picture is the alternating tree and to the left we display the positions of edges in the tree in the bipartite graph. Gray edges are A-edges and white are B-edges.

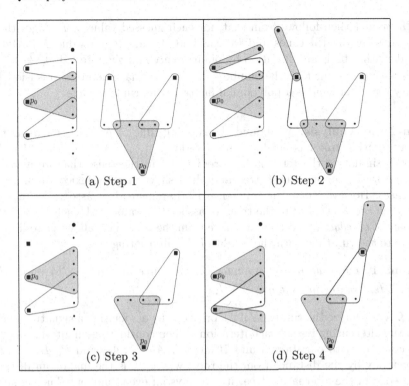

Fig. 1. Alternating tree visualization. The right part of every picture is the alternating tree and to the left we display the positions of edges in the tree in the bipartite graph. Gray edges are in the set A and white edges are in the set B.

In Figure 1(a) we start by adding an A-edge to the tree. There are 2 edges in the matching intersecting this edge, so we add them as blocking edges. Then in Figure 1(b) we add a fat edge that has no blockers, so we add it to the matching and thus remove one blocking edge, as we can see in Figure 1(c). Then in Figure 1(d) we add a thin edge which has no blockers. Now the A and B edges form an alternating path, so by swapping them we increase the size of the matching and the algorithm terminates.

Note that the fat edge in step 2 is added before the thin edge from step 4, because it has shorter distance from the root p_0. Recall that the distance of an edge e is the number of thin edges between e and the root including e, thus the distance of the fat edge is 2 and the distance of the thin edge is 3.

3.3   Analysis of Algorithm

Let the parameter α of the algorithm equal $4 + \epsilon$ for some $\epsilon \in (0, 1]$. We first prove that Algorithm 1 terminates in time $n^{O\left(\frac{1}{\epsilon} \log n\right)}$ where $n = |\mathcal{P}| + |\mathcal{R}|$ and, in the following subsection, we show that it returns a matching of increased size if $CLP(T)$ is feasible.

Theorem 1 then follows from that, for each guessed value of T, Algorithm 1 is at most invoked n times and we can find the maximum value T for which our algorithm finds an allocation by binary search on the interval $[0, \frac{1}{|\mathcal{P}|} \sum_i v_i]$. Since we can assume that the numbers in the input have bounded precision, the binary search only adds a polynomial factor to the running time.

Run-Time Analysis. We bound the running time of Algorithm 1 using that the alternating tree has height at most $O(\log_{(\alpha-1)/3} |\mathcal{P}|) = O\left(\frac{1}{\epsilon} \log |\mathcal{P}|\right)$. The proof is similar to the termination proof in [1] in the sense that we associate a signature vector with each tree and then show that its lexicographic value decreases. However, one key difference is that instead of associating a value with *each edge* of type A in the tree, we associate a value with each "layer" that consists of *all edges* of a certain distance from the root. This allows us to directly associate the run-time with the height of the alternating tree.

Lemma 1. *For a desired approximation guarantee of $1/\alpha = 1/(4 + \epsilon)$, Algorithm 1 terminates in time $n^{O\left(\frac{1}{\epsilon} \log n\right)}$.*

Proof. We analyze the run-time of Algorithm 1 by associating a signature vector with the alternating tree of each iteration. When considering an alternating tree it is convenient to partition A and B into $A_0, A_1, \ldots, A_{2k}$ and $B_0, B_1, \ldots, B_{2k}$ respectively by the distance from the root, where $2k$ is the maximum distance of an edge in the alternating tree (it is always an even number). The signature vector of an alternating tree is then defined to be

$$(-|A_0|, |B_0|, -|A_1|, |B_1|, \ldots, -|A_{2k}|, |B_{2k}|, \infty).$$

We prove that each addition of an edge decreases the lexicographic value of the signature or increases the size of the matching.

On the one hand, if we add an edge with no blocking edges, we either completely collapse the alternating tree or collapse only a part of it and change the signature to $(-|A_0|, |B_0|, \ldots, -|A_{2\ell}|, |B_{2\ell}| - 1, \infty)$ for some $0 \le \ell \le k$ as the algorithm drops all edges farther away or in the same distance from the root as the edge last added to the matching. Thus we either increase the size of the matching or decrease the signature of the alternating tree.

On the other hand, if the added edge e has blocking edges, there are two cases. We either open new layers $A_{2k+1} = \{e\}$ and B_{2k+2} where e is a thin edge and the signature gets smaller, since $-|A_{2k+1}| < \infty$. If we do not open a new layer, we increase the size of some A_ℓ and $-(|A_\ell| + 1) < -|A_\ell|$, so in this case the signature decreases too.

The algorithm only runs as long as the height of the alternating tree is at most $O(\log_{(\alpha-1)/3} |\mathcal{P}|) = O(\log_{1+\epsilon/3} |\mathcal{P}|)$. This can be rewritten as $O\left(\frac{\log |\mathcal{P}|}{\log(1+\epsilon/3)}\right) = O\left(\frac{\log |\mathcal{P}|}{\epsilon}\right)$ where the equality follows from $x \le 2\log(1 + x)$ for $x \in (0, 1]$ and we only consider $\epsilon \in (0, 1]$. There are at most $|\mathcal{P}|$ possible values for each position in a signature, so the total number of signatures encountered during the execution of Algorithm 1 is $|\mathcal{P}|^{O\left(\frac{1}{\epsilon}\log|\mathcal{P}|\right)}$. As adding an edge happens in polynomial time in $n = |\mathcal{P}| + |\mathcal{R}|$, we conclude that Algorithm 1 terminates in time $n^{O\left(\frac{1}{\epsilon}\log n\right)}$. □

Correctness of Algorithm 1. We show that Algorithm 1 is correct, i.e., that it returns an increased matching if $CLP(T)$ is feasible.

We have already proved that the algorithm terminates in Lemma 1. The statement therefore follows from proving that the condition of the while loop always is satisfied assuming that the configuration LP is feasible. In other words, we will prove that there always is an addable edge within the required distance from the root. This strengthens the analogous statement of [1] that states that there always is an addable edge (but without the restriction on the search space that is crucial for our run-time analysis). We shall do so by proving that the number of thin blocking edges increases quickly with respect to the height of the alternating tree and, as there cannot be more than $|\mathcal{P}|$ blocking edges, this in turn bounds the height of the tree.

For this purpose, let us again partition A and B into $A_0, A_1, \ldots, A_{2k}$ and $B_0, B_1, \ldots, B_{2k}$ respectively by the distance from the root. Note that B_i is empty for all odd i. Also, A_{2i} contains only fat edges and A_{2i+1} only thin edges. For a set of edges F denote by F^t all the thin edges in F and by F^f all the fat edges in F. We also use $\mathcal{R}^t$ to denote thin resources and $\mathcal{R}^f$ to denote fat resources.

We are now ready to state the key insight behind the analysis that shows that the number of blocking edges increases as a function of α and the height of the alternating tree.

Lemma 2. *Let $\alpha > 4$. Assuming that $CLP(T)$ is feasible, if there is no addable edge e within distance $2D + 1$ from the root for some integer D, then*

$$\frac{\alpha - 4}{3} \sum_{i=1}^{D} |B_{2i}^t| < |B_{2D+2}^t|.$$

Before giving the proof of Lemma 2, let us see how it implies that there always is an addable edge within distance $2\log_{(\alpha-1)/3}(|\mathcal{P}|) + 1$ from the root assuming the configuration LP is feasible, which in turn implies the correctness of Algorithm 1.

Corollary 1. *If $\alpha > 4$ and $CLP(T)$ is feasible, then there is always an addable edge within distance $2D + 1$ from the root, where $D = \log_{(\alpha-1)/3} |\mathcal{P}|$.*

The proof of the corollary follows intuitively from that Lemma 2 says that the number of blocking edges increases exponentially in terms of the height of the tree and therefore, as there are at most $|\mathcal{P}|$ blocking edges, the height must be $O_\alpha(\log|\mathcal{P}|)$. The detailed proof can be found in the full version of this paper. We complete the correctness analysis of the algorithm by presenting a proof sketch of the key lemma.

Proof (Lemma 2). We give an overview of the proof of Lemma 2 (the complete proof can be found in the full version of this paper).

Suppose toward contradiction that there is no addable edge within distance $2D + 1$ and

$$\frac{\alpha - 4}{3} \sum_{i=1}^{D} |B_{2i}^t| \geq |B_{2D+2}^t|.$$

We show that this implies that the dual of the configuration LP is unbounded, which in turn contradicts the assumption that the primal is feasible. Recall that the objective function of the dual is $\max \sum_{i \in \mathcal{P}} y_i - \sum_{j \in \mathcal{R}} z_j$. Furthermore, as each solution (y, z) of the dual can be scaled by a scalar c to obtain a new solution $(c \cdot y, c \cdot z)$, any solution with positive objective implies unboundedness. We proceed by defining such solution (y^*, z^*), that is determined by the alternating tree. More precisely, we take

$$y_i^* = \begin{cases} \frac{\alpha - 1}{\alpha} & \text{if } i \in \mathcal{P} \text{ is within distance } 2D \text{ from the root,} \\ 0 & \text{otherwise,} \end{cases}$$

and

$$z_j^* = \begin{cases} (\alpha - 1)/\alpha & \text{if } j \in \mathcal{R} \text{ is fat and within distance } 2D \text{ from the root,} \\ v_j/T & \text{if } j \in \mathcal{R} \text{ is thin and within distance } 2D + 2 \text{ from the root,} \\ 0 & \text{otherwise.} \end{cases}$$

It can be shown that (y^*, z^*) is a feasible solution provided that there is no addable edge within distance $2D + 1$.

The proof is completed by showing that

$$\sum_{j \in \mathcal{R}} z_j \leq \frac{\alpha - 1}{\alpha} \sum_{i=0}^{D} |B_{2i}^f| + \frac{\alpha - 1}{\alpha} \sum_{i=1}^{D} |B_{2i}^t| < \frac{\alpha - 1}{\alpha} \left(1 + \sum_{i=0}^{D} |B_{2i}|\right) = \sum_{i \in \mathcal{P}} y_i,$$

so the dual is unbounded, which contradicts the assumption that the primal is feasible. □

4 Conclusions

Asadpour et al. [1] raised as an open question whether their local search (or a modified variant) can be shown to run in polynomial time. We made progress toward proving this statement by showing that a modified local search procedure finds a solution in quasi-polynomial time. Moreover, based on our findings, we conjecture the stronger statement that there is a local search algorithm that does not use the LP solution, i.e., it is combinatorial, and it finds a $1/(4 + \epsilon)$-approximate solution in polynomial time for any fixed $\epsilon > 0$.

References

1. Asadpour, A., Feige, U., Saberi, A.: Santa claus meets hypergraph matchings. In: Proceedings of the 11th International Workshop and 12th International Workshop on Approximation, Randomization and Combinatorial Optimization, pp. 10–20 (2008); see authors' homepages for the lower bound of 1/4 instead of the claimed 1/5 in the conference version

2. Asadpour, A., Saberi, A.: An approximation algorithm for max-min fair allocation of indivisible goods. In: Proceedings of the Thirty-Ninth Annual ACM Symposium on Theory of Computing, STOC 2007, pp. 114–121. ACM, New York (2007)

3. Bansal, N., Sviridenko, M.: The santa claus problem. In: Proceedings of the Thirty-Eighth Annual ACM Symposium on Theory of Computing, STOC 2006, pp. 31–40. ACM Press, New York (2006)

4. Bateni, M., Charikar, M., Guruswami, V.: Maxmin allocation via degree lower-bounded arborescences. In: Proceedings of the 41st Annual ACM Symposium on Theory of Computing, STOC 2009, pp. 543–552. ACM, New York (2009)

5. Bezáková, I., Dani, V.: Allocating indivisible goods. SIGecom. Exch. 5, 11–18 (2005)

6. Chakrabarty, D., Chuzhoy, J., Khanna, S.: On allocating goods to maximize fairness. In: Proceedings of the 2009 50th Annual IEEE Symposium on Foundations of Computer Science FOCS 2009, pp. 107–116. IEEE Computer Society Press, Washington, DC (2009)

7. Dobzinski, S., Schapira, M.: An improved approximation algorithm for combinatorial auctions with submodular bidders. In: Proceedings of the Seventeenth Annual ACM-SIAM Symposium on Discrete Algorithm, SODA 2006, pp. 1064–1073. ACM, New York (2006)

8. Feige, U.: On maximizing welfare when utility functions are subadditive. In: Proceedings of the Thirty-Eighth Annual ACM Symposium on Theory of Computing, STOC 2006, pp. 41–50. ACM, New York (2006)

9. Feige, U.: On allocations that maximize fairness. In: Proceedings of the Nineteenth Annual ACM-SIAM Symposium on Discrete Algorithms, SODA 2008, pp. 287–293. Society for Industrial and Applied Mathematics, Philadelphia (2008)

10. Feige, U.: On estimation algorithms vs approximation algorithms. In: Hariharan, R., Mukund, M., Vinay, V. (eds.) IARCS Annual Conference on Foundations of Software Technology and Theoretical Computer Science (FSTTCS 2008). Leibniz International Proceedings in Informatics (LIPIcs), vol. 2, pp. 357–363. Schloss Dagstuhl–Leibniz-Zentrum fuer Informatik, Dagstuhl, Germany (2008)

11. Feige, U., Vondrak, J.: Approximation algorithms for allocation problems: Improving the factor of 1 - 1/e. In: 47th Annual IEEE Symposium on Foundations of Computer Science, FOCS 2006, pp. 667–676 (October 2006)

12. Haeupler, B., Saha, B., Srinivasan, A.: New constructive aspects of the lovasz local lemma. In: 2010 51st Annual IEEE Symposium on Foundations of Computer Science (FOCS), pp. 397–406 (October 2010)

13. Lenstra, J.K., Shmoys, D.B., Tardos, É.: Approximation algorithms for scheduling unrelated parallel machines. Math. Program. 46, 259–271 (1990)

14. Svensson, O.: Santa claus schedules jobs on unrelated machines. In: Proceedings of the 43rd Annual ACM Symposium on Theory of Computing, STOC 2011, pp. 617–626. ACM, New York (2011)

15. Vondrak, J.: Optimal approximation for the submodular welfare problem in the value oracle model. In: Proceedings of the 40th Annual ACM Symposium on Theory of Computing, STOC 2008, pp. 67–74. ACM, New York (2008)

Strictly-Black-Box Zero-Knowledge and Efficient Validation of Financial Transactions

Michael O. Rabin[1], Yishay Mansour[2,*], S. Muthukrishnan[3], and Moti Yung[4]

[1] Harvard University, Hebrew University
[2] Tel-Aviv University
[3] Google Inc. and Rutgers University
[4] Google Inc. and Columbia University

Abstract. Zero Knowledge Proofs (ZKPs) are one of the most striking innovations in theoretical computer science. In practice, the prevalent ZKP methods are, at times, too complicated to be useful for real-life applications. In this paper we present a practically efficient method for ZKPs which has a wide range applications. Specifically, motivated by the need to provide an upon-demand efficient validation of various financial transactions (e.g., the high-volume Internet auctions), we have developed a novel secure and highly efficient method for validating correctness of the output of a transaction while keeping input values secret. The method applies to input values which are publicly committed to by employing generic commitment functions (even input values submitted using tamper-proof hardware solely with input/ output access can be used.) We call these: strictly black box [SBB] commitments. Hence these commitments are typically much faster than public-key ones, and are the only cryptographic/ security tool we give the poly-time players, throughout. The general problem we solve in this work is: Let SLC be a publicly known staight line computation on n input values taken from a finite field and having k output values. The inputs are publicly committed to in a SBB manner. An Evaluator performs the SLC on the inputs and announces the output values. Upon demand the Evaluator, or a Prover acting on his behalf, can present to a Verifier a proof of correctness of the announced output values. This is done in a manner that (1) The input values as well as all intermediate values of the SLC remain information theoretically secret. (2) The probability that the Verifier will accept a false claim of correctness of the output values can be made exponentially small. (3) The Prover can supply any required number of proofs of correctness to multiple Verifiers. (4) The method is highly efficient. The application to financial processes is straight forward. To this end (1) we first use a novel technique for representation of values from a finite field which we call "split representation", the two coordinates of the split representation are generically committed to; (2) next, the SLC is augmented by the Prover into a "translation" which is presented to the Verifier as a

* This research was supported in by The Israeli Centers of Research Excellence (I-CORE) program, (Center No. 4/11), by a grant from the Israel Science Foundation (ISF), by a grant from United States-Israel Binational Science Foundation (BSF), and by a grant from the Israeli Ministry of Science (MoS).

A. Czumaj et al. (Eds.): ICALP 2012, Part I, LNCS 7391, pp. 738–749, 2012.

sequence of generically committed split representations of values; (3) using the translation, the Prover and Verifier conduct a secrecy preserving proof of correctness of the announced SLC output values; (4) in order to exponentially reduce the probability of cheating by the Prover and also to enable multiple proofs, a novel highly efficient method for preparation of any number of committed-to split representations of the n input values is employed. The extreme efficiency of these ZK methods is of decisive importance for large volume applications. Secrecy preserving validation of announced results of Vickrey auctions is our demonstrative example.

1   Introduction

Many current methods for validation of auctions employ "additive homomorphic encryptions." The Paillier encryption [8], in particular, employs a public integer $n = p \cdot q$ where p, q, are large primes constituting the private key. A value $x \in [0, n - 1]$ is encrypted, using n, as $C = E(x, r)$ where r is a help value. Given two ciphers $C_i = E(x_i, r_i), i = 1, 2$, then $(C_1 \cdot C_2) \bmod n^2 = E((x_1 + x_2) \bmod n, (r_1 \cdot r_2) \bmod n)$. In [1-7] numerous applications of Paillier encryption to secure auctions are given. In [11] Paillier encryption is applied to combinatorial clock-proxy auctions. There are drawbacks to the use of homomorphic encryptions for verifying financial processes. Practicality, [10] employs Paillier encryption for providing secrecy preserving proofs of correctness of Vickery auctions. A proof of correctness of a 100 bidder auction required 800 minutes.

This work uses values $x \in F_p$ where p is a prime. A 128-bit prime is adequate for all applications. A value x is randomly represented by a vector $X = (u, v)$ such that $(u + v) \bmod p = x$, where $u \in F_p$ is randomly chosen. A value x appearing in a proof of correctness of outputs of a straight line computation is represented by a vector X and is committed to by $COM(X) = (COM(u), COM(v))$ where COM is any generic information- theoretic hiding and computationally binding commitment function. In proofs of correctness, the Prover never opens/reveals both coordinates of a commitment $COM(X)$ to a representation of a value x appearing the proof.

The method of using split representations of values and generic commitments appears in Kilian's [19] who credits it to [20]. However, Kilian only treats values in F_2, i.e. bits, and implements correctness proofs for boolean operations on bits. Also, he deals with ZKPs by a Prover without inputs from others, such as bidders in the case of auctions.

In [9] there appears the first use of random split representations of values $x \in F_p$ for a general prime p as well as information theoretic value hiding proofs for straight line computations involving $(x + y) \bmod p = z$, $(x \cdot y) \bmod p = z$, and the predicate $x \leq y$. The method of [9] allows a Prover to provide only a single ZK proof of correctness. Once this is done and commitments to components values are opened, another proof is impossible. A 100 bidder Vickery auction is ZK verified by this method in 4 minutes.

The present work greatly improves over [9] in several ways. The ZK proofs of correctness are considerably simpler and faster. The probability of a Verifier accepting a false claim by the Prover is smaller and better analyzed. Most importantly, the Prover can supply to different Verifiers any required number of verifications. A 100 bidder Vickery auction is ZK verified by this new method in 2 ms. This new work enables repeated ZK verifications of large and large volume auctions and other financial processes.

The full version of the paper will include a full discussion of the SBB ZK proofs, as well as all the omitted proofs.

2   Representation, Commitment, and Translation

2.1   Validation Domain: The Financial Application Domain Settings

Consider as an example a Vickrey auction. An Auctioneer AU receives sealed bids $x_1, \cdots, x_n$ from bidders $P_1, \cdots, P_n$. At closing time the bid values are revealed to AU and he determines that, say, P_1 is the winner and he should pay x_2. Proving correctness of the announced result involves proving $x_1 > x_2$ and $x_2 \geq x_3, \cdots, x_2 \geq x_n$. After announcement of the result, the AU acting as a Prover, wants to prove correctness of the result to a Verifier, say one or all of the bidders, or to a judge, and this without revealing any of the bid values.

In general, the above example and many other real life situations are captured by a model where an Evaluator Prover EP who receives inputs from participants (i.e., they post commitments to the values and decommit privately to EP). EP, in turn, computes certain output values from these inputs (the computational procedure is efficient and public), and announces these "outputs." Later on, possibly upon demand, the EP provides a ZKP of correctness of the output values (given the public commitments) to a Verifier.

2.2   Inputs and Straight Line Computations

In our setting we assume that all inputs, constants, intermediate values and outputs of the EP's calculations are values smaller than $p/32$ where p is a known prime number, say $p \sim 2^{128}$. Our computations are in the finite field F_p so that for $x, y \in F_p$, $x + y$ and $x \cdot y$ are abbreviations for $(x + y) \mod p$ and $x \cdot y \mod p$. For financial applications the range of values $0 \leq x < 2^{128}$ is adequate.

Definition 1. *A straight line computation (SLC) on inputs $x_1, \ldots, x_n$ in F_p with k outputs $x_{L+1}, \ldots, x_{L+k}$ is a sequence*

$$\text{SLC} = x_1, \ldots, x_n, x_{n+1}, \ldots, x_{L+1}, \ldots, x_{L+k} \tag{1}$$

where for all $m > n$ there exist $i, j < m, L$ such that

- $x_m = x_i + x_j$, *or*
- $x_m = x_i \cdot x_j$, *or*
- $x_m = x_i$, *or*
- $x_m = TruthValue(x_i \leq x_j)$. *These x_m are restricted to output values.*

An example of the SLC for the output $x_1 + \ldots + x_n$ is: $x_1, \ldots, x_n, x_{n+1}, \ldots, x_{2n-1}$, where

$$x_{n+1} = x_1 + x_2, x_{n+2} = x_{n+1} + x_3, \ldots \qquad (2)$$

We now come to the main construct for enabling ZKP's for the correctness of the results $x_{L+1}, \ldots, x_{L+k}$ of the SLC in the above definition (with the strictly-black-box restriction as a SBB proof). It also presents a key component of our overall input representation.

Definition 2. *Let $x \in F_p$ be a value. A random representation $RR(x)$ of x is a vector $X = (u, v)$ where $u, v \in F_p, u$ was chosen randomly (notation $u \leftarrow F_p$) and $x = (u + v) \bmod p$. For a vector $X = (u, v)$ we denote $\mathtt{val}(X) = (u + v) \bmod p$.*

The method for creating a $RR(x) = (u, v)$ of x is to randomly chose $u \leftarrow F_p$ and set $v = (x - u) \bmod p$. Note that from u (or v) by itself, no information about x can be deduced.

2.3   Generic Commitment Schemes

We can use any commitment scheme we wish (i.e., a generic commitment). We can use physically secure and physically binding scheme (based on physical assumptions) or any of the two kinds of commitments. Let us define one for concreteness: We express our work in terms of a generic commitment function COM, which is information theoretic hiding and computationally binding and is used as a "black box".

Definition 3. *An information theoretic hiding and computationally binding generic (black-box) commitment for values $u \in F_p$, is a function $COM : F_p \times [0, m-1] \longrightarrow R$, where R is a set of m elements, such that:*

For any fixed $u \in F_p$, $\{COM(u, r) | r \in [0, m-1]\} = R$. I.e. for a fixed u, the mapping $COM(u, r)$ is 1-1 from $[0, m-1]$ onto R.

It is assumed that COM is computationally collision-free. I.e. finding two different pairs $(u_1, r_1), (u_2, r_2)$ such that $COM(u_1, r_1) = COM(u_2, r_2)$ is not possible by a polynomial-time algorithm.

To commit to a value $u \in F_p$, a committer Alice randomly selects a random help value $r \in [0, m-1]$, obtains from the Black Box the commitment value $c = COM(u, r)$ and submits c to the receiver Bob or posts it.

To de-commit c, Alice submits to Bob, or posts, the pair (u, r). Bob, or anyone else, has the result COM(r, u) computed by the Black Box on the decommitted values and verifies the equality of the commitment value c to the newly obtained value $COM(u, r)$.

Because $m = |R|$, and the above 1-1 property for fixed $u \in F_p$, this commitment is clearly information theoretic hiding.

Remark: Note that if the scheme is implemented via a physical envelope it is also information theoretically binding (the only way to get the value is to open the envelope). The literature often employs the highly structured Discrete-Log-based Pedersen commitment function [18].

Definition 4. *Let $X = (u, v)$ be a representation of $x = (u + v) \bmod p$, then a commitment to X is defined as $\mathrm{COM}(X) = (COM(u, r_1), COM(v, r_2))$, where , $r_1, , r_2$ are randomly chosen help values.*

2.4   The Main Theorem: SBB ZK Arguments for SLC

Theorem 1. *Let EP be computationally bounded prover. Having posted generic black-box information-theoretic hiding "split commitments"*

$$\mathrm{COM}(X_1), \cdots, \mathrm{COM}(X_n), \tag{$*$}$$

of representations $X_1, \cdots X_n$ of values $x_1, \cdots, x_n$ in F_p , the EP can create a translation

$$TR = COM(X_1), \cdots, \mathrm{COM}(X_n), \mathrm{COM}(Y_{n+1}), \cdots, \mathrm{COM}(Y_M), x_{L+1}, \cdots, x_{L+k}, \tag{$**$}$$

of the public SLC (1) so that:

1. *Using the translation TR, the EP can conduct a two round interactive Zero Knowledge Argument for the statement that $x_{L+1}, \cdots, x_{L+k}$, are the correct output values of the publicly known SLC (1) [namely, completeness holds].*
2. *The proof is information theoretic hiding [i.e., there is a ZK simulator in the SBB model, and in fact if run over commitment of canonical input the visible transcript has the same distribution, i.e., the proof is Witness Indistinguishable].*
3. *The probability of EP cheating is at most 3/4 [i.e., this is the soundness error, and it is implies validity as in proof of correctness (probability of extracting is bounded by this value)].*
4. *Finally, the length TR is at most $11 \cdot L$.*

3   Proving Correctness of Additions and Equalities

We now show how the EP can prove to a Verifier correctness of an equation (3) for posted commitments (4). Let $X = (u_1, v_1), Y = (u_2, v_2)$, and $Z = (u_3, v_3)$, be random representations of the values x, y, and z. Note that

$$val(X) + val(Y) = val(Z) \tag{3}$$

if and only if there exists a value w such that $X + Y = Z + (w, -w)$.

The EP has prepared commitments

$$
\begin{aligned}
COM(X) &= [COM(u_1, r_1), COM(v_1, s_1)], COM(Y) \\
&= [COM(u_2, r_2), COM(v_2, s_2)], COM(Z) \\
&= [COM(u_3, r_3), COM(v_3, s_3)]
\end{aligned} \tag{4}
$$

The EP posts the commitments (4) or sends them to the Verifier and claims that the hidden vectors X, Y, Z, satisfy the equation (3).

When challenged to prove this claim, the EP posts or sends to Verifier the above value w. The Verifier now presents to EP a randomly chosen challenge $c \leftarrow \{1, 2\}$.

Assume that c = 1. The EP de-commits /reveals to Verifier $u_j, r_j, j = 1, 2, 3$. The Verifier checks the commitments, i.e. computes $\text{COM}(u_j, r_j), j = 1,2,3$ and compares to the posted first coordinates of $\text{COM}(X), \text{COM}(Y), COM(Z)$.

The Verifier next checks that $u_1 + u_2 = u_3 + w$. If c = 2 was chosen, then the Verifier checks that $v_1 + v_2 = v_3 - w$. The following two theorems are immediately obvious.

Theorem 2. *If the equation (3) is not true for the vectors committed in* $\text{COM}(X)$, $\text{COM}(Y)$, $COM(Z)$, *then Verifier will accept with probability at most* $1/2$ *the claim that (3) holds.*

Proof. Under our assumption about the COM function being computationally binding, the EP can open the commitments for $u_j, v_j, j = 1, 2, 3$, in only one way. Now, if (3) does not hold then at least one of the equations $u_1 + u_2 = u_3 + w$, or $v_1 + v_2 = v_3 - w$ is not true. So the probability that a random challenge $c \leftarrow \{1, 2\}$ will not uncover the falsity of the claim (3) is less than $1/2$.

Theorem 3. *The above interactive proof between EP and Verifier reveals nothing about the values* $\text{val}(X)$, $\text{val}(Y), \text{val}(Z)$ *beyond, if successful, that the claim that (3) is (actually may be) true.*

Proof. We note that the interactive proof involves only the revelation of either all the first coordinates of the "split representation" based commitments, or of all the second coordinates, of X, Y, Z. Assume that Verifier's challenge was c = 1. The only revealed values were random u_1, u_2, u_3, w which satisfy $u_1 + u_2 = u_3 + w$. Because the commitment function $C(\, , \,)$ is information theoretically hiding, the un-opened second coordinates in the commitments (4) of $\text{COM}(X), \text{COM}(Y), \text{COM}(Z)$, are consistent with any three values $v_{1,1}, v_{2,2}, , v_{3,3}$, satisfying $v_{1,1} + v_{2,2} = v_{3,3} - w$. Thus the interactive proof is consistent with any three vectors X_1, Y_1, Z_1 satisfying the sum equality (3)– any consistent triple can serve as an alternative input as in Witness Indistinguishable (WI) proofs.

It is clear how to similarly create and conduct ZK Arguments for the correctness of a claim $\text{val}(X) = \text{val}(Y)$, for given commitments $\text{COM}(X), \text{COM}(Y)$.

4   Proving Correctness of Multiplications

For proving correctness of the operations of multiplication $x_m = x_i \cdot x_j$ in the SLC, the EP will have presented to Verifier commitments $\text{COM}(X_m)$, $\text{COM}(X_i)$, $\text{COM}(X_j)$ for random representations of the values x_m, x_i, x_j. The EP has to prove to Verifier that

$$\text{val}(X_i) \cdot val(X_j) = val(X_m) \tag{5}$$

Let $X_i = (u_1, v_1)$, $X_j = (u_2, v_2)$, and $X_m = (u_3, v_3)$. The EP prepares auxiliary vectors $Z_0 = (u_1 \cdot u_2, v_1 \cdot v_2)$, $Z_1 = (u_1 \cdot v_2 + w_1, p - w_1)$, $Z_2 = (u_2 \cdot v_1 + w_2, p - w_2)$, where w_1, w_2 are randomly chosen values. The EP augments the commitments presented to Verifier into:

$$\text{COM}(X_m), \text{COM}(X_i), \text{COM}(X_j), \text{COM}(Z_0), \text{COM}(Z_1), \text{COM}(Z_2) \tag{6}$$

Clearly (5) holds if the following Aspects 0-4 hold true for the vectors committed in (6):

- Aspect 0: Z_0 is $(u_1 \cdot u_2, v_1 \cdot v_2)$.
- Aspect 1: $\text{val}(Z_1) = u_1 \cdot v_2$.
- Aspect 2: $\text{val}(Z_2) = u_2 \cdot v_1$.
- Aspect 4: $\text{val}(X_m) = val(Z_0) + val(Z_1) + val(Z_2)$.

In the interactive proof/verification either Aspects 0 and 4 are checked together, or Aspect 1, or Aspect 2 are separately checked. The Verifier randomly chooses with probability $1/2$ to verify Aspect 0 and the addition in Aspect 4. He randomly chooses $c \leftarrow \{1, 2\}$. Say $c = 1$. The EP reveals the first coordinates of X_m, X_i, X_j and Z_0. Aspect 0 is verified. Aspect 4 is verified in the manner of verification of addition, see Section 3. If the EP's claim is false with respect to Aspect 0 or Aspect 4, then the probability of Verifier accepting is at most $3/4 = 1 - (1/2) \cdot (1/2)$.

The Verifier chooses to check either Aspect 1 or Aspect 2, each with probability $1/4$. Say Aspect 1 was chosen by Verifier. The EP reveals the first coordinate u_1 of X_i and the second coordinate v_2 of X_j and both coordinates of Z_1 and checks the equality of Aspect 1. Note that if Aspect 1 is false and is chosen for verification then Verifier will never accept. Similarly for Aspect 2. Consequently, if (5) is false and the proof of correctness (5) presented by EP to Verifier is false in Aspect 1, or Aspect 2, then the probability that Verifier will accept is at most $3/4$.

Altogether we have:

Theorem 4. *If the product claim is false then the probability that the Verifier will accept EP s proof of correctness is at most 3/4.*

To achieve the information-theoretic ZK property of the above interactive proof of correctness we require an additional step in EP's construction of the posted proof (6). We note that the same x_i can appear in the SLC (1) as left factor and as right factor. One example arises if the SLC has an operation $x_m = x_i \cdot x_i$. In this case verifying Aspect 1 will reveal both coordinates of X_i and hence reveal the value $x_i = val(X_i)$.

When preparing a proof of correctness of SLC the EP creates for every x_i involved in multiplications two random vector representations XL_i and XR_i.

The proof of correctness of the multiplication $x_m = x_i \cdot x_j$ will be:

$$\text{COM}(X_m), \text{COM}(XL_i), \text{COM}(XR_j), \text{COM}(Z_0), \text{COM}(Z_1), \text{COM}(Z_2),$$

where now $XL_i = (u_1, v_1)$, $XR_j = (u_2, v_2)$. It is clear that even if $i = j$, and Aspect 1 is checked, u_1 and v_2 are independent random values from F_p. Similarly if SLC contains another multiplication $x_k = x_s \cdot x_i$ it as well as $x_m = x_i \cdot x_j$ are verified wrt Aspect 1. For the first multiplication XR_i will be employed, for the second multiplication XL_i will be used. Thus again independent random first coordinate of XR_i and and second coordinate of XL_i are revealed. These considerations lead to a proof of:

Theorem 5. *If the* SLC *comprises only the operations* $+$ *and* $\cdot$ *(and no comparisons* $TruthValue(x_i \leq x_j)$*) then the* EP *can prepare a proof of correctness that is information-theoretically hiding and, by Theorem 3, if false will be accepted by Verifier with probability at most* $3/4$.

Proof: Once the commitments (*) to the representations of the input values $x_1, \cdots, x_n$ are posted or sent to the Verifier, the EP prepares a translation TR for the SLC (1) as follows. Successively, after $X_1, \cdots, X_{m-1}$ were created, if $x_m = x_i + x_j$ then EP creates a $\text{RR}(x_m) = X_m$. Thus, by induction, $\text{val}(X_i) + \text{val}(X_j) = \text{val}(X_m)$. If $x_m = x_i \cdot x_j$ then EP creates the vectors Z_0, Z_1, Z_2, X_m as in the proof of correctness of multiplications, Section 4. Now $\text{val}(X_i) \cdot \text{val}(X_j) = \text{val}(X_m)$. In addition EP creates for every x_i appearing in the SLC (1) as a first and second factor in a multiplication, once X_i was created, another $\text{RR}(x_i)$.

The EP now has a translation TR for the SLC (1). He now creates commitments to all the vectors beyond the already posted commitments (*) and posts or sends those to the Verifier.

In the interactive proof of correctness, the Verifier chooses with probability $1/2$ to simultaneously verify all additions, equalities, and Aspect 0 for all multiplications. Verifier randomly chooses $c \leftarrow \{1, 2\}$. Say $c = 1$. The EP reveals the first coordinates of all the X_i, and of all the Z_0, Z_1, Z_2 and all the w values required for proving correctness of additions. Using these first coordinates and w values the Verifier checks all equations. Similarly if $c = 2$. If the TR is false with respect to any addition, equality, or Aspect 0 of any multiplication, then the probability of Verifier accepting is at most $3/4$. The Verifier chooses with probability $1/4$ to simultaneously verify all Aspects 1 of all multiplications, and with probability $1/4$ to simultaneously verify all Aspects 2 of all multiplications. If the TR is false in Aspect 1 or Aspect 2 for any multiplication then the probability of Verifier accepting the correctness of TR and hence the correctness of the results of the SLC (1) is at most $3/4$. The above arguments lead to proving completeness, validity and statistical ZK/WI.

5   Proving Inequalities $x \leq y$, When $x, y < p/32$

Let $b^2 < p/32$ be an explicit bound on all values x_i, x_j in the SLC(1) for which $x_i \leq x_j$ needs to be proved. In the application to auctions, where the inputs $x_1, \cdots, x_n$ are bids, it is required that all bids are bounded by $p/32$.

We note that for integers $x, y < p/2$ we have $x \leq y$ iff $(y - x) \bmod p < p/2$. Thus if the EP proves to a verifier these three inequalities for split representations X, Y, Z of the values x, y, z, then he has proved that as integers $x \leq y$. Following [15], given $0 \leq z \leq b$ the EP can supply within the framework of SLC proofs of correctness, a proof that $(b) \leq z \leq 2b$ (i.e., as an integer $p - b \leq z < p$ or $0 \leq z \leq 2b$). Such a proof verification can be made part of Aspect 4 of the verification. All such inequalities can be simultaneously proved.

How do we get rid of the $p - b \leq z < p$ possibility?

Lagrange proved that every integer x is the sum of four squares of integers, $x = z_1^2 + z_2^2 + z_3^2 + z_4^2$. Rabin in 1977 MIT lectures and [17] gave an efficient polynomial-time algorithm for computing such a representation. For numbers $x \leq 2^{32}$, Schorn's Python implementation computed $60,000$ representations in 1 second.

[16] proposed using Lagrange in the context of proving range statements for encrypted numbers.

We apply Lagrange and [17] in our context of SLCs.

Given $0 \leq x \leq b^2 < p/32$, the EP computes $z_1, \ldots, z_4$ such that $x = z_1^2 + z_2^2 + z_3^2 + z_4^2$. Each z_i is between 0 and b. The numbers $x, z_1, \ldots, z_4$ are represented as usual in a translation TR by pairs $X, Z_1, \ldots, Z_4$.

EP incorporates in the SLC steps for enabling verification that $-b \leq \mathtt{val}(Z_i) \leq 2b$ and that $\mathtt{val}(X) = \mathtt{val}(Z_1)^2 + \ldots + \mathtt{val}(Z_4)^2$. This implies $0 \leq x \leq 16b^2$. Now $32b^2 < p$, i.e. $16b^2 < p/2$.

Theorem 6. *Given a* SLC *including inequalities, the* EP *can create a proof of correctness of the whole* SLC *and present it to* V. *The verification by* V *information-theoretically hides all input and intermediate values. If the proof is false then the probability that* V *will accept is at most* $3/4$.

Proof: For every x_i appearing in an inequality

$$x_m = TruthValue(x_i \leq x_j)$$

and every difference $x_j - x_i$ linked to such an inequality, the EP calculates the Lagrange sum of 4 squares representation. For each such sum of 4 squares $x = z_1^2 + z_2^2 + z_3^2 + z_4^2$ the EP creates random vector representation Z_j of $z_j, 1 \leq j \leq 4$ as well as random representations S_j of $(z_j)^2, 1 \leq j \leq 4$. The proof of correctness of the SLC now reduces to proof of correctness of a SLC involving only the operations $+$ and $\cdot$, so that Theorem 4 applies.

This establishes Main Theorem 1, but to use the result with negligible probabilities we need amplification, thus copying!

6   Exponential Reduction of Probability of Cheating and Multiple Proofs of Correctness

The use of a single translation of a SLC in a proof of correctness allows a probability of $3/4$ for the EP to cheat the Verifier. This soundness probability is unacceptable in real-life applications. The way to reduce the margin of uncertainty is for the EP to add redundancy and present to the Verifier m translations $TR_1, \ldots, TR_m$ of the SLC. The Verifier randomly and independently challenges the EP for each T_j to verify Aspect 4 with probability $1/2$ or one of Aspects 1, 2, each with probability $1/4$. Recall that the EP 's construction of the whole translation for the SLC starts with commitments $\mathrm{COM}(X_1), \ldots, \mathrm{COM}(X_n)$, where X_j is a random representation of the input value $x_j, 1 \leq j \leq n$. As explained in the Overview, $\mathrm{COM}(X_1), \ldots, \mathrm{COM}(X_n)$, were submitted by $P_1, \ldots, P_n$. To reduce the probability of cheating and to allow multiple proofs of correctness, each participant P_j has to submit multiple random representations (i.e., "the redundant split commitment") of his input value x_j. The whole protocol proceeds as follows. $P_1, \ldots, P_n$ submit input values $x_1, \ldots, x_n$ to EP : $P_i, 1 \leq i \leq n$, prepares $3k$ random representations $Y_1^{(i)}, \ldots, Y_{3k}^{(i)}$ of his value x_i. P_i submits commitments $\mathrm{COM}(Y_1^{(i)}), \ldots, \mathrm{COM}(Y_{3k}^{(i)})$ to the EP . EP posts all commitments from all $P_i, 1 \leq i \leq n$, denoted $\mathcal{Y}$:

$$\mathrm{COM}(Y_1^{(1)}), \mathrm{COM}(Y_2^{(1)}), \ldots, \mathrm{COM}(Y_{3k}^{(1)})$$
$$\mathrm{COM}(Y_1^{(2)}), \mathrm{COM}(Y_2^{(2)}), \ldots, \mathrm{COM}(Y_{3k}^{(2)})$$
$$\ldots$$
$$\mathrm{COM}(Y_1^{(n)}), \mathrm{COM}(Y_2^{(n)}), \ldots, \mathrm{COM}(Y_{3k}^{(n)})$$

EP creates additional random representations of input values: Every P_i opens (reveals) $Y_1^{(i)}, \ldots, Y_{3k}^{(i)}$ to EP . The EP chooses L (say $L = 20$) and constructs and posts additional $10kL = m$ columns, denoted $\mathcal{X}$:

$$\mathrm{COM}(X_1^{(1)}), \mathrm{COM}(X_2^{(1)}), \ldots, \mathrm{COM}(X_m^{(1)})$$
$$\mathrm{COM}(X_1^{(2)}), \mathrm{COM}(X_2^{(2)}), \ldots, \mathrm{COM}(X_m^{(2)})$$
$$\ldots$$
$$\mathrm{COM}(X_1^{(n)}), \mathrm{COM}(X_2^{(2)}), \ldots, \mathrm{COM}(X_m^{(n)})$$

Definition 5. *We call two sequences* $X^{(1)}, \ldots, X^{(n)}$ *and* $Y^{(1)}, \ldots, Y^{(n)}$ *value consistent if* $\mathrm{val}(X^{(j)}) = \mathrm{val}(Y^{(j)}), 1 \leq j \leq n$.

We have by the following a way to replicate values consistently:

Theorem 7. *Given commitments* $\mathrm{COM}(X^{(j)}), \mathrm{COM}(Y^{(j)}), 1 \leq j \leq n$, *to two sequences of representations for which the EP claims value consistency. The EP can give a ZKP for this claim such that if the claim is false then the probability that V will accept is at most $1/2$.*

Using this Theorem, we derive the following which will imply SBB proof of correctness with validity probability negligible and statistical security, resulting in our second major Theorem:

Theorem 8. *Interactively with* V, EP *can provide an SBB ZK (WI) proof of knowledge with probability of cheating at most* $(1/2 + 1/e^2)^k + (1/2 + 1/e^2)^{3k}$ *that*

1. *In the* $n \times 3k$ *posted matrix* $\mathcal{Y}$ *of representation of input values, at least* $2k$ *columns are pair-wise value consistent. By definition, the common* $2k$ *majority of values in row* i *is* $P_i's$ *input* x_i.
2. *In the* $n \times m$ *matrix* $\mathcal{X}$ *at least* $(1 - 1/L)m$ *columns are pair-wise value consistent with the majority values of the input matrix.*
3. *The interactive proof involves all input representations of matrix* $\mathcal{Y}$ *and* $6kL$ *columns of the matrix* $\mathcal{X}$. *The remaining untouched* $4kL$ *columns of the matrix* $\mathcal{X}$ *may be used by* EP *to construct* $4L$ *proofs of correctness of announced* SLC *results.*

Proof: In the interactive proof/verification, the Verifier randomly chooses for each of the $3k$ columns C_i of the inputs matrix $\mathcal{Y}$, $2L$ columns of the matrix of the matrix $\mathcal{X}$ constructed by the EP. The EP interactively proves that C_i is value-consistent with each of the $2L$ correspondingly chosen columns. The Verifier accepts only if all those verifications are successful. The proof for claimed probability of soundness will be given in the full version of the paper.

6.1   Putting It All Together

As a consequence of Theorem 8 the EP and V now have available $4kL$ unused columns of the matrix 7 and with high probability at least $(1 - 1/L)4kL$ of these columns are value consistent with the input values $x_1, \ldots, x_n$.

For the interactive ZKP of the correctness of the outputs of the SLC, V randomly chooses k of these columns and presents this choice to the EP . The EP extends each of the k columns (representing commitments to the n inputs to the SLC) to a full proof of correctness according to Theorem 5. The probability of V accepting such a proof for a single translation is $1/L + 3/4$. The $1/L$ terms bounds the probability that the chosen column is not value consistent with the inputs. The probability that the outputs are incorrect and yet V will accept is at most $(1/2 + 1/e^2)^k + (1/2 + 1/e^2)^{3k} + (1/L + 3/4)^k$. This replication of commitment is also the crux of the ability to repeat the proof process.

References

1. Abe, M., Suzuki, K.: M+1-st Price Auction Using Homomorphic Encryption. In: Naccache, D., Paillier, P. (eds.) PKC 2002. LNCS, vol. 2274, pp. 115–124. Springer, Heidelberg (2002)
2. Brandt, F.: How to obtain full privacy in auctions. Tech. rep., Carnegie Mellon University (2005) (online)
3. Burmaster, M., Magkos, E., Chrissikopoulos, V.: Uncoercible e-bidding games. Electrocronic Commerce Research 4(1-2), 113–125 (2004)

4. Chen, X., Kim, K., Lee, B.: Receipt-Free Electronic Auction Schemes Using Homomorphic Encryption. In: Lim, J.-I., Lee, D.-H. (eds.) ICISC 2003. LNCS, vol. 2971, pp. 259–273. Springer, Heidelberg (2004)
5. Damgard, I., Jurik, M.: A generalisation, a simplification and some applications of P probabilistic public-key system. In: PKC 2001 (2001)
6. Jurik, M.J.: Extensions to the paillier cryptosystem with applications to cryptological protocols. Ph.D. thesis, University of Arhus (2003)
7. Lipmaa, H., Asokan, N., Niemi, V.: Secure Vickrey Auctions Without Threshold Trust. In: Blaze, M. (ed.) FC 2002. LNCS, vol. 2357, pp. 87–101. Springer, Heidelberg (2003)
8. Paillier, P.: Public-Key Cryptosystems Based on Composite Degree Residuosity Classes. In: Stern, J. (ed.) EUROCRYPT 1999. LNCS, vol. 1592, pp. 223–239. Springer, Heidelberg (1999)
9. Rabin, M.O., Servedio, R.A., Thorpe, C.: Highly efficient secrecy-preserving proofs of correctness of computations and applications. In: Proc. IEEE Symposium on Logic in Computer Science (2007)
10. Parkes, D.C., Rabin, M.O., Shieber, S.M., Thorpe, C.A.: Practical secrecy-preserving, verifiably correct and trustworthy auctions. In: Proceedings of the 8th International Conference on Electronic Commerce (ICEC), pp. 70–81 (2006)
11. Thorpe, C., Parkes, D.C.: Cryptographic Combinatorial Securities Exchanges. In: Dingledine, R., Golle, P. (eds.) FC 2009. LNCS, vol. 5628, pp. 285–304. Springer, Heidelberg (2009)
12. Bogetoft, P., Christensen, D.L., Damgård, I., Geisler, M., Jakobsen, T., Krøigaard, M., Nielsen, J.D., Nielsen, J.B., Nielsen, K., Pagter, J., Schwartzbach, M., Toft, T.: Secure Multiparty Computation Goes Live. In: Dingledine, R., Golle, P. (eds.) FC 2009. LNCS, vol. 5628, pp. 325–343. Springer, Heidelberg (2009)
13. Goldwasser, S., Micali, S., Rackoff, C.: The Knowledge Complexity of Interactive Proof Systems. SIAM Journal on Computing 18(1), 186–208 (1989)
14. Goldreich, O., Micali, S., Wigderson, A.: Proofs that Yield Nothing but Their Validity, or all Languages in NP have ZKP systems. Journal of the ACM 38(3), 691–729 (1991)
15. Brickell, E.F., Chaum, D., Damgård, I.B., van de Graaf, J.: Gradual and Verifiable Release of a Secret. In: Pomerance, C. (ed.) CRYPTO 1987. LNCS, vol. 293, pp. 156–166. Springer, Heidelberg (1988)
16. Camenisch, J.L., Shoup, V.: Practical Verifiable Encryption and Decryption of Discrete Logarithms. In: Boneh, D. (ed.) CRYPTO 2003. LNCS, vol. 2729, pp. 126–144. Springer, Heidelberg (2003)
17. Rabin, M.O., Shallit, J.O.: Randomized Algorithms in Number Theory. Comm. Pure and Applied Mathematics 39, 239–256 (1986)
18. Pedersen, T.P.: Non-interactive and Information-Theoretic Secure Verifiable Secret Sharing. In: Feigenbaum, J. (ed.) CRYPTO 1991. LNCS, vol. 576, pp. 129–140. Springer, Heidelberg (1992)
19. Kilian, J.: A note on efficient zero-knowledge proofs and arguments. In: Proceedings of STOC 1992, pp. 723–732 (1992)
20. Brassard, G., Chaum, D., Crepeau, C.: Minimum disclosure proofs of knowledge. Journal of Computer and System Sciences 37, 156–189 (1988)

Parameterized Tractability of Multiway Cut
with Parity Constraints

Daniel Lokshtanov[1] and M.S. Ramanujan[2]

[1] University of California, San Diego, USA
daniello@ii.uib.no
[2] The Institute of Mathematical Sciences, Chennai, India
msramanujan@imsc.res.in

Abstract. In this paper, we study a parity based generalization of the classical MULTIWAY CUT problem. Formally, we study the PARITY MULTIWAY CUT
problem, where the input is a graph G, vertex subsets T_e and T_o ($T = T_e \cup T_o$)
called terminals, a positive integer k and the objective is to test whether there
exists a k-sized vertex subset S such that S intersects all odd paths from $v \in T_o$
to $T \setminus \{v\}$ and all even paths from $v \in T_e$ to $T \setminus \{v\}$. When $T_e = T_o$, this
is precisely the classical MULTIWAY CUT problem. If $T_o = \emptyset$ then this is the
EVEN MULTIWAY CUT problem and if $T_e = \emptyset$ then this is the ODD MULTI-
WAY CUT problem. We remark that even the problem of deciding whether there
is a set of at most k vertices that intersects all odd paths between a pair of vertices s and t is NP-complete. Our primary motivation for studying this problem
is the recently initiated parameterized study of parity versions of graphs minors
(Kawarabayashi, Reed and Wollan, FOCS 2011) and separation problems similar
to MULTIWAY CUT. The area of design of parameterized algorithms for graph
separation problems has seen a lot of recent activity, which includes algorithms
for MULTI-CUT on undirected graphs (Marx and Razgon, STOC 2011, Bousquet,
Daligault and Thomassé, STOC 2011), k-WAY CUT (Kawarabayashi and Thorup,
FOCS 2011), and MULTIWAY CUT on directed graphs (Chitnis, Hajiaghayi and
Marx, SODA 2012). A second motivation is that this problem serves as a good
example to illustrate the application of a generalization of important separators
which we introduce, and can be applied even when most of the recently develped
tools fail to apply. We believe that this could be a useful tool for several other separation problems as well. We obtain this generalization by dividing the graph into
slices with small boundaries and applying a divide and conquer paradigm over
these slices. We show that PARITY MULTIWAY CUT is fixed parameter tractable
(**FPT**) by giving an algorithm that runs in time $f(k)n^{\mathcal{O}(1)}$. More precisely, we
show that instances of this problem with solutions of size $\mathcal{O}(\log \log n)$ can be
solved in polynomial time. Along with this new notion of generalized important
separators, our algorithm also combines several ideas used in previous parameterized algorithms for graph separation problems including the notion of important separators and randomized selection of important sets to simplify the input
instance.

1 Introduction

A fundamental min-max theorem about connectivity in graphs is Menger's Theorem,
which states that the maximum number of vertex disjoint paths between two vertices

A. Czumaj et al. (Eds.): ICALP 2012, Part I, LNCS 7391, pp. 750–761, 2012.
© Springer-Verlag Berlin Heidelberg 2012

s and t, is equal to the minimum number of vertices whose removal separates these two vertices. Indeed, a maximum set of vertex disjoint paths between s, t and a minimum size set of vertices separating these two vertices can be computed in polynomial time. A known generalization of this theorem, commonly known as Mader's T-path Theorem [18] states that, given a graph G and a subset T of vertices, there are either k vertex disjoint paths with only the end points in T (such paths are called T-paths and if their length is odd (even), then *odd (even) T-paths*), or there is a vertex set of size at most $2k$ which intersects every T-path. Although computing a maximum set of vertex disjoint T-paths can be done in polynomial time through matching techniques, the decision version of the dual problem of finding a minimum set of vertices that intersects every T-path is NP-complete for $|T| > 2$. Formally, this problem is the classical MULTIWAY CUT problem, where the input is a graph G, a subset of vertices T called terminals, a positive integer k and the objective is to test whether there exists a k-sized vertex subset that intersects every T-path. This is a very well studied problem in terms of approximation, as well as parameterized algorithms [2,8,19]. In this paper we study a generalization of this classical MULTIWAY CUT problem to a parity version. Formally, we study the PARITY MULTIWAY CUT problem which is defined as follows.

PARITY MULTIWAY CUT (PMWC)
    Instance: A graph $G = (V, E)$, vertex subsets T_e and T_o ($T = T_e \cup T_o$), integer k.
    Parameter: k
    Question: Is there a vertex set S of size at most k which intersects
        1. all odd paths from a vertex $v \in T_o$ to some other vertex $u \in T \setminus \{v\}$,
        2. all even paths from a vertex $v \in T_e$ to some other vertex $u \in T \setminus \{v\}$?

When $T_e = T_o$, this is precisely the classical MULTIWAY CUT problem. If $T_o = \emptyset$ then this is the EVEN MULTIWAY CUT (EMWC) problem and if $T_e = \emptyset$ then this is the ODD MULTIWAY CUT (OMWC) problem.

Our main motivation for studying this particular generalization is the recently initiated parameterized study of parity versions of graphs minors by Kawarabayashi, Reed and Wollan [14] and separation problems similar to MULTIWAY CUT [1,4,20]. The area of design of parameterized algorithms for graph separation problems has seen a lot of recent activity, which includes algorithms for MULTI-CUT on undirected graphs [20,1], k-WAY CUT [15] and MULTIWAY CUT on directed graphs [4]. Furthermore, recently, Geelen, Gerards, Reed, Seymour and Vetta [9] proved an odd variant of Mader's T-path Theorem. They showed that, given a graph G and a subset T of vertices, there are either k vertex disjoint odd T-paths, or there is a vertex set of size at most $2k$ which intersects every odd T-path. This result has already turned out to be useful in graph theory [9,17], as well as in the design of parameterized algorithms [10,12,13]. This result was crucial in settling the parameterized complexity of finding k vertex disjoint odd length cycles in a graph [13]. Observe that, this odd variant of Mader's T-path Theorem naturally gives rise to OMWC, a special case of PMWC.

The goal of parameterized complexity is to find ways of solving NP-hard problems more efficiently than by brute force. Here, the aim is to restrict the combinatorial explosion of computational difficulty to a parameter that is hopefully much smaller than the input size. Formally, a *parameterization* of a problem is the assignment of an integer

k to each input instance and we say that a parameterized problem is *fixed-parameter tractable* (FPT) if there is an algorithm that solves the problem in time $f(k) \cdot |I|^{\mathcal{O}(1)}$, where $|I|$ is the size of the input instance and f is an arbitrary computable function depending only on the parameter k. For more background, the reader is referred to the monographs [6,7,21].

Unlike MULTIWAY CUT, PMWC is already NP-complete for the case when $|T| = 2$. Indeed, consider the following reduction from VERTEX COVER to PMWC. Given an instance $(G = (V, E), k)$ of VERTEX COVER, add two new vertices t_1 and t_2, make them both adjacent to every vertex in V, and set $T_o = \{t_1, t_2\}$ and $T_e = \emptyset$. Call this new graph G'. It is easy to see that G has a vertex cover of size at most k if and only if G' has k-sized vertex subset that intersects every odd T_o-path. In fact, our argument shows that OMWC is NP-complete for the case when $|T| = 2$. One can similarly show that EMWC is NP-complete for the case when $|T| = 2$.

Marx [19] was the first to consider cut problems in the context of parameterized complexity. He gave an algorithm for MULTIWAY CUT with running time $\mathcal{O}(4^{k^3} n^{\mathcal{O}(1)})$ with the current fastest algorithm running in time $\mathcal{O}(2^k n^{\mathcal{O}(1)})$ [5]. Even the recent developments in techniques to solve graph separation problems [20] and parity based graph problems [16], do not seem to apply to PMWC, a natural companion of these problems. In this paper, we introduce a new notion of *generalized important separators*, which along with the tools used to solve parameterized cut problems like MULTIWAY CUT and MULTI-CUT, allows us to design an FPT algorithm for PMWC. In general, this notion seems to allow us to bring a number of problems under a single umbrella, and in particular we demonstrate its application to PMWC. The main result of this paper is the following.

Theorem 1. PMWC *can be solved in time* $2^{2^{\mathcal{O}(k)}} n^{\mathcal{O}(1)}$ *time.*

Our algorithm combines the key idea of generalization of separators with several ideas used in previous parameterized algorithms for graph separation problems The algorithm for PMWC has three phases, in the first phase using the well-known technique of iterative compression, we bound the number of even terminals by a linear function of k. In the second phase we remove even terminals using the notion of generalized important separators that we define in this paper and obtain $f(k)$ instances of OMWC. We obtain the generalized important separators by dividing the graph into slices with small boundaries and applying a divide and conquer paradigm over these slices. In the final phase we solve these instances of OMWC by designing an FPT algorithm for OMWC. More precisely, we have the following result.

Lemma 1. OMWC *can be solved in time* $2^{2^{\mathcal{O}(k)}} n^{\mathcal{O}(1)}$ *time.*

We note that OMWC can be shown to be FPT be a simple reduction to the SUBSET ODD CYCLE TRANSVERSAL (SUBSET OCT) problem which was shown to be FPT in [16]. However, such an algorithm for OMWC would have a much worse dependence on the parameter k when compared to the algorithm we present in this paper. Moreover, we would like to point out that the EMWC problem does not seem to reduce to SUBSET OCT. We also note that in the case of the EMWC problem with two terminals, we may subdivide the edges incident on one of them, thus converting all even paths between these terminals into odd paths and vice versa. This reduction shows that OMWC is equivalent to EMWC in the case of two terminals and hence Lemma 1 immediately gives an FPT algorithm for EMWC in the case of two terminals. We also consider the edge version of PMWC, the EDGE PARITY MULTIWAY CUT (EPMWC) problem, where the input is a graph G, a subset of vertices $T = T_e \cup T_o$, a positive integer k and the objective is to determine whether there exists a k-sized edge subset that intersects every even path from a vertex $v \in T_e$ to $T \setminus \{v\}$ and every odd path from a vertex $v \in T_o$ to $T \setminus \{v\}$. We show that this problem is also FPT by establishing a parameter preserving reduction from EPMWC to PMWC.

Related Work. Parity problems hold a lot of promise and remain hitherto unexplored in the light of parameterized complexity, with exceptions that are few and far between. The first parameterized algorithm for ODD CYCLE TRANSVERSAL, finding a k sized vertex subset that intersect all odd cycles only appeared in 2004 [24]. Recently, Kawarabayashi and Reed [12] obtained a faster algorithm for ODD CYCLE TRANSVERSAL that runs in *almost linear time*. Kawarabayashi and Reed [13] settled the parameterized complexity of ODD CYCLE PACKING, finding k vertex disjoint odd cycles in a graph, by showing it to be FPT. The parameterized complexity of ODD CYCLE PACKING was a long standing open problem and is much more general problem than the famous DISJOINT PATHS problem, finding vertex disjoint paths between given pairs of vertices. Recently, Kawarabayashi, Reed and Wollan [14] initiated the parameterized study of parity versions of graphs minors and gave an algorithm to find odd minors. Other studies include finding odd subdivision, parity paths passing through specific vertices [11,10]. On the cut side, as we mentioned before, the area was initiated by the paper of Marx [19]. The notions used in this paper has been useful in settling parameterized complexity of variety of problems including DIRECTED FEEDBACK VERTEX SET [3], ALMOST 2 SAT [23] and ABOVE GUARANTEE VERTEX COVER [23,22]. Recently, Marx and Razgon [20] and Bousquet, Daligault and Thomassé [1] independently showed that MULTI-CUT, finding k vertices to disconnect given pairs of terminals is FPT. Continuing this line of study, Chitnis, Hajiaghayi and Marx studied MULTIWAY CUT on directed graphs and showed it to be FPT [4].

2 Preliminaries

Given a graph $G = (V, E)$ and $T \subseteq V$, paths with only the end points in T are called *T-paths* and if their length is odd (even), then *odd (even) T*-paths. In an instance $(G, T_e \cup T_o, k)$ of PMWC, the vertices in T_e are called *even* terminals and those in T_o are called *odd* terminals. Vertices in $T_e \setminus T_o$ are called *purely even* terminals and those in $T_o \setminus T_e$ are called *purely odd* terminals.

3 PMWC **Parameterized by the Solution Size**

The algorithm for PMWC has three phases; in the first phase, using the well-known technique of iterative compression, we bound the number of even terminals by $7k$, the second phase consists of removing even terminals using the notion of generalized important separators that we define in this section and obtain $f(k)$ instances of OMWC and the final phase consists of solving the instances of OMWC. In this section, we outline the first two phases of the algorithm. Before moving on to the first phase, we make the following observation, which will be used in the description this first phase.

Observation 2. [1] *Let $(G, T = T_e \cup T_o, k)$ be an instance of PMWC and let S be a solution to this instance.*
(a) Any connected component of $G \setminus S$ with at least two terminals contains terminals from exactly one of $T_o \setminus T_e$ or $T_e \setminus T_o$.
(b) Any connected component of $G \setminus S$ contains at most 2 vertices from T_e.

3.1 Bounding the Number of Even Terminals

We will first describe a way to reduce the given instance of PMWC to multiple (but bounded number of) instances, each with a bounded number of even terminals, such that solving these instances will lead to a solution for the input instance. To this end we will use the technique of iterative compression. In this technique, we assume that a solution of size $k + 1$ is part of the input, and attempt to compress it to a solution of size k. The method adopted usually is to begin with a subgraph that trivially admits a $(k + 1)$-sized solution and then expand it iteratively.

Given an instance $(G = (V, E), T = T_e \cup T_o, k)$ of PMWC, where $V = \{v_1, \ldots, v_n\}$, we define a graph $G_i = G[V_i]$ where $V_i = \{v_1, \ldots, v_i\}$. We iterate through the instances $(G_i, T_i = (T_e \cap V_i) \cup (T_o \cap V_i), k)$ starting from $i = k + 1$ and for the ith instance, with the help of a *known* solution S_i of size at most $k + 1$ we try to find a solution $\hat{S}_i$ of size at most k. Formally, the *compression* problem we address is following.

PMWC COMPRESSION
   Instance: $(G = (V, E), T = T_e \cup T_o, k, S)$ where G is an undirected graph, T_e, T_o are
            vertex sets, k a postive integer and S, a PMWC of size at most $k + 1$.
   Parameter: k
   Question: Does there exist a PMWC of size at most k for this instance?

We will reduce the PMWC problem to $n - k$ instances of the PMWC COMPRESSION problem as follows. Let $I_i = (G_i, (T_e \cap V_i) \cup (T_o \cap V_i), S_i, k)$ be the ith instance of PMWC COMPRESSION. Clearly, the set V_{k+1} is a solution of size at most $k + 1$ for the instance I_{k+1}. It is also easy to see that if $\hat{S}_{i-1}$ is a solution of size at most k for instance I_{i-1}, then the set $\hat{S}_{i-1} \cup \{v_i\}$ is a solution of size at most $k + 1$ for the instance I_i. We use these two observations to start off the iteration with the instance $(G_{k+1}, (T_e \cap V_{k+1}) \cup (T_o \cap V_{k+1}), k, S_{k+1} = V_{k+1})$ and try to compute a solution of size at most k for this instance. If there is such a solution $\hat{S}_{k+1}$, we set $S_{k+2} = \hat{S}_{k+1} \cup \{v_{k+1}\}$ and try to compute a solution of size at most k for the instance I_{k+2} and so on. If, during any iteration, the corresponding instance does not have a solution of the required size, it implies that the original instance is also a NO instance. Finally the solution for the original input instance will be $\hat{S}_n$. Since there can be at most n iterations, the total time taken is bounded by n times the time required to solve the PMWC COMPRESSION problem.

---

[1] Proofs not appearing in the main text are given in the full version.

We will now describe a way to bound the number of terminals in an instance of PMWC COMPRESSION. Let $(G, T = T_e \cup T_o, k, S)$ be an instance of PMWC COMPRESSION. Fix a hypothetical solution $\hat{S}$ for this instance. We first guess the set $Y = S \cap \hat{S}$. There are 2^{k+1} such possibilities. For each guess of Y, we delete it from the instance, and also delete vertices which are no longer relevant for the instance. This results in an instance of PMWC which has a solution $N = S \setminus Y$ and we are required to find a solution of size at most $k - |Y|$ which is disjoint from N. Now, we show that if the resulting instance has such a solution, then it must be the case that the number of even terminals in this instance is bounded.

Suppose that the resulting instance indeed has such a solution. Fix such a solution S'. We call a component of $G \setminus N$ *affected* if it contains some vertex of S' and *unaffected* otherwise. Clearly, there can be at most k affected components. Now, consider the unaffected components which contain even terminals. We claim that the number of such components cannot be more that $2k$. Suppose this was not the case. Then there must exist three unaffected components which contain even terminals and share a neighbor in N. But this implies that there will be an even path between atleast two of these terminals which is disjoint from the new solution S', a contradiction. Hence, the number of components of N which contain even terminals is at most $3k$. By Observation 2, any component can contain at most 2 even terminals. Since N contains at most k even terminals, the number of even terminals in the instance is bounded by $7k$. Hence, if the number of terminals in the instance after removing the guess Y exceeds $7k$, we can reject this guess right away. Note that, even if we compute a solution of the required size which is not disjoint from the set N, we can use it to continue the iteration. Hence, once we have an instance with a bounded number of even terminals, we ignore the fact that there is a solution disjoint from N and just compute any solution of the required size for the corresponding PMWC instance. Since we only need to deal with PMWC instances arising from instances of PMWC COMPRESSION, henceforth we will assume that the given instance of PMWC contains at most $7k$ even terminals.

3.2 Removing Even Terminals

We now describe a way to separate and remove the even terminals from the instance. We initially perform the following preprocessing step on the given instance $(G, T = T_e \cup T_o, k)$ of PMWC. For every purely odd terminal $t_i \in T \setminus T_e$, we add $2(k+1)$ new vertices $T_i = \{t_i^1, \ldots, t_i^{k+1}\}$ and $\hat{T}_i = \{\hat{t}_i^1, \ldots, \hat{t}_i^{k+1}\}$ and make t_i adjacent to every vertex in T_i. Finally, we make a complete bipartition between the sets T_i and $\hat{T}_i$. We now define a new set of purely odd terminals $T_o' = \bigcup_{t_i \in T \setminus T_e} \hat{T}_i$. That is, for every t_i in $T \setminus T_e$, we replace t_i with the $k + 1$ vertices $\hat{t}_i^j$ in the set of purely odd terminals. We note that the resulting instance is indeed equivalent to the input instance.

Lemma 2. *Given an instance* $(G, T = T_e \cup T_o, k)$ *of* PMWC, *let* $(G', T' = T_e \cup T_o', k)$ *be the instance obtained as a result of the terminal transformation described above. Then,* (G, T, k) *is a* YES *instance if and only if* (G', T', k) *is a* YES *instance.*

Due to Lemma 2, henceforth, we will assume that the given input instance is already of the form described above. This also allows us to assume that the solution will be disjoint from the set of purely odd terminals. We will now describe a procedure to reduce this instance of PMWC to an instance with no even terminals, thereby resulting in an instance of OMWC.

We fix a hypothetical solution for the PMWC instance and work with this solution. We first guess the intersection of the hypothetical solution with the set of even terminals and delete these vertices from the graph. Let S be the subset of the solution left after this step, that is S is a solution for the remaining instance. We then guess the way S partitions the even terminals into different connected components in the graph $G \setminus S$. There are at most $2^{|T_e|}$ possible intersections and $|T_e|^{|T_e|} = 2^{\mathcal{O}(k \log k)}$ partitions for the even terminals. Hence, the total number of possible guesses is $2^{\mathcal{O}(k \log k)}$. We say that S *conflicts* with a partition if there is a component of $G \setminus S$ containing terminals from two distinct sets of the partition. For each guess of the partition, we attempt to find a solution which partitions the even terminals in a way which does not conflict with the guess. We fix one such guess of the partition, say $\mathcal{P}$ and work with this partition for the rest of the section. In addition, note that, we can now assume that the solution S is disjoint from the entire set of terminals. This is because we have already guessed (and deleted) the intersection with even terminals, and it is already disjoint from the purely odd terminals.

3.3 Important Separators

The notion of important separators was formally introduced in [19] to handle the MULTIWAY CUT problem and the same concept was used implicitly in [2] to give an improved algorithm for the same problem. In this subsection, we recall some definitions related to important separators and a few lemmas which will be required for our algorithm.

Definition 1. *Let $G = (V, E)$ be an undirected graph, let $X, S \subseteq V$ be vertex subsets. We denote by $R_G(X, S)$ the set of vertices of G reachable from X in the graph $G \setminus S$ and we denote by $N R_G(X, S)$ the set of vertices of $G \setminus S$ which are not reachable from X in the graph $G \setminus S$. We drop the subscript G if it is clear from the context.*

Definition 2. *Let $G = (V, E)$ be an undirected graph and let $X, Y \subset V$ be two disjoint vertex sets. A subset $S \subseteq V \setminus (X \cup Y)$ is called an X-Y separator in G if $R_G(X, S) \cap Y = \emptyset$ or in other words there is no path from X to Y in the graph $G \setminus S$. We denote by $\lambda_G(X, Y)$ the size of the smallest X-Y separator in G. An X-Y separator S_1 is said to **cover** an X-Y separator S with respect to X if $R(X, S_1) \supset R(X, S)$ and S_1 is said to **dominate** S if it covers S and $|S_1| \leq |S|$. If the set X is clear from the context, we just say that S_1 dominates S. An X-Y separator is said to be inclusion wise minimal if none of its proper subsets is an X-Y separator.*

Definition 3. *Two X-Y separators are said to be **incomparable** if neither covers the other.*

Observation 3. *Let S_1 and S_2 be two incomparable X-Y separators. Then, $R(X, S_1) \cap S_2 \neq \emptyset$ and $R(X, S_2) \cap S_1 \neq \emptyset$. That is, there is a vertex of S_1 reachable from X in the graph $G \setminus S_2$ and a vertex of S_2 reachable from X in the graph $G \setminus S_1$. Also,*

$NR(X, S_1) \cap S_2 \neq \emptyset$ and $NR(X, S_2) \cap S_1 \neq \emptyset$. That is, there is a vertex of S_1 separated from X in the graph $G \setminus S_2$ and a vertex of S_2 separated from X in the graph $G \setminus S_1$.

Definition 4. Let $G = (V, E)$ be an undirected graph, $X, Y \subset V$ be vertex sets and $S \subseteq V$ be an X-Y separator in G. We say that S is an **important** X-Y separator if it is inclusionwise minimal and there does not exist another X-Y separator S_1 such that S_1 dominates S with respect to X.

Lemma 3. ([19]) Let $G = (V, E)$ be an undirected graph, $X, Y \subset V$ be disjoint vertex sets. There exists a unique important X-Y separator S^* of size $\lambda_G(X, Y)$ and it can be computed in polynomial time.

3.4   Important Separator Sequences and a Generalization of Important Separators

In this subsection we will define the notion of an important separator sequence and use it in the context of PMWC to define generalized vertex separators with the properties we require.

Definition 5. Let $G = (V, E)$ be a graph and let $X, Y \subseteq V$ be disjoint vertex sets. We define an important X-Y separator of order i, S^i to be the unique smallest important X-S^{i-1} separator in G, where $S^0 = Y$.

By Lemma 3, for every i, an important X-Y separator of order i is unique and can be computed in polynomial time.

Definition 6. We define a smallest X-Y separator sequence $\mathcal{I}$ to be a set $\mathcal{I} = \{S^i | 1 \le i \le l\}$, where S^i is an important X-Y separator of order i, for every $1 \le i, j \le l$, $|S^i| = |S^j|$, and $\lambda(X, S^l) > \lambda(X, Y)$, that is there is no X-S^l cut of size $|S^l|$.

Observation 4. Given two X-Y separators S_1 and S_2, we say that $S_1 \preceq S_2$ if S_2 covers S_1. Then, $(\mathcal{I}, \preceq)$ forms a total order where $\mathcal{I}$ is a smallest X-Y separator sequence.

Note that a smallest X-Y separator sequence is unique and can be computed in polynomial time. Observation 4 is the reason we refer to the set $\mathcal{I}$ as a *sequence*.
The key consequence of the definition of the smallest separator sequence is that it defines a natural partition of the graph into slices with *small boundaries*. Using this, we may restrict our search to local parts of the graph, in which case finding separators with certain properties becomes easier. We will now describe how this concept is applied in the context the PMWC problem.

Definition 7. Given sets X and Y and a minimal X-Y separator S, let l be the size of a minimum EMWC of the set X in the graph $G \setminus S$. We say that a minimal X-Y separator S' **well dominates** S (with respect to X) if S' dominates S with respect to X and the size of a minimum EMWC of X in the graph $G \setminus S'$ is at most l.

Note that any X-Y separator well dominates itself. Now, let T_1 be any set in the partition $\mathcal{P}$, and let $\hat{S}$ be a minimal part of S separating T_1 from $T \setminus T_1$. Recall that T_1 is a set of even terminals and by Observation 2, we can assume that T_1 contains at most 2 even terminals. We will first show that for any separator which well dominates $\hat{S}$, there is a solution for the PMWC instance containing this separator. Following that, we will describe an algorithm to compute a T_1-$T \setminus T_1$ separator that well dominates $\hat{S}$.

Lemma 4. *Let $(G, T = T_e \cup T_o, k)$ be an instance of PMWC, let S be a solution for this instance, and T_1 be a set in $\mathcal{P}$, with $T_2 = T \setminus T_1$. Let $\hat{S}$ be a minimal part of S separating T_1 and T_2. Let $\hat{S}_1$ be a T_1-T_2 separator which well dominates $\hat{S}$. Then, there is also a solution for the instance which contains $\hat{S}_1$.*

Proof. Let $\hat{K} \subseteq S \setminus \hat{S}$ be a minimum EMWC of T_1 in the graph $G \setminus \hat{S}$ and let $\hat{K}_1$ be a minimum EMWC of T_1 in the graph $G \setminus \hat{S}_1$. We know that $|\hat{S}| \geq |\hat{S}_1|$ and $|\hat{K}| \geq |\hat{K}_1|$. Now, consider the set $S' = (S \setminus (\hat{S} \cup \hat{K})) \cup (\hat{S}_1 \cup \hat{K}_1)$. We claim that S' is also a solution for the given instance. It is clear that the size of S' is at most that of S. Hence, it remains to show that S' is indeed a PMWC for the given instance.

Suppose that this is not the case and let t_i and t_j be two terminals such that there is a path P of forbidden parity between t_i and t_j in the graph $G \setminus S'$. Then, there must be a vertex $v \in \hat{S} \cup \hat{K}$ such that the path P intersects v. Since $\hat{S}_1$ dominates $\hat{S}$, this vertex must be reachable from T_1 in the graph $G \setminus \hat{S}_1$. But, $\hat{S}_1$ is a T_1-T_2 separator. Hence, if t_i or t_j is in T_2, then the path P must intersect $\hat{S}_1$ and hence also intersects S', a contradiction. Therefore, it must be the case that t_i and t_j are precisely the vertices in T_1. Now, since this path P lies entirely inside the component containing T_1 in the graph $G \setminus \hat{S}_1$, it must be the case that this path intersects $\hat{K}_1$ and this in turn implies that the path P intersects S', a contradiction. This completes the proof of the lemma. $\square$

Lemma 5. *Let $(G, T = T_e \cup T_o, k)$ be an instance of PMWC with a solution S, $\mathcal{P}$ be a partition of T_e such that S does not conflict with $\mathcal{P}$. Let T_1 be a set in $\mathcal{P}$ and T_2 be the set $T \setminus T_1$. Let X be a minimal part of S separating T_1 from T_2. Then, there is an algorithm which runs in time $2^{2^{\mathcal{O}(k)}} n^{\mathcal{O}(1)}$ and returns a set of at most $2^{\mathcal{O}(k^3)}$ T_1-T_2 separators of which at least one separator well dominates X (X is also called the target separator for this instance).*

Proof. For a given subset of vertices, the algorithm computes (if there is one) a T_1-T_2 separator of size at most k, which is contained in the given subset, and well dominates X. Initially, and also when the subset is not explicitly given, we allow this subset to be the entire vertex set of the current graph, and as we prune our search, we will define the subset accordingly. We first fix a hypothetical minimum EMWC of T_1 in the graph $G \setminus X$, say K and guess the size of this set, say l.

Description of Algorithm. We first check if there is a T_1-T_2 separator of size at most k within the given subset Z. If not, we return NO. If there is no path from T_1 to T_2, then we return $\emptyset$. Otherwise, we compute the smallest T_1-T_2 separator sequence, $\mathcal{I}$ comprising only of the vertices of Z. We call a T_1-T_2 separator S' *good* if the size of the minimum EMWC of T_1 in the graph $G \setminus S'$ is at most l and we call it *bad* otherwise. The following observation plays a crucial role in allowing us to ignore (potentially) large parts of the graph during our search.

Observation 5. *If a T_1-T_2 separator is* good, *all T_1-T_2 separators covered by this separator are also* good *and if a T_1-T_2 separator is* bad, *all T_1-T_2 separators which cover this separator are* bad.

For each T_1-T_2 separator in $\mathcal{I}$, we now determine if the separator is good or bad. Since $|T_1| \leq 2$, by Lemma 1, this step takes $2^{2^{O(k)}} n^{O(1)}$ time. Let P_1 be the maximal element of $\mathcal{I}$ which is good and let P_2 be the minimal element of $\mathcal{I}$, which is bad. That is, P_1 is good and every separator in $\mathcal{I} \setminus \{P_1\}$ which covers P_1 is bad, P_2 is bad and every separator in $\mathcal{I} \setminus \{P_2\}$ covered by P_2 is good. If all the separators in $\mathcal{I}$ are good, then P_2 is defined as T_2 and if all separators in $\mathcal{I}$ are bad, then P_1 is defined as T_1. We also have a final base case. If $k = 1$ and neither of the other base case conditions apply, then P_1 is the required separator and we simply return P_1. Otherwise, we will create a number of sub-instances, recurse on each of these sub-instances and finally return the *union* of the sets returned by these recursive calls. The sub-instances are created by exhaustive branching according to the following case analysis on the "relative position" of the target separator with P_1 and P_2.

1. P_1 covers the target separator X or $P_1 = X$. In this case, P_1 itself is a separator which well dominates the target separator and we have indeed found a separator of the required kind. Hence, we return P_1.

2. The target separator X covers P_1, but is itself covered by P_2. Let $\tilde{S}_1$ be the intersection of X with P_1 and $\tilde{S}_2$ be the intersection of X with P_2. We first guess the set $\tilde{S}_1$. If this set is non empty, then we delete it from the graph G, and recursively compute a T_1-T_2 separator of size at most $k - |\tilde{S}_1|$ in the graph $G \setminus \tilde{S}_1$, which lies in the set $NR_G(T_1, P_1)$, and well dominates $X \setminus \tilde{S}_1$ in the graph $G \setminus \tilde{S}_1$. If the set $\tilde{S}_1$ is empty and $P_2 \neq T_2$, then we guess the set $\tilde{S}_2$. If this set is non empty, then we delete it from the graph and recursively compute a set containing a T_1-T_2 separator of size at most $k - |\tilde{S}_2|$ in the graph $G \setminus \tilde{S}_2$, which lies in the set $NR_G(T_1, P_1) \cap R_G(T_1, P_2)$, which also well dominates $X \setminus \tilde{S}_2$ in the graph $G \setminus \tilde{S}_2$. Finally, if the set $\tilde{S}_2$ is also empty, then we recursively compute a set containing a T_1-T_2 separator of size at most k which is contained in the set $NR_G(T_1, P_1) \cap R_G(T_1, P_2)$ and well dominates X in the graph G, and return this set.

3. The target separator X is incomparable with P_1. Let $\tilde{S}_1$ be the intersection of X with P_1, P_1^r be the intersection of P_1 with $R_G(T_1, X)$, P_1^{nr} be the rest of P_1. Also, let X^r be the intersection of X with $R_G(T_1, P_1)$ and let X^{nr} be the rest of X. Since X is incomparable with P_1, by Observation 3, P_1^r, X^r, P_1^{nr} and X^{nr} are non empty. We first guess the set $\tilde{S}_1$. If it is non empty, then we delete it from the graph and recursively compute a set containing a T_1-T_2 separator of size at most $k - |\tilde{S}_1|$ in the graph $G \setminus \tilde{S}_1$ which well dominates $X \setminus \tilde{S}_1$. If it is empty, then we guess the sets P_1^r and P_1^{nr} and also the sizes of the sets X^r and X^{nr}.

We now construct a graph G' as follows. Initially, we set G' as the subgraph of G induced on the set $R_G(T_1, P_1)$. For every vertex in P_1^r, we guess if it is in the set K, in which case, we delete it from the graph G'. From the remaining vertices of P_1^r, for every pair of vertices, we guess if there is an odd (respectively even) path between them in the graph $G \setminus S$, with the internal vertices disjoint from the vertices of G' and add an edge (respectively subdivided edge) between these vertices. We note that it is possible to add both an edge and a subdivided edge between a pair of vertices. This completes the construction of G'. Now, we recursively compute a set containing a T_1-P_1^{nr} separator in the graph G', which well dominates X^r in this graph. Once we compute this set, for each separator X' in the set, we delete it from the graph G and in the resulting graph, recursively compute a set containing a T_1-T_2 separator which lies in the set $NR_G(T_1, P_1)$ and well dominates X^{nr} in $G \setminus X'$. Finally, we construct a new set by pairing up each separator from the first set, with the corresponding separators in the second set, and return this new set.

4. The target separator is incomparable with P_2. This case is analogous to case 3.

We note that the target separator is distinct from P_2 and cannot cover P_2, due to Observation 5 and hence this case need not be taken into consideration. Proof of correctness and the running time analysis appear in the full version.   □

Proof of Theorem 1. Given Lemma 4 and Lemma 5, we do the following. Pick a set T_1 in $\mathcal{P}$ and guess a T_1-$T \setminus T_1$ separator well dominating the minimal part of the solution separating T_1 and $T \setminus T_1$. Once T_1 has been separated from the rest of the terminals, we use Lemma 1 to compute a minimum EMWC of T_1 in the component containing T_1. Following this, we pick another set from $\mathcal{P}$, and repeat. At the end of this procedure, we will be left with an instance of PMWC with no even terminals, resulting in an instance of the OMWC problem. In each step, we either pick a vertex in the solution or discard an even terminal. Hence the number of steps is bounded by $8k$ and by Lemma 5, in $2^{2^{\mathcal{O}(k)}} n^{\mathcal{O}(1)}$ time, we obtain $2^{\mathcal{O}(k^4)}$ instances of OMWC such that the given instance of PMWC is a YES instance if and only if one of these instances of OMWC is a YES instance. This, combined with Lemma 1 proves Theorem 1.   □

We can also give a parameter preserving reduction from the edge version of PMWC (EDGE PARITY MULTIWAY CUT) to PMWC and thus we have the following theorem.

Theorem 6. EDGE PARITY MULTIWAY CUT *can be solved in time* $2^{2^{\mathcal{O}(k)}} n^{\mathcal{O}(1)}$.

4   Conclusion

In this paper, we introduce a notion of generalized important separators and by supplementing this idea with randomized selection of important components, along with the iterative compression technique, we give an FPT algorithm for a parity based generalization of the classical MULTIWAY CUT problem, the PARITY MULTIWAY CUT problem. The design of improved FPT algorithms for this problem, as well as FPT algorithms for other parity based separation problems like the parity version of MULTI-CUT remains an interesting open problems, as does the kernelization complexity of these problems.

Acknowledgements. We would like to thank Saket Saurabh for the insightful discussions on graph separation problems.

References

1. Bousquet, N., Daligault, J., Thomassé, S.: Multicut is fpt. In: STOC, pp. 459–468 (2011)
2. Chen, J., Liu, Y., Lu, S.: An improved parameterized algorithm for the minimum node multiway cut problem. Algorithmica 55(1), 1–13 (2009)
3. Chen, J., Liu, Y., Lu, S., O'Sullivan, B., Razgon, I.: A fixed-parameter algorithm for the directed feedback vertex set problem. J. ACM 55(5) (2008)
4. Chitnis, R.H., Hajiaghayi, M., Marx, D.: Fixed-parameter tractability of directed multiway cut parameterized by the size of the cutset. In: SODA, pp. 1713–1725 (2012)
5. Cygan, M., Pilipczuk, M., Pilipczuk, M., Wojtaszczyk, J.O.: On Multiway Cut Parameterized above Lower Bounds. In: Marx, D., Rossmanith, P. (eds.) IPEC 2011. LNCS, vol. 7112, pp. 1–12. Springer, Heidelberg (2012)
6. Downey, R.G., Fellows, M.R.: Parameterized Complexity. Springer, New York (1999)
7. Flum, J., Grohe, M.: Parameterized Complexity Theory. Texts in Theoretical Computer Science. An EATCS Series. Springer, Berlin (2006)
8. Garg, N., Vazirani, V.V., Yannakakis, M.: Multiway Cuts in Directed and Node Weighted Graphs. In: Shamir, E., Abiteboul, S. (eds.) ICALP 1994. LNCS, vol. 820, pp. 487–498. Springer, Heidelberg (1994)
9. Geelen, J., Gerards, B., Reed, B.A., Seymour, P.D., Vetta, A.: On the odd-minor variant of hadwiger's conjecture. J. Comb. Theory, Ser. B 99(1), 20–29 (2009)
10. Kawarabayashi, K., Li, Z., Reed, B.A.: Recognizing a totally odd k_4-subdivision, parity 2-disjoint rooted paths and a parity cycle through specified elements. In: SODA, pp. 318–328 (2010)
11. Kawarabayashi, K., Reed, B.A.: A nearly linear time algorithm for the half integral parity disjoint paths packing problem. In: SODA, pp. 1183–1192 (2009)
12. Kawarabayashi, K., Reed, B.A.: An (almost) linear time algorithm for odd cyles transversal. In: SODA, pp. 365–378 (2010)
13. Kawarabayashi, K., Reed, B.A.: Odd cycle packing. In: STOC, pp. 695–704 (2010)
14. Kawarabayashi, K., Reed, B.A., Wollan, P.: The graph minor algorithm with parity conditions. In: FOCS, pp. 27–36 (2011)
15. Kawarabayashi, K., Thorup, M.: The minimum k-way cut of bounded size is fixed-parameter tractable. In: FOCS, pp. 160–169 (2011)
16. Kakimura, N., Kawarabayashi, K., Kobayashi, Y.: Erdös-pósa property and its algorithmic applications: parity constraints, subset feedback set, and subset packing. In: SODA, pp. 1726–1736 (2012)
17. Kakimura, N., Kawarabayashi, K., Marx, D.: Packing cycles through prescribed vertices. J. Comb. Theory, Ser. B 101(5), 378–381 (2011)
18. Mader, W.: Über die Maximalzahl kreuzungsfreier H-Wege. Arch. Math. (Basel) 31(4), 387–402 (1978)
19. Marx, D.: Parameterized graph separation problems. Theoret. Comput. Sci. 351(3), 394–406 (2006)
20. Marx, D., Razgon, I.: Fixed-parameter tractability of multicut parameterized by the size of the cutset. In: STOC, pp. 469–478 (2011)
21. Niedermeier, R.: Invitation to Fixed-Parameter Algorithms. Oxford Lecture Series in Mathematics and its Applications, vol. 31. Oxford University Press, Oxford (2006)
22. Raman, V., Ramanujan, M.S., Saurabh, S.: Paths, Flowers and Vertex Cover. In: Demetrescu, C., Halldórsson, M.M. (eds.) ESA 2011. LNCS, vol. 6942, pp. 382–393. Springer, Heidelberg (2011)
23. Razgon, I., O'Sullivan, B.: Almost 2-sat is fixed-parameter tractable. J. Comput. Syst. Sci. 75(8), 435–450 (2009)
24. Reed, B.A., Smith, K., Vetta, A.: Finding odd cycle transversals. Oper. Res. Lett. 32(4), 299–301 (2004)

Set Cover Revisited: Hypergraph Cover with Hard Capacities[*]

Barna Saha[1] and Samir Khuller[2]

[1] AT&T Shannon Research Laboratory
barna@research.att.com
[2] University of Maryland College Park
samir@cs.umd.edu

Abstract. In this paper, we consider generalizations of classical covering problems to handle hard capacities. In the hard capacitated set cover problem, additionally each set has a covering capacity which we are not allowed to exceed. In other words, after picking a set, we may cover at most a specified number of elements. Based on the classical results by Wolsey, an $O(\log n)$ approximation follows for this problem.

Chuzhoy and Naor [FOCS 2002], first studied the special case of unweighted vertex cover with hard capacities and developed an elegant 3 approximation for it based on rounding a natural LP relaxation. This was subsequently improved to a 2 approximation by Gandhi et al. [ICALP 2003]. These results are surprising in light of the fact that for weighted vertex cover with hard capacities, the problem is at least as hard as set cover to approximate. Hence this separates the unweighted problem from the weighted version.

The set cover hardness precludes the possibility of a constant factor approximation for the hard-capacitated vertex cover problem on weighted graphs. However, it was not known whether a better than logarithmic approximation is possible on unweighted *multigraphs*, i.e., graphs that may contain parallel edges. Neither the approach of Chuzhoy and Naor, nor the follow-up work of Gandhi et al. can handle the case of multigraphs. In fact, achieving a constant factor approximation for hard-capacitated vertex cover problem on unweighted multigraphs was posed as an open question in Chuzhoy and Naor's work. In this paper, we resolve this question by providing the first constant factor approximation algorithm for the vertex cover problem with hard capacities on unweighted multigraphs. Previous works cannot handle hypergraphs which is analogous to consider set systems where elements belong to at most f sets. In this paper, we give an $O(f)$ approximation algorithm for this problem. Further, we extend these works to consider partial covers.

1 Introduction

Covering problems have been widely studied in computer science and operations research, starting from the early work on set-cover [11, 15, 18]. In addition, the vertex cover problem has been extremely well studied as well – this is a special case of set

---

[*] Research supported by NSF CCF-0728839, NSF CCF-0937865 and a Google Research Award.

A. Czumaj et al. (Eds.): ICALP 2012, Part I, LNCS 7391, pp. 762–773, 2012.

cover, where each element belongs to exactly two sets [2, 10]. Both these problems have played a central role in the development of many important ideas in algorithms – greedy algorithms, LP rounding, randomized algorithms, primal-dual methods, and have been the vehicle to convey many central ideas in combinatorial optimization.

In this paper, we consider covering problems with hard capacity constraints. In other words, if a set is chosen, it cannot cover all its elements, but there is an upper bound on the number of elements that the set can cover. More formally, consider a ground set of elements $\mathcal{U} = \{a_1, a_2, \ldots, a_n\}$ and a collection of subsets of $\mathcal{U}$, $\mathcal{S} = \{S_1, S_2, \ldots, S_m\}$. Each set $S \in \mathcal{S}$ has a positive integral capacity $k(S) \in \mathbb{N}$ and has an upper bound (denoted by $m(S)$) on the number of copies. In addition, each set can have arbitrary non-negative weight $\tilde{w} : \mathcal{S} \to \mathbb{R}^+$. A solution for capacitated covering problem contains each set $S \in \mathcal{S}$, $x(S)$ times where $x(S) = \{0, 1, 2, \ldots, m(S)\}$ such that there is an assignment of at most $x(S)k(S)$ elements to set S and all the elements are covered by the assignment. The goal is to minimize $\sum_{S \in \mathcal{S}} \tilde{w}(S)x(S)$. Using Wolsey's greedy algorithm [18], we can easily derive a $O(\log n)$ approximation for the capacitated set cover problem with hard capacities.

Approximation algorithms for vertex cover with (soft) capacities were developed by Guha et al [9]. In the soft capacitated covering problem there is no bound on the number of copies of each set (vertex) that can be chosen. In [9], a primal dual algorithm was developed to give a 2 approximation. This algorithm can be extended easily to handle vertex cover with (soft) capacities in hypergraphs. In other words, if we have a hyper graph with hyper edges of size at most f (set cover problem where each element belongs to at most f sets), then we can easily get an f approximation [9]. On the other hand, the case of hard capacities is quite difficult. In a surprising result, Chuzhoy and Naor [4] showed that the weighted vertex cover problem with hard capacities is set-cover hard and showed that for *unweighted* graphs a randomized rounding algorithm can give a 3 approximation. This was subsequently improved to a 2 approximation [7]. Vertex cover is a special case of set cover problem where $f = 2$. This naturally raises the question whether it is possible to obtain an f approximation for the unweighted set cover problem with hard capacities, where each element belongs to at most f sets. The approaches of [4, 7] do not extend to case when $f > 2$. Moreover, the results of [4, 7] only hold for *simple* graphs. *Obtaining a constant factor approximation algorithm for the hard-capacitated vertex cover problem for unweighted multigraphs was posed as an open question in [4]. In this paper, we resolve that question, and extending our approach we also obtain an $O(f)$-approximation for the unweighted set cover problem with hard capacities.* Further, we also provide an $O(f)$ approximation algorithm for partial cover problem with hard capacities. Partial cover is a natural generalization of covering problems where only a desired number of elements need to be covered [8]. While the works of [3, 17] extended the vertex cover with soft capacities to consider partial cover, nothing prior to our work was known in the case of hard capacities.

The notion of capacities is also natural in the context of facility location problems, as well as clustering problems and has been widely studied. Capacitated facility location and k-median problems have been an active area of research [1, 5, 16] and frequently appear in applications involving placement of warehouses, web caches and as a subroutine in several network design protocols. Non-metric capacitated facility location problem is a generalization of hard-capacitated set cover problem for which Bar-Ilan et al. [1] gave an $O(\log n + \log m)$-approximation. In this problem, there are m facilities and n clients; there is a cost associated for opening each facility and each client connects to one of the open facility paying a connection cost while the number of clients that can be assigned to an open facility remains bounded by its capacity. When, the connection costs are either 0 or ∞, we get the set cover problem with hard capacities.

In several set cover applications, an element only belongs to a few sets. This is especially true in the context of scheduling. One such example is the work of Khuller, Li and Saha [12] where they study a scheduling algorithm to allocate jobs to machines in data centers such that the minimum number of machines are activated. The goal is to minimize the energy to run machines while maintaining the makespan (maximum sum of processing times on any machine). In data centers, each data is replicated a *small* number of times (typically 3 copies). Thus a job needed to access specific data can be run on one of a small number of machines. In [12], a $(\ln n + 1)$ approximation algorithm is provided that violates the makespan by a factor of 2. However, it does not consider the fact that each job can be scheduled only on f (here $f \approx 3$) machines. Incorporating this, and in addition, considering that jobs have some fixed processing time, we obtain the hard-capacitated set cover problem with elements belonging to at most f sets. The scheduling model of [6] can also be seen as a hard-capacitated set covering instance with *multiple* capacity constraints.

Our algorithms for the hard-capacitated versions of both vertex cover and set cover are based on rounding linear programming (LP) relaxations. In the following subsection, we outline the main reasons why the previous approaches fail and provide a sketch of our algorithms.

1.1   Our Approach and Contributions

The works of [4,7] cannot handle the hard-capacitated vertex cover problem on multigraphs, neither do their approaches extend to hypergraphs or set systems with elements belonging to at most f sets. The algorithms in both of these works are based on LP rounding and involve three major steps. First, they pick all vertices with fractional values above a desired threshold. Next, a randomized rounding step is performed to choose some additional vertices. If even after step two, there are edges with unsatisfied fractional coverage, an alteration step is performed, in which vertices are chosen as long as all the edges are not fractionally fully covered maintaining the capacity constraints. Finally, the fractional edge assignment variables are rounded through a flow computation. While, the expected cost of selecting vertices in the first two steps can be easily bounded within a small factor of the optimal LP cost, the main crux of the argument relies in showing that with high probability the alteration cost can also be charged within a small factor of the cost incurred in the first two steps. When the graph does not contain any parallel edge, the random variables required to prove such a statement are all independent and thus strong concentration inequalities can be employed for the analysis. However, the presence of parallel edges (or having hypergraphs) make these random variables *positively correlated*. This hinders the application of required concentration inequalities and the analysis breaks down.

We utilize the LP-structure to decompose the problem into two simpler instances. Instead of consolidating the variables corresponding to sets (vertices), we modify the variables associated with assignment of elements (edges) to sets (vertices). Viewing the LP solution as a bipartite graph between elements and sets, the graph is decomposed into a forest (H_1) and an additional subgraph (H_2) such that elements entirely covered by either one of these can be rounded without much loss in the approximation. There may be elements that are partially covered (fractionally) by sets in both H_1 and H_2. We further modify the remaining fractional solution to recast the capacitated covering problem on these unsatisfied elements as a multiset multicover (MM) problem *without* any capacity constraints.

We show that the partially rounded solution is feasible for the natural linear programming relaxation for MM. However the natural LP relaxation for MM has an unbounded integrality gap. Using a stronger LP relaxation, it is possible to give $\log n$-approximation algorithm for MM [14], but our fractional solution may not be feasible for such stronger relaxations. Moreover, a $\log n$ approximation for MM is not sufficient for our purpose. Instead, we show that it is possible to charge the cost of the obtained solution to a constant factor of LP cost for MM and the number of elements in the set system, and this suffices to ensure a constant approximation. Our algorithm for MM follows the paradigm of grouping and scaling used for *column restricted* (each set has same multiplicity for all elements) packing and covering problems [13]. However, our set system is not column restricted. We still can group the elements into *small* and *big* based on the extent of coverage these elements get from sets with relatively lower or higher multiplicities compared to their demands. By scaling the fractional variables and doing randomized rounding, we can satisfy the requirements of small elements, but big elements may still have residual demands left. Satisfying the requirements of big elements need a further step of careful rounding. Details are described in Section 2.2.

Our main contributions are as follows.

- We obtain an $O(1)$ approximation algorithm for the vertex cover problem with hard capacities on unweighted multigraphs for the unit multiplicity case, i.e., when all $m(v) = 1$.

- We show an $O(f)$-approximation algorithm for the unweighted set cover problem with hard capacities where each element belongs to at most f sets.
  As a corollary, we obtain an $O(1)$ approximation for the hard-capacitated vertex cover problem on unweighted multigraphs for arbitrary multiplicities.

- We consider partial covering problem with hard capacities. We give $O(1)$ approximation for partial vertex cover with hard capacities and $O(f)$ approximation for partial set cover problem with hard capacities.

In the following section, we describe a constant factor approximation algorithm for the hard-capacitated vertex cover problem on multigraphs with unit multiplicity ($m(v) = 1, \forall v \in V(G)$). The algorithm and the analysis contain the main technical ingredients which are later used to obtain $O(f)$ approximation algorithms for the set cover and partial cover problems with hard capacities and arbitrary multiplicities. For lack of space, the latter two results appear in the full version of the paper.

2   Vertex Cover on Multigraphs with Hard Capacities

We start with the following linear programming relaxation for hard-capacitated vertex cover with unit multiplicities.

$$\text{minimize} \sum_{v \in V} x(v) \qquad\qquad\qquad\qquad\qquad\qquad\qquad (\text{LP}_{\text{VC}})$$

subject to

$$y(e, u) + y(e, v) = 1 \qquad\qquad \forall e = (u, v) \in E, \qquad (1)$$

$$y(e, v) \le x(v), \; y(e, u) \le x(u) \qquad \forall e = (u, v) \in E, \qquad (2)$$

$$\sum_{e = (u, v)} y(e, v) \le k(v) x(v) \qquad\qquad \forall v \in V, \qquad (3)$$

$$0 \le x(v), y(e, v), y(e, u) \le 1 \quad \forall v \in V, \forall e = (u, v) \in E. \qquad (4)$$

Here $x(v)$ is an indicator variable, which is 1 if vertex v is chosen and 0 otherwise. Variables $y(e, u)$ and $y(e, v)$ are associated with edge $e = (u, v)$. $y(e, u) = 1$ ($y(e, v) = 1$) indicates edge e is assigned to vertex u (v). Constraints (1) ensure each edge is covered by at least one of its end-vertices. Constraints (2) imply an edge cannot be covered by a vertex v, if v is not chosen in the solution. The total number of edges covered by a vertex v is at most $k(v)$ if v is chosen and 0 otherwise (constraints (3)). We relax the variables $x(v), y(e, v)$ to take value in $[0, 1]$ in order to obtain the desired LP-relaxation. The optimal solution of LP_{VC} denoted by $\text{LP}_{\text{VC}}(\text{OPT})$ clearly is a lower bound on the actual optimal cost OPT.

2.1   Rounding Algorithm

Let (x^*, y^*) denote an optimal fractional solution of LP_{VC}. We create a bipartite graph $H = (A, B, E(H))$, where A represents the vertices of G, B represents the *edges* of G [1] and the links $E(H)$ correspond to the (e, v) variables $e \in B$, $v \in A$ with non-zero y^* value [2]. Each $v \in A(H)$ is assigned a weight of $x^*(v)$. Each link (e, v) is assigned a weight of $y^*(e, v)$. We now modify the link weights in a suitable manner to decompose the link sets of H into two graphs H_1 and H_2. Special structures of H_1 and H_2 make rounding relatively simpler on them.

- H_1 *is a forest.* For each node $v \in A(H_1)$ and link $(e, v) \in E(H_1)$, $y^*(e, v) < x^*(v)$.
- *In* H_2, *if* $(e, v) \in E(H_2)$, *then weight of link* (e, v) *is equal to the weight of* v. Thus, for each node $v \in A(H_2)$ and link $(e, v) \in E(H_2)$, $y^*(e, v) = x^*(v)$.

A moment's reflection shows the usefulness of such a property, essentially, in H_2, we can ignore the hard capacity constraints altogether.

---

[1] We often refer a vertex in $B(H)$ by edge-vertex to indicate it belongs to $E(G)$.
[2] in order to avoid confusion between edges of G with edges of H, we refer to edges of H by links.

The decomposition procedure is based on iteratively breaking cycles. We now explain the rounding algorithms on each of H_1 and H_2.

Rounding on H_2.

We discard all isolated vertices from H_2. Let $\eta \geq 2$ be the desired approximation factor. We select all vertices in $A(H_2)$ with value of x^* at least $\frac{1}{\eta}$. Let us denote the chosen vertices by $\mathcal{D}$. Then,

$$\mathcal{D} = \{v \mid v \in A(H_2), x^*(v) \geq \frac{1}{\eta}\}.$$

For every edge-vertex $e = (u, v) \in B(H_2)$, if v (or u) is in $\mathcal{D}$, and $(e, v) \in E(H_2)$ (or $(e, u) \in E(H_2)$), then we set $y^*(e, v) = 1$ (or $y^*(e, u) = 1$). That is, we assign e to v, if the link (e, v) is in $E(H_2)$ and v is in $\mathcal{D}$, else if $u \in \mathcal{D}$ and $(e, u) \in E(H_2)$, the edge e is assigned to u.

Observation 1. *From constraints (3),* $\sum_{e=(u,v)} y(e, v) \leq x(v)k(v)$. *Therefore,* $\sum_{e=(u,v)} \frac{y(e,v)}{x(v)} \leq k(v)$, *and hence in* H_2, *after the assignment of edges to vertices in* $\mathcal{D}$, *all vertices maintain their capacity.*

In fact, in H_2, capacity constraints become irrelevant. *Whenever, we decide to pick a vertex in* $A(H_2)$, *we can immediately cover all the links in* $E(H_2)$ *incident on it.*

All edges with both links in $E(H_2)$ get covered at this stage. In addition, if $e \in B(H_2)$ has only one link $(e, v) \in E(H_2)$, but $x^*(v) = y^*(e, v) \geq \frac{1}{\eta}$, then since $v \in \mathcal{D}$, e gets covered. Therefore, the uncovered edges after this step either have no link in $E(H_2)$ or are fractionally covered to an extent less than $\frac{1}{\eta}$ in H_2.

Rounding on H_1.

H_1 is a forest; edge-vertices in H_1 either have both or one link in $E(H_1)$. While the vertices of H_1 and H_2 may overlap, the link sets are disjoint. Edge-vertices in $B(H_1)$ with only one link in H_1 are called *dangling* edges. We root H_1 arbitrarily to some

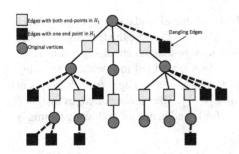

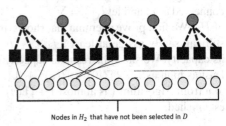

Fig. 1. Structure of H_1, dangling edges are colored black and connected by dashed lines, edges with both end-points in H_1 are colored white and connected by solid lines.

Fig. 2. Structure of H_1 after the edges with two end points in H_1 have been assigned.

node of $A(H_1)$. This naturally defines a parent-child relationship. Figure (1a) depicts the structure of H_1. Dangling edges are shown by dashed lines.

Rounding edges with both links in H_1.

Algorithm (1) describes the procedure to assign edge-vertices that have both links in $E(H_1)$.

Algorithm 1. Assigning edges with two links in H_1

1: let $\mathcal{D}' = \{v \in A(H_1) \mid x^*(v) \geq \frac{1}{\eta}\}$, select all the vertices in $\mathcal{D}'$.
2: **for each** edge-vertex e with two links in H_1 **do**
3:     **if** the child vertex of e is selected in $\mathcal{D}'$ **then**
4:         assign e to the selected child vertex.
5:     **end if**
6: **end for**
7: let $T(v)$ denote the set of unassigned children edge-vertices incident on $v \in A(H_1)$ with both links in H_1.
8: **select** any $t(v) = \lceil \sum_{e=(u,v) \in T(v)} y^*(e, u) \rceil$ vertices from the children of the edge-vertices in $T(v)$, and **assign** the corresponding $t(v)$ edge-vertices in $T(v)$ to these selected children vertices. If v' is a newly selected vertex in this step and there are edges that have links incident on v' in $E(H_2)$, then assign those edges to v' as well.
9: **assign** the remaining edge-vertices from $T(v)$ to v.

We first select a collection of $\mathcal{D}'$ vertices from $A(H_1) \setminus \mathcal{D}$ with x^* value at least $\frac{1}{\eta}$. Any edge-vertex in $B(H_1)$ that has a child vertex chosen in $\mathcal{D}'$ gets assigned to its child. For each vertex $v \in A(H_1)$, we use $T(v)$ to denote the set of children edge-vertices that are not assigned in step (4). We select $t(v) = \lceil \sum_{e=(u,v) \in T(v)} y^*(e, u) \rceil$ vertices from the children of the edge-vertices in $T(v)$. We assign the corresponding $t(v)$ edge-vertices in $T(v)$ to these newly selected children vertices. Rest of the edges in $T(v)$ are assigned to v.

Rounding dangling edges, i.e., with one link in H_1.

After Algorithm 1 finishes, let $L(v)$ denote the set of unassigned dangling edge-vertices connected to v, and let $l(v) = \sum_{e=(u,v), e \in L(v)} y^*(e, u)$. $L(v)$ are the leaf edge-vertices of H_1. We first prove a lemma that shows after the edge-assignment in Algorithm 1, we still can safely assign at least $|L(v)| - \lceil l(v) \rceil$ edges from $L(v)$ to v without violating its capacity. We show the residual capacity of v after assigning edges from $E(H_2)$ is at least as high as $1 + |T(v)| - \lceil t(v) \rceil + |L(v)| - \lceil l(v) \rceil$. The number of edges assigned to v from Algorithm 1 is at most $1 + |T(v)| - \lceil t(v) \rceil$ and hence the following lemma is established.

Lemma 1. *Each vertex $v \in A(H_1)$ can be assigned $|L(v)| - \lceil l(v) \rceil$ leaf edges-vertices without violating its capacity.*

The edge-vertices in $L(v)$ are leaves of H_1, they are connected to v and have their other link in $E(H_2)$. We first pick *one vertex* from $A(H_2)$ such that it covers at least one edge from $L(v)$. Let us denote this vertex by $h2(v)$ and let it cover $p2(v) \geq 1$ parallel edges $(v, h2(v))$. If $l(v) \leq p2(v)$, then following Lemma 1, the rest of the edge-vertices of $L(v)$ can be assigned to v, and we do so.

If $l(v) > p2(v)$. Let $R(v)$ denote the vertices of $A(H_2) \setminus h2(v)$ that are end-points of edges in $L(v)$. If we pick enough vertices from $R(v)$ such that they cover at least $l'(v) = l(v) - p2(v) + 1$ leaf-edges, then again from Lemma 1, rest of the edges from $L(v)$ can be assigned to v.

We scale up all the x^* variables of $\bigcup_{v \in A(H_1)} R(v)$ by a factor of $\frac{1}{1-\frac{1}{\eta}}$. We also scale up the corresponding y^* link variables by a factor of $\frac{1}{1-\frac{1}{\eta}}$. Let $(\bar{x}, \bar{y})$ denote the scaled up variables. Then, $\sum_{\substack{e=(u,v) \in \\ L(v) \setminus (v, h2(v))}} \bar{y}(e, u) = \frac{(l(v)-p2(v)x^*(h2(v)))}{(1-\frac{1}{\eta})} \geq \frac{(l(v)-\frac{p2(v)}{\eta})}{(1-\frac{1}{\eta})} >$
$l(v) - p2(v) + 1 = l'(v)$, where the last inequality follows from the fact that $l(v) > p2(v) \geq 1$. We let $l'(v) = 0$, if $l(v) \leq p2(v)$. We now have the following multi-set multi-cover problem (MM).

For each $v \in A(H_1)$ with $l'(v) > 0$, we create an element $a(v)$. For each vertex $u \in \bigcup_{v \in A(H_1)} R(v)$, we create a multi-set $S(u)$. If there are $d(v, u)$ leaf edge-vertices in $L(v) \setminus (v, h2(v))$ incident upon u, then we include $a(v)$ in $S(u)$, $d(v, u)$ times. Each element $a(v)$ has a requirement of $r(a(v)) = \lfloor l'(v) \rfloor$. The goal is to pick minimum number of sets such that each element $a(v)$ is covered $\lfloor l'(v) \rfloor$ times counting multiplicities.

Note that, since the original graph is a multigraph, $d(v, u)$ can be greater than 1.

Lemma 2. *If we set $z(S(u)) = \bar{x}_u, \forall u \in \bigcup_{v \in A(H_1)} R(v)$, then z is a feasible fractional solution for the above stated multi-set multi-cover problem.*

As described in Section 1.1, existing approaches are not sufficient to obtain an integral solution for the above MM problem that will ensure a constant approximation. We instead, obtain an algorithm where the total number of sets picked is close to $s + \sum_{u \in \bigcup_{v \in A(H_1)} R(v)} \bar{x}_u$, where s is the number of vertices in $A(H_1)$ with $l'(v) > 0$. In Section 2.2, we prove the following theorem.

Theorem 3. *Given any feasible fractional solution $\bar{x}$ with cost F for multi-set multi-cover problem with N elements, there is a polynomial time randomized rounding algorithm that rounds the fractional solution to a feasible integral solution with expected cost at most $21N + 32F$.*

The algorithm for assigning the leaf edge-vertices in $L(v)$ is given in Algorithm (2).

Since, each vertex $v \in A(H_1)$ covers at most $|L(v)| - \lceil l(v) \rceil$ leaf edge-vertices, by Lemma 1 the capacity of all the vertices in H_1 are maintained. We now proceed to analyze the cost.

Theorem 2. *There exists a polynomial time algorithm achieving an approximation factor of 34 for the hard-capacitated vertex cover problem with unit multiplicity on unweighted multigraphs.*

2.2 Proof of Theorem 3

In the multi-set multi-cover problem (MM), we are given a ground set of N elements U and a collection of multi-sets $\mathcal{S}$ of U, $\mathcal{S} = \{S_1, S_2, \ldots, S_M\}$. Each multi-set $S \in \mathcal{S}$

Algorithm 2. Assigning edges with only one link in H_1

1: **for each** vertex $v \in A(H_1)$ with $|L(v)| \geq 1$ **do**
2:     **select** the vertex $h2(v)$ that covers at least one edge-vertex from $L(v)$ and assign the corresponding edge-vertices to $h2(v)$.
3: **end for**
4: **for each** vertex $v \in A(H_1)$ with $l(v) \leq p2(v)$ **do**
5:     **assign** all the remaining edge-vertices (at most $|L(v)| - \lceil l(v) \rceil$) to v
6: **end for**
7: **for each** vertex $v \in A(H_1)$ with $l'(v) > 1$ **do**
8:     **scale up** the x^* variables in $\bigcup_{v \in A(H_1)} R(v)$ by a factor of $\frac{1}{1-\frac{1}{\eta}}$ and denote it by $\bar{x}$.
9: **end for**
10: **create** the MM instance $(\{(a(v), d(v))\}, \{S(u)\})$, and round the fractional solution $\bar{x}$ to obtain an integral solution.
11: **for each** u such that $S(u)$ is chosen by MM algorithm **do**
12:     **select** u and **assign** all the leaf-edges incident on u to it.
13: **end for**
14: **for each** $v \in A(H_1)$ with $l'(v) > 1$ **do**
15:     **assign** all the remaining leaf edge-vertices of $L(v)$ (at most $|L(v)| - \lceil l(v) \rceil$) to it.
16: **end for**

---

contains $M(S, e)$ copies of element $a \in U$. Each element a has a demand of $r(a)$ and needs to be covered $r(a)$ times. The objective is to minimize the number of chosen sets that satisfy the demands of all the elements. Here we propose a new algorithm that proves Theorem 3.

The following is a linear program relaxation for MM.

$$\min \sum_{S \in \mathcal{S}} x(S)$$

$$\sum_{a \in S} M(a, S)x(S) \geq r(a) \qquad \forall a \in U$$

$$0 \leq x(S) \leq 1 \qquad \forall S \in \mathcal{S}$$

2.3 Rounding Algorithm for MM

Let $\mathbf{x}^*$ denote the LP optimal solution. The rounding algorithm has several steps.

Step 1. Selecting Sets with High Fractional Value. First, we pick all sets $S \in \mathcal{S}$ such that $x^*(S) \geq \alpha > 0$, where $\frac{1}{\alpha}$ is the desired approximation factor. Denote the chosen sets by $\mathcal{H}$. Each element a now has a residual requirement of $r(a) - \sum_{a \in S, S \in \mathcal{H}} M(S, a)$. Clearly the fractional solution x^* projected on the sets $\mathcal{S} \setminus \mathcal{H}$ is a feasible solution for the residual problem. For each element $a \in U$, let $\bar{r}(a) = r(a) - \sum_{a \in S, S \in \mathcal{H}} M(S, a)$ be the residual requirement. For some $\beta > 0$ (to be set later), let $y(S) = \beta x^*(S)$, for each $S \in \mathcal{S} \setminus \mathcal{H}$. We have for all elements $a \in U$, $\sum_{a \in S, S \in \mathcal{S} \setminus \mathcal{H}} M(S, a)y(S) \geq \beta \bar{r}(a)$.

Note that after this step, we have a fractional solution with cost

$$|H| + \sum_{S \in \mathcal{S} \setminus \mathcal{H}} y(S) \leq \frac{1}{\alpha} \sum_{S \in \mathcal{H}} x^*(S) + \beta \sum_{S \in \mathcal{S} \setminus \mathcal{H}} x^*(S).$$

For notational simplicity, we denote $\mathcal{C} = \mathcal{S} \setminus \mathcal{H}$. Next, we proceed to round the variables $y(S)$ for $S \in \mathcal{C}$.

Step 2. Rounding into Powers of 2. For each multiplicity $M(S,a)$, $\forall S \in \mathcal{C}, a \in U$, we round it to the highest power of 2 lesser than or equal to $M(S,a)$ and denote it by $M^1(S,a)$. For each requirement $\bar{r}(a)$, $\forall a \in U$, consider the lowest power of 2 greater than or equal to $\bar{r}(a)$ and denote it by $\bar{r}^1(a)$. Clearly, if $\sum_{a \in S, S \in \mathcal{C}} M(S,a)y(S) \geq \beta \bar{r}(a)$, then $\sum_{a \in S, S \in \mathcal{C}} M^1(S,a)4y(S) \geq \beta \bar{r}^1(a)$. We denote $\mathbf{y}^1 = 4\mathbf{y}$.

Step 3. Division into Small and Big Elements. First, for each element if there is a set that completely satisfies its requirement, we pick the set. We continue the process as long as no more element can be covered entirely by a single set. Thus after this procedure, for all elements a, and for all sets S, $M^1(S,a) < \bar{r}^1(a)$ and hence $M^1(S,a) \leq \frac{\bar{r}^1(a)}{2}$. Now for each element a, we divide the sets in $\mathcal{C}$ containing a into *big sets* ($Big(a)$) and *small sets* ($Small(a)$). A set $S \in \mathcal{C}$ is said to be a big set for a, if $M^1(S,a) \geq \frac{1}{18 \ln n} \bar{r}^1(a)$, otherwise it is called a small set, i.e.,

$$Big(a) = \{S \in \mathcal{C} \mid M^1(S,a) \geq \frac{1}{18 \ln n} \bar{r}^1(a)\}$$

$$Small(a) = \{S \in \mathcal{C} \mid M^1(S,a) < \frac{1}{18 \ln n} \bar{r}^1(a)\}$$

Now, we decompose elements into *big* and *small*. An element is *small* if it is covered to an extent of $\bar{r}^1(a)$ by the sets in $Small(a)$. Else, the element is covered at least to an extent of $(\beta - 1)\bar{r}^1(a)$ by the sets in $Big(a)$ and we call it a *big* element. This follows from the inequality

$$\sum_{a \in S, S \in \mathcal{C} \cap Big(a)} M^1(S,a)y^1(S) \quad + \sum_{a \in S, S \in \mathcal{C} \cap Small(a)} M^1(S,a)y^1(S) \geq \beta \bar{r}^1(a).$$

Therefore, either the sets in $Small(a)$ cover a to an extent of $\bar{r}^1(a)$, or the sets in $Big(a)$ cover a to an extent of $(\beta - 1)\bar{r}^1(a)$. Let $\beta_1 = \beta - 1$. In the first case, we refer a as a small element, otherwise it is a big element.

Step 4. Covering Small Elements. We employ simple independent randomized rounding for covering small elements. *We pick each set $S \in \mathcal{C}$ with probability γy_S^1, for some* $\gamma \geq 2$.

Lemma 3. *All small elements are covered in Step 4 with probability at least* $\left(1 - \frac{1}{n^{1/3}}\right)$.

Step 5. Covering Big Elements. This is the most crucial ingredient in the algorithm. For each big element, we consider only the big sets containing it. For each such big element and big set we have $\frac{1}{18\ln n}r_a^1 < M^1(S,a) \le \frac{r_a^1}{2}$. Since, multiplicities are powers of 2, there are at most $l = \ln\ln n + 3$ different values of multiplicities of the sets for each element a.

Let $T_1^a, T_2^a, \ldots T_l^a$ denote the collection of these sets with multiplicities $\frac{\bar{r}^1(a)}{2}, \frac{\bar{r}^1(a)}{2^2}, \ldots, \frac{\bar{r}^1(a)}{2^l}$ respectively. That is, $T_i^a = \{S \in Big(a) \mid M(S,a) = \frac{\bar{r}^1(a)}{2^i}\}$. Set $\beta_1 \ge 3$.

For each $i = 1,2,\ldots,l$, *if* $\sum_{S\in T_i^a} y^1(S) > i$ *and the number of sets that have been picked from* T_i^a *in Step 4 is less than* $\dfrac{\sum_{S\in T_i^a} y^1(S)}{(\beta_1-2)}$, *pick new sets from* T_i^a *such that the total number of chosen sets from* T_i^a *is* $\left\lceil \dfrac{\sum_{S\in T_i^a} y^1(S)}{(\beta_1-2)} \right\rceil$.

We now show that each big element gets covered the required number of times and the total cost is bounded by a constant factor of the optimal cost.

Lemma 4. *Each big element* a *is covered* $r(a)$ *times by the chosen sets.*

Lemma 5. *The expected number of sets selected in Step 4 is at most* $21n'$, *where* n' *are the number of big elements that are not covered after Step 5.*

Theorem 3. *The algorithm returns a solution with expected cost at most* $21N + 32F$, *where* $F = \sum_S x^*(S)$, *and covers all the elements with probability at least* $1 - \frac{1}{n^{1/3}}$.

This completes the description of the $O(1)$ approximation algorithm for hard-capacitated vertex cover problem on multigraphs with unit multiplicities. We have not tried to optimize the constants of our approach, but reducing the approximation ratio to 2 or 3 may require significant new ideas. Theorem 3 is also crucially used to obtain an $O(f)$-approximation algorithm for the set cover and partial cover problem with arbitrary multiplicities. The results for set cover and partial cover problem appear in the full version of the paper.

References

1. Bar-Ilan, J., Kortsarz, G., Peleg, D.: Generalized submodular cover problems and applications. Theor. Comput. Sci. 250, 179–200 (2001)
2. Bar-Yehuda, R., Even, S.: A local-ratio theorem for approximating the weighted vertex cover problem. Annals of Discrete Mathematics 25, 27–45 (1985)
3. Bar-Yehuda, R., Flysher, G., Mestre, J., Rawitz, D.: Approximation of Partial Capacitated Vertex Cover. In: Arge, L., Hoffmann, M., Welzl, E. (eds.) ESA 2007. LNCS, vol. 4698, pp. 335–346. Springer, Heidelberg (2007)
4. Chuzhoy, J., Naor (Seffi)., J.: Covering problems with hard capacities. SIAM J. Comput. 36(2), 498–515 (2006)
5. Chuzhoy, J., Rabani, Y.: Approximating k-median with non-uniform capacities. In: Proceedings of the Sixteenth Annual ACM-SIAM Symposium on Discrete Algorithms, SODA 2005, pp. 952–958 (2005)

6. Demaine, E.D., Zadimoghaddam, M.: Scheduling to minimize power consumption using submodular functions. In: Proceedings of the 22nd ACM Symposium on Parallelism in Algorithms and Architectures, SPAA 2010, pp. 21–29 (2010)

7. Gandhi, R., Halperin, E., Khuller, S., Kortsarz, G., Srinivasan, A.: An improved approximation algorithm for vertex cover with hard capacities. J. Comput. Syst. Sci. 72, 16–33 (2006)

8. Gandhi, R., Khuller, S., Srinivasan, A.: Approximation algorithms for partial covering problems. J. Algorithms 53(1), 55–84 (2004)

9. Guha, S., Hassin, R., Khuller, S., Or, E.: Capacitated vertex covering. Journal of Algorithms 48(1), 257–270 (2003)

10. Hochbaum, D.S.: Approximation algorithms for the set covering and vertex cover problems. Siam Journal on Computing 11, 555–556 (1982)

11. Johnson, D.S.: Approximation algorithms for combinatorial problems. J. Comput. Syst. Sci. 9, 256–278 (1974)

12. Khuller, S., Li, J., Saha, B.: Energy efficient scheduling via partial shutdown. In: SODA, pp. 1360–1372 (2010)

13. Kolliopoulos, S.G.: Approximating covering integer programs with multiplicity constraints. Discrete Appl. Math. 129, 461–473 (2003)

14. Kolliopoulos, S.G., Young, N.E.: Tight approximation results for general covering integer programs. In: IEEE Symposium on Foundations of Computer Science, pp. 522–528 (2001)

15. Lovász, L.: On the ratio of optimal integral and fractional covers. Discrete Mathematics 13(4), 383–390 (1975)

16. Mahdian, M., Pal, M.: Universal facility location. In: Proc. of European Symposium of Algorithms 2003, pp. 409–421 (2003)

17. Mestre, J.: A Primal-Dual Approximation Algorithm for Partial Vertex Cover: Making Educated Guesses. In: Chekuri, C., Jansen, K., Rolim, J.D.P., Trevisan, L. (eds.) APPROX 2005 and RANDOM 2005. LNCS, vol. 3624, pp. 182–191. Springer, Heidelberg (2005)

18. Wolsey, L.A.: An analysis of the greedy algorithm for the submodular set covering problem. Combinatorica 2, 385–393 (1982)

On the Limits of Sparsification*

Rahul Santhanam[1,**] and Srikanth Srinivasan[2,***]

[1] University of Edinburgh
rsanthan@inf.ed.ac.uk
[2] DIMACS, Rutgers University
srikanth@dimacs.rutgers.edu

Abstract. Impagliazzo, Paturi and Zane (JCSS 2001) proved a sparsification lemma for k-CNFs: every k-CNF is a sub-exponential size disjunction of k-CNFs with a linear number of clauses. This lemma has subsequently played a key role in the study of the exact complexity of the satisfiability problem. A natural question is whether an analogous structural result holds for CNFs or even for broader non-uniform classes such as constant-depth circuits or Boolean formulae. We prove a very strong negative result in this connection: For every superlinear function $f(n)$, there are CNFs of size $f(n)$ which cannot be written as a disjunction of $2^{n-\varepsilon n}$ CNFs each having a linear number of clauses for any $\varepsilon > 0$. We also give a hierarchy of such non-sparsifiable CNFs: For every k, there is a k' for which there are CNFs of size $n^{k'}$ which cannot be written as a sub-exponential size disjunction of CNFs of size n^k. Furthermore, our lower bounds hold not just against CNFs but against an *arbitrary* family of functions as long as the cardinality of the family is appropriately bounded.

As by-products of our result, we make progress both on questions about circuit lower bounds for depth-3 circuits and satisfiability algorithms for constant-depth circuits. Improving on a result of Impagliazzo, Paturi and Zane, for any $f(n) = \omega(n \log(n))$, we define a pseudo-random function generator with seed length $f(n)$ such that with high probability, a function in the output of this generator does not have depth-3 circuits of size $2^{n-o(n)}$ with bounded bottom fan-in. We show that if we could decrease the seed length of our generator below n, we would get an explicit function which does not have linear-size logarithmic-depth series-parallel circuits, solving a long-standing open question.

Motivated by the question of whether CNFs sparsify into bounded-depth circuits, we show a *simplification* result for bounded-depth circuits: any bounded-depth circuit of linear size can be written as a sub-exponential size disjunction of linear-size constant-width CNFs. As a corollary, we show that if there is an algorithm for CNF satisfiability which runs in time $O(2^{\alpha n})$ for some fixed $\alpha < 1$ on CNFs of linear size, then there is an algorithm for satisfiability of linear-size constant-depth circuits which runs in time $O(2^{(\alpha+o(1))n})$.

---

* This is an extended abstract with some proofs missing. The full version may be found at [11].
** Partially supported by ESPRC Grant EP/H05068X/1.
*** Work partially done as a Member at the Institute of Advanced Study, Princeton.

A. Czumaj et al. (Eds.): ICALP 2012, Part I, LNCS 7391, pp. 774–785, 2012.
© Springer-Verlag Berlin Heidelberg 2012

1   Introduction

The Satisfiability (SAT) problem is of central importance in theoretical computer science. Since SAT is NP-complete, the NP vs P problem reduces to the question of whether SAT has polynomial-time algorithms. We do not believe that SAT has polynomial-time algorithms, however it is still a very interesting question which the best algorithms are for solving SAT in the worst case. Specifically, by how much can we improve over the "naive" brute-force search algorithm for SAT, which enumerates over all possible 2^n assignments for a SAT instance and checks whether any of them are satisfying? A very concrete motivation for this problem is that SAT instances need to be solved in the real world, in a variety of contexts such as verification, automated planning and testing [6].

From a complexity-theoretic point of view, the importance of improving over brute-force search has been illustrated by the recent results of Williams [14] [15]. He shows that even *marginal* improvements over brute-force search for satisfiability of Boolean circuits in a class $\mathcal{C}$ implies that NEXP does not have polynomial-size circuits in the class $\mathcal{C}$, for a range of natural classes $\mathcal{C}$ of circuits. He applies his methodology [15] to obtain a new circuit lower bound, namely that NEXP $\not\subseteq$ ACC^0, by designing an algorithm performing slightly better than brute-force search for ACC^0-SAT. In fact, there are connections between SAT algorithms and lower bounds in the opposite direction as well, as evidenced in recent work using lower bound techniques to design and analyze improved Satisfiability algorithms [10] [3]. This makes the question of understanding the complexity landscape of the SAT problem even more intriguing.

When trying to design an improved algorithm, a natural approach is to find general structural properties of the class of instances which can be exploited algorithmically. Some examples of such properties for SAT are the downward self-reducibility property used to reduce the search problem to the decision version, and the Satisfiability Coding Lemma of Paturi, Pudlak and Zane, which has been used to design and analyze better algorithms for k-SAT as well as to prove depth-3 circuit lower bounds for restricted classes of circuits [9] [8].

Perhaps the most influential such property is that of *sparsifiability*. The Sparsification Lemma of Impagliazzo, Paturi and Zane [5] plays a key role in the study of the exact complexity of SAT. It states that for any constants $\epsilon > 0$ and k a positive integer, any k-CNF on n variables can be written as the disjunction of $2^{\epsilon n}$ *linear-size* CNFs, where the constant factor in the size depends only on k and ϵ.

The Lemma has found many different applications in both algorithmic and lower bound contexts. Impagliazzo, Paturi and Zane [5] used a constructive version of it in their study of sub-exponential reducibilities between NP-complete problems. Their results indicate that the Exponential-Time Hypothesis (ETH), which states that 3-SAT is not solvable in time $2^{o(n)}$, can be used as a unifying hypothesis in the study of exact complexity of NP-hard problems. They prove that, for various problems such as k-SAT (where $k \geq 3$ is a positive integer), k-Colourability, Clique, Vertex Cover, Satisfiability of linear-size Boolean circuits etc., existence of a $2^{o(n)}$ time algorithm is equivalent to ETH. The Lemma has also been used to undertake more refined studies of the complexity of SAT in terms of various parameters such as clause width and clause density [4] [2]. From the point of view of lower bounds, the Lemma has been used to construct a small pseudorandom family of functions such that with high probability, a function in this family does not have depth-3 circuits of size $2^{n-o(n)}$ and bounded bottom fan-in. This is closely related to classical questions about lower bounds for linear-size logarithmic-depth circuits [13].

It is natural to ask whether a similar sparsifiability property holds for broader classes of formulae or circuits, such as CNFs or even constant-depth circuits. Such a result would be useful in getting better algorithmic results and deriving new lower bounds. For example, while k-SAT is solvable in time $2^{n-\Omega(n)}$ for $m = poly(n)$ and constant k, the best known algorithm for SAT on general CNFs runs in time $2^{n-\Omega(n/\log(m/n))}$. A sparsification lemma for CNFs would be an important step towards a $2^{n-\Omega(n)}$ time algorithm for SAT on polynomial-size formulae. Indeed, this has explicitly been posed as an open question by Calabro, Impagliazzo and Paturi [2].

In this paper, we show a strong *negative* answer to the question of whether CNFs (and hence also more general classes of circuits) can be sparsified.

Theorem 1. *Let $f : \mathbb{N} \to \mathbb{N}$ be any function such that $f(n) = \omega(n)$. Then there is a sequence of CNFs $\{\phi_n\}$, where for each n ϕ_n has n variables and has size at most $f(n)$, such that for any constants $\varepsilon \in (0, 1]$ and $c > 0$, for all large enough n ϕ_n cannot be written as the OR of $2^{n-\varepsilon n}$ CNFs of size at most cn. In particular, CNFs are not sparsifiable.*

In fact, what we show is significantly stronger - for any sequence $\{F_n\}$ of families of Boolean functions such that $|F_n| = n^{O(n)}$, there is a sequence of CNFs which are not expressible as a $2^{n-\Omega(n)}$ size disjunction of functions in F_n. Also, the CNFs for which we show this are very natural. The functions they represent are the solution sets of sparse linear equations.

Theorem 1 only rules out "sparsifying" superlinear-size CNFs to linear-size CNFs. It could potentially still be the case that n^3-size CNFs are sparsifiable into n^2-size CNFs. It turns out that the counter-examples of Theorem 1 cannot establish this stronger statement, however by using a different set of counter-examples and a similar argument, we derive a hierarchy of non-sparsifiable CNFs.

Theorem 2. *Let k and $k' > 2k$ be any fixed constants. There is a fixed $\epsilon > 0$ and a sequence of CNFs $\{\phi_n\}$ where ϕ_n has n variables and $|\phi_n| \leq n^{k'}$ such that for large enough n, ϕ_n cannot be written as the OR of $2^{\epsilon n}$ CNFs of size at most n^k.*

The hard CNFs are again natural - they are simply *random* CNFs of a specified width and size. Thus, in a sense, the proof of Theorem 2 shows that CNFs cannot be sparsified even *on average*.

We motivated the question about sparsification by describing the possible applications of a positive result. It turns out that our negative results have a couple of interesting byproducts as well. By itself, the results give some indication of the obstacles to designing better SAT algorithms, as well as what kinds of instances are likely to be hard. For example it is known that in certain contexts, such as for Resolution-based algorithms, instances encoding subspaces or random instances are hard. Our results are in a similar spirit.

More concretely, motivated by Theorem 1, we construct a simple new sub-exponential time reduction from satisfiability on linear-size constant-depth circuits to k-SAT. The motivation is to apply Theorem 1 to show that CNFs cannot in general be sparsified into linear-size constant depth circuits. We cannot simply use the stronger form of Theorem 1 for arbitrary families of functions of small enough cardinality here, as we are unable to bound the number of functions computed by unbounded fan-in linear-size constant-depth circuits by $n^{O(n)}$. Instead, we show a *positive* result that any linear-size constant-depth circuit can be written as an OR of $2^{\epsilon n}$ k-CNFs for any $\epsilon > 0$ and k depending only on ϵ. This decomposition can actually be done constructively, and this gives us the reduction we mentioned before. The decomposition also implies that superlinear-size CNFs cannot be sparsified into linear-size constant-depth circuits.

Theorem 3. *Let $\{f_n\}$ be a sequence of Boolean functions on n bits, such that f_n is computed by linear-size constant-depth circuits. For any constant $\epsilon > 0$, there is a constant k such that f_n is the disjunction of $2^{\epsilon n}$ functions each of which is computed by a k-CNF of linear size.*

Theorem 1 also has an application to circuit lower bounds. Here we are concerned with lower bounds for depth-3 circuits where there is a bound on the bottom fan-in. If we could show that there is an explicit function which does not have size $2^{n/2}$ depth-3 circuits with bottom fan-in $O(1)$, this would be a lower bound breakthrough, as using a connection due to Valiant[13] it would imply a superlinear-size lower bound against logarithmic-depth series-parallel circuits. Valiant argues that the series-parallel restriction on the structure of the circuit is interesting because the best-known circuits for many problems are series-parallel. Impagliazzo, Paturi and Zane [5] make progress on this question by constructing an explicit pseudo-random family of $2^{O(n^2)}$ functions such that most functions in the family do not have size $2^{n-\Omega(n)}$ depth-3 circuits with bottom fan-in $O(1)$. We improve their result by reducing the size of the function family down to $n^{f(n)}$ for any $f(n) = \omega(n)$. We also argue that a further improvement of the family size to 2^{cn} for $c < 1$ would actually imply a breakthrough lower bound for an explicit function.

In the theorem below, a Σ_3 circuit is an unbounded fan-in depth 3 circuit where the top gate is an OR. Note that when trying to prove a lower bound for an explicit function, we can assume wlog that the top gate is an OR.

Theorem 4. *For each $f(n) = \omega(n)$, there is a sequence $\{F_n\}$ of families of Boolean functions on n bits, where F_n has size at most $n^{f(n)}$, such that with probability $1 - o(1)$, a random function from F_n does not have Σ_3^k circuits of size $2^{n-\Omega(n)}$ with bottom fan-in $O(1)$. Moreover, given $i \in [1, n^{f(n)}]$ in binary and $x \in \{0,1\}^n$, there is a polynomial-time algorithm for evaluating the i'th function in F_n on x.*

2   Preliminaries

2.1   Basic Complexity Notions

We assume a basic knowledge of complexity theory. Standard references for this include the book by Arora and Barak [1] and the Complexity Zoo[1].

When discussing sparsification, we find it convenient to talk of non-uniform complexity measures. A non-uniform complexity measure $\mathcal{CSIZE}$ associates with each integer n and size bound s, a class of Boolean functions $\mathcal{CSIZE}(s(n))$ on n bits, such that for any $s' \geq s, \mathcal{CSIZE}(s(n))) \subseteq \mathcal{CSIZE}(s'(n))$. We will be concerned mainly with measures which correspond directly to standard models of computation, such as CNFs, CNFs of constant width (referred to as $O(1)$-CNFs), constant-depth unbounded fan-in circuits (AC^0), Boolean formulae and Boolean circuits.

By the size of a CNF, we will typically mean the number of clauses. If we mean the total number of literal occurrences, we will make this explicit.

As we will be studying lower bounds for depth-3 circuits, we require some notation for such circuits. Define Σ_d^k to be the set of depth d circuits with top gate OR such that each bottom gate has fan-in at most k. It is known that any Σ_3^k circuit for the Parity function or the Majority function requires $\Omega(2^{n/k})$ gates, and such bounds are tight for $k = O(\sqrt{n})$. For $k = 2$, a $2^{n-o(n)}$ size lower bound is known for an explicit function in P, however not even an $\Omega(2^{n/2})$ size lower bound is known for an explicit function for any $k > 2$. Using a connection due to Valiant [13], this question can be related to classical lower bound questions about linear-size logarithmic-depth Boolean circuits. Valiant's results imply that linear-size logarithmic-depth Boolean circuits with bounded fan-in can be computed by depth-3 unbounded fan-in circuits of size $O(2^{n/\log\log n})$ with bottom fan-in limited by n^ε for arbitrarily small ε. If in addition, the graph of connections of the circuit is restricted to be series-parallel, the simulation can be modified to give size $2^{n/2}$ and fan-in $O(1)$.

Given functions $f, g : \mathbb{N} \to \mathbb{R}^{>0}$, we occasionally use $f \ll g$ to denote $f(n) = o(g(n))$. This notation makes the transitivity of the $o(\cdot)$ relation more transparent.

2.2   Sparsification and Simplification

Definition 1. *Given non-uniform complexity measures $\mathcal{CSIZE}$ and $\mathcal{C'SIZE}$, and functions $s, s' : \mathbb{N} \to \mathbb{N}$, we say that there is a $(\mathcal{C}, s, \mathcal{C'}, s')$-sparsification if for any constant $\epsilon > 0$ and any function $f \in \mathcal{CSIZE}(O(s))$, f is the OR of at most $2^{\epsilon n}$ functions each belonging to $\mathcal{CSIZE}(O(s'))$. We say that $\mathcal{C}$ is sparsifiable to $\mathcal{C'}$ if there is a $(\mathcal{C}, n^k, \mathcal{C'}, n)$-sparsification for each k, and we say simply that $\mathcal{C}$ is sparsifiable if $\mathcal{C}$ is sparsifiable to $\mathcal{C}$.*

Definition 2. *Given non-uniform complexity measures $\mathcal{CSIZE}$ and $\mathcal{C'SIZE}$, and function $s : \mathbb{N} \to \mathbb{N}$, we say that there is an OR-simplification of $\mathcal{C}$ to $\mathcal{C'}$*

---

[1] http://qwiki.stanford.edu/index.php/Complexity_Zoo

at size s if there is a $(\mathcal{C}, s, \mathcal{C}', s)$-sparsification. We say that there is an OR-simplification of $\mathcal{C}$ to $\mathcal{C}'$ if there is an OR-simplification of $\mathcal{C}$ to $\mathcal{C}'$ at size n.

The following proposition is immediate since sub-exponential size ORs are closed under composition.

Proposition 1. *If $\mathcal{C}$ is sparsifiable to $\mathcal{C}'$ and there is an OR-simplification of $\mathcal{C}'$ to $\mathcal{C}$, then $\mathcal{C}$ is sparsifiable.*

There are many interesting positive results on sparsification and simplification. Impagliazzo, Paturi and Zane [5] showed that k-CNFs are sparsifiable for any constant k. Improved parameters were obtained by [2].

Lemma 1 (Sparsification Lemma). *[5] [2] Let $k > 0$ be any integer. For any constant $\epsilon > 0$, there exists a constant $c(k, \epsilon)$ such that for large enough n, any k-CNF over n variables can be expressed as the OR of $2^{\epsilon n}$ k-CNFs each of size at most $c(k, \epsilon)n$.*

The original proof of Lemma 1 [5] yielded c doubly exponential in k but this was subsequently improved to singly exponential in k. Using results of Miltersen, Radhakrishnan and Wegener [7], it can be shown that an exponential dependence on k is necessary.

Schuler [12] showed that there is an OR-simplification of CNFs to $O(1)$-CNFs. This follows from the following more general lemma, the proof of which is similar and is deferred to the full version.

Lemma 2. *For any constant $\varepsilon \in (0, 1]$ and function $c : \mathbb{N} \to \mathbb{N}$, every CNF φ with at most cn clauses can be written as the OR of at most $2^{\varepsilon n}$ many k-CNFs with at most cn clauses, where $k = O(\frac{1}{\varepsilon} \log(\frac{c}{\varepsilon}))$.*

Note that when c is a constant in Lemma 2, k is a constant as well.

Corollary 1. *There is an OR-simplification of CNFs to $O(1)$-CNFs.*

3 The Limits of Sparsification

3.1 Non-sparsifiability of CNFs

We will show that there are CNFs of slightly superlinear size that cannot be written as a subexponential OR of CNFs of linear size.

Given $\ell, r \in \mathbb{N}$, let $\mathcal{S}_{\ell,r}$ denote the collection of all r-tuples of subsets of $[n]$ of size ℓ. Given $\overline{S} = (S_1, \ldots, S_r) \in \mathcal{S}_{\ell,r}$, let $\varphi_{\overline{S}}$ denote some CNF for the following function:

$$G_{\overline{S}} = \bigwedge_{i=1}^{r} \neg \bigoplus_{j \in S_i} x_j$$

Though the above function has not been written in CNF form, it is easy to see that for any $\overline{S}$ as above, $\varphi_{\overline{S}}$ can be chosen to be CNFs of size at most $r2^{\ell}$.

Lemma 3. *Fix any $\ell, r : \mathbb{N} \to \mathbb{N}$. Then we have that for any $\overline{S} \in \mathcal{S}_{\ell,r}$, the CNF $\varphi_{\overline{S}}$ has at least 2^{n-r} satisfying assignments.*

Proof. This follows from the fact that any homogeneous system of r linear equations has at least 2^{n-r} solutions over $\mathbb{F}_2$.   □

Now we proceed to the proof of the main lemma. Given a CNF formula φ, let $\mathrm{Sat}(\varphi)$ denote the set of satisfying assignments of φ.

Fix a $T \subseteq [n]$ and assume that $S \in \binom{[n]}{\ell}$ is chosen uniformly at random. Given $\eta \in [0, 1]$, we call S $(1 - \eta)$-*balanced w.r.t.* T if $|S \cap T| \geq (1 - \eta) \mathbf{E}_S[|S \cap T|]$. We call S *balanced w.r.t.* T if S is $1/2$-balanced w.r.t. T. Given $\overline{S} \in \mathcal{S}_{\ell,r}$, we say that $\overline{S}$ is $(1 - \eta)$-*balanced w.r.t.* T (*balanced w.r.t.* T) if at least half the S_i are $(1 - \eta)$-balanced w.r.t. T (respectively, balanced w.r.t. T).

We need the following technical lemma regarding balance.

Lemma 4. *Let* $\varepsilon, \eta \in (0, 1)$ *be constants. Fix* $\ell = \ell(n), r = r(n)$ *such that* $1 \ll \ell(n)$ *and* $n/\ell \ll r(n)$. *Assume* $T \subseteq [n]$ *such that* $|T| \geq \varepsilon n$. *Then for a randomly chosen* $\overline{S} \in \mathcal{S}_{\ell,r}$, *we have* $\mathrm{Pr}_{\overline{S}}[\overline{S} \text{ is not } (1 - \eta)\text{-balanced w.r.t. } T] = \frac{1}{2^{\omega(n)}}$.

Proof. A simple concentration equality tells us that for any $i \in [r]$, $\mathrm{Pr}_{S_i}[S_i \text{ not } (1 - \eta)\text{-balanced}] \leq 2^{-\Omega(\ell)}$. Hence, given a set of $r/2$ many S_i, the probability that *none* of them are balanced w.r.t. T is bounded by $2^{-\Omega(\ell r)} = 2^{-\omega(n+r)}$, where the last equality follows from the fact that $r \gg n/\ell$. By a union bound, it follows that the probability that there *exists* a subset of $\overline{S}$ of size $r/2$ all of whose elements are not $(1 - \eta)$-balanced w.r.t T is at most $\binom{r}{r/2} 2^{-\omega(n)} \leq 2^r 2^{-\omega(n+r)} \leq 2^{-\omega(n)}$. The lemma now follows since this event corresponds precisely to $\overline{S}$ not being balanced w.r.t T.   □

Lemma 5. *Fix constants* $c, \varepsilon > 0$. *Let* $\ell = \ell(n), r = r(n)$ *be parameters such that* $1 \ll \ell = O(\log n), n/\ell \ll r \ll n$. *Fix any collection* $\mathcal{A}$ *of subsets of* $\{0, 1\}^n$ *of size at most* n^{cn} *such that each* $A \in \mathcal{A}$ *has size at least* $2^{\varepsilon n}$. *Then, for a random* $\overline{S} \in \mathcal{S}_{\ell,r}$, *we have* $\mathrm{Pr}_{\overline{S}}[\exists A \in \mathcal{A} : A \subseteq \mathrm{Sat}(\varphi_{\overline{S}})] = o(1)$.

Proof. Fix any $A \in \mathcal{A}$. Since $\mathrm{Sat}(\varphi)$ is a subspace of $\mathbb{F}_2^n$, we see that $A \subseteq \mathrm{Sat}(\varphi)$ iff $\mathrm{Span}(A) \subseteq \mathrm{Sat}(\varphi)$, where $\mathrm{Span}(A)$ is the span of A in $\mathbb{F}_2^n$. Hence, we assume wlog that every $A \in \mathcal{A}$ is actually a subspace of dimension at least εn. Fix such a subspace A. Let $d \geq \varepsilon n$ denote the dimension of A.

By Gaussian elimination, we can choose a $d \times n$ matrix $M(A)$ such that the rows of $M(A)$ generate A and after some column permutations, $M(A) = [I_d \ M']$ where I_d denotes the $d \times d$ identity matrix. Let the variables indexed by the first d columns of $M(A)$ be denoted $S(A)$.

Consider a uniformly random $\overline{S} = (S_1, \ldots, S_r) \in \mathcal{S}_{\ell,r}$. For $i \in [r]$ let χ_i denote the characteristic vector of S_i. It is easily seen that $A \subseteq \mathrm{Sat}(\varphi_{\overline{S}})$ iff each $\chi_i \in A^\perp$, where $A^\perp$ denotes the dual space of A.

We now consider the probability that $\chi_i \in A^\perp$ for any fixed i. This happens iff $M(A)\chi_i = 0$. Note that this event can occur with probability at least $\frac{1}{2^{O(\ell)}}$ if, for example, $M' = 0$ and it happens that $S_i \subseteq [n] \setminus S(A)$. We now show that this probability is much lower if we condition on the event that S_i is balanced w.r.t. $S(A)$.

Say we condition on $|S_i \cap S(A)| = q$, where $q \in [\ell]$. Note that picking a random S_i conditioned on this event is equivalent to picking a random subset

S_i' of $S(A)$ of size q and a random subset S_i'' of $\overline{S(A)}$ of size $\ell - q$ and setting $S_i = S_i' \cup S_i''$. Let χ_i' and χ_i'' denote the characteristic vectors of S_i' and S_i'' respectively. Then, $M(A)\chi_i = 0$ iff $I_d\chi_i' + M'\chi_i'' = 0$ iff $\chi_i' = M'\chi_i''$. For any fixed choice of χ_i'', the probability over the choice of χ_i' that this occurs is at most $1/\binom{d}{q} \leq (q/\varepsilon n)^q \leq \frac{1}{(\varepsilon n)^{\Omega(q)}}$. Hence, conditioned on S_i being balanced w.r.t. $S(A)$, we see that the probability that $M(A)\chi_i = 0$ is at most $\frac{1}{(\varepsilon n)^{\Omega(\varepsilon \ell)}} \leq \frac{1}{n^{\Omega(\ell)}}$. Using the fact that $r = \omega(n/\ell)$, this implies that $\Pr_{\overline{S}}[\forall i \in [r] : M(A)\chi_i = 0 \mid \overline{S} \text{ balanced w.r.t. } S(A)] \leq \left(\frac{1}{n^{\Omega(\ell)}}\right)^{r/2} = \frac{1}{n^{\omega(n)}}$. $\hspace{1cm}$ (*)

We are now ready to bound the probability that there exists a subspace $A \in \mathcal{A}$ that is contained in $\mathrm{Sat}(\varphi_{\overline{S}})$. Let $E_1(A)$ denote the event that $A \subseteq \mathrm{Sat}(\varphi_{\overline{S}})$. Given $T \subseteq [n]$ s.t. $|T| \geq \varepsilon n$, let $E_2(T)$ denote the event that $\overline{S}$ is not balanced w.r.t. T. We have

$$\Pr_{\overline{S}}[\bigvee_A E_1(A)] \leq \Pr_{\overline{S}}[\bigvee_A E_1(A) \vee \bigvee_{T \subseteq [n]: |T| \geq \varepsilon n} E_2(T)]$$

$$= \Pr_{\overline{S}}[\bigvee_T E_2(T)] + \Pr_{\overline{S}}[\bigvee_A E_1(A) \wedge \neg \bigvee_T E_2(T)]$$

$$\leq \sum_T \Pr_{\overline{S}}[E_2(T)] + \sum_A \Pr_{\overline{S}}[E_1(A) \wedge \neg E_2(S(A))]$$

$$\leq \sum_T \Pr_{\overline{S}}[E_2(T)] + \sum_A \Pr_{\overline{S}}[E_1(A) \mid \neg E_2(S(A))]$$

$$\leq 2^n \cdot \frac{1}{2^{\omega(n)}} + n^{cn} \cdot \frac{1}{n^{\omega(n)}} = o(1)$$

where the last inequality follows from Lemma 4 and (*). This concludes the proof of the lemma. $\hspace{1cm} \square$

Theorem 5. *Fix any constants $c > 0$ and $\varepsilon \in (0, 1]$. Say $\overline{S}$ is chosen uniformly at random from $\mathcal{S}_{\ell, r}$, where ℓ, r are as in the statement of Lemma 5. Then, the probability that $\varphi_{\overline{S}}$ can be written as a union of at most $2^{n-\varepsilon n}$ many CNFs of size at most cn is $o(1)$.*

Proof. Assume that for some $\overline{S}$, $\varphi_{\overline{S}}$ can be written as an OR of at most $2^{n-\varepsilon n}$ many CNFs of size at most cn. By Lemma 2, each such CNF can be written as a union of at most $2^{\varepsilon n/2}$ many k-CNFs of size at most cn, where $k = k(c, \varepsilon)$ is a constant. Moreover, Lemma 3 implies that $|\mathrm{Sat}(\varphi_{\overline{S}})| \geq 2^{n-r} = 2^{n-o(n)}$. Hence, it must be the case that there is some k-CNF ψ of size at most cn such that $|\mathrm{Sat}(\psi)| \geq 2^{\varepsilon n/4}$ and $\mathrm{Sat}(\psi) \subseteq \mathrm{Sat}(\varphi_{\overline{S}})$. Let $\mathcal{A} = \{\mathrm{Sat}(\psi) \mid \psi \text{ a } k\text{-CNF}, \mathrm{Size}(\psi) \leq cn, \text{ and } |\mathrm{Sat}(\psi)| \geq 2^{\varepsilon n/4}\}$; clearly, $|\mathcal{A}| \leq \binom{(2n)^k}{cn} \leq n^{kcn}$. We have seen above that if $\varphi_{\overline{S}}$ can be written as an OR of at most $2^{n-\varepsilon n}$ many CNFs of size at most cn, then there must be an $A \in \mathcal{A}$ such that $A \subseteq \mathrm{Sat}(\varphi_{\overline{S}})$. By Lemma 5, the probability that this happens is $o(1)$. Hence, the theorem follows. $\hspace{1cm} \square$

The above easily yields Theorem 1 by choosing $\ell = \omega(1)$ small enough and $r = n/\sqrt{\ell}$ so that $f(n) \geq n2^{\ell}/\sqrt{\ell}$, and then using Theorem 5 to yield existence of CNFs of the desired size which are non-sparsifiable.

3.2   A Hierarchy Theorem for Non-Sparsifiability

Theorem 5 shows the existence of CNFs of slightly super-linear size which cannot be sparsified into linear-size CNFs. A natural question is whether there is a hierarchy of such non-sparsifiable CNFs: is it true that for each k, there is an $k' > k$ such that there are CNFs of size $n^{k'}$ which cannot be sparsified into CNFs of size n^k.

First note that the hard CNFs we're looking for cannot be of the form $\varphi_{\overline{S}}$ for some $\overline{S} \in \mathcal{S}_{\ell,r}$. This is because the corresponding function $G_{\overline{S}}$ trivially has formulae of size $o(n \log(n))$ over the basis $\{\wedge, \vee, \oplus\}$, and so also is sparsifiable into formulae of the same size over this basis. Lemma 5 shows non-sparsifiability into *any* class of functions of small enough cardinality, so we cannot hope to strengthen Lemma 5 to get the desired result for $k > 1$.

Instead, we use a random CNF ψ with a prescribed width and clause density. Fix $n \in \mathbb{N}$ and $\ell : \mathbb{N} \to \mathbb{N}$. We denote by $\Psi_{n,\ell(n)}$ the collection of all CNF formulas on n boolean variables of width exactly $\ell(n)$ with $2^{\ell(n)}$ many clauses (with possible repetitions). To sample a random ψ from $\Psi_{n,\ell(n)}$, we simply sample $2^{\ell(n)}$ random clauses of width $\ell(n)$. We establish the following theorem, whose proof is omitted in this version.

Theorem 6. *Fix constants $c \geq 1, \eta > 0$. Assume $\ell = \ell(n) = (2c + \eta) \log n$. Then, then there exists a fixed $\delta = \delta(\eta, c) > 0$ such that the probability that a random ψ sampled from $\Psi_{n,\ell}$ can be written as an OR of at most $2^{\delta n}$ many CNFs of size at most $O(n^c)$ is at most $3/4 + o(1)$. In particular, there is no (CNF, $n^{2c+\eta}$, CNF, n^c)-sparsification.*

Theorem 6 straightaway implies Theorem 2.

4   Simplifying AC^0 to CNFs

In this section, all AC^0 circuits considered will have AND gates as their output gates. Note that any AC^0 circuit can be converted to this form by adding an additional AND gate at the output, hence increasing the size and depth by 1.

Definition 3. *Given $s, d, k \in \mathbb{N}$, an AC^0 circuit C with an AND gate as its output gate is said to be (s, d, k)-bounded if it has size at most s, depth at most d, and all of its gates except the output gate have fanin bounded by k.*

Fact 7. *For constants $d, k \in \mathbb{N}$ and any $s \in \mathbb{N}$, any (s, d, k)-bounded AC^0 circuit can be written as a CNF of size $O(s)$ and width k^d.*

Definition 4. *Given $N, s, k \in \mathbb{N}$, a set $\mathcal{C}$ of at most N (s, d, k)-bounded AC^0 circuits is said to be an (N, s, d, k)-disjoint system if the set of satisfying assignments of each pair of distinct circuits $C_1 \neq C_2$ from $\mathcal{C}$ are disjoint. The function computed by $\mathcal{C}$ is defined to be $\bigvee_{C \in \mathcal{C}} C$.*

Lemma 6. *Fix constants $c, d \in \mathbb{N}$ such that $d \geq 2$ and $\varepsilon \in (0, 1]$. There exists a $k = k(c, d, \varepsilon)$ and a $c' = c'(c, d, \varepsilon)$ such that for any AC^0 circuit C of depth d and size at most cn on n variables, there is an $(2^{\varepsilon n}, c'n, d, k)$-disjoint system $\mathcal{C}$ that computes the same function as C.*

Proof. The proof is by induction on d. We need a small variant of Lemma 2, which gives us the base case of $d = 2$:

Claim. For any $c \in \mathbb{N}$ and $\varepsilon \in (0, 1]$, there exists a $k = k(c, \varepsilon) \in \mathbb{N}$ such that for any collection $\mathcal{S}$ of at most cn many clauses (respectively, terms), there is a partition of $\{0, 1\}^n$ into at most $2^{\varepsilon n}$ many parts such that in each part, each clause (resp. term) in $\mathcal{S}$ has size at most k. Moreover, each element of the partition is specified by a k-CNF with at most $(c + 1)n$ clauses.

Proof. We prove the result in the case of clauses; the proof for terms is almost identical. Let k be a parameter that we will choose later. As long as there is a clause of width at least k, choose k literals from the clause and split the remainder of the space into two parts depending on whether the disjunction of these literals is satisfied or not. Call the branch where the literals are *not* satisfied the *good* branch. Along the good branch, we can set k variables to some boolean values; along the other branch, we still end up satisfying the clause.

Note that there can be only $cn + n/k$ many steps overall, since every step either satisfies a clause or sets k variables. Moreover, there can be at most n/k many good steps along any branch. This means that the total number of branches is bounded by $\binom{cn+n/k}{n/k} \leq \binom{(c+1)n}{n/k} \leq (ek(c+1))^{n/k} \leq 2^{O(\log(kc)n/k)} \leq 2^{\varepsilon n}$ for large enough k depending on c and ε.

Note, moreover, that inputs corresponding to each branch is given by a k-CNF, where k with at most $cn + n/k \cdot k = (c + 1)n$ many clauses. $\square$

The above claim easily implies that for any CNF φ with at most cn clauses, there is a $(2^{\varepsilon n}, (2c + 1)n, 2, k)$-disjoint system computing the same function as φ, where k is as defined in Claim 4.

Now consider a circuit of depth $d > 2$. Let $C_{<d}$ be the circuit C up to layer $d - 1$, with the layer of height 1 gates being replaced by a new set of variables $y_1, \ldots, y_m$, where $m \leq cn$. By applying the induction hypothesis to $C_{<d}$ with $\varepsilon = \varepsilon/(2c)$, we see that there exist $c_1, k_1 \in \mathbb{N}$ and a $(2^{\varepsilon n/2}, c_1 n, d - 1, k_1)$-disjoint system $\mathcal{C}$ that computes the same function as $C_{<d}$ on inputs coming from $\{0, 1\}^m$.

Moreover, by applying Claim 4 to the AND and OR gates at height 1, there exists $k_2 \in \mathbb{N}$ and a partition $\mathcal{P}$ of $\{0, 1\}^n$ into at most $2^{\varepsilon n/2}$ parts, each of which is specified by a k_2-CNF of size at most $(c+1)n$, such that in each partition, each gate at height 1 depends on at most k_2 variables. For each $P \in \mathcal{P}$, let φ_P denote the k_2-CNF of size at most $(c + 1)n$ that accepts exactly the inputs in P; given any circuit $C' \in \mathcal{C}$, let C_P denote the circuit $C'' \wedge \varphi_P$, where C'' is obtained by substituting for each y_i the corresponding term or clause of width at most k_2 that agrees with the corresponding gate on inputs from the set P of inputs. The set of all such circuits C_P gives us a $(2^{\varepsilon n}, (c_1 + c + 1)n, d, \max\{k_1, k_2\})$-disjoint system that computes the same function as the circuit C. $\square$

Corollary 2. *There is an OR-simplification of* AC^0 *to $O(1)$-CNFs. In particular, we have:*

1. *For any function $f(n) = \omega(n)$ and constants $c, \varepsilon > 0$, there is a sequence of CNFs $\{\varphi_n\}$, where φ_n has n variables and size at most $f(n)$ such that φ_n*

cannot be written as an OR of at most $2^{n-\varepsilon n}$ many AC^0 circuits of depth d and size at most cn.

2. *If satisfiability of linear-size CNFs can be tested in time $2^{\alpha n}$ for some fixed $\alpha < 1$, then satisfiability of linear-size AC^0 circuits can also be tested in time $2^{(\alpha+\varepsilon)n}$, for any fixed $\varepsilon > 0$.*

Proof. That there is an OR-simplification of AC^0 to $O(1)$-CNFs follows directly from Lemma 6 and Fact 7. Item 1 then follows from Theorem 1. Item 2 follows trivially.   □

Theorem 3 follows from Corollary 2.

5   Circuit Lower Bounds for Depth-3 Circuits

Impagliazzo, Paturi and Zane [5] showed that non-sparsifiability is closely connected to lower bounds for depth-3 circuits with bounded bottom fan-in. It is a long-standing open problem to find an explicit Boolean function which requires Σ_3^k circuits of size $2^{\omega(n/k)}$, where k is the bottom fan-in.

It is implicit in [5] that there is no $(AC^0[\oplus], n^2, \mathcal{C}, n)$-sparsification for any complexity measure $\mathcal{CSIZE}$ such that there are at most $n^{O(n)}$ Boolean functions in $\mathcal{CSIZE}(O(n))$. They use this to construct an explicit family of $2^{O(n^2)}$ Boolean functions such that with probability close to 1, a random function from this family does not have Σ_3^k circuits of size $2^{n-o(n)}$ for $k = o(\log \log(n))$. Note that such a lower bound holds for a *purely random* Boolean function using a straightforward counting argument; what their result gives is a pseudo-random function family of significantly smaller size for which the lower bound still holds with high probability. Their result relies on the sparsification lemma first proved in the same paper. Using our result, we can prove Theorem 4, which reduces the size of the family down to $n^{f(n)}$ for any $f(n) = \omega(n)$, which, as we show, is "close" to getting the lower bound for an explicit function.

Proof (of Theorem 4). The function family $\{F_n\}$ we use is simply the set $\{G_{\overline{S}}\}$, where $\overline{S} \in \mathcal{S}_{\ell,r}$, with ℓ and r chosen as in the proof of Theorem 1. The bound on the cardinality of F_n and the polynomial-time evaluability of functions in F_n are clear. We will show that if a function f cannot be written as an OR of $2^{n-\varepsilon n}$ CNFs of linear size for any $\varepsilon > 0$, then it does not have Σ_3^k circuits of size $2^{n-o(n)}$ with bottom fan-in $O(1)$. Thus the theorem follows using Theorem 5.

Suppose, on the contrary, that there is a constant $c < 1$ such that f has Σ_3^k circuits of size 2^{cn} with bottom fan-in $k = O(1)$. Consider the gates with output wires feeding in to the top OR gate. Each such gate computes an $O(1)$-CNF. By Lemma 1, for any $\varepsilon > 0$ each such gate can be written as the OR of $2^{\varepsilon n}$ $O(1)$-CNFs of size $O(n)$. By choosing ε such that $\varepsilon + c < 1$, we get that f is the OR of $2^{c'n}$ functions, each of which has CNFs of size $O(n)$ for some $c' < 1$. This contradicts the assumption on f, hence we are done.   □

Theorem 8. *Suppose there is a sequence $\{F_n\}$ of families of Boolean functions on n bits, where F_n has size at most $2^{n-\Omega(n)}$, such that for large enough n, there* *exists a function $f_n \in F_n$ such that f_n does not have Σ_3^k circuits of size $2^{n-o(n)}$ with bottom fan-in $k(n) = O(1)$ (resp. $n^{o(1)}$). Also assume that given $i \in [1, |F_n|]$ in binary and $x \in \{0,1\}^n$, there is a polynomial-time algorithm for evaluating the i'th function in F_n on x. Then there is a Boolean function $g \in P$ such that g does not have linear-size logarithmic-depth series-parallel circuits (resp. linear-size logarithmic-depth circuits).*

The proof is omitted in this version.

References

1. Arora, S., Barak, B.: Computational Complexity - A Modern Approach. Cambridge University Press (2009)
2. Calabro, C., Impagliazzo, R., Paturi, R.: A duality between clause width and clause density for SAT. In: Proceedings of IEEE Conference on Computational Complexity, pp. 252–260 (2006)
3. Impagliazzo, R., Matthews, W., Paturi, R.: A satisfiability algorithm for AC0. In: Proceedings of Symposium on Discrete Algorithms (to appear, 2012)
4. Impagliazzo, R., Paturi, R.: On the complexity of k-sat. Journal of Computer and System Sciences 63(4), 512–530 (2001)
5. Impagliazzo, R., Paturi, R., Zane, F.: Which problems have strongly exponential complexity? Journal of Computer and System Sciences 62(4), 512–530 (2001)
6. Malik, S., Zhang, L.: Boolean satisfiability from theoretical hardness to practical success. Communications of the ACM 52(8), 76–82 (2009)
7. Miltersen, P.B., Radhakrishnan, J., Wegener, I.: On converting cnf to dnf. Theoretical Computer Science 347(1-2), 325–335 (2005)
8. Paturi, R., Pudlak, P., Saks, M., Zane, F.: An improved exponential-time algorithm for k-sat. In: Proceedings of 39th International Symposium on Foundations of Computer Sciece (FOCS), pp. 628–637 (1998)
9. Paturi, R., Pudlak, P., Zane, F.: Satisfiability coding lemma. In: Proceedings of 38th International Symposium on Foundations of Computer Science (FOCS), pp. 566–574 (1997)
10. Santhanam, R.: Fighting perebor: New and improved algorithms for formula and QBF satisfiability. In: Proceedings of 51st Annual IEEE Symposium on Foundations of Computer Science, pp. 183–192 (2010)
11. Santhanam, R., Srinivasan, S.: On the limits of sparsification. Electronic Colloquium on Computational Complexity (ECCC) 18, 131 (2011)
12. Schuler, R.: An algorithm for the satisfiability problem of formulas in conjunctive normal form. J. Algorithms 54(1), 40–44 (2005)
13. Valiant, L.G.: Graph-Theoretic Arguments in Low-Level Complexity. In: Gruska, J. (ed.) MFCS 1977. LNCS, vol. 53, pp. 162–176. Springer, Heidelberg (1977)
14. Williams, R.: Improving exhaustive search implies superpolynomial lower bounds. In: Proceedings of the 42nd Annual ACM Symposium on Theory of Computing, pp. 231–240 (2010)
15. Williams, R.: Non-uniform ACC circuit lower bounds. In: Proceedings of 26th Annual IEEE Conference on Computational Complexity, pp. 115–125 (2011)

Certifying 3-Connectivity in Linear Time*

Jens M. Schmidt

MPI für Informatik, Saarbrücken, Germany

Abstract. One of the most noted construction methods of 3-vertex-connected graphs is due to Tutte and based on the following fact: Every 3-vertex-connected graph G on more than 4 vertices contains a contractible edge, i. e., an edge whose contraction generates a 3-connected graph. This implies the existence of a sequence of edge contractions from G to K_4 such that every intermediate graph is 3-vertex-connected. A theorem of Barnette and Grünbaum yields a similar sequence using removals of edges instead of contractions.

We show how to compute both sequences in optimal time, improving the previously best known running times of $O(|V|^2)$ to $O(|E|)$. Based on this result, we give a linear-time test of 3-connectivity that is certifying; finding such an algorithm has been a major open problem in the design of certifying algorithms in the last years. The 3-connectivity test is conceptually different from well-known linear-time tests of 3-connectivity; it uses a certificate that is easy to verify in time $O(|E|)$. We also provide an optimal certifying test of 3-edge-connectivity.

1 Introduction

The class of 3-connected (i. e., 3-vertex-connected) graphs has been studied intensively for many reasons in the past 50 years. Besides being a fundamental graph property, 3-connectivity has numerous applications, in particular (but not only) for problems in graph drawing (see [14] for a survey), problems related to planarity and online problems on planar graphs (see [3] for a survey).

We use graph constructions throughout the paper. Let B be a set of graphs, G be a graph and O be a finite set of graph operations. A sequence of operations of O that generates G when applied to a graph of B is called a *construction sequence from B to G (using O)*. We will call B the set of *base graphs*. When B and O are clear from the context, we just refer to a *construction sequence of G*. Such a sequence can also be described by giving the inverse operations from G to a base graph; we call this the *top-down* variant of a construction sequence, as opposed to a *bottom-up* variant.

The importance of construction sequences for this paper stems mainly from the fact that they certify 3-connectivity; it is however the author's belief that the inductive nature of construction sequences is a very useful, yet not fully utilized, framework to solve computational graph problems efficiently.

One of the most noted constructions for 3-connected graphs was given by Tutte [19]: Every 3-connected graph G on more than 4 vertices contains a *contractible* edge, i. e., an edge whose contraction generates a 3-connected graph. Iteratively contracting such

---

* This research was partly supported by the Deutsche Forschungsgemeinschaft within the research training group "Methods for Discrete Structures" (GRK 1408) and FU Berlin.

A. Czumaj et al. (Eds.): ICALP 2012, Part I, LNCS 7391, pp. 786–797, 2012.
© Springer-Verlag Berlin Heidelberg 2012

an edge yields a top-down construction sequence from G to a K_4-multigraph. Unfortunately, also non-3-connected graphs can contain contractible edges, but adding a side condition establishes a full characterization [6]: A graph G on more than 4 vertices is 3-connected if and only if there is a construction sequence from G to a K_4-multigraph using contractions on edges with both end vertices having at least 3 neighbors; we will call this a *sequence of contractions*. In fact, the existence of the bottom-up variant of this sequence is commonly stated as Tutte's famous *wheel theorem* [19].

Barnette and Grünbaum [2] and Tutte [20] prove that every 3-connected graph G on more than 4 vertices contains a *removable* edge, i. e., an edge whose deletion generates a subdivision of a 3-connected graph. Let *removing* an edge e be the operation that deletes e and, for each end vertex v of e with exactly two distinct neighbors x and y in the remaining graph, deletes v and inserts the edge xy. Removing a removable edge leads, similar as in the sequence of contractions, to a top-down construction sequence from G to K_4; we will call this a *sequence of removals*. Again, adding a side condition fully characterizes the 3-connected graphs [18].

Although both existence theorems on contractible and removable edges are used frequently, the first non-trivial computational result to create the corresponding construction sequences was published more than 45 years afterwards: In 2006, Albroscheit [1] showed how to compute a construction sequence for 3-connected graphs in time $O(|V|^2)$. However, in this algorithm, contractions and removals are allowed to intermix. In 2010, two results [13,18] were given that both computed a sequence of contractions in time $O(|V|^2)$. The latter result also gives an algorithm that computes a sequence of removals in $O(|V|^2)$. All mentioned algorithms do not rely on the 3-connectivity test of Hopcroft and Tarjan [9], which runs in linear time but is rather involved. No algorithm is known that computes any of these sequences in subquadratic time.

We give an algorithm that computes a sequence of removals in linear time. This will also imply a linear-time algorithm that computes a sequence of contractions (in the bottom-up and top-down variants) and has a number of consequences.

Certifying 3-Connectivity in Linear Time. Mehlhorn and Näher [12] (see [11] for a survey) introduced the concept of *certifying algorithms*, i. e., algorithms that produce with each output a *certificate* that the particular output has not been compromised by a bug. Such a *certificate* can be any data that allows to check the correctness of the particular output (uniformly using a verifying algorithm), but should allow for an easy verification. Achieving certifying algorithms is a major goal for problems where the fastest solutions known are complicated and difficult to implement. Testing a graph on 3-connectivity is such a problem, but surprisingly little work has been devoted to certify 3-connectivity, although sophisticated linear-time recognition algorithms are known for over 35 years [9,17,21]. However, none of them describes an easy-to-verify certificate.

The currently fastest algorithms that certify 3-connectivity need $\Theta(|V|^2)$ time and use construction sequences as certificates [1,13,18]. Recently, a linear time certifying algorithm for 3-connectivity has been proposed for the subclass of Hamiltonian graphs, when a Hamiltonian cycle is given [6]. In general, finding a certifying algorithm for 3-connectivity in subquadratic time is a major open problem [11, Chapter 5.4] [6].

We give a linear-time certifying algorithm for 3-connectivity that uses a sequence of removals as certificate. This implies a new linear-time 3-connectivity test that neither assumes the graph to be 2-connected nor needs to compute low-points (see [9] for a definition); instead, it uses the structure of 3-connected graphs implicitly by applying simple path-generating rules. This is conceptually different from all previous linear-time 3-connectivity tests. The algorithm has already been implemented and made publicly available [15]; interestingly, it outperforms the test in [9] on no-instances.

Certifying 3-edge-connectivity in linear time. There is no test for 3-edge-connectivity that is certifying and runs in linear time, although many non-certifying linear-time algorithms for this problem are known, the first being [7]. Based on a reduction in [7], we give a linear-time test on 3-edge-connectivity that is certifying.

Applications. Certifying 3-connectivity allows to make many graph algorithms that use the $SPQR$-*tree* data structure [8] certifying (e. g., [3,14]). Moreover, algorithms on *polytopes* can be augmented with a quick and easy check that their input represents indeed a polytope. Applications in communication networks include certificates for their *reliability* and the property to admit a *perfectly secure message transmission* [5].

We use standard graph-theoretic terminology from [4]; let $n = |V|$ and $m = |E|$. Let $v \to_G w$ denote a path P from vertex v to vertex w in a graph G and let $s(P) = v$ and $t(P) = w$ be the *source* and *target* vertex of P, respectively (this imposes an orientation from $s(P)$ to $t(P)$ on P). Every vertex in $P \setminus \{s(P), t(P)\}$ is called an *inner vertex* of P. For $v \in V(G)$, let $N(v) = \{w \mid vw \in E\}$ (possibly $v \in N(v)$) and $deg(v)$ its degree (counting multiedges). Let $\delta(G)$ be the minimum degree in G. Let T be an undirected tree rooted at r. For two vertices x and y in T, let x be an *ancestor* of y and y be a *descendant* of x if $x \in V(r \to_T y)$. If additionally $x \neq y$, x and y are *proper* ancestors and descendants, respectively. Let $T(x)$ be the subtree of T that contains all descendants of x. Let K_2^m be the graph on 2 vertices that contains exactly m parallel edges and no self-loops.

2   BG-Paths

Iteratively removing removable edges in a 3-connected graph G leads to a *sequence of removals* from G to K_4, in which all generated intermediate graphs are 3-connected. However, the intermediate graphs are not necessarily subgraphs of G, which makes a linear-time computation difficult. For that reason, we reduce the computation to a closely related construction sequence [18], which is described next.

A *subdivision* of a graph G is a graph generated from G by replacing each edge of G by a path of length at least one. Let S be a subdivision of either K_2^3 or of a 3-connected graph. Let a vertex v in S be *real* if $deg(v) \geq 3$ and let $V_{real}(S)$ be the set of real vertices in S. Let the *links* of S be the paths in S that have real end vertices but contain no other real vertices. Note that the links of S are in one-to-one correspondence to the edges of the subdivided graph (which is K_2^3 or 3-connected) and thus partition $E(S)$. Let two links be *parallel* if they share the same end vertices.

Definition 1. *A* BG-path *for S is a path $P = x \to_G y$, $x \neq y$, such that*

1. $V(P) \cap V(S) = \{x, y\}$
2. *Every link of S that contains both x and y contains them as end vertices.*

3. *If x and y are inner vertices of distinct links L_x and L_y of S, respectively, and $|V_{real}(S)| \geq 4$, then L_x and L_y are not parallel.*

It was shown in [18] that a graph G without self-loops is 3-connected if and only if $\delta(G) \geq 3$ and G can be constructed from an (arbitrary) K_4-subdivision in G by adding BG-paths. This implies that every 3-connected graph G contains a subdivision of K_4, a result first shown by J. Isbell [2]. For technical reasons, we will use a slightly modified construction that starts with a K_2^3-subdivision and demand that the first BG-path generates a K_4-subdivision. Thus, a *construction sequence using BG-paths* starts with a K_2^3-subdivision of G, adds one BG-path that generates a K_4-subdivision and then adds BG-paths until G is constructed.

Let $S_3, S_4, S_5, \ldots, S_z = G$ be the intermediate graphs that are generated by such a construction. We benefit from two key features: Each S_l, $3 \leq l < z$, is a subdivision of a 3-connected graph and, additionally, a subgraph of S_{l+1} and therefore of G. This does not only yield an easy representation in linear space, it will also allow to compute a next BG-path efficiently by searching the neighborhood of the current subgraph in G. We give old and new results about construction sequences.

Theorem 1. *The following statements are equivalent.*

(1) *A simple graph G is 3-connected*
(2) *There is a sequence of removals from G to K_4 of removable edges $e = xy$ with $|N(x)| \geq 3$, $|N(y)| \geq 3$ and $|N(x) \cup N(y)| \geq 5$ (see [18])*
(3) *There is a sequence of removals from G to K_4 of   edges $e = xy$ with $|N(x)| \geq 3$, $|N(y)| \geq 3$ and $|N(x) \cup N(y)| \geq 5$*
(4) *There is a sequence of contractions from G to a K_4-multigraph of contractible edges $e = xy$ with $|N(x)| \geq 3$ and $|N(y)| \geq 3$ (see [18], chapter 5 in [19])*
(5) *There is a sequence of contractions from G to a K_4-multigraph of edges $e = xy$ with $|N(x)| \geq 3$ and $|N(y)| \geq 3$ (see [6])*
(6) *$\delta(G) \geq 3$ and there is a sequence of BG-paths from a K_2^3-subdivision in G to G such that the first BG-path generates a K_4-subdivision*
(7) *$\delta(G) \geq 3$ and there is a sequence of BG-paths from a K_4-subdivision in G to G [2,18]*
(8) *$\delta(G) \geq 3$ and there is a sequence of BG-paths from each K_4-subdivision in G to G [18]*

Lemma 1. *There are algorithms that transform a sequence of Type (6) to the sequences of each of the Types (2)–(7) in linear time.*

With Lemma 1, we can transform a sequence of Type (6) to every sequence of Theorem 1 efficiently. We will therefore focus on computing this sequence; if not stated otherwise, a *construction sequence* will refer to a sequence of Type (6). The following lemma provides an iterative algorithmic approach to compute it.

Lemma 2 ([18]). *Let G be a 3-connected graph and $H \subset G$ such that H is a subdivision of either K_2^3 or of a 3-connected graph. There is a BG-path for H in G.*

Clearly, every sequence of Type (4) and (5) must contain exactly $n - 4$ contractions. We give a corresponding result for the number of operations in the other sequences.

Lemma 3. *Every sequence of Type* (2)*,* (3) *and* (7) *contains exactly* $m - n - 2$ *operations and every sequence of Type* (6) *contains exactly* $m - n - 1$ *operations.*

3   Chain Decompositions and Certificates

We first describe a decomposition of a simple graph G, which is closely related to *ear* and *open ear* (i. e., no ear is a cycle) decompositions [10]. This decomposition will be the base structure that allows to compute a sequence of Type (6) efficiently. We define the structure algorithmically on a depth-first search (DFS) forest [9]; a similar procedure for the computation of so-called low-points (see [9]) can be found in [17]. Let F be a (rooted) DFS-forest of G. For every backedge e, let $s(e)$ and $t(e)$ denote the two end vertices of e such that $s(e)$ is a proper ancestor of $t(e)$ in F.

We decompose G into a set $C = \{C_1, \ldots, C_{|C|}\}$ of cycles and paths, called *chains*, by applying the following procedure for each vertex v in DFS-order (see also Figure 1): Let T be the tree in F that contains v and let r be the root of T. For every backedge vw with $s(vw) = v$, we traverse the path $w \to_T r$ until a vertex x is found that is either r or already contained in a chain. The traversed subgraph $vw \cup (w \to_T x)$ forms a new *chain* C_i with $s(C_i) = v$ and $t(C_i) = x$. We call C a *chain decomposition*. Let $<$ be the strict total order on C in which the chains were found, i. e., $C_1 < \cdots < C_{|C|}$. Clearly, the decomposition into chains can be computed in time $O(n + m)$.

Interestingly, C is an open ear decomposition if and only if G is 2-connected, an ear decomposition if and only if G is 2-edge-connected and we can test both facts by checking very easy conditions on C in linear time (proofs omitted). Thus, chain decompositions unify existing linear-time tests on 2-connectivity and 2-edge-connectivity without the necessity to compute low-points in advance. If G is not 2-(edge)-connected, a cut vertex (respectively, a bridge) can be found in linear time.

Easy-to-Verify Certificates for Low (edge-)Connectivity. Suppose there is a vertex or edge cut X of size $k - 1 \geq 0$ in a graph G with $n > 1$ (for vertex-connectivity, let $n > k$). Then X would be a straight-forward certificate for G being not k-(edge-)connected. However, certificates should be as easy to check as possible, while the running time for computing them is less important. We thus apply a paradigm of *shifting as much as possible of the checker's work to the computation of the certificate.*

Instead of using X as certificate, we color the vertices of one connected component of $G \setminus X$ red and the vertices of all other connected components of $G \setminus X$ green (we call this a *red-green coloring*). A checker for G being not k-connected then just needs to check that at most $k - 1$ vertices are uncolored, there is at least one red and one green vertex and that no edge joins a red vertex with a green one. For G being not k-edge-connected, it suffices to check that there is a red, a green and no uncolored vertex and that the end vertices of at most $k - 1$ edges differ in color. The certificates need space $O(n)$ and can be checked in time $O(m)$, as $n > m + 1$ proves G to be disconnected.

A certificate for G being connected is given in [11], using an easy numbering scheme on the vertices. Easy-to-verify certificates for 2- and 2-edge-connectivity are open ear decompositions and ear decompositions [10], respectively, as computed by the chain decomposition. For 3-connectivity, we will use a sequence of Type (6) as certificate, which proves G to be 3-connected due to Theorem 1. A simple checker in $O(m)$ time for this sequence is given in [18].

For certifying 3-edge-connectivity, we use a reduction to 3-connectivity due to Galil and Italiano [7]. The reduction modifies the simple input graph G in linear time to a graph with $m + 3n$ vertices and $3m$ edges. First, a graph G' is generated from G by subdividing each edge with one vertex; these vertices are called *arc-vertices*. For each non-arc-vertex w in G', let $v_1, \ldots, v_{deg(w)}$ be the arc-vertices neighboring w. Then the edges $(v_1v_2, v_2v_3, \ldots, v_{deg(w)}v_1)$ are added to G' if not already existent. The graph G is 3-edge-connected if and only if G' is 3-connected [7]. Moreover, every vertex cut of minimal size in G' contains only arc-vertices (Lemma 2.2 in [7]).

We now apply a certifying 3-connectivity test to G'. If G' is not 3-connected, the test on 3-connectivity returns a vertex cut of minimal size in G', which corresponds to an edge cut X of size at most two in G. We can then use a red-green coloring of the connected components of $G \setminus X$ as certificate.

Otherwise, G' is 3-connected and we have a sequence of Type (6) for G'. The certificate consists of this sequence, G' and the injective mapping ϕ from each vertex in G' to its corresponding vertex or edge in G to certify the construction of G'. For a checker, it suffices to test that G' is 3-connected using the given sequence, every vertex in G has a unique preimage in $V(G')$ under ϕ, every non-arc-vertex v in G' is the hub of a wheel graph with $v + 1$ vertices that are all arc-vertices except for v, every two wheels in G' share at most one arc-vertex and every arc-vertex u in G' is incident to exactly two non-arc-vertices v and w such that $\phi(u) = \phi(v)\phi(w)$ and $\phi(u) \in E(G)$. Note that this checker may fail in detecting additional edges (but not additional vertices) in G and that this does not harm the 3-edge-connectivity of G. In both cases, the given certificate needs linear space and can be checked in time $O(m)$.

4   A Certifying Algorithm for 3-Connectivity in Linear Time

Due to space constraints, we give only a high level description of the certifying algorithm. According to Lemma 2, it suffices to add iteratively BG-paths to an arbitrary K_2^3-subdivision S_3 in G to get a sequence of Type (6) from S_3 to $S_z = G$. With Lemma 3, $z = m - n + 2$. Note that we cannot make wrong decisions when choosing a BG-path (except for the first one), as Lemma 2 ensures a completion of the sequence if G is 3-connected. We aim for adding chains as BG-paths, as they can be efficiently computed.

We compute a chain decomposition on a DFS-forest T of G and check $n \geq 4$, $\delta(G) \geq 3$ and 2-connectivity of G as part of this computation. If the test fails, we can easily find a certificate for G being not 3-connected; otherwise, we obtain the K_2^3-subdivision $S_3 = C_1 \cup C_2$. To keep further explanations as simple as possible, we abuse notation and split the cycle C_1 into two paths by setting $C_0 = t(C_2) \to_T r$ and redefining $C_1 = r \to_{C_1 \setminus E(C_0)} t(C_2)$. We will represent the chain decomposition C as $C_0, \ldots, C_{m-n+1}$.

For every chain $C_i \neq C_0$, C_i contains exactly one backedge, namely its first edge, and $s(C_i)$ is a proper ancestor of $t(C_i)$. We define the following necessity for the 3-connectivity of G, which can be checked in linear time, giving a separation pair if violated. Recall that $T(x)$ is the subtree of T that contains all descendants of x.

Property B: For every chain $C_i \in C \setminus \{C_0\}$ that is not a backedge and for its last inner vertex x, G contains a backedge e that enters $T(x)$ such that $s(e)$ is an inner vertex of $t(C_i) \rightarrow_T s(C_i)$.

Until now we only checked necessary properties for the 3-connectivity of G, which we will take for granted for the rest of the paper. Let the *parent of a chain* $C_i \neq C_0$ be the chain C_k that contains the edge from $t(C_i)$ to the parent of $t(C_i)$ in T. Chains admit the following tree structure.

Lemma 4. *The parent relation on C defines a tree U with $V(U) = C$ and root C_0.*

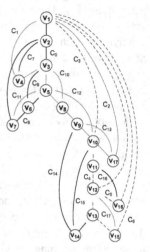

Fig. 1. A chain decomposition. Light solid chains are of *Type 1*, red dashed ones of *Type 2* (*Type 2a*: C_3) and black solid ones of *Type 3* (*Type 3b*: C_{14} and C_{16}, giving the caterpillars $L_{14} = \{C_{14}, C_6, C_4\}$ and $L_{16} = \{C_{16}, C_5\}$).

We assign one of the Types 1, 2a, 2b, 3a and 3b to each chain $C_i \neq C_0$ in ascending order of $<$. Some types will be BG-paths and therefore lead to the next subgraph in the construction sequence. The remaining ones will be grouped into bigger structures that can be decomposed into BG-paths later. Algorithm 1 defines the types in linear time; all chains are unmarked at the beginning. We illustrate the different types in Figure 1.

Algorithm 1 marks every chain of Type 2b. We explain how the algorithm groups chains of certain types. Whenever a chain C_i of Type 3b is found, the path $C_i \rightarrow_U C_0$ is traversed until a chain C_j occurs whose parent is not marked. The chains in $C_i \rightarrow_U C_j$ are stored in a list L_i and unmarked (see Line 15 of Algorithm 1). This way, every chain C_i of Type 3b is associated with a list L_i of chains; we call L_i a *caterpillar*. Property B ensures that caterpillars consist of exactly the chains in C that are of Type 2b and 3b.

In order to decide which chain can be added as BG-path, we want to impose the following structure on every graph S_l, $3 \leq l \leq m - n + 2$.

Definition 2. *Let S_l be* upwards-closed *if, for each vertex v in S_l, the edge from v to its parent in T is contained in S_l. Let S_l be* modular *if S_l is the union of chains.*

Clearly, S_3 is upwards-closed and modular. We would be done if we could restrict every S_l to be upwards-closed and modular, as then every BG-path would be a chain:

Lemma 5. *If S_l and S_{l+1} are upwards-closed and modular, the BG-path P for S_l is a chain.*

Proof. Assume that P is not a chain. Since S_{l+1} is modular, P must be the union of $t > 1$ chains forming a path; let C_i be the first chain in P. Now P cannot start with $t(C_i)$, as Property 1.1 and $s(C_i) \in V(S_l)$ would force C_i to be the only chain in P, contradicting $t > 1$. Thus, P starts with $s(C_i)$ and, for the same reason, $(t(C_i) \rightarrow_T s(C_i)) \not\subseteq S_l$. Since S_{l+1} is upwards-closed, $(t(C_i) \rightarrow_T s(C_i)) \subseteq S_{l+1}$. This contradicts $t > 1$ as well, as a chain in P that contains an edge of $t(C_i) \rightarrow_T s(C_i)$ would induce a vertex of degree at least 3 in the path P, because it contains a backedge. $\square$

Algorithm 1. Classify($C_i \in C \setminus \{C_0\}$, DFS-tree T)   ▷ classifies chains into Types 1,2a,2b,3a,3b

---

1: Let C_k be the parent of C_i in U ▷ $C_k < C_i$
2: **if** $t(C_i) \rightarrow_T s(C_i)$ is contained in C_k **then**   ▷ Type 1
3:      assign Type 1 to C_i
4: **else if** $s(C_i) = s(C_k)$ **then**   ▷ Type 2: $C_k \neq C_0$, $t(C_i)$ is inner vertex of C_k
5:      **if** C_i is a backedge **then**
6:          assign Type 2a to C_i ▷ Type 2a
7:      **else**
8:          assign Type 2b to C_i; mark C_i ▷ Type 2b
9: **else**   ▷ Type 3: $s(C_i) \neq s(C_k)$, $C_k \neq C_0$, $t(C_i)$ is inner vertex of C_k
10:     **if** C_k is not marked **then**
11:         assign Type 3a to C_i ▷ Type 3a
12:     **else**   ▷ C_k is marked
13:         assign Type 3b to C_i; create a list $L_i = \{C_i\}$; $C_j := C_k$ ▷ Type 3b
14:         **while** C_j is marked **do**   ▷ L_i is called a *caterpillar*
15:             unmark C_j; append C_j to L_i; $C_j := parent(C_j)$

Unfortunately, restricting every S_l to be upwards-closed and modular is not possible, as the 3-connected graph in Figure 2 shows: Since every BG-path for S_3 has end vertices x and y, S_4 cannot be modular. We therefore aim to restrict only certain subgraphs. Let a *cluster* be either a caterpillar or a chain of Type 1, 2a or 3a (the *cluster of a chain* is the cluster containing the chain). Instead of adding BG-paths one by one, we will add clusters that can be decomposed into subsequent BG-paths later; we restrict only the subgraph obtained from the last BG-path to be upwards-closed and modular. We list the restrictions for adding a cluster in detail.

Restrictions: We add a cluster to an upwards-closed modular subgraph S_l only if it

(R_1) can be decomposed into as many subsequent BG-paths as it contains chains and creates an upwards-closed and modular subgraph S_{l+t}, $t > 0$, such that

(R_2) no link in S_{l+t} that consists only of tree edges has a parallel link in S_{l+t} (note that $S_{l+t} \neq S_3$).

Finding such a cluster clearly gives the next t BG-path(s) for S_l. In particular, (R_1) ensures that the total number of BG-operations is $|C| - 3 = m - n - 1$, as shown in Lemma 3. Restriction (R_2) implies that the first BG-path generates a K_4-subdivision, as demanded for a construction sequence, and not, e. g., a K_2^4-subdivision. We will assume from now on that S_l was obtained obeying Restrictions (R_1) and (R_2). Let a cluster for S_l that satisfies (R_1) and (R_2) be *addable*. In the following, we investigate how a set of addable clusters for S_l can be obtained.

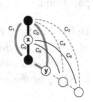

Fig. 2.   No BG-path for S_3 (thick subgraph) preserves modularity.

Definition 3. *For S_l and a chain C_i in S_l, let* Children$_{12}(C_i)$ *be the set of children of C_i of Types 1 and 2 that are not contained in S_l and let* Type$_3(C_i)$ *be the set of chains of Type 3 that start at a vertex in C_i and are not contained in S_l.*

We process chains in the order $<$ of creation, i. e., top-down in the tree U. The key idea for each C_i is to add (among others) the clusters of all chains in *Children$_{12}(C_i)$* and the clusters of all chains in $Type_3(C_i)$. Note that we defined $Children_{12}(C_i)$ to contain only chains of Type 1 and 2. It will not be necessary to consider children of C_i of Type 3, as their clusters will be added before as part of $Type_3(C_j)$ for an ancestor C_j of C_i in U. The sets $Children_{12}(C_i)$ and $Type_3(C_i)$ can be computed efficiently. We give the precise set of clusters we add.

Definition 4. *For S_l and a subset $A \subseteq C$ of chains, let $\mathrm{cl}(A)$ be the set of clusters that contains all (not necessarily proper) ancestors of chains in A that are not contained in S_l. For every C_i, we add the clusters in $\mathrm{cl}(Children_{12}(C_i) \cup Type_3(C_i))$.*

Theorem 2. *Let C_i be a chain in S_l with $Children_{12}(C_j) = Type_3(C_j) = \emptyset$ for every proper ancestor C_j of C_i. If G is 3-connected, there is an order in which the clusters in $\mathrm{cl}(Children_{12}(C_i) \cup Type_3(C_i))$ are successively addable.*

Assume for a moment that G is 3-connected. Clearly, C_0 satisfies the precondition of Theorem 2 for S_3. By induction, let the precondition be true for every $C_j, j \leq i$. Applying Theorem 2 on C_i then generates a subgraph, in which the precondition is satisfied for C_{i+1}. This ensures that iteratively applying Theorem 2 on $C_0, C_1, \ldots, C_{m-n+1}$ constructs G. We obtain the following corollary.

Corollary 1. *For every 3-connected graph G there is a sequence of Type (6) to G that is restricted by (R_1) and (R_2).*

4.1   Reduction to Overlapping Intervals

Theorem 2 provides an algorithmic method to compute a construction sequence: For each $C_i, 0 \leq i \leq m - n + 1$, we add the clusters in $cl(Children_{12}(C_i) \cup Type_3(C_i))$; we say that C_i is *processed*. We describe the processing phase of C_i (see Algorithm 2). Let S_l be the current subgraph. Theorem 2 does not state in which order the clusters are addable; it therefore remains to show how we can compute this order if exists. We first partition the chains in $Type_3(C_i)$ into so-called segments of S_l.

Definition 5. *Let E' be a maximal subset of $E(G) \setminus E(S_l)$ such that every two edges of E' are contained in a path whose inner vertices are disjoint from $V(S_l)$. Then the edge-induced subgraph $G[E']$ is called a* segment *of S_l. Let the* segment *of a chain $C_i \not\subseteq S_l$ be the segment of S_l that contains C_i.*

Note that every segment of S_l is the union of all vertices in a subtree of U, as S_l is upwards-closed and modular. A segment can therefore be represented by the minimal chain it contains. Let X be the subset of chains in $Type_3(C_i)$ whose segments do not contain a chain in $Children_{12}(C_i)$. Then the clusters in $cl(X)$ are successively addable in the order of $<$. We just add them in this order and delete X from $Type_3(C_i)$. For convenience, we abuse notation and let S_l be again the current subgraph.

Let Y be the set of segments that contain a chain in $Children_{12}(C_i)$. Note that every cluster that we still have to add in this processing phase is contained in one segment in Y. For each segment $H \in Y$, let $H \cap S_l$ be the *attachment vertices* of H. It can be deduced from H containing a chain in $Children_{12}(C_i)$ that all attachment vertices of H

Algorithm 2. Certify3Connectivity(Graph G)

1: Compute a DFS-tree T of G, a chain decomposition C, classify the chains and check simple properties
2: Check Property B, Set $S_3 := C_0 \cup C_1 \cup C_2$ ▷ Page 792 and Section 3
3: **for** $i := 0$ to $m - n + 1$ **do**   ▷ process each C_i and add clusters of Theorem 2
4:     Compute the lists $Children_{12}(C_i)$ and $Type_3(C_i)$
5:     Partition $Type_3(C_i)$ into segments
6:     $X :=$ subset of chains in $Type_3(C_i)$ whose segments do not contain a chain of $Children_{12}(C_i)$
7:     Add the clusters in $cl(X)$ successively in the order of $<$; update $Type_3(C_i)$
8:     $Y :=$ set of segments that contain a chain in $Children_{12}(C_i)$
9:     **for each** segment $H \in Y$ **do**
10:        Compute the attachment vertices of H and the dependent path of H
11:        Map H to a set of intervals on C_i ▷ Section 4.1
12:     **if** the merged overlap graph G' of Y is connected **then**
13:        Obtain a proper order σ on Y from G' ▷ Lemma 7
14:        **for each** segment $H \in Y$ in the order of σ **do**   ▷ Add clusters; save construction seq.
15:           Add the clusters in $cl(Type_3(C_i) \cup Children_{12}(C_i))$ that are in H in the order of $<$
16:     **else**
17:        Compute a separation pair   ▷ G' is not 3-connected

---

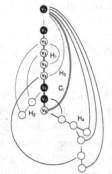

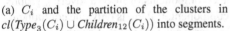

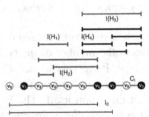

(a) C_i and the partition of the clusters in $cl(Type_3(C_i) \cup Children_{12}(C_i))$ into segments.

(b) For each of the two inner real vertices v_6 and v_7 in C_i, there is one interval in I_0.

Fig. 3. Mapping the segments H_{1-4} to intervals on C_i. Different colors depict different segments.

are contained in C_i (see, e.g., Figure 3(a)). Let the maximal path in C_i that connects two attachment vertices of H be the *dependent path* of H. For example, the dependent path of H_4 in Figure 3(a) is $v_5 \rightarrow_{C_i} v_9$.

Consider a segment $H \in Y$ and its dependent path P. It can be shown that the clusters in $cl(Children_{12}(C_i) \cup Type_3(C_i))$ that are contained in H are successively addable in the order of $<$ if P contains an inner real vertex. Moreover, if P does not contain an inner real vertex, none of these clusters is addable. Whenever we have found a segment with an inner real vertex in its dependent path, we will therefore add all clusters in this segment successively.

Note that adding the clusters of a segment H causes all attachment vertices of H to be real. This might induce new inner real vertices for dependent paths of other segments in Y. It remains to compute in which order the segments of Y can be added such that every dependent path will have an inner real vertex if possible. Let an order σ on Y be *proper* if the dependent path of each segment in σ contains an inner real vertex or an inner vertex that is an attachment vertex of a previous segment in σ. A proper order on Y thus gives the desired order on all clusters in $cl(Children_{12}(C_i) \cup Type_3(C_i))$.

We describe how to compute a proper order σ efficiently, if exists. This is the heart of the reduction. We map each segment H in Y to a set $I(H)$ of intervals on $V(C_i)$: Let $a_1, \ldots, a_k$ be the attachment vertices of H and let $I(H) = \bigcup_{1 < j \le k}\{[a_1, a_j]\} \cup \bigcup_{1 < j < k}\{[a_j, a_k]\}$ (see Figure 3). Additionally, we augment C_i by an artificial vertex v_0 (next to $t(C_i)$) and map the real vertices $b_1, \ldots, b_k$ of C_i to the set of intervals $I_0 = \bigcup_{1 < j < k}\{[v_0, b_j]\}$. The intervals can be efficiently computed; there are at most $|Children_{12}(C_i)| + 2|Type_3(C_i)| + |V_{real}(C_i)| - 2$ intervals for C_i, giving a total of $O(m)$ intervals for all processing phases.

Let two intervals $[a, b]$ and $[c, d]$ overlap if $a < c < b < d$ or $c < a < d < b$. We want to compute a proper order on Y by finding a sequence of overlapping intervals that starts with an interval in I_0. Let the *overlap graph* of Y be the graph with vertex set $I_0 \cup \bigcup_{H \in Y} I(H)$ and an edge between two vertices if and only if the corresponding intervals overlap. Let the *merged overlap graph* of Y be the graph that results from the overlap graph by merging the vertices in I_0 and in $I(H)$, respectively, to one vertex, for every segment $H \in Y$.

Lemma 6. *There is a proper order on the segments in Y if and only if the merged overlap graph G' of Y is connected.*

Clearly, the overlap graph (and the merged overlap graph) can have a quadratic number of edges in the number of intervals, e. g., consider k pairwise distinct intervals of the same length lying very close to each other. Interestingly, the connected components of the merged overlap graph can still be computed in linear time, without the need to construct the graph itself. The key idea is to use a modified variant of a sweep-line algorithm in [16] that computes the connected components of interval overlap graphs by selecting only sparse subgraphs for each component. If there is only one component, a proper order on Y can be obtained from the sparse subgraph of that component.

Lemma 7. *Let k be the number of intervals that have been created for the segments in Y and let G' be the merged overlap graph of Y. There is an algorithm with running time $O(k + |V(C_i)|)$ that computes a proper order σ on Y, if it exists, and that computes the connected components of G', if no proper order on Y exists.*

This computes the desired order. With Lemma 1, we obtain the following theorem.

Theorem 3. *The sequence of each of the types (2)–(7) for a simple 3-connected graph G can be computed in time $O(m)$.*

It is possible to extend the algorithm to certify non-3-connectivity: If the algorithm of Lemma 7 outputs more than one connected component of the merged overlap graph, a separation pair can be computed. This gives our main result.

Theorem 4. *There are certifying algorithms for testing the 3-connectivity and 3-edge-connectivity of graphs G in time $O(n + m)$ using the certificates of Section 3.*

References

1. Albroscheit, S.: Ein Algorithmus zur Konstruktion gegebener 3-zusammenhängender Graphen. Diploma thesis, Freie Universität Berlin (2006)
2. Barnette, D.W., Grünbaum, B.: On Steinitz's theorem concerning convex 3-polytopes and on some properties of 3-connected graphs. In: Many Facets of Graph Theory, pp. 27–40 (1969)
3. Battista, G.D., Tamassia, R.: Online Graph Algorithms With $SPQR$-Trees. In: Paterson, M. (ed.) ICALP 1990. LNCS, vol. 443, pp. 598–611. Springer, Heidelberg (1990)
4. Bondy, J.A., Murty, U.S.R.: Graph Theory. Springer (2008)
5. Dolev, D., Dwork, C., Waarts, O., Yung, M.: Perfectly secure message transmission. J. ACM 40, 17–47 (1993)
6. Elmasry, A., Mehlhorn, K., Schmidt, J.M.: An O(n+m) certifying triconnnectivity algorithm for Hamiltonian graphs. Algorithmica 62(3), 754–766 (2012)
7. Galil, Z., Italiano, G.F.: Reducing edge connectivity to vertex connectivity. SIGACT News 22(1), 57–61 (1991)
8. Gutwenger, C., Mutzel, P.: A Linear Time Implementation of SPQR-Trees. In: Marks, J. (ed.) GD 2000. LNCS, vol. 1984, pp. 77–90. Springer, Heidelberg (2001)
9. Hopcroft, J.E., Tarjan, R.E.: Dividing a graph into triconnected components. SIAM J. Comput. 2(3), 135–158 (1973)
10. Lovász, L.: Computing ears and branchings in parallel. In: Proceedings of the 26th Annual Symposium on Foundations of Computer Science (FOCS 1985), pp. 464–467 (1985)
11. McConnell, R.M., Mehlhorn, K., Näher, S., Schweitzer, P.: Certifying algorithms. Computer Science Review 5(2), 119–161 (2011)
12. Mehlhorn, K., Näher, S.: From Algorithms to Working Programs: On the Use of Program Checking in LEDA. In: Brim, L., Gruska, J., Zlatuška, J. (eds.) MFCS 1998. LNCS, vol. 1450, pp. 84–93. Springer, Heidelberg (1998)
13. Mehlhorn, K., Schweitzer, P.: Progress on Certifying Algorithms. In: Lee, D.-T., Chen, D.Z., Ying, S. (eds.) FAW 2010. LNCS, vol. 6213, pp. 1–5. Springer, Heidelberg (2010)
14. Mutzel, P.: The SPQR-Tree Data Structure in Graph Drawing. In: Baeten, J.C.M., Lenstra, J.K., Parrow, J., Woeginger, G.J. (eds.) ICALP 2003. LNCS, vol. 2719, pp. 34–46. Springer, Heidelberg (2003)
15. Neumann, A.: Implementation of Schmidt's algorithm for certifying triconnectivity testing. Master's thesis, Universität des Saarlandes and Graduate School of CS, Germany (2011)
16. Olariu, S., Zomaya, A.Y.: A time- and cost-optimal algorithm for interlocking sets – With applications. IEEE Trans. Parallel Distrib. Syst. 7(10), 1009–1025 (1996)
17. Ramachandran, V.: Parallel open ear decomposition with applications to graph biconnectivity and triconnectivity. In: Synthesis of Parallel Algorithms, pp. 275–340 (1993)
18. Schmidt, J.M.: Construction sequences and certifying 3-connectedness. In: Proceedings of the 27th Symposium on Theoretical Aspects of Computer Science (STACS 2010), pp. 633–644 (2010)
19. Tutte, W.T.: A theory of 3-connected graphs. Indag. Math. 23, 441–455 (1961)
20. Tutte, W.T.: Connectivity in graphs. In: Mathematical Expositions, vol. 15. University of Toronto Press (1966)
21. Vo, K.-P.: Finding triconnected components of graphs. Linear and Multilinear Algebra 13, 143–165 (1983)

Epsilon-Net Method for Optimizations over Separable States[*]

Yaoyun Shi and Xiaodi Wu

Department of EECS, University of Michigan, Ann Arbor, USA
shiyy@eecs.umich.edu, xiaodiwu@umich.edu

Abstract. We give algorithms for the optimization problem: $\max_\rho \langle Q, \rho \rangle$, where Q is a Hermitian matrix, and the variable ρ is a bipartite *separable* quantum state. This problem lies at the heart of several problems in quantum computation and information, such as the complexity of QMA(2). While the problem is NP-hard, our algorithms are better than brute force for several instances of interest. In particular, they give PSPACE upper bounds on promise problems admitting a QMA(2) protocol in which the verifier performs only logarithmic number of elementary gate on both proofs, as well as the promise problem of deciding if a bipartite local Hamiltonian has large or small ground energy. For $Q \geq 0$, our algorithm runs in time exponential in $\|Q\|_F$. While the existence of such an algorithm was first proved recently by Brandão, Christandl and Yard [*Proceedings of the 43rd annual ACM Symposium on Theory of Computation*, 343–352, 2011], our algorithm is conceptually simpler.

1 Introduction

Entanglement is an essential ingredient in many ingenious applications of quantum information processing. Understanding and exploiting entanglement remains a central theme in quantum information processing research [16]. Denote by SepD $(\mathcal{A}_1 \otimes \mathcal{A}_2)$ the set of separable density operators over the space $\mathcal{A}_1 \otimes \mathcal{A}_2$. The *weak membership* problem for separability that is to decide, given a classical description of $\rho \in$ SepD $(\mathcal{A}_1 \otimes \mathcal{A}_2)$, whether the state ρ is inside or ϵ far away in trace distance from SepD $(\mathcal{A}_1 \otimes \mathcal{A}_2)$, turns out to be NP-hard when ϵ is inverse exponential [14] (or even inverse polynomial [18,12]) in the dimension of $\mathcal{A}_1 \otimes \mathcal{A}_2$.

In this paper we study a closely related problem, namely the linear optimization problem over separable states below where $\langle A, B \rangle$ denotes the Hilbert-Schmidt inner product of A and B.

Problem 1. Given a Hermitian matrix Q over $\mathcal{A}_1 \otimes \mathcal{A}_2$ (of dimension $d \times d$), compute the optimum value, denoted by $\text{OptSep}(Q)$, of the optimization problem

$$\max \langle Q, X \rangle \text{ subject to } X \in \text{SepD} (\mathcal{A}_1 \otimes \mathcal{A}_2).$$

[*] A full version of this paper is available at arXiv:1112.0808. This research was supported in part by National Basic Research Program of China Awards 2011CBA00300 and 2011CBA00301, and by NSF of United States Award 1017335.

A. Czumaj et al. (Eds.): ICALP 2012, Part I, LNCS 7391, pp. 798–809, 2012.
© Springer-Verlag Berlin Heidelberg 2012

It is a standard fact in convex optimization [13,18] that the weak membership problem and the weak linear optimization, a special case of Problem 1, over certain convex set, such as $\text{SepD}(\mathcal{A}_1 \otimes \mathcal{A}_2)$, are equivalent up to polynomial loss in precision and polynomial-time overhead. Thus it is NP-hard to compute $\text{OptSep}(Q)$ with inverse polynomial additive error. Besides the connection mentioned above, Problem 1 can also be understood from various aspects. Firstly, Problem 1 can be viewed as finding the minimum energy of some physical system that is achieved by separable states. Secondly, in the study of the tensor product space [8], the value $\text{OptSep}(Q)$ is precisely the *injective norm* of Q in $\mathcal{L}(\mathcal{A}_1) \otimes \mathcal{L}(\mathcal{A}_2)$, where $\mathcal{L}(\mathcal{A})$ denote the Banach space of operators on $\mathcal{A}$ with the operator norm. Finally, one may be equally motivated from the study in operations research (e.g., "Bi-Quadratic Optimization over Unit Spheres" [23]).

Another motivation to study Problem 1 is the recent interest about the complexity class QMA(2). Originally the class QMA was defined [22] as the quantum counterpart of the complexity class NP. While the extension of NP to allow multiple provers trivially reduces to NP itself, the power of QMA(2), the extension for QMA with multiple *unentangled* provers, remains far from being well understood. The study of the multiple-prover model was initiated in [21], where QMA(k) denotes the complexity class for the *k*-prover case. Much attention was attracted to this model because of the surprising discovery that NP admits *logarithmic*-size unentangled quantum proofs [5], comparing with the fact that single prover quantum logarithm-size proofs only characterize BQP [24]. It seems adding one unentangled prover increases the power of the model substantially. There are several subsequent works on refining the initial protocol either with improved completeness and soundness bounds [3,1] or with less powerful verifiers [7]. Recently it was proved that QMA(2)=QMA(poly) [17] by using the so-called *product test* protocol that determines whether a multipartite state is a product state when two copies of it are given. Also, variants of QMA(2) with restricted verifiers, such as BellQMA and LOCCQMA that refers to restricted verifiers that perform only nonadaptive or adaptive local measurements respectively, were defined in [1] and studied in [6].

Despite much effort, no nontrivial upper bound of QMA(2) is known. The best known upper bound QMA(2)⊆NEXP follows trivially by nondeterministically guessing the two proofs. It would be surprising if QMA(2) = NEXP. Thus it is reasonable to seek a better upper bound like EXP or even PSPACE. It is not hard to see that simulating QMA(2) amounts to distinguishing between two promises of $\text{OptSep}(Q)$, although one is free to choose the appropriate Q.

Our contributions. In this paper we provide efficient algorithms for Problem 1 in either time or space for several Qs of interest. Our idea is to enumerate via epsilon-nets more "cleverly" with the help of certain structure of Q.

Now we briefly describe our strategy of obtaining space-efficient algorithms. When the total number of points to enumerate is not large, one canrepresent and hence enumerate each point in polynomial space. If the additional computation for each point can also be done in polynomial space, one immediately gets a polynomial-space implementation for the whole algorithm by composing those two components naturally. We make use of the relation NC(poly)=PSPACE [4] to obtain space-efficient implementation for the additional computation, which in our case basically includes the following two parts. The first part assures that the enumeration procedure functions correctly because these epsilon-nets of interest are not standard. This part turns out to be a simple application of the so-called *multiplicative matrix weight update* (MMW) method [2,20] to computing a min-max form. The second part only contains fundamental matrix operations, which usually admit efficient parallel algorithms [11]. As a result, both parts of the additional computation admit efficient parallel algorithms, and therefore can be implemented in polynomial space. We summarize below the main results obtained by applying the above ideas.

1. The first property exploited is the so-called *decomposability* of Q which refers to whether Q can be decomposed in the form $Q = \sum_{i=1}^{M} Q_i^1 \otimes Q_i^2$ with small M. Intuitively, if one substitutes this Q's decomposition into $\langle Q, \rho_1 \otimes \rho_2 \rangle$ and treat $\langle Q_1^1, \rho_1 \rangle, \cdots, \langle Q_M^1, \rho_1 \rangle, \langle Q_1^2, \rho_2 \rangle, \cdots, \langle Q_M^2, \rho_2 \rangle$ as variables, the optimization problem becomes quadratic and M is the number of second-order terms in the objective function. If we plug the values of $\langle Q_1^1, \rho_1 \rangle, \cdots, \langle Q_M^1, \rho_1 \rangle$ into the objective function, then the optimization problem reduces to be an efficiently solvable semidefinite program. Hence by enumerating all possible values of $\langle Q_1^1, \rho_1 \rangle, \cdots, \langle Q_M^1, \rho_1 \rangle$ one can efficiently solve the original problem when M is small. Since this approach naturally extends to the k-partite case for $k \geq 2$, we obtain the following general result.

Theorem 1. *Given any Hermitian Q (of dimension d) and its decomposition $Q = \sum_{i=1}^{M} Q_i^1 \otimes \cdots \otimes Q_i^k$, OptSep$(Q)$ can be approximated with additive error δ in quasi-polynomial time[1] in d and $1/\delta$ if kM is $O(ploy\text{-}log(d))$.*

By exploiting the space-efficient algorithm design strategy above, this algorithm can also be made space-efficient. To facilitate the later applications to complexity classes, we choose the input size to be some n such that $d = \exp(\text{poly}(n))$.

Corollary 1. *If $kM/\delta \in O(poly(n))$, the quantity OptSep(Q) can be approximated with additive error δ in PSPACE.*

As a direct application, we prove the following variant of QMA(2) belongs to PSPACE where QMA(2)[poly$(n), O(\log(n))$] refers to the model where the verifier only performs $O(\log(n))$ elementary gates that act on both proofs at the same time and a polynomial number of other elementary gates.

Corollary 2. QMA(2)[*poly*$(n), O(\log(n))$] $\subseteq$ PSPACE.

This result establishes the first PSPACE upper bound for a variant of QMA(2) where the verifier is allowed to generate some quantum entanglement between two proofs. In contrast, previous results are all about variants with nonadaptive or adaptive local measurements, such as BellQMA(2) or LOCCQMA(2).

2. The second structure made use of is the eigenspace of Q of large eigenvalues, where we establish an algorithm in time exponential in $\|Q\|_F$.

---

[1] Quasi-polynomial time is upper bounded by $2^{O((\log n)^c)}$ for some fixed c, where n is the input size.

Theorem 2. *For $Q \geq 0$, OptSep(Q) can be approximated with additive error δ in time $\exp(O(\log(d) + \delta^{-2}\|Q\|_F^2 \ln(\|Q\|_F/\delta)))$.*

A similar running time $\exp(O(\log^2(d)\delta^{-2}\|Q\|_F^2))$ was obtained in [6] using some known results (i.e., the semidefinite programming for finding symmetric extension [9] and an improved quantum de Finetti-type bound) in quantum information theory. In contrast, our algorithm only uses fundamental operations of matrices and epsilon-nets. To approximate with precision δ, it suffices to consider the eigenspace of Q of eigenvalues greater than δ whose dimension is bounded by $\|Q\|_F^2/\delta^2$. Nevertheless, naively enumerating density operators over that subspace does not work since one cannot detect the separability of those density operators. We circumvent this difficulty by making nontrivial use of the Schmidt decomposition of bipartite pure states.

We note, however, that other results in [6] do not follow from our algorithm, and our method cannot be seen as a replacement of the kernel technique therein. Furthermore, our method does not extend to the k-partite case, as there is no Schmidt decomposition in that case.

Organizations: The necessary background knowledge on the epsilon-nets in use is introduced in Section 2. The main algorithm based on the decomposability of Q is illustrated in Section 3, which is followed by the simulation of variants of QMA(2) in Section 4. Finally, the demonstration of an algorithm with running time exponential in $\|Q\|_F$ for Problem 1 can be found in Section 5.

Notations: We assume familiarity with standard concepts from quantum information [22,27]. Let $\mathcal{A}, \mathcal{B}$ denote complex Euclidean spaces and $L(\mathcal{A}), \text{Herm}(\mathcal{A}), D(\mathcal{A})$ denote the linear, Hermitian and density operators over $\mathcal{A}$ respectively. We denote the trace norm of operator Q by $\|Q\|_{\text{tr}}$, i.e. $\|Q\|_{\text{tr}} = \text{Tr}(Q^*Q)^{1/2}$ where Q^* stands for the conjugate transpose of Q. The Frobenius norm is denoted by $\|Q\|_F$ and the operator norm is denoted by $\|Q\|_{\text{op}}$. The ℓ_1 norm of vector $x \in \mathbb{C}^n$ is denoted by $\|x\|_1 = \sum_{i=1}^{n} |x_i|$ and its ℓ_∞ norm is denoted by $\|x\|_\infty = \max_{i=1,\cdots,n} |x_i|$. We use $\|x\|$ to denote the Euclidean norm. The unit ball of $\mathbb{C}^n$ under certain norm $\|\cdot\|$ is denoted by $\mathbf{B}(\mathbb{C}^n, \|\cdot\|)$.

Due to space limit, all proofs except for Theorem 4 and Corollary 4 are omitted and can be found in the full version.

2 Epsilon Net

Definition 1 (ϵ-net). *Let (X, d) [2] be any metric space and let $\epsilon > 0$. A subset $\mathcal{N}_\epsilon$ is called an ϵ-net of X if for each $x \in X$, there exists $y \in \mathcal{N}_\epsilon$ with $d(x, y) \leq \epsilon$.*

Now we turn to the particular ϵ-net in this paper. Let $\mathcal{H}$ be any Hilbert space of dimension d and $\mathcal{Q} = \mathcal{Q}(M, w) = (Q_1, Q_2, \cdots Q_M)$ be a sequence of operators on $\mathcal{H}$ with $\|Q_i\|_{\text{op}} \leq w$, for all i. Define the $\mathcal{Q}$-space, denoted by SP($\mathcal{Q}$), as

$$\text{SP}(\mathcal{Q}) = \{(\langle Q_1, \rho \rangle, \langle Q_2, \rho \rangle, \cdots, \langle Q_M, \rho \rangle) : \rho \in D(\mathcal{H})\} \subseteq \mathbb{C}^M.$$

---

[2] We will abuse the notation later where the metric d is replaced by the norm from which the metric is induced.

The set is convex and compact, and a (possibly proper) subset of Raw-$(M, w) = \{(q_1, q_2, \cdots, q_M) : \forall i, q_i \in \mathbb{C}, \|q_i\| \leq w\}$. In the following, we construct an ϵ-net of the metric space $(\mathrm{SP}(\mathcal{Q}), \ell_1)$ by first generating an ϵ-net of (Raw-$(M, w), \ell_1$) via a standard procedure and then selecting those points close to $\mathcal{Q}$-space.

Selection Process

The selection process determines if some point $\boldsymbol{p}$ in Raw-(M, w) is close to $\mathrm{SP}(\mathcal{Q})$. Denote by dis$(\boldsymbol{p})$ the distance of $\boldsymbol{p} \in \mathbb{C}^M$ to $\mathrm{SP}(\mathcal{Q})$, i.e.,

$$\mathrm{dis}(\boldsymbol{p}) = \min_{\boldsymbol{q} \in \mathrm{SP}(\mathcal{Q})} \|\boldsymbol{p} - \boldsymbol{q}\|_1.$$

The distance dis$(\boldsymbol{p})$ can be efficiently computed in time by casting the problem as a semidefinite program. However, it is unknown whether the time-efficient solutions for SDPs can be made space-efficient in general. In our case where $\mathcal{Q}$ has a concise description, space-efficient solutions correspond to PSPACE upper bound. Thus we need to develop our own space-efficient algorithm for this problem. Due to the duality of the ℓ_1 norm, one has

$$\mathrm{dis}(\boldsymbol{p}) = \min_{\rho \in \mathrm{D}(\mathcal{H})} \max_{\boldsymbol{z} \in \mathbf{B}(\mathbb{C}^M, \|\cdot\|_\infty)} \mathrm{Re} \langle \boldsymbol{p} - \boldsymbol{q}(\rho), \boldsymbol{z} \rangle,$$

where $\boldsymbol{q}(\rho) = (\langle Q_1, \rho \rangle, \langle Q_2, \rho \rangle, \cdots, \langle Q_M, \rho \rangle) \in \mathbb{C}^M$. By rephrasing dis$(\boldsymbol{p})$ in the above form, one shows the quantity dis$(\boldsymbol{p})$ is actually an equilibrium value. This follows from the well-known extensions of von' Neumann's Min-Max Theorem [26,10]. One can easily verify that the density operator set $\mathrm{D}(\mathcal{H})$ and the unit ball of $\mathbb{C}^M$ under ℓ_∞ norm are convex and compact sets and the objective function is a bilinear form over the two sets.

$$\min_{\rho \in \mathrm{D}(\mathcal{H})} \max_{\boldsymbol{z} \in \mathbf{B}(\mathbb{C}^M, \|\cdot\|_\infty)} \mathrm{Re} \langle \boldsymbol{p} - \boldsymbol{q}(\rho), \boldsymbol{z} \rangle = \max_{\boldsymbol{z} \in \mathbf{B}(\mathbb{C}^M, \|\cdot\|_\infty)} \min_{\rho \in \mathrm{D}(\mathcal{H})} \mathrm{Re} \langle \boldsymbol{p} - \boldsymbol{q}(\rho), \boldsymbol{z} \rangle.$$

$$(1)$$

Fortunately, there is an efficient algorithm in both time and space (in terms of $d, M, w, 1/\epsilon$) to approximate dis$(\boldsymbol{p})$ with additive error ϵ. The main tool used here is the so-called matrix multiplicative weight update method [2,20]. Similar min-max forms also appeared before in a series of work on quantum complexity [19,28,15].

Lemma 1. *Given any point $\boldsymbol{p} \in Raw\text{-}(M, w)$ and $\epsilon > 0$, there is an algorithm that approximates dis$(\boldsymbol{p})$ with additive error ϵ in poly$(d, M, w, 1/\epsilon)$ time. Furthermore, if d is considered as the input size and $M, w, 1/\epsilon \in O(poly\text{-}log(d))$, this algorithm is also efficient in parallel, namely, it is inside* NC.

Construction of the ϵ-Net

Given any $\mathcal{Q}(M, w)$ and $\epsilon > 0$, we construct the ϵ-net of $\mathrm{SP}(\mathcal{Q})$ as follows.

- Construct the ϵ-net of the set Raw-(M, w) with the metric induced from the ℓ_1 norm. Denote such an ϵ-net by $\mathcal{R}_\epsilon$.

– For each point $p \in \mathcal{R}_\epsilon$, determine dis($p$) and select it to $\mathcal{N}_\epsilon$ if dis(p) $\leq \epsilon$. We claim $\mathcal{N}_\epsilon$ is the ϵ-net of $(\mathrm{SP}(\mathcal{Q}), \ell_1)$.

The construction for the first step is rather routine. Creating an ϵ'-net T'_ϵ over a bounded complex region $\{z \in \mathbb{C} : \|z\| \leq w\}$ is simple: we can place a 2D grid over the complex plane to cover the disk $\|z\| \leq w$. Simple argument shows $|T'_\epsilon| \in O(\frac{w^2}{\epsilon'^2})$. Then $\mathcal{R}_\epsilon$ can be obtained by the M times cross-product $T'_\epsilon \times \cdots \times T'_\epsilon$. To ensure the closeness in the ℓ_1 norm, we will choose $\epsilon' = \epsilon/M$.

Theorem 3. *The $\mathcal{N}_\epsilon$ constructed above is indeed an ϵ-net of $(\mathrm{SP}(\mathcal{Q}), \ell_1)$ with cardinality at most $O((\frac{w^2 M^2}{\epsilon^2})^M)$. For any point $n \in \mathcal{N}_\epsilon$, we have dis($n$) $\leq \epsilon$.*

3   The Main Algorithm

Without loss of generality, we assume $\mathcal{A}_1, \mathcal{A}_2$ are identical, and of dimension d in Problem 1. Moreover, our algorithm will deal with the set of product states rather than separable states. Namely, we consider the following problem.

$$\text{max:} \quad \langle Q, \rho \rangle, \tag{2}$$
$$\text{subject to:} \quad \rho = \rho_1 \otimes \rho_2, \rho_1 \in \mathrm{D}(\mathcal{A}_1), \rho_2 \in \mathrm{D}(\mathcal{A}_2).$$

It is easy to see these two optimization problems are equivalent since product states are extreme points of the set of separable states. Our algorithm works for both maximization and minimization of the objective function and can be extended naturally to the k-partite version.

Problem 2 (k-partite version). Given any Hermitian matrix Q over $\mathcal{A}_1 \otimes \cdots \otimes \mathcal{A}_k$ ($k \geq 2$), compute the optimum value OptSep(Q) with additive error δ.

$$\text{max:} \quad \langle Q, \rho \rangle, \tag{3}$$
$$\text{subject to:} \quad \rho = \rho_1 \otimes \cdots \otimes \rho_k, \forall i, \rho_i \in \mathrm{D}(\mathcal{A}_i).$$

Before describing the algorithm, we need some terminology about the *decomposability* of a multi-partite operator. Any Hermitian operator Q is called *M-decomposable* if there exists $(Q_1^t, Q_2^t, \cdots, Q_M^t) \in \mathrm{L}(\mathcal{A}_t)^M$ for each t such that

$$Q = \sum_{i=1}^{M} Q_i^1 \otimes Q_i^2 \otimes \cdots \otimes Q_i^{k-1} \otimes Q_i^k.$$

To facilitate the use of ϵ-net, we adopt a slight variation of the decomposability above. Let $w \in \mathbb{R}_+^k$ denote the widths of operators over each $\mathcal{A}_i$. Any Q is called (M, w)-*decomposable* if Q is M-decomposable and the widths of those operators in the decomposition are bounded in the sense that $\max_i \|Q_i^t\|_{op} \leq w_t$ for each t. It is noteworthy to mention that the decomposability defined above is related to the concept tensor rank. However, given the representation Q as input, it is hard in general to compute the tensor rank of Q or its corresponding decomposition. Therefore, for any (M, w)-decomposable Q we assume its corresponding decomposition is also a part of the input to our algorithm.

1. Let $Q_t(M, w_t) = (Q_1^t, Q_2^t, \cdots, Q_M^t)$ for t=1,..., k-1. Let $W = \Pi_{i=1}^k w_i$. Generate the ϵ_t-net (by Theorem 3) of $(SP(Q_t), \ell_1)$ for each t=1,..., k-1 with $\epsilon_t = w_t \delta/(k-1)W$ and denote such a set by $\mathcal{N}_{\epsilon_t}^t$. Also let OPT store the optimum value.

2. For each point $\boldsymbol{q} = (\boldsymbol{q}^1, \boldsymbol{q}^2, \cdots \boldsymbol{q}^{k-1}) \in \mathcal{N}_{\epsilon_1}^1 \times \mathcal{N}_{\epsilon_2}^2 \times \cdots \times \mathcal{N}_{\epsilon_{k-1}}^{k-1}$, let Q^k be

$$Q^k = \sum_{i=1}^M q_i^1 q_i^2 \cdots q_i^{k-1} Q_i^k,$$

and calculate $\tilde{Q}^k = \frac{1}{2}(Q^k + Q^{k*})$. Then compute the maximum eigenvalue of $\tilde{Q}^k$, denoted by $\lambda_{\max}(\boldsymbol{q})$. Update OPT as follows: OPT = $\max\{$OPT, $\lambda_{\max}(\boldsymbol{q})\}$.

3. Return OPT.

Fig. 1. The main algorithm with precision δ

Theorem 4. *Let Q be some $(M, \boldsymbol{w})$-decomposable Hermitian over $\mathcal{A}_1 \otimes \cdots \otimes \mathcal{A}_k$ (each $\mathcal{A}_i$ is of dimension d) and $\delta > 0$. Also let $(Q_1^t, Q_2^t, \cdots, Q_M^t), t = 1, \cdots, k$ be the operators in the corresponding decomposition of Q. The algorithm shown in Fig. 1 approximates $OptSep(Q)$ of Problem 2 with additive error δ in $O((\frac{(k-1)^2 W^2 M^2}{\delta^2})^{(k-1)M}) \times poly(d, M, k, W, 1/\delta)$ time where $W = \Pi_{i=1}^k w_i$.*

Proof. By substituting the identity $Q = \sum_{i=1}^M Q_i^1 \otimes Q_i^2 \otimes \cdots \otimes Q_i^{k-1}$, the optimization problem becomes

$$\text{max:} \quad \left\langle \sum_{i=1}^M p_i^1 p_i^2 \cdots p_i^{k-1} Q_i^k, \rho_k \right\rangle$$

$$\text{subject to:} \quad \forall t \in \{1, \cdots, k-1\}, \boldsymbol{p}_t \in SP(Q_t(M, w_t)), \text{ and } \rho_k \in D(\mathcal{A}_k).$$

Thus, solving the optimization problem amounts to first enumerating $\boldsymbol{p}_t \in SP(Q_t(M, w_1))$ for each t, and then solving the optimization problem over $D(\mathcal{A}_k)$.

Consider any point $\boldsymbol{p} = (\boldsymbol{p}^1, \boldsymbol{p}^2, \cdots, \boldsymbol{p}^{k-1}) \in SP(Q_i)^{k-1}$ where $SP(Q_i)^{k-1}$ denotes $SP(Q_1) \times \cdots \times SP(Q_{k-1})$. Due to Theorem 3, there is at least one point $\boldsymbol{q} = (\boldsymbol{q}^1, \boldsymbol{q}^2, \cdots \boldsymbol{q}^{k-1}) \in \{\mathcal{N}_{\epsilon_i}^i\}^{k-1}$ where $\{\mathcal{N}_{\epsilon_i}^i\}^{k-1}$ denotes $\mathcal{N}_{\epsilon_1}^1 \times \mathcal{N}_{\epsilon_2}^2 \times \cdots \times \mathcal{N}_{\epsilon_{k-1}}^{k-1}$ such that $\|\boldsymbol{q}^t - \boldsymbol{p}^t\|_1 \le \epsilon_t$ for t=1,..,k-1. The choice of $\tilde{Q}^k$ is to symmetrize Q^k. With $\tilde{Q}^k$ being Hermitian, it is clear that $\lambda_{\max}(\boldsymbol{q}) = \max_{\rho_k \in D(\mathcal{A}_k)} \left\langle \tilde{Q}^k, \rho_k \right\rangle$. Now let's analyze how much error will be induced in this process.

Let $P^k(\boldsymbol{p}) = \sum_{i=1}^M p_i^1 p_i^2 \cdots p_i^{k-1} Q_i^k$ and $\tilde{P}^k = \frac{1}{2}(P^k + P^{k*})$. It is not hard to see that $P^k = \tilde{P}^k$. The error bound is achieved by applying a chain of triangle inequalities as follows. Firstly, one has

$$\|\tilde{P}^k - \tilde{Q}^k\|_{\text{op}} = \|\frac{1}{2}(P^k - Q^k) + \frac{1}{2}(P^{k*} - Q^{k*})\|_{\text{op}} \le \|P^k - Q^k\|_{\text{op}}.$$

Substitute the expressions for P^k, Q^k and apply the standard hybrid argument.

$$\|P^k - Q^k\|_{op} = \|\sum_{i=1}^{M}(p_i^1 p_i^2 \cdots p_i^{k-1} - q_i^1 q_i^2 \cdots q_i^{k-1})Q_i^k\|_{op}$$

$$= \|\sum_{i=1}^{M}\sum_{t=1}^{k-1}(q_i^1 \cdots q_i^{t-1} p_i^t p_i^{t+1} \cdots p_i^{k-1} - q_i^1 \cdots q_i^{t-1} q_i^t p_i^{t+1} \cdots p_i^{k-1})Q_i^k\|_{op},$$

which is immediately upper bounded by the sum of the following terms,

$$\sum_{i=1}^{M}|p_i^1 - q_i^1||p_i^2 \cdots p_i^{k-1}|\|Q_i^k\|_{op,}, \cdots, \sum_{i=1}^{M}|q_i^1 \cdots q_i^{k-2}||p_i^{k-1} - q_i^{k-1}|\|Q_i^k\|_{op}.$$

As the tth term above can be upper bounded by $\epsilon_t W/w_t$ for each t, we have,

$$\|\tilde{P}^k - \tilde{Q}^k\|_{op} \le \epsilon_1 W/w_1 + \epsilon_2 W/w_2 + \cdots + \epsilon_{k-1} W/w_{k-1} = \underbrace{\frac{\delta}{k-1} + \cdots + \frac{\delta}{k-1}}_{\text{k-1 terms}} = \delta.$$

Hence the optimum value for any fixed p won't differ too much from the one for its approximation q in the ϵ-net. This is because

$$\max_{\rho_k \in D(\mathcal{A}_k)} \left\langle \tilde{P}^k, \rho_k \right\rangle = \max_{\rho_k \in D(\mathcal{A}_k)} \left\langle \tilde{Q}^k, \rho_k \right\rangle + \left\langle \tilde{P}^k - \tilde{Q}^k, \rho_k \right\rangle.$$

By Hölder Inequalities we have $|\left\langle \tilde{P}^k - \tilde{Q}^k, \rho_k \right\rangle| \le \|\tilde{P}^k - \tilde{Q}^k\|_{op}\|\rho_k\|_{tr} \le \delta,$

$$\lambda_{\max}(q) - \delta \le \max_{\rho_k \in D(\mathcal{A}_k)} \left\langle \tilde{P}^k(p), \rho_k \right\rangle \le \lambda_{\max}(q) + \delta.$$

We now optimize p over $\text{SP}(\mathcal{Q}_i)^{k-1}$ and the corresponding q will run over the ϵ-net $\{\mathcal{N}_{\epsilon_i}^i\}^{k-1}$. As every point $q \in \{\mathcal{N}_{\epsilon_i}^i\}^{k-1}$ is also close to $\text{SP}(\mathcal{Q}_i)^{k-1}$ in the sense that $\text{dis}(q^t) \le \epsilon_t$ for each t, we have

$$\max_{q \in \{\mathcal{N}_{\epsilon_i}^i\}^{k-1}} \lambda_{\max}(q) - \delta \le \max_{p \in \text{SP}(\mathcal{Q}_i)^{k-1}} \max_{\rho_k \in D(\mathcal{A}_k)} \left\langle \tilde{P}^k(p), \rho_k \right\rangle \le \max_{q \in \{\mathcal{N}_{\epsilon_i}^i\}^{k-1}} \lambda_{\max}(q) + \delta.$$

Finally, it is not hard to see that $\text{OPT} = \max_{q \in \{\mathcal{N}_{\epsilon_i}^i\}^{k-1}} \lambda_{\max}(q)$ and therefore

$$\text{OPT} - \delta \le \text{OptSep}(Q) \le \text{OPT} + \delta.$$

Now let us analyze the efficiency of this algorithm. The total number of points in the ϵ-net $\{\mathcal{N}_{\epsilon_i}^i\}^{k-1}$ is upper bounded by $O((\frac{(k-1)^2 W^2 M^2}{\delta^2})^{(k-1)M})$. The generation of each point q will cost time polynomial in $d, M, W, 1/\delta$ (See Lemma 1.). Afterward, one needs to calculate $\tilde{Q}^k$ and its maximum eigenvalue for each point, which can be done in time polynomial in d, k, M. Thus, the total running time is bounded by $O((\frac{(k-1)^2 W^2 M^2}{\delta^2})^{(k-1)M}) \times \text{poly}(d, M, k, W, 1/\delta)$.

Remarks. All operations in the algorithm described in Fig. 1 can be implemented efficiently in parallel in some situation. This is because fundamental operations of matrices can be done in NC and the calculation of dis(p) can be done in NC (See Lemma 1) when $M, W, k, 1/\delta$ are in nice forms of d.

Corollary 3. *Let n be the input size such that $d = \exp(poly(n))$, if $W/\delta \in O(poly(n))$, $kM \in O(poly(n))$, then $OptSep(Q)$ can be approximated with additive error δ in PSPACE.*

4 Simulation of Several Variants of QMA(2)

This section illustrates the use of the algorithm shown in Section 3 to simulate some variants of the complexity class QMA(2). The idea is to show for those variants, the corresponding POVM matrices of acceptance are $(M, \boldsymbol{w})$-decomposable with small Ms. Recall the definition of the complexity class QMA(2).

Definition 2. *A language $\mathcal{L}$ is in $QMA(2)_{m,c,s}$ if there exists a polynomial-time generated family of quantum verification circuits $Q = \{Q_n | n \in \mathbb{N}\}$ such that for any input x of size n, the circuit Q_n implements a two-outcome measurement $\{Q_x^{acc}, \mathbb{I} - Q_x^{acc}\}$. Furthermore,*

- *Completeness: If $x \in \mathcal{L}$, there exist $|\psi_1\rangle \in \mathcal{A}_1, |\psi_2\rangle \in \mathcal{A}_2$, each of m qubits,*

$$\langle Q_x^{acc}, |\psi_1\rangle\langle\psi_1| \otimes |\psi_2\rangle\langle\psi_2|\rangle \geq c.$$

- *Soundness: If $x \notin \mathcal{L}$, then for any states $|\psi_1\rangle \in \mathcal{A}_1, |\psi_2\rangle \in \mathcal{A}_2$,*

$$\langle Q_x^{acc}, |\psi_1\rangle\langle\psi_1| \otimes |\psi_2\rangle\langle\psi_2|\rangle \leq s.$$

We call $QMA(2)=QMA(2)_{poly(n),2/3,1/3}$. It is easy to see that simulating the complexity class QMA(2) amounts to distinguishing between the two promises of the maximum acceptance probability (i.e. $OptSep(Q_x^{acc})$).

The first example is the variant with only logarithm-size proofs, namely $QMA(2)_{O(\log(n)),2/3,1/3}$. It is not hard to find out the corresponding POVMs of acceptance (i.e. Q_x^{acc}) need to be $(poly(n), \boldsymbol{w})$-decomposable where $\boldsymbol{w} = (1, 1)$ since $\mathcal{A}_1, \mathcal{A}_2$ are only of polynomial dimension. Thus, it follows directly from Corollary 3 that $OptSep(Q_x^{acc})$ can be approximated in polynomial space. Namely,

$$QMA(2)_{O(\log(n)),2/3,1/3} \subseteq PSPACE.$$

The next example is slightly less trivial. Before moving on, we need some terminology about the quantum verification circuits Q. Assume the input x is fixed from now on. Let $\mathcal{A}_1, \mathcal{A}_2$ be the Hilbert space of size $d_{\mathcal{A}}$ for the two proofs and let $\mathcal{V}$ be the ancillary space of size $d_{\mathcal{V}}$. Then the quantum verification process will be carried out on the space $\mathcal{A}_1 \otimes \mathcal{A}_2 \otimes \mathcal{V}$ with some initial state $|\psi_1\rangle \otimes |\psi_2\rangle \otimes |0\rangle$ where $|\psi_1\rangle, |\psi_2\rangle$ are provided by the provers. The verification process is also efficient in the sense that the whole circuit only consists of polynomial elementary gates. Without loss of generality, we can fix one universal gate set for the verification circuits. Particularly, we choose the universal gate set to be single qubit gates plus the CNOT gates. One can also choose other universal gate sets without any change of the main result.

We categorize all elementary gates in the verification circuits into two types. A gate is of *type-I* if it only affects the qubits within the same space (i.e, $\mathcal{A}_1, \mathcal{A}_2$, or, $\mathcal{V}$). Otherwise, this gate is of *type-II*. It is easy to see single qubit gates are always type-I gates. The only type-II gates are CNOT gates whose control qubit and target qubit sit in different spaces. Let $p, r : \mathbb{N} \to \mathbb{N}$ be polynomial-bounded functions. A polynomial-time generated family of quantum verification circuits Q is called $Q[p, r]$ if each Q_n only contains $p(n)$ type-I elementary gates and $r(n)$ type-II elementary gates.

Definition 3. *A language $\mathcal{L}$ is in* $\mathrm{QMA}(2)_{m,c,s}[p, r]$ *if $\mathcal{L}$ is in* $\mathrm{QMA}(2)_{m,c,s}$ *with some $Q[p, r]$ verification circuit family.*

It is easy to see that $\mathrm{QMA}(2) = \mathrm{QMA}(2)[\text{poly}, \text{poly}]$ from our definition.

Lemma 2. *For any family of verification circuits $Q[p, r]$, the POVM Q_x^{acc} is $(4^{r(n)}, (1, 1))$-decomposable for any $n \in \mathbb{N}$ and input x. Moreover, this decomposition can be calculated in parallel with $O(t(n)4^{r(n)}) \times \text{poly}(n)$ time.*

We will show that when the number of type-II gates is relatively small, one can simulate this complexity model efficiently by the algorithm in Fig. 1.

Corollary 4. $\mathrm{QMA}(2)[\text{poly}(n), O(\log(n))] \subseteq \mathrm{PSPACE}$.

Proof. This is a simple consequence of Lemma 2 and Corollary 3. For any fixed x of length n, one can first compute the decomposition of Q_x^{acc} in parallel with $O(t(n)4^{r(n)}) \times \text{poly}(n)$ time, which is parallel polynomial time in n when $r(n) = O(\log(n))$ and $t(n) \in \text{poly}(n)$. Hence the first step can be done in polynomial space via the relation $\mathrm{NC}(\text{poly}) = \mathrm{PSPACE}$ [4].

Then one can invoke the parallel algorithm in Corollary 3 to approximate $\mathrm{OptSep}(Q_x^{acc})$ to sufficient precision δ such that one can distinguish between the two promises. Precisely in this case, we choose those parameters as follows,

$$k = 2, W = 1, M = 4^{O(\log(n))} = \text{poly}(n), 1/\delta = \text{poly}(n).$$

Thus the whole algorithm can be done in polynomial space.

Remarks. Although the proof of the result is not too technical, it establishes the first non-trivial upper bound (PSPACE in this case) for variants of QMA(2) that allow quantum operations acting on both proofs at the same time.

However, our results are hard to extend to the most general case of QMA(2). This is because SWAP-test operation uses many more type-II gates than what is allowed in our method. And SWAP-test seems to be inevitable if one wants to fully characterize the power of QMA(2).

1. Compute the spectral decomposition of $Q = \sum_t \lambda_t |\Psi_t\rangle\langle\Psi_t|$. Choose $\epsilon = \delta/2$ and $\Gamma_\epsilon = \{t : \lambda_t \geq \epsilon\}$. Also let OPT store the optimum value.
2. Generate the ε-net of the unit ball of $\mathbb{C}^{|\Gamma_\epsilon|}$ under the Euclidean norm with $\varepsilon = \frac{\delta}{4\|Q\|_F}$. Denote such set by $\mathcal{N}_\varepsilon$. Then for each point $\alpha \in \mathcal{N}_\varepsilon$,
   (a) Compute $|\phi_\alpha\rangle = \sum_{t \in \Gamma_\epsilon} \alpha_t^* \sqrt{\lambda_t} |\Psi_t\rangle$ and compute the Schmidt decomposition of $|\phi_\alpha\rangle$, i.e.

   $$|\phi_\alpha\rangle = \sum_i \mu_i |u_i\rangle |v_i\rangle,$$

   where $\mu_1 \geq \mu_2 \geq \cdots$ and $\{u_i\}, \{v_i\}$ are orthogonal bases.
   (b) Update OPT as follows: OPT=max{OPT,μ_1}.
3. Return OPT.

Fig. 2. The algorithm runs in time exponential in $\|Q\|_F/\delta$.

5   Exponential Running Time Algorithm in $\|Q\|_F$

In this section we demonstrate another application of the simple idea "enumeration". As a result, we obtained an algorithm with running time exponential in $\|Q\|_F$ (or $\|Q\|_{LOCC}$ [25][3]) for computing OptSep(Q) with additive error δ. A similar running time $\exp(O(\log^2(d)\delta^{-2}\|Q\|_F^2))$ was obtained in [6].

Theorem 5. *Given any positive semidefinite Q over $\mathcal{A}_1 \otimes \mathcal{A}_2$ (of dimension $d \times d$) and $\delta > 0$, the algorithm in Fig. 2 approximates OptSep(Q) with additive error δ with running time $\exp(O(\log(d) + \delta^{-2}\|Q\|_F^2 \ln(\|Q\|_F/\delta)))$.*

Acknowledgement. We thank Zhengfeng Ji and John Watrous for helpful discussions and anonymous reviewers for helpful comments on the manuscript.

References

1. Aaronson, S., Beigi, S., Drucker, A., Fefferman, B., Shor, P.: The Power of Unentanglement. Theory of Computing 5, 1–42 (2009)
2. Arora, S., Hazan, E., Kale, S.: The multiplicative weights update method: a meta algorithm and applications (2005)
3. Beigi, S.: NP vs QMALog(2). Quantum Information and Computation 54(1&2), 0141–0151 (2010)
4. Borodin, A.: On relating time and space to size and depth. SIAM Journal on Computing 6(4), 733–744 (1977)
5. Blier, H., Tapp, A.: All languages in NP have very short quantum proofs. In: Proceedings of the ICQNM, pp. 34–37 (2009)

---

[3] This follows easily from the fact $\|Q\|_F = O(\|Q\|_{LOCC})$ [25] where $\|Q\|_{LOCC}$ stands for the LOCC norm of the operator Q.

6. Brandão, F., Christandl, M., Yard, J.: A quasipolynomial-time algorithm for the quantum separability problem. In: Proceedings of the 43rd Annual ACM Symposium on Theory of Computation (STOC 2011), p. 343 (2011)
7. Chen, J., Drucker, A.: Short multi-prover quantum proofs for SAT without entangled measurements. arXiv:1011.0716 (2010)
8. Defant, A., Floret, K.: Tensor norms and operator ideals. North Holland (1992)
9. Doherty, A., Parrilo, P., Spedalieri, F.: A complete family of separability criteria. Physical Review A 69, 022308 (2004)
10. Fan, K.: Minimax theorems. Proceedings of the National Academy of Sciences 39, 42–47 (1953)
11. Gathen, J.: Parallel linear algebra. In: Synthesis of Parallel Algorithms. Morgan Kaufmann Publishers (1993)
12. Gharibian, S.: Strong NP-hardness of the quantum separability problem. Quantum Information and Computation 10, 343 (2010)
13. Grötschel, M., Lovász, L., Schrijver, A.: Geometric algorithms and combinatorial optimization. Springer (1993)
14. Gurvits, L.: Classical complexity and quantum entanglement. Journal of Computer and System Sciences 69, 448 (2004)
15. Gutoski, G., Wu, X.: Parallel approximation of min-max problems with applications to classical and quantum zero-sum games. In: Proceedings of the 27rd Annual IEEE Conference on Computational Complexity (to appear, 2012)
16. Horodecki, R., Horodecki, P., Horodecki, M., Horodecki, K.: Quantum entanglement. Review Modern Physics 81, 865 (2009)
17. Harrow, A., Montanaro, A.: An efficient test for product states, with applications to quantum Merlin-Arthur games. In: Proceedings of IEEE 51st Annual Symposium on Foundations of Computer Science (FOCS 2010), p. 633 (2010)
18. Ioannou, L.: Computational complexity of the quantum separability problem. Quantum Information and Computation 7, 335 (2007)
19. Jain, R., Watrous, J.: Parallel approximation of non-interactive zero-sum quantum games. In: Proceedings of the 24th IEEE Conference on Computational Complexity, pp. 243–253 (2009)
20. Kale, S.: Efficient algorithms using the multiplicative weights update method. PhD thesis. Princeton University (2007)
21. Kobayashi, H., Matsumoto, K., Yamakami, T.: Quantum Certificate Verification: Single versus Multiple Quantum Certificates, quant-ph/0110006 (2001)
22. Kitaev, A., Shen, A., Vyalyi, M.: Classical and Quantum Computation. American Mathematical Society (2002)
23. Ling, C., Qi, L., Nie, J., Ye, Y.: Bi-Quadratic Optimization over Unit Spheres and Semidefinite Programming Relaxations. SIAM Journal on Optimization 20(3), 1286–1310 (2009)
24. Marriott, C., Watrous, J.: Quantum Arthur-Merlin Games. Computational Complexity 14(2), 122–152 (2005)
25. Matthews, W., Wehner, S., Winter, A.: Distinguishability of quantum states under restricted families of measurements with an application to quantum data hiding. Comm. Math. Phys., 291 (2009)
26. Neumann, J.: Zur theorie der gesellschaftsspiele. Mathematische Annalen 100, 295–320 (1928)
27. Watrous, J.: Lecture Notes on Theory of Quantum Information (2008)
28. Wu, X.: Equilibrium value method for the proof of QIP=PSPACE. arXiv:1004.0264v4 (2010)

Faster Algorithms for Privately Releasing Marginals*

Justin Thaler**, Jonathan Ullman***, and Salil Vadhan†

School of Engineering and Applied Sciences &
Center for Research on Computation and Society
Harvard University, Cambridge, MA
{jthaler,jullman,salil}@seas.harvard.edu
http://seas.harvard.edu/~jthaler
http://seas.harvard.edu/~jullman
http://seas.harvard.edu/~salil

Abstract. We study the problem of releasing k-way marginals of a database $D \in (\{0,1\}^d)^n$, while preserving differential privacy. The answer to a k-way marginal query is the fraction of D's records $x \in \{0,1\}^d$ with a given value in each of a given set of up to k columns. Marginal queries enable a rich class of statistical analyses of a dataset, and designing efficient algorithms for privately releasing marginal queries has been identified as an important open problem in private data analysis (cf. Barak et. al., PODS '07).

We give an algorithm that runs in time $d^{O(\sqrt{k})}$ and releases a private summary capable of answering any k-way marginal query with at most $\pm.01$ error on every query as long as $n \geq d^{O(\sqrt{k})}$. To our knowledge, ours is the first algorithm capable of privately releasing marginal queries with non-trivial worst-case accuracy guarantees in time substantially smaller than the number of k-way marginal queries, which is $d^{\Theta(k)}$ (for $k \ll d$).

1 Introduction

Consider a database $D \in (\{0,1\}^d)^n$ in which each of the $n = |D|$ rows corresponds to an individual's record, and each record consists of d binary attributes. The goal of privacy-preserving data analysis is to enable rich statistical analyses on the database while protecting the privacy of the individuals. In this work, we seek to achieve *differential privacy* [6], which guarantees that no individual's data has a significant influence on the information released about the database.

One of the most important classes of statistics on a dataset is its *marginals*. A marginal query is specified by a set $S \subseteq [d]$ and a pattern $t \in \{0,1\}^{|S|}$. The query asks, "What fraction of the individual records in D has each of the attributes

* A full version of this paper appears on the authors' websites.
** Supported by the Department of Defense (DoD) through the National Defense Science & Engineering Graduate Fellowship (NDSEG) Program, and in part by NSF grants CCF-0915922 and IIS-0964473.
*** Supported by NSF grant CNS-0831289 and a gift from Google, Inc.
† Supported by NSF grant CNS-0831289 and a gift from Google, Inc.

A. Czumaj et al. (Eds.): ICALP 2012, Part I, LNCS 7391, pp. 810–821, 2012.
© Springer-Verlag Berlin Heidelberg 2012

$j \in S$ set to t_j?" A major open problem in privacy-preserving data analysis is to *efficiently* create a differentially private summary of the database that enables analysts to answer each of the 3^d marginal queries. A natural subclass of marginals are *k-way marginals*, the subset of marginals specified by sets $S \subseteq [d]$ such that $|S| \leq k$.

Privately answering marginal queries is a special case of the more general problem of privately answering *counting queries* on the database, which are queries of the form, "What fraction of individual records in D satisfy some property q?" Early work in differential privacy [5,2,6] showed how to approximately answer any set of of counting queries $\mathcal{Q}$ by perturbing the answers with appropriately calibrated noise, providing good accuracy (say, within $\pm.01$ of the true answer) as long as $|D| \gtrsim |\mathcal{Q}|^{1/2}$.

In a setting where the queries arrive online, or are known in advance, it may be reasonable to assume that $|D| \gtrsim |\mathcal{Q}|^{1/2}$. However, many situations necessitate a non-interactive data release, where the data owner computes and publishes a single differentially private summary of the database that enables analysts to answer a large class of queries, say all k-way marginals for a suitable choice of k. In this case $|\mathcal{Q}| = d^{\Theta(k)}$, and it may be impractical to collect enough data to ensure $|D| \gtrsim |\mathcal{Q}|^{1/2}$. Fortunately, the remarkable work of Blum et. al. [3] and subsequent refinements [7,9,17,13,12,11], have shown how to privately release approximate answers to any set of counting queries, even when $|\mathcal{Q}|$ is *exponentially larger* than $|D|$. For example, these algorithms can release all k-way marginals as long as $|D| \geq \tilde{\Theta}(k\sqrt{d})$. Unfortunately, all of these algorithms have running time at least 2^d, even when $|\mathcal{Q}|$ is the set of 2-way marginals (and this is inherent for algorithms that produce "synthetic data" [19]; as discussed below).

Given this state of affairs, it is natural to seek *efficient* algorithms capable of privately releasing approximate answers to marginal queries even when $|D| \ll d^k$. A recent series of works [10,4,14] have shown how to privately release answers to k-way marginal queries with small *average error* (over various distributions on the queries) with both running time and minimum database size much smaller than d^k (e.g. $d^{O(1)}$ for product distributions [10,4] and $\min\{d^{O(\sqrt{k})}, d^{O(d^{1/3})}\}$ for arbitrary distributions [14]). Hardt et. al. [14] also gave an algorithm for privately releasing k-way marginal queries with small *worst-case error* and minimum database size much smaller than d^k. However the running time of their algorithm is still $d^{\Theta(k)}$, which is polynomial in the number of queries.

In this paper, we give the first algorithms capable of releasing k-way marginals up to small worst-case error, with both running time and minimum database size substantially smaller than d^k. Specifically, we show how to create a private summary in time $d^{O(\sqrt{k})}$ that gives approximate answers to all k-way marginals as long as $|D|$ is at least $d^{O(\sqrt{k})}$. When $k = d$, our algorithm runs in time $2^{\tilde{O}(\sqrt{d})}$, and is the first algorithm for releasing *all* marginals in time $2^{o(d)}$.

1.1 Our Results and Techniques

In this paper, we give faster algorithms for releasing marginals and other classes of counting queries.

Theorem 1 (Releasing Marginals). *There exists a constant C such that for every $k, d, n \in \mathbb{N}$ with $k \leq d$, every $\alpha \in (0, 1]$, and every $\varepsilon > 0$, there is an ε-differentially private sanitizer that, on input a database $D \in (\{0, 1\}^d)^n$, runs in time $|D| \cdot d^{C\sqrt{k}\log(1/\alpha)}$ and releases a summary that enables computing each of the k-way marginal queries on D up to an additive error of at most α, provided that $|D| \geq d^{C\sqrt{k}\log(1/\alpha)}/\varepsilon$.*

For notational convenience, we focus on *monotone k-way disjunction queries*. However, our results extend straightforwardly to general non-monotone k-way disjunction queries (see Section 4.1), which are equivalent to k-way marginals. A monotone k-way disjunction is specified by a set $S \subseteq [d]$ of size k and asks what fraction of records in D have at least one of the attributes in S set to 1.

Our algorithm is inspired by a series of works reducing the problem of private query release to various problems in learning theory. One ingredient in this line of work is a shift in perspective introduced by Gupta, Hardt, Roth, and Ullman [10]. Instead of viewing disjunction queries as a set of functions on the database, they view the database as a function $f_D: \{0, 1\}^d \to [0, 1]$, in which each vector $s \in \{0, 1\}^d$ is interpreted as the indicator vector of a set $S \subseteq [d]$, and $f_D(s)$ equals the evaluation of the disjunction specified by S on the database D. They use the structure of the functions f_D to privately learn an approximation g_D that has small *average error* over any product distribution on disjunctions.[1]

Cheraghchi, Klivans, Kothari, and Lee [4] observed that the functions f_D can be approximated by a low-degree polynomial with small average error over the uniform distribution on disjunctions. They then use a private learning algorithm for low-degree polynomials to release an approximation to f_D; and thereby obtain an improved dependence on the accuracy parameter, as compared to [10].

Hardt, Rothblum, and Servedio [14] observe that f_D is itself an average of disjunctions (each row of D specifies a disjunction of bits in the indicator vector $s \in \{0, 1\}^d$ of the query), and thus develop private learning algorithms for threshold of sums of disjunctions. These learning algorithms are also based on low-degree approximations of sums of disjunctions. They show how to use their private learning algorithms to obtain a sanitizer with small average error over *arbitrary distributions* with running time and minimum database size $d^{O(\sqrt{k})}$. They then are able to apply the private boosting technique of Dwork, Rothblum, and Vadhan [9] to obtain worst-case accuracy guarantees. Unfortunately, the boosting step incurs a blowup of d^k in the running time.

We improve the above results by showing how to *directly* compute (a noisy version of) a polynomial p_D that is privacy-preserving and still approximates f_D on *all* k-way disjunctions, as long as $|D|$ is sufficiently large. Specifically, the running time and the database size requirement of our algorithm are both polynomial in the number of monomials in p_D, which is $d^{O(\sqrt{k})}$. By "directly", we mean that we compute p_D from the database D itself and perturb its coefficients,

---

[1] In their learning algorithm, privacy is defined with respect to the rows of the database D that defines f_D, not with respect to the examples given to the learning algorithm (unlike earlier works on "private learning" [15]).

rather than using a learning algorithm. Our construction of the polynomial p_D uses the same low-degree approximations exploited by Hardt et. al. in the development of their private learning algorithms.

In summary, the main difference between prior work and ours is that prior work used learning algorithms that have restricted access to the database, and released the hypothesis output by the learning algorithm. In contrast, we do not make use of any learning algorithms, and give our release algorithm direct access to the database. This enables our algorithm to achieve a worst-case error guarantee while maintaining a minimal database size and running time much smaller than the size of the query set. Our algorithm is also substantially simpler than that of Hardt et. al.

We also consider other families of counting queries. We define the class of r-of-k queries. Like a monotone k-way disjunction, an r-of-k query is defined by a set $S \subseteq [d]$ such that $|S| \leq k$. The query asks what fraction of the rows of D have at least r of the attributes in S set to 1. For $r = 1$, these queries are exactly monotone k-way disjunctions, and r-of-k queries are a strict generalization.

Theorem 2 (Releasing r-of-k Queries). *For every $r, k, d, n \in \mathbb{N}$ with $r \leq k \leq d$, every $\alpha \in (0, 1]$, and every $\varepsilon > 0$ there is an ε-differentially private sanitizer that, on input a database $D \in (\{0,1\}^d)^n$, runs in time $|D| \cdot d^{\tilde{O}\left(\sqrt{rk \log(1/\alpha)}\right)}$ and releases a summary that enables computing each of the r-of-k queries on D up to an additive error of at most α, provided that $|D| \geq d^{\tilde{O}\left(\sqrt{rk \log(1/\alpha)}\right)}/\varepsilon$.*

Note that monotone k-way disjunctions are just r-of-k queries where $r = 1$, thus Theorem 2 implies a release algorithm for disjunctions with quadratically better dependence on $\log(1/\alpha)$, at the cost of slightly worse dependence on k (implicit in the switch from $O(\cdot)$ to $\tilde{O}(\cdot)$).

Finally, we present a sanitizer for privately releasing databases in which the *rows* of the database are interpreted as decision lists, and the *queries* are inputs to the decision lists. That is, instead of each record in D being a string of d attributes, each record is an element of the set $\mathrm{DL}_{k,m}$, which consists of all length-k decision lists over m input variables. (See Section 4.3 for a precise definition.) A query is specified by a string $y \in \{0,1\}^d$ and asks "What fraction of database participants would make a certain decision based on the input y?"

As an example application, consider a database that allows high school students to express their preferences for colleges in the form of a decision list. For example, a student may say, "If the school is ranked in the top ten nationwide, I am willing to apply to it. Otherwise, if the school is rural, I am unwilling to apply. Otherwise, if the school has a good basketball team then I am willing to apply to it." And so on. Each student is allowed to use up to k attributes out of a set of m binary attributes. Our sanitizer allows any college (represented by its m binary attributes) to determine the fraction of students willing to apply.

Theorem 3 (Releasing Decision Lists). *For any $k, m \in \mathbb{N}$ s.t. $k \leq m$, any $\alpha \in (0, 1]$, and any $\varepsilon > 1/n$, there is an ε-differentially private sanitizer with running time $m^{\tilde{O}(\sqrt{k} \log(1/\alpha))}$ that, on input a database $D \in (\mathrm{DL}_{k,m})^n$, releases* *a summary that enables computing any length-k decision list query up to an additive error of at most α on every query, provided that $|D| \geq m^{\tilde{O}(\sqrt{k}\log(1/\alpha))}/\varepsilon$.*

For comparison, we note that all the results on releasing k-way disjunctions (including ours) also apply to a dual setting where the database *records* specify a k-way disjunction over m bits and the *queries* are m-bit strings (in this setting m plays the role of d). Theorem 3 generalizes this dual version of Theorem 1, as length-k decision lists are a strict generalization of k-way disjunctions.

We prove the latter two results (Theorems 2 and 3) using the same approach outlined for marginals (Theorem 1), but with different low-degree polynomial approximations appropriate for the different types of queries.

Table 1. Summary of prior results on differentially private release of k-way marginals. The database size column indicates the minimum database size required to release answers to k-way marginals up to an additive error of α. For clarity, we ignore the dependence on the privacy parameters and the failure probability of the algorithms. Notice that this paper contains the first algorithm capable of releasing k-way marginals with running time and worst-case error substantially smaller than the number of queries.

Paper	Running Time	Database Size	Error Type[a]	Synthetic Data?
[5,8,2,6]	$d^{O(k)}$	$O(d^{k/2}/\alpha)$	Worst case	N
[1]	$2^{O(d)}$	$O(d^{k/2}/\alpha)$	Worst case	Y
[3,7,9,12]	$2^{O(d)}$	$\tilde{O}(k\sqrt{d}/\alpha^2)$	Worst case	Y
[10]	$d^{\tilde{O}(1/\alpha^2)}$	$d^{\tilde{O}(1/\alpha^2)}$	Product Dists.	N
[4]	$d^{O(\log(1/\alpha))}$	$d^{O(\log(1/\alpha))}$	Uniform Dist.[b]	N
[14]	$d^{O(d^{1/3}\log(1/\alpha))}$	$d^{O(d^{1/3}\log(1/\alpha))}$	Any Dist.	N
[14]	$d^{O(k)}$	$d^{O(d^{1/3}\log(1/\alpha))}$	Worst case	N
[14]	$d^{O(\sqrt{k}\log(1/\alpha))}$	$d^{O(\sqrt{k}\log(1/\alpha))}$	Any Dist.	N
[14]	$d^{O(k)}$	$d^{O(\sqrt{k}\log(1/\alpha))}$	Worst case	N
This paper	$d^{O(\sqrt{k}\log(1/\alpha))}$	$d^{O(\sqrt{k}\log(1/\alpha))}$	Worst case	N

[a] *Worst case* error indicates that the accuracy guarantee holds for every marginal. The other types of error indicate that accuracy holds for random marginals over a given distribution from a particular class of distributions (e.g. product distributions).
[b] The results of [4] apply only to the uniform distribution over *all* marginals.

On Synthetic Data. An attractive type of summary is a *synthetic database*. A synthetic database is a new database $\widehat{D} \in (\{0,1\}^d)^{\hat{n}}$ whose rows are "fake", but such that $\widehat{D}$ approximately preserves many of the statistical properties of the database D (e.g. all the marginals). Some of the previous work on counting query release has provided synthetic data, starting with Barak et. al. [1] and including [3,7,9,12].

Unfortunately, Ullman and Vadhan [19] (building on [7]) have shown that no differentially private sanitizer with running time $d^{O(1)}$ can take a database $D \in (\{0,1\}^d)^n$ and output a private synthetic database $\widehat{D}$, all of whose 2-way marginals are approximately equal to those of D (assuming the existence of one-way functions). They also showed that there is a constant $k \in \mathbb{N}$ such that no differentially private sanitizer with running time $2^{d^{1-\Omega(1)}}$ can output a private synthetic database, all of whose k-way marginals are approximately equal to those of D (under stronger cryptographic assumptions).

When $k = d$, our sanitizer runs in time $2^{\tilde{O}(\sqrt{d})}$ and releases a private summary that enables an analyst to approximately answer any marginal query on D. Prior to our work it was not known how to release *any* summary enabling approximate answers to all marginals in time $2^{d^{1-\Omega(1)}}$. Thus, our results show that releasing a private summary for all marginal queries can be done considerably more efficiently if we do not require the summary to be a synthetic database (under the hardness assumptions made in [19]).

2 Preliminaries

2.1 Differentially Private Sanitizers

Let a *database* $D \in \mathcal{X}^n$ be a collection of n rows $x^{(1)}, \ldots, x^{(n)}$ from a *data universe* $\mathcal{X}$. We say that two databases $D_1, D_2 \in \mathcal{X}^n$ are *adjacent* if they differ only on a single row, and we denote this by $D_1 \sim D_2$.

A *sanitizer* $\mathcal{A} : \mathcal{X}^n \to \mathcal{R}$ takes a database as input and outputs some data structure in $\mathcal{R}$. We are interested in sanitizers that satisfy *differential privacy*.

Definition 4 (Differential Privacy [6]). *A sanitizer* $\mathcal{A}: \mathcal{X}^n \to \mathcal{R}$ *is* (ε, δ)-*differentially private if for every two adjacent databases* $D, D' \in \mathcal{X}^n$ *and every subset* $S \subseteq \mathcal{R}$, $\Pr[\mathcal{A}(D) \in S] \leq e^\varepsilon \Pr[\mathcal{A}(D') \in S] + \delta$. *In the case where* $\delta = 0$ *we say that* $\mathcal{A}$ *is* ε-*differentially private.*

Since a sanitizer that always outputs $\perp$ satisfies Definition 4, we also need to define what it means for a sanitizer to be accurate. In particular, we are interested in sanitizers that give accurate answers to *counting queries*. A counting query is defined by a boolean predicate $q: \mathcal{X} \to \{0, 1\}$. We define the evaluation of the query q on a database $D \in \mathcal{X}^n$ to be $q(D) = \frac{1}{n} \sum_{i=1}^{n} q(x^{(i)})$. We use $\mathcal{Q}$ to denote a set of counting queries.

Since $\mathcal{A}$ may output an arbitrary data structure, we must specify how to answer queries in $\mathcal{Q}$ from the output $\mathcal{A}(D)$. Hence, we require that there is an *evaluator* $\mathcal{E}: \mathcal{R} \times \mathcal{Q} \to \mathbb{R}$ that estimates $q(D)$ from the output of $\mathcal{A}(D)$. For example, if $\mathcal{A}$ outputs a vector of "noisy answers" $Z = (q(D) + Z_q)_{q \in \mathcal{Q}}$, where Z_q is a random variable for each $q \in \mathcal{Q}$, then $\mathcal{R} = \mathbb{R}^{\mathcal{Q}}$ and $\mathcal{E}(Z, q)$ is the q-th component of Z. Abusing notation, we write $q(Z)$ and $q(\mathcal{A}(D))$ as shorthand for $\mathcal{E}(Z, q)$ and $\mathcal{E}(\mathcal{A}(D), q)$, respectively. Since we are interested in the efficiency of the sanitization process as a whole, when we refer to the running time of $\mathcal{A}$, we also include the running time of the evaluator $\mathcal{E}$. We say that $\mathcal{A}$ is "accurate" for the query set $\mathcal{Q}$ if the values $q(\mathcal{A}(D))$ are close to the answers $q(D)$. Formally,

Definition 5 (Accuracy). *An output* Z *of a sanitizer* $\mathcal{A}(D)$ *is* α-*accurate for the query set* $\mathcal{Q}$ *if* $|q(Z) - q(D)| \leq \alpha$ *for every* $q \in \mathcal{Q}$. *A sanitizer is* (α, β)-*accurate for the query set* $\mathcal{Q}$ *if for every database* D,

$$\Pr\left[\forall q \in \mathcal{Q}, |q(\mathcal{A}(D)) - q(D)| \leq \alpha\right] \geq 1 - \beta,$$

where the probability is taken over the coins of $\mathcal{A}$.

We will make use of the *Laplace mechanism*. Let $\mathrm{Lap}^k(\sigma)$ denote a draw from the random variable over $\mathbb{R}^k$ in which each coordinate is chosen independently according to the density function $\mathrm{Lap}_\sigma(x) \propto e^{-|x|/\sigma}$. Let $D \in \mathcal{X}^n$ be a database and $g : \mathcal{X}^n \to \mathbb{R}^k$ be a function such that for every pair of adjacent databases $D \sim D'$, $\|g(D) - g(D')\|_\infty \leq \Delta$. Then we have the following two theorems:

Lemma 6 (Laplace Mechanism, ε-Differential Privacy [6]). *For D, g, k, Δ as above, the mechanism $\mathcal{A}(D) = g(D) + \mathrm{Lap}^k(\Delta k/\varepsilon)$ satisfies ε-differential privacy. Furthermore, for any $\beta > 0$, $\Pr_{\mathcal{A}}\left[\|g(D) - \mathcal{A}(D)\|_1 \leq \alpha\right] \geq 1 - \beta$, for $\alpha = 2\Delta k^2 \log(k/\beta)/\varepsilon$.*

The choice of the L_1 norm in the accuracy guarantee of the lemma is for convenience, and doesn't matter for the parameters of Theorems 1-3 (except for the hidden constants).

2.2   Query Function Families

We take the approach of Gupta et. al. [10] and think of the database D as specifying a function f_D mapping queries q to their answers $q(D)$, which we call the $\mathcal{Q}$-*representation of D*. We now describe this transformation more formally:

Definition 7 ($\mathcal{Q}$-Function Family). *Let $\mathcal{Q} = \{q_y\}_{y \in Y_{\mathcal{Q}} \subseteq \{0,1\}^m}$ be a set of counting queries on a data universe $\mathcal{X}$, where each query is indexed by an m-bit string. We define the* index set *of $\mathcal{Q}$ to be the set $Y_{\mathcal{Q}} = \{y \in \{0,1\}^m \mid q_y \in \mathcal{Q}\}$.*
We define the $\mathcal{Q}$-function family $\mathcal{F}_{\mathcal{Q}} = \{f_x : \{0,1\}^m \to \{0,1\}\}_{x \in \mathcal{X}}$ as follows: For every possible database row $x \in \mathcal{X}$, the function $f_{\mathcal{Q},x} : \{0,1\}^m \to \{0,1\}$ is defined as $f_{\mathcal{Q},x}(y) = q_y(x)$. Given a database $D \in \mathcal{X}^n$ we define the function $f_{\mathcal{Q},D} : \{0,1\}^m \to [0,1]$ where $f_{\mathcal{Q},D}(q) = \frac{1}{n}\sum_{i=1}^n f_{\mathcal{Q},x^{(i)}}(q)$. When $\mathcal{Q}$ is clear from context we will drop the subscript $\mathcal{Q}$ and simply write f_x, f_D, and $\mathcal{F}$.

For some intuition about this transformation, when the queries are monotone k-way disjunctions on a database $D \in (\{0,1\}^d)^n$, the queries are defined by sets $S \subseteq [d]$, $|S| \leq k$. In this case each query can be represented by the d-bit indicator vector of the set S, with at most k non-zero entries. Thus we can take $m = d$ and $Y_{\mathcal{Q}} = \left\{y \in \{0,1\}^d \mid \sum_{j=1}^d y_j \leq k\right\}$.

2.3   Polynomial Approximations

An m-variate real polynomial $p \in \mathbb{R}[y_1, \ldots, y_m]$ of *degree* t and (L_∞) *norm* T can be written as $p(y) = \sum_{\substack{j_1,\ldots,j_m \geq 0 \\ j_1 + \cdots + j_m \leq t}} c_{j_1,\ldots,j_m} \prod_{\ell=1}^m y_\ell^{j_\ell}$ where $|c_{j_1,\ldots,j_m}| \leq T$ for every $j_1, \ldots, j_m$. Recall that there are at most $\binom{m+t}{t}$ coefficients in an m-variate polynomial of total degree t. Often we will want to associate a polynomial p of degree t and norm T with its coefficient vector $\boldsymbol{p} \in [-T, T]^{\binom{m+t}{t}}$. Specifically,

$p = (c_{j_1,...,j_m})_{\substack{j_1,...,j_m \geq 0 \\ j_1 + \cdots + j_m \leq t}}$. Given a vector p and a point $y \in \{0,1\}^m$ we use $p(y)$ to indicate the evaluation of the polynomial described by the vector p at the point y. Observe this is equivalent to computing $p \cdot y$ where $y \in \{0,1\}^{\binom{m+t}{t}}$ is defined as $y_{j_1,...,j_m} = \prod_{\ell=1}^m y_\ell^{j_\ell}$ for every $j_1,...,j_m \geq 0$, $j_1 + \cdots + j_m \leq t$.

Let $\mathcal{P}_{t,T}$ be the family of all m-variate real polynomials of degree t and norm T. In many cases, the functions $f_{Q,x} : \{0,1\}^m \to \{0,1\}$ can be approximated well on all the indices in Y_Q by a family of polynomials $\mathcal{P}_{t,T}$ with low degree and small norm. Formally:

Definition 8 (Uniform Approximation by Polynomials). *Given a family of m-variate functions $\mathcal{F} = \{f_x\}_{x \in \mathcal{X}}$ and a set $Y \subseteq \{0,1\}^m$, we say that the family $\mathcal{P}_{t,T}$ uniformly γ-approximates $\mathcal{F}$ on Y if for every $x \in \mathcal{X}$, there exists $p_x \in \mathcal{P}_{t,T}$ such that $\max_{y \in Y} |f_x(y) - p_x(y)| \leq \gamma$.*

We say that $\mathcal{P}_{t,T}$ efficiently and uniformly γ-approximates $\mathcal{F}$ if there is an algorithm $\mathcal{P_F}$ that takes $x \in \mathcal{X}$ as input, runs in time $\text{poly}(\log |\mathcal{X}|, \binom{m+t}{t}, \log T)$, and outputs a coefficient vector p_x such that $\max_{y \in Y} |f_x(y) - p_x(y)| \leq \gamma$.

3 From Polynomial Approximations to Data Release Algorithms

In this section we present an algorithm for privately releasing any family of counting queries Q such that $\mathcal{F}_Q$ that can be efficiently and uniformly approximated by polynomials. The algorithm will take an n-row database D and, for each row $x \in D$, constructs a polynomial p_x that uniformly approximates the function $f_{Q,x}$ (recall that $f_{Q,x}(q) = q(x)$, for each $q \in Q$). From these, it constructs a polynomial $p_D = \frac{1}{n} \sum_{x \in D} p_x$ that uniformly approximates $f_{Q,D}$. The final step is to perturb each of the coefficients of p_D using noise from a Laplace distribution (Theorem 6) and bound the error introduced from the perturbation.

Theorem 9 (Releasing Polynomials). *Let $Q = \{q_y\}_{y \in Y_Q \subseteq \{0,1\}^m}$ be a set of counting queries over $\{0,1\}^d$, and $\mathcal{F}_Q$ be the Q function family (Definition 7). Assume that $\mathcal{P}_{t,T}$ efficiently and uniformly γ-approximates $\mathcal{F}_Q$ on Y_Q (Definition 8). Then there is a sanitizer $\mathcal{A}: (\{0,1\}^d)^n \to \mathbb{R}^{\binom{m+t}{t}}$ that*

1. *is ε-differentially private,*
2. *runs in time $\text{poly}(n, d, \binom{m+t}{t}, \log T, \log(1/\varepsilon))$, and*
3. *is (α, β)-accurate for Q for $\alpha = \gamma + \frac{4T\binom{m+t}{t}^2 \log(\binom{m+t}{t}/\beta)}{\varepsilon n}$.*

Proof. First we construct the sanitizer $\mathcal{A}$. See the relevant codebox below.

Privacy. We establish that $\mathcal{A}$ is ε-differentially private. This follows from the observation that for any two adjacent $D \sim D'$ that differ only on row i^*,

$$\|p_D - p_{D'}\|_\infty = \left\| \frac{1}{n} \sum_{i=1}^n p_{x^{(i)}} - \frac{1}{n} \sum_{i=1}^n p_{x'^{(i)}} \right\|_\infty = \frac{1}{n} \|p_{x^{(i^*)}} - p_{x'^{(i^*)}}\|_\infty \leq \frac{2T}{n}.$$

The Sanitizer $\mathcal{A}$

Input: A database $D \in (\{0,1\}^d)^n$, an explicit family of polynomials $\mathcal{P}$, and a parameter $\varepsilon > 0$.

For $i = 1, \ldots, n$

Using efficient approximation of $\mathcal{F}$ by $\mathcal{P}$, compute a polynomial $\boldsymbol{p}_{x^{(i)}} = \mathcal{P}_{\mathcal{F}}(x^{(i)})$ that γ-approximates $f_{x^{(i)}}$ on $Y_{\mathcal{Q}}$.

Let $\boldsymbol{p}_D = \frac{1}{n} \sum_{i=1}^n \boldsymbol{p}_{x^{(i)}}$, where the sum denotes standard entry-wise vector addition.

Let $\tilde{\boldsymbol{p}}_D = \boldsymbol{p}_D + Z$, where Z is drawn from an $\binom{m+t}{t}$-variate Laplace distribution with parameter $2T/\varepsilon n$ (Section 2.1).

Output: $\tilde{\boldsymbol{p}}_D$.

The last inequality is from the fact that for every x, $\boldsymbol{p}_x$ is a vector of L_∞ norm at most T. Part 1 of the Theorem now follows directly from the properties of the Laplace Mechanism (Theorem 6). Now we construct the evaluator $\mathcal{E}$.

The Evaluator $\mathcal{E}$ for the Sanitizer $\mathcal{A}$

Input: A vector $\tilde{\boldsymbol{p}} \in \mathbb{R}^{\binom{m+t}{t}}$ and the description of a query $y \in \{0,1\}^m$.

Output: $\tilde{\boldsymbol{p}}(y)$. Recall that we view $\tilde{\boldsymbol{p}}$ as an m-variate polynomial, p, and $\tilde{\boldsymbol{p}}(y)$ is the evaluation of p on the point y.

Efficiency. Next, we show that $\mathcal{A}$ runs in time $\text{poly}(n, d, \binom{m+t}{t}, \log T, \log(1/\varepsilon))$. Recall that we assumed the polynomial construction algorithm $\mathcal{P}$ runs in time $\text{poly}(d, \binom{m+t}{t}, \log T)$. The algorithm $\mathcal{A}$ needs to run $\mathcal{P}_{\mathcal{F}}$ on each of the n rows, and then it needs to generate $\binom{m+t}{t}$ samples from a univariate Laplace distribution with magnitude $\text{poly}(T, \binom{m+t}{t}, 1/n, 1/\varepsilon)$, which can also be done in time $\text{poly}(\binom{m+t}{t}, \log T, \log n, \log(1/\varepsilon))$. We also establish that $\mathcal{E}$ runs in time $\text{poly}(\binom{m+t}{t}, \log T, \log n, \log(1/\varepsilon))$, observe that $\mathcal{E}$ needs to expand the input into an appropriate vector of dimension $\binom{m+t}{t}$ and take the inner product with the vector $\tilde{\boldsymbol{p}}$, whose entries have magnitude $\text{poly}(\binom{m+t}{t}, T, 1/n, 1/\varepsilon)$. These observations establish Part 2 of the Theorem.

Accuracy. Finally, we analyze the accuracy of the sanitizer $\mathcal{A}$. First, by the assumption that $\mathcal{P}_{t,T}$ uniformly γ-approximates $\mathcal{F}$ on $Y \subseteq \{0,1\}^m$, we have

$$\max_{y \in Y} |f_D(y) - \boldsymbol{p}_D(y)| \le \frac{1}{n} \sum_{i=1}^n \max_{y \in Y} |f_{x^{(i)}}(y) - \boldsymbol{p}_{x^{(i)}}(y)| \le \gamma.$$

Now we want to establish that $\Pr\left[\max_{y \in \{0,1\}^m} |\tilde{\boldsymbol{p}}_D(y) - \boldsymbol{p}_D(y)| \le \alpha'\right] \ge 1 - \beta$ for $\alpha' = 4T\binom{m+t}{t}^2 \log\left(\binom{m+t}{t}/\beta\right)/\varepsilon n$, where the probability is taken over the coins of $\mathcal{A}$. Part (3) of the Theorem will then follow by the triangle inequality.

To see that the above statement is true, observe that by the properties of the Laplace mechanism (Theorem 6), we have $\Pr\left[\|\tilde{\boldsymbol{p}}_D - \boldsymbol{p}_D\|_1 \le \alpha'\right] \ge 1 - \beta$, where the probability is taken over the coins of $\mathcal{A}$. Given that $\|\tilde{p}_D - p_D\|_1 \leq \alpha'$, it holds that for every $y \in \{0,1\}^m$,

$$|\tilde{p}_D(y) - p_D(y)| = |(\tilde{p}_D - p_D)(y)| \leq \|\tilde{p}_D - p_D\|_1 \leq \alpha'.$$

The first inequality follows from the fact that every monomial evaluates to 0 or 1 at the point y. This completes the proof of the theorem.

4 Applications

In this section we establish the existence of explicit families of low-degree polynomials approximating the families $\mathcal{F}_Q$ for some interesting query sets.

4.1 Releasing Monotone Disjunctions

We define the class of monotone k-way disjunctions as follows:

Definition 10 (Monotone k-Way Disjunctions). *Let $\mathcal{X} = \{0,1\}^d$. The query set $\mathcal{Q}_{\mathrm{Disj},k} = \{q_y\}_{y \in Y_k \subseteq \{0,1\}^d}$ of monotone k-way disjunctions over $\{0,1\}^d$ contains a query q_y for every $y \in Y_k = \{y \in \{0,1\}^d \mid |y| \leq k\}$. Each query is defined as $q_y(x_1,\ldots,x_d) = \bigvee_{j=1}^d y_j x_j$. The $\mathcal{Q}_{\mathrm{Disj},k}$ function family $\mathcal{F}_{\mathcal{Q}_{\mathrm{Disj},k}} = \{f_x\}_{x \in \{0,1\}^d}$ contains a function $f_x(y_1,\ldots,y_d) = \bigvee_{j=1}^d y_j x_j$ for every $x \in \{0,1\}^d$.*

Thus the family $\mathcal{F}_{\mathcal{Q}_{\mathrm{Disj},k}}$ consists of all disjunctions, and the index set, Y_k, consists of all vectors $y \in \{0,1\}^d$ with at most k non-zero entries.

The next lemma shows that $\mathcal{F}_{\mathcal{Q}_{\mathrm{Disj},k}}$ can be efficiently and uniformly approximated by polynomials of low degree and low norm. The statement is a well-known application of Chebyshev polynomials, and a similar statement appears in [14] but without bounding the running time of the construction or a bound on the norm of the polynomials.

Lemma 11 (Approximating $\mathcal{F}_{\mathcal{Q}_{\mathrm{Disj},k}}$ by polynomials, similar to [14]). *For every $k, d \in \mathbb{N}$ such that $k \leq d$ and every $\gamma > 0$, the family $\mathcal{P}_{t,T}$ of d-variate real polynomials of degree $t = O(\sqrt{k}\log(1/\gamma))$ and norm $T = d^{O(\sqrt{k}\log(1/\gamma))}$ efficiently and uniformly γ-approximates the family $\mathcal{F}_{\mathcal{Q}_{\mathrm{Disj},k}}$ on the set Y_k.*

Theorem 1 in the introduction follows by combining Theorems 9 and 11.

4.2 Releasing Monotone r-of-k Queries

We define the class of monotone r-of-k queries as follows:

Definition 12 (Monotone r-of-k Queries). *Let $\mathcal{X} = \{0,1\}^d$ and $r, k \in \mathbb{N}$ such that $r \leq k \leq d$. The query set $\mathcal{Q}_{r,k} = \{q_y\}_{y \in Y_k \subseteq \{0,1\}^d}$ of monotone r-of-k queries over $\{0,1\}^d$ contains a query q_y for every $y \in Y_k = \{y \in \{0,1\}^d \mid |y| \leq k\}$. Each query is defined as $q_y(x_1,\ldots,x_d) = 1_{\sum_{j=1}^d y_j x_j \geq r}$. The $\mathcal{Q}_{r,k}$ function family $\mathcal{F}_{\mathcal{Q}_{r,k}} = \{f_x\}_{x \in \{0,1\}^d}$ contains a function $f_x(y_1,\ldots,y_d) = 1_{\sum_{j=1}^d y_j x_j \geq r}$ for every $x \in \{0,1\}^d$.*

The next lemma shows that $\mathcal{F}_{\mathcal{Q}_{r,k}}$ can be efficiently and uniformly approximated over Y_k by low-degree polynomials. The statement is based on approximation-theoretic results of Sherstov [18, Lemma 3.11].

Lemma 13 (Approximating $\mathcal{F}_{\mathcal{Q}_{r,k}}$ on Y_k). *For every $r, k, d \in \mathbb{N}$ such that $r \leq k \leq d$ and every $\gamma > 0$, the family $\mathcal{P}_{t,T}$ of d-variate real polynomials of degree $t = \tilde{O}(\sqrt{kr \log(1/\gamma)})$ and norm $T = d^{\tilde{O}(\sqrt{kr \log(1/\gamma)})}$ efficiently and uniformly γ-approximates the family $\mathcal{F}_{\mathcal{Q}_{r,k}}$ on the set Y_k.*

Remark 1. Using the principle of inclusion-exclusion, the answer to a monotone r-of-k query can be written as a linear combination of the answers to $k^{O(r)}$ monotone k-way disjunctions. Thus, a sanitizer that is $(\alpha/k^{O(r)}, \beta)$-accurate for monotone k-way disjunctions implies a sanitizer that is (α, β)-accurate for monotone r-of-k queries. However, combining this implication with Theorem 1 yields a sanitizer with running time $d^{O(r\sqrt{k}\log(k/\beta))}$, which has a worse dependence on r than what we achieve in Theorem 2.

4.3   Releasing Decision Lists

A *length-k decision list* over m variables is a function which can be written in the form "if ℓ_1 then output b_1 else $\cdots$ else if ℓ_k then output b_k else output b_{k+1}," where each ℓ_i is a boolean literal in $\{x_1, \ldots, x_m\}$, and each b_i is an output bit in $\{0, 1\}$. Note that decision lists of length-k strictly generalize k-way disjunctions and conjunctions. We use $\mathrm{DL}_{k,m}$ to denote the set of all length-k decision lists over m binary input variables.

Definition 14 (Evaluations of Length-k Decision Lists). *Let $k, m \in \mathbb{N}$ such that $k \leq m$ and $\mathcal{X} = \mathrm{DL}_{k,m}$. The query set $\mathcal{Q}_{\mathrm{DL}_{k,m}} = \{q_y\}_{y \in \{0,1\}^m}$ of evaluations of length-k decision lists contains a query q_y for every $y \in \{0,1\}^m$. Each query is defined as $q_y(x) = x(y)$ where $x \in \mathrm{DL}_{k,m}$ is a length-k decision list over m variables. The $\mathcal{Q}_{\mathrm{DL}_{k,m}}$ function family $\mathcal{F}_{\mathcal{Q}_{\mathrm{DL}_{k,m}}} = \{f_x\}_{x \in \mathrm{DL}_{k,m}}$ contains functions $f_x(y) = x(y)$ for every $x \in \mathrm{DL}_{k,m}$. That is, $\mathcal{F}_{\mathcal{Q}_{\mathrm{DL}_{k,m}}} = \mathrm{DL}_{k,m}$.*

We clarify that in this setting, the records in the database are length-k decision lists over $\{0,1\}^m$ and the queries inputs in $\{0,1\}^m$. Thus $|\mathcal{X}| = |\mathrm{DL}_{k,m}| = m^{O(k)}$ and $|\mathcal{Q}| = 2^m$. Alternatively, $\mathcal{X} = \{0,1\}^d$ for $d = k(\log m + 2) + 1$, since a length-k decision list can be described using this many bits.

Lemma 15 ([16]). *For every $k, m \in \mathbb{N}$ such that $k \leq m$ and every $\gamma > 0$, the family $\mathcal{P}_{t,T}$ of m-variate real polynomials of degree $\tilde{O}\left(\sqrt{k}\log(1/\gamma)\right)$ and norm $T = m^{\tilde{O}(\sqrt{k}\log(1/\gamma))}$ efficiently and uniformly γ-approximates the family $\mathcal{F}_{\mathcal{Q}_{\mathrm{DL}_{k,m}}} = \mathrm{DL}_{k,m}$ on all of $\{0,1\}^m$.*

We obtain Theorem 3 of the introduction by combining Theorems 9 and 15.

Acknowledgements. We thank Moritz Hardt, Varun Kanade, Aaron Roth, Guy Rothblum, and Li-Yang Tan for helpful discussions.

References

1. Barak, B., Chaudhuri, K., Dwork, C., Kale, S., McSherry, F., Talwar, K.: Privacy, accuracy, and consistency too: a holistic solution to contingency table release. In: Libkin, L. (ed.) PODS, pp. 273–282. ACM (2007)
2. Blum, A., Dwork, C., McSherry, F., Nissim, K.: Practical privacy: the SuLQ framework. In: Li, C. (ed.) PODS, pp. 128–138. ACM (2005)
3. Blum, A., Ligett, K., Roth, A.: A learning theory approach to non-interactive database privacy. In: Dwork, C. (ed.) STOC, pp. 609–618. ACM (2008)
4. Cheraghchi, M., Klivans, A., Kothari, P., Lee, H.K.: Submodular functions are noise stable. In: Randall, D. (ed.) SODA, pp. 1586–1592. SIAM (2012)
5. Dinur, I., Nissim, K.: Revealing information while preserving privacy. In: PODS, pp. 202–210. ACM (2003)
6. Dwork, C., McSherry, F., Nissim, K., Smith, A.: Calibrating Noise to Sensitivity in Private Data Analysis. In: Halevi, S., Rabin, T. (eds.) TCC 2006. LNCS, vol. 3876, pp. 265–284. Springer, Heidelberg (2006)
7. Dwork, C., Naor, M., Reingold, O., Rothblum, G.N., Vadhan, S.P.: On the complexity of differentially private data release: efficient algorithms and hardness results. In: STOC 2009, pp. 381–390 (2009)
8. Dwork, C., Nissim, K.: Privacy-Preserving Datamining on Vertically Partitioned Databases. In: Franklin, M. (ed.) CRYPTO 2004. LNCS, vol. 3152, pp. 528–544. Springer, Heidelberg (2004)
9. Dwork, C., Rothblum, G.N., Vadhan, S.P.: Boosting and differential privacy. In: FOCS, pp. 51–60. IEEE Computer Society (2010)
10. Gupta, A., Hardt, M., Roth, A., Ullman, J.: Privately releasing conjunctions and the statistical query barrier. In: STOC 2011, pp. 803–812 (2011)
11. Gupta, A., Roth, A., Ullman, J.: Iterative Constructions and Private Data Release. In: Cramer, R. (ed.) TCC 2012. LNCS, vol. 7194, pp. 339–356. Springer, Heidelberg (2012)
12. Hardt, M., Ligett, K., McSherry, F.: A simple and practical algorithm for differentially private data release. CoRR abs/1012.4763 (2010)
13. Hardt, M., Rothblum, G.N.: A multiplicative weights mechanism for privacy-preserving data analysis. In: FOCS, pp. 61–70. IEEE Computer Society (2010)
14. Hardt, M., Rothblum, G.N., Servedio, R.A.: Private data release via learning thresholds. In: Randall, D. (ed.) SODA, pp. 168–187. SIAM (2012)
15. Kasiviswanathan, S.P., Lee, H.K., Nissim, K., Raskhodnikova, S., Smith, A.: What can we learn privately? SIAM J. Comput. 40(3), 793–826 (2011)
16. Klivans, A.R., Servedio, R.A.: Toward Attribute Efficient Learning of Decision Lists and Parities. In: Shawe-Taylor, J., Singer, Y. (eds.) COLT 2004. LNCS (LNAI), vol. 3120, pp. 224–238. Springer, Heidelberg (2004)
17. Roth, A., Roughgarden, T.: Interactive privacy via the median mechanism. In: STOC 2010, pp. 765–774 (2010)
18. Sherstov, A.A.: Approximate inclusion-exclusion for arbitrary symmetric functions. Computational Complexity 18(2), 219–247 (2009)
19. Ullman, J., Vadhan, S.: PCPs and the Hardness of Generating Private Synthetic Data. In: Ishai, Y. (ed.) TCC 2011. LNCS, vol. 6597, pp. 400–416. Springer, Heidelberg (2011)

Stochastic Matching with Commitment[*]

Kevin P. Costello[2], Prasad Tetali[1], and Pushkar Tripathi[1]

[1] Georgia Institute of Technology
[2] University of California at Riverside

Abstract. We consider the following stochastic optimization problem first introduced by Chen et al. in [7]. We are given a vertex set of a random graph where each possible edge is present with probability p_e. We do not know which edges are actually present unless we scan/probe an edge. However whenever we probe an edge and find it to be present, we are constrained to picking the edge and both its end points are deleted from the graph. We wish to find the maximum matching in this model. We compare our results against the optimal omniscient algorithm that knows the edges of the graph and present a 0.573 factor algorithm using a novel sampling technique. We also prove that no algorithm can attain a factor better than 0.898 in this model.

1 Introduction

The matching problem has been a corner-stone of combinatorial optimization and has received considerable attention starting from the work of Jack Edmonds [9]. There has been recent interest in studying stochastic versions of the problem due to its applications to online advertising and several barter exchange settings [22,18,23]. Much of the recent research focused on studying matchings on *bipartite* graphs. In this paper we study a recently introduced variant on the stochastic online matching problem [7] on general graphs as described below.

For p a probability vector indexed by pairs of vertices from a vertex set V, let $G(V, p)$ denote an undirected Erdős-Rényi graph on V. That is, for any $(u, v) \in V \times V$, $p_{uv} = p_{vu}$ denotes the (known) probability that there is an edge connecting u and v in G. For every pair $(u, v) \in V \times V$ we are *not* told a priori whether there is an edge connecting these vertices, until we *probe/scan* this pair. If we scan a pair of vertices and find that there is an edge connecting them we are constrained to *pick* this edge and in this case both u and v are removed from the graph. However, if we find that u and v are not connected by an edge, they continue to be available (to others) in the future. The goal is to maximize the number of vertices that get matched.

We will refer to the above as the Stochastic Matching with Commitment Problem (SMCP), since whenever we probe a pair of adjacent vertices, we are committed to picking them. The performance of our algorithm is compared against the optimal offline algorithm that knows the underlying graph for each

---

[*] The full version is available under the same name at the arxiv.org

A. Czumaj et al. (Eds.): ICALP 2012, Part I, LNCS 7391, pp. 822–833, 2012.

instantiation of the problem and finds the maximum matching in it. Note that since the input is itself random, the average performance of the optimal offline-algorithm is the expected size of the maximum matching in this random graph. We use the somewhat non-standard notation of $G(V, p)$, rather than $G_{n,p}$, since we will need to refer to the (fixed) vertex set V and also since p is a vector with typically different entries.

1.1 Our Results

It is easy to see that the simple greedy algorithm, which probes pairs in an arbitrary order, would return a *maximal* matching in every instance of the problem and is therefore a factor 0.5 approximation algorithm. We give a sampling based algorithm for this problem that does better than this:

Theorem 1. *There exists a randomized algorithm that attains a competitive ratio of at least 0.573 for the Stochastic Matching with Commitment Problem that runs in time $\tilde{O}(n^4)$ for a graph with n vertices. Furthermore, the running time can be reduced to $\tilde{O}(n^3)$ in the case where the expected size of the optimal matching is a positive fraction of the number of vertices in the graph.*

Our algorithm uses *offline simulations* to determine the relative importance of edges to decide the order in which to scan them. It is based on a novel sampling lemma that might be of independent interest in tackling online optimization settings, wherein an algorithm needs to make irrevocable online decisions with limited stochastic knowledge. This sampling trick is explained in section 2.3. Even though the proof for our sampling lemma is based on solving an exponentially large linear program, we also give a fast combinatorial algorithm for it.

On the hardness front, we prove the following theorem, using rigorous analysis of the performance of the *optimal* online algorithm for a carefully chosen graph.

Theorem 2. *No algorithm can attain a competitive ratio better than 0.898 for the SMCP.*

1.2 Previous Work

The problem has similar flavor to several well known stochastic optimization problems such as the stochastic knapsack [8] and the shortest-path [19,20]; refer to [24] for a detailed discussion on these problems. We will now present more explicit connections between SMCP and several previously studied models of matching with limited information.

Stochastic Matching Problems: Chen et al. [7] considered a more general model for stochastic matching than the one presented above. In their model every vertex $v \in V$ had a *patience parameter* $t(v)$ indicating the maximum number of failed probes v is willing to participate in. After $t(v)$ failed attempts, vertex

v would leave the system, and would not be considered for any further matches. Our model can be viewed as a special case of their setting where $t(v) = n$ for every vertex. However Chen et al., and subsequently Bansal et al. [4], compared their performance against the optimal online algorithm. This was necessary because if we consider the case of the star graph, where each edge has a probability of $1/n$ and $t(v) = 1$ for every vertex v, then any online algorithm can match the center of the star with probability at most $1/n$, while there exists an edge incident on the center with probability $1 - 1/e$. In contrast, our results are against the strongest adversary, i.e., the optimal offline omniscient algorithm. Clearly the performance of the optimal online algorithm can be no better than that of such an omniscient algorithm.

In their model [7], Chen et al. presented a 1/4 competitive algorithm which was later improved to a 0.5 factor algorithm by Adamczyk [1]. The results in [7] were extended to the weighted setting by Bansal et al. [4] who gave a 1/4 competitive algorithm for the general case, and a 1/3 competitive algorithm for the special case of bipartite graphs.

Online Bipartite Matching Problems: Online bipartite matching was first introduced by Karp et al. in [15]. Here one side of a bipartite graph is known in advance and the other side arrives online. For each arriving vertex we are revealed its neighbors in the given side. The task is to match the maximum number arriving vertices. In [15], the authors gave a tight $1 - 1/e$ factor algorithm for this problem. This barrier of $1 - 1/e$ has been breached for various stochastic variants of this problem [11,5,17], by assuming prior stochastic knowledge about the arriving vertices. Goel and Mehta [13] studied the random order arrival model where the vertices of the streaming side are presented in a random order and showed that the greedy algorithm attains a factor of $1 - 1/e$. This was later improved to a 0.69 competitive algorithm by [16,14].

Remark 1. The algorithm in [16,14] can be thought of as the following randomized algorithm for finding a large matching in a given bipartite graph - randomly permute one of the sides and consider the vertices of the other side also in a (uniformly) random order. Match every vertex to the first available neighbor (according to the permutation) on the other side. It can be viewed as an oblivious algorithm that ignores the edge structure of the graph and can therefore be simulated in our setting. This yields a 0.69 competitive algorithm for the SMCP restricted to bipartite graphs.

Randomized Algorithms for Maximum Matchings: Fast randomized algorithms for finding maximum matchings have been studied particularly in the context of Erdős-Rényi graphs [3,12,6] starting from the work of Karp and Sipser [21]. However all these algorithms explicitly exploit the edge structure of the graph and are not applicable in our setting. In [2], Aronson et al. analysed the following simple algorithm for finding a matching in a general graph - consider the vertices of the graph in a random order and match each vertex to a randomly chosen neighbor that is unmatched. This algorithm was shown to achieve a factor of 0.50000025.

Remark 2. To the best of our knowledge, the algorithm in [2] is the only non-trivial algorithm for finding a large matching in a general graph that works without looking at the edge structure. Since this algorithm works for *arbitrary* graphs, it can be simulated in our setting and yields a 0.50000025 factor algorithm for SMCP for general graphs. However we manage to improve the factor by exploiting the additional stochastic information available to us in our model.

1.3   Informal Description of the Proof Technique

Observe that the simple algorithm, which weighs (or probes) an edge e according to probability p_e, is not necessarily the best way to proceed. Consider a path having 3 edges such that the middle edge is present with probability 1 whereas the other two edges are each present with probability 0.9. Even though the middle edge is always present, it is unlikely to be involved in any maximum matching. Conversely, the outer edges will always be a part of some maximum matching when they appear.

To determine the relative importance of edges, our algorithm relies on offline simulations. We sample from the given distribution to obtain a collection of representative graphs. We use maximum matchings from these graphs to estimate the probability (denoted by q_e^*) that a given candidate edge e belongs to the maximum matching. Note that this is done as a preprocessing step *without* probing any of the edges in the given graph (a necessary requirement, as probing an edge could lead to unwanted commitments). Clearly the probability that a vertex would get matched in the optimal solution is the sum of q_e^* for all edges incident on it and this gives us a way to approximate the optimal solution.

Similarly we can also calculate the conditional probability that an edge belongs to the maximum matching, given that it is present in the underlying graph. We use this as a measure of the importance of the edge. Observe that it is safe to probe edges where this conditional probability is large, since we are unlikely to make a mistake on such edges. After we are done probing these edges we are left with a residual graph where this conditional probability is small for every edge.

Ideally at this point what we would like to do is to simulate the fractional matching given by the q_e^*, i.e., include every edge with probability q_e^*. However, this is made impossible by the combination of our lack of knowledge of the graph and the commitments we are forced to make as we scan edges to obtain information about the graph. To overcome these limitations, we devise a novel sampling technique, described in section 2.3, that gives us a partial simulation. This sampling algorithm outputs a (randomized) ordering to scan the edges incident to a given vertex, so as to ensure that edge e is included with probability at least some large positive fraction of q_e^*.

2 Preliminaries

2.1 The Model

We are given a set of vertices V, and for every unordered pair of vertices $u, v \in V$, we have a (known) probability p_{uv} of the edge (u, v) being present. These probabilities are independent over the edges. Let $\mathcal{D}$ denote this distribution over all graphs defined by p. Let $G(V, E)$ be a graph drawn from $\mathcal{D}$. We are given only the vertex set V of G, but the edge-set E is not revealed to us unless we *scan* an unordered pair of vertices. A pair $(u, v) \in V \times V$ may be scanned to check if they are adjacent and if so then they are matched and removed from the graph. The objective is to maximize the expected number of vertices that get matched.

We compare our performance to the optimal off-line algorithm that knows the edges before hand, and reports the maximum matching in the underlying graph. We say an online algorithm $\mathcal{A}$ attains a competitive ratio of γ for the SMCP if, for every problem instance $\mathcal{I} = (G(V, .), p)$, the expected size of the matching returned by $\mathcal{A}$ is at least γ times the expected size of the optimal matching in the Erdős Rényi graph $G(V, p)$. That is, $\gamma = \min\limits_{\mathcal{I}=(G(V,.),p)} \left\{ \dfrac{E\left[\mathcal{A}(\mathcal{I})\right]}{E\left[\text{max matching in } G(V,p)\right]} \right\}.$

2.2 Definitions

For any graph H drawn from $\mathcal{D}$, let $M(H)$ be an arbitrarily chosen maximum matching on H. We define $q^*_{uv} = \Pr\limits_{H \leftarrow \mathcal{D}}\left(u \sim v \text{ in } M(H)\right).$

Clearly $q^*_{uv} \leq p_{uv}$, since an edge cannot be part of a maximal matching unless it is actually in the graph. In general, the ratio q^*_{uv}/p_{uv} can be thought of as the conditional probability that an edge is in the matching, given that it appears in the graph. For a given vertex u, the probability that u is matched in M is exactly $Q_u(G) := \sum_v q^*_{uv}$.

This of course is at most 1. We will compare the performance of our algorithm against the expected size of a maximum matching (denoted by OPT) for a graph drawn from $\mathcal{D}$. Thus we have, $\mathbf{E}[|\text{OPT}|] = \mathbf{E}\left[|M(H)|\right] = \frac{1}{2}\sum_u Q_u(G) = \sum_{(u,v)} q^*_{uv}$, where the last sum is taken over unordered pairs. Finally define an unordered pair (u, v) to be a *candidate edge* if both u and v are still unmatched and (u, v) is yet to be scanned. At any stage let $F(G) \subseteq V \times V$ be the set of candidate edges, and for any $u \in V$, let $N(u, G) = \{v \mid uv \in F(G)\}$. A vertex u is defined to be *alive* if $|N(u, G)| > 0$.

2.3 Sampling Technique

In this section we will describe a sampling technique that will be an important component of our algorithm. A curious reader may directly read Corollary 1 and proceed to Section 3 to see an application of this technique. Frequently over the course of our algorithm we will encounter the following framework: We have a vertex v, whose incident edges have known probabilities p_{uv} of being connected to v. We would like to choose an ordering on the incident edges to probe accordingly so that each edge is included(scanned and found to be present) with some target probability of at least r_{uv} (which may depend on u).

Clearly there are some restrictions on the r_{uv} in order for this to be feasible; for example the situation is clearly hopeless if $r_{uv} > p_{uv}$. More generally, for each subset S of the neighborhood of v, it must be the case that the sum of the target probabilities of vertices in S (the desired probability of choosing some member of S) is at most the probability that at least one vertex of S is adjacent to v. As it turns out, these are the *only* necessary restrictions.

Lemma 1. *Let $A_1, A_2, \ldots A_k$ be independent events having probabilities $p_1, \ldots, p_k$. Let $r_1, \ldots, r_k$ be fixed non-negative constants such that for every $S \subseteq \{1, \ldots, k\}$ we have*

$$\sum_{i \in S} r_i \leq 1 - \prod_{i \in S}(1 - p_i). \tag{1}$$

Then there is a probability distribution over permutations π of $\{1, 2, \ldots, k\}$ such that for each i, we have

$$\mathbf{P}(A_i \text{ is the earliest occurring event in } \pi) \geq r_i. \tag{2}$$

Proof. By the Theorem of the Alternative from Linear Duality [10], it suffices to show that the following system of $n! + 1$ inequalities in $n + 1$ variables $\{x_1, \ldots, x_n, y\}$ does not have a non-negative solution:

$$y - \sum_k x_k r_k < 0 \tag{3}$$

$$\forall \pi \in S_n, \quad y - \sum_k x_k p_k \prod_{\substack{j < k \\ \text{in} \pi}}(1 - p_j) \geq 0 \tag{4}$$

Assume such a solution exists. Without loss of generality we may assume $x_1 \geq x_2 \geq \cdots \geq x_n \geq 0$. Combining the first inequality with the inequality from the identity permutation, we have

$$\sum_{i=1}^{n} x_i p_i \prod_{j=1}^{i-1}(1 - p_j) < \sum_{k=1}^{n} x_k r_k. \tag{5}$$

On the other hand, we have for each k by applying (1) to $S = \{1, 2, \ldots k\}$ that $\sum_{i=1}^{k} r_i \leq 1 - \prod_{j=1}^{k}(1 - p_j)$. By weighting each of these equations by $(x_i - x_{i+1})$ and treating $x_{n+1} = 0$ (note that each of these weights are nonnegative by assumption) and adding, we obtain

$$\sum_{k=1}^{n} x_k r_k \leq \sum_{k=1}^{n}(x_k - x_{k+1})[1 - \prod_{j=1}^{k}(1 - p_j)]. \tag{6}$$

It can be checked directly that both the left side of (5) and the right hand side of (6) are equal to

$$\sum_{\substack{S \subseteq \{1, 2, \ldots n\} \\ S \neq \emptyset}} (-1)^{|S|-1} x_{\max(S)} \prod_{i \in S} p_i,$$

implying that the two equations contradict each other. Therefore no such solution to the dual system can exist, so the original system must have been feasible.

In theory, it is possible to find the desired distribution π using linear programming. However, it turns out there is a faster constructive combinatorial algorithm details of which appear in the full version of this paper.

Lemma 2. *A probability distribution π on permutations solving the program (2) can be constructively found in time $O(n^2)$.*

Corollary 1. *Given a graph $G(V,E)$ and $u \in V$, such that $q_{uv}^*/p_{uv} < \alpha < 1$ for every $v \in N(u, G)$, there exists a randomized algorithm for scanning the edges in $\{uv \mid v \in N(u,G)\}$ such any edge uv, $v \in N(u,G)$, is included in the matching with probability at least $\delta(u, G)q_{uv}^*$, where*

$$\delta(u, G) = \frac{1 - \exp(-\sum_{v \in N(u,G)} q_{uv}^*/\alpha)}{\sum_{v \in N(u,G)} q_{uv}^*}$$

Proof. Note that for any $u \in V$, and $S \subseteq N(u,G)$, $1 - \prod_{v \in S}(1 - p_{uv}) \geq \sum_{v \in S} q_{uv}^*$, since the right side represents the probability u is matched to S in our chosen maximal matching and the left side the probability that there is at least one edge connecting u to S. Thus (p, q^*) satisfy the condition for Lemma 1. However, we can do better. For any given S, if we scale each q_e by $\left(1 - \prod_{v \in S}(1 - p_{uv})\right)/\sum_{v \in S} q_{uv}^*$, the above condition still remains satisfied for that S. Since $q_e^*/p_e < \alpha$ we have

$$\frac{1 - \prod_{v \in S}(1 - p_{uv})}{\sum_{v \in S} q_{uv}^*} \geq \frac{1 - \exp(-\sum_{v \in S} p_{uv})}{\sum_{v \in S} q_{uv}^*} \geq \frac{1 - \exp(-\sum_{v \in S} q_{uv}^*/\alpha)}{\sum_{v \in S} q_{uv}^*}$$

$$\tag{7a}$$

$$\geq \frac{1 - \exp(-\sum_{v \in N(u,G)} q_{uv}^*/\alpha)}{\sum_{v \in N(u,G)} q_{uv}^*} = \delta(u, G), \tag{7b}$$

and (7b) follows since $1 - \exp(-\sum_{v \in S} q_{uv}^*/\alpha)/\sum_{v \in S} q_{uv}^*$ is a decreasing function in $\sum_{v \in S} q_{uv}^*$, thus achieving its minimum value at $S = N(u, G)$. Therefore we can replace our q^* by $\delta(u, G)q^*$ and still have the conditions of Lemma 1 hold.

3   Matching Algorithm on Unweighted Erdős-Rényi graphs

Our algorithm can be divided into two stages. The first stage involves several iterations each consisting of two steps - Estimation and Pruning. The parameters α and C will be determined in Section 4.

- **Step 1 (Estimation):** Generate samples $H_1, H_2, \ldots H_C$ of the Erdős-Rényi graph by sampling from $\mathcal{D}$. For each sample, generate the corresponding maximum matching $M(H_j)$. For every prospective edge (u, v), let q_{uv} be the proportion of samples in which the edge (u, v) is contained in $M(H_j)$.

- **Step 2 (Pruning):** Let (u,v) be an edge having maximum (finite) ratio q_{uv}/p_{uv}. If this ratio is less than α, end Stage 1. Otherwise, scan (u,v). If (u,v) is present, add it to the partial matching; remove u and v from V, and return to Step 1; otherwise set p_{uv} to 0 and return to Step 1.

We recompute q_{uv} every time we scan an edge. Stage 1 ends when the maximum (finite) value of q_e/p_e falls below α. Note that at this point we stop recomputing q, and these values of q will remain the same for each pair of vertices for the remainder of the algorithm. We now describe the second stage of the algorithm.

The second stage also has several iterations each consisting of two steps. At the start of this stage define $X = V$. The algorithm terminates when X becomes empty.

- **Step 1 (Random Bipartition):** Randomly partition X into two equal sized sets L and R and let B be the *bipartite graph* induced by L and R.
- **Step 2 (Sample and Match):** Iterate through the vertices in L in an arbitrary order, and for each vertex $u \in L$ sample a neighbor in $N(u, B)$ by choosing a vertex in R using the sampling technique described in Corollary 1[1]. At the end redefine X to be the set of alive vertices in R and discard the unmatched vertices in L. Recall that a vertex was defined to be alive if it is still unmatched and it has at least one candidate edge incident on it.

4 Analysis

In this section we will analyze the competitive ratio for the algorithm described earlier. We begin by analyzing Stage 1 of the algorithm. For each iteration in Stage 1, define the residual graph at the start of the ith iteration to be G_i starting with $G_1 = G$. We denote by $q_{e,i}^*$ the actual probability that e is contained in the maximal matching on G_i and $q_{e,i}$ as our estimate calculated in Step 1. We define $\epsilon_e := \max_i |q_{e,i} - q_{e,i}^*|$

Let the total number of iterations in this stage be k and let $G' = G_k$. Let ALG_1 be the set of edges that are matched in Stage 1 and let $OPT(G_i)$ be the optimal solution in the residual graph at the start of the ith iteration.

Lemma 3. $\mathbf{E}[|OPT| - |OPT(G')|] \leq (2 - \alpha)\mathbf{E}[|ALG_1|] + \sum_e \epsilon_e$

Proof. For $i \in [k]$, let $Gain(i)$ be 1 if the edge scanned in the ith iteration is present, and 0 otherwise. We will first show that $E[|OPT(G_i)| - |OPT(G_{i+1})|] \leq (2 - \alpha)\mathbf{E}[Gain(i)]$. Three cases may arise during the ith iteration.

- **Case 1:** The edge scanned in the ith iteration is not present. Then $OPT(G_i) = OPT(G_{i+1})$ and $Gain(i) = 0$ thus, $|OPT(G_i)| - |OPT(G_{i+1})| = Gain(i) = 0$.

---

[1] The algorithm described in Corollary 1 requires the exact estimates for q_e^*. However we will show in our analysis that for large enough samples C, q_e defined above is a good estimate of q_e^*.

- **Case 2.1:** The edge scanned in the ith iteration is present but does not belong to $OPT(G_{i+1})$. This happens with probability $p_e - q^*_{e,i}$. Then $|OPT(G_i)| - |OPT(G_{i+1})| = 2$ and $Gain(i) = 1$.
- **Case 2.2:** The edge scanned in the ith iteration is present and belongs to $OPT(G_i)$. This happens with probability $q^*_{e,i}$. Then $|OPT(G_i)| - |OPT(G_{i+1})| = 1$ and $Gain(i) = 1$.

Summing over all three cases, we see that

$$\mathbf{E}[|OPT(G_i)| - |OPT(G_{i+1})|] = 2(p_e - q^*_{e,i}) + q^*_{e,i} \leq p_e(2 - \alpha) + \epsilon_e \ ,$$

while the expected gain from scanning the edge is simply p_e. The result follows from adding over all scanned edges, and noting for the additive factor that each edge is scanned at most once in the first stage (and indeed in the whole algorithm).

Analysis of Stage 2: Let us begin by analyzing the first iteration of the second stage of the algorithm. The analysis for the subsequent iterations would follow along similar lines. Let G' be the residual graph at the start of the second stage, where $q_e/p_e < \alpha$ for every candidate edge e, and $1/2 \sum_u Q_u(G') = \mathbf{E}[|OPT(G')|]$. The following lemma bounds the performance of the first iteration of Stage 2 on G'. We defer the proof to the full version of the paper.

Lemma 4. *The expected number of edges that are matched in the first iteration of Stage 2 of the algorithm is at least* $\left(1 - \frac{1}{e}\right)\left(1 - e^{-1/2\alpha}\right)|OPT(G')| - \sum_e \epsilon_e$.

For ease of notation, let $\phi = (1 - 1/e)\left(1 - e^{-1/2\alpha}\right)$. Let ALG_2 be the set of edges that get matched in Stage 2 of the algorithm. Next we lower bound $\mathbf{E}[|ALG_2|]$.

Lemma 5. $\mathbf{E}[|ALG_2|] \geq \mathbf{E}[|OPT(G')|]\left[\frac{\phi}{1 - (\frac{1-\phi}{2})^2}\right] - \sum_e \epsilon_e$

Proof. Observe that not all candidate edges in G' have been considered during the first iteration of Stage 2. In particular, candidate edges with both end points in R are yet to be considered. For analyzing the subsequent iterations in Stage 2, we will consider only these candidate edges. Clearly this only lower bounds the performance of the algorithm.

We can infact prove something slightly stronger than Lemma 4 (refer to the full version for details), i.e., we can show that every vertex $v \in R$ is chosen with probability at least $\phi Q_v(G')$. By slightly altering the algorithm it is easy to ensure that for every $v \in R$, it is chosen with exactly this probability. Thus any vertex in R survives the first iteration with probability $1 - \phi Q_v(G') > 1 - \phi$. Since the partitions L and R are chosen at random, the probability that a vertex is in R and unmatched after the first iteration is at least $\mu = (1 - \phi)/2$. Continuing this argument further, the probability that an ordered pair (u, v) is a candidate edge at the start of the ith iteration is the probability that both u and v have always been in R in all previous iterations, and are still unmatched; this probability is at least $\mu^{2(i-1)}$.

Let G_i' be the residual graph at the start of the ith iteration in Stage 2, with $G_1' = G'$. By the above observation and using linearity of expectation, the expected sum of q_e's on candidate edges in G_i' is lower bounded by $\mu^{2i-2} \sum_{e \in F(G')} q_e$. Observe that the expected size of the matching returned by the ith iteration in the second stage is at least $\phi \mu^{2i-2} \sum_{e \in F(G')} q_e$. Summing over all iterations in Stage 2 we have,

$$\mathbf{E}[|ALG_2|] \geq \sum_i \phi \mu^{2i-2} \sum_{e \in F(G')} q_e \geq \phi \sum_{e \in F(G')} q_e \sum_{i=1} \mu^{2i-2} \tag{8a}$$

$$= \phi \sum_{e \in F(G')} q_e \frac{1}{1 - \mu^2} \quad = \quad \frac{1}{2} \sum_{u \in G'} Q_u(G') \frac{\phi}{1 - \mu^2} \tag{8b}$$

$$= |OPT(G')| \frac{\phi}{1 - \mu^2} \quad = \quad |OPT(G')| \frac{\phi}{1 - \left(\frac{1-\phi}{2}\right)^2} . \tag{8c}$$

Now all that is left is to balance the factors for both the stages and set the optimal value of α. In the subsequent theorem we find the optimal value of α.

Theorem 3. *The above algorithm attains a factor of at least* $0.573 - 2\gamma$ *where* $\sum_e \epsilon_e \leq \gamma \mathbf{E}(|OPT|)$.

Proof. By Lemma 3, $\mathbf{E}[|OPT| - |OPT(G')|] \leq 2/(1+\alpha)\mathbf{E}[|ALG_1|] + \sum \epsilon_e$. Also by Lemma 5, $\mathbf{E}[|OPT(G')|] \leq \frac{1 - (\frac{1-\phi}{2})^2}{\phi} \mathbf{E}[|ALG_2|] + \sum \epsilon_e$. Combining these two and substituting $\alpha = 0.255$ and $\phi = (1 - 1/e)(1 - e^{-1/2\alpha}) = 0.543$ we have,

$$\mathbf{E}[|OPT|] \leq (2 - \alpha)\mathbf{E}[|ALG_1|] + \frac{1 - (\frac{1-\phi}{2})^2}{\phi} \mathbf{E}[|ALG_2|] + 2 \sum_e \epsilon_e \tag{9a}$$

$$= 1.74(\mathbf{E}[|ALG_1|] + \mathbf{E}[|ALG_2|]) + 2 \sum_e \epsilon_e \tag{9b}$$

$$= 1.74 \mathbf{E}[|ALG|] + 2 \sum_e \epsilon_e \tag{9c}$$

Thus $\mathbf{E}[|ALG|] \geq 0.573 \, \mathbf{E}[|OPT|]$.

The above algorithm can be implemented in $\tilde{O}(n^4)$ time. Also, if the optimal matching is a non-negligible fraction of the vertices the running time can be reduced to $\tilde{O}(n^3)$. We defer the analysis to the full version of the paper.

On the hardness front, we prove the following theorem.

Theorem 4. *No randomized algorithm can attain a competitive ratio better than* 0.898 *for the SMCP.*

Proof. The proof of this theorem relies on a full analysis of the performance of the optimal online theorem on the Erdos-Renyi graph $G(4, p)$ with probability $0 < p < 1$ (identical across all edges) to be chosen later. We provide a brief overview of the proof in this extended abstract.

The performance of an algorithm on this graph is $(P_1 + P_2)/(Q_1 + Q_2)$, where P_i denotes the probability the algorithm finds a matching with at least i edges, and Q_i denotes the probability a matching with at least i edges is present.

We know $P_1 = Q_1 = 1 - (1 - p)^6$, the probability $G(p)$ contains an edge. For P_2 and Q_2, we think of the complete graph K_4 as being divided into 3 pairs of opposite edges $(a_1, a_1'), (a_2, a_2'), (a_3, a_3')$. Call an edge "Type 1" if it is the first edge of its pair scanned. An algorithm only finds a matching of size 2 if some type-1 edge is present, and the mate of the *first* type-1 edge found is present. This occurs with probability at most $(1 - (1 - p)^3)p$, an upper bound on P_2. Conversely Q_2 is the probability both edges from *any* pair are present, $1 - (1 - p^2)^3$. Plugging these values into our ratio and optimizing over p, we see we can take $p = 0.638$ to bound the factor by 0.8972.

References

1. Adamczyk, M.: Improved analysis of the greedy algorithm for stochastic matching. Inf. Process. Lett. 111(15), 731–737 (2011)
2. Aronson, J., Dyer, M., Frieze, A., Suen, S.: Randomized greedy matching. ii. Random Struct. Algorithms 6, 55–73 (1995)
3. Aronson, J., Frieze, A., Pittel, B.G.: Maximum matchings in sparse random graphs: Karp-sipser revisited. Random Struct. Algorithms 12, 111–177 (1998)
4. Bansal, N., Gupta, A., Li, J., Mestre, J., Nagarajan, V., Rudra, A.: When LP Is the Cure for Your Matching Woes: Improved Bounds for Stochastic Matchings. In: de Berg, M., Meyer, U. (eds.) ESA 2010, Part II. LNCS, vol. 6347, pp. 218–229. Springer, Heidelberg (2010)
5. Birnbaum, B., Mathieu, C.: On-line bipartite matching made simple. SIGACT News 39, 80–87 (2008)
6. Chebolu, P., Frieze, A., Melsted, P.: Finding a maximum matching in a sparse random graph in o(n) expected time. J. ACM 57, 24:1–24:27 (2010)
7. Chen, N., Immorlica, N., Karlin, A.R., Mahdian, M., Rudra, A.: Approximating Matches Made in Heaven. In: Albers, S., Marchetti-Spaccamela, A., Matias, Y., Nikoletseas, S., Thomas, W. (eds.) ICALP 2009, Part I. LNCS, vol. 5555, pp. 266–278. Springer, Heidelberg (2009)
8. Dean, B.C., Goemans, M.X., Vondrák, J.: Adaptivity and approximation for stochastic packing problems. In: Proceedings of the Sixteenth Annual ACM-SIAM Symposium on Discrete Algorithms, SODA 2005, pp. 395–404. Society for Industrial and Applied Mathematics, Philadelphia (2005)
9. Edmonds, J.: Paths, trees, and flowers. Canadian Journal of Mathematics 17, 449–467 (1965)
10. Farkas, J.G.: Uber die theorie der einfachen ungleichungen. Journal fur die Reine und Angewandte Mathematik 124, 1–27 (1902)
11. Feldman, J., Mehta, A., Mirrokni, V.S., Muthukrishnan, S.: Online stochastic matching: Beating 1-1/e. In: FOCS, pp. 117–126 (2009)
12. Frieze, A., Pittel, B.: Perfect matchings in random graphs with prescribed minimal degree. In: Proceedings of the Fourteenth Annual ACM-SIAM Symposium on Discrete Algorithms, SODA 2003, pp. 148–157. Society for Industrial and Applied Mathematics, Philadelphia (2003)
13. Goel, G., Mehta, A.: Online budgeted matching in random input models with applications to adwords. In: SODA, pp. 982–991 (2008)

14. Karande, C., Mehta, A., Tripathi, P.: Online bipartite matching in the unknown distributional model. In: STOC, pp. 106–117 (2011)
15. Karp, R.M., Vazirani, U.V., Vazirani, V.V.: An optimal algorithm for online bipartite matching. In: Proceedings of the 22nd Annual ACM Symposium on Theory of Computing (1990)
16. Mahdian, M., Yan, Q.: Online bipartite matching with random arrivals: An approach based on strongly factor-revealing lps. In: STOC, pp. 117–126 (2011)
17. Manshadi, V.H., Oveis-Gharan, S., Saberi, A.: Online stochastic matching: Online actions based on offline statistics. In: SODA (2011)
18. Mehta, A., Mirrokni, V.: Online ad serving: Theory and practice. Tutorial (2011)
19. Nikolova, E., Kelner, J.A., Brand, M., Mitzenmacher, M.: Stochastic Shortest Paths Via Quasi-convex Maximization. In: Azar, Y., Erlebach, T. (eds.) ESA 2006. LNCS, vol. 4168, pp. 552–563. Springer, Heidelberg (2006)
20. Papadimitriou, C.H., Yannakakis, M.: Shortest paths without a map. Theor. Comput. Sci. 84, 127–150 (1991)
21. Karp, R.M., Sipser, M.: Maximum matching in sparse random graphs. In: FOCS, pp. 364–375 (1981)
22. Ross, L.F., Rubin, D.T., Siegler, M., Josephson, M.A., Thistlethwaite, J.R., Woodle, E.S.: The case for a living emotionally related international kidney donor exchange registry. Transplantation Proceedings 18, 5–9 (1986)
23. Ross, L.F., Rubin, D.T., Siegler, M., Josephson, M.A., Thistlethwaite, J.R., Woodle, E.S.: Ethics of a paired-kidney-exchange program. The New England Journal of Medicine 336, 1752–1755 (1997)
24. Shmoys, D.B., Swamy, C.: An approximation scheme for stochastic linear programming and its application to stochastic integer programs. J. ACM 53, 978–1012 (2006)

Rademacher-Sketch: A Dimensionality-Reducing Embedding for Sum-Product Norms, with an Application to Earth-Mover Distance

Elad Verbin[1] and Qin Zhang[2]

[1] Aarhus University: MADALGO* and CTIC**
elad.verbin@gmail.com
[2] MADALGO, Aarhus University
qinzhang@cs.au.dk

Abstract. Consider a *sum-product* normed space, i.e. a space of the form $Y = \ell_1^n \otimes X$, where X is another normed space. Each element in Y consists of a length-n vector of elements in X, and the norm of an element in Y is the sum of the norms of its coordinates. In this paper we show a constant-distortion embedding from the normed space $\ell_1^n \otimes X$ into a lower-dimensional normed space $\ell_1^{n'} \otimes X$, where $n' \ll n$ is some value that depends on the properties of the normed space X (namely, on its *Rademacher dimension*). In particular, composing this embedding with another well-known embedding of Indyk [18], we get an $O(1/\epsilon)$-distortion embedding from the earth-mover metric EMD_Δ on the grid $[\Delta]^2$ to $\ell_1^{\Delta^{O(\epsilon)}} \otimes \mathrm{EEMD}_{\Delta^\epsilon}$ (where EEMD is a norm that generalizes earth-mover distance). This embedding is stronger (and simpler) than the sketching algorithm of Andoni et al [4], which maps EMD_Δ with $O(1/\epsilon)$ approximation into sketches of size $\Delta^{O(\epsilon)}$.

1 Introduction

Sum-product norms. A *normed space* $(X, \|\cdot\|_X)$ consists of a linear space X and a norm $\|\cdot\|_X$ (i.e. a positive function from X to the reals, which satisfies the triangle inequality and where for $c \in \mathbb{R}, x \in X$ it holds that $\|c \cdot x\|_X = |c| \cdot \|x\|_X$). A *sum-product normed space* is a normed space of the form $Y = \ell_1^n \otimes X$, where X is another normed space. Each element y in Y consists of a length-n vector $y = (x_1, \ldots, x_n)$ of elements in X, and the norm of y is the sum of the norms of its coordinates, namely, $\|y\|_Y = \sum_{i=1}^n \|x_i\|_X$. Sum-product normed spaces have arisen in the literature on streaming and sketching algorithms. In particular, in 2009 Andoni et. al. [6] used product normed space to overcome the ℓ_1 non-embeddability barrier for the Ulam metric. A year later Andoni and Nguyen [8] used sum-product of Ulam metrics to obtain faster

---

* MADALGO is the Center for Massive Data Algorithmics, a center of the Danish National Research Foundation.
** The first author acknowledges support from the Danish National Research Foundation and The National Science Foundation of China (under the grant 61061130540) for the Sino-Danish Center for the Theory of Interactive Computation, within which part of this work was performed.

A. Czumaj et al. (Eds.): ICALP 2012, Part I, LNCS 7391, pp. 834–845, 2012.

approximation algorithms for computing the Ulam distance between two non-repetitive strings. Sum-product metrics have also been used by Andoni and Onak [9] to compute the edit distance between two strings in near-linear time.

Given two normed spaces Y and Y', an *embedding* (also called *strong embedding*) of Y into Y' is a function $\phi : Y \to Y'$. The *distortion* of ϕ is analogous to the "approximation ratio" achieved by Y' as an approximation of Y. Specifically, the distortion of ϕ is the value

$$\max_{y \in Y} (\|y\|_Y / \|\phi(y)\|_{Y'}) \cdot \max_{y \in Y} (\|\phi(y)\|_{Y'} / \|y\|_Y) .$$

Efficiently-computable embeddings with small distortion have been of much recent interest in theoretical computer science and various branches of mathematics, see, e.g., [25,19]. In particular, if there is an efficient algorithm for computing the norm in the space Y', then the norm in Y can be computed by first applying the embedding and then performing the computation in Y'; the approximation factor of this algorithm is equal to the distortion. Similar approaches were used when designing sketches and data structures: rather than design a data structure for Y from scratch, simply embed Y in an efficient way into a normed space for which good data structures are already known. Applications of this approach are too numerous to cite, see e.g. Indyk's survey [16].

In this paper we show dimensionality-reducing embeddings in sum-product normed spaces: the goal is, given a normed space $Y = \ell_1^n \otimes X$, to find a small-distortion embedding of Y into a smaller-dimensional sum-product normed space $Y' = \ell_1^{n'} \otimes X$. Our embeddings are *generic*, in the sense that their general structures do not depend on the properties of X. This is the first such generic dimension-reduction work that we know of for sum-product spaces. Previous literature has considered dimension-reduction for particular spaces, such as ℓ_1^n or ℓ_2^n. For literature on dimension reduction in ℓ_1^n, see for example the paper of Andoni et al. [5] as well as the references therein. For dimension reduction in ℓ_2^n, consider the classical Johnson-Lindenstrauss lemma [21].

1.1 Our Results

We first define a central concept in this paper: the *Rademacher dimension* of a normed space. As far as we know, this definition was never used or given in the literature; it is somewhat related to the property of being a Rademacher type p metric for $p > 0$ (see e.g. [26], also see the recent paper by Andoni et. al. [7]) but it is not the same.

Definition 1. *A normed space X has* Rademacher dimension α *if for any natural number s, and for any $x_0, x_1, \ldots, x_{s-1} \in X$ with $\|x_i\|_X \leq T$, we have with probability at least $1 - 1/\alpha^c$ (for some universal constant c) that*

$$\left\| \sum_{i \in [s]} \varepsilon_i x_i \right\|_X \leq \alpha \cdot \sqrt{s} \cdot T.$$

Here, $\varepsilon_0, \varepsilon_1, \ldots, \varepsilon_{s-1}$ are (± 1)-valued random variables such that $\mathbf{Pr}[\varepsilon_i = +1] = \mathbf{Pr}[\varepsilon_i = -1] = 1/2$ for all $i \in [s]$, and the probability is taken over the sample space defined by these variables. If there is no real number α which this holds, then we say that the Rademacher dimension of the space is ∞.

As an illustrative example, it is easy to see that the normed space $(\mathbb{R}^d, \|\cdot\|_1)$ has Rademacher dimension $O(d^2)$. The proof of this fact follows from Hoeffding's inequality, in a similar way as in Lemma 1 below.

For $x = \{x_0, x_1, \ldots, x_{n-1}\} \in \ell_1^n \otimes X$, we denote $\|x\|_{1,X} = \sum_{i \in [n]} \|x_i\|_X$. Our main theorem states that any sum-product normed space $\ell_1^n \otimes X$ can be weakly-embedded with distortion $O(1)$ into $\ell_1^{n'} \otimes X$, where $n' \approx \alpha$ is roughly the Rademacher dimension of X:

Theorem 1. *Let X be a normed space with Rademacher dimension α. Let $\lambda = \max\{\alpha, \log^3 n\}$. Then there exists a distribution over linear mappings $\mu : \ell_1^n \otimes X \to \ell_1^{\lambda^{O(1)}} \otimes X$, such that for any $x \in \ell_1^n \otimes X$ we have*

- $\|\mu(x)\|_{1,X} \geq \Omega(\|x\|_{1,X})$ *with probability* $1 - 1/\lambda^{\Omega(1)}$.
- $\|\mu(x)\|_{1,X} \leq O(\|x\|_{1,X})$ *with probability* 0.99.

Remarks:

1. The embedding is *linear*. This is an important property, since it allows efficient updating of the sketch given updates in a streaming way, as well as computing the associated distance function (the distance function associated with the normed space X is the function $d(x, y) = \|x - y\|_X$).

2. The above embedding is a *weak embedding*, in the sense that for each vector, the norm of its embedded representation is good with constant probability. Thus, for each particular instantiation of the random variable μ, we would expect a constant fraction of the vectors in the source space to embed to vectors that are too large in the target space. This is as opposed to a *strong embedding*, that would be good for all of the vectors simultaneously.

   In another dimension-reduction paper, Indyk [17] showed a weak dimension reduction in ℓ_1, which was sufficient for applications such as norm estimation in data streams and approximate nearest neighbor search. In general, weak embeddings seem applicable for most of the purposes where strong embeddings are used, and they might not encounter the same barriers as strong embeddings: our embedding is in fact a good example of this, as we explain next.

3. The last theorem states that $\ell_1^n \otimes X$ can be weakly-embedded into $\ell_1^{n'} \otimes X$ with constant distortion, where $n' \approx \alpha$. It is natural to ask whether there exists a strong embedding with similar properties. The answer is a resounding "no": Even in the special case when X is simply $\ell_1^1 = \mathbb{R}$, a result by Brinkman and Charikar [10] shows that an n point subset of ℓ_1 cannot be embedded into $\ell_1^{n^{\Omega(1/D^2)}}$ with distortion $o(D)$. Thus, if we require a strong embedding with constant distortion, the dimension can be reduced by no more than constant factors.

4. Also, it is interesting to note that Theorem 1 works when X is a *normed space*, but if X was a *metric space* (i.e. a space where we have a measure for the distance between any two points, but not necessarily a norm for each point) then it is not clear whether any similar result can be obtained. Our embeddings inherently rely on the properties of normed space: in particular, we need the ability to sum elements of the space, which is not available in metric spaces.

What happens when the underlying space X has bad, or even infinite, Rademacher dimension? We can still achieve dimensionality reduction, but this time the more we want to reduce the dimension, the larger the distortion will be. Specifically, to reduce from dimension n to dimension n^ϵ, the distortion will be $O(1/\epsilon)$:

Theorem 2. *For any normed space X and any $\lambda \geq \log^3 n$, there exists a distribution over linear mappings $\mu : \ell_1^n \otimes X \to \ell_1^{\lambda^{O(1)}} \otimes X$, such that for any $x \in \ell_1^n \otimes X$, we have*

- $\|\mu(x)\|_{1,X} \geq \Omega(\|x\|_{1,X})$ *with probability* $1 - 1/\lambda^{\Omega(1)}$.
- $\|\mu(x)\|_{1,X} \leq O(\log_\lambda n \cdot \|x\|_{1,X})$ *with probability* 0.99.

The results of this theorem are easier to achieve, and might be folklore in the field. Theorem 2 can be obtained from a similar embedding as we use for proving Theorem 1 and with a simpler proof, so for most of the paper we concentrate on proving Theorem 1, and where appropriate we discuss Theorem 2.

2   Earth-Mover Distance

2.1   Introduction to Earth-Mover Distance

Earth-Mover Distance. Denote $[n] = \{0, 1, \ldots, n - 1\}$. Given two multisets A, B in the grid $[\Delta]^2$ with $|A| = |B| = N$, the *earth-mover distance* is defined as the minimum cost of a perfect matching between points in A and B, where the cost of two matched points $a \in A$ and $b \in B$ is the ℓ_1 distance between them. Namely,

$$\mathrm{EMD}(A, B) = \min_{\pi : A \to B \text{ a bijection}} \sum_{a \in A} \|a - \pi(a)\|_1 .$$

Earth-Mover distance (EMD) is a natural metric that measures the difference between two images: If one, for example, thinks of pixels of a certain color laid out in two images, then the distance between the two images can be defined as the minimum amount of work to move one set of pixels to match the other. EMD has been extensively used in image retrieval and experiments show that it outperform many other similarity measures in various aspects [31,15,14,12,30].

Historically, EMD is a special case of the *Kantorovich metric*, which is proposed by L. V. Kantorovich in an article in 1942 [22]. This metric has numerous applications in probabilistic concurrency, image retrieval, data mining and bioinformatics. One can refer to [13] for a detailed survey. Other equivalent formulations of EMD including Transportation distance, Wasserstein distance and Mallows distance [24].

The general (non-planar) EMD can be solved by the classical *Hungarian method* [23] in time $O(N^3)$. However, this approach is too expensive to scale to perform retrievals from large databases. A number of (approximation) algorithms designed for the planar case are proposed in literature [32,1,33,11,20,2,18]. In particular, Indyk [18] proposed a constant approximation algorithm with running time $O(N \log^{O(1)} N)$, which is almost linear.

Recently, there has been an increasing interest in designing sketching algorithms for EMD. A good sketch can lead to space/time-efficient streaming algorithms and nearest neighbor algorithms [4]. Searching for good sketching algorithms for EMD is considered to be a major open problem in the data stream community [27].

For a multiset A in the grid $[\Delta]^2$, a sketching algorithm defines a mapping f that maps A into a host space (such as the space of short bit-strings $\{0,1\}^S$). The sketching algorithm must satisfy the property that for any two multisets A and B, the earth-mover distance $\mathsf{EMD}(A,B)$ can be approximately reconstructed from the two sketches $f(A)$ and $f(B)$. The sketching algorithm and the reconstruction algorithm should be *space-efficient* (and for practical considerations sometimes also time-efficient) in order to get efficient data structures and streaming algorithms. An embedding can thus be seen as a special type of sketch, where the reconstruction algorithm consists simply of computing the norm of the difference $f(A) - f(B)$ in the host space. Some embeddings of EMD into ℓ_1 space were proposed in [11,20]. However, in [28] the authors showed that it is impossible to embed EMD_Δ into ℓ_1 with distortion $o(\sqrt{\log \Delta})$. Therefore to get constant-approximation algorithms we need to investigate other, probably more sophisticated host spaces.

Recently, Andoni et. al. [4] obtained a sketch algorithm for planar EMD with $O(1/\epsilon)$ approximation ratio and Δ^ϵ space for any $0 \le \epsilon \le 1$. This is the first sublinear sketching algorithm for EMD achieving constant approximation ratio. Their sketching algorithm is not an embedding since their reconstruction algorithm involves operations such as binary decisions which are not metric operations. It remains an interesting open problem to embed EMD into simple normed spaces or products of simple normed spaces with constant distortion.

2.2   Applying our Results to Earth-Mover Distance

We now introduce the metric EEMD, which is an extension of EMD to any multisets $A, B \subseteq [\Delta]^2$ not necessary having the same size. It is defined as follows:

$$\mathsf{EEMD}(A, B) = \min_{S \subseteq A, S' \subseteq B, |S| = |S'|} [\mathsf{EMD}(S, S') + \Delta(|A - S| + |B - S'|)].$$

It is easy to see that when $|A| = |B|$, we have $\mathsf{EEMD}(A,B) = \mathsf{EMD}(A,B)$.

EEMD can be further extended to a *norm*: For a multiset $A \subseteq [\Delta]^2$, let $x(A) \in \mathbb{R}^{\Delta^2}$ be the characteristic vector of A. We next define the norm EEMD such that for any multiset $A, B \subseteq [\Delta]^2$, we have $\mathsf{EEMD}(A, B) = \|x(A) - x(B)\|_{\mathsf{EEMD}}$. The norm $\|\cdot\|_{\mathsf{EEMD}}$ is defined as follows: for each $x \in \mathbb{Z}^d$, let x^+ contain only the positive entries in x, that is, $x^+ = (|x| + x)/2$, and let $x^- = x - x^+$. And then we define $\|x\|_{\mathsf{EEMD}} = \mathsf{EEMD}(x^+, x^-)$. One can easily verify that this norm is well-defined. This definition can also be easily extended to $x \in \mathbb{R}^d$ by an appropriate weighting.

Let EEMD_Δ denote the normed space $(\mathbb{R}^{\Delta^2}, \|\cdot\|_{\mathsf{EEMD}})$.

Lemma 1. *EEMD$_\Delta$ has Rademacher dimension Δ^4.*

Proof. For any $x_0, \ldots, x_{s-1} \in \mathbb{R}^{\Delta^2}$ with $\|x_i\|_{\mathsf{EEMD}} \le T$ for all $i \in [s]$, let x_i^d ($d \in [\Delta^2]$) be the d-th coordinates of x_i. By the triangle inequality and the definition of

EEMD, $\left\|\sum_{i\in[s]}\varepsilon_i x_i\right\|_{\text{EEMD}} \leq \sum_{d\in[\Delta]^2}\left(\Delta\left|\sum_{i\in[s]}\varepsilon_i x_i^d\right|\right)$, where $\varepsilon_0,\ldots,\varepsilon_{s-1}$ are (± 1)-valued random variables. Thus we only need to bound $\left|\sum_{i\in[s]}\varepsilon_i x_i^d\right|$ for each $d\in[\Delta^2]$.

Fix a $d\in[\Delta^2]$. Since for each x_i ($i\in[s]$), we have that $\left|x_i^d\right|\leq T$. By Hoeffding's inequality we have that $\mathbf{Pr}\left[\left|\sum_{i\in[s]}\varepsilon_i x_i^d\right|\geq \Delta\sqrt{s}T\right] \leq 2e^{-\frac{2(\Delta\sqrt{s}T)^2}{s\cdot(2T)^2}} = e^{-\Omega(\Delta^2)}$. Therefore with probability at least $1 - \Delta^2\cdot e^{-\Omega(\Delta^2)} \geq 1 - 1/\Delta^{\Omega(1)}$, we have that $\left\|\sum_{i\in[s]}\varepsilon_i x_i\right\|_{\text{EEMD}} \leq \Delta^4\sqrt{s}T$.

The following fact is shown by Indyk [18].

Fact 1. *([18]) For any $\epsilon \in (0,1)$, there exists a distribution over linear mappings $F = \langle F_0,\ldots,F_{n-1}\rangle$ with $F_i : \text{EEMD}_\Delta \to \text{EEMD}_{\Delta^\epsilon}$ for all $i = 0,\ldots,n-1$, such that for any $x \in \text{EEMD}_\Delta$ we have*

- $\|x\|_{\text{EEMD}} \leq \sum_{i\in[n]}\|F_i(x)\|_{\text{EEMD}}$ *with probability 1.*
- $\sum_{i\in[n]}\|F_i(x)\|_{\text{EEMD}} \leq O(1/\epsilon)\|x\|_{\text{EEMD}}$ *with probability 0.95.*

Moreover, $n = \Delta^{O(1)}$.

Combining Theorem 1, Lemma 1 and Fact 1, we have the following.

Theorem 3. *For any $\epsilon \in \left[\frac{\log\log\Delta}{\log\Delta},1\right]$, there exists a distribution over linear mappings $\nu : \text{EEMD}_\Delta \to \ell_1^{\Delta^{O(\epsilon)}} \otimes \text{EEMD}_{\Delta^\epsilon}$, such that for any two $A, B \subseteq [\Delta]^2$ of equal size, we have*

- $\|\nu(x(A) - x(B))\|_{1,\text{EEMD}} \geq \Omega(\text{EMD}(A,B))$ *with probability $1 - 1/\Delta^{\Omega(\epsilon)}$.*
- $\|\nu(x(A) - x(B))\|_{1,\text{EEMD}} \leq O(1/\epsilon \cdot \text{EMD}(A,B))$ *with probability 0.9.*

The embedding given by this theorem can also serve as an alternative to the sketching algorithm of Andoni et al. [4]; it is simpler so its actual performance might be better. Furthermore, there might be additional advantages to having an embedding rather than a sketching algorithm (e.g., if there exists a good nearest neighbor data structure for $\ell_1^{\Delta^{O(\epsilon)}} \otimes \text{EEMD}_{\Delta^\epsilon}$, then we can use it to answer nearest neighbor queries for EMD_Δ).

3 The Embedding

In this section we construct the random linear mapping μ from Theorem 1. The random linear mapping for Theorem 2 is the same, and its analysis is simpler; we shall address the differences in Section 4.3.

Before giving the embedding, we first introduce a few definitions. Let $x = (x_0,\ldots,x_{n-1}) \in \ell_1^n \otimes X$ be the vector that we want to embed. The embedding will work in ℓ levels, where $\ell = \lceil\log_\lambda(4\lambda n)\rceil$. Note that $\ell \leq \lambda^{1/3}$ since $\lambda \geq \log^3 n$. At each level $k \in [\ell]$ we define a parameter $p_k = \lambda^{-k}$. For each level k we define a subsampled set, a hash function, and a series of (± 1)-valued random variables, all of them random and independent. The subsampled set is a set $I_k \subseteq [n]$ such that each $i \in [n]$ is placed in I_k

with probability p_k; the hash function is a random function $h_k : [n] \to [t]$ where $t = \lambda^5$; the (± 1)-valued variables are $\varepsilon_{k,1}, \ldots, \varepsilon_{k,n}$, each of them is $+1$ with probability $1/2$ and -1 with probability $1/2$. All the random choices are independent.

We denote $\chi[E] = 1$ if event E is true and $\chi[E] = 0$ if it is false.

The Embedding μ. For each level $k \in [\ell]$ and for each value $v \in [t]$ of the hash function h_k, compute

$$Z_k^v = \sum_{i \in [n]} \chi[i \in I_k] \cdot \chi[h_k(i) = v] \cdot \varepsilon_{k,i} \cdot x_i \cdot 1/p_k \,.$$

We see that the embedded vector $\mu(x) \in \ell_1^{t \cdot \ell} \otimes X$ consists of all the values Z_k^v, one after another. These are $t \cdot l = \lambda^{O(1)}$ cells (=coordinates), each of which contains an element from X.

Remarks: The use of ± 1 random variables, also known in this context as Rademacher random variables, is superficially similar to usage in the seminal paper of Alon et. al. [3]. However, these variables are used here for an entirely different purpose. In [3] and other related work, these variables are used to decrease the variance of a random variable that estimates the second frequency moment of a steam of items. In our algorithm they are used for a different purpose: roughly speaking, they are used to isolate a class of items with norms in a certain range from items with much smaller norms by making the variables with smaller norm cancel with one another.

For the purpose of proving Theorem 2, the ± 1 random variables are not needed, and it is enough to define $Z_k^v = \sum_{i \in [n]} \chi[i \in I_k] \cdot \chi[h_k(i) = v] \cdot x_i \cdot 1/p_k$.

Also note that to use the above embedding as a sketching algorithm, it is necessary to remember all the random choices we made. This amount of space is huge: much more than n. However, this is not actually necessary. A standard approach using pseudo-random generators allows to decrease the amount of random bits to $\lambda^{O(1)}$, thus giving the "correct" space complexity. These random bits can be generated by Nisan's pseudo-random generator [29]. See the similar discussions in [17,4].

4   Analysis

We first introduce a few more definitions. Let $M = \|x\|_{1,X}$. Let $T_j = M/\lambda^j$ ($j = 0, 1, \ldots$). Let $S_j = \{i \in [n] \mid \|x_i\|_X \in (T_j/\lambda, T_j]\}$ and let $s_j = |S_j|$. We say x_i is in class j if $i \in S_j$.

It is easy to see that we only need to consider classes up to $\ell - 1$ since elements from classes $j \geq \ell$ contribute at most $n \cdot M/\lambda^\ell \cdot \ell \leq M/4$ to all the levels. Therefore for simplicity we assume that all elements belong to classes $\{0, 1, \ldots, \ell - 1\}$.

Let $\beta = 1/100\ell$. We say class $j \in [\ell]$ is *important* if elements from S_j contribute significantly to the sum M, that is, $\sum_{i \in S_j} \|x_i\|_X \geq \beta M$. Thus for an important class j we have $s_j \geq \beta \cdot M/T_j = \beta\lambda^j$. Also note that $s_j \leq M/(T_j/\lambda) = \lambda^{j+1}$ for all $j \in [\ell]$ by definition. Therefore $s_j \in [\beta\lambda^j, \lambda^{j+1}]$ for each important class j. Let J denote the set of all important classes.

During the analysis when we say an event holds with high probability we mean that the probability is at least $1 - 1/\lambda^{\Omega(1)}$.

A Few More Notations. Before the analysis, we would like to introduce a few more notations to facilitate our exposition. For item class $j \in [\ell]$, sample level $k \in [\ell]$ and cell $v \in [t]$, we define the following random variables.

- Let $S_{j,k}$ be the set of elements in class j that are sampled at sample level k. That is, $S_{j,k} = S_j \cap I_k$. Let $s_{j,k} = |S_{j,k}|$.
- Let $S_{j,k}^v$ be the set of elements in class j that are sampled at level k and hashed to cell v. That is, $S_{j,k}^v = \{i \in S_{j,k} \mid h_k(i) = v\}$. Let $s_{j,k}^v = \left| S_{j,k}^v \right|$.
- For each class j, let $W(S_j) = \sum_{i \in S_j} \|x_i\|_X$. And for each class j and sample level k, let $W(S_{j,k}) = \sum_{i \in S_{j,k}} \|x_i\|_X \cdot 1/p_k$.
- For each class j, sample level k and cell v, let $Z_{j,k}^v = \sum_{i \in S_{j,k}^v} \varepsilon_{k,i} \cdot x_i \cdot 1/p_k$. Note that $\|Z_k^v\|_X \leq \sum_{j \in [\ell]} \left\| Z_{j,k}^v \right\|_X$ by the triangle inequality.
- For each class j and sample level k, let $C_{j,k} = \{v \mid \max\{i \mid i \in S_{j,k}^v\} = j\}$. We also say each cell $v \in C_{j,k}$ a j-dominated cell at level k.
- Let $W(C_{j,k}) = \sum_{v \in C_{j,k}} \|Z_k^v\|_X$. That is, the sum of X-norms of class j elements in those j-dominated cells at level k. Moreover, let $W(C_{j,k}, j') = \sum_{v \in C_{j,k}} \left\| Z_{j',k}^v \right\|_X$ and $W(C_{j,k}, \geq j') = \sum_{j'' \geq j'} W(j, k, j'')$. Note that by the triangle inequality we have $W(C_{j,k}, j) - W(C_{j,k}, \geq j+1) \leq W(C_{j,k}) \leq W(C_{j,k}, j) + W(C_{j,k}, \geq j+1)$.

We need the following tool (c.f. [4]).

Lemma 2. *(A variant of Hoeffding bound) Let $Y_0, Y_1, \ldots, Y_{n-1}$ be n independent random variables such that $Y_i \in [0, T]$ for some $T > 0$. Let $\mu = \mathbf{E}[\sum_i Y_i]$. Then for any $a > 0$, we have $Pr\left[\sum_{i \in [n]} Y_i > a\right] \leq e^{-(a-2\mu)/T}$.*

Now we prove Theorem 1. We accomplish it by two steps.

4.1 No Underestimation

In this section we show that $\|\mu(x)\|_{1,X} \geq \Omega(M)$ with probability $1 - 1/\lambda^{\Omega(1)}$. To show this we first prove the following lemma,

Lemma 3. *For each important class $j \in J$, we have $W(C_{j,j-1}) \geq W(S_j)/2$ for all $j \geq 1$ and $W(C_{0,0}) \geq W(S_0)/2$ for $j = 0$ with probability at least $1 - 1/\lambda^{\Omega(1)}$.*

The proof of the lemma is essentially that for each important class $j \in J$, at sample level $\max\{j - 1, 0\}$, the scaled contribution of elements in class j is close to $W(S_j)$ and the noise from other classes is small. The intuition of the later is simply that the range of each hash function is large enough such that:

1. There is no collision between elements in S_j and elements in $\bigcup_{j'>j} S_{j'}$ with high probability.
2. The noise from $\bigcup_{j'>j} S_{j'}$ is small since only a small fraction of elements from each $S_{j'}$ ($j' < j$) will collide with elements in S_j and the X-norm of each item in $S_{j'}$ ($j' < j$) is much smaller compared with those in S_j.

Proof. (of Lemma 3.) For notational convenience we set $k = j - 1$ in this proof. For each important class $j \geq 1$, we have $\mathbf{E}[|S_{j,k}|] = s_j p_k \in [\beta\lambda, \lambda^2]$. By Chernoff bound we have that with probability at least $1 - e^{-\Omega(\beta\lambda)}$, $\beta\lambda/2 \leq |S_{j,k}| \leq 2\lambda^2$. Similarly, we have that $|S_{j',k}| \leq 2\lambda$ for all class $j' \leq j-1$ with probability $1 - \ell \cdot e^{-\Omega(\lambda)}$. Conditioned on these, the probability that any two of $\bigcup_{j' \leq j} S_{j',k}$ hash into the same cell is at most $\binom{2\lambda^2 + \ell \cdot 2\lambda}{2}/t \leq O(1/\lambda)$. That is, with probability at least $1 - O(1/\lambda) - \ell \cdot e^{-\Omega(\lambda)} = 1 - 1/\lambda^{\Omega(1)}$, there is no collision between elements in class $0, 1, \ldots, j$ at sample level k. In particular, we have $|C_{j,k}| = |S_{j,k}| \in [\beta\lambda/2, 2\lambda^2]$ and $W(C_{j,k}, j) = W(S_{j,k})$ for each class $j \geq 1$ with high probability.

With similar arguments we can show that $W(C_{0,0}, 0) = W(S_{0,0})$ with high probability if $0 \in J$.

Now we show that $W(S_{j,k}) \geq \frac{2}{3}W(S_j)$ with high probability for each important class $j \geq 1$. We define for each $i \in S_j$ a random variable $Y_i = \chi[i \in S_{j,k}] \cdot \|x_i\|_X /T_j$. Note that $0 \leq Y_i \leq 1$ for all $i \in S_j$ and $W(S_{j,k}) = (\sum_{i \in S_j} Y_i) \cdot T_j \cdot 1/p_k$. We also have that

$$\mu = \mathbf{E}[\textstyle\sum_{i \in S_j} Y_i] = W(S_j)p_k/T_j = \Omega(\beta M/T_j \cdot \lambda^{-(j-1)}) = \Omega(\beta\lambda^j \cdot \lambda^{-(j-1)}) = \Omega(\beta\lambda).$$

By Chernoff bound we have $\mathbf{Pr}\left[\left|\sum_{i \in S_j} Y_i - \mu\right| \geq \mu/3\right] \leq 2e^{-\Omega(\mu)} \leq e^{-\Omega(\beta\lambda)}$.

Therefore with probability at least $1 - \ell \cdot e^{-\Omega(\beta\lambda)} \geq 1 - 1/\lambda^{\Omega(1)}$, we have $W(S_{j,k}) \geq \frac{2}{3}W(S_j)$ for all important classes $j \geq 1$. Consequently, we have $W(C_{j,k}, j) \geq \frac{2}{3}W(S_j)$ for all important $j \geq 1$ with high probability.

Moreover, note that $W(S_{0,0}) = W(S_0)$ is trivial since we pick each item at sample level 0. Therefore it also holds that with high probability, $W(C_{0,0}, 0) = W(S_{0,0}) = W(S_0)$ if $0 \in J$.

Next we bound $W(C_{j,k}, \geq j+1)$ for each important class $j \geq 1$ and $W(C_{0,0}, \geq 1)$ if $0 \in J$. We first bound the former. For each class $j' \geq j+1$, let $w = |C_{j,k}|$, we have $\mathbf{E}\left[\sum_{i \in S_{j',k}} \chi[h_k(i) \in C_{j,k}]\right] = s_{j'} p_k \cdot w/t$. By Lemma 2 we have that $\mathbf{Pr}\left[\sum_{i \in S_{j',k}} \chi[h_k(i) \in C_{j,k}] \geq 2 \cdot s_{j'} p_k \cdot w/t + \lambda\right] \leq e^{-\Omega(\lambda)}$. Summing up for all classes $j' \geq j+1$, with probability at least $1 - \ell \cdot e^{-\Omega(\lambda)}$, we have

$$\begin{aligned}
W(C_{j,k}, \geq j+1) &\leq \textstyle\sum_{j' \geq j+1} \sum_{i \in S_{j',k}} \chi[h_k(i) \in C_{j,k}] \cdot T_{j'} \cdot 1/p_k \\
&\leq \textstyle\sum_{j' \geq j+1} (2 \cdot s_{j'} p_k \cdot w/t + \lambda) \cdot T_{j'} \cdot 1/p_k \\
&\leq \ell \cdot (2 \cdot M \cdot w/\lambda^5 + T_j \cdot 1/p_k) \\
&\leq o(\beta M) \leq o(W(S_j))
\end{aligned}$$

The second to the last inequality holds since $w \leq 2\lambda^2$ with probability at least $1 - e^{-\Omega(\lambda)}$. Therefore by a union bound over $j \in [\ell]$ we have that with probability at least $1 - \ell^2 \cdot e^{-\Omega(\lambda)} \geq 1 - 1/\lambda^{\Omega(1)}$, $W(C_{j,k}, \geq j+1) = o(W(S_j))$ for all important $j \geq 1$.

Similarly, we can show that $W(C_{0,0}, \geq 1) = o(W(S_0))$ with high probability if $0 \in J$.

Finally, for each important class $j \geq 1$, we have $W(C_{j,k}) \geq W(C_{j,k}, j) - W(C_{j,k}, \geq j+1)$. Therefore $W(C_{j,k}) \geq \frac{2}{3}W(S_j) - o(W(S_j)) \geq W(S_j)/2$ for all important $j \geq 1$ with high probability. Similarly we can show that $W(C_{0,0}) \geq W(S_0)/2$ with high probability if $0 \in J$.

Note that Lemma 3 immediately gives the following. With probability at least $1 - 1/\lambda^{\Omega(1)}$, we have

$$
\begin{aligned}
\|\mu(x)\|_{1,X} &\geq \sum_{j \in J: j \geq 1} W(C_{j,j-1}) + \chi[0 \in J] \cdot W(C_{0,0}) \\
&\geq \sum_{j \in J} W(S_j)/2 = \left(M - \sum_{j \notin J} W(S_j) \right)/2 \\
&\geq (M - \ell \cdot \beta M)/2 \geq \Omega(M).
\end{aligned}
$$

4.2 No Overestimation

In this section we show that $\|\mu(x)\|_{1,X} \leq O(M)$ with probability 0.99. The general idea is the following:

1. Elements from S_j contribute little to all levels $k < j - 9$. This is because in each of such levels k and each cell $v \in [t]$ at that level, many elements from S_j are sampled and hashed into v and they will cancel with each other heavily due to the random variables $\varepsilon_{k,i} \in \{+1, -1\}$ multiplied.
2. On the other hand, elements from S_j will not be sampled at all levels $k > j + 1$ with high probability according to the sample ratios we choose at each level.

Now we prove the second part of Theorem 1. By the triangle inequality and the fact that $\sum_{v \in [t]} \left\| Z_{j,k}^v \right\|_X \leq W(S_{j,k})$ we have

$$
\begin{aligned}
\|\mu(x)\|_{1,X} &\leq \sum_{j \in [\ell]} \sum_{k \in [\ell]} \sum_{v \in [t]} \left\| Z_{j,k}^v \right\|_X \\
&\leq \sum_{j \in [\ell]} \sum_{k=j+2}^{\ell-1} W(S_{j,k}) + \sum_{j \in [\ell]} \sum_{k=j-9}^{j+1} W(S_{j,k}) + \sum_{j \in [\ell]} \sum_{k=0}^{j-10} \sum_{v \in [t]} \left\| Z_{j,k}^v \right\|_X.
\end{aligned}
$$

We bound the three terms of (1) separately. For the first term, the probability that there exists an element in class j that is sampled at level higher than $j + 2$ is at most $\ell \cdot s_j p_{j+2} \leq O(\ell/\lambda)$. Union bound over all class $j \in [\ell]$ we have that with probability at least $1 - O(\ell^2/\lambda) \geq 1 - 1/\lambda^{\Omega(1)}$, the first term is 0.

For the second term, since $\mathbf{E}[W(S_{j,k})] = W(S_j)$ for each $k \in [\ell]$, by the linearity of expectation and Markov inequality we obtain that with probability at least 0.991,

$$
\sum_{j \in [\ell]} \sum_{k=j-9}^{j+1} W(S_{j,k}) \leq 2000 \cdot \sum_{j \in [\ell]} W(S_j) = 2000M.
$$

Now we try to bound the third term. We know by lemma 2 that $s_{j,k} \leq 2s_j p_k + \lambda$ with probability at least $1 - e^{-\Omega(\lambda)}$. By the assumption X has Rademacher dimension λ we have that for each j, k such that $j \geq k + 10$,

$$\sum_{v \in [t]} \left\| Z_{j,k}^v \right\|_X \leq \sum_{v \in [t]} \lambda T_j \sqrt{s_{j,k}^v} \cdot 1/p_k$$

$$\leq \lambda T_j \cdot t \sqrt{s_{j,k}/t} \cdot \lambda^k$$

$$\leq \lambda M/\lambda^j \cdot \lambda^5 \sqrt{2\lambda^{j+1}\lambda^{-k}/\lambda^5 + \lambda} \cdot \lambda^k$$

$$\leq 2M \cdot \lambda^{1+5/2-\frac{i-k-1}{2}}$$

$$\leq 2M\lambda^{1+5/2-9/2} = 2M/\lambda.$$

Summing over all $j \in [\ell]$ and all $k \in [0, \ldots, j-10]$ we have that the third term is at most $O(2M\ell^2/\lambda) = o(M)$ with probability at least $1 - \ell^2 \cdot e^{-\Omega(\lambda)} \geq 1 - 1/\lambda^{\Omega(1)}$.

To sum up, with probability at least $1 - 1/\lambda^{\Omega(1)} - (1 - 0.991) \geq 0.99$ we have $\|\mu(x)\|_{1,X} \leq 2000M + o(M) = O(M)$.

4.3 Proof for Theorem 2

We can also prove Theorem 2 by two steps. The proof for the first part of the theorem (i.e., no underestimate) is exactly the same as that in Theorem 1 since in that proof we do not use any property of Rademacher dimension. For the second part, just notice that $\|\mu(x)\|_{1,X} \leq \sum_{k \in [\ell]} \sum_{j \in [\ell]} W(S_{j,k})$ where $\ell = O(\log_\lambda n)$, and $\mathbf{E}[W(S_{j,k})] = W(S_j)$ for all $j, k \in [\ell]$. Thus $\mathbf{E}[\|\mu(x)\|_{1,X}] \leq O(\log_\lambda n) \cdot \sum_{j \in [\ell]} W(S_j) = O(\log_\lambda n \cdot M)$. Directly applying Markov inequality gives the result.

References

1. Agarwal, P.K., Efrat, A., Sharir, M.: Vertical decomposition of shallow levels in 3-dimensional arrangements and its applications. In: SoCG, pp. 39–50 (1995)
2. Agarwal, P.K., Varadarajan, K.R.: A near-linear constant-factor approximation for euclidean bipartite matching? In: Symposium on Computational Geometry, pp. 247–252 (2004)
3. Alon, N., Matias, Y., Szegedy, M.: The space complexity of approximating the frequency moments. J. Comput. Syst. Sci. 58, 137–147 (1999)
4. Andoni, A., Ba, K.D., Indyk, P., Woodruff, D.: Efficient sketches for earth-mover distance, with applications. In: FOCS (2009)
5. Andoni, A., Charikar, M.S., Neiman, O., Nguyen, H.L.: Near linear lower bound for dimension reduction in ℓ_1. In: IEEE Symposium on Foundations of Computer Science (2011)
6. Andoni, A., Indyk, P., Krauthgamer, R.: Overcoming the ℓ_1 non-embeddability barrier: algorithms for product metrics. In: SODA, pp. 865–874 (2009)
7. Andoni, A., Krauthgamer, R., Onak, K.: Streaming algorithms via precision sampling. In: FOCS, pp. 363–372 (2011)
8. Andoni, A., Nguyen, H.L.: Near-optimal sublinear time algorithms for Ulam distance. In: Proceedings of the Twenty-First Annual ACM-SIAM Symposium on Discrete Algorithms, SODA 2010, pp. 76–86 (2010)
9. Andoni, A., Onak, K.: Approximating edit distance in near-linear time. In: STOC, pp. 199–204 (2009)
10. Brinkman, B., Charikar, M.: On the impossibility of dimension reduction in ℓ_1. J. ACM 52, 766–788 (2005)

11. Charikar, M.S.: Similarity estimation techniques from rounding algorithms. In: STOC, pp. 380–388 (2002)
12. Chefd'hotel, C., Bousquet, G.: Intensity-based image registration using earth mover's distance. In: SPIE (2007)
13. Deng, Y., Du, W.: The Kantorovich metric in computer science: A brief survey. Electr. Notes Theor. Comput. Sci. 253(3), 73–82 (2009)
14. Grauman, K., Darrell, T.: Fast contour matching using approximate earth movers distance. In: CVPR, pp. 220–227 (2004)
15. Holmes, A.S., Rose, C.J., Taylor, C.J.: Transforming pixel signatures into an improved metric space. Image Vision Comput 20(9-10), 701–707 (2002)
16. Indyk, P.: Algorithmic aspects of geometric embeddings. In: IEEE Symposium on Foundations of Computer Science (2001)
17. Indyk, P.: Stable distributions, pseudorandom generators, embeddings, and data stream computation. J. ACM 53, 307–323 (2006)
18. Indyk, P.: A near linear time constant factor approximation for euclidean bichromatic matching (cost). In: SODA, pp. 39–42 (2007)
19. Indyk, P., Matousek, J.: Low-distortion embeddings of finite metric spaces. In: Handbook of Discrete and Computational Geometry, pp. 177–196. CRC Press (2004)
20. Indyk, P., Thaper, N.: Fast color image retrieval via embeddings. In: Workshop on Statistical and Computational Theories of Vision, at ICCV (2003)
21. Johnson, W., Lindenstrauss, J.: Extensions of Lipschitz mappings into a Hilbert space. In: Conference in modern analysis and probability (New Haven, Conn., 1982). Contemporary Mathematics, vol. 26, pp. 189–206. American Mathematical Society (1984)
22. Kantorovich, L.V.: On the translocation of masses. Dokl. Akad. Nauk SSSR 37(7-8), 227–229 (1942)
23. Lawler, E.: Combinatorial optimization - networks and matroids. Holt, Rinehart and Winston, New York (1976)
24. Levina, E., Bickel, P.J.: The earth mover's distance is the mallows distance: Some insights from statistics. In: ICCV, pp. 251–256 (2001)
25. Linial, N.: Finite metric spaces - combinatorics, geometry and algorithms. In: Proceedings of the International Congress of Mathematicians III, pp. 573–586 (2002)
26. Maurey, B.: Type, cotype and k-convexity. In: Handbook of the Geometry of Banach Spaces, vol. 2, pp. 1299–1332. North-Holland (2003)
27. McGregor, A.: Open problems in data streams, property testing, and related topics (2011), http://www.cs.umass.edu/~mcgregor/papers/11-openproblems.pdf
28. Naor, A., Schechtman, G.: Planar earthmover is not in ℓ_1. SIAM J. Comput. 37(3), 804–826 (2007)
29. Nisan, N.: Pseudorandom generators for space-bounded computations. In: Proceedings of the Twenty-Second Annual ACM Symposium on Theory of Computing, STOC 1990, pp. 204–212 (1990)
30. Puzicha, J., Buhmann, J.M., Rubner, Y., Tomasi, C.: Empirical evaluation of dissimilarity measures for color and texture. In: ICCV, pp. 1165–1173 (1999)
31. Rubner, Y., Tomasi, C., Guibas, L.J.: The earth movers distance as a metric for image retrieval. International Journal of Computer Vision 40 (2000)
32. Vaidya, P.M.: Geometry helps in matching. SIAM J. Comput. 18, 1201–1225 (1989)
33. Varadarajan, K.R., Agarwal, P.K.: Approximation algorithms for bipartite and non-bipartite matching in the plane. In: SODA, pp. 805–814 (1999)

A Matrix Hyperbolic Cosine Algorithm and Applications*

Anastasios Zouzias

Department of Computer Science
University of Toronto, Canada

Abstract. In this paper, we generalize Spencer's hyperbolic cosine algorithm to the matrix-valued setting. We apply the proposed algorithm to several problems by analyzing its computational efficiency under two special cases of matrices; one in which the matrices have a group structure and an other in which they have rank-one. As an application of the former case, we present a deterministic algorithm that, given the multiplication table of a finite group of size n, it constructs an expanding Cayley graph of logarithmic degree in near-optimal $\mathcal{O}(n^2 \log^3 n)$ time. For the latter case, we present a fast deterministic algorithm for spectral sparsification of positive semi-definite matrices, which implies an improved deterministic algorithm for spectral graph sparsification of dense graphs. In addition, we give an elementary connection between spectral sparsification of positive semi-definite matrices and element-wise matrix sparsification. As a consequence, we obtain improved element-wise sparsification algorithms for diagonally dominant-like matrices.

1 Introduction

A non-trivial generalization of Chernoff bound type inequalities for matrix-valued random variables was introduced by Ahlswede and Winter [2]. In parallel, Vershynin and Rudelson introduced similar matrix-valued concentration inequalities using different machinery [27,28]. Following these two seminal papers, many variants have been proposed in the literature [26]; see [39] for more. Such inequalities, similarly to their real-valued ancestors, provide powerful tools to analyze probabilistic constructions and the performance of randomized algorithms. There is a rapidly growing line of research exploiting the power of these inequalities including new proofs of probabilistic constructions of expander graphs [3,21,23], matrix approximation by element-wise sparsification [13], graph approximation via edge sparsification [35], analysis of algorithms for matrix completion and decomposition of low rank matrices [26,22], semi-definite relaxation and rounding of quadratic maximization problems [29].

In many settings, it is desirable to convert the above probabilistic proofs into *efficient* deterministic procedures. That is, to derandomize the proofs. Wigderson and Xiao presented an efficient derandomization of the matrix Chernoff bound by generalizing Raghavan's method of pessimistic estimators to the

---

* A full version of this paper can be found at http://arxiv.org/abs/1103.2793

A. Czumaj et al. (Eds.): ICALP 2012, Part I, LNCS 7391, pp. 846–858, 2012.

matrix-valued setting [41]. In this paper, we generalize Spencer's hyperbolic cosine algorithm to the matrix-valued setting [30]. In an earlier, preliminary version of our paper [42] the generalization of Spencer's hyperbolic cosine algorithm was also based on the method of pessimistic estimators. However, here we present a proof which is based on a simple averaging argument. Next, we carefully analyze two special cases of matrices; one in which the matrices have a group structure and the other in which they have rank-one. We apply our main result to the following problems: deterministically constructing Alon-Roichman expanding Cayley graphs, approximating graphs via edge sparsification and approximating matrices via element-wise sparsification.

The Alon-Roichman theorem asserts that Cayley graphs obtained by choosing a logarithmic number of group elements independently and uniformly at random are expanders [3]. The original proof of Alon and Roichman is based on Wigner's trace method, whereas recent proofs rely on matrix-valued deviation bounds [21]. Wigderson and Xiao's derandomization of the matrix Chernoff bound implies a deterministic $\mathcal{O}(n^4 \log n)$ time algorithm for constructing Alon-Roichman graphs. Independently, Arora and Kale generalized the multiplicative weights update (MWU) method to the matrix-valued setting and, among other interesting implications, they improved the running time to $\mathcal{O}(n^3 \text{polylog}(n))$ [19]. Here we further improve the running time to $\mathcal{O}(n^2 \log^3 n)$ by exploiting the group structure of the problem. In addition, our algorithm is combinatorial in the sense that it only requires counting the number of all closed (even) paths of size at most $\mathcal{O}(\log n)$ in Cayley graphs. All previous algorithms involve numerical matrix computations such as eigenvalue decompositions and matrix exponentiation.

The second problem that we study is the graph sparsification problem. This problem poses the question whether any dense graph can be approximated by a sparse graph under different notions of approximation. Given any undirected graph, the most well-studied notions of approximation by a sparse graph include approximating, *all* pairwise distances up to an additive error [25], every cut to an arbitrarily small multiplicative error [9] and every eigenvalue of the difference of their Laplacian matrices to an arbitrarily small relative error [34]; the resulting graphs are usually called *graph spanners*, *cut sparsifiers* and *spectral sparsifiers*, respectively. Given that the notion of spectral sparsification is stronger than cut sparsification, so we focus on spectral sparsifiers. An efficient randomized algorithm to construct an $(1+\varepsilon)$-spectral sparsifier with $\mathcal{O}(n \log n/\varepsilon^2)$ edges was given in [35]. Furthermore, an $(1+\varepsilon)$-spectral sparsifier with $\mathcal{O}(n/\varepsilon^2)$ edges can be computed in $\mathcal{O}(mn^3/\varepsilon^2)$ deterministic time [8]. The latter result is a direct corollary of the spectral sparsification of positive semi-definite (psd) matrices problem as defined in [36]; see also [24] for more applications. Here we present a fast deterministic algorithm for spectral sparsification of psd matrices and, as a consequence, we obtain an improved deterministic spectral graph sparsification algorithm for the case of dense graphs.

The last problem that we analyze is the element-wise matrix sparsification problem. This problem was first introduced by Achlioptas and McSherry in [1].

They described sampling-based algorithms that select a small number of entries from an input matrix A, forming a sparse matrix $\tilde{A}$, which is close to A in the operator norm sense. The motivation to study this problem lies on the need to speed up several matrix computations including approximate eigenvector computations [1] and semi-definite programming solvers [4,7]. Recently, there are many follow-up results on this problem [5,13]. To the best of our knowledge, all known algorithms for this problem are randomized (see Table 1 of [13]). In this paper we present the first deterministic algorithm and strong sparsification bounds for symmetric matrices that have an approximate diagonally dominant[1] property. Diagonally dominant matrices arise in many applications such as the solution of certain elliptic differential equations via the finite element method [11], several optimization problems in computer vision [20] and computer graphics [18], to name a few.

Organization of the Paper. The paper is organized as follows. In § 2, we present the matrix hyperbolic cosine algorithm (Algorithm 1). We apply the matrix hyperbolic cosine algorithm to derive improved deterministic algorithms for the construction of Alon-Roichman expanding Cayley graphs in § 3, spectral sparsification of psd matrices in § 4 and element-wise matrix sparsification. Due to space constraints, the element-wise matrix sparsification section and all proofs have been deferred to the full version of the paper.

Our Results

The main contribution of this paper is a generalization of Spencer's hyperbolic cosine algorithm to the matrix-valued setting [30], [33, Lecture 4], see Algorithm 1. As mentioned in the introduction, our main result has connections with a recent derandomization of matrix concentration inequalities [41]. We should highlight a few advantages of our result compared to [41]. First, our construction does not rely on composing two separate estimators (or potential functions) to achieve operator norm bounds and does not require knowledge of the sampling probabilities of the matrix samples as in [41]. In addition, the algorithm of [41] requires computations of matrix expectations with matrix exponentials which are computationally expensive, see [41, Footnote 6, p. 63]. In this paper, we demostrate that overcoming these limitations leads to faster and in some cases simpler algorithms.

Next, we demonstrate the usefulness of the main result by analyzing its computational efficiency under two special cases of matrices. We begin by presenting the following result

Theorem 1 (Restatement of Theorem 5). *There is a deterministic algorithm that, given the multiplication table of a group G of size n, constructs an Alon-Roichman expanding Cayley graph of logarithmic degree in $\mathcal{O}(n^2 \log^3 n)$*

---

[1] A symmetric matrix A of size n is called *diagonally dominant* if $|A_{ii}| \geq \sum_{j \neq i} |A_{ij}|$ for every $i \in [n]$.

time. Moreover, the algorithm performs only group algebra operations that correspond to counting closed paths in Cayley graphs.

To the best of our knowledge, the above theorem improves the running time of all previously known deterministic constructions of Alon-Roichman Cayley graphs [6,41,19]. Moreover, notice that the running time of the above algorithm is optimal up-to poly-logarithmic factors since the size of the multiplication table of a finite group of size n is $\mathcal{O}(n^2)$.

In addition, we study the computational efficiency of the matrix hyperbolic cosine algorithm on the case of matrix samples with rank-one. The motivation for studying this special setting is its connection with problems such as graph approximation via edge sparsification as was shown in [8,36] and matrix approximation via element-wise sparsification as we will see later in this paper. The main result for this setting can be summarized in the following theorem (see § 4), which improves the $\mathcal{O}(mn^3/\varepsilon^2)$ running time of [36] when, say, $m = \Omega(n^2)$ and ε is a constant.

Theorem 2. *Suppose $0 < \varepsilon < 1$ and $\mathsf{A} = \sum_{i=1}^{m} v_i \otimes v_i$ are given, with column vectors $v_i \in \mathbb{R}^n$. Then there are non-negative real weights $\{s_i\}_{i \leq m}$, at most $\lceil n/\varepsilon^2 \rceil$ of which are non-zero, such that*

$$(1 - \varepsilon)^3 \mathsf{A} \preceq \widetilde{\mathsf{A}} \preceq (1 + \varepsilon)^3 \mathsf{A},$$

where $\widetilde{\mathsf{A}} = \sum_{i=1}^{m} s_i v_i \otimes v_i$. Moreover, there is a deterministic algorithm which computes the weights s_i in[2] $\widetilde{\mathcal{O}}(mn^2 \log^3 n/\varepsilon^2 + n^4 \log n/\varepsilon^4)$ time.

First, as we have already mentioned the graph sparsification problem can be reduced to spectral sparsification of positive semi-definite matrix. Hence as a corollary of the above theorem (proof omitted, see [36] for details), we obtain a fast deterministic algorithm for sparsifying dense graphs, which improves the currently best known $\mathcal{O}(n^5/\varepsilon^2)$ running time for this problem.

Corollary 1. *Given a weighted dense graph $H = (V, E)$ on n vertices with positive weights and $0 < \varepsilon < 1$, there is a deterministic algorithm that returns an $(1+\varepsilon)$-spectral sparsifier with $\mathcal{O}(n/\varepsilon^2)$ edges in $\widetilde{\mathcal{O}}(n^4 \log n/\varepsilon^2 \max\{\log^2 n, 1/\varepsilon^2\})$ time.*

Second, we give an elementary connection between element-wise matrix sparsification and spectral sparsification of psd matrices. A direct application of this connection implies strong sparsification bounds for symmetric matrices that are close to being *diagonally dominant*. More precisely, we give two element-wise sparsification algorithms for symmetric and diagonally dominant-like matrices; in its randomized and the other in its derandomized version (see Table 1 of [13] for comparison). Here, for the sake of presentation, we state our results for diagonally dominant matrices, although the results hold under a more general setting (see full version for details).

---

[2] The $\widetilde{\mathcal{O}}(\cdot)$ notation hides $\log \log n$ and $\log \log(1/\varepsilon)$ factors throughout the paper.

Theorem 3. *Let* A *be any symmetric and diagonally dominant matrix of size* n *and* $0 < \varepsilon < 1$. *Assume for normalization that* $\|A\| = 1$.

(a) *There is a randomized linear time algorithm that outputs a matrix* $\tilde{A} \in \mathbb{R}^{n \times n}$ *with at most* $\mathcal{O}(n \log n / \varepsilon^2)$ *non-zero entries such that, with probability at least* $1 - 1/n$, $\left\| A - \tilde{A} \right\| \le \varepsilon$.

(b) *There is a deterministic* $\tilde{\mathcal{O}}(\varepsilon^{-2} \mathbf{nnz}(A) n^2 \log n \max\{\log^2 n, 1/\varepsilon^2\})$ *time algorithm that outputs a matrix* $\tilde{A} \in \mathbb{R}^{n \times n}$ *with at most* $\mathcal{O}(n/\varepsilon^2)$ *non-zero entries such that* $\left\| A - \tilde{A} \right\| \le \varepsilon$.

Preliminaries. The next discussion reviews several definitions and facts from linear algebra; for more details, see [10]. By $[n]$ to be the set $\{1, 2, \ldots, n\}$. We denote by $\mathcal{S}^{n \times n}$ the set of symmetric matrices of size n. Let $\mathbf{x} \in \mathbb{R}^n$, we denote by $\mathbf{diag}(\mathbf{x})$ the diagonal matrix containing $x_1, x_2, \ldots, x_n$. For a square matrix M, we also write $\mathbf{diag}(M)$ to denote the diagonal matrix that contains the diagonal entries of M. Let A be an $m \times n$ matrix. $A^{(j)}$ will denote the j-th column of A and $A_{(i)}$ the i-th row of A. We denote $\|A\| = \max\{\|A\mathbf{x}\| \mid \|\mathbf{x}\| = 1\}$, $\|A\|_\infty = \max_{i \in [m]} \sum_{j \in [n]} |A_{ij}|$ and by $\|A\|_F = \sqrt{\sum_{i,j} A_{ij}^2}$ the Frobenius norm of A. Also $\mathbf{sr}(A) := \|A\|_F^2 / \|A\|^2$ is the *stable rank* of A and by $\mathbf{nnz}(A)$ the number of its non-zero entries. The trace of a square matrix B is denoted as $\mathbf{tr}(B)$. We write J_n for the all-ones square matrices of size n. For two symmetric matrices X, Y, we say that $Y \succeq X$ if and only if $Y - X$ is a positive semi-definite (psd) matrix. Let $\mathbf{x} \in \mathbb{R}^n$, then $\mathbf{x} \otimes \mathbf{x}$ is the $n \times n$ matrix such that $(\mathbf{x} \otimes \mathbf{x})_{i,j} = x_i x_j$. Given any matrix A, its *dilation* is defined as $\mathcal{D}(A) = \begin{bmatrix} 0 & A \\ A^\top & 0 \end{bmatrix}$. It is easy to see that $\lambda_{\max}(\mathcal{D}(A)) = \|A\|$, see e.g. [37, Theorem 4.2].

Functions of Matrices. Here we review some basic facts about the matrix exponential and the hyperbolic cosine function, for more details see [17]. All proofs of this section have been deferred to the appendix. The matrix exponential of a symmetric matrix A is defined as $\mathbf{exp}[A] = \mathbf{I} + \sum_{k=1}^\infty \frac{A^k}{k!}$. Let $A = Q$ $matLamQ^\top$ be the eigendecomposition of A. It is easy to see that $\mathbf{exp}[A] = Q\mathbf{exp}[\Lambda] Q^\top$. For any real square matrices A and B of the same size that commute, i.e., $AB = BA$, we have that $\mathbf{exp}[A + B] = \mathbf{exp}[A]\mathbf{exp}[B]$. In general, when A and B do not commute, the following estimate is known for symmetric matrices.

Lemma 1. *[15,38] For any symmetric matrices* A *and* B, $\mathbf{tr}(\mathbf{exp}[A + B]) \le \mathbf{tr}(\mathbf{exp}[A]\mathbf{exp}[B])$.

We will also need the following fact about matrix exponential for rank one matrices.

Lemma 2. *Let* $\mathbf{x}$ *be a non-zero vector in* $\mathbb{R}^n$. *Then* $\mathbf{exp}[\mathbf{x} \otimes \mathbf{x}] = \mathbf{I}_n + \frac{e^{\|\mathbf{x}\|^2} - 1}{\|\mathbf{x}\|^2} \mathbf{x} \otimes \mathbf{x}$. *Similarly,* $\mathbf{exp}[-\mathbf{x} \otimes \mathbf{x}] = \mathbf{I}_n - \frac{1 - e^{-\|\mathbf{x}\|^2}}{\|\mathbf{x}\|^2} \mathbf{x} \otimes \mathbf{x}$.

Let us define the *matrix hyperbolic cosine* function of a symmetric matrix A as $\cosh[A] := (\exp[A] + \exp[-A])/2$. Next, we state a few properties of the matrix hyperbolic cosine.

Lemma 3. *Let A be a symmetric matrix. Then* $tr(exp[\mathcal{D}(A)]) = 2tr(cosh[A])$.

Lemma 4. *Let A be a symmetric matrix and P be a projector matrix that commutes with A, i.e., $PA = AP$. Then* $cosh[PA] = P\,cosh[A] + I - P$.

Lemma 5. *[40, Lemma 2.2] For any positive semi-definite symmetric matrix A of size n and any two symmetric matrices B, C of size n, $B \preceq C$ implies* $tr(AB) \leq tr(AC)$.

2   Balancing Matrices: A Matrix Hyperbolic Cosine Algorithm

We briefly describe Spencer's balancing vectors game and then generalize it to the matrix-valued setting [33, Lecture 4]. Let a two-player perfect information game between Alice and Bob. The game consists of n rounds. On the i-th round, Alice sends a vector v_i with $\|v_i\|_\infty \leq 1$ to Bob, and Bob has to decide on a sign $s_i \in \{\pm 1\}$ knowing only his previous choices of signs and $\{v_k\}_{k<i}$. At the end of the game, Bob pays Alice $\|\sum_{i=1}^n s_i v_i\|_\infty$. We call the latter quantity, the *value* of the game.

It has been shown in [32] that, in the above limited online variant, Spencer's six standard deviations bound [31] does not hold and the best value that we can hope for is $\Omega(\sqrt{n \ln n})$. Such a bound is easy to obtain by picking the signs $\{s_i\}$ uniformly at random. Indeed, a direct application of Azuma's inequality to each coordinate of the random vector $\sum_{i=1}^n s_i v_i$ together with a union bound over all the coordinates gives a bound of $\mathcal{O}(\sqrt{n \ln n})$.

Now, we generalize the balancing vectors game to the matrix-valued setting. That is, Alice now sends to Bob a sequence $\{M_i\}$ of symmetric matrices of size n with[3] $\|M_i\| \leq 1$, and Bob has to pick a sequence of signs $\{s_i\}$ so that, at the end of the game, the quantity $\|\sum_{i=1}^n s_i M_i\|$ is as small as possible. Notice that the balancing vectors game is a restriction of the balancing matrices game in which Alice is allowed to send only diagonal matrices with entries bounded in absolute value by one. Similarly to the balancing vectors game, using matrix-valued concentration inequalities, one can prove that Bob has a randomized strategy that achieves at most $\mathcal{O}(\sqrt{n \ln n})$ w.p. at least $1/2$. Indeed,

Lemma 6. *Let $M_i \in \mathcal{S}^{n \times n}$, $\|M_i\| \leq 1$, $1 \leq i \leq n$. Pick $s_i^* \in \{\pm 1\}$ uniformly at random for every $i \in [n]$. Then* $\|\sum_{i=1}^n s_i^* M_i\| = \mathcal{O}(\sqrt{n \ln n})$ *w.p. at least* $1/2$.

---

[3] A curious reader may ask him/her-self why the operator norm is the right choice. It turns out the the operator norm is the correct matrix-norm analog of the ℓ_∞ vector-norm, viewed as the *infinity* Schatten norm on the space of matrices.

Now, let's assume that Bob wants to achieve the above probabilistic guarantees using a *deterministic* strategy. Is it possible? We answer this question in the affirmative by generalizing Spencer's hyperbolic cosine algorithm (and its proof) to the matrix-valued setting. We call the resulting algorithm *matrix hyperbolic cosine* (Algorithm 1). It is clear that this simple greedy algorithm implies a deterministic strategy for Bob that achieves the probabilistic guarantees of Lemma 6 (set $f_j \sim s_j \mathsf{M}_j$, $t = n$ and $\varepsilon = \mathcal{O}(\sqrt{\ln n / n})$ and notice that γ, ρ^2 are at most one).

Algorithm 1 requires an extra assumption on its random matrices compared to Spencer's original algorithm. That is, we assume that our random matrices have uniformly bounded their "matrix variance", denoted by ρ^2. This requirement is motivated by the fact that in the applications that are studied in this paper such an assumption translates bounds that depend quadratically on the matrix dimensions to bounds that depend linearly on the dimensions.

We will need the following technical lemma for proving the main result of this section, which is a Bernstein type argument generalized to the matrix-valued setting [39].

Lemma 7. *Let $f : [m] \to \mathcal{S}^{n \times n}$ with $\|f(i)\| \le \gamma$ for all $i \in [m]$. Let X be a random variable over $[m]$ such that $\mathbb{E} f(X) = \mathbf{0}$ and $\|\mathbb{E} f(X)^2\| \le \rho^2$. Then, for any $\theta > 0$, $\|\mathbb{E}[exp [\mathcal{D} (\theta f(X))]]\| \le \exp \left(\rho^2 (e^{\theta \gamma} - 1 - \theta \gamma) / \gamma^2 \right)$. In particular, for any $0 < \varepsilon < 1$, setting $\theta = \varepsilon / \gamma$ implies that $\mathbb{E}[exp [\mathcal{D} (\varepsilon f(X) / \gamma)]] \preceq e^{\varepsilon^2 \rho^2 / \gamma^2} \mathbf{I}_{2n}$.*

Now we are ready to prove the correctness of the matrix hyperbolic cosine algorithm.

Algorithm 1. Matrix Hyperbolic Cosine

1: **procedure** MATRIX-HYPERBOLIC($\{f_j\}$, ε, t)   ▷ $f_j : [m] \to \mathcal{S}^{n \times n}$ as in Theorem 4,
       $0 < \varepsilon < 1$.
2:     Set $\theta = \varepsilon / \gamma$
3:     **for** $i = 1$ to t **do**
4:         Compute $x_i^* \in [m]$: $x_i^* = \arg \min_{k \in [m]} \mathrm{tr} \left(\mathbf{cosh} \left[\theta \sum_{j=1}^{i-1} f_j(x_j^*) + \theta f_i(k) \right] \right)$
5:     **end for**
6:     **Output:** t indices $x_1^*, x_2^*, \dots, x_t^*$ such that $\left\| \frac{1}{t} \sum_{j=1}^{t} f_j(x_j^*) \right\| \le \frac{\gamma \ln(2n)}{t \varepsilon} + \frac{\varepsilon \rho^2}{\gamma}$
7: **end procedure**

Theorem 4. *Let $f_j : [m] \to \mathcal{S}^{n \times n}$ with $\|f_j(i)\| \le \gamma$ for all $i \in [m]$ and $j = 1, 2, \dots$. Suppose that there exists independent random variables $X_1, X_2, \dots$ over $[m]$ such that $\mathbb{E} f_j(X_j) = \mathbf{0}$ and $\|\mathbb{E} f_j(X_j)^2\| \le \rho^2$. Algorithm 1 with input $\{f_j\}, \varepsilon, t$ outputs a set of indices $\{x_j^*\}_{j \in [t]}$ over $[m]$ such that $\left\| \frac{1}{t} \sum_{j=1}^{t} f_j(x_j^*) \right\| \le \frac{\gamma \ln(2n)}{t \varepsilon} + \frac{\varepsilon \rho^2}{\gamma}$.*

We conclude with an open question related to Spencer's six standard deviation bound [31]. Does Spencer's six standard deviation bound holds under the matrix setting? More formally, given any sequence of n symmetric matrices $\{M_i\}$ with $\|M_i\| \leq 1$, does there exist a set of signs $\{s_i\}$ so that $\|\sum_{i=1}^{n} s_i M_i\| = \mathcal{O}(\sqrt{n})$?

3 Alon-Roichman Expanding Cayley Graphs

We start by describing expander graphs. Given a connected undirected d-regular graph $H = (V, E)$ on n vertices, let A be its adjacency matrix, i.e., $A_{ij} = w_{ij}$ where w_{ij} is the number of edges between vertices i and j. Moreover, let $\widehat{A} = \frac{1}{d}A$ be its normalized adjacency matrix. We allow self-loops and multiple edges. Let $\lambda_1(\widehat{A}), \ldots, \lambda_n(\widehat{A})$ be its eigenvalues in decreasing order. We have that $\lambda_1(\widehat{A}) = 1$ with corresponding eigenvector $1/\sqrt{n}$, where 1 is the all-one vector. The graph H is called a spectral expander if $\lambda(\widehat{A}) := \max_{2 \leq j}\{|\lambda_j(\widehat{A})|\} \leq \varepsilon$ for some positive constant $\varepsilon < 1$.

Denote by $m_k = m_k(H) := \mathbf{tr}\left(A^k\right)$. By definition, m_k is equal to the number of self-returning walks of length k of the graph H. A graph-spectrum-based invariant, recently proposed by Estrada is defined as $EE(A) := \mathbf{tr}\left(\exp[A]\right)$ [14], which also equals to $\sum_{k=0}^{\infty} m_k/k!$. For $\theta > 0$, we define the $even$ θ-$Estrada$ $index$ by $EE_{\text{even}}(A, \theta) := \sum_{k=0}^{\infty} m_{2k}(\theta A)/(2k)!$.

Now let G be any finite group of order n with identity element id. Let S be a multi-set of elements of G, we denote by $S \sqcup S^{-1}$ the symmetric closure of S, namely the number of occurrences of s and s^{-1} in $S \sqcup S^{-1}$ equals the number of occurrences of $s \in S$. Let R be the right regular representation[4], i.e., $(R(g_1)\phi)(g_2) = \phi(g_1 g_2)$ for every $\phi : G \to \mathbb{R}$ and $g_1, g_2 \in G$. The Cayley graph $\text{Cay}(G; S)$ on a group G with respect to the mutli-set $S \subset G$ is the graph whose vertex set is G, and where g_1 and g_2 are connected by an edge if there exists $s \in S$ such that $g_2 = g_1 s$ (allowing multiple edges for multiple elements in S). In this section we prove the correctness of the following greedy algorithm for constructing expanding Cayley graphs.

Theorem 5. *Algorithm 2, given the multiplication table of a finite group G of size n and $0 < \varepsilon < 1$, outputs a (symmetric) multi-set $S \subset G$ of size $\mathcal{O}(\log n/\varepsilon^2)$ such that $\lambda(\text{Cay}(G; S)) \leq \varepsilon$ in $\mathcal{O}(n^2 \log^3 n/\varepsilon^3)$ time. Moreover, the algorithm performs only group algebra operations that correspond to counting closed paths in Cayley graphs.*

Let $\widehat{A}$ be the normalized adjacency matrix of $\text{Cay}\left(G; S \sqcup S^{-1}\right)$ for some $S \subset G$. It is not hard to see that $\widehat{A} = \frac{1}{2|S|} \sum_{s \in S}(R(s) + R(s^{-1}))$. We want to bound $\lambda(A)$. Notice that $\lambda(A) = \|(I - J/n)A\|$. Since we want to analyze the second-largest eigenvalue (in absolute value), we consider $(I - J/n)A =$

---

[4] In other words, represent each group algebra element with a permutation matrix of size n that preserves the group structure (i.e., it is a group homeomorphism). This is always possible due to Cayley's theorem.

Algorithm 2. Expander Cayley Graph via even Estrada Index Minimization

1: **procedure** GREEDYESTRADAMIN(G, ε)   ▷ Multiplication table of G, $0 < \varepsilon < 1$
2:     Set $S^{(0)} = \emptyset$ and $t = \mathcal{O}(\log n / \varepsilon^2)$
3:     **for** $i = 1, \ldots t$ **do**
4:         Let $g_* \in G$ that (approximately) min. the even $\varepsilon/2$-Estrada index of Cay $\left(G; S^{(i-1)} \cup g \cup g^{-1}\right)$ over all $g \in G$ ▷ Use Lemma 9
5:         Set $S^{(i)} = S^{(i-1)} \cup g_* \cup g_*^{-1}$
6:     **end for**
7:     **Output:** A multi-set $S := S^{(t)}$ of size $2t$ such that $\lambda(\text{Cay}\,(G; S)) \leq \varepsilon$
8: **end procedure**

$\frac{1}{|S|}\sum_{s \in S}(R(s) + R(s^{-1}))/2 - \mathbf{J}/n$. Based on the above calculation, we define our matrix-valued function as

$$f(g) := (R(g) + R(g^{-1}))/2 - \mathbf{J}/n \tag{1}$$

for every $g \in G$. The following lemma connects the potential function that is used in Theorem 4 and the even Estrada index.

Lemma 8. *Let $S \subset G$ and A be the adjacency matrix of* Cay $\left(G; S \sqcup S^{-1}\right)$*. For any $\theta > 0$, $\mathbf{tr}\left(\mathbf{cosh}\left[\theta \sum_{s \in S} f(s)\right]\right) = EE_{even}(A, \theta/2) + 1 - \cosh(\theta|S|)$.*

The following lemma indicates that it is possible to efficiently compute the (even) Estrada index for Cayley graphs with small generating set.

Lemma 9. *Let $S \subset G$, $\theta, \delta > 0$, and A be the adjacency matrix of* Cay $(G; S)$*. There is an algorithm that, given S, computes an additive δ approximation to $EE(\theta A)$ or $EE_{even}(A, \theta)$ in $\mathcal{O}(n|S| \max\{\log(n/\delta), 2e^2|S|\theta\})$ time.*

Proof. (of Theorem 5) By Lemma 8, minimizing the even $\varepsilon/2$-Estrada index in the i-th iteration is equivalent to minimizing $\mathbf{tr}\left(\mathbf{cosh}\left[\theta \sum_{s \in S^{(i-1)}} f(s) + \theta f(g)\right]\right)$ over all $g \in G$ with $\theta = \varepsilon$. Notice that $f(g) \in \mathcal{S}^{n \times n}$ for $g \in G$, $\mathbb{E}_{g \in_R G} f(g) = \mathbf{0}_n$ since $\sum_{g \in G} R(g) = \mathbf{J}$. It is easy to see that $\|f(g)\| \leq 2$ and moreover a calculation implies that $\left\|\mathbb{E}_{g \in_R G} f(g)^2\right\| \leq 2$ as well. Theorem 4 implies that we get a multi-set S of size t such that $\lambda(\text{Cay}\,(G; S \sqcup S^{-1})) = \left\|\frac{1}{|S|}\sum_{s \in S} f(s)\right\| \leq \varepsilon$.

The moreover part follows from Lemma 9 with $\delta = \frac{e^{\varepsilon^2}}{n^c}$ for a sufficient large constant $c > 0$. Indeed, in total we incur (following the proof of Theorem 4) at most an additive $\ln(\delta n e^{\varepsilon^2 t})/\varepsilon$ error which is bounded by ε.

4   Fast Isotropic Sparsification and Spectral Sparsification

Let A be an $m \times n$ matrix with $m \gg n$ whose columns are in isotropic position, i.e., $A^\top A = I_n$. For $0 < \varepsilon < 1$, consider the problem of finding a small subset of (rescaled) rows of A forming a matrix $\widetilde{A}$ such that $\left\|\widetilde{A}^\top \widetilde{A} - I\right\| \leq \varepsilon$. The matrix Bernstein inequality (see [39]) tells us that there exists such a set with size

$\mathcal{O}(n \log n / \varepsilon^2)$. Indeed, set $f(i) = \mathsf{A}_{(i)} \otimes \mathsf{A}_{(i)}/p_i - \mathbf{I}_n$ where $p_i = \left\| \mathsf{A}_{(i)} \right\|^2 / \|\mathsf{A}\|_{\mathrm{F}}^2$. A calculation shows that γ and ρ^2 are $\mathcal{O}(n)$. Moreover, Algorithm 1 implies an $\mathcal{O}(mn^4 \log n / \varepsilon^2)$ time algorithm for finding such a set. The running time of Algorithm 1 for rank-one matrix samples can be improved to $\mathcal{O}(mn^3 \mathrm{polylog}\,(n)/\varepsilon^2)$ by exploiting their rank-one structure. More precisely, using fast algorithms for computing all the eigenvalues of matrices after rank-one updates [16]. Next we show that we can further improve the running time by a more careful analysis.

We show how to improve the running time of Algorithm 1 to $\mathcal{O}(\frac{mn^2}{\varepsilon^2} \mathrm{polylog}\,(n, \frac{1}{\varepsilon}))$ utilizing results from numerical linear algebra including the Fast Multipole Method [12] (FMM) and ideas from [16]. The main idea behind the improvement is that the trace is invariant under any change of basis. At each iteration, we perform a change of basis so that the matrix corresponding to the previous choices of the algorithm is diagonal. Now, Step 4 of Algorithm 1 corresponds to computing all the eigenvalues of m different eigensystems with special structure, i.e., diagonal plus a rank-one matrix. Such eigensystem can be solved in $\mathcal{O}(n \mathrm{polylog}\,(n))$ time using the FMM as was observed in [16]. However, the problem now, is that at each iteration we have to represent all the vectors $\mathsf{A}_{(i)}$ in the new basis, which may cost $\mathcal{O}(mn^2)$. The key observation is that the change of basis matrix at each iteration is a Cauchy matrix (see the full version). It is known that matrix-vector multiplication with Cauchy matrices can be performed efficiently and numerically stable using FMM. Therefore, at each iteration, we can perform the change of basis in $\mathcal{O}(mn \mathrm{polylog}\,(n))$ and m eigenvalue computations in $\mathcal{O}(mn \mathrm{polylog}\,(n))$ time. The next theorem states that the resulting algorithm runs in $\mathcal{O}(mn^2 \mathrm{polylog}\,(n))$ time (see Appendix for proof).

Theorem 6. *Let A be an $m \times n$ matrix with $\mathsf{A}^\top \mathsf{A} = \mathbf{I}_n$, $m \geq n$ and $0 < \varepsilon < 1$. Algorithm 3 returns at most $t = \mathcal{O}(n \ln n / \varepsilon^2)$ indices $x_1^*, x_2^*, \ldots x_t^*$ over $[m]$ with corresponding scalars $s_1, s_2, \ldots, s_t$ using $\tilde{\mathcal{O}}(mn^2 \log^3 n / \varepsilon^2)$ operations such that*

$$\left\| \sum_{i=1}^{t} s_i \mathsf{A}_{(x_i^*)} \otimes \mathsf{A}_{(x_i^*)} - \mathbf{I}_n \right\| \leq \varepsilon. \tag{2}$$

Next, we show that Algorithm 3 can be used as a bootstrapping procedure to improve the time complexity of [36, Theorem 3.1], see also [8, Theorem 3.1]. Such an improvement implies faster algorithms for constructing graph sparsifiers and, as we will see in element-wise matrix sparsification section (see full version).

Theorem 7. *Suppose $0 < \varepsilon < 1$ and $\mathsf{A} = \sum_{i=1}^{m} v_i \otimes v_i$ are given, with column vectors $v_i \in \mathbb{R}^n$ and $m \geq n$. Then there are non-negative weights $\{s_i\}_{i \leq m}$, at most $\lceil n/\varepsilon^2 \rceil$ of which are non-zero, such that*

$$(1 - \varepsilon)^3 \mathsf{A} \preceq \tilde{\mathsf{A}} \preceq (1 + \varepsilon)^3 \mathsf{A}, \tag{3}$$

where $\tilde{\mathsf{A}} = \sum_{i=1}^{m} s_i v_i \otimes v_i$. Moreover, there is an algorithm that computes the weights $\{s_i\}_{i \leq m}$ in deterministic $\tilde{\mathcal{O}}(mn^2 \log^3 n / \varepsilon^2 + n^4 \log n / \varepsilon^4)$ time.

Algorithm 3. Fast Isotropic Sparsification

1: **procedure** ISOTROP(A, ε) $\triangleright$ A $\in \mathbb{R}^{m \times n}$, $\sum_{k=1}^{m}$ A$_{(k)} \otimes$ A$_{(k)} = I_n$ and $0 < \varepsilon < 1$
2:     Set $\theta = \varepsilon/n$, $t = \mathcal{O}(n \ln n/\varepsilon^2)$, and A$_{(k)} \leftarrow$ A$_{(k)}/\sqrt{p_k}$ for every $k \in [m]$, where
    $p_k = \|$A$_{(k)}\|^2/n$
3:     Set $\Lambda_{\{0\}} = 0_n$ and $Z = \sqrt{\theta}$ A
4:     **for** $i = 1$ to t **do**
5:         $x_i^* = \arg\min_{k \in [m]}$ **tr** $\left(\exp\left[\Lambda_{\{i-1\}} + Z_{(k)} \otimes Z_{(k)}\right] e^{-\theta i} + \exp\left[-\Lambda_{\{i-1\}} - Z_{(k)} \otimes Z_{(k)}\right] e^{\theta i}\right)$
        $\triangleright$ Apply m times Lemma 12 (see full version)
6:         $[\Lambda_{\{i\}}, U_{\{i\}}] =$ **eigs**$(\Lambda_{\{i-1\}} + Z_{(x_i^*)} \otimes Z_{(x_i^*)})$ $\triangleright$ **eigs** computes eigensystem
7:         $Z = ZU_{\{i\}}$ $\triangleright$ Apply fast matrix-vector multiplication
8:     **end for**
9:     **Output:** t indices $x_1^*, x_2^*, \ldots, x_t^*$, $x_i^* \in [m]$ s.t. $\left\|\sum_{k=1}^{t} \frac{A_{(x_k^*)} \otimes A_{(x_k^*)}}{t p_{x_k^*}} - I_n\right\| \leq \varepsilon$
10: **end procedure**

Acknowledgements. The author would like to thank Mark Braverman for several interesting discussions and comments about this work.

References

1. Achlioptas, D., McSherry, F.: Fast Computation of Low-rank Matrix Approximations. SIAM J. Comput. 54(2), 9 (2007)
2. Ahlswede, R., Winter, A.: Strong Converse for Identification via Quantum Channels. IEEE Transactions on Information Theory 48(3), 569–579 (2002)
3. Alon, N., Roichman, Y.: Random Cayley Graphs and Expanders. Random Struct. Algorithms 5, 271–284 (1994)
4. Arora, S., Hazan, E., Kale, S.: Fast Algorithms for Approximate Semidefinite Programming using the Multiplicative Weights Update Method. In: Proceedings of the Symposium on Foundations of Computer Science (FOCS), pp. 339–348 (2005)
5. Arora, S., Hazan, E., Kale, S.: A Fast Random Sampling Algorithm for Sparsifying Matrices. In: Díaz, J., Jansen, K., Rolim, J.D.P., Zwick, U. (eds.) APPROX 2006 and RANDOM 2006. LNCS, vol. 4110, pp. 272–279. Springer, Heidelberg (2006)
6. Arora, S., Kale, S.: A Combinatorial, Primal-Dual Approach to Semidefinite Programs. In: Proceedings of the Symposium on Theory of Computing (STOC), pp. 227–236 (2007)
7. d'Aspremont, A.: Subsampling Algorithms for Semidefinite Programming. In: Stochastic Systems, pp. 274–305 (2011)
8. Batson, J.D., Spielman, D.A., Srivastava, N.: Twice-ramanujan sparsifiers. In: Proceedings of the Symposium on Theory of Computing (STOC), pp. 255–262 (2009)
9. Benczúr, A.A., Karger, D.R.: Approximating s-t Minimum Cuts in $\widetilde{\mathcal{O}}(n^2)$ Time. In: Proceedings of the Symposium on Theory of Computing (STOC) (1996)
10. Bhatia, R.: Matrix Analysis, 1st edn. Graduate Texts in Mathematics, vol. 169. Springer, Heidelberg (1996)
11. Boman, E.G., Hendrickson, B., Vavasis, S.: Solving Elliptic Finite Element Systems in Near-Linear Time with Support Preconditioners. SIAM J. on Numerical Analysis 46(6), 3264–3284 (2008)
12. Carrier, J., Greengard, L., Rokhlin, V.: A Fast Adaptive Multipole Algorithm for Particle Simulations. SIAM J. on Scientific and Statistical Computing 9(4), 669–686 (1988)

13. Drineas, P., Zouzias, A.: A note on Element-wise Matrix Sparsification via a Matrix-valued Bernstein Inequality. Information Processing Letters 111(8), 385–389 (2011)
14. Estrada, E., Rodríguez-Velázquez, J.A.: Subgraph Centrality in Complex Networks. Phys. Rev. E 71 (May 2005)
15. Golden, S.: Lower Bounds for the Helmholtz Function. Phys. Rev. 137(4B), B1127–B1128 (1965)
16. Gu, M., Eisenstat, S.C.: A Stable and Efficient Algorithm for the Rank-One Modification of the Symmetric Eigenproblem. SIAM J. Matrix Anal. Appl. 15 (1994)
17. Higham, N.J.: Functions of Matrices: Theory and Computation. Society for Industrial and Applied Mathematics (SIAM) (2008)
18. Joshi, P., Meyer, M., DeRose, T., Green, B., Sanocki, T.: Harmonic Coordinates for Character Articulation. ACM Trans. Graph. 26 (2007)
19. Kale, S.: Efficient Algorithms Using the Multiplicative Weights Update Method. PhD in Computer Science, Princeton University (2007)
20. Koutis, I., Miller, G.L., Tolliver, D.: Combinatorial Preconditioners and Multilevel Solvers for Problems in Computer Vision and Image Processing. In: Bebis, G., Boyle, R., Parvin, B., Koracin, D., Kuno, Y., Wang, J., Wang, J.-X., Wang, J., Pajarola, R., Lindstrom, P., Hinkenjann, A., Encarnação, M.L., Silva, C.T., Coming, D. (eds.) ISVC 2009. LNCS, vol. 5875, pp. 1067–1078. Springer, Heidelberg (2009)
21. Landau, Z., Russell, A.: Random Cayley Graphs are Expanders: a simplified proof of the Alon-Roichman theorem. The Electronic J. of Combinatorics 11(1) (2004)
22. Magen, A., Zouzias, A.: Low Rank Matrix-Valued Chernoff Bounds and Approximate Matrix Multiplication. In: Proceedings of the ACM-SIAM Symposium on Discrete Algorithms (SODA), pp. 1422–1436 (2011)
23. Naor, A.: On the Banach Space Valued Azuma Inequality and Small set Isoperimetry of Alon-Roichman Graphs. To Appear in Combinatorics, Probability and Computing (September 2010), arxiv:1009.5695
24. Naor, A.: Sparse Quadratic Forms and their Geometric Applications (after Batson, Spielman and Srivastava) (January 2011), arxiv:1101.4324
25. Peleg, D., Schäffer, A.A.: Graph Spanners. J. of Graph Theory 13(1), 99–116 (1989)
26. Recht, B.: A Simpler Approach to Matrix Completion. J. of Machine Learning Research, 3413–3430 (December 2011)
27. Rudelson, M.: Random Vectors in the Isotropic Position. J. Funct. Anal. 164(1), 60–72 (1999)
28. Rudelson, M., Vershynin, R.: Sampling from Large Matrices: An Approach through Geometric Functional Analysis. J. ACM 54(4), 21 (2007)
29. So, A.M.C.: Moment Inequalities for sums of Random Matrices and their Applications in Optimization. Mathematical Programming, pp. 1–27 (2009)
30. Spencer, J.: Balancing Games. J. Comb. Theory, Ser. B 23(1), 68–74 (1977)
31. Spencer, J.: Six Standard Deviations Suffice. Transactions of The American Mathematical Society 289, 679–679 (1985)
32. Spencer, J.: Balancing Vectors in the max Norm. Combinatorica 6, 55–65 (1986)
33. Spencer, J.: Ten Lectures on the Probabilistic Method, 2nd edn. Society for Industrial and Applied Mathematics, SIAM (1994)
34. Spielman, D.A.: Algorithms, Graph Theory, and Linear Equations in Laplacian Matrices. In: Proceedings of the International Congress of Mathematicians, vol. IV, pp. 2698–2722 (2010)
35. Spielman, D.A., Srivastava, N.: Graph Sparsification by Effective Resistances. In: Proceedings of the Symposium on Theory of Computing, STOC (2008)

36. Srivastava, N.: Spectral Sparsification and Restricted Invertibility. PhD in Computer Science, Yale University (2010)
37. Stewart, G.W., Sun, J.G.: Matrix Perturbation Theory (Computer Science and Scientific Computing). Academic Press, London (1990)
38. Thompson, C.J.: Inequality with Applications in Statistical Mechanics. J. of Mathematical Physics 6(11), 1812–1813 (1965)
39. Tropp, J.A.: User-Friendly Tail Bounds for Sums of Random Matrices. Foundations of Computational Mathematics, pp. 1–46 (2011)
40. Tsuda, K., Rätsch, G., Warmuth, M.K.: Matrix Exponentiated Gradient Updates for on-line Learning and Bregman Projections. JMLR 6, 995–1018 (2005)
41. Wigderson, A., Xiao, D.: Derandomizing the Ahlswede-Winter Matrix-valued Chernoff Bound using Pessimistic Estimators, and Applications. Theory of Computing 4(1), 53–76 (2008)
42. Zouzias, A.: A Matrix Hyperbolic Cosine Algorithm and Applications. Ver. 1. (March 2011), arxiv:1103.2793

Author Index